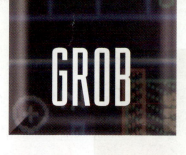

GROB

BASIC ELECTRONICS

EIGHTH EDITION

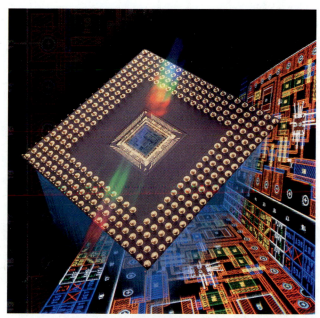

Bernard Grob

Glencoe
McGraw-Hill

New York, New York   Columbus, Ohio   Woodland Hills, California   Peoria, Illinois

IN MEMORY OF RUTH

Photo credits appear on page 1010, which is hereby made part of this copyright page.

Library of Congress Cataloging-in-Publication Data
Grob, Bernard.
    Basic electronics / Bernard Grob.—8th ed.
        p.   cm.
    Includes bibliographical references and index.
    ISBN 0-02-802253-X
    1. Electronics.   I. Title.
TK7816.G75   1997
621.3—dc21   96-39669
                                                          CIP

Glencoe/McGraw-Hill

A Division of The **McGraw·Hill** *Companies*

Basic Electronics, Eighth Edition

Copyright © 1997 by The McGraw-Hill Companies, Inc. All rights reserved. Copyright © 1992, 1989, 1984, 1977, 1971, 1965, 1959 by The McGraw-Hill Companies, Inc. All rights reserved. Printed in the United States of America. Except as permitted under the United States Copyright Act of 1976, no part of this publication may be reproduced or distributed in any form or by any means, or stored in a data base or retrieval system, without the prior written permission of the publisher.

Send all inquiries to:
Glencoe/McGraw-Hill
936 Eastwind Drive
Westerville, Ohio 43081

ISBN 0-02-802253-X   (student edition)
ISBN 0-02-802260-2   (instructor's annotated edition)

2 3 4 5 6 7 8 9   027   01 00 99 98

CONTENTS

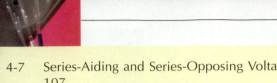

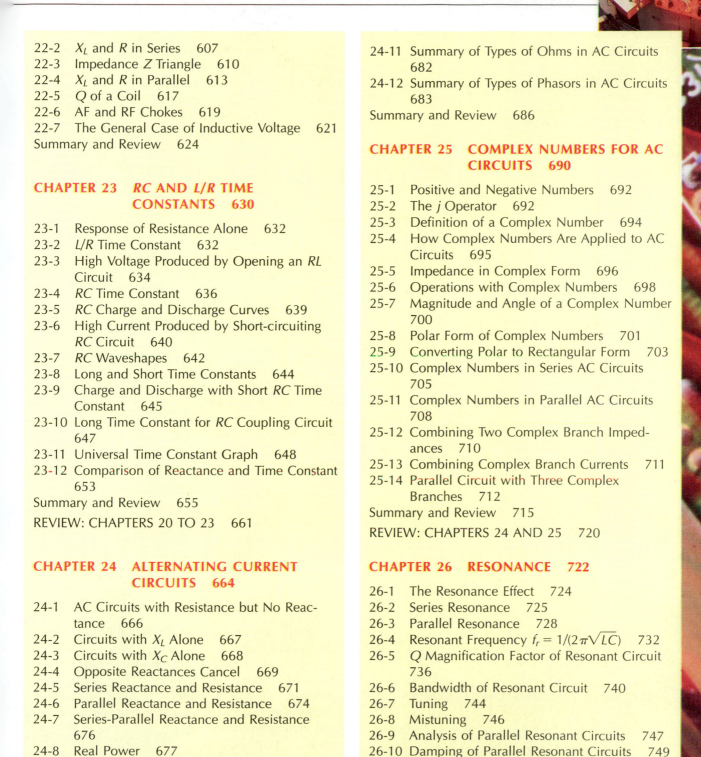

PREFACE

Basic Electronics, now in its eighth edition, is intended for students taking their first course in the fundamentals of electricity and electronics. The book is written for the beginning student, who is assumed to have no prior knowledge of the technical aspects of the subject. The prerequisites for using this book include an understanding of basic algebra and some trigonometry. In many schools, it will be possible to take a basic algebra-trigonometry course concurrently with the use of this book in a course covering the fundamental concepts of dc and ac theory.

The primary focus of this revision was a very careful review of its content and layout. Many additions, deletions, and reordering of topics have occurred as a direct result of an extensive survey sent to electronics instructors all across the country. For current users of the book, the additions and changes will be easy to identify throughout the book.

ORGANIZATION   The book begins with a chapter entitled "Survey of Electronics." This chapter provides a brief overview of the history of the development of electronics, describes a variety of career opportunities available in electronics, explains the most common components used in electronics, and identifies some of the most common types of equipment used by professionals in the electronics field. Following the "Survey of Electronics" chapter, the book provides complete and comprehensive coverage of the subjects which form the real fundamentals of basic electronics. Beginning with the atomic nature of electricity in Chapter 1, the topics progress through a study of resistors, Ohm's law, series and parallel circuits, series-parallel circuits, voltage and current dividers, dc meters, Kirchhoff's laws and network theorems, conductors and insulators, batteries, magnetism, magnetic units, electromagnetic induction, alternating voltage and current, capacitance, capacitive reactance, capacitive circuits, inductance, inductive reactance, inductive circuits, *RC* and *L/R* time constants, ac circuits, complex numbers, resonance, and filters.

Current users of the book will notice that the chapters on capacitance, capacitive reactance, and capacitive circuits now precede the chapters on inductance, inductive reactance, and inductive circuits. This change has been made as a result of the previously mentioned survey that was sent to electronics instructors. If they wish, instructors may choose to cover the chapters on inductance prior to those on capacitance.

In this edition, several changes have been made in Chapter 2, "Resistors." There is expanded coverage of both carbon and metal film resistors and the five-band resistor color code. Also new to this chapter is coverage of surface-mount resistors and zero-ohm resistors. These additions reflect the most state-of-the-art coverage available on resistors in any basic textbook covering dc and ac theory.

In Chapter 11, "Conductors and Insulators," more information on switches has been added. In Chapter 15, "Electromagnetic Induction," a new section on electromechanical relays has been added.

In Chapter 17, "Capacitance," new and updated material regarding the coding system used with a wide variety of types of capacitors has been added. Also new to this chapter is the coverage of surface-mount capacitors and the coding systems used with them. In Chapter 20, "Inductance," new material covers impedance matching and the many ratings associated with transformers. In Chapter 23, "*RC* and *L/R* Time Constants," new information on differentiation and integration is included. In Chapter 27, "Filters," a wealth of new information, on phase angles, calculating cutoff frequency and output voltage, decibels, and frequency response curves, now appears.

The last five chapters of the book provide a basic introduction to semiconductor theory, diodes, transistors, amplifiers, oscillators, modulation, rectifier circuits, circuit configurations, class of operation, troubleshooting, number systems, basic logic gates, Boolean algebra, flip-flops, counters, op amp characteristics, and op amp circuits. The coverage of op amp circuits has been expanded in response to the survey.

Following the text chapters are four appendixes: Appendix A, "Electrical Symbols and Abbreviations," Appendix B, "Solder and the Soldering Process," Appendix C, "Schematic Symbols," and Appendix D, "Using the Oscilloscope." The appendixes are followed by a glossary, answers to self-tests, answers to odd-numbered chapter problems and critical thinking problems, and an index.

CHAPTER LAYOUT   Each chapter begins with a brief introduction of the topic, a list of important terms, chapter objectives (new to this edition), and a list of the sections appearing within the chapter. Within each chapter, test-point questions are given at the end of each section. This provides the student with a quick means of checking his or her understanding of the material in that section. At the end of each chapter are the following items: summary, self-test, questions, problems, and critical thinking questions. Like the chapter objectives, the critical thinking questions are new to the eighth edition. It should be noted that new problems have been added to each chapter. The answers to the test-point questions appear at the end of each chapter.

Step-by-step solutions of typical problems dealing with a particular concept are generously provided in every chapter of the book. Where appropriate, typical calculator keystroke routines are provided as an additional aid to the student. The illustrative examples are highlighted so that students can access them more readily.

ANCILLARY PACKAGE   The following supplements are available to adopters of Grob *Basic Electronics:*

- *Problems in Grob Basic Electronics:* This book, written by Mitchel E. Schultz, provides students and instructors with a source of hundreds of practical problems for self-study, homework assignments, tests, and review. Each chapter contains a number of solved illustrative problems demonstrating, step-by-step, how representative problems on a particular topic are solved. Following the solved problems are sets of problems for the students to solve.

- *Experiments in Grob Basic Electronics:* This book, written by Frank Pugh and Wes Ponick, provides students and instructors with 67 easy-to-follow laboratory experiments. The experiments range from an introduction to laboratory equipment to an experiment on operational amplifiers. All experiments have been student-tested to ensure their effectiveness.

- *Mathematics for Grob Basic Electronics:* This book, written by Bernard Grob, provides students with the basic math skills needed to solve problems in the text, *Grob Basic Electronics.* Included are chapters on algebra, trigonometry, the basics of computer mathematics, and a new chapter on complex numbers for ac circuits.

- *Instructor's Productivity Center for Grob Basic Electronics, Eighth Edition:* This package includes a Windows-based test generator, a math tutorial, and a Power Point presentation for every chapter of the text. It also includes a graphics file of the circuits in the text. These files can be used for tests or presentations. The optional Group Instruction software, developed by HyperGraphics, can be accessed directly from the IPC. An optional Electronics Workbench file of the circuits in *Basic Electronics* is also available.

- *Instructor's Annotated Edition for Grob Basic Electronics, Eighth Edition:* This book includes teaching hints and scheduling suggestions for the instructor and career information for students. Much of this material has been given in the margin of this text for the instructor's ease of reference while teaching a class. Answers to test-point questions, which appear in the student's text, are also included in the margin of this version, produced especially for the instructor.

- *Instructor's Manual for Grob Basic Electronics, Eighth Edition:* This book provides the instructor with answers to all the questions and problems in the text and in its supplements, *Problems in Grob Basic Electronics, Experiments in Grob Basic Electronics,* and *Mathematics for Grob Basic Electronics.*

Bernard Grob

CREDITS AND ACKNOWLEDGMENTS

The author would like to thank those individuals who responded to the survey which was sent out long before this book was revised. Their comments and suggestions provided the information needed to make this the most up-to-date book available on electricity and electronics. The author would also like to thank the reviewers listed below who painstakingly examined every sentence, example, and problem for accuracy prior to the publication of the eighth edition.

In addition, the author would like to thank the highly professional staff at Glencoe in Columbus, Ohio—especially Brian Mackin for his patience, hard work, and understanding during the long period of the manuscript preparation. My thanks also go to Mitchel Schultz for his help on this project. Finally, it is a pleasure to thank my wife, Sylvia, for her help in preparing the manuscript.

Tim Beecher
Wisconsin Indianhead Technical College
Superior, WI

Jack Berger
ITT Technical Institute
Murray, UT

B. J. Tobias Boydell
Seva Electronics
Burford, Ontario, Canada

Patrick J. Chalmers
ITT Technical Institute
Matteson, IL

Michael Fairbanks
ITT Technical Institute
Nashville, TN

Richard L. Green
ITT Technical Institute
Maitland, FL

Rich Hassler
ITT Technical Institute
Youngstown, OH

William M. Hessmiller
QRS Corporation
Dunmore, PA

Barry Hoy
ITT Technical Institute
Norfolk, VA

Arnold Kroeger
Hillsborough Community College
Tampa, FL

James A. McQuoid
ITT Technical Institute
San Antonio, TX

Jim Myers
Wallace Community College
Selma, AL

Tony Richardson
ITT Technical Institute
Strongsville, OH

T. Randall Riggs
ITT Technical Institute
Tampa, FL

Gregg Richley
ITT Technical Institute
Youngstown, OH

Victor Rozeboom
Burlington, NC

John Ryan
ITT Technical Institute
Knoxville, TN

Dan Siddall
ITT Technical Institute
Boise, ID

Mike Siemion
Madison Area Technical College
Madison, WI

William H. Sims III
ITT Technical Institute
Jacksonville, FL

Pat Thomason
Patterson State Technical
College
Montgomery, AL

Bryon K. Van Beek
Courtesy Communications
Spokane, WA

SURVEY

SURVEY OF ELECTRONICS

The broad field of electronics encompasses radio, audio, television, computers, fiber optics, medical, industrial, automotive, and aerospace electronics. Although each of these areas represents a remarkable advance in technology, electronics can still be considered the practical applications of the general principles of electricity. For example, the same electric current produced by a battery to power a flashlight can also be used to power a calculator, a computer, a television, a videocassette recorder (VCR), or even an orbiting satellite. All applications of electronics are based on the fundamental laws of electricity and magnetism. An electric current will always produce magnetism. Furthermore, magnetism will always be produced by an electric current.

The study of the technology of electricity and electronics is, to say the least, very exciting and challenging. Whether your goal is to become an electronics technician or engineer, you will need a thorough understanding of the fundamental concepts presented in this book. Before we analyze the fundamentals of electricity, though, it can be interesting and helpful to look at the history of the development of electronics. We shall also take a look at some of the available career opportunities in electronics. And, finally, we will introduce some of the most common electronic components and test instruments used by electronics technicians and engineers.

CHAPTER OBJECTIVES

Upon completion of this chapter, you should be able to:

- *List* some milestones in the history of the development of electronics.
- *List* the names of famous scientists who contributed to the early development of electronics.
- *List* a variety of different career opportunities available in electronics.
- *List* the names of the most common electronic components and *describe* their main function.
- *List* the names of the most common pieces of electronic test equipment used by technicians and engineers and *describe* their main function.

IMPORTANT TERMS IN THIS CHAPTER

alternating current	DMM	power supply
capacitors	electronics	resistors
communications electronics	function generator	surface-mount technology
digital electronics	integrated circuit	transistors
direct current	oscilloscope	waveform

TOPICS COVERED IN THIS CHAPTER

S-1 A BRIEF HISTORY OF THE DEVELOPMENT OF ELECTRONICS

Modern electronics began with the pioneer days of radio communications. The development of radio, however, was preceded by earlier experiments in electricity and magnetism. The beginning of wireless transmission for radio communications arose from work done by Heinrich Hertz, a German physicist. In 1887 he demonstrated the effects of electromagnetic radiation through space. Even though the distance of transmission was only a few feet, Hertz's experiment proved that radio waves could travel from one place to another without the need for any connecting wires between the transmitting and receiving equipment. The experiment also proved that radio waves, although invisible, travel with the same velocity as light waves. In fact, radio waves and light waves are both examples of *electromagnetic radiation.* This form of energy combines the effects of electricity and magnetism, whereby an electromagnetic wave transmits electric energy through space.

The work of Hertz actually followed even earlier experiments. In 1820, a Danish physicist, Hans Christian Oersted, proved that an electric current produces magnetic effects. Then, in 1831, Michael Faraday, a British physicist, discovered that a magnet in motion can generate electricity. The motion meets the requirement of a change in the magnetic field. In 1864, another British physicist, James Clerk Maxwell, on the basis of earlier work in electricity and magnetism, predicted the electromagnetic waves demonstrated later by Hertz.

The importance of the work done by Hertz, Maxwell, Oersted, and Faraday can be judged by the fact that their names are used as the basic units of measurement in electricity. For example, the maxwell (Mx) and the oerstead (Oe) are basic units of magnetism. Similarly, the hertz (Hz) is the basic unit of frequency, equal to one complete cycle of alternating voltage or alternating current (ac) each second. The farad (F), which is named after Michael Faraday, is the basic unit of capacitance, which indicates how much electric charge is stored in a component known as a capacitor. Other famous scientists for which electrical units are named include André Ampère, Georg Simon Ohm, Joseph Henry, James Prescott Joule, Wilhelm Eduard Weber, James Watt, and Werner von Siemens, to name a few. The ampere is the basic unit of electric current, the ohm is the basic unit of electrical resistance, the joule is the basic unit of electric energy, the weber is a unit of magnetic flux, the henry is the basic unit of inductance, the watt is the basic unit of electric power, and the siemens is the basic unit of conductance.

In 1895, Guglielmo Marconi used a long wire as an antenna in developing a practical radio system for very long distances. (An antenna is needed to provide efficient radiation of the radio waves.) Marconi succeeded in producing wireless communications across the Atlantic Ocean in 1901. Electronic communication remained limited, however, until Lee DeForest invented the audion tube in 1906. Further development of vacuum tubes allowed the field of electronic communications to develop rapidly. In 1920, regularly scheduled radio programs were broadcast by station KDKA. The first radio broadcasts used what is called amplitude modulation (AM) to transmit its program information. The first frequency-modulated (FM) broadcast occurred in 1939, and the first FM stereo broadcast became available to listeners in 1961. Commercial television broadcasting began officially in 1941 but did not become popular until 1945. Our present color television system was adopted in 1953.

Perhaps one of the most important milestones in the history of electronics was the invention of the transistor in 1948 by John Bardeen, Walter Brattain, and William Schockley, all of whom worked for Bell Telephone Laboratories at the time. The transistor (also called a solid-state device or semiconductor) eventually replaced the vacuum tube in radios and televisions. The transistor is much more efficient than the vacuum tube, since it does not have a heater. Other solid-state devices include diodes and integrated circuits. Figure S-1 shows several different solid-state devices. The development of solid-state devices has led to the continued miniaturization of all types of electronic equipment. Each month we learn of new technological advances that produce increasingly sophisticated circuitry in ever smaller integrated circuits. The primary reasons for this trend are improved reliability, lower costs, and increased speed due to the reduction in the length of the interconnecting wires.

FIG. S-1 Semiconductor devices for electronic equipment (not shown to actual size). (*a*) Transistor amplifiers. (*b*) Diode rectifiers. (*c*) Integrated circuit (IC) packages.

S-2 RADIO AND TELEVISION BROADCASTING

The word *broadcast* means to send out in all directions. In a radio or television broadcasting system the transmitter sends out electromagnetic radio waves from its antenna. At the receiving end, an antenna picks up or intercepts the transmitted signal. The receiver reproduces the transmitted information. Because there are many radio signals traveling in space from different transmitters, a receiver must be able to select or *tune* to the frequency of the station desired. All other stations must be tuned out. Figure S-2 illustrates a basic radio broadcasting system. Notice that it consists of a transmitter, a receiver, and two antennas. The distance of transmission for a wireless communication system can be less than a mile or as much as 5000 miles, depending on the type of service. There are many different services, such as the radio broadcasting of voice and music, television broadcasting, amateur ("ham") radio, and citizen's band (CB) radio. Other examples of radio communication include police and aircraft radios, cordless and cellular telephones, and satellite communications. All radio communication services in the United States are regulated by the Federal Communications Commission (FCC). The FCC assigns carrier frequencies to each type of service and also regulates the lawful use of each assigned service.

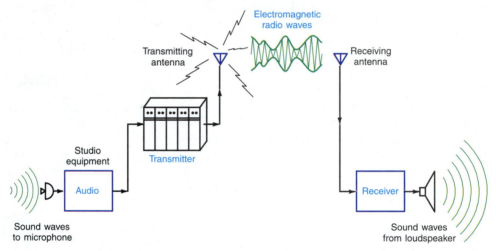

FIG. S-2   Radio broadcasting between transmitter and receiver. An amplitude-modulated (AM) carrier wave is shown here.

AM AND FM BROADCAST BANDS   The AM broadcast band, which ranges from 540 to 1600 kilohertz (kHz), was the first system to be used for public broadcasting of program information such as news, weather, and music. Each station is assigned a carrier frequency by the FCC and is allotted a bandwidth of 10 kHz.

The FM broadcast band extends from 88 to 108 megahertz (MHz), with each station occupying a 200-kHz bandwidth. The FM system has several advantages over the AM system. Most important is the fact that it is inherently less noisy and is also capable of broadcasting a broader range of audio frequencies for better fidelity.

TELEVISION BROADCASTING   *Television* (TV) means to see at a distance. Television is just another form of radio communication, but with video or picture information in addition to sound. Two separate carrier waves are used in the transmission of a television signal: one for the picture and one for the sound. The picture carrier signal is an AM signal, whereas the sound carrier signal is an FM signal. Each television station occupies a bandwidth of 6 MHz, which is 600 times more than an AM broadcast station and 30 times more than an FM broadcast station.

AMATEUR RADIO   Amateur radio is one of the largest noncommercial radio services in the world. Amateur operators, also called "hams," communicate with each other using Morse code, voice, radioteletype, slow-scan television, and digital communications. Public service is a large part of amateur radio. Hams often provide emergency communications in times of national disaster such as flooding, tornadoes, hurricanes, and earthquakes. Amateur radio operators have formed an organization known as the American Radio Relay League (ARRL). The ARRL was founded in 1914 and is headquartered in Newington, Connecticut.

CITIZEN'S BAND RADIO   Citizen's band (CB) radio is another noncommercial radio service which can be used by the general public. The CB band has forty 10-kHz channels extending from 26.965 MHz up to 27.405 MHz. The CB band allows a maximum power output of 4 watts (W). Unlike amateur radio, no license is required to operate a CB radio.

S-3 CAREER OPPORTUNITIES IN ELECTRONICS

Career opportunities in electronics continue to expand and become more varied with each passing year. It seems as if the career opportunities in electronics are almost limitless. The purpose of this section is to provide you, the newcomer to electronics, with some insight into the many possible career opportunities available to you.

CONSUMER ELECTRONICS TECHNICIAN   A consumer electronics technician is a person who repairs home entertainment equipment, such as stereos, televisions, VCRs, CD players, and camcorders. Electronic technicians who service these types of equipment can expect to encounter a wide variety of electronic circuits and mechanical devices. In this field, technicians often troubleshoot down to the component level rather than just the board level. A strong understanding of electronic theory and circuits is a must to be successful in this branch of electronics. With high-definition television (HDTV) and interactive television becoming more and more common, the opportunities awaiting the consumer electronics technician continue to look promising. One major advantage in choosing a career as a consumer electronics technician is that employment can be obtained just about anywhere in the United States.

COMPUTER REPAIR TECHNICIAN   With the use and applications of computers continuing to grow, the need for computer repair technicians is at an all-time high. To be a technician in this branch of electronics, you must have a thorough understanding of both the hardware and the software aspects of computer operation. The electronic, magnetic, and mechanical devices of a computer are referred to as the *hardware,* whereas the programs, which tell the computer what to do, are called the *software.*

    Because computer technology is advancing so rapidly, it is hard to keep up with the latest developments, which may change on almost a week-to-week

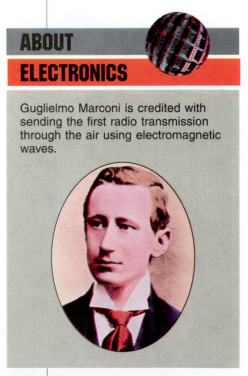

ABOUT ELECTRONICS

Guglielmo Marconi is credited with sending the first radio transmission through the air using electromagnetic waves.

basis. Most computer repair technicians attend training schools on a regular basis in order to keep up to date.

INDUSTRIAL ELECTRONICS TECHNICIAN

An industrial electronics technician is a person who maintains, troubleshoots, and repairs a variety of industrial equipment, such as motors and motor controls, programmable logic controllers (PLCs), robots, and computer-controlled machine tools and fluid power systems. The industrial electronics technician may also be involved in the development and installation of equipment. With manufacturing on the rise in the United States, the need for industrial electronic technicians will continue to grow.

MEDICAL ELECTRONICS TECHNICIAN

A medical electronics technician is a person who works closely with nurses and physicians and other hospital staff to advise about and help maintain the wide range of electronic equipment used in medicine. As an electronic technician in this field you can expect to maintain, troubleshoot, and repair a variety of different types of medical equipment, such as computerized axial tomography (CAT) and magnetic resonance imaging (MRI) systems, surgical lasers, heart-lung bypass machines, dialysis machines, and electronic thermometers. Medical equipment manufacturers often provide training in the maintenance and repair of their products. As a technician in this branch of electronics you can expect to be under some stress. On the plus side, however, medical electronic technicians often make good to excellent wages and benefits, and the job is usually considered quite prestigious.

COMMUNICATION ELECTRONICS TECHNICIAN

A communications electronics technician is a person who is trained to install, maintain, troubleshoot, and repair a variety of electronic communication equipment, such as FM communications transceivers, cellular phones, pagers, repeater systems, antenna systems, microwave systems, and AM and FM broadcasting equipment. Employment opportunities are exploding all across the country as a result of the rapid growth in this area. Wages and benefits are typically very good. The physical demands of this kind of work may range from climbing a 200-ft tower to install antennas, to sitting in an air-conditioned room troubleshooting equipment.

CABLE TELEVISION TECHNICIAN

A cable television technician is a person who is trained to install, test, maintain, troubleshoot, and repair cable television systems. In this branch of electronics an in-depth knowledge of electronic circuitry is helpful but not a necessity. This type of work is more likely to entail installing connectors on coaxial cable, testing coaxial lines for faults, and connecting and disconnecting customers to the cable system. Although this type of work often provides good wages, the physical demands of the job often make it less appealing than other career opportunities available in electronics.

ABOUT ELECTRONICS

The unit of measure for capacitance, the farad, was named for Michael Faraday, an English chemist and physicist who discovered the principle of induction (1 F is the unit of capacitance that will store 1 C of charge when 1 V is applied).

OFFICE EQUIPMENT TECHNICIAN   An office equipment technician is a person who maintains and repairs office equipment such as xerographic copiers, facsimile (FAX) machines, and laser printers. In this type of work, customer relation skills are extremely important, since the technician almost always travels to the customer's location. In many cases the job is more mechanical than it is electrical, since the moving parts within the equipment eventually wear out. Wages in this area are moderate to good, and jobs are plentiful.

S-4 A PREVIEW OF THE MOST COMMON ELECTRONIC COMPONENTS

Although there are an almost unlimited number of applications for electronics, all applications use basically the same types of electronic components. This section introduces the most common types found in all applications of electronics.

Electronic components are categorized as being either *passive* or *active*. Resistors, capacitors, and inductors are examples of passive components, whereas diodes, transistors, and vacuum tubes are examples of active components. Active components are capable of rectifying, amplifying, or changing energy from one form to another. Passive components, on the other hand, can control energy, but they cannot amplify or modify it. Both passive and active components generally are used together to form complete circuits in all types of electronic equipment. For the passive components, it may be surprising to learn that all electronic circuits use only three basic types: resistors, capacitors, and inductors.

RESISTORS   Resistors are perhaps the most common component in electronic circuits. Their main function is to limit current flow or reduce the voltage in a circuit. The resistance may be either fixed or variable. Figure S-3 shows several different types of both fixed and variable resistors. Some fixed resistors are color coded to indicate their resistance value, while others have their resistance values printed right on the body. The basic unit of resistance is the ohm (Ω). Variable resistors usually have their maximum resistance stamped on them. Table S-1

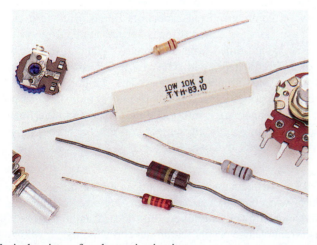

FIG. S-3   Typical resistors for electronic circuits.

TABLE S-1 SCHEMATIC SYMBOLS FOR FIXED AND VARIABLE RESISTORS

TYPE	SCHEMATIC SYMBOL	NOTES
Fixed	*R* ⏦⏦⏦	Limits current and reduces voltage
Variable	*R*	Has three external connections Varies voltage
		Has two external connections Varies current

shows the schematic symbols for both fixed and variable resistors. All fixed resistors use the same symbol, but the symbol for a variable resistor may vary depending on how it is used in a circuit. Resistors are covered in Chap. 2.

CAPACITORS   A capacitor is a component that is able to hold or store an electric charge. Its physical construction consists of two metal plates separated by an insulator. In general, capacitors are used to block direct current (dc) but pass alternating current (ac). The basic unit of capacitance is the farad (F).

Like resistors, capacitors may be either fixed in value or variable. Figure S-4 shows several commonly used capacitors found in modern electronic circuitry. Some capacitors are coded using a special numbering system, whereas others have their value printed right on them. As you can see, capacitors come in a variety of different shapes and sizes. Table S-2 shows the schematic symbols for both fixed and variable capacitors. The symbol for an electrolytic capacitor (the one with polarity markings on its plates) indicates that polarity must be observed when connecting it into a circuit. Capacitors other than electrolytics can be connected into a circuit without concern for proper polarity. Capacitors are covered in detail in Chap. 17.

INDUCTORS   An inductor is a component whose physical construction is simply a coil of wire. For this reason inductors are often called *coils*. The function of an inductor is to provide opposition to a changing or varying current. Whenever a current change occurs within the coil windings, a voltage is induced across the ends of the coil. The polarity of the induced voltage is such that it

DID YOU KNOW?

A deadly ant species originating in South America was found in the United States in 1995. Attracted to buried cables, these "fire ants" can chew through electrical lines and stop telephone service, cut electrical power, and start fires.

FIG. S-4 Typical small capacitors for electronic circuits. Length is ¼ to 1 in. without leads.

opposes the current change occurring within the coil. In more general terms, inductors are the opposite of capacitors in that they are able to pass a dc current and block an ac current. The basic unit of inductance is the henry (H).

Like resistors and capacitors, inductors may be either fixed or variable. Several different types are shown in Fig. S-5. When inductors are color coded, they can be distinguished from resistors by the fact that the first color band is always a wide silver band. The variable inductors frequently used in modern electronic equipment are adjusted by tuning a magnetic slug in and out of its core. The schematic symbols for both fixed and variable inductors are shown in Table S-3.

TABLE S-2   SCHEMATIC SYMBOLS FOR FIXED AND VARIABLE CAPACITORS

TYPE	SCHEMATIC SYMBOL	NOTES
Fixed	*C*	Stores electric charge Blocks dc, passes ac
Electrolytic	*C*	Has large capacitance Must be connected with proper polarity Fixed values only
Variable	*C*	Used for tuning radios, TVs, and other electronic communications equipment

FIG. S-5   Typical inductors or coils for electronic circuits.

A transformer is another component that can technically be defined as a inductor, because it is constructed by placing two coils in close proximity to each other. One of the main applications of a transformer is to step up or step down an ac voltage. Transformers cannot step up or step down a dc voltage. Figure S-6 shows several modern transformers, whereas Table S-4 shows a typical transformer schematic symbol. Inductors and transformers are covered in Chap. 20.

DIODES   A diode is an electronic component which allows current to flow through it in one direction but not the other. A diode's main function is to change an ac voltage into a dc voltage. Figure S-7 shows a variety of different diodes found in modern electronic equipment. The end that is closest to the gray band

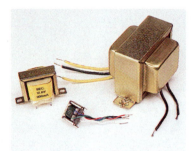

FIG. S-6   Typical modern transformers.

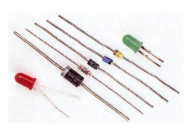

FIG. S-7   Typical diodes.

TABLE S-3	SCHEMATIC SYMBOLS FOR FIXED AND VARIABLE INDUCTORS	
TYPE	**SCHEMATIC SYMBOL**	**NOTES**
Air	L	Induces voltage when current changes Passes dc, limits ac
Iron core	L	Used for low frequencies
Variable	L	Used for tuning radios, TVs, and other electronic communications equipment

TABLE S-4 SCHEMATIC SYMBOL FOR TRANSFORMER

TYPE	SCHEMATIC SYMBOL	NOTES
Iron-core transformer	*T* Primary Secondary	Steps up or steps down ac voltages

identifies the lead known as the *cathode*. The other lead is called the *anode*. Although there are a wide variety of different diode types, Table S-5 shows the schematic symbols seen most frequently.

TABLE S-5 SCHEMATIC SYMBOLS FOR DIODES

TYPE	SCHEMATIC SYMBOL	NOTES
Semiconductor diode	Anode Cathode *D*	Passes current in only one direction
Zener diode	Anode Cathode	Regulates voltage

TRANSISTORS   A transistor is an electronic component that can be used to amplify small ac signals or switch a dc voltage. A wide variety of different types are available, so it is not practical to explain them all. However, Fig. S-8 shows some typical transistors, while Table S-6 shows the most common schematic symbols.

SURFACE-MOUNTED DEVICES   Surface-mounted devices (SMDs), also called *chip components,* have been in use for many years now, primarily where portability and compactness are needed or desired. All electronic components are available as SMDs. The widespread use of the SMD in portable electronic equipment such as camcorders, pagers, and hand-held communications transceivers has significantly reduced the cost of SMDs. Additional cost savings result from the fact that they require much less space than conventional components, and also that no holes need to be drilled to mount them. SMDs are soldered directly to the copper traces of a printed circuit board. SMDs do more than just save money and reduce the size of electronic equipment, however. They also improve a circuit's performance. This is due to the fact that with SMDs the components are all much closer together, which means shorter, more direct electric

FIG. S-8   Typical transistors.

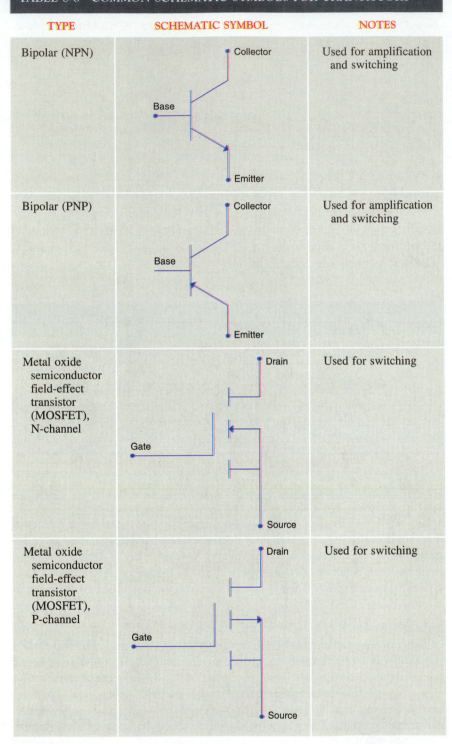

TABLE S-6   COMMON SCHEMATIC SYMBOLS FOR TRANSISTORS

TYPE	SCHEMATIC SYMBOL	NOTES
Bipolar (NPN)	Collector / Base / Emitter	Used for amplification and switching
Bipolar (PNP)	Collector / Base / Emitter	Used for amplification and switching
Metal oxide semiconductor field-effect transistor (MOSFET), N-channel	Drain / Gate / Source	Used for switching
Metal oxide semiconductor field-effect transistor (MOSFET), P-channel	Drain / Gate / Source	Used for switching

connections, which in turn reduce the circuit's susceptibility to noise pickup. Figure S-9 shows a variety of different SMDs. In this book you will be introduced to passive surface-mount devices.

S-5 TOOLS OF THE TRADE

As an electronic technician or engineer, you can expect to encounter a wide variety of equipment when maintaining or troubleshooting electronic equipment. However, nearly all technicians use a digital multimeter (DMM), oscilloscope, power supply, and function generator. This section provides some basic information about the functions of each of these pieces of test equipment.

DIGITAL MULTIMETER   A DMM is used to measure the voltage, current, or resistance in an electronic circuit. Most DMMs also have provisions for testing capacitors, diodes, and transistors. The meter may be either a hand-held or bench-top unit. Both types are shown in Fig. S-10. All digital meters have numerical readouts which indicate directly the value of voltage, current, or resistance being measured. Years ago, analog meters were used to make voltage, current, and resistance measurements. An analog meter uses a moving pointer and a printed scale. DMMs have replaced the older analog meters in almost all applications.

OSCILLOSCOPE   The oscilloscope is a technician's most versatile piece of test equipment. Its basic function is to view and measure ac waveforms. Most oscilloscopes have dual-trace capabilities, which means they can display two waveforms at the same time. Figure S-11 shows a modern dual-trace oscilloscope.

POWER SUPPLY   A power supply is a unit capable of supplying dc voltage and current to electronic circuits under test. Modern power supplies have regulated outputs. This means that their output voltage does not fluctuate as the load current varies. In most power supplies, a red jack is used for the positive terminal and a black jack is used for the negative terminal.

FUNCTION GENERATOR   A function generator is a piece of test equipment capable of producing a number of different output waveforms. Function generators are used when designing, developing, and troubleshooting electronic equipment. All function generators have controls to adjust the amplitude, frequency, and shape of the output waveform.

(a)

(b)

FIG. S-10   Typical multimeters. (a) Hand-held. (b) Bench-top.

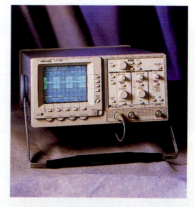

FIG. S-11   Dual-trace oscilloscope.

CHAPTER 1

ELECTRICITY

We see applications of electricity all around us, especially in electronics, but they are derived from more fundamental effects. Basically, electricity is an invisible force that can produce heat, light, and motion, as examples. The force for motion is an attraction or repulsion between electric charges. Specifically, electricity can be explained in terms of electric charge, current, and voltage.

Electric charge is the basic form for a quantity of electricity. The charge has an invisible field of force, extending outward in all directions, that can do the work of moving another charge.

Charge in motion provides an electric current. The practical unit of current is the ampere.

Two charges that are different in polarity provide a difference of potential with a net field of force that can move charges to produce current. This potential difference is indicated by the volt unit.

Electricity is often described as voltage or current but both are usually necessary and they are very different from each other. The closed path for the movement of charges is called a circuit. The circuit usually has resistance also, as a component to limit the amount of current. In short, the practical applications of electricity require the analysis of charge, current, and voltage with the components that make up the circuit.

CHAPTER OBJECTIVES

Upon completion of this chapter, you should be able to:

- *List* the two basic particles of electric charge.
- *Describe* the basic structure of the atom.
- *Define* the terms *conductor, insulator,* and *semiconductor* and give examples of each.
- *Define* the coulomb unit of electric charge.
- *Define* potential difference and voltage and list the unit of each.
- *Define* current and list its unit of measure.
- *Describe* the difference between voltage and current.
- *Define* resistance and conductance and list the unit of each.
- *List* three important characteristics of an electric circuit.
- *Define* the difference between electron flow and conventional current.
- *Describe* the difference between direct and alternating current.

IMPORTANT TERMS IN THIS CHAPTER

alternating current	electric field	open circuit
ampere unit	electron flow	polarity
circuit	hole charge	potential difference
conventional current	ion charge	proton
coulomb unit	insulator	short circuit
conductance	magnetic field	siemens unit
conductor	mho unit	static electricity
direct current	ohm unit	volt unit

TOPICS COVERED IN THIS CHAPTER

1-1 NEGATIVE AND POSITIVE POLARITIES

We see the effects of electricity in a battery, static charge, lightning, radio, television, and many other applications. What do they all have in common that is electrical in nature? The answer is basic particles of electric charge with opposite polarities. All the materials we know, including solids, liquids, and gases, contain two basic particles of electric charge: the *electron* and the *proton*. An electron is the smallest amount of electric charge having the characteristic called *negative polarity*. The proton is a basic particle with *positive polarity*.

Actually, the negative and positive polarities indicate two opposite characteristics that seem to be fundamental in all physical applications. Just as magnets have north and south poles, electric charges have the opposite polarities labeled negative and positive. The opposing characteristics provide a method of balancing one against the other to explain different physical effects.

It is the arrangement of electrons and protons as basic particles of electricity that determines the electrical characteristics of all substances. As an example, this paper has electrons and protons in it. There is no evidence of electricity, though, because the number of electrons equals the number of protons. In that case the opposite electrical forces cancel, making the paper electrically neutral. The neutral condition means that opposing forces are exactly balanced, without any net effect either way.

When we want to use the electrical forces associated with the negative and positive charges in all matter, work must be done to separate the electrons and protons. Changing the balance of forces produces evidence of electricity. A battery, for instance, can do electrical work because its chemical energy separates electric charges to produce an excess of electrons at its negative terminal and an excess of protons at its positive terminal. With separate and opposite charges at the two terminals, electric energy can be supplied to a circuit connected to the battery. Figure 1-1 shows a battery with its negative ($-$) and positive ($+$) terminals marked to emphasize the two opposite polarities.

FIG. 1-1 Positive and negative polarities for the voltage output of a typical battery.

TEST-POINT QUESTION 1-1

Answers at end of chapter.
a. Is the charge of an electron positive or negative?
b. Is the charge of a proton positive or negative?
c. Is it true or false that the neutral condition means equal positive and negative charges?

1-2 ELECTRONS AND PROTONS IN THE ATOM

Although there are any number of possible methods by which electrons and protons might be grouped, they assemble in specific atomic combinations for a stable arrangement. (An atom is the smallest particle of the basic elements which forms the physical substances we know as solids, liquids, and gases.) Each stable combination of electrons and protons makes one particular type of atom. For example, Fig. 1-2 illustrates the electron and proton structure of one atom of the gas hydrogen. This atom consists of a central mass called the *nucleus* and 1 electron outside. The proton in the nucleus makes it the massive and stable part of the atom because a proton is 1840 times heavier than an electron.

In Fig. 1-2, the 1 electron in the hydrogen atom is shown in an orbital ring around the nucleus. In order to account for the electrical stability of the atom, we can consider the electron as spinning around the nucleus, as planets revolve around the sun. Then the electrical force attracting the electrons in toward the nucleus is balanced by the mechanical force outward on the rotating electron. As a result, the electron stays in its orbit around the nucleus.

In an atom that has more electrons and protons than hydrogen, all the protons are in the nucleus, while all the electrons are in one or more outside rings. For example, the carbon atom illustrated in Fig. 1-3a has 6 protons in the nucleus and 6 electrons in two outside rings. The total number of electrons in the outside rings must equal the number of protons in the nucleus in a neutral atom.

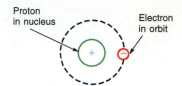

FIG. 1-2 Electron and proton in hydrogen (H) atom.

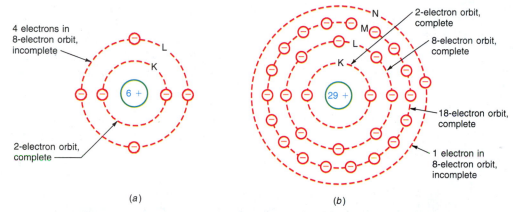

(a)

(b)

FIG. 1-3 Atomic structure showing the nucleus and its orbital rings of electrons. (*a*) Carbon (C) atom has 6 orbital electrons to balance 6 protons in nucleus. (*b*) Copper (Cu) atom with 29 protons in nucleus and 29 orbital electrons.

The distribution of electrons in the orbital rings determines the atom's electrical stability. Especially important is the number of electrons in the ring farthest from the nucleus. This outermost ring requires 8 electrons for stability, except when there is only one ring, which has a maximum of 2 electrons.

In the carbon atom in Fig. 1-3a, with 6 electrons, there are just 2 electrons in the first ring because 2 is its maximum number. The remaining 4 electrons are in the second ring, which can have a maximum of 8 electrons.

As another example, the copper atom in Fig. 1-3b has only 1 electron in the last ring, which can include 8 electrons. Therefore, the outside ring of the copper atom is less stable than the outside ring of the carbon atom.

When there are many atoms close together in a copper wire, the outermost orbital electrons are not sure which atoms they belong to. They can migrate easily from one atom to another at random. Such electrons that can move freely from one atom to the next are often called *free electrons*. This freedom accounts for the ability of copper to conduct electricity very easily. It is the movement of free electrons that provides electric current in a metal conductor.

The net effect in the wire itself without any applied voltage, however, is zero because of the random motion of the free electrons. When voltage is applied, it forces all the free electrons to move in the same direction to produce electron flow, which is an electric current.

CONDUCTORS, INSULATORS, AND SEMICONDUCTORS When electrons can move easily from atom to atom in a material, it is a *conductor.* In general, all the metals are good conductors, with silver the best and copper second. Their atomic structure allows free movement of the outermost orbital electrons. Copper wire is generally used for practical conductors because it costs much less than silver. The purpose of using conductors is to allow electric current to flow with minimum opposition.

The wire conductor is used only as a means of delivering current produced by the voltage source to a device that needs the current in order to function. As an example, a bulb lights only when current is made to flow through the filament.

A material with atoms in which the electrons tend to stay in their own orbits is an *insulator* because it cannot conduct electricity very easily. However, the insulators are able to hold or store electricity better than the conductors. An insulating material, such as glass, plastic, rubber, paper, air, or mica, is also called a *dielectric,* meaning it can store electric charge.

Insulators can be useful when it is necessary to prevent current flow. In addition, for applications requiring the storage of electric charge, as in capacitors, a dielectric material must be used because a good conductor cannot store any charge.

Carbon can be considered a semiconductor, conducting less than the metal conductors but more than the insulators. In the same group are germanium and silicon, which are commonly used for transistors and other semiconductor components. Practically all transistors are made of silicon.

ELEMENTS The combinations of electrons and protons forming stable atomic structures result in different kinds of elementary substances having specific characteristics. A few familiar examples are the elements hydrogen, oxygen, carbon, copper, and iron. An *element* is defined as a substance that cannot be decomposed any further by chemical action. The atom is the smallest particle of an element that still has the same characteristics as the element. *Atom* itself is a Greek word meaning a particle too small to be subdivided. As an example of the fact that atoms are too small to be visible, a particle of carbon the size of a pinpoint contains many billions of atoms. The electrons and protons within the atom are even smaller.

Table 1-1 lists some more examples of elements. These are just a few out of a total of 112. Notice how the elements are grouped. The metals listed across the top row are all good conductors of electricity. Each has an atomic structure with an unstable outside ring that allows many free electrons.

The semiconductors have 4 electrons in the outermost ring. This means they neither gain nor lose electrons but share them with similar atoms. The reason is that 4 is exactly halfway to the stable condition of 8 electrons in the outside ring.

The inert gas neon has a complete outside ring of 8 electrons, which makes it chemically inactive. Remember that 8 electrons in the outside ring is a stable structure.

MOLECULES AND COMPOUNDS A group of two or more atoms forms a molecule. For instance, two atoms of hydrogen (H) form a hydrogen molecule (H_2). When hydrogen unites chemically with oxygen, the result is water (H_2O),

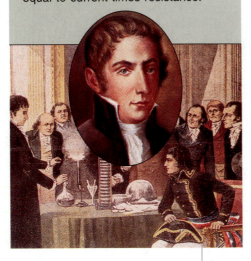

TABLE 1-1 EXAMPLES OF THE CHEMICAL ELEMENTS

GROUP	ELEMENT	SYMBOL	ATOMIC NUMBER	ELECTRON VALENCE
Metal conductors, in order of conductance	Silver	Ag	47	+1
	Copper	Cu	29	+1*
	Gold	Au	79	+1*
	Aluminum	Al	13	+3
	Iron	Fe	26	+2*
Semiconductors	Carbon	C	6	±4
	Silicon	Si	14	±4
	Germanium	Ge	32	±4
Active gases	Hydrogen	H	1	±1
	Oxygen	O	8	−2
Inert gases	Helium	He	2	0
	Neon	Ne	10	0

*Some metals have more than one valence number in forming chemical compounds. Examples are cuprous or cupric copper, ferrous or ferric iron, and aurous or auric gold.

which is a compound. A compound, then, consists of two or more elements. The molecule is the smallest unit of a compound with the same chemical characteristics. We can have molecules for either elements or compounds. However, atoms exist only for the elements.

TEST-POINT QUESTION 1-2

Answers at end of chapter.
a. Which have more free electrons: metals or insulators?
b. Which is the best conductor: silver, carbon, or iron?
c. Which is a semiconductor: copper, silicon, or neon?

1-3 STRUCTURE OF THE ATOM

Although nobody has ever seen an atom, its hypothetical structure fits experimental evidence that has been measured very exactly. The size and electric charge of the invisible particles in the atom are indicated by how much they are deflected by known forces. Our present planetary model of the atom was proposed by Niels Bohr in 1913. His contribution was joining the new ideas of a nuclear atom developed by Lord Rutherford (1871–1937) with the quantum theory of radiation developed by Max Planck (1858–1947) and Albert Einstein (1879–1955).

As illustrated in Figs. 1-2 and 1-3, the nucleus contains protons for all the positive charge in the atom. The number of protons in the nucleus is equal to the number of planetary electrons. Then the positive and negative charges are balanced, as the proton and electron have equal and opposite charges. The orbits for the planetary electrons are also called *shells* or *energy levels*.

ATOMIC NUMBER   This gives the number of protons or electrons required in the atom for each element. For the hydrogen atom in Fig. 1-2, the atomic number is 1, which means the nucleus has 1 proton balanced by 1 orbital electron. Similarly, the carbon atom in Fig. 1-3 with atomic number 6 has 6 protons in the nucleus and 6 orbital electrons. Also, the copper atom has 29 protons and 29 electrons because its atomic number is 29. The atomic number is listed for each of the elements in Table 1-1 to indicate the atomic structure.

ORBITAL RINGS   The planetary electrons are in successive shells called K, L, M, N, O, P, and Q at increasing distances outward from the nucleus. Each shell has a maximum number of electrons for stability. As indicated in Table 1-2, these stable shells correspond to the inert gases, like helium and neon.

TABLE 1-2	SHELLS OF ORBITAL ELECTRONS IN THE ATOM	
SHELL	**MAXIMUM ELECTRONS**	**INERT GAS**
K	2	Helium
L	8	Neon
M	8 (up to calcium) or 18	Argon
N	8, 18, or 32	Krypton
O	8 or 18	Xenon
P	8 or 18	Radon
Q	8	—

   The K shell, closest to the nucleus, is stable with 2 electrons, corresponding to the atomic structure for the inert gas helium. Once the stable number of electrons has filled a shell, it cannot take any more electrons. The atomic structure with all its shells filled to the maximum number for stability corresponds to an inert gas.

   Elements with a higher atomic number have more planetary electrons. These are in successive shells, tending to form the structure of the next inert gas in the periodic table. (The periodic table is a very useful grouping of all elements according to their chemical properties.) After the K shell has been filled with 2 electrons, the L shell can take up to 8 electrons. Ten electrons filling the K and L shells is the atomic structure for the inert gas neon.

   The maximum number of electrons in the remaining shells can be 8, 18, or 32 for different elements, depending on their place in the periodic table. The maximum for an outermost shell, though, is always 8.

   To illustrate these rules, we can use the copper atom in Fig. 1-3*b* as an example. There are 29 protons in the nucleus balanced by 29 planetary electrons. This number of electrons fills the K shell with 2 electrons, corresponding to the helium atom, and the L shell with 8 electrons. The 10 electrons in these two shells correspond to the neon atom, which has an atomic number of 10. The remaining 19 electrons for the copper atom then fill the M shell with 18 electrons and 1 electron in the outermost N shell. These values can be summarized as follows:

$$K \text{ shell} = 2 \text{ electrons}$$
$$L \text{ shell} = 8 \text{ electrons}$$
$$M \text{ shell} = 18 \text{ electrons}$$
$$N \text{ shell} = 1 \text{ electron}$$
$$\text{Total} = 29 \text{ electrons}$$

For most elements, we can use the rule that the maximum number of electrons in a filled inner shell equals $2n^2$, where n is the shell number in sequential order outward from the nucleus. Then the maximum number of electrons in the first shell is $2 \times 1 = 2$; for the second shell $2 \times 2^2 = 8$, for the third shell $2 \times 3^2 = 18$, and for the fourth shell $2 \times 4^2 = 32$. These values apply only to an inner shell that is filled with its maximum number of electrons.

ELECTRON VALENCE This value is the number of electrons in an incomplete outermost shell. A completed outer shell has a valence of zero. Copper, for instance, has a valence of 1, as there is 1 electron in the last shell, after the inner shells have been completed with their stable number. Similarly, hydrogen has a valence of 1, and carbon has a valence of 4. The number of outer electrons is considered positive valence, as these electrons are in addition to the stable shells.

Except for H and He, the goal of valence is 8 for all the atoms, as each tends to form the stable structure of 8 electrons in the outside ring. For this reason, valence can also be considered as the number of electrons in the outside ring needed to make 8. This value is the negative valence. As examples, the valence of copper can be considered +1 or −7; carbon has the valence of ±4. The inert gases have 0 valence, as they all have a complete outer shell.

The valence indicates how easily the atom can gain or lose electrons. For instance, atoms with a valence of +1 can lose this 1 outside electron, especially to atoms with a valence of +7 or −1, which need 1 electron to complete the outside shell with 8 electrons.

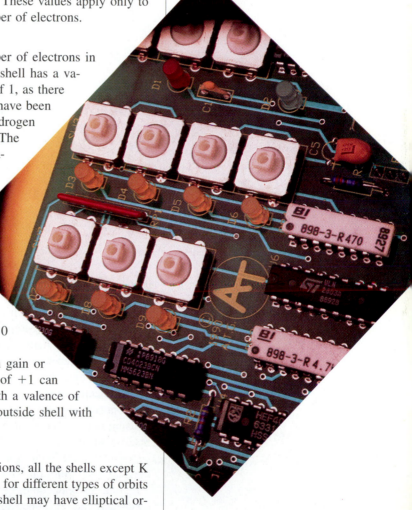

SUBSHELLS Although not shown in the illustrations, all the shells except K are divided into subshells. This subdivision accounts for different types of orbits in the same shell. For instance, electrons in one subshell may have elliptical orbits, while other electrons in the same main shell have circular orbits. The subshells indicate magnetic properties of the atom.

PARTICLES IN THE NUCLEUS A stable nucleus, meaning it is not radioactive, contains protons and neutrons. The neutron is electrically neutral without any net charge. Its mass is almost the same as a proton.

A proton has the positive charge of a hydrogen nucleus. The charge is the same amount as that of an orbital electron but of opposite polarity. There are no electrons in the nucleus. Table 1-3 lists the charge and mass for these three basic particles in all atoms. The C in the charge column is for coulombs.

TABLE 1-3 STABLE PARTICLES IN THE ATOM		
PARTICLE	CHARGE	MASS
Electron, in orbital shells	0.16×10^{-18} C, negative	9.108×10^{-28} g
Proton, in nucleus	0.16×10^{-18} C, positive	1.672×10^{-24} g
Neutron, in nucleus	None	1.675×10^{-24} g

TEST-POINT QUESTION 1-3

Answers at end of chapter.

a. An element with 14 protons and 14 electrons has what atomic number?

b. What is the electron valence of an element with atomic number 3?

c. The atomic number 32 indicates how many planetary electrons?

1-4 THE COULOMB UNIT OF ELECTRIC CHARGE

If you rub a hard rubber pen or comb on a sheet of paper, the rubber will attract a corner of the paper if it is free to move easily. The paper and rubber then give evidence of a static electric charge. The work of rubbing resulted in separating electrons and protons to produce a charge of excess electrons on the surface of the rubber and a charge of excess protons on the paper.

Because paper and rubber are dielectric materials, they hold their extra electrons or protons. As a result, the paper and rubber are no longer neutral, but each has an electric charge. The resultant electric charges provide the force of attraction between the rubber and the paper. This mechanical force of attraction or repulsion between charges is the fundamental method by which electricity makes itself evident.

Any charge is an example of *static electricity* because the electrons or protons are not in motion. There are many examples. When you walk across a wool rug, your body becomes charged with an excess of electrons. Similarly, silk, fur, and glass can be rubbed to produce a static charge. This effect is more evident in dry weather, because a moist dielectric does not hold its charge so well. Also, plastic materials can be charged easily, which is why thin, lightweight plastics seem to stick to everything.

The charge of many billions of electrons or protons is necessary for common applications of electricity. Therefore, it is convenient to define a practical unit called the *coulomb* (C) as equal to the charge of 6.25×10^{18} electrons or protons stored in a dielectric (see Fig. 1-4). The analysis of static charges and their forces is called *electrostatics*.

The symbol for electric charge is Q or q, standing for quantity. For instance, a charge of 6.25×10^{18} electrons* is stated as $Q = 1$ C. This unit is named after Charles A. Coulomb (1736–1806), a French physicist, who measured the force between charges.

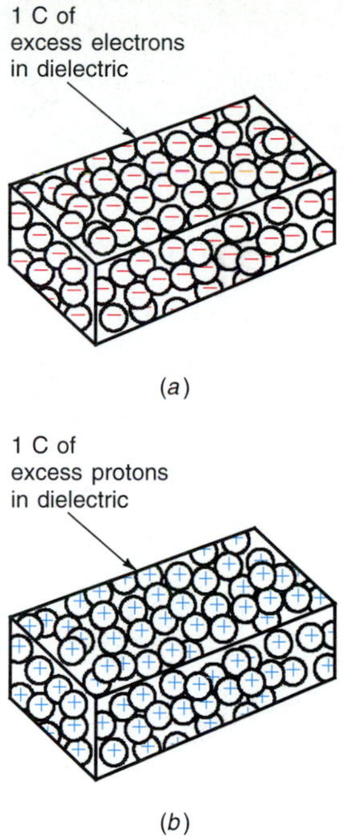

NEGATIVE AND POSITIVE POLARITIES   Historically, the negative polarity has been assigned to the static charge produced on rubber, amber, and resinous materials in general. Positive polarity refers to the static charge produced on glass and other vitreous materials. On this basis, the electrons in all atoms are basic particles of negative charge because their polarity is the same as the charge on rubber. Protons have positive charge because the polarity is the same as the charge on glass.

CHARGES OF OPPOSITE POLARITY ATTRACT   If two small charged bodies of light weight are mounted so that they are free to move easily and are placed close to each other, one can be attracted to the other when the two charges have opposite polarity (Fig. 1-5a). In terms of electrons and protons, they tend to be attracted to each other by the force of attraction between opposite charges. Furthermore, the weight of an electron is only about $\frac{1}{1840}$ the weight of a proton. As a result, the force of attraction tends to make electrons move to protons.

CHARGES OF THE SAME POLARITY REPEL   In Fig. 1-5b and c, it is shown that when the two bodies have an equal amount of charge with the same polarity, they repel each other. The two negative charges repel in Fig. 1-5b, while two positive charges of the same value repel each other in Fig. 1-5c.

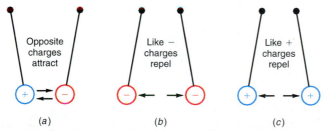

(a)   (b)   (c)

FIG. 1-5   Physical force between electric charges. (*a*) Opposite charges attract. (*b*) Two negative charges repel each other. (*c*) Two positive charges repel.

POLARITY OF A CHARGE   An electric charge must have either negative or positive polarity, labeled $-Q$ or $+Q$, with an excess of either electrons or protons. A neutral condition is considered zero charge. On this basis, consider the

---

*For an explanation of how to use powers of 10, see B. Grob, *Mathematics for Basic Electronics*, Glencoe/McGraw-Hill, Columbus, Ohio.

1 C of excess electrons in dielectric

(a)

1 C of excess protons in dielectric

(b)

FIG. 1-4   The coulomb (C) unit of electric charge. (*a*) Quantity of 6.25×10^{18} excess electrons for a negative charge of 1 C. (*b*) Same amount of protons for a positive charge of 1 C, caused by removing electrons from neutral atoms.

following examples, remembering that the electron is the basic particle of charge and the proton has exactly the same amount, although of opposite polarity.

Example

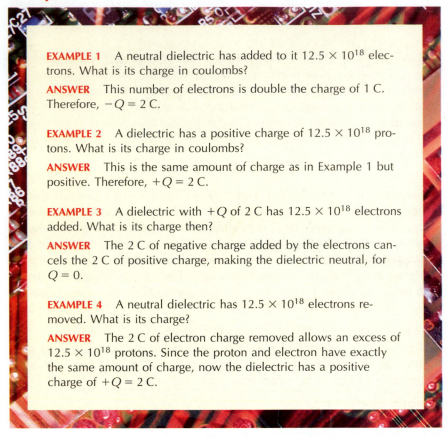

EXAMPLE 1   A neutral dielectric has added to it 12.5×10^{18} electrons. What is its charge in coulombs?

ANSWER   This number of electrons is double the charge of 1 C. Therefore, $-Q = 2$ C.

EXAMPLE 2   A dielectric has a positive charge of 12.5×10^{18} protons. What is its charge in coulombs?

ANSWER   This is the same amount of charge as in Example 1 but positive. Therefore, $+Q = 2$ C.

EXAMPLE 3   A dielectric with $+Q$ of 2 C has 12.5×10^{18} electrons added. What is its charge then?

ANSWER   The 2 C of negative charge added by the electrons cancels the 2 C of positive charge, making the dielectric neutral, for $Q = 0$.

EXAMPLE 4   A neutral dielectric has 12.5×10^{18} electrons removed. What is its charge?

ANSWER   The 2 C of electron charge removed allows an excess of 12.5×10^{18} protons. Since the proton and electron have exactly the same amount of charge, now the dielectric has a positive charge of $+Q = 2$ C.

Note that we generally consider the electrons moving, rather than the heavier protons. However, a loss of a given number of electrons is equivalent to a gain of the same number of protons.

CHARGE OF AN ELECTRON   Fundamentally, the quantity of any charge is measured by its force of attraction or repulsion. The extremely small force of an electron or proton was measured by Millikan* in experiments done from 1908 to 1917. Very briefly, the method consisted of measuring the charge on vaporized droplets of oil, by balancing the gravitational force against an electrical force that could be measured very precisely.

A small drop of oil sprayed from an atomizer becomes charged by friction. Furthermore, the charges can be increased or decreased slightly by radiation. These very small changes in the amount of charge were measured. The three

*Robert A. Millikan (1868–1953), an American physicist. Millikan received the Nobel prize in physics for this oil-drop experiment.

smallest values were 0.16×10^{-18} C, 0.32×10^{-18} C, and 0.48×10^{-18} C. These values are multiples of 0.16. In fact, all the charges measured were multiples of 0.16×10^{-18} C. Therefore, we conclude that 0.16×10^{-18} C is the basic charge from which all other values are derived. This ultimate charge of 0.16×10^{-18} C is the charge of 1 electron or 1 proton. Then

▶　　　1 electron or $Q_e = 0.16 \times 10^{-18}$ C

The reciprocal of 0.16×10^{-18} gives the number of electrons or protons in 1 C. Then

▶　　　1 C $= 6.25 \times 10^{18}$ electrons

Note that the factor 6.25 equals exactly $\frac{1}{0.16}$ and the 10^{18} is the reciprocal of 10^{-18}.

THE ELECTRIC FIELD OF A STATIC CHARGE　　The ability of an electric charge to attract or repel another charge is actually a physical force. To help visualize this effect, lines of force are used, as shown in Fig. 1-6. All the lines form the electric field. The lines and the field are imaginary, since they cannot be seen. Just as the field of the force of gravity is not visible, however, the resulting physical effects prove the field is there.

　　Each line of force in Fig. 1-6 is directed outward to indicate repulsion of another charge in the field with the same polarity as Q, either positive or negative. The lines are shorter further away from Q to indicate that the force decreases inversely as the square of the distance. The larger the charge, the greater is the force. These relations describe Coulomb's law of electrostatics.

　　The electric field in the dielectric between two plates with opposite charges is the basis for the ability of a capacitor to store electric charge. More details are explained in Chap. 17, "Capacitance." In general, any charged insulator has capacitance. Practical capacitors, though, are constructed in a form that concentrates the electric field.

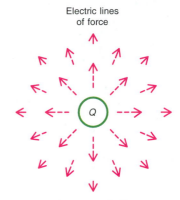

Electric lines of force

FIG. 1-6　Arrows to indicate electric field around a stationary charge Q.

TEST-POINT QUESTION 1-4

Answers at end of chapter.
a. How many electron charges are there in the practical unit of one coulomb?
b. How much is the charge in coulombs for a surplus of 25×10^{18} electrons?
c. Do opposite electric charges attract or repel each other?

1-5 THE VOLT UNIT OF POTENTIAL DIFFERENCE

Potential refers to the possibility of doing work. Any charge has the potential to do the work of moving another charge, by either attraction or repulsion. When we consider two unlike charges, they have a *difference of potential*.

A charge is the result of work done in separating electrons and protons. Because of the separation, there is stress and strain associated with opposite charges, since normally they would be balancing each other to produce a neutral condition. We could consider that the accumulated electrons are drawn tight and are straining themselves to be attracted toward protons in order to return to the neutral condition. Similarly, the work of producing the charge causes a condition of stress in the protons, which are trying to attract electrons and return to the neutral condition. Because of these forces, the charge of electrons or protons has potential, as it is ready to give back the work put into producing the charge. The force between charges is in the electric field.

POTENTIAL BETWEEN DIFFERENT CHARGES When one charge is different from the other, there must be a difference of potential between them. For instance, consider a positive charge of 3 C, shown at the right in Fig. 1-7a. The charge has a certain amount of potential, corresponding to the amount of work this charge can do. The work to be done is moving some electrons, as illustrated.

Assume a charge of 1 C can move 3 electrons. Then the charge of +3 C can attract 9 electrons toward the right. However, the charge of +1 C at the opposite side can attract 3 electrons toward the left. The net result, then, is that 6 electrons can be moved toward the right to the more positive charge.

In Fig. 1-7b, one charge is 2 C, while the other charge is neutral with 0 C. For the difference of 2 C, again 2 × 3 or 6 electrons can be attracted to the positive side.

In Fig. 1-7c, the difference between the charges is still 2 C. The +1 C attracts 3 electrons to the right side. The −1 C repels 3 electrons to the right side also. This effect is really the same as attracting 6 electrons.

Therefore, the net number of electrons moved in the direction of the more positive charge depends on the difference of potential between the two charges. This difference corresponds to 2 C for all three cases in Fig. 1-7. Potential difference is often abbreviated PD.

The only case without any potential difference between charges is where they both have the same polarity and are equal in amount. Then the repelling and attracting forces cancel, and no work can be done in moving electrons between the two identical charges.

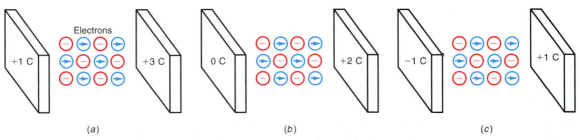

(a)   (b)   (c)

FIG. 1-7 The amount of work required to move electrons between two charges depends on their difference of potential. This potential difference (PD) is equivalent to the charge of 2 C for the examples in (a), (b), and (c).

THE VOLT UNIT OF POTENTIAL DIFFERENCE   This unit is named after Alessandro Volta (1754–1827). Fundamentally, the volt is a measure of the work needed to move an electric charge. When 0.7376 foot-pound (ft · lb) of work is required to move 6.25×10^{18} electrons between two points, each with its own charge, the potential difference is 1 V.

Note that 6.25×10^{18} electrons make up one coulomb. Therefore the definition of a volt is for a coulomb of charge.

Also, 0.7376 ft · lb of work is the same as 1 joule (J), which is the practical metric unit of work or energy. Therefore, we can say briefly that *one volt equals one joule of work per coulomb of charge,* or

▶ $$1 \text{ V} = \frac{1 \text{ J}}{1 \text{ C}}$$

The symbol for potential difference is *V* for voltage. In fact, the volt unit is used so often that potential difference is often called voltage. Remember, though, that voltage is the potential difference between two points. Two terminals are necessary to measure a potential difference.

Consider the 2.2-V lead-acid cell in Fig. 1-8. Its output of 2.2 V means that this is the amount of potential difference between the two terminals. The cell then is a voltage source, or a source of electromotive force (*emf*).

Sometimes the symbol *E* is used for emf, but the standard symbol is *V* for any potential difference. This applies either to the voltage generated by a source or to the voltage drop across a passive component, such as a resistor.

In a practical circuit, the voltage determines how much current can be produced.

(*a*)

$V = 2.2$ V

(*b*)

FIG. 1-8   Chemical cell as a voltage source. (*a*) Voltage output is the potential difference between the two terminals. (*b*) Schematic symbol of any dc voltage source with constant polarity. Longer line indicates positive side.

TEST-POINT QUESTION 1-5

Answers at end of chapter.
  a. How much potential difference is there between two identical charges?
  b. Which applies a greater PD, a 1.5-V battery or a 12-V battery?

1-6 CHARGE IN MOTION IS CURRENT

When the potential difference between two charges forces a third charge to move, the charge in motion is an *electric current.* To produce current, therefore, charge must be moved by a potential difference.

In solid materials, such as copper wire, the free electrons are charges that can be forced to move with relative ease by a potential difference, since they require relatively little work to be moved. As illustrated in Fig. 1-9, if a potential difference is connected across two ends of a copper wire, the applied voltage forces the free electrons to move. This current is a drift of electrons, from the point of negative charge at one end, moving through the wire, and returning to the positive charge at the other end.

To illustrate the drift of free electrons through the wire shown in Fig. 1-9, each electron in the middle row is numbered, corresponding to a copper atom to

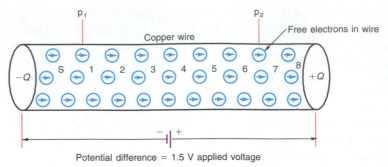

FIG. 1-9 Potential difference across two ends of wire conductor causes drift of free electrons throughout the wire to produce electric current.

which the free electron belongs. The electron at the left is labeled S to indicate that it comes from the negative charge of the source of potential difference. This one electron S is repelled from the negative charge $-Q$ at the left and is attracted by the positive charge $+Q$ at the right. Therefore, the potential difference of the voltage source can make electron S move toward atom 1. Now atom 1 has an extra electron. As a result, the free electron of atom 1 can then move to atom 2. In this way, there is a drift of free electrons from atom to atom. The final result is that the one free electron labeled 8 at the extreme right in Fig. 1-9 moves out from the wire to return to the positive charge of the voltage source.

Considering this case of just one electron moving, note that the electron returning to the positive side of the voltage source is not the electron labeled S that left the negative side. All electrons are the same, however, and have the same charge. Therefore, the drift of free electrons resulted in the charge of one electron moving through the wire. This charge in motion is the current. With more electrons drifting through the wire, the charge of many electrons moves, resulting in more current.

The current is a continuous flow of electrons. Only the electrons move, not the potential difference. For ordinary applications, where the wires are not long lines, the potential difference produces current instantaneously through the entire length of wire. Furthermore, the current must be the same at all points of the wire at any time.

POTENTIAL DIFFERENCE IS NECESSARY TO PRODUCE CURRENT The number of free electrons that can be forced to drift through the wire to produce the moving charge depends upon the amount of potential difference across the wire. With more applied voltage, the forces of attraction and repulsion can make more free electrons drift, producing more charge in motion. A larger amount of charge moving during a given period of time means a higher value of current. Less applied voltage across the same wire results in a smaller amount of charge in motion, which is a smaller value of current. With zero potential difference across the wire, there is no current.

Two cases of zero potential difference and no current can be considered in order to emphasize the important fact that potential difference is needed to pro-

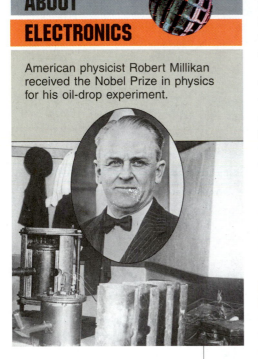

duce current. Assume the copper wire to be by itself, not connected to any voltage source, so that there is no potential difference across the wire. The free electrons in the wire can move from atom to atom, but this motion is random, without any organized drift through the wire. If the wire is considered as a whole, from one end to the other, the current is zero.

As another example, suppose that the two ends of the wire have the same potential. Then free electrons cannot move to either end, because both ends have the same force, and there is no current through the wire. A practical example of this case of zero potential difference would be to connect both ends of the wire to just one terminal of a battery. Each end of the wire would have the same potential, and there would be no current. The conclusion, therefore, is that two connections to two points at different potentials are needed in order to produce the current.

THE AMPERE OF CURRENT   Since current is the movement of charge, the unit for stating the amount of current is defined in rate of flow of charge. When the charge moves at the rate of 6.25×10^{18} electrons flowing past a given point per second, the value of the current is one *ampere* (A). This is the same as one coulomb of charge per second. The ampere unit of current is named after André M. Ampère (1775–1836).

Referring back to Fig. 1-9, note that if 6.25×10^{18} free electrons move past p_1 in 1 second (s), the current is 1 A. Similarly, the current is 1 A at p_2 because the electron drift is the same throughout the wire. If twice as many electrons moved past either point in 1 s, the current would be 2 A.

The symbol for current is I or i for intensity, since the current is a measure of how intense or concentrated the electron flow is. Two amperes of current in a copper wire is a higher intensity than 1 A; a greater concentration of moving electrons results because of more electrons in motion. Sometimes current is called *amperage*. However, the current in electronic circuits is usually in smaller units, milliamperes and microamperes.

HOW CURRENT DIFFERS FROM CHARGE   Charge is a quantity of electricity accumulated in a dielectric, which is an insulator. The charge is static electricity, at rest, without any motion. When the charge moves, usually in a conductor, the current I indicates the intensity of the electricity in motion. This characteristic is a fundamental definition of current:

$$I = \frac{Q}{T} \tag{1-1}$$

where I is the current in amperes, Q is in coulombs, and the time T is in seconds. It does not matter whether the moving charge is positive or negative. The only question is how much charge moves and what its rate of motion is.

In terms of practical units,

$$1\ A = \frac{1\ C}{1\ s} \tag{1-2}$$

One ampere of current results when one coulomb of charge moves past a given point in 1 s.

Example

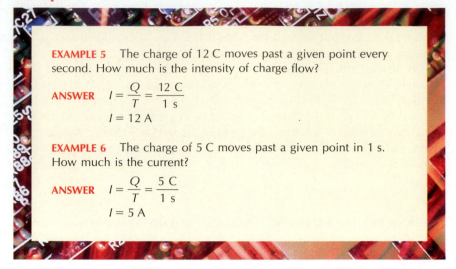

EXAMPLE 5   The charge of 12 C moves past a given point every second. How much is the intensity of charge flow?

ANSWER $I = \dfrac{Q}{T} = \dfrac{12 \text{ C}}{1 \text{ s}}$

$I = 12 \text{ A}$

EXAMPLE 6   The charge of 5 C moves past a given point in 1 s. How much is the current?

ANSWER $I = \dfrac{Q}{T} = \dfrac{5 \text{ C}}{1 \text{ s}}$

$I = 5 \text{ A}$

The fundamental definition of current can also be used to consider the charge as equal to the product of the current multiplied by the time. Or

▶ $$Q = I \times T \tag{1-3}$$

In terms of practical units,

▶ $$1 \text{ C} = 1 \text{ A} \times 1 \text{ s} \tag{1-4}$$

One coulomb of charge results when one ampere of current accumulates charge during the time of one second. The charge is generally accumulated in the dielectric of a capacitor, or at the electrodes of a battery.

For instance, we can have a dielectric connected to conductors with a current of 0.4 A. If the current can deposit electrons for the time of 0.2 s, the accumulated charge in the dielectric will be

$$Q = I \times T = 0.4 \text{ A} \times 0.2 \text{ s}$$
$$Q = 0.08 \text{ C}$$

The formulas $Q = IT$ for charge and $I = Q/T$ for current illustrate the fundamental nature of Q as an accumulation of static charge in an insulator, while I measures the intensity of moving charges in a conductor. Furthermore, current I is different from voltage V. You can have V without I, but you cannot have current without an applied voltage.

THE GENERAL NATURE OF CURRENT   The moving charges that provide current in metal conductors like a copper wire are the free electrons of the copper atoms. In this case, the moving charges have negative polarity. The direction of motion between two terminals for this *electron current,* therefore, is toward the more positive end. It is important to note, however, that there are examples of positive charges in motion. Common applications include current in liquids, gases, and semiconductors. For the case of current resulting from the motion of positive charges, its direction is opposite from the direction of electron flow. Whether negative or positive charges move, though, the current is still defined fundamentally as Q/T. Note also that the current is provided by free charges, which are easily moved by an applied voltage.

TYPES OF ELECTRIC CHARGES FOR CURRENT See Table 1-4. The most common charge is the electron. In metal conductors and solid materials in general, free electrons in the atoms can be forced to move by applying a potential difference. Therefore current is produced. The direction of electron flow is from the negative terminal of the voltage source, through the external circuit, and returning to the positive source terminal. Also, electrons are released by thermionic emission from the heated cathode in a vacuum tube. Finally, N-type semiconductors such as silicon and germanium have unbound electrons as a result of doping with impurity elements that can provide valence electrons from the added atoms.

For P-type semiconductors, the silicon and germanium are doped with impurity elements that cause a deficiency of electrons in the bonds between atoms. Each vacant space where an electron is missing is called a *hole charge*. The polarity is positive, opposite from the electron, but the amount of charge is exactly the same. In short, a hole charge is a deficiency of one valence electron in semiconductors. When hole charges move in a P-type semiconductor, they provide hole current. The direction of flow for the positive charges is from the positive terminal of the voltage source, through the external circuit, and returning to the negative source terminal.

An ion is an atom that has either lost or gained one or more valence electrons to become electrically charged. Therefore, the ion charge may be either positive or negative. The amount may be the charge of 1 electron Q_e, $2Q_e$, $3Q_e$, etc. Ions can be produced by applying voltage to liquids and gases to produce ionization of the atoms. The ions are much less mobile than electrons or hole charges because an ion includes a complex atom with its nucleus.

Note that protons are not included as charge carriers for current in Table 1-4. The reason is that the protons are bound in the nucleus. They cannot be released except by nuclear forces. Therefore, a current of positive charges is a flow of either hole charges or positive ions. The hole charge has the same amount of charge as the proton, which is the same amount as an electron. However, the positive hole charge is in the valence structure of the atoms, not in the nucleus.

TABLE 1-4 TYPES OF ELECTRIC CHARGES FOR CURRENT				
TYPE OF CHARGE	**AMOUNT OF CHARGE**	**POLARITY**	**TYPE OF CURRENT**	**APPLICATIONS**
Electron	$Q_e = 0.16 \times 10^{-18}$ C	Negative	Electron flow	In wire conductors, vacuum tubes, and N-type semiconductors
Ion	Q_e or multiples of Q_e	Positive or negative	Ion current	In liquids and gases
Hole	$Q_e = 0.16 \times 10^{-18}$ C	Positive	Hole current	In P-type semiconductors

MAGNETIC FIELD AROUND AN ELECTRIC CURRENT   When any current flows, it has an associated magnetic field. Figure 1-10 shows how iron filings line up in a circular field pattern corresponding to the magnetic lines of force. The magnetic field is in a plane perpendicular to the current. It should be noted that the iron filings are just a method of making the imaginary lines of force visible. The filings become magnetized by the magnetic field. Both magnetic and electric fields can do the physical work of attraction or repulsion.

The magnetic field of any current is the basis for many electromagnetic applications, including magnets, relays, loudspeakers, transformers, and coils in general. Winding the conductor in the form of a coil concentrates the magnetic field. More details of electromagnetism and magnetic fields are explained in Chap. 15, "Electromagnetic Induction."

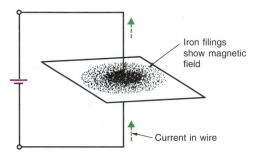

Iron filings show magnetic field

Current in wire

FIG. 1-10   Magnetic field around an electric current. Arrow for current indicates direction of electron flow.

TEST-POINT QUESTION 1-6

Answers at end of chapter.
a. The flow of 2 C/s of electron charges is how many amperes of current?
b. The flow of 2 C/s of hole charges is how many amperes of current?
c. How much is the current with zero potential difference?

1-7 RESISTANCE IS OPPOSITION TO CURRENT

The fact that a wire conducting current can become hot is evidence of the fact that the work done by the applied voltage in producing current must be accomplished against some form of opposition. This opposition, which limits the amount of current that can be produced by the applied voltage, is called *resistance.* Conductors have very little resistance; insulators have a large amount of resistance.

The atoms of a copper wire have a large number of free electrons, which can be moved easily by a potential difference. Therefore, the copper wire has little opposition to the flow of free electrons when voltage is applied, corresponding to a low value of resistance.

Carbon, however, has fewer free electrons than copper. When the same amount of voltage is applied to the carbon as to the copper, fewer electrons will flow. It should be noted that just as much current can be produced in the carbon by applying more voltage. For the same current, though, the higher applied voltage means that more work is necessary, causing more heat. Carbon opposes the current more than copper, therefore, and has a higher value of resistance.

THE OHM   The practical unit of resistance is the *ohm*. A resistance that develops 0.24 calorie of heat when one ampere of current flows through it for one second has one ohm of opposition. As an example of a low resistance, a good conductor like copper wire can have a resistance of 0.01 Ω for a 1-ft length. The resistance-wire heating element in a 600-W 120-V toaster has a resistance of 24 Ω, and the tungsten filament in a 100-W 120-V light bulb has a resistance of 144 Ω. The ohm unit is named after Georg Simon Ohm (1747–1854), a German physicist.

Figure 1-11*a* shows a wire-wound resistor. Resistors also are made with powdered carbon. They can be manufactured with a value from a few ohms to millions of ohms.

The symbol for resistance is *R*. The abbreviation used for the ohm unit is the Greek letter *omega,* written as Ω. In diagrams, resistance is indicated by a zigzag line, as shown by *R* in Fig. 1-11*b*.

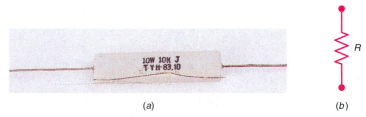

(a)   (b)

FIG. 1-11   (*a*) Wire-wound type of resistor with cement coating for insulation. (*b*) Schematic symbol for any type of fixed resistor.

CONDUCTANCE   The opposite of resistance is conductance. The lower the resistance, the higher the conductance. Its symbol is *G*, and the unit is the *siemens* (S), named after Ernst von Siemens, a European inventor. (The old unit name for conductance is *mho,* which is *ohm* spelled backward.)

Specifically, *G* is the reciprocal of *R*, or *G* = 1/*R*. For example, 5 Ω of resistance is equal to ⅕ or 0.2 S of conductance. Also, *R* = 1/*G*. Therefore,

$$\blacktriangleright \quad R = \frac{1}{0.2 \text{ S}} = 5 \ \Omega$$

Whether to use *R* or *G* for components is usually a matter of convenience. In general, *R* is easier to use in series circuits, because the series voltages are proportional to the resistances. However, *G* can be more convenient in parallel circuits, because the parallel currents are proportional to the conductances. (Series and parallel circuits are explained in Chaps. 4 and 5.)

Answers at end of chapter.

a. Which has more resistance, carbon or copper?

b. With the same voltage applied, which resistance will allow more current, 4.7 Ω or 5000 Ω?

c. What is the conductance value in siemens units for a 10-Ω R?

1-8 THE CLOSED CIRCUIT

In applications requiring the use of current, the components are arranged in the form of a circuit, as shown in Fig. 1-12. A circuit can be defined as a path for current flow. The purpose of this circuit is to light the incandescent bulb. The bulb lights when the tungsten-filament wire inside is white hot, producing an incandescent glow.

By itself the tungsten filament cannot produce current. A source of potential difference is necessary. Since the battery produces a potential difference of 1.5 V across its two output terminals, this voltage is connected across the filament of the bulb by means of the two wires so that the applied voltage can produce current through the filament.

In Fig. 1-12b the schematic diagram of the circuit is shown. Here the components are represented by shorthand symbols. Note the symbols for the battery and resistance. The connecting wires are shown simply as straight lines because their resistance is small enough to be neglected. A resistance of less than 0.01 Ω for the wire is practically zero compared with the 300-Ω resistance of the bulb. If the resistance of the wire must be considered, the schematic diagram includes it as additional resistance in the same current path.

It should be noted that the schematic diagram does not look like the physical layout of the circuit. The schematic shows only the symbols for the components and their electrical connections.

Any electric circuit has three important characteristics:

1. There must be a source of potential difference. Without the applied voltage, current cannot flow.

2. There must be a complete path for current flow, from one side of the applied voltage source, through the external circuit, and returning to the other side of the voltage source.

3. The current path normally has resistance. The resistance is in the circuit for the purpose of either generating heat or limiting the amount of current.

HOW THE VOLTAGE IS DIFFERENT FROM THE CURRENT It is the current that moves through the circuit. The potential difference does not move.

In Fig. 1-12 the voltage across the filament resistance makes electrons flow from one side to the other. While the current is flowing around the circuit, however, the potential difference remains across the filament to do the work of moving electrons through the resistance of the filament.

The circuit is redrawn in Fig. 1-13 to emphasize the comparison between V and I. The voltage is the potential difference across the two ends of the resistance. If you want to measure the PD, just connect the two leads of the voltmeter across the

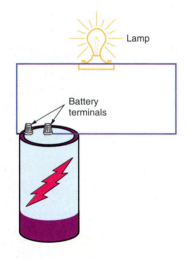

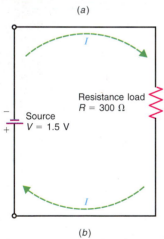

(a)

(b)

FIG. 1-12 Example of an electric circuit with battery as voltage source connected to a light bulb as a resistance. (a) Wiring diagram of the closed path for current. (b) Schematic diagram of the circuit.

resistor. However, the current is the intensity of the electron flow past any one point in the circuit. Measuring the current is not as easy. You would have to break open the path, at any point, and then insert the current meter to complete the circuit.

The word *across* is used with voltage because it is the potential difference between two points. There cannot be a PD at one point. However, current can be considered at one point, as the motion of charges through that point.

To illustrate the difference between *V* and *I* another way, suppose the circuit in Fig. 1-12 is opened by disconnecting the bulb. Now no current can flow because there is no closed path. Still, the battery has its potential difference. If you measure across the two terminals, the voltmeter will read 1.5 V even though the current is zero.

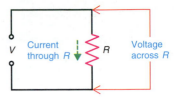

FIG. 1-13 Comparison of voltage (*V*) across a resistance and the current (*I*) through *R*.

THE VOLTAGE SOURCE MAINTAINS THE CURRENT
As current flows in the circuit, electrons leave the negative terminal of the cell or battery in Fig. 1-12, and the same number of free electrons in the conductor are returned to the positive terminal. As electrons are lost from the negative charge and gained by the positive charge, the two charges would tend to neutralize each other. The chemical action inside the battery, however, continuously separates electrons and protons to maintain the negative and positive charges on the outside terminals that provide the potential difference. Otherwise, the current would neutralize the charges, resulting in no potential difference, and the current would stop. Therefore, the battery keeps the current flowing by maintaining the potential difference across the circuit. The battery is the voltage source for the circuit.

THE CIRCUIT IS A LOAD ON THE VOLTAGE SOURCE
We can consider the circuit as a means whereby the energy of the voltage source is carried by means of the current through the filament of the bulb, where the electric energy is used in producing heat energy. On this basis, the battery is the *source* in the circuit, since its voltage output represents the potential energy to be used. The part of the circuit connected to the voltage source is the *load resistance,* since it determines how much work the source will supply. In this case, the bulb's filament is the load resistance for the battery.

The resistance of the filament determines how much current the 1.5-V source will produce. Specifically, the current here is 0.005 A, equal to 1.5 V divided by 300 Ω. With more opposition, the same voltage will produce less current. For the opposite case, less opposition allows more current.

The current that flows through the load resistance is the *load current.* Note that a lower value of ohms for the load resistance corresponds to a higher load current. Unless noted otherwise, the term *load* by itself can be assumed generally to mean the load current. Therefore, a heavy or big load electrically means a high value of load current, corresponding to a large amount of work supplied by the source.

In summary, we can say that the closed circuit, normal circuit, or just a circuit is a closed path that has V to produce I with R to limit the amount of current. The circuit provides a means of using the energy of the battery as a voltage source. The battery has its potential difference V with or without the circuit. However, the battery alone is not doing any work in producing load current. The bulb alone has its resistance, but without current the bulb does not light. With the circuit, the voltage source is used for the purpose of producing current to light the bulb.

OPEN CIRCUIT When any part of the path is open or broken, the circuit is open because there is no continuity in the conducting path. The open circuit can be in the connecting wires or in the bulb's filament as the load resistance. The resistance of an open circuit is infinitely high. The result is no current in an open circuit.

SHORT CIRCUIT In this case, the voltage source has a closed path across its terminals, but the resistance is practically zero. The result is too much current in a short circuit. Usually, the short circuit is a bypass across the load resistance. For instance, a short across the conducting wires for a bulb produces too much current in the wires but no current through the bulb. Then the bulb is shorted out. The bulb is not damaged, but the wires can become hot enough to burn unless the line has a fuse as a safety precaution against too much current.

<hr>

TEST-POINT QUESTION 1-8

Answers at end of chapter.

Answer True or False for the circuit in Fig. 1-12.
a. The bulb has a PD of 1.5 V across its filament only when connected to the voltage source.
b. The battery has a PD of 1.5 V across its terminals only when connected to the bulb.
c. The battery by itself, without the wires and the bulb, has a PD of 1.5 V.

1-9 DIRECTION OF THE CURRENT

Just as a voltage source has polarity, the current has a direction. The reference is with respect to the positive and negative terminals of the voltage source. What the direction is for the current depends on whether we consider the flow of negative electrons or the motion of positive charges in the opposite direction.

ELECTRON FLOW As shown in Fig. 1-14*a*, the direction of electron drift for the current I is out from the negative side of the voltage source. The I flows through the external circuit with R and returns to the positive side of V. Note that this direction from the negative terminal applies to the external circuit connected to the output terminals of the voltage source. Electron flow is also shown in Fig. 1-14*c* with reversed polarity for V.

Inside the battery, the electrons move to the negative terminal because this is how the voltage source produces its potential difference. The battery is doing the work of separating charges, accumulating electrons at the negative terminal and protons at the positive terminal. Then the potential difference across the two output terminals can do the work of moving electrons around the external circuit. For the circuit outside the voltage source, however, the direction of the electron flow is from a point of negative potential to a point of positive potential.

CONVENTIONAL CURRENT   The motion of positive charges, in the opposite direction from electron flow, is considered as conventional current. This direction is generally used for analyzing circuits in electrical engineering. The reason is based on some traditional definitions in the science of physics. By the definitions of force and work with positive values, a positive potential is considered above a negative potential. Then conventional current corresponds to a motion of positive charges "falling downhill" from a positive to a negative potential. The conventional current, therefore, is in the direction of positive charges in motion. An example is shown in Fig. 1-14b. The conventional I is out from the positive side of the voltage source, flows through the external circuit and returns to the negative side of V. Conventional current is also shown in Fig. 1-14d, with the voltage source in reverse polarity.

An example of positive charges in motion for conventional current is the current of hole charges in P-type semiconductors. Also a current of positive ions in liquids and gases moves in the opposite direction from electron flow. For instance, the current through the electrolyte in a battery is ionization current.

It should be noted that with semiconductors, the arrow symbols for I are shown in the direction of conventional current. This method is standardized by the Electronic Industries Association (EIA) for transistors, diodes, and all semiconductor devices.

Actually, either a negative or positive potential of the same value can do the same amount of work in moving a charge. Any circuit, therefore, can be analyzed either with electron flow for I or by conventional current. The two opposite directions with two opposite polarities for the voltage source are summarized by the four diagrams in Fig. 1-14.

In this book, the current is considered as electron flow in the applications where electrons are the moving charges. A dotted or dashed arrow, as in Fig. 1-14a and c, is used to indicate the direction of electron flow for I. In Fig. 14b and d, the solid arrow means the direction of conventional current. These arrows are used for the unidirectional current in dc circuits. For ac circuits, the direction of current can be considered either way because I reverses direction every half-cycle with the reversals in polarity for V.

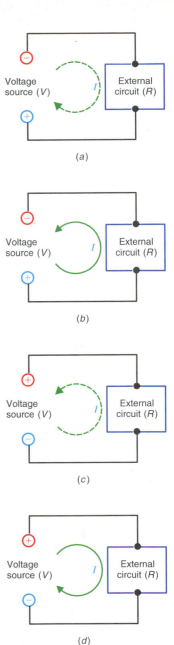

FIG. 1-14   Direction of I in a closed circuit, shown for electron flow and conventional current. The circuit works the same way no matter which direction you consider. (a) Electron flow indicated with dashed arrow in diagram. (b) Conventional current indicated with solid arrow. (c) Electron flow as in (a) but with reversed polarity of voltage source. (d) Conventional I as in (b) but reversed polarity for V.

TEST-POINT QUESTION 1-9

Answers at end of chapter.
  a. Is electron flow out from the positive or negative terminal of the voltage source?
  b. Does conventional current return to the positive or negative terminal of the voltage source?
  c. Is it true or false that electron flow and conventional current are in opposite directions?

1-10 DIRECT CURRENT (DC) AND ALTERNATING CURRENT (AC)

The electron flow illustrated for the circuit with a bulb in Fig. 1-12 is direct current because it has just one direction. The reason for the unidirectional current is that the battery maintains the same polarity of output voltage across its two terminals.

It is the flow of charges in one direction and the fixed polarity of applied voltage that are the characteristics of a dc circuit. Actually, the current can be a flow of positive charges, rather than electrons but the conventional direction of current does not change the fact that the charges are moving only one way.

Furthermore, the dc voltage source can change the amount of its output voltage but, with the same polarity, direct current still flows only in one direction. This type of source provides a fluctuating or pulsating dc voltage. A battery is a steady dc voltage source because it has fixed polarity and its output voltage is a steady value.

An alternating voltage source periodically reverses or alternates in polarity. The resulting alternating current, therefore, periodically reverses in direction. In terms of electron flow, the current always flows from the negative terminal of the voltage source, through the circuit, and back to the positive terminal, but when the generator alternates in polarity, the current must reverse its direction. The 60-cycle ac power line used in most homes is a common example. This frequency means that the voltage polarity and current direction go through 60 cycles of reversal per second.

The unit for 1 cycle per second is 1 hertz (Hz). Therefore 60 cycles per second is a frequency of 60 Hz.

The details of ac circuits are explained in Chap. 16. Direct-current circuits are analyzed first because they usually are simpler. However, the principles of dc circuits also apply to ac circuits. Both types are important, as most electronic circuits include ac voltages and dc voltages. A comparison of dc and ac voltages and their waveforms is illustrated in Figs. 1-15 and 1-16. Their uses are compared in Table 1-5. Note that transistors require dc electrode voltages in order to amplify an ac signal voltage.

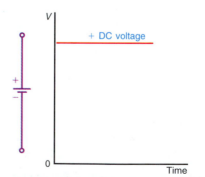

FIG. 1-15   Steady dc voltage of fixed polarity, such as the output of a battery. Note schematic symbol at left.

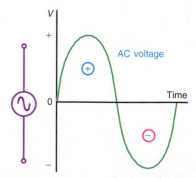

FIG. 1-16   Sine-wave ac voltage with alternating polarity, such as from an ac generator. Note schematic symbol at left. The ac line voltage in your home has this waveform.

TABLE 1-5   COMPARISON OF DC VOLTAGE AND AC VOLTAGE

DC VOLTAGE	AC VOLTAGE
Fixed polarity	Reverses in polarity
Can be steady or vary in magnitude	Varies between reversals in polarity
Steady value cannot be stepped up or down by a transformer	Can be stepped up or down for electric power distribution
Electrode voltages for transistor amplifiers	Signal input and output for amplifiers
Easier to measure	Easier to amplify
Heating effect is the same for direct or alternating current	

TEST-POINT QUESTION 1-10

Answers at end of chapter.

Answer True or False.

a. When the polarity of the applied voltage reverses, the direction of current flow also reverses.

b. A battery is a dc voltage source because it cannot reverse the polarity across its output terminals.

1-11 SOURCES OF ELECTRICITY

There are electrons and protons in the atoms of all materials, but to do useful work the charges must be separated to produce a potential difference that can make current flow. Some of the more common methods of providing electrical effects are listed here.

STATIC ELECTRICITY BY FRICTION   In this method, electrons in an insulator can be separated by the work of rubbing to produce opposite charges that remain in the dielectric.

CONVERSION OF CHEMICAL ENERGY   Wet or dry cells and batteries are the applications. Here a chemical reaction produces opposite charges on two dissimilar metals, which serve as the negative and positive terminals.

ELECTROMAGNETISM   Electricity and magnetism are closely related. Any moving charge has an associated magnetic field; also, any changing magnetic field can produce current. A motor is an example of how current can react with a magnetic field to produce motion; a generator produces voltage by means of a conductor rotating in a magnetic field.

DID YOU KNOW?

Electricity enables people to have healthier and longer lives, but more than half of available commercial energy is used by only a quarter of the world's population. In some developing nations, people without electricity lack basic services. Without access to health services, housing, clean water, and sewage treatment, life expectancy is low and infant mortality high.

PHOTOELECTRICITY   Some materials are photoelectric, meaning they can emit electrons when light strikes the surface. The element cesium is often used as a source of *photoelectrons.* Also, photovoltaic cells or solar cells use silicon to generate output voltage from the light input. In another effect, the resistance of the element selenium changes with light. When this is combined with a fixed voltage source, wide variations between *dark current* and *light current* can be produced. Such characteristics are the basis of many photoelectric devices, including television camera tubes, photoelectric cells, and phototransistors.

THERMAL EMISSION   Some materials when heated can "boil off" electrons from the surface. Then these emitted electrons can be controlled to provide useful applications of electric current. The emitting electrode is called a *cathode,* while an *anode* is used to collect the emitted electrons.

TEST-POINT QUESTION 1-11

Answers at end of chapter.
  a. The excess charges at the negative terminal of a battery are _____.
  b. The charges emitted from a heated cathode are _____.
  c. The charges in a P-type semiconductor are _____.
  d. In a liquid that is conducting current, the moving charges are _____.

1   SUMMARY AND REVIEW

- Electricity is present in all matter in the form of electrons and protons.
- The electron is the basic quantity of negative electricity, the proton of positive electricity. Both have the same amount of charge but opposite polarities. The charge of 6.25×10^{18} electrons or protons equals the practical unit of one coulomb.
- Charges of the same polarity tend to repel each other; charges of opposite polarities attract. There must be a difference of charges for any force of attraction or repulsion.
- Electrons tend to move toward protons because an electron has $\frac{1}{1840}$ the weight of a proton. Electrons in motion provide an electron current.
- The atomic number of an element gives the number of protons in the nucleus of its atom, balanced by an equal number of orbital electrons.
- The number of electrons in the outermost orbit is the valence of the element.
- Types of negative charges include electrons and negative ions. Types of positive charges include protons, positive ions, and hole charges.
- An electric circuit is a closed path for electron flow. Potential difference must be connected across the circuit to produce current. In the external circuit outside the voltage source, electrons flow from the negative terminal toward the positive terminal.
- Direct current has just one direction, as the dc voltage source has a fixed polarity. Alternating current periodically reverses in direction as the ac voltage source reverses its polarity.
- Table 1-6 summarizes the main features of electric circuits.

TABLE 1-6   ELECTRICAL CHARACTERISTICS

CHARACTERISTIC	SYMBOL	UNIT	DESCRIPTION
Charge	Q or q*	Coulomb (C)	Quantity of electrons or protons; $Q = I \times T$
Current	I or i*	Ampere (A)	Charge in motion; $I = Q/T$
Voltage	V or v*,†	Volt (V)	Potential difference between two unlike charges; makes charge move to produce I
Resistance	R or r‡	Ohm (Ω)	Opposition that reduces amount of current; $R = 1/G$
Conductance	G or g‡	Siemens (S)	Reciprocal of R, or $G = 1/R$

*Small letter q, i, or v is used for an instantaneous value of a varying charge, current, or voltage.
†E or e is sometimes used for a generated emf, but the standard symbol for any potential difference is V or v in the international system of units (SI).
‡Small letter r or g is used for internal resistance or conductance of transistors and tubes.

SELF-TEST

ANSWERS AT BACK OF BOOK.

Answer True or False.

1. All matter has electricity in the form of electrons and protons in the atom.
2. The electron is the basic unit of negative charge.
3. A proton has the same amount of charge as the electron but opposite polarity.
4. Electrons are repelled from other electrons but are attracted to protons.
5. The force of attraction or repulsion between charges is in their electric field.
6. The nucleus is the massive stable part of an atom, with positive charge.
7. Neutrons add to the weight of the atom's nucleus but not to its electric charge.
8. An element with atomic number 12 has 12 orbital electrons.
9. The element in question 8 has an electron valence of +2.
10. To produce current in a circuit, potential difference is connected across a closed path.
11. A dc voltage has fixed polarity, while ac voltage periodically reverses its polarity.
12. The coulomb is a measure of the quantity of stored charge.
13. If a dielectric has 2 C of excess electrons, removing 3 C of electrons will leave the dielectric with the positive charge of 1 C.
14. A charge of 5 C flowing past a point each second is a current of 5 A.
15. A 7-A current charging a dielectric will accumulate a charge of 14 C after a period of 2 s.
16. A voltage source has two terminals with different charges.
17. An ion is a charged atom.
18. The resistance of a few feet of copper wire is practically zero.
19. The resistance of the rubber or plastic insulation on wire is practically zero.
20. A resistance of 600 Ω has a conductance of 6 S.

QUESTIONS

1. Briefly define each of the following, giving its unit and symbol: charge, potential difference, current, resistance, and conductance.
2. Name two good conductors, two good insulators, and two semiconductors.
3. Explain briefly why there is no current in a light bulb unless it is connected across a source of applied voltage.
4. Give three differences between voltage and current.
5. In any circuit: (**a**) state two requirements for producing current; (**b**) give the direction of electron flow.

6. Show the atomic structure of the element sodium (Na), with atomic number 11. What is its electron valence?
7. Make up your own name for direct current and dc voltage to indicate how it differs from alternating voltage and current.
8. State the formulas for each of the following two statements: (a) Current is the time rate of change of charge. (b) Charge is current accumulated over a period of time.
9. Refer to Table 1-4 on page 31. (a) Name two types of moving charges that provide current in the direction of electron flow. (b) Name two types that provide conventional current.
10. Why is it that protons are not considered a source of moving charges for current flow?
11. Give one difference and one similarity in comparing electric and magnetic fields.
12. Give three methods of providing electric charges, and give their practical applications.
13. Give one way in which electricity and electronics are similar and one way in which they are different.
14. What kind of a meter would you use to measure the potential difference of a battery?
15. Give at least two voltage sources you have used and classify them as either dc or ac sources.
16. Compare the functions of a good conductor like copper wire with an insulator like paper, air, and vacuum.
17. How would you define *electromagnetism?*
18. Why is the characteristic of frequency used with alternating current and voltage?

PROBLEMS

ANSWERS TO ODD-NUMBERED PROBLEMS AT BACK OF BOOK.

1. A charge of 10 C flows past a given point every 2 s. How much is the current flow in amperes?
2. A charge of 5 C flows past a given point every 0.5 s. How much is the current flow in amperes?
3. A current of 3 A charges an insulator for 3 s. How much charge is accumulated?
4. If 31.25×10^{18} electrons are removed from a neutral dielectric, how much charge, in coulombs, is being stored?
5. If 18.75×10^{18} electrons are added to a neutral dielectric, how much charge, in coulombs, is being stored?
6. A dielectric has a charge Q of 6 C. To how many electrons does this correspond?
7. Show the atomic structure and determine the electron valence for the element silicon (Si), which has an atomic number of 14.
8. How long will it take a neutral dielectric to obtain a charge of $+4$ C if the charging current is 0.5 A?
9. How long will it take a neutral dielectric to obtain a charge of -20 C if the charging current is 2.5 A?

10. Calculate the resistance value in ohms for each of the following conductance values: (a) 0.001 S; (b) 0.01 S; (c) 0.1 S; (d) 1 S.

11. Calculate the resistance value in ohms for each of the following conductance values: (a) 0.002 S; (b) 0.004 S; (c) 0.00833 S; (d) 0.25 S.

12. Calculate the conductance value in siemens for each of the following resistance values: (a) 200 Ω; (b) 100 Ω; (c) 50 Ω; (d) 25 Ω.

13. Calculate the conductance value in siemens for each of the following resistance values: (a) 1 Ω; (b) 10,000 Ω; (c) 40 Ω; (d) 0.5 Ω.

14. Connect the components in Fig. 1-17 to form a complete electric circuit. Label the source voltage, with polarity, and the load resistance, R_L.

15. Calculate the voltage output of a battery if 15 J of energy is expended in moving 1.25 C of charge.

16. Calculate the voltage output of a battery if 3.6 J of energy is expended in moving 0.4 C of charge.

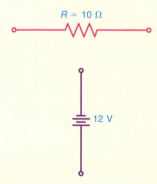

$R = 10\ \Omega$

12 V

FIG. 1-17 Circuit components for Prob. 14.

CRITICAL THINKING

1. Suppose that 1000 electrons are removed from a neutral dielectric. How much charge, in coulombs, is stored in the dielectric?

2. How long will it take an insulator that has a charge of +5 C to charge to +30 C if the charging current is 2 A?

3. Assume that 6.25×10^{15} electrons flow past a given point in a conductor every 10 s. Calculate the current, I, in amperes.

4. The conductance of a wire at 100°C is one-tenth its value at 25°C. If the wire resistance equals 10 Ω at 25°C, calculate the resistance of the wire at 100°C.

1-1 a. negative
     b. positive
     c. true

1-2 a. metals
     b. silver
     c. silicon

1-3 a. 14
     b. 1
     c. 32

1-4 a. 6.25×10^{18}
     b. 4 C
     c. attract

1-5 a. zero
     b. 12 V

1-6 a. 2 A
     b. 2 A
     c. zero

1-7 a. carbon
     b. 4.7 Ω
     c. $\frac{1}{10}$ S

1-8 a. T
     b. F
     c. T

1-9 a. negative
     b. negative
     c. true

1-10 a. T
     b. T

1-11 a. electrons
     b. electrons
     c. holes
     d. ions

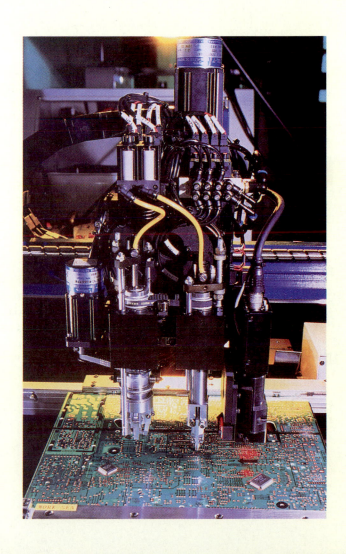

CHAPTER 2

RESISTORS

Resistors are perhaps the most common component in electronic circuits. Their main function is to reduce the current I to the desired value, or to provide the desired voltage V in a circuit. A resistor is manufactured to have a specific value in ohms for its resistance R. The physical size of a resistor determines how much power it can dissipate in the form of heat. However, there is no direct correlation between a resistor's physical size and its resistance value. Resistors are manufactured in a variety of standard values and power ratings.

Resistors can be either fixed or variable in value. A fixed resistor has a resistance value that does not change, whereas a variable resistor has a resistance that can be varied over a range of values. Most fixed resistors have a four- or five-band color code that indicates their resistance value in ohms, but on larger resistors the resistance value may be printed right on the body. One important feature of resistance in general is that its effect is the same for both dc and ac circuits.

CHAPTER OBJECTIVES

Upon completion of this chapter, you should be able to:

- *List* five different types of resistors and describe the makeup of each type.
- *Interpret* the resistor color code to determine the resistance and tolerance of a resistor.
- *Explain* the difference between a potentiometer and a rheostat.
- *Explain* the significance of a resistor's power rating.
- *List* the most common troubles with resistors.
- *Explain* the precautions that must be observed when measuring a resistor with an ohmmeter.

IMPORTANT TERMS IN THIS CHAPTER

carbon-composition resistors	metal-film resistors	surface-mount resistors
carbon-film resistors	noisy controls	tapered control
chip resistor	potentiometer	tolerance
color code	preferred value	wire-wound resistors
decade resistance box	rheostat	zero-ohm resistor

TOPICS COVERED IN THIS CHAPTER

2-1 TYPES OF RESISTORS

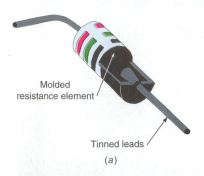

Molded
resistance element

Tinned leads

(a)

(b)

FIG. 2-1 Carbon-composition resistors. (a) Internal construction. Length is ¾ in. without leads for 1-W power rating. Color stripes give R in ohms. Tinned leads have coating of solder. (b) Resistors mounted on printed-circuit (PC) board.

(a)

(b)

FIG. 2-3 Large wire-wound resistors with 50-W power rating. (a) Fixed R, length of 5 in. (b) Variable R, diameter of 2 in.

The two main characteristics of a resistor are its resistance R in ohms and its power rating W in watts. Resistors are available in a very wide range of R values, from a fraction of an ohm to many kilohms (kΩ) and megohms (MΩ). One kilohm is 1000 Ω, and one megohm is 1,000,000 Ω. More details of very small and large units are given in Chap. 3. The power rating for resistors may be as high as several hundred watts or as low as ¹⁄₁₀ W.

The R is the resistance value required to provide the desired current I or voltage. Also important is the wattage rating, because it specifies the maximum power the resistor can dissipate without excessive heat. Dissipation means that the power is wasted, since the resultant heat is not used. Too much heat can make the resistor burn. The wattage rating of the resistor is generally more than the actual power dissipation, as a safety factor.

Most common in electronic equipment are carbon resistors with a power rating of 1 W or less. The construction is illustrated in Fig. 2-1, while Fig. 2-2 shows a group of resistors mounted on a printed-circuit (PC) board. The resistors can be inserted automatically by machine.

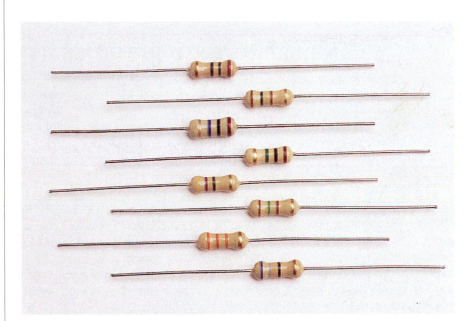

FIG. 2-2 Carbon resistors commonly used on PC board. Typically, leads are cut and formed for insertion into holes with 0.5-in. spacing.

Resistors with higher R values usually have lower wattage ratings because they have less current. As an example, a common value is 1 MΩ at ¼ W, for a resistor only ½ in. long. The lower the power rating, the smaller the actual physical size of the resistor. However, the resistance value is not related to physical size.

WIRE-WOUND RESISTORS In this construction, a special type of wire called *resistance wire* is wrapped around an insulating core, as shown in Fig. 2-3.

The length of wire and its specific resistivity determine the R of the unit. Types of resistance wire include tungsten and manganin, as explained in Chap. 11, "Conductors and Insulators." The insulated core is commonly porcelain, cement, or just plain pressed paper. Bare wire is used, but the entire unit is generally encased in an insulating material.

Since they are generally for high-current applications with low resistance and appreciable power, wire-wound resistors are available in wattage ratings from 5 W up to 100 W or more. The resistance can be less than 1 Ω up to several thousand ohms.

In addition, wire-wound resistors are used where accurate, stable resistance values are necessary. Examples are precision resistors for the function of an ammeter shunt or a precision potentiometer to adjust for an exact amount of R.

For 2 W or less, carbon resistors are preferable because they are small and cost less. Between 2 and 5 W, combinations of carbon resistors can be used. Also, small wire-wound resistors are available in a 3- or 4-W rating.

CARBON-COMPOSITION RESISTORS   This type of resistor is made of finely divided carbon or graphite mixed with a powdered insulating material as a binder, in the proportions needed for the desired R value. As shown in Fig. 2-1a, the resistor element is enclosed in a plastic case for insulation and mechanical strength. Joined to the two ends of the carbon resistance element are metal caps with leads of tinned copper wire for soldering the connections into a circuit. These are called *axial leads* because they come straight out from the ends.

Carbon resistors are commonly available in R values of 1 Ω to 20 MΩ. Examples are 10 Ω, 220 Ω, 4.7 kΩ, and 68 kΩ. The power rating is generally $\frac{1}{10}$, $\frac{1}{8}$, $\frac{1}{4}$, $\frac{1}{2}$, 1, or 2 W.

FILM-TYPE RESISTORS   There are two kinds of film-type resistors: carbon-film and metal-film resistors. The carbon-film resistor, whose construction is shown in Fig. 2-4, is made by depositing a thin layer of carbon on an insulated substrate. The carbon film is then cut in the form of a spiral to form the resistive element. The resistance value is controlled by varying the proportion of carbon to insulator. As compared to carbon-composition resistors, carbon-film resistors have the following advantages: tighter tolerances, less sensitivity to temperature changes and aging, and less noise generated internally.

Metal-film resistors are constructed in a manner similar to the carbon-film type. However, in a metal-film resistor a thin film of metal is sprayed onto a ceramic substrate and then cut in the form of a spiral. The construction of a metal-film resistor is shown in Fig. 2-5. The length, thickness, and width of the metal spiral determine the exact resistance value. Metal-film resistors have even more precise R values than do carbon-film resistors. Like carbon-film resistors, metal-film resistors are affected very little by temperature changes and aging. They also generate very little noise internally. In terms of overall performance, metal-film resistors are the best, carbon-film resistors are next, and carbon-composition resistors are last. Both carbon- and metal-film resistors can be distinguished from carbon-composition resistors by the fact that the diameters of the ends are a little larger than that of the body.

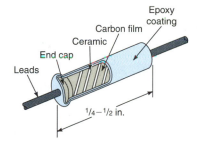

FIG. 2-4   Construction of carbon-film resistor.

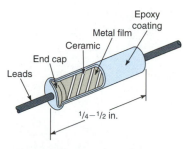

FIG. 2-5   Construction of metal-film resistor.

FIG. 2-6 Typical chip resistors.

SURFACE-MOUNT RESISTORS Surface-mount resistors, also called *chip resistors,* are constructed by depositing a thick carbon film on a ceramic base. The exact resistance value is determined by the composition of the carbon itself, as well as by the amount of trimming done to the carbon deposit. The resistance can vary from a fraction of an ohm to well over a million ohms. Power dissipation ratings are typically ⅛ to ¼ W. Figure 2-6 shows typical chip resistors. Electric connection to the resistive element is made via two leadless solder end electrodes (terminals). The end electrodes are C-shaped. The physical dimensions of a ⅛-W chip resistor are: 0.125 in. long by 0.063 in. wide and approximately 0.028 in. thick. This is many times smaller than a conventional resistor having axial leads. Chip resistors are very temperature-stable and also very rugged. The end electrodes are soldered directly to the copper traces of a circuit board, hence the name *surface-mount.*

FUSIBLE RESISTORS This type is a wire-wound resistor made to burn open easily when the power rating is exceeded. It then serves the dual functions of a fuse and a resistor to limit the current.

TEST-POINT QUESTION 2-1

Answers at end of chapter.

Answer True or False.

a. An R of 10 Ω with a 10-W rating would be a wire-wound resistor.
b. A resistance of 10,000 Ω is the same as 10 kΩ.
c. Axial leads are not used for carbon resistors.
d. Which is more temperature stable, a carbon-composition or a metal-film resistor?
e. Which is larger physically, a 1000-Ω ½ W or a 1000-Ω 1 W carbon-film resistor?

2-2 RESISTOR COLOR CODING

Because carbon resistors are small physically, they are color-coded to mark their R value in ohms. The basis of this system is the use of colors for numerical values, as listed in Table 2-1. In memorizing the colors, note that the darkest col-

TABLE 2-1	COLOR CODE		
COLOR	**VALUE**	**COLOR**	**VALUE**
Black	0	Green	5
Brown	1	Blue	6
Red	2	Violet	7
Orange	3	Gray	8
Yellow	4	White	9

ors, black and brown, are for the lowest numbers, zero and one, whereas white is for nine. The color coding is standardized by the Electronic Industries Association (EIA).

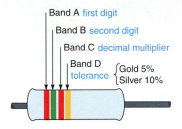

FIG. 2-7 How to read color stripes on carbon resistors for R in ohms.

RESISTANCE COLOR STRIPES   The use of bands or stripes is the most common system for color-coding carbon resistors, as shown in Fig. 2-7. Color stripes are printed at one end of the insulating body, which is usually tan. Reading from left to right, the first band closest to the edge gives the first digit in the numerical value of R. The next band marks the second digit. The third band is the decimal multiplier, which gives the number of zeroes after the two digits.

In Fig. 2-8a, the first stripe is red for 2 and the next stripe is green for 5. The red multiplier in the third stripe means add two zeroes to 25, or "this multiplier is 10^2." The result can be illustrated as follows:

$$
\begin{array}{ccccc}
\text{Red} & \text{Green} & & \text{Red} & \\
\downarrow & \downarrow & & \downarrow & \\
2 & 5 & \times & 100 & = 2500
\end{array}
$$

Therefore, this R value is 2500 Ω.

The example in Fig. 2-8b illustrates that black for the third stripe just means "do not add any zeroes to the first two digits." Since this resistor has red, green, and black stripes, the R value is 25 Ω.

RESISTORS UNDER 10 Ω   For these values, the third stripe is either gold or silver, indicating a fractional decimal multiplier. When the third stripe is gold, multiply the first two digits by 0.1. In Fig. 2-8c, the R value is

$$25 \times 0.1 = 2.5 \ \Omega$$

Silver means a multiplier of 0.01. If the third band in Fig. 2-8c were silver, the R value would be

$$25 \times 0.01 = 0.25 \ \Omega$$

It is important to realize that the gold and silver colors are used as decimal multipliers only in the third stripe. However, gold and silver are used most often as a fourth stripe to indicate how accurate the R value is.

RESISTOR TOLERANCE   The amount by which the actual R can be different from the color-coded value is the *tolerance,* usually given in percent. For instance, a 2000-Ω resistor with ± 10 percent tolerance can have resistance 10 percent above or below the coded value. This R, therefore, is between 1800 and 2200 Ω. The calculations are as follows:

$$10 \text{ percent of } 2000 \text{ is } 0.1 \times 2000 = 200$$

For $+10$ percent, the value is

$$2000 + 200 = 2200 \ \Omega$$

For -10 percent, the value is

$$2000 - 200 = 1800 \ \Omega$$

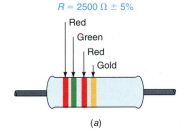

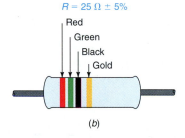

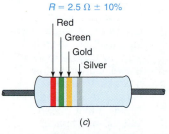

FIG. 2-8 Examples of color-coded R values, with percent tolerance.

As illustrated in Fig. 2-7, silver in the fourth band indicates a tolerance of ±10 percent; gold indicates ±5 percent. If there is no color band for tolerance, it is ±20 percent. The inexact value of carbon-composition resistors is a disadvantage of their economical construction. They usually cost only a few cents each, or less in larger quantities. In most circuits, though, a small difference in resistance can be tolerated.

FIVE-BAND CODE   Precision resistors often use a five-band code rather than the four-band code shown in Fig. 2-7. The purpose is to obtain more precise R values. With the five-band code, the first three color stripes indicate the first three digits, followed by the decimal multiplier in the fourth stripe and the tolerance in the fifth stripe. In the fifth stripe, the colors brown, red, green, blue, and violet represent the following tolerances:

Brown	±1%
Red	±2%
Green	±0.5%
Blue	±0.25%
Violet	±0.1%

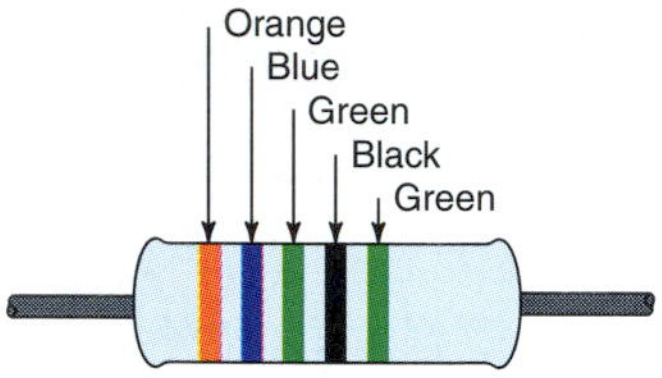

FIG. 2-9   Color-coded resistor.

Orange
Blue
Green
Black
Green

Example

EXAMPLE 1   What is the resistance indicated by the five-band color code in Fig. 2-9? Also, what ohmic range is permissible for the specified tolerance?

ANSWER   The first stripe is orange for the number 3, the second stripe is blue for the number 6, and the third stripe is green for the number 5. Therefore, the first three digits of the resistance are 3, 6, and 5, respectively. The fourth stripe, which is the multiplier, is black, which means add no zeros. The fifth stripe, which indicates the resistor tolerance, is green for ±0.5%. Therefore $R = 365 \, \Omega$ ±0.5%. The permissible ohmic range is calculated as $365 \times 0.005 = ±1.825 \, \Omega$, or 363.175 to 366.825 Ω.

WIRE-WOUND-RESISTOR MARKING   Usually, wire-wound resistors are big enough physically to have the R value printed on the insulating case. The tolerance is generally ±5 percent, except for precision resistors, which have a tolerance of ±1 percent or less.

Some small wire-wound resistors may be color-coded with stripes, however, like carbon resistors. In this case, the first stripe is double the width of the others to indicate a wire-wound resistor. This type may have a wattage rating of 3 or 4 W.

PREFERRED RESISTANCE VALUES   In order to minimize the problem of manufacturing different R values for an almost unlimited variety of circuits, specific values are made in large quantities so that they are cheaper and more easily available than unusual sizes. For resistors of ±10 percent, the preferred values are 10, 12, 15, 18, 22, 27, 33, 39, 47, 56, 68, and 82 with their decimal multiples. As examples, 47, 470, 4700, and 47,000 are preferred values. In this way, there is a preferred value available within 10 percent of any R value needed in a circuit. See Appendix A for a listing of preferred resistance values for tolerances of ±20%, ±10%, and ±5%.

ZERO-OHM RESISTORS   Believe it or not, there is such a thing as a zero-ohm resistor. In fact, zero-ohm resistors are quite common. The zero-ohm value is denoted by the use of a single black band around the center of the resistor body, as shown in Fig. 2-10. Zero-ohm resistors are available in ⅛- or ¼-W sizes. The actual resistance of a so-called ⅛-W zero-ohm resistor is about 0.004 Ω, whereas a ¼-W zero-ohm resistor has a resistance of approximately 0.003 Ω.

But why are zero-ohm resistors used in the first place? The reason is that for most printed-circuit boards, the components are inserted by automatic insertion machines (robots) rather than by human hands. In some instances it may be necessary to short two points on the printed-circuit board, in which case a piece of wire has to be placed between the two points. Because the robot can handle only components such as resistors, and not wires, zero-ohm resistors are used. Before zero-ohm resistors were developed, jumpers had to be installed by hand, which was time-consuming and expensive. Zero-ohm resistors may need to be used as a result of an after-the-fact design change which requires new point connections in a circuit.

CHIP RESISTOR CODING SYSTEM   A chip resistor, shown in Fig. 2-11a, has the following identifiable features.

Body color: white or off-white
Dark film on one side only (usually black, but may also be dark gray or green)
End electrodes (terminals) are C-shaped
Three-digit marking on either the film or the body side (usually the film)

The resistance value of a chip resistor is determined from the three-digit number printed on the film or body side of the component. The three digits provide the same information as the first three color stripes on a four-band resistor. This is shown in Fig. 2-11b. The first two digits indicate the first two numbers in the numerical value of the resistance; the third digit indicates the multiplier. If a four-digit number is used, the first three digits indicate the first three numbers in the numerical value of the resistance, while the fourth digit indicates the multiplier. The letter R is used to signify a decimal point for values between 1 and 10 ohms as in 2R7 = 2.7 Ω. Figure 2-11c shows the symbol used to denote a zero-ohm chip resistor.

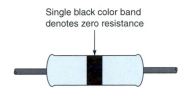

Single black color band denotes zero resistance

FIG. 2-10   A zero-ohm resistor is indicated by a single black color band around the body of the resistor.

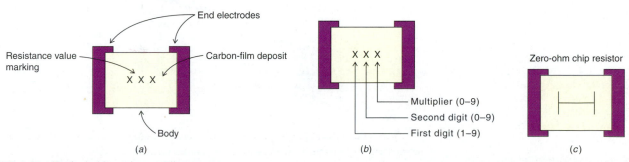

(a)   (b)   (c)

FIG. 2-11   Typical chip resistor coding system.

FIG. 2-12 Chip resistor with number coding.

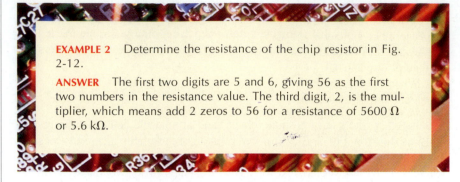

EXAMPLE 2 Determine the resistance of the chip resistor in Fig. 2-12.

ANSWER The first two digits are 5 and 6, giving 56 as the first two numbers in the resistance value. The third digit, 2, is the multiplier, which means add 2 zeros to 56 for a resistance of 5600 Ω or 5.6 kΩ.

TEST-POINT QUESTION 2-2

Answers at end of chapter.
a. Give the color for the number 4.
b. What tolerance does a silver stripe represent?
c. Give the multiplier for red in the third stripe.
d. Give *R* and the tolerance for a resistor coded with yellow, violet, brown, and gold stripes.
e. Assume that the chip resistor in Fig. 2-12 is marked 333. What is its resistance value in ohms?

2-3 VARIABLE RESISTORS

Variable resistors can be wire-wound, as in Fig. 2-3*b*, or the carbon type, illustrated in Fig. 2-13. Inside the metal case of Fig. 2-13*a*, the control has a circular disk, shown in Fig. 2-13*b*, that is the carbon-composition resistance element. It can be a thin coating on pressed paper or a molded carbon disk. Joined to the

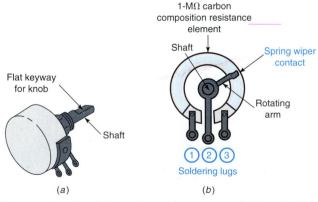

FIG. 2-13 Construction of variable carbon resistance control. Diameter is ¾ in. (*a*) External view. (*b*) Internal view of circular resistance element.

two ends are the external soldering-lug terminals 1 and 3. The middle terminal is connected to the variable arm that contacts the resistor element by a metal spring wiper. As the shaft of the control is turned, the variable arm moves the wiper to make contact at different points on the resistor element. The same idea applies to the slide control in Fig. 2-14, except that the resistor element is straight instead of circular.

When the contact moves closer to one end, the R decreases between this terminal and the variable arm. Between the two ends, however, R is not variable but always has the maximum resistance of the control.

Carbon controls are available with a total R from 1000 Ω to 5 MΩ, approximately. Their power rating is usually ½ to 2 W.

FIG. 2-14 Slide control for variable R. Length is 2 in.

TAPERED CONTROLS The way R varies with shaft rotation is called the *taper* of the control. With a linear taper, one-half rotation changes R by one-half the maximum value. Similarly, all values of R change in direct proportion to rotation. For a nonlinear taper, though, R can change more gradually at one end, with bigger changes at the opposite end. This effect is accomplished by different densities of carbon in the resistance element. For the example of a volume control, its audio taper allows smaller changes in R at low settings. Then it is easier to make changes without having the volume too loud or too low.

DECADE RESISTANCE BOX As shown in Fig. 2-15, the decade resistance box is a convenient unit for providing any one R within a wide range of values. It can be considered as test equipment for trying different R values in a circuit. Inside the box are six series strings of resistors, with one string for each dial switch.

FIG. 2-15 Decade resistance box for a wide range of R values.

The first dial connects in an R of 0 to 9 Ω. It is the *units* or $R \times 1$ dial.
The second dial has units of 10 from 0 to 90 Ω. It is the *tens* or $R \times 10$ dial.
The hundreds or $R \times 100$ dial has an R of 0 to 900 Ω.
The thousands or $R \times 1k$ dial has an R of 0 to 9000 Ω.
The ten-thousandths or $R \times 10k$ dial provides R values of 0 to 90,000 Ω.
The one-hundred thousandths or $R \times 100k$ dial provides R values of 0 to 900,000 Ω.

The six dial sections are connected internally so that their values add to one another. Then any value from 0 to 999,999 Ω can be obtained. Note the exact values that are possible. As an example, when all six dials are on 2, the total R equals $2 + 20 + 200 + 2,000 + 20,000 + 200,000 = 222,222$ Ω.

TEST-POINT QUESTION 2-3

Answers at end of chapter.
a. In Fig. 2-13, which terminal provides variable R?
b. Is an audio taper linear or nonlinear?
c. In Fig. 2-15, how much is the total R if the $R \times 100k$ and $R \times 10k$ dials are set to 4 and 7, respectively, and all other dials are set to zero?

2-4 RHEOSTATS AND POTENTIOMETERS

These are variable resistances, either carbon or wire-wound, used to vary the amount of current or voltage for a circuit. The controls can be used in either dc or ac applications.

A rheostat is a variable R with two terminals connected in series with a load. The purpose is to vary the amount of current.

A potentiometer, generally called a *pot* for short, has three terminals. The fixed maximum R across the two ends is connected across a voltage source. Then the variable arm is used to vary the voltage division between the center terminal and the ends. This function of a potentiometer is compared with a rheostat in Table 2-2.

TABLE 2-2 POTENTIOMETERS AND RHEOSTATS	
RHEOSTAT	**POTENTIOMETER**
Two terminals	Three terminals
In series with load and V source	Ends are connected across V source
Varies the I	Taps off part of V

RHEOSTAT CIRCUIT   The function of the rheostat R_2 in Fig. 2-16 is to vary the amount of current through R_1. For instance, R_1 can be a small light bulb that requires a specified I. Therefore, the two terminals of the rheostat R_2 are connected in series with R_1 and the source V in order to vary the total resistance R_T in the circuit. When R_T changes, the I changes, as read by the meter.

In Fig. 2-16b, R_1 is 5 Ω and the rheostat R_2 varies from 0 to 5 Ω. With R_2 at its maximum of 5 Ω, then R_T equals $5 + 5 = 10$ Ω. The I equals 0.15 A or 150 mA. (The method for calculating I given R and V is covered in Chapter 3, "Ohm's Law.")

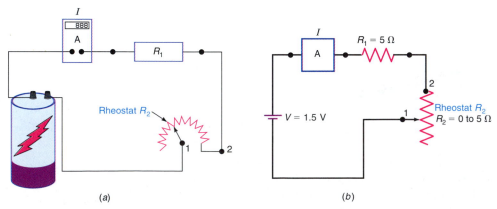

FIG. 2-16   Rheostat connected in series circuit to vary the current I. Symbol for current meter is A, for amperes. (*a*) Wiring diagram with digital meter for I. (*b*) Schematic diagram.

When R_2 is at its minimum value of 0 Ω, the R_T equals 5 Ω. Then I is 0.3 A or 300 mA for the maximum current. As a result, varying the rheostat changes the circuit resistance to vary the current through R_1. The I increases as R decreases.

It is important that the rheostat have a wattage rating high enough for the maximum I when R is minimum. Rheostats are often wire-wound variable resistors used to control relatively large values of current in low-resistance circuits for ac power applications.

POTENTIOMETER CIRCUIT   The purpose of the circuit in Fig. 2-17 is to tap off a variable part of the 100-V from the source. Consider this circuit in two parts:

1.   The applied V is input to the two end terminals of the potentiometer.
2.   The variable V is output between the variable arm and an end terminal.

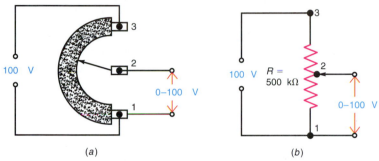

(a)   (b)

FIG. 2-17   Potentiometer connected across voltage source to function as a voltage divider. (a) Wiring diagram. (b) Schematic diagram.

Two pairs of connections to the three terminals are necessary, with one terminal common to the input and output. One pair connects the source V to the end terminals 1 and 3. The other pair of connections is between the variable arm at the center terminal and one end. This end has double connections for input and output. The other end has only an input connection.

When the variable arm is at the middle value of the 500-kΩ R in Fig. 2-17, the 50 V is tapped off between terminals 2 and 1 as one-half the 100-V input. The other 50 V is between terminals 2 and 3. However, this voltage is not used for output.

As the control is turned up to move the variable arm closer to terminal 3, more of the input voltage is available between 2 and 1. With the control at its maximum R, the voltage between 2 and 1 is the entire 100 V. Actually, terminal 2 then is the same as 3.

When the variable arm is at minimum R, rotated to terminal 1, the output between 2 and 1 is zero. Now all the applied voltage is across 2 and 3, with no output for the variable arm. It is important to note that the source voltage is not short-circuited. The reason is that the maximum R of the pot is always across the applied V, regardless of where the variable arm is set. Typical examples of small potentiometers used in electronic circuits are shown in Fig. 2-18.

FIG. 2-18   Small potentiometers and trimmers often used for variable controls in electronic circuits. Terminal leads formed for insertion into a PC board.

POTENTIOMETER USED AS A RHEOSTAT Commercial rheostats are generally wire-wound high-wattage resistors for power applications. However, a small low-wattage rheostat is often needed in electronic circuits. One example is a continuous tone control in a receiver. The control requires the variable series resistance of a rheostat but dissipates very little power.

A method of wiring a potentiometer as a rheostat is to connect just one end of the control and the variable arm, using only two terminals. The third terminal is open, or floating, not connected to anything.

Another method is to wire the unused terminal to the center terminal. When the variable arm is rotated, different amounts of resistance are short-circuited. This method is preferable because there is no floating resistance.

Either end of the potentiometer can be used for the rheostat. The direction of increasing R with shaft rotation reverses, though, for connections at opposite ends. Also, the taper is reversed on a nonlinear control.

TEST-POINT QUESTION 2-4

Answers at end of chapter.
a. How many circuit connections to a potentiometer are needed?
b. How many circuit connections to a rheostat are needed?
c. In Fig. 2-17, with a 500-kΩ linear potentiometer, how much is the output voltage with 400 kΩ between terminals 1 and 2?

2-5 POWER RATING OF RESISTORS

In addition to having the required ohms value, a resistor should have a wattage rating high enough to dissipate the power produced by the current flowing through the resistance, without becoming too hot. Carbon resistors in normal operation are often quite warm, up to a maximum temperature of about 85°C, which is close to the 100°C boiling point of water. Carbon resistors should not be so hot, however, that they "sweat" beads of liquid on the insulating case. Wire-wound resistors operate at very high temperatures, a typical value being 300°C for the maximum temperature. If a resistor becomes too hot because of excessive power dissipation, it can change appreciably in resistance value or burn open.

The power rating is a physical property that depends on the resistor construction, especially physical size. Note the following:

1. A larger physical size indicates a higher power rating.
2. Higher-wattage resistors can operate at higher temperatures.
3. Wire-wound resistors are physically larger with higher wattage ratings than carbon resistors.

For approximate sizes, a 2-W carbon resistor is about 1 in. long with ¼ in. diameter; a ¼-W resistor is about 0.35 in. long with diameter of 0.1 in.

For both types, a higher power rating allows a higher voltage rating. This rating gives the highest voltage that may be applied across the resistor without internal arcing. As examples for carbon resistors, the maximum voltage is 500 V for a 1-W rating, 250 V for ¼-W, and 150 V for ⅛-W. In wire-wound resistors, excessive voltage can produce an arc between turns; in carbon resistors, the arc is between carbon granules.

SHELF LIFE Resistors keep their characteristics almost indefinitely when not used. Without any current in a circuit to heat the resistor, it has practically no change with age. The shelf life of resistors is usually no problem, therefore.

TEST-POINT QUESTION 2-5

Answers at end of chapter.

Answer True or False.
a. The shelf life of a resistor is about 6 months.
b. Resistors should not operate above a temperature of 0°C.
c. Would a 10-W resistor be the wire-wound or carbon type?

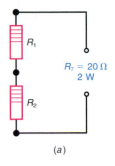

(a)

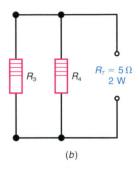

(b)

2-6 SERIES AND PARALLEL COMBINATIONS OF RESISTORS

In some cases two or more resistors are combined to obtain a desired R value with a higher wattage rating. Several examples are shown in Fig. 2-19.

The total resistance R_T depends on the series or parallel connections. However, the combination has a power rating equal to the sum of the individual values, whether the resistors are in series or in parallel. The reason is that the total physical size increases with each added resistor. Equal resistors are generally used in order to have equal distribution of I, V, and P.

In general, series resistors add for a higher R_T. With parallel resistors, R_{EQ} is reduced. More details of series and parallel circuits are explained in Chaps. 4 and 5.

In Fig. 2-19a, the two equal resistors in series double the resistance for R_T. Also, the power rating of the combination is twice the rating for one resistor.

In Fig. 2-19b, the two equal resistors in parallel have one-half the resistance for R_T. However, the combined power rating is still twice the rating for one resistor.

In Fig. 2-19c, the series-parallel combination of four equal resistors makes R_T the same as each R. The total power rating, though, is four times the rating for one resistor.

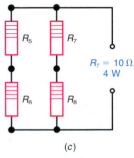

(c)

FIG. 2-19 Total resistance R_T and wattage rating for series and parallel circuits. Each R is 10 Ω with a 1-W rating. (a) For R_1 and R_2 in series, add wattage ratings; R_T is $2 \times R$. (b) For R_3 and R_4 in parallel, add wattage ratings; R_T is ½ R. (c) For this series-parallel combination, add all wattage ratings; R_T is equal to R.

TEST-POINT QUESTION 2-6

Answers at end of chapter.

Give R_T and the combined power rating of two 1-kΩ resistors rated at 2 W connected:
a. In series
b. In parallel
c. Do both methods provide the same power rating?

2-7 RESISTOR TROUBLES

The most common trouble in resistors is an open. When the open resistor is a series component, there is no current in the entire series path.

NOISY CONTROLS   In applications such as volume and tone controls, carbon controls are preferred because the smoother change in resistance results in less noise when the variable arm is rotated. With use, however, the resistance element becomes worn by the wiper contact, making the control noisy. When a volume or tone control makes a scratchy noise as the shaft is rotated, it indicates a worn-out resistance element.

CHECKING RESISTORS WITH AN OHMMETER   Resistance measurements are made with an ohmmeter. The ohmmeter has its own voltage source so that it is always used without any external power applied to the resistance being measured. Separate the resistance from its circuit by disconnecting one lead of the resistor. Then connect the ohmmeter leads across the resistance to be measured.

An open resistor reads infinitely high ohms. For some reason, infinite ohms is often confused with zero ohms. Remember, though, that infinite ohms means an open circuit. The current is zero, but the resistance is infinitely high. Furthermore, it is practically impossible for a resistor to become short-circuited in itself. The resistor may be short-circuited by some other part of the circuit. However, the construction of resistors is such that the trouble they develop is an open circuit, with infinitely high ohms.

The ohmmeter must have an ohms scale capable of reading the resistance value, or the resistor cannot be checked. In checking a 10-MΩ resistor, for instance, if the highest R the ohmmeter can read is 1 MΩ, it will indicate infinite resistance, even if the resistor has its normal value of 10 MΩ. An ohms scale of 100 MΩ or more should be used for checking such high resistances.

To check resistors of less than 10 Ω, a low-ohms scale of about 100 Ω or less is necessary. Center scale should be 6 Ω or less. Otherwise, the ohmmeter will read a normally low resistance value as zero ohms.

When checking resistance in a circuit, it is important to be sure there are no parallel resistance paths. Otherwise, the measured resistance can be much lower than the actual resistor value, as illustrated in Fig. 2-20a. Here, the ohmmeter reads the resistance of R_2 in parallel with R_1. To check across R_2 alone, one end is disconnected, as in Fig. 2-20b.

For very high resistances, it is important not to touch the ohmmeter leads. There is no danger of shock, but the body resistance of about 50,000 Ω as a parallel path will lower the ohmmeter reading.

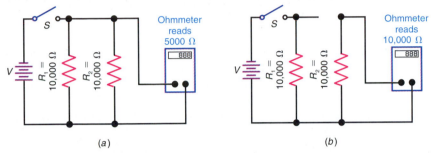

(a)   (b)

FIG. 2-20   Parallel R_1 can lower the ohmmeter reading for testing R_2. (a) The two resistances R_1 and R_2 are in parallel. (b) R_2 is isolated by disconnecting one end of R_1.

CHANGED VALUE OF R   In many cases, the value of a carbon-composition resistor can exceed its allowed tolerance, due to normal resistor heating over a long period of time. In most instances the value change is seen as an increase in R. This is known as *aging*. As you know, carbon-film and metal-film resistors age very little. Surface-mount resistors should never be rubbed or scraped, as this will remove some of the carbon deposit and cause its resistance to change.

TEST-POINT QUESTION 2-7

Answers at end of chapter.
a. What is the ohmmeter reading for a short circuit?
b. What is the ohmmeter reading for an open resistor?
c. Which has a higher R, an open or a short circuit?
d. Which is more likely to change in R value after many years of use, a metal-film or a carbon-composition resistor?

2 SUMMARY AND REVIEW

- The most common types of resistors include carbon-composition, carbon-film, metal-film, wire-wound, and surface-mount or chip resistors. Carbon-film and metal-film resistors are better than carbon-composition resistors because they have tighter tolerances, are less affected by temperature and aging, and generate less noise internally.
- Wire-wound resistors are typically used in high-current applications. Wire-wound resistors are available with wattage ratings of about 5 to 100 W.
- Resistors are usually color-coded to indicate their resistance value in ohms. Either a four-band or a five-band code is used. The five-band code is used for more precise R values. Chip resistors use a three- or four-digit code to indicate their resistance value.
- Zero-ohm resistors are used with automatic insertion machines when it is desired to short two points on a printed-circuit board. Zero-ohm resistors are available in $\frac{1}{8}$- or $\frac{1}{4}$-W ratings.
- A potentiometer is a variable resistor with three terminals. It is used to vary the voltage in a circuit. A rheostat is a variable resistor with two terminals. It is used to vary the current in a circuit.
- The physical size of a resistor determines its wattage rating: the larger the physical size, the larger the wattage rating. There is no correlation between a resistor's physical size and its resistance value.
- Resistors can be combined in series, parallel, or series-parallel to obtain a higher overall power rating. The total power rating equals the sum of the individual wattage ratings of the resistors.
- The most common trouble in resistors is an open. An ohmmeter across the leads of an open resistor will read infinite, assuming there is no other parallel path across the resistor.

SELF-TEST

ANSWERS AT BACK OF BOOK.

Choose (*a*), (*b*), (*c*), or (*d*).

1. A carbon composition resistor having only three color stripes has a tolerance of (*a*) ±5%; (*b*) ±20%; (*c*) ±10%; (*d*) ±100%.
2. A resistor with a power rating of 25 W is most likely a (*a*) carbon-composition resistor; (*b*) metal-film resistor; (*c*) surface-mount resistor; (*d*) wire-wound resistor
3. When checked with an ohmmeter, an open resistor measures (*a*) infinite resistance; (*b*) its color-coded value; (*c*) zero resistance; (*d*) less than its color-coded value.
4. One precaution to observe when checking resistors with an ohmmeter is to (*a*) check high resistances on the lowest ohms range; (*b*) check low resistances on the highest ohms range; (*c*) disconnect all parallel paths; (*d*) make sure your fingers are touching each test lead.

5. A chip resistor is marked 394. Its resistance value is (*a*) 39.4 Ω; (*b*) 394 Ω; (*c*) 390,000 Ω; (*d*) 39,000 Ω.

6. A carbon-film resistor is color-coded with red, violet, black, and gold stripes. What are its resistance and tolerance? (*a*) 27 Ω ± 5%; (*b*) 270 Ω ± 5%; (*c*) 270 Ω ± 10%; (*d*) 27 Ω ± 10%.

7. A potentiometer is a (*a*) three-terminal device used to vary the voltage in a circuit; (*b*) two-terminal device used to vary the current in a circuit; (*c*) fixed resistor; (*d*) two-terminal device used to vary the voltage in a circuit.

8. A metal-film resistor is color-coded with brown, green, red, brown, and blue stripes. What are its resistance and tolerance? (*a*) 1500 Ω ± 1.25%; (*b*) 152 Ω ± 1%; (*c*) 1521 Ω ± 0.5%; (*d*) 1520 Ω ± 0.25%.

9. Which of the following resistors has the smallest physical size? (*a*) wire-wound resistors; (*b*) carbon-composition resistors; (*c*) surface-mount resistors; (*d*) potentiometers.

10. Which of the following statements is true? (*a*) Resistors always have axial leads. (*b*) Resistors are always made from carbon. (*c*) There is no correlation between the physical size of a resistor and its resistance value. (*d*) The shelf life of a resistor is about 1 year.

QUESTIONS

1. List five different types of fixed resistors.
2. List the advantages of using a metal-film resistor versus a carbon-composition resistor.
3. Draw the schematic symbols for a (*a*) fixed resistor; (*b*) potentiometer; (*c*) rheostat.
4. List the colors corresponding to the decimal digits 0 through 9.
5. How can a technician identify a wire-wound resistor that is color-coded?
6. Explain how the resistance of a carbon-film resistor can be controlled during the manufacturing process.
7. Explain an application where it may be desired to use a decade resistance box.
8. List the differences between a potentiometer and a rheostat.
9. For resistors using the four-band code, what are the values for gold and silver as decimal multipliers in the third band?
10. Give the color code for the following resistors using the four-band code: 1.2 Ω; 330 Ω; 2.2 MΩ; 680 kΩ; 1 kΩ; 10 kΩ; and 47 Ω. Each resistor has a tolerance of ±5%.
11. Give the color code for the following resistors using the five-band code: 3320 Ω ± 1%; 7.23 kΩ ± 0.1%; 333 kΩ ± 2%; and 47.5 Ω + 0.25%.
12. Briefly describe how you would check to see whether a 1-MΩ resistor is open or not. Give two precautions to make sure the test is not misleading.
13. Give at least two examples of variable resistors you have seen used for controls in electronic circuits.
14. Show how to connect three 1-kΩ resistors for a total R_T of 1.5 kΩ.
15. Show how to connect resistors for the following examples: (*a*) two 20-kΩ 1-W resistors for a total R_T of 10 kΩ with a power rating of 2 W; (*b*) two 20-kΩ 1-W resistors for an R_T of 40 kΩ and a power rating of 2 W; (*c*) four 20-kΩ 1-W resistors for an R_T of 20 kΩ and a power rating of 4 W.

PROBLEMS

ANSWERS TO ODD-NUMBERED PROBLEMS AT BACK OF BOOK.

1. Indicate the resistance and tolerance for each resistor shown in Fig. 2-21.
2. Indicate the resistance and tolerance for each resistor shown in Fig. 2-22.
3. Indicate the resistance for each chip resistor shown in Fig. 2-23.
4. Calculate the permissible ohmic range of a resistor whose resistance value and tolerance are (*a*) 3.9 kΩ ± 5%; (*b*) 100 Ω ± 10%; (*c*) 120 kΩ ± 2%; (*d*) 2.2 Ω ± 5%; (*e*) 75 Ω ± 1%.
5. Using the four-band code, indicate the colors of the bands for each of the following resistors: (*a*) 10 kΩ ± 5%; (*b*) 2.7 Ω ± 5%; (*c*) 5.6 kΩ ± 10%; (*d*) 1.5 MΩ ± 5%; (*e*) 0.22 Ω ± 5%.
6. Using the five-band code, indicate the colors of the bands for each of the following resistors: (*a*) 110 Ω ± 1%; (*b*) 34 kΩ ± 0.5%; (*c*) 82.5 kΩ ± 2%; (*d*) 62.6 Ω ± 1%; (*e*) 105 kΩ ± 0.1%.
7. Refer to Fig. 2-15 on page 55. Indicate the total resistance R_T for each of the different dial settings in Table 2-3.

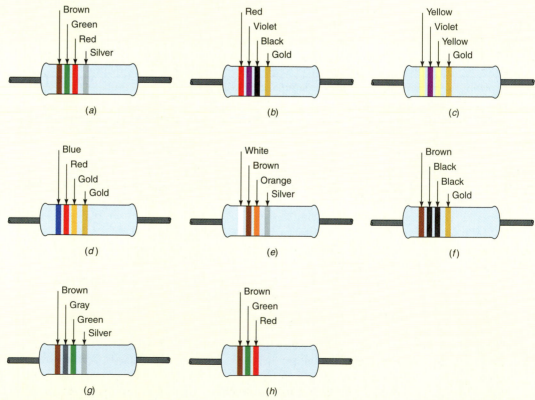

FIG. 2-21 Resistors for Prob. 1.

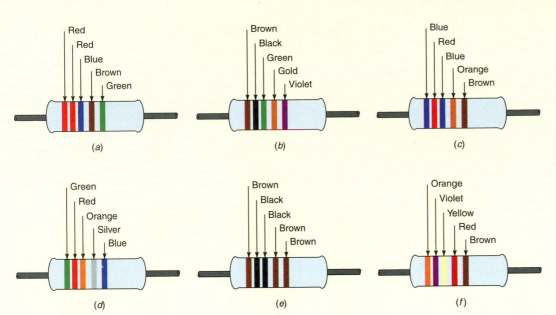

FIG. 2-22 Resistors for Prob. 2.

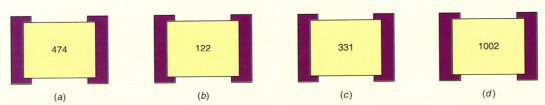

FIG. 2-23 Chip resistors for Prob. 3.

TABLE 2-3 DECADE RESISTANCE BOX (FIG. 2-15) DIAL SETTINGS FOR PROB. 7					
	R × 100k	R × 10k	R × 1k	R × 100	R × 10	R × 1
(a)	6	8	0	2	2	5
(b)	0	0	8	2	5	0
(c)	0	1	8	5	0	3
(d)	2	7	5	0	6	0
(e)	0	6	2	9	8	4

8. Show how to wire two 100-Ω 5-W resistors for an R_T of 200 Ω and a total power rating of 10 W.

9. Show how to wire the resistors in Prob. 8 for an R_T of 50 Ω and a total power rating of 10 W.

10. Show two different ways to wire a potentiometer so it will work as a rheostat.

CRITICAL THINKING

1. A manufacturer of carbon-film resistors specifies a maximum working voltage of 250 V for all its ¼-W resistors. Exceeding 250 V causes internal arcing within the resistor. Above what minimum resistance will the maximum working voltage be exceeded before its ¼-W power dissipation rating is exceeded? **Hint:** The maximum voltage which produces the rated power dissipation can be calculated as $V_{max} = \sqrt{P \times R}$.

ANSWERS TO TEST-POINT QUESTIONS

2-1 a. T
b. T
c. F
d. metal film
e. 1000 Ω 1 W

2-2 a. yellow
b. ±10 percent
c. 100
d. 470 Ω ±5 percent
e. 33,000 Ω or 33 kΩ

2-3 a. terminal 2
b. nonlinear
c. 470,000 Ω or 470 kΩ

2-4 a. four connections to three terminals
b. two
c. 80 V

2-5 a. F
b. F
c. wire-wound

2-6 a. 2 kΩ 4 W
b. 0.5 kΩ 4 W
c. yes

2-7 a. 0 Ω
b. infinite ohms
c. open circuit
d. carbon-composition resistor

CHAPTER 3

OHM'S LAW

When we have a circuit with voltage V applied and resistance R in the closed path, it is important to know how much the current I is in the circuit. This chapter explains how the amount of I increases with more applied voltage but is less with more resistance. Specifically, $I = V/R$, as determined in 1828 by the experiments of Georg Simon Ohm. If we know any two of the factors R, I, and V, the third can be calculated. As an example, when 6 V is applied across an R of 2 Ω, the I is 6 V/2 Ω = 3 A.

Ohm's law is also used to determine the amount of electric power P in the circuit. The P can be calculated as $V \times I$ or as $I^2 \times R$. All these relations derived from Ohm's law apply to both dc and ac circuits.

CHAPTER OBJECTIVES

Upon completion of this chapter, you should be able to:

- *List* the three forms of Ohm's law.
- *Use* Ohm's law to calculate the current, voltage, or resistance in a circuit.
- *List* the multiple and submultiple units of voltage, current, and resistance.
- *Explain* the difference between a linear and a nonlinear resistance.
- *Explain* the difference between work and power and list the units of each.
- *Calculate* the power in a circuit when the voltage and current, current and resistance, or voltage and resistance are known.
- *Understand* the shock hazards associated with working with electricity.
- *Explain* the difference between an open circuit and short circuit.

IMPORTANT TERMS IN THIS CHAPTER

volt	joule	mega
ampere	milli	linear graph
ohm	micro	power
watt	kilo	volt-ampere characteristic

TOPICS COVERED IN THIS CHAPTER

3-1 THE CURRENT *I* = *V/R*

If we keep the same resistance in a circuit but vary the voltage, the current will vary. The circuit in Fig. 3-1 demonstrates this idea. The applied voltage *V* can be varied from 0 to 12 V, as an example. The bulb has a 12-V filament, which requires this much voltage for its normal current to light with normal intensity. The meter *I* indicates the amount of current in the circuit for the bulb.

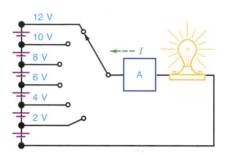

FIG. 3-1 Increasing the applied voltage *V* produces more current *I* to light the bulb with more intensity.

With 12 V applied, the bulb lights, indicating normal current. When *V* is reduced to 10 V, there is less light because of less *I*. As *V* decreases, the bulb becomes dimmer. For zero volts applied there is no current and the bulb cannot

light. In summary, the changing brilliance of the bulb shows that the current is varying with the changes in applied voltage.

For the general case of any V and R, Ohm's law is

▶ $$I = \frac{V}{R}$$ (3-1)

where I is the amount of current through the resistance R connected across the source of potential difference V. With volts as the practical unit for V and ohms for R, the amount of current I is in amperes. Therefore,

▶ $$\text{Amperes} = \frac{\text{volts}}{\text{ohms}}$$

This formula says to simply divide the voltage across R by the ohms of resistance between the two points of potential difference to calculate the amperes of current through R. In Fig. 3-2, for instance, with 6 V applied across a 3-Ω resistance, by Ohm's law the amount of current I equals $\frac{6}{3}$ or 2 A.

HIGH VOLTAGE BUT LOW CURRENT   It is important to realize that with high voltage, the current can have a low value when there is a very high resistance in the circuit. For example, 1000 V applied across 1,000,000 Ω results in a current of only $\frac{1}{1000}$ A. By Ohm's law,

$$I = \frac{V}{R}$$
$$= \frac{1000 \text{ V}}{1,000,000 \ \Omega} = \frac{1}{1000}$$
$$I = 0.001 \text{ A}$$

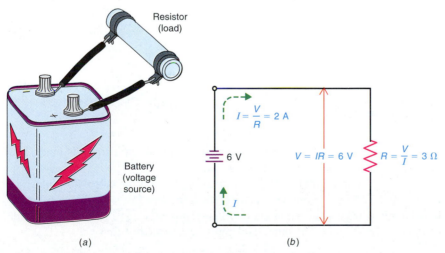

(a)   (b)

FIG. 3-2   Example of using Ohm's law. (*a*) Wiring diagram of a circuit with a 6-V battery for V applied across a load R. (*b*) Schematic diagram of the circuit with values for I and R calculated by Ohm's law.

The practical fact is that high-voltage circuits usually do have a small value of current in electronic equipment. Otherwise, tremendous amounts of power would be necessary.

LOW VOLTAGE BUT HIGH CURRENT   At the opposite extreme, a low value of voltage in a very low resistance circuit can produce a very high current. A 6-V battery connected across a resistance of 0.01 Ω produces 600 A of current:

$$I = \frac{V}{R}$$

$$= \frac{6\ V}{0.01\ \Omega}$$

$$I = 600\ A$$

LESS *I* WITH MORE *R*   Note the values of *I* in the following two examples also:

Example

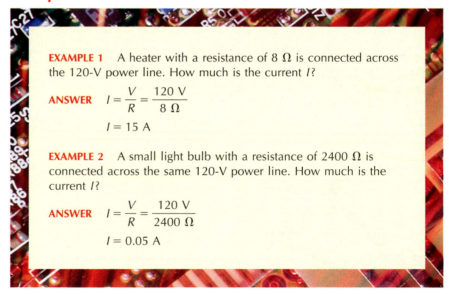

EXAMPLE 1   A heater with a resistance of 8 Ω is connected across the 120-V power line. How much is the current *I*?

ANSWER $I = \dfrac{V}{R} = \dfrac{120\ V}{8\ \Omega}$

$\qquad I = 15\ A$

EXAMPLE 2   A small light bulb with a resistance of 2400 Ω is connected across the same 120-V power line. How much is the current *I*?

ANSWER $I = \dfrac{V}{R} = \dfrac{120\ V}{2400\ \Omega}$

$\qquad I = 0.05\ A$

Although both cases have the same 120 V applied, the current is much less in Example 2 because of the higher resistance.

To do a division problem like *V/R* in Example 2 on the calculator, punch in the number 120 for the numerator, then press the key ⊕ for division before punching in 2400 for the denominator. Finally, press the ⊜ key for the answer of 0.05 on the display. The numerator must be punched in first.

TYPICAL *V* AND *I*   Transistors and integrated circuits generally operate with a dc supply of 5, 6, 9, 12, 15, 24, or 50 V. The current is usually in millionths or thousandths of one ampere up to about 5 A.

3-2 THE VOLTAGE $V = IR$

Referring back to Fig. 3-2, the voltage across R must be the same as the source V because the resistance is connected directly across the battery. The numerical value of this V is equal to the product $I \times R$. For instance, the IR voltage in Fig. 3-2 is 2 A × 3 Ω, which equals the 6 V of the applied voltage. The formula is

▶ $V = IR$ (3-2)

With I in ampere units and R in ohms, their product V is in volts. Actually, this must be so because the I value equal to V/R is the amount that allows the IR product to be the same as the voltage across R.

To do a multiplication problem like $I \times R$ for this example on the calculator, punch in the factor 2, then press the ⊗ key for multiplication before punching in 3 for the other factor. Finally, press the ⊜ key for the answer of 6 on the display. The factors can be multiplied in any order.

Besides the numerical calculations possible with the IR formula, it is useful to consider that the IR product means voltage. Whenever there is current through a resistance, it must have a potential difference across its two ends equal to the IR product. If there were no potential difference, no electrons could flow to produce the current.

3-3 THE RESISTANCE $R = V/I$

As the third and final version of Ohm's law, the three factors V, I, and R are related by the formula

▶ $R = \dfrac{V}{I}$ (3-3)

In Fig. 3-2, R is 3 Ω because 6 V applied across the resistance produces 2 A through it. Whenever V and I are known, the resistance can be calculated as the voltage across R divided by the current through it.

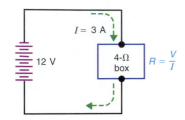

FIG. 3-3 The resistance R of any component is its V/I ratio.

Physically, a resistance can be considered as some material with elements having an atomic structure that allows free electrons to drift through it with more or less force applied. Electrically, though, a more practical way of considering resistance is simply as a V/I ratio. Anything that allows 1 A of current with 10 V applied has a resistance of 10 Ω. This V/I ratio of 10 Ω is its characteristic. If the voltage is doubled to 20 V, the current will also double to 2 A, providing the same V/I ratio of a 10-Ω resistance.

Furthermore, we do not need to know the physical construction of a resistance to analyze its effect in a circuit, so long as we know its V/I ratio. This idea is illustrated in Fig. 3-3. Here, a box with some unknown material in it is connected into a circuit where we can measure the 12 V applied across the box and the 3 A of current through it. The resistance is 12 V/3 A, or 4 Ω. There may be liquid, gas, metal, powder, or any other material in the box, but electrically it is just a 4-Ω resistance because its V/I ratio is 4.

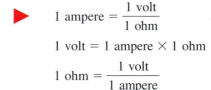

TEST-POINT QUESTION 3-3

Answers at end of chapter.
  a. Calculate R for 12 V with 0.003 A.
  b. Calculate R for 12 V with 0.006 A.
  c. Calculate R for 12 V with 0.001 A.

3-4 PRACTICAL UNITS

The three forms of Ohm's law can be used to define the practical units of current, potential difference, and resistance as follows:

▶ $$1 \text{ ampere} = \frac{1 \text{ volt}}{1 \text{ ohm}}$$

$$1 \text{ volt} = 1 \text{ ampere} \times 1 \text{ ohm}$$

$$1 \text{ ohm} = \frac{1 \text{ volt}}{1 \text{ ampere}}$$

One ampere is the amount of current through a one-ohm resistance that has one volt of potential difference applied across it.

One volt is the potential difference across a one-ohm resistance that has one ampere of current through it.

One ohm is the amount of opposition in a resistance that has a V/I ratio of 1, allowing one ampere of current with one volt applied.

In summary, the circle diagram in Fig. 3-4 for $V = IR$ can be helpful in using Ohm's law. Put your finger on the unknown quantity and the desired formula remains. The three possibilities are

Cover V and you have IR.
Cover I and you have V/R.
Cover R and you have V/I.

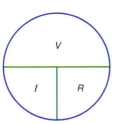

FIG. 3-4 A circle diagram to help in memorizing the Ohm's law formulas $V = IR$, $I = V/R$, and $R = V/I$. The V is always at the top.

Answers at end of chapter.

a. Calculate V for 0.007 A through 5000 Ω.

b. Calculate the amount of I for 12,000 V across 6,000,000 Ω.

c. Calculate R for 8 V with 0.004 A.

3-5 MULTIPLE UNITS

The basic units—ampere, volt, and ohm—are practical values in most electric power circuits, but in many electronics applications these units are either too small or too big. As examples, resistances can be a few million ohms, the output of a high-voltage supply in a television receiver is about 20,000 V, and current through tubes and transistors is generally thousandths or millionths of an ampere.

In such cases, it is helpful to use multiples of the basic units. These multiples are based on the decimal system of tens, hundreds, thousands, etc. The common conversions for V, I, and R are summarized at the end of this chapter, but a complete listing of all the prefixes is in App. A. Note that capital M is used for 10^6 to distinguish it from small m for 10^{-3}.*

Example

EXAMPLE 3   The I of 8 mA flows through a 5-kΩ R. How much is the IR voltage?

ANSWER $V = IR = 8 \times 10^{-3} \times 5 \times 10^3 = 8 \times 5$

$V = 40$ V

In general, milliamperes multiplied by kilohms results in volts for the answer, as 10^{-3} and 10^3 cancel.

EXAMPLE 4   How much current is produced by 60 V across 12 kΩ?

ANSWER $I = \dfrac{V}{R} = \dfrac{60}{12 \times 10^3} = 5 \times 10^{-3} = 5$ mA

Note that volts across kilohms produces milliamperes of current. Similarly, volts across megohms produces microamperes.

---

*The notations 10^6 and 10^{-3} are known as powers of 10. A detailed explanation of the uses and operations involving powers of 10 may be found in either B. Grob, *Mathematics for Basic Electronics,* or M. Schultz, *Problems in Basic Electronics*, Glencoe/McGraw-Hill, Columbus, Ohio.

In summary, common combinations to calculate the current I are

$$\frac{V}{k\Omega} = mA \quad \text{and} \quad \frac{V}{M\Omega} = \mu A$$

Also, common combinations to calculate IR voltage are

$$mA \times k\Omega = V$$
$$\mu A \times M\Omega = V$$

These relationships occur often in electronic circuits because the current is generally in units of milliamperes or microamperes. A useful relationship to remember is that 1 mA is equal to 1000 μA.

TEST-POINT QUESTION 3-5

Answers at end of chapter.
a. Change the following to basic units with powers of 10 instead of metric prefixes: 6 mA, 5 kΩ, and 3 μA.
b. Change the following powers of 10 to units with metric prefixes: 6×10^{-3} A, 5×10^3 Ω, and 3×10^{-6} A.
c. Which is larger, 2 mA or 20 μA?

3-6 THE LINEAR PROPORTION BETWEEN V AND I

The Ohm's law formula $I = V/R$ states that V and I are directly proportional for any one value of R. This relation between V and I can be analyzed by using a fixed resistance of 2 Ω for R_L, as in Fig. 3-5. Then when V is varied, the meter shows I values directly proportional to V. For instance, with 12 V, I equals 6 A; for 10 V, the current is 5 A; an 8-V potential difference produces 4 A.

All the values of V and I are listed in the table in Fig. 3-5b and plotted in the graph in Fig. 3-5c. The I values are one-half the V values because R is 2 Ω. However, I is zero with zero volts applied.

PLOTTING THE GRAPH The voltage values for V are marked on the horizontal axis, called the x axis or *abscissa*. The current values I are on the vertical axis, called the y axis or *ordinate*.

Because the values for V and I depend on each other, they are variable factors. The independent variable here is V because we assign values of voltage and note the resulting current. Generally, the independent variable is plotted on the x axis, which is why the V values are shown here horizontally while the I values are on the ordinate.

The two scales need not be the same. The only requirement is that equal distances on each scale represent equal changes in magnitude. On the x axis here

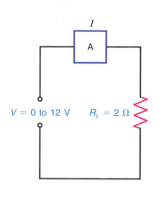

Volts V	Ohms Ω	Amperes A
0	2	0
2	2	1
4	2	2
6	2	3
8	2	4
10	2	5
12	2	6

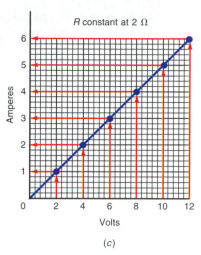

(a) (b) (c)

FIG. 3-5   Experiment to show that I increases in direct proportion to V with the same R. (a) Circuit with variable V but constant R. (b) Table of increasing I for higher V. (c) Graph of V and I values. This is a linear volt-ampere characteristic. It shows a direct proportion between V and I.

2-V steps are chosen, while the y axis has 1-A scale divisions. The zero point at the origin is the reference.

The plotted points in the graph show the values in the table. For instance, the lowest point is 2 V horizontally from the origin, and 1 A up. Similarly, the next point is at the intersection of the 4-V mark and the 2-A mark.

A line joining these plotted points includes all values of I, for any value of V, with R constant at 2 Ω. This also applies to values not listed in the table. For instance, if we take the value of 7 V, up to the straight line and over to the I axis, the graph shows 3.5 A for I.

VOLT-AMPERE CHARACTERISTIC   The graph in Fig. 3-5c is called the volt-ampere characteristic of R. It shows how much current the resistor allows for different voltages. Multiple and submultiple units of V and I can be used, though. For transistors and tubes the units of I are often milliamperes or microamperes.

LINEAR RESISTANCE   The straight-line graph in Fig. 3-5 shows that R is a linear resistor. A linear resistance has a constant value of ohms. Its R does not change with the applied voltage. Then V and I are directly proportional. Doubling the value of V from 4 to 8 V results in twice the current, from 2 to 4 A. Similarly, three or four times the value of V will produce three or four times I, for a proportional increase in current.

NONLINEAR RESISTANCE   This type has a nonlinear volt-ampere characteristic. As an example, the resistance of the tungsten filament in a light bulb is nonlinear. The reason is that R increases with more current

ABOUT ELECTRONICS

The unit of measure for resistance (ohm) is named for German physicist Georg Simon Ohm. Ohm is also known for his development of Ohm's law: voltage = current × resistance.

as the filament becomes hotter. Increasing the applied voltage does produce more current, but I does not increase in the same proportion as the increase in V.

INVERSE RELATION BETWEEN I AND R   Whether R is linear or not, the current I is less for more R, with the applied voltage constant. This is an inverse relation, meaning that I goes down as R goes up. Remember that in the formula $I = V/R$, the resistance is in the denominator. A higher value of R actually lowers the value of the complete fraction.

As an example, let V be constant at 1 V. Then I is equal to the fraction $1/R$. As R increases, the values of I decrease. For R of 2 Ω, I is ½ or 0.5 A. For a higher R of 10 Ω, I will be lower at $\frac{1}{10}$ or 0.1 A.

TEST-POINT QUESTION 3-6

Answers at end of chapter.

Refer to the graph in Fig. 3-5c.
a. Are the values of I on the y or x axis?
b. Is this R linear or nonlinear?
c. If the voltage across a 5-Ω resistor increases from 10 V to 20 V, what happens to I?
d. The voltage across a 5-Ω resistor is 10 V. If R is doubled to 10 Ω, what happens to I?

3-7 ELECTRIC POWER

The unit of electric power is the *watt* (W), named after James Watt (1736–1819). One watt of power equals the work done in one second by one volt of potential difference in moving one coulomb of charge.

Remember that one coulomb per second is an ampere. Therefore power in watts equals the product of volts times amperes.

▶   Power in watts = volts × amperes
$$P = V \times I$$
(3-4)

When a 6-V battery produces 2 A in a circuit, for example, the battery is generating 12 W of power.

The power formula can be used in three ways:

▶ $P = V \times I$

$I = P \div V$ or $\dfrac{P}{V}$

$V = P \div I$ or $\dfrac{P}{I}$

Which formula to use depends on whether you want to calculate P, I, or V. Note the following examples:

Example

EXAMPLE 5 A toaster takes 10 A from the 120-V power line. How much power is used?

ANSWER $P = V \times I = 120 \text{ V} \times 10 \text{ A}$

$P = 1200 \text{ W}$

EXAMPLE 6 How much current flows in the filament of a 300-W bulb connected to the 120-V power line?

ANSWER $I = \dfrac{P}{V} = \dfrac{300 \text{ W}}{120 \text{ V}}$

$I = 2.5 \text{ A}$

EXAMPLE 7 How much current flows in the filament of a 60-W bulb connected to the 120-V power line?

ANSWER $I = \dfrac{P}{V} = \dfrac{60 \text{ W}}{120 \text{ V}}$

$I = 0.5 \text{ A}$

Note that the lower-wattage bulb uses less current.

WORK AND POWER   Work and energy are essentially the same with identical units. Power is different, however, because it is the time rate of doing work.

As an example of work, if you move 100 lb a distance of 10 ft, the work is 100 lb × 10 ft or 1000 ft · lb, regardless of how fast or how slowly the work is done. Note that the unit of work is foot-pounds, without any reference to time.

However, power equals the work divided by the time it takes to do the work. If it takes 1 s, the power in this example is 1000 ft · lb/s; if the work takes 2 s, the power is 1000 ft · lb in 2 s, or 500 ft · lb/s.

Similarly, electric power is the time rate at which charge is forced to move by voltage. This is why the power in watts is the product of volts and amperes. The voltage states the amount of work per unit of charge; the current value includes the time rate at which the charge is moved.

WATTS AND HORSEPOWER UNITS   A further example of how electric power corresponds to mechanical power is the fact that

▶ $746 \text{ W} = 1 \text{ hp} = 550 \text{ ft} \cdot \text{lb/s}$

This relation can be remembered more easily as 1 hp equals approximately ¾ kilowatt (kW). One kilowatt = 1000 W.

PRACTICAL UNITS OF POWER AND WORK   Starting with the watt, we can develop several other important units. The fundamental

principle to remember is that power is the time rate of doing work, while work is power used during a period of time. The formulas are

$$\text{Power} = \frac{\text{work}}{\text{time}} \qquad (3\text{-}5)$$

and

$$\text{Work} = \text{power} \times \text{time} \qquad (3\text{-}6)$$

With the watt unit for power, one watt used during one second equals the work of one joule. Or one watt is one joule per second. Therefore, 1 W = 1 J/s. The joule is a basic practical unit of work or energy.

To summarize these practical definitions,

> 1 joule = 1 watt · second
> 1 watt = 1 joule/second

In terms of charge and current,

> 1 joule = 1 volt · coulomb
> 1 watt = 1 volt · ampere

Remember that the ampere unit includes time in the denominator, since the formula is 1 ampere = 1 coulomb/second.

ELECTRON VOLT (eV)   This unit of work can be used for an individual electron, rather than the large quantity of electrons in a coulomb. An electron is charge, while the volt is potential difference. Therefore, 1 eV is the amount of work required to move an electron between two points that have a potential difference of one volt.

The number of electrons in one coulomb for the joule unit equals 6.25×10^{18}. Also, the work of one joule is a volt-coulomb. Therefore, the number of electron volts equal to one joule must be 6.25×10^{18}. As a formula,

> $1 \text{ J} = 6.25 \times 10^{18} \text{ eV}$

Either the electron volt or joule unit of work is the product of charge times voltage, but the watt unit of power is the product of voltage times current. The division by time to convert work to power corresponds to the division by time that converts charge to current.

KILOWATTHOURS   This is a unit commonly used for large amounts of electrical work or energy. The amount is calculated simply as the product of the power in kilowatts multiplied by the time in hours during which the power is used. As an example, if a light bulb uses 300 W or 0.3 kW for 4 hours (h), the amount of energy is 0.3 × 4, which equals 1.2 kWh.

We pay for electricity in kilowatthours of energy. The power-line voltage is constant at 120 V. However, more appliances and light bulbs require more current because they all add in the main line to increase the power.

Suppose the total load current in the main line equals 20 A. Then the power in watts from the 120-V line is

$$P = 120 \text{ V} \times 20 \text{ A}$$
$$P = 2400 \text{ W or } 2.4 \text{ kW}$$

If this power is used for 5 h, then the energy or work supplied equals $2.4 \times 5 = 12$ kWh. If the cost of electricity is 6 cents/kWh, then 12 kWh of electricity will cost $0.06 \times 12 = 0.72$ or 72 cents. This charge is for a 20-A load current from the 120-V line during the time of 5 h.

<hr>

TEST-POINT QUESTION 3-7

Answers at end of chapter.
a. An electric heater takes 15 A from the 120-V power line. Calculate the amount of power used.
b. How much is the load current for a 100-W bulb connected to the 120-V power line?
c. How many watts is the power of 200 J/s equal to?
d. How much will it cost to operate a 300-W light bulb for 48 h if the cost of electricity is 7 cents/kWh?

3-8 POWER DISSIPATION IN RESISTANCE

When current flows in a resistance, heat is produced because friction between the moving free electrons and the atoms obstructs the path of electron flow. The heat is evidence that power is used in producing current. This is how a fuse opens, as heat resulting from excessive current melts the metal link in the fuse.

The power is generated by the source of applied voltage and consumed in the resistance in the form of heat. As much power as the resistance dissipates in heat must be supplied by the voltage source; otherwise, it cannot maintain the potential difference required to produce the current.

The correspondence between electric power and heat is indicated by the fact that 1 W used during the time of 1 s is equivalent to 0.24 calorie of heat energy. The electric energy converted to heat is considered to be dissipated or used up because the calories of heat cannot be returned to the circuit as electric energy.

Since power is dissipated in the resistance of a circuit, it is convenient to express the power in terms of the resistance R. The formula $P = V \times I$ can be rearranged as follows:

Substituting IR for V,

▶
$$P = V \times I = IR \times I$$
$$P = I^2R \tag{3-7}$$

This is a common form for the power formula because of the heat produced by current in a resistance.

For another form, substitute V/R for I. Then

$$P = V \times I = V \times \frac{V}{R}$$

$$P = \frac{V^2}{R} \qquad\qquad\qquad \textbf{(3-8)}$$

In all the formulas, V is the voltage across R in ohms, producing the current I in amperes, for power in watts.

Any one of the three formulas (3-4), (3-7), and (3-8) can be used to calculate the power dissipated in a resistance. The one to be used is just a matter of convenience, depending on which factors are known.

In Fig. 3-6, for example, the power dissipated with 2 A through the resistance and 12 V across it is $2 \times 12 = 24$ W.

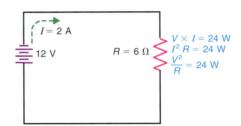

FIG. 3-6 Calculating the electric power in a circuit as $P = V \times I$, $P = I^2R$, or $P = V^2/R$.

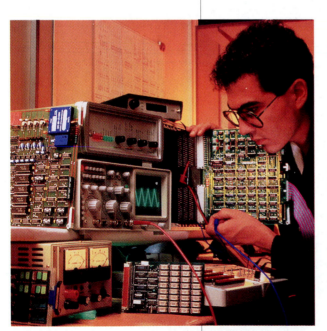

Or, calculating in terms of just the current and resistance, the power is the product of 2 squared, or 4, times 6, which equals 24 W.

Using the voltage and resistance, the power can be calculated as 12 squared, or 144, divided by 6, which also equals 24 W.

No matter which formula is used, 24 W of power is dissipated, in the form of heat. This amount of power must be generated continuously by the battery in order to maintain the potential difference of 12 V that produces the 2-A current against the opposition of 6 Ω.

In some applications, the electric power dissipation is desirable because the component must produce heat in order to do its job. For instance, a 600-W toaster must dissipate this amount of power to produce the necessary amount of heat. Similarly, a 300-W light bulb must dissipate this power to make the filament white-hot so that it will have the incandescent glow that furnishes the light. In other applications, however, the heat may be just an undesirable byproduct of the need to provide current through the resistance in a circuit. In any case, though, whenever there is current I in a resistance R, it dissipates the amount of power P equal to I^2R.

Example

EXAMPLE 8 Calculate the power in a circuit where the source of 100 V produces 2 A in a 50-Ω R.

ANSWER $P = I^2R = 2 \times 2 \times 50 = 4 \times 50$

$P = 200$ W

This means the source delivers 200 W of power to the resistance while the resistance dissipates 200 W in the form of heat.

EXAMPLE 9 Calculate the power in a circuit where the same source of 100 V produces 4 A in a 25-Ω R.

ANSWER $P = I^2R = 4^2 \times 25 = 16 \times 25$

$P = 400$ W

Note the higher power in Example 9 because of more I, even though R is less than in Example 8.

To use the calculator for a problem like Example 8 where I must be squared for $I^2 \times R$, use the following procedure:

Punch in the value of 2 for I.
Press the key marked (x^2) for the square of 2 equal to 4 on the display.
Next press the multiplication $(\times)$ key.
Punch in the value of 50 for R.
Finally, press the $(=)$ key for the answer of 200 on the display.

Be sure to square only the I value before multiplying by the R value.

Components that utilize the power dissipated in their resistance, such as light bulbs and toasters, are generally rated in terms of power. The power rating is given at normal applied voltage, which is usually the 120 V of the power line. For instance, a 600-W 120-V toaster has this rating because it dissipates 600 W in the resistance of the heating element when connected across 120 V.

Note this interesting point about the power relations. The lower the source voltage, the higher the current required for the same power. The reason is that $P = V \times I$. For instance, an electric heater rated at 240 W from the 120-V power line takes 240 W/120 V = 2 A of current from the source. However, the same 240 W from a 12-V source, as in a car or boat, requires 240 W/12 V = 20 A. More current must be supplied by a source with lower voltage, to provide a specified amount of power.

TEST-POINT QUESTION 3-8

Answers at end of chapter.
a. Current I is 2 A in a 5-Ω R. Calculate P.
b. Voltage V is 10 V across a 5-Ω R. Calculate P.
c. Resistance R has 10 V with 2 A. Calculate the values for P and R.

3-9 POWER FORMULAS

In order to calculate I or R for components rated in terms of power at a specified voltage, it may be convenient to use the power formulas in different forms. There are three basic power formulas, but each can be in three forms for nine combinations, as follows:

▶

$$P = VI \qquad\qquad P = I^2R \qquad\qquad P = \frac{V^2}{R}$$

$$\text{or}\quad I = \frac{P}{V} \qquad \text{or}\quad R = \frac{P}{I^2} \qquad \text{or}\quad R = \frac{V^2}{P}$$

$$\text{or}\quad V = \frac{P}{I} \qquad \text{or}\quad I = \sqrt{\frac{P}{R}} \qquad \text{or}\quad V = \sqrt{PR}$$

Example

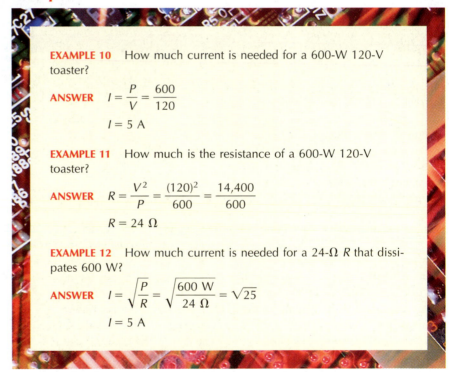

EXAMPLE 10   How much current is needed for a 600-W 120-V toaster?

ANSWER $I = \dfrac{P}{V} = \dfrac{600}{120}$

$I = 5\ A$

EXAMPLE 11   How much is the resistance of a 600-W 120-V toaster?

ANSWER $R = \dfrac{V^2}{P} = \dfrac{(120)^2}{600} = \dfrac{14,400}{600}$

$R = 24\ \Omega$

EXAMPLE 12   How much current is needed for a 24-Ω R that dissipates 600 W?

ANSWER $I = \sqrt{\dfrac{P}{R}} = \sqrt{\dfrac{600\ W}{24\ \Omega}} = \sqrt{25}$

$I = 5\ A$

Note that all these formulas are based on Ohm's law $V = IR$ and the power formula $P = VI$. The following example with a 300-W bulb also illustrates this idea. The bulb is connected across the 120-V line. Its 300-W filament requires current of 2.5 A, equal to P/V. These calculations are

$$I = \frac{P}{V} = \frac{300\ W}{120\ V} = 2.5\ A$$

The proof is that the VI product then is 120×2.5, which equals 300 W.

Furthermore, the resistance of the filament, equal to *V/I,* is 48 Ω. These calculations are

$$R = \frac{V}{I} = \frac{120\ \text{V}}{2.5\ \text{A}} = 48\ \Omega$$

If we use the power formula $R = V^2/P$, the answer is the same 48 Ω. These calculations are

$$R = \frac{V^2}{P} = \frac{120^2}{300}$$

$$R = \frac{14{,}400}{300} = 48\ \Omega$$

In any case, when this bulb is connected across 120 V so that it can dissipate its rated power, the bulb draws 2.5 A from the power line and the resistance of the white-hot filament is 48 Ω.

Furthermore, it is important to note that the Ohm's-law calculations can be used for just about all types of circuits. As an example, Fig. 3-7 shows three resistors in a series-parallel circuit. However, if we consider just R_3, the *I*, *V*, and *P* values can be calculated as shown. Actually, these values are the same as for the one resistor in Fig. 3-6. As far as this resistor is concerned, it could be in either circuit and not know any difference.

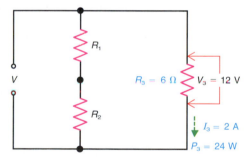

FIG. 3-7 Ohm's law can be applied to the entire circuit or to any one component such as R_1, R_2, or R_3.

To use the calculator for a problem like Example 11 that involves a square and division for V^2/R, use the following procedure:

Punch in the *V* value of 120.
Press the key marked $\boxed{x^2}$ for the square of 120, equal to 14,400 on the display.
Next press the division $\boxed{\div}$ key.
Punch in the value of 600 for *R.*
Finally press the $\boxed{=}$ key for the answer of 24 on the display. Be sure to square only the numerator before dividing.

For Example 12 with a square root and division, be sure to divide first, so that the square root is taken for the quotient, as follows:

Punch in the P of 600.
Press the division $\div$ key.
Punch in 24 for R.
Press the $=$ key for the quotient of 25.

Then press the $\sqrt{}$ key for the square root. This key may be a second function of the same key for squares. If so, press the key marked $\boxed{\text{2ndF}}$ or $\boxed{\text{SHIFT}}$ before pressing the $\sqrt{}$ key. As a result, the square root equal to 5 appears on the display. You do not need the $=$ key for this answer. In general, the $=$ key is pressed only for the multiplication, division, addition, and subtraction operations.

TEST-POINT QUESTION 3-9

Answers at end of chapter.
a. How much is the R of a 100-W 120-V light bulb?
b. How much power is dissipated by a 2-Ω R with 10 V across it?
c. Calculate P for 2 A of I through a 2-Ω resistor.

3-10 CHOOSING THE RESISTOR FOR A CIRCUIT

In deciding what size resistor to use, the first requirement is to have the amount of resistance needed. Consider the example in Fig. 3-8a. Resistor R_2 is to be combined with R_1 for the purpose of limiting the current to 0.1 A with a 100-V source.* The combined R equals 1000 Ω. This calculation is

$$R = \frac{V}{I} = \frac{100 \text{ V}}{0.1 \text{ A}} = 1000 \ \Omega$$

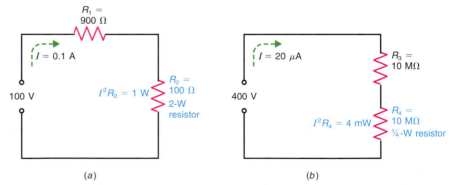

(a)   (b)

FIG. 3-8 Examples of power rating for a resistor. (a) Resistance R_2 dissipates 1 W, but a 2-W resistor is used for a safety factor of 2. (b) Resistance R_4 dissipates only 0.004 W or 4 mW, but a ¼-W resistor is used.

---

*Circuits like this with multiple resistors in one path are explained in Chap. 4, "Series Circuits."

If the combined R_T equals $R_1 + R_2$ and $R_1 = 900\ \Omega$, then the required value of R_2 is $1000 - 900 = 100\ \Omega$.

The I^2R power dissipated in R_2 equals 1 W. This calculation is

$$P = (0.1\ \text{A})^2 \times 100\ \Omega = 0.01 \times 100$$
$$P = 1\ \text{W}$$

However, a 2-W resistor would normally be used. This safety factor of 2 in the power rating is common practice with carbon resistors, so that they will not become too hot in normal operation.

A resistor with a higher wattage rating but the same R could also be used, if there is space on the circuit board for the larger size. The higher power rating would allow the circuit to operate normally and last longer without breaking down from excessive heat. However, a higher wattage rating is inconvenient when the next larger size is wire-wound and physically bigger.

Wire-wound resistors can operate closer to their power rating, assuming adequate ventilation. They have a higher rating for maximum operating temperature, compared with carbon resistors.

Another example of choosing a carbon resistor for a circuit is shown in Fig. 3-8b. The 10 MΩ of R_4 is used with the 10 MΩ of R_3 to provide an IR voltage drop of 200 V, equal to one-half the source of 400 V. Since the two resistances are equal, they divide the applied voltage into two equal parts of 200 V. This IR voltage drop can be calculated from the current. I is equal to 400 V/20 MΩ = 20 μA. Then the voltage drops are

$$V_{R_3} = 20\ \mu\text{A} \times 10\ \text{M}\Omega = 200\ \text{V}$$
$$V_{R_4} = 20\ \mu\text{A} \times 10\ \text{M}\Omega = 200\ \text{V}$$

The I^2R power dissipated in R_4 is 4 mW. This value can be calculated as

$$P = (20\ \mu\text{A})^2 \times 10\ \text{M}\Omega$$
$$= 400 \times 10^{-12} \times 10 \times 10^6$$
$$P = 4000\ \mu\text{W} = 0.004\ \text{W, or 4 mW}$$

However, the wattage rating used here is ¼ W, which equals 250 mW. In this case, the wattage rating is much higher than the actual amount of power dissipated in the resistor. A ¼-W power rating is used to provide a high enough voltage rating.

Notice the small amount of power dissipated in this circuit with a relatively high applied voltage of 400 V. The reason is that the very high resistance limits the current to a low value.

In general, using a resistor with a suitable power rating provides the required voltage 'rating. The exception, however, is a low-current, high-voltage circuit where the applied voltage is of the order of kilovolts (kV).

TEST-POINT QUESTION 3-10

Answers at end of chapter.
a. In Fig. 3-8a, calculate the power dissipation in R_2 as $V_2 \times I$.
b. In Fig. 3-8b, calculate the power dissipation in R_4 as $V_4 \times I$.
c. In Fig. 3-8a, calculate the total power dissipation as I^2R_T.

3-11 ELECTRIC SHOCK

While you are working on electric circuits, there is often the possibility of receiving an electric shock by touching the "live" conductors when the power is on. The shock is a sudden involuntary contraction of the muscles, with a feeling of pain, caused by current through the body. If severe enough, the shock can be fatal. Safety first, therefore, should always be the rule.

The greatest shock hazard is from high-voltage circuits that can supply appreciable amounts of power. The resistance of the human body is also an important factor. If you hold a conducting wire in each hand, the resistance of the body across the conductors is about 10,000 to 50,000 Ω. Holding the conductors tighter lowers the resistance. If you hold only one conductor, your resistance is much higher. It follows that the higher the body resistance, the smaller the current that can flow through you.

A safety rule, therefore, is to work with only one hand if the power is on. Also, keep yourself insulated from earth ground when working on power-line circuits, since one side of the line is usually connected to earth. The final and best safety rule is to work on the circuits with the power disconnected if at all possible and make resistance tests.

Note that it is current through the body, not through the circuit, which causes the electric shock. This is why high-voltage circuits are most important, since sufficient potential difference can produce a dangerous amount of current through the relatively high resistance of the body. For instance, 500 V across a body resistance of 25,000 Ω produces 0.02 A, or 20 mA, which can be fatal. As little as 10 μA through the body can cause an electric shock. In an experiment on electric shock to determine the current at which a person could release the live conductor, this value of "let-go" current was about 9 mA for men and 6 mA for women.

In addition to high voltage, the other important consideration in how dangerous the shock can be is the amount of power the source can supply. The current of 0.02 A through 25,000 Ω means the body resistance dissipates 10 W. If the source cannot supply 10 W, its output voltage drops with the excessive current load. Then the current is reduced to the amount corresponding to how much power the source can produce.

In summary, then, the greatest danger is from a source having an output of more than about 30 V with enough power to maintain the load current through the body when it is connected across the applied voltage. In general, components that can supply high power are physically big because of the need for dissipating heat.

TEST-POINT QUESTION 3-11

Answers at end of chapter.

Answer True or False.

a. The potential difference of 120 V is more dangerous than 12 V for electric shock.

b. Resistance should be measured with power off in the circuit.

3-12 OPEN-CIRCUIT AND SHORT-CIRCUIT TROUBLES

Ohm's law is useful for calculating I, V, and R in a closed circuit with normal values. However, an open circuit or a short circuit causes troubles that can be summarized as follows: An open circuit has zero I because R is infinitely high. It does not matter how much the V is. A short circuit has zero R, which causes excessively high I in the short-circuit path because of no resistance. These opposite troubles are compared in Figs. 3-9 and 3-10.

In Fig. 3-9a, the circuit is normal with I of 2 A produced by 10 V applied across R of 5 Ω. However, the resistor is shown open in Fig. 3-9b. Then the path for current has infinitely high resistance and there is no current in any part of the circuit. The trouble can be caused by an internal open in the resistor or a break in the wire conductors.

In Fig. 3-10a, the same normal circuit is shown with I of 2 A. In Fig. 3-10b, however, there is a short circuit path across R with zero resistance. The result is excessively high current in the short-circuit path, including the wire conductors. It may be surprising, but there is no current in the resistor itself because all the current is in the zero-resistance path around it.

Theoretically, the amount of current could be infinitely high with no R, but the voltage source can only supply a limited amount of I before it loses its ability to provide voltage output. Also, the wire conductors may become hot enough to burn open, which would open the circuit. Also, if there is any fuse in the circuit, it will open because of the excessive current produced by the short circuit.

It should be noted that the resistor itself is not likely to develop a short circuit, because of the nature of its construction. However, the wire conductors may touch or some other component in a circuit connected across the resistor may become short-circuited.

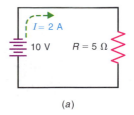

(a)

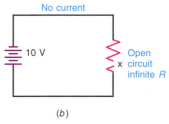

(b)

FIG. 3-9 Effect of an open circuit. (*a*) Normal circuit with current of 2 A for 10 V across 5 Ω. (*b*) Open circuit with no current and infinitely high resistance.

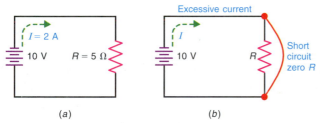

(a)   (b)

FIG. 3-10 Effect of a short circuit. (*a*) Normal circuit with current of 2 A for 10 V across 5 Ω. (*b*) Short circuit with zero resistance and excessively high current.

TEST-POINT QUESTION 3-12

Answers at end of chapter.

Answer True or False.

a. An open circuit has a zero current.
b. A short circuit has excessive current.
c. An open circuit and a short circuit have opposite effects on resistance and current.

3 SUMMARY AND REVIEW

- The three forms of Ohm's law are $I = V/R$, $V = IR$, and $R = V/I$.
- One ampere is the amount of current produced by one volt of potential difference across one ohm of resistance. This current of 1 A is the same as 1 C/s.
- With R constant, the amount of I increases in direct proportion as V increases. This linear relation between V and I is shown by the graph in Fig. 3-5.
- With V constant, the current I decreases as R increases. This is an inverse relation.
- Power is the time rate of doing work or using energy. The unit is the watt. One watt equals $1 \text{ V} \times 1 \text{ A}$. Also, watts = joules per second.
- The unit of work or energy is the joule. One joule equals $1 \text{ W} \times 1 \text{ s}$.
- The most common multiples and submultiples of the practical units are listed in Table 3-1.

TABLE 3-1 CONVERSION FACTORS

PREFIX	SYMBOL	RELATION TO BASIC UNIT	EXAMPLES
mega	M	1,000,000 or 1×10^6	5 MΩ (megohms) = 5,000,000 ohms = 5×10^6 ohms
kilo	k	1000 or 1×10^3	18 kV (kilovolts) = 18,000 volts = 18×10^3 volts
milli	m	0.001 or 1×10^{-3}	48 mA (milliamperes) = 48×10^{-3} ampere = 0.048 ampere
micro	μ	0.000 001 or 1×10^{-6}	15 μV (microvolts) = 15×10^{-6} volt = 0.000 015 volt

- Voltage applied across your body can produce a dangerous electric shock. Whenever possible, shut off the power and make resistance tests. If the power must be on, use only one hand. Do not let the other hand rest on a conductor.
- Table 3-2 summarizes the practical units used with Ohm's law.

TABLE 3-2 PRACTICAL UNITS OF ELECTRICITY

COULOMB	AMPERE	VOLT	WATT	OHM	SIEMENS
6.25×10^{18} electrons	$\dfrac{\text{coulomb}}{\text{second}}$	$\dfrac{\text{joule}}{\text{coulomb}}$	$\dfrac{\text{joule}}{\text{second}}$	$\dfrac{\text{volt}}{\text{ampere}}$	$\dfrac{\text{ampere}}{\text{volt}}$

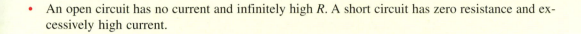

- An open circuit has no current and infinitely high R. A short circuit has zero resistance and excessively high current.

ANSWERS AT BACK OF BOOK.

Fill in the missing answers.

1. With 10 V across 5-Ω R, the current I is _____ A.
2. When 10 V produces 2.5 A, R is _____ Ω.
3. With 8 A through a 2-Ω R, the IR voltage is _____ V.
4. The resistance of 500,000 Ω is _____ MΩ.
5. With 10 V across 5000-Ω R, the current I is _____ mA.
6. The power of 50 W = 2 A $\times$ _____ V.
7. The energy of 50 J = 2 C $\times$ _____ V.
8. The current drawn from the 120-V power line by a 1200-W toaster = _____ A.
9. The current of 400 μA = _____ mA.
10. With 12 V across a 2-Ω R, its power dissipation = _____ W.
11. A circuit has a 4-A I. If V is doubled and R is the same, I = _____ A.
12. A circuit has a 4-A I. If R is doubled and V is the same, I = _____ A.
13. A television receiver using 240 W from the 120-V power line draws current I = _____ A.
14. The rated current for a 100-W 120-V bulb = _____ A.
15. The resistance of the bulb in question 14 is _____ Ω.
16. The energy of 12.5×10^{18} eV = _____ J.
17. The current of 1200 mA = _____ A.
18. In an amplifier, the load resistor R_L of 5 kΩ has 15 V across it. Through R_L, then, the current = _____ mA.
19. In a transistor circuit, a 1-kΩ resistor R_1 has 200 μA through it. Across R_1, then, its voltage = _____ V.
20. In a transistor circuit, a 50-kΩ resistor R_2 has 6 V across it. Through R_2, then, its current = _____ mA.
21. A resistor is to be connected across a 45-V battery to provide 1 mA of current. The required resistance with a suitable wattage rating is (*a*) 4.5 Ω 1 W; (*b*) 45 Ω 10 W; (*c*) 450 Ω 2 W; (*d*) 45,000 Ω ¼ W.

1. State the three forms of Ohm's law relating V, I, and R.
2. (**a**) Why does higher applied voltage with the same resistance result in more current? (**b**) Why does more resistance with the same applied voltage result in less current?

3. Calculate the resistance of a 300-W bulb connected across the 120-V power line, using two different methods to arrive at the same answer.

4. State which unit in each of the following pairs is larger: (**a**) volt or kilovolt; (**b**) ampere or milliampere; (**c**) ohm or megohm; (**d**) volt or microvolt; (**e**) siemens or microsiemens; (**f**) electron volt or joule; (**g**) watt or kilowatt; (**h**) kilowatthour or joule; (**i**) volt or millivolt; (**j**) megohm or kilohm.

5. State two safety precautions to follow when working on electric circuits.

6. Referring back to the resistor shown in Fig. 1-11 in Chap. 1, suppose that it is not marked. How could you determine its resistance by Ohm's law? Show your calculations that result in the V/I ratio of 600 Ω. However, do not exceed the power rating of 10 W.

7. Give three formulas for electric power.

8. What is the difference between work and power? Give two units for each.

9. Prove that 1 kWh is equal to 3.6×10^6 J.

10. Give the metric prefixes for 10^{-6}, 10^{-3}, 10^3, and 10^6.

11. Which two units in Table 3-2 are reciprocals of each other?

12. A circuit has a constant R of 5000 Ω, while V is varied from 0 to 50 V in 10-V steps. Make a table listing the values of I for each value of V. Then draw a graph plotting these values of milliamperes vs. volts. (This graph should be similar to Fig. 3-5c.)

13. Give the voltage and power rating for at least two types of electrical equipment.

14. Which uses more current from the 120-V power line, a 600-W toaster or a 200-W television receiver?

15. Refer to the two resistors in series with each other in Fig. 4-1 in Chapter 4. How much would you guess is the current through R_2?

16. Consider the two resistors R_1 and R_2 shown before in Fig. 3-7. Would you say they are in one series path or separate parallel paths?

17. Give a definition for a short circuit and for an open circuit.

18. Compare the R of zero ohms and infinite ohms.

19. Derive the formula $P = I^2R$ from $P = IV$ by using an Ohm's-law formula.

20. Give two safety factors to consider for preventing an electric shock.

PROBLEMS

ANSWERS TO ODD-NUMBERED PROBLEMS AT BACK OF BOOK.

1. A 15-V dc source is connected across a 1-kΩ resistance. (**a**) Draw a schematic diagram. (**b**) Calculate the current I through the resistance. (**c**) How much current flows through the voltage source? (**d**) If the voltage of the source is doubled, how much is the current I in the circuit?

2. A 12-V battery is connected across a 3-Ω resistance. (*a*) Draw a schematic diagram. (*b*) Calculate the current I in the circuit. (*c*) How much power is dissipated in the resistance? (*d*) If the resistance is doubled to 6 Ω, how much power is dissipated by R?

3. A 240-Ω resistor is dissipating 5.4 W. Calculate: (*a*) the voltage across the resistor; (*b*) the current through the resistor.

4. Convert the following units: (*a*) 20 mA to amperes; (*b*) 3500 V to kilovolts; (*c*) 0.4 A to milliamperes; (*d*) 1.5 mA to microamperes; (*e*) 47,000 Ω to kilohms; (*f*) 2.2 MΩ to kilohms; (*g*) 2500 μA to milliamperes; (*h*) 16 kV to volts.

5. A current of 2 A flows through a 12-Ω resistance connected across a battery. (*a*) How much is the battery voltage? (*b*) How much power is dissipated by the resistance? (*c*) How much power is supplied by the battery?

6. A 560-kΩ resistor has a current of 30 μA. (*a*) How much voltage is across the resistor? (*b*) How much power is dissipated by the resistor?

7. Calculate the current *I*, in ampere (A) units, for each of the following examples: (*a*) 66 V across 220 kΩ; (*b*) 15 mV across 3 kΩ; (*c*) 24 V across 120 Ω; (*d*) 3 V across 10 kΩ; (*e*) 18 V across 18 kΩ; (*f*) 15 kV across 1 MΩ.

8. Calculate the *IR* voltage drop for each of the following examples: (*a*) 2 mA through 1 kΩ; (*b*) 60 μA through 120 kΩ; (*c*) 250 mA through 10 kΩ; (*d*) 330 μA through 2.2 MΩ; (*e*) 20 mA through 6.8 kΩ; (*f*) 0.4 mA through 12 kΩ.

9. Calculate the resistance *R*, in ohms, for each of the following examples: (*a*) 1 mA drawn from a 12-V source; (*b*) 4 mA drawn from a 15-V source; (*c*) 150 mW dissipated with 36 V applied; (*d*) 16.2 W dissipated with a current of 30 mA.

10. A 4-Ω resistor dissipates 16 W of power. (*a*) How much voltage is across the load? (*b*) How much current is flowing through the load?

11. A 1000-ft length of wire has a conductance *G* of 0.4 S. If the wire is carrying a current *I* of 10 A, calculate: (*a*) the resistance of the wire; (*b*) the *IR* voltage drop across the wire; (*c*) the power dissipated by the length of the wire.

12. Refer to Fig. 3-11. Draw a graph of the *I* and *V* values if (*a*) *R* = 2.5 Ω; (*b*) *R* = 5 Ω; (*c*) *R* = 10 Ω. In each case, the voltage source is to be varied in 5-V steps from 0 to 30 V.

13. Refer to Fig. 3-12. Draw a graph of the *I* and *R* values when *R* is varied in 2-Ω steps from 2 to 12 Ω. (*V* is constant at 12 V.)

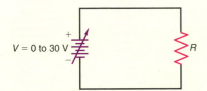

FIG. 3-11 Circuit diagram for Prob. 12.

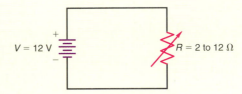

FIG. 3-12 Circuit diagram for Prob. 13.

14. How much will it cost to operate a 1500-W quartz heater for 36 h if the cost of electricity is 7 cents/kWh?

15. Calculate the hot resistance of a 60-W light bulb connected to the 120-V power line.

16. What is the maximum voltage that can be applied across a ½-W 10-kΩ resistor without exceeding its power dissipation rating?

17. What is the maximum current that can be allowed to flow through a 680-Ω ¼-W resistor without exceeding its power dissipation rating?

18. A 50-Ω load dissipates 100 W of power. Calculate: (*a*) the voltage across the load; (*b*) the current through the load.

19. How much current is drawn by a 2-W electric clock connected to the 120-V power line?
20. A 2.2-kΩ resistor is connected across a 9-V dc source. How much power is dissipated by the resistance?
21. A 1.8-kΩ resistor has a current of 25 mA. How much power is dissipated by the resistor?
22. What is the resistance of a 120-V 30-W soldering iron?
23. Determine the resistance and wattage ratings of a resistor to meet the following requirements: 5-V *IR* drop with a current *I* of 25 mA. When choosing the wattage rating, use a safety factor of 2.
24. How much current is drawn by a 6-W 12-V lamp?
25. What value of resistance dissipates 250 mW of power when its current equals 50 mA?

CRITICAL THINKING

1. The percent efficiency of a motor can be calculated as

$$\% \text{ efficiency} = \frac{\text{power out}}{\text{power in}} \times 100$$

where power out represents horsepower (hp). Calculate the current drawn by a 5-hp 240-V motor that is 72 percent efficient.
2. A ½-hp, 120-V motor draws 4.67 A when it is running. Calculate the motor's efficiency.
3. A ¾-hp motor with an efficiency of 75 percent runs 20 percent of the time over a 30-day period. If the cost of electricity is 7 cents/kwh, how much will it cost the user?
4. An appliance uses 14.4×10^6 J of energy for 1 day. How much will this cost the user if the cost of electricity is 6.5 cents/kwh?
5. A certain 1-kΩ resistor has a power rating of ½ W for temperatures up to 70°C. Above 70°C, however, the power rating must be reduced by a factor of 6.25 mW/°C. Calculate the maximum current that the resistor can allow at 120°C without exceeding its power dissipation rating at this temperature.

3-1 **a.** 3 A
    b. 1.5 A
    c. 2 A
    d. 6 A

3-2 **a.** 2 V
    b. 4 V
    c. 4 V

3-3 **a.** 4000 Ω
    b. 2000 Ω
    c. 12,000 Ω

3-4 **a.** 35 V
    b. 0.002 A
    c. 2000 Ω

3-5 **a.** See Prob. **b**
    b. See Prob. **a**
    c. 2 mA

3-6 **a.** y axis
    b. linear
    c. I doubles
    d. I is halved

3-7 **a.** 1.8 kW
    b. 0.83 A
    c. 200 W
    d. $1.01 (approx.)

3-8 **a.** 20 W
    b. 20 W
    c. 20 W and 5 Ω

3-9 **a.** 144 Ω
    b. 50 W
    c. 8 W

3-10 **a.** 1 W
    b. 4 mW
    c. 10 W

3-11 **a.** T
    b. T

3-12 **a.** T
    b. T
    c. T

CHAPTER 4

SERIES CIRCUITS

If two or more components are used with the voltage source, different types of circuits are possible. When the components are connected with only one path for current through all of them, the result is a series circuit. The series path is made by connecting an end of each component to an end of the next.

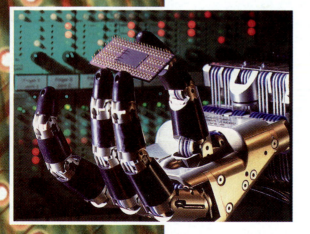

Which end comes first does not matter with resistors. Since there is only one path for electron flow, the current I must be the same in all the series components.

The purpose of a series circuit is to connect different components that need the same current. However, the individual voltages across each of the series components can have different values. These principles of series connections apply to dc and ac circuits.

CHAPTER OBJECTIVES

Upon completion of this chapter, you should be able to:

- *Explain* why the current is the same in all parts of a series circuit.
- *Determine* the total resistance of a series circuit.
- *Determine* the individual resistor voltage drops in a series circuit.
- *Determine* the polarity of a resistor's *IR* voltage drop.
- *Calculate* the total power dissipated in a series circuit.
- *Determine* the net voltage of series-aiding and series-opposing voltage sources.
- *Solve* for the voltage, current, resistance, and power in a series circuit having random unknowns.
- *Describe* the effect of an open in a series circuit.
- *Describe* how series-connected switches can be used to describe the AND logic function.

IMPORTANT TERMS IN THIS CHAPTER

aiding voltages	open circuit	series string
AND gate	opposing voltages	total power
applied voltage	positive potential	total resistance
chassis ground	potential difference	voltage drop
IR drop	proportional parts	voltage polarity
negative potential	series components	

TOPICS COVERED IN THIS CHAPTER

4-1 WHY *I* IS THE SAME IN ALL PARTS OF A SERIES CIRCUIT

An electric current is a movement of charges between two points, produced by the applied voltage. When components are connected in successive order as in Fig. 4-1, they form a series circuit. The resistors R_1 and R_2 are in series with each other and the battery.

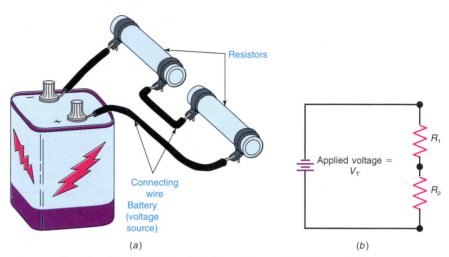
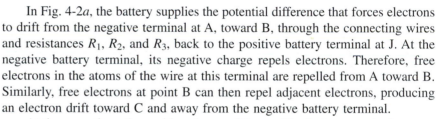

FIG. 4-1 A series circuit. (*a*) Pictorial wiring diagram. (*b*) Schematic diagram.

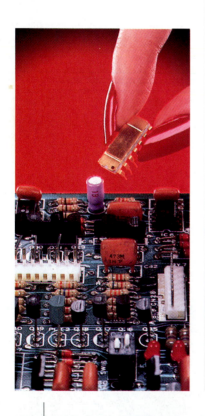

In Fig. 4-2*a*, the battery supplies the potential difference that forces electrons to drift from the negative terminal at A, toward B, through the connecting wires and resistances R_1, R_2, and R_3, back to the positive battery terminal at J. At the negative battery terminal, its negative charge repels electrons. Therefore, free electrons in the atoms of the wire at this terminal are repelled from A toward B. Similarly, free electrons at point B can then repel adjacent electrons, producing an electron drift toward C and away from the negative battery terminal.

At the same time, the positive charge of the positive battery terminal attracts free electrons, causing electrons to drift toward I and J. As a result, the free electrons in R_1, R_2, and R_3 are forced to drift toward the positive terminal.

The positive terminal of the battery attracts electrons just as much as the negative side of the battery repels electrons. Therefore, the motion of free electrons in the circuit starts at the same time and at the same speed in all parts of the circuit.

The electrons returning to the positive battery terminal are not the same electrons as those leaving the negative terminal. Free electrons in the wire are forced to move to the positive terminal because of the potential difference of the battery.

The free electrons moving away from one point are continuously replaced by free electrons flowing from an adjacent point in the series circuit. All electrons have the same speed as those leaving the battery. In all parts of the circuit, therefore, the electron drift is the same, with an equal number of electrons moving at one time with the same speed. That is why the current is the same in all parts of the series circuit.

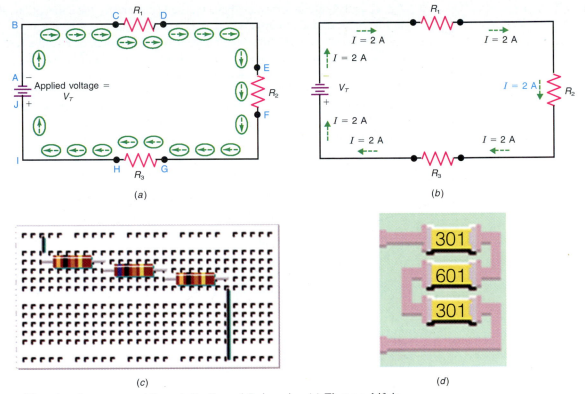

FIG. 4-2 There is only one current through R_1, R_2, and R_3 in series. (a) Electron drift is the same in all parts of a series circuit. (b) Current I is the same at all points in a series circuit. (c) Series circuit assembled on a lab prototype board, using axial-lead resistors. (d) Series circuit using surface-mount resistors assembled on a printed circuit board.

In Fig. 4-2b, when the current is 2 A, for example, this is the value of the current through R_1, R_2, R_3, and the battery at the same instant. Not only is the amount of current the same throughout, but in all parts of a series circuit the current cannot differ in any way because there is just one current path for the entire circuit. Figure 4-2c shows how to assemble resistors on a lab prototype board to form a series circuit. Figure 4-2d shows a series circuit using surface-mount resistors assembled on a printed circuit board.

The order in which components are connected in series does not affect the current. In Fig. 4-3b, resistances R_1 and R_2 are connected in reverse order compared with Fig. 4-3a, but in both cases they are in series. The current through each is the same because there is only one path for the electron flow. Similarly,

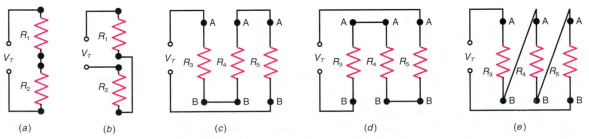

FIG. 4-3 Examples of series connections: R_1 and R_2 are in series in both (a) and (b); also, R_3, R_4, and R_5 are in series in (c), (d), and (e).

R_3, R_4, and R_5 are in series and have the same current for the connections shown in Fig. 4-3c, d, and e. Furthermore, the resistances need not be equal.

The question of whether a component is first, second, or last in a series circuit has no meaning in terms of current. The reason is that I is the same amount at the same time in all the series components.

In fact, series components can be defined as those in the same current path. The path is from one side of the voltage source, through the series components, and back to the other side of the applied voltage. However, the series path must not have any point where the current can branch off to another path in parallel.

TEST-POINT QUESTION 4-1

Answers at end of chapter.
a. In Fig. 4-2b, name five parts that have the I of 2 A.
b. In Fig. 4-3e, when I in R_5 is 5 A, then I in R_3 is _____ A.
c. In Fig. 4-4b, how much is the I in R_2?

4-2 TOTAL R EQUALS THE SUM OF ALL SERIES RESISTANCES

When a series circuit is connected across a voltage source, as shown in Fig. 4-3, the free electrons forming the current must drift through all the series resistances. This path is the only way the electrons can return to the battery. With two or more resistances in the same current path, therefore, the total resistance across the voltage source is the opposition of all the resistances.

Specifically, the total resistance R_T of a series string is equal to the sum of the individual resistances. This rule is illustrated in Fig. 4-4. In Fig. 4-4b, 2 Ω is added in series with the 3 Ω of Fig. 4-4a, producing the total resistance of 5 Ω. The total opposition of R_1 and R_2 limiting the amount of current is the same as though a 5-Ω resistance were used, as shown in the equivalent circuit in Fig. 4-4c.

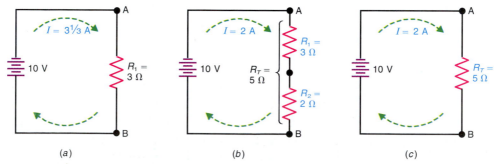

FIG. 4-4   Series resistances are added for the total R_T. (a) R_1 alone is 3 Ω. (b) R_1 and R_2 in series total 5 Ω. (c) The R_T of 5 Ω is the same as one resistance of 5 Ω between points A and B.

SERIES STRING   A combination of series resistances is often called a *string*. The string resistance equals the sum of the individual resistances. For instance, R_1 and R_2 in Fig. 4-4 form a series string having the R_T of 5 Ω. A string can have two or more resistors.

By Ohm's law, the amount of current between two points in a circuit equals the potential difference divided by the resistance between these points. As the entire string is connected across the voltage source, the current equals the voltage applied across the entire string divided by the total series resistance of the string. Between points A and B in Fig. 4-4, for example, 10 V is applied across 5 Ω in Fig. 4-4b and c to produce 2 A. This current flows through R_1 and R_2 in one series path.

SERIES RESISTANCE FORMULA   In summary, the total resistance of a series string equals the sum of the individual resistances. The formula is

▶ $$R_T = R_1 + R_2 + R_3 + \cdots + \text{etc.} \qquad \textbf{(4-1)}$$

where R_T is the total resistance and R_1, R_2, and R_3 are individual series resistances.

This formula applies to any number of resistances, whether equal or not, as long as they are in the same series string. Note that R_T is the resistance to use in calculating the current in a series string. Then Ohm's law is

▶ $$I = \frac{V_T}{R_T} \qquad \textbf{(4-2)}$$

where R_T is the sum of all the resistances, V_T is the voltage applied across the total resistance, and I is the current in all parts of the string.

Example

EXAMPLE 1   Two resistances R_1 and R_2 of 5 Ω each and R_3 of 10 Ω are in series. How much is R_T?

ANSWER $R_T = R_1 + R_2 + R_3 = 5 + 5 + 10$

   $R_T = 20\ \Omega$

EXAMPLE 2   With 80 V applied across the series string of Example 1, how much is the current in R_3?

ANSWER $I = \dfrac{V_T}{R_T} = \dfrac{80\ \text{V}}{20\ \Omega}$

   $I = 4\ \text{A}$

This 4-A current is the same in R_3, R_2, R_1, or any part of the series circuit.

Note that adding series resistance reduces the current. In Fig. 4-4a the 3-Ω R_1 allows 10 V to produce $3\frac{1}{3}$ A. However, I is reduced to 2 A when the 2-Ω R_2 is added for a total series resistance of 5 Ω opposing the 10-V source.

4-3 SERIES *IR* VOLTAGE DROPS

With current I through a resistance, by Ohm's law the voltage across R is equal to $I \times R$. This rule is illustrated in Fig. 4-5 for a string of two resistors. In this circuit, I is 1 A because the applied V_T of 10 V is across the total R_T of 10 Ω, equal to the 4-Ω R_1 plus the 6-Ω R_2. Then I is 10 V/10 Ω = 1 A.

For each *IR* voltage in Fig. 4-5, multiply each R by the 1 A of current in the series circuit. Then

$$V_1 = IR_1 = 1\text{ A} \times 4\ \Omega = 4\text{ V}$$
$$V_2 = IR_2 = 1\text{ A} \times 6\ \Omega = 6\text{ V}$$

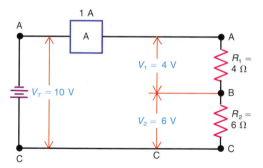

FIG. 4-5   An example of *IR* voltage drops V_1 and V_2 in a series circuit.

The V_1 of 4 V is across the 4 Ω of R_1. Also, the V_2 of 6 V is across the 6 Ω of R_2. The two voltages V_1 and V_2 are in series.

The *IR* voltage across each resistance is called an *IR drop*, or a *voltage drop*, because it reduces the potential difference available for the remaining resistance in the series circuit. Note that the symbols V_1 and V_2 are used for the voltage drops across each resistor to distinguish them from the source V_T applied across both resistors.

In Fig. 4-5, the V_T of 10 V is applied across the total series resistance of R_1 and R_2. However, because of the *IR* voltage drop of 4 V across R_1, the potential difference across R_2 is only 6 V. The negative potential drops from 10 V at point A, with respect to the common reference point at C, down to 6 V at point B with reference to point C. The potential difference of 6 V between B and the reference at C is the voltage across R_2.

Similarly, there is an *IR* voltage drop of 6 V across R_2. The negative potential drops from 6 V at point B with respect to point C, down to 0 V at point C with respect to itself. The potential difference between any two points on the return line to the battery must be zero because the wire has practically zero resistance and therefore no *IR* drop.

It should be noted that voltage must be applied by a source of potential difference such as the battery in order to produce current and have an *IR* voltage drop across the resistance. With no current through a resistor, the resistor has resistance only. There is no potential difference across the two ends of the resistor.

The *IR* drop of 4 V across R_1 in Fig. 4-5 represents that part of the applied voltage used to produce the current of 1 A through the 4-Ω resistance. Also, across R_2 the *IR* drop is 6 V because this much voltage allows 1 A in the 6-Ω resistance. The *IR* drop is more in R_2 because more potential difference is necessary to produce the same amount of current in the higher resistance. For series circuits, in general, the highest *R* has the largest *IR* voltage drop across it.

ABOUT ELECTRONICS

Among his many contributions to modern society is Benjamin Franklin's development of the concept of two types of electrical energy: positive and negative. Franklin concluded that like charged items repel one another and opposite types attract.

TEST-POINT QUESTION 4-3

Answers at end of chapter.

Refer to Fig. 4-5.
a. How much is the sum of V_1 and V_2?
b. Calculate *I* as V_T/R_T.
c. How much is *I* through R_1?
d. How much is *I* through R_2?

4-4 THE SUM OF SERIES *IR* DROPS EQUALS THE APPLIED V_T

The whole applied voltage is equal to the sum of its parts. For example, in Fig. 4-5, the individual voltage drops of 4 V and 6 V total the same 10 V produced by the battery. This relation for series circuits is given by the formula

$$V_T = V_1 + V_2 + V_3 + \cdots + \text{etc.} \qquad \text{(4-3)}$$

where V_T is the applied voltage and V_1, V_2, V_3, . . . are the individual *IR* drops.

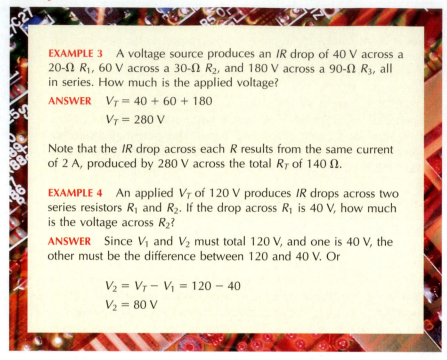

EXAMPLE 3   A voltage source produces an *IR* drop of 40 V across a 20-Ω R_1, 60 V across a 30-Ω R_2, and 180 V across a 90-Ω R_3, all in series. How much is the applied voltage?

ANSWER $V_T = 40 + 60 + 180$

$V_T = 280$ V

Note that the *IR* drop across each *R* results from the same current of 2 A, produced by 280 V across the total R_T of 140 Ω.

EXAMPLE 4   An applied V_T of 120 V produces *IR* drops across two series resistors R_1 and R_2. If the drop across R_1 is 40 V, how much is the voltage across R_2?

ANSWER   Since V_1 and V_2 must total 120 V, and one is 40 V, the other must be the difference between 120 and 40 V. Or

$V_2 = V_T - V_1 = 120 - 40$

$V_2 = 80$ V

It really is logical that V_T is the sum of the series *IR* drops. The current *I* is the same in all the series components. For this reason, the total of all the series voltages V_T is needed to produce the same *I* in the total of all the series resistances R_T as the *I* that each resistor voltage produces in its *R*.

A practical application of voltages in a series circuit is illustrated in Fig. 4-6. In this circuit, two 120-V light bulbs are operated from a 240-V line. If one bulb were connected to 240 V, the filament would burn out. With the two bulbs in series, however, each has 120 V for proper operation. The two 120-V drops across the bulbs in series add to equal the applied voltage of 240 V.

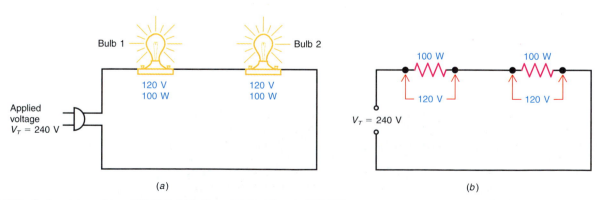

FIG. 4-6   Series string of two 120-V light bulbs operating from a 240-V line. (*a*) Wiring diagram. (*b*) Schematic diagram.

Answers at end of chapter.

a. A series circuit has *IR* drops of 10 V, 20 V, and 30 V. How much is the applied voltage V_T of the source?

b. A 100-V source is applied across R_1 and R_2 in series. If V_1 is 25 V, how much is V_2?

c. A 120-V source is applied across three equal resistances in series. How much is each *IR* drop?

4-5 POLARITY OF *IR* VOLTAGE DROPS

When an *IR* voltage drop exists across a resistance, one end must be either more positive or more negative than the other end. Otherwise, without a potential difference no current could flow through the resistance to produce the *IR* drop. The polarity of this *IR* voltage can be associated with the direction of *I* through *R*. In brief, electrons flow into the negative side of the *IR* voltage and out the positive side (Fig. 4-7*a*).

If we want to consider conventional current, with positive charges moving in the opposite direction from electron flow, the rule is reversed for the positive charges. See Fig. 4-7*b*. Here the positive charges for *I* are moving into the positive side of the *IR* voltage.

However, for either electron flow or conventional current the actual polarity of the *IR* drop is the same. In both *a* and *b* of Fig. 4-7, the top end of *R* in the diagrams is negative since this is the negative terminal of the source producing the current. After all, the resistor does not know which direction of current we are thinking of.

A series circuit with two *IR* voltage drops is shown in Fig. 4-8. We can analyze these polarities in terms of electron flow. The electrons move from the negative terminal of the source V_T through R_1 from point C to D. Electrons move into C and out from D. Therefore C is the negative side of the voltage drop across R_1. Similarly, for the *IR* voltage drop across R_2, point E is the negative side, compared with point F.

A more fundamental way to consider the polarity of *IR* voltage drops in a circuit is the fact that between any two points the one nearer to the positive terminal of the voltage source is more positive; also, the point nearer to the negative terminal of the applied voltage is more negative. A point nearer to the terminal means there is less resistance in its path.

In Fig. 4-8 point C is nearer to the negative battery terminal than point D. The reason is that C has no resistance to A, while the path from D to A includes the resistance of R_1. Similarly, point F is nearer to the positive battery terminal than point E, which makes F more positive than E.

Notice that points D and E in Fig. 4-8 are marked with both plus and minus polarities. The plus polarity at D indicates that it is more positive than C. This polarity, however, is shown just for the voltage across R_1. Point D cannot be more positive than points F and B. The positive terminal of the applied voltage must be the most positive point because the battery is generating the positive potential for the entire circuit.

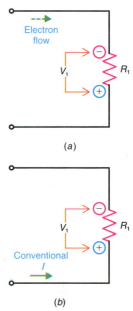

(a)

(b)

FIG. 4-7 Polarity of *IR* voltage drops. (*a*) Electrons flow into the negative side of V_1 across R_1. (*b*) Same polarity of V_1 with positive charges into positive side.

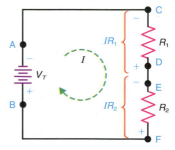

FIG. 4-8 Example of two *IR* voltage drops in series. Electron flow shown for direction of *I*.

Similarly, points A and C must have the most negative potential in the entire string, since point A is the negative terminal of the applied voltage. Actually, the plus polarity marked at D only means that this end of R_1 is less negative than C by the amount of voltage drop across R_1.

Consider the potential difference between E and D in Fig. 4-8, which is only a piece of wire. This voltage is zero because there is no resistance between these two points. Without any resistance here, the current cannot produce the IR drop necessary for a difference in potential. Points E and D are, therefore, the same electrically since they have the same potential.

When we go around the external circuit from the negative terminal of V_T, with electron flow, the voltage drops are drops in negative potential. For the opposite direction, starting from the positive terminal of V_T, the voltage drops are drops in positive potential. Either way, the voltage drop of each series R is its proportional part of the V_T needed for the one value of current in all the resistances.

4-6 TOTAL POWER IN A SERIES CIRCUIT

The power needed to produce current in each series resistor is used up in the form of heat. Therefore, the total power used is the sum of the individual values of power dissipated in each part of the circuit. As a formula,

$$P_T = P_1 + P_2 + P_3 + \cdots + \text{etc.} \qquad (4\text{-}4)$$

As an example, in Fig. 4-9, R_1 dissipates 40 W for P_1, equal to 20 V × 2 A for the VI product. Or, the P_1 calculated as I^2R is $(2 \times 2) \times 10 = 40$ W. Also, the P_1 is V^2/R, or $(20 \times 20)/10 = 40$ W.

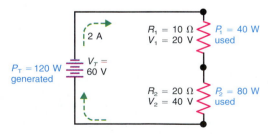

FIG. 4-9 The sum of the individual powers P_1 and P_2 used in each resistance equals the total power P_T produced by the source.

Similarly, P_2 for R_2 is 80 W. This value is 40×2 for VI, $(2 \times 2) \times 20$ for I^2R, or $(40 \times 40)/20$ for V^2/R. The P_2 must be more than P_1 because R_2 is more than R_1 with the same current.

The total power dissipated by R_1 and R_2, then, is $40 + 80 = 120$ W. This power is generated by the source of applied voltage.

The total power can also be calculated as $V_T \times I$. The reason is that V_T is the sum of all the series voltages and I is the same in all the series components. In this case, then, $P_T = V_T \times I = 60 \times 2 = 120$ W.

The total power here is 120 W, calculated either from the total voltage or from the sum of P_1 and P_2. This is the amount of power produced by the battery. The voltage source produces this power, equal to the amount used by the resistors.

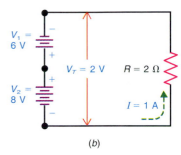

(a)

(b)

FIG. 4-10 Example of voltage sources V_1 and V_2 in series. (a) Note the connections for series-aiding polarities. Here $8\text{ V} + 6\text{ V} = 14\text{ V}$ for the total V_T. (b) Connections for series-opposing polarities. Now $8\text{ V} - 6\text{ V} = 2\text{ V}$ for V_T.

TEST-POINT QUESTION 4-6

Answers at end of chapter.
a. Each of three equal resistances dissipates 2 W. How much P_T is supplied by the source?
b. A 1-kΩ R_1 and 40-kΩ R_2 are in series with a 50-V source. Which R dissipates more power?

4-7 SERIES-AIDING AND SERIES-OPPOSING VOLTAGES

Series-aiding voltages are connected with polarities that allow current in the same direction. In Fig. 4-10a, the 6 V of V_1 alone could produce 3 A electron flow from the negative terminal, with the 2-Ω R. Also, the 8 V of V_2 could produce 4 A in the same direction. The total I then is 7 A.

Instead of adding the currents, however, the voltages V_1 and V_2 can be added, for a V_T of $6 + 8 = 14$ V. This 14 V produces 7 A in all parts of the series circuit with a resistance of 2 Ω. Then I is $14/2 = 7$ A.

Voltages are connected series-aiding when the plus terminal of one is connected to the negative terminal of the next. They can be added for a total equivalent voltage. This idea applies in the same way to voltage sources, such as batteries, and to voltage drops across resistances. Any number of voltages can be added, as long as they are connected with series-aiding polarities.

Series-opposing voltages are subtracted, as shown in Fig. 4-10b. Notice here that the positive terminals of V_1 and V_2 are connected. Subtract the smaller from the larger value, and give the net V the polarity of the larger voltage. In this example, V_T is $8 - 6 = 2$ V. The polarity of V_T is the same as V_2 because its voltage is higher than V_1.

If two series-opposing voltages are equal, the net voltage will be zero. In effect, one voltage balances out the other. The current I also is zero, without any net potential difference.

Answers at end of chapter.

a. Voltage V_1 of 40 V is series-aiding with V_2 of 60 V. How much is V_T?

b. The same V_1 and V_2 are connected series-opposing. How much is V_T?

4-8 ANALYZING SERIES CIRCUITS

FIG. 4-11 Analyzing a series circuit to find I, V_1, V_2, P_1, and P_2. See text for solution.

Refer to Fig. 4-11. Suppose that the source V_T of 50 V is known, with the 14-Ω R_1 and 6-Ω R_2. The problem is to find R_T, I, the individual voltage drops V_1 and V_2 across each resistor, and the power dissipated.

We must know the total resistance R_T to calculate I because the total applied voltage V_T is given. This V_T is applied across the total resistance R_T. In this example, R_T is $14 + 6 = 20 \ \Omega$.

Now I can be calculated as V/R_T, or 50/20, which equals 2.5 A. This 2.5-A I flows through R_1 and R_2.

The individual voltage drops are

$$V_1 = IR_1 = 2.5 \times 14 = 35 \text{ V}$$
$$V_2 = IR_2 = 2.5 \times 6 = 15 \text{ V}$$

Note that V_1 and V_2 total 50 V, equal to the applied V_T.

The calculations to find the power dissipated in each resistor are as follows:

$$P_1 = V_1 \times I = 35 \times 2.5 = 87.5 \text{ W}$$
$$P_2 = V_2 \times I = 15 \times 2.5 = 37.5 \text{ W}$$

These two values of dissipated power total 125 W. The power generated by the source equals $V_T \times I$ or 50×2.5, which is also 125 W.

GENERAL METHODS FOR SERIES CIRCUITS   For other types of problems with series circuits it is useful to remember the following:

1. When you know the I for one component, use this value for I in all the components, as the current is the same in all parts of a series circuit.

2. To calculate I, the total V_T can be divided by the total R_T, or an individual IR drop can be divided by its R. For instance, the current in Fig. 4-11 could be calculated as V_2/R_2 or 15/6, which equals the same 2.5 A for I. However, do not mix a total value for the entire circuit with an individual value for only part of the circuit.

3. When you know the individual voltage drops around the circuit, these can be added to equal the applied V_T. This also means that a known voltage drop can be subtracted from the total V_T to find the remaining voltage drop.

DID YOU KNOW?

Most electric cars need recharging, and this entails a risk of electrocution. A new GM car uses an inductive method that changes the way the batteries are charged. The driver parks the car and inserts a plastic card into the charge slot. An external charger generates a 120-kW magnetic field, energizing the car's converter to produce direct current for the batteries.

These principles are illustrated by the problem in Fig. 4-12. In this circuit R_1 and R_2 are known but not R_3. However, the current through R_3 is given as 3 mA.

With just this information, all values in this circuit can be calculated. The I of 3 mA is the same in all three series resistances. Therefore,

$$V_1 = 3 \text{ mA} \times 10 \text{ k}\Omega = 30 \text{ V}$$
$$V_2 = 3 \text{ mA} \times 30 \text{ k}\Omega = 90 \text{ V}$$

The sum of V_1 and V_2 is $30 + 90 = 120$ V. This 120 V plus V_3 must total 180 V. Therefore, V_3 is $180 - 120 = 60$ V.

With 60 V for V_3, equal to IR_3, then R_3 must be 60/0.003, equal to 20,000 Ω or 20 kΩ. The total circuit resistance is 60 kΩ, which results in the current of 3 mA with 180 V applied, as specified in the circuit.

Another way of doing this problem is to find R_T first. The equation $I = V_T/R_T$ can be inverted to calculate R_T.

$$R_T = \frac{V_T}{I}$$

With a 3-mA I and 180 V for V_T, the value of R_T must be 180 V/3 mA = 60 kΩ. Then R_3 is 60 kΩ − 40 kΩ = 20 kΩ.

The power dissipated in each resistance is 90 mW in R_1, 270 mW in R_2, and 180 mW in R_3. The total power is $90 + 270 + 180 = 540$ mW.

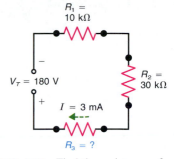

FIG. 4-12   Find the resistance of R_3. See text for analysis of this series circuit.

SERIES VOLTAGE-DROPPING RESISTORS

A common application of series circuits is to use a resistance to drop the voltage from the source V_T to a lower value, as in Fig. 4-13. The load R_L here represents a transistor radio that operates normally with a 9-V battery. When the radio is on, the dc load current with 9 V applied is 18 mA. Therefore, the requirements are 9 V, at 18 mA as the load.

To operate this radio from 12.6 V, the voltage-dropping resistor R_S is inserted in series to provide a voltage drop V_S that will make V_L equal to 9 V. The required voltage drop across V_S is the difference between V_L and the higher V_T. As a formula,

$$V_S = V_T - V_L = 12.6 - 9 = 3.6 \text{ V}$$

Furthermore, this voltage drop of 3.6 V must be provided with a current of 18 mA, as the current is the same through R_S and R_L. To calculate R_S,

$$R_S = \frac{3.6 \text{ V}}{18 \text{ mA}} = 0.2 \text{ k}\Omega = 200 \ \Omega$$

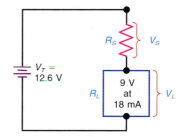

FIG. 4-13   Example of a series voltage-dropping resistor R_S used to drop V_T of 12.6 V to 9 V for R_L. See text for calculations.

CIRCUIT WITH VOLTAGE SOURCES IN SERIES

See Fig. 4-14. Note that V_1 and V_2 are series-opposing, with + to + through R_1. Their net effect, then, is 0 V. Therefore, V_T consists only of V_3, equal to 4.5 V. The total R is $2 + 1 + 2 = 5$ kΩ for R_T. Finally, I is V_T/R_T or 4.5 V/5 kΩ, which is equal to 0.9 mA, or 900 μA.

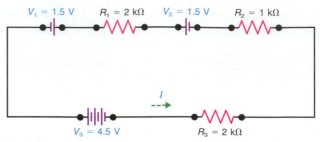

FIG. 4-14 Finding the I for this series circuit with three voltage sources. See text for solution.

Refer to Fig. 4-12.
a. Calculate V_1 across R_1.
b. Calculate V_2 across R_2.
c. How much is V_3?

4-9 EFFECT OF AN OPEN CIRCUIT IN A SERIES PATH

An open circuit is a break in the current path. The resistance of the open path is very high because an insulator like air takes the place of a conducting part of the circuit. Remember that the current is the same in all parts of a series circuit. Therefore, an open in any part results in no current for the entire circuit. As illustrated in Fig. 4-15, the circuit is normal in Fig. 4-15a, but in Fig. 4-15b there is no current in R_1, R_2, or R_3 because of the open in the series path.

The open between P1 and P2, or at any other point in the circuit, has practically infinite resistance because its opposition to electron flow is so great

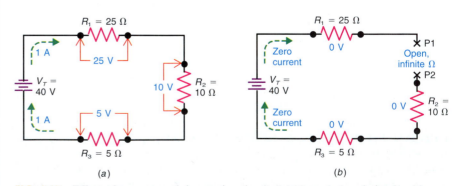

(a)   (b)

FIG. 4-15 Effect of an open path in a series circuit. (a) Normal closed circuit with current of 1 A. (b) Open path in any part of the series path results in no current in the entire circuit.

compared with the resistance of R_1, R_2, and R_3. Therefore, the value of current is practically zero, although the battery produces its normal applied voltage of 40 V.

To take an example, suppose that the open circuit between P1 and P2 has a resistance of 40×10^9 Ω (40 gigaohms). The resistance of the entire circuit is essentially 40×10^9 Ω, since the resistance of R_1, R_2, and R_3 can then be neglected compared with the resistance of the open. Such a high resistance is practically infinite ohms.

By Ohm's law, the current that results from 40 V applied across 40×10^9 Ω is 1×10^{-9} A, which is practically zero. This is the value of current in all parts of the series circuit. With practically no current, the *IR* voltage drop is practically zero across the 25 Ω of R_1, the 10 Ω of R_2, and the 5 Ω of R_3.

In summary, with an open in any part of a series circuit, the current is zero in the entire circuit. There is no *IR* voltage drop across any of the series resistances, although the voltage source still maintains its output voltage.

THE CASE OF ZERO *IR* DROP   In Fig. 4-15*b*, each of the resistors in the open circuit has an *IR* drop of zero. The reason is that current of practically zero is the value in all the series components. Each *R* still has its resistance. However, with zero current the *IR* voltage is zero.

THE SOURCE VOLTAGE V_T IS STILL PRESENT WITH ZERO *I*
The open circuit in Fig. 4-15*b* illustrates another example of how *V* and *I* are different forms of electricity. There is no current with the open circuit because there is no complete path outside the battery between its two terminals. However, the battery has a potential difference across the positive and negative terminals. This source voltage is present with or without current in the external circuit. If you measure V_T, the meter will read 40 V with the circuit closed or open.

The same idea applies to the 120-Vac voltage from the power line in the home. The 120-V potential difference is across the two terminals of the wall outlet. If you connect a lamp or appliance, current will flow in the circuit. When nothing is connected, though, the 120-V potential difference is still there at the outlet. If you should touch it, you would get an electric shock. The power company is maintaining the 120 V at the outlets as a source to produce current in any circuits that are plugged in.

THE APPLIED VOLTAGE IS ACROSS THE OPEN TERMINALS   It is useful to note that the entire applied voltage is present across the open circuit. Between P1 and P2 in Fig. 4-15*b*, there is 40 V. The reason is that essentially all the resistance of the series circuit is between P1 and P2. Therefore, the resistance of the open circuit develops all the *IR* voltage drop.

The extremely small current of one-billionth of an ampere is not enough to develop any appreciable *IR* drop across R_1, R_2, and R_3. Across the open, however, the resistance is 40×10^9 Ω. Therefore, the *IR* voltage across the open here is one-billionth of an ampere multiplied by 40×10^9 Ω, which equals 40 V.

We could also consider the open circuit as a proportional voltage divider. Since practically all the series resistance is between P1 and P2, all the applied voltage is across the open terminals.

The fact that the open terminals have the entire applied voltage indicates a good way to find an open component in a series string. If you measure the voltage across each good component, zero voltage will be normal. However, the component that has the full source voltage is the one that is open.

A practical example is a fuse in a circuit. A good fuse has the normal current, practically zero resistance, and zero voltage. If the fuse blows, the result is no current in the circuit but the applied voltage is across the open fuse.

Another example is illustrated in Fig. 4-16 to emphasize again the difference between voltage and current in a circuit. In Fig. 4-16a, the circuit is normal, with the switch closed to light the bulb. The 120 V of the source is across the bulb to produce the normal current of 0.833 A for the 100-W bulb. Across the switch, the voltage is zero because it has practically no resistance in the closed position. When the switch is opened in Fig. 4-16b, however, the results are: (1) There is no current to light the bulb; (2) there is no voltage across the bulb because the IR drop is zero without any current; and (3) the applied voltage of 120 V is across the open switch.

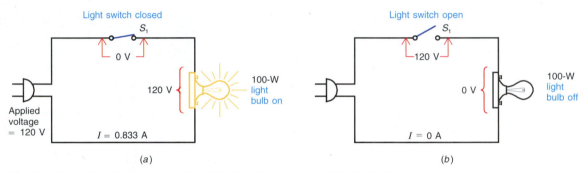

FIG. 4-16 Switching a light bulb on and off. (*a*) Switch S_1 is closed to light the bulb. The bulb has the applied voltage and normal current *I*. (*b*) Switch S_1 open. Now *I* is zero and the applied voltage is across the open switch.

TEST-POINT QUESTION 4-9

Answers at end of chapter.
a. Which component has 120 V in Fig. 4-16a?
b. Which component has 120 V in Fig. 4-16b?
c. How much is the current in the bulb in Fig. 4-16b?

4-10 SERIES SWITCHES REPRESENT THE AND LOGIC FUNCTION

In Fig. 4-17a, two series-connected switches, A and B, are used to control whether a lamp is on or off. If either or both switches A and B are open, the lamp will not light. The reason is that the open switches provide a *break* or *open* in the current path. In order for the lamp to light, both switches A and B must be closed.

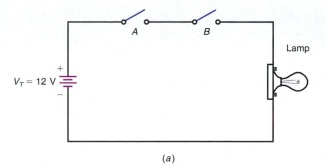

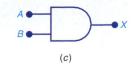

Switches		
A	B	Lamp
Open	Open	OFF
Open	Closed	OFF
Closed	Open	OFF
Closed	Closed	ON

$$A \cdot B = X$$

(a)　　　　　　(b)　　　　　　(c)　　　(d)

FIG. 4-17 Comparison of the AND logic gate to a series circuit and switches. (a) Series circuit with two switches, A and B. (b) Truth table with four possibilities for the switches. (c) Schematic symbol for the AND logic gate. (d) Boolean expression for the AND gate.

With both switches closed, current flows to light the lamp. Figure 4-17b is a table listing all of the possible combinations for switches A and B. As you can see from the table, both switches A and B must be closed in order for the lamp to be on. For the other three switch combinations the lamp is off. The table in Fig. 4-17b is often called a *truth table*.

The operation of the circuit in Fig. 4-17a is similar to that of the digital electronics circuit known as the AND logic gate. (It is important to note that the series circuit in Fig. 4-17a only illustrates the idea of the AND logic function; it is not the actual type of circuit used in digital electronic circuits.) With the actual digital AND logic gate, both inputs must have high voltages to produce a high voltage at the output. In digital circuits, 0 V and +5 V are representative of typical low and high voltages, respectively.

The schematic symbol for an AND logic gate is shown in Fig. 4-17c. The inputs are represented as A and B, whereas the output is represented as X. Figure 4-17d shows the mathematical expression which describes the operation of the AND logic gate. The dot between A and B represents the AND operation. The expression is read "A AND B equals X" rather than "A times B equals X." The expression can also be written as $AB = X$. It should be noted that an AND logic gate can have more than two inputs. Regardless of the number of inputs, however, the output is high only when all the inputs are high.

TEST-POINT QUESTION 4-10

Answers at end of chapter.

Answer True or False.
a. In Fig. 4-17a, the lamp is on only when switches A and B are both closed.
b. The mathematical expression for an AND gate with two inputs is $A \cdot B = X$.
c. When switches A and B are closed in Fig. 4-17a, the voltage across the lamp is 12 V.

4 SUMMARY AND REVIEW

- There is only one current I in a series circuit. $I = V_T/R_T$, where V_T is the voltage applied across the total series resistance R_T. This I is the same in all the series components.
- The total resistance R_T of a series string is the sum of the individual resistances.
- The applied voltage V_T equals the sum of the series IR voltage drops.
- The negative side of an IR voltage drop is where electrons flow in, attracted to the positive side at the opposite end.
- The sum of the individual values of power used in the individual resistances equals the total power supplied by the source.
- Series-aiding voltages are added; series-opposing voltages are subtracted.
- An open circuit results in no current in all parts of the series circuit.
- In an open circuit, the voltage across the two open terminals is equal to the applied voltage.
- The AND gate function corresponds to switches in series.

SELF-TEST

ANSWERS AT BACK OF BOOK.

Choose (a), (b), (c), or (d).

1. When two resistances are connected in series, (a) they must both have the same resistance value; (b) the voltage across each must be the same; (c) they must have different resistance values; (d) there is only one path for current through both resistances.
2. In Fig. 4-3c, if the current through R_5 is 1 A, then the current through R_3 must be (a) ⅓ A; (b) ½ A; (c) 1 A; (d) 3 A.
3. With a 10-kΩ resistance in series with a 2-kΩ resistance, the total R_T equals (a) 2 kΩ; (b) 8 kΩ; (c) 10 kΩ; (d) 12 kΩ.
4. With two 45-kΩ resistances in series across a 90-V battery, the voltage across each resistance equals (a) 30 V; (b) 45 V; (c) 90 V; (d) 180 V.
5. The sum of series IR voltage drops (a) is less than the smallest voltage drop; (b) equals the average value of all the voltage drops; (c) equals the applied voltage; (d) is usually more than the applied voltage.
6. Resistances R_1 and R_2 are in series with 90 V applied. If V_1 is 30 V, then V_2 must be (a) 30 V; (b) 90 V; (c) 45 V; (d) 60 V.
7. With a 4-Ω resistance and a 2-Ω resistance in series across a 6-V battery, the current (a) in the larger resistance is 1⅓ A; (b) in the smaller resistance is 3 A; (c) in both resistances is 1 A; (d) in both resistances is 2 A.

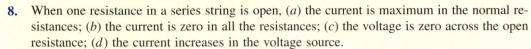

8. When one resistance in a series string is open, (*a*) the current is maximum in the normal re-sistances; (*b*) the current is zero in all the resistances; (*c*) the voltage is zero across the open resistance; (*d*) the current increases in the voltage source.
9. The resistance of an open series string is (*a*) zero; (*b*) infinite; (*c*) equal to the normal resis-tance of the string; (*d*) about double the normal resistance of the string.
10. A source of 100 V is applied across a 20-Ω R_1 and 30-Ω R_2 in series, and V_1 is 40 V. The current in R_2 is (*a*) 5 A; (*b*) 3⅓ A; (*c*) 1⅓ A; (*d*) 2 A.

QUESTIONS

1. Show how to connect two resistances in series with each other across a voltage source.
2. State three rules for the current, voltage, and resistance in a series circuit.
3. For a given amount of current, why does higher resistance have a larger voltage drop across it?
4. Two 300-W 120-V light bulbs are connected in series across a 240-V line. If the filament of one bulb burns open, will the other bulb light? Why? With the open circuit, how much is the voltage across the source and across each bulb?
5. Prove that if $V_T = V_1 + V_2 + V_3$, then $R_T = R_1 + R_2 + R_3$.
6. State briefly a rule for determining polarity of the voltage drop across each resistor in a se-ries circuit.
7. Redraw the circuit in Fig. 4-12, marking the polarity of V_1, V_2, and V_3.
8. State briefly a rule to determine when voltages are series-aiding.
9. Derive the formula $P_T = P_1 + P_2 + P_3$ from the fact that $V_T = V_1 + V_2 + V_3$.
10. In a series string, why does the largest R dissipate the most power?
11. Give one application of series circuits.
12. Describe the function of the AND gate.

PROBLEMS

ANSWERS TO ODD-NUMBERED PROBLEMS AT BACK OF BOOK.

1. A circuit has 20 V applied across a 10-Ω resistance R_1. How much is the current in the cir-cuit? How much resistance R_2 must be added in series with R_1 to reduce the current one-half? Show the schematic diagram of the circuit with R_1 and R_2.
2. Draw the schematic diagram of 20-, 30-, and 40-Ω resistances in series. (**a**) How much is the total resistance of the entire series string? (**b**) How much current flows in each resistance, with a voltage of 18 V applied across the series string? (**c**) Find the voltage drop across each resistance. (**d**) Find the power dissipated in each resistance.
3. An R_1 of 90 kΩ and an R_2 of 10 kΩ are in series across a 12-V source. (**a**) Draw the schematic diagram. (**b**) How much is V_2?

4. Draw a schematic diagram showing two resistances R_1 and R_2 in series across a 100-V source. (a) If the IR voltage drop across R_1 is 60 V, how much is the IR voltage drop across R_2? (b) Label the polarity of the voltage drops across R_1 and R_2. (c) If the current is 1 A through R_1, how much is the current through R_2? (d) How much is the resistance of R_1 and R_2? How much is the total resistance across the voltage source? (e) If the voltage source is disconnected, how much is the voltage across R_1 and across R_2?

5. In Fig. 4-18, calculate I, V_1, V_2, P_1, P_2, and P_T. (Note that R_1 and R_2 are in series even though V_T is shown at the right instead of the left.)

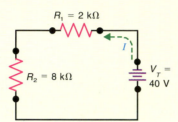

$R_1 = 2$ kΩ

$R_2 = 8$ kΩ

$V_T = 40$ V

FIG. 4-18   Circuit diagram for Probs. 5 and 6.

6. If R_1 is increased to 8 kΩ in Fig. 4-18, what will be the new value of I?
7. Find the total R_T of the following resistances in series: 2 MΩ; 0.5 MΩ; 68 kΩ; 5 kΩ; and 470 Ω.
8. Referring to Fig. 4-13, calculate the power dissipated in the voltage-dropping resistor R_S.
9. Draw the circuit with values for three equal series resistances across a 90-V source, where each R has one-third the applied voltage and the current in the circuit is 5 mA.
10. A 100-W bulb normally takes 0.833 A, and a 200-W bulb takes 1.666 A from the 120-V power line. If these two bulbs were connected in series across a 240-V power line, prove that the current would be 1.111 A in both bulbs, assuming the resistances remain constant.
11. How much voltage V is needed for a 1.8-mA current through resistances R_1 of 4.7 kΩ and R_2 of 6.8 kΩ in series?
12. Three 10-Ω resistances are in series across a voltage source. Show the schematic diagram. If the voltage across each resistor is 10 V, how much is the applied voltage? How much is the current in each resistance?
13. How much resistance R_1 must be added in series with a 100-Ω R_2 to limit the current to 0.3 A with 120 V applied? Show the schematic diagram. How much power is dissipated in each resistance?
14. In Fig. 4-19, find resistance R_1. Why is the electron flow for I in the direction shown?
15. In Fig. 4-20, find the value of R_2.
16. Figure 4-21 shows the circuit for keeping a 12.6-V car battery charged from a 14.8-Vdc generator. Calculate current I and show the direction of electron flow for current between points A and B.
17. In Fig. 4-22, find voltage V_2. Show polarities for the voltage drops.
18. In Fig 4-23, calculate voltage V_T. Show polarities for the voltage drops and V_T.
19. Three resistors of 100, 200, and 300 Ω are in series across a 24-V source. (*a*) If the 200-Ω resistor shorts, how much voltage is across the 300-Ω resistor? (*b*) If the 300-Ω resistor opens, how much voltage is across the 100- and 200-Ω resistors?
20. A 1.5-kΩ resistor is in series with an unknown resistance. The applied voltage equals 36 V and the series current is 14.4 mA. Calculate the value of the unknown resistor.

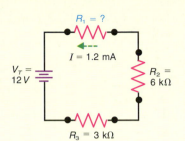

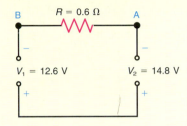

FIG. 4-19 Circuit diagram for Prob. 14.

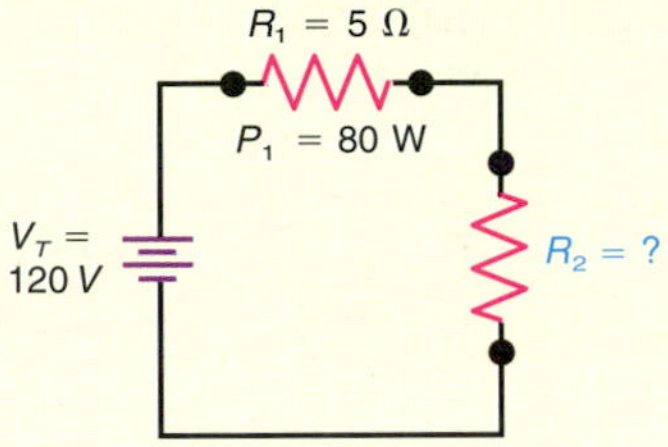

FIG. 4-20 Circuit diagram for Prob. 15.

FIG. 4-21 Circuit diagram for Prob. 16.

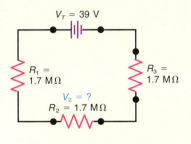

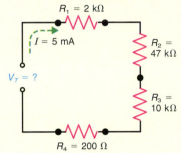

FIG. 4-22 Circuit diagram for Prob. 17.

FIG. 4-23 Circuit diagram for Prob. 18.

21. A 120-Ω resistor is in series with a resistor whose value is unknown. The voltage drop across the unknown resistance is 12 V, and the power dissipated by the 120-Ω resistor is 4.8 W. Calculate the value of the unknown resistance.

22. In Fig. 4-24, solve for the following unknown values: R_T, I, V_1, V_2, V_3, V_4, P_T, P_1, P_2, P_3, and P_4.

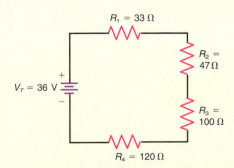

FIG. 4-24 Circuit diagram for Probs. 22 and 23.

23. In Fig. 4-24, assume R_2 opens. What are the values for V_1, V_2, V_3, and V_4?

24. In Fig. 4-25, solve for the following unknown values: R_T, I, V_1, V_2, V_3, V_4, V_5, P_T, P_1, P_2, P_3, P_4, and P_5.

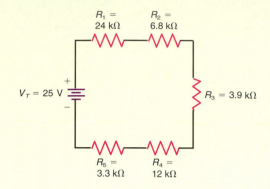

FIG. 4-25   Circuit diagram for Prob. 24.

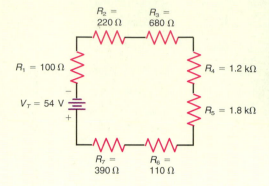

FIG. 4-26   Circuit diagram for Prob. 25.

25. In Fig. 4-26, solve for the following unknown values: R_T, I, V_1, V_2, V_3, V_4, V_5, V_6, V_7, P_T, P_1, P_2, P_3, P_4, P_5, P_6, and P_7.

26. In Fig. 4-27, solve for R_T, V_1, V_3, V_4, R_2, R_3, P_T, P_1, P_2, P_3, and P_4.

27. In Fig. 4-28, solve for I, V_1, V_2, V_3, V_T, R_3, P_T, P_2, and P_3.

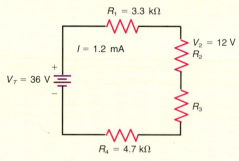

FIG. 4-27   Circuit diagram for Prob. 26.

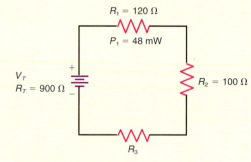

FIG. 4-28   Circuit diagram for Prob. 27.

CRITICAL THINKING

1. Three resistors in series have a total resistance R_T of 2.7 kΩ. If R_2 is twice the value of R_1 and R_3 is three times the value of R_2, what are the values of R_1, R_2, and R_3?

2. Three resistors in series have an R_T of 7 kΩ. If R_3 is 2.2 times larger than R_1 and 1.5 times larger than R_2, what are the values of R_1, R_2, and R_3?

3. A 100-Ω ⅛-W resistor is in series with a 330-Ω ½-W resistor. What is the maximum series current this circuit can handle without exceeding the wattage rating of either resistor?

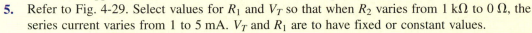

4. A 1.5-kΩ ½-W resistor is in series with a 470-Ω ¼-W resistor. What is the maximum voltage that can be applied to this series circuit without exceeding the wattage rating of either resistor?

5. Refer to Fig. 4-29. Select values for R_1 and V_T so that when R_2 varies from 1 kΩ to 0 Ω, the series current varies from 1 to 5 mA. V_T and R_1 are to have fixed or constant values.

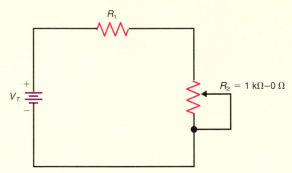

R_1

$R_2 = 1\ k\Omega{-}0\ \Omega$

V_T

FIG. 4-29 Circuit diagram for Critical Thinking Problem 5.

ANSWERS TO TEST-POINT QUESTIONS

4-1 a. R_1, R_2, R_3, V_T and the wires
b. $I = 5$ A
c. 2 A

4-2 a. $I = 2$ mA
b. $R = 10$ kΩ
c. $I = 1$ mA

4-3 a. 10 V
b. $I = 1$ A
c. $I = 1$ A
d. $I = 1$ A

4-4 a. 60 V
b. 75 V
c. 40 V

4-5 a. point A or C
b. point B or F
c. point D

4-6 a. $P = 6$ W
b. R_2

4-7 a. 100 V
b. 20 V

4-8 a. $V_1 = 30$ V
b. $V_2 = 90$ V
c. $V_3 = 60$ V

4-9 a. bulb
b. switch
c. zero

4-10 a. T
b. T
c. T

CHAPTER 5

PARALLEL CIRCUITS

Components in series have one common current, but parallel circuits are used where it is necessary to have one common voltage across all the components. A typical application is for the lights and appliances in a house, where they all need the same 120 V from the ac power line. As a definition, a parallel circuit is formed when two or more components are connected across one voltage source. The polarity of the connections does not matter for resistors. Each parallel path is then a branch circuit, with its own individual current.

Parallel circuits, therefore, have one common voltage across all the branches but separate branch currents that can be different. These characteristics are opposite from series circuits that have one common current but individual voltage drops. The principles of parallel or series connections apply to dc and ac circuits.

CHAPTER OBJECTIVES

Upon completion of this chapter, you should be able to:

- *Explain* why voltage is the same across all branches in a parallel circuit.
- *Calculate* the individual branch currents in a parallel circuit.
- *Calculate* the total current in a parallel circuit.
- *Calculate* the equivalent resistance of two or more resistors in parallel.
- *Explain* why the combined resistance of a parallel circuit is always less than the smallest branch resistance.
- *Calculate* the total conductance of a parallel circuit.
- *Calculate* the total power in a parallel circuit.
- *Solve* for the voltage, current, power, and resistance in a parallel circuit having random unknowns.
- *Describe* the effects of an open and short in a parallel circuit.
- *Describe* how parallel connected switches can be used to describe the OR logic function.

IMPORTANT TERMS IN THIS CHAPTER

inverse relation
main-line current
open branch
OR gate

parallel bank of resistors
parallel branch currents
parallel conductances
reciprocal resistance formula

short circuit
total power

TOPICS COVERED IN THIS CHAPTER

5-1 THE APPLIED VOLTAGE V_A IS THE SAME ACROSS PARALLEL BRANCHES

A parallel circuit is formed when two or more components are connected across a voltage source, as shown in Fig. 5-1. In this figure, R_1 and R_2 are in parallel with each other and a 1.5-V battery. In Fig. 5-1b, the points A, B, C, and E are really equivalent to a direct connection at the negative terminal of the battery because the connecting wires have practically no resistance. Similarly, points H, G, D, and F are the same as a direct connection at the positive battery terminal. Since R_1 and R_2 are directly connected across the two terminals of the battery, both resistances must have the same potential difference as the battery. It follows that the voltage is the same across components connected in parallel. The parallel circuit arrangement is used, therefore, to connect components that require the same voltage.

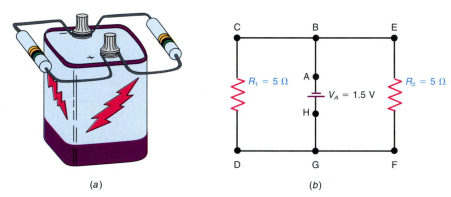

(a)　　　　　　(b)

FIG. 5-1 Example of a parallel circuit with two resistors. (*a*) Wiring diagram. (*b*) Schematic diagram.

A common application of parallel circuits is typical house wiring to the power line, with many lights and appliances connected across the 120-V source (Fig. 5-2). The wall receptacle has the potential difference of 120 V across each pair of terminals. Therefore, any resistance connected to an outlet has the applied

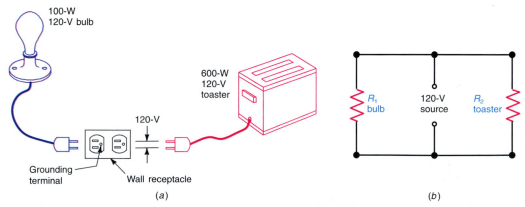

(a)　　　　　　(b)

FIG. 5-2 Light bulb and toaster connected in parallel with the 120-V line. (*a*) Wiring diagram. (*b*) Schematic diagram.

voltage of 120 V. The light bulb is connected to one outlet and the toaster to another outlet, but both have the same applied voltage of 120 V. Therefore, each operates independently of any other appliance, with all the individual branch circuits connected across the 120-V line.

TEST-POINT QUESTION 5-1

Answers at end of chapter.

a. In Fig. 5-1, how much is the common voltage across R_1 and R_2?
b. In Fig. 5-2, how much is the common voltage across the bulb and the toaster?
c. How many parallel branch circuits are connected across the voltage source in Figs. 5-1 and 5-2?

5-2 EACH BRANCH *I* EQUALS V_A/R

In applying Ohm's law, it is important to note that the current equals the voltage applied across the circuit divided by the resistance between the two points where that voltage is applied. In Fig. 5-3*a*, 10 V is applied across the 5 Ω of R_2, resulting in the current of 2 A between points E and F through R_2. The battery

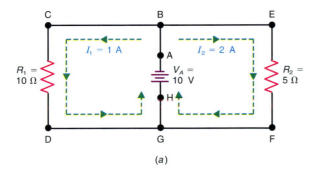

(a)

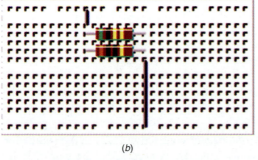

(b)

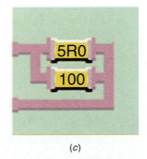

(c)

FIG. 5-3 Parallel circuit. (*a*) The current in each parallel branch equals the applied voltage V_A divided by each branch resistance R. (*b*) Axial-lead resistors assembled on a lab prototype board, forming a parallel circuit. (*c*) Surface-mount resistors assembled on a printed circuit board, forming a parallel circuit. Note: 5R0 equals 5.0 Ω and 100 equals 10 Ω.

voltage is also applied across the parallel resistance of R_1, applying 10 V across 10 Ω. Through R_1, therefore, the current is 1 A between points C and D. The current has a different value through R_1, with the same applied voltage, because the resistance is different. These values are calculated as follows:

$$I_1 = \frac{V_A}{R_1} = \frac{10}{10} = 1 \text{ A}$$

$$I_2 = \frac{V_A}{R_2} = \frac{10}{5} = 2 \text{ A}$$

Figure 5-3*b* shows how to assemble axial-lead resistors on a lab prototype board to form a parallel circuit. Figure 5-3*c* shows surface-mount resistors assembled on a printed circuit board to form a parallel circuit.

Just as in a circuit with just one resistance, any branch that has less R allows more I. If R_1 and R_2 were equal, however, the two branch currents would have the same value. For instance, in Fig. 5-1*b* each branch has its own current equal to 1.5 V/5 Ω = 0.3 A.

The I can be different in parallel circuits that have different R because V is the same across all the branches. Any voltage source generates a potential difference across its two terminals. This voltage does not move. Only I flows around the circuit. The source voltage is available to make electrons move around any closed path connected to the terminals of the source. How much I is in each separate path depends on the amount of R in each branch.

TEST-POINT QUESTION 5-2

Answers at end of chapter.

Refer to Fig. 5-3.
a. How much is the voltage across R_1?
b. How much is I_1 through R_1?
c. How much is the voltage across R_2?
d. How much is I_2 through R_2?

5-3 THE MAIN-LINE I_T EQUALS THE SUM OF THE BRANCH CURRENTS

Components to be connected in parallel are usually wired directly across each other, with the entire parallel combination connected to the voltage source, as illustrated in Fig. 5-4. This circuit is equivalent to wiring each parallel branch directly to the voltage source, as shown in Fig. 5-1, when the connecting wires have essentially zero resistance.

The advantage of having only one pair of connecting leads to the source for all the parallel branches is that usually less wire is necessary. The pair of leads connecting all the branches to the terminals of the voltage source is the *main line*. In Fig. 5-4, the wires from G to A on the negative side and from B to F in the return path form the main line.

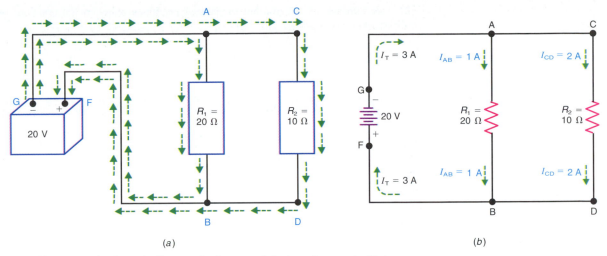

(a)

(b)

FIG. 5-4 The current in the main line equals the sum of the branch currents. Note that from G to A at the top of this diagram is the negative side of the main line, and from B to F at the bottom is the positive side. (a) Wiring diagram. Arrows inside the lines indicate current in the main line for R_1; arrows outside indicate current for R_2. (b) Schematic diagram. I_T is the total line current for both R_1 and R_2.

In Fig. 5-4b, with 20 Ω of resistance for R_1 connected across the 20-V battery, the current through R_1 must be 20 V/20 Ω = 1 A. This current is electron flow from the negative terminal of the source, through R_1, and back to the positive battery terminal. Similarly, the R_2 branch of 10 Ω across the battery has its own branch current of 20 V/10 Ω = 2 A. This current flows from the negative terminal of the source, through R_2, and back to the positive terminal, since it is a separate path for electron flow.

All the current in the circuit, however, must come from one side of the voltage source and return to the opposite side for a complete path. In the main line, therefore, the amount of current is equal to the total of the branch currents.

For example, in Fig. 5-4b, the total current in the line from point G to point A is 3 A. The total current at branch point A subdivides into its component branch currents for each of the branch resistances. Through the path of R_1 from A to B the current is 1 A. The other branch path ACDB through R_2 has a current of 2 A. At the branch point B, the electron flow from both parallel branches combines, so that the current in the main-line return path from B to F has the same value of 3 A as in the other side of the main line.

The formula for the total current I_T in the main line is

$$I_T = I_1 + I_2 + I_3 + \cdots + \text{etc.} \qquad \textbf{(5-1)}$$

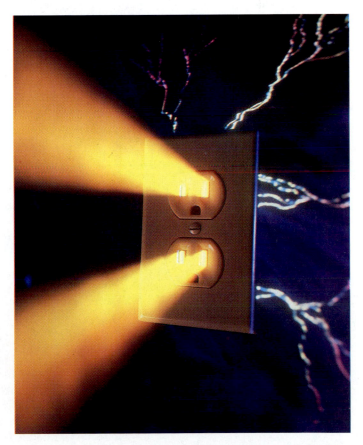

This rule applies for any number of parallel branches, whether the resistances in the branches are equal or unequal.

Example

EXAMPLE 1 An R_1 of 20 Ω, an R_2 of 40 Ω, and an R_3 of 60 Ω are connected in parallel across the 120-V power line. How much is the total line current I_T?

ANSWER Current I_1 for the R_1 branch is 120/20 or 6 A. Similarly, I_2 is 120/40 or 3 A, and I_3 is 120/60 or 2 A. The total current in the main line is

$$I_T = I_1 + I_2 + I_3 = 6 + 3 + 2$$
$$I_T = 11 \text{ A}$$

EXAMPLE 2 Two branches R_1 and R_2 across the 120-V power line draw a total line current I_T of 15 A. The R_1 branch takes 10 A. How much is the current I_2 in the R_2 branch?

ANSWER $I_2 = I_T - I_1 = 15 - 10$
$$I_2 = 5 \text{ A}$$

With two branch currents, one must equal the difference between I_T and the other branch current.

EXAMPLE 3 Three parallel branch currents are 0.1 A, 500 mA, and 800 μA. Calculate I_T.

ANSWER All values must be in the same units in order to be added. In this case, all units will be converted to milliamperes: 0.1 A = 100 mA and 800 μA = 0.8 mA. Then

$$I_T = 100 + 500 + 0.8$$
$$I_T = 600.8 \text{ mA}$$

You can convert the currents to A, mA, or μA units, as long as the same unit is used for adding all the currents.

TEST-POINT QUESTION 5-3

Answers at end of chapter.
a. Branch currents in a parallel circuit are 1 A for I_1, 2 A for I_2, and 3 A for I_3. How much is I_T?
b. Assume $I_T = 6$ A for three branch currents; I_1 is 1 A, and I_2 is 2 A. How much is I_3?
c. Branch currents in a parallel circuit are 1 A for I_1 and 200 mA for I_2. How much is I_T?

5-4 RESISTANCES IN PARALLEL

The combined equivalent resistance across the main line in a parallel circuit can be found by Ohm's law: *Divide the common voltage across the parallel resistances by the total current of all the branches.* Referring to Fig. 5-5a, note that the parallel resistance of R_1 with R_2, indicated by the equivalent resistance R_{EQ}, is the opposition to the total current in the main line. In this example, V_A/I_T is 60 V/3 A = 20 Ω for R_{EQ}.

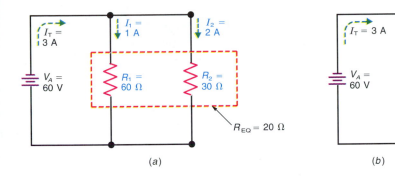

(a)
(b)

FIG. 5-5 Resistances in parallel. (a) Combination of R_1 and R_2 is the total R_{EQ} for the main line. (b) Equivalent circuit showing R_{EQ} drawing the same 3-A I_T as the parallel combination of R_1 and R_2 in (a).

The total load connected to the source voltage is the same as though one equivalent resistance of 20 Ω were connected across the main line. This is illustrated by the equivalent circuit in Fig. 5-5b. For any number of parallel resistances of any value, therefore,

▶ $$R_{EQ} = \frac{V_A}{I_T} \qquad (5\text{-}2)$$

where I_T is the sum of all the branch currents and R_{EQ} is the equivalent resistance of all the parallel branches across the applied voltage source V_A.

The first step in solving for R_{EQ} is to add all the parallel branch currents to find the I_T being delivered by the voltage source. The voltage source thinks it is connected to a single resistance whose value allows I_T to flow in the circuit according to Ohm's law. This single resistance is R_{EQ}. An illustrative example of a circuit with two parallel branches will be used to show how R_{EQ} is calculated.

Example

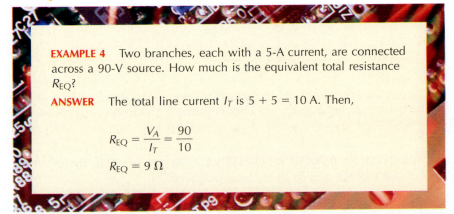

EXAMPLE 4   Two branches, each with a 5-A current, are connected across a 90-V source. How much is the equivalent total resistance R_{EQ}?

ANSWER   The total line current I_T is 5 + 5 = 10 A. Then,

$$R_{EQ} = \frac{V_A}{I_T} = \frac{90}{10}$$

$$R_{EQ} = 9 \ \Omega$$

PARALLEL BANK A combination of parallel branches is often called a *bank*. In Fig. 5-5, the bank consists of the 60-Ω R_1 and 30-Ω R_2 in parallel. Their combined parallel resistance R_{EQ} is the bank resistance, equal to 20 Ω in this example. A bank can have two or more parallel resistors.

When a circuit has more current with the same applied voltage, this greater value of I corresponds to less R because of their inverse relation. Therefore, the combination of parallel resistances R_{EQ} for the bank is always less than the smallest individual branch resistance. The reason is that I_T must be more than any one branch current.

WHY R_{EQ} IS LESS THAN ANY BRANCH R It may seem unusual at first that putting more resistance into a circuit lowers the equivalent resistance. This feature of parallel circuits is illustrated in Fig. 5-6. Note that equal resistances of 30 Ω each are added across the source voltage, one branch at a time. The circuit in Fig. 5-6a has just R_1, which allows 2 A with 60 V applied. In Fig. 5-6b the R_2 branch is added across the same V_A. This branch also has 2 A. Now the parallel circuit has a 4-A total line current because of $I_1 + I_2$. Then the third branch, which also takes 2 A for I_3, is added in Fig. 5-6c. The combined circuit with three branches therefore requires a total load current of 6 A, which is supplied by the voltage source.

The combined resistance across the source, then, is V_A/I_T, which is 60/6 or 10 Ω. This equivalent resistance R_{EQ}, representing the entire load on the voltage source, is shown in Fig. 5-6d. More resistance branches reduce the combined resistance of the parallel circuit because more current is required from the same voltage source.

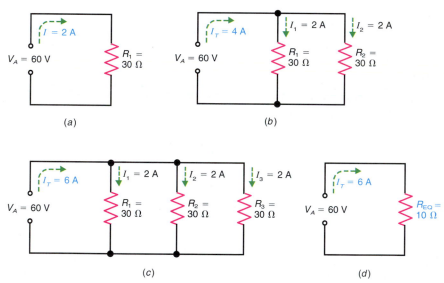

FIG. 5-6 How adding parallel branches of resistors increases I_T but decreases R_{EQ}. (a) One resistor. (b) Two branches. (c) Three branches. (d) Equivalent circuit of the three branches in (c).

RECIPROCAL RESISTANCE FORMULA We can derive this formula from the fact that I_T is the sum of all the branch currents, or,

$$I_T = I_1 + I_2 + I_3 + \cdots + \text{etc.}$$

However, I_T is V/R_{EQ}. Also, each I is V/R. Substituting V/R_{EQ} for I_T on the left side of the formula and V/R for each branch I on the right side, the result is

$$\frac{V}{R_{EQ}} = \frac{V}{R_1} + \frac{V}{R_2} + \frac{V}{R_3} + \cdots + \text{etc.}$$

Dividing by V because the voltage is the same across all the resistances,

▶ $$\frac{1}{R_{EQ}} = \frac{1}{R_1} + \frac{1}{R_2} + \frac{1}{R_3} + \cdots + \text{etc.} \qquad (5\text{-}3)$$

This reciprocal formula applies to any number of parallel resistances of any value. Using the values in Fig. 5-7a as an example,

$$\frac{1}{R_{EQ}} = \frac{1}{20} + \frac{1}{10} + \frac{1}{10} = \frac{1}{20} + \frac{2}{20} + \frac{2}{20} = \frac{5}{20}$$

$$R_{EQ} = \frac{20}{5} = 4\ \Omega$$

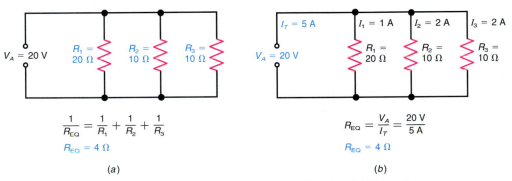

$$\frac{1}{R_{EQ}} = \frac{1}{R_1} + \frac{1}{R_2} + \frac{1}{R_3}$$

$$R_{EQ} = 4\ \Omega$$

(a)

$$R_{EQ} = \frac{V_A}{I_T} = \frac{20\ V}{5\ A}$$

$$R_{EQ} = 4\ \Omega$$

(b)

FIG. 5-7 Two methods of combining parallel resistances to find R_{EQ}. (a) Using the reciprocal resistance formula to calculate R_{EQ} as 4 Ω. (b) Using the total line current method with an assumed line voltage of 20 V gives the same 4 Ω for R_{EQ}.

Notice that the value for $1/R_{EQ}$ must be inverted to obtain R_{EQ} when using Formula (5-3) because the formula gives the reciprocal of R_{EQ}.

When using the calculator to find a reciprocal such as $1/R$, choose either of two methods. Either divide the number 1 by the value of R, or use the reciprocal key labeled $(1/x)$. As an example, to find the reciprocal of $R = 20\ \Omega$ by division:

First punch in the number 1 on the key pad.
Then press the division $(\div)$ key.
Punch in 20 for the value of R.
Finally, press the equal $(=)$ key for the quotient of 0.05 on the display.

To use the reciprocal key, first punch in 20 for R. Then press the $(1/x)$ key. This may be a second function on some calculators, requiring that you push the $(2^{nd}F)$ or $(SHIFT)$ key before pressing $(1/x)$. The reciprocal equal to 0.05 is displayed without the need for the $(=)$ key.

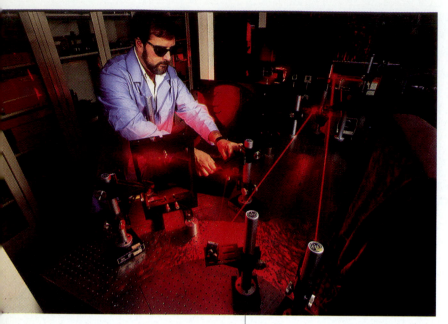

TOTAL-CURRENT METHOD Figure 5-7b shows how this same problem can be calculated in terms of total current instead of by the reciprocal formula, if it is easier to work without fractions. Although the applied voltage is not always known, any convenient value can be assumed because it cancels in the calculations. It is usually simplest to assume an applied voltage of the same numerical value as the highest resistance. Then one assumed branch current will automatically be 1 A and the other branch currents will be more, eliminating fractions less than 1 in the calculations.

In Fig. 5-7b, the highest branch R is 20 Ω. Therefore, assume 20 V for the applied voltage. Then the branch currents are 1 A in R_1, 2 A in R_2, and 2 A in R_3. Their sum is $1 + 2 + 2 = 5$ A for I_T. The combined resistance R_{EQ} across the main line is V_A/I_T, or 20 V/5 A = 4 Ω. This is the same value calculated with the reciprocal resistance formula.

SPECIAL CASE OF EQUAL R IN ALL BRANCHES If R is equal in all branches, the combined R_{EQ} equals the value of one branch resistance divided by the number of branches.

▶ $$R_{EQ} = \frac{R}{n}$$

where R is the resistance in one branch and n is the number of branches.

This rule is illustrated in Fig. 5-8, where three 60-kΩ resistances in parallel equal 20 kΩ.

The rule applies to any number of parallel resistances, but they must all be equal. As another example, five 60-Ω resistances in parallel have the combined resistance of 60/5, or 12 Ω. A common application is two equal resistors wired in a parallel bank for R_{EQ} equal to one-half R.

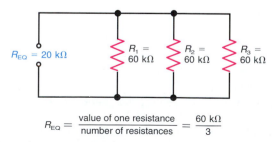

$$R_{EQ} = \frac{\text{value of one resistance}}{\text{number of resistances}} = \frac{60 \text{ k}\Omega}{3}$$

FIG. 5-8 For the special case of all branches having the same resistance, just divide R by the number of branches to find R_{EQ}. Here, $R_{EQ} = 60 \text{ k}\Omega/3 = 20 \text{ k}\Omega$.

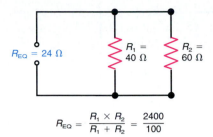

$$R_{EQ} = \frac{R_1 \times R_2}{R_1 + R_2} = \frac{2400}{100}$$

FIG. 5-9 For the special case of only two branch resistances, of any values, R_{EQ} equals their product divided by the sum. Here, $R_{EQ} = 2400/100 = 24\ \Omega$.

SPECIAL CASE OF ONLY TWO BRANCHES When there are two parallel resistances and they are not equal, it is usually quicker to calculate the combined resistance by the method shown in Fig. 5-9. This rule says that the combination of two parallel resistances is their product divided by their sum.

$$\blacktriangleright \qquad R_{EQ} = \frac{R_1 \times R_2}{R_1 + R_2} \qquad\qquad\qquad (5\text{-}4)$$

where R_{EQ} is in the same units as all the individual resistances. For the example in Fig. 5-9,

$$R_{EQ} = \frac{R_1 \times R_2}{R_1 + R_2} = \frac{40 \times 60}{40 + 60} = \frac{2400}{100}$$

$$R_{EQ} = 24\ \Omega$$

Each R can have any value, but there must be only two resistances. Note that this method gives R_{EQ} directly, not its reciprocal. If you use the reciprocal formula for this example, the answer will be $1/R_{EQ} = \frac{1}{24}$, which is the same value as $R_{EQ} = 24\ \Omega$.

    Formula (5-4) states a product over a sum. When using a calculator, group the R values in parentheses before adding. The reason is that the division bar is a mathematical sign of grouping for terms* to be added or subtracted. You must add $R_1 + R_2$ before dividing. By grouping R_1, and R_2 within parentheses, the addition will be done first before the division. The complete process is as follows. Multiply the R values in the numerator. Press the divide ⊘ key and then the left (or open) parentheses ⦅ key. Add the R values, $R_1 + R_2$, and press the right (or close) parentheses ⦆ key. Then press the equal ⊜ key for R_{EQ} on the display. Using the values in Fig. 5-9 as an example, multiply 40×60; press divide ⊘ and left parentheses ⦅ then 40 ⊕ 60 and the right parentheses ⦆. Finally, press ⊜ to display 24 as the answer.

SHORT-CUT CALCULATIONS Figure 5-10 shows how these special rules can help in reducing parallel branches to a simpler equivalent circuit. In Fig. 5-10a, the 60-Ω R_1 and R_4 are equal and in parallel. Therefore, they are equivalent to the 30-Ω R_{14} in Fig. 5-10b. Similarly, the 20-Ω R_2 and R_3 are equivalent to the 10 Ω of R_{23}. The circuit in Fig. 5-10a is equivalent to the

---

*For definitions of terms and factors, see B. Grob, *Mathematics for Basic Electronics*, Glencoe/McGraw-Hill, Columbus, Ohio.

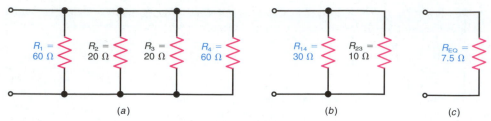

FIG. 5-10 An example of parallel resistance calculations with four branches. (*a*) Original circuit. (*b*) Resistors combined into two branches. (*c*) Equivalent circuit reduces to one R_{EQ} for all the branches.

simpler circuit in Fig. 5-10*b* with just the two parallel resistances of 30 and 10 Ω.

Finally, the combined resistance for these two equals their product divided by their sum, which is 300/40 or 7.5 Ω, as shown in Fig. 5-10*c*. This value of R_{EQ} in Fig. 5-10*c* is equivalent to the combination of the four branches in Fig. 5-10*a*. If you connect a voltage source across either circuit, the current in the main line will be the same for both cases.

The order of connections for parallel resistances does not matter in determining R_{EQ}. There is no question as to which is first or last because they are all across the same voltage source and receive their current at the same time.

FINDING AN UNKNOWN BRANCH RESISTANCE In some cases with two parallel resistors, it is useful to be able to determine what size R_X to connect in parallel with a known R in order to obtain a required value of R_{EQ}. Then the factors can be transposed as follows:

$$\blacktriangleright \qquad R_X = \frac{R \times R_{EQ}}{R - R_{EQ}} \qquad\qquad\qquad \textbf{(5-5)}$$

This formula is just another way of writing Formula (5-4).

Example

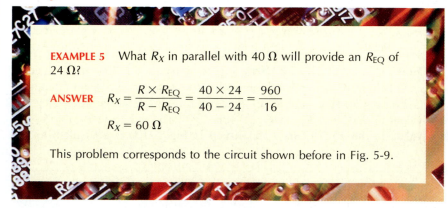

EXAMPLE 5 What R_X in parallel with 40 Ω will provide an R_{EQ} of 24 Ω?

ANSWER $R_X = \dfrac{R \times R_{EQ}}{R - R_{EQ}} = \dfrac{40 \times 24}{40 - 24} = \dfrac{960}{16}$

$R_X = 60\ \Omega$

This problem corresponds to the circuit shown before in Fig. 5-9.

Note that Formula (5-5) for R_X has a product over a difference. The R_{EQ} is subtracted because it is the smallest R. Remember that both Formulas (5-4) and (5-5) can be used with only two parallel branches.

Example

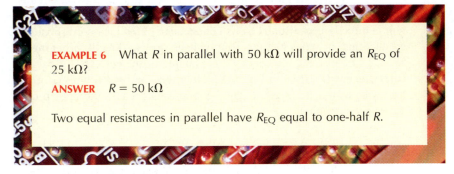

EXAMPLE 6   What R in parallel with 50 kΩ will provide an R_{EQ} of 25 kΩ?

ANSWER $R = 50 \text{ k}\Omega$

Two equal resistances in parallel have R_{EQ} equal to one-half R.

TEST-POINT QUESTION 5-4

Answers at end of chapter.
a. Find R_{EQ} for three 4.7-MΩ resistances in parallel.
b. Find R_{EQ} for 3 MΩ in parallel with 2 MΩ.
c. Find R_{EQ} for two parallel 20-Ω resistances in parallel with 10 Ω.

5-5 CONDUCTANCES IN PARALLEL

Since conductance G is equal to $1/R$, the reciprocal resistance Formula (5-3) can be stated for conductance as

$$G_T = G_1 + G_2 + G_3 + \cdots + \text{etc.} \qquad \text{(5-6)}$$

With R in ohms, G is in siemens. For the example in Fig. 5-11, G_1 is $1/20 = 0.05$, G_2 is $1/5 = 0.2$, and G_3 is $1/2 = 0.5$. Then

$$G_T = 0.05 + 0.2 + 0.5 = 0.75 \text{ S}$$

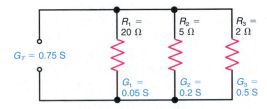

FIG. 5-11   Conductances G_1, G_2, and G_3 in parallel are added for the total G_T.

Notice that adding the conductances does not require reciprocals. Actually, each value of G is the reciprocal of R.

The reason why parallel conductances are added directly can be illustrated by assuming a 1-V source across all the branches. Then calculating the values of $1/R$ for the conductances gives the same values as calculating the branch currents. These values are added for the total I_T or G_T.

Working with G may be more convenient than R in parallel circuits, since it avoids use of the reciprocal formula for R_{EQ}. Each branch current is directly proportional to its conductance. This idea corresponds to the fact that, in series circuits, each voltage drop is directly proportional to each of the series resistances. An example of the currents for parallel conductances is shown in Fig. 5-12. Note that the branch with G of 4 S has twice as much current as the 2-S branches because the branch conductance is doubled.

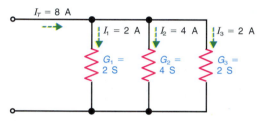

FIG. 5-12 Example of how parallel branch currents are directly proportional to each branch conductance G.

TEST-POINT QUESTION 5-5

Answers at end of chapter.
  a. If G_1 is 2 S and G_2 in parallel is 4 S, calculate G_T.
  b. If G_1 is 0.05 μS, G_2 is 0.2 μS, and G_3 is 0.5 μS, all in parallel, find G_T and its equivalent R_{EQ}.
  c. If G_T is 4 μS for a parallel circuit, how much is R_{EQ}?

5-6 TOTAL POWER IN PARALLEL CIRCUITS

Since the power dissipated in the branch resistances must come from the voltage source, the total power equals the sum of the individual values of power in each branch. This rule is illustrated in Fig. 5-13. We can also use this circuit as an example of how to apply the rules of current, voltage, and resistance for a parallel circuit.

The applied 10 V is across the 10-Ω R_1 and 5-Ω R_2 in Fig. 5-13. The branch current I_1 then is V_A/R_1 or 10/10, which equals 1 A. Similarly, I_2 is 10/5, or 2 A. The total I_T is $1 + 2 = 3$ A. If we want to find R_{EQ}, it equals V_A/I_T or 10/3, which is 3⅓ Ω.

The power dissipated in each branch R is $V_A \times I$. In the R_1 branch, I_1 is 10/10 = 1 A. Then P_1 is $V_A \times I_1$ or $10 \times 1 = 10$ W.

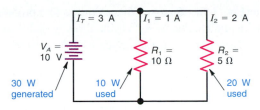

FIG. 5-13 The sum of the power values P_1 and P_2 used in each branch equals the total power P_T produced by the source.

For the R_2 branch, I_2 is 10/5 = 2 A. Then P_2 is $V_A \times I_2$ or $10 \times 2 = 20$ W.

Adding P_1 and P_2, the answer is $10 + 20 = 30$ W. This P_T is the total power dissipated in both branches.

This value of 30 W for P_T is also the total power supplied by the voltage source by means of its total line current I_T. With this method, the total power is $V_A \times I_T$ or $10 \times 3 = 30$ W for P_T. The 30 W of power supplied by the voltage source is dissipated or used up in the branch resistances.

Note that in both parallel and series circuits the sum of the individual values of power dissipated in the circuit equals the total power generated by the source. This can be stated as a formula

$$P_T = P_1 + P_2 + P_3 + \cdots + \text{etc.} \tag{5-7}$$

The series or parallel connections can alter the distribution of voltage or current, but power is the rate at which energy is supplied. The circuit arrangement cannot change the fact that all the energy in the circuit comes from the source.

TEST-POINT QUESTION 5-6

Answers at end of chapter.
a. Two parallel branches each have 2 A at 120 V. How much is P_T?
b. Three parallel branches of 10, 20, and 30 Ω have 60 V applied. How much is P_T?
c. Two parallel branches dissipate power of 15 W each. How much is P_T?

5-7 ANALYZING PARALLEL CIRCUITS

For many types of problems with parallel circuits it is useful to remember the following points:

1. When you know the voltage across one branch, this voltage is across all the branches. There can be only one voltage across branch points with the same potential difference.
2. If you know I_T and one of the branch currents I_1, you can find I_2 by subtracting I_1 from I_T.

The circuit in Fig. 5-14 illustrates these points. The problem is to find the applied voltage V_A and the value of R_3. Of the three branch resistances, only R_1 and R_2 are known. However, since I_2 is given as 2 A, the I_2R_2 voltage must be $2 \times 60 = 120$ V.

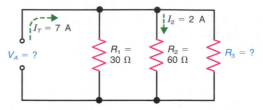

FIG. 5-14   Analyzing a parallel circuit. What are the values for V_A and R_3? See solution in text.

Although the applied voltage is not given, this must also be 120 V. The voltage across all the parallel branches is the same 120 V that is across the R_2 branch.

Now I_1 can be calculated as V_A/R_1. This is 120/30 = 4 A for I_1.

Current I_T is given as 7 A. The two branches take $2 + 4 = 6$ A. The third branch current through R_3 must be $7 - 6 = 1$ A for I_3.

Now R_3 can be calculated as V_A/I_3. This is 120/1 = 120 Ω for R_3.

5-8 EFFECT OF AN OPEN BRANCH IN PARALLEL CIRCUITS

An open in any circuit is an infinite resistance that results in no current. However, in parallel circuits there is a difference between an open circuit in the main line and an open circuit in a parallel branch. These two cases are illustrated in Fig. 5-15. In Fig. 5-15*a* the open circuit in the main line prevents any electron flow in the line to all the branches. The current is zero in every branch, therefore, and none of the bulbs can light.

However, in Fig. 5-15*b* the open is in the branch circuit for bulb 1. The open branch circuit has no current, then, and this bulb cannot light. The current in all the other parallel branches is normal, though, because each is connected to the voltage source. Therefore, the other bulbs light.

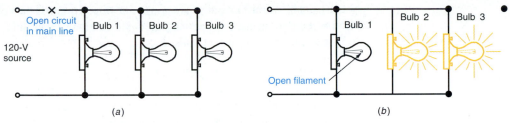

FIG. 5-15 Effect of an open in a parallel circuit. (*a*) Open path in the main line—
no current and no light for all the bulbs. (*b*) Open path in any branch—bulb for that
branch does not light, but the other two bulbs operate normally.

These circuits show the advantage of wiring components in parallel. An open
in one component opens only one branch, while the other parallel branches have
their normal voltage and current.

5-9 EFFECT OF A SHORT CIRCUIT ACROSS PARALLEL BRANCHES

A short circuit has practically zero resistance. Its effect, therefore, is to allow excessive current in the shorted circuit. Consider the example in Fig. 5-16. In Fig.
5-16*a*, the circuit is normal, with 1 A in each branch and 2 A for the total line
current. However, suppose the conducting wire at point H should accidentally
contact the wire at point G, as shown in Fig. 5-16*b*. Since the wire is an excellent conductor, the short circuit results in practically zero resistance between

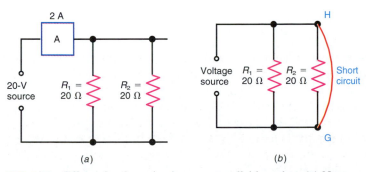

FIG. 5-16 Effect of a short circuit across parallel branches. (*a*) Normal circuit. (*b*) Short circuit across points H and G shorts out all the
branches.

points H and G. These two points are connected directly across the voltage source. With no opposition, the applied voltage could produce an infinitely high value of current through this current path.

THE SHORT-CIRCUIT CURRENT   Practically, the amount of current is limited by the small resistance of the wire. Also, the source usually cannot maintain its output voltage while supplying much more than its rated load current. Still, the amount of current can be dangerously high. For instance, the short-circuit current might be more than 100 A instead of the normal line current of 2 A in Fig. 5-16a. Because of the short circuit, excessive current flows in the voltage source, in the line to the short circuit at point H, through the short circuit, and in the line returning to the source from G. Because of the large amount of current, the wires can become hot enough to ignite and burn the insulation covering the wire. There should be a fuse that would open if there is too much current in the main line because of a short circuit across any of the branches.

THE SHORT-CIRCUITED COMPONENTS HAVE NO CURRENT   For the short circuit in Fig. 5-16b, the I is 0 A in the parallel resistors R_1 and R_2. The reason is that the short circuit is a parallel path with practically zero resistance. Then all the current flows in this path, bypassing the resistors R_1 and R_2. Therefore R_1 and R_2 are short-circuited or *shorted out* of the circuit. They cannot function without their normal current. If they were filament resistances of light bulbs or heaters, they would not light without any current.

The short-circuited components are not damaged, however. They do not even have any current passing through them. Assuming that the short circuit has not damaged the voltage source and the wiring for the circuit, the components can operate again when the circuit is restored to normal by removing the short circuit.

ALL THE PARALLEL BRANCHES ARE SHORT-CIRCUITED   If there were only one R in Fig. 5-16 or any number of parallel components, they would all be shorted out by the short circuit across points H and G. Therefore, a short circuit across one branch in a parallel circuit shorts out all the parallel branches.

This idea also applies to a short circuit across the voltage source in any type of circuit. Then the entire circuit is shorted out.

TEST-POINT QUESTION 5-9

Answers at end of chapter.

Refer to Fig. 5-16.
a. How much is the R of the short circuit between H and G?
b. How much is I_1 in R_1 with the short circuit across R_2?

5-10 PARALLEL SWITCHES REPRESENT THE OR LOGIC FUNCTION

Another common logic circuit found in digital electronics is the OR gate, which in one respect is opposite to the AND logic function described in Chap. 4. The parallel connected switches *A* and *B* in Fig. 5-17*a* illustrate the basic function of an OR gate. If either switch *A* or switch *B* or both is closed, the lamp will be on. The lamp is off only when both switches *A* and *B* are open. Figure 5-17*b* shows the truth table listing all possible combinations for switches *A* and *B*. With both switches open there is no complete current path and the lamp cannot light. With either or both switches closed, current flows to light the lamp.

It should be noted that the parallel-connected switches in Fig. 5-17*a* do not represent the actual type of circuit used to perform the OR logic function in digital circuits. However, the analogy of operation of an OR gate is identical to that of the parallel switches in Fig. 5-17*a*. Figure 5-17*c* shows the symbol used to represent the OR logic function in digital circuits. The inputs to the OR gate are applied to *A* and *B* and the output is taken from *X*. (An OR gate may have more than two inputs.) The mathematical expression describing the operation of the OR gate is shown in Fig. 5-17*d*. The statement is read "*A* or *B* equals *X*," rather than "*A* plus *B* equals *X*."

TEST POINT QUESTION 5-10

Answers at end of chapter.

Answer True or False.
a. In Fig. 5-17*a*, the lamp is on when both switches are open.
b. In Fig. 5-17*a*, the voltage across switch *B* is 12 V if switch *A* is closed and switch *B* is open.
c. The mathematical expression used to describe the circuit in Fig. 5-17*a* is $A + B = X$.
d. An OR gate can have more than two inputs.

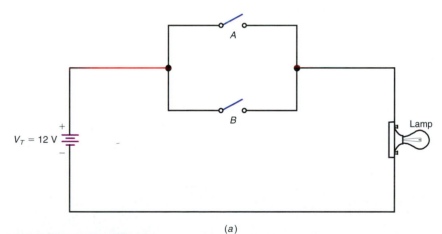

(a)

Switches		
A	*B*	Lamp
Open	Open	OFF
Open	Closed	ON
Closed	Open	ON
Closed	Closed	ON

(b)

(c)

$A + B = X$

(d)

FIG. 5-17 Comparison of the OR logic gate to a parallel circuit and switches. (*a*) Parallel circuit with two switches, *A* and *B*. (*b*) Truth table with four possibilities for the switches. (*c*) Schematic symbol for the OR logic gate. (*d*) Boolean expression for the OR gate.

- There is only one voltage V_A across all components in parallel.
- The current in each branch I_b equals the voltage V_A across the branch divided by the branch resistance R_b. Or $I_b = V_A/R_b$.
- The total line current equals the sum of all the branch currents. Or $I_T = I_1 + I_2 + I_3 + \cdots + $ etc.
- The equivalent resistance R_{EQ} of parallel branches is less than the smallest branch resistance, since all the branches must take more current from the source than any one branch.
- For only *two* parallel resistances of any value, $R_{EQ} = R_1 R_2/(R_1 + R_2)$.
- For any number of *equal* parallel resistances, R_{EQ} is the value of one resistance divided by the number of resistances.
- For the general case of any number of branches, calculate R_{EQ} as V_A/I_T or use the reciprocal resistance formula:

$$1/R_{EQ} = 1/R_1 + 1/R_2 + 1/R_3 + \cdots + \text{etc.}$$

- For any number of conductances in parallel, their values are added for G_T, in the same way as parallel branch currents are added.
- The sum of the individual values of power dissipated in parallel resistances equals the total power produced by the source.
- An open circuit in one branch results in no current through that branch, but the other branches can have their normal current. However, an open circuit in the main line results in no current for any of the branches.
- A short circuit has zero resistance, resulting in excessive current. When one branch is short-circuited, all the parallel paths are also short-circuited. The entire current is in the short circuit and no current is in the short-circuited branches.
- The OR gate function corresponds to switches in parallel.

SELF-TEST

ANSWERS AT BACK OF BOOK.

Choose (*a*), (*b*), (*c*), or (*d*).

1. With two resistances connected in parallel: (*a*) the current through each must be the same; (*b*) the voltage across each must be the same; (*c*) their combined resistance equals the sum of the individual values; (*d*) each must have the same resistance value.
2. With 100 V applied across ten 50-Ω resistances in parallel, the current through each resistance equals (*a*) 2 A; (*b*) 10 A; (*c*) 50 A; (*d*) 100 A.

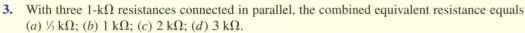

3. With three 1-kΩ resistances connected in parallel, the combined equivalent resistance equals (*a*) ⅓ kΩ; (*b*) 1 kΩ; (*c*) 2 kΩ; (*d*) 3 kΩ.

4. A 1-Ω resistance in parallel with a 2-Ω resistance provides a combined equivalent resistance of (*a*) 3 Ω; (*b*) 1 Ω; (*c*) 2 Ω; (*d*) ⅔ Ω.

5. With resistances of 100, 200, 300, 400, and 500 Ω in parallel, R_{EQ} is (*a*) less than 100 Ω; (*b*) more than 1 MΩ; (*c*) about 500 Ω; (*d*) about 1 kΩ.

6. With two resistances connected in parallel, if each dissipates 10 W, the total power supplied by the voltage source equals (*a*) 5 W; (*b*) 10 W; (*c*) 20 W; (*d*) 100 W.

7. With eight 10-MΩ resistances connected in parallel across a 10-V source, the main-line current equals (*a*) 0.1 μA; (*b*) ⅛ μA; (*c*) 8 μA; (*d*) 10 μA.

8. A parallel circuit with 20 V applied across two branches has a total line current of 5 A. One branch resistance equals 5 Ω. The other branch resistance equals (*a*) 5 Ω; (*b*) 20 Ω; (*c*) 25 Ω; (*d*) 100 Ω.

9. Three 100-W light bulbs are connected in parallel across the 120-V power line. If one bulb opens, how many bulbs can light? (*a*) None; (*b*) one; (*c*) two; (*d*) all.

10. If a parallel circuit is open in the main line, the current (*a*) increases in each branch; (*b*) is zero in all the branches; (*c*) is zero only in the branch that has highest resistance; (*d*) increases in the branch that has lowest resistance.

QUESTIONS

1. Draw a wiring diagram showing three resistances connected in parallel across a battery. Indicate each branch and the main line.

2. State two rules for the voltage and current values in a parallel circuit.

3. Explain briefly why the current is the same in both sides of the main line that connects the voltage source to the parallel branches.

4. (**a**) Show how to connect three equal resistances for a combined equivalent resistance one-third the value of one resistance. (**b**) Show how to connect three equal resistances for a combined equivalent resistance three times the value of one resistance.

5. Why can the current in parallel branches be different when they all have the same applied voltage?

6. Why does the current increase in the voltage source as more parallel branches are added to the circuit?

7. Show how the formula

$$R_{EQ} = R_1 R_2 / (R_1 + R_2)$$

is derived from the reciprocal formula

$$\frac{1}{R_{EQ}} = \frac{1}{R_1} + \frac{1}{R_2}$$

8. Redraw Fig. 5-16 with five parallel resistors R_1 to R_5 and explain why they all would be shorted out with a short circuit across R_3.

9. State briefly why the total power equals the sum of the individual values of power, whether a series circuit or a parallel circuit is used.

10. Explain why an open in the main line disables all the branches, but an open in one branch affects only that branch current.

11. Give two differences between an open circuit and a short circuit.

12. List as many differences as you can in comparing series circuits with parallel circuits.

13. Why are household appliances connected to the 120-V power line in parallel instead of being in series?

14. Give one advantage and one disadvantage of parallel connections.

15. Describe the function of the OR gate.

PROBLEMS

ANSWERS TO ODD-NUMBERED PROBLEMS AT BACK OF BOOK.

1. A 6-Ω R_1 and a 12-Ω R_2 are connected in parallel across a 12-V battery. (**a**) Draw the schematic diagram. (**b**) How much is the voltage across R_1 and R_2? (**c**) How much is the current in R_1 and R_2? (**d**) How much is the main-line current? (**e**) Calculate R_{EQ}.

2. For the circuit in question 1, how much is the total power supplied by the battery?

3. A parallel circuit has three branch resistances of 20, 10, and 5 Ω for R_1, R_2, and R_3. The current through the 20-Ω branch is 1 A. (**a**) Draw the schematic diagram. (**b**) How much is the voltage applied across all the branches? (**c**) Find the current through the 10-Ω branch and the 5-Ω branch.

4. (**a**) Draw the schematic diagram of a parallel circuit with three branch resistances, each having 28 V applied and a 2-A branch current. (**b**) How much is I_T? (**c**) How much is R_{EQ}?

5. Referring to Fig. 5-13, assume that R_2 opens. (**a**) How much is the current in the R_2 branch? (**b**) How much is the current in the R_1 branch? (**c**) How much is the line current? (**d**) How much is the R_{EQ} of the circuit? (**e**) How much power is generated by the battery?

6. Two resistances R_1 and R_2 are in parallel across a 120-V source. The total line current is 10 A. The current I_1 through R_1 is 4 A. Draw a schematic diagram of the circuit, giving the values of currents I_1 and I_2 and resistances R_1 and R_2 in both branches. How much is the R_{EQ} of both branches across the voltage source?

7. Find the R_{EQ} for the following groups of branch resistances: (**a**) 10 Ω and 25 Ω; (**b**) five 10-kΩ resistances; (**c**) two 500-Ω resistances; (**d**) 100 Ω, 200 Ω, and 300 Ω; (**e**) two 5-kΩ and two 2-kΩ resistances; (**f**) four 40-kΩ and two 20-kΩ resistances.

8. Find R_3 in Fig. 5-18.

9. Find the total conductance in siemens for the following branches: $G_1 = 9000$ μS; $G_2 = 8000$ μS; $G_3 = 22,000$ μS.

10. Referring to Fig. 5-11, calculate R_{EQ} by combining resistances. Show that this R_{EQ} equals $1/G_T$, where G_T is 0.75 S.

11. How much parallel R_X must be connected across a 100-kΩ resistance to reduce R_{EQ} to (**a**) 50 kΩ; (**b**) 25 kΩ; (**c**) 10 kΩ?

12. In Fig. 5-19, (**a**) find each branch current and show the direction of electron flow; (**b**) calculate I_T; (**c**) calculate R_{EQ}; (**d**) calculate P_1, P_2, P_3, and P_T.

13. A 2.2-kΩ resistor R_1 and 3.9-kΩ resistor R_2 are in parallel. The current I_1 is 4 mA. Calculate V_1, V_2, and I_2.

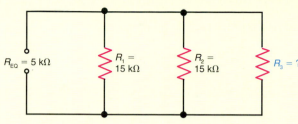

FIG. 5-18 Circuit diagram for Prob. 8.

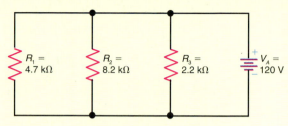

FIG. 5-19 Circuit diagram for Prob. 12.

14. Compare the combined resistance of a 5600-Ω R_1 and 8200-Ω R_2 connected in series and in parallel.

15. A 10-Ω R_1 is in parallel with a branch that has G_2 of 0.4 S. Calculate G_T and R_{EQ} for the two parallel branches.

16. What value of G_3 must be connected in parallel with G_1 of 1.62 S and G_2 of 1.34 S for a G_T of 4 S?

17. In Fig. 5-20, solve for I_1, I_2, I_3, I_T, R_{EQ}, P_1, P_2, P_3, and P_T.

18. In Fig. 5-20, what is R_{EQ} if R_2: (a) opens; (b) shorts?

19. In Fig. 5-21, solve for I_1, I_2, I_3, I_T, R_{EQ}, P_1, P_2, P_3, and P_T.

20. In Fig. 5-21, what is R_{EQ} if a 15-Ω resistor, R_4, is added in parallel with the other three resistors?

21. In Fig. 5-21, how is the branch current, I_1, affected if R_3 opens?

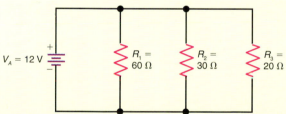

FIG. 5-20 Circuit diagram for Probs. 17 and 18.

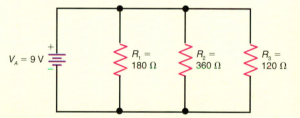

FIG. 5-21 Circuit diagram for Probs. 19, 20, and 21.

22. In Fig. 5-22, solve for I_1, I_2, I_3, I_4, I_T, R_{EQ}, P_1, P_2, P_3, P_4, and P_T.

23. In Fig. 5-22, calculate R_{EQ} if V_A is doubled to 48 V.

24. In Fig. 5-23, calculate I_1, I_2, I_3, I_4, I_T, and R_{EQ}. Also, list the currents which will be indicated by meters M_1 and M_2.

25. In Fig. 5-23, how much current will be indicated by M_2 if: (a) R_4 opens; (b) R_2 opens?

26. In Fig. 5-24, calculate I_1, I_2, I_3, I_4, I_T, and R_{EQ}.

27. In Fig. 5-24, how much current will be indicated by M_1 if: (a) R_1 and R_4 open; (b) R_3 opens?

28. In Fig. 5-24, how much current will flow in M_2 if a 200-Ω resistor is added between points A and B?

29. In Fig. 5-25, find I_T, I_1, I_2, R_1, R_2, R_3, P_T, P_2, and P_3.

30. In Fig. 5-26, find I_T, I_1, I_2, I_4, R_3, R_4, P_T, P_1, P_2, P_3, and P_4.

31. In Fig. 5-27, find R_{EQ}, I_T, I_2, I_3, I_4, R_1, R_3, P_T, P_1, P_2, P_3, and P_4.

32. In Fig. 5-28, find V_A, I_T, I_1, I_2, R_3, P_T, P_1, P_2, and P_3.

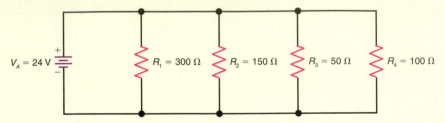

FIG. 5-22 Circuit diagram for Probs. 22 and 23.

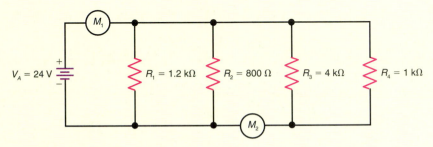

FIG. 5-23 Circuit diagram for Probs. 24 and 25.

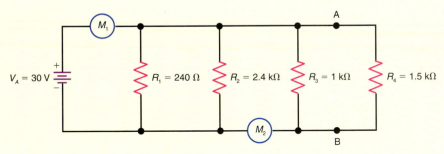

FIG. 5-24 Circuit diagram for Probs. 26, 27, and 28.

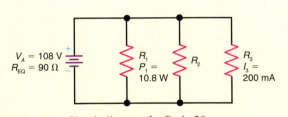

FIG. 5-25 Circuit diagram for Prob. 29.

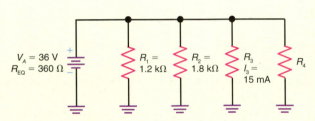

FIG. 5-26 Circuit diagram for Prob. 30.

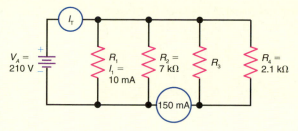

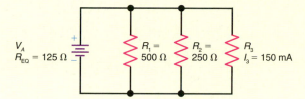

FIG. 5-27 Circuit diagram for Prob. 31.

FIG. 5-28 Circuit diagram for Prob. 32.

CRITICAL THINKING

1. A 180-Ω ¼-W resistor is in parallel with a 1-kΩ ½-W and a 12-kΩ 2-W resistor. What is the maximum total current, I_T, that this parallel combination can have before the wattage rating of any resistor is exceeded?

2. A 470-Ω ⅛-W resistor is in parallel with a 1-kΩ ¼-W and a 1.5-kΩ ½-W resistor. What is the maximum voltage, V, that can be applied to this circuit without exceeding the wattage rating of any resistor?

3. Three resistors in parallel have a combined equivalent resistance, R_{EQ}, of 1 kΩ. If R_2 is twice the value of R_3 and three times the value of R_1, what are the values for R_1, R_2, and R_3?

4. Three resistors in parallel have a combined equivalent resistance, R_{EQ}, of 4 Ω. If the conductance, G_1, is one-fourth that of G_2 and one-fifth that of G_3, what are the values of R_1, R_2, and R_3?

5. A voltage source is connected across four resistors, R_1, R_2, R_3, and R_4, in parallel. The currents are labeled I_1, I_2, I_3, and I_4, respectively. If $I_2 = 2I_1$, $I_3 = 2I_2$, and $I_4 = 2I_3$, calculate the values for R_1, R_2, R_3, and R_4 if $R_{EQ} = 1$ kΩ.

ANSWERS TO TEST-POINT QUESTIONS

5-1 a. 1.5 V
     b. 120 V
     c. Two

5-2 a. 10 V
     b. 1 A
     c. 10 V
     d. 2 A

5-3 a. $I_T = 6$ A
     b. $I_3 = 3$ A
     c. $I_T = 1.2$ A

5-4 a. $R_{EQ} = 1.57$ MΩ
     b. $R_{EQ} = 1.2$ MΩ
     c. $R_{EQ} = 5$ Ω

5-5 a. $G_T = 6$ S
     b. $G_T = 0.75$ μS
        $R_{EQ} = 1.33$ MΩ
     c. $R_{EQ} = 0.25$ MΩ

5-6 a. 480 W
     b. 660 W
     c. 30 W

5-7 a. 120 V
     b. $I_1 = 4$ A
     c. 7 A

5-8 a. infinite ohms
     b. bulbs 1 and 2
     c. all

5-9 a. 0 Ω
     b. $I_1 = 0$ A

5-10 a. F
     b. F
     c. T
     d. T

SERIES-PARALLEL CIRCUITS

Components in parallel with the voltage source V_A all have the same voltage, but what if they need a voltage less than V_A? The solution is to insert a voltage-dropping resistor in series with the parallel combination. The result is one type of series-parallel circuit. As another example, components in series have the same current but another component may need the same voltage that exists across one of the series components. Just connect the components in parallel for the same voltage for another type of series-parallel circuit.

In general, a series-parallel circuit is used when it is necessary to provide different amounts of voltage and current for the components using one source of applied voltage.

These principles are described for dc circuits in this chapter but they also apply to ac circuits, since resistors have the same value for direct or alternating current.

CHAPTER OBJECTIVES

Upon completion of this chapter, you should be able to:

- *Determine* the total resistance of a series-parallel circuit.
- *Calculate* the voltage, current, resistance, and power in a series-parallel circuit.
- *Calculate* the voltage, current, resistance, and power in a series-parallel circuit having random unknowns.
- *Explain* how a Wheatstone bridge can be used to determine the resistance of an unknown resistor.
- *Define* the term *chassis ground.*
- *Calculate* the voltage at a given point in a circuit with respect to chassis ground.
- *Describe* the effects of opens and shorts in series-parallel circuits.
- *Describe* how series-parallel switches can be used to describe the operation of a logic circuit containing AND and OR gates.

IMPORTANT TERMS IN THIS CHAPTER

balanced bridge	grounded tap	voltage divider
banks in series	strings in parallel	Wheatstone bridge
chassis-ground connections	voltage to chassis ground	

TOPICS COVERED IN THIS CHAPTER

6-1 FINDING R_T FOR SERIES-PARALLEL RESISTANCES

In Fig. 6-1, R_1 is in series with R_2. Also, R_3 is in parallel with R_4. However, R_2 is *not* in series with R_3 and R_4. The reason is the branch point A where the current through R_2 divides for R_3 and R_4. As a result, the current through R_3 must be less than the current through R_2. Therefore, R_2 and R_3 cannot be in series because they do not have the same current. For the same reason, R_4 also cannot be in series with R_2.

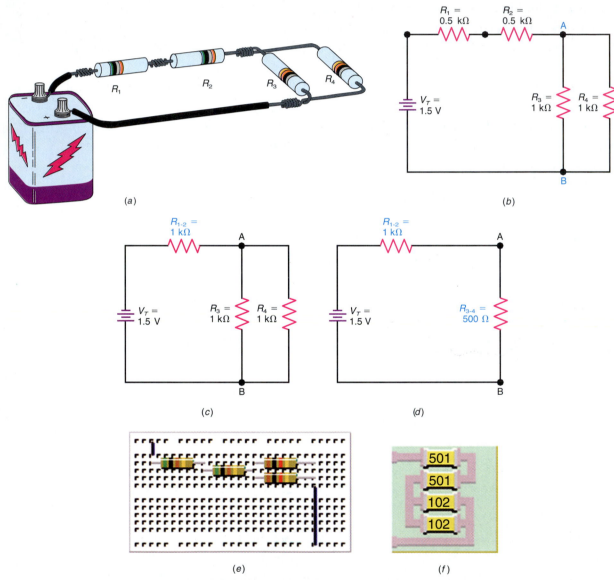

FIG. 6-1 Example of series-parallel circuit. (*a*) Wiring of a series-parallel circuit. (*b*) Schematic diagram of a series-parallel circuit. (*c*) Schematic with R_1 and R_2 in series added for $R_{1\text{-}2}$. (*d*) Schematic with R_3 and R_4 in parallel combined for $R_{3\text{-}4}$. (*e*) Axial-lead resistors assembled on a lab prototype board to form the series-parallel circuit shown in parts *a* and *b*. (*f*) Surface-mount resistors assembled on a printed circuit board to form the series-parallel circuit shown in parts *a* and *b*.

The wiring is shown in Fig. 6-1a with the schematic diagram in Fig. 6-1b. To find R_T, we add the series resistances and combine the parallel resistances.

In Fig. 6-1c, the 0.5-kΩ R_1 and 0.5-kΩ R_2 in series total 1 kΩ for $R_{1\text{-}2}$. The calculations are

$$0.5 \text{ k}\Omega + 0.5 \text{ k}\Omega = 1 \text{ k}\Omega$$

Also, the 1-kΩ R_3 in parallel with the 1-kΩ R_4 can be combined, for an equivalent resistance of 0.5 kΩ for $R_{3\text{-}4}$, as in Fig. 6-1d. The calculations are

$$\frac{1 \text{ } k\Omega}{2} = 0.5 \text{ } k\Omega$$

This parallel $R_{3\text{-}4}$ combination of 0.5 kΩ is then added to the series $R_{1\text{-}2}$ combination for the final R_T value of 1.5 kΩ. The calculations are

   $0.5 \text{ k}\Omega + 1 \text{ k}\Omega = 1.5 \text{ k}\Omega$

The 1.5 kΩ is the R_T of the entire circuit connected across the V_T of 1.5 V.

With R_T known to be 1.5 kΩ, we can find I_T in the main line produced by 1.5 V. Then

$$I_T = \frac{V_T}{R_T} = \frac{1.5 \text{ V}}{1.5 \text{ k}\Omega} = 1 \text{ mA}$$

This 1-mA I_T is the current through resistors R_1 and R_2 in Fig. 6-1a and b or $R_{1\text{-}2}$ in Fig. 6-1c.

At branch point A, at the top of the diagram in Fig. 6-1b, the 1 mA of electron flow for I_T divides into two branch currents for R_3 and R_4. Since these two branch resistances are equal, I_T divides into two equal parts of 0.5 mA each. At branch point B at the bottom of the diagram, the two 0.5-mA branch currents combine to equal the 1-mA I_T in the main line, returning to the source V_T.

Figure 6-1e shows axial-lead resistors assembled on a lab prototype board to form the series-parallel circuit shown in Fig. 6-1a and b. Figure 6-1f shows surface-mount resistors assembled on a printed circuit board to form the series-parallel circuit shown in Fig. 6-1b.

TEST-POINT QUESTION 6-1

Answers at end of chapter.

Refer to Fig. 6-1b.
a. Calculate the series R of R_1 and R_2.
b. Calculate the parallel R of R_3 and R_4.
c. Calculate R_T across the source V_T.

6-2 RESISTANCE STRINGS IN PARALLEL

More details about the voltages and currents in a series-parallel circuit are illustrated by the example in Fig. 6-2. Suppose there are four 120-V 100-W light bulbs to be wired, with a voltage source that produces 240 V. Each bulb needs

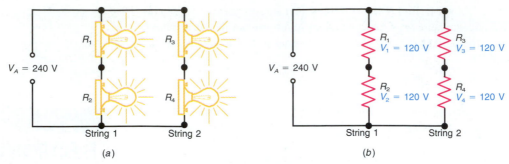

FIG. 6-2 Two identical series strings in parallel. All bulbs have 120-V 100-W rating. (*a*) Wiring diagram. (*b*) Schematic diagram.

120 V for normal brilliance. If the bulbs were connected across the source, each would have the applied voltage of 240 V. This would cause excessive current in all the bulbs that could result in burned-out filaments.

If the four bulbs were connected in series, each would have a potential difference of 60 V, or one-fourth the applied voltage. With too low a voltage, there would be insufficient current for normal operation and the bulbs would not operate at normal brilliance.

However, two bulbs in series across the 240-V line provide 120 V for each filament, which is the normal operating voltage. Therefore, the four bulbs are wired in strings of two in series, with the two strings in parallel across the 240-V source. Both strings have 240 V applied. In each string, two series bulbs divide the 240 V equally to provide the required 120 V for normal operation.

Another example is illustrated in Fig. 6-3. This circuit has just two parallel branches where one branch includes R_1 in series with R_2. The other branch has just the one resistance R_3. Ohm's law can be applied to each branch.

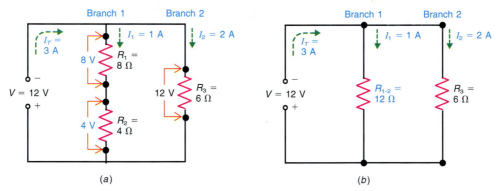

FIG. 6-3 Series string in parallel with another branch. (*a*) Schematic diagram. (*b*) Equivalent circuit.

BRANCH CURRENTS I_1 AND I_2 Each branch current equals the voltage applied across the branch divided by the total resistance in the branch. In branch 1, R_1 and R_2 total $8 + 4 = 12\ \Omega$. With 12 V applied, this branch current I_1 is $12/12 = 1$ A. Branch 2 has only the 6-Ω R_3. Then I_2 in this branch is $12/6 = 2$ A.

SERIES VOLTAGE DROPS IN A BRANCH For any one resistance in a string, the current in the string multiplied by the resistance equals the *IR* voltage drop across that particular resistance. Also, the sum of the series *IR* drops in the string equals the voltage across the entire string.

Branch 1 is a string with R_1 and R_2 in series. The I_1R_1 drop equals 8 V, while the I_1R_2 drop is 4 V. These drops of 8 and 4 V add to equal the 12 V applied. The voltage across the R_3 branch is also the same 12 V.

CALCULATING I_T The total line current equals the sum of the branch currents for all the parallel strings. Here I_T is 3 A, equal to the sum of 1 A in branch 1 and 2 A in branch 2.

CALCULATING R_T The resistance of the total series-parallel circuit across the voltage source equals the applied voltage divided by the total line current. In Fig. 6-3, R_T equals 12 V/3 A, or 4 Ω. This resistance can also be calculated as 12 Ω in parallel with 6 Ω. Using the product divided by the sum formula, 72/18 = 4 Ω for the equivalent combined R_T.

APPLYING OHM'S LAW There can be any number of parallel strings and more than two series resistances in a string. Still, Ohm's law can be used in the same way for the series and parallel parts of the circuit. The series parts have the same current. The parallel parts have the same voltage. Remember that for *V/R* the *R* must include all the resistance across the two terminals of *V*.

TEST-POINT QUESTION 6-2

Answers at end of chapter.

Refer to Fig. 6-3*a*.
a. How much is the voltage across R_3?
b. If *I* in R_2 were 6 A, what would *I* in R_1 be?
c. If the source voltage were 18 V, what would V_3 be across R_3?

6-3 RESISTANCE BANKS IN SERIES

In Fig. 6-4*a*, the group of parallel resistances R_2 and R_3 is a bank. This is in series with R_1 because the total current of the bank must go through R_1.

The circuit here has R_2 and R_3 in parallel in one bank so that these two resistances will have the same potential difference of 20 V across them. The source applies 24 V, but there is a 4-V drop across R_1.

The two series voltage drops of 4 V across R_1 and 20 V across the bank add to equal the applied voltage of 24 V. The purpose of a circuit like this is to pro-

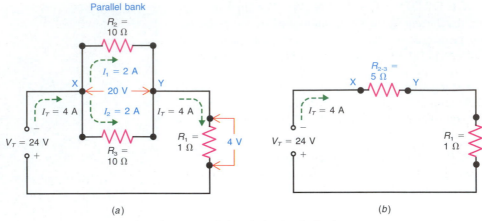

FIG. 6-4   Parallel bank of R_2 and R_3 in series with R_1. (a) Original circuit. (b) Equivalent circuit.

vide the same voltage for two or more resistances in a bank, where the bank voltage must be less than the applied voltage by the amount of IR drop across any series resistance.

To find the resistance of the entire circuit, combine the parallel resistances in each bank and add the series resistance. As shown in Fig. 6-4b, the two 10-Ω resistances R_2 and R_3 in parallel are equivalent to 5 Ω. Since the bank resistance of 5 Ω is in series with 1 Ω for R_1, the total resistance is 6 Ω across the 24-V source. Therefore, the main-line current is 24 V/6 Ω, which equals 4 A.

The total line current of 4 A divides into two parts of 2 A each in the parallel resistances R_2 and R_3. Note that each branch current equals the bank voltage divided by the branch resistance. For this bank, 20/10 = 2 A for each branch.

The branch currents are combined in the line to provide the total 4 A in R_1. This is the same total current flowing in the main line, in the source, into the bank, and out of the bank.

There can be more than two parallel resistances in a bank and any number of banks in series. Still, Ohm's law can be applied the same way to the series and parallel parts of the circuit. The general procedure for circuits of this type is to find the equivalent resistance of each bank and then add all the series resistances.

TEST-POINT QUESTION 6-3

Answers at end of chapter.

Refer to Fig. 6-4a.
a. If V_2 across R_2 were 40 V, what would V_3 across R_3 be?
b. If I in R_2 were 4 A, with 4 A in R_3, what would I in R_1 be?
c. How much is V_1 across R_1 in Fig. 6-4b?

6-4 RESISTANCE BANKS AND STRINGS IN SERIES-PARALLEL

In the solution of such circuits, the most important fact to know is which components are in series with each other and what parts of the circuit are parallel branches. The series components must be in one current path without any branch points. A branch point such as point A or B in Fig. 6-5 is common to two or more current paths. For instance, R_1 and R_6 are *not* in series with each other. They do not have the same current, because the current in R_1 divides at point A into its two component branch currents. Similarly, R_5 is not in series with R_2, because of the branch point B.

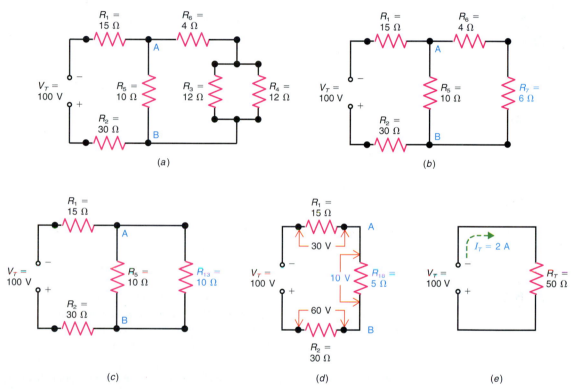

FIG. 6-5 Reducing a series-parallel circuit to an equivalent series circuit to find the R_T. (a) Actual circuit. (b) R_3 and R_4 in parallel combined for the equivalent R_7. (c) R_7 and R_6 in series added for R_{13}. (d) R_{13} and R_5 in parallel combined for R_{18}. (e) The R_{18}, R_1, and R_2 in series are added for the total resistance of 50 Ω for R_T.

To find the currents and voltages in Fig. 6-5, first find R_T in order to calculate the main-line current I_T as V_T/R_T. In calculating R_T, start reducing the branch farthest from the source and work toward the applied voltage. The reason for following this order is that you cannot tell how much resistance is in series with R_1 and R_2 until the parallel branches are reduced to their equivalent resistance. If no source voltage is shown, R_T can still be calculated from the outside in toward the open terminals where a source would be connected.

To calculate R_T in Fig. 6-5, the steps are as follows:

1. The bank of the 12-Ω R_3 and 12-Ω R_4 in parallel in Fig. 6-5*a* is equal to the 6-Ω R_7 in Fig. 6-5*b*.
2. The 6-Ω R_7 and 4-Ω R_6 in series in the same current path total 10 Ω for R_{13} in Fig. 6-5*c*.
3. The 10-Ω R_{13} is in parallel with the 10-Ω R_5, across the branch points A and B. Their equivalent resistance, then, is the 5-Ω R_{18} in Fig. 6-5*d*.
4. Now the circuit in Fig. 6-5*d* has just the 15-Ω R_1, 5-Ω R_{18}, and 30-Ω R_2 in series. These resistances total 50 Ω for R_T, as shown in Fig. 6-5*e*.
5. With a 50-Ω R_T across the 100-V source, the line current I_T is equal to $100/50 = 2$ A.

To see the individual currents and voltages, we can use the I_T of 2 A for the equivalent circuit in Fig. 6-5*d*. Now we work from the source V out toward the branches. The reason is that I_T can be used to find the voltage drops in the main line. The *IR* voltage drops here are

$$V_1 = I_T R_1 = 2 \times 15 = 30 \text{ V}$$
$$V_{18} = I_T R_{18} = 2 \times 5 = 10 \text{ V}$$
$$V_2 = I_T R_2 = 2 \times 30 = 60 \text{ V}$$

The 10-V drop across R_{18} is actually the potential difference between branch points A and B. This means 10 V across R_5 and R_{13} in Fig. 6-5*c*. The 10 V produces 1 A in the 10-Ω R_5 branch. The same 10 V is also across the R_{13} branch.

Remember that the R_{13} branch is actually the string of R_6 in series with the $R_3 R_4$ bank. Since this branch resistance is 10 Ω, with 10 V across it, the branch current here is 1 A. The 1 A through the 4 Ω of R_6 produces a voltage drop of 4 V. The remaining 6-V *IR* drop is across the $R_3 R_4$ bank. With 6 V across the 12-Ω R_3, its current is ½ A; the current is also ½ A in R_4.

Tracing all the current paths from the source, the main-line current through R_1 is 2 A. At the branch point A, this current divides into 1 A for R_5 and 1 A for the string with R_6. There is a 1-A branch current in R_6, but it subdivides in the bank, with ½ A in R_3 and ½ A in R_4. At the branch point B, the total bank current of 1 A combines with the 1 A through the R_5 branch, resulting in a 2-A total line current through R_2, the same as through R_1 in the opposite side of the line.

TEST-POINT QUESTION 6-4

Answers at end of chapter.

Refer to Fig. 6-5*a*.
a. Which R is in series with R_2?
b. Which R is in parallel with R_3?
c. Which R is in series with the $R_3 R_4$ bank?

6-5 ANALYZING SERIES-PARALLEL CIRCUITS

The circuits in Figs. 6-6 to 6-9 will be solved now. The following principles are illustrated:

1. With parallel strings across the main line, the branch currents and I_T can be found without R_T (see Figs. 6-6 and 6-7).
2. When parallel strings have series resistance in the main line, R_T must be calculated to find I_T, assuming no branch currents are known (see Fig. 6-9).
3. The source voltage is applied across the R_T of the entire circuit, producing an I_T that flows only in the main line.
4. Any individual series R has its own IR drop that must be less than the total V_T. In addition, any individual branch current must be less than I_T.

SOLUTION FOR FIG. 6-6   The problem here is to calculate the branch currents I_1 and I_{2-3}, total line current I_T, and the voltage drops V_1, V_2, and V_3. This order will be used for the calculations, because we can find the branch currents from the 90 V across the known branch resistances.

In the 30-Ω branch of R_1, the branch current is 90/30 = 3 A for I_1. The other branch resistance, with a 20-Ω R_2 and a 25-Ω R_3, totals 45 Ω. This branch current then is 90/45 = 2 A for I_{2-3}. In the main line, I_T is 3 A + 2 A, which is equal to 5 A.

For the branch voltages, V_1 must be the same as V_A, equal to 90 V. Or $V_1 = I_1R_1$, which is 3 × 30 = 90 V.

In the other branch, the 2-A I_{2-3} flows through the 20-Ω R_2 and the 25-Ω R_3. Therefore, V_2 is 2 × 20 = 40 V. Also, V_3 is 2 × 25 = 50 V. Note that these 40-V and 50-V series IR drops in one branch add to equal the 90-V source.

If we want to know R_T, it can be calculated as V_A/I_T. Then 90 V/5 A equals 18 Ω. Or R_T can be calculated by combining the branch resistances of 30 Ω in parallel with 45 Ω. Then, using the product-divided-by-sum formula, R_T is (30 × 45)/(30 + 45) or is 1350/75, which equals the same value of 18 Ω for R_T.

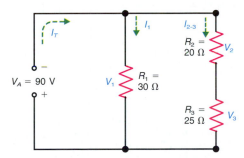

FIG. 6-6   Finding all the currents and voltages by calculating the branch currents first. See text for solution.

SOLUTION FOR FIG. 6-7   To find the applied voltage first, the I_1 branch current is given. This 3-A current through the 10-Ω R_1 produces a 30-V drop V_1 across R_1. The same 3-A current through the 20-Ω R_2 produces 60 V for V_2 across

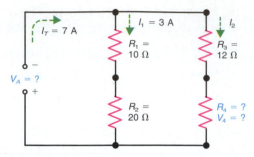

FIG. 6-7 Finding the applied voltage V_A, then V_4 and R_4 from I_2 and the branch voltages. See text for the calculations.

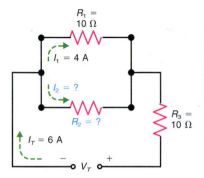

FIG. 6-8 Finding R_2 in the parallel bank and its I_2. See text for solution.

R_2. The 30-V and 60-V drops are in series with each other across the applied voltage. Therefore, V_A equals the sum of $30 + 60$, or 90 V. This 90 V is also across the other branch combining R_3 and R_4 in series.

The other branch current I_2 in Fig. 6-7 must be 4 A, equal to the 7-A I_T minus the 3-A I_1. With 4 A for I_2, the voltage drop across the 12-Ω R_3 equals 48 V for V_3. Then the voltage across R_4 is $90 - 48$, or 42 V for V_4, as the sum of V_3 and V_4 must equal the applied 90 V.

Finally, with 42 V across R_4 and 4 A through it, this resistance equals 42/4, or 10.5 Ω. Note that 10.5 Ω for R_4 added to the 12 Ω of R_3 equals 22.5 Ω, which allows 90/22.5 or a 4-A branch current for I_2.

SOLUTION FOR FIG. 6-8 The division of branch currents also applies to Fig. 6-8, but the main principle here is that the voltage must be the same across R_1 and R_2 in parallel. For the branch currents, I_2 is 2 A, equal to the 6-A I_T minus the 4-A I_1. The voltage across the 10-Ω R_1 is 4×10, or 40 V. This same voltage is also across R_2. With 40 V across R_2 and 2 A through it, R_2 equals 40/2 or 20 Ω.

We can also find V_T in Fig. 6-8 from R_1, R_2, and R_3. The 6-A I_T through R_3 produces a voltage drop of 60 V for V_3. Also, the voltage across the parallel bank with R_1 and R_2 has been calculated as 40 V. This 40 V across the bank in series with 60 V across R_3 totals 100 V for the applied voltage.

SOLUTION FOR FIG. 6-9 In order to find all the currents and voltage drops, we need R_T to calculate I_T through R_6 in the main line. Combining resis-

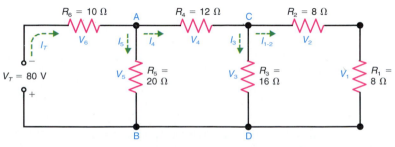

FIG. 6-9 Finding all currents and voltages by calculating R_T and then I_T to find V_6 across R_6 in the main line.

tances for R_T, we start with R_1 and R_2 and work in toward the source. Add the 8-Ω R_1 and 8-Ω R_2 in series with each other for 16 Ω. This 16 Ω combined with the 16-Ω R_3 in parallel equals 8 Ω between points C and D. Add this 8 Ω to the series 12-Ω R_4 for 20 Ω. This 20 Ω with the parallel 20-Ω R_5 equals 10 Ω between points A and B. Add this 10 Ω in series with the 10-Ω R_6, to make R_T of 20 Ω for the entire series-parallel circuit.

Current I_T in the main line is V_T/R_T, or 80/20, which equals 4 A. This 4-A I_T flows through the 10-Ω R_6, producing a 40-V IR drop for V_6.

Now that we know I_T and V_6 in the main line, we use these values to calculate all the other voltages and currents. Start from the main line, where we know the current, and work outward from the source. To find V_5, the IR drop of 40 V for V_6 in the main line is subtracted from the source voltage. The reason is that V_5 and V_6 must add to equal the 80 V of V_T. Then V_5 is $80 - 40 = 40$ V.

Voltages V_5 and V_6 happen to be equal at 40 V each. They split the 80 V in half because the 10-Ω R_6 equals the combined resistance of 10 Ω between branch points A and B.

With V_5 known to be 40 V, then I_5 through the 20-Ω R_5 is $40/20 = 2$ A. Since I_5 is 2 A and I_T is 4 A, I_4 must be 2 A also, equal to the difference between I_T and I_5. At the branch point A, the 4-A I_T divides into 2 A through R_5 and 2 A through R_4.

The 2-A I_4 through the 12-Ω R_4 produces an IR drop equal to $2 \times 12 = 24$ V for V_4. It should be noted now that V_4 and V_3 must add to equal V_5. The reason is that both V_5 and the path with V_4 and V_3 are across the same two points AB or AD. Since the potential difference across any two points is the same regardless of the path, $V_5 = V_4 + V_3$. To find V_3 now, we can subtract the 24 V of V_4 from the 40 V of V_5. Then $40 - 24 = 16$ V for V_3.

With 16 V for V_3 across the 16-Ω R_3, its current I_3 is 1 A. Also I_{1-2} in the branch with R_1 and R_2 is equal to 1 A. The 2-A I_4 into branch point C divides into the two equal branch currents of 1 A each because of the equal branch resistances.

Finally, with 1 A through the 8-Ω R_2 and 8-Ω R_1, their voltage drops are $V_2 = 8$ V and $V_1 = 8$ V. Note that the 8 V of V_1 in series with the 8 V of V_2 add to equal the 16-V potential difference V_3 between points C and D.

All the answers for the solution of Fig. 6-9 are summarized below:

$$
\begin{array}{lll}
R_T = 20\ \Omega & I_T = 4\ \text{A} & V_6 = 40\ \text{V} \\
V_5 = 40\ \text{V} & I_5 = 2\ \text{A} & I_4 = 2\ \text{A} \\
V_4 = 24\ \text{V} & V_3 = 16\ \text{V} & I_3 = 1\ \text{A} \\
I_{1-2} = 1\ \text{A} & V_2 = 8\ \text{V} & V_1 = 8\ \text{V}
\end{array}
$$

Answers at end of chapter.

a. In Fig. 6-6, which R is in series with R_2?
b. In Fig. 6-6, which R is across V_A?
c. In Fig. 6-7, how much is I_2?
d. In Fig. 6-8, how much is V_3?

6-6 WHEATSTONE BRIDGE

A bridge circuit has four terminals, two for input voltage and two for output. The purpose is to have a circuit where the voltage drops can be balanced to provide zero voltage across the output terminals, with voltage applied across the input. In Fig. 6-10a the input terminals are C and D, with output terminals A and B.

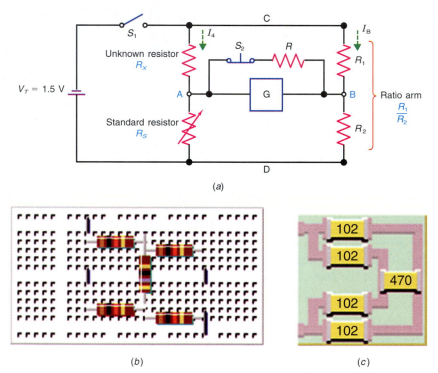

(a)

(b)   (c)

FIG. 6-10 Wheatstone-bridge circuit. (a) Schematic diagram of a Wheatstone-bridge circuit. (b) A balanced Wheatstone bridge assembled using fixed resistors on a lab prototype board. (c) A balanced Wheatstone bridge assembled using surface-mount resistors on a printed circuit board.

The bridge circuit has many uses for comparison measurements. In the Wheatstone* bridge, an unknown resistance R_X is balanced against a standard accurate resistor R_S for precise measurement of resistance.

---

*Sir Charles Wheatstone (1802–1875), English physicist and inventor.

In Fig. 6-10a, S_1 applies battery voltage to the four resistors in the bridge. To balance the bridge, the value of R_S is varied. Balance is indicated by zero current in the galvanometer G. Finally, S_2 is a spring switch that is opened to disconnect R and increase the sensitivity of the meter for the bridge.

The reason for zero current in the meter can be seen by analysis of the voltage drops across the resistors. R_S in series with R_X divides the voltage across V_T; the parallel string of R_1 in series with R_2 is also a voltage divider across the same source. When the voltage divides in the same ratio for both strings, the voltage drop across R_S equals the voltage across R_2. Also, the voltage across R_X then equals the voltage across R_1. In this case, points A and B must be at the same potential. The difference of potential across the meter then must be zero, and there is no deflection of the meter's pointer.

At balance, the equal voltage ratios in the two branches of the Wheatstone bridge can be stated as

▶ $$\frac{I_A R_X}{I_A R_S} = \frac{I_B R_1}{I_B R_2} \quad \text{or} \quad \frac{R_X}{R_S} = \frac{R_1}{R_2}$$

Note that I_A and I_B cancel. Now, multiplying both sides of the equation by R_S,

▶ $$R_X = R_S \times \frac{R_1}{R_2} \tag{6-1}$$

Usually, the total resistance of R_1 and R_2 is fixed, but any desired ratio can be chosen by moving point B on the ratio arm. The bridge is balanced by varying R_S for zero current in the meter. At balance, then, the value of R_X can be determined by multiplying R_S by the ratio of R_1/R_2. As an example, if the ratio is $\frac{1}{100}$ or 0.01 and R_S is 248 Ω, the value of R_X equals $248 \times 0.01 = 2.48$ Ω.

The balanced bridge circuit can be analyzed as simply two series resistance strings in parallel when the current is zero through the meter. Without any current between A and B, this path is effectively open. When current flows through the meter path, however, the bridge circuit must be analyzed by other methods as described in Chaps. 9 and 10.

Figure 6-10b shows a balanced Wheatstone bridge circuit assembled using fixed resistors on a lab prototype board. Figure 6-10c shows a balanced Wheatstone bridge circuit assembled using surface-mount resistors on a printed circuit board.

TEST-POINT QUESTION 6-6

Answers at end of chapter.
a. A bridge circuit has how many *pairs* of terminals?
b. In Fig. 6-10a, how much is V_{AB} at balance?

6-7 CHASSIS-GROUND CONNECTIONS

In the wiring of practical circuits, one side of the voltage source is usually grounded. For the 120-Vac power line in residential wiring, the ground is actually the earth. The ground connection is usually made by driving copper rods into the ground and connecting the ground wire of the electrical system to these rods.

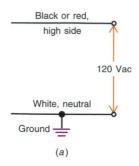

(a)

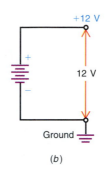

(b)

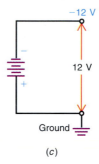

(c)

FIG. 6-11 Example of grounding one side of the voltage source. (a) The ac power line. (b) Negative side of a battery connected to a chassis ground. (c) Positive side of a battery connected to chassis ground.

In some cases, the incoming water pipe is used as the ground connection. For electronic equipment, the ground just indicates a metal chassis, which is used as a common return for connections to the source. With printed wiring on a plastic board instead of a metal chassis, a conducting path around the entire board is used as a common return for chassis ground. The chassis ground may or may not be connected to earth ground. In either case the grounded side is called the "cold side" or "low side" of the applied voltage, while the ungrounded side is the "hot side" or "high side."

GROUNDING ONE SIDE OF THE SOURCE VOLTAGE   Three examples are shown in Fig. 6-11. In Fig. 6-11a, one side of the 120-Vac power line is grounded. Note the symbol for earth ground. This symbol also indicates a chassis ground that is connected to one side of the voltage source. In electronic equipment, black wire is generally used for chassis ground returns and red wire for the high side of the voltage source. In residential wiring, green insulation denotes the grounding wire, whereas white insulation denotes the wire connected to the system's ground.

In Fig. 6-11b and c the 12-V battery is used as an example of a voltage source connected to chassis ground but not to earth. For instance, in an automobile one side of the battery is connected to the metal frame of the car. In Fig. 6-11b, the negative side is grounded, while in Fig. 6-11c the positive side is grounded. Some people have the idea that ground must always be negative, but this is not necessarily so.

The reason for connecting one side of the 120-Vac power line to earth ground is to reduce the possibility of electric shock. However, chassis ground in electronic equipment is mainly a common-return connection. Where the equipment operates from the power line, the metal chassis should be at ground potential, not connected to the hot side of the ac outlet. This connection reduces the possibility of electric shock from the chassis. Also, hum from the power line is reduced in audio, radio, and television equipment.

TEST-POINT QUESTION 6-7

Answers at end of chapter.
a. In Fig. 6-11b, give the voltage to ground with polarity.
b. Do the same for Fig. 6-11c.

6-8 VOLTAGES MEASURED TO CHASSIS GROUND

When a circuit has the chassis as a common return, we generally measure the voltages with respect to chassis. The circuit in Fig. 6-12a is called a voltage divider. Let us consider this circuit without any ground, and then analyze the effect of grounding different points on the divider. It is important to realize that this circuit operates the same way with or without the ground. The only factor that changes is the reference point for measuring the voltages.

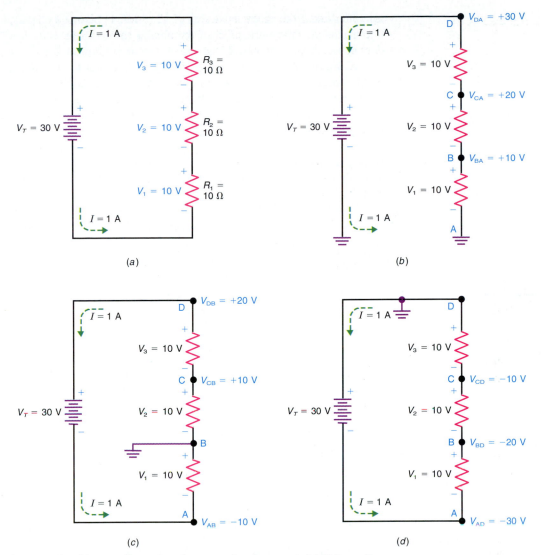

FIG. 6-12 An example of how to figure dc voltages to chassis ground. (*a*) Voltage divider without any ground connection. (*b*) Negative side of source V_T grounded to make all voltages positive with respect to ground. (*c*) Positive and negative voltages in the divider with respect to ground at point B. (*d*) With positive side of source grounded, all voltages are negative to chassis ground.

In Fig. 6-12*a*, the three 10-Ω resistances R_1, R_2, and R_3 divide the 30-V source equally. Then each voltage drop is 30/3 = 10 V for V_1, V_2, and V_3. The polarity is positive at the top and negative at the bottom, the same as V_T.

If we want to consider the current, I is 30/30 = 1 A. Each IR drop is 1 × 10 = 10 V for V_1, V_2, and V_3.

POSITIVE VOLTAGES TO NEGATIVE GROUND

In Fig. 6-12*b*, the negative side of V_T is grounded and the bottom end of R_1 is also grounded to complete the circuit. The ground is at point A. Note that the individual voltages V_1, V_2, and V_3 are still 10 V each. Also the current is still 1 A. The direction is also the same, from the negative side of V_T, through the metal chassis, to the bottom end of R_1. The only effect of the chassis ground here is to provide a conducting path from one side of the source to one side of the load.

With the ground in Fig. 6-12b, though, it is useful to consider the voltages with respect to chassis ground. In other words, the ground at point A will now be the reference for all voltages. When a voltage is indicated for only one point in a circuit, generally the other point is assumed to be chassis ground. We must have two points for a potential difference.

Let us consider the voltages at points B, C, and D. The voltage at B to ground is V_{BA}. This double subscript notation shows that we measure at B with respect to A. In general, the first letter indicates the point of measurement and the second letter is the reference point.

Then V_{BA} is +10 V. The positive sign is used here to emphasize the polarity. The value of 10 V for V_{BA} is the same as V_1 across R_1 because points B and A are across R_1. However, V_1 as the voltage across R_1 really cannot be given any polarity without a reference point.

When we consider the voltage at C, then, V_{CA} is +20 V. This voltage equals $V_1 + V_2$. Also, for point D at the top, V_{DA} is +30 V for $V_1 + V_2 + V_3$.

POSITIVE AND NEGATIVE VOLTAGES TO A GROUNDED TAP

In Fig. 6-12c point B in the divider is grounded. The purpose is to have the divider supply negative and positive voltages with respect to chassis ground. The negative voltage here is V_{AB}, which equals −10 V. This value is the same 10 V as V_1, but V_{AB} is the voltage at the negative end A with respect to the positive end B. The other voltages in the divider are $V_{CB} = +10$ V and $V_{DB} = +20$ V.

We can consider the ground at B as a dividing point for positive and negative voltages. For all points toward the positive side of V_T, any voltage is positive to ground. Going the other way, at all points toward the negative side of V_T, any voltage is negative to ground.

NEGATIVE VOLTAGES TO POSITIVE GROUND

In Fig. 6-12d, point D at the top of the divider is grounded, which is the same as grounding the positive side of the source V_T. The voltage source here is *inverted,* compared with Fig. 6-12b, as the opposite side is grounded. In Fig. 6-12d, all the voltages on the divider are negative to ground. Here, $V_{CD} = −10$ V, while $V_{BD} = −20$ V and $V_{AD} = −30$ V. Any point in the circuit must be more negative than the positive terminal of the source, even when this terminal is grounded.

TEST-POINT QUESTION 6-8

Answers at end of chapter.

Refer to Fig. 6-12c and give the voltage and polarity for
a. A to ground.
b. B to ground.
c. D to ground.
d. V_{DA} across V_T.

6-9 OPENS AND SHORTS IN SERIES-PARALLEL CIRCUITS

A short circuit has practically zero resistance. Its effect, therefore, is to allow excessive current. An open circuit has the opposite effect because an open circuit has infinitely high resistance with practically zero current. Furthermore, in series-parallel circuits an open or short circuit in one path changes the circuit for the other resistances. For example, in Fig. 6-13, the series-parallel circuit in Fig. 6-13a becomes a series circuit with only R_1 when there is a short circuit between terminals A and B. As an example of an open circuit, the series-parallel circuit in Fig. 6-14a becomes a series circuit with just R_1 and R_2 when there is an open circuit between terminals C and D.

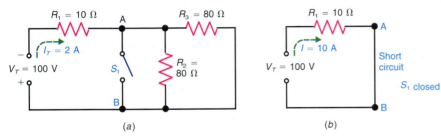

FIG. 6-13 Effect of a short circuit with series-parallel connections. (a) Normal circuit with S_1 open. (b) Circuit with short between points A and B when S_1 is closed; now R_2 and R_3 are short-circuited.

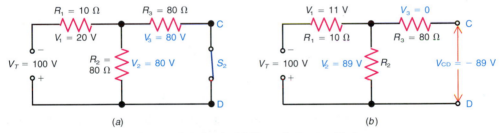

FIG. 6-14 Effect of an open path in a series-parallel circuit. (a) Normal circuit with S_2 closed. (b) Series circuit with R_1 and R_2 when S_2 is open. Now R_3 in the open path has no current and zero IR voltage drop.

EFFECT OF A SHORT CIRCUIT We can solve the series-parallel circuit in Fig. 6-13a in order to see the effect of the short circuit. For the normal circuit, with S_1 open, R_2 and R_3 are in parallel. Although R_3 is drawn horizontally, both ends are across R_2. The switch S_1 has no effect as a parallel branch here because it is open.

The combined resistance of the 80-Ω R_2 in parallel with the 80-Ω R_3 is equivalent to 40 Ω. This 40 Ω for the bank resistance is in series with the 10-Ω R_1. Then R_T is 40 + 10 = 50 Ω.

In the main line, I_T is 100/50 = 2 A. Then V_1 across the 10-Ω R_1 in the main line is 2 × 10 = 20 V. The remaining 80 V is across R_2 and R_3 as a parallel bank. As a result, $V_2 = 80$ V and $V_3 = 80$ V.

Now consider the effect of closing switch S_1. A closed switch has zero resistance. Not only is R_2 short-circuited, but R_3 in the bank with R_2 is also short-circuited. The closed switch short-circuits everything connected between terminals A and B. The result is the series circuit shown in Fig. 6-13b.

Now the 10-Ω R_1 is the only opposition to current. I equals V/R_1, which is $100/10 = 10$ A. This 10 A flows through R_1, the closed switch, and the source. With 10 A through R_1, instead of its normal 2 A, the excessive current can cause excessive heat in R_1. There is no current through R_2 and R_3, as they are short-circuited out of the path for current.

EFFECT OF AN OPEN CIRCUIT Figure 6-14a shows the same series-parallel circuit as Fig. 6-13a, except that switch S_2 is used now to connect R_3 in parallel with R_2. With S_2 closed for normal operation, all currents and voltages have the values calculated for the series-parallel circuit. However, let us consider the effect of opening S_2, as shown in Fig. 6-14b. An open switch has infinitely high resistance. Now there is an open circuit between terminals C and D. Furthermore, because R_3 is in the open path, its 80 Ω cannot be considered in parallel with R_2.

The circuit with S_2 open in Fig. 6-14b is really the same as having just R_1 and R_2 in series with the 100-V source. The open path with R_3 has no effect as a parallel branch. The reason is that no current flows through R_3.

We can consider R_1 and R_2 in series as a voltage divider, where each IR drop is proportional to its resistance. The total series R is $80 + 10 = 90$ Ω. The 10-Ω R_1 is 10/90 or $\frac{1}{9}$ of the total R and the applied V_T. Then V_1 is $\frac{1}{9} \times 100$ V = 11 V and V_2 is $\frac{8}{9} \times 100$ V = 89 V, approximately. The 11-V drop for V_1 and 89-V drop for V_2 add to equal the 100 V of the applied voltage.

Note that V_3 is zero. Without any current through R_3, it cannot have any voltage drop.

Furthermore, the voltage across the open terminals C and D is the same 89 V as the potential difference V_2 across R_2. Since there is no voltage drop across R_3, terminal C has the same potential as the top terminal of R_2. Terminal D is directly connected to the bottom end of resistor R_2. Therefore, the potential difference from terminal C to terminal D is the same 89 V that appears across resistor R_2.

TEST-POINT QUESTION 6-9

Answers at end of chapter.
  a. In Fig. 6-13, the short circuit increases I_T from 2 A to what value?
  b. In Fig. 6-14, the open branch reduces I_T from 2 A to what value?

6-10 SERIES-PARALLEL SWITCHES COMBINE THE AND AND OR LOGIC FUNCTIONS

Figure 6-15a shows how series-parallel switches can be connected to combine the AND and OR logic functions. If both switches A and B are closed or if switch C is closed, a complete path for current flow is provided and the lamp will be

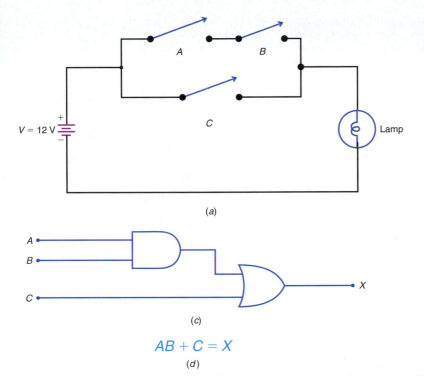

	Switches			
A	B	C		Lamp
Open	Open	Open		OFF
Open	Open	Closed		ON
Open	Closed	Open		OFF
Open	Closed	Closed		ON
Closed	Open	Open		OFF
Closed	Open	Closed		ON
Closed	Closed	Open		ON
Closed	Closed	Closed		ON

(a)

(b)

(c)

$$AB + C = X$$

(d)

FIG. 6-15 Series-parallel switches combine the AND and OR logic functions. (*a*) Circuit representing the logic function $AB + C = X$. (*b*) Truth table showing the eight possible combinations for switches *A*, *B*, and *C*. (*c*) Equivalent logic circuit using the AND and OR logic symbols. (*d*) Mathematical expression describing the AND and OR logic functions.

ON. However, if either switch *A* or switch *B* is open at the same time switch *C* is open, the lamp will be OFF. Figure 6-15*b* shows the truth table listing all of the possible combinations for switches *A*, *B*, and *C*. Notice in the last row of the truth table that the lamp is on when all three switches are closed.

Figure 6-15*c* shows the logic circuit corresponding to the series-parallel connection of switches in Fig. 6-15*a*. The AND logic symbol is used to represent the series switches *A* and *B*, whereas the OR logic symbol is used to represent the parallel connection of switch *C*. Figure 6-15*d* shows the mathematical expression describing both Figs. 6-15*a* and 6-15*c*. The statement is read: "*A* and *B* or *C* equals *X*."

TEST-POINT QUESTION 6-10

Answers at end of chapter.

Answer True or False.
a. In Fig. 6-15*a*, the lamp is ON when switches *A* and *B* are both open at the same time as switch *C* is closed.
b. In Fig. 6-15*a*, the lamp is OFF if either switch *A* or switch *B* is open at the same time switch *C* is open.

6   SUMMARY AND REVIEW

- Table 6-1 summarizes the main characteristics of series and parallel circuits. In circuits combining series and parallel connections, the components in one current path without any branch points are in series; the parts of the circuit connected across the same two branch points are in parallel.
- To calculate R_T in a series-parallel circuit with R in the main line, combine resistances from the outside back toward the source.
- Chassis ground is commonly used as a return connection to one side of the source voltage. Voltages measured to chassis ground can have either negative or positive polarity.
- When the potential is the same at the two ends of a resistance, its voltage is zero. If no current flows through a resistance, it cannot have any IR voltage drop.
- A Wheatstone bridge circuit has two input terminals and two output terminals. When balanced, the Wheatstone bridge can be analyzed as simply two series strings in parallel. The Wheatstone bridge finds many uses in applications where comparison measurements are needed.

TABLE 6-1   COMPARISON OF SERIES AND PARALLEL CIRCUITS

SERIES CIRCUIT	PARALLEL CIRCUIT
Current the same in all components	Voltage the same across all branches
V across each series R is $I \times R$	I in each branch R is V/R
$V_T = V_1 + V_2 + V_3 + \cdots + $ etc.	$I_T = I_1 + I_2 + I_3 + \cdots + $ etc.
$R_T = R_1 + R_2 + R_3 + \cdots + $ etc.	$G_T = G_1 + G_2 + G_3 + \cdots + $ etc.
R_T must be more than the largest individual R	R_{EQ} must be less than the smallest branch R
$P_T = P_1 + P_2 + P_3 + \cdots + $ etc.	$P_T = P_1 + P_2 + P_3 + \cdots + $ etc.
Applied voltage is divided into IR voltage drops	Main-line current is divided into branch currents
The largest IR drop is across the largest series R	The largest branch I is in the smallest parallel R
Open in one component causes entire circuit to be open	Open in one branch does not prevent I in other branches

ANSWERS AT BACK OF BOOK.

Choose (*a*), (*b*), (*c*), or (*d*).

1. In the series-parallel circuit in Fig. 6-1*b*: (*a*) R_1 is in series with R_3; (*b*) R_2 is in series with R_3; (*c*) R_4 is in parallel with R_3; (*d*) R_1 is in parallel with R_3.
2. In the series-parallel circuit in Fig. 6-2*b*: (*a*) R_1 is in parallel with R_3; (*b*) R_2 is in parallel with R_4; (*c*) R_1 is in series with R_2; (*d*) R_2 is in series with R_4.
3. In the series-parallel circuit in Fig. 6-5, the total of all the branch currents into branch point A and out of branch point B equals (*a*) ½ A; (*b*) 1 A; (*c*) 2 A; (*d*) 4 A.
4. In the circuit in Fig. 6-2 with four 120-V 100-W light bulbs, the resistance of one bulb equals (*a*) 72 Ω; (*b*) 100 Ω; (*c*) 144 Ω; (*d*) 120 Ω.
5. In the series-parallel circuit in Fig. 6-4*a*: (*a*) R_2 is in series with R_3; (*b*) R_1 is in series with R_3; (*c*) the equivalent resistance of the R_2R_3 bank is in parallel with R_1; (*d*) the equivalent resistance of the R_2R_3 bank is in series with R_1.
6. In a series circuit with unequal resistances: (*a*) the lowest R has the highest V; (*b*) the highest R has the highest V; (*c*) the lowest R has the most I; (*d*) the highest R has the most I.
7. In a parallel bank with unequal branch resistances: (*a*) the current is highest in the highest R; (*b*) the current is equal in all the branches; (*c*) the voltage is highest across the lowest R; (*d*) the current is highest in the lowest R.
8. In Fig. 6-14, with S_2 open, R_T equals (*a*) 90 Ω; (*b*) 100 Ω; (*c*) 50 Ω; (*d*) 10 Ω.
9. In Fig. 6-12*c*, V_{DA} equals (*a*) +10 V; (*b*) −20 V; (*c*) −30 V; (*d*) +30 V.
10. In the Wheatstone bridge of Fig. 6-10, at balance: (*a*) $I_A = 0$; (*b*) $I_B = 0$; (*c*) $V_2 = 0$; (*d*) $V_{AB} = 0$.

QUESTIONS

1. In a series-parallel circuit, how can you tell which resistances are in series with each other and which are in parallel?
2. Draw a schematic diagram showing two resistances in a bank that is in series with one resistance.
3. Draw a diagram showing how to connect three resistances of equal value so that the combined resistance will be 1½ times the resistance of one unit.
4. Draw a diagram showing two strings in parallel across a voltage source, where each string has three series resistances.
5. Explain why components are connected in series-parallel, showing a circuit as an example of your explanation.
6. Give two differences between a short circuit and an open circuit.
7. Explain the difference between voltage division and current division.
8. Where would a voltage be negative with respect to chassis ground.

9. Draw a circuit with nine 40-V 100-W bulbs connected to a 120-V source.
10. (a) Two 10-Ω resistors are in series with a 100-V source. If a third 10-Ω R is added in series, explain why I will decrease. (b) The same two 10-Ω resistors are in parallel with the 100-V source. If a third 10-Ω R is added in parallel, explain why I_T will increase.

PROBLEMS

ANSWERS TO ODD-NUMBERED PROBLEMS AT BACK OF BOOK.

1. Refer to Fig. 6-1. (a) Calculate the circuit's total resistance if all resistances are 10 Ω. (b) What is the main-line current if V_T equals 100 V?
2. In Fig. 6-2, calculate the total power supplied by the source for the four 100-W bulbs.
3. Refer to the diagram in Fig. 6-16. (a) Why is R_1 in series with R_3 but not with R_2? (b) Find the total circuit resistance across the battery.
4. Two 60-Ω resistances R_1 and R_2 in parallel require 60 V across the bank with 1 A through each branch. Show how to connect a series resistance R_3 in the main line to drop an applied voltage of 100 V to 60 V across the bank. (a) How much is the required voltage across R_3? (b) How much is the required current through R_3? (c) How much is the required resistance of R_3? (d) If R_3 opens, how much is the voltage across R_1 and R_2? (e) If R_1 opens, what are the voltages across R_2 and R_3?
5. Refer to the diagram in Fig. 6-17. (a) Calculate R across points AD. (b) How much is R across points AD with R_4 open?
6. Show how to connect four 100-Ω resistances in a series-parallel circuit with a combined resistance equal to 100 Ω. (a) If the combination is connected across a 100-V source, how much power is supplied by the source? (b) How much power is dissipated in each resistance?
7. In the Wheatstone-bridge circuit of Fig. 6-18, find each voltage, label the polarity, and calculate R_X. The bridge is balanced.
8. In Fig. 6-19, find each voltage V and current I for the four resistors with 30 V applied.
9. Calculate R_T for the series-parallel circuit in Fig. 6-20.
10. For the series-parallel circuit in Fig. 6-21, with 10 V applied, find voltage V_6 for the end resistor R_6.
11. Refer to Fig. 6-22. (a) Calculate V_2. (b) Find V_2 when R_3 is open.

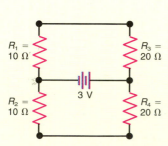

FIG. 6-16 Circuit diagram for Prob. 3.

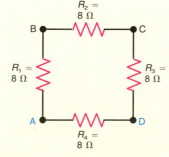

FIG. 6-17 Diagram for Prob. 5.

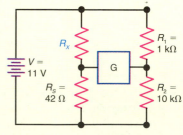

FIG. 6-18 Circuit diagram for Prob. 7.

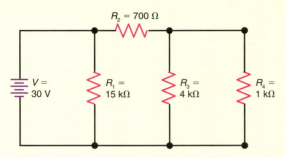

FIG. 6-19 Circuit diagram for Prob. 8.

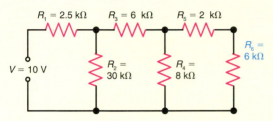

FIG. 6-21 Circuit diagram for Prob. 10.

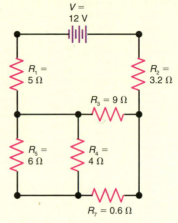

FIG. 6-20 Circuit diagram for Prob. 9.

12. Refer to Fig. 6-23 with 12 V applied. (**a**) Find V_1, V_2, V_3, I_1, I_2, I_3, and I_T in the circuit with three resistors as shown. (**b**) Now connect point G to ground. Give the voltages, with polarity, at points A, B, and C with respect to ground. (**c**) Give the values of I_1, I_2, I_3, and I_T with point G grounded.

13. In Fig. 6-24, find I and V for the five resistors and calculate V_T. All the R values are given and the current I is 8 mA through R_1.

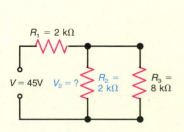

FIG. 6-22 Circuit diagram for Prob. 11.

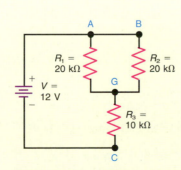

FIG. 6-23 Circuit diagram for Prob. 12.

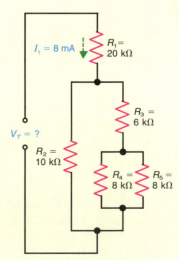

FIG. 6-24 Circuit diagram for Prob. 13.

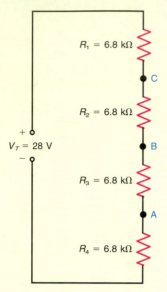

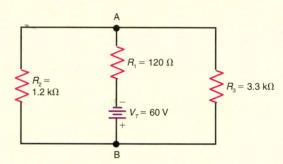

FIG. 6-25   Circuit diagram for Prob. 14.

FIG. 6-26   Circuit diagram for Prob. 15.

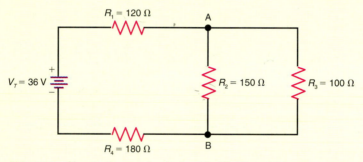

FIG. 6-27   Circuit diagram for Probs. 16, 17, 18, and 19.

14. For the voltage divider in Fig. 6-25 with 28 V applied, give the voltages at points A, B, and C with polarity to ground when (a) point A is grounded; (b) point B is grounded; (c) point C is grounded.

15. In Fig. 6-26, calculate R_T, I_1, I_2, I_3, V_1, V_2, V_3, and V_{AB}.

16. In Fig. 6-27, calculate R_T, I_T, V_{R_1}, V_{AB}, V_{R_4}, I_2, and I_3.

17. Refer to Fig. 6-27. If R_3 opens, does V_{AB} increase, decrease, or stay the same? What about the currents, I_T, I_2, and I_3?

18. Refer to Fig. 6-27. If R_2 decreases to 25 Ω, does V_{AB} increase, decrease, or stay the same? What about the currents, I_T, I_2, and I_3?

19. Refer to Fig. 6-27. If R_3 shorts, does V_{AB} increase, decrease, or stay the same? What about the currents, I_T, I_2, and I_3?

20. In Fig. 6-28, solve for the following unknowns: R_T, I_1, I_2, I_3, I_4, I_5, I_6, I_7, V_1, V_2, V_3, V_4, V_5, V_6, and V_7.

21. In Fig. 6-29, calculate: R_T, I_1, I_2, I_3, I_4, I_5, I_6, I_7, I_8, I_9, V_1, V_2, V_3, V_4, V_5, V_6, V_7, V_8, and V_9.

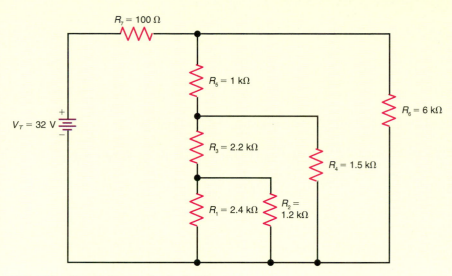

FIG. 6-28   Circuit diagram for Prob. 20.

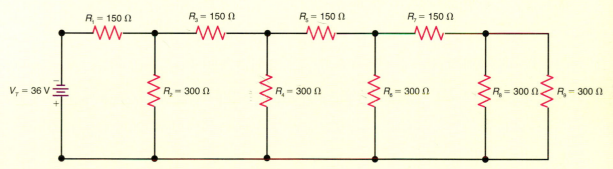

FIG. 6-29   Circuit diagram for Prob. 21.

CRITICAL THINKING

1. In Fig. 6-30, bulbs *A* and *B* each have an operating voltage of 28 V. If the wattage ratings for bulbs *A* and *B* are 1.12 W and 2.8 W respectively, calculate: (**a**) the required resistance of R_1; (**b**) the recommended wattage rating of R_1; (**c**) the total resistance R_T.

2. Refer to Fig. 6-31. How much voltage will be indicated by the voltmeter when the wiper arm of the linear potentiometer R_2 is set: (**a**) to point A; (**b**) to point B; (**c**) midway between points A and B.

3. Three switches, *A*, *B*, and *C*, are used to control whether a lamp is ON or OFF. The lamp is to be on only when switch *A* is closed at the same time as switch *B* or switch *C* (or both) is closed. (**a**) Draw the schematic diagram for the voltage source, three switches, and the lamp. (**b**) Write the mathematical expression which describes the operation of the series-parallel switches in controlling the lamp. (**c**) Draw the logic circuit, using the AND and OR logic symbols, which represents the operation of the series-parallel switches.

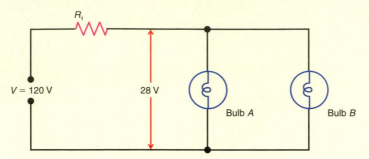

FIG. 6-30 Circuit diagram for Critical Thinking Prob. 1.

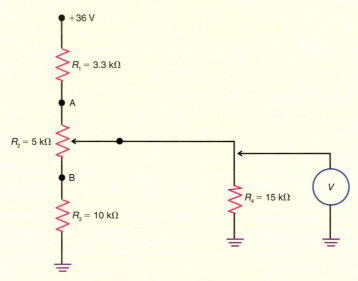

FIG. 6-31 Circuit diagram for Critical Thinking Prob. 2.

ANSWERS TO TEST-POINT QUESTIONS

6-1 **a.** $R = 1\ k\Omega$
    b. $R = 0.5\ k\Omega$
    c. $R_T = 1.5\ k\Omega$

6-2 **a.** 12 V
    b. $I = 6\ A$
    c. $V_3 = 18\ V$

6-3 **a.** $V_3 = 40\ V$
    b. $I = 8\ A$
    c. $V_1 = 4\ V$

6-4 **a.** R_1
    b. R_4
    c. R_6

6-5 **a.** R_3
    b. R_1
    c. $I_2 = 4\ A$
    d. $V_3 = 60\ V$

6-6 **a.** Two
    b. 0 V

6-7 **a.** $+12\ V$
    b. $-12\ V$

6-8 **a.** $-10\ V$
    b. 0 V
    c. $+20\ V$
    d. $+30\ V$

6-9 **a.** $I = 10\ A$
    b. $I = 1.1\ A$

6-10 **a.** T
    b. T

REVIEW: CHAPTERS 1 TO 6

SUMMARY

- The electron is the basic quantity of negative electricity; the proton is the basic quantity of positive electricity. Both have the same charge, but they have opposite polarities.
- A quantity of electrons is a negative charge; a deficiency of electrons is a positive charge. Like charges repel each other; unlike charges attract.
- Charge, Q, is measured in coulombs; 6.25×10^{18} electrons equals one coulomb of charge. Charge in motion is current. One coulomb of charge flowing past a given point each second equals one ampere of current.
- Potential difference, PD, is measured in volts. One volt produces one ampere of current against the opposition of one ohm of resistance.
- The main types of resistors are carbon-composition, carbon-film, metal-film, wire-wound, and surface-mount. Wire-wound resistors are used when the resistance must dissipate a lot of power, such as 5 W or more. Carbon-film and metal-film resistors are better than the older carbon-composition type because they have tighter tolerances, are less sensitive to temperature changes and aging, and generate less noise internally.
- Resistors having a power rating less than 2 W are often color-coded to indicate their resistance value. To review the color code, refer to Table 2-1 and Fig. 2-7.
- Surface-mount resistors (also called chip resistors) use a three-digit number printed on the body to indicate the resistance value in ohms. The first two digits indicate the first two digits in the numerical value of the resistance; the third digit is the multiplier. For example, a surface-mount resistor which is marked 103 has a resistance value of 10,000 Ω or 10 kΩ.
- A potentiometer is a variable resistor that has three terminals. It is used as a variable voltage divider. A rheostat is a variable resistor that has only two terminals. It is used to vary the current in a circuit.
- The most common trouble with resistors is that they develop an open and thus have an infinitely high resistance.
- The three forms of Ohm's law are $I = V/R$, $V = IR$, and $R = V/I$.
- The three power formulas are $P = VI$, $P = I^2R$, and $P = V^2/R$.
- The most common multiple and submultiples of the practical units are mega (M) for 10^6, micro (μ) for 10^{-6}, kilo (k) for 10^3, and milli (m) for 10^{-3}.
- For series resistances: (**1**) the current is the same in all resistances; (**2**) the IR drops can be different with unequal resistances; (**3**) the applied voltage equals the sum of the series IR drops; (**4**) the total resistance equals the sum of the individual resistances; (**5**) an open circuit in one resistance results in no current through the entire series circuit.

- For parallel resistances: (**1**) the voltage is the same across all resistances; (**2**) the branch currents can be different with unequal resistances; (**3**) the total line current equals the sum of the parallel branch currents; (**4**) the combined equivalent resistance, R_{EQ}, of parallel branches is less than the smallest resistance as determined by Formula (5-3); (**5**) an open circuit in one branch does not create an open in the other branches; (**6**) a short circuit across one branch short-circuits all the branches.
- In series-parallel circuits, the resistances in one current path without any branch points are in series; all the rules of series resistances apply. The resistances across the same two branch points are in parallel; all the rules of parallel resistances apply.
- Series-connected switches represent the AND logic function, whereas parallel-connected switches represent the OR logic function. Switches connected in a series-parallel manner combine the AND and OR logic functions.

REVIEW SELF-TEST

ANSWERS AT BACK OF BOOK.

Choose (*a*), (*b*), (*c*), or (*d*).

1. A carbon resistor is color-coded with brown, green, red, and gold stripes from left to right. Its value is (*a*) 1500 Ω ± 5 percent; (*b*) 6800 Ω ± 5 percent; (*c*) 10,000 Ω ± 10 percent; (*d*) 500,000 Ω ± 5 percent.
2. A metal-film resistor is color-coded with orange, orange, orange, red, and green stripes, reading from left to right. Its value is: (*a*) 3.3 kΩ ± 5 percent; (*b*) 333 kΩ ± 5 percent; (*c*) 33.3 kΩ ± 0.5 percent; (*d*) 333 Ω ± 0.5 percent.
3. With 30 V applied across two equal resistors in series, 10 mA of current flows. Typical values for each resistor to be used here are (*a*) 10 Ω 10 W; (*b*) 1500 Ω ½ W; (*c*) 3000 Ω 10 W; (*d*) 30 MΩ 2 W.
4. In which of the following circuits will the voltage source produce the most current? (*a*) 10 V across a 10-Ω resistance; (*b*) 10 V across two 10-Ω resistances in series; (*c*) 10 V across two 10-Ω resistances in parallel; (*d*) 1000 V across a 1-MΩ resistance.
5. Three 120-V 100-W bulbs are in parallel across the 120-V power line. If one bulb burns open: (*a*) the other two bulbs cannot light; (*b*) all three bulbs light; (*c*) the other two bulbs can light; (*d*) there is excessive current in the main line.
6. A circuit allows 1 mA of current to flow with 1 V applied. The conductance of the circuit equals (*a*) 0.002 Ω; (*b*) 0.005 μS; (*c*) 1000 μS; (*d*) 1 S.
7. If 2 A of current is allowed to accumulate charge for 5 s, the resultant charge equals (*a*) 2 C; (*b*) 10 C; (*c*) 5 A; (*d*) 10 A.
8. A potential difference applied across a 1-MΩ resistor produces 1 mA of current. The applied voltage equals (*a*) 1 μV; (*b*) 1 mV; (*c*) 1 kV; (*d*) 1,000,000 V.
9. A string of two 1000-Ω resistances is in series with a parallel bank of two 1000-Ω resistances. The total resistance of the series-parallel circuit equals (*a*) 250 Ω; (*b*) 2500 Ω; (*c*) 3000 Ω; (*d*) 4000 Ω.

10. In the circuit of Question 9, one of the resistances in the series string opens. Then the current in the parallel bank (*a*) increases slightly in both branches; (*b*) equals zero in one branch but is maximum in the other branch; (*c*) is maximum in both branches; (*d*) equals zero in both branches.

11. With 100 V applied across a 10,000-Ω resistance, the power dissipation equals (*a*) 1 mW; (*b*) 1 W; (*c*) 100 W; (*d*) 1 kW.

12. A source of 10 V is applied across R_1, R_2, and R_3 in series, producing 1 A in the series circuit. R_1 equals 6 Ω and R_2 equals 2 Ω. Therefore, R_3 equals (*a*) 2 Ω; (*b*) 4 Ω; (*c*) 10 Ω; (*d*) 12 Ω.

13. A 5-V source and a 3-V source are connected with series-opposing polarities. The combined voltage across both sources equals (*a*) 5 V; (*b*) 3 V; (*c*) 2 V; (*d*) 8 V.

14. In a circuit with three parallel branches, if one branch opens, the main-line current will be (*a*) more; (*b*) less; (*c*) the same; (*d*) infinite.

15. A 10-Ω R_1 and a 20-Ω R_2 are in series with a 30-V source. If R_1 opens, the voltage drop across R_2 will be (*a*) zero; (*b*) 20 V; (*c*) 30 V; (*d*) infinite.

16. A voltage V_1 of 40 V is connected series-opposing with V_2 of 50 V. The total voltage across both components is: (*a*) 10 V; (*b*) 40 V; (*c*) 50 V; (*d*) 90 V.

17. Two series voltage drops V_1 and V_2 total 100 V for V_T. When V_1 is 60 V, then V_2 must equal: (*a*) 40 V; (*b*) 60 V; (*c*) 100 V; (*d*) 160 V.

18. Two parallel branch currents I_1 and I_2 total 100 mA for I_T. When I_1 is 60 mA, then I_2 must equal: (*a*) 40 mA; (*b*) 60 mA; (*c*) 100 mA; (*d*) 160 mA.

19. A surface-mount resistor is marked 224. Its resistance is: (*a*) 224 Ω; (*b*) 220 kΩ; (*c*) 224 kΩ; (*d*) 22 kΩ.

20. If a variable voltage is connected across a fixed resistance: (*a*) *I* and *V* will vary in direct proportion; (*b*) *I* and *V* will be inversely proportional; (*c*) *I* will remain constant as *V* is varied; (*d*) none of the above.

21. If a fixed value of voltage is connected across a variable resistance: (*a*) *I* will vary in direct proportion to *R*; (*b*) *I* will be inversely proportional to *R*; (*c*) *I* will remain constant as *R* is varied; (*d*) none of the above.

REFERENCES

Cooke, N. M., H. F. R. Adams, and P. B. Dell: *Basic Mathematics for Electronics,* Glencoe/McGraw-Hill, Columbus, Ohio.

Grob, B.: *Mathematics for Basic Electronics,* Glencoe/McGraw-Hill, Columbus, Ohio.

Periodic Chart of the Atoms

Understanding Calculator Math, Texas Instruments Incorporated, Dallas.

Calculator instruction booklets.

CHAPTER 7

VOLTAGE DIVIDERS AND CURRENT DIVIDERS

Any series circuit is a voltage divider. The *IR* voltage drops are proportional parts of the applied voltage. Also, any parallel circuit is a current divider. Each branch current is part of the total line current, but in inverse proportion to the branch resistance.

Special formulas can be used for the voltage and current division as short cuts in the calculations. The voltage division formula gives the series voltages even when the current is not known. Also, the current division formula gives the branch currents even when the branch voltage is not known. Finally, we consider a series voltage divider with parallel branches that have load currents. The design of such a loaded voltage divider can be applied to the important case of different voltage taps from the power supply in electronic equipment.

CHAPTER OBJECTIVES

Upon completion of this chapter, you should be able to:

- *Calculate* the voltage drops in an unloaded voltage divider.
- *Explain* why resistor voltage drops are proportional to the resistor values in a series circuit.
- *Calculate* the branch currents in a parallel circuit.
- *Explain* why the branch currents are inversely proportional to the branch resistances in a parallel circuit.
- *Define* what is meant by the term *loaded voltage divider*.
- *Calculate* the voltage, current, and power values in a loaded voltage divider.

IMPORTANT TERMS IN THIS CHAPTER

bleeder current
current divider

load currents
loaded voltage

voltage divider

TOPICS COVERED IN THIS CHAPTER

7-1 SERIES VOLTAGE DIVIDERS

The current is the same in all the resistances in a series circuit. Also, the voltage drops equal the product of I times R. Therefore, the IR voltages are proportional to the series resistances. A higher resistance has a greater IR voltage than a lower resistance in the same series circuit; equal resistances have the same amount of IR drop. If R_1 is double R_2, then V_1 will be double V_2.

The series string can be considered as a *voltage divider*. Each resistance provides an IR drop V equal to its proportional part of the applied voltage. Stated as a formula,

$$\blacktriangleright \qquad V = \frac{R}{R_T} \times V_T \qquad\qquad (7\text{-}1)$$

Example

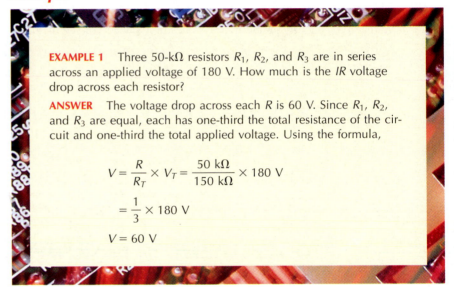

EXAMPLE 1 Three 50-kΩ resistors R_1, R_2, and R_3 are in series across an applied voltage of 180 V. How much is the IR voltage drop across each resistor?

ANSWER The voltage drop across each R is 60 V. Since R_1, R_2, and R_3 are equal, each has one-third the total resistance of the circuit and one-third the total applied voltage. Using the formula,

$$V = \frac{R}{R_T} \times V_T = \frac{50 \text{ k}\Omega}{150 \text{ k}\Omega} \times 180 \text{ V}$$

$$= \frac{1}{3} \times 180 \text{ V}$$

$$V = 60 \text{ V}$$

Note that R and R_T must be in the same units for the proportion. Then V is in the same units as V_T.

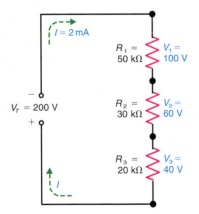

FIG. 7-1 Series string of resistors as a proportional voltage divider. Each V_R is R/R_T fraction of the total source voltage V_T.

TYPICAL CIRCUIT Figure 7-1 illustrates another example of a proportional voltage divider. Let the problem be to find the voltage across R_3. We can either calculate this voltage V_3 as IR_3 or determine its proportional part of the total applied voltage V_T. The answer is the same both ways. Note that R_T is $20 + 30 + 50 = 100$ kΩ.

PROPORTIONAL VOLTAGE METHOD Using Formula (7-1), V_3 equals 20/100 of the 200 V applied for V_T because R_3 is 20 kΩ and R_T is 100 kΩ. Then V_3 is 20/100 of 200 or $\frac{1}{5}$ of 200, which is equal to 40 V. The calculations are

$$V_3 = \frac{R_3}{R_T} \times V_T = \frac{20}{100} \times 200 \text{ V}$$
$$V_3 = 40 \text{ V}$$

In the same way, V_2 is 60 V. The calculations are

$$V_2 = \frac{R_2}{R_T} \times V_T = \frac{30}{100} \times 200 \text{ V}$$
$$V_2 = 60 \text{ V}$$

Also, V_1 is 100 V. The calculations are

$$V_1 = \frac{R_1}{R_T} \times V_T = \frac{50}{100} \times 200 \text{ V}$$
$$V_1 = 100 \text{ V}$$

The sum of V_1, V_2, and V_3 in series is $100 + 60 + 40 = 200$ V, which is equal to V_T.

To do a problem like this on the calculator, you can divide R_3 by R_T first and then multiply by V_T. For the values here, to find V_3, the procedure can be as follows:

Punch in the number 20 for R_3.
Push the $\div$ key, then 100 for R_T and press the $\times$ key for the quotient of 0.2 on the display.
Push the $\times$ key then 200 for V_T and press the $=$ key to display the quotient of 40 as the answer for V_3.

As another method, you can multiply R_3 by V_T first and then divide by R_T, the answers will be the same for either method.

METHOD OF _IR_ DROPS   If we want to solve for the current in Fig. 7-1, the I is V_T/R_T or 200 V/100 kΩ = 2 mA. This I flows through R_1, R_2, and R_3 in series. The IR drops are

$$V_1 = I \times R_1 = 2 \text{ mA} \times 50 \text{ kΩ} = 100 \text{ V}$$
$$V_2 = I \times R_2 = 2 \text{ mA} \times 30 \text{ kΩ} = 60 \text{ V}$$
$$V_3 = I \times R_3 = 2 \text{ mA} \times 20 \text{ kΩ} = 40 \text{ V}$$

These voltages are the same values calculated by Formula (7-1) for proportional voltage dividers.

TWO VOLTAGE DROPS IN SERIES   For this case, it is not necessary to calculate both voltages. After you find one, subtract it from V_T to find the other.

As an example, assume V_T is 48 V across two series resistors R_1 and R_2. If V_1 is 18 V, then V_2 must be $48 - 18 = 30$ V.

THE LARGEST SERIES _R_ HAS THE MOST _V_   The fact that series voltage drops are proportional to the resistances means that a very small R in series

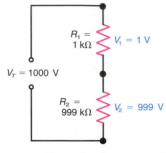

FIG. 7-2 Example of a very small R_1 in series with a large R_2; V_2 is almost equal to the whole V_T.

with a much larger R has a negligible IR drop. An example is shown in Fig. 7-2. Here the 1 kΩ of R_1 is in series with the much larger 999 kΩ of R_2. The V_T is 1000 V.

The voltages across R_1 and R_2 in Fig. 7-2 can be calculated by the voltage divider formula. Note that R_T is $1 + 999 = 1000$ kΩ.

▶ $$V_1 = \frac{R_1}{R_T} \times V_T = \frac{1}{1000} \times 1000 \text{ V} = 1 \text{ V}$$

$$V_2 = \frac{R_2}{R_T} \times V_T = \frac{999}{1000} \times 1000 \text{ V} = 999 \text{ V}$$

The 999 V across R_2 is practically the entire applied voltage. Also, the very high series resistance dissipates almost all the power.

Furthermore, the current of 1 mA through R_1 and R_2 in Fig. 7-2 is determined almost entirely by the 999 kΩ of R_2. The I for R_T is 1000 V/1000 kΩ, which equals 1 mA. However, the 999 kΩ of R_2 alone would allow 1.001 mA of current, which differs very little from the original I of 1 mA.

ADVANTAGE OF THE VOLTAGE DIVIDER METHOD   Using Formula (7-1), we can find the proportional voltage drops from V_T and the series resistances without knowing the amount of I. For odd values of R, calculating the I may be more troublesome than finding the proportional voltages directly. Also, in many cases we can approximate the voltage division without the need for any written calculations.

TEST-POINT QUESTION 7-1

Answers at end of chapter.

Refer to Fig. 7-1.
a. How much is R_T?
b. What fraction of the applied voltage is V_3?
c. If each resistance is doubled in value, how much is V_1?

7-2 CURRENT DIVIDER WITH TWO PARALLEL RESISTANCES

It is often necessary to find the individual branch currents in a bank from the resistances and I_T, but without knowing the voltage across the bank. This problem can be solved by using the fact that currents divide inversely as the branch resistances. An example is shown in Fig. 7-3. The formulas for the two branch currents are

▶ $$I_1 = \frac{R_2}{R_1 + R_2} \times I_T \qquad\qquad (7\text{-}2)$$

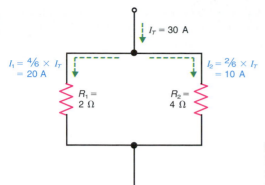

FIG. 7-3 Current divider with two branch resistances. Each branch I is inversely proportional to its R. The smaller R has more I.

and

$$I_2 = \frac{R_1}{R_1 + R_2} \times I_T$$

Notice that the formula for each branch I has the opposite R in the numerator. The reason is that each branch current is inversely proportional to the branch resistance. The denominator is the same in both formulas, equal to the sum of the two branch resistances.

To calculate the currents in Fig. 7-3, with a 30-A I_T, a 2-Ω R_1, and a 4-Ω R_2,

$$I_1 = \frac{4}{2 + 4} \times 30 = \frac{4}{6} \times 30$$
$$I_1 = 20 \text{ A}$$

For the other branch,

$$I_2 = \frac{2}{2 + 4} \times 30 = \frac{2}{6} \times 30$$
$$I_2 = 10 \text{ A}$$

With all the resistances in the same units, the branch currents are in the units of I_T. For instance, kilohms of R and milliamperes of I can be used.

Actually, it is not necessary to calculate both currents. After one I is calculated, the other can be found by subtracting from I_T.

To use the calculator for a problem like this with current division between two branch resistances, as in Formula (7-2), there are several points to note. The numerator has the R of the branch opposite from the desired I. In adding R_1 and R_2, the parens keys ⒪ and ⒪ should be used. The reason is that both terms in the denominator must be added before the division. The procedure for calculating I_1 in Fig. 7-3 can be as follows:

Punch in 4 for R_2. Press the ⊙ key followed by the open parens key ⒪. Punch in 2 ⊕ 4 for R_1 and R_2 followed by the close parens key ⒪. The sum of 6 will be displayed. Press ⊗ and 30, then press ⊜ to display the answer 20 for I_1.

ABOUT
ELECTRONICS

A new "smart" car has an electronic steering wheel that interacts with a driver's palm perspiration and measures alcohol content. The car won't start if the driver is intoxicated or wearing gloves.

Notice that the division of branch currents in a parallel bank is opposite from the voltage division of resistance in a series string. With series resistances, a higher resistance develops a higher IR voltage proportional to its R; with parallel branches, a lower resistance takes more branch current, equal to V/R.

In Fig. 7-3, the 20-A I_1 is double the 10-A I_2 because the 2-Ω R_1 is one-half the 4-Ω R_2. This is an inverse relationship between I and R.

The inverse relation between I and R in a parallel bank means that a very large R has little effect with a much smaller R in parallel. As an example, Fig. 7-4 shows a 999-kΩ R_2 in parallel with a 1-kΩ R_1 dividing the I_T of 1000 mA. The branch currents are calculated as follows:

$$I_1 = \frac{999}{1000} \times 1000 \text{ mA} = 999 \text{ mA}$$

$$I_2 = \frac{1}{1000} \times 1000 \text{ mA} = 1 \text{ mA}$$

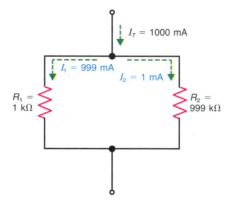

FIG. 7-4 Example of a very large R_2 in parallel with a small R_1. For the branch currents, the small R_1 has almost the entire total line current, I_T.

The 999 mA for I_1 is almost the entire line current of 1000 mA because R_1 is so small compared with R_2. Also, the smallest branch R dissipates the most power because it has the most I.

The current divider Formula (7-2) can be used only for two branch resistances. The reason is the inverse relation between each branch I and its R. In comparison, the voltage divider Formula (7-1) can be used for any number of series resistances because of the direct proportion between each voltage drop V and its R.

For more branches, it is possible to combine the branches in order to work with only two divided currents at a time. However, a better method is to use parallel conductances, because I and G are directly proportional, as explained in the next section.

TEST-POINT QUESTIONS 7-2

Answers at end of chapter.

Refer to Fig. 7-3.
a. What is the ratio of R_2 to R_1?
b. What is the ratio of I_2 to I_1?

7-3 CURRENT DIVISION BY PARALLEL CONDUCTANCES

Remember that the conductance G is $1/R$. Therefore, conductance and current are directly proportional. More conductance allows more current, for the same V. With any number of parallel branches, each branch current is

▶ $$I = \frac{G}{G_T} \times I_T \qquad\qquad (7\text{-}3)$$

where G is the conductance of one branch and G_T is the sum of all the parallel conductances. The unit for G is the siemens (S).

Note that Formula (7-3), for dividing branch currents in proportion to G, has the same form as Formula (7-1), for dividing series voltages in proportion to R. The reason is that both formulas specify a direct proportion.

TWO BRANCHES   As an example of using Formula (7-3), we can go back to Fig. 7-3 and find the branch currents with G instead of R. For the 2 Ω of R_1, the G_1 is $1/2 = 0.5$ S. Also, the 4 Ω of R_2 has G_2 of $1/4 = 0.25$ S. Then G_T is $0.5 + 0.25 = 0.75$ S.

The I_T is 30 A in Fig. 7-3. For the branch currents,

$$I_1 = \frac{G_1}{G_T} \times I_T = \frac{0.50}{0.75} \times 30 \text{ A}$$

$$I_1 = 20 \text{ A}$$

This 20 A is the same I_1 calculated before.

For the other branch, I_2 is $30 - 20 = 10$ A. Also, I_2 can be calculated as 0.25/0.75 or ⅓ of I_T for the same 10-A value.

To calculate I_1 with a calculator, use the same procedure described before with Formula (7-1) for voltage dividers. Using Formula (7-3) for a proportional current divider with conductance, first divide G by G_T and then multiply the answer by I_T. The key strokes will be as follows: G ⊘ G_T ⊗ I_T ⊜. Remember that both Formulas (7-1) and (7-3) are mathematically similar as a direct proportion.

THREE BRANCHES   A circuit with three branch currents is shown in Fig. 7-5. We can find G for the 10-Ω R_1, 2-Ω R_2, and 5-Ω R_3 as follows:

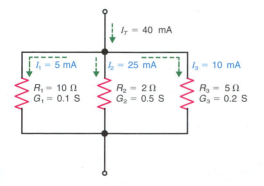

FIG. 7-5   Current divider with branch conductances G_1, G_2, and G_3, each equal to $1/R$. Note that S is the siemens unit for conductance. With conductance values, each branch I is directly proportional to the branch G.

$$G_1 = \frac{1}{R_1} = \frac{1}{10\ \Omega} = 0.1\ \text{S}$$

$$G_2 = \frac{1}{R_2} = \frac{1}{2\ \Omega} = 0.5\ \text{S}$$

$$G_3 = \frac{1}{R_3} = \frac{1}{5\ \Omega} = 0.2\ \text{S}$$

Remember that the siemens (S) unit is the reciprocal of the ohm (Ω) unit. The total conductance then is

$$G_T = G_1 + G_2 + G_3$$
$$= 0.1 + 0.5 + 0.2$$
$$G_T = 0.8\ \text{S}$$

The I_T is 40 mA in Fig. 7-5. To calculate the branch currents with Formula (7-3),

$$I_1 = 0.1/0.8 \times 40\ \text{mA} = 5\ \text{mA}$$
$$I_2 = 0.5/0.8 \times 40\ \text{mA} = 25\ \text{mA}$$
$$I_3 = 0.2/0.8 \times 40\ \text{mA} = 10\ \text{mA}$$

The sum is $5 + 25 + 10 = 40$ mA for I_T.

Although three branches are shown here, Formula (7-3) can be used to find the currents for any number of parallel conductances because of the direct proportion between I and G. The method of conductances is usually easier to use than the method of resistances for three or more branches.

TEST-POINT QUESTIONS 7-3

Answers at end of chapter.

Refer to Fig. 7-5.
a. What is the ratio of G_3 to G_1?
b. What is the ratio of I_3 to I_1?

7-4 SERIES VOLTAGE DIVIDER WITH PARALLEL LOAD CURRENT

The voltage dividers shown so far illustrate just a series string without any branch currents. Actually, though, a voltage divider is often used to tap off part of the applied voltage V_T for a load that needs less voltage than V_T. Then the added load is a parallel branch across part of the divider, as shown in Fig. 7-6. This example shows how the loaded voltage at the tap F is reduced below the potential it would have without the branch current for R_L.

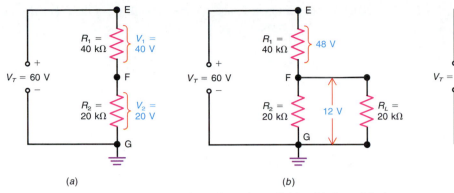

FIG. 7-6 Effect of a parallel load in part of a series voltage divider. (*a*) R_1 and R_2 in series without any branch current. (*b*) Reduced voltage across R_2 and its parallel load R_L. (*c*) Equivalent circuit of the loaded voltage divider.

WHY THE LOADED VOLTAGE DECREASES   We can start with Fig. 7-6*a*, which shows an R_1R_2 voltage divider alone. Resistances R_1 and R_2 in series simply form a proportional divider across the 60-V source for V_T.

For the resistances, R_1 is 40 kΩ and R_2 is 20 kΩ, making R_T equal to 60 kΩ. Also, the current I is V_T/R_T or 60 V/60 kΩ = 1 mA. For the divided voltages in Fig. 7-6*a*,

$$V_1 = \frac{40}{60} \times 60 \text{ V} = 40 \text{ V}$$

$$V_2 = \frac{20}{60} \times 60 \text{ V} = 20 \text{ V}$$

Note that $V_1 + V_2$ is 40 + 20 = 60 V, which is the total applied voltage.

However, in Fig. 7-6*b* the 20-kΩ branch of R_L changes the equivalent resistance at tap F to ground. This change in the proportions of R changes the voltage division. Now the resistance from F to G is 10 kΩ, equal to the 20-kΩ R_2 and R_L in parallel. This equivalent bank resistance is shown as the 10-kΩ R_E in Fig. 7-6*c*.

Resistance R_1 is still the same 40 kΩ because it has no parallel branch. The new R_T for the divider in Fig. 7-6*c* is 40 kΩ + 10 kΩ = 50 kΩ. As a result, V_E from F to G in Fig. 7-6*c* becomes

$$V_E = \frac{R_E}{R_T} \times V_T = \frac{10}{50} \times 60 \text{ V}$$

$$V_E = 12 \text{ V}$$

Therefore, the voltage across the parallel R_2 and R_L in Fig. 7-6*b* is reduced to 12 V. This voltage is at the tap F for R_L.

Note that V_1 across R_1 increases to 48 V in Fig. 7-6*c*. Now V_1 is 40/50 × 60 V = 48 V. The V_1 increases here because there is more current through R_1.

The sum of $V_1 + V_E$ in Fig. 7-6*c* is 12 + 48 = 60 V. The IR drops still add to equal the applied voltage.

PATH OF CURRENT FOR R_L   All the current in the circuit must come from the source V_T. Trace the electron flow for R_L. It starts from the negative side of

V_T, through R_L, to the tap at F, and returns through R_1 in the divider to the positive side of V_T. This current I_L goes through R_1 but not R_2.

BLEEDER CURRENT In addition, both R_1 and R_2 have their own current from the source. This current through all the resistances in the divider is bleeder current I_B. The electron flow for I_B is from the negative side of V_T, through R_2 and R_1, and back to the positive side of V_T.

The bleeder current is a steady drain on the source. However, I_B has the advantage of reducing variations in the total current in the voltage source for different values of load current.

In summary, then, for the three resistances in Fig. 7-6b, note the following currents:

1. Resistance R_L has just its load current I_L.
2. Resistance R_2 has only the bleeder current I_B.
3. Resistance R_1 has both I_L and I_B.

Note that only R_1 is in the path for both the bleeder current and the load current.

TEST-POINT QUESTION 7-4

Answers at end of chapter.

Refer to Fig. 7-6.
a. What is the proportion of R_2/R_T in Fig. 7-6a?
b. What is the proportion of R_E/R_T in Fig. 7-6c?

7-5 DESIGN OF A LOADED VOLTAGE DIVIDER

These principles can be applied to the design of a practical voltage divider, as shown in Fig. 7-7. This type of circuit is used for the output of a power supply in electronic equipment to supply different voltages at the taps, with different load currents. For instance, load D can represent the collector-emitter circuit for one or more power transistors that need +100 V for the collector supply. Also, the tap at E can be the 40-V collector supply for medium-power transistors. Finally, the 18-V tap at F can be for base-emitter bias current in the power transistors and collector voltage for smaller transistors.

Note the load specifications in Fig. 7-7. Load F needs 18 V from point F to chassis ground. When the 18 V is supplied by this part of the divider, a 36-mA branch current will flow through the load. Similarly, 40 V is needed at tap E for 54 mA of I_E in load E. Also, 100 V is available at D with a load current I_D of 180 mA. The total load current here is $36 + 54 + 180 = 270$ mA.

In addition, the bleeder current I_B through the entire divider is generally specified at about 10 percent of the load current. For the example here, I_B is taken as 30 mA to make a total line current I_T of $270 + 30 = 300$ mA from the source. Remember that the 30-mA I_B flows through R_1, R_2, and R_3.

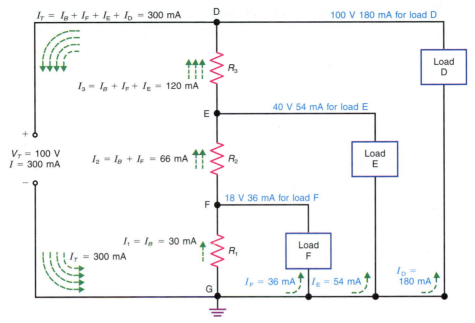

FIG. 7-7 Voltage divider for different voltages and currents from the source V_T. See text for design calculations to find the values of R_1, R_2, and R_3.

The design problem in Fig. 7-7 is to find the values of R_1, R_2, and R_3 needed to provide the specified voltages. Each R is calculated as its ratio of V/I. However, the question is what are the correct values of V and I to use for each part of the divider.

FIND THE CURRENT IN EACH R   We start with R_1 because its current is only the 30-mA bleeder current I_B. No load current flows through R_1. Therefore I_1 through R_1 equals 30 mA.

The 36-mA current I_F for load F returns to the source through R_2 and R_3. Considering just R_2 now, its current is the I_F load current and the 30-mA bleeder current I_B. Therefore, I_2 through R_2 is $36 + 30 = 66$ mA.

The 54-mA current I_E for load E returns to the source through R_3 alone. However, R_3 also has the 36-mA I_F and the 30-mA I_B. Therefore I_3 through R_3 is $54 + 36 + 30 = 120$ mA. The values for I_1, I_2, and I_3 are given in Table 7-1.

Note that the load current I_D for load D at the top of the diagram does not flow through R_3 or any of the resistors in the divider. However, the I_D of 180 mA is the main load current through the source of applied voltage. The 120 mA of bleeder and load currents plus the 180-mA I_D load add to equal 300 mA for I_T in the main line of the power supply.

TABLE 7-1   DESIGN VALUES FOR VOLTAGE DIVIDER IN FIG. 7-7

	CURRENT, mA	VOLTAGE, V	RESISTANCE, Ω
R_1	30	18	600
R_2	66	22	333
R_3	120	60	500

CALCULATE THE VOLTAGE ACROSS EACH R   The voltages at the taps in Fig. 7-7 give the potential to chassis ground. However, we need the voltage across the two ends of each R. For R_1, the voltage V_1 is the indicated 18 V to ground because one end of R_1 is grounded. However, across R_2 the voltage is the difference between the 40-V potential at point E and the 18 V at F. Therefore V_2 is $40 - 18 = 22$ V. Similarly, V_3 is calculated as 100 V at point D minus the 40 V at E, or, V_3 is $100 - 40 = 60$ V. These values for V_1, V_2, and V_3 are summarized in Table 7-1.

CALCULATING EACH R   Now we can calculate the resistance of R_1, R_2, and R_3 as each V/I ratio. For the values listed in Table 7-1,

$$R_1 = \frac{V_1}{I_1} = \frac{18 \text{ V}}{30 \text{ mA}} = 0.6 \text{ k}\Omega = 600 \ \Omega$$

$$R_2 = \frac{V_2}{I_2} = \frac{22 \text{ V}}{66 \text{ mA}} = 0.333 \text{ k}\Omega = 333 \ \Omega$$

$$R_3 = \frac{V_3}{I_3} = \frac{60 \text{ V}}{120 \text{ mA}} = 0.5 \text{ k}\Omega = 500 \ \Omega$$

When these values are used for R_1, R_2, and R_3 and connected in a voltage divider across the source of 100 V, as in Fig. 7-7, each load will have the specified voltage at its rated current.

<hr>

TEST-POINT QUESTION 7-5

Answers at end of chapter.

Refer to Fig. 7-7.
a. How much is the bleeder current I_B through R_1, R_2, and R_3?
b. How much is the voltage for load E at tap E to ground?
c. How much is V_2 across R_2?
d. If load D opens, how much voltage will be measured at tap F to ground?

7   SUMMARY AND REVIEW

- In a series circuit V_T is divided into IR voltage drops proportional to the resistances. Each $V_R = (R/R_T) \times V_T$, for any number of series resistances. The largest series R has the largest voltage drop.
- In a parallel circuit, I_T is divided into branch currents. Each I is inversely proportional to the branch R. The inverse division of branch currents is given by Formula (7-2), for two resistances only. The smaller branch R has the larger branch current.
- For any number of parallel branches, I_T is divided into branch currents directly proportional to each conductance G. Each $I = (G/G_T) \times I_T$.
- A series voltage divider is often tapped for a parallel load, as in Fig. 7-6. Then the voltage at the tap is reduced because of the load current.
- The design of a loaded voltage divider, as in Fig. 7-7, involves calculating each R. Find the I and potential difference V for each R. Then $R = V/I$.

SELF-TEST

ANSWERS AT BACK OF BOOK.

Answer True or False.

1. In a series voltage divider, each IR voltage is proportional to its R.
2. With parallel branches, each branch I is inversely proportional to its R.
3. With parallel branches, each branch I is directly proportional to its G.
4. Formula (7-2) for parallel current dividers can be used for three or more resistances.
5. Formula (7-3) for parallel current dividers can be used for five or more branch conductances.
6. In the series voltage divider of Fig. 7-1, V_1 is 2.5 times V_3 because R_1 is 2.5 times R_3.
7. In the parallel current divider of Fig. 7-3, I_1 is double I_2 because R_1 is one-half R_2.
8. In the parallel current divider of Fig. 7-5, I_2 is five times I_1 because G_2 is five times G_1.
9. In Fig. 7-6b, the branch current I_L flows through R_L, R_2, and R_1.
10. In Fig. 7-7, the bleeder current I_B flows through R_1, R_2, and R_3.

QUESTIONS

1. Define a series voltage divider.
2. Define a parallel current divider.
3. Give two differences between a series voltage divider and a parallel current divider.
4. Give three differences between Formula (7-2) for branch resistances and Formula (7-3) for branch conductances.
5. Define bleeder current.
6. What is the main difference between the circuits in Fig. 7-6a and b?
7. Referring to Fig. 7-1, why is V_1 series-aiding with V_2 and V_3 but in series opposition to V_T? Show polarity of each IR drop.
8. Show the derivation of Formula (7-2) for each branch current in a parallel bank of two resistances. [Hint: The voltage across the bank is $I_T \times R_{EQ}$ and R_{EQ} is $R_1R_2/(R_1 + R_2)$.]

PROBLEMS

ANSWERS TO ODD-NUMBERED PROBLEMS AT BACK OF BOOK.

1. A 100-Ω R_1 is in series with a 200-Ω R_2 and a 300-Ω R_3. The applied voltage is 18 V. Calculate V_1, V_2, and V_3.
2. A 10-kΩ R_1 is in series with a 12-kΩ R_2, a 4.7-kΩ R_3, and a 3.3-kΩ R_4. The applied voltage is 36 V. Calculate V_1, V_2, V_3, and V_4.
3. A 120-Ω R_1 is in parallel with a 60-Ω R_2. The total current I_T is 3 A. Calculate the branch currents I_1 and I_2.
4. A 10-Ω R_1 is in parallel with a 30-Ω R_2, a 60-Ω R_3, and a 20-Ω R_4. I_T = 120 mA. Calculate the branch currents I_1, I_2, I_3, and I_4.
5. In Fig. 7-8, calculate V_1, V_2, V_3, V_4, V_{CG}, V_{BG}, and V_{AG}.
6. In Fig. 7-9, calculate V_1, V_2, V_3, V_4, V_{DG}, V_{CG}, V_{BG}, and V_{AG}.
7. In Fig. 7-10, calculate I_1 and I_2.
8. In Fig. 7-11, calculate I_1, I_2, and I_3.
9. In Fig. 7-12, calculate I_1, I_2, I_3, and I_4.
10. In Fig. 7-12, find the voltage drop across the parallel bank.
11. In Fig. 7-13, calculate the required values for R_1, R_2, and R_3 if the bleeder current I_B (I_3) is chosen to equal 6 mA. Also, calculate the power dissipated by R_1, R_2, and R_3.
12. In Fig. 7-13, calculate the resistances of loads A, B, and C.
13. In Fig. 7-14, calculate the required values for R_1, R_2, and R_3 if I_B = 15 mA. Also, calculate the power dissipated by R_1, R_2, and R_3.
14. In Fig. 7-13, calculate the voltage across load B if load C develops an open.
15. In Fig. 7-13, calculate the voltage across load C if load B develops an open.
16. Select the values for R_1, R_2, R_3, R_4, and R_5 that will provide the specified voltages at points A, B, C, and D in Fig. 7-15. Use a series current I of 10 mA.

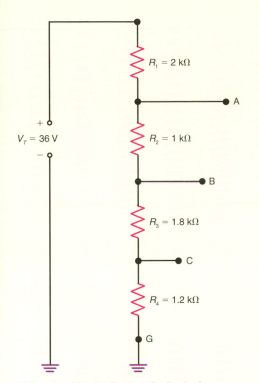

FIG. 7-8   Circuit diagram for Prob. 5.

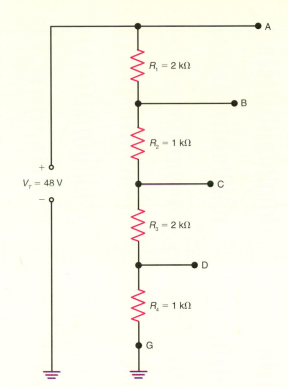

FIG. 7-9   Circuit diagram for Prob. 6.

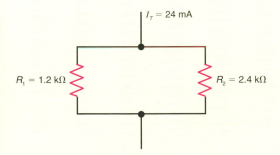

FIG. 7-10   Circuit diagram for Prob. 7.

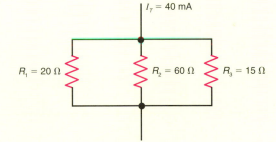

FIG. 7-11   Circuit diagram for Prob. 8.

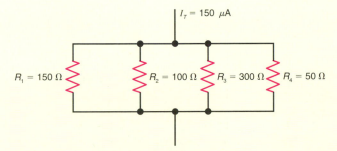

FIG. 7-12   Circuit diagram for Probs. 9 and 10.

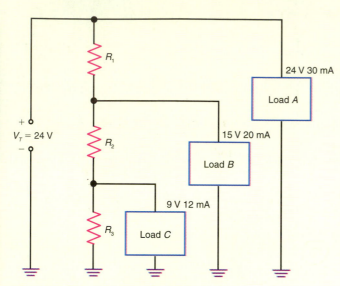

FIG. 7-13 Circuit diagram for Probs. 11, 12, 14, and 15.

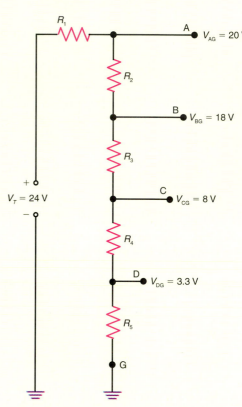

FIG. 7-15 Circuit diagram for Prob. 16.

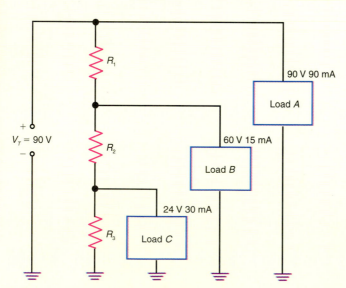

FIG. 7-14 Circuit diagram for Prob. 13.

CRITICAL THINKING

1. Refer to Fig. 7-16. Select values for R_1 and R_3 that will allow the output voltage to vary between 6 V and 15 V.
2. Design a loaded voltage divider, using a 25-V supply, to meet the following requirements: load $A = 25$ V 10 mA; load $B = 15$ V 45 mA; load $C = 6$ V 5 mA; $I_B = 10$ percent of total load current. Draw the schematic diagram including all values.
3. Design a loaded voltage divider, using a 24-V supply, to meet the following requirements: load $A = 18$ V 10 mA; load $B = 12$ V 30 mA; load $C = 5$ V 6 mA; $I_B = 10$ percent of total load current. Draw the schematic diagram including all values.
4. Design a loaded voltage divider, using a 25-V supply, to meet the following requirements: load $A = 20$ V 25 mA; load $B = 12$ V 10 mA; load $C = -5$ V 10 mA; total current $I_T = 40$ mA. Draw the schematic diagram including all values.

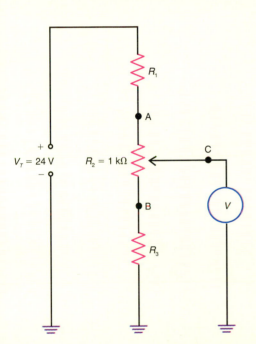

FIG. 7-16   Circuit diagram for Critical Thinking Prob. 1.

ANSWERS TO TEST-POINT QUESTIONS

7-1 a. $R_T = 100$ kΩ
  b. $V_3 = (2/10) \times V_T$
  c. $V_1 = 100$ V

7-2 a. 2 to 1
  b. 1 to 2

7-3 a. 2 to 1
  b. 2 to 1

7-4 a. 1/3
  b. 1/5

7-5 a. $I_B = 30$ mA
  b. $V_E = 40$ V
  c. $V_2 = 22$ V
  d. 18 V

DIRECT-CURRENT METERS

Voltage, resistance, and current measurements are generally made with a combination volt-ohm-milliammeter (VOM) or the digital multimeter (DMM). They both have the same applications in measuring V, R, and I.

Meters with a printed scale and a moving pointer are called *analog* meters. Meters with numerical readouts are called *digital* meters.

To measure voltage, the test leads of the voltmeter are connected in parallel across the two points of potential difference. For dc voltages, observe the correct polarity. Otherwise, the pointer on the meter scale will move to the left instead of up-scale to the right.

Similarly, to measure R in ohms, the two leads of the ohmmeter are connected across the resistance to be measured, but power is off. No power is needed in the circuit being tested because the ohmmeter has its own internal battery. Polarity does not matter in checking the R of a resistor.

To measure current, make the meter a series component in the circuit. For direct current, the meter must be connected in the correct polarity to read up-scale. The circuit usually must be opened temporarily in order to insert the current meter in series. Thus, current measurements are not as convenient as testing V and R in troubleshooting.

Often, meters like the VOM and DMM are called *multimeters*. A multimeter is probably the most important type of test equipment in troubleshooting electronic equipment because it can check V, R, and I. The nondigital multitesters are basically dc meters. However, an internal rectifier is included to convert ac values to dc values for the meter measurement.

CHAPTER OBJECTIVES

Upon completion of this chapter, you should be able to:

- *Explain* the difference between an analog meter and a digital meter.
- *Explain* the construction and operation of a basic moving-coil meter.
- *Explain* how to use a meter to measure the current in a circuit.
- *Calculate* the value of shunt resistance required to extend the current range of a basic moving-coil meter.
- *Explain* how to use a meter to measure the voltage in a circuit.
- *Calculate* the value of multiplier resistance required to make a basic moving-coil meter capable of measuring voltage.
- *Explain* the importance of the ohms-per-volt rating of a voltmeter.
- *Explain* what is meant by *voltmeter loading*.
- *Explain* how a basic moving-coil meter can be used with a battery to construct an ohmmeter.
- *List* the main features of a digital multimeter.
- *Explain* how an ohmmeter can be used to check continuity.

IMPORTANT TERMS IN THIS CHAPTER

back-off ohmmeter scale	diode test	ohms-per-volt sensitivity
clamp probe	galvanometer	taut-band meter
continuity testing	loading effect of voltmeter	volt-ohm-milliammeter
D'Arsonval meter	meter shunt	(VOM)
decibel scale	multiplier resistor	zero-ohms adjustment
digital multimeter (DMM)	ohmmeter	

TOPICS COVERED IN THIS CHAPTER

8-1 MOVING-COIL METER

Figure 8-1 shows two very typical types of multimeters used by electronics technicians. Figure 8-1*a* shows an analog volt-ohm-milliammeter, whereas Fig. 8-1*b* shows a digital multimeter. Both types are capable of measuring voltage, current, and resistance. It should be noted that a typical multimeter can measure dc or ac voltages but only dc values of current.

A moving coil meter movement, shown in Fig. 8-2, is generally used in an analog VOM. The construction consists essentially of a coil of fine wire on a drum mounted between the poles of a permanent magnet. When direct current flows in the coil, the magnetic field of the current reacts with the field of the magnet. The resultant force turns the drum with its pointer, winding up the restoring spring. When the current is removed, the pointer returns to zero. The amount of deflection indicates the amount of current in the coil. When the polarity is connected correctly, the pointer will read up-scale, to the right; the incorrect polarity forces the pointer off-scale, to the left.

The pointer deflection is directly proportional to the amount of current in the coil. If 100 μA is the current needed for full-scale deflection, 50 μA in the coil will produce half-scale deflection. The accuracy of the moving-coil meter mechanism is 0.1 to 2 percent.

The moving-coil principle is applied in several meter types which have different names. A *galvanometer* is an extremely sensitive instrument for measur-

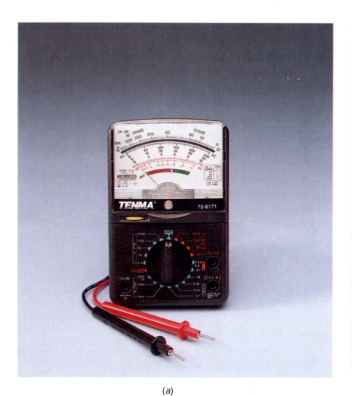

(a)

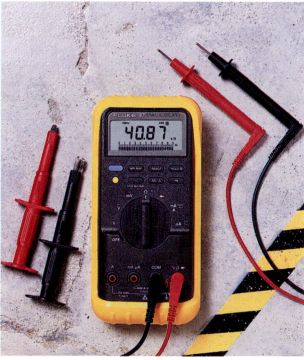

(b)

FIG. 8-1 Typical multimeters used for measuring *V*, *I*, and *R*. (*a*) Analog VOM. (*b*) DMM.

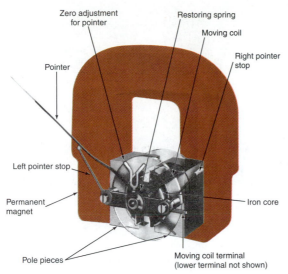

FIG. 8-2   Construction of a moving-coil meter. The diameter of the coil can be ½ to 1 in.

ing very small values of current. Laboratory-type galvanometers, which include a suspended moving coil with an optical system to magnify small deflection, can measure a small fraction of one microampere. A *ballistic galvanometer* is used for reading the value of a small momentary current, to measure electric charge. The moving-coil arrangement is often called a *D'Arsonval movement,* after its inventor, who patented this meter movement in 1881.

VALUES OF I_M   The full-scale deflection current I_M is the amount needed to deflect the pointer all the way to the right to the last mark on the printed scale. Typical values of I_M are from about 10 μA to 30 mA. In a VOM, the I_M is typically either 50 μA or 1 mA.

Refer to the analog VOM in Fig. 8-1*a*. Notice the mirror along the scale to eliminate reading errors. The meter is read when the pointer and its mirror reflection appear as one. This eliminates the optical error called *parallax* caused by looking at the meter from the side.

VALUES OF r_M   This is the internal resistance of the wire of the moving coil. Typical values range from 1.2 Ω for a 30-mA movement to 2000 Ω for a 50-μA movement. A movement with a smaller I_M has a higher r_M because many turns of fine wire are needed. An average value of r_M for a 1-mA movement is about 50 Ω. Figure 8-3 provides a close-up view of the basic components contained within a D'Arsonval meter movement. Notice that the moving coil is wound around a drum which will rotate when dc current flows through the wire of the moving coil.

TAUT-BAND METERS   The meter movement can be constructed with the moving coil and pointer suspended by a metal band, instead of the pivot and jewel design with a restoring spring. Both types of movements have similar operating characteristics. However, taut-band meters generally have lower values of r_M because a smaller coil can be used to force the pointer up-scale.

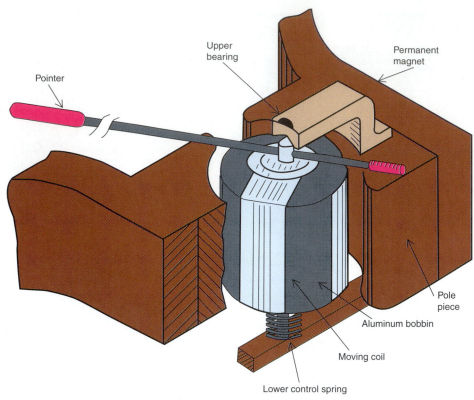

Labels on figure:

Pointer

Upper bearing

Permanent magnet

Pole piece

Aluminum bobbin

Moving coil

Lower control spring

FIG. 8-3   Close-up view of a D'Arsonval moving coil meter movement.

TEST-POINT QUESTION 8-1

Answers at end of chapter.
a. Is a voltmeter connected in parallel or in series?
b. Is a milliammeter connected in parallel or in series?

8-2 MEASUREMENT OF CURRENT

Whether we are measuring amperes, milliamperes, or microamperes, two important facts to remember are:

1. The current meter must be in series in the circuit where the current is to be measured. The amount of deflection depends on the current through the meter. In a series circuit, the current is the same through all series components. Therefore, the current to be measured must be made to flow through the meter as a series component in the circuit.
2. A dc meter must be connected in the correct polarity for the meter to read up-scale. Reversed polarity makes the meter read down-scale, forcing the pointer against the stop at the left, which can bend the pointer.

HOW TO CONNECT A CURRENT METER IN SERIES

As illustrated in Fig. 8-4, the circuit must be opened at one point in order to insert the current meter in series in the circuit. Since R_1, R_2, R_3, and the meter are all in series, the current is the same in each and the meter reads the current in any part of the series circuit.

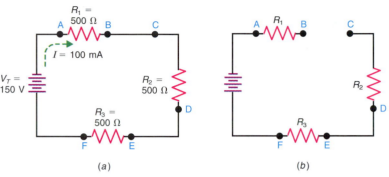

(a)   (b)   (c)

FIG. 8-4 Inserting a current meter in series. (a) Circuit without meter. (b) Connections opened between points B and C for the meter. (c) The meter is connected between R_1 and R_2 in series to read the common current I in the circuit.

In Fig. 8-4, V_T is 150 V with a total series resistance of 1500 Ω for the current of 0.1 A, or 100 mA. This value is the current in R_1, R_2, R_3, and the battery, as shown in Fig. 8-4a. Note that in Fig. 8-4b, the circuit is opened at the junction of R_1 and R_2 for insertion of the meter. In Fig. 8-4c, the meter completes the series circuit to read the current of 100 mA. The meter inserted in series at any point in the circuit would read the same current. Turn off the power to connect the current meter and then put the power back on to read the current.

HOW TO CONNECT A DC METER IN THE CORRECT POLARITY

A dc meter has its terminals marked for polarity, either with + and − signs or red for plus and black for minus. Electrons must flow into the negative side through the movement and out from the positive side for the meter to read upscale.

To have the meter polarity correct, always connect the negative terminal to the point in the circuit that has a path back to the negative side of the voltage source, *without going through the meter.* Similarly, the positive terminal of the meter returns to the positive terminal of the voltage source (Fig. 8-5). Here the negative terminal of the meter is joined to R_2 because this path with R_1 connects to the negative terminal of the battery. The positive meter terminal is connected to R_3. Electrons in the circuit will flow through R_1 and R_2 into the negative side of the meter, through the movement, and out from the meter and return through R_3 to the positive battery terminal.

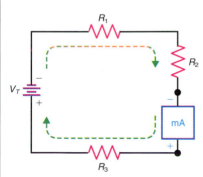

FIG. 8-5 Correct polarity for a dc meter to read current.

A CURRENT METER SHOULD HAVE VERY LOW RESISTANCE

Referring back to Fig. 8-4a, the milliammeter reads 100 mA because its resistance is negligible compared with the total series R of 1500 Ω. Then I is the same with or without the meter.

In general, a current meter should have very low R compared with the circuit where the current is being measured. We take an arbitrary figure of $\frac{1}{100}$. For the circuit in Fig. 8-4, then, the meter resistance should be less than $1500/100 = 15\ \Omega$. Actually, a meter for 100 mA would have an internal R of about $1\ \Omega$ or less because of its internal shunt resistor. The higher the current range of the meter, the smaller its resistance.

An extreme case of a current meter with too much R is shown in Fig. 8-6. Here the series R_T is doubled when the meter is inserted in the circuit. The result is one-half the actual I in the circuit without the meter.

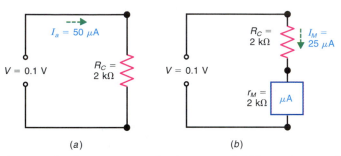

(a)   (b)

FIG. 8-6 An example of a current meter having too much R for accurate readings. (a) Circuit without meter has I of 50 μA. (b) Meter resistance reduces I to 25 μA.

TEST-POINT QUESTION 8-2

Answers at end of chapter.
  a. In Fig. 8-4, how much will the milliammeter read when inserted at point A?
  b. In Fig. 8-5, which R is connected to the positive side of the meter to make it read up-scale?
  c. Should a current meter have very low or very high resistance?
  d. Should a voltmeter have very high or very low resistance?

8-3 METER SHUNTS

A meter shunt is a precision resistor connected across the meter movement for the purpose of shunting, or bypassing, a specific fraction of the circuit's current around the meter movement. The combination then provides a current meter with an extended range. The shunts are usually inside the meter case. However, the schematic symbol for the current meter usually does not show the shunt.

In current measurements, the parallel bank of the movement with its shunt is still connected as a current meter in series in the circuit (Fig. 8-7). It should be noted that a meter with an internal shunt has the scale calibrated to take into account the current through both the shunt and the meter movement. Therefore, the scale reads total circuit current.

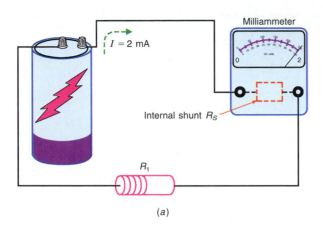

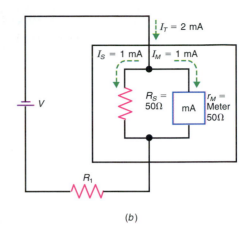

(a)   (b)

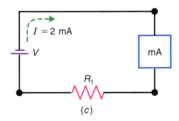

(c)

FIG. 8-7   Example of meter shunt R_S in bypassing current around the movement to extend range from 1 to 2 mA. (*a*) Wiring diagram. (*b*) Schematic diagram showing effect of shunt. With $R_S = r_M$ the current range is doubled. (*c*) Circuit with 2-mA meter to read the current.

RESISTANCE OF THE METER SHUNT   In Fig. 8-7*b*, the 1-mA movement has a resistance of 50 Ω, which is the resistance of the moving coil r_M. To double the range, the shunt resistance R_S is made equal to the 50 Ω of the movement. When the meter is connected in series in a circuit where the current is 2 mA, this total current into one terminal of the meter divides equally between the shunt and the meter movement. At the opposite meter terminal, these two branch currents combine to provide the 2 mA of the circuit current.

Inside the meter, the current is 1 mA through the shunt and 1 mA through the moving coil. Since it is a 1-mA movement, this current produces full-scale deflection. The scale is doubled, however, reading 2 mA, to account for the additional 1 mA through the shunt. Therefore, the scale reading indicates total current at the meter terminals, not just coil current. The movement with its shunt, then, is a 2-mA meter. Its internal resistance is $50 \times \frac{1}{2} = 25$ Ω.

Another example is shown in Fig. 8-8. In general, the shunt resistance for any range can be calculated with Ohm's law from the formula

▶ $$R_S = \frac{V_M}{I_S}$$ (8-1)

where R_S is the resistance of the shunt and I_S is the current through it.

Voltage V_M is equal to $I_M \times r_M$. This is the voltage across both the shunt and the meter movement, which are in parallel.

CALCULATING I_S   This current through the shunt alone is the difference between the total current I_T through the meter and the divided current I_M through the movement. Or

▶ $$I_S = I_T - I_M$$ (8-2)

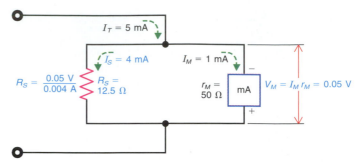

FIG. 8-8 Calculating the resistance of a meter shunt. R_S is equal to V_M/I_S. See text for calculations.

Use the values of current for full-scale deflection, as these are known. In Fig. 8-8,

$$I_S = 5 - 1 = 4 \text{ mA}, \quad \text{or} \quad 0.004 \text{ A}$$

CALCULATING R_S The complete procedure for using the formula $R_S = V_M/I_S$ can be as follows:

1. Find V_M. Calculate this for full-scale deflection as $I_M \times r_M$. In Fig. 8-8, with a 1-mA full-scale current through the 50-Ω movement,

$$V_M = 0.001 \times 50 = 0.05 \text{ V}$$

2. Find I_S. For the values that are shown in Fig. 8-8,

$$I_S = 5 - 1 = 4 \text{ mA} = 0.004 \text{ A}$$

3. Divide V_M by I_S to find R_S. For the final result,

$$R_S = 0.05/0.004 = 12.5 \ \Omega$$

This shunt enables the 1-mA movement to be used for the extended range of 0 to 5 mA.

Note that R_S and r_M are inversely proportional to their full-scale currents. The 12.5 Ω for R_S equals one-fourth the 50 Ω of r_M because the shunt current of 4 mA is four times the 1 mA through the movement for full-scale deflection.

Example

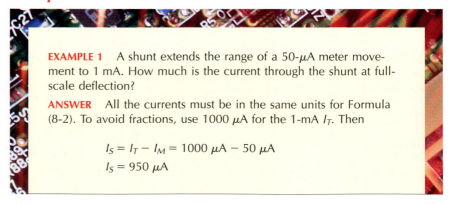

EXAMPLE 1 A shunt extends the range of a 50-μA meter movement to 1 mA. How much is the current through the shunt at full-scale deflection?

ANSWER All the currents must be in the same units for Formula (8-2). To avoid fractions, use 1000 μA for the 1-mA I_T. Then

$$I_S = I_T - I_M = 1000 \ \mu\text{A} - 50 \ \mu\text{A}$$
$$I_S = 950 \ \mu\text{A}$$

EXAMPLE 2 A 50-μA meter movement has r_M of 1000 Ω. What R_S is needed to extend the range to 500 μA?

ANSWER The shunt current I_S is 500 − 50, or 450 μA. Then

$$R_S = \frac{V_M}{I_S}$$

$$= \frac{50 \times 10^{-6} \text{ A} \times 10^3 \text{ } \Omega}{450 \times 10^{-6} \text{ A}} = \frac{50,000}{450}$$

$$R_S = 111.1 \text{ } \Omega$$

The shunts usually are precision wire-wound resistors. For very low values, a short wire of precise size can be used.

TEST-POINT QUESTION 8-3

Answers at end of chapter.

A 50-μA movement with a 900-Ω r_M has a shunt R_S for the range of 500 μA.
a. How much is I_S?
b. How much is V_M?
c. What is the size of R_S?

8-4 VOLTMETERS

Although a meter movement responds only to current in the moving coil, it is commonly used for measuring voltage by the addition of a high resistance in series with the movement (Fig. 8-9). The series resistance must be much higher than the coil resistance in order to limit the current through the coil. The combination of the meter movement with this added series resistance then forms a voltmeter. The series resistor, called a *multiplier,* is usually connected inside the voltmeter case.

Since a voltmeter has high resistance, it must be connected in parallel to measure the potential difference across two points in a circuit. Otherwise, the high-resistance multiplier would add so much series resistance that the current in the circuit would be reduced to a very low value. Connected in parallel, though, the high resistance of the voltmeter is an advantage. The higher the voltmeter resistance, the smaller the effect of its parallel connection on the circuit being tested.

The circuit is not opened to connect the voltmeter in parallel. Because of this convenience, it is common practice to make voltmeter tests in troubleshooting. The voltage measurements apply the same way to an *IR* drop or a generated emf.

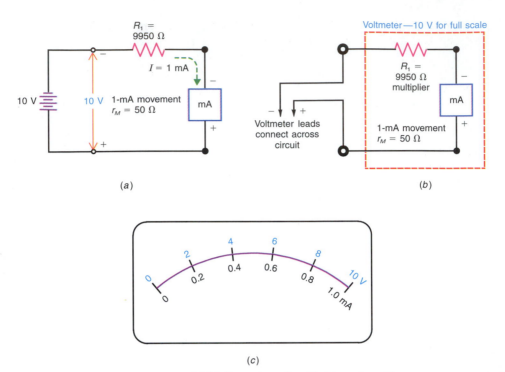

(a)

(b)

(c)

FIG. 8-9 Multiplier resistor R_1 added in series with meter movement to form a voltmeter. (*a*) Resistance of R_1 allows 1 mA for full-scale deflection on 1-mA movement with 10 V applied. (*b*) Internal multiplier R_1 forms a voltmeter. The test leads can be connected across a potential difference to measure 0 to 10 V. (*c*) 10-V scale of voltmeter and corresponding 1-mA scale of meter movement.

The correct polarity must be observed in using a dc voltmeter. Connect the negative voltmeter lead to the negative side of the potential difference being measured and the positive lead to the positive side.

MULTIPLIER RESISTANCE   Figure 8-9 illustrates how the meter movement and its multiplier R_1 form a voltmeter. With 10 V applied by the battery in Fig. 8-9*a*, there must be 10,000 Ω of resistance to limit the current to 1 mA for full-scale deflection of the meter movement. Since the movement has a 50-Ω resistance, 9950 Ω is added in series, resulting in a 10,000-Ω total resistance. Then I is 10 V/10 kΩ = 1 mA.

With 1 mA in the movement, the full-scale deflection can be calibrated as 10 V on the meter scale, as long as the 9950-Ω multiplier is included in series with the movement. It doesn't matter to which side of the movement the multiplier is connected.

If the battery is taken away, as in Fig. 8-9*b*, the movement with its multiplier forms a voltmeter that can indicate a potential difference of 0 to 10 V applied across its terminals. When the voltmeter leads are connected across a potential difference of 10 V in a dc circuit, the resulting 1-mA current through the meter movement produces full-scale deflection, and the reading is 10 V. In Fig. 8-9*c* the 10-V scale is shown corresponding to the 1-mA range of the movement.

If the voltmeter is connected across a 5-V potential difference, the current in the movement is ½ mA, the deflection is one-half of full scale, and the reading is 5 V. Zero voltage across the terminals means no current in the movement, and the voltmeter reads zero. In summary, then, any potential difference up to 10 V, whether an *IR* voltage drop or a generated emf, can be applied across the meter terminals. The meter will indicate less than 10 V in the same ratio that the meter current is less than 1 mA.

The resistance of a multiplier can be calculated from the formula

$$R_{\text{mult}} = \frac{\text{full-scale } V}{\text{full-scale } I} - r_M \qquad (8\text{-}3)$$

Applying this formula to the example of R_1 in Fig. 8-9 gives

$$R_{\text{mult}} = \frac{10 \text{ V}}{0.001 \text{ A}} - 50 \ \Omega = 10,000 - 50$$

$$R_{\text{mult}} = 9950 \ \Omega$$

We can take another example for the same 10-V scale but with a 50-μA meter movement, which is commonly used. Now the multiplier resistance is much higher, though, because less *I* is needed for full-scale deflection. Let the resistance of the 50-μA movement be 2000 Ω. Then

$$R_{\text{mult}} = \frac{10 \text{ V}}{0.000 \ 050 \text{ A}} - 2000 \ \Omega = 200,000 - 2000$$

$$R_{\text{mult}} = 198,000 \ \Omega$$

MULTIPLE VOLTMETER RANGES   Voltmeters often have several multipliers which are used with one meter movement. A range switch selects one multiplier for the required scale. The higher the voltage range is, the higher the multiplier resistance, in essentially the same proportion as the ranges.

Figure 8-10 illustrates two ranges. When the switch is on the 10-V range, multiplier R_1 is connected in series with the 1-mA movement. Then you read the 10-V scale on the meter face. With the range switch on 25 V, R_2 is the multiplier, and the measured voltage is read on the 25-V scale.

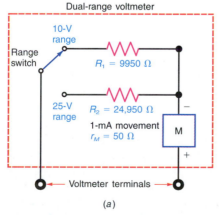
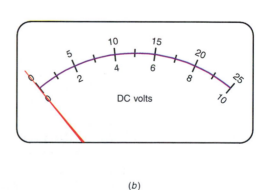

(a) (b)

FIG. 8-10   Voltmeter with range of either 10 or 25 V. (*a*) Range switch selects scale by connecting either R_1 or R_2 as the series multiplier. (*b*) Both voltage scales on the meter.

Several examples of using these two scales are listed in Table 8-1. Note that voltages less than 10 V can be read on either scale. It is preferable, however, to have the pointer read on the middle third of the scale. That is why the scales are usually multiples of 10 and 2.5 or 3.

TABLE 8-1   MULTIPLE VOLTAGE-SCALE READINGS FOR FIG. 8-10

10-V SCALE, R_V* = 10,000 Ω			25-V SCALE, R_V* = 25,000 Ω		
METER, mA	DEFLECTION	SCALE READING, V	METER, mA	DEFLECTION	SCALE READING, V
0	0	0	0	0	0
0.5	½	5	0.2	$^2/_{10}$	5
1.0	Full scale	10	0.4	$^4/_{10}$	10
			0.5	½	12.5
			1.0	Full scale	25

*R_V is total voltmeter resistance of multiplier and meter movement.

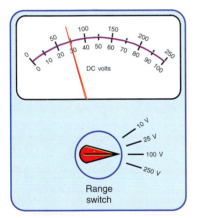

FIG. 8-11   The range switch selects a multiplier for the voltage that can produce full-scale deflection of the meter movement. The reading shown here is 30 V on the 100-V range.

RANGE SWITCH   With multiple ranges, the setting of the selector switch is the voltage that produces full-scale deflection (Fig. 8-11). One scale is generally used for ranges that are multiples of 10. If the range switch is set for 250 V in Fig. 8-11, read the top scale as is. With the range switch at 25 V, however, the readings on the 250-V scale are divided by 10.

Similarly, the 100-V scale is used for the 100-V range and the 10-V range. In Fig. 8-11 the pointer indicates 30 V when the switch is on the 100-V range; this reading on the 10-V range is 3 V.

TYPICAL MULTIPLE VOLTMETER CIRCUIT   Another example of multiple voltage ranges is shown in Fig. 8-12, with a typical switching arrangement. Resistance R_1 is the series multiplier for the lowest voltage range of 2.5 V. When higher resistance is needed for the higher ranges, the switch adds the required series resistors.

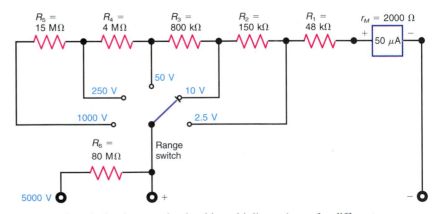

FIG. 8-12   A typical voltmeter circuit with multiplier resistors for different ranges.

The meter in Fig. 8-12 requires 50 μA for full-scale deflection. For the 2.5-V range, a series resistance of $2.5/(50 \times 10^{-6})$, or 50,000 Ω, is needed. Since r_M is 2000 Ω, the value of R_1 is 50,000 − 2000, which equals 48,000 Ω or 48 kΩ.

For the 10-V range, a resistance of $10/(50 \times 10^{-6})$, or 200,000 Ω, is the value needed. Since $R_1 + r_M$ provides 50,000 Ω, R_2 is made 150,000 Ω, for a total of 200,000 Ω series resistance on the 10-V range. Similarly, additional resistors are switched in to increase the multiplier resistance for the higher voltage ranges. Note the separate jack and extra multiplier R_6 on the highest range for 5000 V. This method of adding series multipliers for higher voltage ranges is the circuit generally used in commercial multimeters.

VOLTMETER RESISTANCE The high resistance of a voltmeter with a multiplier is essentially the value of the multiplier resistance. Since the multiplier is changed for each range, the voltmeter resistance changes.

Table 8-2 shows how the voltmeter resistance increases for the higher ranges. The middle column lists the total internal resistance R_V, including R_{mult} and r_M, for the voltmeter circuit in Fig. 8-12. With a 50-μA movement, R_V increases from 50 kΩ on the 2.5-V range to 20 MΩ on the 1000-V range. It should be noted that R_V has these values on each range whether you read full-scale or not.

TABLE 8-2	CHARACTERISTICS OF A VOLTMETER USING A 50-μA MOVEMENT	
FULL-SCALE VOLTAGE V_F	**$R_V = R_{mult} + r_M$**	**OHMS PER VOLT = R_V/V_F**
2.5	50 kΩ	20,000 Ω/V
10	200 kΩ	20,000 Ω/V
50	1 MΩ	20,000 Ω/V
250	5 MΩ	20,000 Ω/V
1000	20 MΩ	20,000 Ω/V

OHMS-PER-VOLT RATING To indicate the voltmeter's resistance independently of the range, voltmeters are generally rated in ohms of resistance needed for 1 V of deflection. This value is the ohms-per-volt rating of the voltmeter. As an example, see the last column in Table 8-2. The values in the top row show that this meter needs 50,000 Ω R_V for 2.5 V of full-scale deflection. The resistance per 1 V of deflection then is 50,000/2.5, which equals 20,000 Ω/V.

The ohms-per-volt value is the same for all ranges. The reason is that this characteristic is determined by the full-scale current I_M of the meter movement. To calculate the ohms-per-volt rating, take the reciprocal of I_M in ampere units. For example, a 1-mA movement results in 1/0.001 or 1000 Ω/V; a 50-μA movement allows 20,000 Ω/V, and a 20-μA movement allows 50,000 Ω/V. The ohms-per-volt rating is also called the *sensitivity* of the voltmeter.

A high ohms-per-volt value means a high voltmeter resistance R_V. In fact R_V can be calculated as the product of the ohms-per-volt rating and the full-scale voltage of each range. For instance, across the second row in Table 8-2, on the 10-V range with a 20,000 Ω/V rating,

$$R_V = 10 \text{ V} \times \frac{20,000 \text{ }\Omega}{\text{volts}}$$

$$R_V = 200,000 \text{ }\Omega$$

These values are for dc volts only. The sensitivity for ac voltage is made lower, generally, to prevent erratic meter deflection produced by stray magnetic fields before the meter leads are connected into the circuit. Usually the ohms-per-volt rating of a voltmeter is printed on the meter face.

The sensitivity of 1000 Ω/V with a 1-mA movement used to be common for dc voltmeters, but 20,000 Ω/V with a 50-μA movement is generally used now. Higher sensitivity is an advantage, not only for less voltmeter loading, but because lower voltage ranges and higher ohmmeter ranges can be obtained.

TEST-POINT QUESTION 8-4

Answers at end of chapter.

Refer to Fig. 8-12.
a. Calculate the voltmeter resistance R_V on the 2.5-V range.
b. Calculate the voltmeter resistance R_V on the 50-V range.
c. Is the voltmeter multiplier resistor in series or parallel with the meter movement?
d. Is a voltmeter connected in series or parallel with the potential difference to be measured?
e. How much is the total R of a voltmeter with a sensitivity of 20,000 Ω/V on the 25-V scale?

8-5 LOADING EFFECT OF A VOLTMETER

When the voltmeter resistance is not high enough, connecting it across a circuit can reduce the measured voltage, compared with the voltage present without the voltmeter. This effect is called *loading down* the circuit, since the measured voltage decreases because of the additional load current for the meter.

LOADING EFFECT Voltmeter loading can be appreciable in high-resistance circuits, as shown in Fig. 8-13. In Fig. 8-13*a*, without the voltmeter, R_1 and R_2 form a voltage divider across the applied voltage of 120 V. The two equal resistances of 100 kΩ each divide the applied voltage equally, with 60 V across each.

When the voltmeter in Fig. 8-13*b* is connected across R_2 to measure its potential difference, however, the voltage division changes. The voltmeter resistance R_V of 100 kΩ is the value for a 1000-ohms-per-volt meter on the 100-V range. Now the voltmeter in parallel with R_2 draws additional current and the equivalent resistance between the measured points 1 and 2 is reduced from 100,000 to 50,000 Ω. This resistance is one-third the total circuit resistance, and the measured voltage across points 1 and 2 drops to 40 V, as shown in Fig. 8-13*c*.

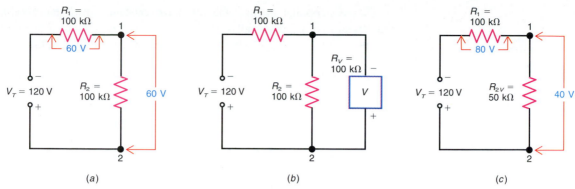

FIG. 8-13 How loading effect of the voltmeter can reduce the voltage reading. (a) High-resistance series circuit without voltmeter. (b) Connecting voltmeter across one of the series resistances. (c) Reduced R and V between points 1 and 2 caused by the voltmeter as a parallel branch across R_2. The R_{2V} is the equivalent of R_2 and R_V in parallel.

As additional current drawn by the voltmeter flows through the other series resistance R_1, this voltage goes up to 80 V.

Similarly, if the voltmeter were connected across R_1, this voltage would go down to 40 V, with the voltage across R_2 rising to 80 V. When the voltmeter is disconnected, the circuit returns to the condition in Fig. 8-13a, with 60 V across both R_1 and R_2.

The loading effect is minimized by using a voltmeter with a resistance much greater than the resistance across which the voltage is measured. As shown in Fig. 8-14, with a voltmeter resistance of 10 MΩ, the loading effect is negligible. Because R_V is so high, it does not change the voltage division in the circuit. The 10 MΩ of the meter in parallel with the 100,000 Ω for R_2 results in an equivalent resistance practically equal to 100,000 Ω.

With multiple ranges on a VOM, the voltmeter resistance changes with the range selected. Higher ranges require more multiplier resistance, increasing the voltmeter resistance for less loading. As examples, a 20,000-ohms-per-volt meter on the 250-V range has an internal resistance R_V of 20,000 × 250, or 5 MΩ. However, on the 2.5-V range the same meter has an R_V of 20,000 × 2.5, which is only 50,000 Ω.

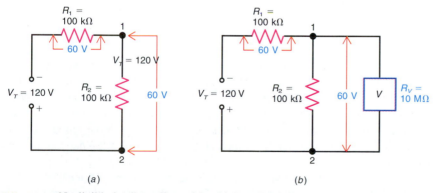

FIG. 8-14 Negligible loading effect with a high-resistance voltmeter. (a) High-resistance series circuit without voltmeter, as in Fig. 8-13a. (b) Same voltages in circuit with voltmeter connected, because R_V is so high.

On any one range, though, the voltmeter resistance is constant whether you read full-scale or less than full-scale deflection. The reason is that the multiplier resistance set by the range switch is the same for any reading on that range.

CORRECTION FOR LOADING EFFECT   The following formula can be used:

▶   *Actual reading + correction*
            ↓   ↓

$$V = V_M + \frac{R_1R_2}{R_V(R_1 + R_2)}V_M \qquad \text{(8-4)}$$

Voltage V is the corrected reading the voltmeter would show if it had infinitely high resistance. Voltage V_M is the actual voltage reading. Resistances R_1 and R_2 are the voltage-dividing resistances in the circuit without the voltmeter resistance R_V. As an example, in Fig. 8-13,

$$V = 40 \text{ V} + \frac{100 \text{ k}\Omega \times 100 \text{ k}\Omega}{100 \text{ k}\Omega \times 200 \text{ k}\Omega} \times 40 \text{ V} = 40 + \frac{1}{2} \times 40 = 40 + 20$$

$$V = 60 \text{ V}$$

The loading effect of a voltmeter causes the voltage reading to be too low because R_V is too low as a parallel resistance. This corresponds to the case of a current meter reading too low because r_M is too high as a series resistance. Both of these effects illustrate the general problem of trying to make any measurement without changing the circuit being measured.

It should be noted that the digital multimeter (DMM) has practically no loading effect as a voltmeter. The input resistance is usually 10 MΩ or 20 MΩ, the same on all ranges.

TEST-POINT QUESTION 8-5

Answers at end of chapter.

With the voltmeter across R_2 in Fig. 8-13, what is the value for
a. V_1?
b. V_2?

8-6 OHMMETERS

Basically, an ohmmeter consists of an internal battery, the meter movement, and a current-limiting resistance, as illustrated in Fig. 8-15. For measuring resistance, the ohmmeter leads are connected across the external resistance to be measured. Power in the circuit being tested must be off. Then only the ohmmeter battery produces current for deflecting the meter movement. Since the amount of current through the meter depends on the external resistance, the scale can be calibrated in ohms.

The amount of deflection on the ohms scale indicates the measured resistance directly. The ohmmeter reads up-scale regardless of the polarity of the leads because the polarity of the internal battery determines the direction of current through the meter movement.

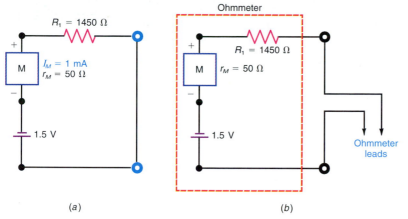

FIG. 8-15 How meter movement M can be used as an ohmmeter with a 1.5-V battery. (*a*) Equivalent closed circuit with R_1 and the battery when ohmmeter leads are short-circuited for zero ohms of external R. (*b*) Internal ohmmeter circuit with test leads open, ready to measure an external resistance.

SERIES OHMMETER CIRCUIT

In Fig. 8-15*a*, the circuit has 1500 Ω for $(R_1 + r_M)$. Then the 1.5-V cell produces 1 mA, deflecting the moving coil full scale. When these components are enclosed in a case, as in Fig. 8-15*b*, the series circuit forms an ohmmeter. Note that M indicates the meter movement.

If the leads are short-circuited together or connected across a short circuit, 1 mA flows. The meter movement is deflected full scale to the right. This ohmmeter reading is 0 Ω.

When the ohmmeter leads are open, not touching each other, the current is zero. The ohmmeter indicates infinitely high resistance or an open circuit across its terminals.

Therefore, the meter face can be marked zero ohms at the right for full-scale deflection and infinite ohms at the left for no deflection. In-between values of resistance result when less than 1 mA flows through the meter movement. The corresponding deflection on the ohms scale indicates how much resistance is across the ohmmeter terminals.

BACK-OFF OHMMETER SCALE

Table 8-3 and Fig. 8-16 illustrate the calibration of an ohmmeter scale in terms of meter current. The current equals V/R_T. Voltage V is the fixed applied voltage of 1.5 V supplied by the internal battery.

TABLE 8-3   CALIBRATION OF OHMMETER IN FIG. 8-16

EXTERNAL R_X, Ω	INTERNAL $R_i = R_1 + r_M$, Ω	$R_T = R_X + R_i$, Ω	$I = V/R_T$, mA	DEFLECTION	SCALE READING, Ω
0	1500	1500	1	Full scale	0
750	1500	2250	⅔ = 0.67	⅔ scale	750
1500	1500	3000	½ = 0.5	½ scale	1500
3000	1500	4500	⅓ = 0.33	⅓ scale	3000
150,000	1500	151,500	0.01	¹⁄₁₀₀ scale	150,000
500,000	1500	501,500	0	None	∞

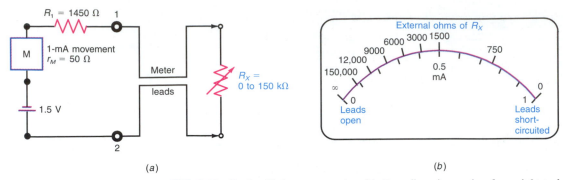

(a)

(b)

FIG. 8-16 Back-off ohmmeter scale with R readings increasing from right to left.
(a) Series ohmmeter circuit for the unknown external resistor R_X to be measured.
(b) Ohms scale has higher R readings to the left of the scale as more R_X decreases I_M.
The R and I values are listed in Table 8-3.

Resistance R_T is the total resistance of R_X and the ohmmeter's internal resistance. Note that R_X is the external resistance to be measured.

The ohmmeter's internal resistance R_i is constant at $50 + 1450$, or $1500\ \Omega$ here. If R_X also equals $1500\ \Omega$, for example, R_T equals $3000\ \Omega$. The current then is $1.5\ \text{V}/3000\ \Omega$, or $0.5\ \text{mA}$, resulting in half-scale deflection for the 1-mA movement. Therefore, the center of the ohms scale is marked for $1500\ \Omega$. Similarly, the amount of current and meter deflection can be calculated for any value of the external resistance R_X.

Note that the ohms scale increases from right to left. This arrangement is called a *back-off scale,* with ohms values increasing to the left as the current backs off from full-scale deflection. The back-off scale is a characteristic of any ohmmeter where the internal battery is in series with the meter movement. Then more external R_X decreases the meter current.

A back-off ohmmeter scale is expanded at the right near zero ohms and crowded at the left near infinite ohms. This nonlinear scale results from the relation of $I = V/R$ with V constant at 1.5 V. Specifically, the back-off ohms scale represents the graph of a hyperbolic curve for a reciprocal function $y = 1/x$, where y is I and x is R.

The highest resistance that can be indicated by the ohmmeter is about 100 times its total internal resistance. Therefore, the infinity mark on the ohms scale, or the "lazy eight" symbol ∞ for infinity, is only relative. It just means that the measured resistance is infinitely greater than the ohmmeter resistance.

For instance, if a $500{,}000\text{-}\Omega$ resistor in good condition were measured with the ohmmeter in Fig. 8-16, it would indicate infinite resistance because this ohmmeter cannot measure as high as $500{,}000\ \Omega$. To read higher values of resistance, the battery voltage can be increased to provide more current, or a more sensitive meter movement is necessary to provide deflection with less current.

MULTIPLE OHMMETER RANGES Commercial multimeters provide for resistance measurements from less than $1\ \Omega$ up to many megohms, in several ranges. The range switch in Fig. 8-17 shows the multiplying factors for the ohms scale. On the $R \times 1$ range, for low-resistance measurements, read the ohms scale directly. In the example here, the pointer indicates $12\ \Omega$. When the range switch

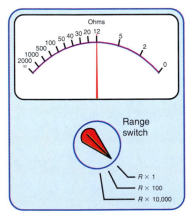

FIG. 8-17 Multiple ohmmeter ranges with just one ohms scale. The ohms reading is multiplied by the factor set on the range switch.

is on $R \times 100$, multiply the scale reading by 100; this reading would then be 12×100 or $1200 \ \Omega$. On the $R \times 10,000$ range, the pointer would indicate $120,000 \ \Omega$.

A multiplying factor, instead of full-scale resistance, is given for each ohms range because the highest resistance is infinite on all the ohms ranges. This method for ohms should not be confused with the full-scale values for voltage ranges. For the ohmmeter ranges, always multiply the scale reading by the $R \times$ factor. On voltage ranges, you may have to multiply or divide the scale reading to match the full-scale voltage with the value on the range switch.

TYPICAL OHMMETER CIRCUIT   For high-ohms ranges a sensitive meter is necessary to read the low values of I with the high values of R_X. For the case of low ohms, however, less sensitivity is needed for the higher currents. These opposite requirements are solved by using a meter shunt across the meter movement and changing the shunt resistance for the multiple ohmmeter ranges. In Fig. 8-18, R_S is the meter shunt.

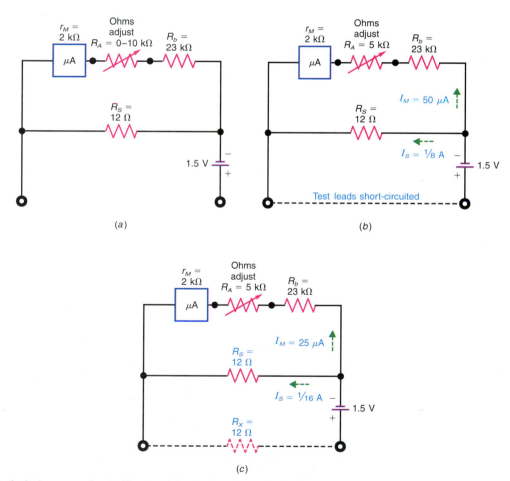

FIG. 8-18   Typical ohmmeter circuit, illustrated for $R \times 1$ range. (*a*) Circuit with test leads open. (*b*) Test leads short-circuited to adjust for zero ohms. (*c*) Ohmmeter measurement with test leads across external R_X of $12 \ \Omega$. This value results in half-scale deflection, as shown on the ohms scale in Fig. 8-17.

To analyze the ohmmeter circuit in Fig. 8-18, three conditions are shown. All are for the $R \times 1$ range with 12-Ω R_S. Figure 8-18a shows the internal circuit before the ohmmeter is adjusted for zero ohms. In Fig. 8-18b the test leads are short-circuited. Then there are two paths for branch current produced by the battery V. One branch is R_S. The other branch includes R_b, R_A, and the meter movement. The 1.5 V is across both branches.

To allow 50 μA through the meter, R_A is adjusted to 5000 Ω. Then the total resistance in this branch is 23 kΩ + 5 kΩ + 2 kΩ, which equals 30 kΩ. With 30 kΩ across 1.5 V, I_M equals 50 μA. Therefore, R_A is adjusted for full-scale deflection to read zero ohms with the test leads short-circuited.

In Fig. 8-18c assume that a resistance R_X being measured is 12 Ω, equal to R_S. Then the meter current is practically 25 μA for half-scale deflection. The center ohms reading on the $R \times 1$ scale, therefore, is 12 Ω. For higher values of R_X, the meter current decreases to indicate higher resistances on the back-off ohms scale.

For higher ohms ranges, the resistance of the R_S branch is increased. The half-scale ohms reading on each range is equal to the resistance of the R_S branch. A higher battery voltage can also be used for the highest ohms range.

On any range, R_A is adjusted for full-scale deflection to read zero ohms with the test leads short-circuited. This variable resistor is the *ohms adjust* or *zero-ohms* adjustment.

ZERO-OHMS ADJUSTMENT To compensate for lower voltage output as the internal battery ages, an ohmmeter includes a variable resistor such as R_A in Fig. 8-18, to calibrate the ohms scale. A back-off ohmmeter is always adjusted for zero ohms. With the test leads short-circuited, vary the ZERO OHMS control on the front panel of the meter until the pointer is exactly on zero at the right edge of the ohms scale. Then the ohms readings are correct for the entire scale.

This type of ohmmeter must be zeroed again every time you change the range. The reason is that the internal circuit is changed.

When the adjustment cannot deflect the pointer all the way to zero at the right edge, it usually means the battery voltage is too low and the internal dry cells must be replaced. Usually, this trouble shows up first on the $R \times 1$ range, which takes the most current from the battery.

TEST-POINT QUESTION 8-6

Answers at end of chapter.
a. The ohmmeter reads 40 Ω on the $R \times 10$ range. How much is R_X?
b. A voltmeter reads 40 V on the 300-V scale, but with the range switch on 30 V. How much is the measured voltage?

8-7 MULTIMETERS

Multimeters are also called *multitesters,* and they are used to measure voltage, current, or resistance. Table 8-4 compares the features of the main types of mul-

TABLE 8-4   COMPARISON OF VOM WITH DMM

VOM	DMM
Analog pointer reading	Digital readout
DC voltmeter R_V changes with range	R_V is 10 or 22 MΩ, the same on all ranges
Zero-ohms adjustment changed for each range	No zero-ohms adjustment
Ohms ranges up to $R \times 10,000 \ \Omega$, as a multiplying factor	Ohms ranges up to 20 MΩ; each range is the maximum

timeters, which are the volt-ohm-milliammeter (VOM), in Fig. 8-19, and the digital multimeter (DMM), in Fig. 8-20. The DMM is explained in more detail in the next section.

Because it is simple, compact, and portable, the VOM is a very common basic instrument. The cost of a typical VOM is less than for a DMM.

Besides its digital readout, an advantage of the DMM is its high input resistance R_V, as a dc voltmeter. The R_V is usually 10 MΩ, the same on all ranges, which is high enough to prevent any loading effect by the voltmeter in most circuits. Some types have R_V of 22 MΩ. Many modern DMMs are autoranging; that is, the internal circuitry selects the proper range for the meter and indicates the range as a readout.

For either a VOM or a DMM, it is important to have a low-voltage dc scale with resolution good enough to read 0.2 V or less. The range of 0.2 to 0.6 V, or 200 to 600 mV, is needed for measuring dc bias voltages in transistor circuits.

LOW-POWER OHMS (LPΩ)   Another feature needed for transistor measurements is an ohmmeter that does not have enough battery voltage to bias a semiconductor junction into the ON or conducting condition. The limit is 0.2 V or less. The purpose is to prevent any parallel conduction paths in the transistor amplifier circuit, which can lower the ohmmeter reading.

DECIBEL SCALE   Most analog multimeters have an ac voltage scale calibrated in decibel (dB) units, for measuring ac signals. The decibel is a logarithmic unit used for comparisons of power levels or voltage levels. The mark of 0 dB on the scale indicates the reference level, which is usually 0.775 V for 1 mW across 600 Ω. Positive decibel values, above the zero mark, indicate ac voltages above the reference of 0.775 V; negative decibel values are less than the reference level.

FIG. 8-19   Analog VOM that combines a function selector and range switch.

FIG. 8-20   Portable digital multimeter (DMM).

AMP-CLAMP PROBE The problem of opening a circuit to measure *I* can be eliminated by using a probe with a clamp that fits around the current-carrying wire. Its magnetic field is used to indicate the amount of current. An example is shown in Fig. 8-21. The clamp probe measures just ac amperes, generally for the 60-Hz power line.

HIGH-VOLTAGE PROBE An accessory probe can be used with a multimeter to measure dc voltages up to 30 kV. This probe is often referred to as a high-voltage probe. One application is measuring the high voltage of 20 to 30 kV at the anode of the color picture tube in a television receiver. The probe is basically just an external multiplier resistance for the dc voltmeter. The required *R* for a 30-kV probe is 580 MΩ with a 20-kΩ/V meter on the 1000-V range.

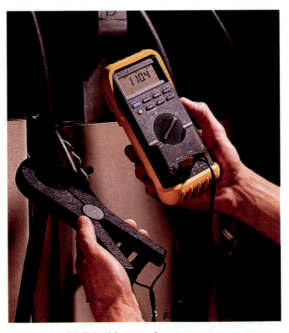

FIG. 8-21   DMM with amp clamp accessory.

Answers at end of chapter.
a. How much is R_V on the 1-V range for a VOM with a sensitivity of 20 kΩ/V?
b. If R_V is 10 MΩ for a DMM on the 100-V range, how much is R_V on the 200-mV range?
c. The low-power ohms function (LPΩ) does not require an internal battery. True or false?

8-8 DIGITAL MULTIMETER (DMM)

The digital multimeter has become a very popular test instrument because the digital value of the measurement is displayed automatically with decimal point, polarity, and the unit for V, A, or Ω. Digital meters are generally easier to use because they eliminate the human error that often occurs in reading different scales on an analog meter with a pointer. Examples of the portable DMM are shown in Figs. 8-20 and 8-22.

The basis of the DMM operation is the use of an analog-to-digital (A/D) converter circuit. It converts analog voltage values in the input to an equivalent binary form. These values are processed by digital circuits to be shown on a liquid-crystal display (LCD) with decimal values.

VOLTAGE MEASUREMENTS   The A/D converter requires a specific range of voltage input, typical values being −200 mV to +200 mV. For DMM input voltages that are higher, the voltage is divided down. When the input voltage is too low, it is increased by a dc amplifier circuit. The measured voltage can then be compared to a fixed reference voltage in the meter by a comparator circuit. Actually, all the functions in the DMM depend on the voltage measurements by the converter and comparator circuits.

The input resistance of the DMM is in the range of 10 to 20 MΩ, shunted by 50 pF of capacitance. This R is high enough to eliminate the problem of voltmeter loading in most transistor circuits. Not only does the DMM have high input resistance, but the R is the same on all ranges.

With ac measurements, the ac input is converted to dc voltage for the A/D converter. The DMM has an internal diode rectifier that serves as an ac converter.

R MEASUREMENT   As an ohmmeter, the internal battery supplies I through the measured R for an IR drop measured by the DMM. The battery is usually the small 9-V type commonly used in portable equipment. A wide range of R values can be measured, from a fraction of an ohm to more than 30 MΩ. Remember that power must be off in the circuit being tested with an ohmmeter.

A DMM ohmmeter usually has an open-circuit voltage across the meter leads, which is much too low to turn on a semiconductor junction. The result is low-power ohms operation.

FIG. 8-22 Typical digital multimeter (DMM) described in text.

I MEASUREMENTS To measure current, internal resistors provide a proportional *IR* voltage. The display shows the *I* values. Note that the DMM still must be connected as a series component in the circuit when current is measured.

DIODE TEST The DMM usually has a setting for testing semiconductor diodes, either silicon or germanium. Current is supplied by the DMM for the diode to test the voltage across its junction. Normal values are 0.7 V for silicon or 0.3 V for germanium. A short-circuited junction will read 0 V. The voltage across an open diode reads much too high. Most diodes are silicon.

RESOLUTION This term for a DMM specifies how many places can be used to display the digits 0 to 9, regardless of the decimal point. For example, 9.99 V is a three-digit display. Also 9.999 V would be a four-digit display. Most portable units, however, compromise with a 3½-digit display. This means that the fourth digit at the left for the most significant place can only be a 1. If not, then the display has three digits. As examples, a 3½ digit display can show 19.99 V but 29.99 V would be read as 30.0 V. It should be noted that better resolution with more digits can be obtained with more expensive meters, especially the larger DMM units for bench mounting. Actually, though, 3½ digit resolution is enough for practically all measurements made in troubleshooting electronic equipment.

RANGE OVERLOAD The DMM selector switch has specific ranges. Any value higher than the selected range is an overload. An indicator on the display warns that the value shown is not correct. Then a higher range is selected. Some units have an *autorange function*, where the meter automatically shifts to a higher range as soon as an overload is indicated.

TYPICAL DMM The unit in Fig. 8-22 can be used as an example. On the front panel, the two jacks at the bottom right are for the test leads. The lower jack is the common, with a black lead, used for all measurements. Above is the jack for the "hot" lead, usually red, used for most measurements of *V*, *R*, and *I*, either dc or ac values. The two jacks at the bottom left side are for the red lead with very high *I* values up to 10 A or very low values below 300 mA.

Consider each function of the large selector switch at the center in Fig. 8-22. The first position at the top, after the switch is turned clockwise from the OFF position, is used to measure ac volts, as indicated by the sine wave. No ranges are given as this meter has autorange function. In operation, the meter has the ranges of 0 to 3.2 V, 32 V, 320 V, and, as a maximum, 750 V.

If the autorange function is not desired, press the range button at the top, to the right of the front panel, to hold the range. Each touch of the button will change the range. Hold the button down to return to autorange operation.

The next position on the function switch is for dc volts. Polarity can be either positive or negative as indicated by the solid and dashed lines next to the *V*. The ranges of dc voltages that can be measured are 0 to 3.2 V, 32 V, 320 V, and 1000 V as a maximum. Values of 300 mV or less are measured on this position of the function switch.

For an ohmmeter, the function switch is set to the position with the Ω symbol. The ohms values are from 0 to 32 MΩ in six ranges. Remember that power must be off in the circuit being measured, or the ohmmeter will read the wrong value.

ABOUT ELECTRONICS

On-board car electronics can make hotel reservations, get you into your locked-out car, give you sports scores, video directions, and traffic reports, or notify emergency service if you are having car trouble.

Next on the function switch is the position for testing semiconductor diodes, as shown by the diode symbol. The lines next to the symbol indicate the meter produces a beep tone. Maximum diode test voltage is 2 V.

The last two positions on the function switch are for current measurements, in the range of 0 to 320 mA. For larger or smaller current values, the special jacks at the lower left are used.

In measuring ac values, either for V or I, the frequency range of the meter is limited to 45 to 1000 Hz, approximately. For amplitudes at higher frequencies, such as RF measurements, special meters are necessary. However, this meter can be used for V and I at the 60-Hz power-line frequency and the 400-Hz test frequency often used for audio equipment.

ANALOG DISPLAY   The bar at the bottom of the display in Fig. 8-22 is used only to show the relative magnitude of the input compared to the full-scale value of the range in use. This function is convenient when adjusting a circuit for a peak value or a minimum (null). The operation is comparable to watching the needle on a VOM for either a maximum or a null adjustment.

TEST-POINT QUESTION 8-8

Answers at end of chapter.

Answer True or False.
a. Typical resistance of a DMM voltmeter is 10 MΩ.
b. The ohmmeter on a portable DMM does not need an internal battery.
c. A DMM voltmeter with 3½ digit resolution can read the value of 3.5555 V.

8-9 METER APPLICATIONS

Table 8-5 summarizes the main points to remember in using a voltmeter, ohmmeter, or milliammeter. These rules apply whether the meter is a single unit or one function on a multimeter. Also, the voltage and current tests apply to either dc or ac circuits.

TABLE 8-5   DIRECT-CURRENT METERS

VOLTMETER	MILLIAMMETER OR AMMETER	OHMMETER
Power on in circuit	Power on in circuit	Power off in circuit
Connect in parallel	Connect in series	Connect in parallel
High internal R	Low internal R	Has internal battery
Has internal series multipliers; higher R for higher ranges	Has internal shunts; lower resistance for higher current ranges	Higher battery voltage and more sensitive meter for higher ohms ranges

To avoid excessive current through the meter, it is good practice to start on a high range when measuring an unknown value of voltage or current. It is very important not to make the mistake of connecting a current meter in parallel, because usually this mistake ruins the meter. The mistake of connecting a voltmeter in series does not damage the meter, but the reading will be wrong.

If the ohmmeter is connected to a circuit where power is on, the meter can be damaged, besides giving the wrong reading. An ohmmeter has its own internal battery, and the power must be off in the circuit being tested. When testing R with an ohmmeter, it may be necessary to disconnect one end of R from the circuit, to eliminate parallel paths.

CONNECTING A CURRENT METER IN THE CIRCUIT   In a series-parallel circuit, the current meter must be inserted in a branch to read branch current. In the main line, the meter reads the total current. These different connections are illustrated in Fig. 8-23. The meters are shown by dashed lines to illustrate the different points at which a meter could be connected to read the respective currents.

If the circuit is opened at point A to insert the meter in series in the main line here, the meter will read total line current I_T through R_1. A meter at B or C will read the same line current.

In order to read the branch current through R_2, this R must be disconnected from its junction with the main line at either end. A meter inserted at D or E, therefore, will read the R_2 branch current I_2. Similarly, a meter at F or G will read the R_3 branch current I_3.

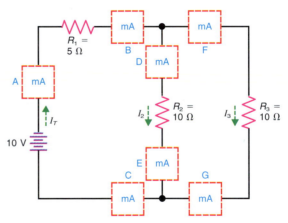

FIG. 8-23   How to insert a current meter in different parts of a series-parallel circuit to read the desired current I. At point A, B, or C the meter reads I_T; at D or E the meter reads I_2; at F or G the meter reads I_3.

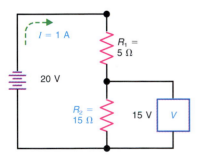

FIG. 8-24   With 15 V measured across a known R of 15 Ω, the I can be calculated as V/R or 15 V/15 Ω = 1 A.

CALCULATING *I* FROM MEASURED VOLTAGE   The inconvenience of opening the circuit to measure current can often be eliminated by the use of Ohm's law. The voltage and resistance can be measured without opening the circuit and the current calculated as V/R. In the example in Fig. 8-24, when the voltage across R_2 is 15 V and its resistance is 15 Ω, the current through R_2 must be

1 A. When values are checked during troubleshooting, if the voltage and resistance are normal, so is the current.

This technique can also be convenient for determining I in low-resistance circuits where the resistance of a microammeter may be too high. Instead of measuring I, measure V and R and calculate I as V/R.

Furthermore, if necessary, we can insert a known resistance R_S in series in the circuit, temporarily, just to measure V_S. Then I is calculated as V_S/R_S. The resistance of R_S, however, must be small enough to have little effect on R_T and I in the series circuit.

This technique is often used with oscilloscopes to produce a voltage waveform of IR which has the same waveform as the current in a resistor. The oscilloscope must be connected as a voltmeter because of its high input resistance.

CHECKING FUSES   Turn the power off or remove the fuse from the circuit to check with an ohmmeter. A good fuse reads 0 Ω. A blown fuse is open, which reads infinity on the ohmmeter.

A fuse can also be checked with the power on in the circuit by using a voltmeter. Connect the voltmeter across the two terminals of the fuse. A good fuse reads 0 V because there is practically no IR drop. With an open fuse, though, the voltmeter reading is equal to the full value of the applied voltage. Having the full applied voltage seems to be a good idea, but it should not be across the fuse.

VOLTAGE TESTS FOR AN OPEN CIRCUIT   Figure 8-25 shows four equal resistors in series with a 100-V source. A ground return is shown here because voltage measurements are usually made to chassis ground. Normally, each resistor would have an IR drop of 25 V. Then, at point B the voltmeter to ground should read $100 - 25 = 75$ V. Also, the voltage at C should be 50 V, with 25 V at D, as shown in Fig. 8-25a.

However, the circuit in Fig. 8-25b has an open in R_3, toward the end of the series string of voltages to ground. Now when you measure at B, the reading is 100 V, equal to the applied voltage. This full voltage at B shows that the series circuit is open, without any IR drop across R_1. The question is, however, which R has the open? Continue the voltage measurements to ground until you find 0 V. In this example, the open is in R_3, between the 100 V at C and 0 V at D.

The points that read the full applied voltage have a path back to the source of voltage. The first point that reads 0 V has no path back to the high side of the source. Therefore, the open circuit must be between points C and D in Fig. 8-25b.

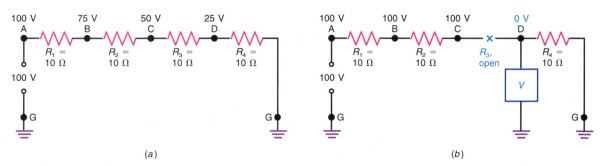

(a)   (b)

FIG. 8-25   Voltage tests to localize an open circuit. (*a*) Normal circuit with voltages to chassis ground. (*b*) Reading of 0 V at point D shows R_3 is open.

Answers at end of chapter.
a. Which type of meter requires an internal battery?
b. How much is the normal voltage across a good fuse?
c. How much is the voltage across R_1 in Fig. 8-25a?
d. How much is the voltage across R_1 in Fig. 8-25b?

8-10 CHECKING CONTINUITY WITH THE OHMMETER

A wire conductor that is continuous without a break has practically zero ohms of resistance. Therefore, the ohmmeter can be useful in testing for the continuity. This test should be done on the lowest ohms range. There are many applications. A wire conductor can have an internal break which is not visible because of the insulated cover, or the wire can have a bad connection at the terminal. Checking for zero ohms between any two points along the conductor tests continuity. A break in the conducting path is evident from a reading of infinite resistance, showing an open circuit.

As another application of checking continuity, suppose there is a cable of wires harnessed together as illustrated in Fig. 8-26, where the individual wires cannot be seen, but it is desired to find the conductor that connects to terminal A. This is done by checking continuity for each conductor to point A. The wire that has zero ohms to A is the one connected to this terminal. Often the individual wires are color-coded, but it may be necessary to check the continuity of each lead.

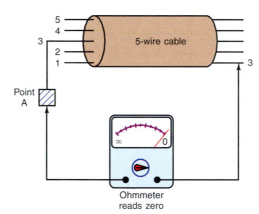

FIG. 8-26 Continuity testing from point A to wire 3 shows this wire is connected.

An additional technique that can be helpful is illustrated in Fig. 8-27. Here it is desired to check the continuity of the two-wire line, but its ends are too far apart for the ohmmeter leads to reach. The two conductors are temporarily short-circuited at one end, however, so that the continuity of both wires can be checked at the other end.

In summary, then, the ohmmeter is helpful in checking the continuity of any wire conductor. This includes resistance-wire heating elements, like the wires in

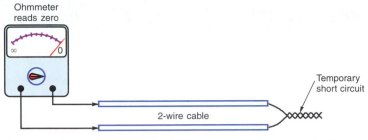

Ohmmeter reads zero

∞   0

Temporary short circuit

2-wire cable

FIG. 8-27   Temporary short circuit at one end of a long two-wire line to check continuity from the opposite end.

a toaster or the filament of an incandescent bulb. Their cold resistance is normally just a few ohms. Infinite resistance means that the wire element is open. Similarly, a good fuse has practically zero resistance; a burned-out fuse has infinite resistance, meaning it is open. Also, any coil for a transformer, solenoid, or motor will have infinite resistance if the winding is open.

TEST-POINT QUESTION 8-10

Answers at end of chapter.
a. On a back-off ohmmeter, is zero ohms at the left or the right end of the scale?
b. What is the ohmmeter reading for an open circuit?

8 SUMMARY AND REVIEW

- Direct current in a moving-coil meter deflects the coil in proportion to the amount of current.
- A current meter is a low-resistance meter connected in series to read the amount of current in the circuit.
- A meter shunt R_S in parallel with the meter movement extends the range of a current meter [see Formula (8-1)].
- A voltmeter consists of the meter movement in series with a high-resistance multiplier. The voltmeter with its multiplier is connected across two points to measure their potential difference in volts. The multiplier R can be calculated from Formula (8-3).
- The ohms-per-volt rating of a voltmeter with series multipliers specifies the sensitivity on all voltage ranges. It equals the reciprocal of the full-scale deflection current of the meter. A typical value is 20,000 Ω/V for a voltmeter using a 50-μA movement. The higher the ohms-per-volt rating, the better.
- Voltmeter resistance R_V is higher for higher ranges because of higher-resistance multipliers. Multiply the ohms-per-volt rating by the voltage range to calculate the R_V for each range.
- An ohmmeter consists of an internal battery in series with the meter movement. Power must be off in the circuit being checked with an ohmmeter. The series ohmmeter has a back-off scale with zero ohms at the right edge and infinity at the left. Adjust for zero ohms with the leads short-circuited each time the ohms range is changed.
- The VOM is a portable multimeter that measures volts, ohms, and milliamperes.
- The digital multimeter generally has an input resistance of 10 MΩ on all dc voltage ranges.
- In checking wire conductors, the ohmmeter reads 0 Ω or very low R for normal continuity and infinite ohms for an open.

SELF-TEST

ANSWERS AT BACK OF BOOK.

Choose (a), (b), (c), or (d).

1. To connect a current meter in series: (a) open the circuit at one point and use the meter to complete the circuit; (b) open the circuit at the positive and negative terminals of the voltage source; (c) short-circuit the resistance to be checked and connect the meter across it; (d) open the circuit at one point and connect the meter to one end.
2. To connect a voltmeter in parallel to read an *IR* drop: (a) open the circuit at one end and use the meter to complete the circuit; (b) open the circuit at two points and connect the meter

across both points; (*c*) allow the circuit to remain as is and connect the meter across the resistance; (*d*) allow the circuit to remain closed but disconnect the voltage source.

3. A shunt for a milliammeter (*a*) extends the range and reduces the meter resistance; (*b*) extends the range and increases the meter resistance; (*c*) decreases the range and the meter resistance; (*d*) decreases the range but increases the meter resistance.

4. For a 50-μA movement with 2000-Ω r_M, the voltage V_M at full-scale deflection is (*a*) 0.1 V; (*b*) 0.2 V; (*c*) 0.5 V; (*d*) 250 μV.

5. A voltmeter using a 20-μA meter movement has a sensitivity of (*a*) 1000 Ω/V; (*b*) 20,000 Ω/V; (*c*) 50,000 Ω/V; (*d*) 11 MΩ/V.

6. When using an ohmmeter, disconnect the applied voltage from the circuit being checked because: (*a*) the voltage source will increase the resistance; (*b*) the current will decrease the resistance; (*c*) the ohmmeter has its own internal battery; (*d*) no current is needed for the meter movement.

7. A multiplier for a voltmeter is (*a*) a high resistance in series with the meter movement; (*b*) a high resistance in parallel with the meter movement; (*c*) usually less than 1 Ω in series with the meter movement; (*d*) usually less than 1 Ω in parallel with the meter movement.

8. To double the current range of a 50-μA 2000-Ω meter movement, the shunt resistance is (*a*) 40 Ω; (*b*) 50 Ω; (*c*) 2000 Ω; (*d*) 18,000 Ω.

9. With a 50-μA movement, a VOM has an input resistance of 6 MΩ on the dc voltage range of (*a*) 3; (*b*) 12; (*c*) 60; (*d*) 300.

10. For a 1-V range, a 50-μA movement with an internal R of 2000 Ω needs a multiplier resistance of (*a*) 1 kΩ; (*b*) 3 kΩ; (*c*) 18 kΩ; (*d*) 50 kΩ.

QUESTIONS

1. (**a**) Why is a milliammeter connected in series in a circuit? (**b**) Why should the milliammeter have low resistance?

2. (**a**) Why is a voltmeter connected in parallel in a circuit? (**b**) Why should the voltmeter have high resistance?

3. A circuit has a battery across two resistances in series. (**a**) Draw a diagram showing how to connect a milliammeter in the correct polarity to read current through the junction of the two resistances. (**b**) Draw a diagram showing how to connect a voltmeter in the correct polarity to read the voltage across one resistance.

4. Explain briefly why a meter shunt equal to the resistance of the moving coil doubles the current range.

5. Describe how to adjust the ZERO OHMS control on a back-off ohmmeter.

6. What is meant by a 3½-digit display on a DMM?

7. Give two advantages of the DMM in Fig. 8-20 compared with the conventional VOM in Fig. 8-19.

8. What does the zero ohms control in the circuit of a back-off ohmmeter do?

9. State two precautions to be observed when you use a milliammeter.

10. State two precautions to be observed when you use an ohmmeter.

11. The resistance of a voltmeter R_V is 300 kΩ on the 300-V range when measuring 300 V. Why is R_V still 300 kΩ when measuring 250 V on the same range?

12. Redraw the schematic diagram in Fig. 6-1b, in Chap. 6, showing a milliammeter to read line current through R_1 and R_2, a meter for R_3 branch current, and a meter for R_4 branch current. Label polarities on each meter.

13. Give a typical value of voltmeter resistance for a DMM.

14. Would you rather use a DMM or VOM in troubleshooting? Why?

PROBLEMS

ANSWERS TO ODD-NUMBERED PROBLEMS AT BACK OF BOOK.

1. Calculate the shunt resistance R_S that is needed to extend the range of a 50-Ω 1-mA movement to: **(a)** 2 mA; **(b)** 10 mA; **(c)** 25 mA; **(d)** 100 mA **(e)** In each case, how much current is indicated by half-scale deflection?

2. Refer to Fig. 8-28. **(a)** Calculate the values for the separate shunt resistances, R_{S_1}, R_{S_2}, and R_{S_3}. **(b)** Calculate the resistance of the meter (R_S in parallel with r_M) for each setting of the range switch.

3. Calculate the multiplier resistance, R_{mult}, for a 50-Ω 1-mA movement if the voltage ranges are: **(a)** 3 V; **(b)** 10 V; **(c)** 30 V; **(d)** 100 V; **(e)** 300 V. In each case, how much voltage is indicated by half-scale deflection?

4. Refer to Fig. 8-29. **(a)** Calculate the values for the multiplier resistors, R_1, R_2, R_3, R_4, R_5, and R_6. **(b)** Calculate the total voltmeter resistance R_V for each setting of the range switch.

5. Refer to Fig. 8-30. **(a)** Calculate the dc voltage that should exist across R_2 without a voltmeter present. **(b)** Calculate the dc voltage that would be measured across R_2 using the volt-

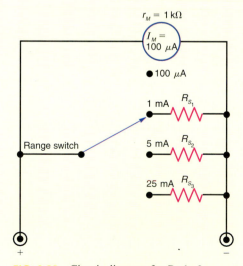

FIG. 8-28 Circuit diagram for Prob. 2.

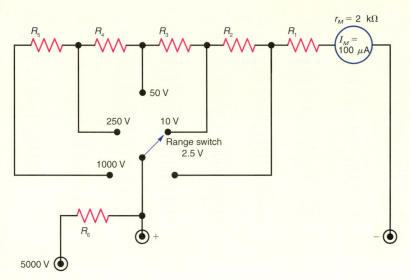

FIG. 8-29 Circuit diagram for Prob. 4.

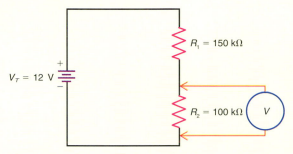

FIG. 8-30 Circuit diagram for Prob. 5.

meter of Fig. 8-29 set to the 10-V range. (**c**) Calculate the dc voltage that would be measured across R_2 using a DMM with an R_V of 10 MΩ on all of its dc voltage ranges.

6. A certain voltmeter has an Ω/V rating of 25 kΩ/V. Calculate the total voltmeter resistance R_V for the following voltmeter ranges: (**a**) 2.5 V; (**b**) 10 V; (**c**) 25 V; (**d**) 100 V; (**e**) 250 V; (**f**) 1000 V; (**g**) 5000 V.

7. Determine the Ω/V rating for the meter in Fig. 8-29.

8. Refer to Fig. 8-30. Suppose a voltmeter having an Ω/V rating of 20 kΩ/V is connected across R_1 to measure its voltage drop. If the voltmeter reads 6 V when set on its 15-V range, use Formula (8-4) to determine the actual value of V_1 without the meter present.

9. Refer to the series ohmmeter in Fig. 8-15*b*. What value of external resistance R_X across the ohmmeter leads will produce: (**a**) full-scale deflection; (**b**) one-quarter full-scale deflection; (**c**) one-third full-scale deflection; (**d**) one-half full-scale deflection; (**e**) two-thirds full-scale deflection; (**f**) three-fourths full-scale deflection?

10. In Fig. 8-15*b*, assume that the battery voltage $V_b = 1.65$ V. What value of external resistance R_X across the ohmmeter leads will produce half-scale deflection?

11. In Fig. 8-15b, assume that the meter movement is shunted to become a 10-mA meter. Calculate the value of R_1 required to produce full-scale deflection when the ohmmeter leads are shorted.

12. Refer to Fig. 8-31. With the ohmmeter leads shorted, to what value must R_A be adjusted to provide full-scale deflection of the meter's pointer?

13. In Fig. 8-31, what value of resistance between the ohmmeter leads will produce half-scale deflection of the meter's pointer?

14. Assume that a 15-Ω resistor is connected across the ohmmeter leads in Fig. 8-31. How much current will flow in the meter movement?

15. In Fig. 8-31, what is the minimum battery voltage that will still provide full-scale deflection of the meter's pointer when the ohmmeter leads are shorted?

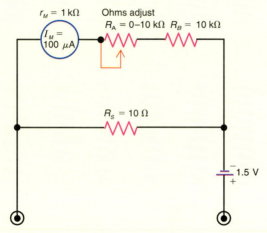

FIG. 8-31   Circuit diagram for Probs. 12, 13, 14, and 15.

CRITICAL THINKING

1. Figure 8-32 shows a universal-shunt current meter. Calculate the values for R_1, R_2, and R_3 which will provide current ranges of 2 mA, 10 mA, and 50 mA.

2. Design a series ohmmeter using a 2-kΩ 50-μA meter movement and a 1.5-V battery. The center-scale ohms reading is to be 150 Ω.

3. The voltmeter across R_2 in Fig. 8-33 shows 20 V. If the voltmeter is set on the 30-V range, calculate the Ω/V rating of the meter.

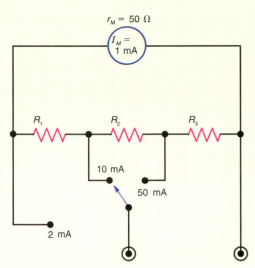

FIG. 8-32 Circuit diagram for Critical Thinking Prob. 1.

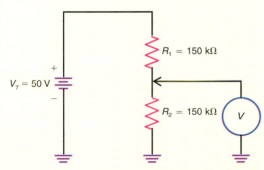

FIG. 8-33 Circuit diagram for Critical Thinking Prob. 3.

ANSWERS TO TEST-POINT QUESTIONS

8-1   **a.** parallel
     b. series

8-2   **a.** $I = 100$ mA
     b. R_3
     c. low
     d. high

8-3   **a.** $I_S = 450$ μA
     b. $V_M = 0.045$ V
     c. $R_S = 100$ Ω

8-4   **a.** $R_V = 50$ kΩ
     b. $R_V = 1$ MΩ
     c. series
     d. parallel
     e. $R = 500$ kΩ

8-5   **a.** $V_1 = 80$ V
     b. $V_2 = 40$ V

8-6   **a.** $R_X = 400$ Ω
     b. $V = 4$ V

8-7   **a.** 20 kΩ
     b. 10 MΩ
     c. F

8-8   **a.** T
     b. F
     c. F

8-9   **a.** ohmmeter
     b. 0 V
     c. 25 V
     d. 0 V

8-10   **a.** right edge
      b. ∞ ohms

REVIEW: CHAPTERS 7 AND 8

SUMMARY

- In a series voltage divider the IR drop across each resistance is proportional to its R. A larger R has a larger voltage drop. Each $V = (R/R_T) \times V_T$. In this way, the series voltage drops can be calculated from V_T without I.
- In a parallel current divider, each branch current is inversely related to its R. A smaller R has more branch current. For only two resistances, we can use the inverse relation

$$I_1 = [R_2/(R_1 + R_2)] \times I_T$$

  In this way, the branch currents can be calculated from I_T without V.
- In a parallel current divider, each branch current is directly proportional to its conductance G. A larger G has more branch current. For any number of parallel resistances, each branch $I = (G/G_T) \times I_T$.
- A milliammeter or ammeter is a low-resistance meter connected in series in a circuit to measure current.
- Different current ranges are obtained by meter shunts in parallel with the meter.
- A voltmeter is a high-resistance meter connected across the voltage to be measured.
- Different voltage ranges are obtained by multipliers in series with the meter.
- An ohmmeter has an internal battery to indicate the resistance of a component across its two terminals, with external power off.
- In making resistance tests, remember that $R = 0\ \Omega$ for continuity or a short circuit, but the resistance of an open circuit is infinitely high.
- Figure 8-1 shows a VOM and DMM. Both types can be used for voltage, current, and resistance measurements.

REVIEW SELF-TEST

ANSWERS AT BACK OF BOOK.

Answer True or False.

1. The internal R of a milliammeter must be low to have minimum effect on I in the circuit.
2. The internal R of a voltmeter must be high to have minimum current through the meter.
3. Power must be off when checking resistance in a circuit because the ohmmeter has its own internal battery.

4. In the series voltage divider in Fig. 8-25, the normal voltage from point B to ground is 75 V.
5. In Fig. 8-25, the normal voltage across R_1, between A and B, is 75 V.
6. The highest ohms range is best for checking continuity with an ohmmeter.
7. With four equal resistors in a series voltage divider with V_T of 44.4 V, each IR drop is 11.1 V.
8. With four equal resistors in parallel with I_T of 44.4 mA, each branch current is 11.1 mA.
9. Series voltage drops divide V_T in direct proportion to each series R.
10. Parallel currents divide I_T in direct proportion to each branch R.
11. The VOM cannot be used to measure current.
12. The DMM can be used as a high-resistance voltmeter.

REFERENCES

Gilmore, C. M.: *Instruments and Measurements,* Glencoe/McGraw-Hill, Columbus, Ohio.

Prensky, S. D., and R. L. Castellucis: *Electronic Instrumentation,* Prentice-Hall Inc., Englewood Cliffs, N.J.

Zbar, P. B., and G. Rockmaker: *Basic Electricity: A Text-Lab Manual,* Glencoe/McGraw-Hill, Columbus, Ohio.

CHAPTER 9

KIRCHHOFF'S LAWS

Many types of circuits have components that are not in series, in parallel, or in series-parallel. For example, a circuit may have two voltages applied in different branches. Another example is an unbalanced bridge circuit. Where the rules of series and parallel circuits cannot be applied, more general methods of analysis become necessary. These methods include the application of Kirchhoff's laws, as described here, and the network theorems explained in Chap. 10.

All circuits can be solved by Kirchhoff's laws because they do not depend on series or parallel connections. Stated in 1847 by the German physicist Gustav R. Kirchhoff, the two basic rules for voltage and current are:

1. *The algebraic sum of the voltage sources and IR voltage drops in any closed path must total zero.*

2. *At any point in a circuit the algebraic sum of the currents directed in and out must total zero.*

Specific methods for applying these basic rules in dc circuits are explained in this chapter.

CHAPTER OBJECTIVES

Upon completion of this chapter, you should be able to:

- *State* Kirchhoff's current and voltage laws.
- *Use* the method of branch currents to solve for all voltages and currents in a circuit containing two or more voltage sources in different branches.
- *Use* node-voltage analysis to solve for the unknown voltages and currents in a circuit containing two or more voltage sources in different branches.
- *Use* the method of mesh currents to solve for the unknown voltages and currents in a circuit containing two or more voltage sources in different branches.

IMPORTANT TERMS IN THIS CHAPTER

current law	network	ΣV
loop	node voltage	voltage law
mesh current	simultaneous equations	

TOPICS COVERED IN THIS CHAPTER

9-1 KIRCHHOFF'S CURRENT LAW (KCL)

The algebraic sum of the currents entering and leaving any point in a circuit must equal zero. Or stated another way: *The algebraic sum of the currents into any point of the circuit must equal the algebraic sum of the currents out of that point.* Otherwise, charge would accumulate at the point, instead of having a conducting path. An algebraic sum means combining positive and negative values.

ALGEBRAIC SIGNS   In using Kirchhoff's laws to solve circuits it is necessary to adopt conventions that determine the algebraic signs for current and voltage terms. A convenient system for currents is: *Consider all currents into a branch point as positive and all currents directed away from that point as negative.*

As an example, in Fig. 9-1 we can write the currents as

$$I_A + I_B - I_C = 0$$

or

$$5\text{ A} + 3\text{ A} - 8\text{ A} = 0$$

Currents I_A and I_B are positive terms because these currents flow into P, but I_C, directed out, is negative.

CURRENT EQUATIONS   For a circuit application, refer to point C at the top of the diagram in Fig. 9-2. The 6-A I_T into point C divides into the 2-A I_3 and 4-A $I_{4\text{-}5}$, both directed out. Note that $I_{4\text{-}5}$ is the current through R_4 and R_5. The algebraic equation is

$$I_T - I_3 - I_{4\text{-}5} = 0$$

Substituting the values for these currents,

$$6\text{ A} - 2\text{ A} - 4\text{ A} = 0$$

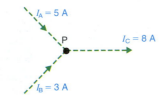

FIG. 9-1   Current I_C out from point P equals 5 A + 3 A into P.

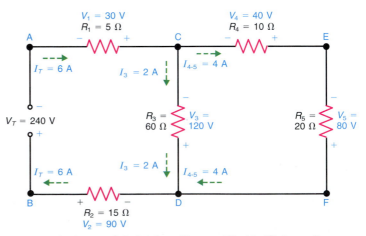

FIG. 9-2   Series-parallel circuit to illustrate Kirchhoff's laws. See text for voltage and current equations.

For the opposite directions, refer to point D at the bottom of Fig. 9-2. Here the branch currents into D combine to equal the main-line current I_T returning to the voltage source. Now I_T is directed out from D, with I_3 and I_{4-5} directed in. The algebraic equation is

$$-I_T + I_3 + I_{4-5} = 0$$
$$-6\,\text{A} + 2\,\text{A} + 4\,\text{A} = 0$$

THE $I_{\text{in}} = I_{\text{out}}$ Note that at either point C or point D in Fig. 9-2, the sum of the 2-A and 4-A branch currents must equal the 6-A total line current. Therefore, Kirchhoff's current law can also be stated as: $I_{\text{in}} = I_{\text{out}}$. For Fig. 9-2, the equations of current can be written:

At point C: $6\,\text{A} = 2\,\text{A} + 4\,\text{A}$
At point D: $2\,\text{A} + 4\,\text{A} = 6\,\text{A}$

Kirchhoff's current law is really the basis for the practical rule in parallel circuits that the total line current must equal the sum of the branch currents.

9-2 KIRCHHOFF'S VOLTAGE LAW (KVL)

The algebraic sum of the voltages around any closed path is zero. If you start from any point at one potential and come back to the same point and the same potential, the difference of potential must be zero.

ALGEBRAIC SIGNS In determining the algebraic signs for voltage terms in a KVL equation, first mark the polarity of each voltage as shown in Fig. 9-2. A convenient system then is: *Go around any closed path and consider any voltage whose negative terminal is reached first as a negative term and any voltage whose positive terminal is reached first as a positive term.* This method applies to *IR* voltage drops and voltage sources. The direction can be clockwise or counterclockwise.

Remember that electrons flowing into a resistor make that end negative with respect to the other end. For a voltage source, the direction of electrons returning to the positive terminal is the normal direction for electron flow, which means the source should be a positive term in the voltage equation.

When you go around the closed path and come back to the starting point, the algebraic sum of all the voltage terms must be zero. There cannot be any potential difference for one point.

If you do not come back to the start, then the algebraic sum is the voltage between the start and finish points.

You can follow any closed path. The reason is that the voltage between any two points in a circuit is the same regardless of the path used in determining the potential difference.

LOOP EQUATIONS   Any closed path is called a *loop*. A loop equation specifies the voltages around the loop.

Figure 9-2 has three loops. The outside loop, starting from point A at the top, through CEFDB, and back to A, includes the voltage drops V_1, V_4, V_5, and V_2 and the source V_T.

The inside loop ACDBA includes V_1, V_3, V_2, and V_T. The other inside loop, CEFDC with V_4, V_5, and V_3, does not include the voltage source.

Consider the voltage equation for the inside loop with V_T. In clockwise direction, starting from point A, the algebraic sum of the voltages is

$$-V_1 - V_3 - V_2 + V_T = 0$$

or

$$-30 \text{ V} - 120 \text{ V} - 90 \text{ V} + 240 \text{ V} = 0$$

Voltages V_1, V_3, and V_2 have the negative sign, because for each of these voltages the negative terminal is reached first. However, the source V_T is a positive term because its plus terminal is reached first, going in the same direction.

For the opposite direction, going counterclockwise in the same loop from point B at the bottom, V_2, V_3, and V_1 have positive values and V_T is negative. Then

$$V_2 + V_3 + V_1 - V_T = 0$$

or

$$90 \text{ V} + 120 \text{ V} + 30 \text{ V} - 240 \text{ V} = 0$$

When we transpose the negative term of -240 V, the equation becomes

$$90 \text{ V} + 120 \text{ V} + 30 \text{ V} = 240 \text{ V}$$

This equation states that the sum of the voltage drops equals the applied voltage.

$\Sigma V = V_T$ The Greek letter Σ means "sum of." In either direction, for any loop, the sum of the *IR* voltage drops must equal the applied voltage V_T. In Fig. 9-2, for the inside loop with the source V_T, going counterclockwise from point B:

$$90 \text{ V} + 120 \text{ V} + 30 \text{ V} = 240 \text{ V}$$

This system does not contradict the rule for algebraic signs. If 240 V were on the left side of the equation, this term would have a negative sign.

Stating a loop equation as $\Sigma V = V_T$ eliminates the step of transposing the negative terms from one side to the other to make them positive. In this form, the loop equations show that Kirchhoff's voltage law is really the basis for the practical rule in series circuits that the sum of the voltage drops must equal the applied voltage.

When a loop does not have any voltage source, the algebraic sum of the *IR* voltage drops alone must total zero. For instance, in Fig. 9-2, for the loop CEFDC without the source V_T, going clockwise from point C, the loop equation of voltages is

$$-V_4 - V_5 + V_3 = 0$$
$$-40 \ V - 80 \ V + 120 \ V = 0$$
$$0 = 0$$

Notice that V_3 is positive now, because its plus terminal is reached first by going clockwise from D to C in this loop.

TEST-POINT QUESTION 9-2

Answers at end of chapter.

Refer to Fig. 9-2.
a. For partial loop CEFD, what is the total voltage across CD with -40 V for V_4 and -80 V for V_5?
b. For loop CEFDC, what is the total voltage with -40 V for V_4, -80 V for V_5, and including 120 V for V_3?

9-3 METHOD OF BRANCH CURRENTS

Now we can use Kirchhoff's laws to analyze the circuit in Fig. 9-3. The problem is to find the currents and voltages for the three resistors.

First, indicate current directions and mark the voltage polarity across each resistor consistent with the assumed current. Remember that electron flow in a resistor produces negative polarity where the current enters. In Fig. 9-3, we

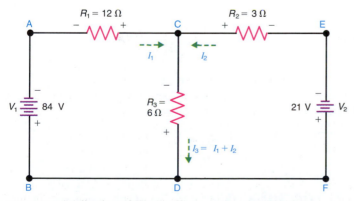

FIG. 9-3 Application of Kirchhoff's laws to a circuit with two sources in different branches. See text for solution by finding the branch currents.

assume that the source V_1 produces electron flow from left to right through R_1, while V_2 produces electron flow from right to left through R_2.

The three different currents in R_1, R_2, and R_3 are indicated as I_1, I_2, and I_3. However, three unknowns would require three equations for the solution. From Kirchhoff's current law, $I_3 = I_1 + I_2$, as the current out of point C must equal the current in. The current through R_3, therefore, can be specified as $I_1 + I_2$.

With two unknowns, two independent equations are needed to solve for I_1 and I_2. These equations are obtained by writing two Kirchhoff's voltage law equations around two loops. There are three loops in Fig. 9-3, the outside loop and two inside loops, but we need only two. The inside loops are used for the solution here.

WRITING THE LOOP EQUATIONS   For the loop with V_1, start at point B, at the bottom left, and go clockwise through V_1, V_{R_1}, and V_{R_3}. This equation for loop 1 is

$$84 - V_{R_1} - V_{R_3} = 0$$

For the loop with V_2, start at point F, at the lower right, and go counterclockwise through V_2, V_{R_2}, and V_{R_3}. This equation for loop 2 is

$$21 - V_{R_2} - V_{R_3} = 0$$

Using the known values of R_1, R_2, and R_3 to specify the IR voltage drops,

$$V_{R_1} = I_1 R_1 = I_1 \times 12 = 12 I_1$$
$$V_{R_2} = I_2 R_2 = I_2 \times 3 = 3 I_2$$
$$V_{R_3} = (I_1 + I_2) R_3 = 6(I_1 + I_2)$$

Substituting these values in the voltage equation for loop 1,

$$84 - 12 I_1 - 6(I_1 + I_2) = 0$$

Also, in loop 2,

$$21 - 3 I_2 - 6(I_1 + I_2) = 0$$

Multiplying $(I_1 + I_2)$ by 6 and combining terms and transposing, the two equations are

$$-18 I_1 - 6 I_2 = -84$$
$$-6 I_1 - 9 I_2 = -21$$

Divide the top equation by -6 and the bottom equation by -3 to reduce the equations to their simplest terms and to have all positive terms. The two equations in their simplest form then become

$$3 I_1 + \ \ I_2 = 14$$
$$2 I_1 + 3 I_2 = 7$$

SOLVING FOR THE CURRENTS   These two equations in the two unknowns I_1 and I_2 contain the solution of the network. It should be noted that the

DID YOU KNOW?

Like gasoline-powered cars, electric cars must have good pickup to be able to merge into traffic. Military ultracapacitors are being remade for this essential driving maneuver, to outfit cars with peak-power storage units.

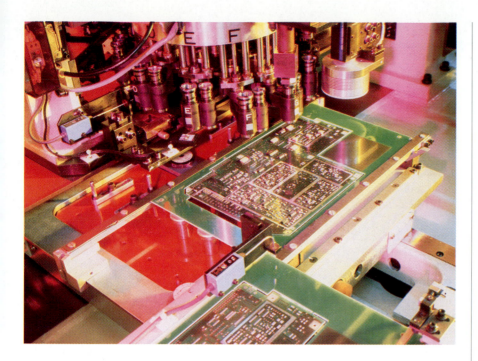

equations include every resistance in the circuit. Currents I_1 and I_2 can be calculated by any of the methods for the solution of simultaneous equations. Using the method of elimination, multiply the top equation by 3 to make the I_2 terms the same in both equations. Then

$$9I_1 + 3I_2 = 42$$
$$2I_1 + 3I_2 = 7$$

Subtract the bottom equation from the top equation, term by term, to eliminate I_2. Then, since the I_2 term becomes zero,

$$7I_1 = 35$$
$$I_1 = 5 \text{ A}$$

The 5-A I_1 is the current through R_1. Its direction is from A to C, as assumed, because the answer for I_1 is positive.

To calculate I_2, substitute 5 for I_1 in either of the two loop equations. Using the bottom equation for the substitution,

$$2(5) + 3I_2 = 7$$
$$3I_2 = 7 - 10$$
$$3I_2 = -3$$
$$I_2 = -1 \text{ A}$$

The negative sign for I_2 means this current is opposite to the assumed direction. Therefore, I_2 flows through R_2 from C to E instead of from E to C as was previously assumed.

WHY THE SOLUTION FOR I_2 IS NEGATIVE   In Fig. 9-3 (page 237), I_2 was assumed to flow from E to C through R_2 because V_2 produces electron flow in this direction. However, the other voltage source V_1 produces electron flow through R_2 in the opposite direction, from point C to E. This solution of -1 A for I_2 shows that the current through R_2 produced by V_1 is more than the current produced by V_2. The net result is 1 A through R_2 from C to E.

The actual direction of I_2 is shown in Fig. 9-4 with all the values for the solution of this circuit. Notice that the polarity of V_{R_2} is reversed from the assumed polarity in Fig. 9-3. Since the net electron flow through R_2 is actually from C to E, the end of R_2 at C is the negative end. However, the polarity of V_2 is the same in both diagrams because it is a voltage source, which generates its own polarity.

To calculate I_3 through R_3,

$$I_3 = I_1 + I_2 = 5 + (-1)$$
$$I_3 = 4 \text{ A}$$

The 4 A for I_3 is in the assumed direction from C to D. Although the negative sign for I_2 only means a reversed direction, its algebraic value of -1 must be used for substitution in the algebraic equations written for the assumed direction.

CALCULATING THE VOLTAGES   With all the currents known, the voltage across each resistor can be calculated as follows:

$$V_{R_1} = I_1 R_1 = 5 \times 12 = 60 \text{ V}$$
$$V_{R_2} = I_2 R_2 = 1 \times\ \ 3 =\ \ 3 \text{ V}$$
$$V_{R_3} = I_3 R_3 = 4 \times\ \ 6 = 24 \text{ V}$$

All the currents are taken as positive, in the correct direction, to calculate the voltages. Then the polarity of each IR drop is determined from the actual direction of current, with electron flow into the negative end (see Fig. 9-4). Notice that V_{R_3} and V_{R_2} have opposing polarities in loop 2. Then the sum of $+3$ V and -24 V equals the -21 V of V_2.

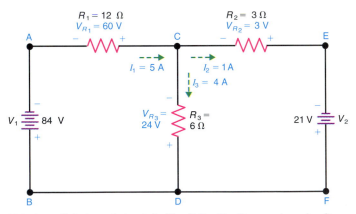

FIG. 9-4   Solution of circuit in Fig. 9-3 with all currents and voltages.

CHECKING THE SOLUTION   As a summary of all the answers for this problem, Fig. 9-4 shows the network with all the currents and voltages. The po-

larity of each V is marked from the known directions. In checking the answers, we can see whether Kirchhoff's current and voltage laws are satisfied:

At point C: $5 \text{ A} = 4 \text{ A} + 1 \text{ A}$
At point D: $4 \text{ A} + 1 \text{ A} = 5 \text{ A}$

Around the loop with V_1 clockwise from B:

$$84 \text{ V} - 60 \text{ V} - 24 \text{ V} = 0$$

Around the loop with V_2 counterclockwise from F:

$$21 \text{ V} + 3 \text{ V} - 24 \text{ V} = 0$$

It should be noted that the circuit has been solved using only the two Kirchhoff laws, without any of the special rules for series and parallel circuits. Any circuit can be solved just by applying Kirchhoff's laws for the voltages around a loop and the currents at a branch point.

TEST-POINT QUESTION 9-3

Answers at end of chapter.

Refer to Fig. 9-4.
a. How much is the voltage around partial loop CEFD?
b. How much is the voltage around loop CEFDC?

9-4 NODE-VOLTAGE ANALYSIS

In the method of branch currents, these currents are used for specifying the voltage drops around the loops. Then loop equations are written to satisfy Kirchhoff's voltage law. Solving the loop equations, we can calculate the unknown branch currents.

Another method uses the voltage drops to specify the currents at a branch point, also called a *node*. Then node equations of currents are written to satisfy Kirchhoff's current law. Solving the node equations, we can calculate the unknown node voltages. This method of node-voltage analysis often is shorter than the method of branch currents.

A node is simply a common connection for two or more components. A *principal node* has three or more connections. In effect, a principal node is just a junction or branch point, where currents can divide or combine. Therefore, we can always write an equation of currents at a principal node. In Fig. 9-5, points N and G are principal nodes.

However, one node must be the reference for specifying the voltage at any other node. In Fig. 9-5, point G connected to chassis ground is the reference node. Therefore, we need only write one current equation for the other node N. In general, the number of current equations required to solve a circuit is one less than the number of principal nodes.

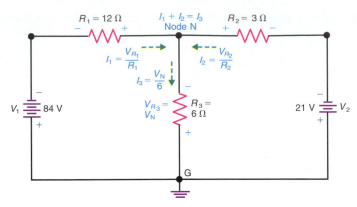

FIG. 9-5 Method of node-voltage analysis for the same circuit as in Fig. 9-3. See text for solution by finding V_N across R_3 from the principal node N to ground.

WRITING THE NODE EQUATIONS

The circuit of Fig. 9-3, earlier solved by the method of branch currents, is redrawn in Fig. 9-5 to be solved now by node-voltage analysis. The problem here is to find the node voltage V_N from N to G. Once this voltage is known, all the other voltages and currents can be determined.

The currents in and out of node N are specified as follows: I_1 is the only current through the 12-Ω R_1. Therefore, I_1 is V_{R_1}/R_1 or $V_{R_1}/12\ \Omega$. Similarly, I_2 is $V_{R_2}/3\ \Omega$. Finally, I_3 is $V_{R_3}/6\ \Omega$.

Note that V_{R_3} is the node voltage V_N that we are to calculate. Therefore, I_3 can also be stated as $V_N/6\ \Omega$. The equation of currents at node N is

$$I_1 + I_2 = I_3$$

or

$$\frac{V_{R_1}}{12} + \frac{V_{R_2}}{3} = \frac{V_N}{6}$$

There are three unknowns here, but V_{R_1} and V_{R_2} can be specified in terms of V_N and the known values of V_1 and V_2. We can use Kirchhoff's voltage law, because the applied voltage V must equal the algebraic sum of the voltage drops. For the loop with V_1 of 84 V,

$$V_{R_1} + V_N = 84 \qquad \text{or} \qquad V_{R_1} = 84 - V_N$$

For the loop with V_2 of 21 V,

$$V_{R_2} + V_N = 21 \qquad \text{or} \qquad V_{R_2} = 21 - V_N$$

Now substitute these values of V_{R_1} and V_{R_2} in the equation of currents:

$$I_1 + I_2 = I_3$$
$$\frac{V_{R_1}}{R_1} + \frac{V_{R_2}}{R_2} = \frac{V_{R_3}}{R_3}$$

Using the value of each V in terms of V_N,

$$\frac{84 - V_N}{12} + \frac{21 - V_N}{3} = \frac{V_N}{6}$$

This equation has only the one unknown, V_N. Clearing fractions by multiplying each term by 12, the equation is

$$(84 - V_N) + 4(21 - V_N) = 2\,V_N$$
$$84 - V_N + 84 - 4\,V_N = 2\,V_N$$
$$-7\,V_N = -168$$
$$V_N = 24\ V$$

This answer of 24 V for V_N is the same as that calculated for V_{R_3} by the method of branch currents. The positive value means the direction of I_3 is correct, making V_N negative at the top of R_3 in Fig. 9-5.

CALCULATING ALL VOLTAGES AND CURRENTS   The reason for finding the voltage at a node, rather than some other voltage, is the fact that a node voltage must be common to two loops. As a result, the node voltage can be used for calculating all the voltages in the loops. In Fig. 9-5, with a V_N of 24 V, then V_{R_1} must be $84 - 24 = 60$ V. Also, I_1 is 60 V/12 Ω, which equals 5 A.

To find V_{R_2}, it must be $21 - 24$, which equals -3 V. The negative answer means that I_2 is opposite to the assumed direction and the polarity of V_{R_2} is the reverse of the signs shown across R_2 in Fig. 9-5. The correct directions are shown in the solution for the circuit in Fig. 9-4. The magnitude of I_2 is 3 V/3 Ω, which equals 1 A.

The following comparisons can be helpful in using node equations and loop equations. A node equation applies Kirchhoff's current law to the currents in and out of a node point. However, the currents are specified as V/R so that the equation of currents can be solved to find a node voltage.

A loop equation applies Kirchhoff's voltage law to the voltages around a closed path. However, the voltages are specified as IR so that the equation of voltages can be solved to find a loop current. This procedure with voltage equations is used for the method of branch currents explained before with Fig. 9-3 and for the method of mesh currents to be described next with Fig. 9-6.

TEST-POINT QUESTION 9-4

Answers at end of chapter.
a. How many principal nodes does Fig. 9-5 have?
b. How many node equations are necessary to solve a circuit with three principal nodes?

A mesh is the simplest possible closed path. The circuit in Fig. 9-6 has the two meshes ACDBA and CEFDC. The outside path ACEFDBA is a loop but not a mesh. Each mesh is like a single window frame. There is only one path without any branches.

A mesh current is assumed to flow around a mesh without dividing. In Fig. 9-6, the mesh current I_A flows through V_1, R_1, and R_3; mesh current I_B flows through V_2, R_2, and R_3. A resistance common to two meshes, such as R_3, has two mesh currents, which are I_A and I_B here.

The fact that a mesh current does not divide at a branch point is the difference between mesh currents and branch currents. A mesh current is an assumed current, while a branch current is the actual current. However, when the mesh currents are known, all the individual currents and voltages can be determined.

As an example, Fig. 9-6, which has the same circuit as Fig. 9-3, will now be solved by using the assumed mesh currents I_A and I_B. The mesh equations are

$$18I_A - 6I_B = 84 \text{ V} \qquad \text{in mesh A}$$
$$-6I_A + 9I_B = -21 \text{ V} \qquad \text{in mesh B}$$

WRITING THE MESH EQUATIONS The number of meshes equals the number of mesh currents, which is the number of equations required. Here two equations are used for I_A and I_B in the two meshes.

The assumed current is usually taken in the same direction around each mesh, in order to be consistent. Generally, the clockwise direction is used, as shown for I_A and I_B in Fig. 9-6.

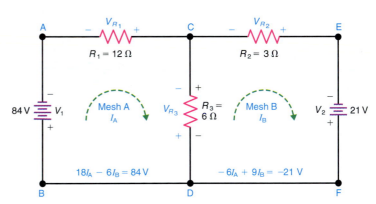

FIG. 9-6 The same circuit as Fig. 9-3 analyzed as two meshes. See text for solution by calculating the assumed mesh currents I_A and I_B.

In each mesh equation, the algebraic sum of the voltage drops equals the applied voltage.

The voltage drops are added going around a mesh in the same direction as its mesh current. Any voltage drop in a mesh produced by its own mesh current is considered positive because it is added in the direction of the mesh current.

Since all the voltage drops of a mesh current in its own mesh must have the same positive sign, they can be written collectively as one voltage drop by adding all the resistances in the mesh. For instance, in the first equation, for mesh A, the total resistance equals $12 + 6$, or $18\ \Omega$. Therefore, the voltage drop for I_A is $18I_A$ in mesh A.

In the second equation, for mesh B, the total resistance is $3 + 6$, or $9\ \Omega$, making the total voltage drop $9I_B$ for I_B in mesh B. You can add all the resistances in a mesh for one R_T, because they can be considered in series for the assumed mesh current.

Any resistance common to two meshes has two opposite mesh currents. In Fig. 9-6, I_A flows down while I_B is up through the common R_3, with both currents clockwise. As a result, a common resistance has two opposing voltage drops. One voltage is positive for the current of the mesh whose equation is being written. The opposing voltage is negative for the current of the adjacent mesh.

In mesh A, the common 6-Ω R_3 has the opposing voltages $6I_A$ and $-6I_B$. The $6I_A$ of R_3 adds to the $12I_A$ of R_1 for the total positive voltage drop of $18I_A$ in mesh A. With the opposing voltage of $-6I_B$, then the equation for mesh A is $18I_A - 6I_B = 84\ \text{V}$.

The same idea applies to mesh B. However, now the voltage $6I_B$ is positive because the equation is for mesh B. The $-6I_A$ voltage is negative here because I_A is for the adjacent mesh. The $6I_B$ adds to the $3I_B$ of R_2 for the total positive voltage drop of $9I_B$ in mesh B. With the opposing voltage of $-6I_A$, the equation for mesh B then is $-6I_A + 9I_B = -21\ \text{V}$.

The algebraic sign of the source voltage in a mesh depends on its polarity. When the assumed mesh current flows into the positive terminal, as for V_1 in Fig. 9-6, it is considered positive for the right-hand side of the mesh equation. This direction of electron flow produces voltage drops that must add to equal the applied voltage.

With the mesh current into the negative terminal, as for V_2 in Fig. 9-6, it is considered negative. This is why V_2 is $-21\ V$ in the equation for mesh B. Then V_2 is actually a load for the larger applied voltage of V_1, instead of V_2 being the source. When a mesh has no source voltage, the algebraic sum of the voltage drops must equal zero.

These rules for the voltage source mean that the direction of electron flow is assumed for the mesh currents. Then electron flow is used to determine the polarity of the voltage drops. Note that considering the voltage source as a positive value with electron flow into the positive terminal corresponds to the normal flow of electron charges. If the solution for a mesh current comes out negative, the actual current for the mesh must be in the opposite direction from the assumed current flow.

SOLVING THE MESH EQUATIONS TO FIND THE MESH CURRENTS

The two equations for the two meshes in Fig. 9-6 are

$$18I_A - 6I_B = 84$$
$$-6I_A + 9I_B = -21$$

These equations have the same coefficients as in the voltage equations written for the branch currents, but the signs are different. The reason is that the directions of the assumed mesh currents are not the same as those of the branch currents.

The solution will give the same answers for either method, but you must be consistent in algebraic signs. Use either the rules for meshes with mesh currents or the rules for loops with branch currents, but do not mix the two methods.

To eliminate I_B and solve for I_A, divide the first equation by 2 and the second equation by 3. Then

$$9I_A - 3I_B = 42$$
$$-2I_A + 3I_B = -7$$

Add the equations, term by term, to eliminate I_B. Then

$$7I_A = 35$$
$$I_A = 5 \text{ A}$$

To calculate I_B, substitute 5 for I_A in the second equation:

$$-2(5) + 3I_B = -7$$
$$3I_B = -7 + 10 = 3$$
$$I_B = 1 \text{ A}$$

The positive solutions mean that the electron flow for both I_A and I_B is actually clockwise, as assumed.

FINDING THE BRANCH CURRENTS AND VOLTAGE DROPS   Referring to Fig. 9-6, the 5-A I_A is the only current through R_1. Therefore, I_A and I_1 are the same. Then V_{R_1} across the 12-Ω R_1 is 5×12, or 60 V. The polarity of V_{R_1} is marked negative at the left, with the electron flow into this side.

Similarly, the 1-A I_B is the only current through R_2. The direction of this electron flow through R_2 is from left to right. Note that this value of 1 A for I_B clockwise is the same as -1 A for I_2, assumed in the opposite direction in Fig. 9-3. Then V_{R_2} across the 3-Ω R_2 is 1×3 or 3 V, with the left side negative.

The current I_3 through R_3, common to both meshes, consists of I_A and I_B. Then I_3 is $5 - 1$ or 4 A. The currents are subtracted because I_A and I_B are in opposing directions through R_3. When all the mesh currents are taken one way, they will always be in opposite directions through any resistance common to two meshes.

The direction of the net 4-A I_3 through R_3 is downward, the same as I_A, because it is more than I_B. Then, V_{R_3} across the 6-Ω R_3 is $4 \times 6 = 24$ V, with the top negative.

THE SET OF MESH EQUATIONS   The system for algebraic signs of the voltages in the mesh equations is different from the method used with branch currents, but the end result is the same. The advantage of mesh currents is the pattern of algebraic signs for the voltages, without the need for tracing any branch currents. This feature is especially helpful in a more elaborate circuit, such as the one in Fig. 9-7 that has three meshes. We can use Fig. 9-7 for more practice in writing mesh equations, without doing the numerical work of solving a set* of three equations. Each R is 2 Ω.

In Fig. 9-7, the mesh currents are shown with solid arrows to indicate conventional current, which is a common way of analyzing these circuits. Also, the volt-

---

*A set with any number of simultaneous linear equations, for any number of meshes, can be solved by determinants. This procedure is explained in B. Grob, *Mathematics for Basic Electronics*, Glencoe/McGraw-Hill, Columbus, Ohio.

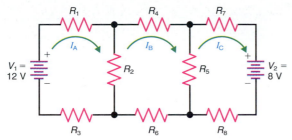

FIG. 9-7   A circuit with three meshes. Each R is 2 Ω.
See text for mesh equations.

age sources V_1 and V_2 have the positive terminal at the top in the diagram. When the direction of conventional current is used, it is important to note that the voltage source is a positive value with mesh current into the negative terminal. This method corresponds to the normal flow of positive charges with conventional current.

For the three mesh equations in Fig. 9-7:

In mesh A: $6I_A - 2I_B + 0 = 12$
In mesh B: $-2I_A + 8I_B - 2I_C = 0$
In mesh C: $0 - 2I_B + 6I_C = -8$

The zero term in equations A and C represents a missing mesh current. Only mesh B has all three mesh currents. However, note that mesh B has a zero term for the voltage source because it is the only mesh with only IR drops.

In summary, the only positive IR voltage in a mesh is for the R_T of each mesh current in its own mesh. All other voltage drops for any adjacent mesh current across a common resistance are always negative. This procedure for assigning algebraic signs to the voltage drops is the same whether the source voltage in the mesh is positive or negative. It also applies even if there is no voltage source in the mesh.

TEST-POINT QUESTION 9-5

Answers at end of chapter.

Answer True or False.
a. A network with four mesh currents needs four mesh equations for a solution.
b. An R common to two meshes has opposing mesh currents.

9   SUMMARY AND REVIEW

- Kirchhoff's voltage law states that the algebraic sum of all voltages around any closed path must equal zero. Stated another way, the sum of the voltage drops equals the applied voltage.
- Kirchhoff's current law states that the algebraic sum of all currents directed in and out at any point in a closed path must equal zero. Stated another way, the current into a point equals the current out of that point.
- A closed path is a loop. The method of using algebraic equations for the voltages around the loops to calculate the branch currents is illustrated in Fig. 9-3.
- A principal node is a branch point where currents divide or combine. The method of using algebraic equations for the currents at a node to calculate each node voltage is illustrated in Fig. 9-5.
- A mesh is the simplest possible loop. A mesh current is assumed to flow around the mesh without branching. The method of using algebraic equations for the voltages around the meshes to calculate the mesh currents is illustrated in Fig. 9-6.

SELF-TEST

ANSWERS AT BACK OF BOOK.

Answer True or False.

1. The algebraic sum of all voltages around any mesh or any loop must equal zero.
2. A mesh with two resistors has two mesh currents.
3. With $I_1 = 3$ A and $I_2 = 2$ A directed into a node, the current I_3 directed out must equal 5 A.
4. In a loop without any voltage source, the algebraic sum of the voltage drops must equal zero.
5. The algebraic sum of $+40$ V and -10 V equals $+30$ V.
6. A principal node is a junction where branch currents can divide or combine.
7. In the node-voltage method, the number of equations of current equals the number of principal nodes.
8. In the mesh-current method, the number of equations of voltage equals the number of meshes.
9. When all mesh currents are clockwise or all counterclockwise, any resistor common to two meshes has two currents in opposite directions.
10. The rules of series voltages and parallel currents are based on Kirchhoff's laws.

1. State Kirchhoff's current law in two ways.
2. State Kirchhoff's voltage law in two ways.
3. What is the difference between a loop and a mesh?
4. What is the difference between a branch current and a mesh current?
5. Define a principal node.
6. Define a node voltage.
7. Use the values in Fig. 9-4 to show that the algebraic sum is zero for all voltages around the outside loop ACEFDBA.
8. Use the values in Fig. 9-4 to show that the algebraic sum is zero for all the currents into and out of node C and node D.

PROBLEMS

ANSWERS TO ODD-NUMBERED PROBLEMS AT BACK OF BOOK.

1. In Fig. 9-8, what is the value of: (a) I_3; (b) I_5; (c) I_6?
2. In Fig. 9-8, use KCL to write an equation for the currents entering and leaving: (a) point A; (b) point B; (c) point C.

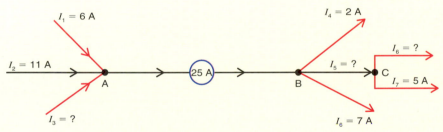

FIG. 9-8 Circuit diagram for Probs. 1 and 2.

3. In Fig. 9-9, calculate: (a) V_T, R_T, I, V_{R_1}, V_{R_2}, and V_{R_3}; (b) V_{AG}, V_{BG}, V_{CG}, and V_{DG}. Use KVL.
4. In Fig. 9-9, assume the polarity of V_2 is reversed. Calculate: (a) V_T, R_T, I, V_{R_1}, V_{R_2}, and V_{R_3}; (b) V_{AG}, V_{BG}, V_{CG}, and V_{DG}.
5. In Fig. 9-10, (a) calculate V_T, R_T, I, V_{R_1}, V_{R_2}, and V_{R_3}. (b) What voltage will be indicated by the voltmeter with the wiper arm of R_2 set to: point A, point B, and midway between points A and B? (Note: R_2 has a linear taper.)
6. In Fig. 9-10, assume R_2 is changed to a 10-kΩ potentiometer. (a) Calculate V_T, R_T, I, V_{R_1}, V_{R_2}, and V_{R_3}. (b) What voltage will be indicated by the voltmeter with the wiper arm of R_2 set to: point A, point B, and midway between points A and B?

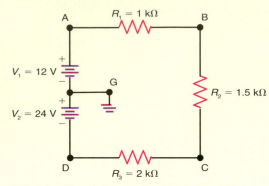

FIG. 9-9 Circuit diagram for Probs. 3 and 4.

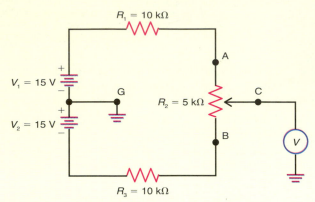

FIG. 9-10 Circuit diagram for Probs. 5 and 6.

7. In Fig. 9-11, use the method of branch currents to solve for I_1, I_2, and I_3. Once the currents have been calculated, determine the values for V_{R_1}, V_{R_2}, and V_{R_3}.

8. In Fig. 9-11, use KVL to prove that the algebraic sum of the voltages is zero in all three closed loops.

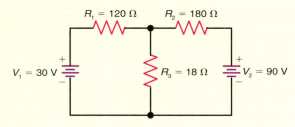

FIG. 9-11 Circuit diagram for Probs. 7, 8, and 9.

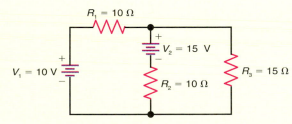

FIG. 9-12 Circuit diagram for Probs. 10, 11, and 12.

9. In Fig. 9-11, use node-voltage analysis to solve for V_{R_1}, V_{R_2}, V_{R_3}, I_1, I_2, and I_3.

10. In Fig. 9-12, use the method of branch currents to solve for I_1, I_2, I_3, V_{R_1}, V_{R_2}, and V_{R_3}.

11. In Fig. 9-12, use KVL to prove that the algebraic sum of the voltages is zero in all three closed loops.

12. In Fig. 9-12, use the method of mesh currents to solve for I_1, I_2, I_3, V_{R_1}, V_{R_2}, and V_{R_3}.

13. In Fig. 9-13, use the method of mesh currents to solve for I_1, I_2, I_3, V_{R_1}, V_{R_2}, and V_{R_3}.

14. In Fig. 9-13, use KVL to prove that the algebraic sum of the voltages is zero in all three closed loops.

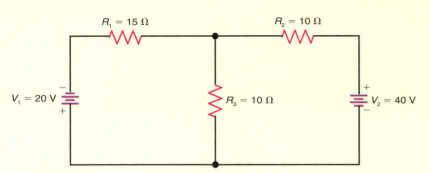

FIG. 9-13 Circuit diagram for Probs. 13 and 14.

CRITICAL THINKING

1. In Fig. 9-14, determine the values for R_1 and R_3 which will allow the output voltage to vary between -5 V and $+5$ V.

2. Refer to Fig. 9-7. If all resistances are 10 Ω, calculate: **(a)** I_A, I_B, and I_C; **(b)** I_1, I_2, I_3, I_4, I_5, I_6, I_7, and I_8.

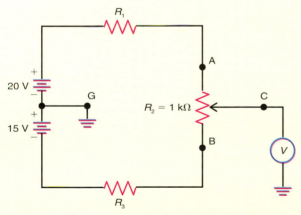

FIG. 9-14 Circuit diagram for Critical Thinking Prob. 1.

ANSWERS TO TEST-POINT QUESTIONS

9-1	**a.** 6 A	9-3	**a.** -24 V	9-5	**a.** T
	b. 4 A		**b.** 0 V		**b.** T
9-2	**a.** -120 V	9-4	**a.** two		
	b. 0 V		**b.** two		

CHAPTER 10

NETWORK THEOREMS

A network is just a combination of components, such as resistances, inter-connected in a way as to achieve a particular end result. However, networks generally need more than the rules of series and parallel circuits for analysis. Kirchhoff's laws can always be applied for any circuit connections. The net-work theorems, though, usually provide shorter methods of solving the circuit.

Some theorems enable us to convert the network into a simpler circuit, equivalent to the original. Then the equivalent circuit can be solved by the rules of series and parallel cir-cuits. Other theorems enable us to convert a given circuit into a form that permits easier solutions.

Only the applications are given here, although all the net-work theorems can be derived from Kirchhoff's laws. It should also be noted that resistance networks with batteries are shown as examples, but the theorems can also be applied to ac net-works.

CHAPTER OBJECTIVES

Upon completion of this chapter, you should be able to:

- *Apply* the superposition theorem to find the voltage across two points in a circuit containing more than one voltage source.
- *State* the requirements for applying the superposition theorem.
- *Determine* the Thevenin and Norton equivalent circuits with respect to any pair of terminals in a complex network.
- *Apply* Thevenin's and Norton's theorems in solving for an unknown voltage or current.
- *Convert* a Thevenin equivalent circuit to a Norton equivalent circuit and vice versa.
- *Apply* Millman's theorem to find a common voltage across any number of parallel branches.
- *Simplify* the analysis of a bridge circuit by using delta to wye conversion formulas.

IMPORTANT TERMS IN THIS CHAPTER

active components
bilateral components
current source
delta (Δ) network
Δ-Y transformations
equivalent circuit
equivalent current source

linear components
Millman's theorem
nortonizing a circuit
Norton's theorem
passive components
pi (π) network
superposition

tee (T) network
Thevenin-Norton conversions
Thevenin's theorem
thevenizing a circuit
voltage source
wye (Y) network

TOPICS COVERED IN THIS CHAPTER

10-1 SUPERPOSITION

The superposition theorem is very useful because it extends the use of Ohm's law to circuits that have more than one source. In brief, we can calculate the effect of one source at a time and then superimpose the results of all the sources. As a definition, the superposition theorem states that: *In a network with two or more sources, the current or voltage for any component is the algebraic sum of the effects produced by each source acting separately.*

In order to use one source at a time, all other sources are "killed" temporarily. This means disabling the source so that it cannot generate voltage or current, without changing the resistance of the circuit. A voltage source such as a battery is killed by assuming a short circuit across its potential difference. The internal resistance remains.

VOLTAGE DIVIDER WITH TWO SOURCES   The problem in Fig. 10-1 is to find the voltage at P to chassis ground for the circuit in Fig. 10-1a. The method is to calculate the voltage at P contributed by each source separately, as in Fig. 10-1b and c, and then superimpose these voltages.

To find the effect of V_1 first, short-circuit V_2 as shown in Fig. 10-1b. Note that the bottom of R_1 then becomes connected to chassis ground because of the short circuit across V_2. As a result, R_2 and R_1 form a series voltage divider for the V_1 source.

Furthermore, the voltage across R_1 becomes the same as the voltage from P to ground. To find this V_{R_1} across R_1 as the contribution of the V_1 source, we use the voltage divider formula:

$$V_{R_1} = \frac{R_1}{R_1 + R_2} \times V_1 = \frac{60 \text{ k}\Omega}{30 \text{ k}\Omega + 60 \text{ k}\Omega} \times 24 \text{ V}$$

$$= \frac{60}{90} \times 24 \text{ V}$$

$$V_{R_1} = 16 \text{ V}$$

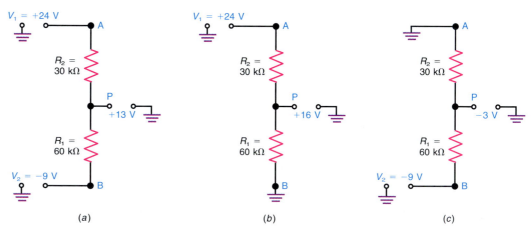

FIG. 10-1   Superposition theorem applied to a voltage divider with two sources V_1 and V_2. (*a*) Actual circuit with +13 V from point P to chassis ground. (*b*) V_1 alone producing +16 V at P. (*c*) V_2 alone producing −3 V at P.

Next find the effect of V_2 alone, with V_1 short-circuited as shown in Fig. 10-1c. Then point A at the top of R_2 becomes grounded. R_1 and R_2 form a series voltage divider again, but here the R_2 voltage is the voltage at P to ground.

With one side of R_2 grounded and the other side to point P, V_{R_2} is the voltage to calculate. Again we have a series divider, but this time for the negative voltage V_2. Using the voltage divider formula for V_{R_2} as the contribution of V_2 to the voltage at P,

$$V_{R_2} = \frac{R_2}{R_1 + R_2} \times V_2 = \frac{30 \text{ k}\Omega}{30 \text{ k}\Omega + 60 \text{ k}\Omega} \times -9 \text{ V}$$

$$= \frac{30}{90} \times -9 \text{ V}$$

$$V_{R_2} = -3 \text{ V}$$

This voltage is negative at P because V_2 is negative.

Finally, the total voltage at P is

$$V_P = V_1 + V_2 = 16 - 3$$

$$V_P = 13 \text{ V}$$

This algebraic sum is positive for the net V_P because the positive V_1 is larger than the negative V_2.

By means of superposition, therefore, this problem was reduced to two series voltage dividers. The same procedure can be used with more than two sources. Also, each voltage divider can have any number of series resistances. Note that in this case we were dealing with ideal voltage sources, that is, sources with zero internal resistance. If the source did have internal resistance, it would have been added in series with R_1 and R_2.

REQUIREMENTS FOR SUPERPOSITION   All the components must be linear and bilateral in order to superimpose currents and voltages. *Linear* means that the current is proportional to the applied voltage. Then the currents calculated for different source voltages can be superimposed.

Bilateral means that the current is the same amount for opposite polarities of the source voltage. Then the values for opposite directions of current can be combined algebraically. Networks with resistors, capacitors, and air-core inductors are generally linear and bilateral. These are also *passive components,* meaning they do not amplify or rectify. *Active components,* such as transistors, semiconductor diodes, and electron tubes, are never bilateral and often are not linear.

Answers at end of chapter.
a. In Fig. 10-1b, which R is shown grounded at one end?
b. In Fig. 10-1c, which R is shown grounded at one end?

10-2 THEVENIN'S THEOREM

Named after M. L. Thevenin, a French engineer, Thevenin's theorem is very useful in simplifying the voltages in a network. By Thevenin's theorem, many sources and components, no matter how they are interconnected, can be represented by an equivalent series circuit with respect to any pair of terminals in the network. In Fig. 10-2 imagine that the block at the left contains a network connected to terminals A and B. Thevenin's theorem states that *the entire* network connected to A and B can be replaced by a single voltage source V_{TH} in series with a single resistance R_{TH}, connected to the same two terminals.

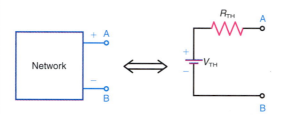

FIG. 10-2 Any network in the block at the left can be reduced to the Thevenin equivalent series circuit at the right.

Voltage V_{TH} is the open-circuit voltage across terminals A and B. This means, find the voltage that the network produces across the two terminals with an open circuit between A and B. The polarity of V_{TH} is such that it will produce current from A to B in the same direction as in the original network.

Resistance R_{TH} is the open-circuit resistance across terminals A and B, but with all the sources killed. This means, find the resistance looking back into the network from terminals A and B. Although the terminals are open, an ohmmeter across AB would read the value of R_{TH} as the resistance of the remaining paths in the network, without any sources operating.

THEVENIZING A CIRCUIT   As an example, refer to Fig. 10-3*a*, where we want to find the voltage V_L across the 2-Ω R_L and its current I_L. To use Thevenin's theorem, mentally disconnect R_L. The two open ends then become terminals A and B. Now we find the Thevenin equivalent of the remainder of the circuit that is still connected to A and B. In general, open the part of the circuit to be analyzed and "thevenize" the remainder of the circuit connected to the two open terminals.

Our only problem now is to find the value of the open-circuit voltage V_{TH} across AB and the equivalent resistance R_{TH}. The Thevenin equivalent always consists of a single voltage source in series with a single resistance, as in Fig. 10-3*d*.

The effect of opening R_L is shown in Fig. 10-3*b*. As a result, the 3-Ω R_1 and 6-Ω R_2 form a series voltage divider, without R_L.

Furthermore, the voltage across R_2 now is the same as the open-circuit voltage across terminals A and B. Therefore V_{R_2} with R_L open is V_{AB}. This is the V_{TH} we need for the Thevenin equivalent circuit. Using the voltage divider formula,

$$V_{R_2} = \frac{6}{9} \times 36 \text{ V} = 24 \text{ V}$$

$$V_{R_2} = V_{AB} = V_{TH} = 24 \text{ V}$$

This voltage is positive at terminal A.

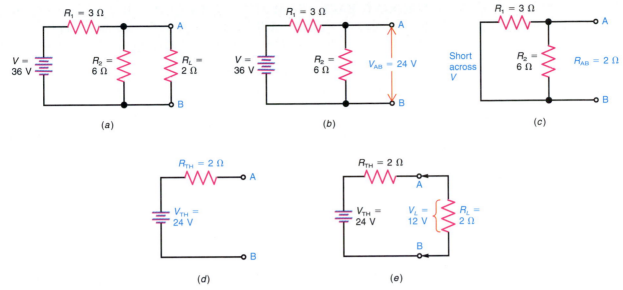

FIG. 10-3 Application of Thevenin's theorem. (*a*) Actual circuit with terminals A and B across R_L. (*b*) Disconnect R_L to find that V_{AB} is 24 V. (*c*) Short-circuit V to find that R_{AB} is 2 Ω. (*d*) Thevenin equivalent circuit. (*e*) Reconnect R_L at terminals A and B to find that V_L is 12 V.

To find R_{TH}, the 2-Ω R_L is still disconnected. However, now the source V is short-circuited. So the circuit looks like Fig. 10-3*c*. The 3-Ω R_1 is now in parallel with the 6-Ω R_2, as they are both connected across the same two points. This combined resistance is

$$R_{TH} = \frac{18}{9} = 2\ \Omega$$

Again, we assume an ideal voltage source whose internal resistance is zero.

As shown in Fig. 10-3*d*, the Thevenin circuit to the left of terminals A and B then consists of the equivalent voltage V_{TH}, equal to 24 V, in series with the equivalent series resistance R_{TH}, equal to 2 Ω. This Thevenin equivalent applies for any value of R_L because R_L was disconnected. We are actually thevenizing the circuit that feeds the open AB terminals.

To find V_L and I_L, we can finally reconnect R_L to terminals A and B of the Thevenin equivalent circuit, as shown in Fig. 10-3*e*. Then R_L is in series with R_{TH} and V_{TH}. Using the voltage divider formula for the 2-Ω R_{TH} and 2-Ω R_L, $V_L = 1/2 \times 24\ \text{V} = 12\ \text{V}$. To find I_L as V_L/R_L, the value is 12 V/2 Ω, which equals 6 A.

These answers of 6 A for I_L and 12 V for V_L apply to R_L in both the original circuit in Fig. 10-3*a* and the equivalent circuit in Fig. 10-3*e*. Note that the 6-A I_L also flows through R_{TH}.

The same answers could be obtained by solving the series-parallel circuit in Fig. 10-3*a*, using Ohm's law. However, the advantage of thevenizing the circuit is that the effect of different values of R_L can be calculated easily. Suppose that R_L were changed to 4 Ω. In the Thevenin circuit, the new value of V_L would be $4/6 \times 24\ \text{V} = 16\ \text{V}$. The new I_L would be 16 V/4 Ω, which equals 4 A. If we used Ohm's law in the original circuit, a complete new solution would be required each time R_L was changed.

LOOKING BACK FROM TERMINALS A AND B   The way we look at the resistance of a series-parallel circuit depends on where the source is connected. In general, we calculate the total resistance from the outside terminals of the circuit in toward the source, as the reference.

When the source is short-circuited for thevenizing a circuit, terminals A and B become the reference. Looking back from A and B to calculate R_{TH}, the situation becomes reversed from the way the circuit was viewed to determine V_{TH}.

For R_{TH}, imagine that a source could be connected across AB, and calculate the total resistance working from the outside in toward terminals A and B. Actually an ohmmeter placed across terminals A and B would read this resistance.

This idea of reversing the reference is illustrated in Fig. 10-4. The circuit in Fig. 10-4a has terminals A and B open, ready to be thevenized. This circuit is similar to Fig. 10-3 but with the 4-Ω R_3 inserted between R_2 and terminal A. The interesting point is that R_3 does not change the value of V_{AB} produced by the source V, but R_3 does increase the value of R_{TH}. When we look back from terminals A and B, the 4 Ω of R_3 is in series with 2 Ω to make R_{TH} 6 Ω, as shown for R_{AB} in Fig. 10-4b and R_{TH} in Fig. 10-4c.

Let us consider why V_{AB} is the same 24 V with or without R_3. Since R_3 is connected to the open terminal A, the source V cannot produce current in R_3. Therefore, R_3 has no IR drop. A voltmeter would read the same 24 V across R_2 and from A to B. Since V_{AB} equals 24 V, this is the value of V_{TH}.

Now consider why R_3 does change the value of R_{TH}. Remember that we must work from the outside in to calculate the total resistance. Then, A and B are like the source terminals. As a result, the 3-Ω R_1 and 6-Ω R_2 are in parallel, for a combined resistance of 2 Ω. Furthermore, this 2 Ω is in series with the 4-Ω R_3 because R_3 is in the main line from terminals A and B. Then R_{TH} is 2 + 4 = 6 Ω. As shown in Fig. 10-4c, the Thevenin equivalent circuit consists of V_{TH} = 24 V and R_{TH} = 6 Ω.

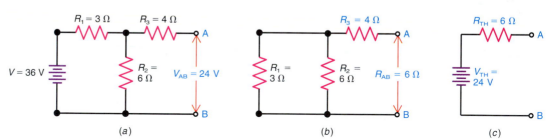

FIG. 10-4   Thevenizing the circuit of Fig. 10-3 but with a 4-Ω R_3 in series with the A terminal. (*a*) V_{AB} is still 24 V. (*b*) Now the R_{AB} is 2 + 4 = 6 Ω. (*c*) Thevenin equivalent circuit.

10-3 THEVENIZING A CIRCUIT WITH TWO VOLTAGE SOURCES

The circuit in Fig. 10-5 has already been solved by Kirchhoff's laws, but we can use Thevenin's theorem to find the current I_3 through the middle resistance R_3. As shown in Fig. 10-5a, first mark the terminals A and B across R_3. In Fig. 10-5b, R_3 is disconnected. To calculate V_{TH}, find V_{AB} across the open terminals.

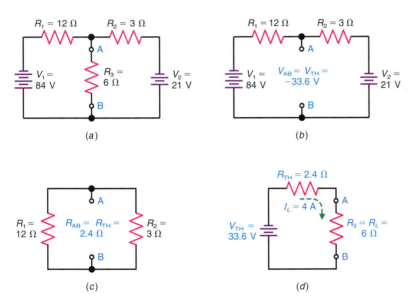

FIG. 10-5   Thevenizing a circuit with two voltage sources V_1 and V_2. (a) Original circuit with terminals A and B across the middle resistor R_3. (b) Disconnect R_3 to find V_{AB} is −33.6 V. (c) Short-circuit V_1 and V_2 to find R_{AB} is 2.4 Ω. (d) Thevenin equivalent with R_L reconnected to terminals A and B.

SUPERPOSITION METHOD

With two sources we can use superposition to calculate V_{AB}. First short-circuit V_2. Then the 84 V of V_1 is divided between R_1 and R_2. The voltage across R_2 is between terminals A and B. To calculate this divided voltage across R_2,

$$V_{R_2} = \frac{R_2}{R_{1\text{-}2}} \times V_1 = \frac{3}{15} \times (-84)$$

$$V_{R_2} = -16.8 \text{ V}$$

This is only the contribution of V_1 to V_{AB}. The polarity is negative at terminal A.

To find the voltage that V_2 produces between A and B, short-circuit V_1. Then the voltage across R_1 is connected from A to B. To calculate this divided voltage across R_1,

$$V_{R_1} = \frac{R_1}{R_{1\text{-}2}} \times V_2 = \frac{12}{15} \times (-21)$$

$$V_{R_1} = -16.8 \text{ V}$$

Both V_1 and V_2 produce −16.8 V across the AB terminals with the same polarity. Therefore, they are added.

DID YOU KNOW?

Blood tests used to require several vials of blood because testing machines couldn't handle small quantities. A new method encapsulates blood within a dime-sized "vial" and moves blood electrically within a computer chip with channels that carry liquid instead of wires. For this procedure, less than a billionth of a liter needs to be sampled.

The resultant value of $V_{AB} = -33.6$ V, shown in Fig. 10-5b, is the value of V_{TH}. The negative polarity means that terminal A is negative with respect to B.

To calculate R_{TH}, short-circuit the sources V_1 and V_2, as shown in Fig. 10-5c. Then the 12-Ω R_1 and 3-Ω R_2 are in parallel across terminals A and B. Their combined resistance is 36/15, or 2.4 Ω, which is the value of R_{TH}.

The final result is the Thevenin equivalent in Fig. 10-5d with an R_{TH} of 2.4 Ω and a V_{TH} of 33.6 V, negative toward terminal A.

In order to find the current through R_3, it is reconnected as a load resistance across terminals A and B. Then V_{TH} produces current through the total resistance of 2.4 Ω for R_{TH} and 6 Ω for R_3:

$$\blacktriangleright \quad I_3 = \frac{V_{TH}}{R_{TH} + R_3} = \frac{33.6}{2.4 + 6} = \frac{33.6}{8.4} = 4 \text{ A}$$

This answer of 4 A for I_3 is the same value calculated before, using Kirchhoff's laws, in Fig. 9-4.

It should be noted that this circuit can be solved by superposition alone, without using Thevenin's theorem, if R_3 is not disconnected. However, opening terminals A and B for the Thevenin equivalent simplifies the superposition, as the circuit then has only series voltage dividers without any parallel current paths. In general, a circuit can often be simplified by disconnecting a component to open terminals A and B for Thevenin's theorem.

TEST-POINT QUESTION 10-3

Answers at end of chapter.

In the Thevenin equivalent circuit in Fig. 10-5d,
a. How much is R_T?
b. How much is V_{R_L}?

10-4 THEVENIZING A BRIDGE CIRCUIT

As another example of Thevenin's theorem, we can find the current through the 2-Ω R_L at the center of the bridge circuit in Fig. 10-6a. When R_L is disconnected to open terminals A and B, the result is as shown in Fig. 10-6b. Notice how the circuit has become simpler because of the open. Instead of the unbalanced bridge in Fig. 10-6a which would require Kirchhoff's laws for a solution, the Thevenin equivalent in Fig. 10-6b consists of just two voltage dividers. Both the R_3R_4 divider and the R_1R_2 divider are across the same 30-V source.

Since the open terminal A is at the junction of R_3 and R_4, this divider can be used to find the potential at point A. Similarly the potential at terminal B can be found from the R_1R_2 divider. Then V_{AB} is the difference between the potentials at terminals A and B.

Note the voltages for the two dividers. In the divider with the 3-Ω R_3 and 6-Ω R_4, the bottom voltage V_{R_4} is 6/9 $\times$ 30 = 20 V. Then V_{R_3} at the top is 10 V because both must add up to equal the 30-V source. The polarities are marked negative at the top, the same as the source voltage V.

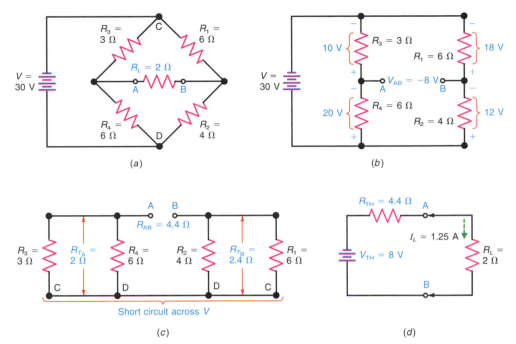

(a)

(b)

Short circuit across V

(c)

(d)

FIG. 10-6 Thevenizing a bridge circuit. (*a*) Original circuit with terminals A and B across middle resistor R_L. (*b*) Disconnect R_L to find V_{AB} of −8 V. (*c*) With source V short-circuited, R_{AB} is 2 + 2.4 = 4.4 Ω. (*d*) Thevenin equivalent with R_L reconnected to terminals A and B.

Similarly, in the divider with the 6-Ω R_1 and 4-Ω R_2, the bottom voltage V_{R_2} is 4/10 × 30 = 12 V. Then V_{R_1} at the top is 18 V, as the two must add up to equal the 30-V source. The polarities are also negative at the top, the same as V.

Now we can determine the potentials at terminals A and B, with respect to a common reference in order to find V_{AB}. Imagine that the positive side of the source V is connected to a chassis ground. Then we would use the bottom line in the diagram as our reference for voltages. Note that V_{R_4} at the bottom of the R_3R_4 divider is the same as the potential of terminal A, with respect to ground. This value is −20 V, with terminal A negative.

Similarly, V_{R_2} in the R_1R_2 divider is the potential at B with respect to ground. This value is −12 V, with terminal B negative. As a result, V_{AB} is the difference between the −20 V at A and the −12 V at B, both with respect to the common ground reference.

The potential difference V_{AB} then equals

$$V_{AB} = -20 - (-12) = -20 + 12 = -8 \text{ V}$$

Terminal A is 8 V more negative than B. Therefore, V_{TH} is 8 V, with the negative side toward terminal A as shown in the Thevenin equivalent in Fig. 10-6*d*.

The potential difference V_{AB} can also be found as the difference between V_{R_3} and V_{R_1} in Fig. 10-6*b*. In this case V_{R_3} is 10 V and V_{R_1} is 18 V, both positive with respect to the top line connected to the negative side of the source V. The potential difference between terminals A and B then is $10 - 18$, which also equals -8 V. Note that V_{AB} must have the same value no matter which path is used to determine the voltage.

To find R_{TH}, the 30-V source is short-circuited while terminals A and B are still open. Then the circuit looks like Fig. 10-6c. Looking back from terminals A and B, the 3-Ω R_3 and 6-Ω R_4 are in parallel, for a combined resistance R_{T_A} of 18/9 or 2 Ω. The reason is that R_3 and R_4 are joined at terminal A, while their opposite ends are connected by the short circuit across the source V. Similarly, the 6-Ω R_1 and 4-Ω R_2 are in parallel, for a combined resistance R_{T_B} of 24/10 = 2.4 Ω. Furthermore, the short circuit across the source now provides a path that connects R_{T_A} and R_{T_B} in series. The entire resistance is $2 + 2.4 = 4.4$ Ω for R_{AB} or R_{TH}.

The Thevenin equivalent in Fig. 10-6d represents the bridge circuit feeding the open terminals A and B, with 8 V for V_{TH} and 4.4 Ω for R_{TH}. Now connect the 2-Ω R_L to terminals A and B in order to calculate I_L. The current is

$$I_L = \frac{V_{TH}}{R_{TH} + R_L} = \frac{8}{4.4 + 2} = \frac{8}{6.4}$$

$$I_L = 1.25 \text{ A}$$

This 1.25 A is the current through the 2-Ω R_L at the center of the unbalanced bridge in Fig. 10-6a. Furthermore, the amount of I_L for any value of R_L in Fig. 10-6a can be calculated from the equivalent circuit in Fig. 10-6d.

TEST-POINT QUESTION 10-4

Answers at end of chapter.

In the Thevenin equivalent circuit in Fig. 10-6d,
a. How much is R_T?
b. How much is V_{R_L}?

10-5 NORTON'S THEOREM

Named after E. L. Norton, a scientist with Bell Telephone Laboratories, Norton's theorem is used for simplifying a network in terms of currents instead of voltages. In many cases, analyzing the division of currents may be easier than voltage analysis. For current analysis, therefore, Norton's theorem can be used to reduce a network to a simple parallel circuit, with a current source. The idea of a current source is that it supplies a total line current to be divided among parallel branches, corresponding to a voltage source applying a total voltage to be divided among series components. This comparison is illustrated in Fig. 10-7.

EXAMPLE OF A CURRENT SOURCE A source of electric energy supplying voltage is often shown with a series resistance which represents the internal resistance of the source, as in Fig. 10-7a. This method corresponds to showing an actual voltage source, such as a battery for dc circuits. However, the source may be represented also as a current source with a parallel resistance, as in Fig. 10-7b. Just as a voltage source is rated at, say, 10 V, a current source may be rated at 2 A. For the purpose of analyzing parallel branches, the concept of a current source may be more convenient than a voltage source.

If the current I in Fig. 10-7b is a 2-A source, it supplies 2 A no matter what is connected across the output terminals A and B. Without anything connected across A and B, all the 2 A flows through the shunt R. When a load resistance R_L is connected across A and B, then the 2-A I divides according to the current division rules for parallel branches.

Remember that parallel currents divide inversely to branch resistances but directly with conductances. For this reason it may be preferable to consider the current source shunted by the conductance G, as shown in Fig. 10-7c. We can always convert between resistance and conductance, because $1/R$ in ohms is equal to G in siemens.

The symbol for a current source is a circle with an arrow inside, as shown in Fig. 10-7b and c, to show the direction of current. This direction must be the same as the current produced by the polarity of the corresponding voltage source. Remember that a source produces electron flow out from the negative terminal.

An important difference between voltage and current sources is that a current source is killed by making it open, compared with short-circuiting a voltage source. Opening a current source kills its ability to supply current without affecting any parallel branches. A voltage source is short-circuited to kill its ability to supply voltage without affecting any series components.

THE NORTON EQUIVALENT CIRCUIT   As illustrated in Fig. 10-8, Norton's theorem states that the entire network connected to terminals A and B can be replaced by a single current source I_N in parallel with a single resistance R_N. The value of I_N is equal to the short-circuit current through the AB terminals. This means, find the current that the network would produce through A and B with a short circuit across these two terminals.

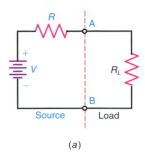

(a)

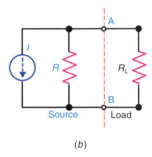

(b)

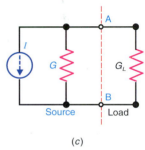

(c)

FIG. 10-7   General forms for a voltage source or current source connected to a load R_L across terminals A and B. (a) Voltage source V with series R. (b) Current source I with parallel R. (c) Current source I with parallel conductance G.

FIG. 10-8   Any network in the block at the left can be reduced to the Norton equivalent parallel circuit at the right.

The value of R_N is the resistance looking back from open terminals A and B. These terminals are not short-circuited for R_N but are open, as in calculating R_{TH} for Thevenin's theorem. Actually, the single resistor is the same for both the Norton and Thevenin equivalent circuits. In the Norton case, this value of R_{AB} is R_N in parallel with the current source; in the Thevenin case, it is R_{TH} in series with the voltage source.

NORTONIZING A CIRCUIT   As an example, let us recalculate the current I_L in Fig. 10-9a, which was solved before by Thevenin's theorem. The first step

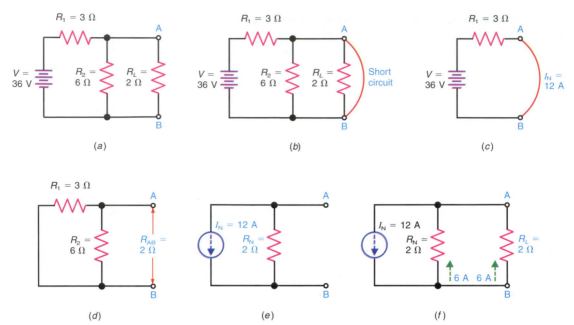

FIG. 10-9 Same circuit as in Fig. 10-3, but solved by Norton's theorem. (*a*) Original circuit. (*b*) Short circuit across terminals A and B. (*c*) The short-circuit current I_N is $36/3 = 12$ A. (*d*) Open terminals A and B but short-circuit V to find R_{AB} is 2 Ω, the same as R_{TH}. (*e*) Norton equivalent circuit. (*f*) R_L reconnected to terminals A and B to find I_L is 6 A.

in applying Norton's theorem is to imagine a short circuit across terminals A and B, as shown in Fig. 10-9*b*. How much current is flowing in the short circuit? Note that a short circuit across AB short-circuits R_L and the parallel R_2. Then the only resistance in the circuit is the 3-Ω R_1 in series with the 36-V source, as shown in Fig. 10-9*c*. The short-circuit current, therefore, is

$$I_N = \frac{36 \text{ V}}{3 \text{ Ω}} = 12 \text{ A}$$

This 12-A I_N is the total current available from the current source in the Norton equivalent in Fig. 10-9*e*.

To find R_N, remove the short circuit across A and B and consider the terminals open, without R_L. Now the source V is considered to be short-circuited. As shown in Fig. 10-9*d*, the resistance seen looking back from terminals A and B is 6 Ω in parallel with 3 Ω, which equals 2 Ω for the value of R_N.

The resultant Norton equivalent is shown in Fig. 10-9*e*. It consists of a 12-A current source I_N shunted by the 2-Ω R_N.

The arrow on the current source shows the direction of electron flow from terminal B to terminal A, as in the original circuit.

Finally, to calculate I_L, replace the 2-Ω R_L between terminals A and B, as shown in Fig. 10-9f. The current source still delivers 12 A, but now that current divides between the two branches of R_N and R_L. Since these two resistances are equal, the 12-A I_N divides into 6 A for each branch, and I_L is equal to 6 A. This value is the same current we calculated in Fig. 10-3, by Thevenin's theorem. Also, V_L can be calculated as $I_L R_L$, or 6 A $\times$ 2 Ω, which equals 12 V.

LOOKING AT THE SHORT-CIRCUIT CURRENT   In some cases, there may be a question of which current is I_N when terminals A and B are short-circuited. Imagine that a wire jumper is connected between A and B to short-circuit these terminals. Then I_N must be the current that flows in this wire between terminals A and B.

Remember that any components directly across these two terminals are also short-circuited by the wire jumper. Then these parallel paths have no effect. However, any components in series with terminal A or terminal B are in series with the wire jumper. Therefore, the short-circuit current I_N also flows through the series components.

An example of a resistor in series with the short circuit across terminals A and B is shown in Fig. 10-10. The idea here is that the short-circuit I_N is a branch current, not the main-line current. Refer to Fig. 10-10a. Here the short circuit connects R_3 across R_2. Also, the short-circuit current I_N is now the same as the current I_3 through R_3. Note that I_3 is only a branch current.

To calculate I_3, the circuit is solved by Ohm's law. The parallel combination of R_2 with R_3 equals 72/18 or 4 Ω. The R_T is 4 + 4 = 8 Ω. As a result, the I_T from the source is 48 V/8 Ω = 6 A.

This I_T of 6 A in the main line divides into 4 A for R_2 and 2 A for R_3. The 2-A I_3 for R_3 flows through short-circuited terminals A and B. Therefore, this current of 2 A is the value of I_N.

To find R_N in Fig. 10-10b, the short circuit is removed from terminals A and B. Now the source V is short-circuited. Looking back from open terminals A and B, the 4-Ω R_1 is in parallel with the 6-Ω R_2. This combination is 24/10 = 2.4 Ω. The 2.4 Ω is in series with the 12-Ω R_3 to make R_{AB} = 2.4 + 12 = 14.4 Ω.

The final Norton equivalent is shown in Fig. 10-10c. Current I_N is 2 A because this branch current in the original circuit is the current that flows through R_3 and short-circuited terminals A and B. Resistance R_N is 14.4 Ω looking back from open terminals A and B with the source V short-circuited the same way as for R_{TH}.

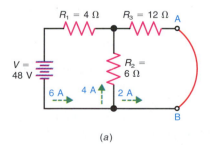
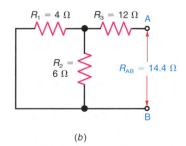
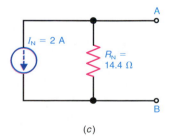

(a)   (b)   (c)

FIG. 10-10   Nortonizing a circuit where the short-circuit current I_N is a branch current. (a) I_N is 2 A through short-circuited terminals A and B and R_3. (b) $R_N = R_{AB}$ = 14.4 Ω. (c) Norton equivalent circuit.

Answers at end of chapter.

Answer True or False.

a. For a Norton equivalent circuit, terminals A and B are short-circuited to find I_N.

b. For a Norton equivalent circuit, terminals A and B are open to find R_N.

10-6 THEVENIN-NORTON CONVERSIONS

Thevenin's theorem says that any network can be represented by a voltage source and series resistance, while Norton's theorem says that the same network can be represented by a current source and shunt resistance. It must be possible, therefore, to convert directly from a Thevenin form to a Norton form and vice versa. Such conversions are often useful.

NORTON FROM THEVENIN   Consider the Thevenin equivalent circuit in Fig. 10-11*a*. What is its Norton equivalent? Just apply Norton's theorem, the same as for any other circuit. The short-circuit current through terminals A and B is

$$I_N = \frac{V_{TH}}{R_{TH}} = \frac{15\ \text{V}}{3\ \Omega} = 5\ \text{A}$$

The resistance, looking back from open terminals A and B with the source V_{TH} short-circuited, is equal to the 3 Ω of R_{TH}. Therefore, the Norton equivalent consists of a current source that supplies the short-circuit current of 5 A, shunted by the same 3-Ω resistance that is in series in the Thevenin circuit. The results are shown in Fig. 10-11*b*.

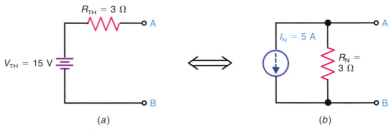

(*a*)   (*b*)

FIG. 10-11   Thevenin equivalent circuit in (*a*) corresponds to the Norton equivalent in (*b*).

THEVENIN FROM NORTON   For the opposite conversion, we can start with the Norton circuit of Fig. 10-11*b* and get back to the original Thevenin circuit. To do this, apply Thevenin's theorem, the same as for any other circuit. First, we find the Thevenin resistance by looking back from open terminals A and B. An important principle here, though, is that while a voltage source is

DID YOU KNOW?

New HID headlights generate light by a gas arc between two electrodes. Compared to halogens, HIDs make three times more light, cover twice as much distance, use two-thirds less energy, and last three times longer. Also, these lights make embedded lane reflectors more visible.

short-circuited to find R_{TH}, a current source is an open circuit. In general, a current source is killed by opening the path between its terminals. Therefore, we have just the 3-Ω R_N, in parallel with the infinite resistance of the open current source. The combined resistance then is 3 Ω.

In general, the resistance R_N always has the same value as R_{TH}. The only difference is that R_N is connected in parallel with I_N, but R_{TH} is in series with V_{TH}.

Now all that is required is to calculate the open-circuit voltage in Fig. 10-11*b* to find the equivalent V_{TH}. Note that with terminals A and B open, all the current of the current source flows through the 3-Ω R_N. Then the open-circuit voltage across the terminals A and B is

$$I_N R_N = 5 \text{ A} \times 3 \ \Omega = 15 \text{ V} = V_{TH}$$

As a result, we have the original Thevenin circuit, which consists of the 15-V source V_{TH} in series with the 3-Ω R_{TH}.

CONVERSION FORMULAS   In summary, the following formulas can be used for these conversions:

Thevenin from Norton:

▶ $R_{TH} = R_N$
  $V_{TH} = I_N \times R_N$

Norton from Thevenin:

▶ $R_N = R_{TH}$
  $I_N = V_{TH}/R_{TH}$

Another example of these conversions is shown in Fig. 10-12.

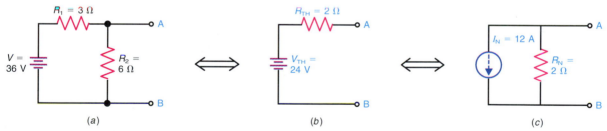

(a)   (b)   (c)

FIG. 10-12   Example of Thevenin-Norton conversions. (*a*) Original circuit, the same as in Figs. 10-3*a* and 10-9*a*. (*b*) Thevenin equivalent. (*c*) Norton equivalent.

TEST-POINT QUESTION 10-6

Answers at end of chapter.

Answer True or False.
a. In Thevenin-Norton conversions, resistances R_N and R_{TH} are equal.
b. In Thevenin-Norton conversions, current I_N is V_{TH}/R_{TH}.
c. In Thevenin-Norton conversions, voltage V_{TH} is $I_N \times R_N$.

10-7 CONVERSION OF VOLTAGE AND CURRENT SOURCES

Norton conversion is a specific example of the general principle that any voltage source with its series resistance can be converted to an equivalent current source with the same resistance in parallel. In Fig. 10-13, the voltage source in Fig. 10-13a is equivalent to the current source in Fig. 10-13b. Just divide the source V by its series R to calculate the value of I for the equivalent current source shunted by the same R. Either source will supply the same current and voltage for any components connected across terminals A and B.

Conversion of voltage and current sources can often simplify circuits, especially those with two or more sources. Current sources are easier for parallel connections, where we can add or divide currents. Voltage sources are easier for series connections, where we can add or divide voltages.

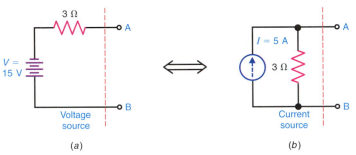

(a)   (b)

FIG. 10-13 The voltage source in (a) corresponds to the current source in (b).

TWO SOURCES IN PARALLEL BRANCHES   Referring to Fig. 10-14a, assume that the problem is to find I_3 through the middle resistor R_3. Note that V_1 with R_1 and V_2 with R_2 are branches in parallel with R_3. All three branches are connected across terminals A and B.

When we convert V_1 and V_2 to current sources in Fig. 10-14b, the circuit has all parallel branches. Current I_1 is $^{84}/_{12}$ or 7 A, while I_2 is $^{21}/_3$, which also happens to be 7 A. Current I_1 has its parallel R of 12 Ω, while I_2 has its parallel R of 3 Ω.

Furthermore, I_1 and I_2 can be combined for the one equivalent current source I_T in Fig. 10-14c. Since both sources produce current in the same direction through R_L, they are added for $I_T = 7 + 7 = 14$ A.

The shunt R for the 14-A combined source is the combined resistance of the 12-Ω R_1 and the 3-Ω R_2 in parallel. This R equals $^{36}/_{15}$ or 2.4 Ω, as shown in Fig. 10-14c.

To find I_L, we can use the current divider formula for the 6- and 2.4-Ω branches, into which the 14-A I_T from the current source was split. Then

$$I_L = \frac{2.4}{2.4 + 6} \times 14 = 4 \text{ A}$$

The voltage V_{R_3} across terminals A and B is $I_L R_L$, which equals $4 \times 6 = 24$ V. These are the same values calculated for V_{R_3} and I_3 by Kirchhoff's laws in Fig. 9-4 and by Thevenin's theorem in Fig. 10-5.

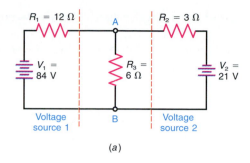

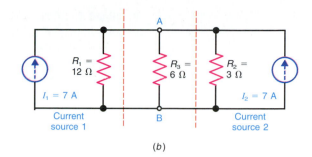

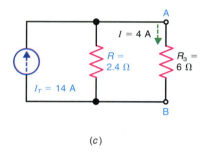

FIG. 10-14 Converting two voltage sources in V_1 and V_2 in parallel branches to current sources I_1 and I_2 that can be combined. (*a*) Original circuit. (*b*) V_1 and V_2 converted to parallel current sources I_1 and I_2. (*c*) Equivalent circuit with one combined current source I_T.

TWO SOURCES IN SERIES Referring to Fig. 10-15, assume that the problem is to find the current I_L through the load resistance R_L between terminals A and B. This circuit has the two current sources I_1 and I_2 in series with each other.

The problem here can be simplified by converting I_1 and I_2 to the series voltage sources V_1 and V_2 shown in Fig. 10-15*b*. The 2-A I_1 with its shunt 4-Ω R_1 is equivalent to 4×2, or 8 V, for V_1 with a 4-Ω series resistance. Similarly, the 5-A I_2 with its shunt 2-Ω R_2 is equivalent to 5×2, or 10 V, for V_2 with a 2-Ω series resistance. The polarities of V_1 and V_2 produce electron flow in the same direction as I_1 and I_2.

The series voltages can now be combined as in Fig. 10-15*c*. The 8 V of V_1 and 10 V of V_2 are added because they are series-aiding, resulting in the total V_T of 18 V. And, the resistances of 4 Ω for R_1 and 2 Ω for R_2 are added, for a

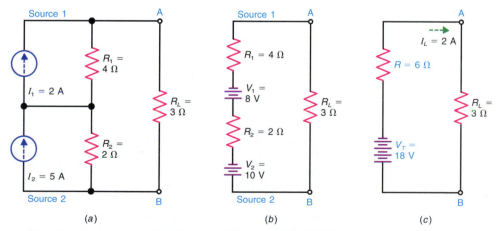

FIG. 10-15 Converting two current sources I_1 and I_2 in series to voltage sources V_1 and V_2 that can be combined. (*a*) Original circuit. (*b*) I_1 and I_2 converted to series voltage sources V_1 and V_2. (*c*) Equivalent circuit with one combined voltage source V_T.

combined R of 6 Ω. This is the series resistance of the 18-V source V_T connected across terminals A and B.

The total resistance of the circuit in Fig. 10-15c is R plus R_L, or $6 + 3 = 9$ Ω. With 18 V applied, $I_L = {}^{18}\!\!/\!_9 = 2$ A through R_L between terminals A and B.

10-8 MILLMAN'S THEOREM

Millman's theorem provides a shortcut for finding the common voltage across any number of parallel branches with different voltage sources. A typical example is shown in Fig. 10-16. For all the branches, the ends at point Y are connected to chassis ground. Furthermore, the opposite ends of all the branches are also connected to the common point X. The voltage V_{XY}, therefore, is the common voltage across all the branches.

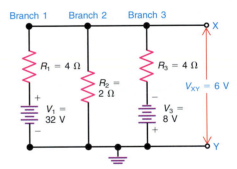

FIG. 10-16 Example of Millman's theorem to find V_{XY}, the common voltage across branches with separate voltage sources.

Finding the value of V_{XY} gives the net effect of all the sources in determining the voltage at X with respect to chassis ground. To calculate this voltage

$$\blacktriangleright \qquad V_{XY} = \frac{V_1/R_1 + V_2/R_2 + V_3/R_3}{1/R_1 + 1/R_2 + 1/R_3} \ldots \text{etc.} \qquad \textbf{(10-1)}$$

This formula is derived from converting the voltage sources to current sources and combining the results. The numerator with V/R terms is the sum of the parallel current sources. The denominator with $1/R$ terms is the sum of the parallel conductances. The net V_{XY} then is the form of I/G or $I \times R$, which is in units of voltage.

CALCULATING V_{XY}   For the values in Fig. 10-16,

$$V_{XY} = \frac{32/4 + 0/2 - 8/4}{1/4 + 1/2 + 1/4}$$

$$= \frac{8 + 0 - 2}{1}$$

$$V_{XY} = 6 \text{ V}$$

Note that in branch 3, V_3 is considered negative because it would make point X negative. However, all the resistances are positive. The positive answer for V_{XY} means that point X is positive with respect to Y.

In branch 2, V_2 is zero because this branch has no voltage source. However, R_2 is still used in the denominator.

This method can be used for any number of branches, but they must all be in parallel, without any series resistances between the branches. In a branch with several resistances, they can be combined as one R_T. When a branch has more than one voltage source, they can be combined algebraically for one V_T.

APPLICATIONS OF MILLMAN'S THEOREM   In many cases, a circuit can be redrawn to show the parallel branches and their common voltage V_{XY}. Then with V_{XY} known the entire circuit can be analyzed quickly. For instance, Fig. 10-17 has been solved before by other methods. For Millman's theorem the common voltage V_{XY} across all the branches is the same as V_3 across R_3. This voltage is calculated with Formula (10-1), as follows:

$$V_{XY} = \frac{-84/12 + 0/6 - 21/3}{1/12 + 1/6 + 1/3} = \frac{-7 + 0 - 7}{7/12}$$

$$= \frac{-14}{7/12} = -14 \times \frac{12}{7}$$

$$V_{XY} = -24 \text{ V} = V_3$$

The negative sign means that point X is the negative side of V_{XY}.

With V_3 known to be 24 V across the 6-Ω R_3, then I_3 must be 24/6 = 4 A. Similarly, all the voltages and currents in this circuit can then be calculated. (See Fig. 9-4 in Chap. 9.)

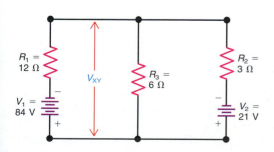

FIG. 10-17   The same circuit as in Fig. 9-4 for Kirchhoff's laws, but shown with parallel branches to calculate V_{XY} by Millman's theorem.

As another application, the example of superposition in Fig. 10-1 has been redrawn in Fig. 10-18 to show the parallel branches with a common voltage V_{XY} to be calculated by Millman's theorem. Then

$$V_{XY} = \frac{24 \text{ V}/30 \text{ k}\Omega - 9 \text{ V}/60 \text{ k}\Omega}{1/(30 \text{ k}\Omega) + 1/(60 \text{ k}\Omega)} = \frac{0.8 \text{ mA} - 0.15 \text{ mA}}{3/(60 \text{ k}\Omega)}$$

$$= 0.65 \times \frac{60}{3} = \frac{39}{3}$$

$$V_{XY} = 13 \text{ V} = V_P$$

The answer of 13 V from point P to ground, using Millman's theorem, is the same value calculated before by superposition.

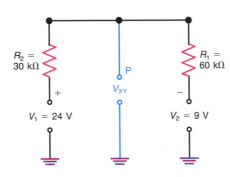

FIG. 10-18 Same circuit as in Fig. 10-1 for superposition, but shown with parallel branches to calculate V_{XY} by Millman's theorem.

TEST-POINT QUESTION 10-8

Answers at end of chapter.

For the example of Millman's theorem in Fig. 10-16,
a. How much is V_{R_2}?
b. How much is V_{R_3}?

10-9 CIRCUITS WITH CURRENT SOURCES

Examples are shown in Figs. 10-19 and 10-20 with multiple energy sources to illustrate how to simplify circuits with current sources. The general principles are as follows:

1. When current sources are in parallel, they can be combined. Add currents that produce I in the same direction through the load; subtract opposite currents.
2. When voltage sources are in series, they can be combined. Add voltages that produce I in the same direction through the load; subtract opposite voltages.
3. When current sources are in series, convert to voltage sources so that they can be combined. The polarity of the voltage must produce I in the same direction as the current source.

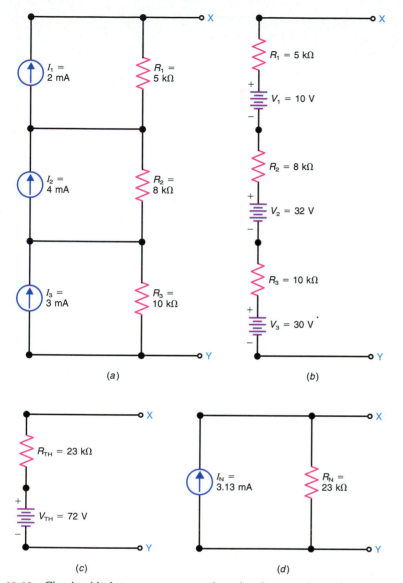

FIG. 10-19 Circuit with three current sources in series. *I* arrows shown for conventional current. See text for analysis that converts circuit to one Thevenin or Norton equivalent. (*a*) Original circuit. (*b*) *I* sources converted to *V* sources. (*c*) Thevenin equivalent. (*d*) Norton equivalent.

4. With voltage sources in parallel, convert to current sources so that they can be combined. The direction of *I* must be in the same direction through the load as would be produced by the voltage source.

The current sources in Figs. 10-19 and 10-20 are shown with solid arrows for the direction of conventional current, opposite from electron flow, although the circuits can be analyzed either way. In conversion, note that conventional *I* is out from the positive terminal of *V* and returns to the negative terminal. Also, these circuits have *I* of milliamperes and *R* of kilohms. Still, the *IR* product is in units of volts.

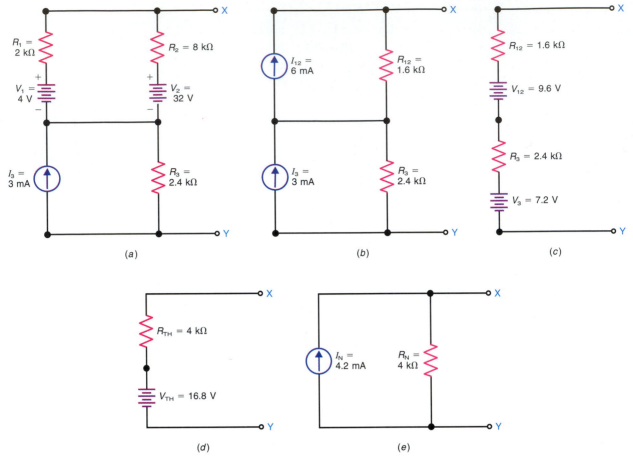

FIG. 10-20 Circuit with series-parallel I and V sources. I arrows shown for conventional current. See text for analysis that converts circuit to one Thevenin or Norton equivalent. (a) Original circuit. (b) V_1 and V_2 converted to one equivalent current source I_{12}. (c) The I_{12} converted to V_{12} and I_3 converted to V_3 for series circuit of voltage sources. (d) Thevenin equivalent. (e) Norton equivalent.

I SOURCES IN SERIES Refer to Fig 10-19a. The three current sources I_1, I_2, and I_3 are in series. They are converted to voltage sources in Fig 10-19b. The calculations are:

$$V_1 = I_1 \times R_1 \qquad\qquad V_2 = I_2 \times R_2 \qquad\qquad V_3 = I_3 \times R_3$$
$$= 2\ \text{mA} \times 5\ \text{k}\Omega \qquad = 4\ \text{mA} \times 8\ \text{k}\Omega \qquad = 3\ \text{mA} \times 10\ \text{k}\Omega$$
$$V_1 = 10\ \text{V} \qquad\qquad V_2 = 32\ \text{V} \qquad\qquad V_3 = 30\ \text{V}$$

The three voltage sources V_1, V_2, and V_3 in series in Fig. 10-19b are combined for one equivalent source V_{TH} in Fig. 10-19c, equal to $10 + 32 + 30 = 72$ V. Also, R_1, R_2, and R_3 in series total 23 kΩ.

The single voltage source of 72 V in Fig. 10-19c with series resistance of 23 kΩ is the Thevenin equivalent of the circuit in Fig. 10-19a. Any load connected across terminals X and Y will have the same V_L and I_L, for either circuit.

Finally, the circuit in Fig. 10-19d is the Norton equivalent of Fig. 10-19c. The I_N is 72 V/23 kΩ, which equals 3.13 mA. Also, R_N is the same 23 kΩ as R_{TH}.

SERIES-PARALLEL ENERGY SOURCES   Refer to Fig. 10-20a. The first step is to convert V_1 and V_2 to current sources so that they can be combined in parallel for the single value of I_{12} in Fig. 10-20b. The calculations are

$$I_1 = V_1/R_1 \qquad\qquad I_2 = V_2/R_2 \qquad\qquad I_{12} = I_1 + I_2$$
$$= 4 \text{ V/2 k}\Omega \qquad\quad = 32 \text{ V/8 k}\Omega \qquad\quad = 2 + 4$$
$$I_1 = 2 \text{ mA} \qquad\qquad I_2 = 4 \text{ mA} \qquad\qquad I_{12} = 6 \text{ mA}$$

The equivalent resistance for R_{12} is the parallel combination of $(2 \times 8)/(2 + 8)$ or 16/10, which equals 1.6 kΩ.

Next, the current sources I_{12} and I_3 in Fig. 10-20b are converted to voltage sources in Fig. 10-20c, because of the series circuit. Then V_{12} becomes 6 mA $\times$ 1.6 kΩ = 9.6 V. The R_{12} is still 1.6 kΩ. Also, the current source I_3 is converted to V_3, equal to 3 mA $\times$ 2.4 kΩ = 7.2 V. The source resistance R_3 is still 2.4 kΩ.

In Fig. 10-20d, the series voltage sources are added for the single value of 9.6 + 7.2 = 16.8 V. This value is V_{TH}. The R_{TH} is 2.4 + 1.6 = 4 kΩ.

For the last step, the Thevenin circuit is converted to a Norton equivalent in Fig. 10-20e. The I_{N} is 16.8 V/4 kΩ, which equals 4.2 mA. Also, R_{N} is the same 4 kΩ as R_{TH}.

TEST-POINT QUESTION 10-9

Answers at end of chapter.

Answer True or False.
a. In Fig. 10-19, the current sources I_1, I_2, and I_3 are converted to voltage sources because they are in series.
b. In Fig. 10-20, the parallel bank of V_1 with R_1 and V_2 with R_2 is equivalent to one source of 9.6 V with series R of 1.6 kΩ.

10-10 T OR Y AND π OR Δ CONNECTIONS

The circuit in Fig. 10-21 is called a T (tee) or Y (wye) network, as suggested by the shape. They are different names for the same network; the only difference is that the R_2 and R_3 legs are shown at an angle in the Y.

The circuit in Fig. 10-22 is called a π (pi) or Δ (delta) network, as the shape is similar to these Greek letters. Both forms are the same network. Actually, the

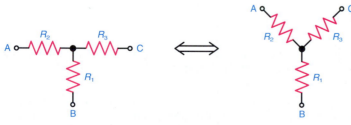

FIG. 10-21   The form of a T or Y network.

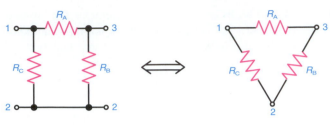

FIG. 10-22 The form of a π or Δ network.

network can have the R_A arm shown at the top or bottom, as long as it is connected between R_B and R_C. In Fig. 10-22, R_A is at the top, as an inverted delta, to look like the π network.

The circuits in Figs. 10-21 and 10-22 are passive networks, without any energy sources. Also, they are three-terminal networks for two pairs of connections for input and output voltages, with one common. In Fig. 10-21, point B is the common terminal and point 2 is common in Fig. 10-22.

The Y and Δ forms are different ways to connect three resistors in a passive network. Note resistors in the Y are labeled with subscripts 1, 2, and 3, while the Δ has subscripts A, B, and C, in order to emphasize the different connections.

CONVERSION FORMULAS   In the analysis of networks, it is often helpful to convert a Δ to Y or vice versa. Either it may be difficult to visualize the circuit without the conversion, or the conversion makes the solution simpler. The formulas for these transformations are given here. All are derived from Kirchhoff's laws. Note that letters are used as subscripts for R_A, R_B, and R_C in the Δ, while the resistances are numbered R_1, R_2, and R_3 in the Y.

▶ Conversions of Y to Δ, or T to π:

$$R_A = \frac{R_1R_2 + R_2R_3 + R_3R_1}{R_1}$$

$$R_B = \frac{R_1R_2 + R_2R_3 + R_3R_1}{R_2}$$

$$R_C = \frac{R_1R_2 + R_2R_3 + R_3R_1}{R_3}$$

(10-2)

or

▶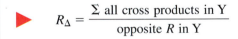

These formulas can be used to convert a Y network to an equivalent Δ, or a T network to π. Both networks will have the same resistance across any pair of terminals.

The three formulas have the same general form, indicated at the bottom as one basic rule. The symbol Σ is the Greek capital letter sigma, meaning "sum of."

For the opposite conversion:

▶ Conversion of Δ to Y or π to T:

$$R_1 = \frac{R_B R_C}{R_A + R_B + R_C}$$

$$R_2 = \frac{R_C R_A}{R_A + R_B + R_C}$$ 　　(10-3)

$$R_3 = \frac{R_A R_B}{R_A + R_B + R_C}$$

or

▶ $$R_Y = \frac{\text{product of two adjacent } R \text{ in } \Delta}{\Sigma \text{ all } R \text{ in } \Delta}$$

As an aid in using these formulas, the following scheme is useful. Place the Y inside the Δ, as shown in Fig. 10-23. Notice that the Δ has three closed sides, while the Y has three open arms. Also note how resistors can be considered opposite each other in the two networks. For instance, the open arm R_1 is opposite the closed side R_A; R_2 is opposite R_B; and R_3 is opposite R_C.

Furthermore, each resistor in an open arm has two adjacent resistors in the closed sides. For R_1, its adjacent resistors are R_B and R_C; also R_C and R_A are adjacent to R_2, while R_A and R_B are adjacent to R_3.

In the formulas for the Y-to-Δ conversion, each side of the delta is found by first taking all possible cross products of the arms of the wye, using two arms at a time. There are three such cross products. The sum of the three cross products is then divided by the opposite arm to find the value of each side of the delta. Note that the numerator remains the same, the sum of the three cross products. However, each side of the delta is calculated by dividing this sum by the opposite arm.

For the case of the Δ-to-Y conversion, each arm of the wye is found by taking the product of the two adjacent sides in the delta and dividing by the sum of the three sides of the delta. The product of two adjacent resistors excludes the opposite resistor. The denominator for the sum of the three sides remains the same in the three formulas. However, each arm is calculated by dividing the sum into each cross product.

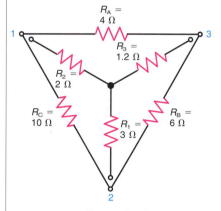

FIG. 10-23 Conversion between Y and Δ networks. See text for conversion formulas.

AN EXAMPLE OF CONVERSION

The values shown for the equivalent Y and Δ in Fig. 10-23 are calculated as follows: Starting with 4, 6, and 10 Ω for sides R_A, R_B, and R_C, respectively, in the delta, the corresponding arms in the wye are

$$R_1 = \frac{R_B R_C}{R_A + R_B + R_C} = \frac{6 \times 10}{4 + 6 + 10} = \frac{60}{20} = 3\ \Omega$$

$$R_2 = \frac{R_C R_A}{20} = \frac{10 \times 4}{20} = \frac{40}{20} = 2\ \Omega$$

$$R_3 = \frac{R_A R_B}{20} = \frac{4 \times 6}{20} = \frac{24}{20} = 1.2\ \Omega$$

As a check on these values, we can calculate the equivalent delta for this wye. Starting with values of 3, 2, and 1.2 Ω for R_1, R_2, and R_3, respectively, in the wye, the corresponding values in the delta are:

$$R_A = \frac{R_1R_2 + R_2R_3 + R_3R_1}{R_1} = \frac{6 + 2.4 + 3.6}{3} = \frac{12}{3} = 4 \ \Omega$$

$$R_B = \frac{12}{R_2} = \frac{12}{2} = 6 \ \Omega$$

$$R_C = \frac{12}{R_3} = \frac{12}{1.2} = 10 \ \Omega$$

These results show that the Y and Δ networks in Fig. 10-23 are equivalent to each other when they have the values obtained with the conversion formulas.

Note that the equivalent R values in the Y are less than in the equivalent Δ network. The reason is that the Y has two legs between the terminals, while the Δ has only one.

SIMPLIFYING A BRIDGE CIRCUIT As an example of the use of such transformations, consider the bridge circuit of Fig. 10-24. The total current I_T from the battery is desired. Therefore, we must find the total resistance R_T.

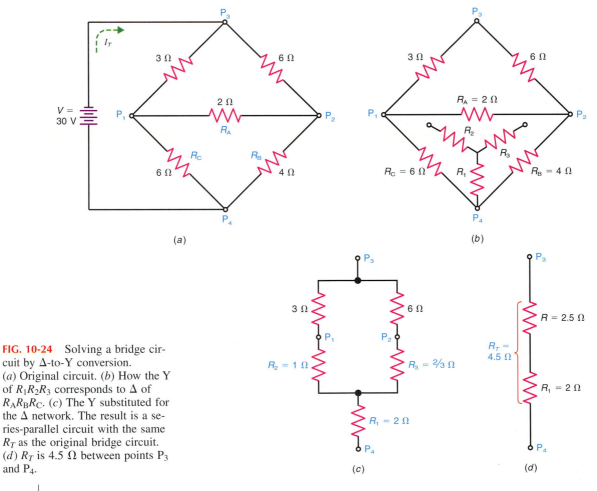

FIG. 10-24 Solving a bridge circuit by Δ-to-Y conversion. (a) Original circuit. (b) How the Y of $R_1R_2R_3$ corresponds to Δ of $R_AR_BR_C$. (c) The Y substituted for the Δ network. The result is a series-parallel circuit with the same R_T as the original bridge circuit. (d) R_T is 4.5 Ω between points P_3 and P_4.

One approach is to note that the bridge in Fig. 10-24a consists of two deltas connected between terminals P_1 and P_2. One of them can be replaced by an equivalent wye. We use the bottom delta with R_A across the top, in the same form as Fig. 10-23. We then replace this delta $R_A R_B R_C$ by an equivalent wye $R_1 R_2 R_3$, as shown in Fig. 10-24b. Using the conversion formulas,

$$\blacktriangleright \quad R_1 = \frac{R_B R_C}{R_A + R_B + R_C} = \frac{24}{12} = 2\ \Omega$$

$$R_2 = \frac{R_C R_A}{12} = \frac{12}{12} = 1\ \Omega$$

$$R_3 = \frac{R_A R_B}{12} = \frac{8}{12} = \frac{2}{3}\ \Omega$$

We next use these values for R_1, R_2, and R_3 in an equivalent wye to replace the original delta. Then the resistances form the series-parallel circuit shown in Fig. 10-24c. The combined resistance of the two parallel branches here is 4×6.67 divided by 10.67, which equals 2.5 Ω for R. Adding this 2.5 Ω to the series R_1 of 2 Ω, the total resistance is 4.5 Ω in Fig. 10-24d.

This 4.5 Ω is the R_T for the entire bridge circuit between P_3 and P_4 connected to source V. Then I_T is 30 V/4.5 Ω, which equals 6.67 A supplied by the source.

Another approach to finding R_T for the bridge circuit in Fig. 10-24a is to recognize that the bridge also consists of two T or Y networks between terminals P_3 and P_4. One of them can be transformed into an equivalent delta. The result is another series-parallel circuit but with the same R_T of 4.5 Ω.

BALANCED NETWORKS   When all the R values are equal in a network, it is balanced. Then the conversion is simplified, as R in the wye network is 1/3 the R in the equivalent delta. As an example, for $R_A = R_B = R_C$ equal to 6 Ω in the delta, the equivalent wye has $R_1 = R_2 = R_3$ equal to 6/3 or 2 Ω. Or, converting the other way, for 2-Ω R values in a balanced wye delta, the equivalent delta network has each R equal to $3 \times 2 = 6\ \Omega$. This example is illustrated in Fig. 10-25.

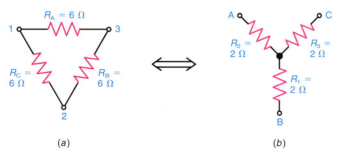

(a)   (b)

FIG. 10-25   Equivalent balanced networks. (a) Delta form. (b) Wye form.

TEST-POINT QUESTION 10-10

Answers at end of chapter.

In the standard form for conversion,
a. Which resistor in the Y is opposite R_A in the Δ?
b. Which two resistors in the Δ are adjacent to R_1 in the Y?

10 SUMMARY AND REVIEW

- *Superposition theorem.* In a linear, bilateral network having more than one source, the current and voltage in any part of the network can be found by adding algebraically the effect of each source separately. All other sources are temporarily killed by short-circuiting voltage sources and opening current sources.
- *Thevenin's theorem.* Any network with two open terminals A and B can be replaced by a single voltage source V_{TH} in series with a single resistance R_{TH} connected to terminals A and B. Voltage V_{TH} is the voltage produced by the network across terminals A and B. Resistance R_{TH} is the resistance across open terminals A and B with all sources short-circuited.
- *Norton's theorem.* Any two-terminal network can be replaced by a single current source I_N in parallel with a single resistance R_N. The value of I_N is the current produced by the network through the short-circuited terminals. R_N is the resistance across the open terminals with all sources short-circuited.
- *Millman's theorem.* The common voltage across parallel branches with different V sources can be determined with Formula (10-1).
- A voltage source V with its series R can be converted to an equivalent current source I with parallel R. Similarly, a current source I with a parallel R can be converted to a voltage source V with a series R. The value of I is V/R, or V is $I \times R$. The value of R is the same for both sources. However, R is in series with V but in parallel with I.
- The comparison between delta and wye networks is illustrated in Fig. 10-23. To convert from one network to the other, Formula (10-2) or (10-3) is used.

SELF-TEST

ANSWERS AT BACK OF BOOK.

Answer True or False.

1. Voltage V_{TH} is an open-circuit voltage.
2. Current I_N is a short-circuit current.
3. Resistances R_{TH} and R_N have the same value.
4. A voltage source has series resistance.
5. A current source has parallel resistance.
6. A voltage source is killed by short-circuiting the terminals.
7. A current source is killed by opening the source.
8. A π network is the same as a T network.

QUESTIONS

1. State the superposition theorem, and discuss how to apply it.
2. State how to calculate V_{TH} and R_{TH} in Thevenin equivalent circuits.
3. State the method of calculating I_N and R_N for a Norton equivalent circuit.
4. How is a voltage source converted to a current source, and vice versa?
5. For what type of circuit is Millman's theorem used?
6. Draw a delta network and a wye network and give the six formulas needed to convert from one to the other.
7. Give two differences between a dc source and an ac source.

PROBLEMS

ANSWERS TO ODD-NUMBERED PROBLEMS AT BACK OF BOOK.

1. Refer to Fig. 10-26. Show the Thevenin equivalent and calculate V_L.
2. Show the Norton equivalent of Fig. 10-26 and calculate I_L.
3. In Fig. 10-26, convert V and R_1 to a current source and calculate I_L.
4. Use Ohm's law to solve Fig. 10-26 as a series-parallel circuit in order to calculate V_L and I_L. (Note: R_L is not opened for Ohm's law.)
5. Why is the value of V_L across terminals A and B in Prob. 4 not the same as V_{AB} for the Thevenin equivalent circuit in Prob. 1?
6. Refer to Fig. 10-27. Determine V_P by superposition.
7. Redraw Fig. 10-27 as two parallel branches to calculate V_P by Millman's theorem.
8. Refer to Fig. 10-28. Calculate V_L across R_L by Millman's theorem and also by superposition.
9. Show the Thevenin equivalent of Fig. 10-29, where terminals A and B are across the middle resistor R_2. Then calculate V_{R_2}.

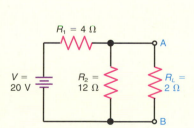

FIG. 10-26 Circuit for Probs. 1, 2, 3, 4, and 5.

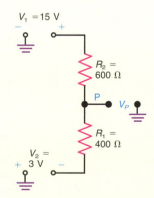

FIG. 10-27 Circuit for Probs. 6 and 7.

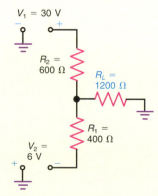

FIG. 10-28 Circuit for Prob. 8.

10. In Fig. 10-29, solve for V_{R_2} by superposition.
11. In Fig. 10-29, solve for V_{R_2} by Millman's theorem.
12. In Fig. 10-30, solve for all the currents by Kirchhoff's laws.
13. In Fig. 10-30, find V_3 by Millman's theorem.
14. Convert the T network in Fig. 10-31 to an equivalent π network.
15. Convert the π network in Fig. 10-32 to an equivalent T network.
16. Show the Thevenin and Norton equivalent circuits for the diagram in Fig. 10-33.
17. In Fig. 10-34, do the following: (**a**) convert the current sources, I_1 and I_2, into equivalent voltage sources, V_1 and V_2, in series with load R_L; (**b**) combine the individual voltage sources into one equivalent voltage source whose voltage and resistance values are identified as V_{TH} and R_{TH}, respectively; (**c**) calculate the load current I_L; (**d**) convert the Thevenin equivalent circuit in part (**b**) into its Norton equivalent. (The arrows in each current source indicate the direction of conventional current flow.)

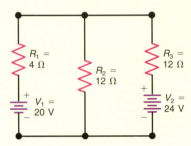

FIG. 10-29   Circuit for Probs. 9, 10, and 11.

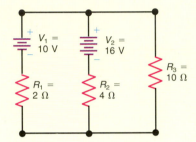

FIG. 10-30   Circuit for Probs. 12 and 13.

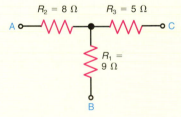

FIG. 10-31   Circuit for Prob. 14.

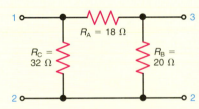

FIG. 10-32   Circuit for Prob. 15.

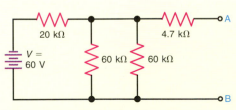

FIG. 10-33   Circuit for Prob. 16.

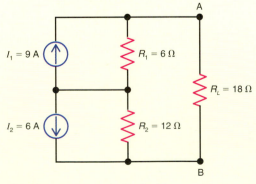

FIG. 10-34   Circuit for Prob. 17.

18. In Fig. 10-35, show the Thevenin equivalent circuit driving terminals A and B. Also, calculate the load current I_L.
19. Show the Thevenin equivalent circuit driving terminals A and B in Fig. 10-36. Also, calculate V_L.
20. In Fig. 10-36, show the Norton equivalent circuit driving terminals A and B. Also, calculate I_L.
21. In Fig. 10-37, show the Thevenin equivalent circuit driving terminals A and B. Also, calculate the values for I_L and V_L.
22. In Fig. 10-38, show the Thevenin equivalent circuit driving terminals A and B. Also, calculate I_L and V_L.
23. In Fig. 10-38, show the Norton equivalent circuit driving terminals A and B. Also, calculate I_L and V_L.

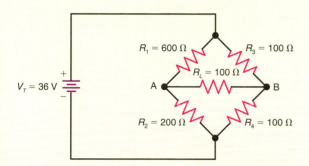

FIG. 10-35 Circuit for Prob. 18.

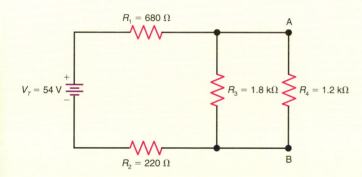

FIG. 10-36 Circuit for Probs. 19 and 20.

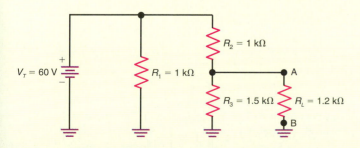

FIG. 10-37 Circuit for Prob. 21.

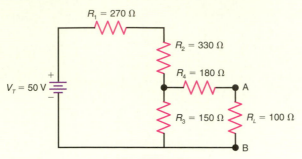

FIG. 10-38   Circuit for Probs. 22 and 23.

24. In Fig. 10-39, show the Thevenin equivalent circuit driving terminals A and B. Also, calculate I_3 and V_{R_3}.

25. In Fig. 10-39, use the superposition theorem to find currents I_1, I_2, I_3 and voltages V_{R_1}, V_{R_2}, and V_{R_3}.

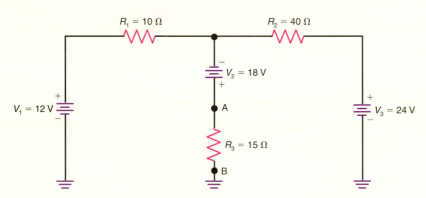

FIG. 10-39   Circuit for Probs. 24 and 25.

CRITICAL THINKING

1. Thevenize the circuit driving terminals A and B in Fig. 10-40. Show the Thevenin equivalent circuit and calculate the values for I_L and V_L.

2. In Fig. 10-41, use the superposition theorem to solve for I_L and V_L.

3. In Fig. 10-41, show the Thevenin equivalent circuit driving terminals A and B.

4. Refer to Fig. 10-35. Remove R_2 from the circuit and show the Thevenin equivalent circuit driving the open terminals. Also, calculate the value of I_2 and V_{R_2}.

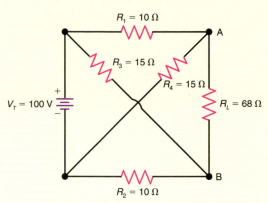

FIG. 10-40 Circuit for Critical Thinking Prob. 1.

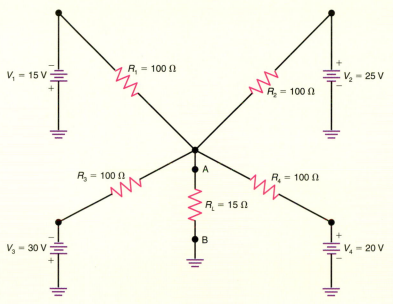

FIG. 10-41 Circuit for Critical Thinking Probs. 2 and 3.

REVIEW: CHAPTERS 9 AND 10

SUMMARY

- Methods of applying Kirchhoff's laws include (**a**) equations of voltages using the branch currents in the loops to specify the voltages; (**b**) equations of currents at a node using the node voltages to specify the node currents; (**c**) equations of voltages using assumed mesh currents to specify the voltages.
- Methods of reducing a network to a simple equivalent circuit include (**a**) the superposition theorem using one source at a time; (**b**) Thevenin's theorem to convert the network to a series circuit with one source; (**c**) Norton's theorem to convert the network to a parallel circuit with one source; (**d**) Millman's theorem to find the common voltage across parallel branches with different sources; (**e**) delta (Δ) wye (Y) conversions to transform a network into a series-parallel circuit.

REVIEW TEST

ANSWERS AT BACK OF BOOK.

Answer True or False.

1. In Fig. 9-3, V_3 can be found by using Kirchhoff's laws with either branch currents or mesh currents.
2. In Fig. 9-3, V_3 can be found by superposition, thevenizing, or using Millman's theorem.
3. In Fig. 10-6, I_L cannot be found by delta-wye conversion because R_L disappears in the transformation.
4. In Fig. 10-6, I_L can be calculated with Kirchhoff's laws, using mesh currents for three meshes.
5. With superposition, we can use Ohm's law for circuits that have more than one source.
6. A Thevenin equivalent is a parallel circuit.
7. A Norton equivalent is a series circuit.
8. Either a Thevenin or a Norton equivalent of a network will produce the same current in any load across terminals A and B.
9. A Thevenin-to-Norton conversion means converting a voltage source to a current source.
10. The volts unit is equal to (volts/ohms) ÷ siemens.
11. A node voltage is a voltage between current nodes.
12. A π network can be converted to an equivalent T network.

13. A 10-V source with 10-Ω series R will supply 5 V to a 10-Ω load R_L.
14. A 10-A source with 10-Ω parallel R will supply 5 A to a 10-Ω load R_L.
15. Current sources in parallel can be added when they supply current in the same direction through R_L.

REFERENCES

Hayt and Kemmerling: *Engineering Circuit Analysis,* McGraw-Hill, New York.

Kaufman and Seldman: *Handbook of Electronics Calculations for Engineers and Technicians,* McGraw-Hill, New York.

Nasar, S.: *3000 Solved Problems in Electric Circuits,* McGraw-Hill, New York.

CHAPTER 11

CONDUCTORS AND INSULATORS

Conductors have very low resistance. An R of 0.1 Ω or less for 10 ft of copper wire is a typical value. The function of the wire conductor is to connect a source of applied voltage to a load with minimum IR voltage drop in the conductor. Then practically all the applied voltage can produce current in the load.

At the opposite extreme, materials having a very high resistance of many megohms are insulators. Some common examples are air, paper, plastics, rubber, mica, glass, cotton, shellac or varnish, and wood. An insulator provides the equivalent of an open circuit with almost infinite R and practically zero I.

Between the extremes of conductors and insulators are semiconductor materials such as carbon (C), silicon (Si), and germanium (Ge). Carbon is used in the manufacture of carbon resistors. Si and Ge are used for transistors. Just about all semiconductor devices are made with silicon. It should be noted, though, that the medium resistance of Si and Ge is not so important as the fact their electrical characteristics can be altered to provide free charges to be controlled by a small applied voltage.

In a class by itself is the category of superconductors. As the name says, these materials have the superb characteristic of zero resistance. However, this very desirable feature can be provided only with special low-temperature conditions or unique ceramic materials.

CHAPTER OBJECTIVES

Upon completion of this chapter, you should be able to:

- *Explain* the main function of a conductor in an electric circuit.
- *Determine* the circular area of round wire when the wire diameter is known.
- *List* the advantages of using stranded wire versus solid wire.
- *List* several common types of connectors used with wire conductors.
- *Define* the terms *pole* and *throw* as they relate to switches.
- *Explain* the difference between a fast-acting and slow-blow fuse.
- *Calculate* the total resistance of a wire conductor whose length, cross-sectional area, and specific resistance are known.
- *Understand* what is meant by the *temperature coefficient of resistance.*
- *Explain* the difference between ion current and electron current.
- *List* three types of semiconductors.
- *Explain* the makeup of N-type and P-type semiconductor materials.
- *Explain* why insulators are sometimes called dielectrics.

IMPORTANT TERMS IN THIS CHAPTER

characteristic impedance	fuse types	RS-232 plug
circuit breaker	hole current	specific resistance
circular mil	hot resistance	superconductor
coaxial cable	ion current	switch types
corona effect	printed wiring	temperature coefficient of *R*
covalent bond	RCA plug	twin lead
cryogenics	resistance wire	wire gage sizes
F connector	ribbon cable	

TOPICS COVERED IN THIS CHAPTER

11-1 FUNCTION OF THE CONDUCTOR

In Fig. 11-1, the resistance of the two 10-ft lengths of copper-wire conductor is 0.08 Ω. This R is negligibly small compared with the 144-Ω R for the tungsten filament in the light bulb. When the current of 0.833 A flows in the bulb and the series conductors, the IR voltage drop of the conductors is only 0.07 V, with 119.93 V across the bulb. Practically all the applied voltage is across the bulb filament. Since the bulb then has its rated voltage of 120 V, approximately, it will dissipate the rated power of 100 W and light with full brilliance.

The current in the wire conductors and the bulb is the same, since they are in series. However, the IR voltage drop in the conductor is practically zero because its R is almost zero.

Also, the I^2R power dissipated in the conductor is negligibly small, allowing the conductor to operate without becoming hot. Therefore, the conductor delivers energy from the source to the load with minimum loss, by means of electron flow in the copper wires.

Although the resistance of wire conductors is very small, for some cases of high current the resultant IR drop can be appreciable. For example, suppose the 120-V power line is supplying 30 A of current to a load through two conductors, each of which has a resistance of 0.2 Ω. In this case, each conductor has an IR drop of 6 V, calculated as 30 A × 0.2 Ω = 6 V. With each conductor dropping 6 V, the load receives a voltage of only 108 V rather than the full 120 V. The lower-than-normal load voltage could result in the load not operating properly. Furthermore, the I^2R power dissipated in each conductor equals 180 W, calculated as 30^2 × 0.2 Ω = 180 W. The I^2R power loss of 180 W in the conductors is considered to be excessively high.

TEST-POINT QUESTION 11-1

Answers at end of chapter.

Refer to Fig. 11-1.
a. How much is R for the 20 ft of copper wire?
b. How much is the IR voltage drop for the wire conductors?
c. The voltage in **b** is what percent of the applied voltage?

11-2 STANDARD WIRE GAGE SIZES

Table 11-1 lists the standard wire sizes in the system known as the American Wire Gage (AWG), or Brown and Sharpe (B&S) gage. The gage numbers specify the size of round wire in terms of its diameter and cross-sectional area. Note the following:

1. As the gage numbers increase from 1 to 40, the diameter and circular area decrease. Higher gage numbers indicate thinner wire sizes.
2. The circular area doubles for every three gage sizes. For example, No. 10 wire has approximately twice the area of No. 13 wire.
3. The higher the gage number and the thinner the wire, the greater the resistance of the wire for any given length.

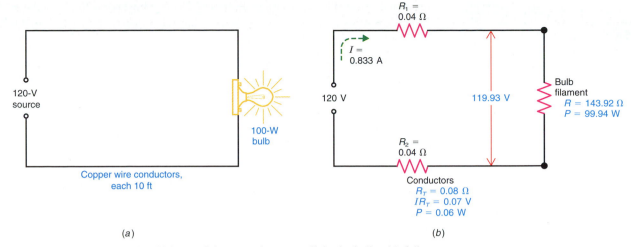

FIG. 11-1 The conductors should have minimum resistance to light the bulb with full brilliance. (*a*) Wiring diagram. (*b*) Schematic diagram. R_1 and R_2 represent the very small resistance of the wire conductors.

In typical applications, hookup wire for electronic circuits with current in the order of milliamperes is generally about No. 22 gage. For this size, 0.5 to 1 A is the maximum current the wire can carry without excessive heating.

House wiring for circuits where the current is 5 to 15 A is usually No. 14 gage. Minimum sizes for house wiring are set by local electrical codes, which

TABLE 11-1 COPPER-WIRE TABLE

GAGE NO.	DIAMETER, MILS	AREA, CIRCULAR-MILS	OHMS PER 1000 FT OF COPPER WIRE AT 25°C*	GAGE NO.	DIAMETER, MILS	AREA, CIRCULAR-MILS	OHMS PER 1000 FT OF COPPER WIRE AT 25°C*
1	289.3	83,690	0.1264	21	28.46	810.1	13.05
2	257.6	66,370	0.1593	22	25.35	642.4	16.46
3	229.4	52,640	0.2009	23	22.57	509.5	20.76
4	204.3	41,740	0.2533	24	20.10	404.0	26.17
5	181.9	33,100	0.3195	25	17.90	320.4	33.00
6	162.0	26,250	0.4028	26	15.94	254.1	41.62
7	144.3	20,820	0.5080	27	14.20	201.5	52.48
8	128.5	16,510	0.6405	28	12.64	159.8	66.17
9	114.4	13,090	0.8077	29	11.26	126.7	83.44
10	101.9	10,380	1.018	30	10.03	100.5	105.2
11	90.74	8234	1.284	31	8.928	79.70	132.7
12	80.81	6530	1.619	32	7.950	63.21	167.3
13	71.96	5178	2.042	33	7.080	50.13	211.0
14	64.08	4107	2.575	34	6.305	39.75	266.0
15	57.07	3257	3.247	35	5.615	31.52	335.0
16	50.82	2583	4.094	36	5.000	25.00	423.0
17	45.26	2048	5.163	37	4.453	19.83	533.4
18	40.30	1624	6.510	38	3.965	15.72	672.6
19	35.89	1288	8.210	39	3.531	12.47	848.1
20	31.96	1022	10.35	40	3.145	9.88	1069

*20° to 25°C or 68° to 77°F is considered average room temperature.

are usually guided by the National Electrical Code published by the National Fire Protection Association. A gage for measuring wire size is shown in Fig. 11-2.

CIRCULAR MILS The cross-sectional area of round wire is measured in circular mils, abbreviated cmil. A mil is one-thousandth of an inch, or 0.001 in. One circular mil is the cross-sectional area of a wire with a diameter of 1 mil. The number of circular mils in any circular area is equal to the square of the diameter in mils.

Example

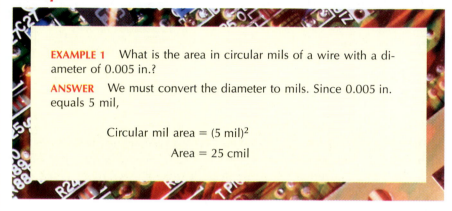

EXAMPLE 1 What is the area in circular mils of a wire with a diameter of 0.005 in.?

ANSWER We must convert the diameter to mils. Since 0.005 in. equals 5 mil,

$$\text{Circular mil area} = (5 \text{ mil})^2$$
$$\text{Area} = 25 \text{ cmil}$$

Note that the circular mil is a unit of area, obtained by squaring the diameter, while the mil is a linear unit of length, equal to one-thousandth of an inch. Therefore, the circular-mil area increases as the square of the diameter. As illustrated in Fig. 11-3, doubling the diameter quadruples the area. Circular mils are convenient for round wire because the cross section is specified without using the formula πr^2 or $\pi d^2/4$ for the area of a circle.

ABOUT
ELECTRONICS

Airplane skin can be "undented" with an electromagnetic device that uses strong fields to unbend the metal.

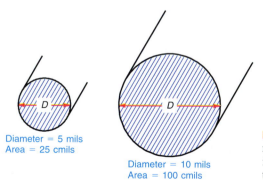

Diameter = 5 mils
Area = 25 cmils

Diameter = 10 mils
Area = 100 cmils

FIG. 11-3 Cross-sectional area for round wire. Doubling the diameter increases the circular area by four times.

TEST-POINT QUESTION 11-2

Answers at end of chapter.
a. How much is R for 1 ft of No. 22 wire?
b. What is the cross-sectional area in circular mils for wire with a diameter of 0.025 in.?
c. What is the wire gage size in Fig. 11-1?

11-3 TYPES OF WIRE CONDUCTORS

Most wire conductors are copper, although aluminum and silver are used sometimes. The copper may be tinned with a thin coating of solder, which gives it a silvery appearance. Tinned wire is easier to solder for connections. The wire can be solid or stranded, as shown in Fig. 11-4*a* and *b*.

Stranded wire is flexible, easier to handle and less likely to develop an open break. Sizes for stranded wire are equivalent to the sum of the areas for the individual strands. For instance, two strands of No. 30 wire correspond to solid No. 27 wire.

Very thin wire, such as No. 30, often has an insulating coating of enamel or shellac. It may look like copper, but the coating must be scraped off the ends to make a good connection. This type of wire is used for small coils.

Heavier wires generally are in an insulating sleeve, which may be rubber or one of many plastic materials. General-purpose wire for connecting electronic components is generally plastic-coated hookup wire of No. 20 gage. Hookup wire that is bare should be enclosed in a hollow insulating sleeve called *spaghetti*.

The braided conductor in Fig. 11-4*c* is used for very low resistance. It is wide for low *R* and thin for flexibility, and the braiding provides many strands. A common application is a grounding connection, which must have very low *R*.

(a)   (b)   (c)

(d)   (e)

FIG. 11-4 Types of wire conductors. (*a*) Solid wire. (*b*) Stranded wire. (*c*) Braided wire for very low *R*. (*d*) Coaxial cable. Note braided wire for shielding the inner conductor. (*e*) Twin-lead cable.

WIRE CABLE   Two or more conductors in a common covering form a cable. Each wire is insulated from the others. Cables often include two, three, ten, or many more pairs of conductors, usually color-coded to help identify the conductors at both ends of a cable.

The ribbon cable in Fig. 11-5 has multiple conductors but not in pairs. This cable is used for multiple connections to a computer and associated equipment.

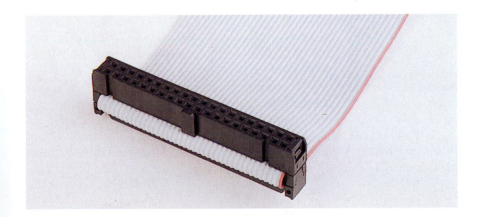

FIG. 11-5   Ribbon cable with multiple conductors.

TRANSMISSION LINES   Constant spacing between two conductors through the entire length provides a transmission line. Common examples are the coaxial cable in Fig. 11-4*d* and the twin lead in Fig 11-4*e*.

Coaxial cable with an outside diameter of ¼ in. is generally used for the signals in cable television. In construction, there is an inner solid wire, insulated from metallic braid that serves as the other conductor. The entire assembly is covered by an outer plastic jacket. In operation, the inner conductor has the desired signal voltage with respect to ground, while the metallic braid is connected to ground to shield the inner conductor against interference. Coaxial cable, therefore, is a shielded type of transmission line.

With twin-lead wire, two conductors are imbedded in plastic to provide constant spacing. This type of line is commonly used in television for connecting the antenna to the receiver. In this application, the spacing is ⅝ in. between wires of No. 20 gage size, approximately. This line is not shielded.

CHARACTERISTIC IMPEDANCE   The coaxial cable and twin-lead transmission lines are so common in television that they can be useful to consider an important characteristic that is usually reserved for more advanced ac theory. Because of the constant spacing, such line has a constant value of impedance between the two conductors. The term *impedance,* with the symbol Z, specifies an opposition to current that can include resistance, inductance, and capacitance.

Any transmission line has a constant value of impedance across the two conductors that is characteristic of the line, depending on the spacing, size of the conductors, and the insulation between the conductors. This value is specified as the *characteristic impedance,* with the symbol Z_0, for that type of transmission line. The Z_0 is constant for any length because it depends on the square root of the ratio for the inductance in the line to the capacitance between the conductors.

For ¼-in. coaxial cable, its characteristic impedance is approximately 75 Ω. This 75-Ω line is used in cable distribution systems for television. For twin-lead wire with ⅜-in. spacing between the two conductors in a plastic ribbon, the characteristic impedance is 300 Ω. This 300-Ω line is generally used for connecting an antenna to the receiver, either TV or FM radio.

The characteristic impedance is the same for any length of line because Z_0 depends mainly on the spacing between conductors and not at all on how long they are.

It is important to realize that Z_0 is an ac value that cannot be measured with an ohmmeter. However, you can check the continuity of each length of conductor with an ohmmeter. The R should be practically zero ohms for the wire in the line. If there is a break in the wire, an ohmmeter reading of infinity will show the open circuit.

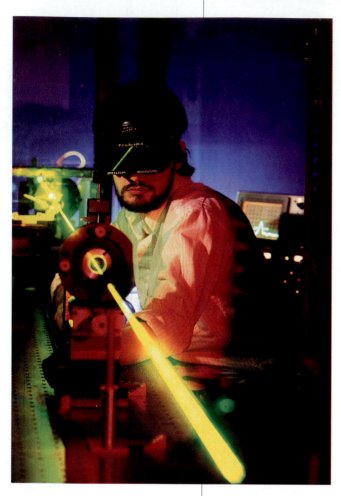

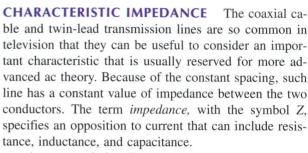

| TEST-POINT QUESTION 11-3 |

Answers at end of chapter.

Answer True or False.
a. The plastic coating on wire conductors has very high R.
b. Coaxial cable is a shielded transmission line.
c. The characteristic impedance of transmission lines cannot be measured with an ohmmeter.

11-4 CONNECTORS

Refer to Fig. 11-6 for different types. The spade lug in Fig. 11-6*a* is often used for screw-type terminals. In Fig. 11-6*b*, the alligator clip is convenient for a temporary connection. Alligator clips come in small and large sizes. The banana pins in Fig. 11-6*c* have spring-type sides that make a tight connection. In Fig. 11-6*d*, the terminal strip provides a block for multiple solder connections.

In Fig. 11-6*e*, the RCA-type plug is commonly used for shielded cables with audio equipment. The inner conductor of the coaxial cable is connected to the center pin of the plug and the cable braid to the shield. Both connections must be soldered.

The phone plug in Fig. 11-6*f* is still used in some applications, but in a smaller size. Note that the ring is insulated from the sleeve, to provide for two connections. There may be a separate tip, ring, and sleeve for three connections. The sleeve is usually the ground side.

The plug in Fig. 11-6*g* is called an *F connector*. It is universally used in cable television because of its convenience. The center conductor of the coaxial cable serves as the center pin of the plug, so that no soldering is needed. Also, the shield on the plug is press-fit onto the braid of the cable, underneath the plastic jacket.

All these are male-type plugs, with pins that fit the corresponding female-type socket. The connector that has the pins is male. However, the plug can be either male or female. The connector for power input usually is a female plug mating with a male socket on the equipment.

The multiple connector in Fig. 11-6*h* is shown with a ribbon cable. It has eight wires in a flat, plastic ribbon. This plug is a female type that mates with a male socket. The ribbon cable often has more than eight wires, as shown in Fig. 11-5*a*.

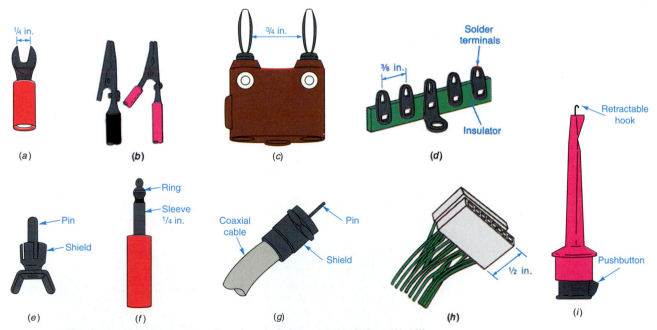

FIG. 11-6 Common types of connectors for wire conductors. (*a*) Spade lug. (*b*) Alligator clips. (*c*) Double banana-pin plug. (*d*) Terminal strip. (*e*) RCA-type plug for audio cables. (*f*) Phone plug. (*g*) F-type plug for cable TV. (*h*) Multiple-pin connector plug. (*i*) Spring-loaded metal hook as grabber for temporary connection in testing circuits.

The RS-232 plug is a standard connector for computer equipment. It has 25 pins. Each is assigned one function for data interface connections, as specified in standards of the Electronic Industries Association (EIA). The male pin plug is normally installed on the computer, the female connector being used for auxiliary equipment.

TEST-POINT QUESTION 11-4

Answers at end of chapter.

Answer True or False.
a. Cable television uses 75-Ω coaxial line.
b. The F-type connector is used with coaxial cable.
c. The RS-232 connector takes two pairs of 300-Ω twin-lead line.

11-5 PRINTED WIRING

Most electronic circuits are mounted on a plastic insulating board with printed wiring, as shown in Fig. 11-7. This is a printed-circuit (PC) or printed-wiring (PW) board. One side has the components, such as resistors, capacitors, coils, transistors, diodes, and integrated-circuit (IC) units. The other side has the conducting paths printed with silver or copper on the board, instead of using wires. On a double-sided board, the component side also has printed wiring. Sockets, small metal eyelets, or just holes in the board are used to connect the components to the wiring.

With a bright light on one side, you can see through to the opposite side to trace the connections. However, the circuit is usually drawn on the PC board.

It is important not to use too much heat in soldering or desoldering. Otherwise the printed wiring can be lifted off the board. Use a small iron of about 25

FIG. 11-7 Printed wiring board. (*a*) Component side with resistors, capacitors, and transistors. (*b*) Side with printed wiring for the circuit.

(a)

(b)

to 35 W rating. When soldering semiconductor diodes and transistors, hold the lead with pliers or connect an alligator clip as a heat sink to conduct heat away from the semiconductor junction.

For desoldering, use a solder-sucker tool, with a soldering iron, to clean each terminal. Another method is to use wire braid. Put the braid on the joint and heat it until the solder runs up into the braid. The terminal must be clean enough to lift out the component easily without damaging the PC board.

A small crack in the printed wiring acts like an open circuit preventing current flow. Cracks can be repaired by soldering a short length of bare wire over the open circuit. If a larger section of printed wiring is open, or if the board is cracked, you can bridge the open circuit with a length of hookup wire soldered at two convenient end terminals of the printed wiring.

Answers at end of chapter.
a. Which is the best size iron to use to solder on a PC board, 25, 100, or 150 W?
b. How much is the resistance of a printed-wire conductor with a crack in the middle?

11-6 SWITCHES

A switch is a component which allows us to control whether the current is ON or OFF in a circuit. A closed switch has practically zero resistance, whereas an open switch has nearly infinite resistance. Figure 11-8 shows a switch in series with a voltage source and a light bulb. With the switch closed, as in Fig. 11-8a, a complete path for current is provided and the light is ON. Since the switch has a very low resistance when it is closed, all of the source voltage is across the load, with 0 V across the closed contacts of the switch. With the switch open, as in Fig. 11-8b, the path for current is interrupted and the bulb does not light. Since the switch has a very high resistance when it is open, all of the source voltage is across the open switch contacts, with 0 V across the load.

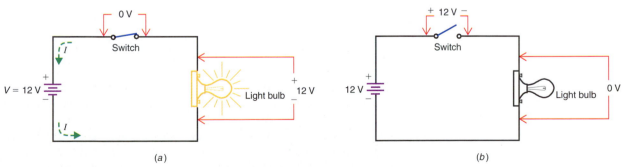

FIG. 11-8   Series switch used to open or close a circuit. (a) With the switch closed, current flows to light the bulb. The voltage drop across the closed switch is 0 V. (b) With the switch open, the light is OFF. The voltage drop across the open switch is 12 V.

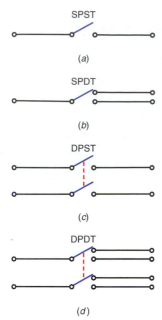

SPST

(a)

SPDT

(b)

DPST

(c)

DPDT

(d)

FIG. 11-9 Switches. *(a)* Single-pole, single-throw (SPST). *(b)* Single-pole, double-throw (SPDT). *(c)* Double-pole, single-throw (DPST). *(d)* Double-pole, double-throw (DPDT).

SWITCH RATINGS   All switches have a current rating and a voltage rating. The current rating corresponds to the maximum allowable current the switch can carry when it is closed. The current rating is based on the physical size of the switch contacts as well as the type of metal used for the contacts. Many switches have gold- or silver-plated contacts to ensure a very low resistance when closed.

The voltage rating of a switch corresponds to the maximum voltage that can safely be applied across the open contacts without internal arcing. The voltage rating does not apply when the switch is closed, since the voltage drop across the closed switch contacts is practically zero.

SWITCH DEFINITIONS   Toggle switches are usually described as having a certain number of poles and throws. For example, the switch in Fig. 11-8 is described as a *single-pole, single-throw* (SPST) switch. Other popular switch types include the *single-pole, double-throw* (SPDT), *double-pole, single-throw* (DPST), and *double-pole, double-throw* (DPDT). The schematic symbols for each type are shown in Fig. 11-9. Notice that the SPST switch has two connecting terminals, whereas the SPDT has three, the DPST has four, and the DPDT has six.

The term *pole* can be defined as the number of completely isolated circuits that can be controlled by the switch. The term *throw* can be defined as the number of closed contact positions that exist per pole. The SPST switch in Fig. 11-8 can only control the current in one circuit and there is only one closed contact position, hence the name single-pole, single-throw.

Figure 11-10 shows a variety of different switch applications. In Fig. 11-10a, a SPDT switch is being used to switch a 12-V dc source between one of two different loads. In Fig. 11-10b, a DPST switch is being used to control two completely separate circuits simultaneously. In Fig. 11-10c, a DPDT switch is being used to reverse the polarity of voltage across the terminals of a dc motor. (Reversing the polarity reverses the direction of the motor.) It is important to note that the dashed lines shown between the poles in Figs. 11-10b and 11-10c indicate that both sets of contacts within the switch are opened and closed simultaneously.

SWITCH TYPES   Figure 11-11 shows a variety of different toggle switches. Although the toggle switch is a very popular type of switch, several other types are found in electronic equipment. Additional types include push-button switches, rocker switches, slide switches, rotary switches, and DIP switches.

Push-button switches are often spring-loaded switches that are either normally open (NO) or normally closed (NC). Figure 11-12 shows the schematic symbols for both types. For the normally open switch in Fig. 11-12a, the switch contacts remain open until the push button is depressed. When the push button is depressed, the switch closes, allowing current to pass. The normally closed switch in Fig. 11-12b operates just the opposite of the normally open switch in Fig. 11-12a. When the push button is depressed, the switch contacts open to interrupt current in the circuit. A typical push-button switch is shown in Fig. 11-13.

Figure 11-14 shows a DIP (dual-inline package) switch. Its construction consists of eight miniature rocker switches, where each switch can be set separately. A DIP switch has pin connections which fit into a standard IC socket.

Figure 11-15 shows another type of switch known as a rotary switch. As shown, its construction consists of three wafers or decks mounted on a common shaft.

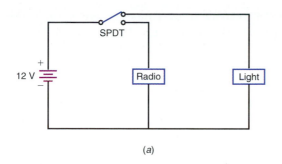

(a)

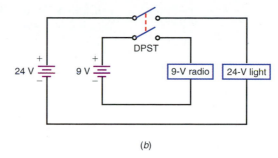

(b)

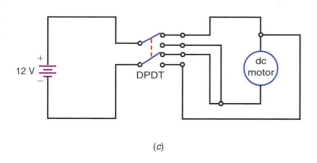

(c)

FIG. 11-10   Switch applications. (*a*) SPDT switch used to switch a 12-V source between one of two different loads. (*b*) DPST switch controlling two completely isolated circuits simultaneously. (*c*) DPDT switch used to reverse the polarity of voltage across a dc motor.

TEST-POINT QUESTION 11-6

Answers at end of chapter.
a. How much is the *IR* voltage drop across a closed switch?
b. How many connections are there on an SPDT switch?
c. What is the resistance across the contacts of an open switch?
d. An SPST switch is rated at 10 A 250 V. Should the switch be used to turn a 120-V 1500-W heater ON and OFF?

FIG. 11-11   A variety of toggle switches.

11-7 FUSES

Many circuits have a fuse in series as a protection against an overload resulting from a short circuit. Excessive current melts the fuse element, blowing the fuse and opening the series circuit. The purpose is to let the fuse blow before the components are damaged. The blown fuse can easily be replaced by a new one, after the overload has been eliminated. A glass-cartridge fuse with holders is shown in Fig. 11-16. This is a type 3AG fuse, with a diameter of ¼ in. and length of 1¼ in. AG is an abbreviation of "automobile glass," since that was one of the first applications of fuses in a glass holder to make the wire link visible.

The metal fuse element may be made of aluminum, tin-coated copper, or nickel. Fuses are available with current ratings from ¹⁄₅₀₀ A to hundreds of amperes. The thinner the wire element in the fuse, the smaller its current rating. For example, a 2-in. length of No. 28 wire can serve as a 2-A fuse. As typical applications, the rating for plug fuses in each branch of house wiring is often 15 A; the high-voltage circuit in a television receiver is usually protected by a glass-

Normally open (NO)   Normally closed (NC)

(a)   (b)

FIG. 11-12   Push-button switch schematic symbols. (*a*) Normally open (NO) push-button switch. (*b*) Normally closed (NC) push-button switch.

FIG. 11-13 Typical push-button switch.

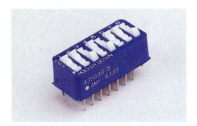

FIG. 11-14 Dual-inline package (DIP) switch.

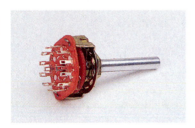

FIG. 11-15 Rotary switch.

cartridge ¼-A fuse. For automobile fuses, the ratings are generally 10 to 30 A because of the higher currents needed with a 12-V source for a given amount of power.

SLOW-BLOW FUSES　These have a coiled construction. They are designed to open only on a continued overload, such as a short circuit. The purpose of coiled construction is to prevent the fuse from blowing on just a temporary current surge. As an example, a slow-blow fuse will hold a 400 percent overload in current for up to 2 s. Typical ratings are shown by the curves in Fig. 11-17. Circuits with an electric motor use slow-blow fuses because the starting current of a motor is much more than its running current.

CIRCUIT BREAKERS　These have a thermal element in the form of a spring. The spring expands with heat and trips open the circuit. The circuit breaker can be reset for normal operation, however, after the short has been eliminated and the thermal element cools down. Large circuit breakers use coils to detect excessive current and open a circuit.

WIRE LINKS　A short length of bare wire is sometimes used as a fuse in television receivers. For instance, a 2-in. length of No. 24 gage wire can hold a 500-mA current but burn open with an overload. The wire link can be mounted between two terminal strips on the chassis. Or, the wire link may be wrapped over a small insulator to make a separate component.

TESTING FUSES　With glass fuses, you can usually see if the wire element inside is burned open. When measured with an ohmmeter, a good fuse has practically zero resistance. An open fuse reads infinite ohms. Power must be off or the fuse must be out of the circuit to test a fuse with an ohmmeter.

When you test with a voltmeter, a good fuse has zero volts across its two terminals (Fig. 11-18a). If you read appreciable voltage across the fuse, this means it is open. In fact, the full applied voltage is across the open fuse in a series circuit, as shown in Fig. 11-18b. This is why fuses also have a voltage rating, which gives the maximum voltage without arcing in the open fuse.

Referring to Fig. 11-18, notice the results when measuring the voltages to ground at the two fuse terminals. In Fig. 11-18a, the voltage is the same 120 V at both ends because there is no voltage drop across the good fuse. In Fig. 11-18b, however, terminal B reads 0 V, as this end is disconnected from V_T because of the open fuse. These tests apply to either dc or ac voltages.

(a)

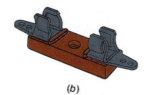

(b)

(c)

FIG. 11-16　(a) Glass cartridge fuse. (b) Fuse holder. (c) Panel-mounted fuse holder.

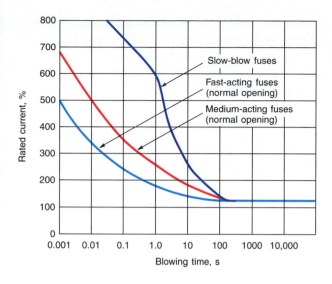

FIG. 11-17 Chart showing percent of rated current to the blowing time for fuses.

TEST-POINT QUESTION 11-7

Answers at end of chapter.
a. How much is the resistance of a good fuse?
b. How much is the *IR* voltage drop across a good fuse?

11-8 WIRE RESISTANCE

The longer a wire, the higher its resistance. More work must be done to make electrons drift from one end to the other. However, the greater the diameter of the wire, the less the resistance, since there are more free electrons in the cross-sectional area. As a formula,

$$R = \rho \frac{l}{A} \qquad (11\text{-}1)$$

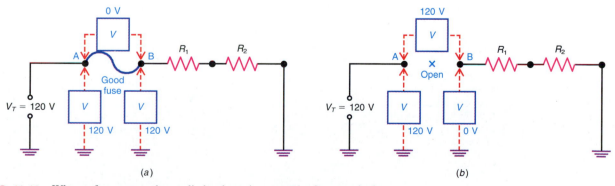

FIG. 11-18 When a fuse opens, the applied voltage is across the fuse terminals. (*a*) Circuit closed with good fuse. Note schematic symbol for any type of fuse. (*b*) Fuse open. Voltage readings are explained in the text.

where R is the total resistance, l the length, A the cross-sectional area, and ρ the specific resistance or *resistivity* of the conductor. The factor ρ then enables the resistance of different materials to be compared according to their nature without regard to different lengths or areas. Higher values of ρ mean more resistance. Note that ρ is the Greek letter *rho,* corresponding to r.

SPECIFIC RESISTANCE   Table 11-2 lists resistance values for different metals having the standard wire size of a 1-ft length with a cross-sectional area of 1 cmil. This rating is the *specific resistance* of the metal, in circular-mil ohms per foot. Since silver, copper, gold, and aluminum are the best conductors, they have the lowest values of specific resistance. Tungsten and iron have a much higher resistance.

Example

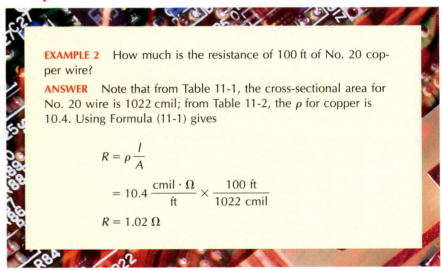

EXAMPLE 2   How much is the resistance of 100 ft of No. 20 copper wire?

ANSWER   Note that from Table 11-1, the cross-sectional area for No. 20 wire is 1022 cmil; from Table 11-2, the ρ for copper is 10.4. Using Formula (11-1) gives

$$R = \rho \frac{l}{A}$$

$$= 10.4 \, \frac{\text{cmil} \cdot \Omega}{\text{ft}} \times \frac{100 \text{ ft}}{1022 \text{ cmil}}$$

$$R = 1.02 \, \Omega$$

TABLE 11-2   PROPERTIES OF CONDUCTING MATERIALS*

MATERIAL	DESCRIPTION AND SYMBOL	SPECIFIC RESISTANCE (ρ) AT 20°C, cmil · Ω/ft	TEMPERATURE COEFFICIENT PER °C, α	MELTING POINT, °C
Aluminum	Element (Al)	17	0.004	660
Carbon	Element (C)	†	−0.0003	3000
Constantan	Alloy, 55% Cu, 45% Ni	295	0 (average)	1210
Copper	Element (Cu)	10.4	0.004	1083
Gold	Element (Au)	14	0.004	1063
Iron	Element (Fe)	58	0.006	1535
Manganin	Alloy, 84% Cu, 12% Mn, 4% Ni	270	0 (average)	910
Nichrome	Alloy 65% Ni, 23% Fe, 12% Cr	676	0.0002	1350
Nickel	Element (Ni)	52	0.005	1452
Silver	Element (Ag)	9.8	0.004	961
Steel	Alloy, 99.5% Fe, 0.5% C	100	0.003	1480
Tungsten	Element (W)	33.8	0.005	3370

*Listings approximate only, since precise values depend on exact composition of material.
†Carbon has about 2500 to 7500 times the resistance of copper. Graphite is a form of carbon.

All the units cancel except the ohms for R. Note that 1.02 Ω for 100 ft is approximately ¹⁄₁₀ the resistance of 10.35 Ω for 1000 ft of No. 20 copper wire listed in Table 11-1, showing that the resistance is proportional to length.

Example

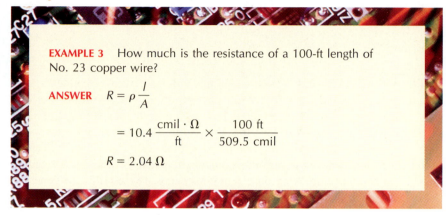

EXAMPLE 3   How much is the resistance of a 100-ft length of No. 23 copper wire?

ANSWER $R = \rho \dfrac{l}{A}$

$$= 10.4 \frac{\text{cmil} \cdot \Omega}{\text{ft}} \times \frac{100 \text{ ft}}{509.5 \text{ cmil}}$$

$$R = 2.04 \ \Omega$$

Note that a wire three gage sizes higher has half the circular area and double the resistance for the same wire length.

UNITS OF OHM-CENTIMETERS FOR ρ   Except for wire conductors, specific resistances are usually compared for the standard size of a 1-cm cube. Then ρ is specified in $\Omega \cdot$ cm for the unit cross-sectional area of 1 cm².

As an example, pure germanium has $\rho = 55 \ \Omega \cdot$ cm, as listed in Table 11-3. This value means that R is 55 Ω for a cube with a cross-sectional area of 1 cm² and length of 1 cm.

For other sizes, use Formula (11-1) with l in cm and A in cm². Then all the units of size cancel to give R in ohms.

Example

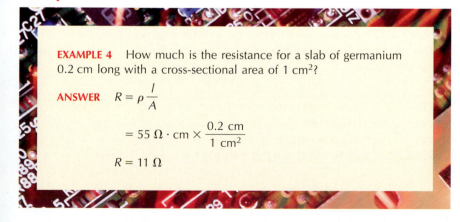

EXAMPLE 4   How much is the resistance for a slab of germanium 0.2 cm long with a cross-sectional area of 1 cm²?

ANSWER $R = \rho \dfrac{l}{A}$

$$= 55 \ \Omega \cdot \text{cm} \times \frac{0.2 \text{ cm}}{1 \text{ cm}^2}$$

$$R = 11 \ \Omega$$

The same size slab of silicon would have R of 11,000 Ω. Note from Table 11-3 that ρ is 1000 times more for silicon than for germanium.

TABLE 11-3 COMPARISON OF RESISTIVITIES		
MATERIAL	**ρ, $\Omega \cdot$ cm, AT 25°C**	**DESCRIPTION**
Silver	1.6×10^{-6}	Conductor
Germanium	55	Semiconductor
Silicon	55,000	Semiconductor
Mica	2×10^{12}	Insulator

TYPES OF RESISTANCE WIRE   For applications in heating elements, such as a toaster, an incandescent light bulb, or a heater, it is necessary to use wire that has more resistance than good conductors like silver, copper, or aluminum. Higher resistance is preferable so that the required amount of I^2R power dissipated as heat in the wire can be obtained without excessive current. Typical materials for resistance wire are the elements tungsten, nickel, or iron and alloys* such as manganin, Nichrome, and constantan. These types are generally called resistance wire because R is greater than for copper wire, for the same length.

TEST-POINT QUESTION 11-8

Answers at end of chapter.
a. Does Nichrome wire have less or more resistance than copper wire?
b. For 100 ft of No. 14 copper wire, R is 0.26 Ω. How much is R for 1000 ft?

11-9 TEMPERATURE COEFFICIENT OF RESISTANCE

This factor with the symbol alpha (α) states how much the resistance changes for a change in temperature. A positive value for α means R increases with temperature; with a negative α, R decreases; zero for α means R is constant. Some typical values of α, for metals and for carbon, are listed in Table 11-2 in the fourth column.

POSITIVE α   All metals in their pure form, such as copper and tungsten, have a positive temperature coefficient. The α for tungsten, for example, is 0.005. Although α is not exactly constant, an increase in wire resistance caused by a rise in temperature can be calculated approximately from the formula

▶ $$R_t = R_0 + R_0(\alpha \Delta t) \qquad \text{(11-2)}$$

---

*An *alloy* is a fusion of elements, without chemical action between them. Metals are commonly alloyed to alter their physical characteristics.

where R_0 is the resistance at 20°C, R_t is the higher resistance at the higher temperature, and Δt is the temperature rise above 20°C.

Example

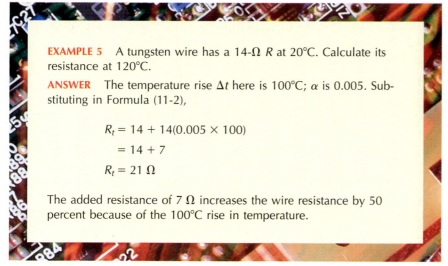

EXAMPLE 5   A tungsten wire has a 14-Ω R at 20°C. Calculate its resistance at 120°C.

ANSWER   The temperature rise Δt here is 100°C; α is 0.005. Substituting in Formula (11-2),

$$R_t = 14 + 14(0.005 \times 100)$$
$$= 14 + 7$$
$$R_t = 21\ \Omega$$

The added resistance of 7 Ω increases the wire resistance by 50 percent because of the 100°C rise in temperature.

In practical terms, a positive α means that heat increases R in wire conductors. Then I is reduced, with a specified applied voltage.

NEGATIVE α   Note that carbon has a negative temperature coefficient. In general, α is negative for all semiconductors, including germanium and silicon. Also, all electrolyte solutions, such as sulfuric acid and water, have a negative α.

A negative value of α means less resistance at higher temperatures. The resistance of semiconductor diodes and transistors, therefore, can be reduced appreciably when they become hot with normal load current.

ZERO α   This means R is constant with changes in temperature. The metal alloys constantan and manganin, for example, have the value of zero for α. They can be used for precision wire-wound resistors, which do not change resistance when the temperature increases.

HOT RESISTANCE   Because resistance wire is made of tungsten, Nichrome, iron, or nickel, there is usually a big difference in the amount of resistance the wire has when hot in normal operation and when cold without its normal load current. The reason is that the resistance increases with higher temperatures, since these materials have a positive temperature coefficient, as shown in Table 11-2.

As an example, the tungsten filament of a 100-W 120-V incandescent bulb has a current of 0.833 A when the bulb lights with normal brilliance at its rated power, since $I = P/V$. By Ohm's law, the hot resistance is V/I, or 120 V/0.833 A, which equals 144 Ω. If, however, the filament resistance is measured with an ohmmeter when the bulb is not lit, the cold resistance is only about 10 Ω.

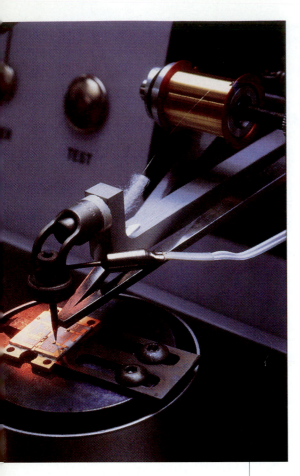

The Nichrome heater elements in appliances and the tungsten heaters in vacuum tubes also become several hundred degrees hotter in normal operation. In these cases, only the cold resistance can be measured with an ohmmeter. The hot resistance must be calculated from voltage and current measurements with the normal value of load current. As a practical rule, the cold resistance is generally about one-tenth the hot resistance. In troubleshooting, however, the problem is usually just to check if the heater element is open. Then it reads infinite ohms on the ohmmeter.

SUPERCONDUCTIVITY   The effect opposite to hot resistance is to cool a metal down to very low temperatures to reduce its resistance. Near absolute zero, 0 K or −273°C, some metals abruptly lose practically all their resistance. As an example, when cooled by liquid helium, the metal tin becomes superconductive at 3.7 K. Tremendous currents can be produced, resulting in very strong electromagnetic fields. Such work at very low temperatures, near absolute zero, is called *cryogenics*.

New types of ceramic materials have been developed and are stimulating great interest in superconductivity because they provide zero resistance at temperatures much above absolute zero. One type is a ceramic pellet, with a 1-in. diameter, that includes yttrium, barium, copper, and oxygen atoms. The superconductivity occurs at a temperature of 93 K, equal to −160°C. This value is still far below room temperature, but the cooling can be done with liquid nitrogen, which is much cheaper than liquid helium. As research continues, it is likely that new materials will be discovered that are superconductive at even higher temperatures.

11-10 ION CURRENT IN LIQUIDS AND GASES

We usually think of metal wire for a conductor, but there are other possibilities. Liquids such as salt water or dilute sulfuric acid can also allow the movement of electric charges. For gases, consider the neon glow lamp, where neon serves as a conductor.

The mechanism may be different for conduction in metal wire, liquids, or gases, but in any case the current is a motion of charges. Furthermore, either positive or negative charges can be the carriers that provide electric current. The amount of current is Q/T. For one coulomb of charge per second, the current is one ampere.

In solid materials like the metals, the atoms are not free to move among each other. Therefore, conduction of electricity must take place by the drift of free electrons. Each atom remains neutral, neither gaining nor losing charge, but the metals are good conductors because they have plenty of free electrons that can be forced to drift through the solid substance.

In liquids and gases, however, each atom is able to move freely among all the other atoms because the substance is not solid. As a result, the atoms can easily take on electrons or lose electrons, particularly the valence electrons in the outside shell. The result is an atom that is no longer electrically neutral. Adding one or more electrons produces a negative charge; the loss of one or more electrons results in a positive charge. The charged atoms are called *ions*. Such charged particles are commonly formed in liquids and gases.

THE ION An ion is an atom, or group of atoms, that has a net electric charge, either positive or negative, resulting from a loss or gain of electrons. In Fig. 11-19*a*, the sodium atom is neutral, with 11 positive charges in the nucleus balanced by 11 electrons in the outside shells. This atom has only 1 electron in the shell farthest from the nucleus. When the sodium is in a liquid solution, this 1 electron can easily leave the atom. The reason may be another atom close by that needs 1 electron in order to have a stable ring of 8 electrons in its outside shell. Notice that if the sodium atom loses 1 valence electron, the atom will still have an outside ring of 8 electrons, as shown in Fig. 11-19*b*. This sodium atom now is a positive ion, with a charge equal to 1 proton. An ion still has the characteristics of the element because the nucleus is not changed.

CURRENT OF IONS Just as in electron flow, opposite ion charges are attracted to each other, while like charges repel. The resultant motion of ions provides electric current. In liquids and gases, therefore, conduction of electricity results mainly from the movement of ions. This motion of ion charges is called *ionization current*. Since an ion includes the nucleus of the atom, the ion charge is much heavier than an electron charge and moves with less velocity. We can say that ion charges are less mobile than electron charges.

The direction of ionization current can be the same as electron flow or the opposite. When negative ions move, they are attracted to the positive terminal of an applied voltage, in the same direction as electron flow. However, when positive ions move, this ionization current is in the opposite direction, toward the negative terminal of an applied voltage.

For either direction, though, the amount of ionization current is determined by the rate at which the charge moves. If 3 C of positive ion charges move past a given point per second, the current is 3 A, the same as 3 C of negative ions or 3 C of electron charges.

IONIZATION IN LIQUIDS Ions are usually formed in liquids when salts or acids are dissolved in water. Salt water is a good conductor because of ionization, but pure distilled water is an insulator. In addition, metals immersed in acids or alkaline solutions produce ionization. Liquids that are good conductors because of ionization are called *electrolytes*. In general, electrolytes have a negative value of α, as more ionization at higher temperatures lowers the resistance.

IONIZATION IN GASES Gases have a minimum striking or ionization potential, which is the lowest applied voltage that will ionize the gas. Before ionization the gas is an insulator, but the ionization current makes the ionized gas a

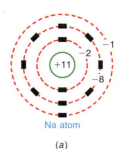
Na atom

(*a*)

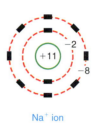
Na⁺ ion

(*b*)

FIG. 11-19 Formation of ions. (*a*) Normal sodium (Na) atom. (*b*) Positively charged ion indicated as Na⁺, missing one free electron.

DID YOU KNOW?

A ring of superconducting material will circulate a current forever, even after the source is removed. For example, high-field magnets using superconducting coils (cooled by helium) maintain themselves without huge amounts of electricity.

low resistance. An ionized gas usually glows. Argon, for instance, emits blue light when the gas is ionized. Ionized neon gas glows red. The amount of voltage needed to reach the striking potential varies with different gases and depends on the gas pressure. For example, a neon glow lamp for use as a night light ionizes at approximately 70 V.

IONIC BONDS   The sodium ion in Fig. 11-20 has a charge of +1 because it is missing 1 electron. If such positive ions are placed near negative ions with a charge of −1, there will be an electrical attraction to form an ionic bond.

A common example is the combination of sodium (Na) ions and chlorine (Cl) ions to form table salt (NaCl), as shown in Fig. 11-20. Notice that the 1 outer electron of the Na atom can fit into the 7-electron shell of the Cl atom. When these two elements are combined, the Na atom gives up 1 electron to form a positive ion, with a stable L shell having 8 electrons; also, the Cl atom adds this 1 electron to form a negative ion, with a stable M shell having 8 electrons. The two opposite types of ions are bound in NaCl because of the strong attractive force between opposite charges close together.

The ions in NaCl can separate in water to make salt water a conductor of electricity, while pure water is not. When current flows in salt water, then, the moving charges must be ions, as another example of ionization current.

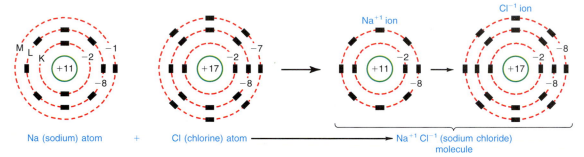

FIG. 11-20   Ionic bond between atoms of sodium (Na) and chlorine (Cl) to form a molecule of sodium chloride (NaCl).

TEST-POINT QUESTION 11-10

Answers at end of chapter.
  a. How much is I for 2 C/s of positive ion charges?
  b. Which have the greatest mobility, positive ions, negative ions, or electrons?
  c. True or False? A dielectric material is a good conductor of electricity.

11-11 ELECTRONS AND HOLE CHARGES IN SEMICONDUCTORS

The semiconductor materials like germanium and silicon are in a class by themselves as conductors, because the charge carriers for current flow are neither ions nor free valence electrons. With a valence of ±4 for these elements, the tendency to gain or lose electrons to form a stable 8 shell is the same either way. As a result, these elements tend to share their outer electrons in pairs of atoms.

An example is illustrated in Fig. 11-21, for two silicon (Si) atoms, each sharing its 4 valence electrons with the other atom to form one Si_2 molecule. This type of combination of atoms sharing their outer electrons to form a stable molecule is called a *covalent bond*.

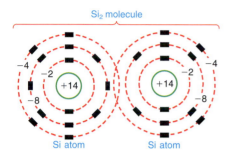

FIG. 11-21 Covalent bond between silicon (Si) atoms.

The covalent-bond structure in germanium and silicon is the basis for their use in transistors. This is because, although the covalent-bond structure is electrically neutral, it permits charges to be added by *doping* the semiconductor with a small amount of impurity atoms.

As a specific example, silicon, with a valence of 4, is combined with phosphorus, with a valence of 5. Then the doped silicon has covalent bonds with an excess of 1 electron for each impurity atom of phosphorus. The result is a negative, or N-type, semiconductor.

For the opposite case, silicon can be doped with aluminum, which has a valence of 3. Then covalent bonds formed with the impurity atoms have 7 outer electrons, instead of the 8 with a pair of silicon atoms.

The 1 missing electron for each covalent bond with an impurity atom corresponds to a positive charge called a *hole*. The amount of charge for each hole is 0.16×10^{-18} C, the same as for an electron, but of opposite polarity. This type of doping results in a P-type semiconductor with positive hole charges.

For either N- or P-type semiconductors the charges can be made to move by an applied voltage that produces current. When electrons move, the current direction is the same as for electron flow. When the positive hole charges move, the direction is opposite from electron current. For either electrons or hole charges, when 1 C moves past a given point in 1 s, the amount of current is 1 A. However, electrons have greater mobility than hole charges.

For semiconductor diodes, the P and N types are combined in a PN junction. It should be noted that this junction is just an electrical boundary between opposite types, but without any physical separation. In a PNP transistor, the N type is between two P types. The opposite case, a P type between two N types, results in the NPN transistor. The NPN and PNP types are junction transistors, with two junctions. More details of semiconductors in general, as important solid-state components, and transistors in particular, as commonly used for amplifier circuits, are given in Chaps. 28 and 29, which explain the operation of semiconductor diodes and transistors. Solid-state integrated circuits are discussed in Chap. 32.

Answers at end of chapter.

a. What is the polarity of the hole charges in P-type doped semiconductors?

b. What is the electron valence of silicon and germanium?

c. What are the charge carriers in N-type semiconductors?

11-12 INSULATORS

Substances that have very high resistance, of the order of many megohms, are classed as insulators. With such high resistance, an insulator cannot conduct appreciable current when voltage is applied. As a result, insulators can have either of two functions. One is to isolate conductors to eliminate conduction between them. The other is to store an electric charge when voltage is applied.

An insulator maintains its charge because electrons cannot flow to neutralize the charge. The insulators are commonly called *dielectric materials,* meaning that they can store a charge.

Among the best insulators, or dielectrics, are air, vacuum, rubber, wax, shellac, glass, mica, porcelain, oil, dry paper, textile fibers, and plastics such as Bakelite, Formica, and polystyrene. Pure water is a good insulator, but salt water is not. Moist earth is a fairly good conductor, while dry, sandy earth is an insulator.

For any insulator, a high enough voltage can be applied to break down the internal structure of the material, forcing the dielectric to conduct. This dielectric breakdown is usually the result of an arc, which ruptures the physical structure of the material, making it useless as an insulator. Table 11-4 compares several insulators in terms of dielectric strength, which is the voltage breakdown rating. The higher the dielectric strength, the better the insulator, since it is less likely to break down with a high value of applied voltage. The breakdown voltages in Table 11-4 are approximate values for the standard thickness of 1 mil, or 0.001 in. More thickness allows a higher breakdown-voltage rating. Note that the value of 20 V/mil for air or vacuum is the same as 20 kV/in.

DID YOU KNOW?

High-temperature superconductors require nitrogen; low-temperature ones require hydrogen. More research is now focused on superconductors that use nitrogen, because it is cheaper, easier to handle, and useful in refrigeration.

TABLE 11-4	VOLTAGE BREAKDOWN OF INSULATORS		
MATERIAL	**DIELECTRIC STRENGTH, V/MIL**	**MATERIAL**	**DIELECTRIC STRENGTH, V/MIL**
Air or vacuum	20	Paraffin wax	200–300
Bakelite	300–550	Phenol, molded	300–700
Fiber	150–180	Polystyrene	500–760
Glass	335–2000	Porcelain	40–150
Mica	600–1500	Rubber, hard	450
Paper	1250	Shellac	900
Paraffin oil	380		

INSULATOR DISCHARGE CURRENT An insulator in contact with a voltage source stores charge, producing a potential on the insulator. The charge tends to remain on the insulator, but it can be discharged by one of the following methods:

1. Conduction through a conducting path. For instance, a wire across the charged insulator provides a discharge path. Then the discharged dielectric has no potential.
2. Brush discharge. As an example, high voltage on a sharp pointed wire can discharge through the surrounding atmosphere by ionization of the air molecules. This may be visible in the dark as a bluish or reddish glow, called the *corona effect*.
3. Spark discharge. This is a result of breakdown in the insulator because of a high potential difference that ruptures the dielectric. The current that flows across the insulator at the instant of breakdown causes the spark.

Corona is undesirable, as it reduces the potential by brush discharge into the surrounding air. In addition, the corona often indicates the beginning of a spark discharge. A potential of the order of kilovolts is usually necessary for corona, as the breakdown voltage for air is approximately 20 kV/in. To reduce the corona effect, conductors that have high voltage should be smooth, rounded, and thick. This equalizes the potential difference from all points on the conductor to the surrounding air. Any sharp point can have a more intense field, making it more susceptible to corona and eventual spark discharge.

TEST-POINT QUESTION 11-12

Answers at end of chapter.
a. Which has a higher voltage breakdown rating, air or mica?
b. Can 30 kV arc across an air gap of 1 in.?

11-13 TROUBLESHOOTING HINTS FOR WIRES AND CONNECTORS

For all types of electronic equipment, a common problem is an open circuit in the wire conductors, the connectors, and the switch contacts.

For conductors, both wires and printed wiring, you can check the continuity with an ohmmeter. A good conductor reads $0\ \Omega$ for continuity. An open reads infinite ohms.

Connectors can also be checked for continuity between the wire and the connector itself. Also, the connector may be tarnished, oxide coated, or rusted. Then it must be cleaned with either fine sandpaper or emery cloth. Sometimes, it helps just to pull out the plug and reinsert it to make sure of tight connections.

With a plug connector for cable, make sure the wires have continuity to the plug. Except for the F-type connector, most plugs require careful soldering to the center pin.

A switch with dirty or pitted contacts can produce intermittent operation. In most cases the switch cannot be disassembled for cleaning. Therefore, the switch must be replaced with a new one.

TEST-POINT QUESTION 11-13

Answers at end of chapter.

Answer True or False.
a. Printed wiring cannot be checked for continuity with an ohmmeter.
b. A tarnished or rusty connection will have a higher than normal resistance.

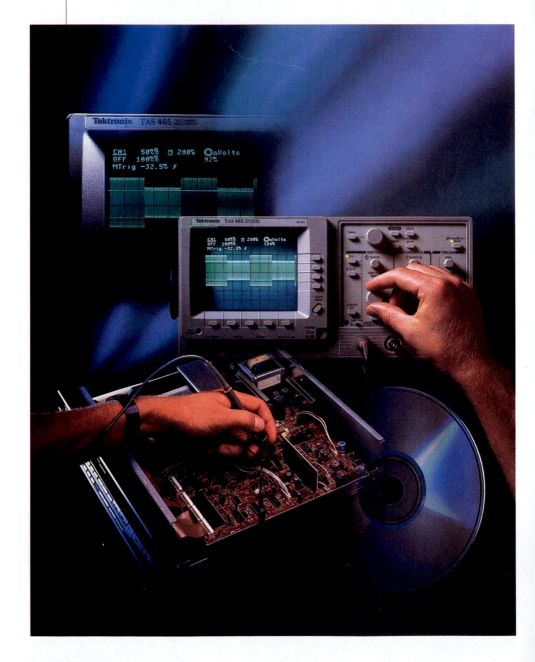

11   SUMMARY AND REVIEW

- A conductor has very low resistance. All the metals are good conductors, the best being silver, copper, and aluminum. Copper is generally used for wire conductors.
- The sizes for copper wire are specified by the American Wire Gage. Higher gage numbers mean thinner wire. Typical sizes are No. 22 gage hookup wire for electronic circuits and No. 12 and No. 14 for house wiring.
- The cross-sectional area of round wire is measured in circular mils. One mil is 0.001 in. The area in circular mils equals the diameter in mils squared.
- The resistance R of a conductor can be found using the formula $R = \rho(l/A)$. The factor ρ is specific resistance, l is length of the conductor, and A is the cross-sectional area of the conductor. Wire resistance increases directly with length l, but decreases inversely with the cross-sectional area A.
- A switch inserted in one side of a circuit opens the entire series circuit. When open, the switch has the applied voltage across it.
- A fuse protects the circuit components against overload, as excessive current melts the fuse element to open the entire series circuit. A good fuse has very low resistance and practically zero voltage across it.
- Ionization in liquids and gases produces atoms that are not electrically neutral. These are ions. Negative ions have an excess of electrons; positive ions have a deficiency of electrons. In liquids and gases, electric current is a result of movement of the ions.
- In the semiconductors, such as germanium and silicon, the charge carriers are electrons in N type and positive hole charges in P type. One hole charge is 0.16×10^{-18} C, the same as one electron, but with positive polarity.
- The resistance of pure metals increases with temperature. For semiconductors and liquid electrolytes, the resistance decreases at higher temperatures.
- An insulator has very high resistance. Common insulating materials are air, vacuum, rubber, paper, glass, porcelain, shellac, and plastics. Insulators are also called dielectrics.
- Superconductors have practically no resistance.
- A common circuit trouble is an open in wire conductors, dirty contacts in switches, and dirt, oxides, and corrosions on connectors and terminals.

ANSWERS AT BACK OF BOOK.

Choose (*a*), (*b*), (*c*), or (*d*).

1. A 10-ft length of copper-wire conductor of No. 20 gage has a total resistance of (*a*) less than 1 Ω; (*b*) 5 Ω; (*c*) 10.4 Ω; (*d*) approximately 1 MΩ.

2. A copper-wire conductor with 0.2 in. diameter has an area of (*a*) 200 cmil; (*b*) 400 cmil; (*c*) 20,000 cmil; (*d*) 40,000 cmil.

3. If a wire conductor of 0.1 Ω resistance is doubled in length, its resistance becomes (*a*) 0.01 Ω; (*b*) 0.02 Ω; (*c*) 0.05 Ω; (*d*) 0.2 Ω.

4. If two similar wire conductors are connected in parallel, their total resistance is (*a*) double the resistance of one wire; (*b*) one-half the resistance of one wire; (*c*) the same as the resistance of one wire; (*d*) two-thirds the resistance of one wire.

5. The hot resistance of the tungsten filament in a bulb is higher than its cold resistance because the filament's temperature coefficient is (*a*) negative; (*b*) positive; (*c*) zero; (*d*) about 10 Ω per degree.

6. A closed switch has a resistance of (*a*) zero; (*b*) infinity; (*c*) about 100 Ω at room temperature; (*d*) at least 1000 Ω.

7. An open fuse has a resistance of (*a*) zero; (*b*) infinity; (*c*) about 100 Ω at room temperature; (*d*) at least 1000 Ω.

8. Insulating materials have the function of (*a*) conducting very large currents; (*b*) preventing an open circuit between the voltage source and the load; (*c*) preventing a short circuit between conducting wires; (*d*) storing very high currents.

9. An ion is (*a*) a free electron; (*b*) a proton; (*c*) an atom with unbalanced charges; (*d*) a nucleus without protons.

10. Ionization current in liquids and gases results from a flow of (*a*) free electrons; (*b*) protons; (*c*) positive or negative ions; (*d*) ions that are lighter in weight than electrons.

QUESTIONS

1. Name three good metal conductors in order of resistance. Describe at least one application.
2. Name four insulators. Give one application.
3. Name two semiconductors. Give one application.
4. Name two types of resistance wire. Give one application.
5. What is meant by the dielectric strength of an insulator?
6. Why does ionization occur more readily in liquids and gases, compared with the solid metals? Give an example of ionization current.

7. Define the following: ion, ionic bond, covalent bond, molecule.
8. Draw a circuit with two bulbs, a battery, and an SPDT switch that determines which bulb lights.
9. Why is it not possible to measure the hot resistance of a filament with an ohmmeter?
10. Give one way in which negative ion charges are similar to electron charges and one way in which they are different.
11. Define the following abbreviations for switches: SPST, SPDT, DPST, DPDT, NO, and NC.
12. Give two common circuit troubles with conductors and connector plugs.

PROBLEMS

ANSWERS TO ODD-NUMBERED PROBLEMS AT BACK OF BOOK.

1. A copper wire has a diameter of 0.032 in. (**a**) How much is its circular-mil area? (**b**) What is its AWG size? (**c**) How much is the resistance of a 100-ft length?
2. Draw the schematic diagram of a resistance in series with an open SPST switch and a 100-V source. (**a**) With the switch open, how much is the voltage across the resistance? Across the open switch? (**b**) With the switch closed, how much is the voltage across the switch and across the resistance? (**c**) Do the voltage drops around the series circuit add to equal the applied voltage in both cases?
3. Draw the schematic diagram of a fuse in series with the resistance of a 100-W 120-V bulb connected to a 120-V source. (**a**) What size fuse can be used? (**b**) How much is the voltage across the good fuse? (**c**) How much is the voltage across the fuse if it is open?
4. Compare the resistance of two conductors: 100 ft of No. 10 gage copper wire and 200 ft of No. 7 gage copper wire.
5. How much is the hot resistance of a 150-W 120-V bulb operating with normal load current?
6. How much is the resistance of a slab of silicon 0.1 cm long with a cross-sectional area of 1 cm^2?
7. A cable with two lengths of No. 10 copper wire is short-circuited at one end. The resistance reading at the open end is 10.35 Ω. What is the cable length in feet? (Temperature is 25°C.)
8. (**a**) How many hole charges are needed to equal 1 C? (**b**) How many electrons? (**c**) How many ions with a negative charge of 1 electron?
9. (**a**) If a copper wire has a resistance of 4 Ω at 25°C, how much is its resistance at 75°C? (**b**) If the wire is No. 10 gage, what is its length in feet?
10. A coil is wound with 1500 turns of No. 20 copper wire. If the average amount of wire in a turn is 4 in., how much is the total resistance of the coil? What will be its resistance if No. 30 wire is used instead? (Temperature is 25°C.)
11. Calculate the voltage drop across 1000 ft of No. 10 gage wire connected to a 3-A load.
12. What is the smallest size of copper wire that will limit the line drop to 5 V, with 120 V applied and a 6-A load? The total line length is 200 ft.

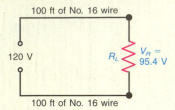

100 ft of No. 16 wire

120 V R_L $V_R = $ 95.4 V

100 ft of No. 16 wire

FIG. 11-22 Circuit diagram for Probs. 13 and 14.

13. Refer to Fig. 11-22. Calculate the load current I for the IR drop of 24.6 V that reduces V_R to 95.4 V with the 120-V supply.

14. From Fig. 11-22, calculate the value of R_L.

15. An extension cord is made up of 26 strands of No. 28 gage copper wire for each of its two conductors. **(a)** What is the equivalent gage size of each conductor in solid wire (approximately)? **(b)** If the extension cord is 50 ft long, what is the total resistance of both conductors? **(c)** What is the resistance of each individual 50-ft strand of No. 28 gage wire?

16. Calculate the resistance of the following conductors: **(a)** 1000 ft of No. 14 gage copper wire; **(b)** 1000 ft of No. 14 gage aluminum wire; **(c)** 1000 ft of No. 14 gage steel wire.

17. A 100-ft pair of No. 12 gage copper wires connects a 15-A load to the 120-V power line. **(a)** What is the total length of No. 12 gage wire? **(b)** What is the resistance of the total length? **(c)** How much voltage is available at the load? **(d)** How much is the I^2R power loss in the conductors?

18. Calculate the resistance of 1000 ft of No. 23 gage aluminum wire. **(a)** What is the resistance if the length is doubled to 2000 ft? **(b)** What is the resistance if the length is cut in half to 500 ft? **(c)** What is the resistance of a 1000-ft length of No. 20 gage aluminum wire?

19. Draw the schematic symbol for a(n): **(a)** single-pole, triple-throw switch; **(b)** triple-pole, triple-throw switch.

20. A No. 16 gage copper wire has 20 Ω of resistance at 65°C. Calculate its length.

CRITICAL THINKING

1. Use two switches having the appropriate number of poles and throws to build a partial decade resistance box. The resistance is to be adjustable in 1-Ω and 10-Ω steps from 0 Ω to 99 Ω across two terminals identified as A and B. Draw the circuit showing all resistance values and switch connections.

2. Show how two SPDT switches can be wired to turn ON and OFF a light from two different locations. The voltage source is a 12-V battery.

11-1 **a.** 0.08 Ω
      b. 0.07 V
      c. 0.06 percent

11-2 **a.** 0.01646 Ω
      b. 625 cmil
      c. No. 16

11-3 **a.** T
      b. T
      c. T

11-4 **a.** T
      b. T
      c. F

11-5 **a.** 25 W
      b. infinite ohms

11-6 **a.** zero
      b. three
      c. infinite
      d. no

11-7 **a.** zero
      b. zero

11-8 **a.** more
      b. 2.6 Ω

11-9 **a.** T
      b. T
      c. T

11-10 **a.** $I = 2$ A
      b. electrons
      c. F

11-11 **a.** positive
      b. four
      c. electrons

11-12 **a.** mica
      b. yes

11-13 **a.** F
      b. T

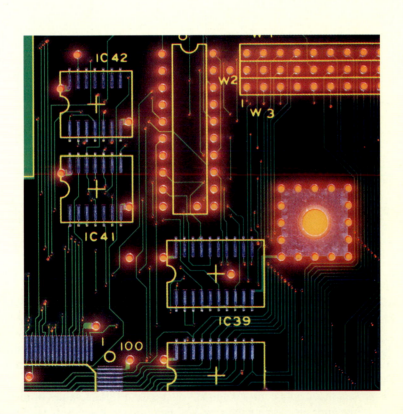

CHAPTER 12

BATTERIES

A battery is a group of cells that generate energy from the internal chemical reaction. The cell itself consists of two different conducting materials as the electrodes that are immersed in an electrolyte. The chemical reaction between the electrodes and the electrolyte results in a separation of electric charges, in the form of ions and free electrons. Then the two electrodes have a difference of potential that provides voltage output from the cell.

The main types are the alkaline cell with an output of 1.5 V and the lead-sulfuric acid wet cell with 2.1 V for output. A common application for dry cells is the 9-V flat battery often used for transistor radios. It has six cells connected in series internally for an output of $6 \times 1.5 = 9$ V. Dry batteries are used for all types of portable electronic equipment, photographic equipment and toys.

The lead-sulfuric acid cell is the type almost always used for automobile batteries. Six cells are connected in series internally for 12-V as the nominal output.

The function of a battery is to provide a source of steady dc voltage of fixed polarity. Furthermore, the battery is a good example of a generator or energy source. The source supplies voltage for a circuit as the load to produce the desired load current. An important factor is the internal resistance r_i of the source, which affects the output voltage when a load is connected. In general, a low r_i means that the source can maintain a constant output voltage for different values of load current. For the opposite case, a high r_i makes the output voltage drop, but a constant value of load current can be maintained.

CHAPTER OBJECTIVES

Upon completion of this chapter, you should be able to:

- *Explain* the difference between a primary cell and a secondary cell.
- *Define* what is meant by the internal resistance of a cell.
- *List* several different types of voltaic cells.
- *Explain* how cells can be connected together to increase the voltage of a battery.
- *Explain* how cells can be connected together to increase the current capacity of a battery.
- *Explain* how cells can be arranged to increase both the voltage and the current ratings of a battery.
- *Explain* why the terminal voltage of a battery drops with more load current.
- *Understand* the difference between a voltage source and a current source.
- *Explain* the concept of maximum power transfer.

IMPORTANT TERMS IN THIS CHAPTER

alkaline cell	internal resistance	open-circuit voltage
ampere-hour capacity	lead-acid cell	primary cell
carbon-zinc cell	Leclanché cell	secondary cell
constant-current generator	lithium cell	solar cell
constant-voltage generator	matching the load	specific gravity
galvanic cell	mercury cell	storage cell
hydrometer	nickel-cadmium cell	

TOPICS COVERED IN THIS CHAPTER

12-1 GENERAL FEATURES OF BATTERIES

A battery is a combination of cells. The chemical battery has always been important as a dc voltage source for the operation of radios and other electronic equipment. The reason is that a transistor amplifier needs dc operating voltages in order to conduct current. With current in the amplifier, the circuit can be used to amplify an ac signal. Originally, all radio receivers used batteries. Then rectifier power supplies were developed to convert the ac power-line voltage to dc output, eliminating the need for batteries. However, now batteries are used more than ever for all types of electronic portable equipment.

From the old days of radio, dry batteries are still called A, B, and C batteries, according to their original functions in vacuum-tube operation. The A battery was used to supply enough current to heat the filament for thermionic emission of electrons from a heated cathode. A typical rating is 4.5 V or 6 V with a load current of 150 mA or more. The C battery was used for a small negative dc bias voltage at the control grid, typically 1.5 V, with practically no current drain.

The A battery is seldom used any more, although a 6-V lantern battery has ratings like an A battery. However, the function of a B battery with medium voltage and current ratings is the same now as it always was. Transistors need a steady dc voltage for the collector electrode, which receives charges from the emitter through the base electrode. For an NPN transistor, positive voltage is needed at the collector or negative voltage at the emitter. With a PNP transistor, the polarities are reversed. The positive dc supply voltage is called B^+ or $+V_{cc}$. In a small transistor radio with a 9-V battery, as an example, the battery is the $+V_{cc}$ supply. The same requirements for dc supply voltage apply to integrated circuits, as the IC chip contains transistor amplifiers.

PRIMARY CELLS   This type cannot be recharged. After it has delivered its rated capacity, the primary cell must be discarded. The reason is that the internal chemical reaction cannot be restored. Figure 12-1 shows a variety of different dry cells and batteries, all of which are of the primary type. In Table 12-1 several different cells are listed by name. Each of the different cells is listed as being of either the primary or the secondary type. Notice the open circuit voltage for each of the cell types listed.

FIG. 12-1   Typical dry cells and batteries. These primary types cannot be recharged.

TABLE 12-1   CELL TYPES AND OPEN-CIRCUIT VOLTAGE

CELL NAME	TYPE	NOMINAL OPEN-CIRCUIT* VOLTAGE, Vdc
Carbon-zinc	Primary	1.5
Zinc chloride	Primary	1.5
Manganese dioxide (alkaline)	Primary or secondary	1.5
Mercuric oxide	Primary	1.35
Silver oxide	Primary	1.5
Lithium	Primary	3.0
Lead-acid	Secondary	2.1
Nickel-cadmium	Secondary	1.25
Nickel-iron (Edison cell)	Secondary	1.2
Silver-zinc	Secondary	1.5
Silver-cadmium	Secondary	1.1

*Open-circuit V is the terminal voltage without a load.

SECONDARY CELLS   This type can be recharged because the chemical action is reversible. When it supplies current to a load resistance, the cell is *discharging*, as the current tends to neutralize the separated charges at the electrodes. For the opposite case, the current can be reversed to re-form the electrodes as the chemical action is reversed. This action is *charging* the cell. The charging current must be supplied by an external dc voltage source, with the cell serving just as a load resistance. The discharging and recharging is called *cycling* of the cell. Since a secondary cell can be recharged, it is also called a *storage cell.* The most common type is the lead-acid cell generally used in automotive batteries (Fig. 12-2). In addition, the list in Table 12-1 indicates which are secondary cells.

FIG. 12-2   Example of a 12-V auto battery using six lead-acid cells in series. This is a secondary type, which can be recharged.

DRY CELLS   What we call a "dry cell" really has moist electrolyte. However, the electrolyte cannot be spilled and the cell can operate in any position.

SEALED RECHARGEABLE CELLS This type is a secondary cell that can be recharged, but it has a sealed electrolyte that cannot be refilled. These cells are capable of charge and discharge in any position.

TEST-POINT QUESTION 12-1

Answers at end of chapter.
a. How much is the output voltage of a carbon-zinc cell?
b. How much is the output voltage of a lead-acid cell?
c. Which type can be recharged, a primary or a secondary cell?

12-2 THE VOLTAIC CELL

When two different conducting materials are immersed in an electrolyte, as illustrated in Fig. 12-3a, the chemical action of forming a new solution results in the separation of charges. This method for converting chemical energy into electric energy is a voltaic cell. It is also called a *galvanic cell,* named after Luigi Galvani (1737–1798).

Referring to Fig. 12-3a, the charged conductors in the electrolyte are the electrodes or plates of the cell. They serve as the terminals for connecting the voltage output to an external circuit, as shown in Fig. 12-3b. Then the potential difference resulting from the separated charges enables the cell to function as a source of applied voltage. The voltage across the cell's terminals forces current to flow in the circuit to light the bulb.

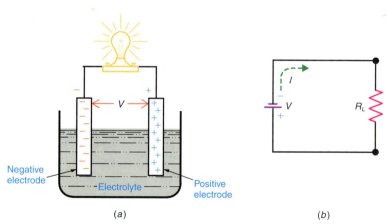

(a)   (b)

FIG. 12-3 How a voltaic cell converts chemical energy into electrical energy. (*a*) Electrodes or plates in liquid electrolyte solution. (*b*) Schematic of a circuit with a voltaic cell as a dc voltage source *V* to produce current in load R_L, which is the light bulb.

CURRENT OUTSIDE THE CELL Electrons from the negative terminal of the cell flow through the external circuit with R_L and return to the positive terminal. The chemical action in the cell separates charges continuously to maintain the terminal voltage that produces current in the circuit.

The current tends to neutralize the charges generated in the cell. For this reason, the process of producing load current is considered discharging of the cell. However, the internal chemical reaction continues to maintain the separation of charges that produces the output voltage.

CURRENT INSIDE THE CELL   The current through the electrolyte is a motion of ion charges. Notice in Fig. 12-3b that the current inside the cell flows from the positive terminal to the negative terminal. This action represents the work being done by the chemical reaction to generate the voltage across the output terminals.

The negative terminal in Fig. 12-3a is considered to be the anode of the cell because it forms positive ions for the electrolyte. The opposite terminal of the cell is its cathode.

INTERNAL RESISTANCE   Any practical voltage source has internal resistance, indicated as r_i, which limits the current it can deliver. For a chemical cell, as in Fig. 12-3, the r_i is mainly the resistance of the electrolyte. For a good cell, r_i is very low, with typical values less than 1 Ω. As the cell deteriorates, though, r_i increases, preventing the cell from producing its normal terminal voltage when load current is flowing. The reason is that the internal voltage drop across r_i opposes the output terminal voltage. This factor is why you can often measure normal voltage on a dry cell with a voltmeter, which drains very little current, but the terminal voltage drops when the load is connected.

The voltage output of a cell depends on the elements used for the electrodes and the electrolyte. The current rating depends mostly on the physical size. Larger batteries can supply more current. Dry cells are generally rated up to 250 mA, while the lead-acid wet cell can supply current up to 300 A or more. Note that a smaller r_i allows a higher current rating.

ELECTROMOTIVE SERIES   The fact that the voltage output of a cell depends on its elements can be seen from Table 12-2. This list, called the *electrochemical series* or *electromotive series,* gives the relative activity in forming ion charges for some of the chemical elements. The potential for each element is the

TABLE 12-2   ELECTROMOTIVE SERIES OF ELEMENTS

ELEMENT	POTENTIAL, V
Lithium	−2.96
Magnesium	−2.40
Aluminum	−1.70
Zinc	−0.76
Cadmium	−0.40
Nickel	−0.23
Lead	−0.13
Hydrogen (reference)	0.00
Copper	+0.35
Mercury	+0.80
Silver	+0.80
Gold	+1.36

voltage with respect to hydrogen as a zero reference. The difference between the potentials for two different elements indicates the voltage of an ideal cell using these electrodes. It should be noted, though, that other factors, such as the electrolyte, cost, stability, and long life, are important for the construction of commercial batteries.

TEST-POINT QUESTION 12-2

Answers at end of chapter.

Answer True or False.
a. The negative terminal of a chemical cell has a charge of excess electrons.
b. The internal resistance of a cell limits the amount of output current.
c. Two electrodes of the same metal provide the highest voltage output.

12-3 CARBON-ZINC DRY CELL

The carbon-zinc dry cell is a very common type because of its low cost. It is also called the *Leclanché cell,* named after the inventor. Examples are shown in Fig. 12-1, while Fig. 12-4 illustrates internal construction for the D-size round cell. Voltage output for the carbon-zinc cell is 1.4 to 1.6 V, with a nominal value of 1.5 V. Suggested current range is up to 150 mA for the D size, which has a height of $2\frac{1}{4}$ in. and volume of 3.18 in.[3]. The C, A, AA, and AAA sizes are smaller, with lower current ratings. The larger No. 6 cell has a height of 6 in., a diameter of $2\frac{1}{2}$ in., and a current range of up to 1500 mA.

The electrochemical system consists of a zinc anode and a manganese dioxide cathode in a moist electrolyte. The electrolyte is a combination of ammonium

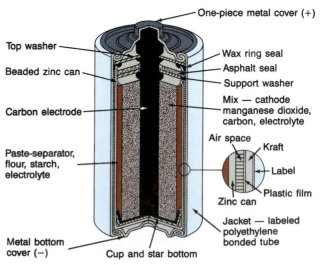

FIG. 12-4 Cutaway view of carbon-zinc dry cell. This is size D, with height of $2\frac{1}{4}$ in.

chloride and zinc chloride dissolved in water. For the round-cell construction, a carbon rod is used down the center, as shown in Fig. 12-4. The rod is chemically inert. However, it serves as a current collector for the positive terminal at the top. The path for current inside the cell includes the carbon rod as the positive terminal, the manganese dioxide, the electrolyte, and the zinc can which is the negative electrode. As additional functions of the carbon rod, it prevents leakage of the electrolyte but is porous to allow the escape of gases which accumulate in the cell.

In operation of the cell, the ammonia releases hydrogen gas which collects around the carbon electrode. This reaction is called *polarization,* and it can reduce the voltage output. However, the manganese dioxide releases oxygen, which combines with the hydrogen to form water. The manganese dioxide functions as a *depolarizer.* Powdered carbon is also added to the depolarizer to improve conductivity and retain moisture.

Carbon-zinc dry cells are generally designed for an operating temperature of 70°F. Higher temperatures will enable the cell to provide greater output. However, temperatures of 125°F or more will cause rapid deterioration of the cell.

The chemical efficiency of the carbon-zinc cell increases with less current drain. Stated another way, the application should allow for the largest battery possible, within practical limits. In addition, performance of the cell is generally better with intermittent operation. The reason is that the cell can recuperate between discharges, probably by the effect of depolarization.

As an example of longer life with intermittent operation, a carbon-zinc D cell may operate for only a few hours with a continuous drain at its rated current. Yet the same cell could be used for a few months or even a year with intermittent operation of less than 1 hour at a time with smaller values of current.

TEST-POINT QUESTION 12-3

Answers at end of chapter.

Answer True or False.
a. A size D cell and a larger No. 6 carbon-zinc cell have the same voltage output of 1.5 V.
b. The zinc can of a carbon-zinc cell is the negative terminal.
c. Polarization at the carbon rod increases the voltage output.

12-4 ALKALINE CELL

Another popular type is the manganese-zinc cell shown in Fig. 12-5, which has an alkaline electrolyte. It is available either as a primary or secondary cell but the primary type is more common. Output is the same 1.5 V as a carbon-zinc cell but the alkaline cell lasts much longer.

The electrochemical system consists of a powdered zinc anode and a manganese dioxide cathode in an alkaline electrolyte. The electrolyte is potassium hydroxide, which is the main difference between the alkaline and Leclanché cells. Hydroxide compounds are alkaline with negative hydroxyl (OH) ions, while an acid electrolyte has positive hydrogen (H) ions. Voltage output from the alkaline cell is 1.5 V.

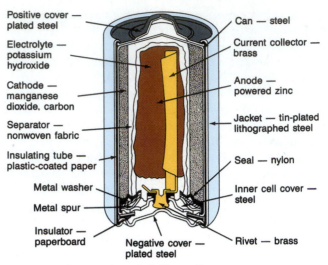

Positive cover — plated steel

Electrolyte — potassium hydroxide

Cathode — manganese dioxide, carbon

Separator — nonwoven fabric

Insulating tube — plastic-coated paper

Metal washer

Metal spur

Insulator — paperboard

Negative cover — plated steel

Can — steel

Current collector — brass

Anode — powered zinc

Jacket — tin-plated lithographed steel

Seal — nylon

Inner cell cover — steel

Rivet — brass

FIG. 12-5 Construction of alkaline cell.

The alkaline cell has many applications because of its ability to work with high efficiency with continuous high discharge rates. Depending on the application, an alkaline cell can provide up to seven times the service of a Leclanché cell. As examples, in a transistor radio an alkaline cell will normally have twice the service life of a general-purpose carbon-zinc cell; in toys the alkaline cell typically provides seven times more service.

The outstanding performance of the alkaline cell is due to its low internal resistance. Its r_i is low because of the dense cathode material, the large surface area of the anode in contact with the electrolyte, and the high conductivity of the electrolyte. In addition, alkaline cells will perform satisfactorily at low temperatures.

ZINC CHLORIDE CELLS   This type is actually a modified carbon-zinc cell with the construction illustrated in Fig. 12-4. However, the electrolyte contains only zinc chloride. The zinc chloride cell is often referred to as the "heavy duty" type. It can normally deliver more current over a longer period of time than the Leclanché cell. Another difference is that the chemical reaction in the zinc chloride cell consumes water along with the chemically active materials, so that the cell is nearly dry at the end of its useful life. As a result, liquid leakage is not a problem.

TEST-POINT QUESTION 12-4

Answers at end of chapter.

Answer True or False.
a. Alkaline cells have the same 1.5-V output as carbon-zinc cells.
b. They are used primarily for their ability to deliver a higher discharge rate than carbon-zinc cells.

12-5 ADDITIONAL TYPES OF PRIMARY CELLS

The miniature button construction shown in Fig. 12-6 is often used for the mercury cell and the silver-oxide cell. Diameter of the cell is ⅜ to 1 in.

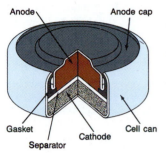

- Anodes are a gelled mixture of amalgamated zinc powder and electrolyte.

- Cathodes
  Silver cells: AgO_2, MnO_2, and conductor
  Mercury cells: HgO and conductor (may contain MnO_2)
  Manganese dioxide cells: MNO_2 and conductor

FIG. 12-6 Construction of miniature button type of primary cell. Diameter is ⅜ to 1 in. Note the chemical symbols AgO_2 for silver oxide, HgO for mercuric oxide, and MnO_2 for manganese dioxide.

MERCURY CELL   The electrochemical system consists of a zinc anode, a mercury compound for the cathode, and an electrolyte of potassium or sodium hydroxide. Mercury cells are available in flat, round cylinder, and miniature button shapes. It should be noted, though, that some round mercury cells have the top button as the negative terminal and the bottom terminal positive. The open-circuit voltage is 1.35 V when the cathode is mercuric oxide (HgO) and 1.4 V or more with mercuric oxide/manganese dioxide. The 1.35-V type is more common.

The mercury cell is used where a relatively flat discharge characteristic is required with high current density. Its internal resistance is low and essentially constant. These cells perform well at elevated temperatures, up to 130°F continuous or 200°F for short periods. One drawback of the mercury cell is its relatively high cost compared with a carbon-zinc cell.

SILVER OXIDE CELL   The electrochemical system consists of a zinc anode, a cathode of silver oxide (AgO_2) with small amounts of manganese dioxide, and an electrolyte of potassium or sodium hydroxide. It is commonly available in the miniature button shape shown in Fig. 12-6. The open-circuit voltage is 1.6 V, but the nominal output with a load is considered to be 1.5 V. Typical applications include hearing aids, cameras, and electronic watches, which use very little current.

TEST-POINT QUESTION 12-5

Answers at end of chapter.

Answer True or False.
a. The mercury cell has output of 2.6 V as a minimum.
b. Most miniature button cells can be recharged.

12-6 SUMMARY OF DRY CELLS

The types of dry cells include carbon-zinc, zinc chloride (heavy-duty), and manganese-zinc (alkaline). Actually, the alkaline cell is better for heavy-duty use than the zinc-chloride type. They are commonly used in the round, cylinder types listed in Table 12-3 for the D, C, AA, and AAA sizes. The small button cells generally use either mercury or silver oxide. All these dry cells are the primary type that cannot be recharged. Each has output of 1.5 V except for the 1.35-V mercury cell.

TABLE 12-3	SIZES FOR POPULAR TYPES OF DRY CELLS*	
SIZE	HEIGHT, IN.	DIAMETER, IN.
D	2¼	1¼
C	1¾	1
AA	1⅞	⁹⁄₁₆
AAA	1¾	⅜

*Cylinder shape shown in Fig. 12-1.

Any dry cell loses its ability to produce output voltage even when it is not being used. The shelf life is about 2 years for the alkaline type, but much less with the carbon-zinc cell, especially for small sizes and partially used cells. The reasons are self-discharge within the cell and loss of moisture in the electrolyte. Therefore, dry cells should be used fresh from the manufacturer.

It should be noted that shelf life can be extended by storing the cell at low temperatures, about 40 to 50°F. Even temperatures below freezing will not harm the cell. However, the cell should be allowed to return to normal room temperature before being used, preferably in its original packaging, to avoid condensation.

The alkaline type of dry cell is probably the most cost-efficient. It costs more but lasts much longer, besides having a longer shelf life. As a comparison with size-D batteries, the alkaline type can last about 10 times longer than the carbon-zinc type in continuous operation, or about 7 times longer for typical intermittent operation. The zinc chloride heavy-duty type can last 2 or 3 times longer than the general-purpose carbon-zinc cell. For low-current applications of about 10 mA or less, however, there is not much difference in the battery life.

TEST-POINT QUESTION 12-6

Answers at end of chapter.
a. Which has longer shelf life, the alkaline or carbon-zinc cell?
b. Which type of dry cell does not have an output of 1.5 V?
c. Which size is physically larger, C or AA?

12-7 LITHIUM CELL

The lithium cell is a relatively new primary cell. However, its high output voltage, long shelf life, low weight, and small volume make the lithium cell an excellent choice for special applications. The open-circuit output voltage is either 2.9 V or 3.7 V, depending on the electrolyte. Note the high potential of lithium in the electromotive list of elements shown before in Table 12-2. Figure 12-7 shows an example of a lithium battery with a 6-V output.

FIG. 12-7 Lithium battery.

A lithium cell can provide at least 10 times more energy than the equivalent carbon-zinc cell. However, lithium is a very active chemical element. Many of the problems in construction have been solved, though, especially for small cells delivering low current. One interesting application is a lithium cell as the dc power source for a cardiac pacemaker. The long service life is important for this use.

Two forms of lithium cells have obtained widespread use. These are the lithium-sulfur dioxide ($LiSO_2$) type and the lithium-thionyl chloride type. Output is approximately 3 V.

In the $LiSO_2$ cell, the sulfur dioxide is kept in a liquid state using a high-pressure container and an organic liquid solvent, usually methyl cyanide. One problem is safe encapsulation of toxic vapor if the container should be punctured or cracked. This problem can be significant for safe disposal of the cells when they are discarded after use.

The shelf life for the lithium cell, 10 years or more, is much longer than that of other types.

TEST-POINT QUESTION 12-7

Answers at end of chapter.

Answer True or False.
a. Shelf life for the lithium cell is much longer than for a carbon-zinc cell.
b. Output voltage for a lithium cell is higher than for a silver-oxide cell.

12-8 LEAD-ACID WET CELL

Where high values of load current are necessary, the lead-acid cell is the type most commonly used. The electrolyte is a dilute solution of sulfuric acid (H_2SO_4). In the application of battery power to start the engine in an automobile, for example, the load current to the starter motor is typically 200 to 400 A. One cell has a nominal output of 2.1 V, but lead-acid cells are often used in a series combination of three for a 6-V battery and six for a 12-V battery. Examples are shown in Figs. 12-2 and 12-8.

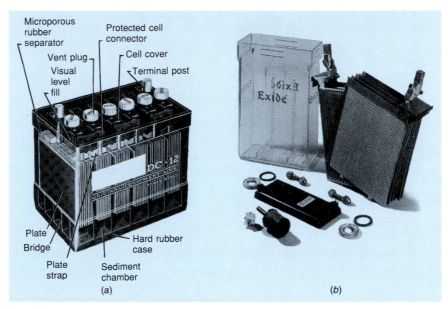

FIG. 12-8 Construction of lead-acid batteries. (*a*) 12-V type. (*b*) Individual plates of 6-V battery.

The lead-acid type is a secondary cell or storage cell, which can be recharged. The charge and discharge cycle can be repeated many times to restore the output voltage, as long as the cell is in good physical condition. However, heat with excessive charge and discharge currents shortens the useful life to about 3 to 5 years for an automobile battery. Of the different types of secondary cells, the lead-acid type has the highest output voltage, which allows less cells for a specified battery voltage.

CONSTRUCTION   Inside a lead-acid battery, the positive and negative electrodes consist of a group of plates welded to a connecting strap. The plates are immersed in the electrolyte, consisting of 8 parts of water to 3 parts of concentrated sulfuric acid. Each plate is a grid or framework, made of a lead-antimony alloy. This construction enables the active material, which is lead oxide, to be pasted into the grid. In manufacture of the cell, a forming charge produces the positive and negative electrodes. In the forming process, the active material in the positive plate is changed to lead peroxide (PbO_2). The negative electrode is spongy lead (Pb).

DID YOU KNOW?

Sony and Nissan are working on a car that runs on a high-current lithium battery that would provide triple the range of traditional batteries.

Automobile batteries are usually shipped dry from the manufacturer. The electrolyte is put in at the time of installation, then the battery is charged to form the plates. With maintenance-free batteries, little or no water need be added in normal service. Some types are sealed, except for a pressure vent, without provision for adding water.

CHEMICAL ACTION   Sulfuric acid is a combination of hydrogen and sulfate ions. When the cell discharges, lead peroxide from the positive electrode combines with hydrogen ions to form water and with sulfate ions to form lead sulfate. The lead sulfate is also produced by combining lead on the negative plate with sulfate ions. Therefore, the net result of discharge is to produce more water, which dilutes the electrolyte, and to form lead sulfate on the plates.

As discharge continues, the sulfate fills the pores of the grids, retarding circulation of acid in the active material. Lead sulfate is the powder often seen on the outside terminals of old batteries. When the combination of weak electrolyte and sulfation on the plate lowers the output of the battery, charging is necessary.

On charge, the external dc source reverses the current in the battery. The reversed direction of ions flowing in the electrolyte results in a reversal of the chemical reactions. Now the lead sulfate on the positive plate reacts with the water and sulfate ions to produce lead peroxide and sulfuric acid. This action re-forms the positive plate and makes the electrolyte stronger by adding sulfuric acid. At the same time, charging enables the lead sulfate on the negative plate to react with hydrogen ions; this also forms sulfuric acid while re-forming lead on the negative electrode.

As a result, the charging current can restore the cell to full output, with lead peroxide on the positive plates, spongy lead on the negative plate, and the required concentration of sulfuric acid in the electrolyte. The chemical equation for the lead-acid cell is

$$Pb + PbO_2 + 2H_2SO_4 \underset{\text{Discharge}}{\overset{\text{Charge}}{\rightleftharpoons}} 2PbSO_4 + 2H_2O$$

On discharge, the Pb and PbO_2 combine with the SO_4 ions at the left side of the equation to form lead sulfate ($PbSO_4$) and water (H_2O) at the right side of the equation.

On charge, with reverse current through the electrolyte, the chemical action is reversed. Then the Pb ions from the lead sulfate on the right side of the equation re-form the lead and lead peroxide electrodes. Also the SO_4 ions combine with H_2 ions from the water to produce more sulfuric acid at the left side of the equation.

CURRENT RATINGS   Lead-acid batteries are generally rated in terms of how much discharge current they can supply for a specified period of time. The output voltage must be maintained above a minimum level, which is 1.5 to 1.8 V per cell. A common rating is ampere-hours (A·h) based on a specific discharge time, which is often 8 h. Typical values for automobile batteries are 100 to 300 A·h.

As an example, a 200 A·h battery can supply a load current of 200/8 or 25 A, based on 8 h discharge. The battery can supply less current for a longer time or more current for a shorter time. Automobile batteries may be rated for "cold cranking power," which is related to the job of starting the engine. A typical rating is 450 A for 30 s at a temperature of 0°F.

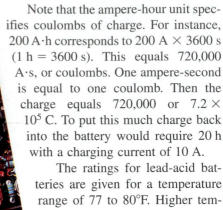

Note that the ampere-hour unit specifies coulombs of charge. For instance, 200 A·h corresponds to 200 A × 3600 s (1 h = 3600 s). This equals 720,000 A·s, or coulombs. One ampere-second is equal to one coulomb. Then the charge equals 720,000 or 7.2×10^5 C. To put this much charge back into the battery would require 20 h with a charging current of 10 A.

The ratings for lead-acid batteries are given for a temperature range of 77 to 80°F. Higher temperatures increase the chemical reaction, but operation above 110°F shortens the battery life. Low temperatures reduce the current capacity and voltage output. The ampere-hour capacity is reduced approximately 0.75 percent for each decrease of 1°F below normal temperature rating. At 0°F the available output is only 60 percent of the ampere-hour battery rating. In cold weather, therefore, it is very important to have an automobile battery up to full charge. In addition, the electrolyte freezes more easily when diluted by water in the discharged condition.

SPECIFIC GRAVITY The state of discharge for a lead-acid cell is generally checked by measuring the specific gravity of the electrolyte. Specific gravity is a ratio comparing the weight of a substance with the weight of water. For instance, concentrated sulfuric acid is 1.835 times as heavy as water for the same volume. Therefore, its specific gravity equals 1.835. The specific gravity of water is 1, since it is the reference.

In a fully charged automotive cell, the mixture of sulfuric acid and water results in a specific gravity of 1.280 at room temperatures of 70 to 80°F. As the cell discharges, more water is formed, lowering the specific gravity. When it is down to about 1.150, the cell is completely discharged.

Specific-gravity readings are taken with a battery hydrometer, such as the one in Fig. 12-9. Note that the calibrated float with the specific gravity marks will rest higher in an electrolyte of higher specific gravity. The decimal point is often omitted for convenience. For example, the value of 1.220 in Fig. 12-9c is simply read "twelve twenty." A hydrometer reading of 1260 to 1280 indicates full charge, approximately 1250 is half charge, and 1150 to 1200 indicates complete discharge.

The importance of the specific gravity can be seen from the fact that the open-circuit voltage of the lead-acid cell is approximately equal to

$$V = \text{specific gravity} + 0.84$$

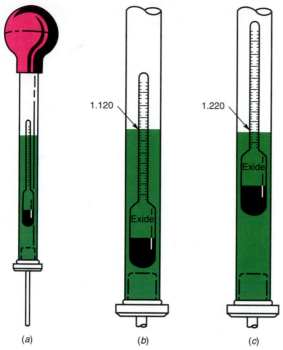

FIG. 12-9 Hydrometer to check specific gravity of lead-acid battery. (*a*) Syringe to suck up electrolyte. (*b*) Float at specific gravity of 1.120. (*c*) Float is higher for higher reading of 1.220.

For the specific gravity of 1.280, the voltage is $1.280 + 0.84 = 2.12$ V, as an example. These values are for a fully charged battery.

CHARGING THE LEAD-ACID BATTERY

The requirements are illustrated in Fig. 12-10. An external dc voltage source is necessary to produce current in one direction. Also, the charging voltage must be more than the battery emf. Approximately 2.5 V per cell is enough to overcome the cell emf so that the charging voltage can produce current opposite to the direction of discharge current.

Note that the reversal of current is obtained just by connecting the battery V_B and charging source V_G with + to + and − to −, as shown in Fig. 12-10*b*. The charging current is reversed because the battery effectively becomes a load resistance for V_G when it is higher than V_B. In this example, the net voltage available to produce charging current is $15 - 12 = 3$ V.

A commercial charger for automobile batteries is shown in Fig. 12-11. This unit can also be used to test batteries and jump-start cars. The charger is essentially a dc power supply, rectifying input from the ac power line to provide dc output for charging batteries.

Float charging refers to a method in which the charger and the battery are always connected to each other for supplying current to the load. In Fig. 12-12, the charger provides current for the load and the current necessary to keep the battery fully charged. The battery here is an auxiliary source for dc power.

It may be of interest to note that an automobile battery is in a floating-charge circuit. The battery charger is an ac generator or alternator with rectifier diodes, driven by a belt from the engine. When you start the car, the battery supplies the

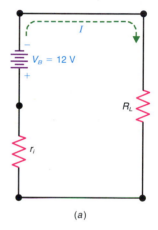

(*a*)

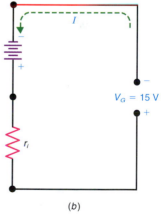

(*b*)

FIG. 12-10 Reversed directions for charge and discharge currents of a battery. The r_i is internal resistance. (*a*) V_B of battery discharges to supply load current for R_L. (*b*) Battery is load resistance for V_G, which is an external source of charging voltage.

FIG. 12-11 Charger for auto batteries.

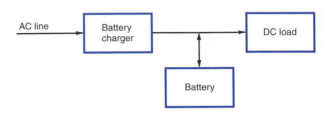

FIG. 12-12 Circuit for battery in float-charge application.

cranking power. Once the engine is running, the alternator charges the battery. It is not necessary for the car to be moving. A voltage regulator is used in this system to maintain the output at approximately 13 to 15 V.

TEST-POINT QUESTION 12-8

Answers at end of chapter.
a. How many lead-acid cells in series are needed for a 12-V battery?
b. A battery is rated for 120 A·h at the 8-h rate at 77°F. How much discharge current can it supply for 8 h?
c. Which of the following is the specific gravity reading for a good lead-acid cell: 1070, 1170, or 1270?

12-9 ADDITIONAL TYPES OF SECONDARY CELLS

A secondary cell is a storage cell that can be recharged by reversing the internal chemical reaction. A primary cell must be discarded after it has been completely discharged. The lead-acid cell is the most common type of storage cell. However, other types of secondary cells are available. Some of these are described next.

NICKEL-CADMIUM (NiCd) CELL   This type is popular because of its ability to deliver high current and to be cycled many times for recharging. Also, the cell can be stored for a long time, even when discharged, without any damage. The NiCd cell is available in both sealed and nonsealed designs, but the sealed construction shown in Fig. 12-13 is common. Nominal output voltage is 1.25 V per cell. Applications include portable power tools, alarm systems, and portable radio or television equipment.

The chemical equation for the NiCd cell can be written as follows:

$$2\ Ni(OH)_3 + Cd \underset{Discharge}{\overset{Charge}{\rightleftarrows}} 2\ Ni(OH)_2 + Cd(OH)_2$$

The electrolyte is potassium hydroxide (KOH), but it does not appear in the chemical equation. The reason is that the function of this electrolyte is just to act as a conductor for the transfer of hydroxyl (OH) ions. Therefore, unlike the lead-acid cell, the specific gravity of the electrolyte in the NiCd cell does not change with the state of charge.

The NiCd cell is a true storage cell with a reversible chemical reaction for recharging that can be cycled up to 1000 times. Maximum charging current is equal to the 10-h discharge rate. It should be noted that a new NiCd battery may need charging before use.

NICKEL-IRON (EDISON) CELL

Developed by Thomas Edison, this cell was once used extensively in industrial truck and railway applications. However, it has been replaced almost entirely by the lead-acid battery. New methods of construction for less weight, though, are making this cell a possible alternative in some applications.

The Edison cell has a positive plate of nickel oxide, a negative plate of iron, and an electrolyte of potassium hydroxide in water with a small amount of lithium hydroxide added. The chemical reaction is reversible for recharging. Nominal output is 1.2 V per cell.

NICKEL-ZINC CELL

This type has been used in limited railway applications. There has been renewed interest in it for use in electric cars, because of its high energy density. However, one drawback is its limited cycle life for recharging. Nominal output is 1.6 V per cell.

ZINC-CHLORINE (HYDRATE) CELL

This cell has been under development for use in electric vehicles. It is sometimes considered a zinc-chloride cell. This type has high energy density with a good cycle life. Nominal output is 2.1 V per cell.

LITHIUM-IRON SULFIDE CELL

This type is under development for commercial energy applications. Nominal output is 1.6 V per cell. The normal operating temperature is 800 to 900°F, which is high compared with the more popular types of cells.

SODIUM-SULFUR CELL

This is another type of cell being developed for electric vehicle applications. It has the potential of long life at low cost with high efficiency. The cell is designed to operate at temperatures between 550 and 650°F. Its most interesting feature is the use of a ceramic electrolyte.

PLASTIC CELLS

A recent development in battery technology is the rechargeable plastic cell made from a conductive polymer, which is a combination of organic chemical compounds. These cells could have ten times the power of the lead-acid type with one-tenth the weight and one-third the volume. In addition,

FIG. 12-13 Examples of nickel-cadmium batteries. The output voltage for each is 1.25 V.

ABOUT
ELECTRONICS

A Nobel prize was earned by researchers J. Bardeen, L. N. Cooper, and J. R. Schrieffer when they learned the secret of superconductivity—that pairs of linked electrons move within mercury without meeting obstacles.

the plastic cell does not require maintenance. One significant application could be for electric vehicles.

A plastic cell consists of an electrolyte between two polymer electrodes. Operation is similar to that of a capacitor. During charge, electrons are transferred from the positive electrode to the negative electrode by a dc source. On discharge, the stored electrons are driven through the external circuit to provide current in the load.

SOLAR CELLS   This type converts the sun's light energy directly into electric energy. The cells are made of semiconductor materials, which generate voltage output with light input. Silicon, with an output of 0.5 V per cell, is mainly used now. Research is continuing, however, on other materials, such as cadmium sulfide and gallium arsenide, that might provide more output. In practice, the cells are arranged in modules that are assembled into a large solar array for the required power.

In most applications, the solar cells are used in combination with a lead-acid cell specifically designed for this use. When there is sunlight, the solar cells charge the battery, as well as supplying power to the load. When there is no light, the battery supplies the required power.

TEST-POINT QUESTION 12-9

Answers at end of chapter.

Answer True or False.
a. The NiCd cell is a primary type.
b. Output of the NiCd cell is 1.25 V.
c. The Edison cell is a storage type.
d. Output of a solar cell is typically 0.5 V.

12-10 SERIES AND PARALLEL CELLS

An applied voltage higher than the emf of one cell can be obtained by connecting cells in series. The total voltage available across the battery of cells is equal to the sum of the individual values for each cell. Parallel cells have the same voltage as one cell but have more current capacity. The combination of cells is a *battery*.

SERIES CONNECTIONS   Figure 12-14 shows series-aiding connections for three dry cells. Here the three 1.5-V cells in series provide a total battery voltage of 4.5 V. Notice that the two end terminals A and B are left open to serve as the plus and minus terminals of the battery. These terminals are used to connect the battery to the load circuit, as shown in Fig. 12-14c.

In the lead-acid battery in Fig. 12-2, short, heavy metal straps connect the cells in series. The current capacity of a battery with cells in series is the same as for one cell because the same current flows through all the series cells.

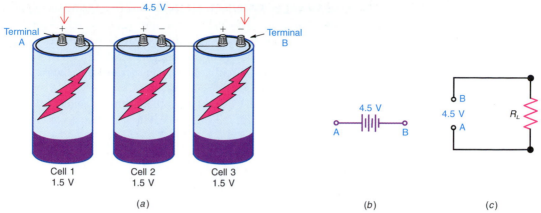

FIG. 12-14   Cells connected in series for higher voltage. Current rating is the same as for one cell. (*a*) Wiring. (*b*) Schematic symbol for battery with three series cells. (*c*) Battery connected to load resistance R_L.

PARALLEL CONNECTIONS   For more current capacity, the battery has cells in parallel, as shown in Fig. 12-15. All the positive terminals are strapped together, as are all the negative terminals. Any point on the positive side can be the plus terminal of the battery, and any point on the negative side can be the negative terminal.

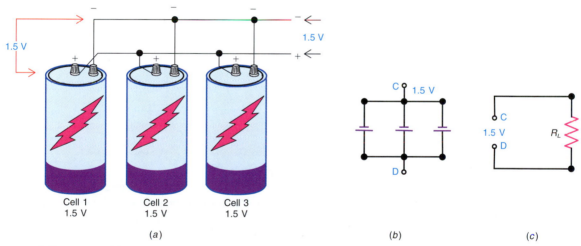

FIG. 12-15   Cells connected in parallel for higher current rating. (*a*) Wiring. (*b*) Schematic symbol for battery with three parallel cells. (*c*) Battery connected to load resistance R_L.

The parallel connection is equivalent to increasing the size of the electrodes and electrolyte, which increases the current capacity. The voltage output of the battery, however, is the same as for one cell.

Identical cells in parallel supply equal parts of the load current. For example, with three identical parallel cells producing a load current of 300 mA, each cell has a drain of 100 mA. Bad cells should not be connected in parallel with good cells, however, since the cells in good condition will supply more current, which may overload the good cells. In addition, a cell with lower output voltage will act as a load resistance, draining excessive current from the cells that have higher output voltage.

SERIES-PARALLEL CONNECTIONS In order to provide higher output and more current capacity, cells can be connected in series-parallel combinations. Figure 12-16 shows four No. 6 cells connected in series-parallel to form a battery that has a 3-V output with a current capacity of ½ A. Two of the 1.5-V cells in series provide 3 V total output voltage. This series string has a current capacity of ¼ A, however, assuming this current rating for one cell.

To double the current capacity, another string is connected in parallel. The two strings in parallel have the same 3-V output as one string, but with a current capacity of ½ A instead of the ¼ A for one string.

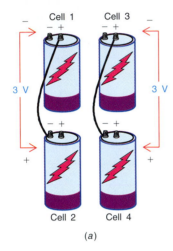

(a)

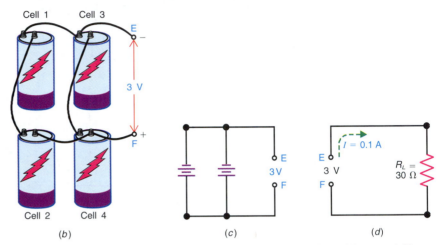

(b)   (c)   (d)

FIG. 12-16 Cells connected in series-parallel combinations. (*a*) Wiring two 3-V strings, each with two 1.5-V cells in series. (*b*) Wiring two 3-V strings in parallel. (*c*) Schematic symbol for the battery in (*b*) with output of 3 V. (*d*) Equivalent battery connected to load resistance R_L.

TEST-POINT QUESTION 12-10

Answers at end of chapter.
a. How many carbon-zinc cells in series are required to obtain a 9-V dc output? How many lead-acid cells are required to obtain 12 Vdc?
b. How many identical cells in parallel would be required to double the current rating of a single cell?
c. How many cells rated 1.5 Vdc 300 mA would be required in a series-parallel combination which would provide a rating of 900 mA at 6 Vdc?

12-11 CURRENT DRAIN DEPENDS ON LOAD RESISTANCE

It is important to note that the current rating of batteries, or any voltage source, is only a guide to typical values permissible for normal service life. The actual amount of current produced when the battery is connected to a load resistance is equal to $I = V/R$, by Ohm's law.

Figure 12-17 illustrates three different cases of using the applied voltage of 1.5 V from a dry cell. In Fig. 12-17a, the load resistance R_1 is 7.5 Ω. Then I is $1.5/7.5 = \frac{1}{5}$ A or 200 mA.

A No. 6 carbon-zinc cell used as the voltage source could supply this load of 200 mA continuously for about 74 h at a temperature of 70°F before dropping to an end voltage of 1.2 V. If an end voltage of 1.0 V could be used, the same load would be served for approximately 170 h.

In Fig. 12-17b, a larger load resistance R_2 is used. The value of 150 Ω limits the current to $1.5/150 = 0.01$ A or 10 mA. Again using the No. 6 carbon-zinc cell at 70°F, the load could be served continuously for 4300 h with an end voltage of 1.2 V. The two principles here are

1. The cell delivers less current for a higher resistance in the load circuit.
2. The cell can deliver a smaller load current for a longer time.

In Fig. 12-17c, the load resistance R_3 is reduced to 2.5 Ω. Then I is $1.5/2.5 = 0.6$ A or 600 mA. The No. 6 cell could serve this load continuously for only 9 h for an end voltage of 1.2 V. The cell could deliver even more load current, but for a shorter time. The relationship between current and time is not linear. For any one example, though, the amount of current is determined by the circuit, not by the current rating of the battery.

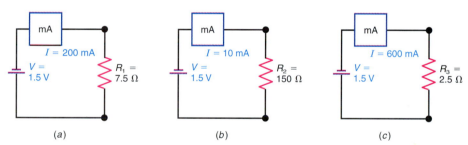

(a)   (b)   (c)

FIG. 12-17 An example of how current drain from a battery used as a voltage source depends on R of the load resistance. Different values of I are shown for the same V of 1.5 V. (a) The V/R_1 equals I of 200 mA. (b) The V/R_2 equals I of 10 mA. (c) The V/R_3 equals I of 600 mA.

TEST-POINT QUESTION 12-11

Answers at end of chapter.

Answer True or False.
a. A cell rated at 250 mA will produce this current for any value of R_L.
b. A higher value of R_L allows the cell to operate with normal voltage for a longer time.

12-12 INTERNAL RESISTANCE OF A GENERATOR

Any source that produces voltage output continuously is a generator. It may be a cell separating charges by chemical action or a rotary generator converting mo-

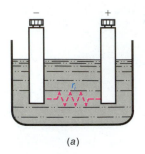

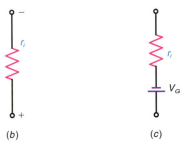

(a)

(b)   *(c)*

FIG. 12-18 Internal resistance r_i is in series with the generator voltage V_G. (*a*) Physical arrangement for a voltage cell. (*b*) Schematic symbol for r_i. (*c*) Equivalent circuit of r_i in series with V_G.

tion and magnetism into voltage output, for common examples. In any case, all generators have internal resistance, which is labeled r_i in Fig. 12-18.

The internal resistance r_i is important when a generator supplies load current because its internal voltage drop Ir_i subtracts from the generated emf, resulting in lower voltage across the output terminals. Physically, r_i may be the resistance of the wire in a rotary generator, or in a chemical cell r_i is the resistance of the electrolyte between electrodes. More generally, the internal resistance r_i is the opposition to load current inside the generator.

Since any current in the generator must flow through the internal resistance, r_i is in series with the generated voltage, as shown in Fig. 12-18c. It may be of interest to note that, with just one load resistance connected across a generator, they are in series with each other because R_L is in series with r_i.

If there is a short circuit across the generator, its r_i prevents the current from becoming infinitely high. As an example, if a 1.5-V cell is temporarily short-circuited, the short-circuit current I_{sc} could be about 15 A. Then r_i is V/I_{sc}, which equals 1.5/15, or 0.1 Ω for the internal resistance. These are typical values for a carbon-zinc D-size cell.

TEST-POINT QUESTION 12-12

Answers at end of chapter.

Answer True or False.
a. The generator's internal resistance r_i is in series with the load.
b. More load current produces a larger voltage drop across r_i.

12-13 WHY THE TERMINAL VOLTAGE DROPS WITH MORE LOAD CURRENT

Figure 12-19 illustrates how the output of a 100-V source can drop to 90 V because of the internal 10-V drop across r_i. In Fig. 12-19a, the voltage across the

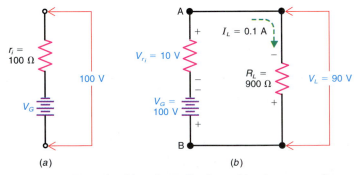

(a)   *(b)*

FIG. 12-19 Example of how internal voltage drop decreases voltage at output terminal of generator. (*a*) Open-circuit voltage output equals V_G of 100 V because there is no load current. (*b*) Terminal voltage V_L between points A and B is reduced to 90 V because of 10-V drop across 100-Ω r_i with 0.1-A I_L.

output terminals is equal to the 100 V of V_G because there is no load current on an open circuit. With no current, the voltage drop across r_i is zero. Then the full generated voltage is available across the output terminals. This value is the generated emf, *open-circuit voltage,* or *no-load voltage.*

We cannot connect the test leads inside the source to measure V_G. However, measuring this no-load voltage without any load current provides a method of determining the internally generated emf. We can assume the voltmeter draws practically no current because of its very high resistance.

In Fig. 12-19b with a load, however, current of 0.1 A flows, to produce a drop of 10 V across the 100 Ω of r_i. Note that R_T is 900 + 100 = 1000 Ω. Then I_L equals 100/1000, which is 0.1 A.

As a result, the voltage output V_L equals 100 − 10 = 90 V. This terminal voltage or load voltage is available across the output terminals when the generator is in a closed circuit with load current. The 10-V internal drop is subtracted from V_G because they are series-opposing voltages.

The graph in Fig. 12-20 shows how the terminal voltage V_L drops with increasing load current I_L. The reason is the greater internal voltage drop across r_i, as shown by the calculated values listed in Table 12-4. For this example, V_G is 100 V and r_i is 100 Ω.

Across the top row, infinite ohms for R_L means an open circuit. Then I_L is zero, there is no internal drop V_i, and V_L is the same 100 V as V_G.

Across the bottom row, zero ohms for R_L means a short circuit. Then the short-circuit current of 1 A results in zero output voltage because the entire generator voltage is dropped across the internal resistance. Or, we can say that with a short circuit of zero ohms across the load, the current is limited to V_G/r_i.

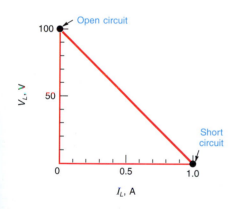

FIG. 12-20 How terminal voltage V_L drops with more load current. Graph is plotted for values in Table 12-4.

$V_G,$ V	$r_i,$ Ω	$R_L,$ Ω	$R_T = R_L + r_i,$ Ω	$I_L = V_G/R_T,$ A	$V_i = I_Lr_i,$ V	$V_L = V_G − V_i,$ V
100	100	∞	∞	0	0	100
100	100	900	1000	0.1	10	90
100	100	600	700	0.143	14.3	85.7
100	100	300	400	0.25	25	75
100	100	100	200	0.5	50	50
100	100	0	100	1.0	100	0

TABLE 12-4 HOW V_L DROPS WITH MORE I_L (FOR FIG. 12-20)

The lower the internal resistance of a generator, the better it is in terms of being able to produce full output voltage when supplying current for a load. For example, the very low r_i, about 0.01 Ω, for a 12-V lead-acid battery is the reason it can supply high values of load current and maintain its output voltage.

For the opposite case, a higher r_i means that the terminal voltage of a generator is much less with load current. As an example, an old dry battery with r_i of 500 Ω would appear normal when measured by a voltmeter but be useless because of low voltage when normal load current flows in an actual circuit.

HOW TO MEASURE r_i The internal resistance of any generator can be measured indirectly by determining how much the output voltage drops for a specified amount of load current. The difference between the no-load voltage and the load voltage is the amount of internal voltage drop $I_L r_i$. Dividing by I_L gives the value of r_i. As a formula,

$$r_i = \frac{V_{NL} - V_L}{I_L} \tag{12-1}$$

Example

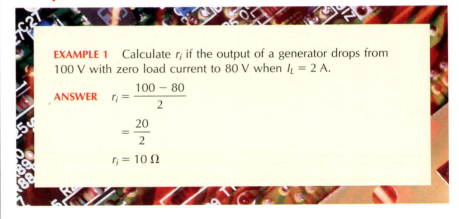

EXAMPLE 1 Calculate r_i if the output of a generator drops from 100 V with zero load current to 80 V when $I_L = 2$ A.

ANSWER
$$r_i = \frac{100 - 80}{2}$$
$$= \frac{20}{2}$$
$$r_i = 10 \ \Omega$$

A convenient technique for measuring r_i is to use a variable load resistance R_L. Vary R_L until the load voltage is one-half the no-load voltage. This value of R_L is also the value of r_i, since they must be equal to divide the generator voltage equally. For the same 100-V generator with the 10-Ω r_i used in Example 1, if a 10-Ω R_L were used, the load voltage would be 50 V, equal to one-half the no-load voltage.

You can solve this circuit by Ohm's law to see that I_L is 5 A with 20 Ω for the combined R_T. Then the two voltage drops of 50 V each add to equal the 100 V of the generator.

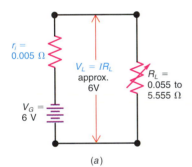

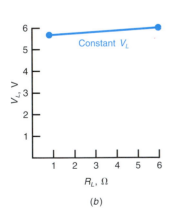

FIG. 12-21 Constant-voltage generator with low r_i. The V_L stays approximately the same 6 V as I varies with R_L. (a) Circuit. (b) Graph for V_L.

Answers at end of chapter.

Answer True or False.
a. For Eq. (12-1), V_L must be more than V_{NL}.
b. For Eq. (12-1), when V_L is one-half V_{NL}, the r_i is equal to R_L.

12-14 CONSTANT-VOLTAGE AND CONSTANT-CURRENT SOURCES

A generator with very low internal resistance is considered to be a constant-voltage source. Then the output voltage remains essentially the same when the load current changes. This idea is illustrated in Fig. 12-21a for a 6-V lead-acid battery with an r_i of 0.005 Ω. If the load current varies over the wide range of 1 to 100 A, for any of these values, the internal Ir_i drop across 0.005 Ω is less than 0.5 V.

CONSTANT-CURRENT GENERATOR   It has very high resistance, compared with the external load resistance, resulting in constant current, although the output voltage varies.

The constant-current generator shown in Fig. 12-22 has such high resistance, with an r_i of 0.9 MΩ, that it is the main factor determining how much current can be produced by V_G. Here R_L varies in a 3:1 range from 50 to 150 kΩ. Since the current is determined by the total resistance of R_L and r_i in series, however, I is essentially constant at 1.05 to 0.95 mA, or approximately 1 mA. This relatively constant I_I is shown by the graph in Fig. 12-22b.

Note that the terminal voltage V_L varies in approximately the same 3:1 range as R_L. Also, the output voltage is much less than the generator voltage because of the high internal resistance compared with R_L. This is a necessary condition, however, in a circuit with a constant-current generator.

A common practice is to insert a series resistance to keep the current constant, as shown in Fig. 12-23a (page 344). Resistance R_1 must be very high compared with R_L. In this example, I_L is 50 μA with 50 V applied, and R_T is practically equal to the 1 MΩ of R_1. The value of R_L can vary over a range as great as 10:1 without changing R_T or I_I appreciably.

The circuit with an equivalent constant-current source is shown in Fig. 12-23b. Note the arrow symbol for a current source. As far as R_L is concerned, its terminals A and B can be considered as receiving either 50 V in series with 1 MΩ or 50 μA in shunt with 1 MΩ.

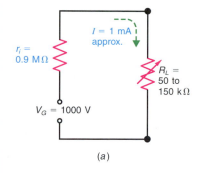

(a)

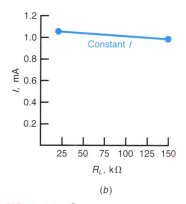

(b)

FIG. 12-22   Constant-current generator with high r_i. The I stays approximately the same 1 mA as V_L varies with R_L. (a) Circuit. (b) Graph for I.

TEST-POINT QUESTION 12-14

Answers at end of chapter.

Is the internal resistance high or low for:
a. A constant-voltage source?
b. A constant-current source?

12-15 MATCHING A LOAD RESISTANCE TO THE GENERATOR r_i

In the diagram in Fig. 12-24 (page 345), when R_L equals r_i the load and generator are matched. The matching is significant because the generator then produces maximum power in R_L, as verified by the values listed in Table 12-5.

MAXIMUM POWER IN R_L   When R_L is 100 Ω to match the 100 Ω of r_i, maximum power is transferred from the generator to the load. With higher resistance for R_L, the output voltage V_L is higher, but the current is reduced. Lower resistance for R_L allows more current, but V_L is less. When r_i and R_L both equal 100 Ω, this combination of current and voltage produces the maximum power of 100 W across R_L.

With generators that have very low resistance, however, matching is often impractical. For example, if a 6-V lead-acid battery with a 0.003-Ω internal resistance were connected to a 0.003-Ω load resistance, the battery could be damaged by excessive current as high as 1000 A.

MAXIMUM VOLTAGE ACROSS R_L   If maximum voltage, rather than power, is desired, the load should have as high a resistance as possible. Note that R_L and r_i form a voltage divider for the generator voltage, as illustrated in Fig. 12-24b. The values for IR_L listed in Table 12-5 show how the output voltage V_L increases with higher values of R_L.

MAXIMUM EFFICIENCY   Note also that the efficiency increases as R_L increases because there is less current, resulting in less power lost in r_i. When R_L equals r_i, the efficiency is only 50 percent, since one-half the total generated power is dissipated in r_i, the internal resistance of the generator. In conclusion, then, matching the load and generator resistances is desirable when the load requires maximum power rather than maximum voltage or efficiency, assuming that the match does not result in excessive current.

TEST-POINT QUESTION 12-15

Answers at end of chapter.

Answer True or False.
a. When $R_L = r_i$, the P_L is maximum.
b. The V_L is maximum when R_L is maximum.

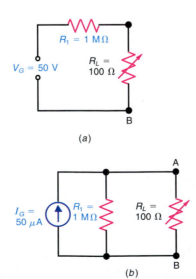

(a)

(b)

FIG. 12-23   Voltage source in (a) equivalent to current source in (b) for load resistance R_L across terminals A and B.

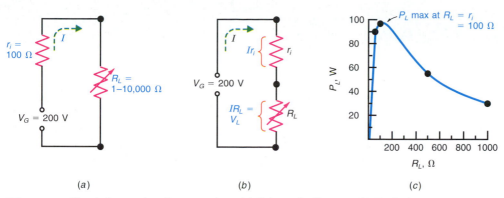

(a)　　　　　　　　　(b)　　　　　　　　　(c)

FIG. 12-24 Circuit for varying R_L to match r_i. (a) Schematic diagram. (b) Equivalent voltage divider for voltage output across R_L. (c) Graph of power output P_L for different values of R_L. All values are listed in Table 12-5.

TABLE 12-5　EFFECT OF LOAD RESISTANCE ON GENERATOR OUTPUT*

	R_L, V	$I = V_G/R_T$, A	Ir_i, V	IR_L, V	P_L, W	P_i, W	P_T, W	EFFICIENCY = P_L/P_T, %
	1	1.98	198	2	4	392	396	1
	50	1.33	133	67	89	178	267	33
$R_L = r_i \rightarrow$	100	1	100	100	100	100	200	50
	500	0.33	33	167	55	11	66	83
	1,000	0.18	18	180	32	3.24	35.24	91
	10,000	0.02	2	198	4	0.04	4.04	99

*Values calculated approximately for circuit in Fig. 12-24, with $V_G = 200$ V and $r_i = 100$ Ω.

- A voltaic cell consists of two different conductors as electrodes immersed in an electrolyte. The voltage output depends only on the chemicals in the cell. The current capacity increases with larger sizes. A primary cell cannot be recharged. A secondary or storage cell can be recharged.
- A battery is a group of cells in series or in parallel. With cells in series, the voltages add, but the current capacity is the same as that of one cell. With cells in parallel, the voltage output is the same as that of one cell, but the total current capacity is the sum of the individual values.
- The carbon-zinc dry cell is the most common type of primary cell. Zinc is the negative electrode; carbon is the positive electrode. Its output voltage is approximately 1.5 V.
- The lead-acid cell is the most common form of storage battery. The positive electrode is lead peroxide; spongy lead is the negative electrode. Both are in a dilute solution of sulfuric acid for the electrolyte. The voltage output is approximately 2.1 V per cell.
- To charge a lead-acid battery, connect it to a dc voltage equal to approximately 2.5 V per cell. Connecting the positive terminal of the battery to the positive side of the charging source and the negative terminal to the negative side results in charging current through the battery.
- The mercury cell is a primary cell with an output of 1.35 or 1.4 V.
- The nickel-cadmium cell is rechargeable, with an output of 1.25 V.
- A constant-voltage generator has very low internal resistance. Output voltage is relatively constant with changing values of load because of the low internal voltage drop.
- A constant-current generator has a very high internal resistance. This determines the constant value of current in the generator circuit relatively independent of the load resistance.
- Any generator has an internal resistance r_i. With load current I_L, the internal $I_L r_i$ drop reduces the voltage across the output terminals. When I_L makes the terminal voltage drop to one-half the no-load voltage, the external R_L equals the internal r_i.
- Matching a load to a generator means making the R_L equal to the generator's r_i. The result is maximum power delivered to the load from the generator.

SELF-TEST

ANSWERS AT BACK OF BOOK.

Choose (*a*), (*b*), (*c*), or (*d*).

1. Which of the following is false? (*a*) A lead-acid cell can be recharged. (*b*) A primary cell has an irreversible chemical reaction. (*c*) A storage cell has a reversible chemical reaction. (*d*) A carbon-zinc cell has unlimited shelf life.
2. The output voltage of a lead-acid cell is (*a*) 1.25 V; (*b*) 1.35 V; (*c*) 2.1 V; (*d*) 6 V.

3. The current in a chemical cell is a movement of (a) positive hole charges; (b) positive and negative ions; (c) positive ions only; (d) negative ions only.

4. Cells are connected in series to (a) increase the voltage output; (b) decrease the voltage output; (c) decrease the internal resistance; (d) increase the current capacity.

5. Cells are connected in parallel to (a) increase the voltage output; (b) increase the internal resistance; (c) decrease the current capacity; (d) increase the current capacity.

6. Which of the following is a sealed secondary cell? (a) Edison cell; (b) carbon-zinc cell; (c) mercury cell; (d) nickel-cadmium cell.

7. When R_L equals the generator r_i, which of the following is maximum? (a) Power in R_L; (b) current; (c) voltage across R_L; (d) efficiency of the circuit.

8. Five carbon-zinc cells in series have an output of (a) 1.5 V; (b) 5.0 V; (c) 7.5 V; (d) 11.0 V.

9. A constant-voltage generator has (a) low internal resistance; (b) high internal resistance; (c) minimum efficiency; (d) minimum current capacity.

10. A generator has an output of 10 V on open circuit, which drops to 5 V with a load current of 50 mA and an R_L of 1000 Ω. The internal resistance r_i equals (a) 25 Ω; (b) 50 Ω; (c) 100 Ω; (d) 1000 Ω.

QUESTIONS

1. Draw a sketch illustrating the construction of a carbon-zinc dry cell. Indicate the negative and positive electrodes and the electrolyte.

2. Draw a sketch illustrating construction of the lead-acid cell. Indicate the negative and positive electrodes and the electrolyte.

3. Show the wiring for the following batteries: (a) six lead-acid cells for a voltage output of approximately 12 V; (b) six standard No. 6 dry cells for a voltage output of 4.5 V with a current capacity of ½ A. Assume a current capacity of ¼ A for one cell.

4. (a) What is the advantage of connecting cells in series? (b) What is connected to the end terminals of the series cells?

5. (a) What is the advantage of connecting cells in parallel? (b) Why can the load be connected across any one of the parallel cells?

6. How many cells are necessary in a battery to double the voltage and current ratings of a single cell? Show the wiring diagram.

7. Draw a diagram showing two 12-V lead-acid batteries being charged by a 15-V source.

8. Why is a generator with very low internal resistance called a constant-voltage source?

9. Why does discharge current lower the specific gravity in a lead-acid cell?

10. Would you consider the lead-acid battery a constant-current source or a constant-voltage source? Why?

11. List five types of chemical cells, giving two features of each.

12. Referring to Fig. 12-21b, draw the corresponding graph that shows how I varies with R_L.

13. Referring to Fig. 12-22b, draw the corresponding graph that shows how V_L varies with R_L.

14. Referring to Fig. 12-24c, draw the corresponding graph that shows how V_L varies with R_L.

1. A 1.5-V No. 6 carbon-zinc dry cell is connected across an R_L of 1000 Ω. How much current flows in the circuit?

2. Draw the wiring diagram for six No. 6 cells providing a 3-V output with a current capacity of ¾ A. Assume a current capacity of ¼ A for one cell. Draw the schematic diagram of this battery connected across a 10-Ω resistance. (**a**) How much current flows in the circuit? (**b**) How much power is dissipated in the resistance? (**c**) How much power is supplied by the battery?

3. A 6-V lead-acid battery has an internal resistance of 0.01 Ω. How much current will flow if the battery has a short circuit?

4. How much is the specific gravity of a solution with equal parts of sulfuric acid and water?

5. A lead-acid battery discharges at the rate of 8 A for 10 h. (**a**) How many coulombs of charge must be put back into the battery to restore the original charge, assuming 100 percent efficiency? (**b**) How long will this recharging take, with a charging current of 2 A?

6. The output voltage of a battery drops from 90 V at no load to 60 V with a load current of 50 mA. (**a**) How much is the internal r_i of the battery? (**b**) How much is R_L for this load current? (**c**) How much R_L reduces the load voltage to one-half the no-load voltage?

7. The output voltage of a source reads 60 V with a DMM. When a meter with 1000 Ω/V sensitivity is used, the reading is 50 V on the 100-V range. How much is the internal resistance of the source?

8. A 100-V source with an internal resistance of 10 kΩ is connected to a variable load resistance R_L. Tabulate I, V_L, and power in R_L for values of 1 kΩ, 5 kΩ, 10 kΩ, 15 kΩ, and 20 kΩ.

9. A generator has an open-circuit emf of 18 V. Its terminal voltage drops to 15 V with an R_L of 30 Ω. Calculate r_i.

10. Referring to Fig. 12-24, calculate P_L when R_L is 200 Ω. Compare this value with the maximum P_L at $R_L = r_i = 100$ Ω.

11. In Fig. 12-25, calculate I, V_L, P_L, P_T, and the percent efficiency $[(P_L/P_T) \times 100]$ for the following values of R_L: 1 Ω, 3 Ω, 5 Ω, 7 Ω, 10 Ω, 15 Ω, 45 Ω, and 100 Ω.

12. In Fig. 12-26, calculate: (**a**) the value of I_L for the following values of R_L: 0 Ω, 1 Ω, 10 Ω, 100 Ω, 1 kΩ, 10 kΩ, 100 kΩ, and 1 MΩ; (**b**) the value of R_L for which maximum power transfer will occur; (**c**) the power transferred to R_L when $R_L = r_i$.

13. Refer to Fig. 12-27. With S_1 in position 1, $V = 50$ V. With S_1 in position 2, $V = 37.5$ V. Calculate r_i.

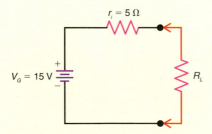

FIG. 12-25 Circuit diagram for Prob. 11.

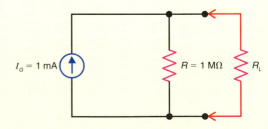

FIG. 12-26 Circuit diagram for Prob. 12.

14. A 1.5-V cell develops a terminal voltage of 1.35 V while delivering 150 mA to a load R_L. Calculate: (**a**) r_i; (**b**) R_L; (**c**) P_L; (**d**) P_T; (**e**) percent efficiency.

15. A 10-V generator has an internal resistance r_i of 50 Ω. Calculate the value of R_L which will provide an efficiency of 75 percent in the transfer of power to R_L.

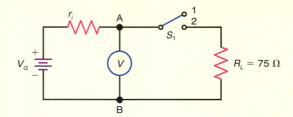

FIG. 12-27   Circuit diagram for Prob. 13.

CRITICAL THINKING

1. In Fig. 12-28, calculate (**a**) V_L, (**b**) I_L, and (**c**) the current supplied to R_L by each separate voltage source.

2. In Fig. 12-29, calculate: (**a**) the value of R_L for which the maximum transfer of power occurs; (**b**) the maximum power delivered to R_L.

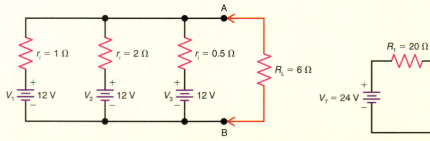

FIG. 12-28   Circuit diagram for Critical Thinking Prob. 1.

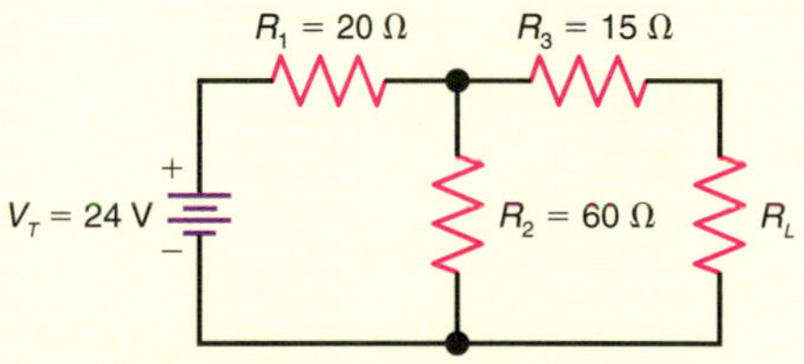

FIG. 12-29   Circuit diagram for Critical Thinking Prob. 2.

ANSWERS TO TEST-POINT QUESTIONS

12-1	**a.** 1.5 V	**12-3**	**a.** T	**12-6**	**a.** alkaline	**12-9**	**a.** F	**12-12** **a.** T
	b. 2.1 V		**b.** T		**b.** mercury		**b.** T	**b.** T
	c. secondary		**c.** F		**c.** size C		**c.** T	**12-13** **a.** F
							d. T	**b.** T
12-2	**a.** T	**12-4**	**a.** T	**12-7**	**a.** T			
	b. T		**b.** T		**b.** T	**12-10**	**a.** six	**12-14** **a.** low
	c. F						**b.** two	**b.** high
		12-5	**a.** F	**12-8**	**a.** six		**c.** twelve	
			b. F		**b.** 15 A			**12-15** **a.** T
					c. 1270	**12-11**	**a.** F	**b.** T
							b. T	

REVIEW: CHAPTERS 11 AND 12

SUMMARY

- A conductor is a material whose resistance is very low. Some examples of good conductors are silver, copper, and aluminum, with copper generally being used for wire. An insulator is a material whose resistance is very high. Some examples of good insulators include air, mica, rubber, porcelain, and plastics.
- The gage sizes for copper wires are listed in Table 11-1. As the gage sizes increase from 1 to 40, the diameter and circular area decrease. Higher gage numbers correspond to thinner wire.
- For switches, the term *pole* refers to the number of completely isolated circuits that can be controlled by the switch. The term *throw* refers to the number of closed-contact positions that exist per pole. A switch can have any number of poles and throws.
- A good fuse has very low resistance, with an *IR* voltage of practically zero. An open fuse has nearly infinite resistance. If an open fuse exists in a series circuit or in the main line of a parallel circuit, its voltage drop equals the applied voltage.
- The resistance of a wire is directly proportional to its length and inversely proportional to its cross-sectional area.
- All metals in their purest form have a positive temperature coefficient, which means that their resistance increases with an increase in temperature. Carbon has a negative temperature coefficient, which means that its resistance decreases as the temperature increases.
- An ion is an atom that has either gained or lost electrons. A negative ion is an atom with more electrons than protons. Conversely, a positive ion is an atom with more protons than electrons. Ions can move to provide electric current in liquids and gases. The motion of ions is called ionization current.
- A battery is a combination of individual voltaic cells. A primary cell cannot be recharged, whereas a secondary cell can be recharged several times. The main types of cells for batteries include alkaline, silver oxide, nickel cadmium, lithium, and lead-acid.
- With individual cells in series, the total battery voltage equals the sum of the individual cell voltages. This is assuming that the cells are connected in a series-aiding manner. The current rating of the series-aiding cells is the same as that for the cell with the lowest current rating.
- With individual cells in parallel, the voltage is the same as that across one cell. However, the current rating of the combination equals the sum of the individual current-rating values. Only cells that have the same voltage should be connected in parallel.
- All types of dc and ac generators have an internal resistance r_i. Physically, the value of r_i may be the resistance of the electrolyte in a battery or the wire in a rotary generator.
- When a generator supplies current to a load, the terminal voltage drops because some voltage is dropped across the internal resistance r_i.

- Matching a load to a generator means making R_L equal to r_i. When $R_L = r_i$, maximum power is delivered from the generator to the load.
- A constant-voltage source has a very low internal r_i, whereas a constant-current source has a very high internal r_i.

REVIEW SELF-TEST

ANSWERS AT BACK OF BOOK.

Choose (a), (b), (c), or (d).

1. Which of the following is the best conductor of electricity? (a) carbon; (b) silicon; (c) rubber; (d) copper.
2. Which of the following wires has the largest cross-sectional area? (a) No. 28 gage; (b) No. 23 gage; (c) No. 12 gage; (d) No. 16 gage.
3. The filament of a light bulb measures 2.5 Ω when cold. With 120 V applied across the filament, the bulb dissipates 75 W of power. What is the hot resistance of the bulb? (a) 192 Ω; (b) 0.625 Ω; (c) 2.5 Ω; (d) 47 Ω.
4. A DPST switch has how many terminal connections for soldering? (a) 3; (b) 1; (c) 4; (d) 6.
5. Which of the following materials has a negative temperature coefficient? (a) steel; (b) carbon; (c) tungsten; (d) nichrome.
6. The IR voltage across a good fuse equals: (a) the applied voltage; (b) one-half the applied voltage; (c) infinity; (d) zero.
7. A battery has a no-load voltage of 9 V. Its terminal voltage drops to 8.25 V when a load current of 200 mA is drawn from the battery. The internal resistance r_i equals: (a) 0.375 Ω; (b) 3.75 Ω; (c) 41.25 Ω; (d) 4.5 Ω.
8. When $R_L = r_i$: (a) maximum voltage is across R_L; (b) maximum power is delivered to R_L; (c) the efficiency is 100 percent; (d) the minimum power is delivered to R_L.
9. A constant-current source has a(n): (a) very high internal resistance; (b) constant output voltage; (c) very low internal resistance; (d) output voltage that is always zero.
10. Cells can be connected in series-parallel to: (a) increase the voltage above that of a single cell; (b) increase the current capacity above that of a single cell; (c) reduce the voltage and current rating below that of a single cell; (d) both (a) and (b).

REFERENCES

Adams, J. A., and G. Rockmaker: *Industrial Electricity—Principles and Practices,* Glencoe/McGraw-Hill, Columbus, Ohio.

Fowler, R. J.: *Electricity—Principles and Applications,* Glencoe/McGraw-Hill, Columbus, Ohio.

Richter, H. P.: *Practical Electrical Wiring,* McGraw-Hill, New York.

CHAPTER 13

MAGNETISM

Electrical effects exist in two forms, voltage and current. In terms of voltage, separated electric charges have the potential to do mechanical work in attracting or repelling charges. Similarly, any electric current has an associated magnetic field that can do the work of attraction or repulsion. Materials made of iron, nickel, and cobalt, particularly, concentrate their magnetic effects at opposite ends, where the magnetic material meets a nonmagnetic medium such as air.

The points of concentrated magnetic strength are called north and south poles. The opposite magnetic poles correspond to the idea of opposite polarities of electric charges. The name *magnetism* is derived from the iron oxide mineral *magnetite.* Ferromagnetism refers specifically to the magnetic properties of iron.

CHAPTER OBJECTIVES

Upon completion of this chapter, you should be able to:

- *Describe* the magnetic field surrounding a magnet.
- *Define* the units of magnetic flux and flux density.
- *Describe* how an iron bar is magnetized by induction.
- *Define* the term *relative permeability*.
- *Explain* the difference between a bar magnet and an electromagnet.
- *List* the three classifications of magnetic materials.
- *Explain* the electrical and magnetic properties of ferrites.
- *Describe* the Hall effect.

IMPORTANT TERMS IN THIS CHAPTER

air gap	gauss unit	permanent magnet
diamagnetic	Hall effect	relative permeability
electromagnet	keeper	shielding
ferrites	magnetic induction	tesla unit
ferromagnetic	magnetic poles	toroid
flux	maxwell unit	weber unit
flux density	paramagnetic	

TOPICS COVERED IN THIS CHAPTER

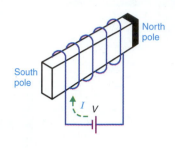

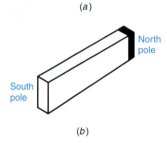

FIG. 13-1 Poles of a magnet. (*a*) Electromagnet (EM) produced by current from a battery. (*b*) Permanent magnet (PM) without any external source of current.

As shown in Fig. 13-1, the north and south poles of a magnet are the points of concentration of magnetic strength. The practical effects of this ferromagnetism result from the magnetic field of force between the two poles at opposite ends of the magnet. Although the magnetic field is invisible, evidence of its force can be seen when small iron filings are sprinkled on a glass or paper sheet placed over a bar magnet (Fig. 13-2*a*). Each iron filing becomes a small bar magnet. If the sheet is tapped gently to overcome friction so that the filings can move, they become aligned by the magnetic field.

Many filings cling to the ends of the magnet, showing that the magnetic field is strongest at the poles. The field exists in all directions but decreases in strength with increasing distance from the poles of the magnet.

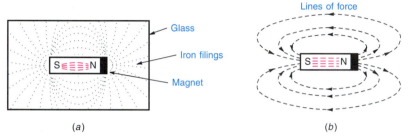

FIG. 13-2 Magnetic field of force around a bar magnet. (*a*) Field outlined by iron filings. (*b*) Field indicated by lines of force.

FIELD LINES In order to visualize the magnetic field without iron filings, we show the field as lines of force, as in Fig. 13-2*b*. The direction of the lines outside the magnet shows the path a north pole would follow in the field, repelled away from the north pole of the magnet and attracted to its south pole. Although we cannot actually have a unit north pole by itself, the field can be explored by noting how the north pole on a small compass needle moves.

The magnet can be considered as the generator for an external magnetic field, provided by the two opposite magnetic poles at the ends. This idea corresponds to the two opposite terminals on a battery as the source for an external electric field provided by opposite charges.

Magnetic field lines are unaffected by nonmagnetic materials such as air, vacuum, paper, glass, wood, or plastics. When these materials are placed in the magnetic field of a magnet, the field lines are the same as though the material were not there.

However, the magnetic field lines become concentrated when a magnetic substance like iron is placed in the field. Inside the iron, the field lines are more dense, compared with the field in air.

NORTH AND SOUTH MAGNETIC POLES The earth itself is a huge natural magnet, with its greatest strength at the north and south poles. Because of the earth's magnetic poles, if a small bar magnet is suspended so that it can turn easily, one end will always point north. This end of the bar magnet is defined as the *north-seeking pole,* as shown in Fig. 13-3. The opposite end is the

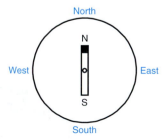

FIG. 13-3 Definition of north and south poles of bar magnet.

south-seeking pole. When polarity is indicated on a magnet, the north-seeking end is the north pole (N) and the opposite end is the south pole (S). It should be noted that the magnetic north pole deviates from true geographic north, the amount depending on location.

Similar to the force between electric charges is a force between magnetic poles causing attraction of opposite poles and repulsion between similar poles:

1. A north pole (N) and a south pole (S) tend to attract each other.
2. A north pole (N) tends to repel another north pole (N), while a south pole (S) tends to repel another south pole (S).

These forces are illustrated by the field of iron filings between opposite poles in Fig. 13-4*a* and between similar poles in Fig. 13-4*b*.

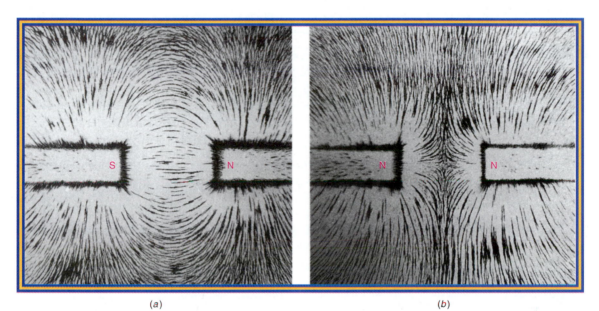

(a)   (b)

FIG. 13-4 Magnetic field patterns produced by iron filings. (*a*) Field between opposite poles. The north and south poles could be reversed. (*b*) Field between similar poles. The two north poles could be south poles.

TEST-POINT QUESTION 13-1

Answers at end of chapter.

Answer True or False.
a. On a magnet, the north-seeking pole is labeled N.
b. Like poles have a force of repulsion.

13-2 MAGNETIC FLUX ϕ

The entire group of magnetic field lines, which can be considered to flow outward from the north pole of a magnet, is called *magnetic flux.* Its symbol is the

Greek letter ϕ (phi). A strong magnetic field has more lines of force and more flux than a weak magnetic field.

THE MAXWELL   One maxwell (Mx) unit equals one magnetic field line. In Fig. 13-5, as an example, the flux illustrated is 6 Mx because there are 6 field lines flowing in or out for each pole. A 1-lb magnet can provide a magnetic flux ϕ of about 5000 Mx. This unit is named for James Clerk Maxwell (1831–1879), an important Scottish mathematical physicist who contributed much to electrical and field theory.

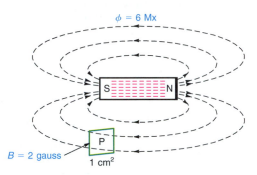

$\phi = 6$ Mx

$B = 2$ gauss

1 cm²

FIG. 13-5   Total flux ϕ is 6 lines or 6 Mx. Flux density B at point P is 2 lines per square centimeter or 2 G.

THE WEBER   This is a larger unit of magnetic flux. One weber (Wb) equals 1×10^8 lines or maxwells. Since the weber is a large unit for typical fields, the microweber unit can be used. Then $1\ \mu\text{Wb} = 10^{-6}$ Wb. This unit is named for Wilhelm Weber (1804–1890), a German physicist.

To convert microwebers to lines, multiply by the conversion factor 10^8 lines per weber, as follows:

$$1\ \mu\text{Wb} = 1 \times 10^{-6}\ \text{Wb} \times 10^8\ \frac{\text{lines}}{\text{Wb}}$$

$$= 1 \times 10^2\ \text{lines}$$

$$1\ \mu\text{Wb} = 100\ \text{lines or Mx}$$

Note that the conversion is arranged to make the weber units cancel, since we want maxwell units in the answer.

Even the microweber unit is larger than the maxwell unit. For the same 1-lb magnet producing the magnetic flux of 5000 Mx, it corresponds to 50 μWb. The calculations for this conversion of units are

$$\frac{5000\ \text{Mx}}{100\ \text{Mx}/\mu\text{Wb}} = 50\ \mu\text{Wb}$$

Note that the maxwell units cancel. Also, the $1/\mu$Wb becomes inverted from the denominator to μWb in the numerator.

SYSTEMS OF MAGNETIC UNITS   The basic units in metric form can be defined in two ways:

1.   The centimeter-gram-second system defines small units. This is the cgs system.

2. The meter-kilogram-second system is for larger units of a more practical size. This is the mks system.

Furthermore, the Système Internationale (SI) units provide a worldwide standard in mks dimensions. They are practical values based on the ampere of current.

With magnetic flux ϕ, the maxwell (Mx) is a cgs unit, while the weber (Wb) is an mks or SI unit. For science and engineering, the SI units are preferred values, but the cgs units are still used in many practical applications of magnetism.

Answers at end of chapter.

The value of 2000 magnetic lines is how much flux in
a. Maxwell units?
b. Microweber units?

13-3 FLUX DENSITY *B*

As shown in Fig. 13-5, the flux density is the number of magnetic field lines per unit area of a section perpendicular to the direction of flux. As a formula,

$$B = \frac{\phi}{A} \qquad \text{(13-1)}$$

where ϕ is the flux through an area A, and the flux density is B.

THE GAUSS   In the cgs system, this unit is one line per square centimeter, or 1 Mx/cm^2. As an example, in Fig. 13-5, the total flux ϕ is 6 lines, or 6 Mx. At point P in this field, however, the flux density B is 2 G because there are 2 lines per cm^2. The flux density has a higher value close to the poles, where the flux lines are more crowded.

As an example of flux density, B for a 1-lb magnet would be 1000 G at the poles. This unit is named for Karl F. Gauss (1777–1855), a German mathematician.

Example

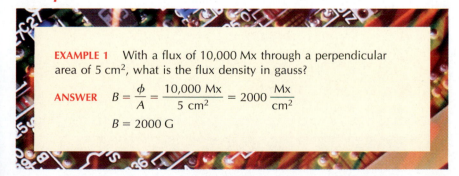

EXAMPLE 1   With a flux of 10,000 Mx through a perpendicular area of 5 cm^2, what is the flux density in gauss?

ANSWER $B = \dfrac{\phi}{A} = \dfrac{10{,}000\ \text{Mx}}{5\ \text{cm}^2} = 2000\ \dfrac{\text{Mx}}{\text{cm}^2}$

$B = 2000\ \text{G}$

ABOUT
ELECTRONICS

Home owners in Los Angeles who have security alarm systems can add a day and night vision camera, no larger than a thumbnail, that will show police via the phone line who has just tripped the alarm.

As typical values, B for the earth's magnetic field can be about 0.2 G; a large laboratory magnet produces B of 50,000 G. Since the gauss is so small, it is often used in kilogauss units, where $1 \text{ kG} = 10^3 \text{ G}$.

THE TESLA In SI, the unit of flux density B is webers per square meter (Wb/m^2). One weber per square meter is called a *tesla,* abbreviated T. This unit is named for Nikola Tesla (1857–1943), a Yugoslav-born American inventor in electricity and magnetism.

When converting between cgs and mks units, note that

$$1 \text{ m} = 100 \text{ cm or } 1 \times 10^2 \text{ cm}$$
$$1 \text{ m}^2 = 10,000 \text{ cm}^2 \text{ or } 10^4 \text{ cm}^2$$

These conversions are from the larger m and m^2 to smaller units of cm and cm^2. To go the opposite way,

$$1 \text{ cm} = 0.01 \text{ m or } 1 \times 10^{-2} \text{ m}$$
$$1 \text{ cm}^2 = 0.0001 \text{ m}^2 \text{ or } 1 \times 10^{-4} \text{ m}^2$$

As an example, 5 cm^2 is equal to 0.0005 m^2 or $5 \times 10^{-4} \text{ m}^2$. The calculations for the conversion are

$$5 \text{ cm}^2 \times \frac{0.0001 \text{ m}^2}{\text{cm}^2} = 0.0005 \text{ m}^2$$

In powers of 10, the conversion is

$$5 \text{ cm}^2 \times \frac{1 \times 10^{-4} \text{ m}^2}{\text{cm}^2} = 5 \times 10^{-4} \text{ m}^2$$

In both cases, note that the units of cm^2 cancel to leave m^2 for the desired unit.

Example

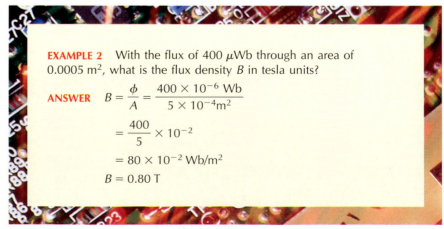

EXAMPLE 2 With the flux of 400 μWb through an area of 0.0005 m^2, what is the flux density B in tesla units?

ANSWER
$$B = \frac{\phi}{A} = \frac{400 \times 10^{-6} \text{ Wb}}{5 \times 10^{-4} \text{ m}^2}$$
$$= \frac{400}{5} \times 10^{-2}$$
$$= 80 \times 10^{-2} \text{ Wb/m}^2$$
$$B = 0.80 \text{ T}$$

The tesla is a larger unit than the gauss, as

▶ $1\text{ T} = 1 \times 10^4\text{ G}$

For example, the flux density of 20,000 G is equal to 2 T. The calculations for this conversion are

$$\frac{20{,}000\text{ G}}{1 \times 10^4\text{ G/T}} = \frac{2 \times 10^4\text{ T}}{1 \times 10^4} = 2\text{ T}$$

Note that the G units cancel to leave T units for the desired answer. Also, the 1/T in the denominator becomes inverted to T units in the numerator.

COMPARISON OF FLUX AND FLUX DENSITY   Remember that the flux ϕ includes total area, while the flux density B is for a specified unit area. The difference between ϕ and B is illustrated in Fig. 13-6 with cgs units. The total area A here is 9 cm², equal to 3 cm × 3 cm. For one unit box of 1 cm², 16 lines are shown. Therefore, the flux density B is 16 lines or maxwells per cm², which equals 16 G. The total area includes nine of these boxes. Therefore, the total flux ϕ is 144 lines or maxwells, equal to 9 × 16 for $B \times A$.

    For the opposite case, if the total flux ϕ is given as 144 lines or maxwells, the flux density is found by dividing 144 by 9 cm². This division of 144/9 equals 16 lines or maxwells per cm², which is 16 G.

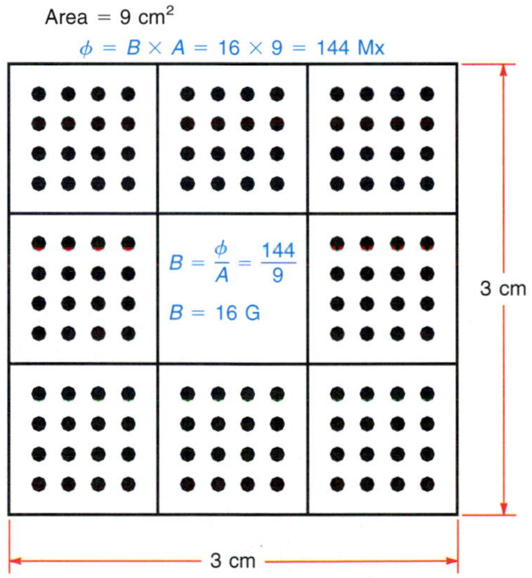

Area = 9 cm²

$\phi = B \times A = 16 \times 9 = 144\text{ Mx}$

$B = \dfrac{\phi}{A} = \dfrac{144}{9}$

$B = 16\text{ G}$

3 cm

3 cm

FIG. 13-6   Comparison of total flux ϕ and flux density B. Total area of 9 cm² has 144 lines or 144 Mx. For 1 cm² the flux density is 144/9 = 16 G.

TEST-POINT QUESTION 13-3

Answers at end of chapter.
a. The ϕ is 9000 Mx through 3 cm². How much is B in gauss units?
b. How much is B in tesla units for ϕ of 90 μWb through 0.0003 m²?

13-4 INDUCTION BY THE MAGNETIC FIELD

The electric or magnetic effect of one body on another without any physical contact between them is called *induction*. For instance, a permanent magnet can induce an unmagnetized iron bar to become a magnet, without the two touching. The iron bar then becomes a magnet, as shown in Fig. 13-7. What happens is that the magnetic lines of force generated by the permanent magnet make the internal molecular magnets in the iron bar line up in the same direction, instead of the random directions in unmagnetized iron. The magnetized iron bar then has magnetic poles at the ends, as a result of the magnetic induction.

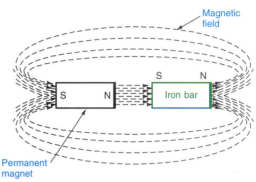

FIG. 13-7   Magnetizing an iron bar by induction.

Note that the induced poles in the iron have opposite polarity from the poles of the magnet. Since opposite poles attract, the iron bar will be attracted. Any magnet attracts to itself all magnetic materials by induction.

Although the two bars in Fig. 13-7 are not touching, the iron bar is in the magnetic flux of the permanent magnet. It is the invisible magnetic field that links the two magnets, enabling one to affect the other. Actually, this idea of magnetic flux extending outward from the magnetic poles is the basis for many inductive effects in ac circuits. More generally, the magnetic field between magnetic poles and the electric field between electric charges form the basis for wireless radio transmission and reception.

POLARITY OF INDUCED POLES   Note that the north pole of the permanent magnet in Fig. 13-7 induces an opposite south pole at this end of the iron bar. If the permanent magnet were reversed, its south pole would induce a north pole. The closest induced pole will always be of opposite polarity. This is the reason why either end of a magnet can attract another magnetic material to itself. No matter which pole is used, it will induce an opposite pole, and the opposite poles are attracted.

RELATIVE PERMEABILITY   Soft iron, as an example, is very effective in concentrating magnetic field lines, by induction in the iron. This ability to concentrate magnetic flux is called *permeability*. Any material that is easily magnetized has high permeability, therefore, as the field lines are concentrated because of induction.

DID YOU KNOW?

Some Chicago buses are emission-free because they run on hydrogen-generated electricity, which operates fuel cells.

Numerical values of permeability for different materials compared with air or vacuum can be assigned. For example, if the flux density in air is 1 G but an iron core in the same position in the same field has a flux density of 200 G, the relative permeability of the iron core equals 200/1, or 200.

The symbol for relative permeability is μ_r (mu), where the subscript r indicates relative permeability. Typical values for μ_r are 100 to 9000 for iron and steel. There are no units, because μ_r is a comparison of two flux densities and the units cancel. The symbol K_m may also be used for relative permeability, to indicate this characteristic of a material for a magnetic field, corresponding to K_ϵ for an electric field.

TEST-POINT QUESTION 13-4

Answers at end of chapter.

Answer True or False.
a. Induced poles always have opposite polarity from the inducing poles.
b. The relative permeability of air or vacuum is approximately 300.

13-5 AIR GAP OF A MAGNET

As shown in Fig. 13-8, the air space between poles of a magnet is its air gap. The shorter the air gap, the stronger the field in the gap for a given pole strength. Since air is not magnetic and cannot concentrate magnetic lines, a larger air gap only provides additional space for the magnetic lines to spread out.

Referring to Fig. 13-8a, note that the horseshoe magnet has more crowded magnetic lines in the air gap, compared with the widely separated lines around the bar magnet in Fig. 13-8b. Actually, the horseshoe magnet can be considered as a bar magnet bent around to place the opposite poles closer. Then the magnetic lines of the poles reinforce each other in the air gap. The purpose of a short air gap is to concentrate the magnetic field outside the magnet, for maximum induction in a magnetic material placed in the gap.

RING MAGNET WITHOUT AIR GAP   When it is desired to concentrate magnetic lines within a magnet, however, the magnet can be formed as a closed magnetic loop. This method is illustrated in Fig. 13-9a by the two permanent horseshoe magnets placed in a closed loop with opposite poles touching. Since the loop has no open ends, there can be no air gap and no poles. The north and south poles of each magnet cancel as opposite poles touch.

Each magnet has its magnetic lines inside, plus the magnetic lines of the other magnet, but outside the magnets the lines cancel because they are in opposite directions. The effect of the closed magnetic loop, therefore, is maximum concentration of magnetic lines in the magnet with minimum lines outside.

The same effect of a closed magnetic loop is obtained with the *toroid* or ring magnet in Fig. 13-9b, made in the form of a doughnut. Iron is often used for the core. This type of electromagnet has maximum strength in the iron ring, with little flux outside. As a result, the toroidal magnet is less sensitive to induction from external magnetic fields and, conversely, has little magnetic effect outside the coil.

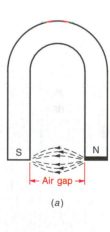

(a)

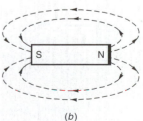

(b)

FIG. 13-8   The horseshoe magnet in (a) has a smaller air gap than the bar magnet in (b).

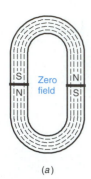

(a)

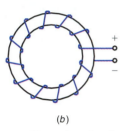

(b)

FIG. 13-9 Examples of a closed magnetic ring without any air gap. (*a*) Two PM horseshoe magnets with opposite poles touching. (*b*) Toroid magnet.

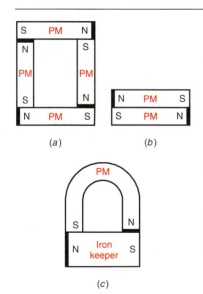

(a)

(b)

(c)

FIG. 13-10 Storing permanent magnets in a closed loop, with opposite poles touching. (*a*) Four bar magnets. (*b*) Two bar magnets. (*c*) Horseshoe magnet with iron keeper across air gap.

It should be noted that, even if the winding is over only a small part of the ring, practically all the flux is in the iron core because its permeability is so much greater than that of air. The small part of the field in the air is called *leakage flux*.

KEEPER FOR A MAGNET The principle of the closed magnetic ring is used to protect permanent magnets in storage. In Fig. 13-10*a*, four permanent-magnet bars are in a closed loop, while Fig. 13-10*b* shows a stacked pair. Additional even pairs can be stacked this way, with opposite poles touching. The closed loop in Fig. 13-10*c* shows one permanent horseshoe magnet with a soft-iron *keeper* across the air gap. The keeper maintains the strength of the permanent magnet as it becomes magnetized by induction to form a closed loop. Then any external magnetic field is concentrated in the closed loop without inducing opposite poles in the permanent magnet. If permanent magnets are not stored this way, the polarity can be reversed with induced poles produced by a strong external field from a dc source; an alternating field can demagnetize the magnet.

TEST-POINT QUESTION 13-5

Answers at end of chapter.

Answer True or False.

a. A short air gap has a stronger field than a large air gap, for the same magnetizing force.
b. A toroid magnet has no air gap.

13-6 TYPES OF MAGNETS

The two broad classes are permanent magnets and electromagnets. An electromagnet needs current from an external source to maintain its magnetic field. With a permanent magnet, not only is its magnetic field present without any external current, but the magnet can maintain its strength indefinitely. Sharp mechanical shock as well as extreme heat, however, can cause demagnetization.

ELECTROMAGNETS Current in a wire conductor has an associated magnetic field. If the wire is wrapped in the form of a coil, as in Fig. 13-11, the current and its magnetic field become concentrated in a smaller space, resulting in a stronger field. With the length much greater than its width, the coil is called a *solenoid*. It acts like a bar magnet, with opposite poles at the ends.

More current and more turns make a stronger magnetic field. Also, the iron core concentrates magnetic lines inside the coil. Soft iron is generally used for the core because it is easily magnetized and demagnetized.

The coil in Fig. 13-11, with the switch closed and current in the coil, is an electromagnet that can pick up the steel nail shown. If the switch is opened, the magnetic field is reduced to zero, and the nail will drop off. This ability of an electromagnet to provide a strong magnetic force of attraction that can be turned on or off easily has many applications in lifting magnets, buzzers, bells or chimes, and relays. A relay is a switch with contacts that are opened or closed by an electromagnet.

Another common application is magnetic tape recording. The tape is coated with fine particles of iron oxide. The recording head is a coil that produces a magnetic field in proportion to the current. As the tape passes through the air gap of the head, small areas of the coating become magnetized by induction. On playback, the moving magnetic tape produces variations in electric current.

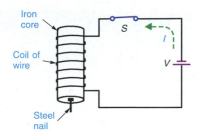

FIG. 13-11 Electromagnet holding nail when switch S is closed for current in coil.

PERMANENT MAGNETS   These are made of hard magnetic materials, such as cobalt steel, magnetized by induction in the manufacturing process. A very strong field is needed for induction in these materials. When the magnetizing field is removed, however, residual induction makes the material a permanent magnet. A common PM material is *alnico,* a commercial alloy of aluminum, nickel, and iron, with cobalt, copper, and titanium added to produce about 12 grades. The Alnico V grade is often used for PM loudspeakers (Fig. 13-12). In this application, a typical size of PM slug for a steady magnetic field is a few ounces to about 5 lb, with a flux ϕ of 500 to 25,000 lines or maxwells. One advantage of a PM loudspeaker is that only two connecting leads are needed for the voice coil, as the steady magnetic field of the PM slug is obtained without any field-coil winding.

Commercial permanent magnets will last indefinitely if they are not subjected to high temperatures, to physical shock, or to a strong demagnetizing field. If the magnet becomes hot, however, the molecular structure can be rearranged, resulting in loss of magnetism that is not recovered after cooling. The point at which a magnetic material loses its ferromagnetic properties is the *Curie temperature.* For iron, this temperature is about 800°C, when the relative permeability drops to unity. A permanent magnet does not become exhausted with use, as its magnetic properties are determined by the structure of the internal atoms and molecules.

CLASSIFICATION OF MAGNETIC MATERIALS   When we consider materials simply as either magnetic or nonmagnetic, this division is really based on the strong magnetic properties of iron. However, weak magnetic materials can be important in some applications. For this reason, a more exact classification includes the following three groups:

1. *Ferromagnetic materials.* These include iron, steel, nickel, cobalt, and commercial alloys such as alnico and Permalloy. They become strongly magnetized in the same direction as the magnetizing field, with high values of permeability from 50 to 5000. Permalloy has μ_r of 100,000 but is easily saturated at relatively low values of flux density.
2. *Paramagnetic materials.* These include aluminum, platinum, manganese, and chromium. The permeability is slightly more than 1. They become weakly magnetized in the same direction as the magnetizing field.
3. *Diamagnetic materials.* These include bismuth, antimony, copper, zinc, mercury, gold, and silver. The permeability is less than 1. They become weakly magnetized, but in the opposite direction from the magnetizing field.

The basis of all magnetic effects is the magnetic field associated with electric charges in motion. Within the atom, the motion of its orbital electrons generates a magnetic field. There are two kinds of electron motion in the atom. First is the electron revolving in its orbit. This motion provides a diamagnetic effect. However, this magnetic effect is weak because thermal agitation at normal room temperature results in random directions of motion that neutralize each other.

FIG. 13-12   Example of a PM loudspeaker.

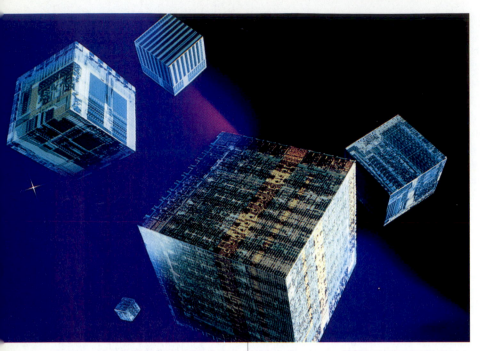

More effective is the magnetic effect from the motion of each electron spinning on its own axis. The spinning electron serves as a tiny permanent magnet. Opposite spins provide opposite polarities. Two electrons spinning in opposite directions form a pair, neutralizing the magnetic fields. In the atoms of ferromagnetic materials, however, there are many unpaired electrons with spins in the same direction, resulting in a strong magnetic effect.

In terms of molecular structure, iron atoms are grouped in microscopically small arrangements called *domains*. Each domain is an elementary *dipole magnet,* with two opposite poles. In crystal form, the iron atoms have domains that are parallel to the axes of the crystal. Still, the domains can point in different directions, because of the different axes. When the material becomes magnetized by an external magnetic field, though, the domains become aligned in the same direction. With PM materials, the alignment remains after the external field is removed.

Answers at end of chapter.

Answer True or False.
a. An electromagnet needs current to maintain its magnetic field.
b. A relay coil is an electromagnet.
c. Steel is a diamagnetic material.

13-7 FERRITES

Ferrites is the name for nonmetallic materials that have the ferromagnetic properties of iron. The ferrites have very high permeability, like iron. However, a ferrite is a nonconducting ceramic material, while iron is a conductor. The permeability of ferrites is in the range of 50 to 3000. The specific resistance is $10^5 \ \Omega \cdot$ cm, which makes the ferrite an insulator.

A common application is a ferrite core, usually adjustable, in the coils for RF transformers. The ferrite core is much more efficient than iron when the current alternates at a high frequency. The reason is that less I^2R power is lost by eddy currents in the core because of its very high resistance.

A ferrite core is used in small coils and transformers for signal frequencies up to 20 MHz, approximately. The high permeability means the transformer can be very small. However, the ferrites are easily saturated at low values of magnetizing current. This disadvantage means the ferrites are not used for power transformers.

Another application is in ferrite beads (Fig. 13-13). A bare wire is used as a string for one or more beads. The bead concentrates the magnetic field of the current in the wire. This construction serves as a simple, economical RF choke, instead of a coil. The purpose of the choke is to reduce the current just for an undesired radio frequency.

FIG. 13-13 Ferrite bead equivalent to coil with 20 μH of inductance at 10 MHz.

13-8 MAGNETIC SHIELDING

The idea of preventing one component from affecting another through their common electric or magnetic field is called *shielding*. Examples are the braided copper-wire shield around the inner conductor of a coaxial cable, a metal shield can that encloses an RF coil, or a shield of magnetic material enclosing a cathode-ray tube.

The problem in shielding is to prevent one component from inducing an effect in the shielded component. The shielding materials are always metals, but there is a difference between using good conductors with low resistance like copper and aluminum and using good magnetic materials like soft iron.

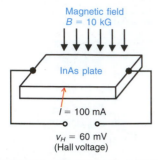

Magnetic field
B = 10 kG

InAs plate

I = 100 mA

v_H = 60 mV
(Hall voltage)

FIG. 13-14 The Hall effect. Hall voltage V_H generated across the element is proportional to the perpendicular flux density B.

A good conductor is best for two shielding functions. One is to prevent induction of static electric charges. The other is to shield against the induction of a varying magnetic field. For static charges, the shield provides opposite induced charges, which prevent induction inside the shield. For a varying magnetic field, the shield has induced currents that oppose the inducing field. Then there is little net field strength to produce induction inside the shield.

The best shield for a steady magnetic field is a good magnetic material of high permeability. A steady field is produced by a permanent magnet, a coil with steady direct current, or the earth's magnetic field. A magnetic shield of high permeability concentrates the magnetic flux. Then there is little flux to induce poles in a component inside the shield. The shield can be considered as a short circuit for the lines of magnetic flux.

TEST-POINT QUESTION 13-8

Answers at end of chapter.

Answer True or False.
a. Magnetic material with high permeability is a good shield for a steady magnetic field.
b. A conductor is a good shield against a varying magnetic field.

13-9 THE HALL EFFECT

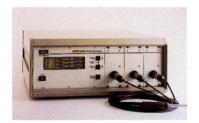

FIG. 13-15 Gaussmeter to measure flux density, with probe containing indium arsenide element.

In 1879, E. H. Hall observed that a small voltage is generated across a conductor carrying current in an external magnetic field. The Hall voltage was very small with typical conductors, and little use was made of this effect. However, with the development of semiconductors, larger values of Hall voltage can be generated. The semiconductor material indium arsenide (InAs) is generally used. As illustrated in Fig. 13-14, the InAs element inserted in the magnetic field can generate 60 mV with B equal to 10 kG and an I of 100 mA. The applied flux must be perpendicular to the direction of current. With current in the direction of the length of conductor, the generated voltage is developed across the width.

The amount of Hall voltage v_H is directly proportional to the value of flux density B. This means that values of B can be measured by means of v_H. As an example, the gaussmeter in Fig. 13-15 uses an InAs probe in the magnetic field to generate a proportional Hall voltage v_H. This value of v_H is then read by the meter, which is calibrated in gauss. The original calibration is made in terms of a reference magnet with a specified flux density.

TEST-POINT QUESTION 13-9

Answers at end of chapter.
a. In Fig. 13-14, how much is the generated Hall voltage?
b. Does the gaussmeter in Fig. 13-15 measure flux or flux density?

13   SUMMARY AND REVIEW

- Iron, nickel, and cobalt are common examples of magnetic materials. Air, paper, wood, and plastics are nonmagnetic.
- The pole of a magnet that seeks the magnetic north pole of the earth is called a north pole; the opposite pole is a south pole.
- Opposite magnetic poles have a force of attraction; similar poles repel.
- An electromagnet needs current from an external source to provide a magnetic field. Permanent magnets retain their magnetism indefinitely.
- Any magnet has an invisible field of force outside the magnet, indicated by magnetic field lines. Their direction is from the north to the south pole.
- The open ends of a magnet where it meets a nonmagnetic material provide magnetic poles. At opposite open ends, the poles have opposite polarity.
- A magnet with an air gap has opposite poles with magnetic lines of force across the gap. A closed magnetic ring has no poles.
- Magnetic induction enables the field of a magnet to induce magnetic poles in a magnetic material without touching.
- Permeability is the ability to concentrate magnetic flux. A good magnetic material has high permeability.
- Magnetic shielding means isolating a component from a magnetic field. The best shield against a steady magnetic field is a material with high permeability.
- The Hall voltage is a small emf generated across the width of a conductor carrying current through its length, when magnetic flux is applied perpendicular to the current. This effect is generally used in the gaussmeter to measure flux density.
- Table 13-1 summarizes the units of magnetic flux ϕ and flux density B.

TABLE 13-1   MAGNETIC FLUX Φ AND FLUX DENSITY B

NAME	SYMBOL	CGS UNITS	MKS OR SI UNITS
Flux, or total lines	$\phi = B \times \text{area}$	1 maxwell (Mx) = 1 line	1 weber (Wb) = 10^8 Mx
Flux density, or lines per unit area	$B = \dfrac{\phi}{\text{area}}$	1 gauss (G) = $\dfrac{1 \text{ Mx}}{\text{cm}^2}$	1 tesla (T) = $\dfrac{1 \text{ Wb}}{\text{m}^2}$

ANSWERS AT BACK OF BOOK.

Answer True or False.

1. Iron and steel are ferromagnetic materials with high permeability.
2. Ferrites are ferromagnetic but have high electrical resistance.
3. Air, vacuum, wood, paper, and plastics have no effect on magnetic flux.
4. Aluminum is ferromagnetic.
5. Magnetic poles exist on opposite sides of an air gap.
6. A closed magnetic ring has no poles and no air gap.
7. A magnet can pick up a steel nail by magnetic induction.
8. Induced poles are always opposite from the original field poles.
9. Soft iron concentrates magnetic flux by means of induction.
10. Without current, an electromagnet has practically no magnetic field.
11. The total flux ϕ of 5000 lines equals 5 Mx.
12. A flux ϕ of 5000 Mx through a cross-sectional area of 5 cm^2 has a flux density B of 1000 G or 1 kG.
13. The flux density B of 1000 G equals 1000 lines/cm^2.
14. A magnetic pole is a terminal where a magnetic material meets a nonmagnetic material.
15. High permeability for magnetic flux corresponds to high resistance for a current conductor.

QUESTIONS

1. Name two magnetic materials and three nonmagnetic materials.
2. Explain the difference between a permanent magnet and an electromagnet.
3. Draw a horseshoe magnet, with its magnetic field. Label the magnetic poles, indicate the air gap, and show the direction of flux.
4. Define: relative permeability, shielding, induction, Hall voltage.
5. Give the symbol, cgs unit, and SI unit for magnetic flux and for flux density.
6. How are the north and south poles of a bar magnet determined with a magnetic compass?
7. Referring to Fig. 13-11, why can either end of the magnet pick up the nail?
8. What is the difference between flux ϕ and flux density B?

PROBLEMS

ANSWERS TO ODD-NUMBERED PROBLEMS AT BACK OF BOOK.

1. A magnet produces 5000 field lines. Find ϕ in maxwells and webers.
2. If the area of the pole in Prob. 1 is 5 cm^2, calculate B in gauss units.

3. Calculate B in tesla units for a 200-μWb flux through an area of 5×10^{-4} m^2.
4. Convert 3000 G to tesla units.
5. For a flux density B of 3 kG at a pole with a cross-sectional area of 8 cm^2, how much is the total flux ϕ in maxwell units?
6. Convert 17,000 Mx to weber units.
7. The flux density is 0.002 T in the air core of an electromagnet. When an iron core is inserted, the flux density in the core is 0.6 T. How much is the relative permeability μ_r of the iron core?
8. Draw the diagram of an electromagnet operated from a 12-V battery, in series with a switch. (a) If the coil resistance is 60 Ω, how much is the current in the coil with the switch closed? (b) Why is the magnetic field reduced to zero when the switch is opened?
9. Derive the conversion of 1 μWb = 100 Mx from the fact that 1 μWb = 10^{-6} Wb and 1 Wb = 10^8 Mx.
10. Derive the relation 1 T = 10^4 G. (Note: 1 m^2 = 10,000 cm^2.)
11. Make the following conversions: (a) 50 T to kG; (b) 6000 μWb to Mx; (c) 2500 G to T; (d) 15,000 Mx to Wb; (e) 0.004 T to G.
12. A permanent magnet produces a flux ϕ of 100,000 Mx. How many weber units does this represent?
13. A flux ϕ of 500 μWb exists in an area A of 0.01 m^2. What is the flux density B in gauss units?
14. The flux density in an iron core is 0.5 T. If the area of the core is 10 cm^2, calculate the total number of flux lines in the core.
15. Calculate the flux density B, in teslas for a flux ϕ of 400 μWb in an area of 0.005 m^2.

CRITICAL THINKING

1. A flux ϕ of 25 μWb exists in an area of 0.25 in^2. What is the flux density B in: (a) gauss units; (b) teslas?
2. At the north pole of an electromagnet, the flux density B equals 5 T. If the area A equals 0.125 in^2, determine the total number of flux lines ϕ in: (a) maxwells; (b) webers.

ANSWERS TO TEST-POINT QUESTIONS

13-1 a. T
     b. T
13-2 a. 2000 Mx
     b. 20 μWb
13-3 a. 3000 G
     b. 0.3 T
13-4 a. T
     b. F
13-5 a. T
     b. T
13-6 a. T
     b. T
     c. F
13-7 a. ferrites
     b. conductor
13-8 a. T
     b. T
13-9 a. 60 mV
     b. flux density

CHAPTER 14

MAGNETIC UNITS

A magnetic field is always associated with charges in motion. Therefore, the magnetic units can be derived from the current that produces the field.

The current in a conductor and its magnetic flux through the medium outside the conductor are related as follows:

1. The current I supplies a magnetizing force, or magneto-motive force (mmf), that increases with the amount of I.
2. The mmf results in a magnetic field intensity H that decreases with the length of conductor, as the field is less concentrated with more length.
3. The field intensity H produces a flux density B that increases with the permeability of the medium.

CHAPTER OBJECTIVES

Upon completion of this chapter, you should be able to:

- *Describe* what is meant by the terms *magnetomotive force* and *field intensity* and *list* the units of each.
- *Define absolute permeability.*
- *Explain* the *B-H* magnetization curve.
- *Define* the term *saturation* as it relates to a magnetic core.
- *Understand* magnetic hysteresis.
- *Write* the three forms of Ohm's law for magnetic circuits.
- *Explain* the difference between electric and magnetic fields.

IMPORTANT TERMS IN THIS CHAPTER

ampere-turns	hysteresis	reluctance
ampere-turns/meter	International System of	saturation
Coulomb's law	Units (SI)	tesla unit
degaussing	magnetomotive force (mmf)	
demagnetization	permeability (μ or mu)	

TOPICS COVERED IN THIS CHAPTER

With a coil magnet, the strength of the magnetic field depends on how much current flows in the turns of the coil. The more current, the stronger the magnetic field. Also, more turns in a specific length concentrate the field. The coil serves as a bar magnet, with opposite poles at the ends, providing a magnetic field proportional to the ampere-turns. As a formula,

▶   Ampere-turns $= I \times N = $ mmf   **(14-1)**

where I is the current in amperes multiplied by the number of turns N. The quantity IN specifies the amount of *magnetizing force or magnetic potential,* which is the *magnetomotive force (mmf)*.

The practical unit is the ampere-turn. The SI abbreviation for ampere-turn is A, the same as for the ampere, since the number of turns in a coil usually is constant but the current can be varied. However, for clarity we shall use the abbreviation A · t.

As shown in Fig. 14-1, a solenoid with 5 turns and 2 amperes has the same magnetizing force as one with 10 turns and 1 ampere, as the product of the amperes and turns is 10 for both cases. With thinner wire, more turns can be used in a given space. The amount of current is determined by the resistance of the wire and the source voltage. How many ampere-turns are necessary depends on the required magnetic field strength.

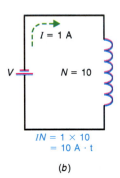

$I = 2$ A

V $N = 5$

$IN = 2 \times 5$
$= 10$ A · t

(a)

$I = 1$ A

V $N = 10$

$IN = 1 \times 10$
$= 10$ A · t

(b)

FIG. 14-1 Two examples of equal ampere-turns for the same mmf. (a) IN is $2 \times 5 = 10$. (b) IN is $1 \times 10 = 10$.

Example

EXAMPLE 1   Calculate the ampere-turns of mmf for a coil with 2000 turns and a 5-mA current.

ANSWER   mmf $= I \times N = 2000 \times 5 \times 10^{-3}$
$$= 10 \text{ A} \cdot \text{t}$$

EXAMPLE 2   A coil with 4 A is to provide the magnetizing force of 600 A · t. How many turns are necessary?

ANSWER $N = \dfrac{A \cdot t}{I} = \dfrac{600}{4}$

$N = 150$ turns

EXAMPLE 3   A coil with 400 turns must provide 800 A · t of magnetizing force. How much current is necessary?

ANSWER $I = \dfrac{A \cdot t}{N} = \dfrac{800}{400}$

$I = 2$ A

EXAMPLE 4 The wire in a solenoid of 250 turns has a resistance of 3 Ω. (**a**) How much is the current with the coil connected to a 6-V battery? (**b**) Calculate the ampere-turns of mmf.

ANSWER

a. $I = \dfrac{V}{R} = \dfrac{6 \text{ V}}{3 \text{ }\Omega}$

 $I = 2 \text{ A}$

b. $\text{mmf} = I \times N = 2 \text{ A} \times 250 \text{ t}$

 $= 500 \text{ A} \cdot \text{t}$

The ampere-turn is an SI unit. It is calculated as *IN*, with the current in amperes.

The cgs unit of mmf is the *gilbert,** abbreviated Gb. One ampere-turn equals 1.26 Gb. The number 1.26 is approximately $4\pi/10$, derived from the surface area of a sphere, which is $4\pi r^2$.

To convert *IN* to gilberts, multiply the ampere-turns by the constant conversion factor 1.26 Gb/1 A · t. As an example, 1000 A · t is the same mmf as 1260 Gb. The calculations are

$$1000 \text{ A} \cdot \text{t} \times 1.26 \, \frac{\text{Gb}}{1 \text{ A} \cdot \text{t}} = 1260 \text{ Gb}$$

Note that the units of A · t cancel in the conversion.

TEST-POINT QUESTION 14-1

Answers at end of chapter.
a. If the mmf is 243 A · t, and *I* is doubled from 2 to 4 A with the same number of turns, how much is the new mmf?
b. Convert 500 A · t to gilberts.

14-2 FIELD INTENSITY (*H*)

The ampere-turns of mmf specify the magnetizing force, but the intensity of the magnetic field depends on how long the coil is. At any point in space, a specific value of ampere-turns must produce less field intensity for a long coil than for a

---

*William Gilbert (1540–1603) was an English scientist who investigated the magnetism of the earth.

short coil that concentrates the same mmf. Specifically, the field intensity H in mks units is

$$H = \frac{\text{ampere-turns of mmf}}{l \text{ meters}} \qquad (14\text{-}2)$$

This formula is for a solenoid. The field intensity H is at the center of an air core. With an iron core, H is the intensity through the entire core. By means of units for H, the magnetic field intensity can be specified for either electromagnets or permanent magnets, since both provide the same kind of magnetic field.

The length in Formula (14-2) is between poles. In Fig. 14-2a, the length is 1 m between the poles at the ends of the coil. In Fig. 14-2b, also, l is 1 m between the ends of the iron core. In Fig. 14-2c, though, l is 2 m between the poles at the ends of the iron core, although the winding is only 1 m long.

The examples in Fig. 14-2 illustrate the following comparisons:

1. In all three cases, the mmf is 1000 A · t, with the same value of IN.
2. In Fig. 14-2a and b, H equals 1000 A · t/m. In a, this H is the intensity at the center of the air core; in b, this H is the intensity through the entire iron core.
3. In Fig. 14-2c, because l is 2 m, H is 1000/2, or 500 A · t/m. This H is the intensity in the entire iron core.

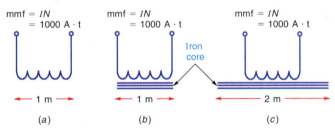

FIG. 14-2 Relation between ampere-turns of mmf and the resultant field intensity H for different cores. Note that $H = $ mmf/length. (a) Intensity H is 1000 A · t/m with an air core. (b) $H = $ 1000 A · t/m in an iron core of the same length as the coil. (c) H is 1000/2 = 500 A · t/m in an iron core twice as long as the coil.

UNITS FOR H The field intensity is basically mmf per unit of length. In practical units, H is ampere-turns per meter. The cgs unit for H is the *oersted,** abbreviated Oe, which equals one gilbert of mmf per centimeter.

CONVERSION OF UNITS To convert SI units of A · t/m to cgs units of Oe, multiply by the conversion factor 0.0126 Oe per 1 A · t/m. As an example, 1000 A · t/m is the same H as 12.6 Oe. The calculations are

$$1000 \, \frac{\text{A} \cdot \text{t}}{\text{m}} \times 0.0126 \, \frac{\text{Oe}}{1 \, \text{A} \cdot \text{t/m}} = 12.6 \, \text{Oe}$$

---

*H. C. Oersted (1777–1851), a Danish physicist, discovered electromagnetism.

Note that the units of A · t and m cancel. The m in the conversion factor becomes inverted to the numerator.

TEST-POINT QUESTION 14-2

Answers at end of chapter.

a. Suppose that H is 250 A · t/m. The length is doubled from 0.2 to 0.4 m for the same IN. How much is the new H?

b. Convert 500 A · t/m to oersted units.

14-3 PERMEABILITY (μ)

Whether we say H is 1000 A · t/m or 12.6 Oe, these units specify how much field intensity is available to produce magnetic flux. However, the amount of flux actually produced by H depends on the material in the field. A good magnetic material with high relative permeability can concentrate flux and produce a large value of flux density B for a specified H. These factors are related by the formula:

$$B = \mu \times H \tag{14-3}$$

or

$$\mu = \frac{B}{H} \tag{14-4}$$

Using SI units, B is the flux density in webers per square meter, or teslas; H is the field intensity in ampere-turns per meter. In the cgs system the units are gauss for B and oersted for H. The factor μ is the absolute permeability, not referred to any other material, in units of B/H.

In the cgs system the units of gauss for B and oersteds for H have been defined to give μ the value of 1 G/Oe, for vacuum, air, or space. This simplification means that B and H have the same numerical values in air and in vacuum. For instance, the field intensity H of 12.6 Oe produces the flux density of 12.6 G, in air.

Furthermore, the values of relative permeability μ_r are the same as those for absolute permeability in B/H units in the cgs system. The reason is that μ is 1 for air or vacuum, used as the reference for the comparison. As an example, if μ_r for an iron sample is 600, the absolute μ is also 600 G/Oe.

In SI, however, the permeability of air or vacuum is not 1. Specifically, this value is $4\pi \times 10^{-7}$, or 1.26×10^{-6}, with the symbol μ_0. Therefore, values of relative permeability μ_r must be multiplied by 1.26×10^{-6} for μ_0 to calculate μ as B/H in SI units.

For an example of $\mu_r = 100$, the SI value of μ can be calculated as follows:

▶

$$\mu = \mu_r \times \mu_0$$

$$= 100 \times 1.26 \times 10^{-6} \, \frac{\text{T}}{\text{A} \cdot \text{t/m}}$$

$$\mu = 126 \times 10^{-6} \, \frac{\text{T}}{\text{A} \cdot \text{t/m}}$$

Example

EXAMPLE 5 A magnetic material has a μ_r of 500. Calculate the absolute μ as B/H (**a**) in cgs units, and (**b**) in SI units.

ANSWER

a. $\mu = \mu_r \times \mu_0$ in cgs units. Then

$$= 500 \times 1 \, \frac{\text{G}}{\text{Oe}}$$

$$\mu = 500 \, \frac{\text{G}}{\text{Oe}}$$

b. $\mu = \mu_r \times \mu_0$ in SI units. Then

$$= 500 \times 1.26 \times 10^{-6} \, \frac{\text{T}}{\text{A} \cdot \text{t/m}}$$

$$\mu = 630 \times 10^{-6} \, \frac{\text{T}}{\text{A} \cdot \text{t/m}}$$

EXAMPLE 6 For this example of $\mu = 630 \times 10^{-6}$ in SI units, calculate the flux density B that will be produced by the field intensity H equal to 1000 A · t/m.

ANSWER $B = \mu H = \left(630 \times 10^{-6} \, \dfrac{\text{T}}{\text{A} \cdot \text{t/m}} \right) \left(1000 \, \dfrac{\text{A} \cdot \text{t}}{\text{m}} \right)$

$$= 630 \times 10^{-3} \, \text{T}$$

$$B = 0.63 \, \text{T}$$

Note that the ampere-turns and meter units cancel, leaving only the tesla unit for the flux density B.

TEST-POINT QUESTION 14-3

Answers at end of chapter.

a. What is the value of μ_r for air, vacuum, or space?

b. An iron core has 200 times more flux density than air for the same field intensity H. How much is μ_r?

c. An iron core produces 200 G of flux density for 1 Oe of field intensity H. How much is μ?

14-4 B-H MAGNETIZATION CURVE

The *B-H* curve in Fig. 14-3 is often used to show how much flux density *B* results from increasing the amount of field intensity *H*. This curve is for soft iron, plotted for the values in Table 14-1, but similar curves can be obtained for all magnetic materials.

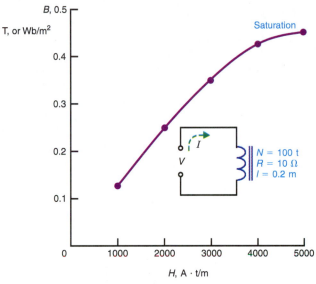

FIG. 14-3 *B-H* magnetization curve for soft iron. No values are shown near zero, where μ may vary with previous magnetization.

CALCULATING *H* AND *B*   The values in Table 14-1 are calculated as follows:

1. The current *I* in the coil equals *V/R*. For a 10-Ω coil resistance with 20 V applied, *I* is 2 A, as listed in the top row of Table 14-1. Increasing values of *V* produce more current in the coil.
2. The ampere-turns *IN* of magnetizing force increase with more current. Since the turns are constant at 100, the values of *IN* increase from 200 for 2 A in the top row to 1000 for 10 A in the bottom row.

TABLE 14-1   B-H VALUES FOR FIG. 14-3

V, VOLTS	R, Ω	I = V/R, AMPERES	N, TURNS	mmf, A·t	l, m	H, A·t/m	μ_R	B = $\mu \times$ H, T
20	10	2	100	200	0.2	1000	100	0.126
40	10	4	100	400	0.2	2000	100	0.252
60	10	6	100	600	0.2	3000	100	0.378
80	10	8	100	800	0.2	4000	85	0.428
100	10	10	100	1000	0.2	5000	70	0.441

3. The field intensity H increases with higher IN. The values of H are in mks units of ampere-turns per meter. These values equal $IN/0.2$, as the length is 0.2 m. Therefore, each IN is just divided by 0.2, or multiplied by 5, for the corresponding values of H. Since H increases in the same proportion as I, sometimes the horizontal axis on a B-H curve is calibrated only in amperes, instead of in H units.

4. The flux density B depends on the field intensity H and permeability of the iron. The values of B in the last column are obtained by multiplying $\mu \times H$. However, with SI units the values of μ_r listed must be multiplied by 1.26×10^{-6} to obtain $\mu \times H$ in teslas.

SATURATION   Note that the permeability decreases for the highest values of H. With less μ, the iron core cannot provide proportional increases in B for increasing values of H. In the graph, for values of H above 4000 A · t/m, approximately, the values of B increase at a much slower rate, making the curve relatively flat at the top. The effect of little change in flux density when the field intensity increases is called *saturation*.

The reason is that the iron becomes saturated with magnetic lines of induction. After most of the molecular dipoles and the magnetic domains are aligned by the magnetizing force, very little additional induction can be produced. When the value of μ is specified for a magnetic material, it is usually the highest value before saturation.

14-5 MAGNETIC HYSTERESIS

Hysteresis means "a lagging behind." With respect to the magnetic flux in an iron core of an electromagnet, the flux lags the increases or decreases in magnetizing force. The hysteresis results from the fact that the magnetic dipoles are not perfectly elastic. Once aligned by an external magnetizing force, the dipoles do not return exactly to their original positions when the force is removed. The effect is the same as if the dipoles were forced to move against an internal friction between molecules. Furthermore, if the magnetizing force is reversed in direction by reversal of the current in an electromagnet, the flux produced in the opposite direction lags behind the reversed magnetizing force.

HYSTERESIS LOSS   When the magnetizing force reverses thousands or millions of times per second, as with rapidly reversing alternating current, the hysteresis can cause a considerable loss of energy. A large part of the magnetizing force is then used just to overcome the internal friction of the molecular dipoles. The work done by the magnetizing force against this internal friction produces heat. This energy wasted in heat as the molecular dipoles lag the magnetizing force is called *hysteresis loss*. For steel and other hard magnetic materials, the hysteresis losses are much higher than in soft magnetic materials like iron.

When the magnetizing force varies at a slow rate, the hysteresis losses can be considered negligible. An example is an electromagnet with direct current that is simply turned on and off, or the magnetizing force of an alternating current that reverses 60 times per second or less. The faster the magnetizing force changes, however, the greater the hysteresis effect.

HYSTERESIS LOOP　　To show the hysteresis characteristics of a magnetic material, its values of flux density B are plotted for a periodically reversing magnetizing force. See Fig. 14-4. This curve is the hysteresis loop of the material. The larger the area enclosed by the curve, the greater the hysteresis loss. The hysteresis loop is actually a B-H curve with an ac magnetizing force.

On the vertical axis, values of flux density B are indicated. The units can be gauss or teslas.

The horizontal axis indicates values of field intensity H. On this axis the units can be oersteds, ampere-turns per meter, ampere-turns, or just magnetizing current, as all factors are constant except I.

Opposite directions of current result in the opposite directions of $+H$ and $-H$ for the field lines. Similarly, opposite polarities are indicated for flux density as $+B$ or $-B$.

The current starts from zero at the center, when the material is unmagnetized. Then positive H values increase B to saturation at $+B_{max}$. Next H decreases to zero, but B drops to the value B_R, instead of to zero, because of hysteresis. When H becomes negative, B drops to zero and continues to $-B_{max}$, which is saturation in the opposite direction from $+B_{max}$ because of the reversed magnetizing current.

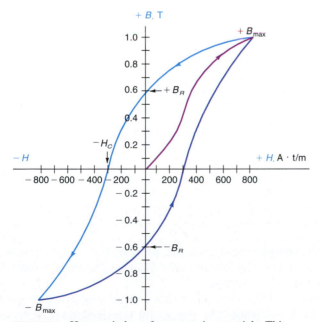

FIG. 14-4　Hysteresis loop for magnetic materials. This graph is a B-H curve like Fig. 14-3, but H alternates in polarity with alternating current.

Then, as the $-H$ values decrease, the flux density is reduced to $-B_R$. Finally, the loop is completed, with positive values of H producing saturation at B_{max} again. The curve does not return to the zero origin at the center because of hysteresis. As the magnetizing force periodically reverses, the values of flux density are repeated to trace out the hysteresis loop.

The value of either $+B_R$ or $-B_R$, which is the flux density remaining after the magnetizing force has been reduced to zero, is the *residual induction* of a magnetic material, also called its *retentivity*. In Fig. 14-4, the residual induction is 0.6 T, in either the positive or negative direction.

The value of $-H_C$, which equals the magnetizing force that must be applied in the reverse direction to reduce the flux density to zero, is the *coercive force* of the material. In Fig. 14-4, the coercive force $-H_C$ is 300 A · t/m.

DEMAGNETIZATION In order to demagnetize a magnetic material completely, the residual induction B_R must be reduced to zero. This usually cannot be accomplished by a reversed dc magnetizing force, because the material then would just become magnetized with opposite polarity. The practical way is to magnetize and demagnetize the material with a continuously decreasing hysteresis loop. This can be done with a magnetic field produced by alternating current. Then as the magnetic field and the material are moved away from each other, or the current amplitude is reduced, the hysteresis loop becomes smaller and smaller. Finally, with the weakest field, the loop collapses practically to zero, resulting in zero residual induction.

This method of demagnetization is also called *degaussing*. One application is degaussing the metal electrodes in a color picture tube, with a deguassing coil providing alternating current from the power line. Another example is erasing the recorded signal on magnetic tape by demagnetizing with an ac bias current. The average level of the erase current is zero, and its frequency is much higher than the recorded signal.

TEST-POINT QUESTION 14-5

Answers at end of chapter.

Answer True or False.
a. Hysteresis loss increases with higher frequencies.
b. Degaussing is done with alternating current.

14-6 OHM'S LAW FOR MAGNETIC CIRCUITS

In comparison with electric circuits, the magnetic flux ϕ corresponds to current. The flux ϕ is produced by ampere-turns IN of magnetomotive force. Therefore, the mmf corresponds to voltage.

Opposition to the production of flux in a material is called its *reluctance*, comparable with resistance. The symbol for reluctance is $\mathcal{R}$. Reluctance is inversely proportional to permeability. Iron has high permeability and low reluctance. Air or vacuum has low permeability and high reluctance.

In Fig. 14-5, the ampere-turns of the coil produce magnetic flux throughout the magnetic path. The reluctance is the total opposition to the flux ϕ. In Fig. 14-5a, there is little reluctance in the closed iron path, and few ampere-turns are necessary. In Fig. 14-5b, however, the air gap has high reluctance, which requires many more ampere-turns for the same flux as in Fig. 14-5a.

The three factors—flux, ampere-turns of mmf, and reluctance—are related as follows:

$$\blacktriangleright \qquad \phi = \frac{\text{mmf}}{\mathcal{R}} \qquad\qquad\qquad (14\text{-}5)$$

which is known as Ohm's law for magnetic circuits, corresponding to $I = V/R$. The mmf is considered to produce flux ϕ in a magnetic material against the opposition of its reluctance $\mathcal{R}$. This relationship corresponds to emf or voltage producing current in a conducting material against the opposition of its resistance.

Remember that the units for the flux ϕ are maxwells and webers. These units measure total lines, as distinguished from flux density B, which equals lines per unit area.

There are no specific units for reluctance, but it can be considered as an mmf/ϕ ratio, just as resistance is a V/I ratio. Then $\mathcal{R}$ is ampere-turns per weber in SI units, or gilberts per maxwell in the cgs system.

The units for mmf are either gilberts in the cgs system or ampere-turns in SI.

Example

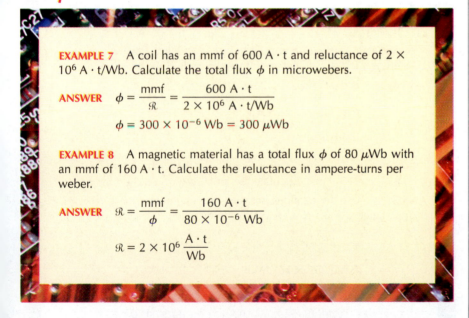

EXAMPLE 7   A coil has an mmf of 600 A · t and reluctance of 2 × 10^6 A · t/Wb. Calculate the total flux ϕ in microwebers.

ANSWER $\phi = \dfrac{\text{mmf}}{\mathcal{R}} = \dfrac{600 \text{ A} \cdot \text{t}}{2 \times 10^6 \text{ A} \cdot \text{t/Wb}}$

$\phi = 300 \times 10^{-6} \text{ Wb} = 300 \ \mu\text{Wb}$

EXAMPLE 8   A magnetic material has a total flux ϕ of 80 μWb with an mmf of 160 A · t. Calculate the reluctance in ampere-turns per weber.

ANSWER $\mathcal{R} = \dfrac{\text{mmf}}{\phi} = \dfrac{160 \text{ A} \cdot \text{t}}{80 \times 10^{-6} \text{ Wb}}$

$\mathcal{R} = 2 \times 10^6 \ \dfrac{\text{A} \cdot \text{t}}{\text{Wb}}$

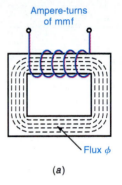

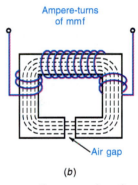

(a)

$\phi = \dfrac{\text{mmf}}{\mathcal{R}}$

(b)

FIG. 14-5   Two examples of a magnetic circuit. (*a*) Closed iron path with low reluctance that requires little mmf. (*b*) Higher-reluctance path with an air gap that requires more mmf.

TEST-POINT QUESTION 14-6

Answers at end of chapter.

Answer True or False.
a. Air has higher reluctance than soft iron.
b. More reluctance means more flux for a specified mmf.

14-7 RELATIONS BETWEEN MAGNETIC UNITS

The following examples show how the values of mmf, H, ϕ, B, and $\mathcal{R}$ depend on each other. These calculations are in SI units, which are generally used for magnetic circuits.

Example

EXAMPLE 9   For a coil having 50 turns and 2 A, how much is the mmf?

ANSWER   mmf $= IN = 2 \times 50$

mmf $= 100\ \text{A} \cdot \text{t}$

The value of $100\ \text{A} \cdot \text{t}$ is the mmf producing the magnetic field, with either an air core or an iron core.

EXAMPLE 10   If this coil is on an iron core with a length of 0.2 m, how much is the field intensity H throughout the iron?

ANSWER $H = \dfrac{\text{mmf}}{l} = \dfrac{100\ \text{A} \cdot \text{t}}{0.2\ \text{m}}$

$H = 500\ \text{A} \cdot \text{t/m}$

This is an example of calculating the field intensity of the external magnetic field from the mmf of the current in the coil.

EXAMPLE 11   If the iron core in Example 10 has a relative permeability μ_r of 200, calculate the flux density B in teslas.

ANSWER $B = \mu H = \mu_r \times 1.26 \times 10^{-6} \times H$

$= 200 \times 1.26 \times 10^{-6}\ \dfrac{\text{T}}{\text{A} \cdot \text{t/m}} \times \dfrac{500\ \text{A} \cdot \text{t}}{\text{m}}$

$B = 0.126\ \text{T}$

EXAMPLE 12   The iron core in Example 11 has a cross-sectional area of $2 \times 10^{-4}\ \text{m}^2$. Calculate the amount of flux ϕ in the core.

ANSWER   Use the relations between flux and density: $\phi = B \times$ Area. Since $B = 0.126\ \text{T}$, or $0.126\ \text{Wb/m}^2$, then

$\phi = B \times \text{Area} = 0.126\ \dfrac{\text{Wb}}{\text{m}^2} \times 2 \times 10^{-4}\ \text{m}^2$

$= 0.252 \times 10^{-4}\ \text{Wb} = 25.2 \times 10^{-6}\ \text{Wb}$

$\phi = 25.2\ \mu\text{Wb}$

EXAMPLE 13   With the mmf of $100\ \text{A} \cdot \text{t}$ for the coil in Fig. 14-5a and a value for ϕ of approximately $25 \times 10^{-6}\ \text{Wb}$ in the iron core, calculate its reluctance $\mathcal{R}$.

ANSWER   Using Ohm's law for magnetic circuits,

$\mathcal{R} = \dfrac{\text{mmf}}{\phi} = \dfrac{100\ \text{A} \cdot \text{t}}{25 \times 10^{-6}\ \text{Wb}}$

$\mathcal{R} = 4 \times 10^6\ \text{A} \cdot \text{t/Wb}$

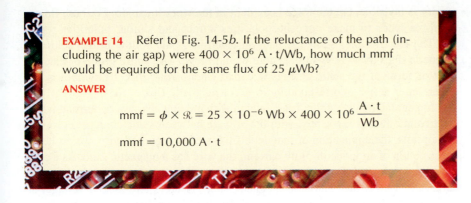

EXAMPLE 14   Refer to Fig. 14-5*b*. If the reluctance of the path (including the air gap) were 400×10^6 A·t/Wb, how much mmf would be required for the same flux of 25 μWb?

ANSWER

$$\text{mmf} = \phi \times \mathcal{R} = 25 \times 10^{-6}\,\text{Wb} \times 400 \times 10^6\,\frac{\text{A·t}}{\text{Wb}}$$

$$\text{mmf} = 10{,}000\,\text{A·t}$$

Notice that the 10,000 A·t of mmf here is 100 times more than the 100 A·t in Example 13, because of the higher reluctance with an air gap. This idea corresponds to the higher voltage needed to produce the same current in a higher resistance.

Answers at end of chapter.

Answer True or False.
a. More *I* in a coil produces more mmf for a specified number of turns.
b. More length for the coil produces more field intensity *H* for a specified mmf.
c. Higher permeability of the core material produces more flux density *B* for a specified *H*.

14-8 COMPARISON OF MAGNETIC AND ELECTRIC FIELDS

As shown in Fig. 14-6*a*, there is an external field of lines of force between two electric charges, similar to the magnetic field between the magnetic poles in Fig. 14-6*b*. We cannot see the force of attraction and repulsion, just as the force of

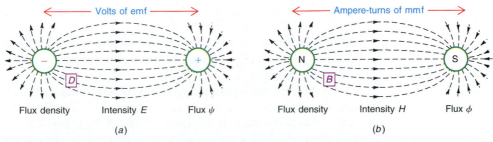

FIG. 14-6   Comparison of electric and magnetic fields. (*a*) Electric field of attraction between static charges of opposite polarity. (*b*) Magnetic field of attraction between opposite magnetic poles.

gravity is invisible, but the force is evident in the work it can do. For both fields, the force tends to make opposite polarities attract and similar polarities repel.

The electric lines show the path an electron would follow in the field; the magnetic lines show how a north pole would move. The entire group of electric lines of force of the static charges is called *electrostatic flux.* Its symbol is the Greek letter ψ (psi), corresponding to ϕ for magnetic flux.

In general, magnetic flux is associated with moving charges, or current, while electrostatic flux is associated with the voltage between static charges. For electric circuits, the application of magnetic flux is often a coil of wire, which is the construction of an inductor. With current, the wire has a magnetic field. As a coil, the wire's magnetic flux is concentrated in the coil. Furthermore, when the magnetic field varies, the change in magnetic flux produces an induced voltage, as explained in Chap. 15, "Electromagnetic Induction," and Chap. 20, "Inductance."

For the case of an electric field, the application is often a dielectric between two conducting plates, which is the construction of a capacitor. With voltage across the dielectric, it has an electric field. As a capacitor, the dielectric's electric field is concentrated between the plates. Furthermore, when the electric field varies, the result is induced current through any conducting path connected to the capacitor. More details are explained in Chap. 17, "Capacitance."

COULOMB'S LAW   The electric lines of force in Fig. 14-6a illustrate the force on an electron in the field. The amount of force between two charges is given by Coulomb's law:

$$F = 9 \times 10^9 \times \frac{q_1 q_2}{r^2} \tag{14-6}$$

where q_1 and q_2 are in coulomb units, F is in newtons, and r is the distance in meters between the charges. The constant factor 9×10^9 converts the values to SI units of newtons for the force in air or vacuum.

Coulomb's law states that the force increases with the amount of charge, but decreases as the square of the distance between charges. Typical values of q are in microcoulombs, since the coulomb is a very large unit of charge.

INTERNATIONAL SYSTEM OF UNITS   In order to provide a closer relation between practical units for electricity and magnetism, these mks units were standardized in 1960 by international agreement. The abbreviation is SI, for *Système International.* Table 14-2 lists the magnetic SI units. The corresponding electrical SI units include the coulomb, which is used for both electric flux and charge, the ampere for current, the volt for potential, and the ohm for resistance. The henry unit for inductance and the farad unit for capacitance are also SI units.

In Table 14-2, note that the reciprocal of reluctance is *permeance,* corresponding to conductance as the reciprocal of resistance. The SI unit for conductance is the siemens (S), equal to $1/\Omega$. This unit is named after Ernst von Siemens, a European inventor.

As another comparison, the permeability μ of a magnetic material with magnetic flux corresponds to the electric *permittivity* ϵ of an insulator with electric flux. Just as permeability is the ability of a magnetic material to concentrate magnetic flux, permittivity is the ability of an insulator to concentrate electric flux. The symbol K_ϵ is used for relative permittivity, corresponding to K_m for relative permeability.

TABLE 14-2   INTERNATIONAL SYSTEM OF MKS UNITS (SI) FOR MAGNETISM

QUANTITY	SYMBOL	UNIT
Flux	ϕ	Weber (Wb)
Flux density	B	Wb/m^2 = tesla (T)
Potential	mmf	Ampere-turn (A · t)
Field intensity	H	Ampere-turn per meter (A · t/m)
Reluctance	$\mathcal{R}$	Ampere-turn per weber (A · t/Wb)
Permeance	$\rho = \dfrac{1}{\mathcal{R}}$	Weber per ampere-turn (Wb/A · t)
Relative μ	μ_r or K_m	None
Permeability	$\mu = \mu_r \times 1.26 \times 10^{-6}$	$\dfrac{B}{H} = \dfrac{\text{tesla (T)}}{\text{ampere-turn per meter (A · t/m)}}$

TEST-POINT QUESTION 14-8

Answers at end of chapter.

Give the SI units for the following:
a. Voltage potential.
b. Magnetic potential.
c. Electric current.
d. Magnetic flux.

14   SUMMARY AND REVIEW

- Table 14-2 summarizes the magnetic units and their definitions.

ANSWERS AT BACK OF BOOK.

Answer True or False.

1. A current of 4 A through 200 turns provides an mmf of $IN = 200$ A · t.
2. For the mmf of 200 A · t with 100 turns, a current of 2 A is necessary.
3. An mmf of 200 A · t across a flux path of 0.1 m provides a field intensity H of 2000 A · t/m.
4. A magnetic material with relative permeability μ_r of 100 has an absolute permeability μ of 126 G/Oe in cgs units.
5. There are no units for relative permeability μ_r.
6. Hysteresis losses are greater in soft iron than in air.
7. Magnetic saturation means that flux density B does not increase in proportion to increases in field intensity H.
8. The units for a B-H curve can be teslas plotted against ampere-turns.
9. In Ohm's law for magnetic circuits, reluctance $\mathcal{R}$ is the opposition to flux ϕ.
10. Ampere-turns of mmf between magnetic poles do not depend on the length of the coil.
11. The ampere-turn is a unit of magnetomotive force.
12. The tesla is an SI unit.

QUESTIONS

1. In Ohm's law for magnetic circuits, what magnetic quantities correspond to V, I, and R?
2. Why can reluctance and permeability be considered opposite characteristics?
3. Give the SI magnetic unit and symbol for each of the following: (**a**) flux; (**b**) flux density; (**c**) field intensity; (**d**) absolute permeability.
4. What cgs units correspond to the following SI units? (**a**) weber; (**b**) tesla; (**c**) ampere-turn; (**d**) ampere-turn per meter.
5. Define the following: (**a**) saturation; (**b**) relative permeability; (**c**) relative permittivity.

6. Explain briefly how to demagnetize a metal object that has become temporarily magnetized.
7. Draw a *B-H* curve with μ, *N*, *I*, and *V* the same as in Fig. 14-3, but with a coil resistance of 5 Ω.
8. Give the formula for Coulomb's law for the force between electrostatic charges, with SI units.
9. What is meant by magnetic hysteresis? Why can hysteresis cause losses in an iron core with a reversing field produced by alternating current?

PROBLEMS

ANSWERS TO ODD-NUMBERED PROBLEMS AT BACK OF BOOK.

1. A coil of 2000 turns with a 100-mA current has a length of 0.4 m. (**a**) Calculate the mmf in ampere-turns. (**b**) Calculate the field intensity *H* in ampere-turns per meter.
2. If the current in the coil of Prob. 1 is increased to 400 mA, calculate the increased values of mmf and *H*.
3. An iron core has a flux density *B* of 3600 G with an *H* of 12 Oe. Calculate (**a**) the permeability μ in cgs units; (**b**) the permeability μ in SI units; (**c**) the relative permeability μ_r of the iron core.
4. A coil of 250 turns with a 400-mA current is 0.2 m long with an iron core of the same length. Calculate the following in mks units: (**a**) mmf; (**b**) *H*; (**c**) *B* in the iron core with a μ_r of 200; (**d**) *B* with an air core instead of the iron core.
5. Referring to the *B-H* curve in Fig. 14-3, calculate the μ in SI units for the iron core at an *H* of: (**a**) 3000 A · t/m; (**b**) 5000 A · t/m.
6. Referring to the hysteresis loop in Fig. 14-4, give the values of (**a**) residual induction B_R, and (**b**) coercive force $-H_C$.
7. A battery is connected across a coil of 100 turns and a 20-Ω *R*, with an iron core 0.2 m long. (**a**) Draw the circuit diagram. (**b**) How much battery voltage is needed for 200 A · t? (**c**) Calculate *H* in the iron core in ampere-turns per meter. (**d**) Calculate *B* in teslas in the iron core if its μ_r is 300. (**e**) Calculate ϕ in webers at each pole with an area of 8×10^{-4} m². (**f**) How much is the reluctance $\Re$ of the iron core, in ampere-turns per weber?
8. In cgs units, how much is the flux density *B* in gauss, for a field intensity *H* of 24 Oe, with μ of 500?
9. Calculate the force, in newtons, between two 4-μC charges separated by 0.1 m in air or vacuum.
10. Make the following conversions: (**a**) 250 A · t to Gb; (**b**) 150 Gb to A · t; (**c**) 400 A · t/m to Oe; (**d**) 1260 Oe to A · t/m.
11. Calculate the relative permeability μ_r of an iron core when a field intensity *H* of 750 A · t/m produces a flux density *B* of 0.126 T.
12. Find *H* in A · t/m units when *B* = 50.4 T and μ_r = 200.
13. A coil with an iron core has a field intensity *H* of 5 A · t/m. If the relative permeability μ_r equals 300, calculate the flux density *B* in teslas.
14. A magnetic material has a total flux ϕ of 300 μWb with an mmf of 12 A · t. Calculate the reluctance in A · t/Wb units.

15. A coil has an mmf of 1000 A · t and a reluctance of 0.1×10^6 A · t/Wb. Calculate the total flux in webers.

16. If the reluctance of a material is 0.5×10^6 A · t/Wb, how much mmf is required to produce a flux ϕ of 600 μWb?

CRITICAL THINKING

1. Derive the value of 1.26×10^{-6} T/(A · t/m) for μ_0 from $\mu = B/H$.
2. What is the relative permeability (μ_r) of a piece of soft iron whose permeability (μ) equals 3.0×10^{-3} T/(A · t/m)?

ANSWERS TO TEST-POINT QUESTIONS

14-1 **a.** 486 A · t
     b. 630 Gb

14-2 **a.** 125 A · t/m
     b. 6.3 Oe

14-3 **a.** 1
     b. 200
     c. 200 G/Oe

14-4 **a.** 0.2 T
     b. 4000 A · t/m

14-5 **a.** T
     b. T

14-6 **a.** T
     b. F

14-7 **a.** T
     b. F
     c. T

14-8 **a.** volt
     b. ampere-turn
     c. ampere
     d. weber

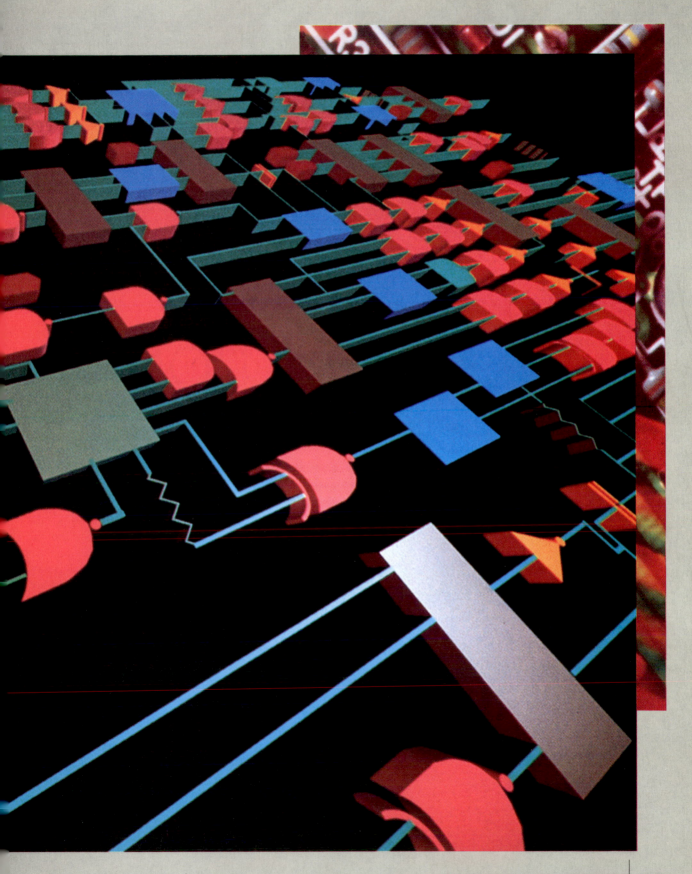

CHAPTER 15

ELECTRO-MAGNETIC INDUCTION

The link between electricity and magnetism was discovered in 1824 by Oersted, who found that current in a wire could move a magnetic compass needle. A few years later the opposite effect was discovered: A magnetic field in motion forces electrons to move, producing current. This important effect was studied by the English physicist and pioneer in electromagnetism, Michael Faraday (1791–1867); the American physicist, Joseph Henry (1797–1878); and the Russian physicist, H. F. E. Lenz (1804–1865).

The henry unit of inductance is named after Joseph Henry. The farad unit of capacitance is named after Michael Faraday, who did important work in electrostatics as well as electromagnetism. There is no unit named after Lenz, but Lenz' law explains many important electromagnetic effects with a varying electric current.

Electromagnetism combines the effect of an electric current and magnetism. Electrons in motion, or any current, always have an associated magnetic field. For the opposite effect, when a magnetic field is in motion, the work put into this action forces electrons to move in any conductor within the field, producing current. These electromagnetic effects have many practical applications that are the basis for motors and generators. More fundamentally, these effects explain why a coil of wire has special characteristics in a circuit when the current is changing in value.

CHAPTER OBJECTIVES

Upon completion of this chapter, you should be able to:

- *Describe* the magnetic field associated with an electric current.
- *Determine* the magnetic polarity of a coil by using the left-hand rule.
- *Explain* what is meant by motor action between two magnetic fields.
- *Understand* how a magnetic flux cutting across a conductor can produce an induced current.
- *State* Lenz' law.
- *Calculate* the amount of induced voltage across the ends of a conductor by using Faraday's law.
- *Understand* the construction and operation of a basic relay.
- *List* and *explain* some important relay ratings.

IMPORTANT TERMS IN THIS CHAPTER

attraction	induced current and voltage	pickup current
chatter	left-hand rule	relay chatter
Faraday's law of induced voltage	Lenz' law	repulsion
field intensity	magnetic flux	solenoid
generator action	magnetic polarity	switching contacts
holding current	make and break	time rate of change
	motor action	torque

TOPICS COVERED IN THIS CHAPTER

In Fig. 15-1, the iron filings aligned in concentric rings around the conductor show the magnetic field of the current in the wire. The iron filings are dense next to the conductor, showing that the field is strongest at this point. Furthermore, the field strength decreases inversely as the square of the distance from the conductor. It is important to note the following two factors about the magnetic lines of force:

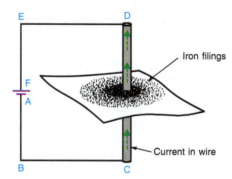

FIG. 15-1 How iron filings can be used to show the invisible magnetic field around the electric current in a wire conductor.

1. The magnetic lines are circular, as the field is symmetrical with respect to the wire in the center.
2. The magnetic field with circular lines of force is in a plane perpendicular to the current in the wire.

From points C to D in the wire, its circular magnetic field is in the horizontal plane because the wire is vertical. Also, the vertical conductor between points EF and AB has the associated magnetic field in the horizontal plane. Where the conductor is horizontal, as from B to C and D to E, the magnetic field is in a vertical plane.

These two requirements of a circular magnetic field in a perpendicular plane apply to any charge in motion. Whether electron flow or a motion of positive charges is considered, the associated magnetic field must be at right angles to the direction of current.

In addition, the current need not be in a wire conductor. As an example, the beam of moving electrons in the vacuum of a cathode-ray tube has an associated magnetic field. In all cases, the magnetic field has circular lines of force in a plane perpendicular to the direction of motion of the electric charges.

CLOCKWISE AND COUNTERCLOCKWISE FIELDS   With circular lines of force, the magnetic field would tend to move a magnetic pole in a circular path. Therefore, the direction of the lines must be considered as either clockwise or counterclockwise. This idea is illustrated in Fig. 15-2, showing how a north pole would move in the circular field.

The directions are tested with a magnetic compass needle. When the compass is in front of the wire, the north pole on the needle points up. On the opposite side, the compass points down. If the compass were placed at the top, its

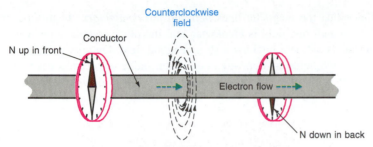

FIG. 15-2 Rule for determining direction of circular field around straight conductor. Field is counterclockwise for direction of electron flow shown here. Circular field is clockwise for conventional current.

needle would point toward the back of the wire; below the wire, the compass would point forward.

When all these directions are combined, the result is the circular magnetic field shown, with counterclockwise lines of force. This field has the magnetic lines upward at the front of the conductor and downward at the back.

Instead of testing every conductor with a magnetic compass, however, we can use the following rule to determine the circular direction of the magnetic field: *If you look along the wire in the direction of electron flow, the magnetic field is counterclockwise.* In Fig. 15-2, the line of electron flow is from left to right. Facing this way, you can assume the circular magnetic flux in a perpendicular plane has lines of force in the counterclockwise direction.

The opposite direction of electron flow produces a reversed field. Then the magnetic lines of force have clockwise rotation. If the charges were moving from right to left in Fig. 15-2, the associated magnetic field would be in the opposite direction, with clockwise lines of force.

FIELDS AIDING OR CANCELING   When the magnetic lines of two fields are in the same direction, the lines of force aid each other, making the field stronger. With magnetic lines in opposite directions, the fields cancel.

In Fig. 15-3 the fields are shown for two conductors with opposite directions of electron flow. The dot in the middle of the field at the left indicates the tip of an arrowhead to show current up from the paper. The cross symbolizes the back of an arrow to indicate electron flow into the paper.

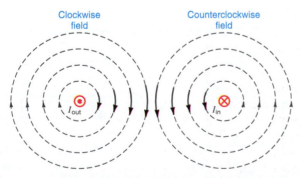

FIG. 15-3 Magnetic fields aiding between parallel conductors with opposite directions of current.

Notice that the magnetic lines *between the conductors* are in the same direction, although one field is clockwise and the other counterclockwise. Therefore, the fields aid here, making a stronger total field. On either side of the conductors, the two fields are opposite in direction and tend to cancel each other. The net result, then, is to strengthen the field in the space between the conductors.

TEST-POINT QUESTION 15-1

Answers at end of chapter.

Answer True or False.
a. Magnetic field lines around a conductor are circular in a perpendicular cross section of the conductor.
b. In Fig. 15-3, the field is strongest between the conductors.

15-2 MAGNETIC POLARITY OF A COIL

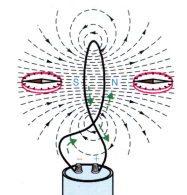

FIG. 15-4 Magnetic poles of a current loop.

Bending a straight conductor into the form of a loop, as shown in Fig. 15-4, has two effects. First, the magnetic field lines are more dense inside the loop. The total number of lines is the same as for the straight conductor, but inside the loop the lines are concentrated in a smaller space. Furthermore, all the lines inside the loop are aiding in the same direction. This makes the loop field effectively the same as a bar magnet with opposite poles at opposite faces of the loop.

SOLENOID AS A BAR MAGNET   A coil of wire conductor with more than one turn is generally called a *solenoid*. An ideal solenoid, however, has a length much greater than its diameter. Like a single loop, the solenoid concentrates the magnetic field inside the coil and provides opposite magnetic poles at the ends. These effects are multiplied, however, by the number of turns as the magnetic field lines aid each other in the same direction inside the coil. Outside the coil, the field corresponds to a bar magnet with north and south poles at opposite ends, as illustrated in Fig. 15-5.

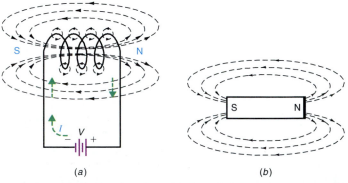

(a)   (b)

FIG. 15-5   Magnetic poles of a solenoid. (*a*) Coil winding. (*b*) Equivalent bar magnet.

MAGNETIC POLARITY To determine the magnetic polarity, use the *left-hand rule* illustrated in Fig. 15-6: *If the coil is grasped with the fingers of the left hand curled around the coil in the direction of electron flow, the thumb points to the north pole of the coil.* The left hand is used here because the current is electron flow.

The solenoid acts like a bar magnet whether it has an iron core or not. Adding an iron core increases the flux density inside the coil. In addition, the field strength then is uniform for the entire length of the core. The polarity is the same, however, for air-core and iron-core coils.

The magnetic polarity depends on the direction of current flow and the direction of winding. The current is determined by the connections to the voltage source. Electron flow is from the negative side of the voltage source, through the coil, and back to the positive terminal.

The direction of winding can be over and under, starting from one end of the coil, or under and over with respect to the same starting point. Reversing either the direction of winding or the direction of current reverses the magnetic poles of the solenoid. See Fig. 15-7. With both reversed, though, the polarity is the same.

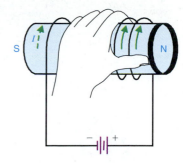

FIG. 15-6 Left-hand rule for north pole of a coil with current *I*. The *I* is electron flow here.

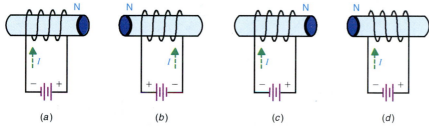

| (a) | (b) | (c) | (d) |

FIG. 15-7 Examples for determining the magnetic polarity of a coil with direct current *I*. The *I* is electron flow here. The polarities are reversed in (*a*) and (*b*) because the battery is reversed to reverse the direction of current. Also, (*d*) is the opposite of (*c*) because of the reversed winding.

TEST-POINT QUESTION 15-2

Answers at end of chapter.
a. In Fig. 15-5, if the battery is reversed, will the north pole be at the left or the right?
b. If one end of a solenoid is a north pole, is the opposite end a north or a south pole?

15-3 MOTOR ACTION BETWEEN TWO MAGNETIC FIELDS

The physical motion resulting from the forces of magnetic fields is called *motor action*. One example is the simple attraction or repulsion between bar magnets.

We know that like poles repel and unlike poles attract. It can also be considered that fields in the same direction repel and opposite fields attract.

Consider the repulsion between two north poles, illustrated in Fig. 15-8. Similar poles have fields in the same direction. Therefore, the similar fields of the two like poles repel each other.

A more fundamental reason for motor action, however, is the fact that the force in a magnetic field tends to produce motion from a stronger field toward a weaker field. In Fig. 15-8, note that the field intensity is greatest in the space between the two north poles. Here the field lines of similar poles in both magnets reinforce in the same direction. Farther away the field intensity is less, for essentially one magnet only. As a result there is a difference in field strength, providing a net force that tends to produce motion. The direction of motion is always toward the weaker field.

To remember the directions, we can consider that the stronger field moves to the weaker field, tending to equalize the field intensity. Otherwise, the motion would make the strong field stronger and the weak field weaker. This must be impossible, because then the magnetic field would multiply its own strength without any work being added.

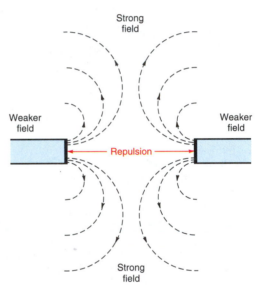

FIG. 15-8 Repulsion between similar poles of two bar magnets. The motion is from the stronger field to the weaker field.

FORCE ON A STRAIGHT CONDUCTOR IN A MAGNETIC FIELD

Current in a conductor has its associated magnetic field. When this conductor is placed in another magnetic field from a separate source, the two fields can react to produce motor action. The conductor must be perpendicular to the magnetic field, however, as shown in Fig. 15-9. This way, the perpendicular magnetic field produced by the current then is in the same plane as the external magnetic field.

Unless the two fields are in the same plane, they cannot affect each other. In the same plane, however, lines of force in the same direction reinforce to make a stronger field, while lines in the opposite direction cancel and result in a weaker field.

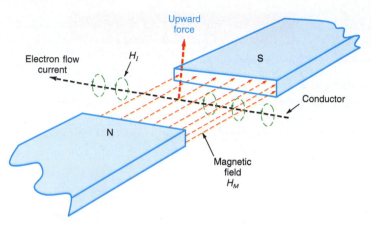

FIG. 15-9 Motor action of current in a straight conductor when it is in an external magnetic field. The H_I is the circular field of the current. The H_M indicates field lines between the north and south poles of the external magnet.

To summarize these directions:

1. With the conductor at 90°, or perpendicular to the external field, the reaction between the two magnetic fields is maximum.
2. With the conductor at 0°, or parallel to the external field, there is no effect between them.
3. When the conductor is at an angle between 0 and 90°, only the perpendicular component is effective.

In Fig. 15-9, electrons flow in the wire conductor in the plane of the paper, from the bottom to the top of the page. This flow provides the counterclockwise field H_I around the wire, in a perpendicular plane cutting through the paper. The external field H_M has lines of force from left to right in the plane of the paper. Then lines of force in the two fields are parallel above and below the wire.

Below the conductor, its field lines are left to right in the same direction as the external field. Therefore, these lines reinforce to produce a stronger field. Above the conductor the lines of the two fields are in opposite directions, causing a weaker field. As a result, the net force of the stronger field makes the conductor move upward out of the page, toward the weaker field.

If electrons flow in the reverse direction in the conductor, or if the external field is reversed, the motor action will be in the opposite direction. Reversing both the field and the current, however, results in the same direction of motion.

ROTATION OF A CURRENT LOOP IN A MAGNETIC FIELD   With a loop of wire in the magnetic field, opposite sides of the loop have current in opposite directions. Then the associated magnetic fields are opposite. The resulting forces are upward on one side of the loop and downward on the other side, making it rotate. This effect of a force in producing rotation is called *torque.*

The principle of motor action between magnetic fields producing rotational torque is the basis of all electric motors. Also, the moving-coil meter described in Sec. 8-1 is a similar application. Since the torque is proportional to current, the amount of rotation indicates how much current flows through the coil.

Answers at end of chapter.

Answer True or False.
a. In Fig. 15-8, the field is strongest between the two north poles.
b. In Fig. 15-9, if both the magnetic field and the current are reversed, the motion will still be upward.

15-4 INDUCED CURRENT

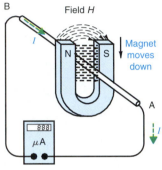

FIG. 15-10 Induced current produced by magnetic flux cutting across a conductor. Direction of *I* here is for electron flow.

Just as electrons in motion provide an associated magnetic field, when magnetic flux moves, the motion of magnetic lines cutting across a conductor forces free electrons in the conductor to move, producing current. This action is called *induction* because there is no physical connection between the magnet and the conductor. The induced current is a result of generator action as the mechanical work put into moving the magnetic field is converted into electric energy when current flows in the conductor.

Referring to Fig. 15-10, let the conductor AB be placed at right angles to the flux in the air gap of the horseshoe magnet. Then, when the magnet is moved up or down, its flux cuts across the conductor. The action of magnetic flux cutting across the conductor generates current. The fact that current flows is indicated by the microammeter.

When the magnet is moved downward, current flows in the direction shown. If the magnet is moved upward, current will flow in the opposite direction. Without motion, there is no current.

DIRECTION OF MOTION   Motion is necessary in order to have the flux lines of the magnetic field cut across the conductor. This cutting can be accomplished by motion of either the field or the conductor. When the conductor is moved upward or downward, it cuts across the flux. The generator action is the same as moving the field, except that the relative motion is opposite. Moving the conductor upward, for instance, corresponds to moving the magnet downward.

CONDUCTOR PERPENDICULAR TO EXTERNAL FLUX   In order to have electromagnetic induction, the conductor and the magnetic lines of flux must be perpendicular to each other. Then the motion makes the flux cut through the cross-sectional area of the conductor. As shown in Fig. 15-10, the conductor is at right angles to the lines of force in the field *H*.

The reason the conductor must be perpendicular is to make its induced current have an associated magnetic field in the same plane as the external flux. If the field of the induced current does not react with the external field, there can be no induced current.

HOW INDUCED CURRENT IS GENERATED   The induced current can be considered the result of motor action between the external field *H* and the magnetic field of free electrons in every cross-sectional area of the wire. Without an external field, the free electrons move at random without any specific direction, and they have no net magnetic field. When the conductor is in the magnetic field H, there still is no induction without relative motion, since the magnetic fields for the free electrons are not disturbed. When the field or conductor moves, however, there must be a reaction opposing the motion. The reaction is a flow of free electrons resulting from motor action on the electrons.

Referring to Fig. 15-10, for example, the induced current must flow in the direction shown because the field is moved downward, pulling the magnet away from the conductor. The induced current of electrons then has a clockwise field, with lines of force aiding H above the conductor and canceling H below. With motor action between the two magnetic fields tending to move the conductor toward the weaker field, the conductor will be forced downward, staying with the magnet to oppose the work of pulling the magnet away from the conductor.

The effect of electromagnetic induction is increased where a coil is used for the conductor. Then the turns concentrate more conductor length in a smaller area. As illustrated in Fig. 15-11, moving the magnet into the coil enables the flux to cut across many turns of conductors.

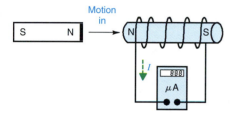

FIG. 15-11 Induced current produced by magnetic flux cutting across turns of wire in a coil. Direction of I here is for electron flow.

TEST-POINT QUESTION 15-4

Answers at end of chapter.

Answer True or False.
a. Refer to Fig. 15-10. If the conductor is moved up, instead of the magnet down, the induced current will flow in the same direction.
b. Refer to Fig. 15-10. The electron flow through the meter is from terminal A to B.

15-5 LENZ' LAW

Lenz' law is the basic principle that is used to determine the direction of an induced voltage or current. Based on the principle of conservation of energy, the law simply states that the direction of the induced current must be such that its own magnetic field will oppose the action that produced the induced current.

In Fig. 15-11, for example, the induced current has the direction that produces a north pole at the left to oppose the motion by repulsion of the north pole being moved in. This is why it takes some work to push the permanent magnet into the coil. The work expended in moving the permanent magnet is the source of energy for the current induced in the coil.

Using Lenz' law, we can start with the fact that the left end of the coil in Fig. 15-11 must be a north pole to oppose the motion. Then the direction of the induced current is determined by the left-hand rule for electron flow. If the fingers coil around the direction of electron flow shown, under and over the winding, the thumb will point to the left for the north pole.

For the opposite case, suppose that the north pole of the permanent magnet in Fig. 15-11 is moved away from the coil. Then the induced pole at the left end of the coil must be a south pole, by Lenz' law. The induced south pole will attract the north pole to oppose the motion of the magnet being moved away. For a south pole at the left end of the coil, then, the electron flow will be reversed from the direction shown in Fig. 15-11. We could actually generate an alternating current in the coil by moving the magnet periodically in and out.

TEST-POINT QUESTION 15-5

Answers at end of chapter.

Refer to Fig. 15-11.
a. If the north end of the magnet is moved away from the coil, will its left side be north or south?
b. If the south end of the magnet is moved in, will the left end of the coil be north or south?

15-6 GENERATING AN INDUCED VOLTAGE

Consider the case of magnetic flux cutting a conductor that is not in a closed circuit, as shown in Fig. 15-12. The motion of flux across the conductor forces free electrons to move, but with an open circuit, the displaced electrons produce opposite electric charges at the two open ends.

For the directions shown, free electrons in the conductor are forced to move to point A. Since the end is open, electrons accumulate here. Point A then develops a negative potential.

At the same time, point B loses electrons and becomes charged positive. The result is a potential difference across the two ends, provided by the separation of electric charges in the conductor.

The potential difference is an electromotive force (emf), generated by the work of cutting across the flux. You can measure this potential difference with a voltmeter. However, a conductor cannot store electric charge. Therefore, the voltage is present only while the motion of flux cutting across the conductor is producing the induced voltage.

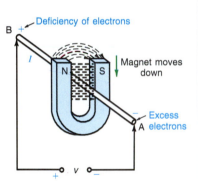

FIG. 15-12 Voltage induced across open ends of conductor cut by magnetic flux.

INDUCED VOLTAGE ACROSS A COIL With a coil, as in Fig. 15-13a, the induced emf is increased by the number of turns. Each turn cut by flux adds to the induced voltage, since they all force free electrons to accumulate at the negative end of the coil, with a deficiency of electrons at the positive end.

The polarity of the induced voltage follows from the direction of induced current. The end of the conductor to which the electrons go and at which they

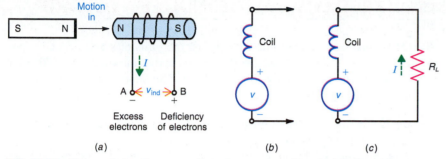

FIG. 15-13   Voltage induced across coil cut by magnetic flux. (*a*) Motion of flux generating voltage across coil. (*b*) Induced voltage acts in series with coil. (*c*) Induced voltage is a source that can produce current in an external load resistor R_L connected across coil.

accumulate is the negative side of the induced voltage. The opposite end, with a deficiency of electrons, is the positive side. The total emf across the coil is the sum of the induced voltages, since all the turns are in series.

Furthermore, the total induced voltage acts in series with the coil, as illustrated by the equivalent circuit in Fig. 15-13*b*, showing the induced voltage as a separate generator. This generator represents a voltage source with a potential difference resulting from the separation of charges produced by electromagnetic induction. The source *v* then can produce current in an external load circuit connected across the negative and positive terminals, as shown in Fig. 15-13*c*.

The induced voltage is in series with the coil because current produced by the generated emf must flow through all the turns. An induced voltage of 10 V, for example, with R_L equal to 5 Ω, results in a current of 2 A, which flows through the coil, the equivalent generator *v*, and the load resistance R_L.

The direction of current in Fig. 15-13*c* shows electron flow around the circuit. Outside the source *v*, the electrons move from its negative terminal, through R_L, and back to the positive terminal of *v* because of its potential difference.

Inside the generator, however, the electron flow is from the + terminal to the − terminal. This direction of electron flow results from the fact that the left end of the coil in Fig. 15-13*a* must be a north pole by Lenz' law, to oppose the north pole being moved in.

Notice how motors and generators are similar in using the motion of a magnetic field, but with opposite applications. In a motor, current is supplied so that an associated magnetic field can react with the external flux to produce motion of the conductor. In a generator, motion must be supplied so that the flux and conductor can cut across each other to induce voltage across the ends of the conductor.

TEST-POINT QUESTION 15-6

Answers at end of chapter.

Refer to Fig. 15-13.
a. Is terminal A or B the negative side of the induced voltage?
b. Is the bottom of R_L the negative side of V_{R_L}?

The voltage induced by magnetic flux cutting the turns of a coil depends upon the number of turns and how fast the flux moves across the conductor. Either the flux or the conductor can move. Specifically, the amount of induced voltage is determined by the following three factors:

1. *Amount of flux.* The more magnetic lines of force that cut across the conductor, the higher the amount of induced voltage.
2. *Number of turns.* The more turns in a coil, the higher the induced voltage. The v_{ind} is the sum of all the individual voltages generated in each turn in series.
3. *Time rate of cutting.* The faster the flux cuts a conductor, the higher the induced voltage. Then more lines of force cut the conductor within a specific period of time.

These factors are of fundamental importance in many applications. Any conductor with current will have voltage induced in it by a change in current and its associated magnetic flux.

The amount of induced voltage can be calculated by Faraday's law:

$$v_{ind} = N \frac{d\phi \text{ (webers)}}{dt \text{ (seconds)}} \qquad \textbf{(15-1)}$$

where N is the number of turns and $d\phi/dt$ specifies how fast the flux ϕ cuts across the conductor. With $d\phi/dt$ in webers per second, the induced voltage is in volts.

As an example, suppose that magnetic flux cuts across 300 turns at the rate of 2 Wb/s.

To calculate the induced voltage,

$$v_{ind} = N \frac{d\phi}{dt}$$
$$= 300 \times 2$$
$$v_{ind} = 600 \text{ V}$$

It is assumed that all the flux links all the turns, which is true with an iron core.

TIME RATE OF CHANGE   The symbol d in $d\phi$ and dt is an abbreviation for *change.* The $d\phi$ means a change in the flux ϕ, and dt means a change in time. In mathematics, dt represents an infinitesimally small change in time, but in this book we are using the d to mean rate of change in general. The results are exactly the same for the practical changes used here because the rate of change is constant.

As an example, if the flux ϕ is 4 Wb one time but then changes to 6 Wb, the change in flux $d\phi$ is 2 Wb. The same idea applies to a decrease as well as an increase. If the flux changed from 6 to 4 Wb, $d\phi$ would still be 2 Wb. However, an increase is usually considered a change in the positive direction, with an upward slope, while a decrease has a negative slope downward.

Similarly, dt means a change in time. If we consider the flux at a time 2 s after the start and at a later time 3 s after the start, the change in time is $3 - 2$, or 1 s for dt. Time always increases in the positive direction.

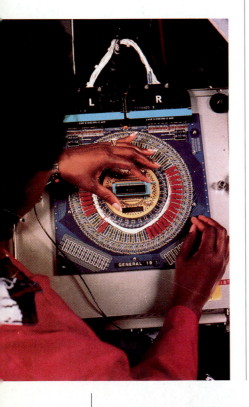

Combining the two factors of $d\phi$ and dt, we can say that for magnetic flux increasing by 2 Wb in 1 s, $d\phi/dt$ equals 2/1, or 2 Wb/s. This states the time rate of change of the magnetic flux.

As another example, suppose that the flux increases by 2 Wb in the time of $\frac{1}{2}$ or 0.5 s. Then

▶ $$\frac{d\phi}{dt} = \frac{2 \text{ Wb}}{0.5 \text{ s}} = 4 \text{ Wb/s}$$

ANALYSIS OF INDUCED VOLTAGE AS $N(d\phi/dt)$ This fundamental concept of voltage induced by a change in flux is illustrated by the graphs in Fig. 15-14, for the values listed in Table 15-1. The linear rise in Fig. 15-14a shows values of flux ϕ increasing at a uniform rate. In this case, the curve goes up 2 Wb for every 1-s interval of time. The slope of this curve, then, equal to $d\phi/dt$, is 2 Wb/s. Note that, although ϕ increases, the rate of change is constant because the linear rise has a constant slope.

For induced voltage, only the $d\phi/dt$ factor is important, not the actual value of flux. To emphasize this basic concept, the graph in Fig. 15-14b shows the $d\phi/dt$ values alone. This graph is just a straight horizontal line for the constant value of 2 Wb/s.

The induced-voltage graph in Fig. 15-14c is also a straight horizontal line. Since $v_{\text{ind}} = N(d\phi/dt)$, the graph of induced voltage is just the $d\phi/dt$ values mul-

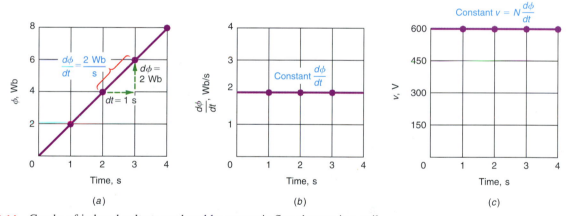

FIG. 15-14   Graphs of induced voltage produced by magnetic flux changes in a coil. (a) Linear increase of flux ϕ. (b) Constant rate of change for $d\phi/dt$ at 2 Wb/s. (c) Constant induced voltage of 600 V, for a coil with 300 turns.

TABLE 15-1   INDUCED-VOLTAGE CALCULATIONS FOR FIG. 15-14

ϕ, Wb	$d\phi$, Wb	t, s	dt, s	$d\phi/dt$, Wb/s	N, TURNS	$N(d\phi/dt)$, v
2	2	1	1	2	300	600
4	2	2	1	2	300	600
6	2	3	1	2	300	600
8	2	4	1	2	300	600

tiplied by the number of turns. The result is a constant 600 V, with 300 turns cut by flux changing at the constant rate of 2 Wb/s.

The example illustrated here can be different in several ways without changing the basic fact that the induced voltage is equal to $N(d\phi/dt)$. First, the number of turns or the $d\phi/dt$ values can be greater than the values assumed here, or less. More turns will provide more induced voltage, while fewer turns mean less voltage. Similarly, a higher value for $d\phi/dt$ results in more induced voltage.

Note that two factors are included in $d\phi/dt$. Its value can be increased by a higher value of $d\phi$ or a smaller value of dt. As an example, the value of 2 Wb/s for $d\phi/dt$ can be doubled by either increasing $d\phi$ to 4 Wb or reducing dt to ½ s. Then $d\phi/dt$ is 4/1 or 2/0.5, which equals 4 Wb/s in either case. The same flux changing within a shorter time means a faster rate of flux cutting the conductor, resulting in a higher value of $d\phi/dt$ and more induced voltage.

For the opposite case, a smaller value of $d\phi/dt$, with less flux or a slower rate of change, results in a lower value of induced voltage. As $d\phi/dt$ decreases, the induced voltage will reverse polarity.

Finally, it should be noted that the $d\phi/dt$ graph in Fig. 15-14b has the constant value of 2 Wb/s because the flux is increasing at a linear rate. However, the flux need not have a uniform rate of change. Then the $d\phi/dt$ values will not be constant. In any case, though, the values of $d\phi/dt$ at all instants of time will determine the values of the induced voltage equal to $N(d\phi/dt)$.

POLARITY OF THE INDUCED VOLTAGE The polarity is determined by Lenz' law. Any induced voltage has the polarity that opposes the change causing the induction. Sometimes this fact is indicated by using a negative sign for v_{ind} in Formula (15-1). However, the absolute polarity depends on whether the flux is increasing or decreasing, the method of winding, and which end of the coil is the reference.

When all these factors are considered, v_{ind} has the polarity such that the current it produces and the associated magnetic field will oppose the change in flux producing the induced voltage. If the external flux increases, the magnetic field of the induced current will be in the opposite direction. If the external field decreases, the magnetic field of the induced current will be in the same direction as the external field to oppose the change by sustaining the flux. In short, the induced voltage has the polarity that opposes the change.

TEST-POINT QUESTION 15-7

Answers at end of chapter.
  a. The magnetic flux of 8 Wb changes to 10 Wb in 1 s. How much is $d\phi/dt$?
  b. The flux of 8 μWb changes to 10 μWb in 1 μs. How much is $d\phi/dt$?

15-8 RELAYS

A relay is an electromechanical device which operates on the basis of electromagnetic induction. It uses either an ac- or a dc-actuated electromagnet to open or close one or more sets of contacts. Relay contacts which are open when the relay is not energized are called *normally open* (NO) contacts. Conversely, relay contacts which are closed when the relay is not energized are called *normally closed* (NC). Relay contacts are held in their resting or normal position either by a spring or by some type of gravity-actuated mechanism. In most cases, an adjustment of the spring tension is provided to set the restraining force on the normally open and normally closed contacts to some desired level based on predetermined circuit conditions.

Figure 15-15 shows the schematic symbols that are commonly used to represent relay contacts. Figure 15-15*a* shows the symbols used to represent normally open contacts, whereas Fig. 15-15*b* shows the symbols used to represent normally closed contacts. When normally open contacts close, they are said to *make*, whereas when normally closed contacts open they are said to *break*. Like mechanical switches, the switching contacts of a relay can have any number of poles and throws.

Figure 15-16 shows the basic parts of an SPDT armature relay. Terminal connections 1 and 2 provide connection to the electromagnet, and terminal connections 3, 4, and 5 provide connections to the SPDT relay contacts which open or close when the relay is energized. A relay is said to be *energized* when NO contacts close and NC contacts open. The movable arm of an electromechanical relay is called the *armature*. The armature is magnetic and has contacts which make or break with other contacts when the relay is energized. For example, when terminals 1 and 2 are connected to a dc source in Fig. 15-16, current flows in the relay coil and an electromagnet is formed. If there is sufficient current in the relay coil, contacts 3 and 4 close (make) and contacts 4 and 5 open (break). The armature is attracted whether the electromagnet produces a north or a south pole on the end adjacent to the armature. Figure 15-17 shows a photo of a typical relay.

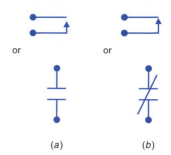

FIG. 15-15 Schematic symbols commonly used to represent relay contacts. (*a*) Symbols used to represent normally open (NO) contacts. (*b*) Symbols used to represent normally closed (NC) contacts.

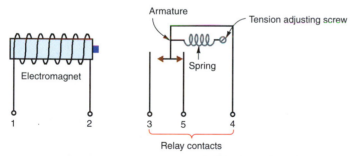

FIG. 15-16 Basic parts of an SPDT armature relay. Terminal connections 1 and 2 provide connection to the electromagnet, and terminal connections 3, 4, and 5 provide connections to the SPDT relay contacts which open or close when the relay is energized.

RELAY SPECIFICATIONS Manufacturers of electromechanical relays always supply a specification sheet for each of their relays. The specification sheet contains voltage and current ratings for both the relay coil and its switching contacts. The specification sheet also includes information regarding the location of the relay coil and switching contact terminals. And finally, the specification sheet will indicate whether the relay can be energized from either an ac or dc source. The following is an explanation of a relay's most important ratings.

FIG. 15-17 Typical relay.

Pickup voltage. The minimum amount of relay coil voltage necessary to energize or operate the relay.

Pickup current. The minimum amount of relay coil current necessary to energize or operate the relay.

Holding current. The minimum amount of current required to keep a relay energized or operating. (The holding current is less than the pickup current.)

Dropout voltage. The maximum relay coil voltage at which the relay is no longer energized.

Contact voltage rating. The maximum voltage the relay contacts are capable of switching safely.

Contact current rating. The maximum current the relay contacts are capable of switching safely.

Contact voltage drop. The voltage drop across the closed contacts of a relay when operating.

Insulation resistance. The resistance measured across the relay contacts in the open position.

RELAY APPLICATIONS Figure 15-18 shows schematic diagrams for two different relay systems. The diagram in Fig. 15-18*a* represents an open-circuit system. With the control switch S_1 open, the SPST relay contacts are open and the load is inoperative. Closing S_1 energizes the relay. This closes the NO relay contacts and makes the load operative.

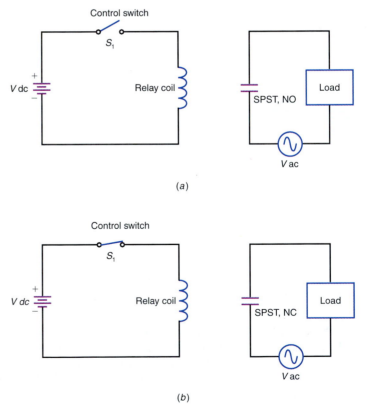

FIG. 15-18 Schematic diagrams for two relay systems. (*a*) Open-circuit system. (*b*) Closed-circuit system.

Figure 15-18*b* represents a closed-circuit system. In this case, the relay is energized by the control switch S_1, which is closed during normal operation. With the relay energized, the normally closed relay contacts are open and the load is inoperative. When it is desired to operate the load, the control switch S_1 is opened. This returns the relay contacts to their normally closed position, thereby activating the load.

It is important to note that a relay can be energized using a low-voltage, low-power source. However, the relay contacts can be used to control a circuit whose load consumes much more power at a much higher voltage than the relay coil circuit. In fact, one of the main advantages of using a relay is its ability to switch or control very high power loads with a relatively low amount of input power. In remote-control applications, a relay can control a high power load a long distance away much more efficiently than a mechanical switch can.

COMMON RELAY TROUBLES   If the relay coil develops an open, the relay cannot be energized. The reason is simple. With an open relay coil the current is zero and no magnetic field is set up by the electromagnet to attract the armature. An ohmmeter can be used to check for the proper relay coil resistance. Since it is usually not practical to repair an open relay coil, the entire relay must be replaced.

A common problem with relays is dirty switching contacts. The switching contacts develop a thin carbon coating after extended use as a result of arcing across the contact terminals when they are opened and closed. Dirty switching contacts usually produce intermittent operation of the load being controlled—for example, a motor. In some cases the relay contacts may chatter (vibrate) if they are dirty.

One final point: The manufacturer of a relay usually indicates its life expectancy in terms of the number of times the relay can be energized (operated). A typical value is 5 million operations.

TEST-POINT QUESTION 15-8

Answers at end of chapter.

Answer True or False.
a. A relay is energized if a set of NC contacts are opened.
b. The pickup current is the minimum relay coil current required to keep a relay energized.
c. The voltage drop across a set of closed relay contacts carrying 1 A of current is very low.
d. An open relay coil measures 0 Ω with an ohmmeter.

15   SUMMARY AND REVIEW

- Current in a straight conductor has an associated magnetic field with circular lines of force in a plane perpendicular to the conductor. The direction of the circular field is counterclockwise when you look along the conductor in the direction of electron flow.
- With two fields in the same plane, produced by either current or a permanent magnet, lines of force in the same direction aid each other to provide a stronger field. Lines of force in opposite directions cancel, resulting in a weaker field.
- A solenoid is a long, narrow coil of wire which concentrates the conductor and its associated magnetic field. Because the fields for all turns aid inside the coil and cancel outside, a solenoid has a resultant electromagnetic field like a bar magnet with north and south poles at opposite ends.
- The left-hand rule for polarity of an electromagnet says that when your fingers curl around the turns in the direction of electron flow, the thumb points to the north pole.
- Motor action is the motion that results from the net force of two fields that can aid or cancel each other. The direction of the resultant force is always from the stronger field to the weaker field.
- Generator action refers to induced voltage. For N turns, $v_{ind} = N(d\phi/dt)$, where $d\phi/dt$ stands for the change in flux (ϕ) with time (t). The change is given in webers per second. There must be a change in the flux to produce induced voltage.
- Lenz' law states that the polarity of the induced voltage will produce I with a magnetic field that opposes the change in magnetic flux causing the induction.
- The faster the flux changes, the higher the induced voltage.
- When the flux changes at a constant rate, the induced voltage is constant.
- The switching contacts of an electromechanical relay may be either normally open (NO) or normally closed (NC). The contacts are held in their normal or resting positions by springs or some gravity-actuated mechanism.
- The movable arm on a relay is called the armature. The armature is magnetic and has contacts which open or close with other contacts when the relay is energized.
- The pickup current of a relay is the minimum amount of relay coil current which will energize the relay. The holding current is the minimum relay coil current required to keep a relay energized.
- The main advantage of a relay over an ordinary mechanical switch is that it can control a high-power load a long distance away with very low I^2R power losses in the wire conductors. The relay is energized using a low-power source. The switching contacts are used to turn on and off the high-power load.

Answer True or False.

1. A vertical wire with electron flow into this page has an associated magnetic field counter-clockwise in the plane of the paper.
2. Lines of force of two magnetic fields in the same direction aid each other to produce a stronger resultant field.
3. Motor action always tends to produce motion toward the weaker field.
4. In Fig. 15-6, if the battery connections are reversed, the magnetic poles of the coil will be reversed.
5. A solenoid is a coil that acts as a bar magnet only when current flows.
6. A torque is a force tending to cause rotation.
7. In Fig. 15-9, if external poles are reversed, motor action will be downward.
8. In Fig. 15-10, if the conductor is moved down, instead of the magnet, the induced current flows in the opposite direction.
9. An induced voltage is produced only while the flux is changing.
10. Faraday's law determines the amount of induced voltage.
11. Lenz' law determines the polarity of an induced voltage.
12. Induced voltage increases with a faster rate of flux cutting.
13. An induced voltage is effectively in series with the turns of the coil in which the voltage is produced.
14. A decrease in flux will induce a voltage of opposite polarity from an increase in flux, with the same direction of field lines in both cases.
15. The flux of 1000 lines increasing to 1001 lines in 1 s produces a flux change $d\phi/dt$ of 1 line per s.
16. The pickup voltage of a relay is the maximum voltage the relay contacts can switch safely.
17. The switching contacts of a relay are always open when the relay is not energized.
18. The movable arm on a relay is called the armature.
19. One difference between a standard mechanical switch and a relay is that relay contacts never get dirty.

QUESTIONS

1. Draw a diagram showing two conductors connecting a battery to a load resistance through a closed switch. (a) Show the magnetic field of the current in the negative side of the line and in the positive side. (b) Where do the two fields aid? Where do they oppose?
2. State the rule for determining the magnetic polarity of a solenoid. (a) How can the polarity be reversed? (b) Why are there no magnetic poles when the current through the coil is zero?
3. Why does the motor action between two magnetic fields result in motion toward the weaker field?

4. Why does current in a conductor perpendicular to this page have a magnetic field in the plane of the paper?

5. Why must the conductor and the external field be perpendicular to each other in order to have motor action or to generate induced voltage?

6. Explain briefly how either motor action or generator action can be obtained with the same conductor in a magnetic field.

7. Assume that a conductor being cut by the flux of an expanding magnetic field has 10 V induced with the top end positive. Now analyze the effect of the following changes: (a) The magnetic flux continues to expand, but at a slower rate. How does this affect the amount of induced voltage and its polarity? (b) The magnetic flux is constant, neither increasing nor decreasing. How much is the induced voltage? (c) The magnetic flux contracts, cutting across the conductor with the opposite direction of motion. How does this affect the polarity of the induced voltage?

8. Redraw the graph in Fig. 15-14c for 500 turns, with all other factors the same.

9. Redraw the circuit with the coil and battery in Fig. 15-6, showing two different ways to reverse the magnetic polarity.

10. Referring to Fig. 15-14, suppose that the flux decreases from 8 Wb to zero at the same rate as the increase. Tabulate all the values as in Table 15-1 and draw the three graphs corresponding to those in Fig. 15-14.

11. Assume you have in your possession a relay whose pickup and holding current values are unknown. Explain how you can determine their values experimentally.

PROBLEMS

ANSWERS TO ODD-NUMBERED PROBLEMS AT BACK OF BOOK.

1. A magnetic flux of 800 Mx cuts across a coil of 1000 turns in 1 μs. How much is the voltage induced in the coil? (1 Mx = 10^{-8} Wb)

2. Refer to Fig. 15-13. (a) Show the induced voltage here connected to a load resistance R_L of 50 Ω. (b) If the induced voltage is 100 V, how much current flows in R_L? (c) Give one way to reverse the polarity of the induced voltage.

3. Calculate the constant rate of flux change with time ($d\phi/dt$) in webers per second for the following: (a) 6 Wb increasing to 8 Wb in 1 s; (b) 8 Wb decreasing to 6 Wb in 1 s.

4. Calculate the voltage induced in 400 turns by each of the flux changes in Prob. 3.

5. Draw a circuit with a 20-V battery connected to a 100-Ω coil of 400 turns with an iron core 0.2 m long. Using SI magnetic units, calculate (a) I; (b) ampere-turns of mmf; (c) field intensity H; (d) flux density B in a core with a μ_r of 500; (e) total flux ϕ at each pole with an area of 6×10^{-4} m^2. (f) Show the direction of winding and magnetic polarity of the coil.

6. For the coil in Prob. 5: (a) If the iron core is removed, how much will the flux be in the air-core coil? (b) How much induced voltage would be produced by this change in flux while the core is being moved out in 1 s? (c) How much is the induced voltage after the core is removed?

7. A magnetic field cuts across a coil, which has 500 turns, at the rate of 2000 μWb/s. Calculate the value of induced voltage, v_{ind}.

8. A coil that has 250 turns has an induced voltage v_{ind} of 10 V. Calculate the rate of flux change, $d\phi/dt$, in Wb/s.

9. A coil has an induced voltage v_{ind} of 40 V when the rate of flux change equals 25,000 μWb/s. How many turns are in the coil?

10. A magnetic field cuts across a coil that has 2000 turns at the rate of 2500 μWb/s. Calculate the induced voltage v_{ind}.

CRITICAL THINKING

1. Refer to Fig. 15-19a. Calculate: (a) the total wire resistance R_W of the No. 12 gage copper wires; (b) the total resistance R_T of the circuit; (c) the voltage available across the load R_L; (d)

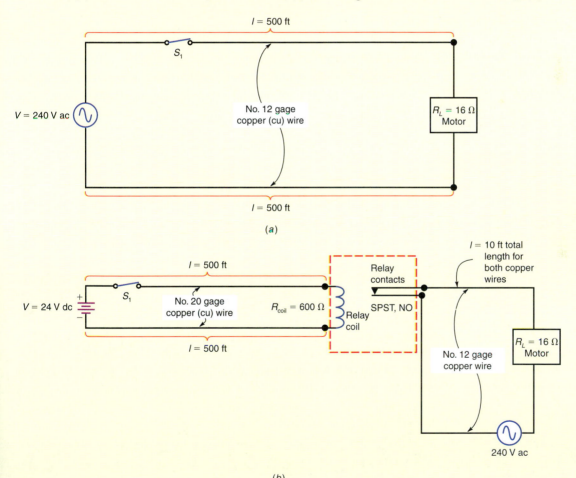

(a)

(b)

FIG. 15-19 Circuit diagram for Critical Thinking Probs. 1 and 2. (a) Mechanical switch controlling a high-power load a long distance away. (b) Relay controlling a high-power load a long distance away.

the I^2R power loss in the wire conductors; (**e**) the load power P_L; (**f**) the total power P_T consumed by the circuit; (**g**) the efficiency of the system calculated as $(P_L/P_T) \times 100$.

2. Refer to Fig. 15-19*b*. Calculate: (**a**) the total wire resistance R_W of the No. 20 gage copper wires; (**b**) the total resistance R_T of the relay coil circuit; (**c**) the voltage across the relay coil; (**d**) the I^2R power loss in the No. 20 gage copper wires in the relay coil circuit; (**e**) the total wire resistance R_W of the 10-ft length of No. 12 gage copper wires which connect the 16-Ω load, R_L to the 240-V_{ac} power line; (**f**) the voltage available across the load R_L; (**g**) the I^2R power loss in the 10-ft length of the No. 12 gage copper wire; (**h**) the load power P_L; (**i**) the total power P_T consumed by the load side of the circuit; (**j**) the percent efficiency of the system calculated as $(P_L/P_T) \times 100$.

3. Explain the advantage of using a relay rather than an ordinary mechanical switch when controlling a high-power load a long distance away. Use your solutions from Critical Thinking Probs. 1 and 2 to support your answer.

ANSWERS TO TEST-POINT QUESTIONS

15-1 a. T
    b. T

15-2 a. left
    b. south pole

15-3 a. T
    b. T

15-4 a. T
    b. T

15-5 a. south
    b. south

15-6 a. A
    b. yes

15-7 a. 2 Wb/s
    b. 2 Wb/s

15-8 a. T
    b. F
    c. T
    d. F

CHAPTER 16

ALTERNATING VOLTAGE AND CURRENT

This chapter begins the analysis of alternating voltage, as used for the 120-Vac power line, and the alternating current that the voltage produces in an ac circuit. Alternating voltage reverses in polarity and amplitude periodically with time. One cycle includes two alternations in polarity. The number of cycles per second is the frequency whose unit is the hertz (Hz). One hertz is equal to one cycle per second (1 Hz = 1 cps). The ac power line frequency is standardized at 60 Hz in the United States.

For an ac voltage:

1. The *V* reverses polarity at a specific rate. Consider one terminal of the ac source positive at a given time, with respect to the other terminal. A little later in time, the positive terminal will become negative to reverse the polarity of the ac output voltage. The polarity reversals are continuously repeated at a regular rate.
2. For either polarity, the ac voltage varies in amplitude. In fact, the voltage must vary from a maximum value to zero in order to be ready for the next polarity reversal.

The alternating current that results has the following features:

1. The *I* reverses in direction with the polarity reversal in *V*.
2. The amplitude of *I* varies with the changing values of voltage.

The ac waveform with its polarity reversals and amplitude variations is very important in electronics because the many audio, radio, and video signals are examples of ac voltages.

CHAPTER OBJECTIVES

Upon completion of this chapter, you should be able to:

- *Understand* how a sine wave of alternating voltage is generated.
- *Calculate* the instantaneous value of a sine wave.
- *Define* the following values for a sine wave: peak, peak-to-peak, root-mean-square, and average.
- *Calculate* the rms, average, and peak-to-peak values of a sine wave when the peak value is known.
- *Define* the terms *frequency* and *period* and list the units of each.
- *Calculate* the wavelength when the frequency is known.
- *Understand* the concept of phase angles.
- *Understand* the makeup of a nonsinusoidal waveform.
- Define the term *harmonics*.
- *Understand* the 60-Hz ac power line and the basics of residential house wiring.

IMPORTANT TERMS IN THIS CHAPTER

alternation	field winding	sawtooth wave
alternator	frequency	sine wave
armature	harmonic	sinusoid
average value	hertz	slip rings
brushes	octave	square wave
commutator	peak value	three-phase power
cycle	phase angle	wavelength
delta connections	phasor	wye connections
effective value	rms value	

TOPICS COVERED IN THIS CHAPTER

16-1 ALTERNATING CURRENT APPLICATIONS

Figure 16-1 shows the output from an ac voltage generator, with the reversals between positive and negative polarities and the variations in amplitude. In Fig. 16-1*a*, the waveform shown simulates an ac voltage as it would appear on the screen of an oscilloscope, which is an important test instrument for ac voltages. The oscilloscope shows a picture of any ac voltage connected to its input terminals, while indicating the amplitude. The details of how to use the oscilloscope for ac voltage measurements are explained in App. D, "Using the Oscilloscope."

In Fig. 16-1*b*, the graph of the ac waveform shows how the output from the generator in Fig. 16-1*c* varies with respect to time. Assume that this graph shows *V* at terminal 2 with respect to terminal 1. Then the voltage at terminal 1 corresponds to the zero axis in the graph as the reference level. At terminal 2, the output voltage has positive amplitude variations from zero up to the peak value and down to zero. All these voltage values are with respect to terminal 1. After a half-cycle, the voltage at terminal 2 becomes negative, still with respect to the other terminal. Then the same voltage variations are repeated at terminal 2, but they have negative polarity compared to the reference level. It should be noted that if we take the voltage at terminal 1 with terminal 2 as the reference, the waveform in Fig. 16-1*b* would have the same shape but be inverted in polarity. The negative half-cycle would come first, but it does not matter which is first or second.

The characteristic of varying values is the reason why ac circuits have so many uses. For instance, a transformer can operate only with alternating current, to step up or step down an ac voltage. The reason is that the changing current produces changes in its associated magnetic field. This application is just an example of inductance *L* in ac circuits, where the changing magnetic flux of a varying current can produce induced voltage. The details of inductance are explained in Chaps. 20, 21, and 22.

A similar but opposite effect in ac circuits is capacitance *C*. The capacitance is important with the changing electric field of a varying voltage. Just as *L* has

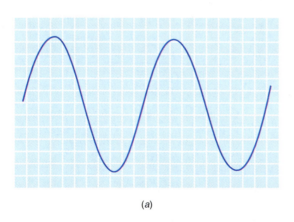

(a)

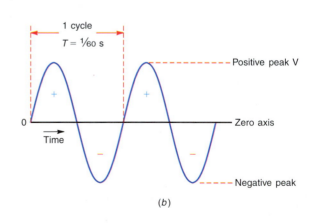

(b)

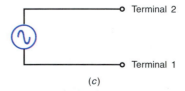

(c)

FIG. 16-1 Waveform of ac power-line voltage with frequency of 60 Hz. Two cycles are shown. (*a*) Oscilloscope readout. (*b*) Details of waveform and alternating polarities. (*c*) Symbol for an ac voltage source.

an effect with alternating current, C has an effect which depends on alternating voltage. The details of capacitance are explained in Chaps. 17, 18, and 19.

The L and C are additional factors, besides resistance R, in the operation of ac circuits. It should be noted that R is the same for either a dc or an ac circuit. However, the effects of L and C depend on having an ac source. The rate at which the ac variations occur, which determines the frequency, allows a greater or lesser reaction by L and C. Therefore, the effect is different for different frequencies. One important application is a resonant circuit with L and C which is tuned to a particular frequency. Tuning in radio and television stations are applications of resonance in an LC circuit.

In general, electronic circuits are combinations of R, L, and C, with both direct current and alternating current. The audio, video, and radio signals are ac voltages and currents. However, the amplifiers that use transistors need dc voltages in order to conduct any current at all. The resulting output of an amplifier circuit, therefore, consists of direct current with a superimposed ac signal. More details of amplifiers are explained in Chap. 29, "Electronic Circuits."

TEST-POINT QUESTION 16-1

Answers at end of chapter.

Answer True or False.
a. An ac voltage varies in magnitude and reverses in polarity.
b. A transformer can operate with either ac or steady dc input.
c. Inductance L and capacitance C are important factors in ac circuits.

16-2 ALTERNATING-VOLTAGE GENERATOR

We can define an ac voltage as one that continuously varies in magnitude and periodically reverses in polarity. In Fig. 16-1, the variations up and down on the waveform show the changes in magnitude. The zero axis is a horizontal line across the center. Then voltages above the center have positive polarity, while the values below center are negative.

Figure 16-2 illustrates how such a voltage waveform is produced by a rotary generator. The conductor loop rotates through the magnetic field to generate the induced ac voltage across its open terminals. The magnetic flux shown here is vertical, with lines of force down in the plane of the paper.

In Fig. 16-2a the loop is in its horizontal starting position in a plane perpendicular to the paper. When the loop rotates counterclockwise, the two longer conductors move around a circle. Note that in the flat position shown, the two long conductors of the loop move vertically up or down but parallel to the vertical flux lines. In this position, motion of the loop does not induce a voltage because the conductors are not cutting across the flux.

When the loop rotates through the upright position in Fig. 16-2b, however, the conductors cut across the flux, producing maximum induced voltage. The shorter connecting wires in the loop do not have any appreciable voltage induced in them.

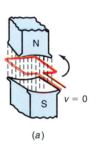

(a)

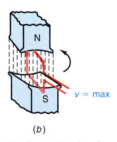

(b)

FIG. 16-2 Loop rotating in magnetic field to produce induced voltage v with alternating polarities. (a) Loop conductors moving parallel to magnetic field results in zero voltage. (b) Loop conductors cutting across magnetic field produce maximum induced voltage.

Each of the longer conductors has opposite polarity of induced voltage because the conductor at the top is moving to the left while the bottom conductor is moving to the right. The amount of voltage varies from zero to maximum as the loop moves from a flat position to upright, where it can cut across the flux. Also, the polarity at the terminals of the loop reverses as the motion of each conductor reverses during each half-revolution.

With one revolution of the loop in a complete circle back to the starting position, therefore, the induced voltage provides a potential difference v across the loop, varying in the same way as the wave of voltage shown in Fig. 16-1. If the loop rotates at the speed of 60 revolutions per second, the ac voltage will have the frequency of 60 Hz.

THE CYCLE   One complete revolution of the loop around the circle is a *cycle*. In Fig. 16-3, the generator loop is shown in its position at each quarter-turn during one complete cycle. The corresponding wave of induced voltage also goes through one cycle. Although not shown, the magnetic field is from top to bottom of the page as in Fig. 16-2.

At position A in Fig. 16-3, the loop is flat and moves parallel to the magnetic field, so that the induced voltage is zero. Counterclockwise rotation of the loop moves the dark conductor to the top at position B, where it cuts across the field to produce maximum induced voltage. The polarity of the induced voltage here makes the open end of the dark conductor positive. This conductor at the top is cutting across the flux from right to left. At the same time, the opposite conductor below is moving from left to right, causing its induced voltage to have opposite polarity. Therefore, maximum induced voltage is produced at this time across the two open ends of the loop. Now the top conductor is positive with respect to the bottom conductor.

In the graph of induced voltage values below the loop in Fig. 16-3, the polarity of the dark conductor is shown with respect to the other conductor. Positive voltage is shown above the zero axis in the graph. As the dark conductor

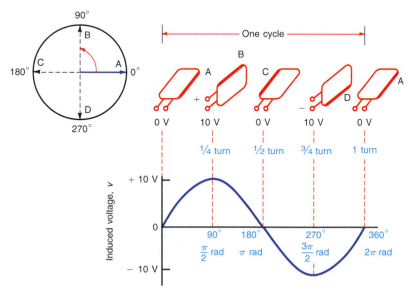

FIG. 16-3 One cycle of alternating voltage generated by rotating loop. Magnetic field, not shown here, is directed from top to bottom, as in Fig. 16-2.

rotates from its starting position parallel to the flux toward the top position, where it cuts maximum flux, more and more induced voltage is produced, with positive polarity.

When the loop rotates through the next quarter-turn, it returns to the flat position shown in C, where it cannot cut across flux. Therefore, the induced voltage values shown in the graph decrease from the maximum value to zero at the half-turn, just as the voltage was zero at the start. The half-cycle of revolution is called an *alternation*.

The next quarter-turn of the loop moves it to the position shown at D in Fig. 16-3, where the loop cuts across the flux again for maximum induced voltage. Note, however, that here the dark conductor is moving left to right at the bottom of the loop. This motion is reversed from the direction it had when it was at the top, moving right to left. Because the direction of motion is reversed during the second half-revolution, the induced voltage has opposite polarity, with the dark conductor negative. This polarity is shown as negative voltage, below the zero axis. The maximum value of induced voltage at the third quarter-turn is the same as at the first quarter-turn but with opposite polarity.

When the loop completes the last quarter-turn in the cycle, the induced voltage returns to zero as the loop returns to its flat position at A, the same as at the start. This cycle of values of induced voltage is repeated as the loop continues to rotate, with one complete cycle of voltage values, as shown, for each circle of revolution.

Note that zero at the start and zero after the half-turn of an alternation are not the same. At the start, the voltage is zero because the loop is flat, but the dark conductor is moving upward in the direction that produces positive voltage. After one half-cycle, the voltage is zero with the loop flat, but the dark conductor is moving downward in the direction that produces negative voltage. After one complete cycle, the loop and its corresponding waveform of induced voltage are the same as at the start. *A cycle can be defined, therefore, as including the variations between two successive points having the same value and varying in the same direction.*

ANGULAR MEASURE   Because the cycle of voltage in Fig. 16-3 corresponds to rotation of the loop around a circle, it is convenient to consider parts of the cycle in angles. The complete circle includes 360°. One half-cycle, or one alternation, is 180° of revolution. A quarter-turn is 90°. The circle next to the loop positions in Fig. 16-3 illustrates the angular rotation of the dark conductor as it rotates counterclockwise from 0 to 90 to 180° for one half-cycle, then to 270° and returning to 360° to complete the cycle. Therefore, one cycle corresponds to 360°.

RADIAN MEASURE   In angular measure it is convenient to use a specific unit angle called the *radian* (abbreviated rad), which is an angle equal to 57.3°. Its convenience is due to the fact that a radian is the angular part of the circle that includes an arc equal to the radius r of the circle, as shown in Fig. 16-4. The circumference around the circle equals $2\pi r$. A circle includes 2π rad, then, as each radian angle includes one length r of the circumference. Therefore, one cycle equals 2π rad.

As shown in the graph in Fig. 16-3, divisions of the cycle can be indicated by angles in either degrees or radians. The comparison between degrees and radians can be summarized as follows:

Zero degrees is also zero radians
$360° = 2\pi$ rad
$180° = \frac{1}{2} \times 2\pi$ rad $= \pi$ rad
$90° = \frac{1}{2} \times \pi$ rad $= \pi/2$ rad
$270° = 180° + 90°$ or π rad $+ \pi/2$ rad $= 3\pi/2$ rad

The constant 2π in circular measure is numerically equal to 6.2832. This is double the value of 3.1416 for π. The Greek letter π (pi) is used to represent the ratio of the circumference to the diameter for any circle, which always has the numerical value of 3.1416. The fact that 2π rad is 360° can be shown as $2 \times 3.1416 \times 57.3° = 360°$ for a complete cycle.

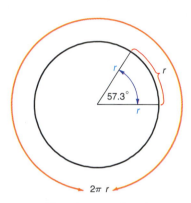

FIG. 16-4   One radian (rad) is the angle equal to 57.3°. The complete circle with 360° includes 2π rad.

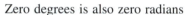

TEST-POINT QUESTION 16-2

Answers at end of chapter.

Refer to Fig. 16-3.
a. How much is the induced voltage at $\pi/2$ rad?
b. How many degrees are in a complete cycle?

16-3 THE SINE WAVE

The voltage waveform in Figs. 16-1 and 16-3 is called a *sine wave, sinusoidal wave,* or *sinusoid* because the amount of induced voltage is proportional to the sine of the angle of rotation in the circular motion producing the voltage. The sine is a trigonometric function* of an angle; it is equal to the ratio of the opposite side to the hypotenuse in a right triangle. This numerical ratio increases from zero for 0° to a maximum value of 1 for 90° as the side opposite the angle becomes larger.

The voltage waveform produced by the circular motion of the loop is a sine wave, because the induced voltage increases to a maximum at 90°, when the loop is vertical, in the same way that the sine of the angle of rotation increases to a maximum at 90°. The induced voltage and sine of the angle correspond for the full 360° of the cycle. Table 16-1 lists the numerical values of the sine for several important angles, to illustrate the specific characteristics of a sine wave.

TABLE 16-1   VALUES IN A SINE WAVE

ANGLE θ		SIN θ	LOOP VOLTAGE
DEGREES	RADIANS		
0	0	0	Zero
30	$\dfrac{\pi}{6}$	0.500	50% of maximum
45	$\dfrac{\pi}{4}$	0.707	70.7% of maximum
60	$\dfrac{\pi}{3}$	0.866	86.6% of maximum
90	$\dfrac{\pi}{2}$	1.000	Positive maximum value
180	π	0	Zero
270	$\dfrac{3\pi}{2}$	−1.000	Negative maximum value
360	2π	0	Zero

Notice that the sine wave reaches ½ its maximum value in 30°, which is only ⅓ of 90°. This fact means that the sine wave has a sharper slope of changing values when the wave is near the zero axis, compared with the more gradual changes near the maximum value.

The instantaneous value of a sine-wave voltage for any angle of rotation is expressed by the formula

▶ $v = V_M \sin \theta$ **(16-1)**

where θ (Greek letter *theta*) is the angle, sin is the abbreviation for its sine, V_M is the maximum voltage value, and v is the instantaneous value of voltage at angle θ.

---

*More details are given in B. Grob, *Mathematics for Basic Electronics*, Glencoe/McGraw-Hill, Columbus, Ohio.

Example

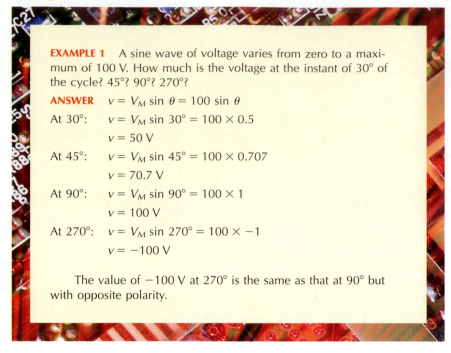

EXAMPLE 1 A sine wave of voltage varies from zero to a maximum of 100 V. How much is the voltage at the instant of 30° of the cycle? 45°? 90°? 270°?

ANSWER $v = V_M \sin \theta = 100 \sin \theta$

At 30°: $v = V_M \sin 30° = 100 \times 0.5$
            $v = 50$ V

At 45°: $v = V_M \sin 45° = 100 \times 0.707$
            $v = 70.7$ V

At 90°: $v = V_M \sin 90° = 100 \times 1$
            $v = 100$ V

At 270°: $v = V_M \sin 270° = 100 \times -1$
            $v = -100$ V

The value of -100 V at 270° is the same as that at 90° but with opposite polarity.

To do the problems in Example 1 you must either refer to a table of trigonometric functions or use a scientific calculator that has trig functions. With the calculator, be sure it is set for degrees, not radians or grad units. To find the value of the sine function, just punch in the number for angle θ in degrees and push the (SIN) key to see the values of sin θ on the display.

Applying this procedure to Formula (16-1), find the value of sin θ and multiply by the peak value V_M. Specifically, for the first problem in Example 1 with V_M of 100 and θ of 30°, first punch in 30 on the calculator. Next press the (SIN) key to see 0.5 on the display, which is sin 30°. Then push the multiplication (×) key, punch in 100 for V_M, and press the (=) key for the final answer of 50. The same method is used for all the other values of angle θ.

Between zero at 0° and maximum at 90° the amplitudes of a sine wave increase exactly as the sine value for the angle of rotation. These values are for the first quadrant in the circle, that is, 0° to 90°. From 90° to 180°, in the second quadrant, the values decrease as a mirror image of the first 90°. The values in the third and fourth quadrants, from 180° to 360°, are exactly the same as 0° to 180° but with opposite sign. At 360° the waveform is back to 0° to repeat its values every 360°.

In summary, the characteristics of the sine-wave ac waveform are:

1. The cycle includes 360° or 2π rad.
2. The polarity reverses each half-cycle.
3. The maximum values are at 90° and 270°.
4. The zero values are at 0° and 180°.
5. The waveform changes its values the fastest when it crosses the zero axis.
6. The waveform changes its values the slowest when it is at its maximum value. The values must stop increasing before they can decrease.

A perfect example of the sine-wave ac waveform is the 60-Hz power-line voltage in Fig. 16-1.

TEST-POINT QUESTION 16-3

Answers at end of chapter.

A sine-wave voltage has a peak value of 170 V. What is its value at
a. 30°?
b. 45°?
c. 90°?

16-4 ALTERNATING CURRENT

When a sine wave of alternating voltage is connected across a load resistance, the current that flows in the circuit is also a sine wave. In Fig. 16-5, let the sine-wave voltage at the left in the diagram be applied across R of 100 Ω. The resulting sine wave of alternating current is shown at the right in the diagram. Note that the frequency is the same for v and i.

During the first alternation of v in Fig. 16-5, terminal 1 is positive with respect to terminal 2. Since the direction of electron flow is from the negative side of v, through R, and back to the positive side of v, current flows in the direction indicated by arrow A for the first half-cycle. This direction is taken as the positive direction of current in the graph for i, corresponding to positive values of v.

The amount of current is equal to v/R. If several instantaneous values are taken, when v is zero, i is zero; when v is 50 V, i equals 50 V/100, or 0.5 A; when v is 100 V, i equals 100 V/100, or 1 A. For all values of applied voltage with positive polarity, therefore, the current is in one direction, increasing to its maximum value and decreasing to zero, just like the voltage.

On the next half-cycle, the polarity of the alternating voltage reverses. Then terminal 1 is negative with respect to terminal 2. With reversed voltage polarity, current flows in the opposite direction. Electron flow is from terminal 1 of the voltage source, which is now the negative side, through R, and back to terminal 2. This direction of current, as indicated by arrow B in Fig. 16-5, is negative.

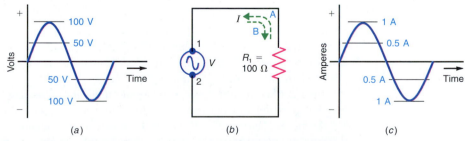

FIG. 16-5 A sine wave of alternating voltage applied across R produces a sine wave of alternating current in the circuit. (*a*) Waveform of applied voltage. (*b*) AC circuit. Note the symbol for sine-wave generator V. (*c*) Waveform of current in the circuit.

The negative values of i in the graph have the same numerical values as the positive values in the first half-cycle, corresponding to the reversed values of applied voltage. As a result, the alternating current in the circuit has sine-wave variations corresponding exactly to the sine-wave alternating voltage.

Only the waveforms for v and i can be compared. There is no comparison between relative values, because the current and voltage are different quantities.

It is important to note that the negative half-cycle of applied voltage is just as useful as the positive half-cycle in producing current. The only difference is that the reversed polarity of voltage produces the opposite direction of current.

Furthermore, the negative half-cycle of current is just as effective as the positive values when heating the filament to light a bulb. With positive values, electrons flow through the filament in one direction. Negative values produce electron flow in the opposite direction. In both cases, electrons flow from the negative side of the voltage source, through the filament, and return to the positive side of the source. For either direction, the current heats the filament. The direction does not matter, since it is just the motion of electrons against resistance that produces power dissipation. In short, resistance R has the same effect in reducing I for either direct current or alternating current.

TEST-POINT QUESTION 16-4

Answers at end of chapter.

Refer to Fig. 16-5.
a. When v is 70.7 V, how much is i?
b. How much is i at 30°?

16-5 VOLTAGE AND CURRENT VALUES FOR A SINE WAVE

Since an alternating sine wave of voltage or current has many instantaneous values through the cycle, it is convenient to define specific magnitudes for comparing one wave with another. The peak, average, and root-mean-square (rms) values can be specified, as indicated in Fig. 16-6. These values can be used for either current or voltage.

PEAK VALUE   This is the maximum value V_M or I_M. For example, specifying that a sine wave has a peak value of 170 V states the highest value the sine wave reaches. All other values during the cycle follow a sine wave. The peak value applies to either the positive or the negative peak.

In order to include both peak amplitudes, the *peak-to-peak* (p-p) *value* may be specified. For the same example, the peak-to-peak value is 340 V, double the peak value of 170 V, since the positive and negative peaks are symmetrical. It should be noted, though, that the two opposite peak values cannot occur at the same time. Furthermore, in some waveforms the two peaks are not equal.

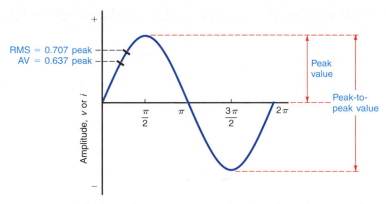

FIG. 16-6 Definitions of important amplitude values for a sine wave of voltage or current.

AVERAGE VALUE   This is an arithmetic average of all the values in a sine wave for one alternation, or half-cycle. The half-cycle is used for the average because over a full cycle the average value is zero, which is useless for comparison purposes. If the sine values for all angles up to 180°, for one alternation, are added and then divided by the number of values, this average equals 0.637. These calculations are shown in Table 16-2.

Since the peak value of the sine function is 1 and the average equals 0.637, then

▶   Average value = 0.637 × peak value   **(16-2)**

TABLE 16-2   DERIVATION OF AVERAGE AND RMS VALUES FOR A SINE-WAVE ALTERNATION

INTERVAL	ANGLE θ	SIN θ	$(\text{SIN } \theta)^2$
1	15°	0.26	0.07
2	30°	0.50	0.25
3	45°	0.71	0.50
4	60°	0.87	0.75
5	75°	0.97	0.93
6	90°	1.00	1.00
7*	105°	0.97	0.93
8	120°	0.87	0.75
9	135°	0.71	0.50
10	150°	0.50	0.25
11	165°	0.26	0.07
12	180°	0.00	0.00
	Total	7.62	6.00
		Average voltage: $\dfrac{7.62}{12} = 0.635$†	RMS value: $\sqrt{6/12} = \sqrt{0.5} = 0.707$

*For angles between 90 and 180°, sin θ = sin (180° − θ).
†More intervals and precise values are needed to get the exact average of 0.637.

With a peak of 170 V, for example, the average value is 0.637×170 V, which equals approximately 108 V.

ROOT-MEAN-SQUARE, OR EFFECTIVE, VALUE   The most common method of specifying the amount of a sine wave of voltage or current is by relating it to dc voltage and current that will produce the same heating effect. This is called its *root-mean-square* value, abbreviated rms. The formula is

> rms value = $0.707 \times$ peak value   **(16-3)**

or

> $V_{\text{rms}} = 0.707 V_{\text{max}}$ and $I_{\text{rms}} = 0.707 I_{\text{max}}$

With a peak of 170 V, for example, the rms value is 0.707×170, or 120 V, approximately. This is a voltage of the commercial ac power line, which is always given in rms value.

It is often necessary to convert from rms to peak value. This can be done by inverting Formula (16-3), as follows:

> Peak $= \dfrac{1}{0.707} \times$ rms $= 1.414 \times$ rms   **(16-4)**

or

> $V_{\text{max}} = 1.414 V_{\text{rms}}$ and $I_{\text{max}} = 1.414 I_{\text{rms}}$

Dividing by 0.707 is the same as multiplying by 1.414.

For example, the commercial power-line voltage with an rms value of 120 V has a peak value of 120×1.414, which equals 170 V, approximately. Its peak-to-peak value is 2×170, or 340 V, which is double the peak value. As a formula,

> Peak-to-peak value = $2.828 \times$ rms value   **(16-5)**

The factor 0.707 for rms value is derived as the square root of the average (mean) of all the squares of the sine values. If we take the sine for each angle in the cycle, square each value, add all the squares, divide by the number of values added to obtain the average square, and then take the square root of this mean value, the answer is 0.707. These calculations are shown in Table 16-2 for one alternation from 0° to 180°. The results are the same for the opposite alternation.

The advantage of the rms value derived in terms of the squares of the voltage or current values is that it provides a measure based on the ability of the sine wave to produce power, which is I^2R or V^2/R. As a result, the rms value of an alternating sine wave corresponds to the same amount of direct current or voltage in heating power. An alternating voltage with an rms value of 120 V, for instance, is just as effective in heating the filament of a light bulb as 120 V from a steady dc voltage source. For this reason, the rms value is also called the *effective* value.

Unless indicated otherwise, all sine-wave ac measurements are in rms values. The capital letters V and I are used, corresponding to the symbols for dc values. As an example, $V = 120$ V for the ac power-line voltage.

The ratio of the rms to average values is the *form factor*. For a sine wave, this ratio is 0.707/0.637 = 1.11.

Note that sine waves can have different amplitudes but still follow the sinusoidal waveform. Figure 16-7 compares a low-amplitude voltage with a high-amplitude voltage. Although different in amplitude, they are both sine waves. In each wave, the rms value = 0.707 × peak value.

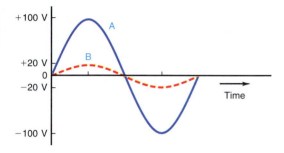

FIG. 16-7   Waveforms A and B have different amplitudes, but they are both sine waves.

TEST-POINT QUESTION 16-5

Answers at end of chapter.
a. Convert 170 V peak to rms value.
b. Convert 10 V rms to peak value.
c. Convert 1 V rms to peak-to-peak value.

16-6 FREQUENCY

The number of cycles per second is the *frequency,* with the symbol *f.* In Fig. 16-3, if the loop rotates through 60 complete revolutions, or cycles, during 1 s, the frequency of the generated voltage is 60 cps, or 60 Hz. You see only one cycle of the sine waveform, instead of 60 cycles, because the time interval shown here is ⅟₆₀ s. Note that the factor of time is involved. More cycles per second means a higher frequency and less time for one cycle, as illustrated in Fig. 16-8. Then the changes in values are faster for higher frequencies.

A complete cycle is measured between two successive points that have the same value and direction. In Fig. 16-8 the cycle is between successive points where the waveform is zero and ready to increase in the positive direction. Or the cycle can be measured between successive peaks.

On the time scale of 1 s, waveform *a* goes through one cycle; waveform *b* has much faster variations, with four complete cycles during 1 s. Both waveforms are sine waves, even though each has a different frequency.

In comparing sine waves, the amplitude has no relation to frequency. Two waveforms can have the same frequency with different amplitudes (Fig. 16-7), the same amplitude but different frequencies (Fig. 16-8), or different amplitudes and frequencies. The amplitude indicates how much the voltage or current is, while the frequency indicates the time rate of change of the amplitude variations, in cycles per second.

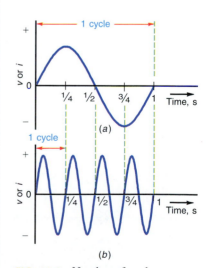

FIG. 16-8   Number of cycles per second is the frequency in hertz (Hz) units. (*a*) *f* = 1 Hz. (*b*) *f* = 4 Hz.

FREQUENCY UNITS   The unit called the *hertz* (Hz), named after H. Hertz, is used for cycles per second. Then 60 cps = 60 Hz. All the metric prefixes can be used. As examples:

$$1 \text{ kilocycle per second} = 1 \times 10^3 \text{ Hz} = 1 \text{ kHz}$$

$$1 \text{ megacycle per second} = 1 \times 10^6 \text{ Hz} = 1 \text{ MHz}$$

$$1 \text{ gigacycle per second} = 1 \times 10^9 \text{ Hz} = 1 \text{ GHz}$$

AUDIO AND RADIO FREQUENCIES   The entire frequency range of alternating voltage or current from 1 Hz to many megahertz can be considered in two broad groups: audio frequencies (AF) and radio frequencies (RF). *Audio* is a Latin word meaning "I hear." The audio range includes frequencies that can be heard in the form of sound waves by the human ear. This range of audible frequencies is approximately 16 to 16,000 Hz.

The higher the frequency, the higher the pitch or tone of the sound. High audio frequencies, about 3000 Hz and above, can be considered to provide *treble* tone. Low audio frequencies, about 300 Hz and below, provide *bass* tone.

Loudness is determined by amplitude. The greater the amplitude of the AF variation, the louder is its corresponding sound.

Alternating current and voltage above the audio range provide RF variations, since electrical variations of high frequency can be transmitted by electromagnetic radio waves. Examples of frequency allocations are given in Table 16-3.

SONIC AND ULTRASONIC FREQUENCIES   These terms refer to sound waves, which are variations in pressure generated by mechanical vibrations, rather than electrical variations. The velocity of transmission for sound waves equals 1130 ft/s, through dry air at 20°C. Sound waves above the audible range of frequencies are called *ultrasonic* waves. The range of frequencies for ultrasonic applications, therefore, is from 16,000 Hz up to several megahertz. Sound waves in the audible range of frequencies below 16,000 Hz can be considered *sonic* or sound frequencies, reserving *audio* for electrical variations that can be heard when converted to sound waves.

TEST-POINT QUESTION 16-6

Answers at end of chapter.
a. What is the frequency of the bottom waveform in Fig. 16-8?
b. Convert 1605 kHz to megahertz.

16-7 PERIOD

The amount of time it takes to go through one cycle is called the *period*. Its symbol is T for time. With a frequency of 60 Hz, as an example, the time for one cycle is $\frac{1}{60}$ s. Therefore, the period is $\frac{1}{60}$ s in this case. The frequency and period are reciprocals of each other:

TABLE 16-3 EXAMPLES OF COMMON FREQUENCIES

FREQUENCY	USE
60 Hz	AC power line
50–15,000 Hz	Audio equipment
535–1605 kHz*	AM radio band
54–60 MHz	TV channel 2
88–108 MHz	FM radio band

*Expanded to 1705 kHz in 1991.

$$T = \frac{1}{f} \quad \text{or} \quad f = \frac{1}{T} \tag{16-6}$$

The higher the frequency, the shorter the period. In Fig. 16-8a, the period for the wave, with a frequency of 1 Hz, is 1 s, while the higher-frequency wave of 4 Hz in Fig. 16-8b has the period of ¼ s for a complete cycle.

UNITS OF TIME   The second is the basic unit, but for higher frequencies and shorter periods, smaller units of time are convenient. Those used most often are:

$$T = 1 \text{ millisecond} = 1 \text{ ms} = 1 \times 10^{-3} \text{ s}$$
$$T = 1 \text{ microsecond} = 1 \ \mu\text{s} = 1 \times 10^{-6} \text{ s}$$
$$T = 1 \text{ nanosecond} = 1 \text{ ns} = 1 \times 10^{-9} \text{ s}$$

These units of time for period are reciprocals of the corresponding units for frequency. The reciprocal of frequency in kilohertz gives the period T in milliseconds; the reciprocal of megahertz is microseconds; the reciprocal of gigahertz is nanoseconds.

Example

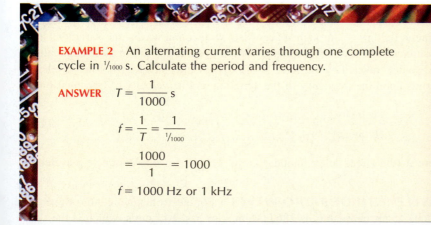

EXAMPLE 2   An alternating current varies through one complete cycle in $\frac{1}{1000}$ s. Calculate the period and frequency.

ANSWER $T = \dfrac{1}{1000} \text{ s}$

$f = \dfrac{1}{T} = \dfrac{1}{\frac{1}{1000}}$

$= \dfrac{1000}{1} = 1000$

$f = 1000 \text{ Hz or } 1 \text{ kHz}$

TEST-POINT QUESTION 16-7

Answers at end of chapter.
a. $T = \frac{1}{400}$ s. Calculate f.
b. $f = 400$ Hz. Calculate T.

16-8 WAVELENGTH

When a periodic variation is considered with respect to distance, one cycle includes the *wavelength,* which is the length of one complete wave or cycle (Fig. 16-9). For example, when a radio wave is transmitted, variations in the electromagnetic field travel through space. Also, with sound waves, the variations in air pressure corresponding to the sound wave move through air. In these applications, the distance traveled by the wave in one cycle is the wavelength. The wavelength depends upon the frequency of the variation and its velocity of transmission:

$$\lambda = \frac{\text{velocity}}{\text{frequency}} \qquad \text{(16-7)}$$

where λ (the Greek letter lambda) is the symbol for one complete wavelength.

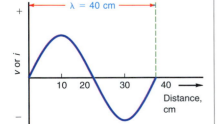

FIG. 16-9   Wavelength λ is the distance traveled by the wave in one cycle.

WAVELENGTH OF RADIO WAVES   For electromagnetic radio waves, the velocity in air or vacuum is 186,000 mi/s, or 3×10^{10} cm/s, which is the speed of light. Therefore,

$$\lambda \text{ (cm)} = \frac{3 \times 10^{10} \text{ cm/s}}{f(\text{Hz})} \qquad \text{(16-8)}$$

Note that the higher the frequency is, the shorter the wavelength. For instance, the short-wave radio broadcast band of 5.95 to 26.1 MHz includes higher frequencies than the standard radio broadcast band of 535 to 1605 kHz.

Example

EXAMPLE 4   Calculate λ for a radio wave with f of 30 GHz.

ANSWER $\lambda = \dfrac{3 \times 10^{10} \text{ cm/s}}{30 \times 10^9 \text{ Hz}} = \dfrac{3}{30} \times 10 \text{ cm}$

$\qquad\qquad = 0.1 \times 10$

$\qquad \lambda = 1 \text{ cm}$

Such short wavelengths are called *microwaves*. This range includes λ of 1 m or less, for frequencies of 300 MHz or more.

EXAMPLE 5   The length of a TV antenna is $\lambda/2$ for radio waves with f of 60 MHz. What is the antenna length in centimeters and feet?

ANSWER

a. $\lambda = \dfrac{3 \times 10^{10} \text{ cm/s}}{60 \times 10^6 \text{ Hz}} = \dfrac{1}{20} \times 10^4 \text{ cm}$

$\qquad = 0.05 \times 10^4$

$\quad \lambda = 500 \text{ cm}$

Then, $\lambda/2 = {}^{500}/_2 = 250$ cm.

b.   Since 2.54 cm = 1 in.,

$\quad \lambda/2 = \dfrac{250 \text{ cm}}{2.54 \text{ cm/in.}} = 98.4 \text{ in.}$

$\quad \lambda/2 = \dfrac{98.4 \text{ in.}}{12 \text{ in./ft}} = 8.2 \text{ ft}$

EXAMPLE 6   For the 6-m band used in amateur radio, what is the corresponding frequency?

ANSWER   The formula $\lambda = v/f$ can be inverted

$$f = \frac{v}{\lambda}$$

Then

$$f = \frac{3 \times 10 \text{ cm/s}}{6 \text{ m}} = \frac{3 \times 10^{10} \text{ cm/s}}{6 \times 10^2 \text{ cm}}$$

$$= \frac{3}{6} \times 10^8 = 0.5 \times 10^8 \text{ Hz}$$

$$f = 50 \times 10^6 \text{ Hz} \qquad \text{or} \qquad 50 \text{ MHz}$$

WAVELENGTH OF SOUND WAVES   The velocity of sound waves is much lower, compared with that of radio waves, because sound waves result from mechanical vibrations rather than electrical variations. For average conditions the velocity of sound waves in air equals 1130 ft/s. To calculate the wavelength, therefore,

▶ $$\lambda = \frac{1130 \text{ ft/s}}{f \text{ Hz}}$$ **(16-9)**

This formula can also be used for ultrasonic waves. Although their frequencies are too high to be audible, ultrasonic waves are still sound waves rather than radio waves.

Example

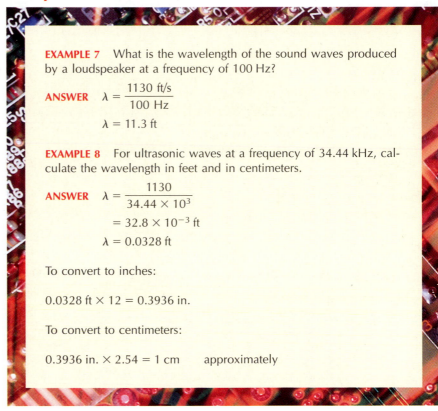

EXAMPLE 7   What is the wavelength of the sound waves produced by a loudspeaker at a frequency of 100 Hz?

ANSWER $\lambda = \dfrac{1130 \text{ ft/s}}{100 \text{ Hz}}$

$\lambda = 11.3 \text{ ft}$

EXAMPLE 8   For ultrasonic waves at a frequency of 34.44 kHz, calculate the wavelength in feet and in centimeters.

ANSWER $\lambda = \dfrac{1130}{34.44 \times 10^3}$

$= 32.8 \times 10^{-3} \text{ ft}$

$\lambda = 0.0328 \text{ ft}$

To convert to inches:

$0.0328 \text{ ft} \times 12 = 0.3936 \text{ in.}$

To convert to centimeters:

$0.3936 \text{ in.} \times 2.54 = 1 \text{ cm}$ approximately

Note that the 34.44 kHz sound waves in this example have the same wavelength (1 cm) as the 30 GHz radio waves in Example 4. The reason is that radio waves have a much higher velocity than sound waves.

TEST-POINT QUESTION 16-8

Answers at end of chapter.

Answer True or False.
a. The higher the frequency, the shorter the wavelength λ.
b. The higher the frequency, the longer the period T.
c. The velocity of propagation for radio waves in free space is 3×10^{10} cm/s.

16-9 PHASE ANGLE

Referring back to Fig. 16-3, suppose that the generator started its cycle at point B, where maximum voltage output is produced, instead of starting at the point of zero output. If we compare the two cases, the two output voltage waves would be as shown in Fig. 16-10. Each is the same waveform of alternating voltage, but wave B starts at maximum, while wave A starts at zero. The complete cycle of wave B through 360° takes it back to the maximum value from which it started. Wave A starts and finishes its cycle at zero. With respect to time, therefore, wave B is ahead of wave A in its values of generated voltage. The amount it leads in time equals one quarter-revolution, which is 90°. This angular difference is the phase angle between waves B and A. Wave B leads wave A by the phase angle of 90°.

The 90° phase angle between waves B and A is maintained throughout the complete cycle and in all successive cycles, as long as they both have the same frequency. At any instant of time, wave B has the value that A will have 90° later. For instance, at 180° wave A is at zero, but B is already at its negative maximum value, where wave A will be later at 270°.

In order to compare the phase angle between two waves, they must have the same frequency. Otherwise, the relative phase keeps changing. Also, they must have sine-wave variations, as this is the only kind of waveform that is measured in angular units of time. The amplitudes can be different for the two waves, although they are shown the same here. We can compare the phase of two voltages, two currents, or a current with a voltage.

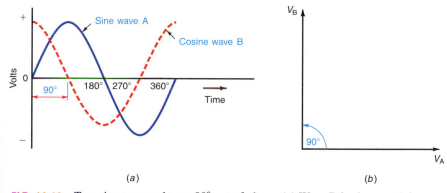

(a)   (b)

FIG. 16-10   Two sine-wave voltages 90° out of phase. (*a*) Wave B leads wave A by 90°. (*b*) Corresponding phasors V_B and V_A for the two sine-wave voltages with phase angle $\theta = 90°$. The right angle shows quadrature phase.

THE 90° PHASE ANGLE   The two waves in Fig. 16-10 represent a sine wave and a cosine wave 90° out of phase with each other. The 90° phase angle means that one has its maximum amplitude when the other is at zero value. Wave A starts at zero, corresponding to the sine of 0°, has its peak amplitude at 90 and 270°, and is back to zero after one cycle of 360°. Wave B starts at its peak value, corresponding to the cosine of 0°, has its zero value at 90 and 270°, and is back to the peak value after one cycle of 360°.

However, wave B can also be considered a sine wave that starts 90° before wave A in time. This phase angle of 90° for current and voltage waveforms has many applications in sine-wave ac circuits with inductance or capacitance.

The sine and cosine waveforms really have the same variations, but displaced by 90°. In fact, both waveforms are called *sinusoids*. The 90° angle is called *quadrature phase*.

PHASE-ANGLE DIAGRAMS   To compare phases of alternating currents and voltages, it is much more convenient to use phasor diagrams corresponding to the voltage and current waveforms, as shown in Fig. 16-10b. The arrows here represent the phasor quantities corresponding to the generator voltage.

A phasor is a quantity that has magnitude and direction. The length of the arrow indicates the magnitude of the alternating voltage, in rms, peak, or any ac value as long as the same measure is used for all the phasors. The angle of the arrow with respect to the horizontal axis indicates the phase angle.

The terms *phasor* and *vector* are used for a quantity that has direction, requiring an angle to specify the value completely. However, a vector quantity has direction in space, while a phasor quantity varies in time. As an example of a vector, a mechanical force can be represented by a vector arrow at a specific angle, with respect to either the horizontal or vertical direction.

For phasor arrows, the angles shown represent differences in time. One sinusoid is chosen as the reference. Then the timing of the variations in another sinusoid can be compared to the reference by means of the angle between the phasor arrows.

The phasor corresponds to the entire cycle of voltage, but is shown only at one angle, such as the starting point, since the complete cycle is known to be a sine wave. Without the extra details of a whole cycle, phasors represent the alternating voltage or current in a compact form that is easier for comparing phase angles.

In Fig. 16-10b, for instance, the phasor V_A represents the voltage wave A, with a phase angle of 0°. This angle can be considered as the plane of the loop in the rotary generator where it starts with zero output voltage. The phasor V_B is vertical to show the phase angle of 90° for this voltage wave, corresponding to the vertical generator loop at the start of its cycle. The angle between the two phasors is the phase angle.

The symbol for a phase angle is θ (the Greek letter theta). In Fig. 16-10, as an example, $\theta = 90°$.

PHASE-ANGLE REFERENCE   The phase angle of one wave can be specified only with respect to another as reference. How the phasors are drawn to show the phase angle depends on which phase is chosen as the reference. Generally, the reference phasor is horizontal, corresponding to 0°. Two possibilities are shown in Fig. 16-11. In Fig. 16-11a the voltage wave A or its phasor V_A is the reference. Then the phasor V_B is 90° counterclockwise. This method is standard practice, using counterclockwise rotation as the positive direction for angles. Also, a leading angle is positive. In this case, then, V_B is 90° counterclockwise from the reference V_A to show that wave B leads wave A by 90°.

However, wave B is shown as the reference in Fig. 16-11b. Now V_B is the horizontal phasor. In order to have the same phase angle, V_A must be 90° clockwise, or −90° from V_B. This arrangement shows that negative angles, clockwise from the 0° reference, are used to show a lagging phase angle. The reference determines whether the phase angle is considered leading or lagging in time.

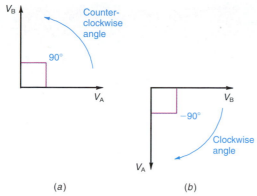

(a)   (b)

FIG. 16-11 Leading and lagging phase angles for 90°. (a) When phasor V_A is the horizontal reference, phasor V_B leads by 90°. (b) When phasor V_B is the horizontal reference, phasor V_A lags by −90°.

The phase is not actually changed by the method of showing it. In Fig. 16-11, V_A and V_B are 90° out of phase, and V_B leads V_A by 90° in time. There is no fundamental difference whether we say V_B is ahead of V_A by +90° or V_A is behind V_B by −90°.

Two waves and their corresponding phasors can be out of phase by any angle, either less or more than 90°. For instance, a phase angle of 60° is shown in Fig. 16-12. For the waveforms in Fig. 16-12a, wave D is behind C by 60° in time. For the phasors in Fig. 16-12b this lag is shown by the phase angle of −60°.

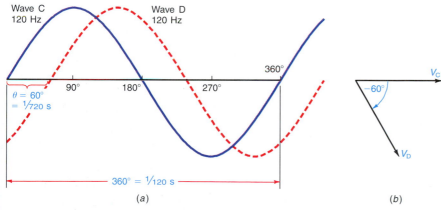

(a)

FIG. 16-12 Phase angle of 60° is the time for 60/360 or 1/6 of the cycle. (a) Waveforms. (b) Phasor diagram.

IN-PHASE WAVEFORMS   A phase angle of 0° means the two waves are in phase (Fig. 16-13).

OUT-OF-PHASE WAVEFORMS   An angle of 180° means opposite phase, or the two waveforms are exactly out of phase (Fig. 16-14). Then the amplitudes are opposing.

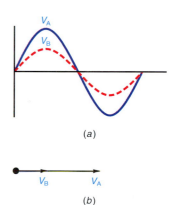

(a)

(b)

FIG. 16-13 Two waveforms in phase, or the phase angle is 0°. (a) Waveforms. (b) Phasor diagram.

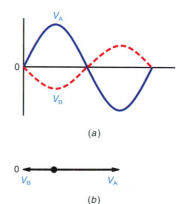

(a)

(b)

FIG. 16-14 Two waveforms out of phase or in opposite phase, with phase angle of 180°. (a) Waveforms. (b) Phasor diagram.

Answers at end of chapter.

Give the phase angle in
a. Fig. 16-10.
b. Fig. 16-12.
c. Fig. 16-13.

16-10 THE TIME FACTOR IN FREQUENCY AND PHASE

It is important to remember that the waveforms we are showing are just graphs drawn on paper. The physical factors represented are variations in amplitude, usually on the vertical scale, with respect to equal intervals on the horizontal scale, which can represent either distance or time. To show wavelength, as in Fig. 16-9, the cycles of amplitude variations are plotted against distance or length units. To show frequency, the cycles of amplitude variations are shown with respect to time in angular measure. The angle of 360° represents the time for one cycle, or the period T.

As an example of how frequency involves time, a waveform with stable frequency is actually used in electronic equipment as a clock reference for very small units of time. Assume a voltage waveform with the frequency of 10 MHz. The period T is 0.1 μs. Every cycle is repeated at 0.1-μs intervals, therefore. When each cycle of voltage variations is used to indicate time, then, the result is effectively a clock that measures 0.1-μs units. Even smaller units of time can be measured with higher frequencies. In everyday applications, an electric clock connected to the power line keeps correct time because it is controlled by the exact frequency of 60 Hz.

Furthermore, the phase angle between two waves of the same frequency indicates a specific difference in time. As an example, Fig. 16-12 shows a phase angle of 60°, with wave C leading wave D. They both have the same frequency of 120 Hz. The period T for each wave then is $\frac{1}{120}$ s. Since 60° is one-sixth of the complete cycle of 360°, this phase angle represents one-sixth of the complete period of $\frac{1}{120}$ s. Multiplying $\frac{1}{6} \times \frac{1}{120}$, the answer is $\frac{1}{720}$ s for the time corresponding to the phase angle of 60°. If we consider wave D lagging wave C by 60°, this lag is a time delay of $\frac{1}{720}$ s.

More generally, the time for a phase angle θ can be calculated as

$$t = \frac{\theta}{360} \times \frac{1}{f} \tag{16-10}$$

where f is in Hz, θ is in degrees, and t is in seconds.

The formula gives the time of the phase angle as its proportional part of the total period of one cycle. For the example of θ equal to 60° with f at 120 Hz,

$$t = \frac{\theta}{360} \times \frac{1}{f}$$

$$= \frac{60}{360} \times \frac{1}{120} = \frac{1}{6} \times \frac{1}{120}$$

$$t = \frac{1}{720} \text{ s}$$

TEST-POINT QUESTION 16-10

Answers at end of chapter.

a. In Fig. 16-12, how much time corresponds to 180°?

b. For two waves with the frequency of 1 MHz, how much time is the phase angle of 36°?

16-11 ALTERNATING CURRENT CIRCUITS WITH RESISTANCE

An ac circuit has an ac voltage source. Note the symbol in Fig. 16-15 used for any source of sine-wave alternating voltage. This voltage connected across an external load resistance produces alternating current of the same waveform, frequency, and phase as the applied voltage.

The amount of current equals V/R by Ohm's law. When V is an rms value, I is also an rms value. For any instantaneous value of V during the cycle, the value of I is for the corresponding instant of time.

In an ac circuit with only resistance, the current variations are in phase with the applied voltage, as shown in Fig. 16-15b. This in-phase relationship between V and I means that such an ac circuit can be analyzed by the same methods used for dc circuits, since there is no phase angle to consider. Circuit components that have R alone include resistors, the filaments of light bulbs, and heating elements.

DID YOU KNOW?

Automatic car airbags work on the closing of a simple electric circuit using magnets and metal balls.

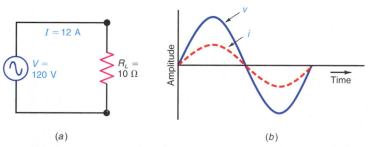

(a)   (b)

FIG. 16-15   An ac circuit with resistance R alone. (a) Schematic diagram. (b) Waveforms.

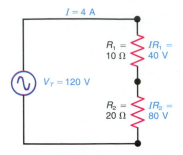

FIG. 16-16 Series ac circuit with resistance only.

The calculations in ac circuits are generally in rms values, unless noted otherwise. In Fig. 16-15a, for example, the 120 V applied across the 10-Ω R_L produces rms current of 12 A. The calculations are

$$I = \frac{V}{R_L} = \frac{120\ V}{10\ \Omega} = 12\ A$$

Furthermore, the rms power dissipation is I^2R, or

$$P = 144 \times 10 = 1440\ W$$

SERIES AC CIRCUIT WITH R   In Fig. 16-16, R_T is 30 Ω, equal to the sum of 10 Ω for R_1 plus 20 Ω for R_2. The current in the series circuit is

$$I = \frac{V}{R_T} = \frac{120\ V}{30\ \Omega} = 4\ A$$

The 4-A current is the same in all parts of the series circuit. This principle applies for either an ac or a dc source.

Next, we can calculate the series voltage drops in Fig. 16-16. With 4 A through the 10-Ω R_1, its IR voltage drop is

$$V_1 = I \times R_1 = 4\ A \times 10\ \Omega = 40\ V$$

The same 4 A through the 20-Ω R_2 produces an IR voltage drop of 80 V. The calculations are

$$V_2 = I \times R_2 = 4\ A \times 20\ \Omega = 80\ V$$

Note that the sum of 40 V for V_1 and 80 V for V_2 in series equals the 120 V applied.

PARALLEL AC CIRCUIT WITH R   In Fig. 16-17, the 10-Ω R_1 and 20-Ω R_2 are in parallel across the 120-V ac source. Therefore, the voltage across the parallel branches is the same as the applied voltage.

Each branch current, then, is equal to 120 V divided by the branch resistance. The branch current for the 10-Ω R_1 is

$$I_1 = \frac{120\ V}{10\ \Omega} = 12\ A$$

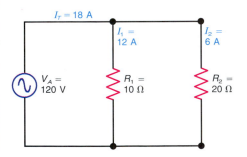

FIG. 16-17   Parallel ac circuit with resistance only.

The same 120 V is across the 20-Ω branch with R_2. Its branch current is

$$\blacktriangleright \quad I_2 = \frac{120 \text{ V}}{20 \text{ }\Omega} = 6 \text{ A}$$

The total line current I_T is $12 + 6 = 18$ A, or the sum of the branch currents.

SERIES-PARALLEL AC CIRCUIT WITH R   See Fig. 16-18. The 20-Ω R_2 and 20-Ω R_3 are in parallel, for an equivalent bank resistance of 20/2 or 10 Ω. This 10-Ω bank is in series with the 20-Ω R_1 in the main line, for a total of 30 Ω for R_T across the 120-V source. Therefore, the main line current produced by the 120-V source is

$$I_T = \frac{V}{R_T} = \frac{120 \text{ V}}{30 \text{ }\Omega} = 4 \text{ A}$$

The voltage drop across R_1 in the main line is calculated as

$$V_1 = I_T \times R_1 = 4 \text{ A} \times 20 \text{ }\Omega = 80 \text{ V}$$

Subtracting this 80-V drop from the 120 V of the source, the remaining 40 V is across the bank of R_2 and R_3 in parallel. Since the branch resistances are equal, the 4-A I_T divides equally, with 2 A in R_2 and 2 A in R_3. The branch currents can be calculated as

$$I_2 = \frac{40 \text{ V}}{20 \text{ }\Omega} = 2 \text{ A}$$

$$I_3 = \frac{40 \text{ V}}{20 \text{ }\Omega} = 2 \text{ A}$$

Note that the 2 A for I_2 and 2 A for I_3 in parallel branches add to equal the 4-A current in the main line.

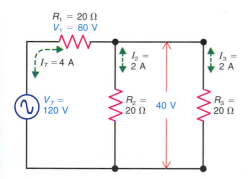

FIG. 16-18   Series-parallel ac circuit with resistance only.

TEST-POINT QUESTION 16-11

Answers at end of chapter.

Calculate R_T in
a. Fig. 16-16.
b. Fig. 16-17.
c. Fig. 16-18.

16-12 NONSINUSOIDAL AC WAVEFORMS

The sine wave is the basic waveform for ac variations for several reasons. This waveform is produced by a rotary generator, as the output is proportional to the angle of rotation. In addition, electronic oscillator circuits with inductance and capacitance naturally produce sine-wave variations.

Because of its derivation from circular motion, any sine wave can be analyzed in terms of angular measure, either in degrees from 0 to 360° or in radians from 0 to 2π rad.

Another feature of a sine wave is its basic simplicity, as the rate of change for the amplitude variations corresponds to a cosine wave which is similar but 90° out of phase. The sine wave is the only waveform that has this characteristic of a rate of change with the same waveform as the original changes in amplitude.

In many electronic applications, however, other waveshapes are important. Any waveform that is not a sine or cosine wave is a *nonsinusoidal waveform.* Common examples are the square wave and sawtooth wave in Fig. 16-19.

With nonsinusoidal waveforms, for either voltage or current, there are important differences and similarities to consider. Note the following comparisons with sine waves.

1. In all cases, the cycle is measured between two points having the same amplitude and varying in the same direction. The period is the time for one cycle. In Fig. 16-19, T for any of the waveforms is 4 μs and the corresponding frequency is $1/T$, equal to ¼ MHz, or 0.25 MHz.
2. Peak amplitude is measured from the zero axis to the maximum positive or negative value. However, peak-to-peak amplitude is better for measuring nonsinusoidal waveshapes because they can have unsymmetrical peaks, as in Fig. 16-19*d.* For all the waveforms shown here, though, the peak-to-peak (p–p) amplitude is 20 V.
3. The rms value 0.707 of maximum applies only to sine waves, as this factor is derived from the sine values in the angular measure used only for the sine waveform.
4. Phase angles apply only to sine waves, as angular measure is used only for sine waves. Note that the horizontal axis for time is divided into angles for the sine wave in Fig. 16-19*a,* but there are no angles shown for the nonsinusoidal waveshapes.
5. All the waveforms represent ac voltages. Positive values are shown above the zero axis, with negative values below the axis.

The sawtooth wave in Fig. 16-19*b* represents a voltage that slowly increases, with a uniform or linear rate of change, to its peak value, and then drops sharply to its starting value. This waveform is also called a *ramp voltage.* It is also often referred to as a *time base* because of its constant rate of change.

Note that one complete cycle includes the slow rise and the fast drop in voltage. In this example, the period T for a complete cycle is 4 μs. Therefore, these sawtooth cycles are repeated at the frequency of ¼ MHz, which equals 0.25 MHz. The sawtooth waveform of voltage or current is often used for horizontal deflection of the electron beam in the cathode-ray tube (CRT) for oscilloscopes and TV receivers.

The square wave in Fig. 16-19*c* represents a switching voltage. First, the 10-V peak is instantaneously applied in positive polarity. This voltage remains

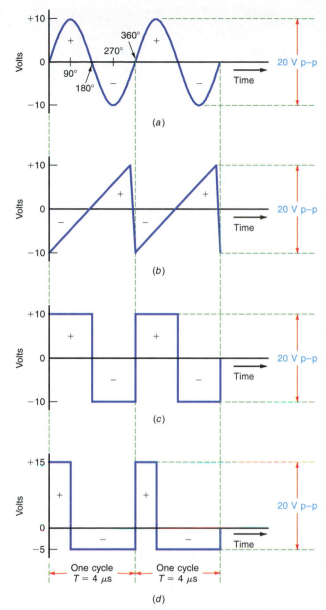

FIG. 16-19 Comparison of sine wave with nonsinusoidal waveforms. Two cycles shown. (*a*) Sine wave. (*b*) Sawtooth wave. (*c*) Symmetrical square wave. (*d*) Unsymmetrical rectangular wave or pulse waveform

on for 2 μs, which is one half-cycle. Then the voltage is instantaneously reduced to zero and applied in reverse polarity for another 2 μs. The complete cycle then takes 4 μs, and the frequency is ¼ MHz.

The rectangular waveshape in Fig. 16-19*d* is similar, but the positive and negative half-cycles are not symmetrical, either in amplitude or in time. However, the frequency is the same 0.25 MHz and the peak-to-peak amplitude is the same 20 V, as in all the waveshapes. This waveform shows pulses of voltage or current, repeated at a regular rate.

Answers at end of chapter.
a. In Fig. 16-19c, for how much time is the waveform at +10 V?
b. In Fig. 16-19d, what voltage is the positive peak amplitude?

16-13 HARMONIC FREQUENCIES

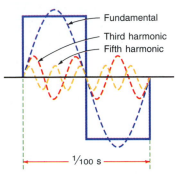

Fundamental
Third harmonic
Fifth harmonic

¹⁄₁₀₀ s

FIG. 16-20 Fundamental and harmonic frequencies for an example of a 100-Hz square wave.

Consider a repetitive nonsinusoidal waveform, such as a 100-Hz square wave. Its fundamental rate of repetition is 100 Hz. Exact multiples of the fundamental frequency are called *harmonic frequencies*. The second harmonic is 200 Hz, the third harmonic is 300 Hz, etc. Even multiples are even harmonics, while odd multiples are odd harmonics.

Harmonics are useful in analyzing distorted sine waves or nonsinusoidal waveforms. Such waveforms consist of a pure sine wave at the fundamental frequency plus harmonic frequency components. For example, Fig. 16-20 illustrates how a square wave corresponds to a fundamental sine wave with odd harmonics. Typical audio waveforms include odd and even harmonics. It is the harmonic components that make one source of sound different from another with the same fundamental frequency.

A common unit for frequency multiples is the *octave*, which is a range of 2:1. Doubling the frequency range—from 100 to 200 Hz, from 200 to 400 Hz, and from 400 to 800 Hz, as examples—raises the frequency by one octave. The reason for this name is that an octave in music includes eight consecutive tones, for double the frequency. One-half the frequency is an octave lower.

Another unit for representing frequency multiples is the decade. A decade corresponds to a 10:1 range in frequencies such as 100 Hz to 1 kHz and 30 kHz to 300 kHz.

Answers at end of chapter.
a. What frequency is the fourth harmonic of 12 MHz?
b. Give the frequency one octave above 220 Hz.

16-14 THE 60-HZ AC POWER LINE

Practically all homes in the United States are supplied alternating voltage between 115 and 125 V rms, at a frequency of 60 Hz. This is a sine-wave voltage produced by a rotary generator. The electricity is distributed by high voltage power lines from the generating station and reduced to the lower voltages used in the home. Here the incoming voltage is wired to all the wall outlets and electrical equipment in parallel. The 120-V source of commercial electricity is the *60-Hz power line* or the *mains,* indicating it is the main line for all the parallel branches.

ADVANTAGES The incoming electric service to residences is normally given as 120 V rms. With an rms value of 120 V, the ac power is equivalent to 120-V dc power in heating effect. If the value were higher, there would be more danger of a fatal electric shock. Lower voltages would be less efficient in supplying power.

Higher voltage can supply electric power with less I^2R loss, since the same power is produced with less I. Note that the I^2R power loss increases as the square of the current. For applications where large amounts of power are used such as central air-conditioners and clothes dryers, a line voltage of 240 V is often used.

The advantage of ac over dc power is greater efficiency in distribution from the generating station. Alternating voltages can easily be stepped up by means of a transformer, with very little loss, but a transformer cannot operate on direct current. The reason is that a transformer needs the varying magnetic field produced by an ac voltage.

Using a transformer, the alternating voltage at the generating station can be stepped up to values as high as 500 kV for high-voltage distribution lines. These high-voltage lines supply large amounts of power with much less current and less I^2R loss, compared with a 120-V line. At the home, the lower voltage required is supplied by a step-down transformer. The step-up and step-down characteristics of a transformer refer to the ratio of voltages across the input and output connections.

The frequency of 60 Hz is convenient for commercial ac power. Much lower frequencies would require much bigger transformers because larger windings would be necessary. Also, too low a frequency for alternating current in a lamp could cause the light to flicker. For the opposite case, too high a frequency results in excessive iron-core heating in the transformer because of eddy currents and hysteresis losses. Based on these factors, 60 Hz is the frequency of the ac power line in the United States. However, the frequency of the ac power mains in England and most European countries is 50 Hz.

THE 60-HZ FREQUENCY REFERENCE All power companies in the United States, except those in Texas, are interconnected in a grid that maintains the ac power-line frequency between 59.98 and 60.02 Hz. The frequency is compared with the time standard provided by the Bureau of Standards radio station WWV at Fort Collins, Colorado. As a result the 60-Hz power-line frequency is maintained accurate to ±0.033 percent. This accuracy makes the power-line voltage a good secondary standard for checking frequencies based on 60 Hz.

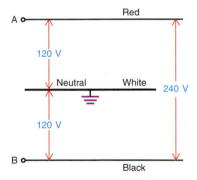

FIG. 16-21 Three-wire, single-phase power lines that can provide either 240 or 120 V.

RESIDENTIAL WIRING Most homes have at the electrical service entrance the three-wire power lines illustrated in Fig. 16-21. The three wires, including the grounded neutral, can be used for either 240 or 120 V single phase. The 240 V at the residence is stepped down from the high-voltage distribution lines.

Note the color coding for the wiring in Fig. 16-21. The grounded neutral is white, or bare wire is used. Each high side can use any color except white or green, but usually they use black* or red. White is reserved for the neutral wire, and green is reserved for grounding.

---

*It should be noted that in electronic equipment black is the color-coded wiring used for chassis-ground returns. However, in electric power work, black wire is used for high-side connections.

From either the red or black high side to the neutral, 120 V is available for separate branch circuits to the lights and outlets. Across the red and black wires, 240 V is available for high-power appliances. This three-wire service with a grounded neutral is called the *Edison system*.

The electrical service is commonly rated for 100 A. At 240 V, then, the power available is $100 \times 240 = 24{,}000$ W, or 24 kW.

The main wires to the service entrance, where the power enters the house, are generally No. 4 to 8 gage. Sizes 6 and heavier are always stranded wire. The 120-V branch circuits, usually rated at 15 A or 20 A, use No. 12 or 14 gage wire. Each branch has its own fuse or circuit breaker. A main switch is usually included to cut off all power from the service entrance.

The neutral wire is grounded at the service entrance to a water pipe or a metal rod driven into the earth, which is *ground*. All 120-V branches must have one side connected to the grounded neutral. White wire is used for these connections. In addition, all the metal boxes for outlets, switches, and lights must have a continuous ground to each other and to the neutral. The wire cable usually has a bare wire for this grounding of the boxes.

Cables commonly used are armored sheath with the trade name BX and non-metallic flexible cable with the trade name Romex. Each has two or more wires for the neutral, high-side connections, and grounding. Both cables contain an extra bare wire for grounding. Rules and regulations for residential wiring are governed by local electrical codes. These are usually based on the National Electrical Code published by the National Fire Protection Association.

GROUNDING   In ac power distribution systems, grounding is the practice of connecting one side of the power line to earth or ground. The purpose is safety, in two ways. First is protection against dangerous electric shock. Also, the power distribution lines are protected against excessive high voltage, particularly from lightning. If the system is struck by lightning, excessive current in the grounding system will energize a cutout device to deenergize the lines.

The grounding in the power distribution system means that it is especially important to have grounding for the electric wiring at the residence. For instance, suppose that an electric appliance such as a clothes dryer does not have its metal case grounded. An accidental short circuit in the equipment can connect the metal frame to the "hot" side of the ac power line. Then the frame has voltage with respect to earth ground. If somebody touches the frame and has a return to ground, the result is a dangerous electric shock. With the case grounded, however, the accidental short circuit blows the fuse or circuit breaker to cut off the power.

In normal operation, the electric circuits function the same way with or without the ground, but the grounding is an important safety precaution. Figure 16-22 shows two types of plug connectors for the ac power line that help in providing protection because they are polarized with respect to the ground connections. Although an ac voltage does not have any fixed polarity, the plugs ensure grounding of the chassis or frame of equipment connected to the power line. In Fig. 16-22a, the plug has two blades for the 120-V line but the wider blade will fit only the side of the outlet that is connected to the neutral wire. This wiring is standard practice. For the three-prong plug in Fig. 16-22b, the rounded pin is for a separate grounding wire, usually color coded green.

In some cases, there may be leakage of current from the "hot" side of the power line to ground. A leakage current of 5 mA or more is considered danger-

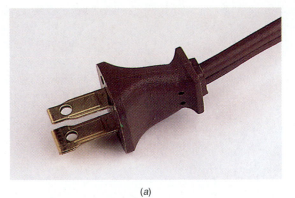 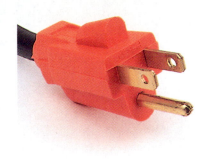

(a)   (b)

FIG. 16-22   Plug connectors polarized for ground connection to an ac power line.
(*a*) Wider blade connects to neutral. (*b*) Rounded pin connects to ground.

ous. The ground-fault circuit interrupter (GCFI) shown in Fig. 16-23 is a device
that can sense excessive leakage current and open the circuit as a protection
against shock hazard.

It may be of interest to note that with high-fidelity audio equipment, the lack
of proper grounding can cause a hum to be heard in the sound. The hum is usually not any safety problem but it still is undesirable.

TEST-POINT QUESTION 16-14

Answers at end of chapter.

Answer True or False.
a. The 120 V of the ac power line is a peak-to-peak value.
b. The frequency of the ac power-line voltage is 60 Hz $\pm$ 0.033 percent.
c. In Fig. 16-21 the voltage between black and white wires is 120 V.
d. The color code for grounding wires is green.

FIG. 16-23   Ground-fault circuit
interrupter (GFCI).

16-15 MOTORS AND GENERATORS

A generator converts mechanical energy into electric energy; a motor does the
opposite, converting electricity into rotary motion. The main parts in the assembly of motors and generators are essentially the same (Fig. 16-24).

ARMATURE   In a generator, the armature connects to the external circuit to
provide the generator output voltage. In a motor, it connects to the electrical
source that drives the motor. The armature is often constructed in the form of a
drum, using many conductor loops for increased output. In Fig. 16-24 the rotating armature is the *rotor* part of the assembly.

FIG. 16-24　Main parts of a dc motor.

FIELD WINDING　This electromagnet provides the flux cut by the rotor. In a motor, current for the field is produced by the same source that supplies the armature. In a generator, the field current may be obtained from a separate exciter source, or from its own armature output. Residual magnetism in the iron yoke of the field allows this *self-excited generator* to start.

The field coil may be connected in series with the armature, in parallel, or in a series-parallel *compound winding*. When the field winding is stationary, it is the *stator* part of the assembly.

SLIP RINGS　In an ac machine, two or more slip rings or *collector rings* enable the rotating loop to be connected to the stationary wire leads for the external circuit.

BRUSHES　These graphite connectors are spring-mounted to brush against the spinning rings on the rotor. The stationary external leads are connected to the brushes for connection to the rotating loop. Constant rubbing slowly wears down the brushes, and they must be replaced after they are worn.

COMMUTATOR　A dc machine has a commutator ring instead of the slip rings. As shown in Fig. 16-24, the commutator ring has segments, with one pair for each loop in the armature. Each of the commutator segments is insulated from the others by mica.

The commutator converts the ac machine to dc operation. In a generator, the commutator segments reverse the loop connections to the brushes every half-cycle to maintain a constant polarity of output voltage. For a dc motor, the commutator segments allow the dc source to produce torque in one direction.

Brushes are necessary with a commutator ring. The two stationary brushes contact opposite segments on the rotating commutator. Graphite brushes are used for very low resistance.

ALTERNATING CURRENT INDUCTION MOTOR   This type, for alternating current only, does not have any brushes. The stator is connected directly to the ac source. Then alternating current in the stator winding induces current in the rotor without any physical connection between them. The magnetic field of the current induced in the rotor reacts with the stator field to produce rotation. Alternating-current induction motors are economical and rugged, without any troublesome brush arcing.

With a single-phase source, however, a starting torque must be provided for an ac induction motor. One method uses a starting capacitor in series with a separate starting coil. The capacitor supplies an out-of-phase current just for starting, and then is switched out. Another method of starting uses shaded poles. A solid copper ring on the main field pole makes the magnetic field unsymmetrical to allow starting.

The rotor of an ac induction motor may be wire-wound or the squirrel-cage type. This rotor is constructed with a frame of metal bars.

UNIVERSAL MOTOR   This type operates on either alternating or direct current because the field and armature are in series. Its construction is like that of a dc motor, with the rotating armature connected to a commutator and brushes. The universal motor is commonly used for small machines such as portable drills and food mixers.

ALTERNATORS   Alternating current generators are alternators. For large power requirements, the alternator usually has a rotating field, while the armature is the stator.

TEST-POINT QUESTION 16-15

Answers at end of chapter.

Answer True or False.
a. In Fig. 16-24 the commutator segments are on the armature.
b. Motor brushes are made of graphite because of its very low resistance.
c. A starting capacitor is used with dc motors that have small brushes.

16-16 THREE-PHASE AC POWER

In an alternator with three generator windings equally spaced around the circle, the windings will produce output voltages 120° out of phase with each other. The three-phase output is illustrated by the sine-wave voltages in Fig. 16-25a and the corresponding phasors in Fig. 16-25b. The advantage of three-phase ac voltage is more efficient distribution of power. Also, ac induction motors are self-start-

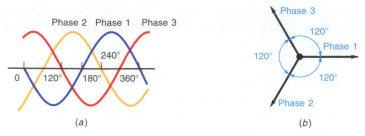

(a)   (b)

FIG. 16-25 Three-phase alternating voltage or current with 120° between each phase. (*a*) Sine waves. (*b*) Phasor diagram.

ing with three-phase alternating current. Finally, the ac ripple is easier to filter in the rectified output of a dc power supply.

In Fig. 16-26*a*, the three windings are in the form of a Y, also called *Wye* or *star* connections. All three coils are joined at one end, with the opposite ends for the output terminals A, B, and C. Note that any pair of terminals is across two coils in series. Each coil has 120 V. The voltage output across any two output terminals is 120 × 1.73 = 208 V, because of the 120° phase angle.

In Fig. 16-26*b*, the three windings are connected in the form of a *delta* (Δ). Any pair of terminals is across one generator winding. The output then is 120 V. However, the other coils are in a parallel branch. Therefore, the current capacity to the line is increased by the factor 1.73.

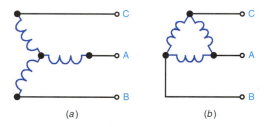

(a)   (b)

FIG. 16-26 Types of connections for three-phase power. (*a*) Wye or Y. (*b*) Delta or Δ.

In Fig. 16-27, the center point of the Y is used for a fourth line, as the neutral wire in the three-phase power distribution system. This way, power is avail-

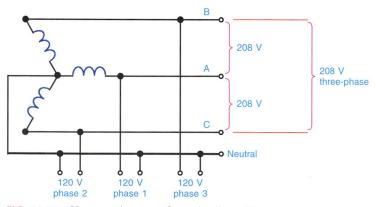

FIG. 16-27 Y connections to a four-wire line with neutral.

able either at 208 V three phase or 120 V single phase. Note that the three-phase voltage is 208 V, not the 240 V in the Edison single-phase system. From terminal A, B, or C to the neutral line in Fig. 16-27, the output is 120 V across one coil. This 120 V single-phase power is used in conventional lighting circuits. However, across terminals AB, BC, or CA, without the neutral, the output is 208 V for three-phase induction motors or other circuits that need three-phase power. Although illustrated here for the 120-V 60-Hz power line, it should be noted that three-phase connections are commonly used for higher voltages.

TEST-POINT QUESTION 16-16

Answers at end of chapter.
a. What is the angle between three-phase voltages?
b. For the Y in Fig. 16-26a how much is V_{AC} or V_{AB}?

16   SUMMARY AND REVIEW

- Alternating voltage continuously varies in magnitude and reverses in polarity. When alternating voltage is applied across a load resistance, the result is alternating current in the circuit.
- A complete set of values repeated periodically is one cycle of the ac waveform. The cycle can be measured from any one point on the wave to the next successive point having the same value and varying in the same direction. One cycle includes 360° in angular measure, or 2π rad.
- The rms value of a sine wave is 0.707 × peak value.
- The peak amplitude, at 90° and 270° in the cycle, is 1.414 × rms value.
- The peak-to-peak value is double the peak amplitude, or 2.828 × rms for a symmetrical ac waveform.
- The average value is 0.637 × peak value.
- The frequency equals the number of cycles per second. One cps is 1 Hz. The audio-frequency (AF) range is 16 to 16,000 Hz. Higher frequencies up to 300,000 MHz are radio frequencies (RF).
- The amount of time for one cycle is the period T. The period and frequency are reciprocals: $T = 1/f$, or $f = 1/T$. The higher the frequency, the shorter the period.
- Wavelength λ is the distance a wave travels in one cycle. The higher the frequency, the shorter the wavelength. The wavelength also depends on the velocity at which the wave travels: $\lambda = v/f$ where v is velocity of the wave and f is the frequency.
- Phase angle is the angular difference in time between corresponding values in the cycles for two waveforms of the same frequency.
- When one sine wave has its maximum value while the other is at zero, the two waves are 90° out of phase. Two waveforms with zero phase angle between them are in phase; a 180° phase angle means opposite phase.
- The length of a phasor arrow indicates amplitude, while the angle corresponds to the phase. Leading phase is shown by counterclockwise angles.
- Sine-wave alternating voltage V applied across a load resistance R produces alternating current I in the circuit. The current has the same waveform, frequency, and phase as the applied voltage because of the resistive load. The amount of $I = V/R$.
- The sawtooth wave and square wave are two common examples of nonsinusoidal waveforms. The amplitudes of these waves are usually measured in peak-to-peak value.
- Harmonic frequencies are exact multiples of the fundamental frequency.
- The ac voltage used in residences range from 115 to 125 V rms with a frequency of 60 Hz. The nominal voltage is usually given as 120 V.
- For residential wiring, the three-wire single-phase Edison system shown in Fig. 16-21 is used to provide either 120 or 240 V.
- In a motor, the rotating armature connects to the power line. The stator field coils provide the magnetic flux cut by the armature as it is forced to rotate. A generator has the opposite effect; it converts mechanical energy into electrical output.

- A dc motor has commutator segments contacted by graphite brushes for the external connections to the power source. An ac induction motor does not have brushes.
- In three-phase power, each phase angle is 120°. For the Y connections in Fig. 16-26a, each pair of output terminals has output of $120 \times 1.73 = 208$ V. This voltage is known as the line-to-line voltage.

SELF-TEST

ANSWERS AT BACK OF BOOK.

Answer True or False.

1. An ac voltage varies in magnitude and reverses in polarity.
2. A dc voltage always has one polarity.
3. Sine-wave alternating current flows in a load resistor with a sine-wave voltage applied.
4. When two waves are 90° out of phase, one has its peak value when the other is at zero.
5. When two waves are in phase, they have their peak values at the same time.
6. The positive peak of a sine wave cannot occur at the same time as the negative peak.
7. The angle of 90° is the same as π rad.
8. A period of 2 μs corresponds to a higher frequency than T of 1 μs.
9. A wavelength of 2 ft corresponds to a lower frequency than a wavelength of 1 ft.
10. When we compare the phase between two waveforms, they must have the same frequency.

Fill in the missing answers.

11. For the rms voltage of 10 V, the peak-to-peak value is _____ V.
12. With 120 V rms across 100 Ω R_L, the rms current equals _____ A.
13. For a peak value of 100 V, the rms value is _____ V.
14. The wavelength of a 1000-kHz radio wave is _____ cm.
15. The period of a 1000-kHz voltage is _____ ms.
16. The period of $\frac{1}{60}$ s corresponds to a frequency of _____ Hz.
17. The frequency of 100 MHz corresponds to a period of _____ μs.
18. The square wave in Fig. 16-19c has the frequency of _____ MHz.
19. The rms voltage for the sine wave in Fig. 16-19a is _____ V.
20. The ac voltage across R_2 in Fig. 16-18 is _____ V.
21. For an audio signal with a T of 0.001 s, its frequency is _____ Hz.
22. For the 60-Hz ac power-line voltage, the third harmonic is _____ Hz.
23. For a 10-V average value, the rms value is _____ V.
24. For a 340-V p–p value, the rms value is _____ V.
25. An audio signal that produces four cycles in the time it takes for one cycle of ac voltage from the power line has the frequency of _____ Hz.
26. In Fig. 16-21, the voltage between the red and black wires is _____ V.
27. In Fig. 16-25, the angle between the three phases is _____ degrees.
28. In Fig. 16-26a, the voltage between terminals B and C is _____ V.

1. **(a)** Define an alternating voltage. **(b)** Define an alternating current. **(c)** Why does ac voltage applied across a load resistance produce alternating current in the circuit?
2. **(a)** State two characteristics of a sine wave of voltage. **(b)** Why does the rms value of $0.707 \times$ peak value apply just to sine waves?
3. Draw two cycles of an ac sawtooth voltage waveform with a peak-to-peak amplitude of 40 V. Do the same for a square wave.
4. Give the angle, in degrees and radians, for each of the following: one cycle, one half-cycle, one quarter-cycle, three quarter-cycles.
5. The peak value of a sine wave is 1 V. How much is its average value? Rms value? Effective value? Peak-to-peak value?
6. State the following ranges in Hz: **(a)** audio frequencies; **(b)** radio frequencies; **(c)** standard AM radio broadcast band; **(d)** FM broadcast band; **(e)** VHF band; **(f)** microwave band.
7. Make a graph with two waves, one with a frequency of 500 kHz and the other with 1000 kHz. Mark the horizontal axis in time, and label each wave.
8. Draw the sine waves and phasor diagrams to show **(a)** two waves 180° out of phase; **(b)** two waves 90° out of phase.
9. Give the voltage value for the 60-Hz ac line voltage with an rms value of 120 V at each of the following times in a cycle: 0°, 30°, 45°, 90°, 180°, 270°, 360°.
10. **(a)** The phase angle of 90° equals how many radians? **(b)** For two sine waves 90° out of phase with each other, compare their amplitudes at 0°, 90°, 180°, 270°, and 360°.
11. Tabulate the sine and cosine values every 30° from 0 to 360° and draw the corresponding sine wave and cosine wave.
12. Draw a graph of the values for $(\sin \theta)^2$ plotted against θ for every 30° from 0 to 360°.
13. Why is the wavelength of an ultrasonic wave at 34.44 kHz the same 1 cm as for the much higher frequency radio wave at 30 GHz?
14. Draw the sine waves and phasors to show wave V_1 leading wave V_2 by 45°.
15. Why are amplitudes for nonsinusoidal waveforms generally measured in peak-to-peak values, rather than rms or average value?
16. Define harmonic frequencies, giving numerical values.
17. Define one octave, with an example of numerical values.
18. Which do you consider more important for applications of alternating current—the polarity reversals or the variations in value?
19. Define the following parts in the assembly of motors: **(a)** armature rotor; **(b)** field stator; **(c)** collector rings; **(d)** commutator segments.
20. Show diagrams of Y and Δ connections for three-phase ac power.

PROBLEMS

ANSWERS TO ODD-NUMBERED PROBLEMS AT BACK OF BOOK.

1. The 60-Hz power-line voltage of 120 V is applied across a resistance of 20 Ω. **(a)** How much is the rms current in the circuit? **(b)** What is the frequency of the current? **(c)** What is the

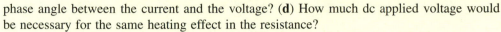

phase angle between the current and the voltage? **(d)** How much dc applied voltage would be necessary for the same heating effect in the resistance?

2. What is the frequency for the following ac variations? **(a)** 10 cycles in 1 s; **(b)** 1 cycle in ⅒ s; **(c)** 50 cycles in 1 s; **(d)** 50 cycles in ½ s; **(e)** 50 cycles in 5 s.

3. Calculate the time delay for a phase angle of 45° at the frequency of **(a)** 500 Hz; **(b)** 2 MHz.

4. Calculate the period T for the following frequencies: **(a)** 500 Hz; **(b)** 5 MHz; **(c)** 5 GHz.

5. Calculate the frequency for the following periods: **(a)** 0.05 s; **(b)** 5 ms; **(c)** 5 μs; **(d)** 5 ns.

6. Refer to Fig. 16-18; calculate the I^2R power dissipated in R_1, R_2, and R_3.

7. Give the plus and minus peak values for each wave in Fig. 16-19*a* to *d*.

8. An ac circuit has a 5-MΩ resistor R_1 in series with a 10-MΩ resistor R_2 across a 200-V source. Calculate I, V_1, V_2, P_1, and P_2.

9. The two resistors in Prob. 8 are connected in parallel. Calculate I_1, I_2, V_1, V_2, P_1, and P_2.

10. A series-parallel ac circuit has two branches across the 60-Hz 120-V power line. One branch has a 10-Ω R_1 in series with a 20-Ω R_2. The other branch has a 10-MΩ R_3 in series with a 20-MΩ R_4. Find V_1, V_2, V_3, and V_4.

11. How much I does a 300-W 120-V bulb take from a 120-V 60-Hz line?

12. In Fig. 16-28, calculate V_{rms}, period T, and frequency f.

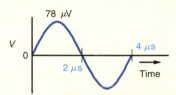

FIG. 16-28 Sine wave for Prob. 12.

13. A sine-wave ac voltage has a rms value of 19.2 V. **(a)** Find the peak value. **(b)** What is the instantaneous value at 50° of the cycle?

14. In Fig. 16-29, calculate I, V_1, V_2, and V_3.

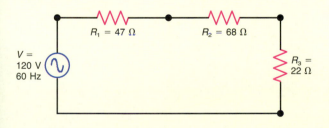

FIG. 16-29 Circuit diagram for Prob. 14.

15. In Fig. 16-30, calculate I_1, I_2, I_3, and I_T.

16. A 5-Ω R in a circuit connected to the ac power line has I of 1.17 A. Calculate the power dissipated in the resistor.

17. Convert to peak-to-peak voltage the following rms values of sine-wave ac signal voltage: **(a)** 164 μV; **(b)** 3.49 mV; **(c)** 12.48 mV.

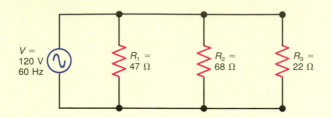

$V =$
120 V
60 Hz

$R_1 =$
47 Ω

$R_2 =$
68 Ω

$R_3 =$
22 Ω

FIG. 16-30   Circuit diagram for Prob. 15.

18. Convert to rms voltage the following peak-to-peak values of sine-wave ac signal voltage: **(a)** 462.5 μV; **(b)** 9.84 mV; **(c)** 35.19 mV.

19. **(a)** What size R_S is needed to drop 120 V from the ac power line to 9 V with a load current of 14 mA? **(b)** Calculate the power dissipated in R_S.

20. Do the same as in Prob. 19 but for a load current of 1.2 A.

21. A sine wave of voltage has an average value of 38.22 V. Calculate the waveform's: **(a)** rms value; **(b)** peak value; **(c)** peak-to-peak value.

22. A sine wave of voltage has an instantaneous value of 24 V at $\theta = 60°$. Calculate the waveform's value at $\theta = 270°$.

23. List the first four harmonics of 7.5 MHz. Also, identify each harmonic as being either an even or odd multiple of the fundamental frequency.

24. Calculate the wavelength λ in meters for the following radio frequencies: **(a)** 1.875 MHz; **(b)** 3.75 MHz; **(c)** 7.5 MHz; **(d)** 15 MHz; **(e)** 20 MHz; **(f)** 30 MHz.

25. Calculate the wavelength λ in feet for the following sound wave frequencies: **(a)** 10 Hz; **(b)** 50 Hz; **(c)** 250 Hz; **(d)** 1 kHz; **(e)** 15 kHz; **(f)** 20 kHz.

26. List the frequency three decades above 100 Hz.

27. Raising the frequency of 400 Hz by two octaves corresponds to what frequency?

28. What is the frequency three octaves below 40 kHz?

29. Two waveforms, A and B, each have a frequency of 10 kHz. If waveform A reaches its maximum positive value 6.25 μs after waveform B, calculate the phase-angle difference between the two waveforms.

30. Draw the phase-angle diagram for Prob. 29. Represent waveform A as V_A and waveform B as V_B. Use V_A as the reference phasor.

31. Calculate the period T of a radio wave whose wavelength λ is 2 m.

CRITICAL THINKING

1. The electrical length of an antenna is to be one-half wavelength long at a frequency f of 7.2 MHz. Calculate the length of the antenna in: **(a)** feet; **(b)** centimeters.

2. A transmission line has a length l of 7.5 m. What is its electrical wavelength at 10 MHz?

3. The total length of an antenna is 120 ft. At what frequency is the antenna one-half wavelength long?

4. A cosine wave of current has an instantaneous amplitude of 45 mA at $\theta = \pi/3$ rad. Calculate the waveform's instantaneous amplitude at $\theta = 3\pi/2$ rad.

16-1   **a.** T
      b. F
      c. T

16-2   **a.** 10 V
      b. 360°

16-3   **a.** 85 V
      b. 120 V
      c. 170 V

16-4   **a.** 0.707 A
      b. 0.5 A

16-5   **a.** 120 V rms
      b. 14.14 V peak
      c. 2.8 V p–p

16-6   **a.** 4 Hz
      b. 1.605 MHz

16-7   **a.** 400 Hz
      b. $\frac{1}{400}$ s

16-8   **a.** T
      b. F
      c. T

16-9   **a.** 90°
      b. 60°
      c. 0°

16-10   **a.** $\frac{1}{240}$ s
      b. 0.1 μs

16-11   **a.** 30 Ω
      b. 6.67 Ω
      c. 30 Ω

16-12   **a.** 2 μs
      b. 15 V

16-13   **a.** 48 MHz
      b. 440 Hz

16-14   **a.** F
      b. T
      c. T
      d. T

16-15   **a.** T
      b. T
      c. F

16-16   **a.** 120°
      b. 208 V

REVIEW: CHAPTERS 13 TO 16

SUMMARY

- Iron, nickel, and cobalt are magnetic materials. Magnets have a north pole and a south pole at opposite ends. Opposite poles attract; like poles repel.
- A magnet has an invisible, external magnetic field. This magnetic flux is indicated by field lines. The direction of field lines outside the magnet is from north pole to south pole.
- An electromagnet has an iron core that becomes magnetized when current flows in the coil winding.
- Magnetic units are defined in Tables 13-1 and 14-2.
- Continuous magnetization and demagnetization of an iron core by means of alternating current causes hysteresis losses, which increase with higher frequencies.
- Current in a conductor has an associated magnetic field with circular lines of force in a plane perpendicular to the wire.
- Motor action results from the net force of two fields that can aid or cancel. The direction of the resultant force is from the stronger field to the weaker.
- The motion of flux cutting across a perpendicular conductor generates an induced emf.
- Faraday's law of induced voltage states that $v = N \, d\phi/dt$.
- Lenz' law states that an induced voltage must have the polarity that opposes the change causing the induction.
- Alternating voltage varies in magnitude and reverses in polarity.
- One cycle includes the values between points having the same value and varying in the same direction. The cycle includes 360°, or 2π rad.
- Frequency f equals the cycles per second (cps). One cps = 1 Hz.
- Period T is the time for one cycle. It equals $1/f$. When f is in cycles per second, T is in seconds.
- Wavelength λ is the distance a wave travels in one cycle. $\lambda = v/f$.
- The rms, or effective value, of a sine wave equals $0.707 \times$ peak value. Or the peak value equals $1.414 \times$ rms value. The average value equals $0.637 \times$ peak value.
- Phase angle θ is the angular difference between corresponding values in the cycles for two sine waves of the same frequency. The angular difference can be expressed in time based on the frequency of the waves.
- Phasors, similar to vectors, indicate the amplitude and phase angle of alternating voltage or current. The length of the phasor is the amplitude, while the angle is the phase.
- The square wave and sawtooth wave are common examples of nonsinusoidal waveforms.
- Direct current motors generally use commutator segments with graphite brushes. Alternating current motors are usually the induction type without brushes.

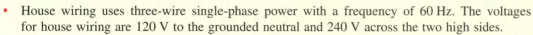

- House wiring uses three-wire single-phase power with a frequency of 60 Hz. The voltages for house wiring are 120 V to the grounded neutral and 240 V across the two high sides.
- Three-phase ac power has three legs 120° out of phase. A Y connection, with 120 V across each phase, has 208 V available across each two legs.

REVIEW SELF-TEST

ANSWERS AT BACK OF BOOK.

Choose (a), (b), (c), or (d).

1. Which of the following statements is true? (a) Alnico is commonly used for electromagnets. (b) Paper cannot affect magnetic flux because it is not a magnetic material. (c) Iron is generally used for permanent magnets. (d) Ferrites have lower permeability than air or vacuum.

2. Hysteresis losses (a) are caused by high-frequency alternating current in a coil with an iron core; (b) generally increase with direct current in a coil; (c) are especially important with permanent magnets that have a steady magnetic field; (d) cannot be produced in an iron core, because it is a conductor.

3. A magnetic flux of 25,000 lines through an area of 5 cm^2 results in (a) 5 lines of flux; (b) 5000 Mx of flux; (c) flux density of 5000 G; (d) flux density corresponding to 25,000 A.

4. If 10 V is applied across a relay coil with 100 turns having 2 Ω of resistance, the total force producing magnetic flux in the circuit is (a) 10 Mx; (b) 50 G; (c) 100 Oe; (d) 500 A · t.

5. The ac power-line voltage of 120 V rms has a peak value of (a) 100 V; (b) 170 V; (c) 240 V; (d) 338 V.

6. Which of the following can produce the most induced voltage? (a) 1-A direct current; (b) 50-A direct current; (c) 1-A 60-Hz alternating current; (d) 1-A 400-Hz alternating current.

7. Which of the following has the highest frequency? (a) $T = \frac{1}{1000}$ s; (b) $T = \frac{1}{60}$ s; (c) $T = 1$ s; (d) $T = 2$ s.

8. Two waves of the same frequency have opposite phase when the phase angle between them is (a) 0°; (b) 90°; (c) 360°; (d) π rad.

9. The 120-V 60-Hz power-line voltage is applied across a 120-Ω resistor. The current equals (a) 1 A, peak value; (b) 120 A, peak value; (c) 1 A, rms value; (d) 5 A, rms value.

10. When an alternating voltage reverses in polarity, the current it produces (a) reverses in direction; (b) has a steady dc value; (c) has a phase angle of 180°; (d) alternates at 1.4 times the frequency of the applied voltage.

REFERENCES

Adams, J. E., and G. Rockmaker: *Industrial Electricity—Principles and Practices,* Glencoe/McGraw-Hill, Columbus, Ohio.

Petruzella, F. D.: *Industrial Electronics,* Glencoe/McGraw-Hill, Columbus, Ohio.

CHAPTER 17

CAPACITANCE

Capacitance is the ability of a dielectric to store electric charge. The more the charge that is stored for a given voltage, the higher the value of capacitance. Its symbol is C and the unit is the farad (F), named after Michael Faraday.

A capacitor consists of an insulator (also called a dielectric) between two conductors. The conductors make it possible to apply voltage across the insulator. Different types of capacitors are manufactured for specific values of C. They are named according to the dielectric. Common types are air, ceramic, mica, paper, film, and electrolytic capacitors. Capacitors used in electronic circuits are small and economical.

The most important property of a capacitor is its ability to block a steady dc voltage, while passing ac signals. The higher the frequency is, the less the opposition for ac voltages.

Capacitors are a common source of troubles because they can either have an open at the conductors or a short circuit through the dielectric. These troubles are described here, including the method of checking a capacitor with an ohmmeter, even though a capacitor is actually an insulator.

CHAPTER OBJECTIVES

Upon completion of this chapter, you should be able to:

- *Understand* how charge is stored in the dielectric of a capacitor.
- *Understand* how a capacitor charges and discharges.
- *Define* the farad unit of capacitance.
- *List* the physical factors affecting the capacitance of a capacitor.
- *List* several types of capacitors and the characteristics of each.
- *Understand* how an electrolytic capacitor is constructed.
- *Understand* how capacitors are coded.
- *Calculate* the total capacitance of parallel connected capacitors.
- *Calculate* the equivalent capacitance of series connected capacitors.
- *Calculate* the energy stored in a capacitor.
- *Understand* how an ohmmeter can be used to test a capacitor.

IMPORTANT TERMS IN THIS CHAPTER

aluminum capacitor	electrostatic induction	shelf life of capacitors
capacitor charge and discharge	farad (F) unit	stored charge
ceramic capacitor	film capacitor	stray capacitance
chip capacitor	ganged capacitors	tantalum capacitor
coding of capacitors	leakage resistance	testing of capacitors
dielectric constant	mica capacitor	tuning capacitor
distributed capacitance	Mylar capacitor	
electrolytic capacitor	paper capacitor	

TOPICS COVERED IN THIS CHAPTER

17-1 HOW CHARGE IS STORED IN THE DIELECTRIC

It is possible for dielectric materials such as air or paper to hold an electric charge because free electrons cannot flow through an insulator. However, the charge must be applied by some source. In Fig. 17-1*a*, the battery can charge the capacitor shown. With the dielectric contacting the two conductors connected to the potential difference *V*, electrons from the voltage source accumulate on the side of the capacitor connected to the negative terminal of *V*. The opposite side of the capacitor connected to the positive terminal of *V* loses electrons.

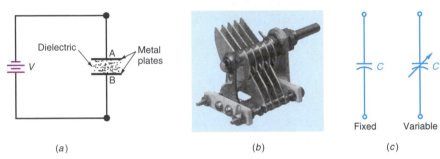

(a) (b) (c)

Fixed Variable

FIG. 17-1 Capacitance stores the charge in the dielectric between two conductors. (*a*) Structure. (*b*) Air-dielectric variable capacitor. Length is 2 in. (*c*) Schematic symbols for fixed and variable capacitors.

As a result, the excess of electrons produces a negative charge on one side of the capacitor, while the opposite side has a positive charge. As an example, if 6.25×10^{18} electrons are accumulated, the negative charge equals 1 coulomb (C). The charge on only one plate need be considered, as the number of electrons accumulated on one plate is exactly the same as the number taken from the opposite plate.

What the voltage source does is simply redistribute some electrons from one side of the capacitor to the other. This process is called *charging* the capacitor. Charging continues until the potential difference across the capacitor is equal to the applied voltage. Without any series resistance, the charging is instantaneous. Practically, however, there is always some series resistance. This charging current is transient, or temporary, as it flows only until the capacitor is charged to the applied voltage. Then there is no current in the circuit.

The result is a device for storing charge in the dielectric. Storage means that the charge remains even after the voltage source is disconnected. The measure of how much charge can be stored is the capacitance *C*. More charge stored for a given amount of applied voltage means more capacitance. Components made to provide a specified amount of capacitance are called *capacitors,* or by their old name *condensers.*

Electrically, then, capacitance is the ability to store charge. Physically, a capacitor consists simply of two conductors separated by an insulator. For example, Fig. 17-1*b* shows a capacitor using air for the dielectric between the metal plates. There are many types with different dielectric materials, including paper, mica, and ceramics, but the schematic symbols shown in Fig. 17-1*c* apply to all capacitors.

ELECTRIC FIELD IN THE DIELECTRIC   Any voltage has a field of electric lines of force between the opposite electric charges. The electric field corresponds to the magnetic lines of force of the magnetic field associated with electric current.* What a capacitor does is concentrate the electric field in the dielectric between the plates. This concentration corresponds to a magnetic field concentrated in the turns of a coil. The only function of the capacitor plates and wire conductors is to connect the voltage source V across the dielectric. Then the electric field is concentrated in the capacitor, instead of being spread out in all directions.

ELECTROSTATIC INDUCTION   The capacitor has opposite charges because of electrostatic induction by the electric field. Electrons that accumulate on the negative side of the capacitor provide electric lines of force that repel electrons from the opposite side. When this side loses electrons, it becomes positively charged. The opposite charges induced by an electric field correspond to the idea of opposite poles induced in magnetic materials by a magnetic field.

TEST-POINT QUESTION 17-1

Answers at end of chapter.
a. In a capacitor, is the electric charge stored in the dielectric or in the metal plates?
b. What is the unit of capacitance?

17-2 CHARGING AND DISCHARGING A CAPACITOR

Charging and discharging are the two main effects of capacitors. Applied voltage puts charge in the capacitor. The accumulation of charge results in a buildup of potential difference across the capacitor plates. When the capacitor voltage equals the applied voltage, there is no more charging. The charge remains in the capacitor, with or without the applied voltage connected.

The capacitor discharges when a conducting path is provided across the plates, without any applied voltage. Actually, it is only necessary that the capacitor voltage be more than the applied voltage. Then the capacitor can serve as voltage source, temporarily, to produce discharge current in the discharge path. The capacitor discharge continues until the capacitor voltage drops to zero or is equal to the applied voltage.

APPLYING THE CHARGE   In Fig. 17-2a, the capacitor is neutral with no charge because it has not been connected to any source of applied voltage and there is no electrostatic field in the dielectric. Closing the switch in Fig. 17-2b, however, allows the negative battery terminal to repel free electrons in the

---

*Electric and magnetic fields are compared in Fig. 14-6 on page 383.

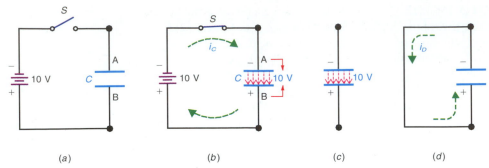

FIG. 17-2 Storing electric charge in a capacitance. (*a*) Capacitor without any charge. (*b*) Battery charges capacitor to applied voltage of 10 V. (*c*) Stored charge remains in capacitor, providing 10 V without the battery. (*d*) Discharging the capacitor.

conductor to plate A. At the same time, the positive terminal attracts free electrons from plate B. The side of the dielectric at plate A accumulates electrons because they cannot flow through the insulator, while plate B has an equal surplus of protons.

Remember that the opposite charges have an associated potential difference, which is the voltage across the capacitor. The charging process continues until the capacitor voltage equals the battery voltage, which is 10 V in this example. Then no further charging is possible because the applied voltage cannot make free electrons flow in the conductors.

Note that the potential difference across the charged capacitor is 10 V between plates A and B. There is no potential difference from each plate to its battery terminal, however, which is the reason why the capacitor stops charging.

STORING THE CHARGE   The negative and positive charges on opposite plates have an associated electric field through the dielectric, as shown by the dotted lines in Figs. 17-2*b* and 17-2*c*. The direction of these electric lines of force is shown repelling electrons from plate B, making this side positive. It is the effect of electric lines of force through the dielectric that results in storage of the charge. The electric field distorts the molecular structure so that the dielectric is no longer neutral. The dielectric is actually stressed by the invisible force of the electric field. As evidence, the dielectric can be ruptured by a very intense field with high voltage across the capacitor.

The result of the electric field, then, is that the dielectric has charge supplied by the voltage source. Since the dielectric is an insulator that cannot conduct, the charge remains in the capacitor even after the voltage source is removed, as illustrated in Fig. 17-2*c*. You can now take this charged capacitor by itself out of the circuit, and it still has 10 V across the two terminals.

DISCHARGING   The action of neutralizing the charge by connecting a conducting path across the dielectric is called *discharging* the capacitor. In Fig. 17-2*d*, the wire between plates A and B is a low-resistance path for discharge current. With the stored charge in the dielectric providing the potential difference, 10 V is available to produce discharge current. The negative plate repels electrons, which are attracted to the positive plate through the wire, until the positive and negative charges are neutralized. Then there is no net charge. The capacitor is completely discharged, the voltage across it equals zero, and there is no discharge current. Now the capacitor is in the same uncharged condition as in Fig. 17-2a. It can be charged again, however, by a source of applied voltage.

NATURE OF THE CAPACITANCE   A capacitor has the ability to store the amount of charge necessary to provide a potential difference equal to the charging voltage. If 100 V were applied in Fig. 17-2, the capacitor would charge to 100 V.

The capacitor charges to the applied voltage because, when the capacitor voltage is less, it takes on more charge. As soon as the capacitor voltage equals the applied voltage, no more charging current can flow. *Note that any charge or discharge current flows through the conducting wires to the plates but not through the dielectric.*

CHARGE AND DISCHARGE CURRENTS   In Fig. 17-2b, i_C is in the opposite direction from i_D in Fig. 17-2d. In both cases the current shown is electron flow. However, i_C is charging current to the capacitor and i_D is discharge current from the capacitor. The charge and discharge currents must always be in opposite directions. In Fig. 17-2b, the negative plate of C accumulates electrons from the voltage source. In Fig. 17-2d, the charged capacitor serves as a voltage source to produce electron flow around the discharge path.

More charge and discharge current result with a higher value of C for a given amount of voltage. Also, more V produces more charge and discharge current with a given amount of capacitance. However, the value of C does not change with the voltage, as the amount of C depends on the physical construction of the capacitor.

TEST-POINT QUESTION 17-2

Answers at end of chapter.

Refer to Fig. 17-2.
a. If the applied voltage were 14.5 V, how much would the voltage be across C after it has charged?
b. How much is the voltage across C after it is completely discharged?
c. Can the capacitor be charged again after it is discharged?

17-3 THE FARAD UNIT OF CAPACITANCE

With more charging voltage, the electric field is stronger and more charge is stored in the dielectric. The amount of charge Q stored in the capacitance is therefore proportional to the applied voltage. Also, a larger capacitance can store more charge. These relations are summarized by the formula

▶ $Q = CV$ coulombs   (17-1)

where Q is the charge stored in the dielectric in coulombs (C), and V is the voltage across the plates of the capacitor, and C is the capacitance in farads.

The C is a physical constant, indicating the capacitance in terms of how much charge can be stored for a given amount of charging voltage. When one coulomb is stored in the dielectric with a potential difference of one volt, the capacitance is one *farad*.

Practical capacitors have sizes in millionths of a farad, or smaller. The reason is that typical capacitors store charge of microcoulombs or less. Therefore, the common units are

$$1 \text{ microfarad} = 1 \ \mu\text{F} = 1 \times 10^{-6} \text{ F}$$
$$1 \text{ nanofarad} = 1 \text{ nF} = 1 \times 10^{-9} \text{ F}$$
$$1 \text{ picofarad} = 1 \text{ pF} = 1 \times 10^{-12} \text{ F}$$

Although traditionally it has not been used, the nanofarad unit of capacitance is gaining acceptance in the electronics industry.

Example

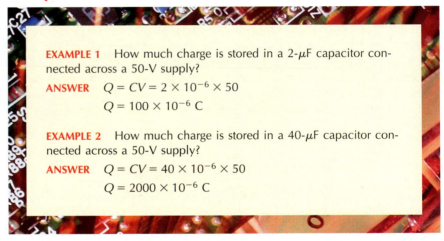

EXAMPLE 1   How much charge is stored in a 2-μF capacitor connected across a 50-V supply?

ANSWER $Q = CV = 2 \times 10^{-6} \times 50$
$Q = 100 \times 10^{-6}$ C

EXAMPLE 2   How much charge is stored in a 40-μF capacitor connected across a 50-V supply?

ANSWER $Q = CV = 40 \times 10^{-6} \times 50$
$Q = 2000 \times 10^{-6}$ C

Note that the larger capacitor stores more charge for the same voltage, in accordance with the definition of capacitance as the ability to store charge.

The factors in $Q = CV$ can be inverted to

$$C = \frac{Q}{V} \tag{17-2}$$

or

$$V = \frac{Q}{C} \tag{17-3}$$

For all three formulas, the basic units are volts for V, coulombs for Q, and farads for C. Note that the formula $C = Q/V$ actually defines one farad of capacitance as one coulomb of charge stored for one volt of potential difference. The letter C (in italic, or slanted, type) is the symbol for capacitance. The same letter C (in roman, or upright, type) is the abbreviation for coulomb unit of charge. The difference between C and C will be made clearer in the examples that follow.

Example

LARGER PLATE AREA INCREASES CAPACITANCE As illustrated in Fig. 17-3, when the area of each plate is doubled, the capacitance in Fig. 17-3b stores twice the charge of Fig. 17-3a. The potential difference in both cases is still 10 V. This voltage produces a given strength of electric field. A larger plate area, however, means that more of the dielectric surface can contact each plate, allowing more lines of force through the dielectric between the plates and less flux leakage outside the dielectric. Then the field can store more charge in the dielectric. The result of larger plate area is more charge stored for the same applied voltage, which means the capacitance is larger.

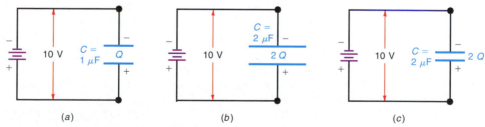

FIG. 17-3 Increasing stored charge and capacitance by increasing the plate area and decreasing the distance between plates. (a) Capacitance of 1 μF. (b) A 2-μF capacitance with twice the plate area and the same distance. (c) A 2-μF capacitance with one-half the distance and the same plate area.

THINNER DIELECTRIC INCREASES CAPACITANCE   As illustrated in Fig. 17-3c, when the distance between plates is reduced one-half, the capacitance stores twice the charge of Fig. 17-3a. The potential difference is still 10 V, but its electric field has greater flux density in the thinner dielectric. Then the field between opposite plates can store more charge in the dielectric. With less distance between the plates, the stored charge is greater for the same applied voltage, which means the capacitance is greater.

DIELECTRIC CONSTANT K_ϵ   This indicates the ability of an insulator to concentrate electric flux. Its numerical value is specified as the ratio of flux in the insulator compared with the flux in air or vacuum. The dielectric constant of air or vacuum is 1, since it is the reference.

Mica, for example, has an average dielectric constant of 6, meaning it can provide a density of electric flux six times as great as that of air or vacuum for the same applied voltage and equal physical size. Insulators generally have a dielectric constant K_ϵ greater than 1, as listed in Table 17-1. Higher values of K_ϵ allow greater values of capacitance.

TABLE 17-1	DIELECTRIC MATERIALS*	
MATERIAL	**DIELECTRIC CONSTANT K_ϵ**	**DIELECTRIC STRENGTH, V/MIL**
Air or vacuum	1	20
Aluminum oxide	7	
Ceramics	80–1200	600–1250
Glass	8	335–2000
Mica	3–8	600–1500
Oil	2–5	275
Paper	2–6	1250
Plastic film	2–3	
Tantalum oxide	25	

*Exact values depend on the specific composition of different types.

It should be noted that the aluminum oxide and tantalum oxide listed in Table 17-1 are used for the dielectric in electrolytic capacitors. Also, plastic film is often used instead of paper for the rolled-foil type of capacitor.

The dielectric constant for an insulator is actually its *relative permittivity*, with the symbol ϵ_r or K_ϵ, indicating the ability to concentrate electric flux. This factor corresponds to relative permeability, with the symbol μ_r or K_m, for magnetic flux. Both ϵ_r and μ_r are pure numbers without units, as they are just ratios.*

---

*The absolute permittivity ϵ_0 is 8.854×10^{-12} F/m, in SI units, for electric flux in air or vacuum. This value corresponds to an absolute permeability μ_0 of $4\pi \times 10^{-7}$ H/m, in SI units, for magnetic flux in air or vacuum.

These physical factors for a parallel-plate capacitor are summarized by the formula

$$\blacktriangleright \quad C = K_\epsilon \times \frac{A}{d} \times 8.85 \times 10^{-12}\,\text{F} \qquad\qquad (17\text{-}4)$$

where A is the area in square meters of either plate, d is the distance in meters between plates, K_ϵ is the dielectric constant, or relative permittivity, as listed in Table 17-1, and C is capacitance in farads. The constant factor 8.85×10^{-12} is the absolute permittivity of air or vacuum, in SI, since the farad is an SI unit.

Example

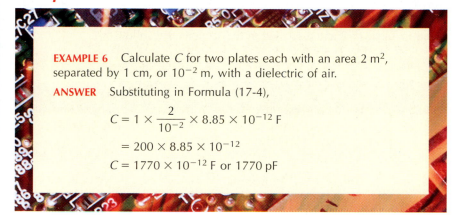

EXAMPLE 6 Calculate C for two plates each with an area 2 m^2, separated by 1 cm, or 10^{-2} m, with a dielectric of air.

ANSWER Substituting in Formula (17-4),

$$C = 1 \times \frac{2}{10^{-2}} \times 8.85 \times 10^{-12}\,\text{F}$$

$$= 200 \times 8.85 \times 10^{-12}$$

$$C = 1770 \times 10^{-12}\,\text{F or 1770 pF}$$

This value means the capacitor can store 1770×10^{-12} C of charge with 1 V. Note the relatively small capacitance, in picofarad units, with the extremely large plates of 2 m^2, which is really the size of a table or a desktop.

If the dielectric used is paper with a dielectric constant of 6, then C will be six times greater. Also, if the spacing between plates is reduced by one-half to 0.5 cm, the capacitance will be doubled. It should be noted that practical capacitors for electronic circuits are much smaller than this parallel-plate capacitor. They use a very thin dielectric, with a high dielectric constant, and the plate area can be concentrated in a small space.

DIELECTRIC STRENGTH Table 17-1 also lists breakdown-voltage ratings for typical dielectrics. Dielectric strength is the ability of a dielectric to withstand a potential difference without arcing across the insulator. This voltage rating is important because rupture of the insulator provides a conducting path through the dielectric. Then it cannot store charge, because the capacitor has been short-circuited. Since the breakdown voltage increases with greater thickness, capacitors with higher voltage ratings have more distance between the plates. This increased distance reduces the capacitance, however, all other factors remaining the same.

Answers at end of chapter.
a. A capacitor charged to 100 V has 1000 μC of charge. How much is C?
b. A mica capacitor and ceramic capacitor have the same physical dimensions. Which has more C?

17-4 TYPICAL CAPACITORS

Commercial capacitors are generally classified according to the dielectric. Most common are air, mica, paper, plastic film, and ceramic capacitors, plus the electrolytic type. Electrolytic capacitors use a molecular-thin oxide film as the dielectric, resulting in large capacitance values in little space. These types are compared in Table 17-2 and discussed in the sections that follow.

TABLE 17-2   TYPES OF CAPACITORS

DIELECTRIC	CONSTRUCTION	CAPACITANCE	BREAKDOWN, V
Air	Meshed plates	10–400 pF	400 (0.02-in. air gap)
Ceramic	Tubular	0.5–1600 pF	500–20,000
	Disk	1 pF to 1 μF	
Electrolytic	Aluminum	1–6800 μF	10–450
	Tantalum	0.047 to 330 μF	6–50
Mica	Stacked sheets	10–5000 pF	500–20,000
Paper	Rolled foil	0.001–1 μF	200–1600
Plastic film	Foil or metalized	100 pF to 100 μF	50–600

Except for electrolytic capacitors, capacitors can be connected to a circuit without regard to polarity, since either side can be the more positive plate. Electrolytic capacitors are marked to indicate the side that must be connected to the positive or negative side of the circuit. *It should be noted that it is the polarity of the charging source that determines the polarity of the capacitor voltage.* Failure to observe the correct polarity can damage the dielectric and lead to the complete destruction of the capacitor.

MICA CAPACITORS   Thin mica sheets as the dielectric are stacked between tinfoil sections for the conducting plates to provide the required capacitance. Alternate strips of tinfoil are connected together and brought out as one terminal for one set of plates, while the opposite terminal connects to the other set of interlaced plates. The construction is shown in Fig. 17-4a. The entire unit is generally in a molded Bakelite case. Mica capacitors are often used for small capacitance values of about 10 to 5000 pF; their length is ¾ in. or less with about ⅛-in. thickness. A typical mica capacitor is shown in Fig. 17-4b.

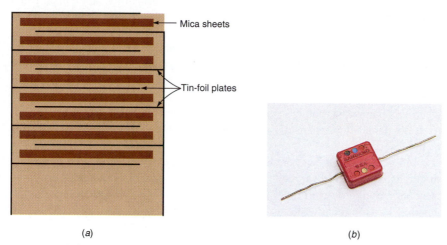

(a)   (b)

FIG. 17-4   Mica capacitor. (a) Physical construction. (b) Example of a mica capacitor.

PAPER CAPACITORS   In this construction, shown in Fig. 17-5a, two rolls of tinfoil conductor separated by a paper dielectric are rolled into a compact cylinder. Each outside lead connects to its roll of tinfoil as a plate. The entire cylinder is generally placed in a cardboard container coated with wax or encased in plastic. Paper capacitors are often used for medium capacitance values of 0.001 to 1.0 μF, approximately. The physical size for 0.05 μF is typically 1 in. long with ⅜-in. diameter. A paper capacitor is shown in Fig. 17-5b.

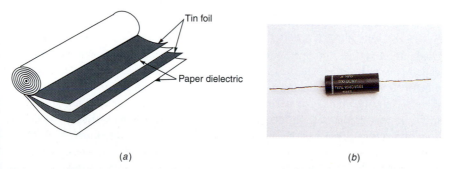

(a)   (b)

FIG. 17-5   Paper capacitor. (a) Physical construction. (b) Example of a paper capacitor.

A black or a white band at one end of a paper capacitor indicates the lead connected to the outside foil. This lead should be used for the ground or low-potential side of the circuit to take advantage of shielding by the outside foil. There is no required polarity, however, since the capacitance is the same no matter which side is grounded. It should also be noted that in the schematic symbol for C the curved line usually indicates the low-potential side of the capacitor.

FILM CAPACITORS Film capacitors are constructed much like paper capacitors except that the paper dielectric is replaced with a plastic film such as polypropylene, polystrene, polycarbonate, or polyethelene terepthalate (Mylar). There are two main types of film capacitors: the foil type and the metallized type. The foil type uses sheets of metal foil, such as aluminum or tin, for its conductive plates. The metallized type is constructed by depositing (spraying) a thin layer of metal, such as aluminum or zinc, on the plastic film. The sprayed-on metal serves as the plates of the capacitor. The advantage of the metallized type over the foil type is that the metallized type is much smaller for a given capacitance value and breakdown voltage rating. The reason is that the metallized type has much thinner plates due to the fact that they are sprayed on. Another advantage of the metallized type is that they are self-healing. This means that if the dielectric is punctured, due to exceeding its breakdown voltage rating, the capacitor is not damaged permanently. Instead the capacitor heals itself. This is not true of the foil type.

Film capacitors are very temperature stable and are therefore used frequently in circuits that require very stable capacitance values. Some examples are radio frequency oscillators and timer circuits. Film capacitors are available with values ranging from about 100 pF to 100 μF. Figure 17-6 shows a typical film capacitor.

FIG. 17-6 Film capacitor.

CERAMIC CAPACITORS The ceramic materials used in ceramic capacitors are made from earth fired under extreme heat. By using titanium dioxide or one of several types of silicates, very high values of dielectric constant, K_E, can be obtained. Most ceramic capacitors come in disk form, as shown in Fig. 17-7. In the disk form, silver is deposited on both sides of the ceramic dielectric to form the capacitor plates. Ceramic capacitors are available with values of 1 pF (or less) up to about 1 μF. The wide range of values is possible because the dielectric constant K_E can be tailored to provide almost any desired value of capacitance.

It should be noted that ceramic capacitors are also available in forms other than disk form. Some ceramic capacitors are available with axial leads and use a color code similar to that of a resistor.

FIG. 17-7 Disk ceramic capacitor.

SURFACE-MOUNT CAPACITORS Like resistors, capacitors are also available as surface-mounted components. Surface-mounted capacitors are often called *chip capacitors*. Chip capacitors are constructed by placing a ceramic dielectric material between layers of conductive film which form the capacitor plates. The capacitance is determined by the dielectric constant K_E and the physical area of the plates. Chip capacitors are available in many different sizes. A common size is 0.125 in. long by 0.063 in. wide in various thicknesses. Another common size is 0.080 in. long by 0.050 in. wide in various thicknesses. Figure 17-8 shows two different sizes of chip capacitors. Like chip resistors, chip capacitors have their end electrodes soldered directly to the copper traces of the printed circuit board. Chip capacitors are available with values ranging from a fraction of a picofarad up to several microfarads.

VARIABLE CAPACITORS Figure 17-1*b* shows a variable air capacitor. In this construction, the fixed metal plates connected together form the *stator*. The movable plates connected together on the shaft form the *rotor*. Capacitance is varied by rotating the shaft to make the rotor plates mesh with the stator plates.

They do not touch, however, since air is the dielectric. Full mesh is maximum capacitance. Moving the rotor completely out of mesh provides minimum capacitance.

A common application is the tuning capacitor in radio receivers. When you tune to different stations, the capacitance varies as the rotor moves in or out of mesh. Combined with an inductance, the variable capacitance then tunes the receiver to a different resonant frequency for each station. Usually two or three capacitor sections are *ganged* on one common shaft.

TEMPERATURE COEFFICIENT Ceramic capacitors are often used for temperature compensation, to increase or decrease capacitance with a rise in temperature. The temperature coefficient is given in parts per million (ppm) per degree Celsius, with a reference of 25°C. As an example, a negative 750 ppm unit is stated as N750. A positive temperature coefficient of the same value would be stated as P750. Units that do not change in capacitance are labeled NPO.

CAPACITANCE TOLERANCE Ceramic disk capacitors for general applications usually have a tolerance of ±20 percent. For closer tolerances, mica or film capacitors are used. These have tolerance values of ±2 to 20 percent. Silver-plated mica capacitors are available with a tolerance of ±1 percent.

The tolerance may be less on the minus side to make sure there is enough capacitance, particularly with electrolytic capacitors, which have a wide tolerance. For instance, a 20-μF electrolytic with a tolerance of −10 percent, +50 percent may have a capacitance of 18 to 30 μF. However, the exact capacitance value is not critical in most applications of capacitors for filtering, ac coupling, and bypassing.

FIG. 17-8 Chip capacitors.

VOLTAGE RATING OF CAPACITORS This rating specifies the maximum potential difference that can be applied across the plates without puncturing the dielectric. Usually the voltage rating is for temperatures up to about 60°C. Higher temperatures result in a lower voltage rating. Voltage ratings for general-purpose paper, mica, and ceramic capacitors are typically 200 to 500 V. Ceramic capacitors with ratings of 1 to 20 kV are also available.

Electrolytic capacitors are typically available in 16-, 35-, and 50-V ratings. For applications where a lower voltage rating is permissible, more capacitance can be obtained in a smaller physical size.

The potential difference across the capacitor depends upon the applied voltage and is not necessarily equal to the voltage rating. A voltage rating higher than the potential difference applied across the capacitor provides a safety factor for long life in service. With electrolytic capacitors, however, the actual capacitor voltage should be close to the rated voltage to produce the oxide film that provides the specified capacitance.

The voltage ratings are for dc voltage applied. The breakdown rating is lower for ac voltage because of the internal heat produced by continuous charge and discharge.

CAPACITOR APPLICATIONS In most electronic circuits, a capacitor has dc voltage applied, combined with a much smaller ac signal voltage. The usual function of the capacitor is to block the dc voltage but pass the ac signal voltage, by means of the charge and discharge current. These applications include coupling, bypassing, and filtering for ac signal.

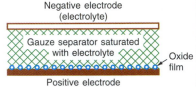

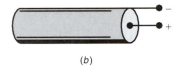

FIG. 17-9 Construction of aluminum electrolytic capacitor. (*a*) Internal electrodes. (*b*) Foil rolled into cartridge. (*c*) Typical capacitor with multiple sections.

17-5 ELECTROLYTIC CAPACITORS

Electrolytic capacitors are commonly used for C values ranging from about 1 to 6800 μF, because electrolytics provide the most capacitance in the smallest space with least cost.

CONSTRUCTION Figure 17-9 shows the aluminum-foil type. The two aluminum electrodes are in an electrolyte of borax, phosphate, or carbonate. Between the two aluminum strips, absorbent gauze soaks up electrolyte to provide the required electrolysis that produces an oxide film. This type is considered a wet electrolytic, but it can be mounted in any position.

When dc voltage is applied to form the capacitance in manufacture, the electrolytic action accumulates a molecular-thin layer of aluminum oxide at the junction between the positive aluminum foil and the electrolyte. The oxide film is an insulator. As a result, capacitance is formed between the positive aluminum electrode and the electrolyte in the gauze separator. The negative aluminum electrode simply provides a connection to the electrolyte. Usually, the metal can itself is the negative terminal of the capacitor, as shown in Fig. 17-9c.

Because of the extremely thin dielectric film, very large C values can be obtained. The area is increased by using long strips of aluminum foil and gauze, which are rolled into a compact cylinder with very high capacitance. For example, an electrolytic capacitor the same size as a 0.1-μF paper capacitor, but rated at 10 V breakdown, may have 1000 μF of capacitance or more. Higher voltage ratings, up to 450 V, are available, with typical C values up to about 6800 μF. The very high C values usually have lower voltage ratings.

POLARITY Electrolytic capacitors are used in circuits that have a combination of dc voltage and ac voltage. The dc voltage maintains the required polarity across the electrolytic capacitor to form the oxide film. A common application is for electrolytic filter capacitors to eliminate the 60- or 120-Hz ac ripple in a dc power supply. Another use is for audio coupling capacitors in transistor amplifiers. In both these applications, for filtering or coupling, electrolytics are needed for large C with a low-frequency ac component, while the circuit has a dc component for the required voltage polarity. Incidentally, the difference between filtering an ac component out or coupling it into a circuit is only a question of parallel or series connections. The filter capacitors for a power supply are typically 100 to 1000 μF. Audio capacitors are usually 10 to 47 μF.

If the electrolytic is connected in opposite polarity, the reversed electrolysis forms gas in the capacitor. It becomes hot and may explode. This is a possibility only with electrolytic capacitors.

LEAKAGE CURRENT   The disadvantage of electrolytics, in addition to the required polarization, is their relatively high leakage current compared with other capacitors, since the oxide film is not a perfect insulator. The problem with leakage current in a capacitor is that it allows part of the dc component to be coupled into the next circuit along with the ac component. In the newer electrolytic capacitors, the leakage current is quite small.

NONPOLARIZED ELECTROLYTICS   This type is available for applications in circuits without any dc polarizing voltage, as in the 60-Hz ac power line. One application is the starting capacitor for ac motors. A nonpolarized electrolytic actually contains two capacitors, connected internally in series-opposing polarity.

TANTALUM CAPACITORS   This is another form of electrolytic capacitor, using tantalum (Ta) instead of aluminum. Titanium (Ti) is also used. Typical tantalum capacitors are shown in Fig. 17-10. They feature:

1. Larger C in a smaller size
2. Longer shelf life
3. Less leakage current

However, tantalum electrolytics cost more than the aluminum type. Methods of construction for tantalum capacitors include the wet-foil type and a solid chip or slug. The solid tantalum is processed in manufacture to have an oxide film as the dielectric. Referring back to Table 17-1, note that tantalum oxide has a dielectric constant of 25, compared with 7 for aluminum oxide.

TEST-POINT QUESTION 17-5

Answers at end of chapter.

Answer True or False.
a. The rating of 1000 μF at 25 V could be for an electrolytic capacitor.
b. Electrolytic capacitors allow more leakage current than a mica capacitor.
c. Tantalum capacitors have a longer shelf life than aluminum electrolytics.

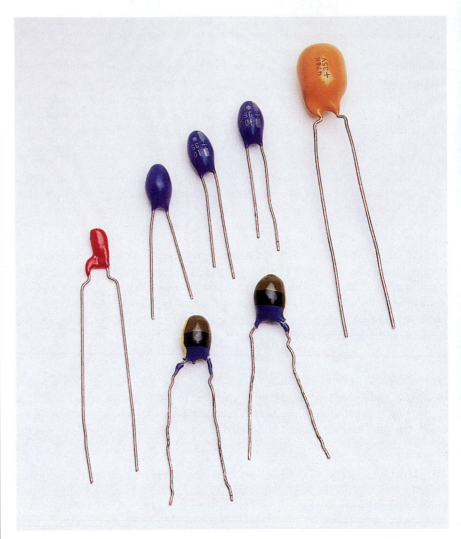

FIG. 17-10 Tantalum capacitors.

17-6 CAPACITOR CODING

The value of a capacitor is always specified in either μF or pF units of capacitance. This is true for all types of capacitors. As a general rule, if a capacitor (other than an electrolytic capacitor) is marked using a whole number such as 33, 220, 680, etc., the capacitance C is in picofarads (pF). Conversely, if a capacitor is labeled using a decimal fraction such as 0.1, 0.047, or 0.0082, the capacitance C is in microfarads (μF). There are a variety of different ways in which a manufacturer may indicate the value of a capacitor. What follows is an explanation of the most frequently encountered coding systems.

FILM-TYPE CAPACITORS   Figure 17-11 shows a popular coding system used with film-type capacitors. The first two numbers printed on the capacitor indicate the first two digits in the numerical value of the capacitance. The third

Film-Type Capacitors

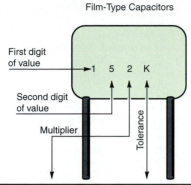

First digit
of value → 1 5 2 K

Second digit
of value

Multiplier

Tolerance

Multiplier			Tolerance of Capacitor	
For the Number	Multiplier	Letter	10 pF or Less	Over 10 pF
0	1	B	±0.1 pF	
1	10	C	±0.25 pF	
2	100	D	±0.5 pF	
3	1,000	F	±1.0 pF	±1%
4	10,000	G	±2.0 pF	±2%
5	100,000	H		±3%
		J		±5%
8	0.01	K		±10%
9	0.1	M		±20%

Examples:
  152K = 15 × 100 = 1500 pF or 0.0015 μF, ±10%
  759J = 75 × 0.1 = 7.5 pF, ±5%

Note: The letter R may be used at times to signify a decimal point, as in
2R2 = 2.2 (pF or μF).

FIG. 17-11   Film capacitor coding system.

number is the *multiplier,* indicating by what factor the first two digits must be multiplied. The letter at the far right indicates the capacitor's tolerance. With this coding system the capacitance is always in pF units. The capacitor's breakdown voltage rating is usually printed on the body directly below the coded value of capacitance.

Example

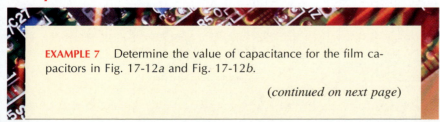

EXAMPLE 7   Determine the value of capacitance for the film capacitors in Fig. 17-12*a* and Fig. 17-12*b*.

(continued on next page)

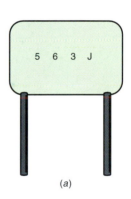

5 6 3 J

(a)

4 7 9 C

(b)

FIG. 17-12   Film capacitors for Example 7.

ANSWER In Fig. 17-12*a*, the first two numbers are 5 and 6, respectively, for 56 as the first two digits in the numerical value of the capacitance. The third number, 3, indicates a multiplier of 1000, or 56 × 1000 = 56,000 pF. The letter J indicates a capacitor tolerance of ±5 percent.

In Fig. 17-12*b*, the first two numbers are 4 and 7, respectively, for 47 as the first two digits in the numerical value of the capacitance. The third number, 9, indicates a fractional multiplier of 0.1, or 47 × 0.1 = 4.7 pF. The letter C indicates a capacitor tolerance of ±0.25 pF.

DISK CERAMIC CAPACITORS Figure 17-13 shows the way in which most disk ceramic capacitors are marked to indicate their capacitance. As you can see, the capacitance is expressed either as a whole number or as a decimal fraction. The type of coding system used depends on the manufacturer. Disk ceramic

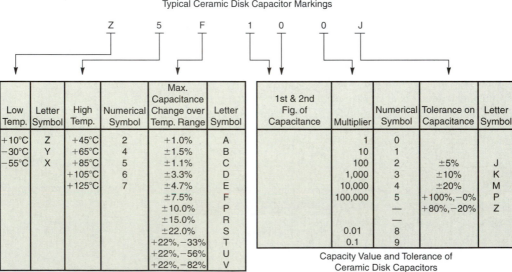

FIG. 17-13 Disk ceramic capacitor coding system.

capacitors are often used for coupling and bypassing ac signals, where it is allowable to have a wide or lopsided tolerance.

Example

FIG. 17-14   Disk ceramic capacitor for Example 8.

EXAMPLE 8   In Fig. 17-14, determine: (**a**) the capacitance value and tolerance; (**b**) the temperature-range identification information.

ANSWER   (**a**) Since the capacitance is expressed as a decimal fraction, its value is in microfarads. In this case, $C = 0.047 \mu F$. The letter Z, to the right of 0.047, indicates a capacitor tolerance of +80 percent, −20 percent. Notice that the actual capacitance value can be as much as 80 percent above its coded value but only 20 percent below its coded value.

(**b**) The alphanumeric code, Z5V, printed below the capacitance value provides additional capacitor information. Referring to Fig. 17-13, note that the letter Z and number 5 indicate the low and high temperatures of +10°C and +85°C, respectively. The letter V indicates that the maximum capacitance change over the specified temperature range (+10°C to +85°C) is +22 percent, −82 percent. With temperature changes less than the range indicated, the percent change in capacitance will be less than that indicated.

MICA CAPACITORS   Mica capacitors are coded using colored dots to indicate the capacitance value in picofarads. Three different coding systems are shown in Fig. 17-16. The color code is best understood through the use of an example.

Example

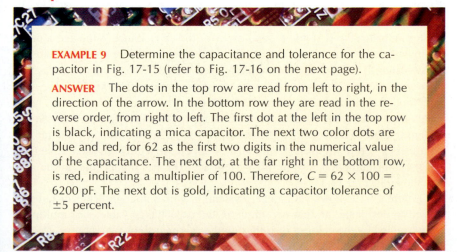

EXAMPLE 9   Determine the capacitance and tolerance for the capacitor in Fig. 17-15 (refer to Fig. 17-16 on the next page).

ANSWER   The dots in the top row are read from left to right, in the direction of the arrow. In the bottom row they are read in the reverse order, from right to left. The first dot at the left in the top row is black, indicating a mica capacitor. The next two color dots are blue and red, for 62 as the first two digits in the numerical value of the capacitance. The next dot, at the far right in the bottom row, is red, indicating a multiplier of 100. Therefore, $C = 62 \times 100 = 6200$ pF. The next dot is gold, indicating a capacitor tolerance of ±5 percent.

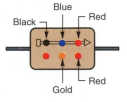

FIG. 17-15   Mica capacitor for Example 9.

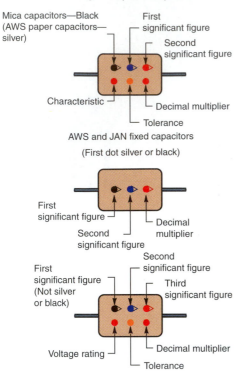

Postage Stamp Mica Capacitors

Mica capacitors—Black (AWS paper capacitors—silver)

First significant figure

Second significant figure

Characteristic

Decimal multiplier

Tolerance

AWS and JAN fixed capacitors

(First dot silver or black)

First significant figure

Second significant figure

Decimal multiplier

First significant figure (Not silver or black)

Second significant figure

Third significant figure

Voltage rating

Decimal multiplier

Tolerance

FIG. 17-16 Three different coding systems used with mica capacitors.

Color	Significant Figure	Multiplier	Tolerance (%)	Voltage Rating
Black	0	1	—	—
Brown	1	10	1	100
Red	2	100	2	200
Orange	3	1,000	3	300
Yellow	4	10,000	4	400
Green	5	100,000	5	500
Blue	6	1,000,000	6	600
Violet	7	10,000,000	7	700
Gray	8	100,000,000	8	800
White	9	1,000,000,000	9	900
Gold	—	0.1	5	1,000
Silver	—	0.01	10	2,000
No color	—	—	20	500

CHIP CAPACITORS Before determining the capacitance value of a chip capacitor, make sure it is a capacitor and not a resistor. Chip capacitors have the following identifiable features:

The body is one solid color, such as off-white, beige, gray, tan, or brown. Also, the end electrodes completely enclose the end of the part.

Three popular coding systems are currently being used by the different manufacturers of chip capacitors. In all three systems, the values represented are in picofarads. One system, shown in Fig. 17-17, uses a two-place system in which a letter indicates the first and second digits of the capacitance value and a number indicates the multiplier (0 to 9). Thirty-three symbols are used to represent the two significant figures. The symbols used include 24 uppercase letters and 9 lowercase letters. In Fig. 17-17, note that J3 represents 22,000 pF.

Another system, shown in Fig. 17-18, also uses two places. In this case, however, values below 100 pF are indicated using two numbers from which the capacitance value is read directly. Values above 100 pF are indicated by a letter and a number as before. In this system, only 24 uppercase letters are used. Also note that the alphanumeric codes in this system are 10 times higher than in the system shown in Fig. 17-17.

Figure 17-19 shows yet another system, in which a single letter or number is used to designate the first two digits in the capacitance value. The multiplier is determined by the color of the letter. In the example shown, an orange-colored W represents a capacitance C of 4.7 pF.

Value (33 Value Symbols)—Upper and Lowercase Letters					Multiplier
A-1.0	H-2.0	b-3.5	f-5.0	X-7.5	0 = × 1.0
B-1.1	J-2.2	P-3.6	T-5.1	t-8.0	1 = × 10
C-1.2	K-2.4	Q-3.9	U-5.6	Y-8.2	2 = × 100
D-1.3	a-2.5	d-4.0	m-6.0	y-9.0	3 = × 1,000
E-1.5	L-2.7	R-4.3	V-6.2	Z-9.1	4 = × 10,000
F-1.6	M-3.0	e-4.5	W-6.8		5 = × 100,000
G-1.8	N-3.3	S-4.7	n-7.0		etc.

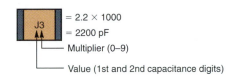

$= 2.2 \times 1000$
$= 2200$ pF
— Multiplier (0–9)
— Value (1st and 2nd capacitance digits)

FIG. 17-17 Chip capacitor coding system.

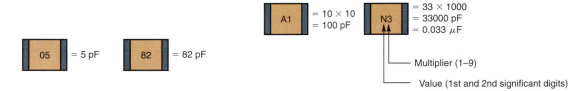

Alternate Two-Place Code
• Values below 100 pF—Value read directly

• Values 100 pF and above—Letter/number code

$= 10 \times 10$
$= 100$ pF

$= 33 \times 1000$
$= 33000$ pF
$= 0.033\ \mu F$

— Multiplier (1–9)
— Value (1st and 2nd significant digits)

$= 5$ pF $= 82$ pF

Value (24 Value Symbols)—Uppercase Letters Only					Multiplier
A-10	F-16	L-27	R-43	W-68	1 = × 10
B-11	G-18	M-30	S-47	X-75	2 = × 100
C-12	H-20	N-33	T-51	Y-82	3 = × 1,000
D-13	J-22	P-36	U-56	Z-91	4 = × 10,000
E-15	K-24	Q-39	V-62		5 = × 100,000 etc.

FIG. 17-18 Chip capacitor coding system.

Standard Single-Place Code
— Orange

$= 4.7 \times 1.0 = 4.7$ pF
— Color multiplier
— Symbol value

Examples: R (Green) = 3.3 × 100 = 330 pF
7 (Blue) = 8.2 × 1000 = 8200 pF

Value (24 Value Symbols)—Uppercase Letters and Numerals					Multiplier (Color)
A-1.0	H-1.6	N-2.7	V-4.3	3-6.8	Orange = × 1.0
B-1.1	I-1.8	O-3.0	W-4.7	4-7.5	Black = × 10
C-1.2	J-2.0	R-3.3	X-5.1	7-8.2	Green = × 100
D-1.3	K-2.2	S-3.6	Y-5.6	9-9.1	Blue = × 1,000
E-1.5	L-2.4	T-3.9	Z-6.2		Violet = × 10,000
					Red = × 100,000

FIG. 17-19 Chip capacitor coding system.

It should be noted that other coding systems are used with chip capacitors; these systems are not covered here. However, the three coding systems shown in this section are the most common systems presently in use.

TANTALUM CAPACITORS   Tantalum capacitors are frequently coded to indicate their capacitance in picofarads. Figure 17-20 shows how to interpret this system.

Dipped Tantalum Capacitors

| Color | Rated Voltage | Capacitance in Picofarads | | Multiplier |
		1st Figure	2nd Figure	
Black	4	0	0	—
Brown	6	1	1	—
Red	10	2	2	—
Orange	15	3	3	—
Yellow	20	4	4	10,000
Green	25	5	5	100,000
Blue	35	6	6	1,000,000
Violet	50	7	7	10,000,000
Gray	—	8	8	—
White	3	9	9	—

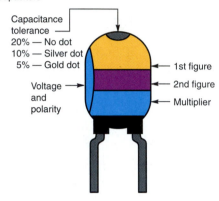

Capacitance tolerance
20% — No dot
10% — Silver dot
5% — Gold dot

Capacitance

1st figure
2nd figure
Multiplier

Voltage and polarity

FIG. 17-20   Tantalum capacitor coding system.

Example

Silver

Yellow
Violet
Blue

Blue

FIG. 17-21   Tantalum capacitor for Example 10.

EXAMPLE 10   For the tantalum capacitor shown in Fig. 17-21, determine the capacitance C in both pF and μF units. Also, determine the voltage rating and tolerance.

ANSWER   Moving from top to bottom, the first two color bands are yellow and violet, which represent the digits 4 and 7, respectively. The third color band is blue, indicating a multiplier of 1,000,000. Therefore the capacitance C is 47 × 1,000,000 = 47,000,000 pF, or 47 μF. The blue color at the left indicates a voltage rating of 35 V. And, finally, the silver dot at the very top indicates a tolerance of ±10 percent.

a. A disk ceramic capacitor that is marked .01 has a capacitance of 0.01 pF.
b. A film capacitor that is marked 224 has a capacitance of 220,000 pF.
c. A chip capacitor has a green letter E marked on it. Its capacitance is 150 pF.
d. A disk ceramic capacitor is marked .001P. Its tolerance is +100 percent, 0 percent.

17-7 PARALLEL CAPACITANCES

Connecting capacitances in parallel is equivalent to adding the plate areas. There-fore, the total capacitance is the sum of the individual capacitances. As illustrated in Fig. 17-22,

▶ $$C_T = C_1 + C_2 + \cdots + \text{etc.} \qquad (17\text{-}5)$$

A 10-μF capacitor in parallel with a 5-μF capacitor, for example, provides a 15-μF capacitance for the parallel combination. The voltage is the same across the parallel capacitors. Note that adding parallel capacitances is opposite to the case of inductances in parallel, and resistances in parallel.

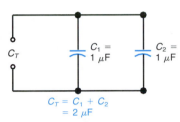

FIG. 17-22 Capacitances in parallel.

a. How much is C_T for 0.01 μF in parallel with 0.02 μF?
b. What C must be connected in parallel with 100 pF to make C_T of 250 pF?

17-8 SERIES CAPACITANCES

Connecting capacitances in series is equivalent to increasing the thickness of the dielectric. Therefore, the combined capacitance is less than the smallest individ-ual value. As shown in Fig. 17-23, the combined equivalent capacitance is cal-culated by the reciprocal formula:

▶ $$\frac{1}{C_{EQ}} = \frac{1}{C_1} + \frac{1}{C_2} + \cdots + \text{etc.} \qquad (17\text{-}6)$$

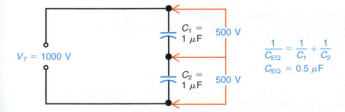

FIG. 17-23 Capacitances in series.

Any of the short-cut calculations for the reciprocal formula apply. For example, the combined capacitance of two equal capacitances of 10 μF in series is 5 μF.

Capacitors are used in series to provide a higher working voltage rating for the combination. For instance, each of three equal capacitances in series has one-third the applied voltage.

DIVISION OF VOLTAGE ACROSS UNEQUAL CAPACITANCES   In series, the voltage across each C is inversely proportional to its capacitance, as illustrated in Fig. 17-24. The smaller capacitance has the larger proportion of the applied voltage. The reason is that the series capacitances all have the same charge because they are in one current path. With equal charge, a smaller capacitance has a greater potential difference.

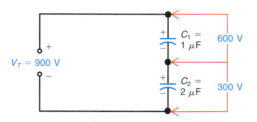

FIG. 17-24   With series capacitors, the smaller C has more voltage for the same charge.

We can consider the amount of charge in the series capacitors in Fig. 17-24. Let the charging current be 600 μA flowing for 1 s. The charge Q equals $I \times t$ or 600 μC. Both C_1 and C_2 have Q equal to 600 μC, as they are in the same series path for charging current.

Although the charge is the same in C_1 and C_2, they have different voltages because of different capacitance values. For each capacitor $V = Q/C$. For the two capacitors in Fig. 17-24, then:

$$V_1 = \frac{Q}{C_1} = \frac{600\ \mu C}{1\ \mu F} = 600\ V$$

$$V_2 = \frac{Q}{C_2} = \frac{600\ \mu F}{2\ \mu F} = 300\ V$$

CHARGING CURRENT FOR SERIES CAPACITANCES   The charging current is the same in all parts of the series path, including the junction between C_1 and C_2, even though this point is separated from the source voltage by two insulators. At the junction, the current is the resultant of electrons repelled by the negative plate of C_2 and attracted by the positive plate of C_1. The amount of current in the circuit is determined by the equivalent capacitance of C_1 and C_2 in series. In Fig. 17-24, the equivalent capacitance is $\frac{2}{3}$ μF.

TEST-POINT QUESTION 17-8

Answers at end of chapter.
a. How much is C_{EQ} for two 0.2-μF capacitors in series?
b. With 50 V applied across both, how much is V_C across each capacitor?
c. How much is C_{EQ} for 100 pF in series with 50 pF?

The study of surface-mount technology is on the rise at technical institutes. Here a student at Texas State Technical College in Harlingen, Texas, uses surface-mount laboratory equipment manufactured by PACE Inc.

17-9 STRAY CAPACITIVE AND INDUCTIVE EFFECTS

Stray capacitive and inductive effects can occur in all circuits with all types of components. A capacitor has a small amount of inductance in the conductors. A coil has some capacitance between windings. A resistor has a small amount of inductance and capacitance. After all, a capacitance physically is simply an insulator between two conductors having a difference of potential. An inductance is basically just a conductor carrying current.

Actually, though, these stray effects are usually quite small, compared with the concentrated or lumped values of capacitance and inductance. Typical values of stray capacitance may be 1 to 10 pF, while stray inductance is usually a fraction of 1 μH. For very high radio frequencies, however, when small values of L and C must be used, the stray effects become important. As another example, any wire cable has capacitance between the conductors.

A practical case of problems caused by stray L and C is the example of a long cable used for RF signals. If the cable is rolled in a coil to save space, a serious change in the electrical characteristics of the line will take place. Specifically, for twin-lead or coaxial cable feeding the antenna input to a television receiver, the line should not be coiled, as the added L or C can affect the signal. Any excess line should be cut off, leaving just the little slack that may be needed. This precaution is not so important with audio cables.

STRAY CIRCUIT CAPACITANCE   The wiring and the components in a circuit have capacitance to the metal chassis. This stray capacitance C_S is typically 5 to 10 pF. To reduce C_S, the wiring should be short, with the leads and components placed high off the chassis. Sometimes, for very high frequencies, the stray capacitance is included as part of the circuit design. Then changing the

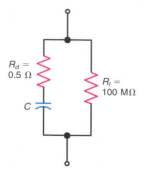

FIG. 17-25 Equivalent circuit of a capacitor; R_l is leakage resistance and R_d is absorption loss dissipated in dielectric.

placement of components or wiring affects the circuit operation. Such critical *lead dress* is usually specified in the manufacturer's service notes.

LEAKAGE RESISTANCE OF A CAPACITOR   Consider a capacitor charged by a dc voltage source. After the charging voltage is removed, a perfect capacitor would keep its charge indefinitely. Because there is no perfect insulator, after a long period of time, however, the charge will be neutralized by a small leakage current through the dielectric and across the insulated case between terminals. For paper, ceramic, and mica capacitors, though, the leakage current is very slight or, inversely, the leakage resistance is very high. As shown in Fig. 17-25, the leakage resistance R_l is indicated by a high resistance in parallel with the capacitance C. For paper, ceramic, or mica capacitors R_l is 100 MΩ or more. However, electrolytic capacitors may have a leakage resistance which is much less.

ABSORPTION LOSSES IN CAPACITORS   With ac voltage applied to a capacitor, the continuous charge, discharge, and reverse charging action cannot be followed instantaneously in the dielectric. This corresponds to hysteresis in magnetic materials. With a high-frequency charging voltage applied to the capacitor, there may be a difference between the amount of ac voltage applied and the ac voltage stored in the dielectric. The difference can be considered *absorption loss* in the dielectric. With higher frequencies, the losses increase. In Fig. 17-25, the small value of 0.5 Ω for R_d indicates a typical value for paper capacitors. For ceramic and mica capacitors, the dielectric losses are even smaller. These losses need not be considered for electrolytic capacitors because they are generally not used for radio frequencies.

POWER FACTOR OF A CAPACITOR   The quality of a capacitor in terms of minimum loss is often indicated by its power factor. The lower the numerical value of the power factor, the better is the quality of the capacitor. Since the losses are in the dielectric, the power factor of the capacitor is essentially the power factor of the dielectric, independent of capacitance value or voltage rating. At radio frequencies, approximate values of power factor are 0.000 for air or vacuum, 0.0004 for mica, about 0.01 for paper, and 0.0001 to 0.03 for ceramics.

The reciprocal of the power factor can be considered the Q of the capacitor, similar to the idea of Q of a coil. For instance, a power factor of 0.001 corresponds to a Q of 1000. A higher Q therefore means better quality for the capacitor.

INDUCTANCE OF A CAPACITOR   Capacitors with a coiled construction, particularly paper and electrolytic capacitors, have some internal inductance. The larger the capacitor, the greater is its series inductance. Mica and ceramic capacitors have very little inductance, however, which is why they are generally used for radio frequencies.

For use above audio frequencies, the rolled-foil type of capacitor must have a noninductive construction. This means the start and finish of the foil winding must not be the terminals of the capacitor. Instead, the foil windings are offset. Then one terminal can contact all layers of one foil at one edge, while the opposite edge of the other foil contacts the second terminal. Most rolled-foil capacitors, including the paper and film types, are constructed this way.

DISTRIBUTED CAPACITANCE OF A COIL As illustrated in Fig. 17-26, a coil has distributed capacitance C_d between turns. Note that each turn is a conductor separated from the next turn by an insulator, which is the definition of capacitance. Furthermore, the potential of each turn is different from the next, providing part of the total voltage as a potential difference to charge C_d. The result then is the equivalent circuit shown for an RF coil. The L is the inductance and R_e its internal effective ac resistance in series with L, while the total distributed capacitance C_d for all the turns is across the entire coil.

Special methods for minimum C_d include *space-wound* coils, where the turns are spaced far apart; the honeycomb or *universal* winding, with the turns crossing each other at right angles; and the *bank winding,* with separate sections called *pies*. These windings are for RF coils. In audio and power transformers, a grounded conductor shield, called a *Faraday screen,* is often placed between windings to reduce capacitive coupling.

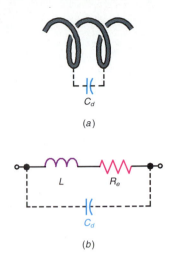

FIG. 17-26 Equivalent circuit of an RF coil. (*a*) Distributed capacitance C_d between turns of wire. (*b*) Equivalent circuit.

REACTIVE EFFECTS IN RESISTORS As illustrated by the high-frequency equivalent circuit in Fig. 17-27, a resistor can include a small amount of inductance and capacitance. For carbon-composition resistors, the inductance is usually negligible. However, approximately 0.5 pF of capacitance across the ends may have an effect, particularly with large resistances used for high radio frequencies. Wire-wound resistors definitely have enough inductance to be evident at radio frequencies. However, special resistors are available with double windings in a noninductive method based on cancellation of opposing magnetic fields.

CAPACITANCE OF AN OPEN CIRCUIT An open switch or a break in a conducting wire has capacitance C_O across the open. The reason is that the open consists of an insulator between two conductors. With a voltage source in the circuit, C_O charges to the applied voltage. Because of the small C_O, in the order of picofarads, the capacitance charges to the source voltage in a short time. This charging of C_O is the reason why an open series circuit has the applied voltage across the open terminals. After a momentary flow of charging current, C_O charges to the applied voltage and stores the charge needed to maintain this voltage.

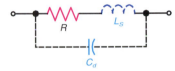

FIG. 17-27 High-frequency equivalent circuit of a resistor.

TEST-POINT QUESTION 17-9

Answers at end of chapter.

Answer True or False.
a. A two-wire cable has distributed C between the conductors.
b. A coil has distributed C between the turns.
c. The leakage resistance of ceramic capacitors is very high.

17-10 ENERGY IN ELECTROSTATIC FIELD OF CAPACITANCE

The electrostatic field of the charge stored in the dielectric has electric energy supplied by the voltage source that charges C. This energy is stored in the dielectric. The proof is the fact that the capacitance can produce discharge current when the voltage source is removed. The electric energy stored is

$$\text{Energy} = \mathcal{E} = \tfrac{1}{2}\,CV^2 \qquad \text{joules} \tag{17-7}$$

where C is the capacitance in farads and V is the voltage across the capacitor, and $\mathcal{E}$ is the electric energy in joules. For example, a 1-μF capacitor charged to 400 V has stored energy equal to

$$\mathcal{E} = \tfrac{1}{2}\,CV^2 = \frac{1 \times 10^{-6} \times (4 \times 10^2)^2}{2}$$

$$= \frac{1 \times 10^{-6} \times (16 \times 10^4)}{2} = 8 \times 10^{-2}$$

$$\mathcal{E} = 0.08 \text{ J}$$

This 0.08 J of energy is supplied by the voltage source that charges the capacitor to 400 V. When the charging circuit is opened, the stored energy remains as charge in the dielectric. With a closed path provided for discharge, the entire 0.08 J is available to produce discharge current. As the capacitor discharges, the energy is used in producing discharge current. When the capacitor is completely discharged, the stored energy is zero.

The stored energy is the reason why a charged capacitor can produce an electric shock, even when not connected into a circuit. When you touch the two leads of the charged capacitor, its voltage produces discharge current through your body. Stored energy greater than 1 J can be dangerous with a capacitor charged to a voltage high enough to produce an electric shock.

Example

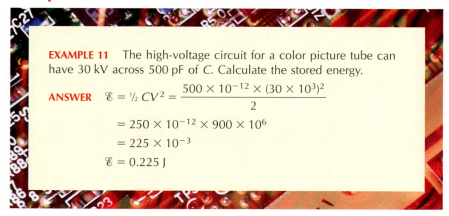

EXAMPLE 11　The high-voltage circuit for a color picture tube can have 30 kV across 500 pF of C. Calculate the stored energy.

ANSWER　$\mathcal{E} = \tfrac{1}{2}\,CV^2 = \dfrac{500 \times 10^{-12} \times (30 \times 10^3)^2}{2}$

$$= 250 \times 10^{-12} \times 900 \times 10^6$$

$$= 225 \times 10^{-3}$$

$$\mathcal{E} = 0.225 \text{ J}$$

Notice that the energy is less, even with 30 kV, because C is so small.

TEST-POINT QUESTION 17-10

Answers at end of chapter.

Answer True or False.
a. The stored energy in C increases with more V.
b. The stored energy decreases with less C.

17-11 TROUBLES IN CAPACITORS

Capacitors can become open or short-circuited. In either case, the capacitor is useless because it cannot store charge. A leaky capacitor is equivalent to a partial short circuit where the dielectric gradually loses its insulating properties under the stress of applied voltage, lowering its resistance. A good capacitor has very high resistance of the order of megohms; a short-circuited capacitor has zero ohms resistance, or continuity; the resistance of a leaky capacitor is lower than normal.

CHECKING CAPACITORS WITH AN OHMMETER   A capacitor usually can be checked with an ohmmeter. The highest ohms range, such as $R \times$ 1 MΩ, is preferable. Also, disconnect one side of the capacitor from the circuit to eliminate any parallel resistance paths that can lower the resistance. Keep your fingers off the connections, since the body resistance lowers the reading.

As illustrated in Fig. 17-28, the ohmmeter leads are connected across the capacitor. For a good capacitor, the meter pointer moves quickly toward the low-resistance side of the scale and then slowly recedes toward infinity. The reading when the pointer stops moving is the insulation resistance of the capacitor, which is normally very high. For paper, mica, and ceramic capacitors, the resistance can be 500 to 1000 MΩ, or more, which is practically infinite resistance. However, electrolytic capacitors will usually measure a much lower resistance of about 500 kΩ to 10 MΩ. In all cases, discharge the capacitor before checking with the ohmmeter.

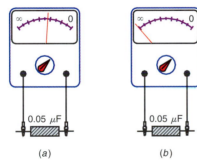

(a)   (b)

FIG. 17-28   Checking a capacitor with an ohmmeter. The R scale is shown right to left, as on a VOM. Use the highest ohms range. (*a*) Capacitor action as needle is moved by the charging current from the battery in the ohmmeter. (*b*) Practically infinite leakage resistance reading after the capacitor has been charged.

When the ohmmeter is initially connected, its battery charges the capacitor. This charging current is the reason the meter pointer moves away from infinity, since more current through the ohmmeter means less resistance. Maximum current flows at the first instant of charge. Then the charging current decreases as the capacitor voltage increases toward the applied voltage; therefore, the needle pointer slowly moves toward infinite resistance. Finally, the capacitor is completely charged to the ohmmeter battery voltage, the charging current is zero, and the ohmmeter reads just the small leakage current through the dielectric. This charging effect, called *capacitor action,* shows that the capacitor can store charge, indicating a normal capacitor. It should be noted that both the rise and fall of the

DID YOU KNOW?

Flashes far above storm clouds are caused by nitrogen molecules becoming agitated by electron collisions. During the process, lightning sends pulses of energy upward and causes slower changes in the atmosphere's electric field.

meter readings are caused by charging. The capacitor discharges when the meter leads are reversed.

OHMMETER READINGS   Troubles in a capacitor are indicated as follows:

1. If an ohmmeter reading immediately goes practically to zero and stays there, the capacitor is short-circuited.
2. If the capacitor shows charging, but the final resistance reading is appreciably less than normal, the capacitor is leaky. Such capacitors are particularly troublesome in high-resistance circuits. When checking electrolytics, reverse the ohmmeter leads and take the higher of the two readings.
3. If the capacitor shows no charging action but just reads very high resistance, it may be open. Some precautions must be remembered, however, since very high resistance is a normal condition for capacitors. Reverse the ohmmeter leads to discharge the capacitor, and check it again. In addition, remember that capacitance values of 100 pF, or less, normally have very little charging current for the low battery voltage of the ohmmeter.

SHORT-CIRCUITED CAPACITORS   In normal service, capacitors can become short-circuited because the dielectric deteriorates with age, usually over a period of years under the stress of charging voltage, especially with higher temperatures. This effect is more common with paper and electrolytic capacitors. The capacitor may become leaky gradually, indicating a partial short circuit, or the dielectric may be punctured, causing a short circuit.

OPEN CAPACITORS   In addition to the possibility of an open connection in any type of capacitor, electrolytics develop high resistance in the electrolyte with age, particularly at high temperatures. After service of a few years, if the electrolyte dries up, the capacitor will be partially open. Much of the capacitor action is gone, and the capacitor should be replaced.

LEAKY CAPACITORS   A leaky capacitor reads R less than normal with an ohmmeter. However, dc voltage tests are more definite. In a circuit, the dc voltage at one terminal of the capacitor should not affect the dc voltage at the other terminal.

SHELF LIFE   Except for electrolytics, capacitors do not deteriorate with age while stored, since there is no applied voltage. Electrolytic capacitors, however, like dry cells, should be used fresh from manufacture. The reason is the wet electrolyte may dry out over a period of time.

CAPACITOR VALUE CHANGE   All capacitors can change value over time, but some are more prone to change than others. Ceramic capacitors often change value by 10 to 15 percent over the first year, as the ceramic material relaxes. Electrolytics change value from simply sitting, because the electrolytic solution dries out.

REPLACEMENT CAPACITORS   Approximately the same *C* and *V* ratings should be used when installing a new capacitor. Except for tuning capacitors, the *C* value is not critical. In most applications the tolerance of capacitors is −20 to +50 percent. Also, a higher voltage rating can be used. An important exception, however, is the electrolytic capacitor. Then the ratings should be close to the original values for two reasons. First, the specified voltage is needed to form the internal oxide film that provides the required capacitance. Also, too much *C* may allow excessive charging current in the circuit that charges the capacitor. Remember that electrolytics generally have large values of capacitance.

TEST-POINT QUESTION 17-11

Answers at end of chapter.
a. What is the ohmmeter reading for a shorted capacitor?
b. Does capacitor action with an ohmmeter show the capacitor is good or bad?
c. Which type of capacitor is more likely to develop trouble, mica or electrolytic?

17 SUMMARY AND REVIEW

- A capacitor consists of two conductors separated by an insulator, or dielectric. Its ability to store charge is the capacitance C. Applying voltage to store charge is called charging the capacitor; short-circuiting the two leads or terminals of the capacitor to neutralize the charge is called discharging the capacitor. Schematic symbols for C are summarized in Fig. 17-29.

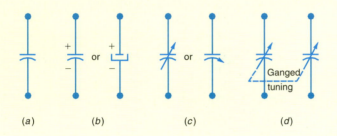

FIG. 17-29 Schematic symbols for types of C. (*a*) Fixed type with air, paper, plastic film, mica, or ceramic dielectric. (*b*) Electrolytic type, which has polarity. (*c*) Variable. (*d*) Ganged variable capacitors on one shaft.

- The unit of capacitance is the farad. One farad of capacitance stores one coulomb of charge with one volt applied. Practical capacitors have much smaller capacitance values, from 1 pF to 1000 μF. A capacitance of 1 pF is 1×10^{-12} F; 1 μF $= 1 \times 10^{-6}$ F; and 1 nF $= 1 \times 10^{-9}$ F.
- $Q = CV$, where Q is the charge in coulombs, C the capacitance in farads, and V is the potential difference across the capacitor in volts.
- Capacitance increases with larger plate area and less distance between the plates.
- The ratio of charge stored in different insulators to the charge stored in air is the dielectric constant K_ϵ of the material. Air or vacuum has a dielectric constant of 1.
- The most common types of commercial capacitors are air, plastic film, paper, mica, ceramic, and electrolytic. Electrolytics are the only capacitors that require observing polarity when connecting to a circuit. The different types are compared in Table 17-2.
- Capacitors are coded to indicate their capacitance, in either microfarads (μF) or picofarads (pF).
- For parallel capacitors, $C_T = C_1 + C_2 + C_3 + \cdots +$ etc.
- For series capacitors, $1/C_{EQ} = 1/C_1 + 1/C_2 + 1/C_3 + \cdots +$ etc.
- The electric field of a capacitance has stored energy $\mathscr{E} = \frac{1}{2} CV^2$, where V is volts, C is in farads, and the electric energy is in joules.
- When checked with an ohmmeter, a good capacitor shows charging current, and then the ohmmeter reading steadies at the insulation resistance. All types except electrolytics normally have a very high insulation resistance of 500 to 1000 MΩ. Electrolytics have more leakage current, with a typical resistance of about 500 kΩ to 10 MΩ.

Choose (*a*), (*b*), (*c*), or (*d*).

1. A capacitor consists of two (*a*) conductors separated by an insulator; (*b*) insulators separated by a conductor; (*c*) conductors alone; (*d*) insulators alone.
2. A capacitance of 0.02 μF equals (*a*) 0.02×10^{-12} F; (*b*) 0.02×10^{-6} F; (*c*) 0.02×10^{6} F; (*d*) 200×10^{-12} F.
3. A 10-μF capacitance charged to 10 V has a stored charge equal to (*a*) 10 μC; (*b*) 100 μC; (*c*) 200 μC; (*d*) 1 C.
4. Capacitance increases with (*a*) larger plate area and greater distance between plates; (*b*) smaller plate area and less distance between plates; (*c*) larger plate area and less distance between plates; (*d*) higher values of applied voltage.
5. Which of the following statements is correct? (*a*) Air capacitors have a black band to indicate the outside foil. (*b*) Mica capacitors are available in capacitance values of 1 to 10 μF. (*c*) Electrolytic capacitors must be connected in the correct polarity. (*d*) Ceramic capacitors must be connected in the correct polarity.
6. Voltage applied across a ceramic dielectric produces an electrostatic field 100 times greater than in air. The dielectric constant K_ϵ of the ceramic equals (*a*) 33⅓; (*b*) 50; (*c*) 100; (*d*) 10,000.
7. A six-dot mica capacitor color-coded white, red, green, brown, red, and yellow has the capacitance value of (*a*) 25 pF; (*b*) 124 pF; (*c*) 250 pF; (*d*) 925 pF.
8. The combination of two 0.02-μF 500-V capacitors in series has a capacitance and a working voltage rating of (*a*) 0.01 μF, 500 V; (*b*) 0.01 μF, 1000 V; (*c*) 0.02 μF, 500 V; (*d*) 0.04 μF, 500 V.
9. The combination of two 0.02-μF 500-V capacitors in parallel has a capacitance and a working voltage rating of (*a*) 0.01 μF, 1000 V; (*b*) 0.02 μF, 500 V; (*c*) 0.04 μF, 500 V; (*d*) 0.04 μF, 1000 V.
10. For a good 0.05-μF paper capacitor, the ohmmeter reading should (*a*) go quickly to 100 Ω, approximately, and remain there; (*b*) show low resistance momentarily and back off to a very high resistance; (*c*) show high resistance momentarily and then a very low resistance; (*d*) not move at all.

QUESTIONS

1. Define capacitance with respect to physical structure and electrical function. Explain how a two-wire conductor has capacitance.
2. **(a)** What is meant by a dielectric material? **(b)** Name five common dielectric materials. **(c)** Define dielectric flux.
3. Explain briefly how to charge a capacitor. How is a charged capacitor discharged?
4. Define 1 F of capacitance. Convert the following into farads using powers of 10: **(a)** 50 pF; **(b)** 0.001 μF; **(c)** 0.047 μF; **(d)** 0.01 μF; **(e)** 10 μF.

5. State the effect on capacitance of (a) larger plate area; (b) thinner dielectric; (c) higher value of dielectric constant.

6. Give one reason for your choice of the type of capacitor to be used in the following applications: (a) 80-μF capacitance for a circuit where one side is positive and the applied voltage never exceeds 150 V; (b) 1.5-pF capacitance for an RF circuit where the required voltage rating is less than 500 V; (c) 5-μF capacitance for an audio circuit where the required voltage rating is less than 25 V.

7. Give the capacitance value of six-dot mica capacitors color-coded as follows: (a) Black, red, green, brown, black, black. (b) White, green, brown, black, silver, brown. (c) Brown, green, black, red, gold, blue.

8. Draw a diagram showing the least number of 400-V 2-μF capacitors needed for a combination rated at 800 V with 2 μF total capacitance.

9. Suppose you are given two identical uncharged capacitors. One is charged to 50 V and connected across the uncharged capacitor. Why will the voltage across both capacitors then be 25 V?

10. Describe briefly how you would check a 0.05-μF capacitor with an ohmmeter. State the ohmmeter indications for the case of the capacitor being good, short-circuited, or open.

11. Define the following: (a) stray circuit capacitance; (b) distributed capacitance of a coil; (c) leakage resistance of a capacitor; (d) power factor and Q of a capacitor.

12. Give two comparisons between the electric field in a capacitor and the magnetic field in a coil.

13. Give three types of troubles in capacitors.

14. When a capacitor discharges, why is its discharge current in the opposite direction from the charging current?

15. Compare the features of aluminum and tantalum electrolytic capacitors.

16. Why can plastic film be used instead of paper for capacitors?

17. What two factors determine the breakdown voltage rating of a capacitor?

PROBLEMS

ANSWERS TO ODD-NUMBERED PROBLEMS AT BACK OF BOOK.

1. How much charge in coulombs is in a 4-μF capacitor charged to 100 V?

2. A 4-μF capacitor has 400 μC of charge. (a) How much voltage is across the capacitor? (b) How much is the voltage across an 8-μF capacitor with the same 400-μC charge?

3. A 2-μF capacitor is charged by a constant 3-μA charging current for 6 s. (a) How much charge is stored in the capacitor? (b) How much is the voltage across the capacitor?

4. A 1-μF capacitor C_1 and a 10-μF capacitor C_2 are in series with a constant 2-mA charging current. (a) After 4 s, how much charge is in C_1 and in C_2? (b) How much is the voltage across capacitor C_1 and across capacitor C_2?

5. Calculate C for a mica capacitor, with $K_\epsilon = 8$, a thickness of 0.02 cm, plates of 6 cm^2, and five sections in parallel. (Hint: 1 cm $= 10^{-2}$ m and 1 cm$^2 = 10^{-4}$ m^2.)

6. How much capacitance stores 6000 μC of charge with 150 V applied? The charge of how many electrons is stored? What type of capacitor is this most likely to be?

7. With 100 V across a capacitor, it stores 100 μC of charge. Then the applied voltage is doubled to 200 V. **(a)** How much is the voltage across the capacitor? **(b)** How much charge is stored? **(c)** How much is its capacitance?

8. Referring to the parallel capacitors in Fig. 17-22, calculate the charge Q_1 in C_1 and Q_2 in C_2 with 50 V. How much is the total charge Q_T in both capacitors? Calculate the total capacitance C_T as Q_T/V.

9. Calculate the energy in joules stored in **(a)** a 500-pF capacitor charged to 10 kV; **(b)** a 1-μF capacitor charged to 5 kV; **(c)** a 40-μF capacitor charged to 400 V.

10. Three capacitors are in series. C_1 is 100 pF, C_2 is 100 pF, and C_3 is 50 pF. Calculate C_{EQ}.

11. Calculate C_T for the series-parallel combination of capacitors in Figs. 17-30a and 17-30b.

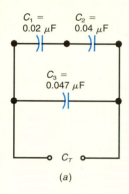

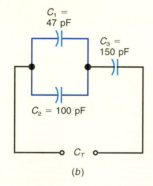

FIG. 17-30 Circuit diagrams for Probs. 11, 13, and 14.

(a) (b)

12. What C must be connected in series with 0.47 μF for an equivalent capacitance of 0.02 μF? (Hint: Use formulas for parallel R).

13. In Fig. 17-30a: **(a)** Change C_3 to 10 pF, and calculate C_T in μF units; **(b)** change C_2 to 10 pF and calculate C_T.

14. In Fig. 17-30b: **(a)** Change C_3 to 100 pF, and calculate C_T in pF units; **(b)** change C_1 to 100 pF and calculate C_T in pF units.

15. Determine the capacitance and tolerance for the capacitors shown in Fig. 17-31.

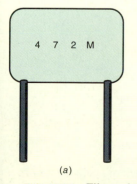
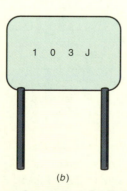
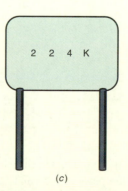
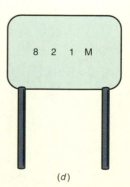

(a) (b) (c) (d)

FIG. 17-31 Film and disk ceramic capacitors for Prob. 15. (*continues on next page*)

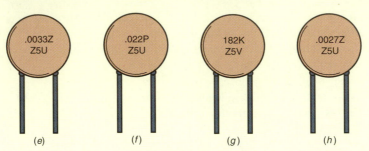

FIG. 17-31   (*continued*) Film and disk ceramic capacitors for Prob. 15.

16.   Determine the capacitance of each chip capacitor shown in Fig. 17-32. Use the coding system shown in Fig. 17-17.

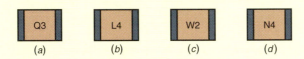

FIG. 17-32   Chip capacitors for Prob. 16.

17.   Determine the capacitance for each chip capacitor shown in Fig. 17-33. Use the coding system shown in Fig. 17-18.

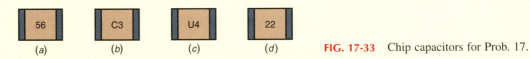

FIG. 17-33   Chip capacitors for Prob. 17.

18.   Determine the capacitance for each chip capacitor in Fig. 17-34.

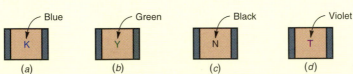

FIG. 17-34   Chip capacitors for Prob. 18.

19.   Determine the capacitance for each tantalum capacitor in Fig. 17-35.
20.   Calculate the permissible capacitance range at 25°C for the disk ceramic capacitor in Fig. 17-31*e*.

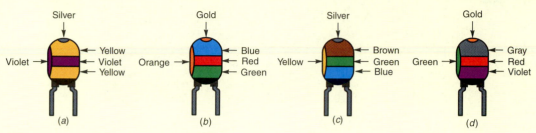

FIG. 17-35 Tantalum capacitors for Prob. 19.

CRITICAL THINKING

1. Three capacitors in series have a combined equivalent capacitance, C_{EQ}, of 1.6 nF. If $C_1 = 4C_2$ and $C_3 = 20C_1$, calculate the values for C_1, C_2, and C_3.

2. A 100-pF ceramic capacitor has a temperature coefficient T_C of N500. Calculate its capacitance at: **(a)** 75°C; **(b)** 125°C; **(c)** −25°C.

3. **(a)** Calculate the energy stored by a 100-μF capacitor charged to 100 V. **(b)** If this capacitor is now connected across another 100-μF capacitor which is uncharged, calculate the total energy stored by both capacitors. **(c)** Is the energy stored by both capacitors in part **(b)** less than the energy stored by the single capacitor in part **(a)**? If yes, where did the energy go?

ANSWERS TO TEST-POINT QUESTIONS

17-1	**a.** dielectric	**17-4**	**a.** T	**17-6**	**a.** F	**17-9**	**a.** T
	b. farad		**b.** F		**b.** T		**b.** T
17-2	**a.** 14.5 V		**c.** T		**c.** T		**c.** T
	b. 0 V		**d.** T		**d.** T	**17-10**	**a.** T
	c. Yes	**17-5**	**a.** T	**17-7**	**a.** 0.03 μF		**b.** T
17-3	**a.** 10 μF		**b.** T		**b.** 150 pF	**17-11**	**a.** 0 Ω
	b. ceramic		**c.** T	**17-8**	**a.** 0.1 μF		**b.** good
					b. 25 V		**c.** electrolytic
					c. 33.3 pF		

CHAPTER 18

CAPACITIVE REACTANCE

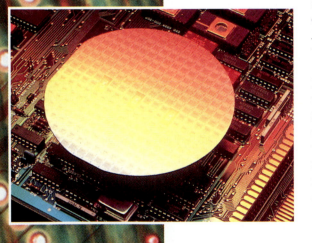

When a capacitor charges and discharges with a varying voltage applied, alternating current can flow. Although there cannot be any current through the dielectric of the capacitor, its charge and discharge produces alternating current in the circuit connected to the capacitor plates. The amount of I that results from the applied sine-wave V depends on the capacitor's capacitive reactance. The symbol for capacitive reactance is X_C and its unit is the ohm. The X in X_C indicates reactance, whereas the subscript C specifies capacitive reactance.

The amount of X_C is a V/I ratio but it can also be calculated as $X_C = 1/(2\pi f C)$ in terms of the value of the capacitance and the frequency of the varying V and I. With f and C in the basic units of the hertz and farad, the X_C is in the basic units of ohms. The reciprocal relation in $1/(2\pi f C)$ means that the ohms of X_C decrease for higher frequencies and with more C. The reason is that more charge and discharge current results either with more capacitance or faster changes in the applied voltage.

CHAPTER OBJECTIVES

Upon completion of this chapter, you should be able to:

- *Explain* how alternating current can flow in a capacitive circuit.
- *Calculate* the reactance of a capacitor when the frequency and capacitance are known.
- *Calculate* the total capacitive reactance of series connected capacitors.
- *Calculate* the equivalent capacitive reactance of parallel connected capacitors.
- *Explain* how Ohm's law can be applied to capacitive reactance.
- *Calculate* the capacitive current when the capacitance and rate of voltage change are known.

IMPORTANT TERMS IN THIS CHAPTER

capacitive reactance	discharge current	phase angle
charge current	inverse relation	series capacitance
dc blocking	parallel capacitance	

TOPICS COVERED IN THIS CHAPTER

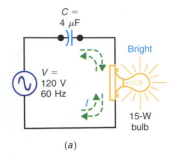

(a)

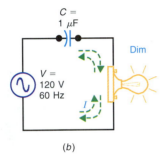

(b)

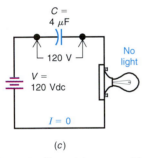

(c)

FIG. 18-1 Current in a capacitive circuit. (a) The 4-μF capacitor allows enough current I to light the bulb brightly. (b) Less current with smaller capacitor causes dim light. (c) Bulb cannot light with dc voltage applied because a capacitor blocks the direct current.

The fact that current flows with ac voltage applied is demonstrated in Fig. 18-1, where the bulb lights in Fig. 18-1a and b because of the capacitor charge and discharge current. There is no current through the dielectric, which is an insulator. While the capacitor is being charged by increasing applied voltage, however, the charging current flows in one direction in the conductors to the plates. While the capacitor is discharging, when the applied voltage decreases, the discharge current flows in the reverse direction. With alternating voltage applied, the capacitor alternately charges and discharges.

First the capacitor is charged in one polarity, and then it discharges; next the capacitor is charged in the opposite polarity, and then it discharges again. The cycles of charge and discharge current provide alternating current in the circuit, at the same frequency as the applied voltage. This is the current that lights the bulb.

In Fig. 18-1a, the 4-μF capacitor provides enough alternating current to light the bulb brightly. In Fig. 18-1b, the 1-μF capacitor has less charge and discharge current because of the smaller capacitance, and the light is not so bright. Therefore, the smaller capacitor has more opposition to alternating current as less current flows with the same applied voltage; that is, it has more reactance for less capacitance.

In Fig. 18-1c, the steady dc voltage will charge the capacitor to 120 V. Because the applied voltage does not change, though, the capacitor will just stay charged. Since the potential difference of 120 V across the charged capacitor is a voltage drop opposing the applied voltage, no current can flow. Therefore, the bulb cannot light. The bulb may flicker on for an instant as charging current flows when voltage is applied, but this current is only temporary until the capacitor is charged. Then the capacitor has the applied voltage of 120 V, but there is zero voltage across the bulb.

As a result, the capacitor is said to *block* direct current or voltage. In other words, after the capacitor has been charged by a steady dc voltage, there is no current in the dc circuit. All the applied dc voltage is across the charged capacitor, with zero voltage across any series resistance.

In summary, then, this demonstration shows the following points:

1. Alternating current flows in a capacitive circuit with ac voltage applied.
2. A smaller capacitance allows less current, which means more X_C with more ohms of opposition.
3. Lower frequencies for the applied voltage result in less current and more X_C. With a steady dc voltage source, which corresponds to a frequency of zero, the opposition of the capacitor is infinite and there is no current. In this case the capacitor is effectively an open circuit.

These effects have almost unlimited applications in practical circuits because X_C depends on frequency. A very common use of a capacitor is to provide little opposition for ac voltage but to block any dc voltage. Another example is to use X_C for less opposition to a high-frequency alternating current, compared with lower frequencies.

CAPACITIVE CURRENT   The reason why a capacitor allows current to flow in an ac circuit is the alternate charge and discharge. If we insert an ammeter

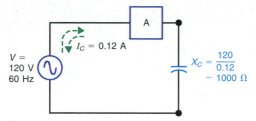

FIG. 18-2 Capacitive reactance X_C is the ratio V_C/I_C.

in the circuit, as shown in Fig. 18-2, the ac meter will read the amount of charge and discharge current. In this example I_C is 0.12 A. This current is the same in the voltage source, the connecting leads, and the plates of the capacitor. However, there is no current through the insulator between the plates of the capacitor.

VALUES FOR X_C   When we consider the ratio of V_C/I_C for the ohms of opposition to the sine-wave current, this value is 120/0.12, which equals 1000 Ω. This 1000 Ω is what we call X_C, to indicate how much current can be produced by sine-wave voltage applied to a capacitor. In terms of current, $X_C = V_C/I_C$. In terms of frequency and capacitance, $X_C = 1/(2\pi fC)$.

   The X_C value depends on the amount of capacitance and the frequency of the applied voltage. If C in Fig. 18-2 were increased, it could take on more charge for more charging current and then produce more discharge current. Then X_C is less for more capacitance. Also, if the frequency in Fig. 18-2 were increased, the capacitor could charge and discharge faster to produce more current. This action also means V_C/I_C would be less, with more current for the same applied voltage. Therefore, X_C is less for higher frequencies. Reactance X_C can actually have almost any value, from practically zero to almost infinite ohms.

TEST-POINT QUESTION 18-1

Answers at end of chapter.
a. Which has more reactance, a 0.1- or a 0.5-μF capacitor, at the same frequency?
b. Which allows more charge and discharge current, a 0.1- or a 0.5-μF capacitor, at the same frequency?

18-2 THE AMOUNT OF X_C EQUALS $1/(2\pi fC)$

The effects of frequency and capacitance are included in the formula for calculating the ohms of reactance. The f is in hertz units and C in farads for X_C in ohms. As an example, we can calculate X_C for C of 2.65 μF and f of 60 Hz. Then

$$X_C = \frac{1}{2\pi fC} \qquad\qquad\qquad\text{(18-1)}$$

$$= \frac{1}{2\pi \times 60 \times 2.65 \times 10^{-6}} = \frac{1}{6.28 \times 159 \times 10^{-6}}$$

$$= 0.00100 \times 10^6$$

$$X_C = 1000\ \Omega$$

The constant factor 2π, equal to 6.28, indicates the circular motion from which a sine wave is derived. Therefore, the formula for X_C applies only to sine-wave circuits. Remember that C must be in farad units for X_C in ohms. Although C values are usually microfarads (10^{-6}) or picofarads (10^{-12}), substitute the value of C in farads with the required negative power of 10.

Example

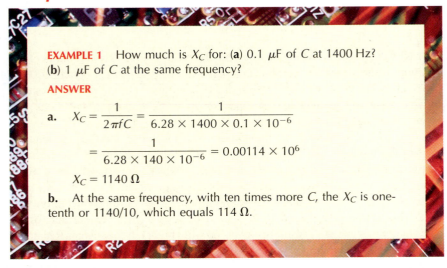

EXAMPLE 1   How much is X_C for: (**a**) 0.1 μF of C at 1400 Hz?
(**b**) 1 μF of C at the same frequency?

ANSWER

a. $X_C = \dfrac{1}{2\pi fC} = \dfrac{1}{6.28 \times 1400 \times 0.1 \times 10^{-6}}$

$$= \frac{1}{6.28 \times 140 \times 10^{-6}} = 0.00114 \times 10^6$$

$$X_C = 1140\ \Omega$$

b. At the same frequency, with ten times more C, the X_C is one-tenth or 1140/10, which equals 114 Ω.

When using Formula (18-1) with a calculator, probably the best method is to multiply all the factors in the denominator and then take the reciprocal of the total product. To save time, memorize 2π as $2 \times 3.14 = 6.28$. If your calculator does not have an (EXP) key, keep the powers of 10 separate. Remember that the negative sign of the exponent becomes positive in the reciprocal value. Specifically, for Example 1, the procedure can be as follows:

1. Punch in 6.28 as the numbers for 2π.
2. Press the ⊗ key and punch in the factor of 1400, then ⊗ and 0.1.
3. Press the ⊜ key to see the total product of 879.2.
4. While 879.2 is on the display, press the reciprocal key (1/x). This may require pushing the (2ndF) key first.
5. The reciprocal value is 0.00114.
6. The reciprocal of 10^{-6} in the denominator becomes 10^6 in the numerator.
7. For the final answer, then, move the decimal point six places to the right, as indicated by 10^6, for the final answer of 1140.

Example

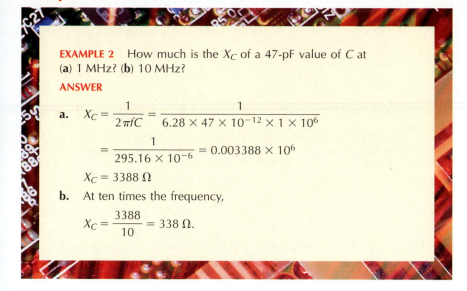

EXAMPLE 2 How much is the X_C of a 47-pF value of C at (a) 1 MHz? (b) 10 MHz?

ANSWER

a. $X_C = \dfrac{1}{2\pi f C} = \dfrac{1}{6.28 \times 47 \times 10^{-12} \times 1 \times 10^{6}}$

$= \dfrac{1}{295.16 \times 10^{-6}} = 0.003388 \times 10^{6}$

$X_C = 3388\ \Omega$

b. At ten times the frequency,

$X_C = \dfrac{3388}{10} = 338\ \Omega.$

Note that X_C in Example 2b is one-tenth the value in Example 2a because the f is 10 times greater.

X_C IS INVERSELY PROPORTIONAL TO CAPACITANCE   This statement means that X_C increases as the capacitance is decreased. In Fig. 18-3, when C is reduced by the factor of 1/10, from 1.0 to 0.1 μF, then X_C increases ten times, from 1000 to 10,000 Ω. Also, decreasing C one-half, from 0.2 to 0.1 μF, doubles X_C from 5000 to 10,000 Ω.

This inverse relation between C and X_C is illustrated by the graph in Fig. 18-3. Note that values of X_C increase downward on the graph, indicating negative reactance that is opposite from inductive reactance. With C increasing to the right, the decreasing values of X_C approach the zero axis of the graph.

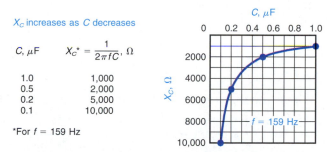

FIG. 18-3   A table of values and a graph to show that capacitive reactance X_C decreases with higher values of C. Frequency is constant at 159 Hz.

X_C IS INVERSELY PROPORTIONAL TO FREQUENCY   Figure 18-4 illustrates the inverse relationship between X_C and f. With f increasing to the right

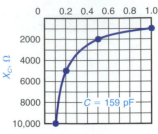

X_C increases as f decreases

f, MHz	$X_C^* = \dfrac{1}{2\pi fC}$, Ω
1.0	1,000
0.5	2,000
0.2	5,000
0.1	10,000

*For C = 159 pF

FIG. 18-4 A table of values and a graph to show that capacitive reactance X_C decreases with higher frequencies. C is constant at 159 pF.

in the graph from 0.1 to 1 MHz, the negative value of X_C for the 159-pF capacitor decreases from 10,000 to 1000 Ω as the X_C curve comes closer to the zero axis.

The graphs are nonlinear because of the inverse relation between X_C and f or C. At one end, the curves approach infinitely high reactance for zero capacitance or zero frequency. At the other end, the curves approach zero reactance for infinitely high capacitance or frequency.

CALCULATING *C* FROM ITS REACTANCE   In some applications, it may be necessary to find the value of capacitance required for a desired amount of X_C. For this case the reactance formula can be inverted to

$$C = \frac{1}{2\pi f X_C}$$ **(18-2)**

The value of 6.28 for 2π is still used. The only change from Formula (18-1) is that the C and X_C values are inverted between denominator and numerator on the left and right side of the equation.

Example

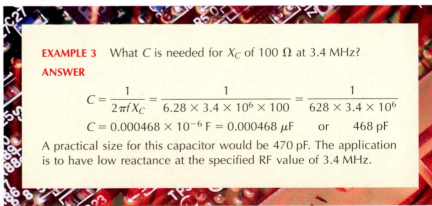

EXAMPLE 3   What C is needed for X_C of 100 Ω at 3.4 MHz?

ANSWER

$$C = \frac{1}{2\pi f X_C} = \frac{1}{6.28 \times 3.4 \times 10^6 \times 100} = \frac{1}{628 \times 3.4 \times 10^6}$$

$$C = 0.000468 \times 10^{-6} \text{ F} = 0.000468 \ \mu\text{F} \quad \text{or} \quad 468 \text{ pF}$$

A practical size for this capacitor would be 470 pF. The application is to have low reactance at the specified RF value of 3.4 MHz.

CALCULATING FREQUENCY FROM THE REACTANCE   Another use is to find the frequency at which a capacitor has a specified amount of X_C. Again, the reactance formula can be inverted to the form shown in Formula (18-3).

$$f = \frac{1}{2\pi C X_C} \qquad\qquad \textbf{(18-3)}$$

The following example illustrates the use of this formula.

Example

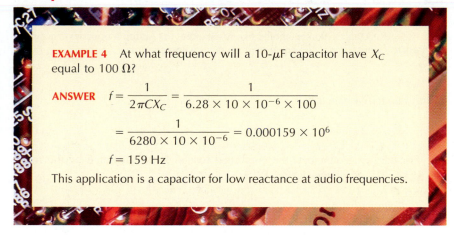

EXAMPLE 4 At what frequency will a 10-μF capacitor have X_C equal to 100 Ω?

ANSWER $\quad f = \dfrac{1}{2\pi C X_C} = \dfrac{1}{6.28 \times 10 \times 10^{-6} \times 100}$

$$= \frac{1}{6280 \times 10 \times 10^{-6}} = 0.000159 \times 10^{6}$$

$$f = 159 \text{ Hz}$$

This application is a capacitor for low reactance at audio frequencies.

SUMMARY OF THE X_C FORMULAS Formula (18-1) is the basic form to calculate X_C when f and C are the known values. As another possibility, the value of X_C can be measured as V_C/I_C.

With X_C known, the value of C can be calculated for a specified f by Formula (18-2). Or the f can be calculated with a known value of C by using Formula (18-3).

TEST-POINT QUESTION 18-2

Answers at end of chapter.

The X_C for a capacitor is 400 Ω at 8 MHz.
a. How much is X_C at 16 MHz?
b. How much is X_C at 4 MHz?
c. Is a smaller or larger C needed for less X_C?

18-3 SERIES OR PARALLEL CAPACITIVE REACTANCES

Because capacitive reactance is an opposition in ohms, series or parallel reactances are combined in the same way as resistances. As shown in Fig. 18-5a, series capacitive reactances are added arithmetically.

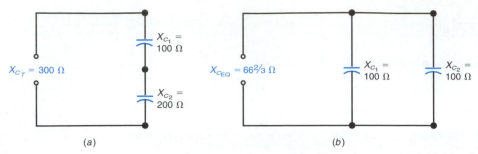

FIG. 18-5 Reactances alone combine like resistances. (*a*) Addition of series reactances. (*b*) Two reactances in parallel equal their product over their sum.

Series capacitive reactance:

▶ $$X_{C_T} = X_{C_1} + X_{C_2} + \cdots + \text{etc.} \tag{18-4}$$

For parallel reactances, the combined reactance is calculated by the reciprocal formula, as shown in Fig. 18-5*b*.

Parallel capacitive reactance:

▶ $$\frac{1}{X_{C_{EQ}}} = \frac{1}{X_{C_1}} + \frac{1}{X_{C_2}} + \cdots + \text{etc.} \tag{18-5}$$

In Fig. 18-5*b* the parallel combination of 100 and 200 Ω is 66⅔ Ω for $X_{C_{EQ}}$. The combined parallel reactance is less than the lowest branch reactance. Any short cuts for combining parallel resistances also apply to parallel reactances.

Combining capacitive reactances is opposite to the way capacitances are combined. The two procedures are compatible, however, because capacitive reactance is inversely proportional to capacitance. The general case is that ohms of opposition add in series but combine by the reciprocal formula in parallel. This rule applies to resistances, to a combination of inductive reactances alone, or to capacitive reactances alone.

TEST-POINT QUESTION 18-3

Answers at end of chapter.
a. How much is X_{C_T} for a 200-Ω X_{C_1} in series with a 300-Ω X_{C_2}?
b. How much is $X_{C_{EQ}}$ for a 200-Ω X_{C_1} in parallel with a 300-Ω X_{C_2}?

18-4 OHM'S LAW APPLIED TO X_C

The current in an ac circuit with X_C alone is equal to the applied voltage divided by the ohms of X_C. Three examples with X_C are illustrated in Fig. 18-6. In Fig. 18-6*a* there is just one reactance of 100 Ω. The current *I* then is equal to V/X_C, or 100 V/100 Ω, which is 1 A.

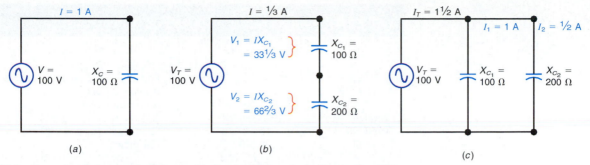

FIG. 18-6   Example of circuit calculations with X_C. (a) With a single X_C, the $I = V/X_C$. (b) Sum of series voltage drops equals the applied voltage V_T. (c) Sum of parallel branch currents equals total line current I_T.

For the series circuit in Fig. 18-6b, the total reactance, equal to the sum of the series reactances, is 300 Ω. Then the current is 100 V/300 Ω, which equals ⅓ A. Furthermore, the voltage across each reactance is equal to its IX_C product. The sum of these series voltage drops equals the applied voltage.

For the parallel circuit in Fig. 18-6c, each parallel reactance has its individual branch current, equal to the applied voltage divided by the branch reactance. The applied voltage is the same across both reactances, since they are all in parallel. In addition, the total line current of 1½ A is equal to the sum of the individual branch currents of 1 and ½ A each. With the applied voltage an rms value, all the calculated currents and voltage drops in Fig. 18-6 are also rms values.

TEST-POINT QUESTION 18-4

Answers at end of chapter.
a. In Fig. 18-6b, how much is X_{C_T}?
b. In Fig. 18-6c, how much is $X_{C_{EQ}}$?

18-5 APPLICATIONS OF CAPACITIVE REACTANCE

The general use of X_C is to block direct current but provide low reactance for alternating current. In this way, a varying ac component can be separated from a steady direct current. Furthermore, a capacitor can have less reactance for alternating current of high frequencies, compared with lower frequencies.

Note the following difference in ohms of R and X_C. Ohms of R remain the same for dc circuits or ac circuits, whereas X_C, depends on the frequency.

If 100 Ω is taken as a desired value of X_C, capacitor values can be calculated for different frequencies, as listed in Table 18-1. The C values indicate typical capacitor sizes for different frequency applications. Note that the required C becomes smaller for higher frequencies.

TABLE 18-1 CAPACITANCE VALUES FOR A REACTANCE OF 100 Ω		
C (APPROX.)	**FREQUENCY**	**REMARKS**
27 μF	60 Hz	Power-line and low audio frequency
1.6 μF	1,000 Hz	Audio frequency
0.16 μF	10,000 Hz	Audio frequency
1600 pF	1,000 kHz (RF)	AM radio
160 pF	10 MHz (HF)	Short-wave radio
16 pF	100 MHz (VHF)	FM radio

The 100 Ω of reactance for Table 18-1 is taken as a low X_C in common applications of C as a coupling capacitor, bypass capacitor, or filter capacitor for ac variations. For all these functions, the X_C must be low compared with the resistance in the circuit. Typical values of C, then, are 16 to 1600 pF for RF signals and 0.16 to 27 μF for AF signals. The power line frequency of 60 Hz, which is a low audio frequency, requires C values of about 27 μF or more.

TEST-POINT QUESTION 18-5

Answers at end of chapter.

A 20-μF C has 100 Ω of X_C at 60 Hz.
a. How much is X_C at 120-Hz?
b. How much is X_C at 6 Hz?

18-6 SINE-WAVE CHARGE AND DISCHARGE CURRENT

In Fig. 18-7, sine-wave voltage applied across a capacitor produces alternating charge and discharge current. The action is considered for each quarter-cycle. Note that the voltage v_C across the capacitor is the same as the applied voltage

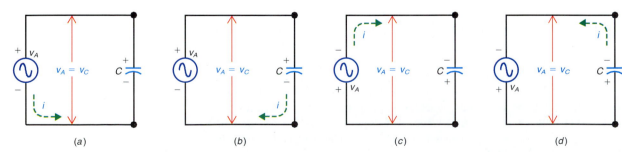

FIG. 18-7 Capacitive charge and discharge currents. (*a*) Voltage V_A increases positive to charge C. (*b*) The C discharges as V_A decreases. (*c*) Voltage V_A increases negative to charge C in opposite polarity. (*d*) The C discharges as reversed V_A decreases.

v_A at all times because they are in parallel. The values of current i, however, depend on the charge and discharge of C. When v_A is increasing, it charges C to keep v_C at the same voltage as v_A; when v_A is decreasing, C discharges to maintain v_C at the same voltage as v_A. When v_A is not changing, there is no charge or discharge current.

During the first quarter-cycle, in Fig. 18-7a, v_A is positive and increasing, charging C in the polarity shown. The electron flow is from the negative terminal of the source voltage, producing charging current in the direction indicated by the arrow for i. Next, when the applied voltage decreases during the second quarter-cycle, v_C also decreases by discharging. The discharge current is from the negative plate of C through the source, and back to the positive plate. Note that the discharge current in Fig. 18-7b has the opposite direction from the charge current in Fig. 18-7a.

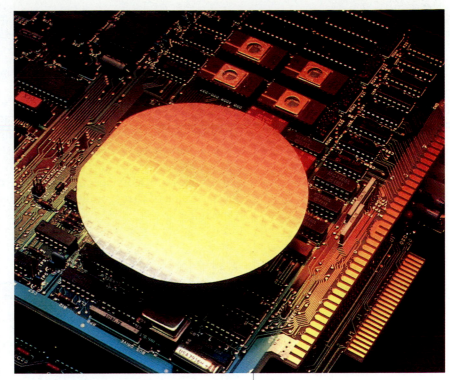

For the third quarter-cycle, in Fig. 18-7c, the applied voltage v_A increases again but in the negative direction. Now C charges again but in reversed polarity. Here the charging current is in the opposite direction from the charge current in Fig. 18-7a but in the same direction as the discharge current in Fig. 18-7b. Finally, the negative applied voltage decreases during the final quarter-cycle in Fig. 18-7d. As a result, C discharges. This discharge current is opposite to the charge current in Fig. 18-7c, but in the same direction as the charge current in Fig. 18-7a.

For the sine wave of applied voltage, therefore, the capacitor provides a cycle of alternating charge and discharge current. Notice that capacitive current flows for either charge or discharge, whenever the voltage changes, for either an increase or decrease. Also, i and v have the same frequency.

CALCULATING THE VALUES OF i_C   The greater the voltage change, the greater is the amount of capacitive current. Furthermore, a larger capacitor can allow more charge current when the applied voltage increases and produce more discharge current. Because of these factors the amount of capacitive current can be calculated as

$$i_C = C\,\frac{dv}{dt} \qquad\qquad\qquad \text{(18-6)}$$

where i is in amperes, C is in farads, and dv/dt is in volts per second. As an example, suppose that the voltage across a 240-pF capacitor changes by 25 V in 1 μs. The amount of capacitive current then is

$$i_C = C\frac{dv}{dt} = 240 \times 10^{-12} \times \frac{25}{1 \times 10^{-6}}$$

$$= 240 \times 25 \times 10^{-6} = 6000 \times 10^{-6}$$

$$i_C = 6 \times 10^{-3} \text{ A or 6 mA}$$

Notice how Formula (18-6) is similar to the capacitor charge formula $Q = CV$. When the voltage changes, this dv/dt factor produces a change in the charge Q. When the charge moves, this dq/dt change is the current i_C. Therefore, dq/dt or i_C is proportional to dv/dt. With the constant factor C, then, i_C becomes equal to $C(dv/dt)$.

By means of Formula (18-6), then, i_C can be calculated to find the instantaneous value of charge or discharge current when the voltage changes across a capacitor.

Example

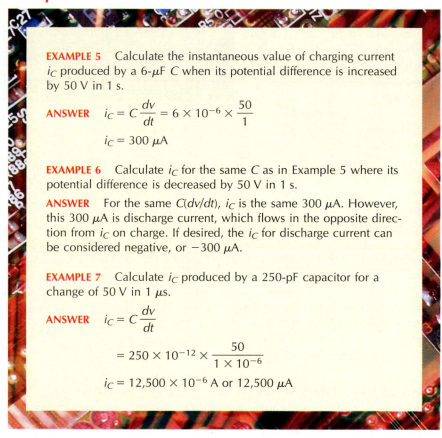

EXAMPLE 5   Calculate the instantaneous value of charging current i_C produced by a 6-μF C when its potential difference is increased by 50 V in 1 s.

ANSWER $i_C = C\dfrac{dv}{dt} = 6 \times 10^{-6} \times \dfrac{50}{1}$

$i_C = 300 \ \mu$A

EXAMPLE 6   Calculate i_C for the same C as in Example 5 where its potential difference is decreased by 50 V in 1 s.

ANSWER   For the same $C(dv/dt)$, i_C is the same 300 μA. However, this 300 μA is discharge current, which flows in the opposite direction from i_C on charge. If desired, the i_C for discharge current can be considered negative, or $-300 \ \mu$A.

EXAMPLE 7   Calculate i_C produced by a 250-pF capacitor for a change of 50 V in 1 μs.

ANSWER $i_C = C\dfrac{dv}{dt}$

$$= 250 \times 10^{-12} \times \frac{50}{1 \times 10^{-6}}$$

$i_C = 12,500 \times 10^{-6}$ A or 12,500 μA

Notice that more i_C is produced in Example 7, although C is smaller than in Example 6, because dv/dt is a much faster voltage change.

WAVESHAPES OF v_C AND i_C   More details of capacitive circuits can be analyzed by plotting the values calculated in Table 18-2. Figure 18-8 shows the waveshapes representing these values. Figure 18-8a shows a sine wave of volt-

TABLE 18-2   VALUES FOR $i_C = C(dv/dt)$ CURVES IN FIG. 18-8

TIME		dt		dv, V	dv/dt, V/μs	C, pF	$i_C = C(dv/dt)$, mA
θ	μs	θ	μs				
30°	2	30°	2	50	25	240	6
60°	4	30°	2	36.6	18.3	240	4.4
90°	6	30°	2	13.4	6.7	240	1.6
120°	8	30°	2	−13.4	−6.7	240	−1.6
150°	10	30°	2	−36.6	−18.3	240	−4.4
180°	12	30°	2	−50	−25	240	−6
210°	14	30°	2	−50	−25	240	−6
240°	16	30°	2	−36.6	−18.3	240	−4.4
270°	18	30°	2	−13.4	−6.7	240	−1.6
300°	20	30°	2	13.4	6.7	240	1.6
330°	22	30°	2	36.6	18.3	240	4.4
360°	24	30°	2	50	25	240	6

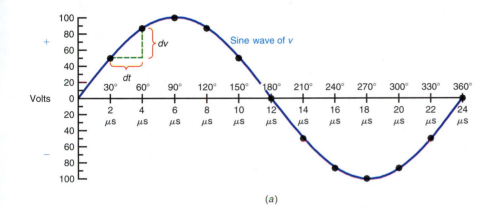

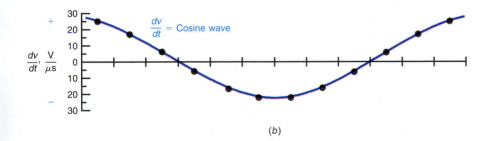

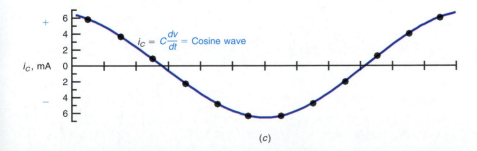

FIG. 18-8 Waveshapes of capacitive circuits. (*a*) Waveshape of sine-wave voltage at top. (*b*) Changes in voltage below causing (*c*) current i_c charge and discharge waveshape. Values plotted are those given in Table 18-2.

age v_C across a 240-pF capacitance C. Since the capacitive current i_C depends on the rate of change of voltage, rather than the absolute value of v, the curve in Fig. 18-8b shows how much the voltage changes. In this curve, the dv/dt values are plotted for every 30° of the cycle.

Figure 18-8c shows the actual capacitive current i_C. This i_C curve is similar to the dv/dt curve because i_C equals the constant C multiplied by dv/dt.

90° PHASE ANGLE The i_C curve at the bottom in Fig. 18-8 has its zero values when the v_C curve at the top is at maximum. This comparison shows that the curves are 90° out of phase, as i_C is a cosine wave of current for the sine wave of voltage v_C. The 90° phase difference results from the fact that i_C depends on the dv/dt rate of change, rather than on v itself. More details of this 90° phase angle for capacitance are explained in the next chapter.

For each of the curves, the period T is 24 μs. Therefore, the frequency is $1/T$ or $\frac{1}{24}$, which equals 41.67 kHz. Each curve has the same frequency, although there is a 90° phase difference between i and v.

OHMS OF X_C The ratio of v_C/i_C actually specifies the capacitive reactance, in ohms. For this comparison, we use the actual value of v_C, which has the peak of 100 V. The rate-of-change factor is included in i_C. Although the peak of i_C at 6 mA is 90° ahead of the peak of v_C at 100 V, we can compare these two peak values. Then v_C/i_C is 100/0.006, which equals 16,667 Ω.

This X_C is only an approximate value because i_C cannot be determined exactly for the large dt changes every 30°. If we used smaller intervals of time, the peak i_C would be 6.28 mA with X_C then 15,900 Ω, the same as $1/(2\pi f C)$ with a 240-pF C and a frequency of 41.67 kHz.

<hr>

TEST-POINT QUESTION 18-6

Answers at end of chapter.

Refer to the curves in Fig. 18-8.
a. At what angle does v have its maximum positive value?
b. At what angle does dv/dt have its maximum positive value?
c. What is the phase angle difference between v_C and i_C?

18   SUMMARY AND REVIEW

- Capacitive reactance, indicated by X_C, is the opposition of a capacitance to the flow of sine-wave alternating current.
- Reactance X_C is measured in ohms because it limits the current to the value V/X_C. With V in volts and X_C in ohms, I is in amperes.
- $X_C = 1/(2\pi fC)$. With f in hertz and C in farads, X_C is in ohms.
- For the same value of capacitance, X_C decreases when the frequency increases.
- For the same frequency, X_C decreases when the capacitance increases.
- With X_C and f known, the capacitance $C = 1/(2\pi fX_C)$.
- With X_C and C known, the frequency $f = 1/(2\pi CX_C)$.
- The total X_C of capacitive reactances in series equals the sum of the individual values, as for series resistances. The series reactances have the same current. The voltage across each reactance is IX_C.
- With parallel capacitive reactances, the combined reactance is calculated by the reciprocal formula, as for parallel resistances. Each branch current is V/X_C. The total line current is the sum of the individual branch currents.
- Table 18-3 summarizes the differences between C and X_C.

TABLE 18-3   COMPARISON OF CAPACITANCE AND CAPACITIVE REACTANCE

CAPACITANCE	CAPACITIVE REACTANCE
Symbol is C	Symbol is X_C
Measured in farad units	Measured in ohm units
Depends on construction of capacitor	Depends on frequency of sine-wave voltage
$C = i_C/(dv/dt)$ or Q/V	$X_C = v_C/i_C$ or $1/(2\pi fC)$

ANSWERS AT BACK OF BOOK.

Choose (*a*), (*b*), (*c*), or (*d*).

1. Alternating current can flow in a capacitive circuit with ac voltage applied because (*a*) of the high peak value; (*b*) varying voltage produces charge and discharge current; (*c*) charging current flows when the voltage decreases; (*d*) discharge current flows when the voltage increases.
2. With higher frequencies, the amount of capacitive reactance (*a*) increases; (*b*) stays the same; (*c*) decreases; (*d*) increases only when the voltage increases.
3. At the same frequency, larger capacitance results in (*a*) more reactance; (*b*) the same reactance; (*c*) less reactance; (*d*) less reactance if the voltage amplitude decreases.
4. The capacitive reactance of a 0.1-μF capacitor at 1000 Hz equals (*a*) 1000 Ω; (*b*) 1600 Ω; (*c*) 2000 Ω; (*d*) 3200 Ω.
5. Two 1000-Ω X_C values in series have a total reactance of (*a*) 500 Ω; (*b*) 1000 Ω; (*c*) 1414 Ω; (*d*) 2000 Ω.
6. Two 1000-Ω X_C values in parallel have a combined reactance of (*a*) 500 Ω; (*b*) 707 Ω; (*c*) 1000 Ω; (*d*) 2000 Ω.
7. With 50 V rms applied across a 100-Ω X_C, the rms current in the circuit equals (*a*) 0.5 A; (*b*) 0.637 A; (*c*) 0.707 A; (*d*) 1.414 A.
8. With steady dc voltage from a battery applied to a capacitance, after it charges to the battery voltage, the current in the circuit (*a*) depends on the current rating of the battery; (*b*) is greater for larger values of capacitance; (*c*) is smaller for larger values of capacitance; (*d*) is zero for any capacitance value.
9. The capacitance needed for a 1000-Ω reactance at 2 MHz is (*a*) 2 pF; (*b*) 80 pF; (*c*) 1000 pF; (*d*) 2000 pF.
10. A 0.2-μF capacitance will have a reactance of 1000 Ω at the frequency of (*a*) 800 Hz; (*b*) 1 kHz; (*c*) 1 MHz; (*d*) 8 MHz.

QUESTIONS

1. Why is capacitive reactance measured in ohms? State two differences between capacitance and capacitive reactance.
2. Explain briefly why the bulb lights in Fig. 18-1*a* but not in Fig. 18-1*c*.
3. Explain briefly what is meant by two factors being inversely proportional. How does this apply to X_C and C? X_C and f?
4. In comparing X_C and R, give two differences and one similarity.
5. Why are the waves in Fig. 18-8*a* and *b* considered to be 90° out of phase, while the waves in Fig. 18-8*b* and *c* have the same phase?
6. Referring to Fig. 18-3, how does this graph show an inverse relation between X_C and C?
7. Referring to Fig. 18-4, how does this graph show an inverse relation between X_C and f?

8. Referring to Fig. 18-8, draw three similar curves but for a sine wave of voltage with a period $T = 12$ μs for the full cycle. Use the same C of 240 pF. Compare the value of X_C obtained as $1/(2\pi f C)$ and v_C/i_C.

9. (a) What is the relationship between charge q and current i? (b) How is this comparison similar to the relation between the two formulas $Q = CV$ and $i = C(dv/dt)$?

PROBLEMS

ANSWERS TO ODD-NUMBERED PROBLEMS AT BACK OF BOOK.

1. Referring to Fig. 18-4, give the values of C needed for 2000 Ω of X_C at the four frequencies listed.

2. What size capacitance is needed for 50-Ω reactance at 100 kHz?

3. A capacitor with an X_C of 2000 Ω is connected across a 9-V 1000-Hz source. (a) Draw the schematic diagram. (b) How much is the current in the circuit? (c) What is the frequency of the current?

4. How much is the capacitance of a capacitor that draws 0.1 A from the 60-Hz 120-V power line?

5. A 1000-Ω X_{C_1} and a 4000-Ω X_{C_2} are in series across a 10-V source. (a) Draw the schematic diagram. (b) Calculate the current in the series circuit. (c) How much is the voltage across X_{C_1}? (d) How much is the voltage across X_{C_2}?

6. The 1000-Ω X_{C_1} and 4000-Ω X_{C_2} in Prob. 5 are in parallel across the 10-V source. (a) Draw the schematic diagram. (b) Calculate the branch current in X_{C_1}. (c) Calculate the branch current in X_{C_2}. (d) Calculate the total line current. (e) How much is the voltage across both reactances?

7. At what frequency will a 0.01-μF capacitor have a reactance of 5000 Ω?

8. Four capacitive reactances of 100, 200, 300, and 400 Ω each are connected in series across a 40-V source. (a) Draw the schematic diagram. (b) How much is the total X_{C_T}? (c) Calculate I. (d) Calculate the voltages across each capacitance. (e) If the frequency of the applied voltage is 1600 kHz, calculate the required value of each capacitance.

9. Three equal capacitive reactances of 600 Ω each are in parallel. (a) How much is the equivalent combined reactance? (b) If the frequency of the applied voltage is 800 kHz, how much is the capacitance of each capacitor and what is the equivalent combined capacitance of the three in parallel?

10. A 2-μF C is in series with a 4-μF C. The frequency is 5 kHz. (a) How much is C_T? (b) Calculate X_{C_T}. (c) Calculate X_{C_1} and X_{C_2} to see if their sum equals X_{C_T}.

11. A capacitor across the 120-V 60-Hz ac power line allows a 0.4-A current. (a) Calculate X_C and C. (b) What size C is needed to double the current?

12. A 0.01-μF capacitor is connected across a 10-V source. Tabulate the values of X_C and current in the circuit at 0 Hz (for steady dc voltage) and at 20 Hz, 60 Hz, 100 Hz, 500 Hz, 5 kHz, 10 kHz, and 455 kHz.

13. Calculate X_C for 470 pF at 1640 kHz.

14. What C is needed for the same X_C in Prob. 13 but at 500 Hz?

15. How much is I with 162 mV applied for the X_C in Probs. 13 and 14?

16. At what frequencies will X_C be 200 Ω for the following capacitors: **(a)** 2 μF; **(b)** 0.1 μF; **(c)** 0.05 μF; **(d)** 0.002 μF; **(e)** 250 pF; **(f)** 100 pF; **(g)** 47 pF?

17. What size C is needed to have X_C the same as the X_L of a 6-mH L at 100 kHz?

18. A capacitor is in series with a 5-kΩ R. At what frequency will X_C equal R for the following values of C: **(a)** 47 pF; **(b)** 500 pF; **(c)** 0.1 pF; **(d)** 10 μF?

19. Find the C needed for X_C of 1274 Ω at the following frequencies: **(a)** 500 kHz; **(b)** 1 MHz; **(c)** 250 kHz; **(d)** 5 MHz; **(e)** 50 kHz?

20. Calculate X_C for a 500 μF capacitor at the frequency of 60 Hz.

21. Calculate the frequency f that will provide X_C of 1274 Ω for the following values of C: **(a)** 250 pF; **(b)** 125 pF; **(c)** 500 pF; **(d)** 25 pF; **(e)** 2500 pF.

22. In Fig. 18-9, calculate X_{C_1}, X_{C_2}, X_{C_3}, X_{C_T}, I, V_{C_1}, V_{C_2}, and V_{C_3}.

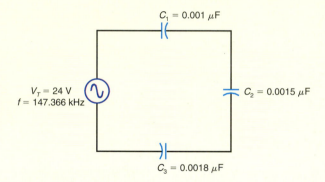

$C_1 = 0.001\ \mu$F

$V_T = 24$ V
$f = 147.366$ kHz

$C_2 = 0.0015\ \mu$F

$C_3 = 0.0018\ \mu$F

FIG. 18-9 Circuit for Probs. 22 and 23.

23. In Fig. 18-9, assume that the frequency f is doubled. What happens to: **(a)** X_{C_T}; **(b)** I?

24. In Fig. 18-10, calculate X_{C_1}, X_{C_2}, X_{C_3}, I_1, I_2, I_3, I_T, and $X_{C_{EQ}}$?

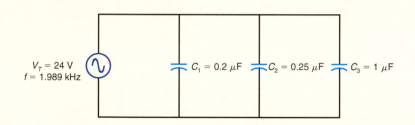

$V_T = 24$ V
$f = 1.989$ kHz

$C_1 = 0.2\ \mu$F $C_2 = 0.25\ \mu$F $C_3 = 1\ \mu$F

FIG. 18-10 Circuit for Probs. 24 and 25.

25. In Fig. 18-10, assume that the frequency f is reduced by one-half. What happens to: **(a)** I_T; **(b)** $X_{C_{EQ}}$?

26. Calculate the charging current i_c for a 0.33-μF capacitor if the voltage across its plates is increasing at the rate of 10 V/1 ms.

27. A capacitor has a discharge current i_c of 15 mA when the voltage across its plates decreases at the rate of 150 V/1 μs. Calculate C.

28. What rate of voltage change will produce a charging current of 25 mA in a 0.01-μF capacitor? Express your answer in volts per second.

1. Explain an experimental procedure for determining the value of an unmarked capacitor. (Assume that a capacitance tester is not available.)

2. In Fig. 18-11, calculate X_{C_T}, X_{C_1}, X_{C_2}, C_1, C_3, V_{C_1}, V_{C_2}, V_{C_3}, I_2, and I_3.

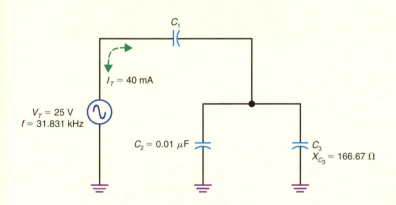

FIG. 18-11 Circuit for Critical Thinking Prob. 2.

18-1 a. 0.1 μF
 b. 0.5 μF

18-2 a. 200 Ω
 b. 800 Ω
 c. larger

18-3 a. 500 Ω
 b. 120 Ω

18-4 a. 300 Ω
 b. 66.7 Ω

18-5 a. 50 Ω
 b. 1000 Ω

18-6 a. 90°
 b. 0 or 360°
 c. 90°

CAPACITIVE CIRCUITS

This chapter analyzes circuits that combine capacitive reactance X_C and resistance R. The main questions are: how do we combine the ohms of opposition, how much current flows, and what is the phase angle? Although X_C and R are both measured in ohms, they have some different characteristics. Specifically, X_C decreases with more C and higher frequencies for sine-wave ac voltage applied, while R is the same for dc and ac circuits. Furthermore, the phase angle for the voltage across X_C is at $-90°$ as measured in the clockwise direction with i_C as the reference at $0°$.

In addition, the practical application of a coupling capacitor shows how a low value of X_C can be used to pass the desired ac signal variations, while blocking the steady dc level of a fluctuating dc voltage. In a coupling circuit with C and R in series, the ac component is across R for the output voltage but the dc component across C is not connected across the output terminals.

Finally, the general case of capacitive charge and discharge current produced when the applied voltage changes is shown with nonsinusoidal voltage variations. In this case, we compare the waveshapes of v_C and i_C. Remember that the $-90°$ angle for an IX_C voltage applies only to sine waves.

CHAPTER OBJECTIVES

Upon completion of this chapter, you should be able to:

- *Explain* why the current leads the voltage by 90° for a capacitor.
- *Define* the term *impedance*.
- *Calculate* the total impedance and phase angle of a series *RC* circuit.
- *Describe* the operation and application of an *RC* phase-shifter circuit.
- *Calculate* the total current, equivalent impedance, and phase angle of a parallel *RC* circuit.
- *Understand* how a capacitor can couple some AC frequencies but not others.
- *Calculate* the individual capacitor voltage drops for capacitors in series.
- *Calculate* the capacitive current that flows with nonsinusoidal waveforms.

IMPORTANT TERMS IN THIS CHAPTER

capacitive voltage divider
coupling capacitor
leading current

phase-shifter circuit
phasor triangle

rectangular waveform
sawtooth waveform

TOPICS COVERED IN THIS CHAPTER

For a sine wave of applied voltage, the capacitor provides a cycle of alternating charge and discharge current, as shown in Fig. 19-1a. In Fig. 19-1b, the wave-shape of this charge and discharge current i_C is compared with the voltage v_C.

Note that the instantaneous value of i_C is zero when v_C is at its maximum value. At either its positive or negative peak, v_C is not changing. For one instant at both peaks, therefore, the voltage must have a static value before changing its direction. Then v is not changing and C is not charging or discharging. The result is zero current at this time.

Also note that i_C is maximum when v_C is zero. When v_C crosses the zero axis, i_C has its maximum value because then the voltage is changing most rapidly.

Therefore, i_C and v_C are 90° out of phase, since the maximum value of one corresponds to the zero value of the other; i_C leads v_C because i_C has its maximum value a quarter-cycle before the time that v_C reaches its peak. The phasors in Fig. 19-1c show i_C leading v_C by the counterclockwise angle of 90°. Here v_C is the horizontal phasor for the reference angle of 0°. In Fig. 19-1d, however, the current i_C is the horizontal phasor for reference. Since i_C must be 90° leading, v_C is shown lagging by the clockwise angle of $-90°$. In series circuits, the current i_C is the reference and then the voltage v_C can be considered to lag i_C by 90°.

The 90° phase angle results because i_C depends on the rate of change of v_C. As shown previously in Fig. 18-8 for a sine wave of v_C, the capacitive charge and discharge current is a cosine wave. This 90° phase between v_C and i_C is true in any sine-wave ac circuit, whether C is in series or parallel and whether C is alone or combined with other components. We can always say that for any X_C its current and voltage are 90° out of phase.

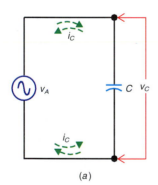

(a)

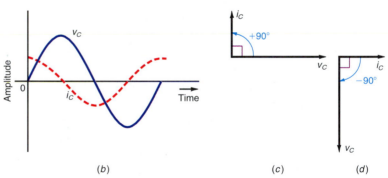

(b)   (c)   (d)

FIG. 19-1 Capacitive current i_C leads v_C by 90°. (a) Circuit with sine-wave V_A across C. (b) Waveshapes of i_C 90° ahead of v_C. (c) Phasor diagram of i_C leading the horizontal reference v_C by a counterclockwise angle of 90°. (d) Phasor diagram with i_C as the reference phasor to show v_C lagging i_C by an angle of $-90°$.

CAPACITIVE CURRENT IS THE SAME IN A SERIES CIRCUIT The leading phase angle of capacitive current is only with respect to the voltage across the capacitor, which does not change the fact that the current is the same in all parts of a series circuit. In Fig. 19-1a, for instance, the current in the generator, the connecting wires, and both plates of the capacitor must be the same because they are all in the same path.

CAPACITIVE VOLTAGE IS THE SAME ACROSS PARALLEL BRANCHES

In Fig. 19-1a, the voltage is the same across the generator and C because they are in parallel. There cannot be any lag or lead in time between these two parallel voltages. At any instant, whatever the voltage value is across the generator at that time, the voltage across C is the same. With respect to the series current, however, both v_A and v_C are 90° out of phase with i_C.

THE FREQUENCY IS THE SAME FOR v_C AND i_C

Although v_C lags i_C by 90°, both waves have the same frequency. For example, if the frequency of the sine wave v_C in Fig. 19-1b is 100 Hz, this is also the frequency of i_C.

TEST-POINT QUESTION 19-1

Answers at end of chapter.

Refer to Fig. 19-1.
a. What is the phase between v_A and v_C?
b. What is the phase between v_C and i_C?
c. Does v_C lead or lag i_C?

19-2 X_C AND R IN SERIES

When resistance is in series with capacitive reactance (Fig. 19-2), both determine the current. Current I is the same in X_C and R since they are in series. Each has its own series voltage drop, equal to IR for the resistance and IX_C for the reactance.

If the capacitive reactance alone is considered, its voltage drop lags the series current I by 90°. The IR voltage has the same phase as I, however, because resistance provides no phase shift. Therefore, R and X_C combined in series must be added by phasors because of the 90° phase angle.

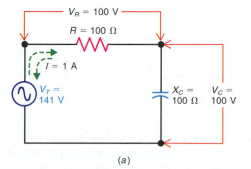
(a)

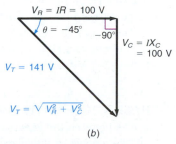
(b)

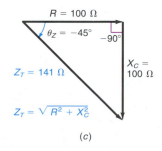
(c)

FIG. 19-2 Circuit with X_C and R in series. (a) Schematic diagram. (b) Phasor triangle of voltages with V_C lagging V_R by −90°. This triangle is used to find the resultant V_T. (c) Similar impedance triangle with X_C lagging R by −90°. This triangle is used to find the resultant Z_T.

PHASOR ADDITION OF V_C AND V_R

In Fig. 19-2b, the current phasor is shown horizontal, as the reference phase, because I is the same in a series circuit. The resistive voltage drop IR has the same phase as I. The capacitor voltage IX_C must be 90° clockwise from I and IR, as the capacitive voltage lags.

The phasor voltages V_R and V_C are 90° out of phase and thus form a right triangle. Therefore

$$V_T = \sqrt{V_R^2 + V_C^2} \qquad\qquad \textbf{(19-1)}$$

This formula applies just to series circuits because then V_C is 90° out of phase with V_R. All the voltages must be in the same units. When V_R and V_C are rms values, then V_T is an rms value.

In calculating the value of V_T, first square V_R and V_C, then add and take the square root. For the example in Fig. 19-2,

$$V_T = \sqrt{100^2 + 100^2} = \sqrt{10{,}000 + 10{,}000} = \sqrt{20{,}000}$$
$$V_T = 141 \text{ V}$$

The two phasor voltages total 141 V instead of 200 V because the 90° phase difference means the peak value of one occurs when the other is at zero.

PHASOR ADDITION OF X_C AND R

Figure 19-2c shows a triangle for R and X_C in series. This corresponds to the voltage triangle in Fig. 19-2b except that the common factor I cancels because the current is the same in both R and X_C. The resultant phasor for R and X_C in series represents the total opposition in ohms offered by the series circuit. The phasor sum of R and X_C is called the *total impedance,* represented by the symbol Z_T. Therefore, the triangle consisting of R, X_C, and Z_T is called the *impedance triangle.* Because the phasors for R and X_C form a right triangle, Z_T is calculated as

$$Z_T = \sqrt{R^2 + X_C^2} \qquad\qquad \textbf{(19-2)}$$

With R and X_C in ohms, Z_T is also in ohms. For the example in Fig. 19-2c, the values are

$$Z_T = \sqrt{100^2 + 100^2} = \sqrt{10{,}000 + 10{,}000} = \sqrt{20{,}000}$$
$$Z_T = 141 \ \Omega$$

In Fig. 19-2a, note that the total impedance of 141 Ω divided into the applied voltage of 141 V produces the current of 1 A in the series circuit. The IR voltage drop is 1×100, or 100 V; the IX_C voltage drop is also 1×100, or 100 V.

The phasor sum of the two series voltage drops of 100 V each equals the applied voltage of 141 V. Also, the applied voltage is equal to $I \times Z_T$, or 1×141, which is 141 V for V_T.

PHASE ANGLE WITH SERIES X_C

As shown in Fig. 19-2b and c, the phase angle θ, between the generator voltage and the series current, can be calculated from the voltage or impedance triangle.

With series X_C, the phase angle is negative, clockwise from the zero reference angle of I, because the X_C voltage lags its current. To indicate the negative

phase angle, therefore, this 90° phasor points downward from the horizontal reference. To calculate the phase angle with series X_C and R,

▶ $$\tan\theta_Z = -\frac{X_C}{R} \tag{19-3}$$

Using this formula for the circuit in Fig. 19-2c,

$$\tan\theta_Z = -\frac{X_C}{R} = -\frac{100}{100}$$
$$= -1$$
$$\theta_Z = \arctan(-1)$$
$$\theta_Z = -45°$$

The negative sign means the angle is clockwise from zero, to indicate that V_T lags behind the leading I.

Example

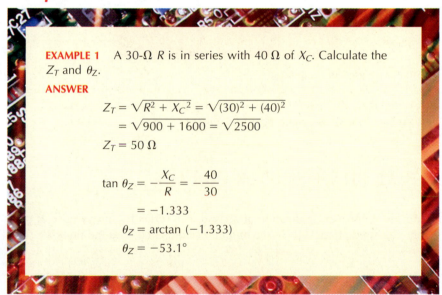

EXAMPLE 1   A 30-Ω R is in series with 40 Ω of X_C. Calculate the Z_T and θ_Z.

ANSWER

$$Z_T = \sqrt{R^2 + X_C^2} = \sqrt{(30)^2 + (40)^2}$$
$$= \sqrt{900 + 1600} = \sqrt{2500}$$
$$Z_T = 50\ \Omega$$

$$\tan\theta_Z = -\frac{X_C}{R} = -\frac{40}{30}$$
$$= -1.333$$
$$\theta_Z = \arctan(-1.333)$$
$$\theta_Z = -53.1°$$

SERIES COMBINATIONS OF X_C AND R   In series, the higher the X_C compared with R, the more capacitive the circuit. There is more voltage drop across the capacitive reactance, and the phase angle increases toward $-90°$. The series X_C always makes the current lead the applied voltage. With all X_C and no R, the entire applied voltage is across X_C, and θ equals $-90°$.

Several combinations of X_C and R in series are listed in Table 19-1, with their resultant impedance values and phase angle. Note that a ratio of 10:1, or more, for X_C/R means the circuit is practically all capacitive. The phase angle of $-84.3°$ is almost $-90°$, and the total impedance Z_T is approximately equal to X_C. The voltage drop across X_C in the series circuit is then practically equal to the applied voltage, with almost none across the R.

TABLE 19-1   SERIES R AND X_C COMBINATIONS

R, Ω	X_C, Ω	Z_T, Ω (APPROX.)	PHASE ANGLE θ_Z
1	10	$\sqrt{101} = 10$	$-84.3°$
10	10	$\sqrt{200} = 14$	$-45°$
10	1	$\sqrt{101} = 10$	$-5.7°$

Note: θ_Z is the phase angle of Z_T or V_T with respect to the reference phasor I in series circuits.

At the opposite extreme, when R is 10 times more than X_C, the series circuit is mainly resistive. The phase angle of $-5.7°$ then means the current is almost in phase with the applied voltage; Z_T is approximately equal to R, and the voltage drop across R is practically equal to the applied voltage with almost none across the X_C.

For the case when X_C and R equal each other, the resultant impedance Z_T is 1.41 times either one. The phase angle then is $-45°$, halfway between 0° for resistance alone and $-90°$ for capacitive reactance alone.

TEST-POINT QUESTION 19-2

Answers at end of chapter.
a. How much is Z_T for a 20-Ω R in series with a 20-Ω X_C?
b. How much is V_T for 20 V across R and 20 V across X_C in series?
c. What is the phase angle θ_Z of this circuit?

19-3 *RC* PHASE-SHIFTER CIRCUIT

Figure 19-3 shows an application of X_C and R in series for the purpose of providing a desired phase shift in the output V_R compared with the input V_T. The

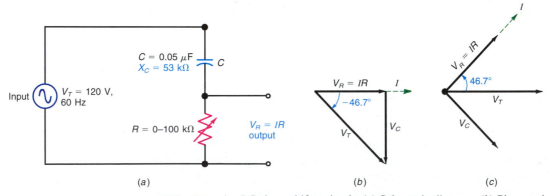

FIG. 19-3   An *RC* phase-shifter circuit. (*a*) Schematic diagram. (*b*) Phasor triangle with *IR*, or V_R, as the horizontal reference. V_R leads V_T by 46.7° with R set at 50 kΩ. (*c*) Phasors shown with V_T as the horizontal reference.

R can be varied up to 100 kΩ to change the phase angle. The C is 0.05 μF here for the 60-Hz ac power line voltage, but a smaller C would be used for a higher frequency. The capacitor must have an appreciable value of reactance for the phase shift.

For the circuit in Fig. 19-3a, assume that R is set for 50 kΩ, at its middle value. The reactance for the 0.05-μF capacitor at 60 Hz is approximately 53 kΩ. For these values of X_C and R, the phase angle of the circuit is $-46.7°$. This angle has the tangent of $-53/50 = -1.06$.

The phasor triangle in Fig. 19-3b shows that IR or V_R is out of phase with V_T by the leading angle or 46.7°. Note that V_C is always 90° lagging V_R in a series circuit. The angle between V_C and V_T then becomes $90° - 46.7° = 43.3°$.

The purpose of this circuit is to provide a phase-shifted voltage V_R in the output, with respect to the input. For this reason, the phasors are redrawn in Fig. 19-3c to show the voltages with the input V_T as the horizontal reference. The conclusion, then, is that the output voltage across R leads the input V_T by 46.7°.

Now let R be varied for a higher value at 90 kΩ, while X_C stays the same. The phase angle becomes $-30.5°$. This angle has the tangent $-53/90 = -0.59$. As a result, V_R leads V_T by 30.5°.

For the opposite case, let R be reduced to 10 kΩ. Then the phase angle becomes $-79.3°$. This angle has the tangent $-53/10 = -5.3$. Then V_R leads V_T by 79.3°. Notice that the phase angle becomes larger as the series circuit becomes more capacitive with less resistance.

A practical application for this circuit is to provide a voltage of variable phase to set the conduction time of semiconductors in power-control circuits. As R is varied, the phase angle of the output V_R is varied with respect to the power-line voltage V_T.

TEST-POINT QUESTION 19-3

Answers at end of chapter.

In Fig. 19-3, give the phase angle between
a. V_R and V_T.
b. V_R and V_C.
c. V_C and V_T.

19-4 X_C AND R IN PARALLEL

The 90° phase angle for X_C in the parallel circuit must be with respect to branch currents instead of voltage drops as in a series circuit. In Fig. 19-4a, the voltage is the same across X_C, R, and the generator, since they are all in parallel. There cannot be any phase difference between the parallel voltages.

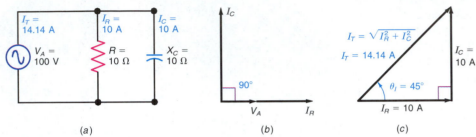

(a)   (b)   (c)

FIG. 19-4 Circuit of X_C and R in parallel, with branch currents I_C and I_R. (a) Schematic diagram. (b) Current phasors showing I_C leading V_A by 90°. (c) Phasor triangle of branch currents I_C and I_R is used to calculate resultant total line current I_T.

Each branch, however, has its individual current. For the resistive branch, I_R is V_A/R; in the capacitive branch, $I_C = V_A/X_C$. These current phasors are shown in Fig. 19-4b.

Note that the phasor diagram has the generator voltage V_A as the reference phasor because it is the same throughout the circuit. The resistive branch current I_R is in phase with V_A, but the capacitive branch current I_C leads V_A by 90°.

The phasor for I_C is up, compared with down for an X_C phasor, because the parallel branch current I_C leads the reference V_A. This I_C phasor for a parallel branch current is opposite from an X_C phasor.

The phasor addition of the branch currents in a parallel RC circuit can be calculated using the phasor triangle for currents shown in Fig. 19-4c. The phasor sum of I_R and I_C equals I_T. As a result, the formula for I_T is

$$\blacktriangleright \quad I_T = \sqrt{I_R^2 + I_C^2} \tag{19-4}$$

In Fig. 19-2c, the phasor sum of 10 A for I_R and 10 A for I_C equals 14.14 A. The branch currents are added by phasors since they are the factors 90° out of phase in a parallel circuit, corresponding to the voltage drops 90° out of phase in a series circuit.

IMPEDANCE OF X_C AND R IN PARALLEL As usual, the impedance of a parallel circuit equals the applied voltage divided by the total line current: $Z_{EQ} = V_A/I_T$. In Fig. 19-4, for example,

$$\blacktriangleright \quad Z_{EQ} = \frac{V_A}{I_T} = \frac{100}{14.14 \text{ A}} = 7.07 \text{ } \Omega$$

which is the opposition in ohms across the generator. This Z_{EQ} of 7.07 Ω is equal to the resistance of 10 Ω in parallel with the reactance of 10 Ω. Notice that the impedance of equal values of R and X_C is not one-half but equals 70.7 percent of either one.

PHASE ANGLE IN PARALLEL CIRCUITS   In Fig. 19-4c, the phase angle θ is 45° because R and X_C are equal, resulting in equal branch currents. The phase angle is between the total current I_T and the generator voltage V_A. However, V_A and I_R are in phase. Therefore θ is also between I_T and I_R.

Using the tangent formula to find θ from the current triangle in Fig. 19-4c gives

▶ $$\tan \theta_I = \frac{I_C}{I_R} \qquad\qquad\qquad (19\text{-}5)$$

The phase angle is positive because the I_C phasor is upward, leading V_A by 90°. This direction is opposite from the lagging phasor of series X_C. The effect of X_C is no different, however. Only the reference is changed for the phase angle.

Note that the phasor triangle of branch currents for parallel circuits gives θ_I as the angle of I_T with respect to the generator voltage V_A. This phase angle for I_T is labeled θ_I with respect to the applied voltage. For the phasor triangle of voltages in a series circuit, the phase angle for Z_T and V_T is labeled θ_Z with respect to the series current.

Example

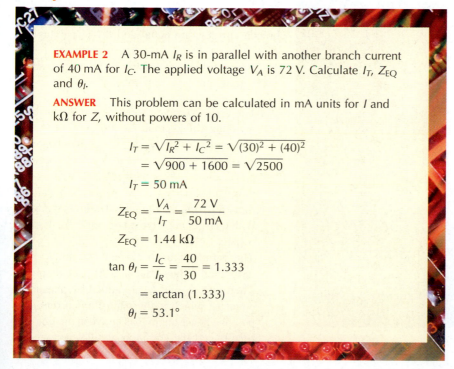

EXAMPLE 2   A 30-mA I_R is in parallel with another branch current of 40 mA for I_C. The applied voltage V_A is 72 V. Calculate I_T, Z_{EQ} and θ_I.

ANSWER   This problem can be calculated in mA units for I and $k\Omega$ for Z, without powers of 10.

$$I_T = \sqrt{I_R^2 + I_C^2} = \sqrt{(30)^2 + (40)^2}$$
$$= \sqrt{900 + 1600} = \sqrt{2500}$$
$$I_T = 50 \text{ mA}$$

$$Z_{EQ} = \frac{V_A}{I_T} = \frac{72 \text{ V}}{50 \text{ mA}}$$
$$Z_{EQ} = 1.44 \text{ k}\Omega$$

$$\tan \theta_I = \frac{I_C}{I_R} = \frac{40}{30} = 1.333$$
$$= \arctan (1.333)$$
$$\theta_I = 53.1°$$

PARALLEL COMBINATIONS OF X_C AND R   In Table 19-2, when X_C is ten times R, the parallel circuit is practically resistive because there is little leading capacitive current in the main line. The small value of I_C results from the high reactance of shunt X_C. Then the total impedance of the parallel circuit is approximately equal to the resistance, since the high value of X_C in a parallel branch has little effect. The phase angle of 5.7° is practically 0° because almost all the line current is resistive.

TABLE 19-2	PARALLEL RESISTANCE AND CAPACITANCE COMBINATIONS*					
R, Ω	X_C, Ω	I_R, A	I_C, A	I_T, A (APPROX.)	Z_{EQ}, Ω (APPROX.)	PHASE ANGLE θ_I
1	10	10	1	$\sqrt{101} = 10$	1	5.7°
10	10	1	1	$\sqrt{2} = 1.4$	7.07	45°
10	1	1	10	$\sqrt{101} = 10$	1	84.3°

*V_A = 10 V. Note that θ_I is the phase angle of I_T with respect to the reference V_A in parallel circuits.

As X_C becomes smaller, it provides more leading capacitive current in the main line. When X_C is $\frac{1}{10}$ R, practically all the line current is the I_C component. Then, the parallel circuit is practically all capacitive, with a total impedance practically equal to X_C. The phase angle of 84.3° is almost 90° because the line current is mostly capacitive. Note that these conditions are opposite to the case of X_C and R in series. With X_C and R equal, their branch currents are equal and the phase angle is 45°.

As additional comparisons between series and parallel circuits, remember that

1. The series voltage drops V_R and V_C have individual values that are 90° out of phase. Therefore, V_R and V_C are added by phasors to equal the applied voltage V_T. The negative phase angle $-\theta_Z$ is between V_T and the common series current I. More series X_C allows more V_C to make the circuit more capacitive, with a larger negative phase angle for V_T with respect to I.

2. The parallel branch currents I_R and I_C have individual values that are 90° out of phase. Therefore, I_R and I_C are added by phasors to equal I_T, which is the main-line current. The positive phase angle θ_I is between the line current I_T and the common parallel voltage V_A. Less parallel X_C allows more I_C to make the circuit more capacitive, with a larger positive phase angle for I_T with respect to V_A.

TEST-POINT QUESTION 19-4

Answers at end of chapter.
a. How much is I_T for branch currents I_R of 2 A and I_C of 2 A?
b. Find the phase angle θ_I between I_T and V_A.

19-5 RF AND AF COUPLING CAPACITORS

In Fig. 19-5, C_C is used in the application of a coupling capacitor. Its low reactance allows practically all the ac signal voltage of the generator to be developed across R. Very little of the ac voltage is across C_C.

The coupling capacitor is used for this application because at lower frequencies it provides more reactance, resulting in less ac voltage coupled across R and more across C_C. For dc voltage, all the voltage is across C with none across R, since the capacitor blocks direct current. As a result, the output signal voltage across R includes the desired higher frequencies but not direct current or very low frequencies. This application of C_C, therefore, is called *ac coupling*.

The dividing line for C_C to be a coupling capacitor at a specific frequency can be taken as X_C one-tenth or less of the series R. Then the series RC circuit is primarily resistive. Practically all the voltage drop of the ac generator is across R, with little across C. In addition, the phase angle is almost 0°.

Typical values of a coupling capacitor for audio or radio frequencies can be calculated if we assume a series resistance of 16,000 Ω. Then X_C must be 1600 Ω or less. Typical values for C_C are listed in Table 19-3. At 100 Hz, a coupling capacitor must be 1 μF to provide 1600 Ω of reactance. Higher frequencies allow a smaller value of C_C for a coupling capacitor having the same reactance. At 100 MHz in the VHF range the required capacitance is only 1 pF.

It should be noted that the C_C values are calculated for each frequency as a lower limit. At higher frequencies, the same size C_C will have less reactance than one-tenth of R, which improves the coupling.

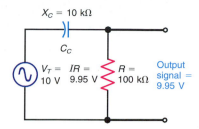

FIG. 19-5 Series circuit for RC coupling. Small X_C compared with R allows practically all the applied voltage across R for the output, with little across C.

f	C_C	REMARKS
TABLE 19-3 COUPLING CAPACITORS WITH A REACTANCE OF 1600 Ω*		
100 Hz	1 μF	Low audio frequencies
1000 Hz	0.1 μF	Audio frequencies
10 kHz	0.01 μF	Audio frequencies
1000 kHz	100 pF	Radio frequencies
100 MHz	1 pF	Very high frequencies

*For an X_C one-tenth of a series R of 16,000 Ω.

CHOOSING A COUPLING CAPACITOR FOR A CIRCUIT As an example of using these calculations, suppose that we have the problem of determining C_C for a transistorized audio amplifier. This application also illustrates the relatively large capacitance needed with low series resistance. The C is to be a coupling capacitor for audio frequencies of 50 Hz and up, with a series R of 4000 Ω. Then the required X_C is 4000/10, or 400 Ω. To find C at 50 Hz,

$$C = \frac{1}{2\pi f X_C} = \frac{1}{6.28 \times 50 \times 400}$$

$$= \frac{1}{125,600} = 0.0000079$$

$$C = 7.9 \times 10^{-6} \quad \text{or} \quad 7.9 \text{ μF}$$

A typical commercial size of low-voltage electrolytic readily available is 10 μF. The slightly higher capacitance value is better for coupling. The voltage rating can be 3 to 10 V, depending on the circuit, with a typical transistor supply voltage of 9 V. Although electrolytic capacitors have a slight leakage current, they can be used for coupling capacitors in this application because of the low series resistance.

TEST-POINT QUESTION 19-5

Answers at end of chapter.
a. The X_C of a coupling capacitor is 70 Ω at 200 Hz. How much is its X_C at 400 Hz?
b. From Table 19-3, what C would be needed for 1600 Ω of X_C at 50 MHz?

19-6 CAPACITIVE VOLTAGE DIVIDERS

When capacitors are connected in series across a voltage source, the series capacitors serve as a voltage divider. Each capacitor has part of the applied voltage, and the sum of all the series voltage drops equals the source voltage.

The amount of voltage across each is inversely proportional to its capacitance. For instance, with 2 μF in series with 1 μF, the smaller capacitor has double the voltage of the larger capacitor. Assuming 120 V applied, one-third of this, or 40 V, is across the 2-μF capacitor, with two-thirds, or 80 V, across the 1-μF capacitor.

The two series voltage drops of 40 and 80 V add to equal the applied voltage of 120 V. The phasor addition is the same as the arithmetic sum of the two voltages because they are in phase. When voltages are out of phase with each other arithmetic addition is not possible and phasor addition becomes necessary.

AC DIVIDER With sine-wave alternating current, the voltage division between series capacitors can be calculated on the basis of reactance. In Fig. 19-6a, the total reactance is 120 Ω across the 120-V source. The current in the series circuit then is 1 A. This current is the same for X_{C_1} and X_{C_2} in series. Therefore, the IX_C voltage across C_1 is 40 V, with 80 V across C_2.

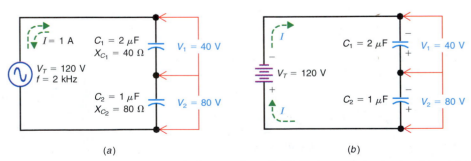

(a)   (b)

FIG. 19-6 Series capacitors divide V_T inversely proportional to each C. The smaller C has more V. (a) An ac divider with more X_C for the smaller C. (b) A dc divider.

The voltage division is proportional to the series reactances, as it is to series resistances. However, reactance is inversely proportional to capacitance. As a result, the smaller capacitance has more reactance and a greater part of the applied voltage.

DC DIVIDER   In Fig. 19-6b, both C_1 and C_2 will be charged by the battery. The voltage across the series combination of C_1 and C_2 must equal V_T. When charging current flows, electrons repelled from the negative battery terminal accumulate on the negative plate of C_1, repelling electrons from its positive plate. These electrons flow through the conductor to the negative plate of C_2. With the positive battery terminal attracting electrons, the charging current from the positive plate of C_2 returns to the positive side of the dc source. Then C_1 and C_2 become charged in the polarity shown.

Since C_1 and C_2 are in the same series path for charging current, both have the same amount of charge. However, the potential difference provided by the equal charges is inversely proportional to capacitance. The reason is that $Q = CV$, or $V = Q/C$. Therefore, the 1-μF capacitor has double the voltage of the 2-μF capacitor, with the same charge in both.

If you measure with a dc voltmeter across C_1, the meter reads 40 V. Across C_2 the dc voltage is 80 V. The measurement from the negative side of C_1 to the positive side of C_2 is the same as the applied battery voltage of 120 V.

If the meter is connected from the positive side of C_1 to the negative plate of C_2, however, the voltage is zero. These plates have the same potential because they are joined by a conductor of zero resistance.

The polarity marks at the junction between C_1 and C_2 indicate the voltage at this point with respect to the opposite plate of each capacitor. This junction is positive compared with the opposite plate of C_1 with a surplus of electrons. However, the same point is negative compared with the opposite plate of C_2, which has a deficiency of electrons.

In general, the following formula can be used for capacitances in series as a voltage divider:

$$\blacktriangleright \qquad V_C = \frac{C_{EQ}}{C} \times V_T \qquad\qquad \textbf{(19-6)}$$

Note that C_{EQ} is in the numerator, since it must be less than the smallest individual C with series capacitances. For the divider examples in Fig. 19-6a and b,

$$V_1 = \frac{C_{EQ}}{C_1} \times 120 = \frac{2/3}{2} \times 120 = 40 \text{ V}$$

$$V_2 = \frac{C_{EQ}}{C_2} \times 120 = \frac{2/3}{1} \times 120 = 80 \text{ V}$$

This method applies to series capacitances as a divider for either dc or ac voltage, as long as there is no series resistance. It should be noted that the case of capacitive dc dividers also applies to pulse circuits. Furthermore, bleeder resistors may be used across each of the capacitors to ensure more exact division.

ABOUT ELECTRONICS

Aura Systems has designed an automobile engine with no camshafts or valve trains. Its electronic valve system opens and shuts by means of an armature disc between two spring-loaded electromagnets. Fuel injection and ignition use sensors and engine-controlled computers.

19-7 THE GENERAL CASE OF CAPACITIVE CURRENT i_C

The capacitive charge and discharge current i_C is always equal to $C(dv/dt)$. A sine wave of voltage variations for v_C produces a cosine wave of current i. This means v_C and i_C have the same waveform, but they are 90° out of phase.

It is usually convenient to use X_C for calculations in sine-wave circuits. Since X_C is $1/(2\pi f C)$, the factors that determine the amount of charge and discharge current are included in f and C. Then I_C equals V_C/X_C. Or, if I_C is known, V_C can be calculated as $I_C \times X_C$.

With a nonsinusoidal waveform for voltage v_C, the concept of reactance cannot be used. Reactance X_C applies only to sine waves. Then i_C must be determined as $C(dv/dt)$. An example is illustrated in Fig. 19-7 to show the change of waveform here, instead of the change of phase angle in sine-wave circuits.

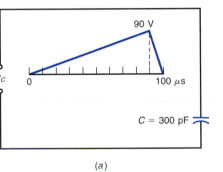

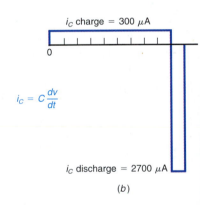

$$i_C = C\frac{dv}{dt}$$

FIG. 19-7 Waveshape of i_C equal to $C(dv/dt)$. (*a*) Sawtooth waveform of V_C. (*b*) Rectangular current waveform of i_C resulting from uniform rate of change in sawtooth waveform of voltage.

Note that the sawtooth waveform of voltage v_C corresponds to a rectangular waveform of current. The linear rise of the sawtooth wave produces a constant amount of charging current i_C because the rate of change is constant for the charging voltage. When the capacitor discharges, v_C drops sharply. Then discharge current is in the opposite direction from charge current. Also, the discharge current has a much larger value because of the faster rate of change in v_C.

19   SUMMARY AND REVIEW

- In a sine-wave ac circuit, the voltage across a capacitance lags its charge and discharge current by 90°.
- Therefore, capacitive reactance X_C is a phasor quantity out of phase with its series resistance by $-90°$ because $i_C = C(dv/dt)$. This fundamental fact is the basis of all the following relations.
- The combination of X_C and R in series is their total impedance Z_T. These three types of ohms of opposition to current are compared in Table 19-4.

R	$\mathbf{X_C = 1/(2\pi fC)}$	$\mathbf{Z_T = \sqrt{R^2 + X_C^2}}$
Ohms unit	Ohms unit	Ohms unit
IR voltage in phase with I	IX_C voltage lags I_C by 90°	IZ_T is the applied voltage
Same ohm value for all f	Ohms value decreases for higher f	Becomes more resistive with more f Becomes more capacitive with less f

TABLE 19-4   COMPARISON OF R, X_C, AND Z

- The opposite characteristics for series and parallel circuits with X_C and R are summarized in Table 19-5.

TABLE 19-5   SERIES AND PARALLEL RC CIRCUITS

$\mathbf{X_C}$ **AND R IN SERIES**	$\mathbf{X_C}$ **AND R IN PARALLEL**
I the same in X_C and R	V the same across X_C and R
$V_T = \sqrt{V_R^2 + V_C^2}$	$I_T = \sqrt{I_R^2 + I_C^2}$
$Z_T = \sqrt{R^2 + X_C^2}$	$Z_{EQ} = \dfrac{V}{I_T}$
V_C lags V_R by 90°	I_C leads I_R by 90°
$\tan \theta_Z = -\dfrac{X_C}{R}$; θ_Z increases as X_C increases, resulting in more V_C	$\tan \theta_I = \dfrac{I_C}{I_R}$; θ_I decreases as X_C increases, resulting in less I_C

- Two or more capacitors in series across a voltage source serve as a voltage divider. The smallest C has the largest part of the applied voltage.
- A coupling capacitor has X_C less than its series resistance by the factor of $\frac{1}{10}$ or less, for the purpose of providing practically all the ac applied voltage across R with little across C.
- In sine-wave circuits, $I_C = V_C/X_C$. Then I_C is out of phase with V_C by $90°$.
- For a circuit with X_C and R in series, $\tan \theta_Z = -(X_C/R)$ while in parallel $\tan \theta_I = I_C/I_R$. See Table 19-5.
- When the voltage is not a sine wave, $i_C = C(dv/dt)$. Then the waveshape of i_C is different from the voltage.

SELF-TEST

ANSWERS AT BACK OF BOOK.

Choose (*a*), (*b*), (*c*), or (*d*).

1. In a capacitive circuit (*a*) a decrease in applied voltage makes a capacitor charge; (*b*) a steady value of applied voltage causes discharge; (*c*) an increase in applied voltage makes a capacitor discharge; (*d*) an increase in applied voltage makes a capacitor charge.

2. In a sine-wave ac circuit with X_C and R in series, the (*a*) phase angle of the circuit is $180°$ with high series resistance; (*b*) voltage across the capacitance must be $90°$ out of phase with its charge and discharge current; (*c*) voltage across the capacitance has the same phase angle as its charge and discharge current; (*d*) charge and discharge current of the capacitor must be $90°$ out of phase with the applied voltage.

3. When v_C across a 1-μF C drops from 43 to 42 V in 1 s, the discharge current i_C equals (*a*) 1 μA; (*b*) 42 μA; (*c*) 43 μA; (*d*) 43 A.

4. In a sine-wave ac circuit with R and C in parallel, (*a*) the voltage across C lags the voltage across R by $90°$; (*b*) resistive I_R is $90°$ out of phase with I_C; (*c*) I_R and I_C are in phase; (*d*) I_R and I_C are $180°$ out of phase.

5. In a sine-wave ac circuit with a 90-Ω R in series with a 90-Ω X_C, the phase angle equals (*a*) $-90°$; (*b*) $-45°$; (*c*) $0°$; (*d*) $90°$.

6. The combined impedance of a 1000-Ω R in parallel with a 1000-Ω X_C equals (*a*) 500 Ω; (*b*) 707 Ω; (*c*) 1000 Ω; (*d*) 2000 Ω.

7. With 100 V applied across two series capacitors of 5 μF each, the voltage across each capacitor will be (*a*) 5 V; (*b*) 33$\frac{1}{3}$ V; (*c*) 50 V; (*d*) 66$\frac{2}{3}$ V.

8. In a sine-wave ac circuit with X_C and R in series, the (*a*) voltages across R and X_C are in phase; (*b*) voltages across R and X_C are $180°$ out of phase; (*c*) voltage across R leads the voltage across X_C by $90°$; (*d*) voltage across R lags the voltage across X_C by $90°$.

9. A 0.01-μF capacitance in series with R is used as a coupling capacitor C_C for 1000 Hz. At 10,000 Hz: (*a*) C_C has too much reactance to be good for coupling; (*b*) C_C has less reactance, which improves the coupling; (*c*) C_C has the same reactance and coupling; (*d*) the voltage across R is reduced by one-tenth.

10. In an RC coupling circuit the phase angle is (*a*) $90°$; (*b*) close to $0°$; (*c*) $-90°$; (*d*) $180°$.

1. (a) Why does a capacitor charge when the applied voltage increases? (b) Why does the capacitor discharge when the applied voltage decreases?
2. A sine wave of voltage V is applied across a capacitor C. (a) Draw the schematic diagram. (b) Draw the sine waves of voltage and current out of phase by 90°. (c) Draw a phasor diagram showing the phase angle of $-90°$ between V and I.
3. Why will a circuit with R and X_C in series be less capacitive as the frequency of the applied voltage is increased?
4. Define the following: coupling capacitor, sawtooth voltage, capacitive voltage divider.
5. Give two comparisons between RC circuits with sine-wave voltage applied and nonsinusoidal voltage applied.
6. State two troubles possible in coupling capacitors and describe briefly how you would check the capacitors with an ohmmeter.
7. Explain the function of R and C in an RC coupling circuit for an ac signal from one transistor amplifier to the next stage.
8. Explain briefly why a capacitor is able to block dc voltage.
9. What is the waveshape of i_C for a sine-wave v_C?
10. Explain why the impedance Z_{EQ} of a parallel RC circuit decreases as the frequency increases.
11. In a series RC circuit explain why θ_Z increases (becomes more negative) as frequency decreases.

ANSWERS TO ODD-NUMBERED PROBLEMS AT BACK OF BOOK.

1. A 40-Ω R is in series with a 30-Ω X_C across a 100-V sine-wave ac source. (a) Draw the schematic diagram. (b) Calculate Z_T. (c) Calculate I. (d) Calculate the voltages across R and C. (e) What is the phase angle of the circuit?
2. A 40-Ω R and a 30-Ω X_C are in parallel across a 100-V sine-wave ac source. (a) Draw the schematic diagram. (b) Calculate each branch current. (c) How much is I_T? (d) Calculate Z_{EQ}. (e) What is the phase angle of the circuit? (f) Compare the phase angle of the voltage across R and X_C.
3. Draw the schematic diagram of a capacitor in series with a 20-kΩ resistance across a 10-Vac source. What size C is needed for equal voltages across R and X_C at frequencies of 100 Hz and 100 kHz?
4. Draw the schematic diagram of two capacitors C_1 and C_2 in series across 10,000 V. The C_1 is 900 pF and has 9000 V across it. (a) How much is the voltage across C_2? (b) How much is the capacitance of C_2?
5. In Fig. 19-2a, how much is C for the X_C value of 100 Ω at frequencies of 60 Hz, 1000 Hz, and 1 MHz?
6. A 1500-Ω R is in series with a 0.01-μF C across a 30-V source with a frequency of 8 kHz. Calculate X_C, Z_T, θ_Z, I, V_R, and V_C.

7. The same R and C as in Prob. 6 are in parallel. Calculate I_C, I_R, I_T, θ_I, Z_{EQ}, V_R, and V_C.

8. A 0.05-μF capacitor is in series with a 50,000-Ω R and a 10-V source. Tabulate the values of X_C, I, V_R, and V_C at the frequencies of 0 (for steady dc voltage), 20, 60, 100, 500, 5000, and 15,000 Hz.

9. A capacitive voltage divider has C_1 of 1 μF, C_2 of 2 μF, and C_3 of 4 μF in series across a 700-V source V_T. (a) Calculate V_1, V_2, and V_3 for a steady dc source. (b) Calculate V_1, V_2, and V_3 for an ac source with a frequency of 400 Hz.

10. (a) A 40-Ω X_C and a 30-Ω R are in series across a 120-V source. Calculate Z_T, I, and θ_Z.
(b) The same X_C and R are in parallel. Calculate I_T, Z_{EQ}, and θ_I.

11. A 500-Ω R is in series with 300-Ω X_C. Find Z_T, I, and θ_Z. $V_T = 120$ V.

12. A 300-Ω R is in series with a 500-Ω X_C. Find Z_T, I, and θ_Z. Compare θ_Z here with Prob. 11, with the same 120 V applied.

13. A 500-Ω R is parallel with a 300-Ω X_C. Find I_T, Z_{EQ}, and θ_I. Compare θ_I here with θ_Z in Prob. 11, with the same 120 V applied.

14. For the waveshape of capacitor voltage v_C in Fig. 19-8, show the corresponding charge and discharge current i_C, with values for a 200-pF capacitance.

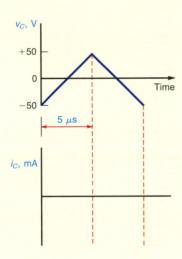

FIG. 19-8 Waveshapes for Prob. 14.

15. Calculate the values needed in Fig. 19-6a for the same voltage division but with a frequency of 60 Hz for V_T.

16. Find the angle θ_Z for X_C and R in series for the following combinations. (a) X_C is 5200 Ω and R is 5200 Ω. (b) X_C is 2600 Ω and R is 5200 Ω. (c) X_C is 520 Ω and R is 5200 Ω. (d) X_C is 52 Ω and R is 5200 Ω. (e) X_C is 5200Ω and R is 2600 Ω.

17. Calculate the angle θ_Z for the RC coupling in Fig. 19-5.

18. Find the angle θ_I for I_C and I_R in parallel with the same combinations as in Prob. 16. Assume V_A of 10 V to determine I_C and I_R in mA units.

19. How much C is needed for X_C of 52 Ω at the frequency of 4 MHz?

20. In Fig. 19-9, calculate X_C, Z_T, I, V_C, V_R, and θ_Z.

21. Repeat Prob. 20 if $f = 500$ Hz.

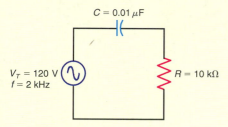

$C = 0.01\ \mu F$

$V_T = 120\ V$
$f = 2\ kHz$

$R = 10\ k\Omega$

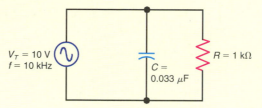

$V_T = 10\ V$
$f = 10\ kHz$

$C = 0.033\ \mu F$

$R = 1\ k\Omega$

FIG. 19-9 Circuit for Probs. 20 and 21.

FIG. 19-10 Circuit for Probs. 22, 23, and 24.

22. In Fig. 19-10, calculate X_C, I_C, I_R, I_T, Z_{EQ}, and θ_I.
23. Repeat Prob. 22 if $f = 5$ kHz.
24. In Fig. 19-10, calculate Z_{EQ} if another 1-kΩ R is added in parallel with the 1-kΩ R shown ($f = 10$ kHz).
25. In Fig. 19-11, calculate the voltages, with respect to ground, at points A, B, and C.
26. In Fig. 19-12, show the corresponding charge and discharge current for the waveshape of capacitor voltage shown.
27. Calculate the minimum coupling capacitance C_C in series with a 10-kΩ R if the frequency of the applied voltage ranges from: (**a**) 100 Hz to 10 kHz; (**b**) 15 kHz to 300 kHz.

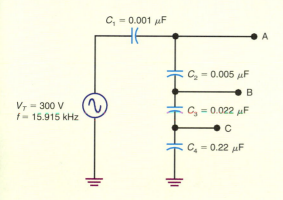

$C_1 = 0.001\ \mu F$

A

$C_2 = 0.005\ \mu F$

B

$V_T = 300\ V$
$f = 15.915\ kHz$

$C_3 = 0.022\ \mu F$

C

$C_4 = 0.22\ \mu F$

FIG. 19-11 Circuit for Prob. 25.

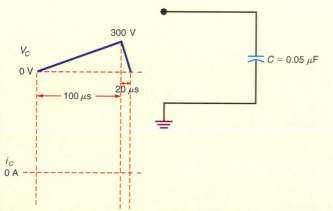

300 V

V_C

0 V

$C = 0.05\ \mu F$

100 μs

20 μs

i_C
0 A

FIG. 19-12 Diagram and circuit for Prob. 26.

CRITICAL THINKING

1. In Fig. 19-13, calculate X_C, Z_T, I, f, V_T, and V_R.
2. In Fig. 19-14, calculate I_C, I_R, V_T, X_C, C, and Z_{EQ}.
3. In Fig. 19-15, calculate I_C, I_R, I_T, X_C, R, and C.

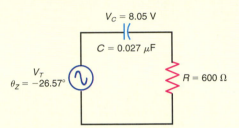

FIG. 19-13 Circuit for Critical Thinking Prob. 1.

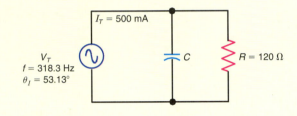

FIG. 19-14 Circuit for Critical Thinking Prob. 2.

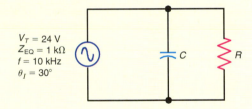

FIG. 19-15 Circuit for Critical Thinking Prob. 3.

ANSWERS TO TEST-POINT QUESTIONS

19-1 **a.** 0°
  b. 90°
  c. lag

19-2 **a.** 28.28 Ω
  b. 28.28 V
  c. $\theta_Z = -45°$

19-3 **a.** 46.7°
  b. 90°
  c. 43.3°

19-4 **a.** 2.828 A
  b. $\theta_I = 45°$

19-5 **a.** 35 Ω
  b. 2 pF

19-6 **a.** $V_1 = 18$ kV
    $V_2 = 2$ kV
  b. $X_{C_T} = 120$ Ω

19-7 **a.** $dv/dt = 1 \times 10^6$ V/s
  b. $i_C = 300$ μA

REVIEW: CHAPTERS 17 TO 19

SUMMARY

- A capacitor consists of two conductors separated by an insulator, which is a dielectric material. With voltage applied to the conductors, charge is stored in the dielectric. One coulomb of charge stored with one volt applied corresponds to one farad of capacitance C. The common units of capacitance are microfarads ($1\ \mu F = 10^{-6}\ F$) or picofarads ($1\ pF = 10^{-12}\ F$).
- Capacitance increases with plate area and larger values of dielectric constant but decreases with increased distance between the plates.
- The most common types of capacitors are air, film, paper, mica, disk ceramic, surface-mount (chip), and electrolytic. Electrolytics must be connected in the correct polarity. The capacitance coding systems for film, disk ceramic, mica, and tantalum capacitors are illustrated in Figs. 17-11, 17-13, 17-16, and 17-20, respectively. The different capacitance coding systems used with chip capacitors are illustrated in Figs. 17-17, 17-18, and 17-19.
- The total capacitance of parallel capacitors is the sum of the individual values; the combined capacitance of series capacitors is found by the reciprocal formula. These rules are opposite from the formulas used for resistors in series or parallel.
- When checking a capacitor with an ohmmeter, a good capacitor shows charging current and then the ohmmeter reads a very high value of ohms equal to the insulation resistance. A short-circuited capacitor reads zero ohms; an open capacitor does not show any charging current.
- $X_C = 1/(2\pi fC)\ \Omega$, where f is in hertz, C is in farads, and X_C is in ohms. The higher the frequency and the greater the capacitance, the smaller X_C is.
- The total X_C of capacitive reactances in series equals the sum of the individual values, just as for series resistances. The series reactances have the same current. The voltage across each X_C equals IX_C.
- With parallel capacitive reactances the combined reactance is calculated using the reciprocal formula, as for parallel resistances. Each branch current equals V_A/X_C. The total current is the sum of the individual branch currents.
- A common application of X_C is in AF or RF coupling capacitors, which have low reactance for higher frequencies but more reactance for lower frequencies.
- Reactance X_C is a phasor quantity where the voltage across the capacitor lags 90° behind its charge and discharge current.
- In a series RC circuit, R and X_C are added by phasors because the voltage drops are 90° out of phase. Therefore, the total impedance Z_T equals $\sqrt{R^2 + X_C^2}$; the current I equals V_T/Z_T.
- For parallel RC circuits, the resistive and capacitive branch currents are added by phasors, $I_T = \sqrt{I_R^2 + I_C^2}$; the impedance $Z_{EQ} = V_A/I_T$.
- Capacitive charge or discharge current i_C is equal to $C(dv/dt)$ for any waveshape of v_C.

- For series capacitors the amount of voltage drop is inversely proportional to its capacitance. That is, the smaller the capacitance, the larger is the voltage drop.

Answer True or False.

1. A capacitor can store charge because it has a dielectric between two conductors.
2. With 100-V applied, a 0.01-μF capacitor stores 1 μC of charge.
3. The smaller the capacitance, the higher is the potential difference across it for a given amount of charge stored in the capacitor.
4. A 250-pF capacitance equals 250×10^{-12} F.
5. The thinner the dielectric, the greater is the capacitance and the lower is the breakdown voltage rating for a capacitor.
6. Larger plate area increases capacitance.
7. Capacitors in series provide less capacitance but a higher breakdown voltage rating for the combination.
8. Capacitors in parallel increase the total capacitance with the same voltage rating.
9. Two 0.01-μF capacitors in parallel have a total C of 0.005 μF.
10. A good 0.1-μF film capacitor will show charging current and read 500 MΩ or more on an ohmmeter.
11. If the capacitance is doubled, the reactance is halved.
12. If the frequency is doubled, the capacitive reactance is doubled.
13. The reactance of a 0.1-μF capacitor at 60 Hz is approximately 60 Ω.
14. In a series RC circuit, the voltage across X_C lags 90° behind the current.
15. The phase angle of a series RC circuit can be any angle between 0° and −90°, depending on the ratio of X_C to R.
16. In a parallel RC circuit, the voltage across X_C lags 90° behind its capacitive branch current.
17. In a parallel circuit of two resistances with 1 A in each branch, the total line current equals 1.414 A.
18. A 1000-Ω X_C in parallel with a 1000-Ω R has a combined Z of 707 Ω.
19. A 1000-Ω X_C in series with a 1000-Ω R has a total Z of 1414 Ω.
20. Neglecting its sign, the phase angle is 45° for both circuits in Probs. 18 and 19.
21. The total impedance of a 1-MΩ R in series with a 5-Ω X_C is approximately 1 MΩ with a phase angle of 0°.
22. The combined impedance of a 5-Ω R in parallel with a 1-MΩ X_C is approximately 5 Ω with a phase angle of 0°.
23. Resistance and impedance are both measured in ohms.
24. The impedance Z of an RC circuit can change with frequency because the circuit includes reactance.
25. Capacitors in series have the same charge and discharge current.
26. Capacitors in parallel have the same voltage.

27. The phasor combination of a 30-Ω R in series with a 40-Ω X_C equals 70 Ω impedance.

28. A film capacitor coded as 103 has a value of 0.001 μF.

29. Capacitive current can be considered leading current in a series circuit.

30. In a series RC circuit, the higher the value of X_C, the greater is its voltage drop compared with the IR drop.

REFERENCES

Bogart, T.: *Electric Circuits,* Glencoe/ McGraw-Hill, Columbus, Ohio.

Schuler, C., and Fowler, R.: *Electric Circuit Analysis,* Glencoe/McGraw-Hill, Columbus, Ohio.

CHAPTER 20

INDUCTANCE

Inductance is the ability of a conductor to produce induced voltage when the current varies. A long wire has more inductance than a short wire, since more conductor length cut by magnetic flux produces more induced voltage. Similarly, a coil has more inductance than the equivalent length of straight wire because the coil concentrates magnetic flux. Components manufactured to have a definite value of inductance are just coils of wire, thus, called *inductors*. The symbol for inductance is L, and the unit is the henry (H).

The wire for a coil can be wound around a hollow, insulating tube. Or the coil can be just the wire itself. This type is an air-core coil, as the magnetic field of current in the coil is in air. With another basic type, the wire is wound on an iron core, in order to concentrate the magnetic flux for more inductance.

Air-core coils are used in RF circuits because higher frequencies need less L for the required inductive effect. Iron-core inductors are used in the audio-frequency range, especially the ac power-line frequency of 60 Hz, and for lower frequencies in general.

CHAPTER OBJECTIVES

Upon completion of this chapter, you should be able to:

- *Explain* the concept of self-inductance.
- *Define* the henry unit of inductance and *define* mutual inductance.
- *Calculate* the inductance when the induced voltage and rate of current change are known.
- *List* the physical factors affecting the inductance of an inductor.
- *Calculate* the induced voltage across an inductor given the inductance and rate of current change.
- *Explain* how induced voltage opposes a change in current.
- *Describe* how a transformer works and *list* important transformer ratings.
- *Calculate* the currents, voltages, and impedances of a transformer circuit.
- *Identify* the different types of transformer cores.
- *Calculate* the total inductance of series connected inductors.
- *Calculate* the equivalent inductance of parallel connected inductors.
- *List* some common troubles with inductors.

IMPORTANT TERMS IN THIS CHAPTER

air-core	henry unit	phasing dots
apparent power	hysteresis	reflected impedance
autotransformer	impedance matching	self-inductance
counter emf (Cemf)	iron-core	stray inductance
coupling coefficient	leakage flux	transformer
eddy current	Lenz' law	turns ratio
efficiency	magnetic coupling	Variac
ferrite	mutual inductance	

TOPICS COVERED IN THIS CHAPTER

Induced voltage is the result of flux cutting across a conductor. This action can be produced by physical motion of either the magnetic field or the conductor. When the current in a conductor varies in amplitude, however, the variations of current and its associated magnetic field are equivalent to motion of the flux. As the current increases in value, the magnetic field expands outward from the conductor. When the current decreases, the field collapses into the conductor. As the field expands and collapses with changes of current, the flux is effectively in motion. Therefore, a varying current can produce induced voltage without the need for motion of the conductor.

Figure 20-1 illustrates the changes in the magnetic field associated with a sine wave of alternating current. Since the alternating current varies in amplitude and reverses in direction, its associated magnetic field has the same variations. At point A, the current is zero and there is no flux. At B, the positive direction of current provides some field lines taken here in the counterclockwise direction. Point C has maximum current and maximum counterclockwise flux.

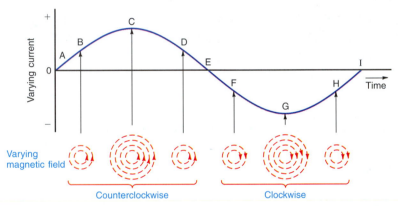

FIG. 20-1 Magnetic field of an alternating current is effectively in motion as it expands and contracts with the current variations.

At D there is less flux than at C. Now the field is collapsing because of the reduced current. At E, with zero current, there is no magnetic flux. The field can be considered as having collapsed into the wire.

The next half-cycle of current allows the field to expand and collapse again, but the directions are reversed. When the flux expands at points F and G, the field lines are clockwise, corresponding to current in the negative direction. From G to H and I, this clockwise field collapses into the wire.

The result of an expanding and collapsing field, then, is the same as that of a field in motion. This moving flux cuts across the conductor that is providing the current, producing induced voltage in the wire itself. Furthermore, any other conductor in the field, whether carrying current or not, also is cut by the varying flux and has induced voltage.

It is important to note that induction by a varying current results from the change in current, not the current value itself. The current must change to provide motion of the flux. A steady direct current of 1000 A, as an example of a large current, cannot produce any induced voltage as long as the current value is constant. A current of 1 μA changing to 2 μA, however, does induce voltage. Also, the faster the current changes, the higher the induced voltage because when the flux moves at a higher speed, it can induce more voltage.

Since inductance is a measure of induced voltage, the amount of inductance has an important effect in any circuit in which the current changes. The inductance is an additional characteristic of the circuit besides its resistance. The characteristics of inductance are important in:

1. *AC circuits.* Here the current is continuously changing and producing induced voltage. Lower frequencies of alternating current require more inductance to produce the same amount of induced voltage as a higher-frequency current. The current can have any waveform, as long as the amplitude is changing.
2. *DC circuits in which the current changes in value.* It is not necessary for the current to reverse direction. One example is a dc circuit being turned on or off. When the direct current is changing between zero and its steady value, the inductance affects the circuit at the time of switching. This effect with a sudden change is called the *transient response.* A steady direct current that does not change in value is not affected by inductance, however, because there can be no induced voltage without a change in current.

TEST-POINT QUESTION 20-1

Answers at end of chapter.
 a. For the same number of turns and frequency, which has more inductance, a coil with an iron core or one without an iron core?
 b. In Fig. 20-1, are the changes of current faster at time B or C?

20-2 SELF-INDUCTANCE *L*

The ability of a conductor to induce voltage in itself when the current changes is its *self-inductance* or simply *inductance.* The symbol for inductance is *L*, for linkages of the magnetic flux, and its unit is the *henry* (H). This unit is named after Joseph Henry (1797–1878).

DEFINITION OF THE HENRY UNIT   As illustrated in Fig. 20-2, one henry is the amount of inductance that allows one volt to be induced when the current changes at the rate of one ampere per second. The formula is

$$L = \frac{v_L}{di/dt} \qquad\qquad (20\text{-}1)$$

where v_L is in volts and di/dt is the current change in amperes per second.

Again the symbol d is used to indicate an infinitesimally small change in current with time. The factor di/dt for the current variation with respect to time

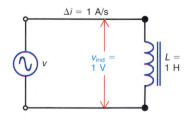

FIG. 20-2   When a current change of 1 A/s induces 1 V across *L*, its inductance equals 1 H.

really specifies how fast the current's associated magnetic flux is cutting the conductor to produce v_L.

Example

EXAMPLE 1   The current in an inductor changes from 12 to 16 A in 1 s. How much is the di/dt rate of current change in amperes per second?

ANSWER   The di is the difference between 16 and 12, or 4 A in 1 s. Then

$$\frac{di}{dt} = 4 \text{ A/s}$$

EXAMPLE 2   The current in an inductor changes by 50 mA in 2 μs. How much is the di/dt rate of current change in amperes per second?

ANSWER $\dfrac{di}{dt} = \dfrac{50 \times 10^{-3}}{2 \times 10^{-6}} = 25 \times 10^3$

$$\frac{di}{dt} = 25,000 \text{ A/s}$$

EXAMPLE 3   How much is the inductance of a coil that induces 40 V when its current changes at the rate of 4 A/s?

ANSWER $L = \dfrac{v_L}{di/dt} = \dfrac{40}{4}$

$L = 10 \text{ H}$

EXAMPLE 4   How much is the inductance of a coil that induces 1000 V when its current changes at the rate of 50 mA in 2 μs?

ANSWER   For this example, the $1/dt$ factor in the denominator of Formula (20-1) can be inverted to the numerator.

$$L = \frac{v_L}{di/dt} = \frac{v_L \times dt}{di}$$

$$= \frac{1 \times 10^3 \times 2 \times 10^{-6}}{50 \times 10^{-3}}$$

$$= \frac{2 \times 10^{-3}}{50 \times 10^{-3}} = \frac{2}{50}$$

$L = 0.04 \text{ H or } 40 \text{ mH}$

Notice that the smaller inductance in Example 4 produces much more v_L than the inductance in Example 3. The very fast current change in Example 4 is equivalent to 25,000 A/s.

INDUCTANCE OF COILS   In terms of physical construction, the inductance depends on how a coil is wound. Note the following factors.

1. A greater number of turns N increases L because more voltage can be induced. Actually L increases in proportion to N^2. Double the number of turns in the same area and length increases the inductance four times.
2. More area A enclosed by each turn increases L. This means a coil with larger turns has more inductance. The L increases in direct proportion to A and as the square of the diameter of each turn.
3. The L increases with the permeability of the core. For an air core μ_r is 1. With a magnetic core, L is increased by the μ_r factor as the magnetic flux is concentrated in the coil.
4. The L decreases with more length for the same number of turns, as the magnetic field then is less concentrated.

These physical characteristics of a coil are illustrated in Fig. 20-3. For a long coil, where the length is at least ten times the diameter, the inductance can be calculated from the formula

▶ $$L = \mu_r \times \frac{N^2 \times A}{l} \times 1.26 \times 10^{-6} \qquad \text{H} \qquad\qquad \textbf{(20-2)}$$

where L is in henrys, l is in meters and A is in square meters. The constant factor 1.26×10^{-6} is the absolute permeability of air or vacuum, in SI units, to calculate L in henrys.

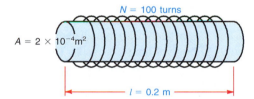

$N = 100$ turns

$A = 2 \times 10^{-4} \text{m}^2$

$l = 0.2$ m

FIG. 20-3 Physical factors for inductance L of a coil. See text for calculating L.

For the air-core coil in Fig. 20-3,

$$L = 1 \times \frac{10^4 \times 2 \times 10^{-4}}{0.2} \times 1.26 \times 10^{-6}$$

$$L = 12.6 \times 10^{-6} \text{ H} = 12.6 \ \mu\text{H}$$

This value means that the coil can produce a self-induced voltage of 12.6 μV when its current changes at the rate of 1 A/s, as $v_L = L(di/dt)$. Furthermore, if the coil has an iron core with $\mu_r = 100$, then L will be 100 times greater.

TYPICAL COIL INDUCTANCE VALUES   Air-core coils for RF applications have L values in millihenrys (mH) and microhenrys (μH). A typical air-core RF inductor (called a *choke*) is shown with its schematic symbol in Fig. 20-4a. Note that

$$1 \text{ mH} = 1 \times 10^{-3} \text{ H}$$
$$1 \ \mu\text{H} = 1 \times 10^{-6} \text{ H}$$

For example, an RF coil for the radio broadcast band of 535 to 1605 kHz may have an inductance L of 250 μH or 0.250 mH.

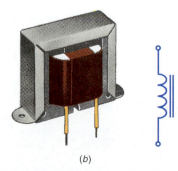

(a)

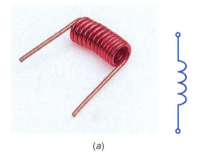

(b)

FIG. 20-4 Typical inductors with symbols. (*a*) Air-core coil used as RF choke. Length is 2 in. (*b*) Iron-core coil used for 60 Hz. Height is 2 in.

Iron-core inductors for the 60-Hz power line and for audio frequencies have inductance values of about 1 to 25 H. An iron-core choke is shown in Fig. 20-4b.

TEST-POINT QUESTION 20-2

Answers at end of chapter.
a. A coil induces 2 V with di/dt of 1 A/s. How much is L?
b. A coil has L of 8 mH with 125 turns. If the number of turns is doubled, how much will L be?

20-3 SELF-INDUCED VOLTAGE v_L

The self-induced voltage across an inductance L produced by a change in current di/dt can be stated as

$$v_L = L\frac{di}{dt} \qquad\qquad (20\text{-}3)$$

where v_L is in volts, L in henrys, and di/dt in amperes per second. This formula is just an inverted version of Formula (20-1) which defines inductance.

Actually both versions are based on Formula (15-1): $v = N(d\phi/dt)$ for magnetism. This gives the voltage in terms of how much magnetic flux is cut by a conductor per second. When the magnetic flux associated with the current varies the same as i, then Formula (20-3) gives the same results for calculating induced voltage. Remember also that the induced voltage across the coil is actually the result of inducing electrons to move in the conductor, so that there is also an induced current. In using Formula (20-3) to calculate v_L, just multiply L by the di/dt factor.

Example

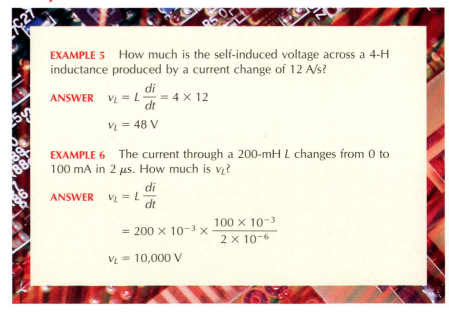

EXAMPLE 5   How much is the self-induced voltage across a 4-H inductance produced by a current change of 12 A/s?

ANSWER $v_L = L\dfrac{di}{dt} = 4 \times 12$

$v_L = 48$ V

EXAMPLE 6   The current through a 200-mH L changes from 0 to 100 mA in 2 μs. How much is v_L?

ANSWER $v_L = L\dfrac{di}{dt}$

$= 200 \times 10^{-3} \times \dfrac{100 \times 10^{-3}}{2 \times 10^{-6}}$

$v_L = 10{,}000$ V

Note the high voltage induced in the 200-mH inductance because of the fast change in current.

The induced voltage is an actual voltage that can be measured, although v_L is produced only while the current is changing. When di/dt is present for only a short time, v_L is in the form of a voltage pulse. With a sine-wave current, which is always changing, v_L is a sinusoidal voltage 90° out of phase with i_L.

20-4 HOW v_L OPPOSES A CHANGE IN CURRENT

By Lenz' law, the induced voltage v_L must produce current with a magnetic field that opposes the change of current that induces v_L. The polarity of v_L, therefore, depends on the direction of the current variation di. When di increases, v_L has the polarity that opposes the increase of current; when di decreases, v_L has the opposite polarity to oppose the decrease of current.

In both cases, the change of current is opposed by the induced voltage. Otherwise, v_L could increase to an unlimited amount without the need for adding any work. *Inductance, therefore, is the characteristic that opposes any change in current*. This is the reason why an induced voltage is often called a *counter emf* or *back emf*.

More details of applying Lenz' law to determine the polarity of v_L in a circuit are illustrated in Fig. 20-5. Note the directions carefully. In Fig. 20-5a, the electron flow is into the top of the coil. This current is increasing. By Lenz' law, v_L must have the polarity needed to oppose the increase. The induced voltage shown with the top side negative opposes the increase in current. The reason is that this polarity of v_L can produce current in the opposite direction, from minus to plus in the external circuit. Note that for this opposing current, v_L is the generator. This action tends to keep the current from increasing.

In Fig. 20-5b, the source is still producing electron flow into the top of the coil, but i is decreasing, because the source voltage is decreasing. By Lenz' law, v_L must have the polarity needed to oppose the decrease in current. The induced voltage shown with the top side positive now opposes the decrease. The reason is that this polarity of v_L can produce current in the same direction, tending to keep the current from decreasing.

In Fig. 20-5c, the voltage source reverses polarity to produce current in the opposite direction, with electron flow into the bottom of the coil. This reversed direction of current is now increasing. The polarity of v_L must oppose the increase. As shown, now the bottom of the coil is made negative by v_L to produce current opposing the source current. Finally, in Fig. 20-5d the reversed current is decreasing. This decrease is opposed by the polarity shown for v_L to keep the current flowing in the same direction as the source current.

DID YOU KNOW?

Certain materials such as mercury, tin, lead, and vanadium become superconductors when cooled by liquid helium to low temperatures.

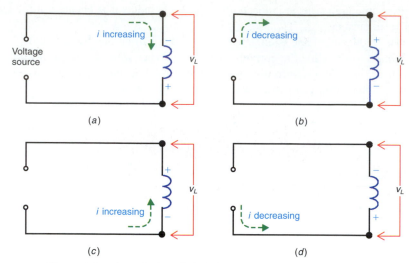

FIG. 20-5 Determining the polarity of V_L that opposes the change in i. (*a*) The i is increasing, and V_L has the polarity that produces an opposing current. (*b*) The i is decreasing, and V_L produces an aiding current. (*c*) The i is increasing but is flowing in the opposite direction. (*d*) The same direction of i as in (*c*), but with decreasing values.

Notice that the polarity of v_L reverses for either a reversal of direction for i or a reversal of change in di between increasing or decreasing values. When both the direction of the current and the direction of change are reversed, as in a comparison of Fig. 20-5*a* and *d*, the polarity of v_L remains unchanged.

Sometimes the formulas for induced voltage are written with a minus sign, in order to indicate the fact that v_L opposes the change, as specified by Lenz' law. However, the negative sign is omitted here so that the actual polarity of the self-induced voltage can be determined in typical circuits.

In summary, Lenz' law states that the reaction v_L opposes its cause, which is the change in i. When i is increasing, v_L produces an opposing current. For the opposite case when i is decreasing, v_L produces an aiding current.

TEST-POINT QUESTION 20-4

Answers at end of chapter.

Answer True or False.
a. In Fig. 20-5*a* and *b* the v_L has opposite polarities.
b. In Fig. 20-5*b* and *c* the polarity of v_L is the same.

20-5 MUTUAL INDUCTANCE L_M

When the current in an inductor changes, the varying flux can cut across any other inductor nearby, producing induced voltage in both inductors. In Fig. 20-6, the coil L_1 is connected to a generator that produces varying current in the turns. The winding L_2 is not connected to L_1, but the turns are linked by the magnetic field.

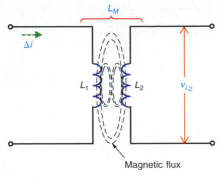

FIG. 20-6 Mutual inductance L_M between L_1 and L_2 linked by magnetic flux.

A varying current in L_1, therefore, induces voltage across L_1 and across L_2. If all the flux of the current in L_1 links all the turns of the coil L_2, each turn in L_2 will have the same amount of induced voltage as each turn in L_1. Furthermore, the induced voltage v_{L_2} can produce current in a load resistance connected across L_2.

When the induced voltage produces current in L_2, its varying magnetic field induces voltage in L_1. The two coils L_1 and L_2 have mutual inductance, therefore, because current in one can induce voltage in the other.

The unit of mutual inductance is the henry, and the symbol is L_M. *Two coils have L_M of 1 H when a current change of 1 A/s in one coil induces 1 V in the other coil.*

The schematic symbol for two coils with mutual inductance is shown in Fig. 20-7a for an air core, and for an iron core in Fig. 20-7b. Iron increases the mutual inductance, since it concentrates magnetic flux. Any magnetic lines that do not link the two coils result in *leakage flux*.

COEFFICIENT OF COUPLING The fraction of total flux from one coil linking another coil is the coefficient of coupling k between the two coils. As examples, if all the flux of L_1 in Fig. 20-6 links L_2, then k equals 1, or unity coupling; if half the flux of one coil links the other, k equals 0.5. Specifically, the coefficient of coupling is

$$k = \frac{\text{flux linkages between } L_1 \text{ and } L_2}{\text{flux produced by } L_1}$$

There are no units for k, as it is just a ratio of two values of magnetic flux. The value of k is generally stated as a decimal fraction, like 0.5, rather than as a percent.

The coefficient of coupling is increased by placing the coils close together, possibly with one wound on top of the other, by placing them parallel rather than perpendicular to each other, or by winding the coils on a common iron core. Several examples are shown in Fig. 20-8.

A high value of k, called *tight coupling,* allows the current in one coil to induce more voltage in the other coil. *Loose coupling,* with a low value of k, has the opposite effect. In the extreme case of zero coefficient of coupling, there is no mutual inductance. Two coils may be placed perpendicular to each other and far apart for essentially zero coupling when it is desired to minimize interaction between the coils.

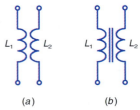

(a)   (b)

FIG. 20-7 Schematic symbols for two coils with mutual inductance. (a) Air core. (b) Iron core.

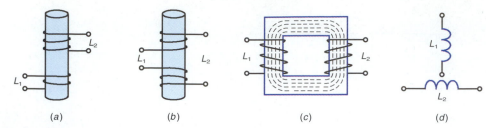

(a) (b) (c) (d)

FIG. 20-8 Examples of coupling between two coils linked by L_M. (a) L_1 or L_2 on paper or plastic form with air core; k is 0.1. (b) L_1 wound over L_2 for tighter coupling; k is 0.3. (c) L_1 and L_2 on the same iron core; k is 1. (d) Zero coupling between perpendicular air-core coils.

Air-core coils wound on one form have values of k equal to 0.05 to 0.3, approximately, corresponding to 5 to 30 percent linkage. Coils on a common iron core can be considered to have practically unity coupling, with k equal to 1. As shown in Fig. 20-8c, for both windings L_1 and L_2 practically all the magnetic flux is in the common iron core. Mutual inductance is also called *mutual coupling*.

Example

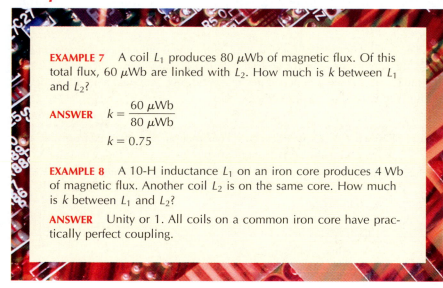

EXAMPLE 7 A coil L_1 produces 80 μWb of magnetic flux. Of this total flux, 60 μWb are linked with L_2. How much is k between L_1 and L_2?

ANSWER $k = \dfrac{60\ \mu\text{Wb}}{80\ \mu\text{Wb}}$

$k = 0.75$

EXAMPLE 8 A 10-H inductance L_1 on an iron core produces 4 Wb of magnetic flux. Another coil L_2 is on the same core. How much is k between L_1 and L_2?

ANSWER Unity or 1. All coils on a common iron core have practically perfect coupling.

CALCULATING L_M The mutual inductance increases with higher values for the primary and secondary inductances and tighter coupling:

$$L_M = k\sqrt{L_1 \times L_2}\qquad \text{H} \qquad\qquad\text{(20-4)}$$

where L_1 and L_2 are the self-inductance values of the two coils, k is the coefficient of coupling, and L_M is the mutual inductance linking L_1 and L_2, in the same units as L_1 and L_2. The k factor is needed to indicate the flux linkages between the two coils.

As an example, suppose that $L_1 = 2$ H and $L_2 = 8$ H, with both coils on an iron core for unity coupling. Then the mutual inductance is

$$L_M = 1 \sqrt{2 \times 8} = \sqrt{16} = 4 \text{ H}$$

The value of 4 H for L_M in this example means that when the current changes at the rate of 1 A/s in either coil, it will induce 4 V in the other coil.

Example

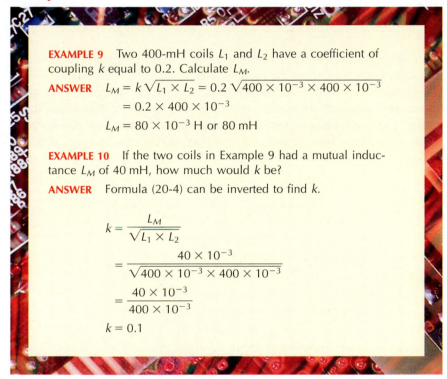

EXAMPLE 9   Two 400-mH coils L_1 and L_2 have a coefficient of coupling k equal to 0.2. Calculate L_M.

ANSWER $L_M = k \sqrt{L_1 \times L_2} = 0.2 \sqrt{400 \times 10^{-3} \times 400 \times 10^{-3}}$

$$= 0.2 \times 400 \times 10^{-3}$$

$$L_M = 80 \times 10^{-3} \text{ H or } 80 \text{ mH}$$

EXAMPLE 10   If the two coils in Example 9 had a mutual inductance L_M of 40 mH, how much would k be?

ANSWER   Formula (20-4) can be inverted to find k.

$$k = \frac{L_M}{\sqrt{L_1 \times L_2}}$$

$$= \frac{40 \times 10^{-3}}{\sqrt{400 \times 10^{-3} \times 400 \times 10^{-3}}}$$

$$= \frac{40 \times 10^{-3}}{400 \times 10^{-3}}$$

$$k = 0.1$$

Notice that the same two coils have one-half the mutual inductance L_M, because the coefficient of coupling k is 0.1 instead of 0.2.

To do Example 9 on a calculator that does not have $\boxed{\text{EXP}}$ key, multiply $L_1 \times L_2$, take the square root of the product, and multiply by k. Keep the powers of 10 separate. Specifically, punch in 400 for L_1, push the $\boxed{\times}$ key, punch in 400 for L_2, and push the $\boxed{=}$ key for the product, 16,000. Press the $\boxed{\sqrt{}}$ key, which is sometimes the $\boxed{\text{2ndF}}$ of the $\boxed{x^2}$ key to get 400. While it is on the display, push the $\boxed{\times}$ key, punch in 0.2, and press the $\boxed{=}$ key for the answer of 80. For the powers of 10, $10^{-3} \times 10^{-3} = 10^{-6}$, and the square root is equal to 10^{-3} for the unit of mH in the answer.

For Example 10, the formula is L_M divided by the reciprocal of $\sqrt{L_1 \times L_2}$. Specifically, punch in 40 for the value in the numerator, press the $\boxed{(}$ key, multiply 400×400, and press the $\boxed{)}$ key, followed by the $\boxed{\sqrt{}}$ and $\boxed{=}$ keys. The display will read 0.1. The powers of 10 cancel with 10^{-3} in the numerator and denominator. Also, there are no units for k, since the units of L cancel.

Answers at end of chapter.

a. All the flux from the current in L_1 links L_2. How much is the coefficient of coupling k?

b. Mutual inductance L_M is 9 mH with k of 0.2. If k is doubled to 0.4, how much will L_M be?

20-6 TRANSFORMERS

The transformer is an important application of mutual inductance. As shown in Fig. 20-9, a transformer has the primary winding inductance L_P connected to a voltage source that produces alternating current, while the secondary winding inductance L_S is connected across the load resistance R_L. The purpose of the transformer is to transfer power from the primary, where the generator is connected, to the secondary, where the induced secondary voltage can produce current in the load resistance that is connected across L_S.

ABOUT ELECTRONICS

The first microprocessor-run grill allows the cookout chef to use a remote control while resting in a hammock. The keypad is solar-powered and sets an exact temperature. It reignites the grill if there's a strong breeze, and a timer says when dinner's ready.

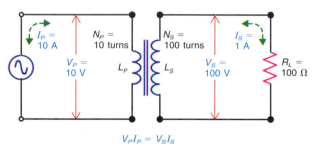

$$V_P I_P = V_S I_S$$

FIG. 20-9 Iron-core transformer with 1:10 turns ratio. Primary current I_P induces secondary voltage V_S, which produces current in secondary load R_L.

Although the primary and secondary are not physically connected to each other, power in the primary is coupled into the secondary by the magnetic field linking the two windings. The transformer is used to provide power for the load resistance R_L, instead of connecting R_L directly across the generator, whenever the load requires an ac voltage higher or lower than the generator voltage. By having more or fewer turns in L_S, compared with L_P, the transformer can step up or step down the generator voltage to provide the required amount of secondary voltage. Typical transformers are shown in Figs. 20-10 and 20-11. It should be noted that a steady dc voltage cannot be stepped up or down by a transformer, because a steady current cannot produce induced voltage.

TURNS RATIO The ratio of the number of turns in the primary to the number in the secondary is the turns ratio of the transformer:

Turns ratio $= \dfrac{N_P}{N_S}$ **(20-5)**

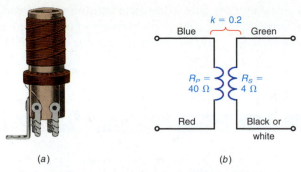

(a)

k = 0.2

Blue   Green

$R_P = 40\ \Omega$ $R_S = 4\ \Omega$

Red   Black or white

(b)

FIG. 20-10 (a) Air-core RF transformer. Height is 2 in. (b) Color code and typical dc resistance of windings.

where N_P = number of turns in the primary and N_S = number of turns in the secondary. For example, 500 turns in the primary and 50 turns in the secondary provide a turns ratio of 500/50, or 10 : 1, which is stated as "ten-to-one."

VOLTAGE RATIO With unity coupling between primary and secondary, the voltage induced in each turn of the secondary is the same as the self-induced voltage of each turn in the primary. Therefore, the voltage ratio is in the same proportion as the turns ratio:

▶
$$\frac{V_P}{V_S} = \frac{N_P}{N_S} \qquad\qquad\qquad (20\text{-}6)$$

When the secondary has more turns than the primary, the secondary voltage is higher than the primary voltage and the primary voltage is said to be stepped up. This principle is illustrated in Fig. 20-9 with a step-up ratio of 10/100, or 1 : 10. When the secondary has fewer turns, the voltage is stepped down.

In either case, the ratio is in terms of the primary voltage, which may be stepped up or down in the secondary winding.

FIG. 20-11 Iron-core power transformer.

These calculations apply only to iron-core transformers with unity coupling. Air-core transformers for RF circuits (as shown in Fig. 20-10*a*) are generally tuned to resonance. In this case, the resonance factor is considered instead of the turns ratio.

Example

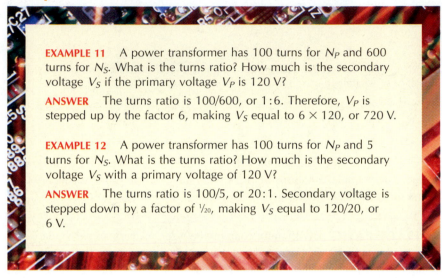

EXAMPLE 11   A power transformer has 100 turns for N_P and 600 turns for N_S. What is the turns ratio? How much is the secondary voltage V_S if the primary voltage V_P is 120 V?

ANSWER   The turns ratio is 100/600, or 1:6. Therefore, V_P is stepped up by the factor 6, making V_S equal to 6 × 120, or 720 V.

EXAMPLE 12   A power transformer has 100 turns for N_P and 5 turns for N_S. What is the turns ratio? How much is the secondary voltage V_S with a primary voltage of 120 V?

ANSWER   The turns ratio is 100/5, or 20:1. Secondary voltage is stepped down by a factor of $\frac{1}{20}$, making V_S equal to 120/20, or 6 V.

SECONDARY CURRENT   By Ohm's law, the amount of secondary current equals the secondary voltage divided by the resistance in the secondary circuit. In Fig. 20-9, with a value of 100 Ω for R_L and negligible coil resistance assumed,

$$I_S = \frac{V_S}{R_L} = \frac{100 \text{ V}}{100 \text{ }\Omega} = 1 \text{ A}$$

POWER IN THE SECONDARY   The power dissipated by R_L in the secondary is $I_S{}^2 \times R_L$ or $V_S \times I_S$, which equals 100 W in this example. The calculations are

$$P = I_S{}^2 \times R_L = 1 \times 100 = 100 \text{ W}$$
$$P = V_S \times I_S = 100 \times 1 = 100 \text{ W}$$

It is important to note that power used by the secondary load, such as R_L in Fig. 20-9, is supplied by the generator in the primary. How the load in the secondary draws power from the generator in the primary can be explained as follows.

With current in the secondary winding, its magnetic field opposes the varying flux of the primary current. The generator must then produce more primary current to maintain the self-induced voltage across L_P and the secondary voltage developed in L_S by mutual induction. If the secondary current doubles, for instance, because the load resistance is reduced one-half, the primary current will also double in value to provide the required power for the secondary. Therefore, the effect of the secondary-load power on the generator is the same as though R_L were in the primary, except that in the secondary the voltage for R_L is stepped up or down by the turns ratio.

CURRENT RATIO With zero losses assumed for the transformer, the power in the secondary equals the power in the primary:

▶ $$V_S I_S = V_P I_P \qquad\qquad (20\text{-}7)$$

or

▶ $$\frac{I_S}{I_P} = \frac{V_P}{V_S} \qquad\qquad (20\text{-}8)$$

The current ratio is the inverse of the voltage ratio; that is, voltage step-up in the secondary means current step-down, and vice versa. The secondary does not generate power but only takes it from the primary. Therefore, the current step-up or step-down is in terms of the secondary current I_S, which is determined by the load resistance across the secondary voltage. These points are illustrated by the following two examples.

Example

EXAMPLE 13 A transformer with a 1:6 turns ratio has 720 V across 7200 Ω in the secondary. **(a)** How much is I_S? **(b)** Calculate the value of I_P.

ANSWER

a. $I_S = \dfrac{V_S}{R_L} = \dfrac{720 \text{ V}}{7200 \text{ Ω}}$

   $I_S = 0.1$ A

b. With a turns ratio of 1:6, the current ratio is 6:1. Therefore,

   $I_P = 6 \times I_S = 6 \times 0.1$

   $I_P = 0.6$ A

EXAMPLE 14 A transformer with a 20:1 voltage step-down ratio has 6 V across 0.6 Ω in the secondary. **(a)** How much is I_S? **(b)** How much is I_P?

ANSWER

a. $I_S = \dfrac{V_S}{R_L} = \dfrac{6 \text{ V}}{0.6 \text{ Ω}}$

   $I_S = 10$ A

b. $I_P = \frac{1}{20} \times I_S = \frac{1}{20} \times 10$

   $I_P = 0.5$ A

As an aid in these calculations, remember that the side with the higher voltage has the lower current. The primary and secondary V and I are in the same proportion as the number of turns in the primary and secondary.

TOTAL SECONDARY POWER EQUALS PRIMARY POWER Figure 20-12 illustrates a power transformer with two secondary windings L_1 and L_2. There can be one, two, or more secondary windings with unity coupling to the primary as long as all the windings are on the same iron core. Each secondary winding has induced voltage in proportion to its turns ratio with the primary winding, which is connected across the 120-V source.

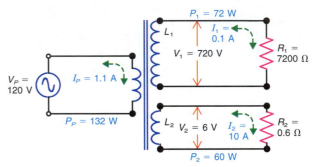

FIG. 20-12 Total power used by two secondary loads R_1 and R_2 is equal to the power supplied by the source in the primary.

The secondary winding L_1 has a voltage step-up of 6:1, providing 720 V. The 7200-Ω load resistance R_1, across L_1, allows the 720 V to produce 0.1 A for I_1 in this secondary circuit. The power here is 720 V $\times$ 0.1 A = 72 W.

The other secondary winding L_2 provides voltage step-down, with the ratio 20:1, resulting in 6 V across R_2. The 0.6-Ω load resistance in this circuit allows 10 A for I_2. Therefore, the power here is 6 V $\times$ 10 A, or 60 W. Since the windings have separate connections, each can have its individual values of voltage and current.

The total power used in the secondary circuits is supplied by the primary. In this example, the total secondary power is 132 W, equal to 72 W for P_1 and 60 W for P_2. The power supplied by the 120-V source in the primary then is 72 + 60 = 132 W.

The primary current I_P equals the primary power P_P divided by the primary voltage V_P. This is 132 W divided by 120 V, which equals 1.1 A for the primary current. The same value can be calculated as the sum of 0.6 A of primary current providing power for L_1 plus 0.5 A of primary current for L_2, resulting in the total of 1.1 A as the value of I_P.

This example shows how to analyze a loaded power transformer. The main idea is that the primary current depends on the secondary load. The calculations can be summarized as follows:

1. Calculate V_S from the turns ratio and V_P.
2. Use V_S to calculate I_S: $I_S = V_S/R_L$.
3. Use I_S to calculate P_S: $P_S = V_S \times I_S$.
4. Use P_S to find P_P: $P_P = P_S$.
5. Finally, I_P can be calculated: $I_P = P_P/V_P$.

With more than one secondary, calculate each I_S and P_S. Then add all the P_S for the total secondary power, which equals the primary power.

AUTOTRANSFORMERS As illustrated in Fig. 20-13, an autotransformer consists of one continuous coil with a tapped connection such as terminal 2 between the ends at terminals 1 and 3. In Fig. 20-13a the autotransformer steps up the generator voltage. Voltage V_P between 1 and 2 is connected across part of the total turns, while V_S is induced across all the turns. With six times the turns for the secondary voltage, V_S also is six times V_P.

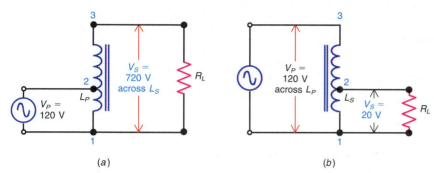

(a)   (b)

FIG. 20-13 Autotransformer with tap at terminal 2 for 10 turns of the complete 60-turn winding. (a) V_P between terminals 1 and 2 stepped up across 1 and 3. (b) V_P between terminals 1 and 3 stepped down across 1 and 2.

In Fig. 20-13b the autotransformer steps down the primary voltage connected across the entire coil. Then the secondary voltage is taken across less than the total turns.

The winding that connects to the voltage source to supply power is the primary, while the secondary is across the load resistance R_L. The turns ratio and voltage ratio apply the same way as in a conventional transformer having an isolated secondary winding.

Autotransformers are used often because they are compact, efficient, and usually cost less since they have only one winding. Note that the autotransformer in Fig. 20-13 has only three leads, compared with four leads for the transformer in Fig. 20-9 with an isolated secondary.

ISOLATION OF THE SECONDARY In a transformer with a separate winding for L_S, as in Fig. 20-9, the secondary load is not connected directly to the ac power line in the primary. This isolation is an advantage in reducing the chance of electric shock. With an autotransformer, as in Fig. 20-13, the secondary is not isolated. Another advantage of an isolated secondary is the fact that any direct current in the primary is blocked from the secondary. Sometimes a transformer with a 1:1 turns ratio is used just for isolation from the ac power line.

TRANSFORMER EFFICIENCY Efficiency is defined as the ratio of power out to power in. Stated as a formula,

$$\text{Efficiency} = \frac{P_{\text{out}}}{P_{\text{in}}} \times 100\% \qquad \text{(20-9)}$$

For example, when the power out in watts equals one-half the power in, the efficiency is one-half, which equals 0.5×100 percent, or 50 percent. In a transformer, power out is secondary power, while power in is primary power.

Assuming zero losses in the transformer, power out equals power in and the efficiency is 100 percent. Actual power transformers, however, have an efficiency slightly less than 100 percent. The efficiency is approximately 80 to 90 percent for transformers that have high power ratings. Transformers for higher power are more efficient because they require heavier wire, which has less resistance. In a transformer that is less than 100 percent efficient, the primary supplies more than the secondary power. The primary power that is lost is dissipated as heat in the transformer, resulting from I^2R in the conductors and certain losses in the core material. The R of the primary winding is generally about 10 Ω or less, for power transformers.

TRANSFORMER COLOR CODES The colors of the leads show the required connections in electronic circuits. For the RF transformer in Fig. 20-10, the leads are:

Blue—Output electrode of transistor amplifier
Red—DC supply voltage for this electrode
Green—Input electrode of next amplifier
Black or white—Return line of secondary winding

This system applies to all coupling transformers between amplifier stages, including iron-core transformers for audio circuits.

For the power transformer in Fig. 20-11, the primary is connected to the ac power line. The leads are:

Black—Primary leads without tap
Black with yellow—Tap on primary
Red—High-voltage secondary to rectifier in power supply
Red with yellow—Tap on high-voltage secondary
Green-yellow—Low-voltage secondary

TEST-POINT QUESTION 20-6

Answers at end of chapter.
a. A transformer connected to the 120-Vac line has a turns ratio of 1:2. Calculate the stepped-up V_S.
b. This V_S is connected across a 2400-Ω R_L. Calculate I_S.
c. An autotransformer has an isolated secondary. True or False?
d. With more I_S for the secondary load, does the I_P increase or decrease?

20-7 TRANSFORMER RATINGS

Like most other components, transformers have voltage, current, and power ratings which must not be exceeded. Exceeding any of these ratings will usually destroy the transformer. What follows is a brief description of the most important transformer ratings.

VOLTAGE RATINGS   Manufacturers of transformers always specify the voltage rating of the primary and secondary windings. Under no circumstances should the primary voltage rating be exceeded. In many cases the rated primary and secondary voltages are printed right on the transformer. For example, consider the transformer shown in Fig. 20-14a. Its rated primary voltage is 120 V, and its secondary voltage is specified as 12.6–0–12.6, which indicates that the secondary is center tapped. The notation 12.6–0–12.6 indicates that 12.6 V is available between the center tap connection and either outside secondary lead. The total secondary voltage available is 2 × 12.6 V or 25.2 V. In Fig. 20-14a, the black leads coming out of the top of the transformer provide connection to the primary winding. The two yellow leads coming out of the bottom of the transformer provide connection to the outer leads of the secondary winding. The bottom middle black lead connects to the center tap on the secondary winding.

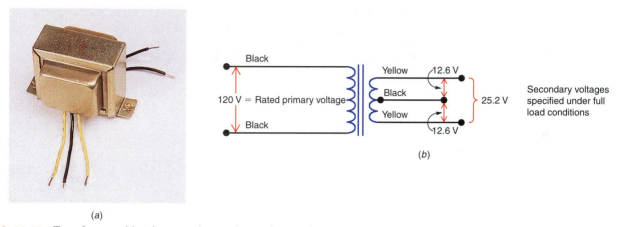

(a)

(b)

FIG. 20-14   Transformer with primary and secondary voltage ratings. (a) Top black leads are primary leads. Yellow and black leads on bottom are secondary leads. (b) Schematic symbol.

It should be noted that manufacturers may specify the secondary voltages of a transformer differently. For example, the secondary in Fig. 20-14a may be specified as 25.2 V CT, where CT indicates a center-tapped secondary. Another way to specify the secondary voltage in Fig. 20-14a would be: 12.6 V each side of center.

Regardless of how the secondary voltage of a transformer is specified, it should be noted that the rated value is always specified under full load conditions with the rated primary voltage applied. A transformer is considered fully loaded when the rated current is drawn from the secondary. When unloaded, the secondary voltage will measure a value which is approximately 5 to 10 percent higher than its rated value. Let's use the transformer in Fig. 20-14a as an example. It has a rated secondary current of 2 A. If 120 V is connected to the primary and no load is connected to the secondary, each half of the secondary will measure somewhere between 13.2 V and 13.9 V approximately. However, with the rated current of 2 A drawn from the secondary, each half of the secondary will measure approximately 12.6 V.

Figure 20-14b shows the schematic diagram for the transformer in Fig. 20-14a. Notice that the colors of each lead are identified for clarification.

As you already know, transformers can have more than one secondary winding. They can also have more than one primary winding. The purpose is to allow the transformer to be used with more than one value of primary voltage. Figure 20-15 shows a transformer that has two separate primaries and a single secondary. This transformer can be wired to work with a primary voltage of either 120 V or 240 V. For either value of primary voltage, the secondary voltage is 24 V. Figure 20-15a shows the individual primary windings with phasing dots to identify those leads with the same instantaneous polarity. Figure 20-15b shows how to connect the primary windings for connection to 240 V. Notice the connections of the leads with the phasing dots. With this connection, each half of the primary voltage is in the proper phase to provide a series-aiding connection of the induced voltages. Furthermore, the series connection of the primary windings provides a turns ratio N_P/N_S of 10:1, thus allowing a secondary voltage of 24 V. Figure 20-15c shows how to connect the primaries for connection to 120 V. Again, notice the connection of the leads with the phasing dots. When the primary windings are in parallel, the total primary current I_P is divided evenly between the windings. The parallel connection also provides a turns ratio N_P/N_S of 5:1, thus allowing a secondary voltage of 24 V.

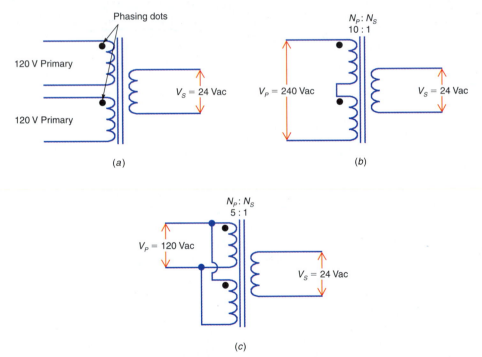

FIG. 20-15 Transformer with multiple primary windings. (a) Phasing dots show primary leads with same instantaneous polarity. (b) Primary windings connected in series to work with a primary voltage of 240 V; $N_P/N_S = 10:1$. (c) Primary windings connected in parallel to work with a primary voltage of 120 V; $N_P/N_S = 5:1$.

Figure 20-16 shows a transformer that can operate with a primary voltage of either 120 V or 440 V. In this case, only one of the primary windings is used with a given primary voltage. For example, if 120 V is applied to the lower pri-

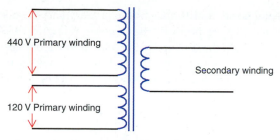

FIG. 20-16 Transformer that has two primaries, which are used separately and never together.

mary, the upper primary winding is not used. Conversely, if 440 V is applied to the upper primary, the lower primary winding is not used.

CURRENT RATINGS   Manufacturers of transformers usually specify the current ratings for secondary windings only. The reason is quite simple. If the secondary current is not exceeded, there is no possible way the primary current can be exceeded. If the secondary current exceeds its rated value, excessive I^2R losses will result in the secondary winding. This will cause the secondary, and perhaps the primary, to overheat, thus eventually destroying the transformer. The IR voltage drop across the secondary windings is the reason why the secondary voltage decreases as the load current increases.

Example

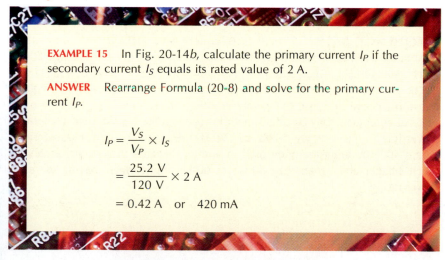

EXAMPLE 15   In Fig. 20-14b, calculate the primary current I_P if the secondary current I_S equals its rated value of 2 A.

ANSWER   Rearrange Formula (20-8) and solve for the primary current I_P.

$$I_P = \frac{V_S}{V_P} \times I_S$$

$$= \frac{25.2 \text{ V}}{120 \text{ V}} \times 2 \text{ A}$$

$$= 0.42 \text{ A} \quad \text{or} \quad 420 \text{ mA}$$

POWER RATINGS   The power rating of a transformer refers to the amount of power the transformer can deliver to a resistive load. The power rating is specified in volt-amperes (VA) rather than watts (W), because the power is not actually dissipated by the transformer. The product VA is called *apparent power,* since it is the power which is *apparently* used by the transformer. The unit of apparent power is VA because the watt unit is reserved for the actual dissipation of power in a resistance.

Assume that a power transformer whose primary and secondary voltage ratings are 120 V and 25 V, respectively, has a power rating of 125 VA. What does this mean? It means that the product of the transformer's primary, or secondary, voltage and current must not exceed 125 VA. If it does, the transformer will overheat and be destroyed. The maximum allowable secondary current for this transformer can be calculated as

$$I_{S(\text{max})} = \frac{125\ \text{VA}}{25\ \text{V}}$$
$$= 5\ \text{A}$$

The maximum allowable primary current can be calculated as

$$I_{P(\text{max})} = \frac{125\ \text{VA}}{120\ \text{V}}$$
$$= 1.04\ \text{A}$$

With multiple secondary windings the VA rating of each individual secondary may be given without any mention of the primary's VA rating. In this case, the sum of all the secondary VA ratings must be divided by the rated primary voltage to determine the maximum allowable primary current.

In summary, you will never overload a transformer or exceed any of its maximum ratings if you obey two fundamental rules:

1. Never apply more than the rated voltage to the primary.
2. Never draw more than the rated current from the secondary.

FREQUENCY RATINGS   All transformers have a frequency rating which must be adhered to. Typical frequency ratings for power transformers are 50 Hz, 60 Hz, and 400 Hz. A power transformer with a frequency rating of 400 Hz cannot be used at 50 Hz or 60 Hz, because it will overheat. However, many power transformers are designed to operate at either 50 Hz or 60 Hz, because many types of equipment may be sold in both Europe and the United States, where the power-line frequencies are 50 Hz and 60 Hz, respectively. Power transformers with a 400-Hz rating are often used in aircraft, because these transformers are much smaller and lighter than 50-Hz or 60-Hz transformers having the same power rating.

TEST-POINT QUESTION 20-7

Answers at end of chapter.

Answer True or False.
a. The measured voltage across an unloaded secondary is usually 5 to 10 percent higher than its rated value.
b. The current rating of a transformer is usually specified only for the secondary windings.
c. A power rating of 300 VA for a transformer means that the transformer secondary must be able to dissipate this amount of power.

20-8 IMPEDANCE TRANSFORMATION

Transformers can be used to change or transform a secondary load impedance to a new value as seen by the primary. The secondary load impedance is said to be reflected back into the primary and is therefore called a *reflected impedance*. The reflected impedance of the secondary may be stepped up or down in accordance with the square of the transformer turns ratio.

By manipulating the relationships between the currents, voltages, and turns ratio in a transformer, an equation for the reflected impedance can be developed. This relationship is

▶ $$Z_P = \left(\frac{N_P}{N_S}\right)^2 \times Z_S \tag{20-10}$$

where Z_P = primary impedance and Z_S = secondary impedance (see Fig. 20-17). If the turns ratio N_P/N_S is greater than 1, Z_S will be stepped up in value. Conversely, if the turns ratio N_P/N_S is less than 1, Z_S will be stepped down in value. It should be noted that the term *impedance* is used rather loosely here, since the primary and secondary impedances may be purely resistive in nature. In the discussions and examples that follow, Z_P and Z_S will be assumed to be purely resistive. The concept of reflected impedance has several practical applications in electronics.

To find the required turns ratio when the impedance ratio is known, rearrange Formula (20-10) as follows:

▶ $$\frac{N_P}{N_S} = \sqrt{\frac{Z_P}{Z_S}} \tag{20-11}$$

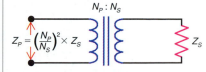

FIG. 20-17 The secondary load impedance Z_S is reflected back into the primary as a new value which is proportional to the square of the turns ratio, N_P/N_S.

Example

EXAMPLE 16 Determine the primary impedance Z_P for the transformer circuit in Fig. 20-18 shown on the next page.

ANSWER Use Formula (20-10). Since $Z_S = R_L$, we have

$$Z_P = \left(\frac{N_P}{N_S}\right)^2 \times R_L$$

$$= \left(\frac{4}{1}\right)^2 \times 8\ \Omega$$

$$= 16 \times 8\ \Omega$$

$$= 128\ \Omega$$

The value of 128 Ω obtained for Z_P using Formula (20-10) can be verified as follows.

(continued on next page)

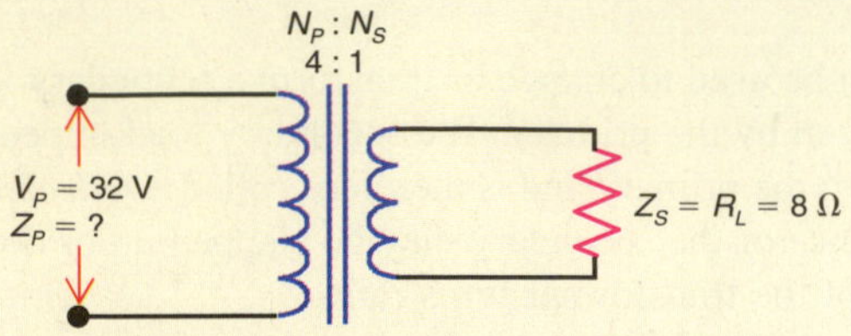

$N_P : N_S$
4 : 1

$V_P = 32$ V
$Z_P = $?

$Z_S = R_L = 8 \, \Omega$

FIG. 20-18 Circuit for Example 16.

$$V_S = \frac{N_S}{N_P} \times V_P$$

$$= \frac{1}{4} \times 32 \text{ V}$$

$$= 8 \text{ V}$$

$$I_S = \frac{V_S}{R_L}$$

$$= \frac{8 \text{ V}}{8 \, \Omega}$$

$$= 1 \text{ A}$$

$$I_P = \frac{V_S}{V_P} \times I_S$$

$$= \frac{8 \text{ V}}{32 \text{ V}} \times 1 \text{ A}$$

$$= 0.25 \text{ A}$$

And finally,

$$Z_P = \frac{V_P}{I_P}$$

$$= \frac{32 \text{ V}}{0.25 \text{ A}}$$

$$= 128 \, \Omega$$

EXAMPLE 17 In Fig. 20-19, calculate the turns ratio N_P/N_S which will produce a reflected primary impedance Z_P of: **(a)** 75 Ω; **(b)** 600 Ω.

ANSWER **(a)** Use Formula (20-11).

$$\frac{N_P}{N_S} = \sqrt{\frac{Z_P}{Z_S}}$$

$$= \sqrt{\frac{75 \, \Omega}{300 \, \Omega}}$$

$$= \sqrt{\frac{1}{4}}$$

$$= \frac{1}{2}$$

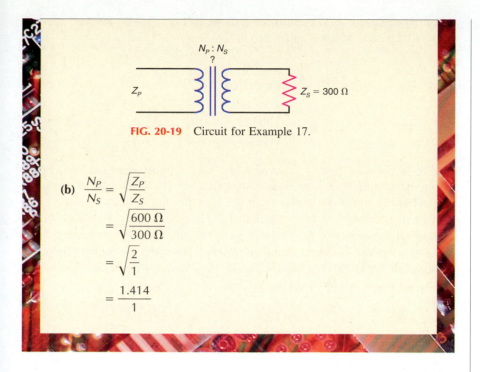

FIG. 20-19 Circuit for Example 17.

(b) $\dfrac{N_P}{N_S} = \sqrt{\dfrac{Z_P}{Z_S}}$

$= \sqrt{\dfrac{600\ \Omega}{300\ \Omega}}$

$= \sqrt{\dfrac{2}{1}}$

$= \dfrac{1.414}{1}$

IMPEDANCE MATCHING FOR MAXIMUM POWER TRANSFER

Transformers are used when it is necessary to achieve maximum transfer of power from a generator to a load when the generator and load impedances are not the same. This application of a transformer is called *impedance matching*.

As an example, consider the amplifier and load in Fig. 20-20a. Notice that the internal resistance r_i of the amplifier is 200 Ω and the load R_L is 8 Ω. If the

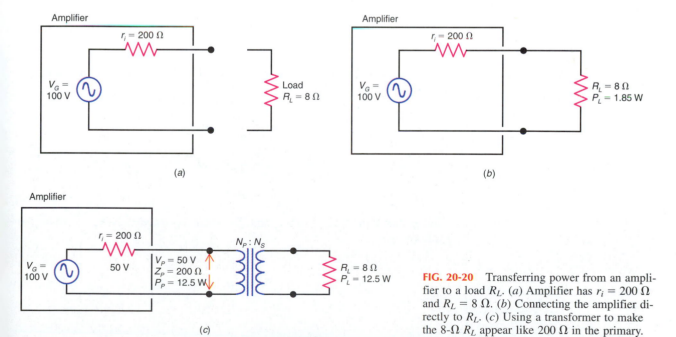

FIG. 20-20 Transferring power from an amplifier to a load R_L. (*a*) Amplifier has $r_i = 200\ \Omega$ and $R_L = 8\ \Omega$. (*b*) Connecting the amplifier directly to R_L. (*c*) Using a transformer to make the 8-Ω R_L appear like 200 Ω in the primary.

amplifier and load are connected directly as in Fig. 20-20b, the load receives 1.85 W of power, which is calculated as

$$P_L = \left(\frac{V_G}{r_i + R_L}\right)^2 \times R_L$$

$$= \left(\frac{100 \text{ V}}{200 \text{ }\Omega + 8 \text{ }\Omega}\right)^2 \times 8 \text{ }\Omega$$

$$= 1.85 \text{ W}$$

To increase the power delivered to the load, a transformer can be used between the amplifier and load. This is shown in Fig. 20-20c. We know that in order to transfer maximum power from the amplifier to the load, R_L must somehow be transformed to a value equaling 200 Ω in the primary. With Z_P equaling r_i, maximum power will be delivered from the amplifier to the primary. Since the primary power P_P must equal the secondary power P_S, maximum power will also be delivered to the load R_L. In Fig. 20-20c, the turns ratio which provides a Z_P of 200 Ω can be calculated as

$$\frac{N_P}{N_S} = \sqrt{\frac{Z_P}{Z_S}}$$

$$= \sqrt{\frac{200 \text{ }\Omega}{8 \text{ }\Omega}}$$

$$= \frac{5}{1}$$

With r_i and Z_P equal, the power delivered to the primary can be calculated as

$$P_P = \left(\frac{V_G}{r_i + Z_P}\right)^2 \times Z_P$$

$$= \left(\frac{100 \text{ V}}{400 \text{ }\Omega}\right)^2 \times 200 \text{ }\Omega$$

$$= 12.5 \text{ W}$$

Since $P_P = P_S$, the load R_L also receives 12.5 W of power. As proof, calculate the secondary voltage.

$$V_S = \frac{N_S}{N_P} \times V_P$$

$$= \frac{1}{5} \times 50 \text{ V}$$

$$= 10 \text{ V}$$

(Notice that V_P is $\frac{1}{2}V_G$, since r_i and Z_P divide V_G evenly.) Next, calculate the load power P_L.

$$P_L = \frac{V_S^2}{R_L}$$

$$= \frac{10^2 \text{ V}}{8 \text{ }\Omega}$$

$$= 12.5 \text{ W}$$

Notice how the transformer has been used as an impedance matching device to obtain the maximum transfer of power from the amplifier to the load. Compare the power dissipated by R_L in Fig. 20-20b to that in Fig. 20-20c. There is a big difference between the load power of 1.85 W in Fig. 20-20b and the load power of 12.5 W in Fig. 20-20c.

TEST-POINT QUESTION 20-8

Answers at end of chapter.

Answer True or False.
  a. The turns ratio will not affect the primary impedance Z_P.
  b. When the turns ratio N_P/N_S is greater than 1, the primary impedance Z_P is less than the value of Z_S.
  c. If the turns ratio N_P/N_S of a transformer is 2/1 and $Z_S = 50\ \Omega$, the primary impedance $Z_P = 200\ \Omega$.

20-9 CORE LOSSES

The fact that the magnetic core can become warm, or even hot, shows that some of the energy supplied to the coil is used up in the core as heat. The two main effects are eddy-current losses and hysteresis losses.

EDDY CURRENTS In any inductance with an iron core, alternating current induces voltage in the core itself. Since it is a conductor, the iron core has current produced by the induced voltage. This current is called an *eddy current* because it flows in a circular path through the cross section of the core, as illustrated in Fig. 20-21.

The eddy currents represent wasted power dissipated as heat in the core. Note in Fig. 20-21 that the eddy-current flux opposes the coil flux, so that more current is required in the coil to maintain its magnetic field. The higher the frequency of the alternating current in the inductance, the greater the eddy-current loss.

Eddy currents can be induced in any conductor near a coil with alternating current, not only in its core. For instance, a coil has eddy-current losses in a metal cover. In fact, the technique of induction heating is an application of heat resulting from induced eddy currents.

RF SHIELDING The reason why a coil may have a metal cover, usually copper or aluminum, is to provide a shield against the varying flux of RF current. In this case, the shielding effect depends on using a good conductor for the eddy currents produced by the varying flux, rather than the magnetic materials used for shielding against static magnetic flux.

The shield cover not only isolates the coil from external varying magnetic fields, but also minimizes the effect of the coil's RF current for external circuits. The reason why the shield helps both ways is the same, as the induced eddy currents have a field that opposes the field that is inducing the current. It should be

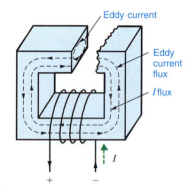

FIG. 20-21 Cross-sectional view of iron core showing eddy currents.

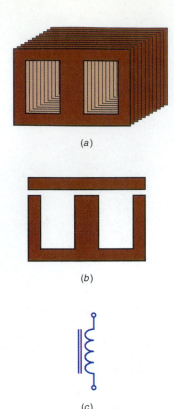

FIG. 20-22 Laminated iron core. (a) Shell-type construction. (b) E- and I-shaped laminations. (c) Symbol for iron core.

FIG. 20-23 RF coils with ferrite core. Width of coil is ½ in. (a) Variable L from 1 to 3 mH. (b) Tuning coil for 40 MHz.

noted that the clearance between the sides of the coil and the metal should be equal to or greater than the coil radius, to minimize the effect of the shield in reducing the inductance.

HYSTERESIS LOSSES Another loss factor present in magnetic cores is hysteresis, although these are not as great as eddy-current losses. The hysteresis losses result from the additional power needed to reverse the magnetic field in magnetic materials in the presence of alternating current. The greater the frequency, the more hysteresis losses.

AIR-CORE COILS It should be noted that air has practically no losses from eddy currents or hysteresis. However, the inductance for small coils with an air core is limited to low values in the microhenry or millihenry range.

Answers at end of chapter.
a. Which has greater eddy-current losses, an iron core or an air core?
b. Which produces more hysteresis losses, 60 Hz or 60 MHz?

20-10 TYPES OF CORES

In order to minimize losses while maintaining high flux density, the core can be made of laminated steel insulated from each other. Insulated powdered-iron granules and ferrite materials can also be used. These core types are illustrated in Figs. 20-22 and 20-23. The purpose is to reduce the amount of eddy currents. The type of steel itself can help reduce hysteresis losses.

LAMINATED CORE Figure 20-22a shows a shell-type core formed with a group of individual laminations. Each laminated section is insulated by a very thin coating of iron oxide, silicon steel, or varnish. The insulating material increases the resistance in the cross section of the core to reduce the eddy currents, but allows a low-reluctance path for high flux density around the core. Transformers for audio frequencies and 60-Hz power are generally made with a laminated iron core.

POWDERED-IRON CORE To reduce eddy currents in the iron core of an inductance for radio frequencies, powdered iron is generally used. It consists of individual insulated granules pressed into one solid form called a *slug*.

FERRITE CORE The ferrites are synthetic ceramic materials that are ferromagnetic. They provide high values of flux density, like iron, but have the advantage of being insulators. Therefore, a ferrite core can be used for high frequencies with minimum eddy-current losses.

This core is usually a slug that can move in or out of the coil to vary L, as in Fig. 20-23a. In Fig. 20-23b, the core has a hole to fit a plastic alignment tool for tuning the coil. Maximum L results with the slug in the coil.

20-11 VARIABLE INDUCTANCE

The inductance of a coil can be varied by one of the methods illustrated in Fig. 20-24. In Fig. 20-24a, more or fewer turns can be used by connection to one of the taps on the coil. Also, in Fig. 20-24b, a slider contacts the coil to vary the number of turns used. These methods are for large coils.

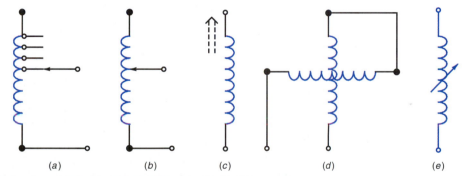

(a)   (b)   (c)   (d)   (e)

FIG. 20-24 Methods of varying inductance. (a) Tapped coil. (b) Slider contact. (c) Adjustable slug. (d) Variometer. (e) Symbol for variable L.

Figure 20-24c shows the schematic symbol for a coil with a slug of powdered iron or ferrite. The dotted lines indicate that the core is not solid iron. The arrow shows that the slug is variable. Usually, an arrow at the top means the adjustment is at the top of the coil. An arrow at the bottom, pointing down, shows the adjustment is at the bottom.

The symbol in Fig. 20-24d is a *variometer*, which is an arrangement for varying the position of one coil within the other. The total inductance of the series-aiding coils is minimum when they are perpendicular.

For any method of varying L, the coil with an arrow in Fig. 20-24e can be used. However, an adjustable slug is usually shown as in Fig. 20-24c.

A practical application of variable inductance is the *Variac*. The Variac is an autotransformer with a variable tap to change the turns ratio. The output voltage

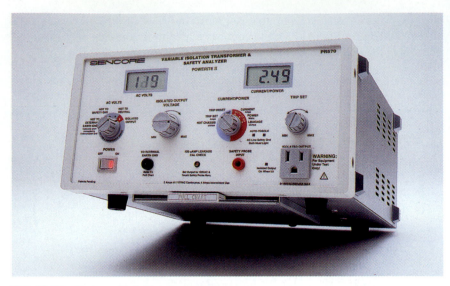

FIG. 20-25 Variac with isolated output.

in the secondary can be varied from 0 to approximately 140 V, with input from the 120-V 60-Hz power line. One use is to test equipment with voltage above or below the normal line voltage.

The Variac is plugged into the power line, and the equipment to be tested is plugged into the Variac. Note that the power rating of the Variac should be equal to or more than the power used by the equipment being tested. Figure 20-25 shows a Variac with an isolated output.

TEST-POINT QUESTION 20-11

Answers at end of chapter.

Answer True or False.
a. The Variac is an autotransformer with a variable tap for the primary.
b. Figure 20-24*c* shows a ferrite or powdered iron core.

20-12 INDUCTANCES IN SERIES OR PARALLEL

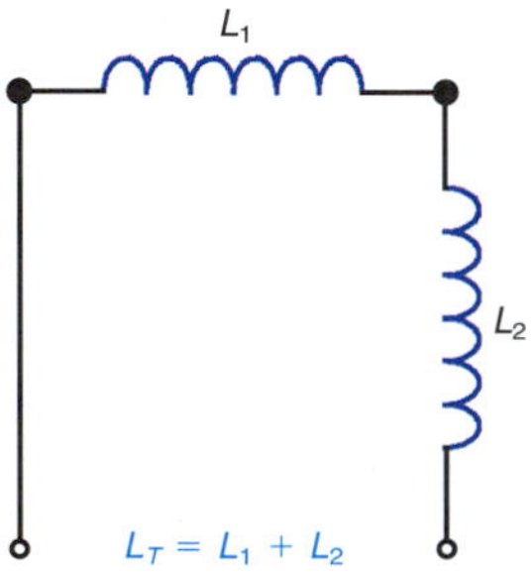

As shown in Fig. 20-26, the total inductance of coils connected in series is the sum of the individual L values, as for series R. Since the series coils have the same current, the total induced voltage is a result of the total number of turns. Therefore, total series inductance is,

$$L_T = L_1 + L_2 + L_3 + \cdots + \text{etc.} \qquad \textbf{(20-12)}$$

where L_T is in the same units of inductance as L_1, L_2, and L_3. This formula assumes no mutual induction between the coils.

L_1

L_2

$L_T = L_1 + L_2$

FIG. 20-26 Inductances L_1 and L_2 in series without mutual coupling.

Example

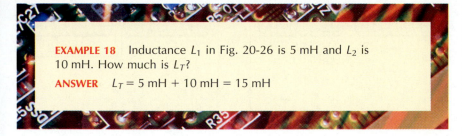

EXAMPLE 18 Inductance L_1 in Fig. 20-26 is 5 mH and L_2 is 10 mH. How much is L_T?

ANSWER $L_T = 5 \text{ mH} + 10 \text{ mH} = 15 \text{ mH}$

With coils connected in parallel, the equivalent inductance is calculated from the reciprocal formula

$$\frac{1}{L_{EQ}} = \frac{1}{L_1} + \frac{1}{L_2} + \frac{1}{L_3} + \cdots + \text{etc.} \qquad \textbf{(20-13)}$$

Again, no mutual induction is assumed, as illustrated in Fig. 20-27.

Example

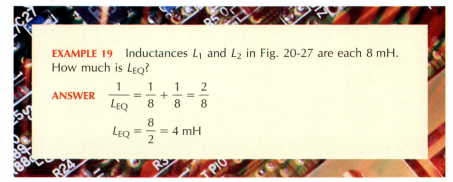

EXAMPLE 19 Inductances L_1 and L_2 in Fig. 20-27 are each 8 mH. How much is L_{EQ}?

ANSWER $\dfrac{1}{L_{EQ}} = \dfrac{1}{8} + \dfrac{1}{8} = \dfrac{2}{8}$

$L_{EQ} = \dfrac{8}{2} = 4 \text{ mH}$

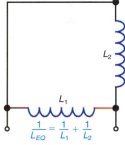

FIG. 20-27 Inductances L_1 and L_2 in parallel without mutual coupling.

All the shortcuts for calculating parallel R can be used with parallel L, since both are based on the reciprocal formula. In this example L_{EQ} is $\frac{1}{2} \times 8 = 4$ mH.

SERIES COILS WITH L_M   This case depends on the amount of mutual coupling and on whether the coils are connected series-aiding or series-opposing. *Series-aiding* means that the common current produces the same direction of magnetic field for the two coils. The *series-opposing* connection results in opposite fields.

The coupling depends on the coil connections and direction of winding. Reversing either one reverses the field. Inductances L_1 and L_2 with the same direction of winding are connected series-aiding in Fig. 20-28*a* on the next page. However, they are series-opposing in Fig. 20-28*b* because L_1 is connected to the opposite end of L_2. To calculate the total inductance of two coils that are series-connected and have mutual inductance,

$$L_T = L_1 + L_2 \pm 2L_M \qquad \textbf{(20-14)}$$

The mutual inductance L_M is plus, increasing the total inductance, when the coils are series-aiding, or minus when they are series-opposing to reduce the total inductance.

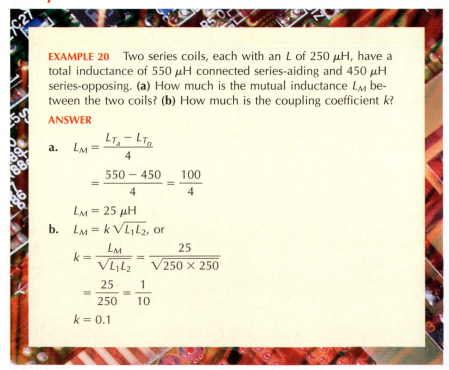

Series-aiding L_M

L_1 L_2

$$L_T = L_1 + L_2 + 2L_M$$

(a)

Series-opposing L_M

L_1 L_2

$$L_T = L_1 + L_2 - 2L_M$$

(b)

FIG. 20-28 Inductances L_1 and L_2 in series but with mutual coupling L_M. (a) Aiding magnetic fields. (b) Opposing magnetic fields.

Note the phasing dots above the coils in Fig. 20-28. Coils with phasing dots at the same end have the same direction of winding. When current enters the dotted ends for two coils, their fields are aiding and L_M has the same sense as L.

HOW TO MEASURE L_M   Formula (20-14) provides a method of determining the mutual inductance between two coils L_1 and L_2 of known inductance. First, the total inductance is measured for the series-aiding connection. Let this be L_{T_a}. Then the connections to one coil are reversed to measure the total inductance for the series-opposing coils. Let this be L_{T_o}. Then

▶ $$L_M = \frac{L_{T_a} - L_{T_o}}{4}$$ (20-15)

When the mutual inductance is known, the coefficient of coupling k can be calculated from the fact that $L_M = k\sqrt{L_1 L_2}$.

Example

EXAMPLE 20   Two series coils, each with an L of 250 μH, have a total inductance of 550 μH connected series-aiding and 450 μH series-opposing. **(a)** How much is the mutual inductance L_M between the two coils? **(b)** How much is the coupling coefficient k?

ANSWER

a. $L_M = \dfrac{L_{T_a} - L_{T_o}}{4}$

$= \dfrac{550 - 450}{4} = \dfrac{100}{4}$

$L_M = 25\ \mu$H

b. $L_M = k\sqrt{L_1 L_2}$, or

$k = \dfrac{L_M}{\sqrt{L_1 L_2}} = \dfrac{25}{\sqrt{250 \times 250}}$

$= \dfrac{25}{250} = \dfrac{1}{10}$

$k = 0.1$

Coils may also be in parallel with mutual coupling. However, the inverse relations with parallel connections and the question of aiding or opposing fields make this case complicated. Actually, it would hardly ever be used.

20-13 STRAY INDUCTANCE

Although practical inductors are generally made as coils, all conductors have inductance. The amount of L is $v_L/(di/dt)$, as with any inductance producing induced voltage when the current changes. The inductance of any wiring not included in the conventional inductors can be considered stray inductance. In most cases, the stray inductance is very small, typical values being less than 1 μH. For high radio frequencies, though, even a small L can have an appreciable inductive effect.

One source of stray inductance is the connecting leads. A wire of 0.04-in. diameter and 4 in. long has an L of approximately 0.1 μH. At low frequencies, this inductance is negligible. However, consider the case of RF current, where i varies from 0 to 20-mA peak value in the short time of 0.025 μs for a quarter-cycle of a 10-MHz sine wave. Then v_L equals 80 mV, which is an appreciable inductive effect. This is one reason why the connecting leads must be very short in RF circuits.

As another example, wire-wound resistors can have appreciable inductance when wound as a straight coil. This is why carbon resistors are preferred for minimum stray inductance in RF circuits. However, noninductive wire-wound resistors can also be used. These are wound in such a way that adjacent turns have current in opposite directions, so that the magnetic fields oppose each other to cancel the inductance. Another application of this technique is twisting a pair of connecting leads to reduce the inductive effect.

20-14 ENERGY IN MAGNETIC FIELD OF INDUCTANCE

Magnetic flux associated with current in an inductance has electric energy supplied by the voltage source producing the current. The energy is stored in the field, since it can do the work of producing induced voltage when the flux moves. The amount of electric energy stored is

▶ $$\text{Energy} = \mathscr{E} = \tfrac{1}{2}LI^2 \quad \text{J} \tag{20-16}$$

The factor of ½ gives the average result of I in producing energy. With L in henrys and I in amperes, the energy is in watt-seconds, or *joules*. For a 10-H L with a 3-A I, the electric energy stored in the magnetic field equals

▶ $$\text{Energy} = \tfrac{1}{2}LI^2 = \frac{10 \times 9}{2} = 45 \text{ J}$$

This 45 J of energy is supplied by the voltage source that produces 3 A in the inductance. When the circuit is opened, the magnetic field collapses. The energy in the collapsing magnetic field is returned to the circuit in the form of induced voltage, which tends to keep the current flowing.

The entire 45 J is available for the work of inducing voltage, since no energy is dissipated by the magnetic field. With resistance in the circuit, however, the I^2R loss with induced current dissipates all the energy after a period of time.

Example

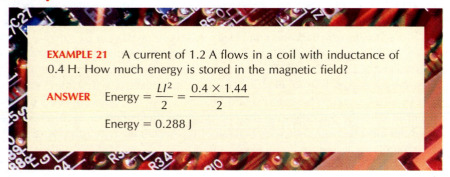

EXAMPLE 21   A current of 1.2 A flows in a coil with inductance of 0.4 H. How much energy is stored in the magnetic field?

ANSWER $\text{Energy} = \dfrac{LI^2}{2} = \dfrac{0.4 \times 1.44}{2} =$

$\text{Energy} = 0.288 \text{ J}$

To do this problem on a calculator, first square the I, multiply by L, and divide by 2. Specifically, punch in 1.2 and push the $\fbox{$x^2$}$ key for 1.44. While this is on the display, push the $\fbox{$\times$}$ key, punch in 0.4 and push the $\fbox{$=$}$ key for 0.576 on the display. Now push the $\fbox{$\div$}$ key, punch in 2, and then push the $\fbox{$=$}$ key to get 0.288 as the answer.

TEST-POINT QUESTION 20-14

Answers at end of chapter.
a. What is the unit of electric energy stored in a magnetic field?
b. Does a 4-H coil store more or less energy than a 2-H coil?

20-15 TROUBLES IN COILS

The most common trouble in coils is an open winding. As illustrated in Fig. 20-29, an ohmmeter connected across the coil reads infinite resistance for the open circuit. It does not matter whether the coil has an air core or an iron core. Since the coil is open, it cannot conduct current and therefore has no inductance, because it cannot produce induced voltage. When the resistance is checked, the coil should be disconnected from the external circuit to eliminate any parallel paths that could affect the resistance readings.

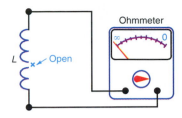

FIG. 20-29 An open coil reads infinite ohms when its continuity is checked with an ohmmeter.

DIRECT CURRENT RESISTANCE OF A COIL A coil has dc resistance equal to the resistance of the wire used in the winding. The amount of resistance is less with heavier wire and fewer turns. For RF coils with inductance values up to several millihenrys, requiring 10 to 100 turns of fine wire, the dc resistance is 1 to 20 Ω, approximately. Inductors for 60 Hz and audio frequencies with several hundred turns may have resistance values of 10 to 500 Ω, depending on the wire size.

As shown in Fig. 20-30, the dc resistance and inductance of a coil are in series, since the same current that induces voltage in the turns must overcome the resistance of the wire. Although resistance has no function in producing induced voltage, it is useful to know the dc coil resistance because if it is normal, usually the inductance can also be assumed to have its normal value.

OPEN COIL An open winding has infinite resistance, as indicated by an ohmmeter reading. With a transformer that has four leads or more, check the resistance across the two leads for the primary, across the two leads for the secondary, and across any other pairs of leads for additional secondary windings. For an autotransformer with three leads, check the resistance from one lead to each of the other two.

When the open circuit is inside the winding, it is usually not practical to repair the coil, and the entire unit is replaced. In some cases, an open connection at the terminals can be resoldered.

VALUE CHANGE The value of an inductor can change over time due to core breakage, windings relaxing, or shorted turns. A coil whose inductance value has changed can be difficult to locate without the proper test equipment. It should be noted that a coil whose inductance value is changed may check okay with an ohmmeter.

FIG. 20-30 The internal dc resistance r_i of a coil is in series with its inductance L.

OPEN PRIMARY WINDING   When the primary of a transformer is open, no primary current can flow and there is no voltage induced in any of the secondary windings.

OPEN SECONDARY WINDING   When the secondary of a transformer is open, it cannot supply power to any load resistance across the open winding. Furthermore, with no current in the secondary, the primary current is also practically zero, as though the primary winding were open. The only primary current needed is the small magnetizing current to sustain the field producing induced voltage across the secondary without any load. If the transformer has several secondary windings, however, an open winding in one secondary does not affect transformer operation for the secondary circuits that are normal.

SHORT ACROSS SECONDARY WINDING   In this case excessive primary current flows, as though it were short-circuited, often burning out the primary winding. The reason is that the large secondary current has a strong field that opposes the flux of the self-induced voltage across the primary, making it draw more current from the generator.

TEST-POINT QUESTION 20-15

Answers at end of chapter.
a. The normal R of a coil is 18 Ω. How much will an ohmmeter read if the coil is open?
b. The primary of a 1:3 step-up autotransformer is connected to the 120-Vac power line. How much will the secondary voltage be if the primary is open?

20 SUMMARY AND REVIEW

- Varying current induces voltage in a conductor, since the expanding and collapsing field of the current is equivalent to flux in motion.
- Lenz' law states that the induced voltage produces I that opposes the change in current causing the induction. Inductance, therefore, tends to keep the current from changing.
- The ability of a conductor to produce induced voltage across itself when the current varies is its self-inductance, or inductance. The symbol is L, and the unit of inductance is the henry. One henry of inductance allows 1 V to be induced when the current changes at the rate of 1 A/s. For smaller units, 1 mH = 1×10^{-3} H and 1 μH = 1×10^{-6} H.
- To calculate the self-induced voltage, $v_L = L(di/dt)$, with v in volts, L in henrys, and di/dt in amperes per second.
- Mutual inductance is the ability of varying current in one conductor to induce voltage in another conductor nearby. Its symbol is L_M, measured in henrys. $L_M = k \sqrt{L_1 L_2}$, where k is the coefficient of coupling between conductors.
- A transformer consists of two or more windings with mutual inductance. The primary winding connects to the source voltage; the load resistance is connected across the secondary winding. A separate winding is an isolated secondary. The transformer is used to step up or step down ac voltage.
- An autotransformer is a tapped coil, used to step up or step down the primary voltage. There are three leads with one connection common to both the primary and secondary.
- A transformer with an iron core has essentially unity coupling. Therefore, the voltage ratio is the same as the turns ratio: $V_P/V_S = N_P/N_S$.
- Assuming 100 percent efficiency for an iron-core power transformer, the power supplied to the primary equals the power used in the secondary.
- The voltage rating of a transformer's secondary is always specified under full load conditions with the rated primary voltage applied. The measured voltage across an unloaded secondary is usually 5 to 10 percent higher than its rated value.
- The current or power rating of a transformer is usually specified only for the secondary windings.
- Transformers can be used to reflect a secondary load impedance back into the primary as a new value which is either larger or smaller than its actual value. The primary impedance Z_P can be determined using Formula (20-10).
- The impedance transforming properties of a transformer make it possible to obtain maximum transfer of power from a generator to a load when the generator and load impedances are not equal. The required turns ratio can be determined using Formula (20-11).
- Eddy currents are induced in the iron core of an inductance, causing wasted power that heats the core. Eddy-current losses increase with higher frequencies of alternating current. To reduce eddy currents, the iron core is laminated. Powdered-iron and ferrite cores have minimum eddy-current losses for radio frequencies. Hysteresis also causes power loss.
- With no mutual coupling, series inductances are added like series resistances. For parallel inductances, the equivalent inductance is calculated by the reciprocal formula, as for parallel resistances.

- The magnetic field of an inductance has stored energy $\mathcal{E} = \frac{1}{2} LI^2$. With I in amperes and L in henrys, the energy $\mathcal{E}$ is in joules.
- In addition to its inductance, a coil has dc resistance equal to the resistance of the wire in the coil. An open coil has infinitely high resistance.
- An open primary in a transformer results in no induced voltage in any of the secondary windings.
- Figure 20-31 summarizes the main types of inductors, or coils, with their schematic symbols.
- Characteristics of inductance and capacitance are compared in Table 20-1.

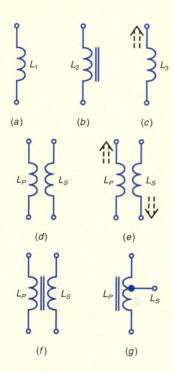

FIG. 20-31 Summary of types of inductors. (*a*) Air-core coil. (*b*) Iron-core coil. (*c*) Adjustable ferrite core. (*d*) Air-core transformer. (*e*) Variable L_P and L_S. (*f*) Iron-core transformer. (*g*) Autotransformer.

TABLE 20-1   COMPARISON OF CAPACITANCE AND INDUCTANCE

CAPACITANCE	INDUCTANCE
Symbol is C	Symbol is L
Unit is the farad (F)	Unit is the henry (H)
Needs dielectric as insulator	Needs conductor for circuit path
More plate area allows more C	More turns allow more L
Dielectric can concentrate electric field for more C	Core can concentrate magnetic field for more L
$\dfrac{1}{C_{EQ}} = \dfrac{1}{C_1} + \dfrac{1}{C_2}$ in series	$L_T = L_1 + L_2$ in series
$C_T = C_1 + C_2$ in parallel	$\dfrac{1}{L_{EQ}} = \dfrac{1}{L_1} + \dfrac{1}{L_2}$ in parallel

Choose (*a*), (*b*), (*c*), or (*d*).

1. Alternating current can induce voltage because alternating current has a (*a*) high peak value; (*b*) varying magnetic field; (*c*) stronger magnetic field than direct current; (*d*) constant magnetic field.

2. When current in a conductor increases, Lenz' law says that the self-induced voltage will (*a*) tend to increase the amount of current; (*b*) aid the applied voltage; (*c*) produce current opposite to the increasing current; (*d*) aid the increasing current.

3. A 1:5 voltage step-up transformer has 120 V across the primary and a 600-Ω resistance across the secondary. Assuming 100 percent efficiency, the primary current equals (*a*) $\frac{1}{5}$ A; (*b*) 600 mA; (*c*) 5 A; (*d*) 10 A.

4. An iron-core transformer with an 1:8 step-up ratio has 120 V applied across the primary. The voltage across the secondary equals (*a*) 15 V; (*b*) 120 V; (*c*) 180 V; (*d*) 960 V.

5. With double the number of turns but the same length and area, the inductance is (*a*) the same; (*b*) double; (*c*) quadruple; (*d*) one-quarter.

6. Current changing from 4 to 6 A in 1 s induces 40 V in a coil. Its inductance equals (*a*) 40 mH; (*b*) 4 H; (*c*) 6 H; (*d*) 20 H.

7. A laminated iron core has reduced eddy-current losses because (*a*) the laminations are stacked vertically; (*b*) the laminations are insulated from each other; (*c*) the magnetic flux is concentrated in the air gap of the core; (*d*) more wire can be used with less dc resistance in the coil.

8. Two 250-μH coils in series without mutual coupling have a total inductance of (*a*) 125 μH; (*b*) 250 μH; (*c*) 400 μH; (*d*) 500 μH.

9. If a transformer has a turns ratio N_P/N_S of 3:1 and $Z_S = 16\ \Omega$, the primary impedance Z_P equals (*a*) 48 Ω; (*b*) 144 Ω; (*c*) 1.78 Ω; (*d*) 288 Ω.

10. An open coil has (*a*) infinite resistance and zero inductance; (*b*) zero resistance and high inductance; (*c*) infinite resistance and normal inductance; (*d*) zero resistance and inductance.

QUESTIONS

1. Define 1 H of self-inductance and 1 H of mutual inductance.
2. State Lenz' law in terms of induced voltage produced by varying current.
3. Refer to Fig. 20-5. Explain why the polarity of v_L is the same for the examples in Fig. 20-5*a* and *d*.
4. Make a schematic diagram showing the primary and secondary of an iron-core transformer with a 1:6 voltage step-up ratio: (**a**) using an autotransformer; (**b**) using a transformer with isolated secondary winding. Then (**c**) with 100 turns in the primary, how many turns are in the secondary for both cases?
5. Define the following: coefficient of coupling, transformer efficiency, stray inductance, and eddy-current losses.

6. Why are eddy-current losses reduced with the following cores: (a) laminated; (b) powdered iron; (c) ferrite?

7. Why is a good conductor used for an RF shield?

8. Show two methods of providing a variable inductance.

9. (a) Why will the primary of a power transformer have excessive current if the secondary is short-circuited? (b) Why is there no voltage across the secondary if the primary is open?

10. (a) Describe briefly how to check a coil for an open winding with an ohmmeter. What ohmmeter range should be used? (b) What leads will be checked on an autotransformer with one secondary and a transformer with two isolated secondary windings?

11. Derive the formula $L_M = (L_{T_a} - L_{T_o})/4$ from the fact that $L_{T_a} = L_1 + L_2 + 2L_M$ while $L_{T_o} = L_1 + L_2 - 2L_M$.

12. Explain how a transformer with a $1:1$ turns ratio and an isolated secondary can be used to reduce the chance of electric shock from the 120 Vac power line.

PROBLEMS

ANSWERS TO ODD-NUMBERED PROBLEMS AT BACK OF BOOK.

1. Convert the following current changes to amperes per second: (a) zero to 3 A in 2 s; (b) zero to 50 mA in 5 μs; (c) 100 to 150 mA in 5 μs; (d) 150 to 100 mA in 5 μs.

2. Convert into henrys using powers of 10: (a) 250 μH; (b) 40 μH; (c) 40 mH; (d) 7 mH; (e) 0.005 H.

3. Calculate the values of v_L across a 5-mH inductance for each of the current variations in Prob. 1.

4. A coil produces a self-induced voltage of 42 mV when i varies at the rate of 19 mA/ms. How much is L?

5. A power transformer with a $1:8$ turns ratio has 60 Hz 120 V across the primary. (a) What is the frequency of the secondary voltage? (b) How much is the secondary voltage? (c) With a load resistance of 10,000 Ω across the secondary, how much is the secondary current? Draw the schematic diagram showing primary and secondary circuits. (d) How much is the primary current? Assume 100 percent efficiency. (Note: The ratio of L_P to L_S is $1:8$.)

6. How much would the primary current be in a power transformer having a primary resistance of 5 Ω if it were connected by mistake to a 120-Vdc line instead of the 120-Vac line?

7. For a 100-μH inductance L_1 and a 200-μH inductance L_2, calculate: (a) the total inductance L_T of L_1 and L_2 in series without mutual coupling; (b) the combined inductance of L_1 and L_2 in parallel without mutual coupling; (c) the L_T of L_1 and L_2 series-aiding, and series-opposing, with 10-μH mutual inductance; (d) the value of the coefficient of coupling k.

8. Calculate the inductance L for the following long coils: (a) air core, 20 turns, area 3.14 cm^2, length 25 cm; (b) same coil as (a) with ferrite core having a μ_r of 5000; (c) air core, 200 turns, area 3.14 cm^2, length 25 cm; (d) air core, 20 turns, area 3.14 cm^2, length 50 cm; (e) air core, 20 turns, diameter 4 cm, length 50 cm. (Note: 1 cm $= 10^{-2}$ m, and 1 cm$^2 = 10^{-4}$ m^2.)

9. Calculate the resistance of the following coil, using Table 11-1: 400 turns, each using 3 in. of No. 30 gage wire.

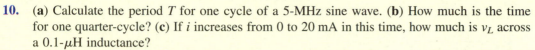

10. **(a)** Calculate the period T for one cycle of a 5-MHz sine wave. **(b)** How much is the time for one quarter-cycle? **(c)** If i increases from 0 to 20 mA in this time, how much is v_L across a 0.1-μH inductance?

11. Calculate the energy in joules stored in the magnetic field of a 60-mH L with a 90-mA I.

12. For a power transformer connected to the 120-Vac line, calculate the turns ratio needed for each of the following secondary voltages: **(a)** 5 V; **(b)** 9 V; **(c)** 24 V; **(d)** 30 V; **(e)** 120 V.

13. **(a)** A transformer delivers 400 W out with 500 W in. **(a)** Calculate the efficiency in percent. **(b)** A transformer with 80 percent efficiency delivers 400 W total secondary power. Calculate the primary power.

14. A 20-mH L and a 40-mH L are connected series-aiding, with $k = 0.4$. Calculate L_T.

15. Calculate the inductance of the coil in Fig. 20-3 with $\mu_r = 100$.

16. An autotransformer has L_S with one-tenth the turns of L_P. When L_P is connected to the 120-Vac power line, calculate: **(a)** secondary voltage V_S; **(b)** Secondary current I_S with 500-Ω R_L; **(c)** P_S in the secondary circuit; **(d)** P_P in the primary circuit, assuming 100 percent efficiency.

17. Do the same as in Prob. 16 but with a 5-kΩ load resistance R_L.

18. A power transformer with a $2:1$ voltage step up is connected to the 120-Vac power line. What is the lowest R_L that can be connected across the secondary without exceeding the power rating of 30 VA?

19. In Fig. 20-32, calculate V_{S_1}, I_{S_1}, P_{S_1}, V_{S_2}, I_{S_2}, P_{S_2}, P_P, and I_P.

20. In Fig. 20-32, calculate the primary current I_P if the load R_{L_1} opens.

21. In Fig. 20-33, the rated secondary current equals 10 A as shown. Calculate: **(a)** the value of secondary load resistance R_L which will draw 10 A of current; **(b)** the primary current I_P when $I_S = 10$ A.

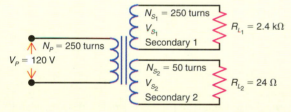

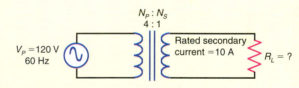

FIG. 20-32 Circuit for Probs. 19, 20, and Critical Thinking Prob. 4.

FIG. 20-33 Circuit for Prob. 21.

22. Refer to Fig. 20-34. Calculate the maximum allowable current for: **(a)** secondary 1; **(b)** secondary 2; **(c)** the primary.

23. In Fig. 20-35, calculate the primary impedance Z_P for a turns ratio N_P/N_S of: **(a)** $2:1$; **(b)** $1:2$; **(c)** $11.18:1$; **(d)** $10:1$; **(e)** $1:3.16$.

24. In Fig. 20-36, calculate the required turns ratio N_P/N_S for: **(a)** $Z_P = 10$ kΩ, $R_L = 75$ Ω; **(b)** $Z_P = 100$ Ω, $R_L = 25$ Ω; **(c)** $Z_P = 100$ Ω, $R_L = 10$ kΩ; **(d)** $Z_P = 1$ kΩ, $R_L = 200$ Ω; **(e)** $Z_P = 200$ Ω, $R_L = 10$ Ω.

25. In Fig. 20-37, what turns ratio N_P/N_S will provide the maximum transfer of power from the amplifier to R_L?

26. Using the turns ratio from Prob. 25 in Fig. 20-37, calculate: **(a)** Z_P; **(b)** P_P; **(c)** P_{RL}.

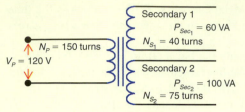

FIG. 20-34   Circuit for Prob. 22.

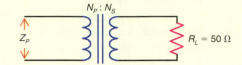

FIG. 20-35   Circuit for Prob. 23.

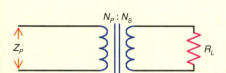

FIG. 20-36   Circuit for Prob. 24.

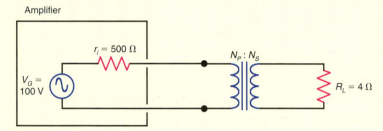

FIG. 20-37   Circuit for Probs. 25 and 26.

CRITICAL THINKING

1. Derive the formula:

$$Z_P = \left(\frac{N_P}{N_S}\right)^2 \times Z_S$$

2. Calculate the primary impedance Z_P in Fig. 20-38.
3. In Fig. 20-39, calculate the primary impedance Z_P across primary leads: (**a**) 1 and 3; (**b**) 1 and 2. (Note: Terminal 2 is a center-tap connection on the transformer primary. Also, the turns ratio of 4:1 is specified using leads 1 and 3 of the primary.)
4. Refer to Fig. 20-32. If the transformer has an efficiency of 80 percent, calculate the primary current I_P.

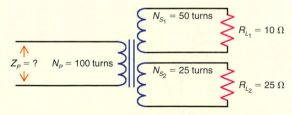

FIG. 20-38   Circuit for Critical Thinking Prob. 2.

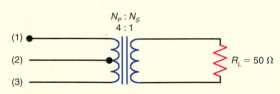

FIG. 20-39   Circuit for Critical Thinking Prob. 3.

20-1   **a.** coil with an iron core
        b. time B

20-2   **a.** $L = 2$ H
        b. $L = 32$ mH

20-3   **a.** $v_L = 2$ V
        b. $v_L = 200$ V

20-4   **a.** T
        b. T

20-5   **a.** $k = 1$
        b. $L_M = 18$ mH

20-6   **a.** $V_S = 240$ V
        b. $I_S = 0.1$ A
        c. F
        d. increase

20-7   **a.** T
        b. T
        c. F

20-8   **a.** F
        b. F
        c. T

20-9   **a.** iron core
        b. 60 MHz

20-10   **a.** T
        b. T
        c. T

20-11   **a.** T
        b. T

20-12   **a.** $L_T = 1.5$ mH
        b. $L_{EQ} = 0.33$ mH

20-13   **a.** T
        b. T

20-14   **a.** joule
        b. more

20-15   **a.** infinite ohms
        b. 0 V

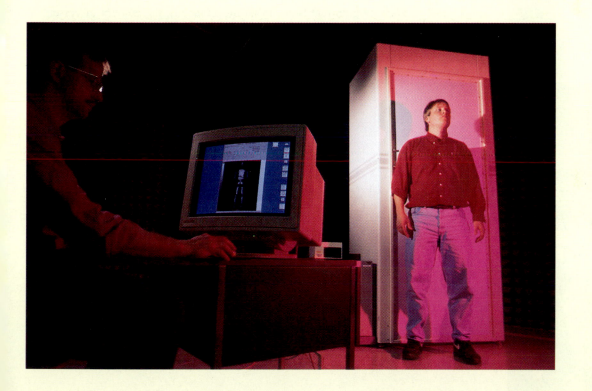

CHAPTER 21

INDUCTIVE REACTANCE

When alternating current flows in an inductance L, the amount of current is much less than the resistance alone would allow. The reason is that the current variations induce a voltage across L that opposes the applied voltage. This additional opposition of an inductance to sine-wave alternating current is specified by the amount of its inductive reactance X_L. It is an opposition to current, measured in ohms. The X_L is the ohms of opposition, therefore, that an inductance L has for sine-wave current.

The amount of X_L equals $2\pi fL$ ohms, with f in hertz and L in henrys. Note that the opposition in ohms of X_L increases for higher frequencies and more inductance. The constant factor 2π indicates sine-wave variations.

The requirements for having X_L correspond to what is needed for producing induced voltage. There must be variations in current and its associated magnetic flux. For a steady direct current without any changes in current, the X_L is zero. However, with sine-wave alternating current, the X_L is the best way to analyze the effect of L.

CHAPTER OBJECTIVES

Upon completion of this chapter, you should be able to:

- *Explain* how inductive reactance reduces the amount of alternating current.
- *Calculate* the reactance of an inductor when the frequency and inductance are known.
- *Calculate* the total reactance of series connected inductors.
- *Calculate* the equivalent reactance of parallel connected inductors.
- *Understand* how Ohm's law can be applied to inductive reactance.
- *Understand* the waveshape of induced voltage produced by sine-wave alternating current.

IMPORTANT TERMS IN THIS CHAPTER

cosine wave
inductive reactance X_L

90° phase angle

quadrature angle

TOPICS COVERED IN THIS CHAPTER

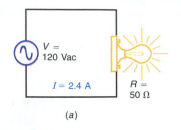

(a)

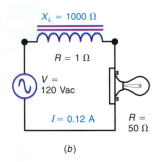

(b)

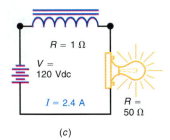

(c)

FIG. 21-1 Illustrating the effect of inductive reactance X_L in reducing the amount of sine-wave alternating current. (a) Bulb lights with 2.4 A. (b) Inserting an X_L of 1000 Ω reduces I to 0.12 A, and the bulb cannot light. (c) With direct current, the coil has no inductive reactance, and the bulb lights.

Figure 21-1 illustrates the effect of X_L in reducing the alternating current for a light bulb. The more ohms of X_L, the less current flows. When X_L reduces I to a very small value, the bulb cannot light.

In Fig. 21-1a, there is no inductance, and the ac voltage source produces a 2.4-A current to light the bulb with full brilliance. This 2.4-A I results from 120 V applied across the 50-Ω R of the bulb's filament.

In Fig. 21-1b, however, a coil is connected in series with the bulb. The coil has a dc resistance of only 1 Ω, which is negligible, but the reactance of the inductance is 1000 Ω. This X_L is a measure of the coil's reaction to sine-wave current in producing a self-induced voltage that opposes the applied voltage and reduces the current. Now I is 120 V/1000 Ω, approximately, which equals 0.12 A. This I is not enough to light the bulb.

Although the dc resistance is only 1 Ω, the X_L of 1000 Ω for the coil limits the amount of alternating current to such a low value that the bulb cannot light. This X_L of 1000 Ω for a 60-Hz current can be obtained with an inductance L of approximately 2.65 H.

In Fig. 21-1c, the coil is also in series with the bulb, but the applied battery voltage produces a steady value of direct current. Without any current variations, the coil cannot induce any voltage and, therefore, it has no reactance. The amount of direct current, then, is practically the same as though the dc voltage source were connected directly across the bulb, and it lights with full brilliance. In this case, the coil is only a length of wire, as there is no induced voltage without current variations. The dc resistance is the resistance of the wire in the coil.

In summary, we can make the following conclusions:

1. An inductance can have appreciable X_L in ac circuits, to reduce the amount of current. Furthermore, the higher the frequency of the alternating current, and the greater the inductance, the higher is the X_L opposition.

2. There is no X_L for steady direct current. In this case, the coil is just a resistance equal to the resistance of the wire.

These effects have almost unlimited applications in practical circuits. Consider how useful ohms of X_L can be for different kinds of current, compared with resistance, which always has the same ohms of opposition. One example is to use X_L where it is desired to have high ohms of opposition to alternating current but little opposition to direct current. Another example is to use X_L for more opposition to a high-frequency alternating current, compared with lower frequencies.

X_L IS AN INDUCTIVE EFFECT An inductance can have X_L to reduce the amount of alternating current because self-induced voltage is produced to oppose the applied voltage. In Fig. 21-2, V_L is the voltage across L, induced by the variations in sine-wave current produced by the applied voltage V_A.

The two voltages V_A and V_L are the same because they are in parallel. However, the current I_L is the amount that allows the self-induced voltage V_L to be equal to V_A. In this example, I is 0.12 A. This value of a 60-Hz current in the inductance produces a V_L of 120 V.

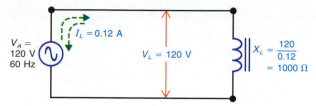

FIG. 21-2 The inductive reactance X_L equals the V_L/I_L ratio in ohms.

THE REACTANCE IS A V/I RATIO   When we consider the V/I ratio for the ohms of opposition to the sine-wave current, this value is 120/0.12, which equals 1000 Ω. This 1000 Ω is what we call X_L, to indicate how much current can be produced by sine-wave voltage across an inductance. The ohms of X_L can be almost any amount, but the 1000 Ω here is a typical example.

THE EFFECT OF L AND f ON X_L   The X_L value depends on the amount of inductance and the frequency of the alternating current. If L in Fig. 21-2 were increased, it could induce the same 120 V for V_L with less current. Then the ratio of V_L/I_L would be greater, meaning more X_L for more inductance.

Also, if the frequency were increased in Fig. 21-2, the current variations would be faster with a higher frequency. Then the same L could produce the 120 V for V_L with less current. For this condition also, the V_L/I_L ratio would be greater because of the smaller current, indicating more X_L for a higher frequency.

TEST-POINT QUESTION 21-1

Answers at end of chapter.
a. For the dc circuit in Fig. 21-1c, how much is X_L?
b. For the ac circuit in Fig. 21-1b, how much is the V/I ratio for X_L?

21-2 $X_L = 2\pi fL$

The formula $X_L = 2\pi fL$ includes the effects of frequency and inductance for calculating the inductive reactance. The frequency is in hertz and L is in henrys for an X_L in ohms. As an example, we can calculate X_L for an inductance of 2.65 H at the frequency of 60 Hz:

▶ $X_L = 2\pi fL$ (21-1)
$\quad = 6.28 \times 60 \times 2.65$
$X_L = 1000\ \Omega$

Note the following factors in the formula $X_L = 2\pi fL$.

1.   The constant factor 2π is always $2 \times 3.14 = 6.28$. It indicates the circular motion from which a sine wave is derived. Therefore, this formula applies

Frequency, Hz	$X_L = 2\pi fL$, Ω
0	0
100	200
200	400
300	600
400	800

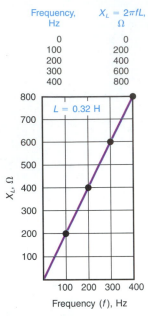

FIG. 21-3 Graph of values to show linear increase of X_L for higher frequencies. The L is constant at 0.32 H.

X_L increases as L increases

Inductance, H	$X_L = 2\pi fL$, Ω
0	0
0.32	200
0.64	400
0.96	600
1.28	800

FIG. 21-4 Graph of values to show linear increase of X_L for higher values of inductance L. The frequency is constant at 100 Hz.

only to sine-wave ac circuits. The 2π is actually 2π rad or 360° for a complete circle or cycle.

2. The frequency f is a time element. Higher frequency means that the current varies at a faster rate. A faster current change can produce more self-induced voltage across a given inductance. The result is more X_L.

3. The inductance L indicates the physical factors of the coil that determine how much voltage it can induce for a given current change.

4. Inductive reactance X_L is in ohms, corresponding to a V_L/I_L ratio for sine-wave ac circuits, to determine how much current L allows for a given applied voltage.

Stating X_L as V_L/I_L and as $2\pi fL$ are two ways of specifying the same value of ohms. The $2\pi fL$ formula gives the effect of L and f on the X_L. The V_L/I_L ratio gives the result of $2\pi fL$ in reducing the amount of I.

The formula $2\pi fL$ shows that X_L is proportional to frequency. When f is doubled, for instance, X_L is doubled. This linear increase of inductive reactance with frequency is illustrated in Fig. 21-3.

The reactance formula also shows that X_L is proportional to the inductance. When the value of henrys for L is doubled, the ohms of X_L is also doubled. This linear increase of inductive reactance with frequency is illustrated in Fig. 21-4.

Example

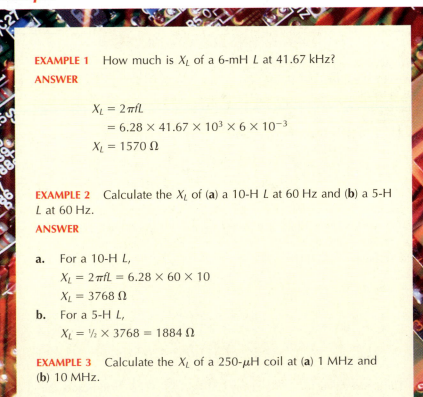

EXAMPLE 1 How much is X_L of a 6-mH L at 41.67 kHz?

ANSWER

$$X_L = 2\pi fL$$
$$= 6.28 \times 41.67 \times 10^3 \times 6 \times 10^{-3}$$
$$X_L = 1570\ \Omega$$

EXAMPLE 2 Calculate the X_L of (**a**) a 10-H L at 60 Hz and (**b**) a 5-H L at 60 Hz.

ANSWER

a. For a 10-H L,
$$X_L = 2\pi fL = 6.28 \times 60 \times 10$$
$$X_L = 3768\ \Omega$$

b. For a 5-H L,
$$X_L = \tfrac{1}{2} \times 3768 = 1884\ \Omega$$

EXAMPLE 3 Calculate the X_L of a 250-μH coil at (**a**) 1 MHz and (**b**) 10 MHz.

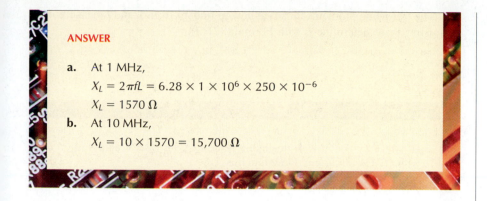

a. At 1 MHz,

$X_L = 2\pi fL = 6.28 \times 1 \times 10^6 \times 250 \times 10^{-6}$

$X_L = 1570\ \Omega$

b. At 10 MHz,

$X_L = 10 \times 1570 = 15{,}700\ \Omega$

The last two examples illustrate the fact that X_L is proportional to frequency and inductance. In Example 2b, X_L is one-half the value in Example 2a because the inductance is one-half. In Example 3b, the X_L is ten times more than in Example 3a because the frequency is ten times higher.

To do a problem like Example 1 with a calculator just requires continued multiplication. Multiply all the factors and then press the ⟨=⟩ key only at the end. If the calculator does not have an ⟨EXP⟩ (exponential) function key, do the powers* of 10 separately without the calculator. Specifically, for this example with $2\pi \times 6 \times 10^{-3} \times 41.67 \times 10^3$, the 10^3 and 10^{-3} cancel. Then calculate $2\pi \times 6 \times 41.67$ as factors. To save time in the calculation, 2π can be memorized as 6.28, since it occurs in many ac formulas. For the multiplication, punch in 6.28 for 2π then push the ⟨×⟩ key, punch in 6 and push the ⟨×⟩ key again, punch in 41.67, and push the ⟨=⟩ key for the total product of 1570 as the final answer. It is not necessary to use the ⟨=⟩ key until the last step for the final product. The factors can be multiplied in any order.

FINDING L FROM X_L   Not only can X_L be calculated from f and L, but if any two factors are known, the third can be found. Very often X_L can be determined from voltage and current measurements. With the frequency known, L can be calculated as

$$L = \frac{X_L}{2\pi f} \tag{21-2}$$

This formula just has the factors inverted from Formula (21-1). Use the basic units with ohms for X_L and hertz for f to calculate L in henrys.

It should be noted that Formula (21-2) can also be stated as

$$L = \frac{1}{2\pi f} \times X_L$$

This form is easier to use with a calculator because $1/2\pi f$ can be found as a reciprocal value and then multiplied by X_L.

---

*For an explanation of powers of 10 see B. Grob, *Mathematics for Basic Electronics*, Glencoe/McGraw-Hill, Columbus, Ohio, and M. Schultz, *Problems in Basic Electronics*, Glencoe/McGraw-Hill, Columbus, Ohio.

CHAPTER 21   INDUCTIVE REACTANCE   589

The following problems illustrate how to find X_L from V and I measurements and using X_L to determine L with Formula (21-2).

Example

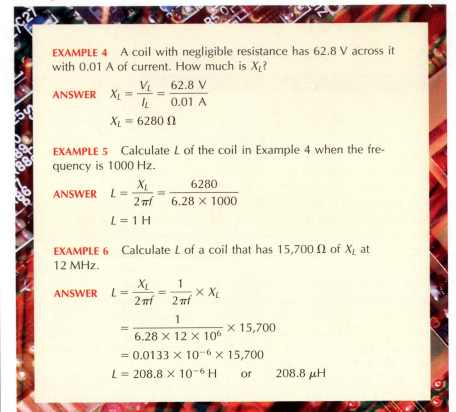

EXAMPLE 4 A coil with negligible resistance has 62.8 V across it with 0.01 A of current. How much is X_L?

ANSWER $X_L = \dfrac{V_L}{I_L} = \dfrac{62.8 \text{ V}}{0.01 \text{ A}}$

$X_L = 6280 \; \Omega$

EXAMPLE 5 Calculate L of the coil in Example 4 when the frequency is 1000 Hz.

ANSWER $L = \dfrac{X_L}{2\pi f} = \dfrac{6280}{6.28 \times 1000}$

$L = 1 \text{ H}$

EXAMPLE 6 Calculate L of a coil that has 15,700 Ω of X_L at 12 MHz.

ANSWER $L = \dfrac{X_L}{2\pi f} = \dfrac{1}{2\pi f} \times X_L$

$= \dfrac{1}{6.28 \times 12 \times 10^6} \times 15{,}700$

$= 0.0133 \times 10^{-6} \times 15{,}700$

$L = 208.8 \times 10^{-6} \text{ H} \quad \text{or} \quad 208.8 \; \mu\text{H}$

To do Example 6 with a calculator, first find the product $2\pi f$ and then take the reciprocal to multiply by 15,700. Note that with powers of 10 a reciprocal value has the sign reversed for the exponent. Specifically, 10^6 in the denominator here becomes 10^{-6} as the reciprocal. To multiply the factors, punch in 6.28 then push the ⊗ key, punch in 12, and push the ⊜ key for the total product of 75.36. Take the reciprocal by using the (1/x) key, while the product is still on the display. This may require pushing the (2ndF) or "shift" key on the calculator. The reciprocal value is 0.0133. Now press the ⊗ key, punch in 15,700, and push the ⊜ key for the answer of 208.8×10^{-6}.

FINDING f FROM X_L For a third version of the inductive reactance formula,

▶ $f = \dfrac{X_L}{2\pi L}$ (21-3)

Use the basic units of ohms for X_L and henrys for L to calculate the frequency in hertz.

Formula 21-3 can also be stated as

$$f = \frac{1}{2\pi L} \times X_L$$

This form is easier to use with a calculator. Find the reciprocal value and multiply by X_L, as explained before with Example 6.

Example

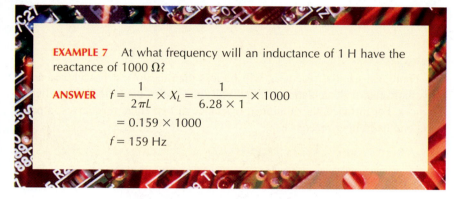

EXAMPLE 7 At what frequency will an inductance of 1 H have the reactance of 1000 Ω?

ANSWER $f = \dfrac{1}{2\pi L} \times X_L = \dfrac{1}{6.28 \times 1} \times 1000$

$$= 0.159 \times 1000$$

$$f = 159 \text{ Hz}$$

TEST-POINT QUESTION 21-2

Answers at end of chapter.

Calculate X_L for the following:
a. L is 1 H and f is 100 Hz.
b. L is 0.5 H and f is 100 Hz.
c. L is 1 H and f is 1000 Hz.

21-3 SERIES OR PARALLEL INDUCTIVE REACTANCES

Since reactance is an opposition in ohms, the values X_L in series or in parallel are combined the same way as ohms of resistance. With series reactances, the total is the sum of the individual values, as shown in Fig. 21-5a. For example, the series reactances of 100 and 200 Ω add to equal 300 Ω of X_L across both reactances. Therefore, in series,

$$X_{L_T} = X_{L_1} + X_{L_2} + X_{L_3} + \cdots + \text{etc.} \qquad (21\text{-}4)$$

For the case of parallel reactances, the combined reactance is calculated by the reciprocal formula. As shown in Fig. 21-5b, in parallel

$$\frac{1}{X_{L_{EQ}}} = \frac{1}{X_{L_1}} + \frac{1}{X_{L_2}} + \frac{1}{X_{L_3}} + \cdots + \text{etc.} \qquad (21\text{-}5)$$

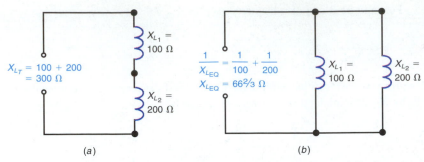

FIG. 21-5 Combining ohms of X_L for inductive reactances. (*a*) X_{L_1} and X_{L_2} in series. (*b*) X_{L_1} and X_{L_2} in parallel.

The combined parallel reactance will be less than the lowest branch reactance. Any short cuts for calculating parallel resistances also apply to the parallel reactances. For instance, the combined reactance of two equal reactances in parallel is one-half either reactance.

Answers at end of chapter.
  a. An X_L of 200 Ω is in series with a 300-Ω X_L. How much is the total X_LT?
  b. An X_L of 200 Ω is in parallel with a 300-Ω X_L. How much is the combined $X_{L_{EQ}}$ in this problem?

21-4 OHM'S LAW APPLIED TO X_L

The amount of current in an ac circuit with just inductive reactance is equal to the applied voltage divided by X_L. Three examples are illustrated in Fig. 21-6. No dc resistance is indicated, since it is assumed to be practically zero for the coils shown. In Fig. 21-6a, there is just one reactance of 100 Ω. Then I equals V/X_L, or 100 V/100 Ω, which is 1 A.

In Fig. 21-6b, the total reactance is the sum of the two individual series reactances of 100 Ω each, for a total of 200 Ω. The current, calculated as V/X_{L_T}, then equals 100 V/200 Ω, which is ½ A or 0.5 A. This current is the same in both series reactances. Therefore, the voltage across each reactance equals its IX_L product. This is 0.5 A × 100 Ω, or 50 V across each X_L.

In Fig. 21-6c, each parallel reactance has its individual branch current, equal to the applied voltage divided by the branch reactance. Then each branch current equals 100 V/100 Ω, which is 1 A. The voltage is the same across both reactances, equal to the generator voltage, since they are all in parallel.

The total line current of 2 A is the sum of the two individual 1-A branch currents. With the rms value for the applied voltage, all the calculated values of currents and voltage drops in Fig. 21-6 are also rms values.

Answers at end of chapter.
a. In Fig. 21-6b, how much is the I through both X_{L_1} and X_{L_2}?
b. In Fig. 21-6c, how much is the V across both X_{L_1} and X_{L_2}?

21-5 APPLICATIONS OF X_L FOR DIFFERENT FREQUENCIES

The general use of inductance is to provide minimum reactance for relatively low frequencies but more for higher frequencies. In this way, the current in an ac circuit can be reduced for higher frequencies because of more X_L. There are many circuits in which voltages of different frequencies are applied to produce current with different frequencies. Then, the general effect of X_L is to allow the most current for direct current and low frequencies, with less current for higher frequencies, as X_L increases.

Compare this frequency factor for ohms of X_L with ohms of resistance. The X_L increases with frequency, but R has the same effect in limiting direct current or alternating current of any frequency.

If 1000 Ω is taken as a suitable value of X_L for many applications, typical inductances can be calculated for different frequencies. These are listed in Table 21-1.

At 60 Hz, for example, the inductance L in the top row of Table 21-1 is 2.65 H for 1000 Ω of X_L. The calculations are

$$L = \frac{X_L}{2\pi f} = \frac{1000}{2\pi \times 60}$$
$$= \frac{1000}{377}$$
$$L = 2.65 \text{ H}$$

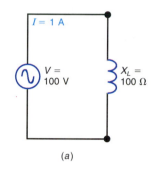

(a)

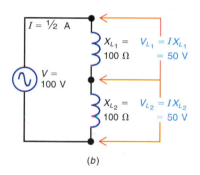

(b)

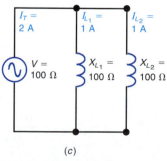

(c)

FIG. 21-6 Circuit calculations with V, I, and ohms of reactance X_L. (a) One reactance. (b) Two series reactances. (c) Two parallel reactances.

TABLE 21-1   VALUES OF INDUCTANCE L FOR X_L OF 1000 Ω

L* (APPROX.)	FREQUENCY	REMARKS
2.65 H	60 Hz	Power-line frequency and low audio frequency
160 mH	1000 Hz	Medium audio frequency
16 mH	10,000 Hz	High audio frequency
160 μH	1000 kHz (RF)	In radio broadcast band
16 μH	10 MHz (HF)	In short-wave radio band
1.6 μH	100 MHz (VHF)	In FM broadcast band

*Calculated as $L = 1000/(2\pi f)$.

For this case, the inductance has practically no reactance for direct current or for very low frequencies below 60 Hz. However, above 60 Hz, the inductive reactance increases to more than 1000 Ω.

To summarize, the effects of increasing frequencies for this 2.65-H inductance are as follows:

Inductive reactance X_L is zero for 0 Hz which corresponds to a steady direct current.
Inductive reactance X_L is less than 1000 Ω for frequencies below 60 Hz.
Inductive reactance X_L equals 1000 Ω at 60 Hz.
Inductive reactance X_L is more than 1000 Ω for frequencies above 60 Hz.

Note that the smaller inductances at the bottom of the first column still have the same X_L of 1000 Ω as the frequency is increased. Typical RF coils, for instance, have an inductance value of the order of 100 to 300 μH. For the very high radio-frequency (VHF) range, only several microhenrys of inductance are needed for an X_L of 1000 Ω.

It is necessary to use smaller inductance values as the frequency is increased because a coil that is too large can have excessive losses at high frequencies. With iron-core coils, particularly, the hysteresis and eddy-current losses increase with frequency.

TEST-POINT QUESTION 21-5

Answers at end of chapter.

Refer to Table 21-1.
a. Which frequency requires the smallest L for 1000 Ω of X_L?
b. How much would X_L be for the 1.6-μH L at 200 MHz?

21-6 WAVESHAPE OF v_L INDUCED BY SINE-WAVE CURRENT

More details of inductive circuits can be analyzed by means of the waveshapes in Fig. 21-7, plotted for the calculated values in Table 21-2. The top curve shows a sine wave of current i_L flowing through a 6-mH inductance L. Since induced voltage depends on rate of change of current rather than the absolute value of i, the curve in Fig. 21-7b shows how much the current changes. In this curve the di/dt values are plotted for the current changes every 30° of the cycle. The bottom curve shows the actual induced voltage v_L. This v_L curve is similar to the di/dt curve because v_L equals the constant factor L multiplied by di/dt. It should be noted that di/dt indicates infinitely small changes in i and t.

90° PHASE ANGLE   The v_L curve at the bottom of Fig. 21-7 has its zero values when the i_L curve at the top is at maximum. This comparison shows that the curves are 90° out of phase. The v_L is a cosine wave of voltage for the sine wave of current i_L.

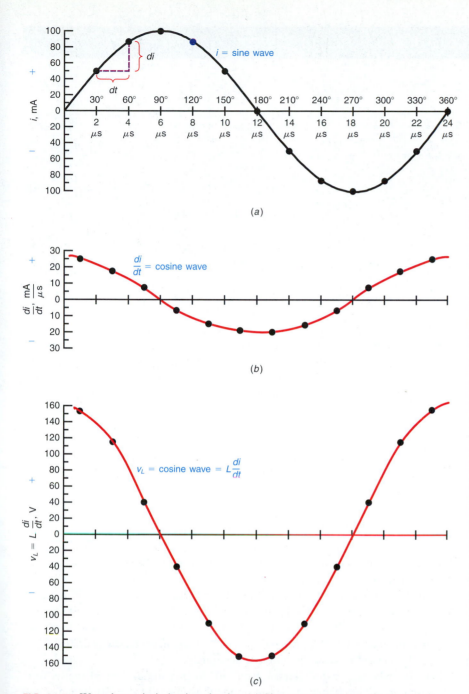

FIG. 21-7 Waveshapes in inductive circuits. (*a*) Sine-wave current, *i*; (*b*) changes in current with time, di/dt; (*c*) induced voltage, v_L.

The 90° phase difference results from the fact that v_L depends on the di/dt rate of change, rather than on i itself. More details of this 90° phase angle between v_L and i_L for inductance are explained in the next chapter.

FREQUENCY For each of the curves, the period T is 24 μs. Therefore, the frequency is $1/T$ or $\frac{1}{24}$ μs, which equals 41.67 kHz. Each curve has the same frequency.

TABLE 21-2   VALUES FOR $v_L = L(di/dt)$ CURVES IN FIG. 21-7

| TIME | | dt | | | di/dt, | | |
θ	μs	θ	μs	di, mA	mA/μs	L, mH	$v_L = L(di/dt)$, V
30°	2	30°	2	50	25	6	150
60°	4	30°	2	36.6	18.3	6	109.8
90°	6	30°	2	13.4	6.7	6	40.2
120°	8	30°	2	−13.4	−6.7	6	−40.2
150°	10	30°	2	−36.6	−18.3	6	−109.8
180°	12	30°	2	−50	−25	6	−150
210°	14	30°	2	−50	−25	6	−150
240°	16	30°	2	−36.6	−18.3	6	−109.8
270°	18	30°	2	−13.4	−6.7	6	−40.2
300°	20	30°	2	13.4	6.7	6	40.2
330°	22	30°	2	36.6	18.3	6	109.8
360°	24	30°	2	50	25	6	150

OHMS OF X_L   The ratio of v_L/i_L actually specifies the inductive reactance in ohms. For this comparison, we use the actual value of i_L, which has a peak value of 100 mA. The rate-of-change factor is included in the induced voltage v_L. Although the peak of v_L at 150 V is 90° before the peak of i_L at 100 mA, we can compare these two peak values. Then v_L/i_L is 150/0.1, which equals 1500 Ω.

This X_L is only an approximate value because v_L cannot be determined exactly for the large dt changes every 30°. If we used smaller intervals of time, the peak v_L would be 157 V. Then X_L would be 1570 Ω, the same as $2\pi fL$ Ω with a 6-mH L and a frequency of 41.67 kHz. This is the same X_L problem as Example 1 on page 588.

THE TABULATED VALUES FROM 0° TO 90°   The numerical values in Table 21-2 are calculated as follows: The i curve is a sine wave. This means it rises to one-half its peak value in 30° and to 0.866 of the peak in 60°, and the peak value is at 90°.

In the di/dt curve the changes in i are plotted. For the first 30° the di is 50 mA; the dt change is 2 μs. Then di/dt is 50/2 or 25 mA/μs. This point is plotted between 0° and 30° to indicate that 25 mA/μs is the rate of change of current for the 2-μs interval between 0° and 30°. If smaller intervals were used, the di/dt values could be determined more accurately.

During the next 2-μs interval, from 30° to 60°, the current increases from 50 to 86.6 mA. The change of current during this time is 86.6 − 50, which equals 36.6 mA. The time is the same 2 μs for all the intervals. Then di/dt for the next plotted point is 36.6/2, or 18.3.

For the final 2-μs change before i reaches its peak at 100 mA, the di value is 100 − 86.6, or 13.4 mA, and the di/dt value is 6.7. All these values are listed in Table 21-2.

Notice that the di/dt curve in Fig. 21-7b has its peak at the zero value of the i curve, while the peak i values correspond to zero on the di/dt curves. These conditions result because the sine wave of i has its sharpest slope at the zero values. The rate of change is greatest when the i curve is going through the zero axis. The i curve flattens near the peaks and has zero rate of change exactly at the peak. The curve must stop going up before it can come down. In summary, then, the *di/dt* curve and the *i* curve are 90° out of phase with each other.

The v_L curve follows the *di/dt* curve exactly, as $v_L = L(di/dt)$. The phase of the v_L curve is exactly the same as that of the *di/dt* curve, 90° out of phase with the *i* curve. For the first plotted point,

$$v_L = L\frac{di}{dt} = 6 \times 10^{-3} \times \frac{50 \times 10^{-3}}{2 \times 10^{-6}}$$

$$v_L = 150 \text{ V}$$

The other v_L values are calculated the same way, multiplying the constant factor of 6 mH by the *di/dt* value for each 2-μs interval.

90° TO 180°   In this quarter-cycle, the sine wave of *i* decreases from its peak of 100 mA at 90° to zero at 180°. This decrease is considered a negative value for *di*, as the slope is negative going downward. Physically, the decrease in current means its associated magnetic flux is collapsing, compared with the expanding flux as the current increases. The opposite motion of the collapsing flux must make v_L of opposite polarity, compared with the induced voltage polarity for increasing flux. This is why the *di* values are negative from 90° to 180°. The *di/dt* values are also negative, and the v_L values are negative.

180° TO 270°   In this quarter-cycle, the current increases in the reverse direction. If the magnetic flux is considered counterclockwise around the conductor with +*i* values, the flux is in the reversed clockwise direction with −*i* values. Any induced voltage produced by expanding flux in one direction will have opposite polarity from voltage induced by expanding flux in the opposite direction. This is why the *di* values are considered negative from 180° to 270°, as in the second quarter-cycle, compared with the positive *di* values from 0° to 90°. Actually, increasing negative values and decreasing positive values

are changing in the same direction. This is why v_L is negative for both the second and third quarter-cycles.

270° TO 360° In the last quarter-cycle, the negative i values are decreasing. Now the effect on polarity is like two negatives making a positive. The current and its magnetic flux have the negative direction. But the flux is collapsing, which induces opposite voltage from increasing flux. Therefore, the di values from 270° to 360° are positive, as are the di/dt values and the induced voltages v_L.

The same action is repeated for each cycle of sine-wave current. Then the current i_L and the induced voltage v_L are 90° out of phase. The reason is that v_L depends on di/dt, not on i alone.

APPLICATION OF THE 90° PHASE ANGLE IN A CIRCUIT The phase angle of 90° between V_L and I will always apply for any L with sine-wave current. Remember, though, that the specific comparison is only between the induced voltage across any one coil and the current flowing in its turns. To emphasize this important principle, Fig. 21-8 shows an ac circuit with a few coils and resistors. The details of this complex circuit are not to be analyzed now. However, for each L in the circuit, the V_L is 90° out of phase with its I. The I lags V_L by 90°, or V_L leads I. For the examples of three coils in Fig. 21-8:

Current I_1 lags V_{L_1} by 90°.
Current I_2 lags V_{L_2} by 90°.
Current I_3 lags V_{L_3} by 90°.
Note that I_3 is also I_T for the series-parallel circuit.

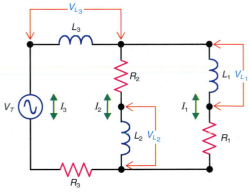

FIG. 21-8 How 90° phase angle for the V_L applies in a complex circuit with more than one inductance. The current I_1 lags V_{L_1} by 90°; I_2 lags V_{L_2} by 90°; and I_3 lags V_{L_3} by 90°.

TEST-POINT QUESTION 21-6

Answers at end of chapter.

Refer to Fig. 21-7.
a. At what angle does i have its maximum positive value?
b. At what angle does v_L have its maximum positive value?
c. What is the phase angle difference between the waveforms for i and v_L?

21   SUMMARY AND REVIEW

- Inductive reactance is the opposition of an inductance to the flow of sine-wave alternating current. The symbol for inductive reactance is X_L.
- Reactance X_L is measured in ohms because it limits the current to the value $I = V/X_L$. With V in volts and X_L in ohms, I is in amperes.
- $X_L = 2\pi fL$, where f is in hertz, L is in henrys, and X_L is in ohms.
- With a constant L, X_L increases proportionately with higher frequencies.
- At a constant frequency, X_L increases proportionately with higher inductances.
- With X_L and f known, the inductance $L = X_L/(2\pi f)$.
- With X_L and L known, the frequency $f = X_L/(2\pi L)$.
- The total X_L of reactances in series is the sum of the individual values, as for series resistances. Series reactances have the same current. The voltage across each inductive reactance is IX_L.
- With parallel reactances, the equivalent reactance is calculated by the reciprocal formula, as for parallel resistances. Each branch current is V/X_L. The total line current is the sum of the individual branch currents.
- Table 21-3 summarizes the differences between L and X_L.

TABLE 21-3   COMPARISON OF INDUCTANCE AND INDUCTIVE REACTANCE

INDUCTANCE	INDUCTIVE REACTANCE
Symbol is L	Symbol is X_L
Measured in henry units	Measured in ohm units
Depends on construction of coil	Depends on frequency and inductance
$L = v_L/(di/dt)$, in H units	$X_L = v_L/i_L$ or $2\pi fL$, in Ω units

- Table 21-4 compares X_L and R.

TABLE 21-4   COMPARISON OF X_L AND R

X_L	R
Ohm unit	Ohm unit
Increases for higher frequencies	Same for all frequencies
Current lags voltage by 90° ($\theta = 90°$)	Current in phase with voltage ($\theta = 0°$)

- Table 21-5 summarizes the differences between capacitive reactance and inductive reactance.

TABLE 21-5   COMPARISON OF CAPACITIVE AND INDUCTIVE REACTANCES

X_C, Ω	X_L, Ω

Decreases with more capacitance C
Decreases with increase in frequency f
Allows less current at lower frequencies;
  blocks direct current.
	Increases with more inductance L
	Increases with increase in frequency f
	Allows more current at lower frequencies;
	  passes direct current.

SELF-TEST

ANSWERS AT BACK OF BOOK.

Choose (a), (b), (c), or (d).

1. Inductive reactance is measured in ohms because it (a) reduces the amplitude of alternating current; (b) increases the amplitude of alternating current; (c) increases the amplitude of direct current; (d) has a back emf opposing a steady direct current.

2. Inductive reactance applies only to sine waves because it (a) increases with lower frequencies; (b) increases with lower inductance; (c) depends on the factor 2π; (d) decreases with higher frequencies.

3. An inductance has a reactance of 10,000 Ω at 10,000 Hz. At 20,000 Hz, its inductive reactance equals (a) 500 Ω; (b) 2000 Ω; (c) 20,000 Ω; (d) 32,000 Ω.

4. A 16-mH inductance has a reactance of 1000 Ω. If two of these are connected in series without any mutual coupling, their total reactance equals (a) 500 Ω; (b) 1000 Ω; (c) 1600 Ω; (d) 2000 Ω.

5. Two 5000-Ω inductive reactances in parallel have an equivalent reactance of (a) 2500 Ω; (b) 5000 Ω; (c) 10,000 Ω; (d) 50,000 Ω.

6. With 10 V applied across an inductive reactance of 100 Ω, the current equals (a) 10 μA; (b) 10 mA; (c) 100 mA; (d) 10 A.

7. A current of 100 mA through an inductive reactance of 100 Ω produces a voltage drop equal to (a) 1 V; (b) 6.28 V; (c) 10 V; (d) 100 V.

8. The inductance required for a 2000-Ω reactance at 20 MHz equals (a) 10 μH; (b) 15.9 μH; (c) 159 μH; (d) 320 μH.

9. A 160-μH inductance will have a 5000-Ω reactance at the frequency of (a) 5 kHz; (b) 200 kHz; (c) 1 MHz; (d) 5 MHz.

10. A coil has an inductive reactance of 1000 Ω. If its inductance is doubled and the frequency is doubled, then the inductive reactance will be (a) 1000 Ω; (b) 2000 Ω; (c) 4000 Ω; (d) 16,000 Ω.

1. Explain briefly why X_L limits the amount of alternating current.
2. Give two differences and one similarity between X_L and R.
3. Explain why X_L increases with higher frequencies and more inductance.
4. Give two differences between inductance L of a coil and its reactance X_L.
5. Why are the waves in Fig. 21-7a and b considered to be 90° out of phase, while the waves in Fig. 21-7b and c have the same phase?
6. Referring to Fig. 21-3, how does this graph show a linear proportion between X_L and frequency?
7. Referring to Fig. 21-4, how does this graph show a linear proportion between X_L and L?
8. Referring to Fig. 21-3, tabulate the values of L that would be needed for each frequency listed but for an X_L of 2000 Ω. (Do not include 0 Hz.)
9. (**a**) Draw the circuit for a 40-Ω R across a 120-V 60-Hz source. (**b**) Draw the circuit for a 40-Ω X_L across a 120-V 60-Hz source. (**c**) Why is I equal to 3 A for both circuits? (**d**) Give two differences between the circuits.
10. Why are coils for RF applications generally smaller than AF coils?

PROBLEMS

ANSWERS TO ODD-NUMBERED PROBLEMS AT BACK OF BOOK.

1. Calculate the X_L of a 0.5-H inductance at 100, 200, and 1000 Hz.
2. How much is the inductance for 628 Ω reactance at 100 Hz? 200 Hz? 1000 Hz? 500 kHz?
3. A coil with an X_L of 748 Ω is connected across a 16-Vac generator. (**a**) Draw the schematic diagram. (**b**) Calculate the current. (**c**) How much is the voltage across the coil?
4. A 20-H coil has 10 V applied, with a frequency of 60 Hz. (**a**) Draw the schematic diagram. (**b**) How much is the inductive reactance of the coil? (**c**) Calculate the current. (**d**) What is the frequency of the current?
5. How much is the inductance of a coil with negligible resistance if the current is 0.1 A when connected across the 60-Hz 120-V power line?
6. Referring to Fig. 21-6b, how much is the inductance of L_T, L_1, and L_2 if the frequency of the source voltage is 400 Hz?
7. How much is the inductance of a coil that has a reactance of 1000 Ω at 1000 Hz? How much will the reactance be for the same coil at 10 kHz?
8. How much is the reactance of a 20-μH inductance at 40 MHz?
9. A 1000-Ω X_{L_1} and a 4000-Ω X_{L_2} are in series across a 10-V 60-Hz source. Draw the schematic diagram and calculate the following: (**a**) total X_L; (**b**) current in X_{L_1} and in X_{L_2}; (**c**) voltage across X_{L_1} and across X_{L_2}; (**d**) L_1 and L_2.
10. The same 1000-Ω X_{L_1} and 4000-Ω X_{L_2} are in parallel across the 10-V 60-Hz source. Draw the schematic diagram and calculate the following: branch currents in X_{L_1} and in X_{L_2}; total current in the generator; voltage across X_{L_1} and across X_{L_2}; inductance of L_1 and L_2.

11. At what frequencies will X_L be 2000 Ω for the following inductors: **(a)** 2 H; **(b)** 250 mH; **(c)** 800 μH; **(d)** 200 μH; **(e)** 20 μH?

12. A 6-mH L_1 is in series with an 8-mH L_2. The frequency is 40 kHz. **(a)** How much is L_T? **(b)** Calculate X_{L_T}. **(c)** Calculate X_{L_1} and X_{L_2}.

13. Calculate X_L of a 2.4-mH coil at 108 kHz.

14. Calculate X_L of a 40-μH coil at 3.2 MHz.

15. Calculate X_L of a 2-H coil at 60 Hz.

16. How much is I when the X_L of Prob. 15 is connected to 120-V 60-Hz?

17. A 250-mH inductor with negligible resistance is connected across a 10-V source. Tabulate the values of X_L and current in the circuit for alternating current at **(a)** 20 Hz; **(b)** 60 Hz; **(c)** 100 Hz; **(d)** 500 Hz; **(e)** 5000 Hz; **(f)** 15,000 Hz.

18. Do the same as in Prob. 17 for an 8-H inductor.

19. What inductance L is needed for X_L of 785 Ω at the following frequencies: **(a)** 500 kHz; **(b)** 1 MHz; **(c)** 250 kHz; **(d)** 5 MHz; **(e)** 50 kHz?

20. What frequency will provide X_L of 785 Ω for the following values of inductance L? **(a)** 250 μH; **(b)** 125 μH; **(c)** 500 μH; **(d)** 25 μH; **(e)** 2.5 mH?

21. In Fig. 21-9, calculate X_{L_1}, X_{L_2}, X_{L_3}, X_{L_T}, I, V_{L_1}, V_{L_2}, and V_{L_3}.

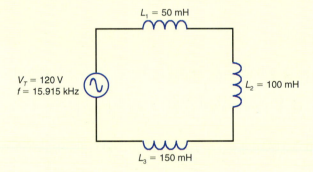

$L_1 = 50$ mH

$V_T = 120$ V
$f = 15.915$ kHz

$L_2 = 100$ mH

$L_3 = 150$ mH

FIG. 21-9 Circuit for Probs. 21 and 22.

22. In Fig. 21-9, what happens to I if the frequency f is doubled?

23. In Fig. 21-10, calculate X_{L_1}, X_{L_2}, X_{L_3}, I_1, I_2, I_3, I_T, and $X_{L_{EQ}}$.

24. In Fig. 21-10, what happens to $X_{L_{EQ}}$ if the frequency f is doubled?

25. Two inductors connected in series have a mutual inductance L_M of 1 mH. Given $L_1 = 5$ mH and $L_2 = 10$ mH, calculate X_{L_T} for a frequency of 50 kHz when: **(a)** L_M is series aiding; **(b)** L_M is series opposing.

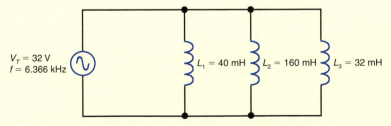

$V_T = 32$ V
$f = 6.366$ kHz

$L_1 = 40$ mH $L_2 = 160$ mH $L_3 = 32$ mH

FIG. 21-10 Circuit for Probs. 23 and 24.

CRITICAL THINKING

1. In Fig. 21-11, calculate L_1, L_2, L_3, L_T, X_{L_1}, X_{L_2}, X_{L_T}, V_{L_1}, V_{L_3}, I_{L_2}, and I_{L_3}.
2. Two inductors in series without L_M have a total inductance L_T of 120 μH. If $L_1/L_2 = 1/20$, what are the values for L_1 and L_2?
3. Three inductors in parallel have an equivalent inductance L_{EQ} of 7.5 mH. If $L_2 = 3L_3$ and $L_3 = 4L_1$, calculate L_1, L_2, and L_3.

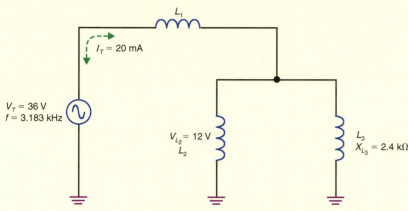

FIG. 21-11 Circuit for Critical Thinking Prob. 1.

ANSWERS TO TEST-POINT QUESTIONS

21-1 a. 0 Ω
    b. 1000 Ω

21-2 a. X_L = 628 Ω
    b. X_L = 314 Ω
    c. X_L = 6280 Ω

21-3 a. X_{L_T} = 500 Ω
    b. X_{L_T} = 120 Ω

21-4 a. 0.5 A
    b. 100 V

21-5 a. 100 MHz
    b. 2000 Ω

21-6 a. 90°
    b. 0° or 360°
    c. 90°

CHAPTER 22

INDUCTIVE CIRCUITS

This chapter analyzes circuits that combine inductive reactance X_L and resistance R. The main questions are: how do we combine the ohms of opposition, how much current flows, and what is the phase angle? Although X_L and R both are measured in ohms, they have some different characteristics. Specifically, X_L increases with more L and higher frequencies, with sine-wave ac voltage applied, while R is the same for dc or ac circuits. Furthermore, the phase angle for the voltage across X_L is at 90° with respect to the current through L.

In addition, the practical application of using a coil as a choke to reduce the current for a specific frequency is explained here. For a circuit with L and R in series, the X_L can be high for an undesired ac signal frequency, while R is the same for either direct current or alternating current.

Finally, the general case of induced voltage produced across L is shown with nonsinusoidal current variations. In this case, we compare the waveshapes of i_L and v_L instead of their phase. Remember that the 90° angle for an IX_L voltage applies only to sine waves.

With nonsinusoidal waveforms, such as pulses of current or voltage, the circuit can be analyzed in terms of its L/R time constant, as explained in Chap. 23.

CHAPTER OBJECTIVES

Upon completion of this chapter, you should be able to:

- *Explain* why the voltage leads the current by 90° for an inductor.
- *Calculate* the total impedance and phase angle of a series *RL* circuit.
- *Calculate* the total current, equivalent impedance, and phase angle of a parallel *RL* circuit.
- *Define* what is meant by the *Q* of a coil.
- *Understand* how an inductor can be used to pass some ac frequencies but block others.
- *Calculate* the induced voltage that is produced by a nonsinusoidal current.

IMPORTANT TERMS IN THIS CHAPTER

| choke | lagging current | sawtooth waveform |
| impedance *Z* | *Q* of a coil | |

TOPICS COVERED IN THIS CHAPTER

22-1 SINE-WAVE i_L LAGS v_L BY 90°

With sine-wave variations of current producing an induced voltage, the current lags its induced voltage by exactly 90°, as shown in Fig. 22-1. The inductive circuit in Fig. 22-1*a* has the current and voltage waveshapes shown in Fig. 22-1*b*. The phasors in Fig. 22-1*c* show the 90° phase angle between i_L and v_L. Therefore, we can say that i_L lags v_L by 90°. Or, v_L leads i_L by 90°.

This 90° phase relationship between i_L and v_L is true in any sine-wave ac circuit, whether L is in series or parallel, and whether L is alone or combined with other components. We can always say that the voltage across any X_L is 90° out of phase with the current through it.

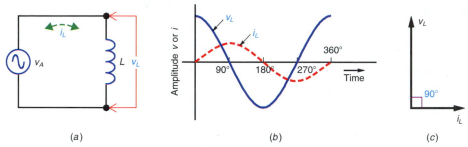

(a)	(b)	(c)

FIG. 22-1 (*a*) Circuit with inductance *L*. (*b*) Sine wave of i_L *lags* v_L by 90°. (*c*) Phasor diagram.

WHY THE PHASE ANGLE IS 90°   This results because v_L depends on the rate of change of i_L. As previously shown in Fig. 21-7 for a sine wave of i_L, the induced voltage is a cosine wave. In other words, v_L has the phase of di/dt, not the phase of i.

WHY i_L LAGS v_L   The 90° difference can be measured between any two points having the same value on the i_L and v_L waves. A convenient point is the positive peak value. Note that the i_L wave does not have its positive peak until 90° after the v_L wave. Therefore, i_L lags v_L by 90°. This 90° lag is in time. The time lag equals one quarter-cycle, which is one-quarter of the time for a complete cycle.

INDUCTIVE CURRENT IS THE SAME IN A SERIES CIRCUIT   The time delay and resultant phase angle for the current in an inductance apply only with respect to the voltage across the inductance. This condition does not change the fact that the current is the same in all parts of a series circuit. In Fig. 22-1*a*, the current in the generator, the connecting wires, and L must be the same because they are in series. At any instant, whatever the current value is at that time, it is the same in all the series components. The time lag is between current and voltage.

INDUCTIVE VOLTAGE IS THE SAME ACROSS PARALLEL BRANCHES
In Fig. 22-1*a*, the voltage across the generator and the voltage across L are the same because they are in parallel. There cannot be any lag or lead in time between these two parallel voltages. At any instant, whatever the voltage value is across the generator at that time, the voltage across L is the same. Considering the parallel voltage v_A or v_L, it is 90° out of phase with the current.

In this circuit the voltage across L is determined by the applied voltage, since they must be the same. The inductive effect here is to make the current have the values that produce $L(di/dt)$ equal to the parallel voltage.

THE FREQUENCY IS THE SAME FOR i_L AND v_L   Although i_L lags v_L by 90°, both waves have the same frequency. The i_L wave reaches it peak values 90° later than the v_L wave, but the complete cycles of variations are repeated at the same rate. As an example, if the frequency of the sine wave v_L in Fig. 22-1*b* is 100 Hz, this is also the frequency for i_L.

TEST-POINT QUESTION 22-1

Answers at end of chapter.

Refer to Fig. 22-1.
a. What is the phase angle between v_A and v_L?
b. What is the phase angle between v_L and i_L?
c. Does i_L lead or lag v_L?

22-2 X_L AND R IN SERIES

When a coil has series resistance, the current is limited by both X_L and R. This current I is the same in X_L and R, since they are in series. Each has its own series voltage drop, equal to IR for the resistance and IX_L for the reactance.

Note the following points about a circuit that combines series X_L and R, as in Fig. 22-2:

1. The current is labeled I, rather than I_L, because I flows through all the series components.
2. The voltage across X_L, labeled V_L, can be considered an IX_L voltage drop, just as we use V_R for an IR voltage drop.
3. The current I through X_L must lag V_L by 90°, as this is the phase angle between current through an inductance and its self-induced voltage.
4. The current I through R and its IR voltage drop are in phase. There is no reactance to sine-wave current in any resistance. Therefore, I and IR have a phase angle of 0°.

Resistance R can be either the internal resistance of the coil or an external series resistance. The I and V values may be rms, peak, or instantaneous, as long as the same measure is applied to all. Peak values are used here for convenience in comparing the waveforms.

PHASE COMPARISONS   Note the following:

1. Voltage V_L is 90° out of phase with I.
2. However, V_R and I are in phase.
3. If I is used as the reference, V_L is 90° out of phase with V_R.

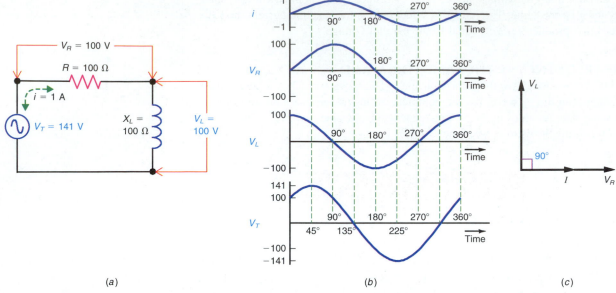

| (a) | (b) | (c) |

FIG. 22-2 Inductive reactance X_L and resistance R in series. (a) Circuit. (b) Waveforms of current and voltages. (c) Phasor diagram.

Specifically, V_R lags V_L by 90°, just as the current I lags V_L. These phase relations are shown by the waveforms in Fig. 22-2b and the phasors in Fig. 22-2c.

COMBINING V_R AND V_L   As shown in Fig. 22-2b, when the V_R voltage wave is combined with the V_L voltage wave, the result is the voltage wave for the applied generator voltage V_T. The voltage drops must add to equal the applied voltage. The 100-V peak values for V_R and for V_L total 141 V, however, instead of 200 V, because of the 90° phase difference.

Consider some instantaneous values to see why the 100-V peak V_R and 100-V peak V_L cannot be added arithmetically. When V_R is at its maximum of 100 V, for instance, V_L is at zero. The total for V_T then is 100 V. Similarly, with V_L at its maximum of 100 V, then V_R is zero and the total V_T is also 100 V.

Actually, V_T has its maximum value of 141 V at the time when V_L and V_R are each 70.7 V. When series voltage drops that are out of phase are combined, therefore, they cannot be added without taking the phase difference into account.

PHASOR-VOLTAGE TRIANGLE   Instead of combining waveforms that are out of phase, we can add them more quickly by using their equivalent phasors, as shown in Fig. 22-3. The phasors in Fig. 22-3a just show the 90° angle without any addition. The method in Fig. 22-3b is to add the tail of one phasor to the arrowhead of the other, using the angle required to show their relative phase. Voltages V_R and V_L are at right angles because they are 90° out of phase. The sum of the phasors is a resultant phasor from the start of one to the end of the other. Since the V_R and V_L phasors form a right angle, the resultant

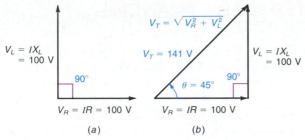

(a)   (b)

FIG. 22-3 Addition of two voltages 90° out of phase. (*a*) Phasors for V_L and V_R are 90° out of phase. (*b*) Resultant of the two phasors is hypotenuse of right triangle for value of V_T.

phasor is the hypotenuse of a right triangle. The hypotenuse is the side opposite the 90° angle.

From the geometry of a right triangle, the pythagorean theorem states that the hypotenuse is equal to the square root of the sum of the squares of the sides. For the voltage triangle in Fig. 22-3*b*, therefore, the resultant is

▶ $$V_T = \sqrt{V_R{}^2 + V_L{}^2}$$ (22-1)

where V_T is the phasor sum of the two voltages V_R and V_L 90° out of phase.

This formula is for V_R and V_L when they are in series, since then they are 90° out of phase. All the voltages must be in the same units. When V_A is an rms value, V_R and V_L are also rms values. For the example in Fig. 22-3,

$$V_T = \sqrt{100^2 + 100^2} = \sqrt{10,000 + 10,000}$$
$$= \sqrt{20,000}$$
$$V_T = 141 \text{ V}$$

To do a problem like this on the calculator, remember that the square root sign is a sign of grouping. All terms within the group must be added before you take the square root. Also, each term must be squared individually before adding for the sum. Specifically for this problem:

1. Punch in 100 and push the (x^2) button for 10,000 as the square. Press $(+)$ and $($ $($ $)$. ·
2. Next punch in 100 and (x^2). Press $($ $)$ $)$ and $(=)$. The display should read 20000.
3. Press $(\sqrt{\ \ })$ to read the answer 141.421.

In some calculators either the (x^2) or the $(\sqrt{\ \ })$ key must be preceded by the second function key (F).

<hr>

TEST-POINT QUESTION 22-2

<hr>

Answers at end of chapter.
a. In a series circuit with X_L and R, what is the phase angle between I and V_R?
b. What is the phase angle between V_R and V_L?

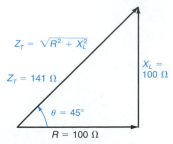

FIG. 22-4 Addition of *R* and X_L 90° out of phase in series circuit, to find the resultant impedance Z_T.

A triangle of *R* and X_L in series corresponds to the voltage triangle, as shown in Fig. 22-4. It is similar to the voltage triangle in Fig. 22-3, but the common factor *I* cancels because the current is the same in X_L and *R*. The resultant of the phasor addition of *R* and X_L is their total opposition in ohms, called *impedance,* with the symbol Z_T.* The *Z* takes into account the 90° phase relation between *R* and X_L.

For the impedance triangle of a series circuit with reactance and resistance,

▶ $$Z_T = \sqrt{R^2 + X_L^2} \tag{22-2}$$

where *R*, X_L, and Z_T are all in ohms. For the example in Fig. 22-4,

$$Z_T = \sqrt{100^2 + 100^2} = \sqrt{10{,}000 + 10{,}000}$$
$$= \sqrt{20{,}000}$$
$$Z_T = 141 \ \Omega$$

Note that the total impedance of 141 Ω divided into the applied voltage of 141 V results in 1 A of current in the series circuit. The *IR* voltage is 1 × 100, or 100 V; the IX_L voltage is also 1 × 100, or 100 V. The total of the series *IR* and IX_L drops of 100 V each added by phasors equals the applied voltage of 141 V. Finally, the applied voltage equals *IZ*, or 1 × 141, which is 141 V.

To summarize the similar phasor triangles for volts and ohms in a series circuit:

1. The phasor for *R*, *IR*, or V_R is used as a reference at 0°.
2. The phasor for X_L, IX_L, or V_L is at 90°.
3. The phasor for *Z*, *IZ*, or V_T has the phase angle θ of the complete circuit.

PHASE ANGLE WITH SERIES X_L   The angle between the generator voltage and its current is the phase angle of the circuit. Its symbol is θ (theta). In Fig. 22-3, the phase angle between V_T and *IR* is 45°. Since *IR* and *I* have the same phase, the angle is also 45° between V_T and *I*.

In the corresponding impedance triangle in Fig. 22-4, the angle between Z_T and *R* is also equal to the phase angle. Therefore, the phase angle can be calculated from the impedance triangle of a series circuit by the formula

▶ $$\tan \theta_Z = \frac{X_L}{R} \tag{22-3}$$

The tangent (tan) is a trigonometric function of any angle, equal to the ratio of the opposite side to the adjacent side of a triangle. In this impedance triangle, X_L is the opposite side and *R* is the adjacent side of the angle. We use the subscript *z* for θ to show that θ_Z is found from the impedance triangle for a series circuit. To calculate this phase angle,

$$\tan \theta_Z = \frac{X_L}{R} = \frac{100}{100} = 1$$

---

*Although the Z_T is a passive component, we consider it as a phasor here because it determines the phase angle of V and I.

The angle that has the tangent equal to 1 is 45°. Therefore, the phase angle is 45° in this example. The numerical values of the trigonometric functions can be found from a table or scientific calculator.

Note that the phase angle of 45° is halfway between 0° and 90° because R and X_L are equal.

Example

EXAMPLE 1   If a 30-Ω R and a 40-Ω X_L are in series with 100 V applied, find the following: Z_T, I, V_R, V_L, and θ_Z. What is the phase angle between V_L and V_R with respect to I? Prove that the sum of the series voltage drops equals the applied voltage V_T.

ANSWER $Z_T = \sqrt{R^2 + X_L^2} = \sqrt{900 + 1600}$

$$= \sqrt{2500}$$

$$Z_T = 50 \ \Omega$$

$$I = \frac{V_T}{Z_T} = \frac{100}{50} = 2 \text{ A}$$

$$V_R = IR = 2 \times 30 = 60 \text{ V}$$

$$V_L = IX_L = 2 \times 40 = 80 \text{ V}$$

$$\tan \theta_Z = \frac{X_L}{R} = \frac{40}{30} = \frac{4}{3} = 1.3333$$

$$\theta_Z = 53.1°$$

Therefore, I lags V_T by 53.1°. Furthermore, I and V_R are in phase, and I lags V_L by 90°. Finally.

$$V_T = \sqrt{V_R^2 + V_L^2} = \sqrt{60^2 + 80^2} = \sqrt{3600 + 6400}$$

$$= \sqrt{10,000}$$

$$V_T = 100 \text{ V}$$

Note that the phasor sum of the voltage drops equals the applied voltage.

To do the trigonometry in Example 1 with a calculator, there are several points to keep in mind:

1. The ratio of X_L/R specifies the angle's tangent function as a numerical value, but this is not the angle θ in degrees. Finding X_L/R is just a division problem.
2. The angle θ itself is an *inverse function* of tan θ that is indicated as arctan θ or $\tan^{-1} \theta$. A scientific calculator can give the trigonometric functions directly from the value of the angle, or inversely show the angle from its trig functions.
3. As a check on your values, note that for tan $\theta = 1$, $\tan^{-1}\theta$ is 45°. Tangent values less than 1 must be for angles smaller than 45°; angles more than 45° must have tangent values higher than 1.

For the values in Example 1, specifically, punch in 40 for X_L, push the $\div$ key, punch in 30 for R, and push the $=$ key for the ratio of 1.3333 on the display. This value is tan θ. While it is on the display, push the $\boxed{\text{TAN}^{-1}}$ key and the answer of 53.1° appears for angle θ. Use of the tan^{-1} key is usually preceded by pressing the function key $\boxed{\text{F}}$ or $\boxed{\text{2ndF}}$.

SERIES COMBINATIONS OF X_L AND R   In a series circuit, the higher the value of X_L compared with R, the more inductive the circuit is. This means there is more voltage drop across the inductive reactance and the phase angle increases toward 90°. The series current lags the applied generator voltage. With all X_L and no R, the entire applied voltage is across X_L and θ_Z equals 90°.

Several combinations of X_L and R in series are listed in Table 22-1 with their resultant impedance and phase angle. Note that a ratio of 10:1 or more for X_L/R means that the circuit is practically all inductive. The phase angle of 84.3° is only slightly less than 90° for the ratio of 10:1, and the total impedance Z_T is approximately equal to X_L. The voltage drop across X_L in the series circuit will be practically equal to the applied voltage, with almost none across R.

TABLE 22-1   SERIES R AND X_L COMBINATIONS

R, Ω	X_L, Ω	Z_T, Ω (APPROX.)	IMPEDANCE ANGLE θ_Z
1	10	$\sqrt{101} = 10$	84.3°
10	10	$\sqrt{200} = 14.1$	45°
10	1	$\sqrt{101} = 10$	5.7°

Note: θ_Z is the angle of Z_T with respect to the reference I in a series circuit.

At the opposite extreme, when R is ten times as large as X_L, the series circuit is mainly resistive. The phase angle of 5.7°, then, means the current is almost in phase with the applied voltage, the total impedance Z_T is approximately equal to R, and the voltage drop across R is practically equal to the applied voltage, with almost none across X_L.

For the case when X_L and R equal each other, their resultant impedance Z_T is 1.41 times the value of either one. The phase angle then is 45°, halfway between 0° for resistance alone and 90° for inductive reactance alone.

TEST-POINT QUESTION 22-3

Answers at end of chapter.
a. How much is Z_T for a 20-Ω R in series with a 20-Ω X_L?
b. How much is V_T for 20 V across R and 20 V across X_L in series?
c. What is the phase angle of the circuit in **a** and **b**?

22-4 X_L AND R IN PARALLEL

For parallel circuits with X_L and R, the 90° phase angle must be considered for each of the branch currents, instead of voltage drops in a series circuit. Remember that any series circuit has different voltage drops but one common current. A parallel circuit has different branch currents but one common voltage.

In the parallel circuit in Fig. 22-5a, the applied voltage V_A is the same across X_L, R, and the generator, since they are all in parallel. There cannot be any phase difference between these voltages. Each branch, however, has its individual current. For the resistive branch, $I_R = V_A/R$; in the inductive branch, $I_L = V_A/X_L$.

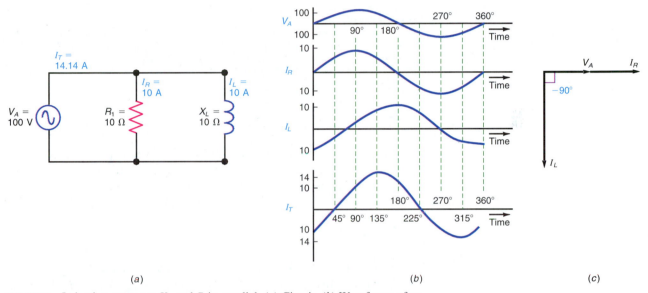

FIG. 22-5   Inductive reactance X_L and R in parallel. (*a*) Circuit. (*b*) Waveforms of applied voltage and branch currents. (*c*) Phasor diagram.

The resistive branch current I_R is in phase with the generator voltage V_A. The inductive branch current I_L lags V_A, however, because the current in an inductance lags the voltage across it by 90°.

The total line current, therefore, consists of I_R and I_L, which are 90° out of phase with each other. The phasor sum of I_R and I_L equals the total line current I_T. These phase relations are shown by the waveforms in Fig. 22-5b, with the phasors in Fig. 22-5c. Either way, the phasor sum of 10 A for I_R and 10 A for I_L is equal to 14.14 A for I_T.

Both methods illustrate the general principle that quadrature components must be combined by phasor addition. The branch currents are added by phasors here because they are the factors that are 90° out of phase in a parallel circuit. This method is similar to combining voltage drops 90° out of phase in a series circuit.

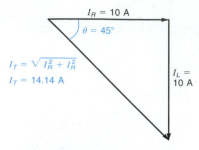

$I_R = 10$ A

$\theta = 45°$

$I_T = \sqrt{I_R^2 + I_R^2}$

$I_T = 14.14$ A

$I_L = 10$ A

FIG. 22-6 Phasor triangle of inductive and resistive branch currents 90° out of phase in a parallel circuit to find resultant I_T.

PHASOR CURRENT TRIANGLE Note that the phasor diagram in Fig. 22-5c has the applied voltage V_A of the generator as the reference phasor. The reason is that V_A is the same throughout the parallel circuit.

The phasor for I_L is down, as compared with up for an X_L phasor. Here the parallel branch current I_L lags the parallel voltage reference V_A. In a series circuit the X_L voltage leads the series current reference I. For this reason the I_L phasor is shown with a negative 90° angle. The −90° means the current I_L lags the reference phasor V_A.

The phasor addition of the branch currents in a parallel circuit can be calculated by the phasor triangle for currents shown in Fig. 22-6. Peak values are used for convenience in this example, but when the applied voltage is an rms value, the calculated currents are also in rms values. To calculate the total line current, we have

$$I_T = \sqrt{I_R^2 + I_L^2}$$ (22-4)

For the values in Fig. 22-6,

$$I_T = \sqrt{10^2 + 10^2} = \sqrt{100 + 100}$$
$$= \sqrt{200}$$
$$I_T = 14.14 \text{ A}$$

IMPEDANCE OF X_L AND R IN PARALLEL A practical approach to the problem of calculating the total impedance of X_L and R in parallel is to calculate the total line current I_T and divide this into the applied voltage:

$$Z_{EQ} = \frac{V_A}{I_T}$$ (22-5)

For example, in Fig. 22-5, V_A is 100 V and the resultant I_T, obtained as the phasor sum of the resistive and reactive branch currents, is equal to 14.14 A. Therefore, we calculate the impedance as

$$Z_{EQ} = \frac{V_A}{I_T} = \frac{100 \text{ V}}{14.14 \text{ A}}$$
$$Z_{EQ} = 7.07 \text{ }\Omega$$

This impedance is the combined opposition in ohms across the generator, equal to the resistance of 10 Ω in parallel with the reactance of 10 Ω.

Note that the impedance for equal values of R and X_L in parallel is not one-half but equals 70.7 percent of either one. Still, the combined value of ohms must be less than the lowest ohms value in the parallel branches.

For the general case of calculating the impedance of X_L and R in parallel, any number can be assumed for the applied voltage because in the calculations for Z in terms of the branch currents the value of V_A cancels. A good value to assume for V_A is the value of either R or X_L, whichever is the higher number. This way there are no fractions smaller than one in calculation of the branch currents.

Example

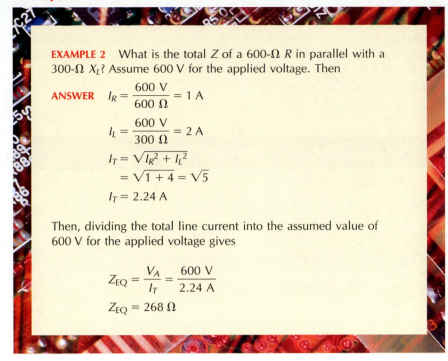

EXAMPLE 2   What is the total Z of a 600-Ω R in parallel with a 300-Ω X_L? Assume 600 V for the applied voltage. Then

ANSWER $I_R = \dfrac{600 \text{ V}}{600 \text{ }\Omega} = 1 \text{ A}$

$I_L = \dfrac{600 \text{ V}}{300 \text{ }\Omega} = 2 \text{ A}$

$I_T = \sqrt{I_R^2 + I_L^2}$

$\quad = \sqrt{1 + 4} = \sqrt{5}$

$I_T = 2.24 \text{ A}$

Then, dividing the total line current into the assumed value of 600 V for the applied voltage gives

$Z_{EQ} = \dfrac{V_A}{I_T} = \dfrac{600 \text{ V}}{2.24 \text{ A}}$

$Z_{EQ} = 268 \text{ }\Omega$

The combined impedance of a 600-Ω R in parallel with a 300-Ω X_L is equal to 268 Ω, no matter how much the applied voltage is.

PHASE ANGLE WITH PARALLEL X_L AND R   In a parallel circuit, the phase angle is between the line current I_T and the common voltage V_A applied across all the branches. However, the resistive branch current I_R has the same phase as V_A. Therefore, the phase of I_R can be substituted for the phase of V_A. This is shown in Fig. 22-5c. The triangle of currents is in Fig. 22-6. To find θ_I from the branch currents, use the tangent formula

$\blacktriangleright \quad \tan \theta_I = -\dfrac{I_L}{I_R}$ \hfill **(22-6)**

We use the subscript I for θ to show that θ_I is found from the triangle of branch currents in a parallel circuit. In Fig. 22-6, θ_I is $-45°$ because I_L and I_R are equal. Then $\tan \theta_I = -1$.

The negative sign is used for this current ratio because I_L is lagging at $-90°$, compared with I_R. The phase angle of $-45°$ here means that I_T lags I_R and V_A by 45°.

Note that the phasor triangle of branch currents gives θ_I as the angle of I_T with respect to the generator voltage V_A. This phase angle for I_T is with respect to the applied voltage as the reference at 0°. For the phasor triangle of voltages in a series circuit, the phase angle θ_Z for Z_T and V_T is with respect to the series current as the reference phasor at 0°.

PARALLEL COMBINATIONS OF X_L AND R Several combinations of X_L and R in parallel are listed in Table 22-2. When X_L is 10 times R, the parallel circuit is practically resistive because there is little inductive current in the line. The small value of I_L results from the high X_L. The total impedance of the parallel circuit is approximately equal to the resistance, then, since the high value of X_L in a parallel branch has little effect. The phase angle of $-5.7°$ is practically $0°$ because almost all the line current is resistive.

TABLE 22-2 PARALLEL RESISTANCE AND INDUCTANCE COMBINATIONS*

R, Ω	X_L, Ω	I_R, A	I_L, A	I_T, A (APPROX.)	$Z_{EQ} = V_A/I_T$, Ω	PHASE ANGLE θ_I
1	10	10	1	$\sqrt{101} = 10$	1	$-5.7°$
10	10	1	1	$\sqrt{2} = 1.4$	7.07	$-45°$
10	1	1	10	$\sqrt{101} = 10$	1	$-84.3°$

*$V_A = 10$ V. Note that θ_I is the angle of I_T with respect to the reference V_A in parallel circuits.

As X_L becomes smaller, it provides more inductive current in the main line. When X_L is $\frac{1}{10} R$, practically all the line current is the I_L component. Then the parallel circuit is practically all inductive, with a total impedance practically equal to X_L. The phase angle of $-84.3°$ is almost $-90°$ because the line current is mostly inductive. Note that these conditions are opposite from the case of X_L and R in series.

When X_L and R are equal, their branch currents are equal and the phase angle is $-45°$. All these phase angles are negative for parallel I_L and I_R.

As additional comparisons between series and parallel circuits, remember that

1. The series voltage drops V_R and V_L have individual values that are $90°$ out of phase. Therefore, V_R and V_L are added by phasors to equal the applied voltage V_T. The phase angle θ_Z is between V_T and the common series current I. More series X_L allows more V_L to make the circuit more inductive with a larger positive phase angle for V_T with respect to I.

2. The parallel branch currents I_R and I_L have individual values that are $90°$ out of phase. Therefore, I_R and I_L are added by phasors to equal I_T, which is the main-line current. The negative phase angle $-\theta_I$ is between the line current I_T and the common parallel voltage V_A. Less parallel X_L allows more I_L to make the circuit more inductive with a larger negative phase angle for I_T with respect to V_A.

TEST-POINT QUESTION 22-4

Answers at end of chapter.
a. How much is I_T for a branch current I_R of 2 A and I_L of 2 A?
b. Find the phase angle θ_I.

22-5 Q OF A COIL

The ability of a coil to produce self-induced voltage is indicated by X_L, since it includes the factors of frequency and inductance. However, a coil has internal resistance equal to the resistance of the wire in the coil. This internal r_i of the coil reduces the current, which means less ability to produce induced voltage. Combining these two factors of X_L and r_i, the *quality* or *merit* of a coil is indicated by

▶ $$Q = \frac{X_L}{r_i} = \frac{2\pi f L}{r_i} \qquad\qquad (22\text{-}7)$$

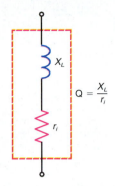

FIG. 22-7   The Q of a coil depends on its inductive reactance X_L and resistance r_i.

As shown in Fig. 22-7, the internal r_i is in series with X_L.

As an example, a coil with X_L of 500 Ω and r_i of 5 Ω has a Q of 500/5 = 100. The Q is a numerical value without any units, since the ohms cancel in the ratio of reactance to resistance. This Q of 100 means that the X_L of the coil is 100 times more than its r_i.

The Q of coils may range in value from less than 10 for a low-Q coil up to 1000 for a very high Q. Radio-frequency coils generally have a Q of about 30 to 300.

At low frequencies, r_i is just the dc resistance of the wire in the coil. However, for RF coils the losses increase with higher frequencies and the effective r_i increases. The increased resistance results from eddy currents and other losses.

Because of these losses, the Q of a coil does not increase without limit as X_L increases for higher frequencies. Generally, the Q can increase by a factor of about 2 for higher frequencies, within the range for which the coil is designed. The highest Q for RF coils generally results with an inductance value that provides an X_L of about 1000 Ω at the operating frequency.

More fundamentally, Q can be defined as the ratio of reactive power in the inductance to the real power dissipated in the resistance. Then

$$Q = \frac{P_L}{P_{r_i}} = \frac{I^2 X_L}{I^2 r_i} = \frac{X_L}{r_i} = \frac{2\pi f L}{r_i}$$

which is the same as Formula (22-7).

SKIN EFFECT   Radio-frequency current tends to flow at the surface of a conductor, at very high frequencies, with little current in the solid core at the center. This skin effect results from the fact that current in the center of the wire encounters slightly more inductance because of the magnetic flux concentrated in the metal, compared with the edges, where part of the flux is in air. For this reason, conductors for VHF currents are often made of hollow tubing. The skin effect increases the effective resistance, as a smaller cross-sectional area is used for the current path in the conductor.

AC EFFECTIVE RESISTANCE   When the power and current applied to a coil are measured for RF applied voltage, the $I^2 R$ loss corresponds to a much higher resistance than the dc resistance measured with an ohmmeter. This higher resistance is the ac effective resistance R_e. Although a result of high-frequency

alternating current, R_e is not a reactance; R_e is a resistive component because it draws in-phase current from the ac voltage source.

The factors that make the R_e of a coil more than its dc resistance include skin effect, eddy currents, and hysteresis losses. Air-core coils have low losses but are limited to small values of inductance.

For a magnetic core in RF coils, a powdered-iron or ferrite slug is generally used. In a powdered-iron slug, the granules of iron are insulated from each other to reduce eddy currents. Ferrite materials have small eddy-current losses, as they are insulators, although magnetic. A ferrite core is easily saturated. Therefore, its use must be limited to coils with low values of current. A common application is the ferrite-core antenna coil in Fig. 22-8.

To reduce the R_e for small RF coils, stranded wire can be made with separate strands insulated from each other and braided so that each strand is as much on the outer surface as all the other strands. This is called *litzendraht* or *litz wire*.

As an example of the total effect of ac losses, assume that an air-core RF coil of 50-μH inductance has a dc resistance of 1 Ω measured with the battery in an ohmmeter. However, in an ac circuit with a 2-MHz current, the effective coil resistance R_e can increase to 12 Ω. The increased resistance reduces the Q of the coil.

Actually, the Q can be used to determine the effective ac resistance. Since Q is X_L/R_e, then R_e equals X_L/Q. For this 50-μH L at 2 MHz, its X_L, equal to $2\pi fL$, is 628 Ω. The Q of the coil can be measured on a Q meter, which operates on the principle of resonance. Let the measured Q be 50. Then $R_e = 628/50$, equal to 12.6 Ω.

Example

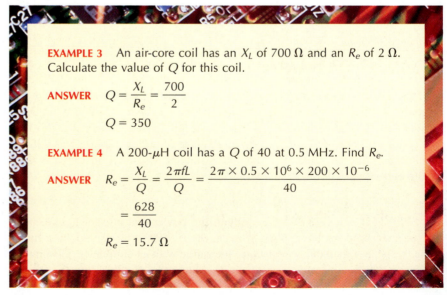

EXAMPLE 3 An air-core coil has an X_L of 700 Ω and an R_e of 2 Ω. Calculate the value of Q for this coil.

ANSWER $Q = \dfrac{X_L}{R_e} = \dfrac{700}{2}$

$Q = 350$

EXAMPLE 4 A 200-μH coil has a Q of 40 at 0.5 MHz. Find R_e.

ANSWER $R_e = \dfrac{X_L}{Q} = \dfrac{2\pi fL}{Q} = \dfrac{2\pi \times 0.5 \times 10^6 \times 200 \times 10^{-6}}{40}$

$= \dfrac{628}{40}$

$R_e = 15.7\ \Omega$

FIG. 22-8 Ferrite coil antenna for a radio receiver.

In general, the lower the internal resistance for a coil, the higher is its Q.

22-6 AF AND RF CHOKES

Inductance has the useful characteristic of providing more ohms of reactance at higher frequencies. Resistance has the same opposition at all frequencies and for direct current. The skin effect for L at very high frequencies is not being considered here. These characteristics of L and R are applied to the circuit in Fig. 22-9, where X_L is much greater than R for the frequency of the ac source V_T. The result is that L has practically all the voltage drop in this series circuit with very little of the applied voltage across R.

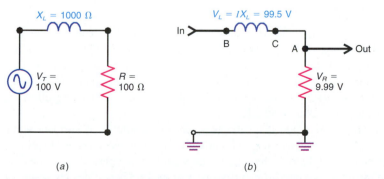

(a)   (b)

FIG. 22-9 Coil used as a choke with X_L at least $10 \times R$. Note that R is an external resistor; V_L across L is practically all of the applied voltage, with very little V_R. (a) Circuit with X_L and R in series. (b) Input and output voltages.

DID YOU KNOW?

In 35 seconds on July 2d, 1996, one tree caused 2 million people to lose power in fifteen U.S. states, two Canadian provinces, and one state in Mexico. The line, carrying 345,000 volts from a Wyoming plant, short-circuited.

The inductance L is used here as a *choke*. Therefore, a choke is an inductance in series with an external R to prevent the ac signal voltage from developing any appreciable output across R, at the frequency of the source.

The dividing line in calculations for a choke can be taken as X_L ten or more times the series R. Then the circuit is primarily inductive. Practically all the ac voltage drop is across L, with little across R. This case also results in θ of practically 90°, but the phase angle is not related to the action of X_L as a choke.

Fig. 22-9b illustrates how a choke is used to prevent ac voltage in the input from developing voltage in the output for the next circuit. Note that the output here is V_R from point A to chassis ground. Practically all the ac input voltage is across X_L between points B and C. However, this voltage is not coupled out because neither B nor C is grounded.

The desired output across R could be direct current from the input side without any ac component. Then X_L has no effect on the steady dc component. Practically all the dc voltage would be across R for the output, but the ac voltage would be just across X_L. The same idea applies to passing an AF signal through to R, while blocking an RF signal as IX_L across the choke because of more X_L at the higher frequency.

CALCULATIONS FOR A CHOKE Typical values for audio or radio frequencies can be calculated if we assume a series resistance of 100 Ω, as an example. Then X_L must be at least 1000 Ω. As listed in Table 22-3, at 100 Hz the relatively large inductance of 1.6 H provides 1000 Ω of X_L. Higher frequencies allow a smaller value of L for a choke with the same reactance. At 100 MHz, in the VHF range, the choke is only 1.6 μH.

TABLE 22-3   TYPICAL CHOKES FOR A REACTANCE OF 1000 Ω*

F	L	REMARKS
100 Hz	1.6 H	Low audio frequency
1000 Hz	0.16 H	Audio frequency
10 kHz	16 mH	Audio frequency
1000 kHz	0.16 mH	Radio frequency
100 MHz	1.6 μH	Very high radio frequency

*For an X_L that is 10 times a series R of 100 Ω.

Some typical chokes are shown in Fig. 22-10. The iron-core choke in Fig. 22-10a is for audio frequencies. The air-core choke in Fig. 22-10b is for radio frequencies. The RF choke in Fig. 22-10c has color coding, which is often used for small coils. The color values are the same as for resistors, except that the values of L are given in microhenrys. As an example, a coil with yellow, red, and black stripes or dots is 42 μH.

It should be noted that inductors are also available as surface-mount components. There are basically two body styles: completely encased and open. The encased body style looks like a thick capacitor with a black body. The open body style inductors are easy to identify because the coil is visible. The value of a surface-mount inductor, if marked, is usually represented using the same three-digit system used for resistors with the value displayed in microhenrys (μH).

(a)

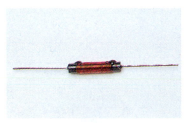

(b)

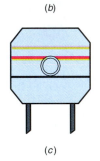

(c)

FIG. 22-10 Typical chokes. (a) Choke for 60 Hz with 8-H inductance and r_i of 350 Ω. Width is 2 in. (b) RF choke with 5 mH of inductance and r_i of 50 Ω. Height is 1 in. (c) Small RF choke encapsulated in plastic with leads for printed circuit board; $L = 42$ μH. Width is ¼ in.

CHOOSING A CHOKE FOR A CIRCUIT As an example of using these calculations, suppose that we have the problem of determining what kind of a coil to use as a choke for the following application. The L is to be an RF choke in series with an external R of 300 Ω, with a current of 90 mA and a frequency of 0.2 MHz. Then X_L must be at least $10 \times 300 = 3000$ Ω. At f of 0.2 MHz,

$$L = \frac{X_L}{2\pi f} = \frac{3{,}000}{2\pi \times 0.2 \times 10^6} = \frac{3 \times 10^3}{1.256 \times 10^6}$$

$$= \frac{3}{1.256} \times 10^{-3}$$

$$L = 2.4 \text{ mH}$$

A typical commercial size easily available is 2.5 mH, with a current rating of 115 mA and an internal resistance of 20 Ω, similar to the RF choke in Fig. 22-10b. Note that the higher current rating is suitable. Also, the internal resistance is negligible compared with the external R. An inductance a little higher than the calculated value will provide more X_L, which is better for a choke.

<hr>

TEST-POINT QUESTION 22-6

Answers at end of chapter.
a. How much is the minimum X_L for a choke in series with R of 80 Ω?
b. If X_L is 800 Ω at 3 MHz, how much will X_L be at 6 MHz for the same coil?

22-7 THE GENERAL CASE OF INDUCTIVE VOLTAGE

The voltage across any inductance in any circuit is always equal to $L(di/dt)$. This formula gives the instantaneous values of v_L based on the self-induced voltage produced by a change in magnetic flux associated with a change in current.

A sine waveform of current i produces a cosine waveform for the induced voltage v_L, equal to $L(di/dt)$. This means v_L has the same waveform as i, but they are 90° out of phase for sine-wave variations.

The inductive voltage can be calculated as IX_L in sine-wave ac circuits. Since X_L is $2\pi fL$, the factors that determine the induced voltage are included in the frequency and inductance. Usually, it is more convenient to work with IX_L for the inductive voltage in sine-wave ac circuits, instead of $L(di/dt)$.

However, with a nonsinusoidal current waveform, the concept of reactance cannot be used. The X_L applies only to sine waves. Then v_L must be calculated as $L(di/dt)$, which applies for any inductive voltage.

An example is illustrated in Fig. 22-11a for sawtooth current. This waveform is often used in the deflection circuits for the picture tube in television receivers. The sawtooth rise is a uniform or linear increase of current from zero to 90 mA in this example. The sharp drop in current is from 90 mA to zero. Note that the rise is relatively slow; it takes 90 μs. This is nine times longer than the fast drop in 10 μs.

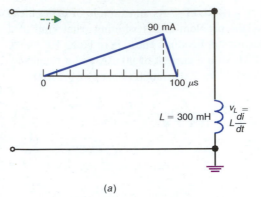
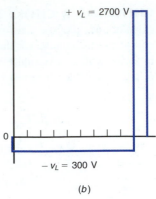

(a)   (b)

FIG. 22-11   Rectangular waveshape of V_L produced by sawtooth current through inductance L. (a) Waveform of current i. (b) Induced voltage V_i equal to $L(di/dt)$.

The complete period of one cycle of this sawtooth wave is 100 μs. A cycle includes the rise of i to the peak value and its drop back to the starting value.

THE SLOPE OF I   The slope of any curve is a measure of how much it changes vertically for each horizontal unit. In Fig. 22-11a the increase of current has a constant slope. Here i increases 90 mA in 90 μs, or 10 mA for every 10 μs of time. Then di/dt is constant at 10 mA/10 μs for the entire rise time of the sawtooth waveform. Actually di/dt is the slope of the i curve. The constant di/dt is why the v_L waveform has a constant value of voltage during the linear rise of i. Remember that the amount of induced voltage depends on the change in current with time.

The drop in i is also linear but much faster. During this time, the slope is 90 mA/10 μs for di/dt.

THE POLARITY OF v_L   In Fig. 22-11, apply Lenz' law to indicate that v_L opposes the change in current. With electron flow into the top of L, the v_L is negative to oppose an increase of current. This polarity opposes the direction of electron flow shown for the current i produced by the source. For the rise time, then, the induced voltage here is labeled $-v_L$.

During the drop of current, the induced voltage has opposite polarity, which is labeled $+v_L$. These voltage polarities are for the top of L with respect to chassis ground.

CALCULATIONS FOR v_L   The values of induced voltage across the 300-mH L are calculated as follows.

For the sawtooth rise:

$$-v_L = L\frac{di}{dt}$$

$$= 300 \times 10^{-3} \times \frac{10 \times 10^{-3}}{10 \times 10^{-6}}$$

$$-v_L = 300 \text{ V}$$

For the sawtooth drop:

$$+v_L = L\,\frac{di}{dt}$$

$$= 300 \times 10^{-3} \times \frac{90 \times 10^{-3}}{10 \times 10^{-6}}$$

$$+v_L = 2700 \text{ V}$$

The decrease in current produces nine times more voltage because the sharp drop in i is nine times faster than the relatively slow rise.

Remember that the di/dt factor can be very large, even with small currents, when the time is short. For instance, a current change of 1 mA in 1 μs is equivalent to the very high di/dt value of 1000 A/s.

An interesting feature of the inductive waveshapes in Fig. 22-11 is that they are the same as the capacitive waveshapes shown before in Fig. 19-7, but with current and voltage waveshapes interchanged. This comparison follows from the fact that both v_L and i_C depend on the rate of change. Then i_C is $C(dv/dt)$, and the v_L is $L(di/dt)$.

It is important to note that v_L and i_L have different waveshapes with nonsinusoidal current. In this case, we compare the waveshapes instead of the phase angle. Common examples of nonsinusoidal waveshapes for either v or i are the sawtooth waveform, square wave, and rectangular pulses. For a sine wave, the $L(di/dt)$ effects result in a cosine wave, as shown before, in Fig. 21-7.

<hr />

TEST-POINT QUESTION 22-7

Answers at end of chapter.

Refer to Fig. 22-11.
a. How much is di/dt in amperes per second for the sawtooth rise of i?
b. How much is di/dt in amperes per second for the drop in i?

22   SUMMARY AND REVIEW

- In a sine-wave ac circuit, the current through an inductance lags 90° behind the voltage across the inductance because $v_L = L(di/dt)$. This fundamental fact is the basis of all the following relations.
- Therefore, inductive reactance X_L is a phasor quantity 90° out of phase with R. The phasor combination of X_L and R is their impedance Z_T.
- These three types of opposition to current are compared in Table 22-4.

TABLE 22-4   COMPARISON OF R, X_L, AND Z_T

R	$X_L = 2\pi f L$	$Z_T = \sqrt{R^2 + X_L^2}$
Ohms unit	Ohms unit	Ohms unit
IR voltage in phase with I	IX_L voltage leads I by 90°	IZ is applied voltage; it leads line I by $\theta°$
Same for all frequencies	Increases as frequency increases	Increases with X_L at higher frequencies

- The phase angle θ is the angle between the applied voltage and its current.
- The opposite characteristics for series and parallel circuits with X_L and R are summarized in Table 22-5.

TABLE 22-5   SERIES AND PARALLEL RL CIRCUITS

X_L AND R IN SERIES	X_L AND R IN PARALLEL
I the same in X_L and R	V_A the same across X_L and R
$V_T = \sqrt{V_R^2 + V_L^2}$	$I_T = \sqrt{I_R^2 + I_L^2}$
$Z_T = \sqrt{R^2 + X_L^2}$	$Z_{EQ} = \dfrac{V_A}{I_T}$
V_L leads V_R by 90°	I_L lags I_R by 90°
$\tan \theta_Z = \dfrac{X_L}{R}$	$\tan \theta_I = -\dfrac{I_L}{I_R}$
The θ_Z increases with more X_L, which means more V_L, thus making the circuit more inductive	The $-\theta_I$ decreases with more X_L, which means less I_L, thus making the circuit less inductive

- The Q of a coil is X_L/r_i, where r_i is the coil's internal resistance.
- A choke is an inductance with X_L greater than the series R by a factor of 10 or more.
- In sine-wave circuits, $V_L = IX_L$. Then V_L is out of phase with I by an angle of 90°.
- For a circuit with X_L and R in in series, $\tan \theta_Z = X_L/R$. When the components are in parallel, $\tan \theta_I = -(I_L/I_R)$. See Table 22-5.
- When the current is not a sine wave, $v_L = L(di/dt)$. Then the waveshape of V_L is different from the waveshape of I.
- Inductors are available as surface-mount components. Surface-mount inductors are available in both completely encased and open body styles.

Choose (a), (b), (c), or (d).

1. In a sine-wave ac circuit with inductive reactance, the (a) phase angle of the circuit is always 45°; (b) voltage across the inductance must be 90° out of phase with the applied voltage; (c) current through the inductance lags its induced voltage by 90°; (d) current through the inductance and voltage across it are 180° out of phase.
2. In a sine-wave ac circuit with X_L and R in series, the (a) voltages across R and X_L are in phase; (b) voltages across R and X_L are 180° out of phase; (c) voltage across R lags the voltage across X_L by 90°; (d) voltage across R leads the voltage across X_L by 90°.
3. In a sine-wave ac circuit with a 40-Ω R in series with a 30-Ω X_L, the total impedance Z_T equals (a) 30 Ω; (b) 40 Ω; (c) 50 Ω; (d) 70 Ω.
4. In a sine-wave ac circuit with a 90-Ω R in series with a 90-Ω X_L, phase angle θ equals (a) 0°; (b) 30°; (c) 45°; (d) 90°.
5. A 250-μH inductance is used as a choke at 10 MHz. At 12 MHz the choke (a) does not have enough inductance; (b) has more reactance; (c) has less reactance; (d) needs more turns.
6. The combined impedance of a 1000-Ω R in parallel with a 1000-Ω X_L equals (a) 500 Ω; (b) 707 Ω; (c) 1000 Ω; (d) 2000 Ω.
7. A coil with a 1000-Ω X_L at 3 MHz and 10 Ω internal resistance has a Q of (a) 3; (b) 10; (c) 100; (d) 1000.
8. With a 2-A I_R and a 2-A I_L in parallel branches, I_T is (a) 1 A; (b) 2 A; (c) 2.8 A; (d) 4 A.
9. In Fig. 22-11 the di/dt for the drop in sawtooth current is (a) 90 mA/s; (b) 100 mA/s; (c) 100 A/s; (d) 9000 A/s.

QUESTIONS

1. What characteristic of the current in an inductance determines the amount of induced voltage? State briefly why.
2. Draw a schematic diagram showing an inductance connected across a sine-wave voltage source and indicate the current and voltage that are 90° out of phase with one another.

3. Why is the voltage across a resistance in phase with the current through the resistance?
4. (a) Draw the sine waveforms for two voltages 90° out of phase, each with a peak value of 100 V. (b) Why does their phasor sum equal 141 V and not 200 V? (c) When will the sum of two 100-V drops in series equal 200 V?
5. (a) Define the phase angle of a sine-wave ac circuit. (b) State the formula for the phase angle in a circuit with X_L and R in series.
6. Define the following: (a) Q of a coil; (b) ac effective resistance; (c) RF choke; (d) sawtooth current.
7. Why do the waveshapes in Fig. 22-2b all have the same frequency?
8. Describe how to check the trouble of an open choke with an ohmmeter.
9. Redraw the circuit and graph in Fig. 22-11 for a sawtooth current with a peak of 30 mA.
10. Why is the R_e of a coil considered resistance rather than reactance?
11. Why are RF chokes usually smaller than AF chokes?
12. What is the waveshape of v_L for a sine-wave i_L?

PROBLEMS

ANSWERS TO ODD-NUMBERED PROBLEMS AT BACK OF BOOK.

1. Draw the schematic diagram of a circuit with X_L and R in series across a 100-V source. Calculate Z_T, I, IR, IX_L, and θ for these values: (a) 100-Ω R, 1-Ω X_L; (b) 1-Ω R, 100-Ω X_L; (c) 50-Ω R, 50-Ω X_L.
2. Draw the schematic diagram of a circuit with X_L and R in parallel across a 100-V source. Calculate I_R, I_L, I_T, and Z_{EQ} for the following values: (a) 100-Ω R, 1-Ω X_L; (b) 1-Ω R, 100-Ω X_L; (c) 50-Ω R, 50-Ω X_L.
3. A coil has an inductance of 1 H and a 100-Ω internal resistance. (a) Draw the equivalent circuit of the coil showing its internal resistance in series with its inductance. (b) How much is the coil's inductive reactance at 60 Hz? (c) How much is the total impedance of the coil at 60 Hz? (d) How much current will flow when the coil is connected across a 120-V source with a frequency of 60 Hz? (e) How much is I with an f of 400 Hz?
4. Calculate the minimum inductance required for a choke in series with a resistance of 100 Ω when the frequency of the current is 5 kHz, 5 MHz, and 50 MHz. Do the same for the case where the series resistance is 10 Ω.
5. How much is the impedance Z of a coil that allows 0.3 A current when connected across a 120-V 60-Hz source? How much is the X_L of the coil if its resistance is 5 Ω? (Hint: $X_L^2 = Z^2 - R^2$.)
6. A 200-Ω R is in series with L across a 141-V 60-Hz generator V_T. The V_R is 100 V. Find L. (Hint: $V_L^2 = V_T^2 - V_R^2$.)
7. A 350-μH L has a Q of 35 at 1.5 MHz. Calculate the effective ac resistance R_e of the coil.
8. How much L is required to produce V_L equal to 9 kV when i_L drops from 300 mA to zero in 8 μs?
9. A 400-Ω R and 400-Ω X_L are in series with a 100-V 400-Hz source. Find Z_T, I, V_L, V_R, and θ_Z.

10. The same R and X_L of Prob. 9 are in parallel. Find I_R, I_L, I_T, Z_{EQ}, and θ_I.

11. The frequency is raised to 800 Hz for the circuit in Prob. 10. Compare the values of I_R, I_L, and θ_I for the two frequencies of 400 and 800 Hz.

12. A 0.4-H L and a 180-Ω R are in series across a 120-V 60-Hz source. Find the current I and θ_Z.

13. An inductance L has 20 V across it at 40 mA. The frequency is 5 kHz. Calculate X_L in ohms and L is in henrys.

14. A 500-Ω R is in series with 300-Ω X_L. Find Z_T, I, and θ_Z. $V_T = 120$ V.

15. A 300-Ω R is in series with a 500-Ω X_L. Find Z_T, I, and θ_Z. Compare θ_Z here with Prob. 14, with the same 120 V applied.

16. A 500-Ω R is in parallel with a 300-Ω X_L. Find I_T, Z_{EQ}, and θ_I. Compare θ_I here with θ_Z in Prob. 14 with the same 120 V applied.

17. The current shown in Fig. 22-12 flows through an 8-mH inductance. Show the corresponding waveform of induced voltage with values.

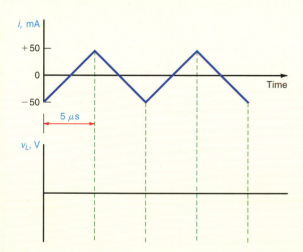

FIG. 22-12 Current waveform for Probs. 17 and 18.

18. Do the same as in Prob. 17 for a 2-mH inductance.

19. Find the angle θ_Z with the following combinations for X_L and R in series: **(a)** X_L is 120 Ω and R is 120 Ω; **(b)** X_L is 240 Ω and R is 120 Ω; **(c)** X_L is 1200 Ω and R is 120 Ω; **(d)** X_L is 120 Ω and R is 60 Ω.

20. What value of L is needed for X_L of 1200 Ω with f of 4 MHz?

21. Calculate the angle θ_I for I_L and I_R in parallel with the same combinations as in Prob. 19. Assume V_A of 1 V to determine I_L and I_R in mA units.

22. What is the frequency for X_L of 1200 Ω with a 50-mH coil?

23. In Fig. 22-13, calculate X_L, Z_T, I, V_L, V_R, and θ_Z.

24. Repeat Prob. 23 if $f = 1.591$ kHz.

25. In Fig. 22-14, calculate X_L, I_L, I_R, I_T, Z_{EQ}, and θ_I.

26. Repeat Prob. 25 if f is reduced to 3.183 kHz.

27. In Fig. 22-15, calculate L, Z_T, I, V_L, V_R, and θ_Z.

28. In Fig. 22-16, calculate X_L, I_L, I_R, R, I_T, and Z_{EQ}.

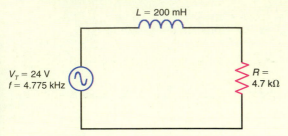

FIG. 22-13 Circuit for Probs. 23 and 24.

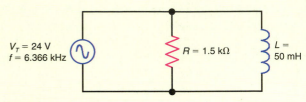

FIG. 22-14 Circuit for Probs. 25 and 26.

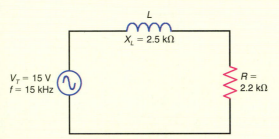

FIG. 22-15 Circuit for Prob. 27.

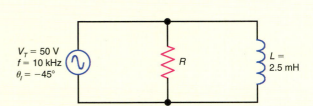

FIG. 22-16 Circuit for Prob. 28.

29. Calculate the minimum inductance L for a choke in series with a resistance of $1\text{ k}\Omega$ if the lowest frequency of the applied voltage is: (a) 2 kHz; (b) 100 kHz.

30. In Fig. 22-17, show the corresponding values of induced voltage for the waveform of current flowing through the 250-mH inductor.

CRITICAL THINKING

1. In Fig. 22-18, calculate X_L, R, L, I, V_L, and V_R.
2. In Fig. 22-19, calculate I_T, I_R, I_L, X_L, R, and L.
3. In Fig. 22-20, calculate V_R, V_{L_1}, X_{L_1}, X_{L_2}, I, Z_T, L_1, L_2, and θ_Z.

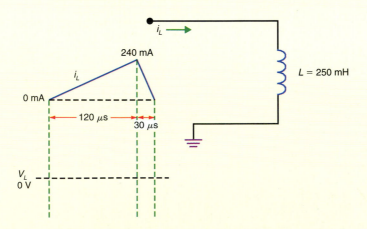

FIG. 22-17 Diagram for Prob. 30.

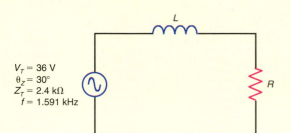

FIG. 22-18   Circuit for Critical Thinking Prob. 1.

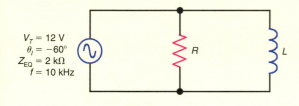

FIG. 22-19   Circuit for Critical Thinking Prob. 2.

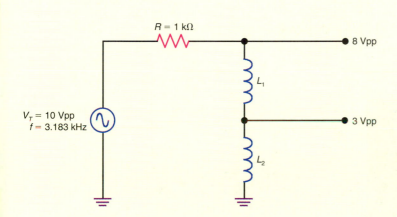

FIG. 22-20   Circuit for Critical Thinking Prob. 3.

ANSWERS TO TEST-POINT QUESTIONS

22-1 a. $0°$
  b. $90°$
  c. lag

22-2 a. $0°$
  b. $90°$

22-3 a. $28.28 \ \Omega$
  b. $28.28 \ V$
  c. $\theta_Z = 45°$

22-4 a. $I_T = 2.828 \ A$
  b. $\theta_I = -45°$

22-5 a. $Q = 75$
  b. $R_e = 10 \ \Omega$

22-6 a. $X_L = 800 \ \Omega$
  b. $X_L = 1600 \ \Omega$

22-7 a. $di/dt = 1000 \ A/s$
  b. $di/dt = 9000 \ A/s$

CHAPTER 23

RC AND *L/R* TIME CONSTANTS

Many applications of inductance are for sine-wave ac circuits, but any time the current changes, *L* has the effect of producing induced voltage. Examples of nonsinusoidal waveshapes include dc voltages that are switched on or off, square waves, sawtooth waves, and rectangular pulses. For capacitance, also, many applications are for sine waves, but any time the voltage changes, *C* produces charge or discharge current.

Actually, *RC* circuits are more common. The reasons are that capacitors are small, economical, and do not have strong magnetic fields.

With nonsinusoidal voltage and current, the effect of *L* or *C* is to produce a change in waveshape. This effect can be analyzed by means of the time constant for capacitive and inductive circuits. The time constant is the time for a change of 63.2 percent in the current through *L* or the voltage across *C*.

CHAPTER OBJECTIVES

Upon completion of this chapter, you should be able to:

- *Define* the term *transient response*.
- *Define* the term *time constant*.
- *Calculate* the time constant of a circuit containing resistance and inductance.
- *Explain* the effect of producing a high voltage when opening an *RL* circuit.
- *Calculate* the time constant of a circuit containing resistance and capacitance.
- *Explain* how capacitance opposes a change in voltage.
- *List* the criteria for proper differentiation and integration.
- *Explain* why a long time constant is required for an *RC* coupling circuit.
- *Use* the universal time constant graph.
- *Explain* the difference between time constants and reactance.

IMPORTANT TERMS IN THIS CHAPTER

collapsing magnetic field	*L/R* time constant	steady-state *I* or *V*
decay	long time constant	transient response
differentiation	rate of charge or discharge	voltage pulses
energy stored	*RC* time constant	waveshapes
integration	short time constant	

TOPICS COVERED IN THIS CHAPTER

23-1 RESPONSE OF RESISTANCE ALONE

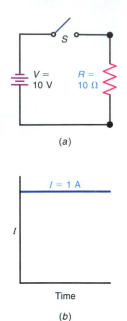

(a)

$I = 1\ A$

I

Time

(b)

FIG. 23-1 Response of circuit with R alone. When switch is closed, current I is 10 V/10 Ω = 1 A. (a) Circuit. (b) Graph of steady I.

In order to emphasize the special features of L and C, the circuit in Fig. 23-1 illustrates how an ordinary resistive circuit behaves. When the switch is closed, the battery supplies 10 V across the 10-Ω R and the resultant I is 1 A. The graph in Fig. 23-1b shows that I changes from 0 to 1 A instantly when the switch is closed. If the applied voltage is changed to 5 V, the current will change instantly to 0.5 A. If the switch is opened, I will immediately drop to zero.

Resistance has only opposition to current; there is no reaction to a change. The reason is that R has no concentrated magnetic field to oppose a change in I, like inductance, and no electric field to store charge that opposes a change in V, like capacitance.

TEST-POINT QUESTION 23-1

Answers at end of chapter.

Answer True or False.
a. Resistance R does not produce induced voltage for a change in I.
b. Resistance R does not produce charge or discharge current for a change in V.

23-2 *L/R* TIME CONSTANT

Consider the circuit in Fig. 23-2 where L is in series with R. When S is closed, the current changes as I increases from zero. Eventually, I will have the steady value of 1 A, equal to the battery voltage of 10 V divided by the circuit

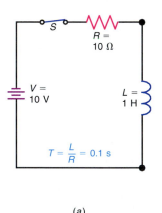

(a)

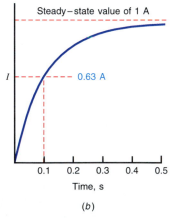

(b)

FIG. 23-2 Transient response of circuit with R and inductance L. When switch is closed, I rises from zero to the steady-state value of 1 A. (a) Circuit with time constant L/R of 1 H/10 Ω = 0.1 s. (b) Graph of I during five time constants. Compare with graph in Fig. 23-1b.

resistance of 10 Ω. While the current is building up from 0 to 1 A, however, *I* is changing and the inductance opposes the change. The action of the *RL* circuit during this time is its *transient response,* meaning a temporary condition existing only until the steady-state current of 1 A is reached. Similarly, when *S* is opened, the transient response of the *RL* circuit opposes the decay of current toward the steady-state value of zero.

The transient response is measured in terms of the ratio *L/R,* which is the time constant of an inductive circuit. To calculate the time constant

▶ $$T = \frac{L}{R} \qquad\qquad\qquad\qquad (23\text{-}1)$$

where *T* is the time constant in seconds, *L* is the inductance in henrys, and *R* is the resistance in ohms. The resistance is in series with *L,* being either the coil resistance, an external resistance, or both in series. In Fig. 23-2,

$$T = \frac{L}{R} = \frac{1}{10} = 0.1 \text{ s}$$

Specifically, the time constant is a measure of how long it takes the current to change by 63.2 percent, or approximately 63 percent. In Fig. 23-2, the current increases from 0 to 0.63 A, which is 63 percent of the steady-state value, in the period of 0.1 s, which is one time constant. In the period of five time constants, the current is practically equal to its steady-state value of 1 A.

The reason why *L/R* equals time can be illustrated as follows: Since induced voltage $V = L\,(di/dt)$, by transposing terms, *L* has the dimensions of $V \times T/I$. Dividing *L* by *R* results in $V \times T/IR$. As the *IR* and *V* factors cancel, *T* remains to indicate the dimension of time for the ratio *L/R.*

Example

EXAMPLE 1   What is the time constant of a 20-H coil having 100 Ω of series resistance?

ANSWER $T = \dfrac{L}{R} = \dfrac{20 \text{ H}}{100 \text{ Ω}}$

$T = 0.2 \text{ s}$

EXAMPLE 2   An applied dc voltage of 10 V will produce a steady-state current of 100 mA in the 100-Ω coil of Example 1. How much is the current after 0.2 s? After 1 s?

ANSWER   Since 0.2 s is one time constant, *I* then is 63 percent of 100 mA, which equals 63 mA. After five time constants, or 1 s (0.2 s × 5), the current will reach its steady-state value of 100 mA and remain at this value as long as the applied voltage stays at 10 V.

(*continued on next page*)

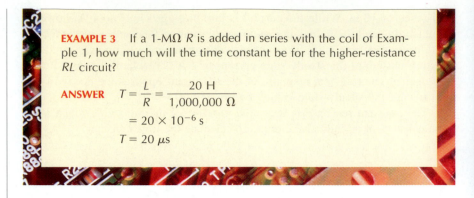

EXAMPLE 3 If a 1-MΩ R is added in series with the coil of Example 1, how much will the time constant be for the higher-resistance RL circuit?

ANSWER
$$T = \frac{L}{R} = \frac{20 \text{ H}}{1{,}000{,}000 \text{ }\Omega}$$
$$= 20 \times 10^{-6} \text{ s}$$
$$T = 20 \text{ }\mu\text{s}$$

The L/R time constant becomes longer with larger values of L. More series R, however, makes the time constant shorter. With more series resistance, the circuit is less inductive and more resistive.

TEST-POINT QUESTION 23-2

Answers at end of chapter.
a. Calculate the time constant for 2 H in series with 100 Ω.
b. Calculate the time constant for 2 H in series with 4000 Ω.

23-3 HIGH VOLTAGE PRODUCED BY OPENING AN *RL* CIRCUIT

When an inductive circuit is opened, the time constant for current decay becomes very short because L/R becomes smaller with the high resistance of the open circuit. Then the current drops toward zero much faster than the rise of current when the switch is closed. The result is a high value of self-induced voltage V_L across a coil whenever an RL circuit is opened. This high voltage can be much greater than the applied voltage.

There is no gain in energy, though, because the high-voltage peak exists only for the short time the current is decreasing at a very fast rate at the start of the decay. Then, as I decays with a slower rate of change, the value of V_L is reduced. After the current has dropped to zero, there is no voltage across L.

This effect can be demonstrated by a neon bulb connected across the coil, as shown in Fig. 23-3. The neon bulb requires 90 V for ionization, at which time it glows. The source here is only 8 V, but when the switch is opened, the self-induced voltage is high enough to light the bulb for an instant. The sharp voltage pulse or spike is more than 90 V just after the switch is opened, when I drops very fast at the start of the decay in current.

Note that the 100-Ω R_1 is the internal resistance of the 2-H coil. This resistance is in series with L whether S is closed or open. The 4-kΩ R_2 across the switch is in the circuit only when S is opened, in order to have a specific

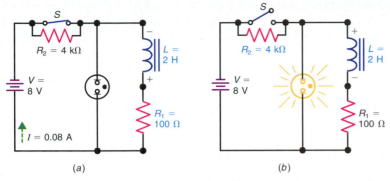

(a)   (b)

FIG. 23-3 Demonstration of high voltage produced by opening inductive circuit. (a) With switch closed, 8 V applied cannot light the 90-V neon bulb. (b) When switch is opened, the short L/R time constant results in high V_L, which lights the bulb.

resistance across the open switch. Since R_2 is much more than R_1, the L/R time constant is much shorter with the switch open.

CLOSING THE CIRCUIT   In Fig. 23-3a, the switch is closed to allow current in L and to store energy in the magnetic field. Since R_2 is short-circuited by the switch, the 100-Ω R_1 is the only resistance. The steady-state I is $V/R_1 = 8/100 = 0.08$ A. This value of I is reached after five time constants.

One time constant is $L/R = 2/100 = 0.02$ s. Five time constants equal $5 \times 0.02 = 0.1$ s. Therefore, I is 0.08 A after 0.1 s, or 100 ms. The energy stored in the magnetic field is 64×10^{-4} J, equal to $\frac{1}{2}LI^2$.

OPENING THE CIRCUIT   When the switch is opened in Fig. 23-3b, R_2 is in series with L, making the total resistance 4100 Ω, or approximately 4 kΩ. The result is a much shorter time constant for current decay. Then L/R is 2/4000, or 0.5 ms. The current decays practically to zero in five time constants, or 2.5 ms.

This rapid drop in current results in a magnetic field collapsing at a fast rate, inducing a high voltage across L. The peak v_L in this example is 320 V. Then v_L serves as the voltage source for the bulb connected across the coil. As a result, the neon bulb becomes ionized, and it lights for an instant. One problem produced when an inductive circuit is opened is arcing. Arcing can destroy contact points and under certain conditions cause fires or explosions.

CALCULATING THE PEAK OF v_L   The value of 320 V for the peak induced voltage when S is opened in Fig. 23-3 can be determined as follows: With the switch closed, I is 0.08 A in all parts of the series circuit. The instant S is opened, R_2 is added in series with L and R_1. The energy stored in the magnetic field maintains I at 0.08 A for an instant before the current decays. With 0.08 A in the 4-kΩ R_2 its potential difference is $0.08 \times 4000 = 320$ V. The collapsing magnetic field induces this 320-V pulse to allow an I of 0.08 A at the instant the switch is opened.

THE di/dt FOR v_L   The required rate of change in current is 160 A/s for the v_L of 320 V induced by the L of 2 H. Since $v_L = L\,(di/dt)$, this formula can be transposed to specify di/dt as equal to v_L/L. Then di/dt corresponds to 320 V/2 H, or 160 A/s. This value is the actual di/dt at the start of the decay in current when the switch is opened in Fig. 23-3b, as a result of the short time constant.*

APPLICATIONS OF INDUCTIVE VOLTAGE PULSES   There are many uses of the high voltage generated by opening an inductive circuit. One example is the high voltage produced for the ignition system in an automobile. Here the circuit of the battery in series with a high-inductance spark coil is opened by the breaker points of the distributor to produce the high voltage needed for each spark plug. By opening an inductive circuit very rapidly, 10,000 V can easily be produced. Another important application is the high voltage of 10 to 30 kV for the anode of the picture tube in television receivers.

TEST-POINT QUESTION 23-3

Answers at end of chapter.
a. Is the L/R time constant longer or shorter in Fig. 23-3 when S is opened?
b. Which produces more v_L, a faster di/dt or a slower di/dt?

23-4 *RC* TIME CONSTANT

For capacitive circuits, the transient response is measured in terms of the product $R \times C$. To calculate the time constant,

▶ $$T = R \times C \qquad\qquad (23\text{-}2)$$

where R is in ohms, C is in farads, and T is in seconds. In Fig. 23-4, for example, with an R of 3 MΩ and a C of 1 μF,

$$T = 3 \times 10^6 \times 1 \times 10^{-6}$$
$$T = 3 \text{ s}$$

Note that the 10^6 for megohms and the 10^{-6} for microfarads cancel. Therefore, multiplying the units of M$\Omega \times \mu$F gives the RC product in seconds.

Common combinations of units for the RC time constant are

$$\text{M}\Omega \times \mu\text{F} = \text{s}$$
$$\text{k}\Omega \times \mu\text{F} = \text{ms}$$
$$\text{M}\Omega \times \text{pF} = \mu\text{s}$$

The reason why the RC product corresponds to time can be illustrated as follows: $C = Q/V$. The charge Q is the product of $I \times T$. The factor V is IR. Therefore, RC is equivalent to $(R \times Q)/V$, or $(R \times IT)/IR$. Since I and R cancel, T remains to indicate the dimension of time.

---

*The di/dt value can be calculated from the slope at the start of decay, shown by the dashed line for curve b in Fig. 23-9.

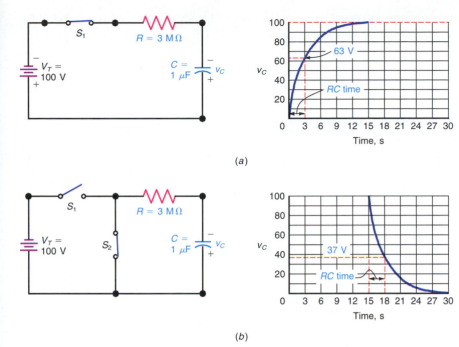

(a)

(b)

FIG. 23-4 Details of how a capacitor charges and discharges in an *RC* circuit. (*a*) With S_1 closed, *C* charges through *R* to 63 percent of V_T in one *RC* time constant of 3 s and is almost completely charged in five time constants. (*b*) With S_1 opened to disconnect the battery and S_2 closed for *C* to discharge through *R*, V_C drops to 37 percent of its initial voltage in one time constant of 3 s and is almost completely discharged in five time constants.

THE TIME CONSTANT INDICATES THE RATE OF CHARGE OR DISCHARGE On charge, *RC* specifies the time it takes *C* to charge to 63 percent of the charging voltage. Similarly, on discharge, *RC* specifies the time it takes *C* to discharge 63 percent of the way down, to the value equal to 37 percent of the initial voltage across *C* at the start of discharge.

In Fig. 23-4*a*, for example, the time constant on charge is 3 s. Therefore, in 3 s, *C* charges to 63 percent of the 100 V applied, reaching 63 V in *RC* time. After five time constants, which is 15 s here, *C* is almost completely charged to the full 100 V applied. If *C* discharges after being charged to 100 V, then *C* will discharge down to 36.8 V or approximately 37 V in 3 s. After five time constants, *C* discharges down to zero.

A shorter time constant allows the capacitor to charge or discharge faster. If the *RC* product in Fig. 23-4 is 1 s, then *C* will charge to 63 V in 1 s instead of 3 s. Also, v_C will reach the full applied voltage of 100 V in 5 s instead of 15 s. Charging to the same voltage in less time means a faster charge.

On discharge also, the shorter time constant will allow *C* to discharge from 100 to 37 V in 1 s instead of 3 s. Also, v_C will be down to zero in the time of 5 s instead of 15 s.

For the opposite case, a longer time constant means slower charge or discharge of the capacitor. More *R* or *C* results in a longer time constant.

RC APPLICATIONS   Several examples are given here to illustrate how the time constant can be applied to *RC* circuits.

Example

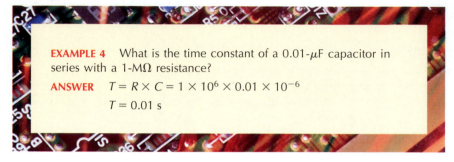

EXAMPLE 4   What is the time constant of a 0.01-μF capacitor in series with a 1-MΩ resistance?

ANSWER $T = R \times C = 1 \times 10^6 \times 0.01 \times 10^{-6}$

$\qquad\qquad T = 0.01$ s

This is the time constant for charging or discharging, assuming the series resistance is the same for charge or discharge.

Example

EXAMPLE 5   With a dc voltage of 300 V applied, how much is the voltage across *C* in Example 4 after 0.01 s of charging? After 0.05 s? After 2 hours? After 2 days?

ANSWER   Since 0.01 s is one time constant, the voltage across *C* then is 63 percent of 300 V, which equals 189 V. After five time constants, or 0.05 s, *C* will be charged practically to the applied voltage of 300 V. After 2 hours or 2 days *C* will still be charged to 300 V if the applied voltage is still connected.

EXAMPLE 6   If the capacitor in Example 5 is allowed to charge to 300 V and then discharged, how much is the capacitor voltage 0.01 s after the start of discharge? The series resistance is the same on discharge as on charge.

ANSWER   In one time constant *C* discharges to 37 percent of its initial voltage, or 0.37 × 300 V, which equals 111 V.

EXAMPLE 7   If the capacitor in Example 5 is made to discharge after being charged to 200 V, how much will the voltage across *C* be 0.01 s later? The series resistance is the same on discharge as on charge.

ANSWER   In one time constant *C* discharges to 37 percent of its initial voltage, or 0.37 × 200, which equals 74 V.

This example shows that the capacitor can charge or discharge from any voltage value. The rate at which it charges or discharges is determined by *RC* counting from the time the charge or discharge starts.

The RC time constant becomes longer with larger values of R and C. More capacitance means that the capacitor can store more charge. Therefore, it takes longer to store the charge needed to provide a potential difference equal to 63 percent of the applied voltage. More resistance reduces the charging current, requiring more time for charging the capacitor.

It should be noted that the RC time constant specifies just a rate. The actual amount of voltage across C depends upon the applied voltage as well as upon the RC time constant.

The capacitor takes on charge whenever its voltage is less than the applied voltage. The charging continues at the RC rate until either the capacitor is completely charged, the applied voltage decreases, or the voltage is disconnected.

The capacitor discharges whenever its voltage is more than the applied voltage. The discharge continues at the RC rate until either the capacitor is completely discharged, the applied voltage increases, or the load is disconnected.

To summarize these two important principles:

1. Capacitor C charges when the net charging voltage is more than v_C.
2. Capacitor C discharges when v_C is more than the net charging voltage.

The net charging voltage equals the difference between v_C and the applied voltage.

TEST-POINT QUESTION 23-4

Answers at end of chapter.
a. How much is the RC time constant for 470 pF in series with 2 MΩ on charge?
b. How much is the RC time constant for 470 pF in series with 1 kΩ on discharge?

23-5 *RC* CHARGE AND DISCHARGE CURVES

In Fig. 23-4, the RC charge curve has the rise shown because the charging is fastest at the start, then tapers off as C takes on additional charge at a slower rate. As C charges, its potential difference increases. Then the difference in voltage between V_T and v_C is reduced. Less potential difference reduces the current that puts the charge in C. The more C charges, the more slowly it takes on additional charge.

Similarly, on discharge, C loses its charge at a slower rate. At the start of discharge, v_C has its highest value and can produce maximum discharge current. With the discharge continuing, v_C goes down and there is less discharge current. The more C discharges, the more slowly it can lose the remainder of its charge.

CHARGE AND DISCHARGE CURRENT   There is often the question of how current can flow in a capacitive circuit with a battery as the dc source. The answer is that current flows any time there is a change in voltage. When V_T is connected, the applied voltage changes from zero. Then charging current flows to charge C to the applied voltage. After v_C equals V_T, there is no net charging voltage and I is zero.

Similarly, C can produce discharge current any time v_C is greater than V_T. When V_T is disconnected, v_C can discharge down to zero, producing discharge current in the opposite direction from the charging current. After v_C equals zero, there is no current.

CAPACITANCE OPPOSES VOLTAGE CHANGES ACROSS ITSELF
This ability corresponds to the ability of inductance to oppose a change of current. In terms of the RC circuit, when the applied voltage increases, the voltage across the capacitance cannot increase until the charging current has stored enough charge in C. The increase in applied voltage is present across the resistance in series with C until the capacitor has charged to the higher applied voltage. When the applied voltage decreases, the voltage across the capacitor cannot go down immediately because the series resistance limits the discharge current.

The voltage across the capacitance in an RC circuit, therefore, cannot follow instantaneously the changes in applied voltage. As a result, the capacitance is able to oppose changes in voltage across itself. The instantaneous variations in V_T are present across the series resistance, however, since the series voltage drops must add to equal the applied voltage at all times.

TEST-POINT QUESTION 23-5

Answers at end of chapter.
a. From the curve in Fig. 23-4a, how much is v_C after 3 s of charge?
b. From the curve in Fig. 23-4b, how much is v_C after 3 s of discharge?

23-6 HIGH CURRENT PRODUCED BY SHORT-CIRCUITING RC CIRCUIT

Specifically, a capacitor can be charged slowly with a small charging current through a high resistance and then discharged fast through a low resistance to obtain a momentary surge, or pulse, of discharge current. This idea corresponds to the pulse of high voltage obtained by opening an inductive circuit.

The circuit in Fig. 23-5 illustrates the application of a battery-capacitor (BC) unit to fire a flash bulb for cameras. The flash bulb needs 5 A to ignite, but this

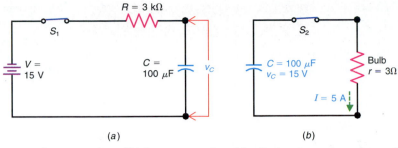

FIG. 23-5 Demonstration of high current produced by discharging a charged capacitor through a low resistance. (a) When S_1 is closed, C charges to 15 V through 3 kΩ. (b) Without the battery, S_2 is closed to allow V_C to produce the peak discharge current of 5 A through the 3-Ω bulb. V_C in (b) is across the same C used in (a).

is too much load current for the small 15-V battery that has a rating of 30 mA for normal load current. Instead of using the bulb as a load for the battery, though, the 100-μF capacitor is charged by the battery through the 3-kΩ R in Fig. 23-5a, and then the capacitor is discharged through the bulb in Fig. 23-5b.

CHARGING THE CAPACITOR

In Fig. 23-5a, S_1 is closed to charge C through the 3-kΩ R without the bulb. The time constant of the RC charging circuit is 0.3 s.

After five time constants, or 1.5 s, C is charged to the 15 V of the battery. The peak charging current, at the first instant of charge, is V/R or 15 V/3 kΩ, which equals 5 mA. This value is an easy load current for the battery.

DISCHARGING THE CAPACITOR

In Fig. 23-5b, v_C is 15 V without the battery. Now S_2 is closed, and C discharges through the 3-Ω resistance of the bulb. The time constant for discharge with the lower r of the bulb is $3 \times 100 \times 10^{-6}$, which equals 300 μs. At the first instant of discharge, when v_C is 15 V, the peak discharge current is 15/3, which equals 5 A. This current is enough to fire the bulb.

ENERGY STORED IN C

When the 100-μF C is charged to 15 V by the battery, the energy stored in the electric field is $CV^2/2$, which equals 0.01 J, approximately. This energy is available to maintain v_C at 15 V for an instant when the switch is closed. The result is the 5-A I through the 3-Ω r of the bulb at the start of the decay. Then v_C and i_C drop to zero in five time constants.

THE dv/dt FOR i_C

The required rate of change in voltage is 0.05×10^6 V/s for the discharge current i_C of 5 A produced by the C of 100 μF. Since $i_C = C(dv/dt)$, this formula can be transposed to specify dv/dt as equal to i_C/C. Then dv/dt corresponds to 5 A/100 μF, or 0.05×10^6 V/s. This value is the actual dv/dt at the start of discharge when the switch is closed in Fig. 23-5b. The dv/dt is high because of the short RC time constant.*

*See footnote on p. 636.

Answers at end of chapter.

a. Is the RC time constant longer or shorter in Fig. 23-5b compared with Fig. 23-5a?

b. Which produces more i_C, a faster dv/dt or a slower dv/dt?

23-7 *RC* WAVESHAPES

The voltage and current waveshapes in an RC circuit are shown in Fig. 23-6 for the case where a capacitor is allowed to charge through a resistance for RC time and then discharge through the same resistance for the same amount of time. It should be noted that this particular case is not typical of practical RC circuits, but the waveshapes show some useful details about the voltage and current for charging and discharging. The RC time constant here equals 0.1 s to simplify the calculations.

SQUARE WAVE OF APPLIED VOLTAGE   The idea of closing S_1 to apply 100 V and then opening it to disconnect V_T at a regular rate corresponds to a square wave of applied voltage, as shown by the waveform in Fig. 23-6a. When S_1 is closed for charge, S_2 is open; when S_1 is open, S_2 is closed for discharge. Here the voltage is on for the RC time of 0.1 s and off for the same time of 0.1 s. The period of the square wave is 0.2 s, and f is 1/0.2 s, which equals 5 Hz for the frequency.

CAPACITOR VOLTAGE v_C   As shown in Fig. 23-6b, the capacitor charges to 63 V, equal to 63 percent of the charging voltage, in the RC time of 0.1 s. Then the capacitor discharges because the applied V_T drops to zero. As a result, v_C drops to 37 percent of 63 V, or 23.3 V in RC time.

The next charge cycle begins with v_C at 23.3 V. The net charging voltage now is $100 - 23.3 = 76.7$ V. The capacitor voltage increases by 63 percent of 76.7 V, or 48.3 V. Adding 48.3 V to 23.3 V, then v_C rises to 71.6 V. On discharge, after 0.3 s, v_C drops to 37 percent of 71.6 V, or to 26.5 V.

CHARGE AND DISCHARGE CURRENT   As shown in Fig. 23-6c, the current i has its positive peak at the start of charge and its negative peak at the start of discharge. On charge, i is calculated as the net charging voltage, which is $(V_T - v_C)$, divided by R. On discharge, i always equals the value of v_C/R.

At the start of charge, i is maximum because the net charging voltage is maximum before C charges. Similarly, the peak i for discharge occurs at the start when v_C is maximum before C discharges.

Note that i is actually an ac waveform around the zero axis, since the charge and discharge currents are in opposite directions. We are arbitrarily taking the charging current as positive values for i.

DID YOU KNOW?

Electric power is provided to homes from a group of interconnected regional plants. This makes it necessary that the 60-cycle frequency of all the generators be synchronized. Your power may arrive at your home via a loop hundreds of miles long rather than by the shortest distance.

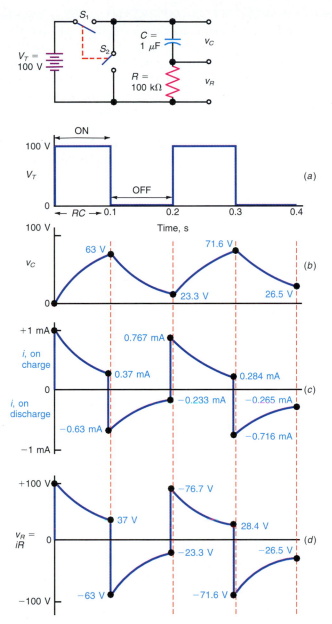

FIG. 23-6 Waveshapes for the charge and discharge of an RC circuit in RC time. Circuit above with S_1 and S_2 provides the square wave of applied voltage.

RESISTOR VOLTAGE v_R   This waveshape in Fig. 23-6d follows the waveshape of current, as v_R is $i \times R$. Because of the opposite directions of charge and discharge current, the iR waveshape is an ac voltage.

Note that on charge v_R must always be equal to $V_T - v_C$ because of the series circuit.

On discharge v_R has the same values as v_C because they are parallel, without V_T. Then S_2 is closed to connect R across C.

WHY THE i_C WAVESHAPE IS IMPORTANT   The v_C waveshape of capacitor voltage in Fig. 23-6 shows the charge and discharge directly, but the i_C waveshape is very interesting. First, the voltage waveshape across R is the same as the i_C waveshape. Also, whether C is charging or discharging, the i_C waveshape is really the same except for the reversed polarity. We can see the i_C waveshape as the voltage across R. It generally is better to connect an oscilloscope for voltage waveshapes across R, especially with one side grounded.

Finally, we can tell what v_C is from the v_R waveshape. The reason is that at any instant of time V_T must equal the sum of v_R and v_C. Therefore v_C is equal to $V_T - v_R$, when V_T is charging C. For the case when C is discharging, there is no V_T. Then v_R is the same as v_C.

TEST POINT QUESTION 23-7

Answers at end of chapter.

Refer to Fig. 23-6.
a. When v_C is 63 V, how much is v_R?
b. When v_R is 76.7 V, how much is v_C?

23-8 LONG AND SHORT TIME CONSTANTS

Useful waveshapes can be obtained by using RC circuits with the required time constant. In practical applications, RC circuits are used more than RL circuits because almost any value of an RC time constant can be obtained easily. With coils, the internal series resistance cannot be short-circuited and the distributed capacitance often causes resonance effects.

LONG RC TIME   Whether an RC time constant is long or short depends on the pulse width of the applied voltage. We can arbitrarily define a long time constant as at least five times longer than the pulse width, in time, for the applied voltage. As a result, C takes on very little charge. The time constant is too long for v_C to rise appreciably before the applied voltage drops to zero and C must discharge. On discharge also, with a long time constant, C discharges very little before the applied voltage rises to make C charge again.

SHORT RC TIME   A short time constant can be defined as no more than one-fifth the pulse width, in time, for the applied voltage V_T. Then V_T is applied for a period of at least five time constants, allowing C to become completely charged. After C is charged, v_C remains at the value of V_T, while the voltage is applied. When V_T drops to zero, C discharges completely in five time constants and remains at zero while there is no applied voltage. On the next cycle, C charges and discharges completely again.

DIFFERENTIATION   The voltage across R in an RC circuit is called a *differentiated output* because v_R can change instantaneously. A short time constant is always used for differentiating circuits to provide sharp pulses of v_R.

INTEGRATION   The voltage across C is called an *integrated output* because it must accumulate over a period of time. A medium or long time constant is always used for integrating circuits.

23-9 CHARGE AND DISCHARGE WITH SHORT *RC* TIME CONSTANT

Usually, the time constant is made much shorter or longer than the factor of 5, to obtain better waveshapes. In Fig. 23-7, RC is 0.1 ms. The frequency for the square wave is 25 Hz, with a period of 0.04 s, or 40 ms. One-half this period is the time V_T is applied. Therefore, the applied voltage is on for 20 ms and off for 20 ms. The RC time constant of 0.1 ms is shorter than the pulse width of 20 ms by a factor of $\frac{1}{200}$. Note that the time axis of all the waveshapes is calibrated in seconds for the period of V_T, not in RC time constants.

SQUARE WAVE OF V_T IS ACROSS C   The waveshape of v_C in Fig. 23-7b is essentially the same as the square wave of applied voltage. The reason is that the short time constant allows C to charge or discharge completely very soon after V_T is applied or removed. The charge or discharge time of five time constants is much less than the pulse width.

SHARP PULSES OF i   The waveshape of i shows sharp peaks for the charge or discharge current. Each current peak is $V_T/R = 1$ mA, decaying to zero in five RC time constants. These pulses coincide with the leading and trailing edges of the square wave of V_T.

   Actually, the pulses are much sharper than shown. They are not to scale horizontally in order to indicate the charge and discharge action. Also, v_C is actually a square wave like the applied voltage but with slightly rounded corners for the charge and discharge.

SHARP PULSES OF v_R   The waveshape of voltage across the resistor follows the current waveshape, as $v_R = iR$. Each current pulse of 1 mA across the 100-kΩ R results in a voltage pulse of 100 V.

   More fundamentally, the peaks of v_R equal the applied voltage V_T before C charges. Then v_R drops to zero as v_C rises to the value of V_T.

   On discharge, $v_R = v_C$, which is 100 V at the start of discharge. Then the pulse drops to zero in five time constants. The pulses of v_R in Fig. 23-7 are

ABOUT ELECTRONICS

In teleassistance, robots learn how to perform tasks as they "watch" and match the movements of a person wearing a virtual-reality glove.

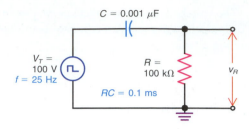

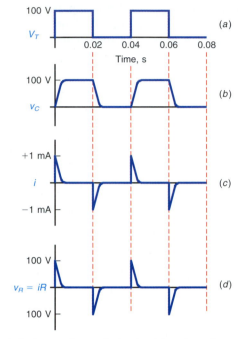

FIG. 23-7 Charge and discharge of an *RC* circuit with a short time constant. Note that the waveshape of V_R in (*d*) has sharp voltage peaks for the leading and trailing edges of the square-wave applied voltage.

useful as timing pulses that match the edges of the square-wave applied voltage V_T. Either the positive or the negative pulses can be used.

The *RC* circuit in Fig. 23-7*a* is a good example of an *RC* differentiator. With the *RC* time constant much shorter than the pulse width of V_T, the voltage V_R follows instantaneously the changes in the applied voltage. Keep in mind that a differentiator must have a short time constant with respect to the pulse width of V_T to provide good differentiation.

TEST-POINT QUESTION 23-9

Answers at end of chapter.

Refer to Fig. 23-7.
a. Is the time constant here short or long?
b. Is the square wave of applied voltage across *C* or *R*?

23-10 LONG TIME CONSTANT FOR *RC* COUPLING CIRCUIT

The *RC* circuit in Fig. 23-8 is the same as in Fig. 23-7, but now the *RC* time constant is long because of the higher frequency of the applied voltage. Specifically, the *RC* time of 0.1 ms is 200 times longer than the 0.5-μs pulse width of V_T with a frequency of 1 MHz. Note that the time axis is calibrated in microseconds for the period of V_T, not in *RC* time constants.

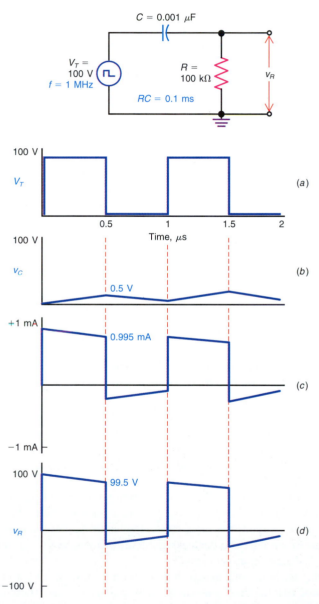

FIG. 23-8 Charge and discharge of an *RC* circuit with a long time constant. Note that the waveshape of v_R in (*d*) has essentially the same waveform as the applied voltage.

VERY LITTLE OF V_T IS ACROSS C   The waveshape of v_C in Fig. 23-8b shows very little voltage rise because of the long time constant. During the 0.5 μs that V_T is applied, C charges to only $\frac{1}{200}$ of the charging voltage. On discharge, also, v_C drops very little.

SQUARE WAVE OF i   The waveshape of i stays close to the 1-mA peak at the start of charge. The reason is that v_C does not increase much, allowing V_T to maintain the charging current. On discharge, the reverse i for discharge current is very small because v_C is low.

SQUARE WAVE OF V_T IS ACROSS R   The waveshape of v_R is the same square wave as i, as $v_R = iR$. Actually, the waveshapes of i and v_R are essentially the same as the square-wave V_T applied. They are not shown to scale vertically in order to indicate the slight charge and discharge action.

Eventually, v_C will climb to the average dc value of 50 V, i will vary ± 0.5 mA above and below zero, while v_R will vary ± 50 V above and below zero. This application is an RC coupling circuit to block the average value of the varying dc voltage V_T as the capacitive voltage v_C, while v_R provides an ac voltage output having the same variations as V_T.

If the output is taken across C rather than R in Fig. 23-8a, the circuit is classified as an RC integrator. Looking at Fig. 23-8b, it can be seen that C combines, or integrates its original voltage with the new change in voltage. Eventually, however, the voltage across C will reach a steady-state value of 50 V after the input waveform has been applied for approximately five RC time constants. Keep in mind that an integrator must have a long time constant with respect to the pulse width of V_T to provide good integration.

TEST-POINT QUESTION 23-10

Answers at end of chapter.

Refer to Fig. 23-8.
a. Is the RC time constant here short or long?
b. Is the square wave of applied voltage across R or C?

23-11 UNIVERSAL TIME CONSTANT GRAPH

We can determine transient voltage and current values for any amount of time, with the curves in Fig. 23-9. The rising curve a shows how v_C builds up as C charges in an RC circuit; the same curve applies to i_L, increasing in the inductance for an RL circuit. The decreasing curve b shows how v_C drops as C discharges or i_L decays in an inductance.

Note that the horizontal axis is in units of time constants rather than absolute time. Suppose that the time constant of an RC circuit is 5 μs. Therefore, one RC time unit = 5 μs, two RC units = 10 μs, three RC units = 15 μs, four RC units = 20 μs, and five RC units = 25 μs.

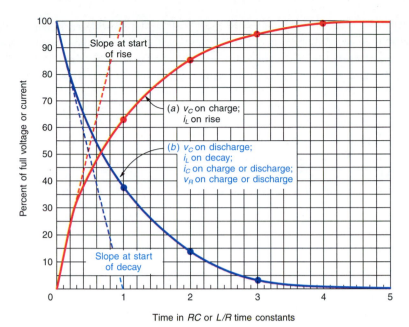

FIG. 23-9 Universal time-constant chart for *RC* and *RL* circuits. The rise or fall changes by 63 percent in one time constant.

As an example, to find v_C after 10 μs of charging, we can take the value of curve *a* in Fig. 23-9 at two *RC*. This point is at 86 percent amplitude. Therefore, we can say that in this *RC* circuit with a time constant of 5 μs, v_C charges to 86 percent of the applied V_T, after 10 μs. Similarly, some important values that can be read from the curve are listed in Table 23-1.

If we consider curve *a* in Fig. 23-9 as an *RC* charge curve, v_C adds 63 percent of the net charging voltage for each additional unit of one time constant, although it may not appear so. For instance, in the second interval of *RC* time, v_C adds 63 percent of the net charging voltage, which is 0.37 V_T. Then 0.63 × 0.37 equals 0.23, which is added to 0.63 to give 0.86, or 86 percent, as the total charge from the start.

TABLE 23-1   TIME CONSTANT FACTORS

FACTOR	AMPLITUDE
0.2 time constant	20%
0.5 time constant	40%
0.7 time constant	50%
1 time constant	63%
2 time constants	86%
3 time constants	96%
4 time constants	98%
5 time constants	99%

SLOPE AT $T = 0$ The curves in Fig. 23-9 can be considered linear for the first 20 percent of change. In 0.1 time constant, for instance, the change in amplitude is 10 percent; in 0.2 time constant, the change is 20 percent. The dashed lines in Fig. 23-9 show that if this constant slope continued, the result would be 100 percent charge in one time constant. This does not happen, though, because the change is opposed by the energy stored in L and C. However, at the first instant of rise or decay, at $t = 0$, the change in v_C or i_L can be calculated from the dotted slope line.

EQUATION OF THE DECAY CURVE   The rising curve a in Fig. 23-9 may seem more interesting because it describes the buildup of v_C or i_L, but the decaying curve b is more useful. For RC circuits, curve b can be applied to

1. v_C on discharge
2. i and v_R on charge or discharge

If we use curve b for the voltage in RC circuits, the equation of this decay curve can be written as

▶ $$v = V \times \epsilon^{-t/RC} \qquad\qquad (23\text{-}3)$$

where V is the voltage at the start of decay and v is the instantaneous voltage after the time t. Specifically, v can be v_R on charge and discharge, or v_C only on discharge.

The constant ϵ is the base 2.718 for natural logarithms.* The negative exponent $-t/RC$ indicates a declining exponential or logarithmic curve. The value of t/RC is the ratio of actual time of decline t to the RC time constant.

This equation can be converted to common logarithms for easier calculations. Since the natural base ϵ is 2.718, its logarithm to base 10 equals 0.434. Therefore, the equation becomes

▶ $$v = \text{antilog} \left(\log V - 0.434 \times \frac{t}{RC} \right) \qquad\qquad (23\text{-}4)$$

CALCULATIONS FOR v_R As an example, let us calculate v_R dropping from 100 V, after RC time. Then the factor t/RC is 1. Substituting these values,

$$v_R = \text{antilog} \,(\log 100 - 0.434 \times 1)$$
$$= \text{antilog} \,(2 - 0.434)$$
$$= \text{antilog} \, 1.566$$
$$v_R = 37 \text{ V}$$

All these logs are to base 10. Note that log 100 is taken first so that 0.434 can be subtracted from 2 before the antilog of the difference is found. The antilog of 1.566 is 37.

We can also use V_R to find V_C, which is $V_T - V_R$. Then $100 - 37 = 63$ V for V_C. These answers agree with the fact that in one time constant, the V_R drops 63 percent while V_C rises 63 percent.

---

*For an explanation of logarithms, see Grob, B.: *Mathematics for Basic Electronics*, Glencoe/McGraw-Hill, Columbus, Ohio.

Figure 23-10 illustrates how the voltages across R and C in series must add to equal the applied voltage V_T. The four examples with 100 V applied are:

1. At time zero, at the start of charging, V_R is 100 V and V_C is 0 V. Then $100 + 0 = 100$ V.
2. After one time constant, V_R is 37 V and V_C is 63 V. Then $37 + 63 = 100$ V.
3. After two time constants, V_R is 14 V and V_C is 86 V. Then $14 + 86 = 100$ V.
4. After five time constants, V_R is 0 V and V_C is 100 V, approximately. Then $0 + 100 = 100$ V

It should be emphasized that Formulas (23-3) and (23-4) can be used to calculate any decaying value on curve b in Fig. 23-9. These applications for an RC circuit include V_R on charge or discharge, i on charge or discharge, and V_C only on discharge. For an RC circuit in which C is charging, Formula (23-5) can be used to calculate the capacitor voltage v_C at any point along curve a in Fig. 23-9:

▶ $$v_C = V(1 - \epsilon^{-t/RC}) \qquad \textbf{(23-5)}$$

In Formula (23-5), V represents the maximum voltage to which C can charge, whereas v_C is the instantaneous capacitor voltage after the time t. Formula (23-5) is derived from the fact that v_C must equal $V_T - V_R$ while C is charging.

Example

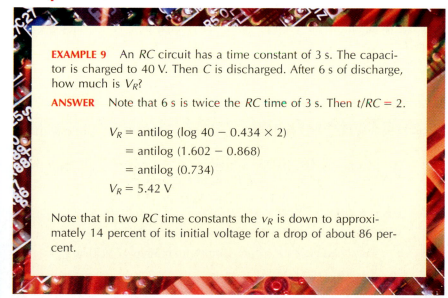

EXAMPLE 9 An RC circuit has a time constant of 3 s. The capacitor is charged to 40 V. Then C is discharged. After 6 s of discharge, how much is V_R?

ANSWER Note that 6 s is twice the RC time of 3 s. Then $t/RC = 2$.

$$V_R = \text{antilog } (\log 40 - 0.434 \times 2)$$
$$= \text{antilog } (1.602 - 0.868)$$
$$= \text{antilog } (0.734)$$
$$V_R = 5.42 \text{ V}$$

Note that in two RC time constants the v_R is down to approximately 14 percent of its initial voltage for a drop of about 86 percent.

To do this problem on a calculator, the steps can be as follows:

1. Find log 40. Punch in 40 on the keyboard and press the (log) key for 1.60 on the display. Be sure not to use the (ln) key for natural logs. The $\log_{10}$ key may require pushing the (2ndF) or shift key first.
2. With 1.60 on the display, push the (−) key, punch in ( ) then 0.434 (×) 2 ( ) and press the (=) key for 0.734 on the display.

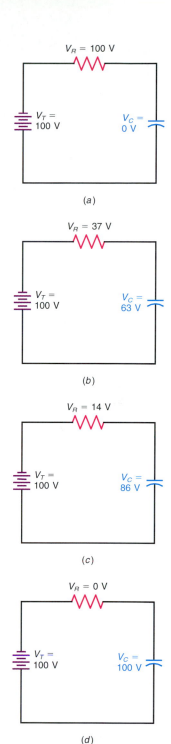

$V_R = 100$ V

$V_T = 100$ V $V_C = 0$ V

(a)

$V_R = 37$ V

$V_T = 100$ V $V_C = 63$ V

(b)

$V_R = 14$ V

$V_T = 100$ V $V_C = 86$ V

(c)

$V_R = 0$ V

$V_T = 100$ V $V_C = 100$ V

(d)

FIG. 23-10 How v_C and v_R add to equal the applied voltage V_T of 100 V. (a) Zero time at the start of charging. (b) After one RC time constant. (c) After two RC time constants. (d) After five or more RC time constants.

3. For the antilog of 0.734, use the $\boxed{10^x}$ key, which is usually a second function, for the answer of 5.42 on the display. This key is for the antilog of any logarithm to base 10.

CALCULATIONS FOR T   Furthermore, Formula (23-4) can be transposed to find the time t for a specific voltage decay. Then

$$\blacktriangleright \quad t = 2.3\, RC \log \frac{V}{v} \qquad\qquad\qquad \text{(23-6)}$$

where V is the higher voltage at the start and v is the lower voltage at the finish. The factor 2.3 is 1/0.434.

As an example, let RC be 1 s. How long will it take for v_R to drop from 100 to 50 V? The required time for this decay is

$$t = 2.3 \times 1 \times \log \frac{100}{50} = 2.3 \times 1 \times \log 2$$

$$= 2.3 \times 1 \times 0.3$$

$$t = 0.7 \text{ s} \qquad \text{approximately}$$

This answer agrees with the fact that the time for a drop of 50 percent takes 0.7 time constant. Formula (23-6) can also be used to calculate the time for any decay of v_C or v_R.

Formula (23-6) cannot be used for a rise in v_C. However, if you convert this rise to an equivalent drop in v_R, the calculated time is the same for both cases.

Example

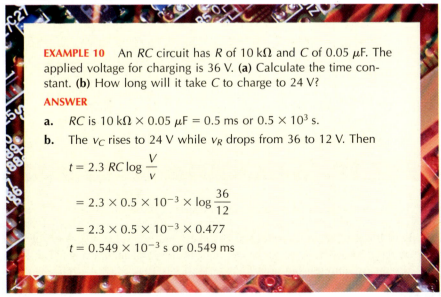

EXAMPLE 10   An *RC* circuit has *R* of 10 kΩ and *C* of 0.05 μF. The applied voltage for charging is 36 V. **(a)** Calculate the time constant. **(b)** How long will it take *C* to charge to 24 V?

ANSWER

a. RC is 10 kΩ × 0.05 μF = 0.5 ms or 0.5×10^3 s.

b.   The v_C rises to 24 V while v_R drops from 36 to 12 V. Then

$$t = 2.3\, RC \log \frac{V}{v}$$

$$= 2.3 \times 0.5 \times 10^{-3} \times \log \frac{36}{12}$$

$$= 2.3 \times 0.5 \times 10^{-3} \times 0.477$$

$$t = 0.549 \times 10^{-3} \text{ s or } 0.549 \text{ ms}$$

Answers at end of chapter.

Answer True or False for the universal curves in Fig. 23-9.
a. Curve *a* applies to v_C on charge.
b. Curve *b* applies to v_C on discharge.
c. Curve *b* applies to v_R when *C* charges or discharges.

23-12 COMPARISON OF REACTANCE AND TIME CONSTANT

The formula for capacitive reactance includes the factor of time in terms of frequency as $X_C = 1/(2\pi fC)$. Therefore, X_C and the *RC* time constant are both measures of the reaction of *C* to a change in voltage. The reactance X_C is a special case but a very important one that applies only to sine waves. The *RC* time constant can be applied to square waves and rectangular pulses.

PHASE ANGLE OF REACTANCE The capacitive charge and discharge current i_C is always equal to $C(dv/dt)$. A sine wave of voltage variations for v_C produces a cosine wave of current i_C. This means v_C and i_C are both sinusoids, but 90° out of phase.

In this case, it is usually more convenient to use X_C for calculations in sine-wave ac circuits to determine *Z*, *I*, and the phase angle θ. Then $I_C = V_C/X_C$. Moreover, if I_C is known, $V_C = I_C \times X_C$. The phase angle of the circuit depends on the amount of X_C compared with the resistance *R*.

CHANGES IN WAVESHAPE With nonsinusoidal voltage applied, X_C cannot be used. Then i_C must be calculated as $C(dv/dt)$. In this comparison of i_C and v_C, their waveshapes can be different, instead of the change in phase angle for sine waves. The waveshapes of v_C and i_C depend on the *RC* time constant.

COUPLING CAPACITORS If we consider the application of a coupling capacitor, X_C must be one-tenth or less of its series *R* at the desired frequency. This condition is equivalent to having an *RC* time constant that is long compared with the period of one cycle. In terms of X_C, the *C* has little IX_C voltage, with practically all the applied voltage across the series *R*. In terms of a long *RC* time constant, *C* cannot take on much charge. Practically all the applied voltage is developed as $v_R = iR$ across the series resistance by the charge and discharge current. These comparisons are summarized in Table 23-2.

INDUCTIVE CIRCUITS Similar comparisons can be made between $X_L = 2\pi fL$ for sine waves and the *L/R* time constant. The voltage across any inductance is $v_L = L(di/dt)$. Sine-wave variations for i_L produce a cosine wave of voltage v_L, 90° out of phase.

In this case X_L can be used to determine *Z*, *I*, and the phase angle θ. Then $I_L = V_L/X_L$. Furthermore, if I_L is known, $V_L = I_L \times X_L$. The phase angle of the circuit depends on the amount of X_L compared with *R*.

DID YOU KNOW?

Researchers have designed a car engine that expands to accelerate and shrinks for high-speed cruises. There are no rotating parts. The car goes from zero to 60 miles an hour in 7 seconds and gets 80 miles per gallon; its lifetime is 500,000 miles.

TABLE 23-2 COMPARISON OF REACTANCE X_C AND RC TIME CONSTANT

SINE-WAVE VOLTAGE	NONSINUSOIDAL VOLTAGE
Examples are 60-Hz power line, AF signal voltage, RF signal voltage	Examples are dc circuit turned on and off, square waves, rectangular pulses
Reactance $X_C = \dfrac{1}{2\pi fC}$	Time constant $T = RC$
Larger C results in smaller reactance X_C	Larger C results in longer time constant
Higher frequency results in smaller X_C	Shorter pulse width corresponds to longer time constant
$I_C = \dfrac{V_C}{X_C}$	$i_C = C\dfrac{dv}{dt}$
X_C makes I_C and V_C 90° out of phase	Waveshape changes between i_C and v_C

With nonsinusoidal voltage, however, X_L cannot be used. Then v_L must be calculated as $L(di/dt)$. In this comparison, i_L and v_L can have different waveshapes, depending on the L/R time constant.

CHOKE COILS   For this application, the idea is to have almost all the applied ac voltage across L. The condition of X_L being at least 10 times R corresponds to having a long time constant. The high value of X_L means practically all the applied ac voltage is across X_L as IX_L, with little IR voltage.

The long L/R time constant means i_L cannot rise appreciably, resulting in little v_R voltage across the resistor. The waveform for i_L and v_R in an inductive circuit corresponds to v_C in a capacitive circuit.

WHEN DO WE USE THE TIME CONSTANT?   In electronic circuits, the time constant is useful in analyzing the effect of L or C on the waveshape of nonsinusoidal voltages, particularly rectangular pulses. Another application is the transient response when a dc voltage is turned on or off. The 63 percent change in one time constant is a natural characteristic of v or i, where the magnitude of one is proportional to the rate of change of the other.

WHEN DO WE USE REACTANCE?   The X_L and X_C are generally used for sine-wave V or I. We can determine Z, I, voltage drops, and phase angles. The phase angle of 90° is a natural characteristic of a cosine wave where its magnitude is proportional to the rate of change in a sine wave.

TEST-POINT QUESTION 23-12

Answers at end of chapter.
a. Does an RC coupling circuit have a small or large X_C compared with R?
b. Does an RC coupling circuit have a long or short time constant for the frequency of the applied voltage?

- The transient response of an inductive circuit with nonsinusoidal current is indicated by the time constant L/R. With L in henrys and R in ohms, T is the time in seconds for the current i_L to change by 63 percent. In five time constants, i_L reaches the steady value of V_T/R.

- At the instant an inductive circuit is opened, high voltage is generated across L because of the fast current decay with a short time constant. The induced voltage $v_L = L(di/dt)$. The di is the change in i_L.

- The transient response of a capacitive circuit with nonsinusoidal voltage is indicated by the time constant RC. With C in farads and R in ohms, T is the time in seconds for the voltage across the capacitor v_C to change by 63 percent. In five time constants, v_C reaches the steady value of V_T.

- At the instant a charged capacitor is discharged through a low resistance, a high value of discharge current can be produced. The discharge current $i_C = C(dv/dt)$ can be large because of the fast discharge with a short time constant. The dv is the change in v_C.

- The waveshapes of v_C and i_L correspond, as both rise relatively slowly to the steady-state value. This is an integrated output.

- Also i_C and v_L correspond, as they are the waveforms that can change instantaneously. This is a differentiated output.

- For both RC and RL circuits the resistor voltage $v_R = iR$.

- A short time constant is one-fifth or less of the pulse width, in time, for the applied voltage.

- A long time constant is greater than the pulse width, in time, for the applied voltage by a factor of 5 or more.

- An RC circuit with a short time constant produces sharp voltage spikes for v_R at the leading and trailing edges of a square-wave applied voltage. The waveshape of voltage V_T is across the capacitor as v_C. See Fig. 23-7.

- An RC circuit with a long time constant allows v_R to be essentially the same as the variations in applied voltage V_T, while the average dc value of V_T is blocked as v_C. See Fig. 23-8.

- The universal rise and decay curves in Fig. 23-9 can be used for current or voltage in RC and RL circuits for any time up to five time constants.

- The concept of reactance is useful for sine-wave ac circuits with L and C.

- The time constant method is used with L or C to analyze nonsinusoidal waveforms.

Choose (*a*), (*b*), (*c*), or (*d*).

1. A 250-μH L is in series with a 50-Ω R. The time constant is (*a*) 5 μs; (*b*) 25 μs; (*c*) 50 μs; (*d*) 250 μs.
2. If V_T is 500 mV in the preceding circuit, after 5 μs I rises to the value of (*a*) 3.7 mA; (*b*) 5 mA; (*c*) 6.3 mA; (*d*) 10 mA.
3. In the preceding circuit, I will have the steady-state value of 10 mA after (*a*) 5 μs; (*b*) 6.3 μs; (*c*) 10 μs; (*d*) 25 μs.
4. The arc across a switch when it opens an *RL* circuit is a result of the (*a*) long time constant; (*b*) large self-induced voltage across L; (*c*) low resistance of the open switch; (*d*) surge of resistance.
5. A 250-pF C is in series with a 1-MΩ R. The time constant is (*a*) 63 μs; (*b*) 100 μs; (*c*) 200 μs; (*d*) 250 μs.
6. If V_T is 100 V in the preceding circuit, after 250 μs, v_C rises to the value of (*a*) 37 V; (*b*) 50 V; (*c*) 63 V; (*d*) 100 V.
7. In the preceding circuit, v_C will have the steady-state value of 100 V after (*a*) 250 μs; (*b*) 630 μs; (*c*) 1000 μs or 1 ms; (*d*) 1.25 ms.
8. In the preceding circuit, after 3 hours v_C will be (*a*) zero; (*b*) 63 V; (*c*) 100 V; (*d*) 200 V.
9. For a square-wave applied voltage with the frequency of 500 Hz, a long time constant is (*a*) 1 ms; (*b*) 2 ms; (*c*) 3.7 ms; (*d*) 5 ms.
10. An *RC* circuit has a 2-μF C in series with a 1-MΩ R. The time of 6 s equals how many time constants? (*a*) one; (*b*) two; (*c*) three; (*d*) six.

QUESTIONS

1. Give the formula, with units, for calculating the time constant of an *RL* circuit.
2. Give the formula, with units, for calculating the time constant of an *RC* circuit.
3. Redraw the *RL* circuit and graph in Fig. 23-2 for a 2-H L and a 100-Ω R.
4. Redraw the graphs in Fig. 23-4 to fit the circuit in Fig. 23-5 with a 100-μF C. Use a 3000-Ω R for charge but a 3-Ω R for discharge.
5. List two comparisons of *RC* and *RL* circuits for nonsinusoidal voltage.
6. List two comparisons between *RC* circuits with nonsinusoidal voltage and sine-wave voltage applied.
7. Define the following: (**a**) a long time constant; (**b**) a short time constant; (**c**) an *RC* differentiating circuit; (**d**) an *RC* integrating circuit.
8. Redraw the horizontal time axis of the universal curve in Fig. 23-9, calibrated in absolute time units of milliseconds for an *RC* circuit with a time constant equal to 2.3 ms.
9. Redraw the circuit and graphs in Fig. 23-7 with everything the same except that R is 20 kΩ, making the *RC* time constant shorter.

10. Redraw the circuit and graphs in Fig. 23-8 with everything the same except that R is 500 kΩ, making the RC time constant longer.
11. Invert the equation $T = RC$, in two forms, to find R or C from the time constant.
12. Show three types of nonsinusoidal waveforms.
13. Give an application in electronic circuits for an RC circuit with a long time constant and with a short time constant.
14. Why can arcing voltage be a problem with coils used in switching circuits?

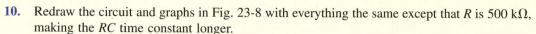

PROBLEMS

ANSWERS TO ODD-NUMBERED PROBLEMS AT BACK OF BOOK.

1. Calculate the time constant of the following inductive circuits: (**a**) L is 20 H and R is 400 Ω; (**b**) L is 20 μH and R is 400 Ω; (**c**) L is 50 mH and R is 50 Ω; (**d**) L is 40 μH and R is 2 Ω.
2. Calculate the time constant of the following capacitive circuits: (**a**) C is 0.001 μF and R is 1 MΩ; (**b**) C is 1 μF and R is 1000 Ω; (**c**) C is 0.05 μF and R is 250 kΩ; (**d**) C is 100 pF and R is 10 kΩ.
3. A 100-V source is in series with a 2-MΩ R and a 2-μF C. (**a**) How much time is required for v_C to be 63 V? (**b**) How much is v_C after 20 s?
4. The C in Prob. 3 is allowed to charge for 4 s and then made to discharge for 8 s. How much is v_C?
5. A 100-V source is applied in series with a 1-MΩ R and a 4-μF C that has already been charged to 63 V. How much is v_C after 4 s?
6. What value of R is needed with a 0.05-μF C for an RC time constant of 0.02 s? For 1 ms?
7. An RC circuit has a time constant of 1 ms. V_T applied is 20 V. How much is v_C on charge after 1.4 ms?
8. A 0.05-μF C charges through a 0.5-MΩ R but discharges through a 2-kΩ R. Calculate the time constants for charge and discharge. Why will the capacitor discharge faster than charge?
9. A 0.05-μF C is charged to 264 V. It discharges through a 40-kΩ R. How much is the time for v_C to discharge down to 132 V?
10. Referring to Fig. 23-6b, calculate the value of v_C on the next charge, starting from 26.5 V.
11. Determine whether 75 μs will be a long or short time constant for applied signal voltage with the following frequencies: (**a**) 60 Hz; (**b**) 1000 Hz; (**c**) 4 MHz.
12. What R is needed with C of 0.001 μF for a time constant of 75 μs?
13. Calculate the C needed for a time constant of 50 ms with R of 5 kΩ.
14. Determine the frequency of a square-wave signal that will have voltage applied for one time constant with RC of 50 ms.
15. An RC circuit has a time constant of 68 μs. The capacitor is charged to 14 V. How much is v_R after 136 μs of discharge?
16. For an RC circuit with R of 1 MΩ and C of 68 pF, 9 V is applied. How long will it take C to charge to 5 V?
17. Use the slope line in Fig. 23-9b to calculate dv/dt at the start of the decay in v_C for the circuit in Fig. 23-5b.

18. Use the slope line in Fig. 23-9 to calculate di/dt at the start of the decay in i_L for the circuit in Fig. 23-3b. (Hint: You can ignore the steady 8 V and 100-Ω R_1 because they do not change the di/dt value.)

19. In Fig. 23-11, draw the waveform you would expect to measure across the 10-kΩ R. Indicate the resistor voltage V_R at the beginning and end of each 1-ms pulse interval. Draw the V_R waveform in the proper time relationship with respect to V_{in}.

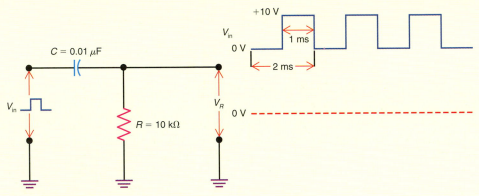

FIG. 23-11 Diagram for Prob. 19.

20. In Fig. 23-12, draw the waveform you would expect to measure across the 0.01-μF C. Indicate the capacitor voltage V_C at the beginning and end of each 27.3-μs pulse interval. Draw the V_C waveform in the proper time relationship with respect to V_{in}. (C is initially uncharged.)

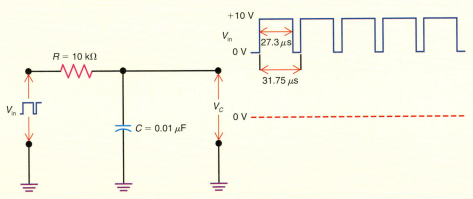

FIG. 23-12 Diagram for Prob. 20.

21. In Fig. 23-13, the capacitor is initially charged to 5 V. If switch S_1 is closed, calculate V_C and V_R after: **(a)** 0 s; **(b)** 0.693 s; **(c)** 1 s; **(d)** 2 s; **(e)** 3.5 s; **(f)** 5 s.

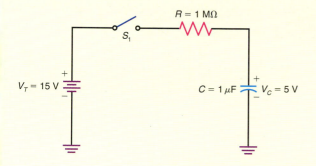

FIG. 23-13   Circuit for Prob. 21.

CRITICAL THINKING

1. Refer to Fig. 23-14. (a) If S_1 is closed long enough for the capacitor C to become fully charged, what voltage is across C? (b) With C fully charged, how long will it take C to fully discharge when S_1 is opened?

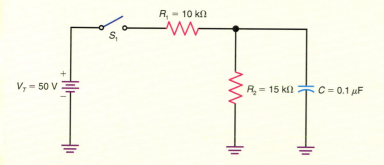

FIG. 23-14   Circuit for Critical Thinking Probs. 1 and 2.

2. Refer to Fig. 23-14. (a) How long will it take C to fully charge after S_1 is initially closed? (b) What is V_C 1 ms after S_1 is initially closed? (c) What is V_C 415.8 μs after S_1 is initially closed? (d) What is V_C 1.5 ms after S_1 is initially closed?

3. Refer to Fig. 23-15. Assume C is allowed to charge fully and then the polarity of V_T is suddenly reversed. What is the capacitor voltage v_C for the following time intervals after the reversal of V_T: (a) 0 s; (b) 6.93 ms; (c) 10 ms; (d) 15 ms; (e) 30 ms?

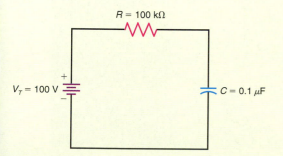

FIG. 23-15   Circuit for Critical Thinking Prob. 3.

23-1 a. T
          b. T

23-2 a. 0.02 s
          b. 0.5 ms

23-3 a. shorter
          b. faster

23-4 a. 940 μs
          b. 470 ns

23-5 a. 63.2 V
          b. 36.8 V

23-6 a. shorter
          b. faster

23-7 a. $v_R = 37$ V
          b. $v_C = 23.3$ V

23-8 a. short
          b. long

23-9 a. short
          b. across C

23-10 a. long
          b. across R

23-11 a. T
          b. T
          c. T

23-12 a. small X_C
          b. long time constant

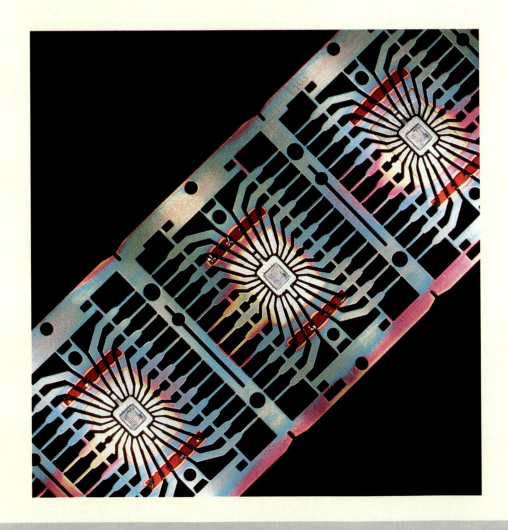

REVIEW: CHAPTERS 20 TO 23

SUMMARY

- The ability of a conductor to produce induced voltage across itself when the current changes is its self-inductance, or inductance. The symbol is L, and the unit is the henry. One henry allows 1 V to be induced when the current changes at the rate of 1 A/s.
- The polarity of the induced voltage always opposes the change in current that is causing the induced voltage. This is Lenz' law.
- Mutual inductance is the ability of varying current in one coil to induce voltage in another coil nearby, without any connection between them. Its symbol is L_M, and the unit is also the henry.
- A transformer consists of two or more windings with mutual inductance. The primary connects to the source voltage, the secondary to the load. With an iron core, the voltage ratio between primary and secondary equals the turns ratio.
- The efficiency of a transformer equals the ratio of power output from the secondary to power input to the primary × 100 percent.
- Eddy currents are induced in the iron core of an inductance, causing I^2R losses that increase with higher frequencies. Laminated iron, powdered-iron, or ferrite cores have minimum eddy-current losses. Hysteresis also increases the losses.
- Series inductances without mutual coupling add like series resistances. With parallel inductances, the combined inductance is calculated by the reciprocal formula, as with parallel resistances.
- Inductive reactance X_L equals $2\pi fL\ \Omega$, where f is in hertz and L is in henrys. Reactance X_L increases with more inductance and higher frequencies.
- A common application of X_L is an AF or RF choke, which has high reactance for one group of frequencies but less reactance for lower frequencies.
- Reactance X_L is a phasor quantity that has its current lagging 90° behind its induced voltage. In series circuits, R and X_L are added by phasors because their voltage drops are 90° out of phase. In parallel circuits, the resistive and inductive branch currents are 90° out of phase.
- Impedance Z, in ohms, is the total opposition of an ac circuit with resistance and reactance. For series circuits, $Z_T = \sqrt{R^2 + X_L^2}$ and $I = V_T/Z_T$. For parallel circuits, $I_T = \sqrt{I_R^2 + I_L^2}$ and $Z_{EQ} = V_A/I_T$.
- The Q of a coil is X_L/r_i.
- Energy stored by an inductance is $\frac{1}{2}LI^2$, where I is in amperes, L in henrys, and the energy is in joules.
- The voltage across L is always equal to $L(di/dt)$ for any waveshape of current.
- The transient response of a circuit refers to the temporary condition which exists until the circuit's current or voltage reaches its steady-state value. The transient response of a circuit is

measured in time constants, where one time constant is defined as the length of time over which a 63.2 percent change in current or voltage occurs.

- For an inductive circuit, one time constant is the time in seconds for the current to change by 63.2 percent. For inductive circuits, one time constant equals L/R; that is, $T = L/R$, where L is in henries, R is in ohms, and T is in seconds. The current reaches its steady-state value after a time of five L/R time constants has elapsed.
- For a capacitive circuit, one time constant is the time in seconds for the capacitor voltage to change by 63.2 percent. For capacitive circuits, one time constant equals RC; that is, $T = RC$, where R is in ohms, C is in farads, and T is in seconds. The capacitor voltage reaches its steady-state value after a time of five RC time constants has elapsed.
- When the input voltage to an inductive or capacitive circuit is nonsinusoidal in nature, time constants rather than reactances are used to determine the circuit's voltage and current values.
- Whether an L/R or RC time constant is considered short or long depends on its relationship to the pulse width of the applied voltage. In general, a short time constant is considered to be one which is one-fifth or less the time of the pulse width of the applied voltage. Conversely, a long time constant is generally considered to be one which is five or more times longer than the pulse width of the applied voltage.
- To calculate the voltage across a capacitor during charge, use curve a in Fig. 23-9 or use Formula (23-5). To calculate the voltage across a resistor during charge, use curve b in Fig. 23-9 or Formula (23-4). To calculate the voltage across a capacitor or resistor during discharge, use curve b in Fig. 23-9 or Formula (23-4).

REVIEW SELF-TEST

ANSWERS AT BACK OF BOOK.

Choose (a), (b), (c), or (d).

1. A coil induces 200 mV when the current changes at the rate of 1 A/s. The inductance L is: (a) 1 mH; (b) 2 mH; (c) 200 mH; (d) 100 mH.
2. Alternating current in an inductance produces maximum induced voltage when the current has its (a) maximum value; (b) maximum change in magnetic flux; (c) minimum change in magnetic flux; (d) rms value of $0.707 \times$ peak.
3. An iron-core transformer connected to the 120-V 60-Hz power line has a turns ratio of $1:20$. The voltage across the secondary equals (a) 20 V; (b) 60 V; (c) 120 V; (d) 2400 V.
4. Two 250-mH chokes in series have a total inductance of (a) 60 mH; (b) 125 mH; (c) 250 mH; (d) 500 mH.
5. Which of the following will have minimum eddy-current losses? (a) Solid iron core; (b) laminated iron core; (c) powdered-iron core; (d) air core.
6. Which of the following will have maximum inductive reactance? (a) 2-H inductance at 60 Hz; (b) 2-mH inductance at 60 kHz; (c) 5-mH inductance at 60 kHz; (d) 5-mH inductance at 100 kHz.
7. A 100-Ω R is in series with 100 Ω of X_L. The total impedance Z equals (a) 70.7 Ω; (b) 100 Ω; (c) 141 Ω; (d) 200 Ω.

8. A 100-Ω R is in parallel with 100 Ω of X_L. The total impedance Z equals (a) 70.7 Ω; (b) 100 Ω; (c) 141 Ω; (d) 200 Ω.

9. If two waves have the frequency of 1000 Hz and one is at the maximum value when the other is at zero, the phase angle between them is (a) 0°; (b) 90°; (c) 180°; (d) 360°.

10. If an ohmmeter check on a 50-μH choke reads 3 Ω, the coil is probably (a) open; (b) defective; (c) normal; (d) partially open.

11. An inductive circuit with $L = 100$ mH and $R = 10$ kΩ has a time constant of: (a) 1 μs; (b) 100 μs; (c) 10 μs; (d) 1000 μs.

12. A capacitive circuit with $R = 1.5$ kΩ and $C = 0.01$ μF has a time constant of: (a) 15 μs; (b) 1.5 ms; (c) 150 μs; (d) 150 s.

13. With respect to the pulse width of the applied voltage, the time constant of an RC integrator should be: (a) short; (b) the same as the pulse width of V_T; (c) long; (d) shorter than the pulse width of V_T.

14. With respect to the pulse width of the applied voltage, the time constant of an RC differentiator should be: (a) long; (b) the same as the pulse width of V_T; (c) longer than the pulse width of V_T; (d) short.

15. The current rating of a transformer is usually specified for: (a) the primary windings only; (b) the secondary windings only; (c) both the primary and secondary windings; (d) the core only.

16. The secondary of a transformer is connected to a 15-Ω resistor. If the turns ratio $N_P/N_S = 3:1$, the primary impedance Z_P equals: (a) 135 Ω; (b) 45 Ω; (c) 5 Ω; (d) none of the above.

REFERENCES

Bogart, Theodore F., Jr.: *Electric Circuits*, Glencoe/McGraw-Hill, Columbus, Ohio.

Lister, Eugene C.: *Electric Circuits and Machines*, Glencoe/McGraw-Hill, Columbus, Ohio.

CHAPTER 24

ALTERNATING CURRENT CIRCUITS

This chapter shows how to analyze sine-wave ac circuits that have R, X_L, and X_C. How do we combine these three types of ohms of opposition, how much current flows, and what is the phase angle? These questions are answered for both series and parallel circuits.

The problems are simplified by the fact that in series circuits X_L is at 90° and X_C at −90°, which are opposite phase angles. Then all of one reactance can be canceled by part of the other reactance, resulting in only a single net reactance.

Similarly, in parallel circuits, I_L and I_C have opposite phase angles. These phasor currents oppose each other and result in a single net reactive line current.

Finally, the idea of how ac power and dc power can differ because of ac reactance is explained. Also, types of ac current meters are described including the wattmeter.

CHAPTER OBJECTIVES

Upon completion of this chapter, you should be able to:

- *Explain* why opposite reactances in series cancel.
- *Determine* the total impedance and phase angle of a series circuit containing resistance, capacitance, and inductance.
- *Determine* the total current, equivalent impedance, and phase angle of a parallel circuit containing resistance, capacitance, and inductance.
- *Define* the terms *real power, apparent power, volt-ampere reactive,* and *power factor.*
- *Calculate* the power factor of a circuit.

IMPORTANT TERMS IN THIS CHAPTER

apparent power	real power	voltampere unit
double-subscript notation	VAR unit	wattmeter
power factor		

TOPICS COVERED IN THIS CHAPTER

24-1 AC CIRCUITS WITH RESISTANCE BUT NO REACTANCE

Combinations of series and parallel resistances are shown in Fig. 24-1. In Fig. 24-1a and b, all voltages and currents throughout the resistive circuit are in phase. There is no reactance to cause a lead or lag in either current or voltage.

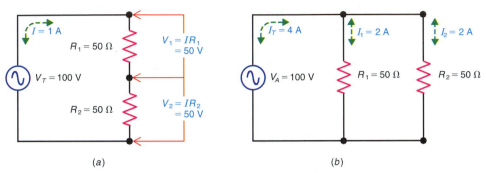

FIG. 24-1   Alternating-current circuits with resistance but no reactance. (a) Resistances R_1 and R_2 in series. (b) Resistances R_1 and R_2 in parallel.

SERIES RESISTANCES   For the circuit in Fig. 24-1a, with two 50-Ω resistances in series across the 100-V source, the calculations are as follows:

$$R_T = R_1 + R_2 = 50 + 50 = 100 \; \Omega$$
$$I = \frac{V_T}{R_T} = \frac{100}{100} = 1 \; \text{A}$$
$$V_1 = IR_1 = 1 \times 50 = 50 \; \text{V}$$
$$V_2 = IR_2 = 1 \times 50 = 50 \; \text{V}$$

Note that the series resistances R_1 and R_2 serve as a voltage divider, as in dc circuits. Each R has one-half the applied voltage for one-half the total series resistance.

The voltage drops V_1 and V_2 are both in phase with the series current I, which is the common reference. Also I is in phase with the applied voltage V_T because there is no reactance.

PARALLEL RESISTANCES   For the circuit in Fig. 24-1b, with two 50-Ω resistances in parallel across the 100-V source, the calculations are

$$I_1 = \frac{V_A}{R_1} = \frac{100}{50} = 2 \; \text{A}$$
$$I_2 = \frac{V_A}{R_2} = \frac{100}{50} = 2 \; \text{A}$$
$$I_T = I_1 + I_2 = 2 + 2 = 4 \; \text{A}$$

With a total current of 4 A in the main line from the 100-V source, the combined parallel resistance is 25 Ω. This R_{EQ} equals 100 V/4 A for the two 50-Ω branches.

Each branch current has the same phase as the applied voltage. Voltage V_A is the reference because it is common to both branches.

24-2 CIRCUITS WITH X_L ALONE

The circuits with X_L in Figs. 24-2 and 24-3 correspond to the series and parallel circuits in Fig. 24-1, with ohms of X_L equal to the R values. Since the applied voltage is the same, the values of current correspond because ohms of X_L are just as effective as ohms of R in limiting the current or producing a voltage drop.

Although X_L is a phasor quantity with a 90° phase angle, all the ohms of opposition are the same kind of reactance in this example. Therefore, without any R or X_C, the series ohms of X_L can be combined directly. Similarly, the parallel I_L currents can be added.

DID YOU KNOW?

Half of all available commercial energy is used by industrialized countries, which constitute only a quarter of the world's population.

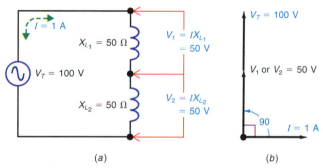

(a)   (b)

FIG. 24-2 Series circuit with X_L alone. (*a*) Schematic diagram. (*b*) Phasor diagram of voltages and line current.

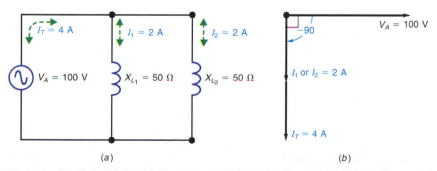

(a)   (b)

FIG. 24-3 Parallel circuit with X_L alone. (*a*) Schematic diagram. (*b*) Phasor diagram of branch and line currents and the applied voltage.

X_L VALUES IN SERIES For Fig. 24-2a, the calculations are

$$X_{L_T} = X_{L_1} + X_{L_2} = 50 + 50 = 100 \ \Omega$$

$$I = \frac{V_T}{X_{L_T}} = \frac{100}{100} = 1 \ A$$

$$V_1 = IX_{L_1} = 1 \times 50 = 50 \ V$$

$$V_2 = IX_{L_2} = 1 \times 50 = 50 \ V$$

Note that the two series voltage drops of 50 V each add to equal the total applied voltage of 100 V.

With regard to the phase angle for the inductive reactance, the voltage across any X_L always leads the current through it by 90°. In Fig. 24-2b, I is the reference phasor because it is common to all the series components. Therefore, the voltage phasors for V_1 and V_2 across either reactance, or V_T across both reactances, are shown leading I by 90°.

I_L VALUES IN PARALLEL For Fig. 24-3a the calculations are

$$I_1 = \frac{V_A}{X_{L_1}} = \frac{100}{50} = 2 \ A$$

$$I_2 = \frac{V_A}{X_{L_2}} = \frac{100}{50} = 2 \ A$$

$$I_T = I_1 + I_2 = 2 + 2 = 4 \ A$$

These two branch currents can be added because they both have the same phase. This angle is 90° lagging the voltage reference phasor as shown in Fig. 24-3b.

Since the voltage V_A is common to the branches, this voltage is across X_{L_1} and X_{L_2}. Therefore V_A is the reference phasor for parallel circuits.

Note that there is no fundamental change between Fig. 24-2b, which shows each X_L voltage leading its current by 90°, and Fig. 24-3b, showing each X_L current lagging its voltage by −90°. The phase angle between the inductive current and voltage is still the same 90°.

TEST-POINT QUESTION 24-2

Answers at end of chapter.
a. In Fig. 24-2, what is the phase angle of V_T with respect to I?
b. In Fig. 24-3, what is the phase angle of I_T with respect to V_A?

24-3 CIRCUITS WITH X_C ALONE

Again, reactances are shown in Figs. 24-4 and 24-5 but with X_C values of 50 Ω. Since there is no R or X_L, the series ohms of X_C can be combined directly. Also the parallel I_C currents can be added.

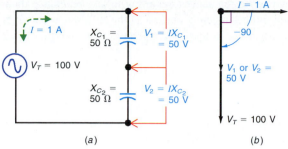

(a)   (b)

FIG. 24-4 Series circuit with X_C alone. (a) Schematic diagram. (b) Phasor diagram of voltages and line current.

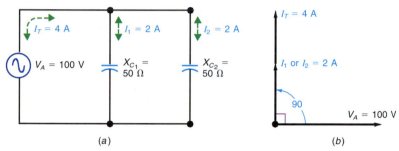

(a)   (b)

FIG. 24-5 Parallel circuit with X_C alone. (a) Schematic diagram. (b) Phasor diagram of branch and line currents and the applied voltage.

X_C VALUES IN SERIES For Fig. 24-4a, the calculations for V_1 and V_2 are the same as before. These two series voltage drops of 50 V each add to equal the total applied voltage.

With regard to the phase angle for the capacitive reactance, the voltage across any X_C always lags its capacitive charge and discharge current I by 90°. For the series circuit in Fig. 24-4, I is the reference phasor. The capacitive current leads by 90°. Or, we can say that each voltage lags I by −90°.

I_C VALUES IN PARALLEL For Fig. 24-5, V_A is the reference phasor. The calculations for I_1 and I_2 are the same as before. However, now each of the capacitive branch currents or the I_T leads V_A by 90°.

TEST-POINT QUESTION 24-3

Answers at end of chapter.
a. In Fig. 24-4, what is the phase angle of V_T with respect to I?
b. In Fig. 24-5, what is the phase angle of I_T with respect to V_A?

24-4 OPPOSITE REACTANCES CANCEL

In a circuit with both X_L and X_C, the opposite phase angles enable one to offset the effect of the other. For X_L and X_C in series, the net reactance is the difference between the two series reactances, resulting in less reactance than either one. In parallel circuits, the net reactive current is the difference between the I_L and I_C branch currents resulting in less total line current than either branch current.

X_L AND X_C IN SERIES   For the example in Fig. 24-6, the series combination of a 60-Ω X_L and a 40-Ω X_C in Fig. 24-6a and b is equivalent to the net reactance of the 20-Ω X_L shown in Fig. 24-6c. Then, with 20 Ω as the net reactance across the 120-V source, the current is 6 A. This current lags the applied voltage V_T by 90° because the net reactance is inductive.

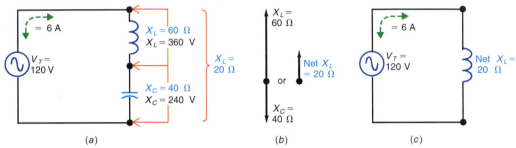

FIG. 24-6   When X_L and X_C are in series, their ohms of reactance subtract. (a) Series circuit with 60-Ω X_L and 40-Ω X_C. (b) Phasor diagram. (c) Equivalent circuit with net value of 20 Ω of X_L for the total reactance.

For the two series reactances in Fig. 24-6a, the current is the same through both X_L and X_C. Therefore, the voltage drops can be calculated as

$$V_L \text{ or } IX_L = 6 \text{ A} \times 60 \text{ }\Omega = 360 \text{ V}$$
$$V_C \text{ or } IX_C = 6 \text{ A} \times 40 \text{ }\Omega = 240 \text{ V}$$

Note that each individual reactive voltage drop can be more than the applied voltage. The phasor sum of the series voltage drops still is 120 V, however, equal to the applied voltage. This results because the IX_L and IX_C voltages are opposite. The IX_L voltage leads the series current by 90°; the IX_C voltage lags the same current by 90°. Therefore, IX_L and IX_C are 180° out of phase with each other, which means they are of opposite polarity and offset one another. Then the total voltage across the two in series is 360 V minus 240 V, which equals the applied voltage of 120 V.

If the values in Fig. 24-6 were reversed, with an X_C of 60 Ω and an X_L of 40 Ω, the net reactance would be a 20-Ω X_C. The current would be 6 A again, but with a lagging phase angle of −90° for the capacitive voltage. The IX_C voltage would then be greater at 360 V, than an IX_L value of 240 V, but the difference still equals the applied voltage of 120 V.

X_L AND X_C IN PARALLEL   In Fig. 24-7, the 60-Ω X_L and 40-Ω X_C are in parallel across the 120-V source. Then the 60-Ω X_L branch current I_L is 2 A, and the 40-Ω X_C branch current I_C is 3 A. The X_C branch has more current because its reactance is less than X_L.

In terms of phase angle, I_L lags the parallel voltage V_A by 90°, while I_C leads the same voltage by 90°. Therefore, the opposite reactive branch currents are 180°

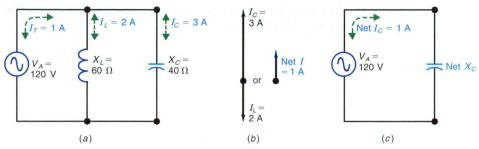

FIG. 24-7 When X_L and X_C are in parallel, their branch currents subtract. (*a*) Parallel circuit with 3-A I_C and 2-A I_L. (*b*) Phasor diagram. (*c*) Equivalent circuit with net value of 1 A of I_C for the total line current.

out of phase with each other. The net line current then is the difference between 3 A for I_C and 2 A for I_L, which equals the net value of 1 A. This resultant current leads V_A by 90° because it is capacitive current.

If the values in Fig. 24-7 were reversed, with an X_C of 60 Ω and an X_L of 40 Ω, I_L would be larger. The I_L then equals 3 A, with an I_C of 2 A. The net line current is 1 A again but inductive, with a net I_L.

TEST-POINT QUESTION 24-4

Answers at end of chapter.
a. In Fig. 24-6, how much is the net X_L?
b. In Fig. 24-7, how much is the net I_C?

24-5 SERIES REACTANCE AND RESISTANCE

In the case of series reactance and resistance, the resistive and reactive effects must be combined by phasors. For series circuits, the ohms of opposition are added to find Z_T. First add all the series resistances for one total R. Also combine all the series reactances, adding all the X_Ls and all the X_Cs and finding the net X by subtraction. The result is one net reactance. It may be either capacitive or inductive, depending on which kind of reactance is larger. Then the total R and net X can be added by phasors to find the total ohms of opposition for the entire series circuit.

MAGNITUDE OF Z_T   After the total R and net reactance X are found, they can be combined by the formula

▶ $$Z_T = \sqrt{R^2 + X^2} \qquad\qquad\qquad \textbf{(24-1)}$$

The circuit's total impedance Z_T is the phasor sum of the series resistance and reactance. Whether the net X is at +90° for X_L or −90° for X_C does not matter in calculating the magnitude of Z_T.

An example is illustrated in Fig. 24-8. Here the net series reactance in Fig. 24-8*b* is a 30-Ω X_C. This value is equal to a 60-Ω X_L subtracted from a

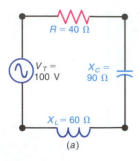

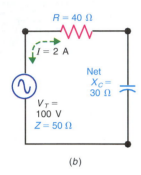

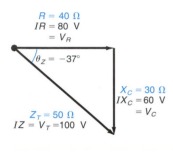

FIG. 24-8 Impedance Z_T of series circuit with resistance and reactance. (*a*) Circuit with R, X_L, and X_C in series. (*b*) Equivalent circuit with one net reactance. (*c*) Phasor diagram. The voltage triangle of phasors is equivalent to an impedance triangle for series circuits.

90-Ω X_C as shown in Fig. 24-8*a*. The net 30-Ω X_C in Fig. 24-8*b* is in series with a 40-Ω R. Therefore

$$Z_T = \sqrt{R^2 + X^2} = \sqrt{(40)^2 + (30)^2} = \sqrt{1600 + 900}$$
$$= \sqrt{2500}$$
$$Z_T = 50 \ \Omega$$

I = V/Z_T The current is 100 V/50 Ω in this example, or 2 A. This value is the magnitude, without considering the phase angle.

SERIES VOLTAGE DROPS All the series components have the same 2-A current. Therefore, the individual drops in Fig. 24-8*a* are

$$V_R = IR = 2 \times 40 = 80 \text{ V}$$
$$V_C = IX_C = 2 \times 90 = 180 \text{ V}$$
$$V_L = IX_L = 2 \times 60 = 120 \text{ V}$$

Since IX_C and IX_L are voltages of opposite polarity, the net reactive voltage is 180 V minus 120 V, which equals 60 V. The phasor sum of IR at 80 V and the net reactive voltage IX of 60 V equals the applied voltage V_T of 100 V.

ANGLE OF Z_T The impedance angle of the series circuit is the angle whose tangent equals X/R. This angle is negative for X_C but positive for X_L.

In this example, X is the net reactance of 30 Ω for X_C and R is 40 Ω. Then $\tan \theta_Z = -0.75$ and θ_Z is $-37°$, approximately.

The negative angle for Z indicates a net capacitive reactance for the series circuit. If the values of X_L and X_C were reversed, the θ_Z would be $+37°$, instead of $-37°$, because of the net X_L. However, the magnitude of Z would still be the same.

Example

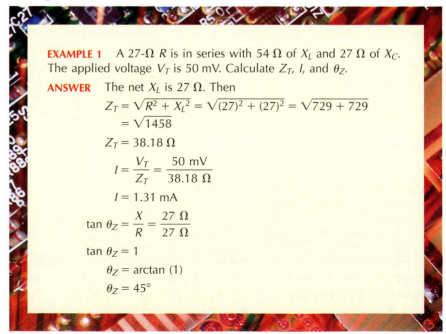

EXAMPLE 1 A 27-Ω R is in series with 54 Ω of X_L and 27 Ω of X_C. The applied voltage V_T is 50 mV. Calculate Z_T, I, and θ_Z.

ANSWER The net X_L is 27 Ω. Then

$$Z_T = \sqrt{R^2 + X_L^2} = \sqrt{(27)^2 + (27)^2} = \sqrt{729 + 729}$$
$$= \sqrt{1458}$$
$$Z_T = 38.18 \ \Omega$$
$$I = \frac{V_T}{Z_T} = \frac{50 \text{ mV}}{38.18 \ \Omega}$$
$$I = 1.31 \text{ mA}$$
$$\tan \theta_Z = \frac{X}{R} = \frac{27 \ \Omega}{27 \ \Omega}$$
$$\tan \theta_Z = 1$$
$$\theta_Z = \arctan (1)$$
$$\theta_Z = 45°$$

In general, when the series resistance and reactance are equal, Z_T is 1.414 times either value. Here, Z_T is $1.414 \times 27 = 38.18 \ \Omega$. Also, $\tan \theta$ must be 1 and the angle is $45°$ for equal sides in a right triangle. To find Z_T on a calculator, see the procedure described on page 609 for the square root of the sum of two squares.

MORE SERIES COMPONENTS   How to combine any number of series resistances and reactances is illustrated by Fig. 24-9. Here the total series R of 40 Ω is the sum of 30 Ω for R_1 and 10 Ω for R_2. Note that the order of connection does not matter, since the current is the same in all series components.

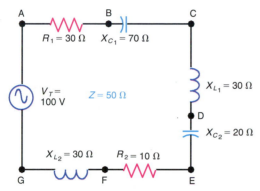

FIG. 24-9   Series circuit with more components than Fig. 24-8 but the same Z_T, I, and θ_Z.

The total series X_C is 90 Ω, equal to the sum of 70 Ω for X_{C_1} and 20 Ω for X_{C_2}. Similarly, the total series X_L is 60 Ω. This value is equal to the sum of 30 Ω for X_{L_1} and 30 Ω for X_{L_2}.

The net reactance X equals 30 Ω, which is 90 Ω of X_C minus 60 Ω of X_L. Since X_C is larger than X_L, the net reactance is capacitive. The circuit in Fig. 24-9 is equivalent to Fig. 24-8, therefore, since a 40-Ω R is in series with a net X_C of 30 Ω.

DOUBLE-SUBSCRIPT NOTATION   This method for specifying ac and dc voltages is useful to indicate the polarity or phase. For instance, in Fig. 24-9 the voltage across R_2 can be taken as either V_{EF} or V_{FE}. With opposite subscripts, these two voltages are $180°$ out of phase. In using double subscripts, note that the first letter in the subscript is the point of measurement with respect to the second letter.

TEST-POINT QUESTION 24-5

Answers at end of chapter.
a. In Fig. 24-8, how much is the net reactance?
b. In Fig. 24-9, how much is the net reactance?
c. In Fig. 24-9, give the phase difference between V_{CD} and V_{DC}.

24-6 PARALLEL REACTANCE AND RESISTANCE

With parallel circuits, the branch currents for resistance and reactance are added by phasors. Then the total line current is found by the formula

$$I_T = \sqrt{I_R^2 + I_X^2} \qquad (24\text{-}2)$$

CALCULATING I_T   As an example, Fig. 24-10a shows a circuit with three branches. Since the voltage across all the parallel branches is the applied 100 V, the individual branch currents are

$$I_R = \frac{V_A}{R} = \frac{100\ V}{25\ \Omega} = 4\ A$$

$$I_L = \frac{V_A}{X_L} = \frac{100\ V}{25\ \Omega} = 4\ A$$

$$I_C = \frac{V_A}{X_C} = \frac{100\ V}{100\ \Omega} = 1\ A$$

The net reactive branch current I_X is 3 A, then, equal to the difference between the 4-A I_L and the 1-A I_C, as shown in Fig. 24-10b.

The next step is to calculate I_T as the phasor sum of I_R and I_X. Then

$$I_T = \sqrt{I_R^2 + I_X^2} = \sqrt{4^2 + 3^2} = \sqrt{16 + 9} = \sqrt{25}$$
$$I_T = 5\ A$$

The phasor diagram for I_T is shown in Fig. 24-10c.

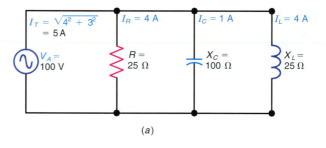

(a)

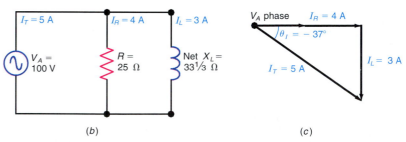

(b)

(c)

FIG. 24-10   Total line current I_T of parallel circuit with resistance and reactance. (a) Parallel branches with I_R, I_C, and I_L. (b) Equivalent circuit with net I_X. (c) Phasor diagram.

$Z_{EQ} = V_A/I_T$ This gives the total impedance of a parallel circuit. In this example, Z_{EQ} is 100 V/5 A, which equals 20 Ω. This value is the equivalent impedance of all three branches in parallel across the source.

PHASE ANGLE   The phase angle of the parallel circuit is found from the branch currents. Now $θ$ is the angle whose tangent equals I_X/I_R.

For this example, I_X is the net inductive current of the 3-A I_L. Also, I_R is 4 A. These phasors are shown in Fig. 24-10c. Then $θ$ is a negative angle with the tangent of $-¾$ or -0.75. This phase angle is $-37°$, approximately.

The negative angle for I_T indicates lagging inductive current. The value of $-37°$ is the phase angle of I_T with respect to the voltage reference V_A.

When Z_{EQ} is calculated as V_A/I_T for a parallel circuit, the phase angle is the same value as for I_T but with opposite sign. In this example, Z_{EQ} is 20 Ω with a phase angle of $+37°$, for an I_T of 5 A with an angle of $-37°$. We can consider that Z_{EQ} has the phase angle of the voltage source with respect to I_T.

Example

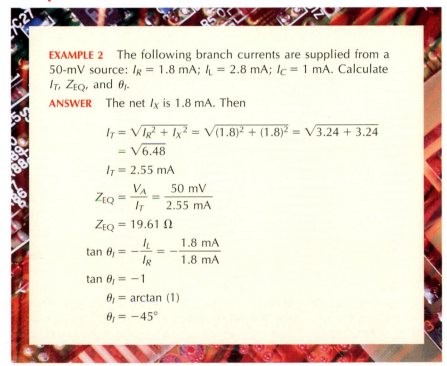

EXAMPLE 2   The following branch currents are supplied from a 50-mV source: $I_R = 1.8$ mA; $I_L = 2.8$ mA; $I_C = 1$ mA. Calculate I_T, Z_{EQ}, and $θ_I$.

ANSWER   The net I_X is 1.8 mA. Then

$$I_T = \sqrt{I_R^2 + I_X^2} = \sqrt{(1.8)^2 + (1.8)^2} = \sqrt{3.24 + 3.24}$$
$$= \sqrt{6.48}$$
$$I_T = 2.55 \text{ mA}$$
$$Z_{EQ} = \frac{V_A}{I_T} = \frac{50 \text{ mV}}{2.55 \text{ mA}}$$
$$Z_{EQ} = 19.61 \text{ Ω}$$
$$\tan θ_I = -\frac{I_L}{I_R} = -\frac{1.8 \text{ mA}}{1.8 \text{ mA}}$$
$$\tan θ_I = -1$$
$$θ_I = \arctan (1)$$
$$θ_I = -45°$$

Note that with equal branch currents, I_T is $1.414 × 1.8 = 2.55$ mA. Also, the phase angle $θ_I$ is negative for inductive branch current.

MORE PARALLEL BRANCHES   Figure 24-11 shows how any number of parallel resistances and reactances can be combined. The total resistive branch current I_R of 4 A is the sum of 2 A each for the R_1 branch and the R_2 branch. Note that the order of connection does not matter, since the parallel branch currents

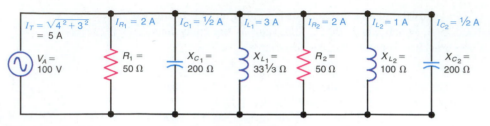

$$I_T = \sqrt{4^2 + 3^2} = 5\text{ A}$$

| $I_{R_1} = 2$ A | $I_{C_1} = \frac{1}{2}$ A | $I_{L_1} = 3$ A | $I_{R_2} = 2$ A | $I_{L_2} = 1$ A | $I_{C_2} = \frac{1}{2}$ A |

$V_A = 100$ V $R_1 = 50\ \Omega$ $X_{C_1} = 200\ \Omega$ $X_{L_1} = 33\frac{1}{3}\ \Omega$ $R_2 = 50\ \Omega$ $X_{L_2} = 100\ \Omega$ $X_{C_2} = 200\ \Omega$

FIG. 24-11 Parallel ac circuit with more components than Fig. 24-10 but the same values of I_T, Z_{EQ}, and θ.

add in the main line. Effectively, two 50-Ω resistances in parallel are equivalent to one 25-Ω resistance.

Similarly, the total inductive branch current I_L is 4 A, equal to 3 A for I_{L_1} and 1 A for I_{L_2}. Also, the total capacitive branch current I_C is 1 A, equal to $\frac{1}{2}$ A each for I_{C_1} and I_{C_2}.

The net reactive branch current I_X is 3 A, then, equal to a 4-A I_L minus a 1-A I_C. Since I_L is larger, the net current is inductive.

The circuit in Fig. 24-11 is equivalent to the circuit in Fig. 24-10, therefore. Both have a 4-A resistive current I_R and a 3-A net reactive current I_X. These values added by phasors make a total of 5 A for I_T in the main line.

24-7 SERIES-PARALLEL REACTANCE AND RESISTANCE

Figure 24-12 shows how a series-parallel circuit can be reduced to a series circuit with just one reactance and one resistance. The method is straightforward as long as resistance and reactance are not combined in one parallel bank or series string.

Working backward toward the generator from the outside branch in Fig. 24-12a, we have an X_{L_1} and an X_{L_2} of 100 Ω each in series, which total 200 Ω. This string in Fig. 24-12a is equivalent to X_{L_5} in Fig. 24-12b.

In the other branch, the net reactance of X_{L_3} and X_C is equal to 600 Ω minus 400 Ω. This is equivalent to the 200 Ω of X_{L_4} in Fig. 24-12b. The X_{L_4} and X_{L_5} of 200 Ω each in parallel are combined for an X_L of 100 Ω.

In Fig. 24-12c, the 100-Ω X_L is in series with the 100-Ω $R_{1\text{-}2}$. This value is for R_1 and R_2 in parallel.

The triangle diagram for the equivalent circuit in Fig. 24-12d shows the total impedance Z of 141 Ω for a 100-Ω R in series with a 100-Ω X_L.

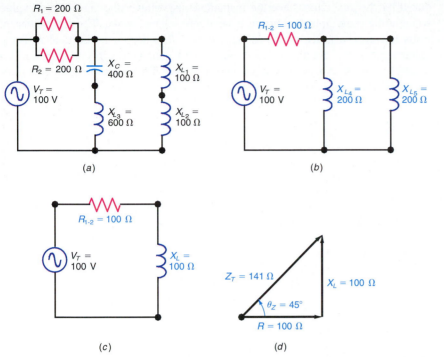

FIG. 24-12   Reducing an ac series-parallel circuit with R, X_L, and X_C to a series circuit with one net resistance and one net reactance. (*a*) Actual circuit. (*b*) Simplified arrangement. (*c*) Equivalent series circuit. (*d*) Impedance triangle with phase angle.

With a 141-Ω impedance across the applied V_T of 100 V, the current in the generator is 0.7 A. The phase angle θ is 45° for this circuit.*

TEST-POINT QUESTION 24-7

Answers at end of chapter.

Refer to Fig. 24-12.
a. How much is $X_{L_1} + X_{L_2}$?
b. How much is $X_{L_3} - X_C$?
c. How much is X_{L_4} in parallel with X_{L_5}?

24-8 REAL POWER

In an ac circuit with reactance, the current I supplied by the generator either leads or lags the generator voltage V. Then the product VI is not the real power pro-

---

*More complicated ac circuits with series-parallel impedances are analyzed with complex numbers, as explained in Chap. 25.

duced by the generator, since the instantaneous voltage may have a high value while at the same time the current is near zero, or vice versa. The real power, in watts however, can always be calculated as $I^2 R$, where R is the total resistive component of the circuit, because current and voltage are in phase in a resistance. To find the corresponding value of power as VI, this product must be multiplied by the cosine of the phase angle θ. Then

▶   Real power $= P = I^2 R$ **(24-3)**

or

▶   Real power $= P = VI \cos \theta$ **(24-4)**

where V and I are in rms values, and P, the real power, is in watts. Multiplying VI by the cosine of the phase angle provides the resistive component for real power equal to $I^2 R$.

For example, the ac circuit in Fig. 24-13 has 2 A through a 100-Ω R in series with the X_L of 173 Ω. Therefore

$$P = I^2 R = 4 \times 100 = 400 \text{ W}$$

Furthermore, in this circuit the phase angle is 60° with a cosine of 0.5. The applied voltage is 400 V. Therefore

$$P = VI \cos \theta = 400 \times 2 \times 0.5 = 400 \text{ W}$$

In both examples, the real power is the same 400 W, because this is the amount of power supplied by the generator and dissipated in the resistance. Either formula can be used for calculating the real power, depending on which is more convenient.

Real power can be considered as resistive power, which is dissipated as heat. A reactance does not dissipate power but stores energy in the electric or magnetic field.

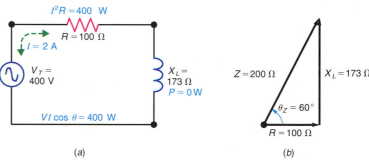

FIG. 24-13   Real power, P, in a series circuit. (*a*) Schematic diagram. (*b*) Impedance triangle with phase angle.

POWER FACTOR   Because it indicates the resistive component, $\cos \theta$ is the power factor of the circuit, converting the VI product to real power. The power factor formulas are

For series circuits:

▶ $$\text{Power factor} = PF = \cos\theta = \frac{R}{Z} \qquad \textbf{(24-5)}$$

For parallel circuits:

▶ $$\text{Power factor} = \cos\theta = \frac{I_R}{I_T} \qquad \textbf{(24-6)}$$

In Fig. 24-13, as an example of a series circuit, we use R and Z for the calculations:

$$PF = \cos\theta = \frac{R}{Z} = \frac{100\ \Omega}{200\ \Omega} = 0.5$$

For the parallel circuit in Fig. 24-10, we use the resistive current I_R and the I_T:

$$PF = \cos\theta = \frac{I_R}{I_T} = \frac{4\ \text{A}}{5\ \text{A}} = 0.8$$

The power factor is not an angular measure but a numerical ratio, with a value between 0 and 1, equal to the cosine of the phase angle.

With all resistance and zero reactance, R and Z are the same for a series circuit, or I_R and I_T are the same for a parallel circuit, and the ratio is 1. Therefore, unity power factor means a resistive circuit. At the opposite extreme, all reactance with zero resistance makes the power factor zero, meaning that the circuit is all reactive. Power factor is frequently given in percent so that unity power factor is 100 percent. To convert from decimal PF to percent PF, merely multiply by 100 percent.

APPARENT POWER   When V and I are out of phase because of reactance, the product of $V \times I$ is called *apparent power.* The unit is *voltamperes* (VA) instead of watts, since the watt is reserved for real power.

For the example in Fig. 24-13, with 400 V and the 2-A I, 60° out of phase, the apparent power is VI, or $400 \times 2 = 800$ VA. Note that apparent power is the VI product alone, without considering the power factor cos θ.

The power factor can be calculated as the ratio of real power to apparent power, as this ratio equals cos θ. As an example, in Fig. 24-13, the real power is 400 W, and the apparent power is 800 VA. The ratio of 400/800 then is 0.5 for the power factor, the same as cos 60°.

THE VAR   This is an abbreviation for voltampere reactive. Specifically, VARs are voltamperes at the angle of 90°.

In general, for any phase angle θ between V and I, multiplying VI by sin θ gives the vertical component at 90° for the value of the VARs. In Fig. 24-13, the value of VI sin 60° is $800 \times 0.866 = 692.8$ VAR.

Note that the factor sin θ for the VARs gives the vertical or reactive component of the apparent power VI. However, multiplying VI by cos θ as the power factor gives the horizontal or resistive component for the real power.

CORRECTING THE POWER FACTOR   In commercial use, the power factor should be close to unity for efficient distribution of electric power. However, the inductive load of motors may result in a power factor of 0.7, as an example, for the phase angle of 45°. To correct for this lagging inductive component of the current in the main line, a capacitor can be connected across the line to draw leading current from the source. To bring the power factor up to 1.0, that is, unity PF, the value of capacitance is calculated to take the same amount of voltamperes as the VARs of the load.

TEST-POINT QUESTION 24-8

Answers at end of chapter.
- **a.** What is the unit for real power?
- **b.** What is the unit for apparent power?
- **c.** Is I^2R real or apparent power?

24-9 AC METERS

The D'Arsonval moving-coil type of meter movement will not read if it is used in an ac circuit because the ac wave is changing polarity too rapidly. Since the two opposite polarities cancel, an alternating current cannot deflect the meter movement either up-scale or down-scale. An ac meter must produce deflection of the meter pointer up-scale regardless of polarity. This deflection is accomplished by one of the following three methods for nonelectronic ac meters.

1. *Thermal type.* In this method, the heating effect of the current, which is independent of polarity, is used to provide meter deflection. Two examples are the thermocouple type and hot-wire meter.

2. *Electromagnetic type.* In this method, the relative magnetic polarity is maintained constant although the current reverses. Examples are the iron-vane meter, dynamometer, and wattmeter.

3. *Rectifier type.* The rectifier changes the ac input to dc output for the meter, which is usually a D'Arsonval movement. This type is the most common for ac voltmeters generally used for audio and radio frequencies.

All analog ac meters (meters with scales and pointers) have scales calibrated in rms values, unless noted otherwise on the meter.

A thermocouple consists of two dissimilar metals joined together at one end but open at the opposite side. Heat at the short-circuited junction produces a small dc voltage across the open ends, which are connected to a dc meter movement. In the hot-wire meter, current heats a wire to make it expand, and this motion is converted into meter deflection. Both types are used as ac meters for radio frequencies.

The iron-vane meter and dynamometer have very low sensitivity, compared with a D'Arsonval movement. They are used in power circuits, for either direct current or 60-Hz alternating current.

TEST-POINT QUESTION 24-9

Answers at end of chapter.

Answer True or False.
a. The iron-vane meter can read alternating current.
b. The D'Arsonval meter movement is for direct current only.

24-10 WATTMETERS

The wattmeter uses fixed coils to measure current in the circuit, while the movable coil measures voltage (Fig. 24-14). The deflection then is proportional to power. Either dc power or real ac power can be read directly by the wattmeter.

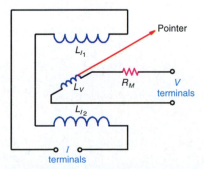

FIG. 24-14 Schematic of voltage and current coils of an analog wattmeter.

In Fig. 24-14, the coils L_{I_1} and L_{I_2} in series are the heavy stationary coils serving as an ammeter to measure current. The two I terminals are connected in one side of the line in series with the load. The movable coil L_V and its multiplier resistance R_M are used as a voltmeter, with the V terminals connected across the line in parallel with the load. Then the current in the fixed coils is proportional to I, while the current in the movable coil is proportional to V. As a result, the deflection is proportional to V and I.

Furthermore, it is the VI product for each instant of time that produces deflection. For instance, if the V value is high when the I value is low, for a phase angle close to 90°, there will be little deflection. The meter deflection is proportional to the watts of real power, therefore, regardless of the power factor in ac circuits. The wattmeter is commonly used to measure power from the 60-Hz power line. For radio frequencies, however, power is generally measured in terms of heat transfer.

Answers at end of chapter.
a. Does a wattmeter measure real or apparent power?
b. In Fig. 24-14, does the movable coil of the wattmeter measure V or I?

24-11 SUMMARY OF TYPES OF OHMS IN AC CIRCUITS

The differences in R, X_L, X_C, and Z_T are listed in Table 24-1, but the following general features should also be noted. Ohms of opposition limit the amount of current in dc circuits or ac circuits. Resistance R is the same for either case. However, ac circuits can have ohms of reactance because of the variations in alternating current or voltage. Reactance X_L is the reactance of an inductance with sine-wave changes in current. Reactance X_C is the reactance of a capacitor with sine-wave changes in voltage.

TABLE 24-1   TYPES OF OHMS IN AC CIRCUITS

	RESISTANCE R, Ω	INDUCTIVE REACTANCE X_L, Ω	CAPACITIVE REACTANCE X_C, Ω	IMPEDANCE Z_T, Ω
Definition	In-phase opposition to alternating or direct current	90° leading opposition to alternating current	90° lagging opposition to alternating current	Combination of resistance and reactance $Z_T = \sqrt{R^2 + X^2}$
Effect of frequency	Same for all frequencies	Increases with higher frequencies	Decreases with higher frequencies	X_L component increases, but X_C decreases with higher frequencies
Phase angle	0°	I_L lags V_L by 90°	I_C leads V_C by 90°	$\tan \theta_Z = \pm\dfrac{X}{R}$ in series, $\tan \theta_I = \pm\dfrac{I_X}{I_R}$ in parallel

Both X_L and X_C are measured in ohms, like R, but reactance has a 90° phase angle, while the phase angle for resistance is 0°. A circuit with steady direct current cannot have any reactance.

Ohms of X_L or X_C are opposite, as X_L has a phase angle of +90°, while X_C has the angle of −90°. Any individual X_L or X_C always has a phase angle that is exactly 90°.

Ohms of impedance Z result from the phasor combination of resistance and reactance. In fact, Z can be considered the general form of any ohms of opposition in ac circuits.

Impedance can have any phase angle, depending on the relative amounts of R and X. When Z consists mostly of R with little reactance, the phase angle of Z is close to 0°. With R and X equal, the phase angle of Z is 45°. Whether the angle is positive or negative depends on whether the net reactance is inductive or capacitive. When Z consists mainly of X with little R, the phase angle of Z is close to 90°.

The phase angle is θ_Z for Z or V_T with respect to the common I in a series circuit. With parallel branch currents, θ_I is for I_T in the main line with respect to the common voltage.

TEST-POINT QUESTION 24-11

Answers at end of chapter.
a. Which of the following does not change with frequency: Z, X_L, X_C, or R?
b. Which has lagging current: R, X_L, or X_C?
c. Which has leading current: R, X_L, or X_C?

24-12 SUMMARY OF TYPES OF PHASORS IN AC CIRCUITS

The phasors for ohms, volts, and amperes are shown in Fig. 24-15. Note the similarities and differences.

SERIES COMPONENTS   In series circuits, ohms and voltage drops have similar phasors. The reason is the common I for all the series components. Therefore:

V_R or IR has the same phase as R.
V_L or IX_L has the same phase as X_L.
V_C or IX_C has the same phase as X_C.

RESISTANCE   The R, V_R, and I_R always have the same phase angle because there is no phase shift in a resistance. This applies to R in either a series or a parallel circuit.

REACTANCE   Reactances X_L and X_C are 90° phasors in opposite directions. The X_L or V_L has the angle of +90° with an upward phasor, while the X_C or V_C has the angle of −90° with a downward phasor.

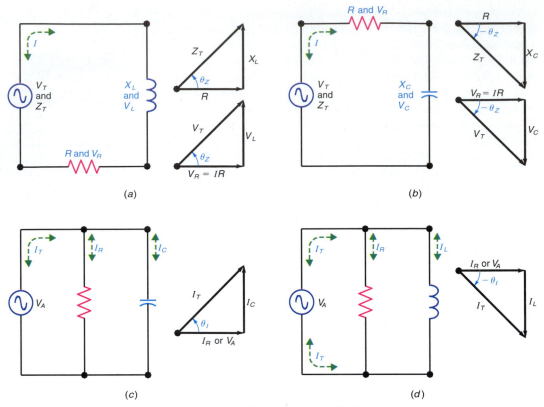

FIG. 24-15 Summary of phasor relations in ac circuits. (*a*) Series *R* and X_L. (*b*) Series *R* and X_C. (*c*) Parallel branches with I_R and I_C. (*d*) Parallel branches with I_R and I_L.

REACTIVE BRANCH CURRENTS   The phasor of a parallel branch current is opposite from its reactance. Therefore, I_C is upward at $+90°$, opposite from X_C downward at $-90°$. Also, I_L is downward at $-90°$, opposite from X_L upward at $+90°$.

   In short, I_C and I_L are opposite from each other, and both are opposite from their corresponding reactances.

ANGLE θ_Z   The phasor resultant for ohms of reactance and resistance is *Z*. The phase angle θ for *Z* can be any angle between $0°$ and $90°$. In a series circuit θ_Z for *Z* is the same as θ for V_T with respect to the common current *I*.

ANGLE θ_I   The phasor resultant of branch currents is the total line current I_T. The phase angle of I_T can be any angle between 0 and $90°$. In a parallel circuit, θ_I is the angle of I_T with respect to the applied voltage V_A.

   Such phasor combinations are necessary in sine-wave ac circuits in order to take into account the effect of reactance. The phasors can be analyzed either graphically, as in Fig. 24-15, or by the shorter technique of complex numbers, with a *j* operator that corresponds to the $90°$ phasor. Complex numbers are explained in the next chapter.

CIRCUIT PHASE ANGLE θ   For all types of sine-wave ac circuits, the phase angle is usually considered as the angle between the current I from the source and its applied voltage as the reference. This angle can be labeled θ, without any subscript. No special identification is necessary because θ is the phase angle of the circuit. Then there are only the two possibilities shown in Fig. 24-16. In Fig. 24-16a, the θ is a counterclockwise angle for a positive value, which means that I leads V. The leading I is in a circuit with series X_C or with I_C in a parallel branch. In Fig. 24-16b, the phase angle is clockwise for $-\theta$, meaning that I lags V. The lagging I is produced in a circuit with series X_L or with I_L in a parallel branch.

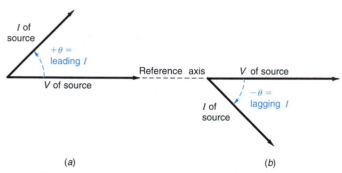

(a)   (b)

FIG. 24-16   Positive and negative values of θ as the phase angle for an ac circuit. (a) Positive θ with I leading V. (b) Negative θ with I lagging V.

It should be noted that, in general, θ is the same as θ_I in parallel branch currents. However, θ has the opposite sign from θ_Z with series reactances.

TEST-POINT QUESTION 24-12

Answers at end of chapter.
a. Of the following phasors, which two are 180° apart: V_L, V_C, or V_R?
b. Of the following phasors, which two are out of phase by 90°: I_R, I_T, or I_L?

24 SUMMARY AND REVIEW

- In ac circuits with resistance alone, the circuit is analyzed the same way as for dc circuits, generally with rms ac values. Without any reactance, the phase angle between V and I is zero.
- When capacitive reactances alone are combined, the X_C values are added in series and combined by the reciprocal formula in parallel, just like ohms of resistance. Similarly, ohms of X_L alone can be added in series or combined by the reciprocal formula in parallel, just like ohms of resistance.
- Since X_C and X_L are opposite reactances, they offset each other. In series, the ohms of X_C and X_L can be subtracted. In parallel, the capacitive and inductive branch currents I_C and I_L can be subtracted.
- In ac circuits, R, X_L, and X_C can be reduced to one equivalent resistance and one net reactance.
- In series, the total R and net X at 90° are combined as $Z_T = \sqrt{R^2 + X^2}$. The phase angle of the series R and X is the angle with tangent $\pm X/R$. To find I, first we calculate Z_T and then divide into V_T.
- For parallel branches, the total I_R and net reactive I_X at 90° are combined as $I_T = \sqrt{I_R^2 + I_X^2}$. The phase angle of the parallel R and X is the angle with tangent $\pm I_X/I_R$. To find Z_T, first we calculate I_T and then divide into V_A.
- The quantities R, X_L, X_C, and Z in ac circuits all are ohms of opposition. The differences with respect to frequency and phase angle are summarized in Table 24-1.
- The phase relations for resistance and reactance are summarized in Fig. 24-15.
- In ac circuits with reactance, the real power, P, in watts equals I^2R, or $VI \cos \theta$, where θ is the phase angle. The real power is the power dissipated as heat in resistance. $\cos \theta$ is the power factor of the circuit.
- The wattmeter measures real ac power or dc power.

SELF-TEST

ANSWERS AT BACK OF BOOK.

Choose (a), (b), (c), or (d).

1. In an ac circuit with resistance but no reactance, (a) two 1000-Ω resistances in series total 1414 Ω; (b) two 1000-Ω resistances in series total 2000 Ω; (c) two 1000-Ω resistances in parallel total 707 Ω; (d) a 1000-Ω R in series with a 400-Ω R totals 600 Ω.
2. An ac circuit has a 100-Ω X_{C_1}, a 50-Ω X_{C_2}, a 40-Ω X_{L_1}, and a 30-Ω X_{L_2}, all in series. The net reactance is equal to (a) an 80-Ω X_L; (b) a 200-Ω X_L; (c) an 80-Ω X_C; (d) a 220-Ω X_C.

3. An ac circuit has a 40-Ω R, a 90-Ω X_L, and a 60-Ω X_C, all in series. The impedance Z equals (a) 50 Ω; (b) 70.7 Ω; (c) 110 Ω; (d) 190 Ω.

4. An ac circuit has a 100-Ω R, a 100-Ω X_L, and a 100-Ω X_C, all in series. The impedance Z of the series combination is equal to (a) 33⅓ Ω; (b) 70.7 Ω; (c) 100 Ω; (d) 300 Ω.

5. An ac circuit has a 100-Ω R, a 300-Ω X_L, and a 200-Ω X_C, all in series. The phase angle θ of the circuit equals (a) 0°; (b) 37°; (c) 45°; (d) 90°.

6. The power factor of an ac circuit equals (a) the cosine of the phase angle; (b) the tangent of the phase angle; (c) zero for a resistive circuit; (d) unity for a reactive circuit.

7. Which phasors in the following combinations are *not* in opposite directions? (a) X_L and X_C; (b) X_L and I_C; (c) I_L and I_C; (d) X_C and I_C.

8. In Fig. 24-8a, the voltage drop across X_L equals (a) 60 V; (b) 66⅔ V; (c) 120 V; (d) 200 V.

9. In Fig. 24-10a, the combined impedance of the parallel circuit equals (a) 5 Ω; (b) 12.5 Ω; (c) 20 Ω; (d) 100 Ω.

10. The wattmeter (a) has voltage and current coils to measure real power; (b) has three connections, two of which are used at a time; (c) measures apparent power because the current is the same in the voltage and current coils; (d) can measure dc power but not 60-Hz ac power.

QUESTIONS

1. Why can series or parallel resistances be combined in ac circuits the same way as in dc circuits?

2. (a) Why do X_L and X_C reactances in series offset each other? (b) With X_L and X_C reactances in parallel, why can their branch currents be subtracted?

3. Give one difference in electrical characteristics comparing R and X_C, R and Z, X_C and C, X_L and L.

4. Name three types of ac meters.

5. Make a diagram showing a resistance R_1 in series with the load resistance R_L, with a wattmeter connected to measure the power in R_L.

6. Make a phasor diagram for the circuit in Fig. 24-8a showing the phase of the voltage drops IR, IX_C, and IX_L with respect to the reference phase of the common current I.

7. Explain briefly why the two opposite phasors at +90° for X_L and −90° for I_L both follow the principle that any self-induced voltage leads the current through the coil by 90°.

8. Why is it that a reactance phasor is always at exactly 90° but an impedance phasor can be less than 90°?

9. Why must the impedance of a series circuit be more than either its X or R?

10. Why must I_T in a parallel circuit be more than either I_R or I_X?

11. Compare real power and apparent power.

12. Define power factor.

13. Make a phasor diagram showing the opposite direction of positive and negative angles.

14. In Fig. 24-15, which circuit has leading current with a positive phase angle θ where I from the source leads the V applied by the source?

PROBLEMS

ANSWERS TO ODD-NUMBERED PROBLEMS AT BACK OF BOOK.

1. For Fig. 24-1a, (**a**) What is the total real power supplied by the source? (**b**) Why is the phase angle zero? (**c**) What is the power factor of the circuit?

2. In a series ac circuit, 2 A flows through a 20-Ω R, a 40-Ω X_L, and a 60-Ω X_C. (**a**) Make a schematic diagram of the series circuit. (**b**) Calculate the voltage drop across each series component. (**c**) How much is the applied voltage? (**d**) Calculate the power factor of the circuit. (**e**) What is the phase angle θ_Z?

3. A parallel circuit has the following five branches: three resistances of 30 Ω each; an X_L of 600 Ω; an X_C of 400 Ω. (**a**) Make a schematic diagram of the circuit. (**b**) If 100 V is applied, how much is I_T? (**c**) What is Z_{EQ} for the circuit? (**d**) What is the phase angle θ_I?

4. Referring to Fig. 24-8, assume that the frequency is doubled from 500 to 1000 Hz. Find X_L, X_C, Z, I, and θ_I for 1000 Hz. Find L and C.

5. A series circuit has a 300-Ω R, a 500-Ω X_{C_1}, a 300-Ω X_{C_2}, an 800-Ω X_{L_1}, and 400-Ω X_{L_2}, all in series with an applied voltage V of 400 V. (**a**) Draw the schematic diagram with all components. (**b**) Draw the equivalent circuit reduced to one resistance and one reactance. (**c**) Calculate Z_T, I, and θ_Z.

6. Repeat Prob. 5 for a circuit with the same components in parallel across the voltage source.

7. A series circuit has a 600-Ω R, a 10-μH inductance L, and a 4-μF capacitance C, all in series with the 60-Hz 120-V power line as applied voltage. (**a**) Find the reactance of L and of C. (**b**) Calculate Z_T, I, and θ_Z.

8. Repeat Prob. 7 for the same circuit, but the 120-V source has $f = 10$ MHz.

9. (**a**) Referring to the series circuit Fig. 24-6, what is the phase angle between the IX_L voltage of 360 V and the IX_C voltage of 240 V? (**b**) Draw the two sine waves for these voltages, showing their relative amplitudes and phase corresponding to the phasor diagram in Fig. 24-6b. Also show the resultant sine wave of voltage across the net X.

10. What resistance dissipates 600 W ac power, with 4.3-A rms current?

11. How much resistance must be inserted in series with a 0.95-H inductance to limit the current to 0.25 A from the 120-V 60-Hz power line?

12. How much resistance is needed in series with a 10-μF capacitance to provide the angle of $-45°$ for θ_Z? The source is the 120-V 60-Hz power line.

13. With the same R as in Prob. 12, what value of C is necessary for θ_Z angle of $-45°$ at the frequency of 2 MHz?

14. A parallel ac circuit has the following branch currents: $I_{R_1} = 4.2$ mA; $I_{R_2} = 2.4$ mA; $I_{L_1} = 7$ mA; $I_{L_2} = 1$ mA; $I_C = 6$ mA. Calculate I_T.

15. What R is needed in series with a 0.01-μF capacitor for θ_Z of $-64°$, with f of 800 Hz?

16. What C is needed with a 5-kΩ R for a phase angle of 45° if $f = 2.5$ MHz?

17. Refer to the series-parallel ac circuit in Fig. 24-17. Calculate Z_T, I, θ_Z, and θ.

18. Calculate the values of L and C for the reactances in Fig. 24-17 with frequency of 8 kHz.

19. (**a**) Double the value of f in Fig. 24-17 to 16 kHz and calculate the values of L and C needed for the reactances given. (**b**) Why are Z_T, I, and θ the same as in Prob. 19?

20. In Fig. 24-18, calculate Z_T, I, V_L, V_C, V_R, θ_Z, real power, apparent power, and power factor (PF).

21. In Fig. 24-19, calculate I_L, I_C, I_R, I_T, Z_{EQ}, θ_I, real power, apparent power, and power factor (PF).

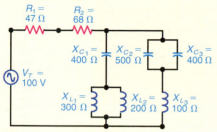

FIG. 24-17   Circuit for Probs. 17, 18, and 19.

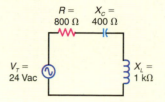

FIG. 24-18   Circuit for Prob. 20.

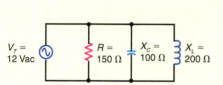

FIG. 24-19   Circuit for Prob. 21.

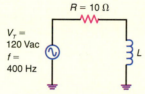

FIG. 24-20   Circuit for Critical Thinking Prob. 1.

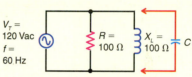

FIG. 24-21   Circuit for Critical Thinking Prob. 2.

CRITICAL THINKING

1. In Fig. 24-20, what value of L will produce a circuit power factor of 0.8?
2. In Fig. 24-21, what value of C in parallel with R and L will produce a power factor of 0.9?

ANSWERS TO TEST-POINT QUESTIONS

24-1 a. 0°
      b. 0°

24-2 a. 90°
      b. −90°

24-3 a. −90°
      b. 90°

24-4 a. 20 Ω
      b. 1 A

24-5 a. $X_C = 30\ \Omega$
      b. $X_C = 30\ \Omega$
      c. 180°

24-6 a. $I_L = 3$ A
      b. $I_L = 3$ A

24-7 a. 200 Ω
      b. 200 Ω
      c. 100 Ω

24-8 a. watt
      b. voltampere
      c. real

24-9 a. T
      b. T

24-10 a. real power
      b. V

24-11 a. R
      b. X_L
      c. X_C

24-12 a. V_L and V_C
      b. I_R and I_L

CHAPTER 25

COMPLEX NUMBERS FOR AC CIRCUITS

Complex numbers refer to a numerical system that includes the phase angle of a quantity, with its magnitude. Therefore, complex numbers are useful in ac circuits when the reactance of X_L or X_C makes it necessary to consider the phase angle. For instance, the complex notation really explains why θ_Z is negative with X_C and θ_I is negative with I_L.

Any type of ac circuit can be analyzed with complex numbers. They are especially convenient for solving series-parallel circuits that have both resistance and reactance in one or more branches. Although graphical analysis with phasor arrows can be used, the method of complex numbers is probably the best way to analyze ac circuits with series-parallel impedances.

CHAPTER OBJECTIVES

Upon completion of this chapter, you should be able to:

- *Explain* the *j* operator.
- *Describe* the makeup of a complex number.
- *Understand* how to add, subtract, multiply, and divide complex numbers.
- *Explain* the difference between the rectangular and polar forms of a complex number.
- *Convert* a complex number from polar to rectangular form and vice versa.
- *Explain* how to use complex numbers to solve series and parallel ac circuits containing resistance, capacitance, and inductance.

IMPORTANT TERMS IN THIS CHAPTER

admittance	*j* operator	rectangular form
complex numbers	polar form	susceptance
imaginary numbers	real numbers	

TOPICS COVERED IN THIS CHAPTER

25-1 POSITIVE AND NEGATIVE NUMBERS

Our common use of numbers as either positive or negative represents only two special cases. In their more general form, numbers have both quantity and phase angle. In Fig. 25-1, positive and negative numbers are shown as corresponding to the phase angles of 0° and 180°, respectively.

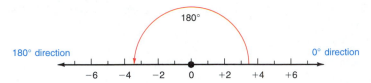

FIG. 25-1 Positive and negative numbers.

For example, the numbers 2, 4, and 6 represent units along the horizontal or *x* axis, extending toward the right along the line of zero phase angle. Therefore, positive numbers really represent units having the phase angle of 0°. Or this phase angle corresponds to the factor of +1. To indicate 6 units with zero phase angle, then, 6 is multiplied by +1 as a factor for the positive number 6. The + sign is often omitted, as it is assumed unless indicated otherwise.

In the opposite direction, negative numbers correspond to 180°. Or, this phase angle corresponds to the factor of −1. Actually, −6 represents the same quantity as 6 but rotated through the phase angle of 180°. The angle of rotation is the *operator* for the number. The operator for −1 is 180°; the operator for +1 is 0°.

25-2 THE *j* OPERATOR

The operator for a number can be any angle between 0° and 360°. Since the angle of 90° is important in ac circuits, the factor *j* is used to indicate 90°. See Fig. 25-2. Here, the number 5 means 5 units at 0°, the number −5 is at 180°, while *j*5 indicates the number 5 at the 90° angle.

The *j* is usually written before the number. The reason is that the *j* sign is a 90° operator, just as the + sign is a 0° operator and the − sign is a 180° operator. Any quantity at right angles to the zero axis, or 90° counterclockwise, is on the +*j* axis.

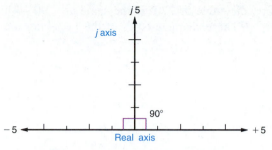

FIG. 25-2 The *j* axis at 90° from the horizontal real axis.

In mathematics, numbers on the horizontal axis are real numbers, including positive and negative values. Numbers on the *j* axis are called *imaginary numbers,* only because they are not on the real axis. In mathematics the abbreviation *i* is used to indicate imaginary numbers. In electricity, however, *j* is used to avoid confusion with *i* as the symbol for current. Furthermore, there is nothing imaginary about electrical quantities on the *j* axis. An electric shock from *j*500 V is just as dangerous as 500 V positive or negative.

More features of the *j* operator are shown in Fig. 25-3. The angle of 180° corresponds to the *j* operation of 90° repeated twice. This angular rotation is indicated by the factor j^2. Note that the *j* operation multiplies itself, instead of adding.

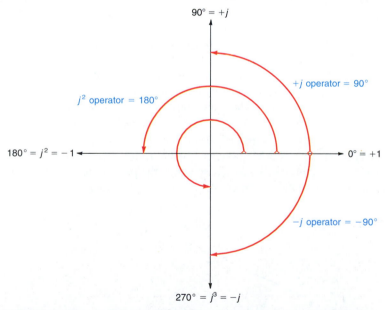

FIG. 25-3 The *j* operator indicates 90° rotation from the real axis; the −*j* operator is −90°; j^2 operation is 180° rotation back to the real axis in a negative direction.

Since j^2 means 180°, which corresponds to the factor of −1, we can say that j^2 is the same as −1. In short, the operator j^2 for a number means multiply by −1. For instance, $j^2 8$ is −8.

Furthermore, the angle of 270° is the same as −90°, which corresponds to the operator −j. These characteristics of the j operator are summarized as follows:

▶ $0° = 1$
$90° = j$
$180° = j^2 = -1$
$270° = j^3 = j^2 \times j = -1 \times j = -j$
$360° = $ same as $0°$

As examples, the number 4 or −4 represents 4 units on the real horizontal axis; $j4$ means 4 units with a leading phase angle of 90°; −$j4$ means 4 units with a lagging phase angle of −90°.

TEST-POINT QUESTION 25-2

Answers at end of chapter.
a. What is the angle for the operator j?
b. What is the angle for the operator −j?

25-3 DEFINITION OF A COMPLEX NUMBER

The combination of a real and an imaginary term is called a *complex number.* Usually, the real number is written first. As an example, $3 + j4$ is a complex number including 3 units on the real axis added to 4 units 90° out of phase on the j axis. Complex numbers must be added as phasors.

Phasors for complex numbers are shown in Fig. 25-4 as typical examples. The $+j$ phasor is up for 90°; the $−j$ phasor is down for −90°. The phasors are shown with the end of one joined to the start of the next, to indicate addition. Graphically, the sum is the hypotenuse of the right triangle formed by the two phasors. Since a number like $3 + j4$ specifies the phasors in rectangular coordinates, this system is the *rectangular form* of complex numbers.

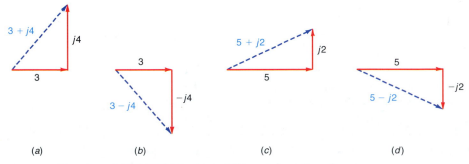

(a)   (b)   (c)   (d)

FIG. 25-4 Phasors corresponding to real terms and imaginary (j) terms, in rectangular coordinates.

Be careful to distinguish a number like $j2$, where 2 is a coefficient, from j^2, where 2 is the exponent. The number $j2$ means 2 units up on the j axis of 90°. However, j^2 is the operator of -1, which is on the real axis in the negative direction.

Another comparison to note is between $j3$ and j^3. The number $j3$ is 3 units up on the j axis, while j^3 is the same as the $-j$ operator, which is down on the $-90°$ axis.

Also note that either the real term or j term can be the larger of the two. When the j term is larger, the angle is more than 45°; when the j term is smaller, the angle is less than 45°. If the j term and the real term are equal, the angle is 45°.

25-4 HOW COMPLEX NUMBERS ARE APPLIED TO AC CIRCUITS

Applications of complex numbers are just a question of using a real term for 0°, $+j$ for 90°, and $-j$ for $-90°$, to denote the phase angles. Figure 25-5 illustrates the following rules:

An *angle of 0°* or a real number without any j operator is used for resistance R. For instance, 3 Ω of R is stated just as 3 Ω.

An *angle of 90° or $+j$* is used for inductive reactance X_L. For instance, a 4-Ω X_L is $j4$ Ω. This rule always applies to X_L, whether it is in series or parallel with R. The reason is the fact that IX_L represents voltage across an inductance, which always leads the current in the inductance by 90°. The $+j$ is also used for V_L.

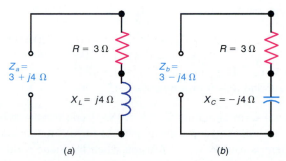

(a)   (b)

FIG. 25-5 Rectangular form of complex numbers for impedances. (*a*) Reactance X_L is $+j$. (*b*) Reactance X_C is $-j$.

An *angle of* $-90°$ *or* $-j$ is used for X_C. For instance, a 4-Ω X_C is $-j4\ \Omega$. This rule always applies to X_C, whether it is in series or parallel with R. The reason is that IX_C is the voltage across a capacitor, which always lags the capacitor's charge and discharge current by $-90°$. The $-j$ is also used for V_C.

With reactive branch currents, the sign for j is reversed, compared with reactive ohms, because of the opposite phase angle. In Fig. 25-6a and b on the next page, $-j$ is used for inductive branch current I_L and $+j$ for capacitive branch current I_C.

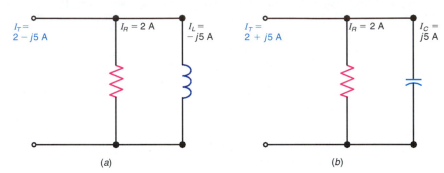

FIG. 25-6 Rectangular form of complex numbers for branch currents. (a) Current I_L is $-j$. (b) Current I_C is $+j$.

TEST-POINT QUESTION 25-4

Answers at end of chapter.
a. Write 3 kΩ of X_L using the j operator.
b. Write 5 mA of I_L using the j operator.

25-5 IMPEDANCE IN COMPLEX FORM

The rectangular form of complex numbers is a convenient way to state the impedance of series resistance and reactance. In Fig. 25-5a, the impedance is $3 + j4$, as Z_a is the phasor sum of a 3-Ω R in series with $j4\ \Omega$ for X_L. Similarly, Z_b is $3 - j4$ for a 3-Ω R in series with $-j4\ \Omega$ for X_C. The minus sign in Z_b results from adding the negative term for $-j$. That is, $3 + (-j4) = 3 - j4$.

For a 4-kΩ R and a 2-kΩ X_L in series: $Z_T = 4000 + j2000\ \Omega$
For a 3-kΩ R and a 9-kΩ X_C in series: $Z_T = 3000 - j9000\ \Omega$
For $R = 0$ and a 7-Ω X_L in series: $Z_T = 0 + j7\ \Omega$
For a 12-Ω R and $X = 0$ in series: $Z_T = 12 + j0$

Note the general form of stating $Z = R \pm j\text{X}$. If one term is zero, substitute 0 for this term, in order to keep Z in its general form. This procedure is not required, but there is usually less confusion when the same form is used for all types of Z.

The advantage of this method is that multiple impedances written as complex numbers can then be calculated as follows:

For series impedances:

▶ $Z_T = Z_1 + Z_2 + Z_3 + \cdots + \text{etc.}$

For parallel impedances:

▶ $\dfrac{1}{Z_T} = \dfrac{1}{Z_1} + \dfrac{1}{Z_2} + \dfrac{1}{Z_3} + \cdots + \text{etc.}$

For two parallel impedances:

▶ $Z_T = \dfrac{Z_1 \times Z_2}{Z_1 + Z_2}$

Examples are shown in Fig. 25-7. The circuit in Fig. 25-7*a* is just a series combination of resistances and reactances. Combining the real terms and *j* terms separately, $Z_T = 12 + j4$. The calculations are $3 + 9 = 12\ \Omega$ for *R* and *j*6 added to $-j2$ equals *j*4 for the net *X*.

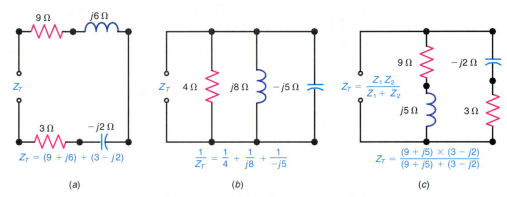

(a) (b) (c)

FIG. 25-7 Reactance X_L is a $+j$ term and X_C is a $-j$ term whether in series or parallel. (*a*) Series circuit. (*b*) Parallel branches. (*c*) Complex branch impedances Z_1 and Z_2 in parallel.

The parallel circuit in Fig. 25-7*b* shows that X_L is $+j$ and X_C is $-j$ even though they are in parallel branches, as they are reactances, not currents.

So far, these types of circuits can be analyzed with or without complex numbers. For the series-parallel circuit in Fig. 25-7*c*, however, the notation of complex numbers is necessary to state the complex impedance Z_T, consisting of branches with reactance and resistance in one or more of the branches. Impedance Z_T is just stated here in its form as a complex impedance. In order to calculate Z_T, some of the rules described in the next section must be used for combining complex numbers.

TEST-POINT QUESTION 25-5

Answers at end of chapter.

Write the following impedances in complex form.
a. X_L of 7 Ω in series with *R* of 4 Ω.
b. X_C of 7 Ω in series with zero *R*.

Real numbers and j terms cannot be combined directly because they are 90° out of phase. The following rules apply:

FOR ADDITION OR SUBTRACTION Add or subtract the real and j terms separately:

$$(9 + j5) + (3 + j2) = 9 + 3 + j5 + j2$$
$$= 12 + j7$$
$$(9 + j5) + (3 - j2) = 9 + 3 + j5 - j2$$
$$= 12 + j3$$
$$(9 + j5) + (3 - j8) = 9 + 3 + j5 - j8$$
$$= 12 - j3$$

The answer should be in the form of $R \pm jX$, where R is the algebraic sum of all the real or resistive terms and X is the algebraic sum of all the imaginary or reactive terms.

TO MULTIPLY OR DIVIDE A j TERM BY A REAL NUMBER Just multiply or divide the numbers. The answer is still a j term. Note the algebraic signs in the following examples. If both factors have the same sign, either + or −, the answer is +; if one factor is negative, the answer is negative.

$$4 \times j3 = j12 \qquad\qquad j12 \div 4 = j3$$
$$j5 \times 6 = j30 \qquad\qquad j30 \div 6 = j5$$
$$j5 \times (-6) = -j30 \qquad -j30 \div (-6) = j5$$
$$-j5 \times 6 = -j30 \qquad -j30 \div 6 = -j5$$
$$-j5 \times (-6) = j30 \qquad j30 \div (-6) = -j5$$

TO MULTIPLY OR DIVIDE A REAL NUMBER BY A REAL NUMBER
Just multiply or divide the real numbers, as in arithmetic. There is no j operation. The answer is still a real number.

TO MULTIPLY A j TERM BY A j TERM Multiply the numbers and the j coefficients to produce a j^2 term. The answer is a real term because j^2 is −1, which is on the real axis. Multiplying two j terms shifts the number 90° from the j axis to the real axis of 180°. As examples:

$$j4 \times j3 = j^2 12 = (-1)(12)$$
$$= -12$$
$$j4 \times (-j3) = -j^2 12 = -(-1)(12)$$
$$= 12$$

TO DIVIDE A _j_ TERM BY A _j_ TERM   Divide the _j_ coefficients to produce a real number; the _j_ factors cancel. For instance:

$$j12 \div j4 = 3 \qquad\qquad -j12 \div j4 = -3$$
$$j30 \div j5 = 6 \qquad\qquad j30 \div (-j6) = -5$$
$$j15 \div j3 = 5 \qquad\qquad -j15 \div (-j3) = 5$$

TO MULTIPLY COMPLEX NUMBERS   Follow the rules of algebra for multiplying two factors, each having two terms:

$$(9 + j5) \times (3 - j2) = 27 - j18 + j15 - j^210$$
$$= 27 - j3 - (-1)10$$
$$= 27 - j3 + 10$$
$$= 37 - j3$$

Note that $-j^210$ equals $+10$ because the operator j^2 is -1 and $-(-1)10$ becomes $+10$.

TO DIVIDE COMPLEX NUMBERS   This process becomes more involved because division of a real number by an imaginary number is not possible. Therefore, the denominator must first be converted to a real number without any _j_ term.

Converting the denominator to a real number without any _j_ term is called _rationalization_ of the fraction. To do this, multiply both numerator and denominator by the _conjugate_ of the denominator. Conjugate complex numbers have equal terms but opposite signs for the _j_ term. For instance, $(1 + j2)$ has the conjugate $(1 - j2)$.

Rationalization is permissible because the value of a fraction is not changed when both numerator and denominator are multiplied by the same factor. This procedure is the same as multiplying by 1. In the following example of division with rationalization the denominator $(1 + j2)$ has the conjugate $(1 - j2)$:

$$\frac{4 - j1}{1 + j2} = \frac{4 - j1}{1 + j2} \times \frac{(1 - j2)}{(1 - j2)} = \frac{4 - j8 - j1 + j^22}{1 - j^2 + j^2 - j^24}$$
$$= \frac{4 - j9 - 2}{1 + 4}$$
$$= \frac{2 - j9}{5}$$
$$= 0.4 - j1.8$$

As a result of the rationalization, $4 - j1$ has been divided by $1 + j2$ to find the quotient that is equal to $0.4 - j1.8$.

Note that the product of a complex number and its conjugate always equals the sum of the squares of the numbers in each term. As another example, the product of $(2 + j3)$ and its conjugate $(2 - j3)$ must be $4 + 9$, which equals 13. Simple numerical examples of division and multiplication are given here because when the required calculations become too long, it is easier to divide and multiply complex numbers in polar form, as explained in Sec. 25-8.

Answers at end of chapter.
a. $(2 + j3) + (3 + j4) = ?$
b. $(2 + j3) \times 2 = ?$

25-7 MAGNITUDE AND ANGLE OF A COMPLEX NUMBER

In electrical terms a complex impedance $(4 + j3)$ means 4 Ω of resistance and 3 Ω of inductive reactance with a leading phase angle of 90°. See Fig. 25-8a. The magnitude of Z is the resultant, equal to $\sqrt{16 + 9} = \sqrt{25} = 5$ Ω. Finding the square root of the sum of the squares is vector or phasor addition of two terms in quadrature, 90° out of phase.

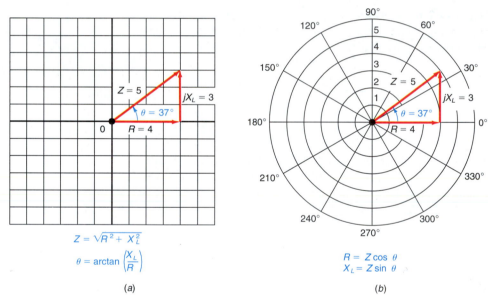

(a) (b)

FIG. 25-8 Magnitude and angle of a complex number. (a) Rectangular form. (b) Polar form.

The phase angle of the resultant is the angle whose tangent is ¾ or 0.75. This angle equals 37°. Therefore, $4 + j3 = 5 \underline{/37°}$.

When calculating the tangent ratio, note that the j term is the numerator and the real term is the denominator because the tangent of an angle is the ratio of the opposite side to the adjacent side. With a negative j term, the tangent is negative, which means a negative angle.

Note the following definitions: $(4 + j3)$ is the complex number in rectangular coordinates. The real term is 4. The imaginary term is $j3$. The resultant 5 is the magnitude, absolute value, or modulus of the complex number. Its phase angle or argument is 37°. The resultant value by itself can be written as $|5|$, with vertical lines to indicate it is the magnitude without the phase angle. The magnitude is the value a meter would read. For instance, with a current of $5\underline{/37°}$ A in a circuit, an ammeter reads 5 A. As additional examples:

$$2 + j4 = \sqrt{4 + 16}\ \underline{/\arctan 2} = 4.47\underline{/63°}$$
$$4 + j2 = \sqrt{16 + 4}\ \underline{/\arctan 0.5} = 4.47\underline{/26.5°}$$
$$8 + j6 = \sqrt{64 + 36}\ \underline{/\arctan 0.75} = 10\underline{/37°}$$
$$8 - j6 = \sqrt{64 + 36}\ \underline{/\arctan -0.75} = 10\underline{/-37°}$$
$$4 + j4 = \sqrt{16 + 16}\ \underline{/\arctan 1} = 5.66\underline{/45°}$$
$$4 - j4 = \sqrt{16 + 16}\ \underline{/\arctan -1} = 5.66\underline{/-45°}$$

Note that arctan 0.75 in the third example means the angle with a tangent equal to 0.75. This value is ⅝ or ¾ for the ratio of the opposite side to the adjacent side. The arctan can also be indicated as $\tan^{-1} 0.75$. In either case, this angle is specified as having 0.75 for its tangent, which makes the angle 36.87°.

Many scientific calculators have keys that can convert from rectangular coordinates to the magnitude-phase angle form (called polar coordinates) directly. See your calculator manual for the particular steps used. If your calculator does not have these keys, the problem can be done in two separate parts: (1) the magnitude as the square root of the sum of two squares, and (2) the angle as the arctan equal to the j term divided by the real term.

Using the calculator for the magnitude in the first example, punch in 2 and then press the $\boxed{x^2}$ key for the square, equal to 4. Press $\boxed{+}$ then $\boxed{(}$ and 4; press $\boxed{x^2}$, $\boxed{)}$, $\boxed{=}$, and $\boxed{\sqrt{\ }}$ in sequence. The display will show 4.47 which is the magnitude.

To find the angle from its tangent value, after the display is cleared, punch in 4 for the opposite side. Then press the $\boxed{÷}$ key, punch in 2 for the adjacent side, and push the $\boxed{=}$ key for the ratio of 2 as the tangent. With 2 as tan θ on the display, press the $\boxed{\text{TAN}^{-1}}$ key, which is usually the second function of the $\boxed{\text{TAN}}$ key. Then 63.4 appears on the display as the angle. Be sure the calculator is set for degrees in the answer, not rad or grad units.

TEST-POINT QUESTION 25-7

Answers at end of chapter.

For the complex impedance $10 + j10$ Ω,
a. Calculate the magnitude.
b. Calculate the phase angle.

25-8 POLAR FORM OF COMPLEX NUMBERS

Calculating the magnitude and phase angle of a complex number is actually converting to an angular form in polar coordinates. As shown in Fig. 25-8, the rectangular form $4 + j3$ is equal to $5\underline{/37°}$ in polar form. In polar coordinates, the distance out from the center is the magnitude of the phasor Z. Its phase angle θ is counterclockwise from the 0° axis.

To convert any complex number to polar form:

1. Find the magnitude by phasor addition of the j term and real term.
2. Find the angle whose tangent is the j term divided by the real term. As examples:

$$2 + j4 = 4.47 \angle 63°$$
$$4 + j2 = 4.47 \angle 26.5°$$
$$8 + j6 = 10 \angle 37°$$
$$8 - j6 = 10 \angle -37°$$
$$4 + j4 = 5.66 \angle 45°$$
$$4 - j4 = 5.66 \angle -45°$$

These examples are the same as those given before for finding the magnitude and phase angle of a complex number.

The magnitude in polar form must be more than either term in rectangular form, but less than their arithmetic sum. For instance, in $8 + j6 = 10 \angle 37°$ the magnitude of 10 is more than 8 or 6 but less than their sum of 14.

Applied to ac circuits with resistance for the real term and reactance for the j term, then, the polar form of a complex number states the resultant impedance and its phase angle. Note the following cases for an impedance where either the resistance or reactance is zero.

$$0 + j5 = 5 \angle 90°$$
$$0 - j5 = 5 \angle -90°$$
$$5 + j0 = 5 \angle 0°$$

The polar form is much more convenient for multiplying or dividing complex numbers. The reason is that multiplication in polar form merely involves multiplying the magnitudes and adding the angles. Division involves dividing the magnitudes and subtracting the angles. The following rules apply.

FOR MULTIPLICATION   Multiply the magnitudes but add the angles algebraically:

$$24 \angle 40° \times 2 \angle 30° = 24 \times 2 \angle 40° + 30° = 48 \angle +70°$$
$$24 \angle 40° \times (-2 \angle 30°) = -48 \angle +70°$$
$$12 \angle -20° \times 3 \angle -50° = 36 \angle -70°$$
$$12 \angle -20° \; 4 \angle 5° = 48 \angle -15°$$

When you multiply by a real number, just multiply the magnitudes:

$$4 \times 2 \angle 30° = 8 \angle 30°$$
$$4 \times 2 \angle -30° = 8 \angle -30°$$
$$-4 \times 2 \angle 30° = -8 \angle 30°$$
$$-4 \times (-2 \angle 30°) = 8 \angle 30°$$

This rule follows from the fact that a real number has an angle of 0°. When you add 0° to any angle, the sum equals the same angle.

FOR DIVISION   Divide the magnitudes and subtract the angles algebraically:

$$24\,\angle 40° \div 2\,\angle 30° = 24 \div 2\,\angle 40° - 30° = 12\,\angle 10°$$
$$12\,\angle 20° \div 3\,\angle 50° = 4\,\angle -30°$$
$$12\,\angle -20° \div 4\,\angle 50° = 3\,\angle -70°$$

To divide by a real number, just divide the magnitudes:

$$12\,\angle 30° \div 2 = 6\,\angle 30°$$
$$12\,\angle -30° \div 2 = 6\,\angle -30°$$

This rule is also a special case that follows from the fact that a real number has a phase angle of 0°. When you subtract 0° from any angle, the remainder equals the same angle.

For the opposite case, however, when you divide a real number by a complex number, the angle of the denominator changes its sign in the answer in the numerator. This rule still follows the procedure of subtracting angles for division, since a real number has a phase angle of 0°. As examples,

$$\frac{10}{5\,\angle 30°} = \frac{10\,\angle 0°}{5\,\angle 30°} = 2\,\angle 0° - 30°$$
$$= 2\,\angle -30°$$

$$\frac{10}{5\,\angle -30°} = \frac{10\,\angle 0°}{5\,\angle -30°} = 2\,\angle 0° - (-30°)$$
$$= 2\,\angle +30°$$

Stated another way, we can say that the reciprocal of an angle is the same angle but with opposite sign. Note that this operation is similar to working with powers of 10. Angles and powers of 10 follow the general rules of exponents.

TEST-POINT QUESTION 25-8

Answers at end of chapter.
a. $6\,\angle 20° \times 2\,\angle 30° = ?$
b. $6\,\angle 20° \div 2\,\angle 30° = ?$

25-9 CONVERTING POLAR TO RECTANGULAR FORM

Complex numbers in polar form are convenient for multiplication and division, but they cannot be added or subtracted if their angles are different. The reason is the real and imaginary parts that make up the magnitude are different. When complex numbers in polar form are to be added or subtracted, therefore, they must be converted back into rectangular form.

Consider the impedance $Z\,\angle\,\theta$ in polar form. Its value is the hypotenuse of a right triangle with sides formed by the real term and j term in rectangular coordinates. See Fig. 25-9. Therefore, the polar form can be converted to rectangular form by finding the horizontal and vertical sides of the right triangle. Specifically:

▶ Real term for $R = Z \cos \theta$

j term for $X = Z \sin \theta$

In Fig. 25-9a, assume that $Z \angle \theta$ in polar form is $5 \angle 37°$. The sine of 37° is 0.6 and its cosine is 0.8.

To convert to rectangular form:

$$R = Z \cos \theta = 5 \times 0.8 = 4$$
$$X = Z \sin \theta = 5 \times 0.6 = 3$$

Therefore,

$$5 \angle 37° = 4 + j3$$

This example is the same as the illustration in Fig. 25-8. The $+$ sign for the j term means it is X_L, not X_C.

In Fig. 25-9b, the values are the same, but the j term is negative when θ is negative. The negative angle has a negative j term because the opposite side is in the fourth quadrant, where the sine is negative. However, the real term is still positive because the cosine is positive.

Note that R for $\cos \theta$ is the horizontal component, which is an adjacent side of the angle. The X for $\sin \theta$ is the vertical component, which is opposite the angle. The $+X$ is X_L; the $-X$ is X_C.

These rules apply for angles in the first or fourth quadrant, from 0 to 90° or from 0 to −90°. As examples:

$$14.14 \angle 45° = 14.14 \cos 45° + 14.14 \sin 45° = 10 + j10$$
$$14.14 \angle -45° = 14.14 \cos (-45°) + 14.14 \sin (-45°) = 10 + j(-10) = 10 - j10$$
$$10 \angle 90° = 0 + j10$$
$$10 \angle -90° = 0 - j10$$
$$100 \angle 30° = 86.6 + j50$$
$$100 \angle -30° = 86.6 - j50$$
$$100 \angle 60° = 50 + j86.6$$
$$100 \angle -60° = 50 - j86.6$$

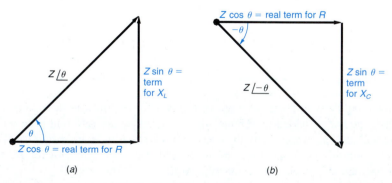

(a)   (b)

FIG. 25-9 Converting polar form of $Z \angle \theta$ to rectangular form of $R \pm jX$. (a) Positive angle θ in first quadrant has $+j$ term. (b) Negative angle $-\theta$ in fourth quadrant has $-j$ term.

When going from one form to the other, keep in mind whether the angle is smaller or greater than 45° and if the j term is smaller or larger than the real term. For angles between 0 and 45°, the opposite side, which is the j term, must be smaller than the real term. For angles between 45 and 90°, the j term must be larger than the real term.

Conversion to rectangular form can be done fast with a calculator. Again, some scientific calculators contain conversion keys that make going from polar coordinates to rectangular coordinates a simple four-key procedure. Check your calculator manual for the exact procedure. If your calculator does not have this capability, use the following routine. Punch in the value of the angle θ in degrees. Make sure the correct sign is used and the calculator is set to handle angles in degrees. Find $\cos \theta$ or $\sin \theta$ and multiply by the magnitude for each term. Remember to use $\cos \theta$ for the real term and $\sin \theta$ for the j term. For the example of $100 \angle 30°$, punch in the number 30 and press the (COS) key for 0.866 as $\cos \theta$. While it is on the display, press the (×) key, punch in 100, and press the (=) key or the answer of 86.6 as the real term. Clear the display for the next operation with $\sin \theta$. Punch in 30, push the (SIN) key for 0.5 as $\sin \theta$, press the (×) key, punch in 100, and push the (=) key for the answer of 50 as the j term.

To summarize how complex numbers are used in ac circuits in rectangular and polar form:

1. For addition or subtraction, complex numbers must be in rectangular form. This procedure applies to the addition of impedances in a series circuit. If the series impedances are in rectangular form, just combine all the real terms and the j terms separately. If the series impedances are in polar form, they must be converted to rectangular form to be added.
2. For multiplication and division, complex numbers are generally used in polar form because the calculations are faster. If the complex number is in rectangular form, convert to polar form. With the complex number available in both forms, then you can quickly add or subtract in rectangular form and multiply or divide in polar form. Sample problems showing how to apply these methods in ac circuits are illustrated in the following sections.

TEST-POINT QUESTION 25-9

Answers at end of chapter.

Convert to rectangular form.
a. $14.14 \angle 45°$.
b. $14.14 \angle -45°$.

25-10 COMPLEX NUMBERS IN SERIES AC CIRCUITS

Refer to Fig. 25-10. Although a circuit like this with only series resistances and reactances can be solved graphically with phasor arrows, the complex numbers show more details of the phase angles.

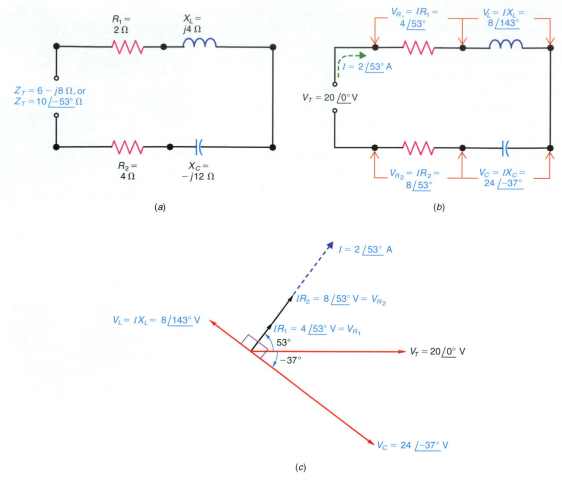

FIG. 25-10 Complex numbers applied to series ac circuits. See text for analysis. (*a*) Circuit with series impedances. (*b*) Current and voltages in the circuit. (*c*) Phasor diagram of current and voltages.

Z_T IN RECTANGULAR FORM

The total Z_T in Fig. 25-10*a* is the sum of the impedances:

$$Z_T = 2 + j4 + 4 - j12$$
$$= 6 - j8$$

The total series impedance then is $6 - j8$. Actually, this amounts to adding all of the series resistances for the real term and finding the algebraic sum of all the series reactances for the j term.

Z_T IN POLAR FORM

We can convert Z_T from rectangular to polar form as follows:

$$Z_T = 6 - j8$$
$$= \sqrt{36 + 64} \; \underline{/\text{arctan} -8/6}$$
$$= \sqrt{100} \; \underline{/\text{arctan} -1.33}$$
$$Z_T = 10\underline{/-53°} \; \Omega$$

The angle of $-53°$ for Z_T means the applied voltage and the current are $53°$ out of phase. Specifically, this angle is θ_Z.

CALCULATING I   The reason for the polar form is to divide Z_T into the applied voltage V_T to calculate the current I. See Fig. 25-10b. Note that the V_T of 20 V is a real number without any j term. Therefore, the applied voltage is $20\angle 0°$. This angle of $0°$ for V_T makes it the reference phase for the following calculations. We can find the current as

$$I = \frac{V_T}{Z_T} = \frac{20\angle 0°}{10\angle -53°} = 2\angle 0° - (-53°)$$
$$I = 2\angle 53° \text{ A}$$

Note that Z_T has the negative angle of $-53°$ but the sign changes to $+53°$ for I because of the division into a quantity with the angle of $0°$. In general, the reciprocal of an angle in polar form is the same angle with opposite sign.

PHASE ANGLE OF THE CIRCUIT   The fact that I has the angle of $+53°$ means it leads V_T. The positive angle for I shows the series circuit is capacitive, with leading current. This angle is more than $45°$ because the net reactance is more than the total resistance, resulting in a tangent function greater than 1.

FINDING EACH IR DROP   To calculate the voltage drops around the circuit, each resistance or reactance can be multiplied by I:

$$V_{R_1} = IR_1 = 2\angle 53° \times 2\angle 0° = 4\angle 53° \text{ V}$$
$$V_L = IX_L = 2\angle 53° \times 4\angle 90° = 8\angle 143° \text{ V}$$
$$V_C = IX_C = 2\angle 53° \times 12\angle -90° = 24\angle -37° \text{ V}$$
$$V_{R_2} = IR_2 = 2\angle 53° \times 4\angle 0° = 8\angle 53° \text{ V}$$

PHASE ANGLE OF EACH VOLTAGE   The phasors for these voltages are in Fig. 25-10c. They show the phase angles using the applied voltage V_T as the zero reference phase.

The angle of $53°$ for V_{R_1} and V_{R_2} shows that the voltage across a resistance has the same phase as I. These voltages lead V_T by $53°$ because of the leading current.

For V_C, its angle of $-37°$ means it lags the generator voltage V_T by this much. However, this voltage across X_C still lags the current by $90°$, which is the difference between $53°$ and $-37°$.

The angle of $143°$ for V_L in the second quadrant is still $90°$ leading the current at $53°$, as $143° - 53° = 90°$. With respect to the generator voltage V_T, though, the phase angle of V_L is $143°$.

TOTAL VOLTAGE V_T EQUALS THE PHASOR SUM OF THE SERIES VOLTAGE DROPS   If we want to add the voltage drops around the circuit to see if they equal the applied voltage, each V must be converted to rectangular form. Then these values can be added. In rectangular form then the individual voltages are

$$
\begin{aligned}
V_{R_1} = 4 \underline{/53°} &= 2.408 + j3.196 \text{ V} \\
V_L = 8 \underline{/143°} &= -6.392 + j4.816 \text{ V} \\
V_C = 24 \underline{/-37°} &= 19.176 - j14.448 \text{ V} \\
V_{R_2} = 8 \underline{/53°} &= 4.816 + j6.392 \text{ V} \\
\text{Total } V &= 20.008 - j0.044 \text{ V}
\end{aligned}
$$

or converting to polar form,

$$V_T = 20 \underline{/0°} \text{ V} \qquad \text{approximately}$$

Note that for $8 \underline{/143°}$ in the second quadrant, the cosine is negative for a negative real term but the sine is positive for a positive j term.*

TEST-POINT QUESTION 25-10

Answers at end of chapter.

Refer to Fig. 25-10.
a. What is the phase angle of I with reference to V_T?
b. What is the phase angle of V_L with reference to V_T?
c. What is the phase angle of V_L with reference to V_R?

25-11 COMPLEX NUMBERS IN PARALLEL AC CIRCUITS

A useful application is converting a parallel circuit to an equivalent series circuit. See Fig. 25-11, with a 10-Ω X_L in parallel with a 10-Ω R. In complex notation, R is $10 + j0$ while X_L is $0 + j10$. Their combined parallel impedance Z_T equals the product over the sum. For Fig. 25-11a, then:

$$Z_T = \frac{(10 + j0) \times (0 + j10)}{(10 + j0) + (0 + j10)} = \frac{10 \times j10}{10 + j10} = \frac{j100}{10 + j10}$$

$$Z_T = \frac{j10}{1 + j1}$$

Converting to polar form for division,

$$Z_T = \frac{j100}{10 + j10} = \frac{100 \underline{/90°}}{} = 7.07 \underline{/45°}$$

---

*For an explanation of quadrants, see B. Grob, *Mathematics for Basic Electronics*, Glencoe/McGraw-Hill, Columbus, Ohio.

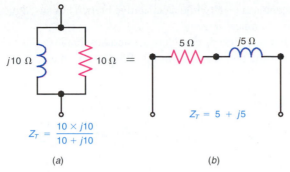

$$Z_T = \frac{10 \times j10}{10 + j10}$$

(a)

$$Z_T = 5 + j5$$

(b)

FIG. 25-11 Complex numbers used for parallel ac circuit to convert a parallel bank to an equivalent series impedance.

Converting the Z_T of $7.07 \underline{/45°}$ into rectangular form to see its resistive and reactive components,

$$\text{Real term} = 7.07 \cos 45°$$
$$= 7.07 \times 0.707 = 5$$
$$j \text{ term} = 7.07 \sin 45°$$
$$= 7.07 \times 0.707 = 5$$

Therefore,

$$Z_T = 7.07 \underline{/45°} \quad \text{in polar form}$$
$$Z_T = 5 + j5 \quad \text{in rectangular form}$$

The rectangular form of Z_T means that 5-Ω R in series with 5-Ω X_L is the equivalent of 10-Ω R in parallel with 10-Ω X_L, as shown in Fig. 25-11*b*.

ADMITTANCE *Y* AND SUSCEPTANCE *B* In parallel circuits, it is usually easier to add branch currents than to combine reciprocal impedances. For this reason, branch conductance G is often used instead of branch resistance, where $G = 1/R$. Similarly, reciprocal terms can be defined for complex impedances. The two main types are *admittance Y,* which is the reciprocal of impedance, and *susceptance B,* which is the reciprocal of reactance. These reciprocals can be summarized as follows:

▶ $$\text{Conductance} = G = \frac{1}{R} \quad \text{S}$$

$$\text{Susceptance} = B = \frac{1}{\pm X} \quad \text{S}$$

$$\text{Admittance} = Y = \frac{1}{Z} \quad \text{S}$$

With R, X, and Z in units of ohms, the reciprocals G, B, and Y are in siemens (S) units.

The phase angle for B or Y is the same as current. Therefore, the sign is opposite from the angle of X or Z because of the reciprocal relation. An inductive

ABOUT ELECTRONICS

Half of the American population doesn't get a good fit with retail clothing. An electronic body scan can fix this by cloning your form for the tailor. In the future, your body "grid" may be modemed to the factory, which will then be able to provide custom-made clothes or shoes.

branch has susceptance $-jB$, while a capacitive branch has susceptance $+jB$, with the same angle as branch current.

With parallel branches of conductance and susceptance the total admittance $Y_T = G \pm jB$. For the two branches in Fig. 25-11a, as an example, G is $\frac{1}{10}$ or 0.1 and B is also 0.1.

In rectangular form:

$$Y_T = 0.1 - j0.1 \text{ S}$$

In polar form:

$$Y_T = 0.14\angle{-45°} \text{ S}$$

This value for Y_T is the same as I_T with 1 V applied across Z_T of $7.07\angle{45°}\ \Omega$.

As another example, suppose that a parallel circuit has 4 Ω for R in one branch and $-j4\ \Omega$ for X_C in the other branch. In rectangular form, then, Y_T is $0.25 + j0.25$ S. Also, the polar form is $Y_T = 0.35\angle{45°}$ S.

TEST-POINT QUESTION 25-11

Answers at end of chapter.
a. A Z of $3 + j4\ \Omega$ is in parallel with an R of 2 Ω. State Z_T in rectangular form.
b. Do the same as in Part **a** for X_C instead of X_L.

25-12 COMBINING TWO COMPLEX BRANCH IMPEDANCES

A common application is a circuit with two branches Z_1 and Z_2, where each is a complex impedance with both reactance and resistance. A circuit, such as the one in Fig. 25-12, can be solved only graphically or by complex numbers. Actually, using complex numbers is the shortest method.

The procedure here is to find Z_T as the product divided by the sum for Z_1 and Z_2. A good way to start is to state each branch impedance in both rectangular and polar forms. Then Z_1 and Z_2 are ready for addition, multiplication, and division. The solution of this circuit is as shown on the next page.

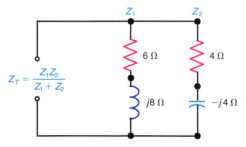

FIG. 25-12 Finding Z_T for any two complex impedances Z_1 and Z_2 in parallel. See text for solution.

$$Z_1 = 6 + j8 = 10\angle 53°$$
$$Z_2 = 4 - j4 = 5.66\angle -45°$$

The combined impedance is

▶ $$Z_T = \frac{Z_1 \times Z_2}{Z_1 + Z_2}$$

Use the polar form of Z_1 and Z_2 to multiply, but add in rectangular form:

$$Z_T = \frac{10\angle 53° \times 5.66\angle -45°}{}$$

$$= \frac{56.6\angle 8°}{}$$

Converting the denominator to polar form for easier division,

$$10 + j4 = 10.8\angle 22°$$

Then

$$Z_T = \frac{56.6\angle 8°}{} = 5.24\angle -14° \ \Omega$$

We can convert Z_T into rectangular form. The R component is $5.24 \times \cos(-14°)$ or $5.24 \times 0.97 = 5.08$. Note that $\cos\theta$ is positive in the first and fourth quadrants. The j component equals $5.24 \times \sin(-14°)$ or $5.24 \times (-0.242) = -1.127$. In rectangular form, then,

$$Z_T = 5.08 - j1.27$$

Therefore, this series-parallel circuit combination is equivalent to 5.08 Ω of R in series with 1.27 Ω of X. Notice that the minus j term means the circuit is capacitive. This problem can also be done in rectangular form by rationalizing the fraction for Z_T.

TEST-POINT QUESTION 25-12

Answers at end of chapter.

Refer to Fig. 25-12.
a. Add $(6 + j8) + (4 - j4)$ for the sum of Z_1 and Z_2.
b. Multiply $10\angle 53° \times 5.66\angle -45°$ for the product of Z_1 and Z_2.

25-13 COMBINING COMPLEX BRANCH CURRENTS

An example with two branches is shown in Fig. 25-13, to find I_T. The branch currents can just be added in rectangular form for the total I_T of parallel branches.

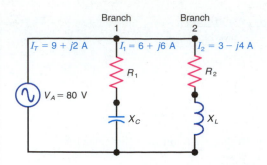

FIG. 25-13 Finding I_T for two branch currents in parallel.

This method corresponds to adding series impedances in rectangular form to find Z_T. The rectangular form is necessary for the addition of phasors.

Adding the branch currents in Fig. 25-13,

$$I_T = I_1 + I_2$$
$$= (6 + j6) + (3 - j4)$$
$$I_T = 9 + j2 \text{ A}$$

Note that I_1 has $+j$ for the $+90°$ of capacitive current, while I_2 has $-j$ for inductive current. These current phasors have the opposite signs from their reactance phasors.

In polar form the I_T of $9 + j2$ A is calculated as the phasor sum of the branch currents.

$$I_T = \sqrt{9^2 + 2^2} = \sqrt{85}$$
$$I_T = 9.22 \text{ A}$$
$$\tan \theta = \frac{2}{9} = 0.222$$
$$\theta_I = \arctan(0.22)$$
$$\theta_I = 12.53°$$

Therefore, I_T is $9 + j2$ A in rectangular form or $9.22 \angle 12.53°$ A in polar form. The complex currents for any number of branches can be added in rectangular form.

TEST-POINT QUESTION 25-13

Answers at end of chapter.
a. Find I_T in rectangular form for I_1 of $0 + j2$ A and I_2 of $4 + j3$ A.
b. Find I_T in rectangular form for I_1 of $6 + j7$ A and I_2 of $3 - j9$ A.

25-14 PARALLEL CIRCUIT WITH THREE COMPLEX BRANCHES

Because the circuit in Fig. 25-14 has more than two complex impedances in parallel, the method of branch currents is used. There will be several conver-

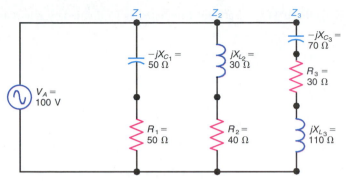

FIG. 25-14 Finding Z_T for any three complex impedances in parallel. See text for solution by means of branch currents.

sions between rectangular and polar form, since addition must be in rectangular form, but division is easier in polar form. The sequence of calculations is:

1. Convert each branch impedance to polar form. This is necessary for dividing into the applied voltage V_A to calculate the individual branch currents. If V_A is not given, any convenient value can be assumed. Note that V_A has a phase angle of $0°$ because it is the reference.
2. Convert the individual branch currents from polar to rectangular form so that they can be added for the total line current. This step is necessary because the resistive and reactive components must be added separately.
3. Convert the total line current from rectangular to polar form for dividing into the applied voltage to calculate Z_T.
4. The total impedance can remain in polar form with its magnitude and phase angle, or can be converted to rectangular form for its resistive and reactive components.

These steps are used in the following calculations to solve the circuit in Fig. 25-14. All the values are in A, V, or Ω units.

BRANCH IMPEDANCES Each Z is converted from rectangular form to polar form:

$$Z_1 = 50 - j50 = 70.7\underline{/-45°}$$
$$Z_2 = 40 + j30 = 50\underline{/+37°}$$
$$Z_3 = 30 + j40 = 50\underline{/+53°}$$

BRANCH CURRENTS Each I is calculated as V_A divided by Z in polar form:

$$I_1 = \frac{V_A}{Z_1} = \frac{100 \angle 0°}{\underline{\qquad}} = 1.414 \angle +45° = 1 + j1$$

$$I_2 = \frac{V_A}{Z_2} = \frac{100 \angle 0°}{\underline{\qquad}} = 2.00 \angle -37° = 1.6 - j1.2$$

$$I_3 = \frac{V_A}{Z_3} = \frac{100 \angle 0°}{\underline{\qquad}} = 2.00 \angle -53° = 1.2 - j1.6$$

The polar form of each I is converted to rectangular form, for addition of the branch currents.

TOTAL LINE CURRENT In rectangular form,

$$\begin{aligned} I_T &= I_1 + I_2 + I_3 \\ &= (1 + j1) + (1.6 - j1.2) + (1.2 - j1.6) \\ &= 1 + 1.6 + 1.2 + j1 - j1.2 - j1.6 \\ I_T &= 3.8 - j1.8 \end{aligned}$$

Converting $3.8 - j1.8$ into polar form,

$$I_T = 4.2 \angle -25.4°$$

TOTAL IMPEDANCE In polar form,

$$Z_T = \frac{V_A}{I_T} = \frac{100 \angle 0°}{\underline{\qquad}}$$

$$Z_T = 23.8 \angle +25.4° \ \Omega$$

Converting $23.8 \angle +25.4°$ into rectangular form,

$$Z_T = 21.5 + j10.2 \ \Omega$$

Therefore, the complex ac circuit in Fig. 25-14 is equivalent to the combination of 21.5 Ω of R in series with 10.2 Ω of X_L. The circuit is inductive.

This problem can also be done by combining Z_1 and Z_2 in parallel as $Z_1 Z_2/(Z_1 + Z_2)$. Then combine this value with Z_3 in parallel to find the total Z_T of the three branches.

TEST-POINT QUESTION 25-14

Answers at end of chapter.

Refer to Fig. 25-14.
a. State Z_2 in rectangular form for branch 2.
b. State Z_2 in polar form.
c. Find I_2.

25 SUMMARY AND REVIEW

- In complex numbers, resistance R is a real term and reactance is a j term. Thus, an 8-Ω R is 8; an 8-Ω X_L is $j8$; an 8-Ω X_C is $-j8$. The general form of a complex impedance with series resistance and reactance then is $Z_T = R \pm jX$, in rectangular form.
- The same notation can be used for series voltages where $V = V_R \pm jV_X$.
- For branch currents $I_T = I_R \pm jI_X$, but the reactive branch currents have signs opposite from impedances. Capacitive branch current is jI_C, while inductive branch current is $-jI_L$.
- The complex branch currents are added in rectangular form for any number of branches to find I_T.
- To convert from rectangular to polar form: $R \pm jX = Z_T \angle\theta$. The angle is θ_Z. The magnitude of Z_T is $\sqrt{R^2 + X^2}$. Also, θ_Z is the angle with $\tan = X/R$.
- To convert from polar to rectangular form, $Z_T \angle\theta_Z = R \pm jX$, where R is $Z_T \cos\theta_Z$ and the j term is $Z_T \sin\theta_Z$. A positive angle has a positive j term; a negative angle has a negative j term. Also, the angle is more than 45° for a j term larger than the real term; the angle is less than 45° for a j term smaller than the real term.
- The rectangular form must be used for addition or subtraction of complex numbers.
- The polar form is usually more convenient in multiplying and dividing complex numbers. For multiplication, multiply the magnitudes and add the angles; for division, divide the magnitudes and subtract the angles.
- To find the total impedance Z_T of a series circuit, add all the resistances for the real term and find the algebraic sum of the reactances for the j term. The result is $Z_T = R \pm jX$. Then convert Z_T to polar form for dividing into the applied voltage to calculate the current.
- To find the total impedance Z_T of two complex branch impedances Z_1 and Z_2 in parallel, Z_T can be calculated as $Z_1Z_2/(Z_1 + Z_2)$.

SELF-TEST

ANSWERS AT BACK OF BOOK.

Match the values in the column at the left with those at the right (*list continues on p. 716*).

1. $24 + j5 + 16 + j10$
2. $24 - j5 + 16 - j10$
3. $j12 \times 4$
4. $j12 \times j4$
5. $j12 \div j3$
6. $(4 + j2) \times (4 - j2)$

a. $14\angle 50°$
b. $7\angle 6°$
c. $1200 - j800\ \Omega$
d. $40 + j15$
e. $90 + j60$ V
f. $45\angle 42°$

7. $1200 \, \Omega$ of $R + 800 \, \Omega$ of X_C **g.** $24\angle -45°$
8. 5 A of $I_R + 7$ A of I_C **h.** 4
9. 90 V of $V_R + 60$ V of V_L **i.** $j48$
10. $14\angle 28° \times \angle 22°$ **j.** -48
11. $14\angle 28° \div 2\angle 22°$ **k.** $5 + j7$ A
12. $15\angle 42° \times 3\angle 0°$ **l.** 20
13. $6\angle -75° \times 4\angle 30°$ **m.** $40 - j15$

QUESTIONS

1. Give the mathematical operator for the angles of 0°, 90°, 180°, 270°, and 360°.
2. Define the sine, cosine, and tangent functions of an angle.
3. How are mathematical operators similar for logarithms, exponents, and angles?
4. Compare the following combinations: resistance R and conductance G; reactance X and susceptance B; impedance Z and admittance Y.
5. What are the units for admittance Y and susceptance B?
6. Why do Z_T and I_T for a circuit have angles with opposite signs?

PROBLEMS

ANSWERS TO ODD-NUMBERED PROBLEMS AT BACK OF BOOK.

1. State Z_T in rectangular form for the following series circuits: **(a)** 4-Ω R and 3-Ω X_C; **(b)** 4-Ω R and 3-Ω X_L; **(c)** 3-Ω R and 6-Ω X_L; **(d)** 3-Ω R and 3-Ω X_C.
2. Draw the schematic diagrams for the impedances in Prob. 1.
3. Convert the following impedances to polar form: **(a)** $4 - j3$; **(b)** $4 + j3$; **(c)** $3 + j$; **(d)** $3 - j3$.
4. Convert the following impedances to rectangular form: **(a)** $5\angle -27°$; **(b)** $5\angle 27°$; **(c)** $6.71\angle 63.4°$; **(d)** $4.24\angle -45°$.
5. Find the total Z_T in rectangular form for the following three series impedances: **(a)** $12\angle 10°$; **(b)** $25\angle 15°$; **(c)** $34\angle 26°$.
6. Multiply the following, in polar form: **(a)** $45\angle 24° \times 10\angle 54°$; **(b)** $45\angle -24° \times 10\angle 54°$; **(c)** $18\angle -64° \times 4\angle 14°$; **(d)** $18\angle -64° \times 4\angle -14°$.
7. Divide the following, in polar form: **(a)** $45\angle 24° \div 10\angle 10°$; **(b)** $45\angle 24° \div 10\angle -10°$; **(c)** $500\angle -72° \div 5\angle 12°$; **(d)** $500\angle -72° \div 5\angle -12°$.
8. Match the four phasor diagrams in Fig. 25-4a, b, c, and d with the four circuits in Figs. 25-5 and 25-6.
9. Find Z_T in polar form for the series circuit in Fig. 25-7a.
10. Find Z_T in polar form for the series-parallel circuit in Fig. 25-7c.
11. In Fig. 25-12, find Z_T in rectangular form by rationalization.
12. Solve the circuit in Fig. 25-12 to find Z_T in polar form, using the method of branch currents. Assume an applied voltage of 56.6 V.

13. Show the equivalent series circuit of Fig. 25-12.

14. Solve the circuit in Fig. 25-14 to find Z_T in polar form, without using branch currents. (Find the Z of two branches in parallel; then combine this Z with the third branch Z.)

15. Show the equivalent series circuit of Fig. 25-14.

16. Refer to Fig. 25-13. **(a)** Find Z_1 and Z_2 for the two branch currents given. **(b)** Calculate the values needed for R_1, R_2, X_C, and X_L for these impedances. **(c)** What are the L and C values for a frequency of 60 Hz?

17. Solve the series ac circuit in Fig. 24-8 in the previous chapter by the use of complex numbers. Find $Z\angle\theta$, $I\angle\theta$, and each $V\angle\theta$. Prove that the sum of the complex voltage drops around the circuit equals the applied voltage V_T. Make a phasor diagram showing all phase angles with respect to V_T.

18. The following components are in series: $L = 100\ \mu H$, $C = 20\ pF$, $R = 2000\ \Omega$. At the frequency of 2 MHz calculate X_L, X_C, Z_T, I, θ_Z, V_R, V_L, and V_C. The applied $V_T = 8$ V.

19. Solve the same circuit as in Prob. 18 for the frequency of 4 MHz. Give three effects of the higher frequency.

20. In Fig. 25-15, show that $Z_T = 4.8\ \Omega$ and $\theta_Z = 36.9°$ by **(a)** the method of branch currents; **(b)** calculating Z_T as $Z_1 Z_2 / (Z_1 + Z_2)$.

21. In Fig. 25-16, find $Z_T\angle\theta$ by calculating Z_{bc} of the parallel bank and combining with the series Z_{ab}.

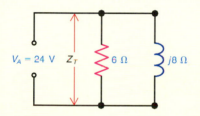

FIG. 25-15 Circuit for Prob. 20.

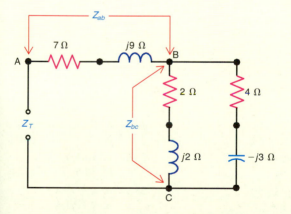

FIG. 25-16 Circuit for Prob. 21.

22. Find Z_T, in polar form, for the series-parallel circuit in Fig. 25-17.

23. Find Z_T and I_T, in polar form, for the series-parallel circuit in Fig. 25-18. Also, find V_{R_1}, V_{C_1}, V_{L_1}, and V_{R_2}. State each voltage drop in polar form.

24. In Fig. 25-19, calculate the output voltage V_{out}, in polar form.

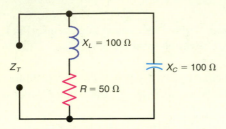

FIG. 25-17   Circuit for Prob. 22.

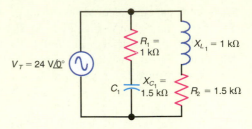

FIG. 25-18   Circuit for Prob. 23.

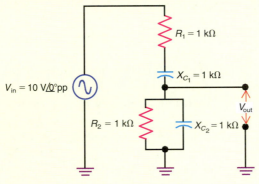

FIG. 25-19   Circuit for Prob. 24.

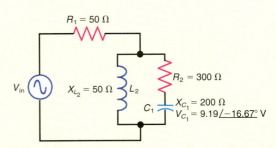

FIG. 25-20   Circuit for Critical Thinking Prob. 1.

CRITICAL THINKING

1. In Fig. 25-20, calculate the input voltage V_{in}, in polar form.

ANSWERS TO TEST-POINT QUESTIONS

25-1 **a.** $0°$
         b. $180°$

25-2 **a.** $90°$
         b. -90 or $270°$

25-3 **a.** T
         b. T

25-4 **a.** $j3$ kΩ
         b. $-j5$ mA

25-5 **a.** $4 + j7$
         b. $0 - j7$

25-6 **a.** $5 + j7$
         b. $4 + j6$

25-7 **a.** $14.14\ \Omega$
         b. $45°$

25-8 **a.** $12 \angle 50°$
         b. $3 \angle -10°$

25-9 **a.** $10 + j10$
         b. $10 - j10$

25-10 **a.** $53°$
          b. $143°$
          c. $90°$

25-11 **a.** $(6 + j8)/(5 + j4)$
          b. $(6 - j8)/(5 - j4)$

25-12 **a.** $10 + j4$
          b. $56.6 \angle 8°$

25-13 **a.** $4 + j5$ A
          b. $9 - j2$ A

25-14 **a.** $40 + j30$
          b. $50 \angle 37°\ \Omega$
          c. $2 \angle -37°$ A

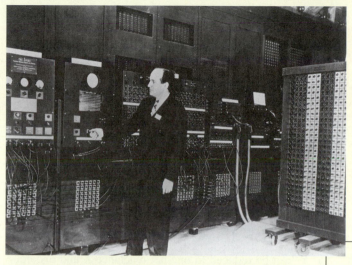

A photographic history of the computer. One of the first computers was the Eniac (*upper left*), developed in the 1940s. The 1970s marked the expanded use of the computer by businesses. The mainframe computer (*upper right*) was the tool of the time. In the 1980s personal computers such as the Apple IIe (*lower left*) brought computers into our homes and schools. Today, personal computers can go anywhere, as laptop computers (*lower right*) increase in popularity.

REVIEW: CHAPTERS 24 AND 25

SUMMARY

- Reactances X_C and X_L are opposite. In series, the ohms of X_C and X_L cancel. In parallel, the branch currents I_C and I_L cancel.
- As a result, circuits with R, X_C, and X_L can be reduced to one net reactance X and one equivalent R.
- In series circuits, the net X is added with the total R by phasors for the impedance: $Z_T = \sqrt{R^2 + X^2}$. Then $I = V_T/Z_T$.
- For the branch currents in parallel circuits, the net I_X is added with I_R by phasors for the total line current: $I_T = \sqrt{I_R^2 + I_X^2}$. Then $Z_{EQ} = V/I_T$.
- The characteristics for ohms of R, X_C, X_L, and Z in ac circuits are compared in Table 24-1.
- In ac circuits with reactance, the real power in watts equals I^2R. This value equals $VI \cos \theta$, where θ is the phase angle of the circuit and $\cos \theta$ is the power factor.
- The wattmeter uses an ac meter movement to read V and I at the same time, measuring watts of real power.
- In complex numbers, R is a real term at $0°$ and reactance is a $\pm j$ term at $\pm 90°$. In rectangular form, $Z_T = R \pm jX$. For example, $10 \, \Omega$ of R in series with $10 \, \Omega$ of X_L is $10 + j10 \, \Omega$.
- The polar form of $10 + j10 \, \Omega$ is $14 \angle 45° \, \Omega$. The angle of $45°$ is arctan X/R. The magnitude of 14 is $\sqrt{R^2 + X^2}$.
- The rectangular form of complex numbers must be used for addition and subtraction. Add or subtract the real terms and the j terms separately.
- The polar form of complex numbers is easier for multiplication and division. For multiplication, multiply the magnitudes and add the angles. For division, divide the magnitudes and subtract the angle of the divisor.
- In double-subscript notation for a voltage, such as V_{BE}, the first letter in the subscript is the point of measurement with respect to the second letter. So V_{BE} is the base voltage with respect to the emitter, in a transistor.

REVIEW SELF-TEST

ANSWERS AT BACK OF BOOK.

Fill in the numerical answer.

1. An ac circuit with 100 Ω R_1 in series with 200 Ω R_2 has R_T of _____ Ω.
2. With 100 Ω X_{L_1} in series with 200 Ω X_{L_2}, the total X_L is _____ Ω.
3. For 200 Ω X_{C_1} in series with 100 Ω X_{C_2}, the total X_C is _____ Ω.
4. Two X_C branches of 500 Ω each in parallel have combined X_C of _____ Ω.
5. Two X_L branches of 500 Ω each in parallel have combined X_L of _____ Ω.
6. A 500-Ω X_L is in series with a 300-Ω X_C. The net X_L is _____ Ω.
7. For 500 Ω X_C in series with 300 Ω X_{L_1}, the net X_C is _____ Ω.
8. A 10-Ω X_L is in series with a 10-Ω R. The total Z_T is _____ Ω.
9. With a 10-Ω X_C in series with a 10-Ω R, the total Z_T is _____ Ω.
10. With 14 V applied across 14 Ω Z_T, the I is _____ A.
11. For 10 Ω X_L and 10 Ω R in series, the phase angle θ is _____ degrees.
12. For 10 Ω X_C and 10 Ω R in series, the phase angle θ is _____ degrees.
13. A 10-Ω X_L and a 10-Ω R are in parallel across 10 V. The amount of each branch I is _____ A.
14. In question 13, the total line current I_T equals _____ A.
15. In questions 13 and 14, Z_T of the parallel branches equals _____ Ω.
16. With 120 V, an I of 10 A, and θ of 60°, a wattmeter reads _____ W.
17. The Z of $4 + j4$ Ω converted to polar form is _____ Ω.
18. The impedance value of $8 \angle 40°/2 \angle 30°$ is equal to _____ Ω.

Answer True or False.

19. In an ac circuit with X_C and R in series, if the frequency is raised, the current will increase.
20. In an ac circuit with X_L and R in series, if the frequency is increased, the current will be reduced.
21. The voltampere is a unit of apparent power.
22. The polar form of complex numbers is best for adding impedance values.

REFERENCES

Bogart, T.: *Electric Circuits,* Glencoe/McGraw-Hill, Columbus, Ohio.

Schuler, C., and R. Fowler: *Electric Circuit Analysis,* Glencoe/McGraw-Hill, Columbus, Ohio.

CHAPTER 26

RESONANCE

This chapter explains how X_L and X_C can be combined to favor one particular frequency, the resonant frequency to which the LC circuit is tuned. The resonance effect occurs when the inductive and capacitive reactances are equal. The main application of resonance is in RF circuits for tuning to an ac signal of the desired frequency. Tuning in radio and television receivers, transmitters, and electronics equipment in general are applications of resonance.

Tuning by means of the resonant effect provides the practical application of selectivity. The resonant circuit can select a particular frequency for the output, with many different frequencies at the input.

CHAPTER OBJECTIVES

Upon completion of this chapter, you should be able to:

- *Define* the term *resonance*.
- *List* four circuit characteristics of a series resonant circuit.
- *List* four circuit characteristics of a parallel resonant circuit.
- *Understand* how the resonant frequency formula is derived.
- *Calculate* the Q of a series and parallel resonant circuit.
- *Calculate* the equivalent impedance of a parallel resonant circuit.
- *Explain* what is meant by the *bandwidth of a resonant circuit*.
- *Calculate* the bandwidth of a series or parallel resonant circuit.
- *Explain* the effect of varying L or C in tuning an LC circuit.
- *Choose* L or C for a resonant circuit.

IMPORTANT TERMS IN THIS CHAPTER

antiresonance
bandwidth
damping
flywheel effect

half-power frequencies
parallel resonance
Q of resonant circuit
ringing

series resonance
tank circuit
tuning

TOPICS COVERED IN THIS CHAPTER

26-1 THE RESONANCE EFFECT

Inductive reactance increases as the frequency is increased, but capacitive reactance decreases with higher frequencies. Because of these opposite characteristics, for any *LC* combination there must be a frequency at which the X_L equals the X_C, as one increases while the other decreases. This case of equal and opposite reactances is called *resonance,* and the ac circuit is then a *resonant circuit.*

Any *LC* circuit can be resonant. It all depends on the frequency. At the resonant frequency, an *LC* combination provides the resonance effect. Off the resonant frequency, either below or above, the *LC* combination is just another ac circuit.

The frequency at which the opposite reactances are equal is the *resonant frequency.* This frequency can be calculated as $f_r = 1/(2\pi\sqrt{LC})$ where L is the inductance in henrys, C is the capacitance in farads, and f_r is the resonant frequency in hertz that makes $X_L = X_C$.

In general, we can say that large values of L and C provide a relatively low resonant frequency. Smaller values of L and C allow higher values for f_r. The resonance effect is most useful for radio frequencies, where the required values of microhenrys for L and picofarads for C are easily obtained.

The most common application of resonance in RF circuits is called *tuning.* In this use, the *LC* circuit provides maximum voltage output at the resonant frequency, compared with the amount of output at any other frequency either below or above resonance. This idea is illustrated in Fig. 26-1, where the *LC* circuit resonant at 1000 kHz magnifies the effect of this particular frequency. The result is maximum output at 1000 kHz, compared with lower or higher frequencies.

Tuning in radio and television are applications of resonance. When you tune a radio to one station, the *LC* circuits are tuned to resonance for that particular carrier frequency. Also, when you tune a television receiver to a particular channel, the *LC* circuits are tuned to resonance for that station. There are almost unlimited uses for resonance in ac circuits.

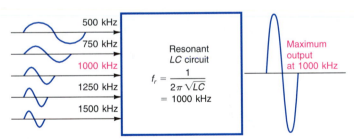

FIG. 26-1   *LC* circuit resonant at f_r of 1000 kHz to provide maximum output at this frequency.

TEST-POINT QUESTION 26-1

Answers at end of chapter.

Refer to Fig. 26-1.
a. Give the resonant frequency.
b. Give the frequency that has maximum output.

26-2 SERIES RESONANCE

In the series ac circuit in Fig. 26-2a, when the frequency of the applied voltage is 1000 kHz, the reactance of the 239-μH inductance equals 1500 Ω. At the same frequency, the reactance of the 106-pF capacitance also is 1500 Ω. Therefore, this LC combination is resonant at 1000 kHz. This is f_r, because the inductive reactance and capacitive reactance are equal at this frequency.

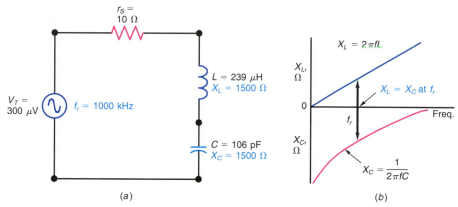

(a)   (b)

FIG. 26-2 Series resonance. (a) Schematic diagram of series r_S, L, and C. (b) Graph to show reactances X_C and X_L are equal and opposite at the resonant frequency f_r. Inductive reactance is shown up for jX_L and capacitive reactance is down for $-jX_C$.

In a series ac circuit, inductive reactance leads by 90°, compared with the zero reference angle of the resistance, while capacitive reactance lags by 90°. Therefore, X_L and X_C are 180° out of phase. The opposite reactances cancel each other completely when they are equal.

Figure 26-2b shows X_L and X_C equal, resulting in a net reactance of zero ohms. The only opposition to current then is the coil resistance r_S, which is the limit on how low the series resistance in the circuit can be. With zero reactance and just the low value of series resistance, the generator voltage produces the greatest amount of current in the series LC circuit at the resonant frequency. The series resistance should be as small as possible for a sharp increase in current at resonance.

MAXIMUM CURRENT AT SERIES RESONANCE

The main characteristic of series resonance is the resonant rise of current to its maximum value of V_T/r_S at the resonant frequency. For the circuit in Fig. 26-2a, the maximum current at series resonance is 30 μA, equal to 300 μV/10 Ω. At any other frequency either below or above the resonant frequency, there is less current in the circuit.

This resonant rise of current to 30 μA at 1000 kHz is illustrated in Fig. 26-3. In Fig. 26-3a, the amount of current is shown as the amplitude of individual cycles of the alternating current produced in the circuit by the ac generator voltage. Whether the amplitude of one ac cycle is considered in terms of peak, rms, or average value, the amount of current is greatest at the resonant frequency. In Fig. 26-3b, the current amplitudes are plotted on a graph for frequencies at and near the resonant frequency, producing a typical *response*

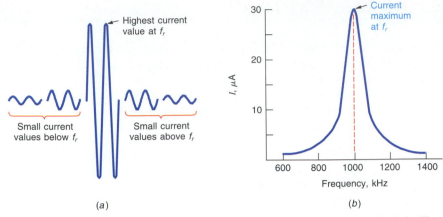

FIG. 26-3   Graphs showing maximum current at resonance for the series circuit in Fig. 26-2. (*a*) Amplitudes of individual cycles. (*b*) Response curve to show amount of *I* below and above resonance. Values of *I* are in Table 26-1.

curve for a series resonant circuit. The response curve in Fig. 26-3*b* can be considered as an outline of the increasing and decreasing amplitudes for the individual cycles shown in Fig. 26-3*a*.

The response curve of the series resonant circuit shows that the current is small below resonance, rises to its maximum value at the resonant frequency, and then drops off to small values above resonance. To prove this fact, Table 26-1 lists the calculated values of impedance and current in the circuit of Fig. 26-2 at the resonant frequency of 1000 kHz and at two frequencies below and two frequencies above resonance.

Below resonance, at 600 kHz, X_C is more than X_L and there is appreciable net reactance, which limits the current to a relatively low value. At the higher frequency of 800 kHz, X_C decreases and X_L increases, making the two reactances closer to the same value. The net reactance is then smaller, allowing more current.

At the resonant frequency, X_L and X_C are equal, the net reactance is zero, and the current has its maximum value equal to V_T/r_S.

Above resonance at 1200 and 1400 kHz, X_L is greater than X_C, providing net reactance that limits the current to much smaller values than at resonance.

TABLE 26-1   SERIES-RESONANCE CALCULATIONS FOR THE CIRCUIT IN FIG. 26-2*

FREQUENCY, kHz	$X_L = 2\pi fL$, Ω	$X_C = 1/(2\pi fC)$, Ω	NET REACTANCE, Ω $X_C - X_L$	$X_L - X_C$	Z_T, Ω†	$I = V_T/Z_T$, μA†	$V_L = IX_L$, μV	$V_C = IX_C$, μV
600	900	2500	1600		1600	0.19	171	475
800	1200	1875	675		675	0.44	528	825
$f_r \rightarrow$ 1000	1500	1500	0	0	10	30	45,000	45,000
1200	1800	1250		550	550	0.55	990	688
1400	2100	1070		1030	1030	0.29	609	310

*$L = 239$ μH, $C = 106$ pF, $V_T = 300$ μV, $r_S = 10$ Ω.
†Z_T and I calculated without r_S when its resistance is very small compared with the net X_L or X_C. Z_T and I are resistive at f_r.

In summary:

1. Below the resonant frequency, X_L is small, but X_C has high values that limit the amount of current.
2. Above the resonant frequency, X_C is small, but X_L has high values that limit the amount of current.
3. At the resonant frequency, X_L equals X_C, and they cancel to allow maximum current.

MINIMUM IMPEDANCE AT SERIES RESONANCE   Since the reactances cancel at the resonant frequency, the impedance of the series circuit is minimum, equal to just the low value of series resistance. This minimum impedance at resonance is resistive, resulting in zero phase angle. At resonance, therefore, the resonant current is in phase with the generator voltage.

RESONANT RISE IN VOLTAGE ACROSS SERIES _L_ OR _C_   The maximum current in a series LC circuit at resonance is useful because it produces maximum voltage across either X_L or X_C at the resonant frequency. As a result, the series resonant circuit can select one frequency by providing much more voltage output at the resonant frequency, compared with frequencies above and below resonance. Figure 26-4 illustrates the resonant rise in voltage across the capacitance in a series ac circuit. At the resonant frequency of 1000 kHz, the voltage across C rises to the value of 45,000 μV, while the input voltage is only 300 μV.

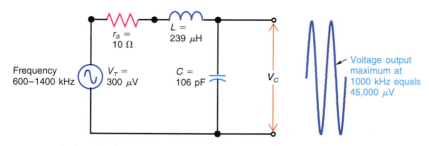

FIG. 26-4   Series circuit selects frequency by producing maximum IX_C voltage output across C at resonance.

In Table 26-1, the voltage across C is calculated as IX_C and across L as IX_L. Below the resonant frequency, X_C has a higher value than at resonance, but the current is small. Similarly, above the resonant frequency, X_L is higher than at resonance, but the current has a low value because of the inductive reactance. At resonance, although X_L and X_C cancel each other to allow maximum current, each reactance by itself has an appreciable value. Since the current is the same in all parts of a series circuit, the maximum current at resonance produces maximum voltage IX_C across C and an equal IX_L voltage across L for the resonant frequency.

Although the voltage across X_C and X_L is reactive, it is an actual voltage that can be measured. In Fig. 26-5, the voltage drops around the series resonant circuit are 45,000 μV across C and 45,000 μV across L, with 300 μV across r_S. The voltage across the resistance is equal to and in phase with the generator voltage.

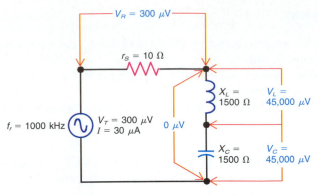

FIG. 26-5 Voltage drops around series resonant circuit.

Across the series combination of both L and C, the voltage is zero because the two series voltage drops are equal and opposite. In order to use the resonant rise of voltage, therefore, the output must be connected across either L or C alone. We can consider the V_L and V_C voltages as similar to the idea of two batteries connected in series opposition. Together, the resultant is zero for the equal and opposite voltages, but each battery still has its own potential difference.

In summary, for a series resonant circuit the main characteristics are:

1. The current I is maximum at the resonant frequency f_r.
2. The current I is in phase with the generator voltage, or the phase angle of the circuit is 0°.
3. The voltage is maximum across either L or C alone.
4. The impedance is minimum at f_r, equal only to the low r_S.

26-3 PARALLEL RESONANCE

With L and C in parallel as shown in Fig. 26-6, when X_L equals X_C, the reactive branch currents are equal and opposite at resonance. Then they cancel each other to produce minimum current in the main line. Since the line current is minimum, the impedance is maximum. These relations are based on r_S being very small compared with X_L at resonance. In this case, the branch currents are practically equal when X_L and X_C are equal.

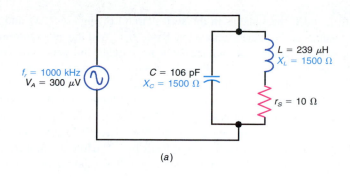

(a)

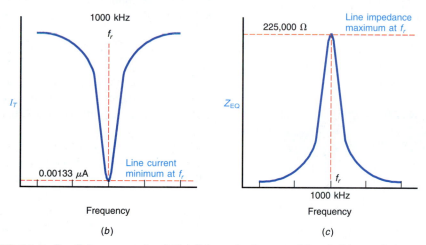

(b)   (c)

FIG. 26-6 Parallel resonant circuit. (*a*) Schematic diagram of *L* and *C* in parallel branches. (*b*) Response curve of I_T shows that the line current dips to minimum at f_r. (*c*) Response curve of Z_{EQ} shows that it rises to maximum at f_r.

MINIMUM LINE CURRENT AT PARALLEL RESONANCE   To show how the current in the main line dips to its minimum value when the parallel *LC* circuit is resonant, Table 26-2 lists the values of branch currents and the total line current for the circuit in Fig. 26-6.

TABLE 26-2   PARALLEL-RESONANCE CALCULATIONS FOR THE CIRCUIT IN FIG. 26-6*

| FREQUENCY, kHz | $X_C = 1/(2\pi f C),$ Ω | $X_L = 2\pi f L,$ Ω | $I_C = V/X_C,$ μA | $I_L = V/X_L,$ $\mu A\dagger$ | NET REACTIVE LINE CURRENT, μA | | $I_T,$ $\mu A\dagger$ | $Z_{EQ} = V_A/I_T,$ $\Omega\dagger$ |
					$I_L - I_C$	$I_C - I_L$		
600	2500	900	0.12	0.33	0.21		0.21	1400
800	1875	1200	0.16	0.25	0.09		0.09	3333
$f_r \rightarrow$ 1000	1500	1500	0.20	0.20	0	0	0.001 33	225,000‡
1200	1250	1800	0.24	0.17		0.07	0.07	3800
1400	1070	2100	0.28	0.14		0.14	0.14	2143

*$L = 239\ \mu H$, $C = 106\ pF$, $V_T = 300\ \mu V$, $r_S = 10\ \Omega$.
†Z_{EQ} and I calculated without r_S when its resistance is very small compared with the net X_L or X_C. Z_{EQ} and I are resistive at f_r.
‡At resonance Z_{EQ} calculated by formula (26-7). Z_{EQ} and I_T are resistive at f_r.

With L and C the same as in the series circuit of Fig. 26-2, X_L and X_C have the same values at the same frequencies. Since L, C, and the generator are in parallel, the voltage applied across the branches equals the generator voltage of 300 μV. Therefore, each reactive branch current is calculated as 300 μV divided by the reactance of the branch.

The values in the top row of Table 26-2 are obtained as follows: At 600 kHz the capacitive branch current equals 300 μV/2500 Ω, or 0.12 μA. The inductive branch current at this frequency is 300 μV/900 Ω, or 0.33 μA. Since this is a parallel ac circuit, the capacitive current leads by 90° while the inductive current lags by 90°, compared with the reference angle of the generator voltage, which is applied across the parallel branches. Therefore, the opposite currents are 180° out of phase. The net current in the line, then, is the difference between 0.33 and 0.12, which equals 0.21 μA.

Following this procedure, the calculations show that as the frequency is increased toward resonance, the capacitive branch current increases because of the lower value of X_C, while the inductive branch current decreases with higher values of X_L. As a result, there is less net line current as the two branch currents become more nearly equal.

At the resonant frequency of 1000 kHz, both reactances are 1500 Ω, and the reactive branch currents are both 0.20 μA, canceling each other completely.

Above the resonant frequency, there is more current in the capacitive branch than in the inductive branch, and the net line current increases above its minimum value at resonance.

The dip in I_T to its minimum value at f_r is shown by the graph in Fig. 26-6b. At parallel resonance, I_T is minimum and Z_{EQ} is maximum.

The in-phase current due to r_S in the inductive branch can be ignored off resonance because it is so small compared with the reactive line current. At the resonant frequency when the reactive currents cancel, however, the resistive component is the entire line current. Its value at resonance equals 0.00133 μA in this example. This small resistive current is the minimum value of the line current at parallel resonance.

MAXIMUM LINE IMPEDANCE AT PARALLEL RESONANCE

The minimum line current resulting from parallel resonance is useful because it corresponds to maximum impedance in the line across the generator. Therefore, an impedance that has a high value for just one frequency but a low impedance for other frequencies, either below or above resonance, can be obtained by using a parallel LC circuit resonant at the desired frequency. This is another method of selecting one frequency by resonance. The response curve in Fig. 26-6c shows how the impedance rises to maximum for parallel resonance.

The main application of parallel resonance is the use of an LC tuned circuit as the load impedance Z_L in the output circuit of RF amplifiers. Because of the high impedance, then, the gain of the amplifier is maximum at f_r. The voltage gain of an amplifier is directly proportional to Z_L. The advantage of a resonant LC circuit is that Z is maximum only for an ac signal at the resonant frequency. Also, L has practically no dc resistance, which means practically no dc voltage drop.

Referring to Table 26-2, the total impedance of the parallel ac circuit is calculated as the generator voltage divided by the total line current. At 600 kHz, for example, Z_{EQ} equals 300 μV/0.21 μA, or 1400 Ω. At 800 kHz, the impedance is higher because there is less line current.

At the resonant frequency of 1000 kHz, the line current is at its minimum value of 0.00133 μA. Then the impedance is maximum and is equal to 300 μV/0.00133 μA, or 225,000 Ω.

Above 1000 kHz, the line current increases, and the impedance decreases from its maximum value.

The idea of how the line current can have a very low value even though the reactive branch currents are appreciable is illustrated in Fig. 26-7. In Fig. 26-7a, the resistive component of the total line current is shown as though it were a separate branch drawing an amount of resistive current from the generator in the main line equal to the current resulting from the coil resistance. Each reactive branch current has its value equal to the generator voltage divided by the reactance. Since they are equal and of opposite phase, however, in any part of the circuit where both reactive currents are present, the net amount of electron flow in one direction at any instant of time corresponds to zero current. The graph in Fig. 26-7b shows how equal and opposite currents for I_L and I_C cancel.

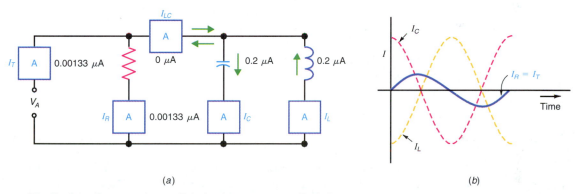

(a)   (b)

FIG. 26-7 Distribution of currents in parallel circuit at resonance. Resistive current shown as an equivalent branch for I_R. (a) Circuit with branch currents for R, L, and C. (b) Graph of equal and opposite reactive currents I_L and I_C.

If a meter is inserted in series with the main line to indicate total line current I_T, it dips sharply to the minimum value of line current at the resonant frequency. With minimum current in the line, the impedance across the line is maximum at the resonant frequency. The maximum impedance at parallel resonance corresponds to a high value of resistance, without reactance, since the line current is then resistive with zero phase angle.

In summary, for a parallel resonant circuit, the main characteristics are

1. The line current I_T is minimum at the resonant frequency.
2. The current I_T is in phase with the generator voltage V_A, or the phase angle of the circuit is 0°.
3. The impedance Z_{EQ}, equal to V_A/I_T, is maximum at f_r because of the minimum I_T.

THE *LC* TANK CIRCUIT   It should be noted that the individual branch currents are appreciable at resonance, although I_T is minimum. For the example in Table 26-2, at f_r either the I_L or I_C equals 0.2 μA. This current is greater than the I_C values below f_r or the I_L values above f_r.

The branch currents cancel in the main line because I_C is at 90° with respect to the source V_A while I_L is at $-90°$, making them opposite with respect to each other.

However, inside the *LC* circuit, I_L and I_C do not cancel because they are in separate branches. Then I_L and I_C provide a circulating current in the *LC* circuit, which equals 0.2 μA in this example. For this reason, a parallel resonant *LC* circuit is often called a *tank circuit*.

Because of the energy stored by L and C, the circulating tank current can provide full sine waves of current and voltage output when the input is only a pulse. The sine-wave output is always at the natural resonant frequency of the *LC* tank circuit. This ability of the *LC* circuit to supply complete sine waves is called the *flywheel effect*. Also, the process of producing sine waves after a pulse of energy has been applied is called *ringing* of the *LC* circuit.

TEST-POINT QUESTION 26-3

Answers at end of chapter.

Answer True or False, for parallel resonance.
a. Currents I_L and I_C are maximum.
b. Currents I_L and I_C are equal.
c. Current I_T is minimum.

26-4 RESONANT FREQUENCY $f_r = 1/(2\pi\sqrt{LC})$

The formula for the resonant frequency is derived from $X_L = X_C$. Using f_r to indicate the resonant frequency in the formulas for X_L and X_C, we have

▶ $$2\pi f_r L = \frac{1}{2\pi f_r C}$$

Inverting the factor f_r gives

$$2\pi L(f_r)^2 = \frac{1}{2\pi C}$$

Inverting the factor $2\pi L$ gives

$$(f_r)^2 = \frac{1}{(2\pi)^2 LC}$$

The square root of both sides is then

▶ $$f_r = \frac{1}{2\pi\sqrt{LC}}$$ (26-1)

where L is in henrys, C is in farads, and the resonant frequency f_r is in hertz (Hz) units. For example, to find the resonant frequency of the LC combination in Fig. 26-2, the values of 239×10^{-6} and 106×10^{-12} are substituted for L and C. Then:

$$f_r = \frac{1}{2\pi\sqrt{LC}} = \frac{1}{2\pi\sqrt{239 \times 10^{-6} \times 106 \times 10^{-12}}}$$

$$= \frac{1}{6.28\sqrt{25,334 \times 10^{-18}}} = \frac{1}{6.28 \times 159.2 \times 10^{-9}} = \frac{1}{1000 \times 10^{-9}}$$

$$f_r = 1 \times 10^6 \text{ Hz} = 1 \text{ MHz} = 1000 \text{ kHz}$$

For any LC circuit, series or parallel, the f_r equal to $1/(2\pi\sqrt{LC})$ is the resonant frequency that makes the inductive and capacitive reactances equal.

To do this problem on a calculator, keep in mind the following points:

1. If your calculator does not have an exponential (EXP) key, work with the powers* of 10 separately, without the calculator. For multiplication, add the exponents. The square root has one-half the exponent, but be sure the exponent is an even number before dividing by 2. The reciprocal has the same exponent but with opposite sign.
2. Multiply the L and C first, take the square root of the product, and multiply by 2π or 6.28. The (√) key is usually a second function of the (x^2) key.
3. After all the multiplications are complete, take the reciprocal of the final product by using the (1/x) key. This operation may require using the (2ndF) key.

For the example just solved with 239 μH for L and 106 pF for C, first punch in 239, push the (×) key, punch in 106 and then press the (=) key for the product of 25,334. While this number is on the display, push the (√) key for 159.2. Keep this display, press the (×) key, punch in 6.28 for 2π and push the (=) key for the total product of approximately 1000 in the denominator.

The powers of 10 in the denominator are $10^{-6} \times 10^{-12} = 10^{-18}$. Its square root is 10^{-9}.

For the reciprocal, while 1000 for the denominator is on the display, press the (1/x) key for the reciprocal, equal to 0.001. The reciprocal of 10^{-9} is 10^9. The answer for f_r then is 0.001×10^9, which equals 1×10^6.

HOW THE f_r VARIES WITH L AND C   It is important to note that higher values of L and C result in lower values of f_r. Either L or C, or both, can be varied. An LC circuit can be resonant at any frequency, from a few hertz to many megahertz.

As examples, an LC combination with the relatively large values of an 8-H inductance and a 20-μF capacitance is resonant at the low audio frequency of 12.6 Hz. For a much higher frequency in the RF range, a small inductance of 2 μH will resonate with the small capacitance of 3 pF for an f_r of 64.9 MHz. These examples are solved in the next two problems for more practice with the resonant frequency formula. Such calculations are often used in practical

DID YOU KNOW?

Nuclear plants in the United States number 110 and meet 20 percent of the country's power needs. Since 1986, no new reactors have been authorized.

---

*For an explanation of powers of 10 see Grob, B.: *Mathematics for Basic Electronics*, Glencoe/McGraw-Hill, Columbus, Ohio.

applications of tuned circuits. Probably the most important feature of any *LC* combination is its resonant frequency, especially in RF circuits. The applications of resonance are mainly for radio frequencies.

Example

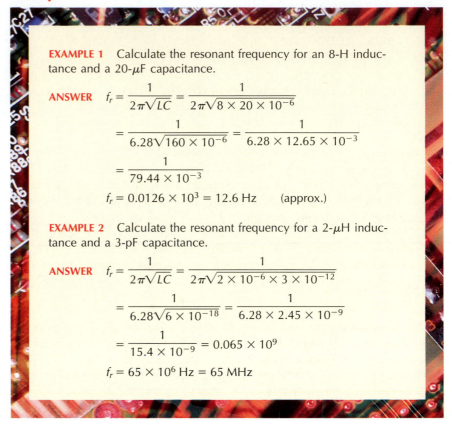

EXAMPLE 1 Calculate the resonant frequency for an 8-H inductance and a 20-μF capacitance.

ANSWER

$$f_r = \frac{1}{2\pi\sqrt{LC}} = \frac{1}{2\pi\sqrt{8 \times 20 \times 10^{-6}}}$$

$$= \frac{1}{6.28\sqrt{160 \times 10^{-6}}} = \frac{1}{6.28 \times 12.65 \times 10^{-3}}$$

$$= \frac{1}{79.44 \times 10^{-3}}$$

$$f_r = 0.0126 \times 10^3 = 12.6 \text{ Hz} \qquad \text{(approx.)}$$

EXAMPLE 2 Calculate the resonant frequency for a 2-μH inductance and a 3-pF capacitance.

ANSWER

$$f_r = \frac{1}{2\pi\sqrt{LC}} = \frac{1}{2\pi\sqrt{2 \times 10^{-6} \times 3 \times 10^{-12}}}$$

$$= \frac{1}{6.28\sqrt{6 \times 10^{-18}}} = \frac{1}{6.28 \times 2.45 \times 10^{-9}}$$

$$= \frac{1}{15.4 \times 10^{-9}} = 0.065 \times 10^9$$

$$f_r = 65 \times 10^6 \text{ Hz} = 65 \text{ MHz}$$

Specifically, because of the square root in the denominator of Formula (26-1), the f_r decreases inversely as the square root of *L* or *C*. For instance, if *L* or *C* is quadrupled, the f_r is reduced one half. The ½ is equal to the square root of ¼.

As a numerical example, suppose f_r is 6 MHz with particular values of *L* and *C*. If either *L* or *C* is made four times larger, then f_r will be reduced to 3 MHz.

Or, to take the opposite case of doubling the frequency from 6 MHz to 12 MHz, the following can be done:

1. Use one-fourth the *L* with the same *C*.
2. Use one-fourth the *C* with the same *L*.
3. Reduced both *L* and *C* by one-half.
4. Use any new combination of *L* and *C* whose product will be one-fourth the original product of *L* and *C*.

LC PRODUCT DETERMINES f_r There are any number of *LC* combinations that can be resonant at one frequency. With more *L*, then less *C* can be used for the same f_r. Or less *L* can be used with more *C*. Table 26-3 lists five

TABLE 26-3 LC COMBINATIONS RESONANT AT 1000 kHz

L, μH	C, pF	L × C LC PRODUCT	X_L, Ω AT 1000 kHz	X_C, Ω AT 1000 kHz
23.9	1060	25,334	150	150
119.5	212	25,334	750	750
239	106	25,334	1,500	1,500
478	53	25,334	3,000	3,000
2390	10.6	25,334	15,000	15,000

possible combinations of L and C resonant at 1000 kHz, just as an example of one f_r. The resonant frequency is the same 1000 kHz here for all five combinations. When either L or C is increased by a factor of 10 or 2, the other is decreased by the same factor, resulting in a constant value for the LC product.

The reactance at resonance changes with different combinations of L and C, but in all five cases X_L and X_C are equal to each other at 1000 kHz. This is the resonant frequency determined by the value of the LC product in $f_r = 1/(2\pi\sqrt{LC})$.

MEASURING L OR C BY RESONANCE   Of the three factors L, C, and f_r in the resonant-frequency formula, any one can be calculated when the other two are known. The resonant frequency of the LC combination can be found experimentally by determining the frequency that produces the resonant response in an LC combination. With a known value of either L or C, and the resonant frequency determined, the third factor can be calculated. This method is commonly used for measuring inductance or capacitance. A test instrument for this purpose is the Q meter, which also measures the Q of a coil.

CALCULATING C FROM f_r   The C can be taken out of the square root sign or radical in the resonance formula, as follows:

$$f_r = \frac{1}{2\pi\sqrt{LC}}$$

Squaring both sides to eliminate the radical gives

$$f_r^2 = \frac{1}{(2\pi)^2 LC}$$

Inverting C and f_r^2 gives

▶ $$C = \frac{1}{4\pi^2 f_r^2 L} \qquad\qquad (26\text{-}2)$$

where f_r is in hertz, C is in farads, and L is in henrys.

CALCULATING L FROM f_r   Similarly, the resonance formula can be transposed to find L. Then

▶ $$L = \frac{1}{4\pi^2 f_r^2 C} \qquad\qquad (26\text{-}3)$$

With Formula (26-3), L is determined by its f_r with a known value of C. Similarly, C is determined from Formula (26-2) by its f_r with a known value of L.

Example

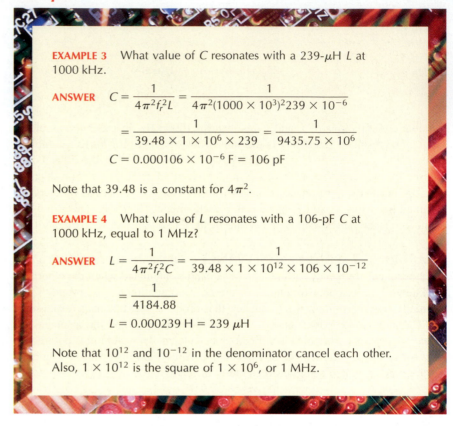

EXAMPLE 3 What value of C resonates with a 239-μH L at 1000 kHz.

ANSWER
$$C = \frac{1}{4\pi^2 f_r^2 L} = \frac{1}{4\pi^2(1000 \times 10^3)^2 239 \times 10^{-6}}$$
$$= \frac{1}{39.48 \times 1 \times 10^6 \times 239} = \frac{1}{9435.75 \times 10^6}$$
$$C = 0.000106 \times 10^{-6} \text{ F} = 106 \text{ pF}$$

Note that 39.48 is a constant for $4\pi^2$.

EXAMPLE 4 What value of L resonates with a 106-pF C at 1000 kHz, equal to 1 MHz?

ANSWER
$$L = \frac{1}{4\pi^2 f_r^2 C} = \frac{1}{39.48 \times 1 \times 10^{12} \times 106 \times 10^{-12}}$$
$$= \frac{1}{4184.88}$$
$$L = 0.000239 \text{ H} = 239 \ \mu\text{H}$$

Note that 10^{12} and 10^{-12} in the denominator cancel each other. Also, 1×10^{12} is the square of 1×10^6, or 1 MHz.

The values in Examples 3 and 4 are from the LC circuit illustrated in Fig. 26-2 for series resonance and Fig. 26-6 for parallel resonance.

TEST-POINT QUESTION 26-4

Answers at end of chapter.
a. To increase f_r, must the C be increased or decreased?
b. If C is increased from 100 to 400 pF, L must be decreased from 800 μH to what value for the same f_r?
c. Give the constant value for $4\pi^2$.

26-5 Q MAGNIFICATION FACTOR OF RESONANT CIRCUIT

The quality, or *figure of merit,* of the resonant circuit, in sharpness of resonance, is indicated by the factor Q. In general, the higher the ratio of the reactance at resonance to the series resistance, the higher is the Q and the sharper the resonance effect.

Q OF SERIES CIRCUIT   In a series resonant circuit we can calculate Q from the following formula:

$$\blacktriangleright \quad Q = \frac{X_L}{r_S} \qquad\qquad\qquad (26\text{-}4)$$

where Q is the figure of merit, X_L is the inductive reactance in ohms at the resonant frequency, and r_S is the resistance in ohms in series with X_L. For the series resonant circuit in Fig. 26-2,

$$Q = \frac{1500\ \Omega}{10\ \Omega} = 150$$

The Q is a numerical factor without any units, because it is a ratio of reactance to resistance and the ohms cancel. Since the series resistance limits the amount of current at resonance, the lower the resistance, the sharper the increase to maximum current at the resonant frequency, and the higher the Q. Also, a higher value of reactance at resonance allows the maximum current to produce a higher value of voltage for the output.

The Q has the same value if it is calculated with X_C instead of X_L, since they are equal at resonance. However, the Q of the circuit is generally considered in terms of X_L, because usually the coil has the series resistance of the circuit. In this case, the Q of the coil and the Q of the series resonant circuit are the same. If extra resistance is added, the Q of the circuit will be less than the Q of the coil. The highest possible Q for the circuit is the Q of the coil.

The value of 150 can be considered as a high Q. Typical values are 50 to 250, approximately. Less than 10 is a low Q; more than 300 is a very high Q.

HIGHER L/C RATIO CAN PROVIDE HIGHER Q   As shown before in Table 26-3, different combinations of L and C can be resonant at the same frequency. However, the amount of reactance at resonance is different. More X_L can be obtained with a higher L and lower C for resonance, although X_L and X_C must be equal at the resonant frequency. Therefore, both X_L and X_C are higher with a higher L/C ratio for resonance.

More X_L can allow a higher Q if the ac resistance does not increase as much as the reactance. With typical RF coils, an approximate rule is that maximum Q can be obtained when X_L is about 1000 Ω. In many cases, though, the minimum C is limited by the stray capacitance in the circuit.

Q RISE IN VOLTAGE ACROSS SERIES L OR C   The Q of the resonant circuit can be considered a magnification factor that determines how much the voltage across L or C is increased by the resonant rise of current in a series circuit. Specifically, the voltage output at series resonance is Q times the generator voltage:

$$\blacktriangleright \quad V_L = V_C = Q \times V_{\text{gen}} \qquad\qquad (26\text{-}5)$$

In Fig. 26-4, for example, the generator voltage is 300 μV and Q is 150. The resonant rise of voltage across either L or C then equals 300 μV $\times$ 150, or 45,000 μV. Note that this is the same value calculated in Table 26-1 for V_C or V_L at resonance.

HOW TO MEASURE Q IN A SERIES RESONANT CIRCUIT The fundamental nature of Q for a series resonant circuit is seen from the fact that the Q can be determined experimentally by measuring the Q rise in voltage across either L or C and comparing this voltage with the generator voltage. As a formula,

$$\blacktriangleright \quad Q = \frac{V_{\text{out}}}{V_{\text{in}}} \qquad\qquad (26\text{-}6)$$

where V_{out} is the ac voltage measured across the coil or capacitor and V_{in} is the generator voltage.

Referring to Fig. 26-5, suppose that you measure with an ac voltmeter across L or C and this voltage equals 45,000 μV at the resonant frequency. Also, measure the generator input of 300 μV. Then

$$Q = \frac{V_{\text{out}}}{V_{\text{in}}}$$

$$= \frac{45,000\ \mu\text{V}}{300\ \mu\text{V}}$$

$$Q = 150$$

This method is better than the X_L/r_S formula for determining Q because r_S is the ac resistance of the coil, which is not so easily measured. Remember that the coil's ac resistance can be more than double the dc resistance measured with an ohmmeter. In fact, measuring Q with Formula (26-6) makes it possible to calculate the ac resistance. These points are illustrated in the following examples.

Example

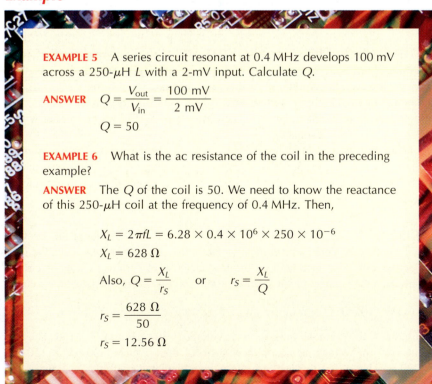

EXAMPLE 5 A series circuit resonant at 0.4 MHz develops 100 mV across a 250-μH L with a 2-mV input. Calculate Q.

ANSWER $\quad Q = \dfrac{V_{\text{out}}}{V_{\text{in}}} = \dfrac{100\ \text{mV}}{2\ \text{mV}}$

$\qquad\qquad Q = 50$

EXAMPLE 6 What is the ac resistance of the coil in the preceding example?

ANSWER The Q of the coil is 50. We need to know the reactance of this 250-μH coil at the frequency of 0.4 MHz. Then,

$$X_L = 2\pi f L = 6.28 \times 0.4 \times 10^6 \times 250 \times 10^{-6}$$

$$X_L = 628\ \Omega$$

$$\text{Also, } Q = \frac{X_L}{r_S} \quad\text{ or }\quad r_S = \frac{X_L}{Q}$$

$$r_S = \frac{628\ \Omega}{50}$$

$$r_S = 12.56\ \Omega$$

Q OF PARALLEL CIRCUIT In a parallel resonant circuit, where r_S is very small compared with X_L, the Q also equals X_L/r_S. Note that r_S is still the resistance of the coil in series with X_L (see Fig. 26-8). The Q of the coil determines the Q of the parallel circuit here because it is less than the Q of the capacitive branch. Capacitors used in tuned circuits generally have a very high Q because of their low losses. In Fig. 26-8, the Q is 1500 Ω/10 Ω, or 150, the same as the series resonant circuit with the same values.

This example assumes that the generator resistance is very high and that there is no other resistance branch shunting the tuned circuit. Then the Q of the parallel resonant circuit is the same as the Q of the coil. Actually, shunt resistance can lower the Q of a parallel resonant circuit, as analyzed in Sec. 26-10.

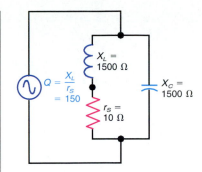

FIG. 26-8 The Q of a parallel resonant circuit in terms of X_L and its series resonance r_S.

Q RISE IN IMPEDANCE ACROSS PARALLEL RESONANT CIRCUIT

For parallel resonance, the Q magnification factor determines by how much the impedance across the parallel LC circuit is increased because of the minimum line current. Specifically, the impedance across the parallel resonant circuit is Q times the inductive reactance at the resonant frequency:

▶ $$Z_{EQ} = Q \times X_L \qquad\qquad\qquad\qquad (26\text{-}7)$$

Referring back to the parallel resonant circuit in Fig. 26-6, as an example, X_L is 1500 Ω and Q is 150. The result is a rise of impedance to the maximum value of 150 × 1500 Ω, or 225,000 Ω, at the resonant frequency.

Since the line current equals V_A/Z_{EQ}, the minimum value of line current is 300 μV/225,000 Ω, which equals 0.00133 μA.

At f_r the minimum line current is $1/Q$ of either branch current. In Fig. 26-7, I_L or I_C is 0.2 μA and Q is 150. Therefore, I_T is 0.2/150, or 0.00133 μA, which is the same answer as V_A/Z_{EQ}. Or, stated another way, the circulating tank current is Q times the minimum I_T.

HOW TO MEASURE Z_{EQ} OF A PARALLEL RESONANT CIRCUIT

Formula (26-7) for Z_{EQ} is also useful in its inverted form as $Q = Z_{EQ}/X_L$. We can measure Z_{EQ} by the method illustrated in Fig. 26-9. Then Q can be calculated from the value of Z_{EQ} and the inductive reactance of the coil.

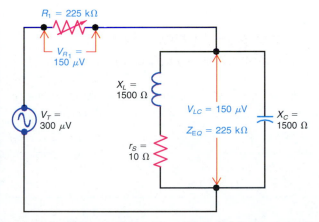

FIG. 26-9 How to measure Z_{EQ} of a parallel resonant circuit. Adjust R_1 to make its V_R equal to V_{LC}. Then $Z_{EQ} = R_1$.

To measure Z_{EQ}, first tune the LC circuit to resonance. Then adjust R_1 in Fig. 26-9 to the resistance that makes its ac voltage equal to the ac voltage across the tuned circuit. With equal voltages, the Z_{EQ} must be the same value as R_1.

For the example here, which corresponds to the parallel resonance shown in Figs. 26-6 and 26-8, the Z_{EQ} is equal to 225,000 Ω. This high value is a result of parallel resonance. The X_L is 1500 Ω. Therefore, to determine Q, the calculations are

$$Q = \frac{Z_{EQ}}{X_L} = \frac{225,000}{1500} = 150$$

Example

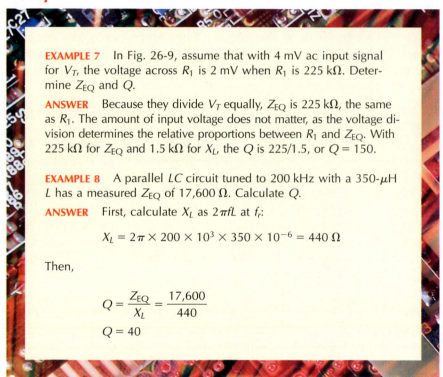

EXAMPLE 7   In Fig. 26-9, assume that with 4 mV ac input signal for V_T, the voltage across R_1 is 2 mV when R_1 is 225 kΩ. Determine Z_{EQ} and Q.

ANSWER   Because they divide V_T equally, Z_{EQ} is 225 kΩ, the same as R_1. The amount of input voltage does not matter, as the voltage division determines the relative proportions between R_1 and Z_{EQ}. With 225 kΩ for Z_{EQ} and 1.5 kΩ for X_L, the Q is 225/1.5, or $Q = 150$.

EXAMPLE 8   A parallel LC circuit tuned to 200 kHz with a 350-μH L has a measured Z_{EQ} of 17,600 Ω. Calculate Q.

ANSWER   First, calculate X_L as $2\pi fL$ at f_r:

$$X_L = 2\pi \times 200 \times 10^3 \times 350 \times 10^{-6} = 440\ \Omega$$

Then,

$$Q = \frac{Z_{EQ}}{X_L} = \frac{17,600}{440}$$

$$Q = 40$$

TEST-POINT QUESTION 26-5

Answers at end of chapter.
a. In a series resonant circuit, V_L is 300 mV with input of 3 mV. Calculate Q.
b. In a parallel resonant circuit, X_L is 500 Ω. With a Q of 50, calculate Z_{EQ}.

26-6 BANDWIDTH OF RESONANT CIRCUIT

When we say an LC circuit is resonant at one frequency, this is true for the maximum resonance effect. However, other frequencies close to f_r also are effective.

For series resonance, frequencies just below and above f_r produce increased current, but a little less than the value at resonance. Similarly, for parallel resonance, frequencies close to f_r can provide a high impedance, although a little less than the maximum Z_{EQ}.

Therefore, any resonant frequency has an associated band of frequencies that provide resonance effects. How wide the band is depends on the Q of the resonant circuit. Actually, it is practically impossible to have an LC circuit with a resonant effect at only one frequency. The width of the resonant band of frequencies centered around f_r is called the *bandwidth* of the tuned circuit.

MEASUREMENT OF BANDWIDTH The group of frequencies with a response 70.7 percent of maximum, or more, is generally considered the bandwidth of the tuned circuit, as shown in Fig. 26-10b. The resonant response here is increasing current for the series circuit in Fig. 26-10a. Therefore, the bandwidth is measured between the two frequencies, f_1 and f_2, producing 70.7 percent of the maximum current at f_r.

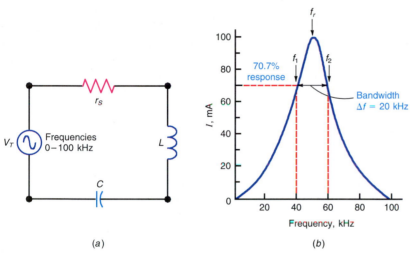

(a)   (b)

FIG. 26-10 Bandwidth of a tuned LC circuit. (a) Series circuit with input of 0 to 100 kHz. (b) Response curve with bandwidth Δf equal to 20 kHz between f_1 and f_2.

For a parallel circuit, the resonant response is increasing impedance Z_{EQ}. Then the bandwidth is measured between the two frequencies allowing 70.7 percent of the maximum Z_{EQ} at f_r.

The bandwidth indicated on the response curve in Fig. 26-10b equals 20 kHz. This value is the difference between f_2 at 60 kHz and f_1 at 40 kHz, both with 70.7 percent response.

Compared with the maximum current of 100 mA for f_r at 50 kHz, f_1 below resonance and f_2 above resonance each allow a rise to 70.7 mA. All frequencies in this band 20 kHz wide allow 70.7 mA, or more, as the resonant response in this example.

BANDWIDTH EQUALS f_r/Q Sharp resonance with high Q means narrow bandwidth. The lower the Q, the broader the resonant response and the greater the bandwidth.

Also, the higher the resonant frequency, the greater is the range of frequency values included in the bandwidth for a given sharpness of resonance. Therefore, the bandwidth of a resonant circuit depends on the factors f_r and Q. The formula is

$$\blacktriangleright \qquad f_2 - f_1 = \Delta f = \frac{f_r}{Q} \qquad\qquad\qquad (26\text{-}8)$$

where Δf is the total bandwidth in the same units as the resonant frequency f_r. The bandwidth Δf can also be abbreviated BW.

For example, a series circuit resonant at 800 kHz with a Q of 100 has a bandwidth of 800/100, or 8 kHz. Then the I is 70.7 percent of maximum, or more, for all frequencies for a band 8 kHz wide. This frequency band is centered around 800 kHz, from 796 to 804 kHz.

With a parallel resonant circuit having a Q higher than 10, Formula (26-8) also can be used for calculating the bandwidth of frequencies which provide 70.7 percent or more of the maximum Z_{EQ}. However, the formula cannot be used for parallel resonant circuits with low Q, as the resonance curve then becomes unsymmetrical.

HIGH Q MEANS NARROW BANDWIDTH   The effect for different values of Q is illustrated in Fig. 26-11. Note that a higher Q for the same resonant frequency results in less bandwidth. The slope is sharper for the sides or *skirts* of the response curve, in addition to its greater amplitude.

High Q is generally desirable for more output from the resonant circuit. However, it must have enough bandwidth to include the desired range of signal frequencies.

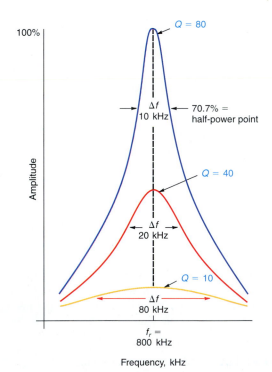

FIG. 26-11   Higher Q provides a sharper resonant response. Amplitude is I for series resonance or Z_{EQ} for parallel resonance. Bandwidth at half-power frequencies is Δf.

THE EDGE FREQUENCIES Both f_1 and f_2 are separated from f_r by one-half of the total bandwidth. For the top curve in Fig. 26-11, as an example, with a Q of 80, Δf is ± 5 kHz centered around 800 kHz for f_r. To determine the edge frequencies:

$$f_1 = f_r - \frac{\Delta f}{2} = 800 - 5 = 795 \text{ kHz}$$

$$f_2 = f_r + \frac{\Delta f}{2} = 800 + 5 = 805 \text{ kHz}$$

These examples assume the resonance curve is symmetrical. This is true for a high-Q parallel resonant circuit and a series resonant circuit with any Q.

Example

EXAMPLE 9 An LC circuit resonant at 2000 kHz has a Q of 100. Find the total bandwidth Δf and the edge frequencies f_1 and f_2.

ANSWER $\Delta f = \dfrac{f_r}{Q} = \dfrac{2000 \text{ kHz}}{100} = 20 \text{ kHz}$

$f_1 = f_r - \dfrac{\Delta f}{2} = 2000 - 10 = 1990 \text{ kHz}$

$f_2 = f_r + \dfrac{\Delta f}{2} = 2000 + 10 = 2010 \text{ kHz}$

EXAMPLE 10 Repeat Example 9 for an f_r equal to 6000 kHz and the same Q of 100.

ANSWER $f = \dfrac{f_r}{Q} = \dfrac{6000 \text{ kHz}}{100} = 60 \text{ kHz}$

$f_1 = 6000 - 30 = 5970 \text{ kHz}$

$f_2 = 6000 + 30 = 6030 \text{ kHz}$

Notice that Δf is three times as wide as Δf in Example 9 for the same Q because f_r is three times higher.

HALF-POWER POINTS It is simply for convenience in calculations that the bandwidth is defined between the two frequencies having 70.7 percent response. At each of these frequencies, the net capacitive or inductive reactance equals the resistance. Then the total impedance of the series reactance and resistance is 1.4 times greater than R. With this much more impedance, the current is reduced to 1/1.414, or 0.707, of its maximum value.

Furthermore, the relative current or voltage value of 70.7 percent corresponds to 50 percent in power, since power is I^2R or V^2/R and the square of 0.707 equals 0.50. Therefore, the bandwidth between frequencies having 70.7 percent response in current or voltage is also the bandwidth in terms of half-power points. Formula (26-8) is derived for Δf between the points with 70.7 percent response on the resonance curve.

MEASURING BANDWIDTH TO CALCULATE Q   The half-power frequencies f_1 and f_2 can be determined experimentally. For series resonance, find the two frequencies at which the current is 70.7 percent of maximum I. Or, for parallel resonance, find the two frequencies that make the impedance 70.7 percent of the maximum Z_{EQ}. The following method uses the circuit in Fig. 26-9 for measuring Z_{EQ}, but with different values to determine its bandwidth and Q.

1.   Tune the circuit to resonance and determine its maximum Z_{EQ} at f_r. In this example, assume that Z_{EQ} is 10,000 Ω at the resonant frequency of 200 kHz.
2.   Keep the same amount of input voltage, but change its frequency slightly below f_r to determine the frequency f_1 which results in a Z_1 equal to 70.7 percent of Z_{EQ}. The required value here is 0.707 × 10,000, or 7070 Ω, for Z_1 at f_1. Assume this frequency f_1 is determined to be 195 kHz.
3.   Similarly, find the frequency f_2 above f_r that results in the impedance Z_2 of 7070 Ω. Assume f_2 is 205 kHz.
4.   The total bandwidth between the half-power frequencies equals $f_2 - f_1$ or 205 − 195. Then the value of $\Delta f = 10$ kHz.
5.   Then $Q = f_r/\Delta f$ or 200 kHz/10 kHz = 20 for the calculated value of Q.

In this way, measuring the bandwidth makes it possible to determine the Q. With Δf and f_r, the Q can be determined for either parallel or series resonance.

TEST-POINT QUESTION 26-6

Answers at end of chapter.
a.   An LC circuit with f_r of 10 MHz has a Q of 40. Calculate the half-power bandwidth.
b.   For an f_r of 500 kHz and bandwidth Δf of 10 kHz, calculate Q.

26-7 TUNING

Tuning means obtaining resonance at different frequencies by varying either L or C. As illustrated in Fig. 26-12, the variable capacitance C can be adjusted to tune the series LC circuit to resonance at any one of the five different frequencies. Each of the voltages V_1 to V_5 indicates an ac input with a specific frequency. Which one is selected for maximum output is determined by the resonant frequency of the LC circuit.

When C is set to 424 pF, for example, the resonant frequency of the LC circuit is 500 kHz for f_{r_1}. The input voltage that has the frequency of 500 kHz then produces a resonant rise of current which results in maximum output voltage across C. At other frequencies, such as 707 kHz, the voltage output is less than the input. With C at 424 pF, therefore, the LC circuit tuned to 500 kHz selects this frequency by providing much more voltage output than other frequencies.

Suppose that we want maximum output for the ac input voltage that has the frequency of 707 kHz. Then C is set at 212 pF to make the LC circuit resonant at 707 kHz for f_{r_2}. Similarly, the tuned circuit can resonate at a different frequency for each input voltage. In this way, the LC circuit is tuned to select the desired frequency.

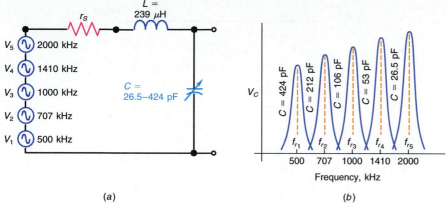

(a)   (b)

FIG. 26-12 Tuning a series *LC* circuit. (*a*) Input voltages at different frequencies. (*b*) Relative response for each frequency when *C* is varied. (Not to scale.)

The variable capacitance *C* can be set at the values listed in Table 26-4 to tune the *LC* circuit to different frequencies. Only five frequencies are listed here, but any one capacitance value between 26.5 and 424 pF can tune the 239-μH coil to resonance at any frequency in the range of 500 to 2000 kHz. It should be noted that a parallel resonant circuit also can be tuned by varying *C* or *L*.

TABLE 26-4	TUNING *LC* CIRCUIT BY VARYING *C*	
L, μH	**C, pF**	**f_r, kHz**
239	424	500
239	212	707
239	106	1000
239	53	1410
239	26.5	2000

TUNING RATIO   When an *LC* circuit is tuned, the change in resonant frequency is inversely proportional to the square root of the change in *L* or *C*. Referring to Table 26-4, notice that when *C* is decreased to one-fourth, from 424 to 106 pF, the resonant frequency doubles from 500 to 1000 kHz. Or the frequency is increased by the factor $1/\sqrt{1/4}$, which equals 2.

Suppose that we want to tune through the whole frequency range of 500 to 2000 kHz. This is a tuning ratio of 4:1 for the highest frequency to the lowest frequency. Then the capacitance must be varied from 424 to 26.5 pF, which is a 16:1 capacitance ratio.

RADIO TUNING DIAL   Figure 26-13 illustrates a typical application of resonant circuits in tuning a receiver to the carrier frequency of a desired station in the band. The tuning is done by the air capacitor *C*, which can be varied from 360 pF with the plates completely in mesh to 40 pF out of mesh. The fixed plates form the *stator*, while the *rotor* has the plates that move in and out.

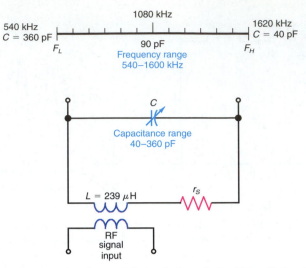

FIG. 26-13 Application of tuning an *LC* circuit through the AM radio band.

Note that the lowest frequency F_L at 540 kHz is tuned in with the highest C at 360 pF. Resonance at the highest frequency F_H at 1620 kHz results with the lowest C at 40 pF.

The capacitance range of 40 to 360 pF tunes through the frequency range from 1620 kHz down to 540 kHz. Frequency F_L is one-third F_H because the maximum C is nine times the minimum C. The tuning dial, in kHz, usually omits the last zero to save space.

The same idea applies to tuning through the commercial FM broadcast band of 88 to 108 MHz, with smaller values of L and C. Also, television receivers are tuned to a specific broadcast channel by resonance at the desired frequencies.

For electronic tuning, the C is varied by a *varactor*. This is a semiconductor diode that varies in capacitance when its voltage is changed. See Chapter 28, "Electronic Devices."

TEST-POINT QUESTION 26-7

Answers at end of chapter.
a. When a tuning capacitor is completely in mesh, is the station tuned in the highest or lowest frequency in the band?
b. A tuning ratio of 2:1 in frequency requires what ratio of variable L or C?

26-8 MISTUNING

Suppose that a series *LC* circuit is tuned to 1000 kHz but the frequency of the input voltage is 17 kHz, completely off resonance. The circuit could provide a Q rise in output voltage for current having the frequency of 1000 kHz, but there is no input voltage and therefore no current at this frequency.

The input voltage produces current that has the frequency of 17 kHz. This frequency cannot produce a resonant rise in current, however, because the current is limited by the net reactance. When the frequency of the input voltage and the resonant frequency of the *LC* circuit are not the same, therefore, the mistuned circuit has very little output compared with the *Q* rise in voltage at resonance.

Similarly, when a parallel circuit is mistuned, it does not have a high value of impedance. Furthermore, the net reactance off resonance makes the *LC* circuit either inductive or capacitive.

SERIES CIRCUIT OFF RESONANCE   When the frequency of the input voltage is lower than the resonant frequency of a series *LC* circuit, the capacitive reactance is greater than the inductive reactance. As a result, there is more voltage across the capacitive reactance than across the inductive reactance. The series *LC* circuit is capacitive below resonance, therefore, with capacitive current leading the generator voltage.

Above the resonant frequency, the inductive reactance is greater than the capacitive reactance. As a result, the circuit is inductive above resonance, with inductive current that lags the generator voltage. In both cases, there is much less output voltage than at resonance.

PARALLEL CIRCUIT OFF RESONANCE   With a parallel *LC* circuit, the smaller amount of inductive reactance below resonance results in more inductive branch current than capacitive branch current. The net line current is inductive, therefore, making the parallel *LC* circuit inductive below resonance, as the line current lags the generator voltage.

Above the resonant frequency, the net line current is capacitive because of the higher value of capacitive branch current. Then the parallel *LC* circuit is capacitive, with line current leading the generator voltage. In both cases the total impedance of the parallel circuit is much less than the maximum impedance at resonance. Note that the capacitive and inductive effects off resonance are opposite for series and parallel *LC* circuits.

TEST-POINT QUESTION 26-8

Answers at end of chapter.
  a. Is a series resonant circuit inductive or capacitive below resonance?
  b. Is a parallel resonant circuit inductive or capacitive below resonance?

26-9 ANALYSIS OF PARALLEL RESONANT CIRCUITS

Parallel resonance is more complex than series resonance because the reactive branch currents are not exactly equal when X_L equals X_C. The reason is that the coil has its series resistance r_S in the X_L branch, while the capacitor has only X_C in its branch.

For high-Q circuits, we consider r_S to be negligible. In low-Q circuits, however, the inductive branch must be analyzed as a complex impedance with X_L and r_S in series. This impedance is in parallel with X_C, as shown in Fig. 26-14. The

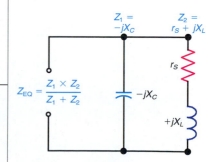

FIG. 26-14   General method of calculating Z_{EQ} for a parallel resonant circuit as $(Z_1 \times Z_2)/(Z_1 + Z_2)$ with complex numbers.

total impedance Z_{EQ} can then be calculated by using complex numbers, as explained in Chap. 25.

HIGH-Q CIRCUIT   We can apply the general method in Fig. 26-14 to the parallel resonant circuit shown before in Fig. 26-6 to see if Z_{EQ} is 225,000 Ω. In this example, X_L and X_C are 1500 Ω and r_S is 10 Ω. The calculations are

$$Z_{EQ} = \frac{Z_1 \times Z_2}{Z_1 + Z_2} = \frac{-j1500 \times (j1500 + 10)}{-j1500 + j1500 + 10}$$

$$= \frac{-j^2 2.25 \times 10^6 - j15{,}000}{10} = -j^2 2.25 \times 10^5 - j1500$$

$$Z_{EQ} = 225{,}000 - j1500 = 225{,}000 \underline{/0°}\ \Omega$$

Note that $-j^2$ is $+1$. Also, the reactive $j1500\ \Omega$ is negligible compared with the resistive 225,000 Ω. This answer for Z_{EQ} is the same as $Q \times X_L$, or $150 \times 1{,}500$, because of the high Q with negligibly small r_S.

LOW-Q CIRCUIT   We can consider a Q less than 10 as low. For the same circuit in Fig. 26-6, if r_S is 300 Ω with an X_L of 1500 Ω, the Q will be 1500/300, which equals 5. For this case of appreciable r_S, the branch currents cannot be equal when X_L and X_C are equal because then the inductive branch will have more impedance and less current.

With a low-Q circuit Z_{EQ} must be calculated in terms of the branch impedances. For this example, the calculations are simpler with all impedances stated in kilohms:

$$Z_{EQ} = \frac{Z_1 \times Z_2}{Z_1 + Z_2} = \frac{-j1.5 \times (j1.5 + 0.3)}{-j1.5 + j1.5 + 0.3} = \frac{-j^2 2.25 - j0.45}{0.3}$$

$$Z_{EQ} = 7.5 - j1.5\ \Omega = 7.65\ \underline{/-11.3°}\ k\Omega = 7650\ \underline{/-11.3°}\ \Omega$$

The phase angle θ is not zero because the reactive branch currents are unequal, even though X_L and X_C are equal. The appreciable value of r_S in the X_L branch makes this branch current smaller than I_C in the X_C branch.

CRITERIA FOR PARALLEL RESONANCE   The frequency f_r that makes $X_L = X_C$ is always $1/(2\pi\sqrt{LC})$. However, for low-Q circuits f_r does not necessarily provide the desired resonance effect. The three main criteria for parallel resonance are

1. Zero phase angle and unity power factor.
2. Maximum impedance and minimum line current.
3. $X_L = X_C$. This is the resonance at $f_r = 1/(2\pi\sqrt{LC})$.

These three effects do not occur at the same frequency in parallel circuits that have a low Q. The condition for unity power factor is often called *antiresonance* in a parallel LC circuit to distinguish it from the case of equal X_L and X_C.

It should be noted that when Q is 10 or higher, though, the parallel branch currents are practically equal when $X_L = X_C$. Then at $f_r = 1/(2\pi\sqrt{LC})$, the line current is minimum with zero phase angle, and the impedance is maximum.

For a series resonant circuit there are no parallel branches to consider. Therefore, the current is maximum at exactly f_r, whether the Q is high or low.

Answers at end of chapter.

a. Is the Q of 8 a high or low value?

b. With this Q, will the I_L be more or less than I_C in the parallel branches when $X_L = X_C$?

26-10 DAMPING OF PARALLEL RESONANT CIRCUITS

In Fig. 26-15a, the shunt R_P across L and C is a damping resistance because it lowers the Q of the tuned circuit. The R_P may represent the resistance of the external source driving the parallel resonant circuit, or R_P can be an actual resistor added for lower Q and greater bandwidth. Using the parallel R_P to reduce Q is better than increasing the series resistance r_S because the resonant response is more symmetrical with shunt damping.

The effect of varying the parallel R_P is opposite from the series r_S. A lower value of R_P lowers the Q and reduces the sharpness of resonance. Remember that less resistance in a parallel branch allows more current. This resistive branch current cannot be canceled at resonance by the reactive currents. Therefore, the resonant dip to minimum line current is less sharp with more resistive line current. Specifically, when Q is determined by parallel resistance

▶ $$Q = \frac{R_P}{X_L}$$ **(26-9)**

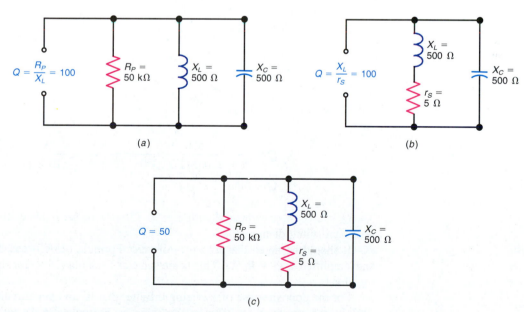

FIG. 26-15   The Q of a parallel resonant circuit in terms of coil resistance r_S and parallel damping resistor R_P. See Formula (26-10) for calculating Q. (a) Parallel R_P but negligible r_S. (b) Series r_S but no R_P branch. (c) Both R_P and r_S.

This relationship with shunt R_P is the reciprocal of the Q formula with series r_S. Reducing R_P decreases Q, but reducing r_S increases Q. The damping can be done by series r_S, parallel R_P, or both.

PARALLEL R_P WITHOUT r_S   In Fig. 26-15a, Q is determined only by the R_P, as no series r_S is shown. We can consider that r_S is zero or very small. Then the Q of the coil is infinite or high enough to be greater than the damped Q of the tuned circuit, by a factor of 10 or more. The Q of the damped resonant circuit here is $R_P/X_L = 50,000/500 = 100$.

SERIES r_S WITHOUT R_P   In Fig. 26-15b, Q is determined only by the coil resistance r_S, as no shunt damping resistance is used. Then $Q = X_L/r_S = 500/5 = 100$. This value is the Q of the coil, which is also the Q of the parallel resonant circuit without shunt damping.

CONVERSION OF r_S OR R_P   For the circuits in both Fig. 26-15a and b, Q is 100 because the 50,000-Ω R_P is equivalent to the 5-Ω r_S as a damping resistance. One value can be converted to the other. Specifically,

$$r_S = \frac{X_L{}^2}{R_P} \quad \text{or} \quad R_P = \frac{X_L{}^2}{r_S}$$

In this example, r_S equals $250,000/50,000 = 5\ \Omega$, or R_P is $250,000/5 = 50,000\ \Omega$.

DAMPING WITH BOTH r_S AND R_P   Figure 26-15c shows the general case of damping where both r_S and R_P must be considered. Then the Q of the circuit can be calculated as

$$Q = \frac{X_L}{r_S + X_L{}^2/R_P} \tag{26-10}$$

For the values in Fig. 26-15c,

$$Q = \frac{500}{5 + 250,000/50,000} = \frac{500}{5 + 5} = \frac{500}{10}$$

$$Q = 50$$

The Q is lower here compared with Fig. 26-15a or b because this circuit has both series and shunt damping.

It should be noted that for an r_S of zero, Formula (26-10) can be inverted and simplified to $Q = R_P/X_L$. This is the same as Formula (26-9) for shunt damping alone.

For the opposite case of R_P being infinite, that is, an open circuit, Formula (26-10) reduces to X_L/r_S. This is the same as Formula (26-4) without shunt damping.

Answers at end of chapter.
a. A parallel resonant circuit has an X_L of 1000 Ω and an r_S of 20 Ω, without any shunt damping. Calculate Q.
b. A parallel resonant circuit has an X_L of 1000 Ω, negligible r_S, and shunt R_P of 50 kΩ. Calculate Q.
c. How much is Z_{EQ} at f_r for the circuits in (**a**) and (**b**)?

26-11 CHOOSING *L* AND *C* FOR A RESONANT CIRCUIT

The following example illustrates how resonance is really just an application of X_L and X_C. Suppose that we have the problem of determining the inductance and capacitance for a circuit to be resonant at 159 kHz. First, we need a known value for either L or C, in order to calculate the other. Which one to choose depends on the application. In some cases, particularly at very high frequencies, C must be the minimum possible value, which might be about 10 pF. At medium frequencies, though, we can choose L for the general case where an X_L of 1000 Ω is desirable and can be obtained. Then the inductance of the required L, equal to $X_L/2\pi f$, is 0.001 H or 1 mH, for the inductive reactance of 1000 Ω.

For resonance at 159 kHz with a 1-mH L, the required C is 0.001 μF or 1000 pF. This value of C can be calculated for an X_C of 1000 Ω, equal to X_L at the f_r of 159 kHz, or from Formula (26-2). In either case, if you substitute 1×10^{-9} F for C and 1×10^{-3} H for L in the resonant frequency formula, f_r will be 159 kHz.

This combination is resonant at 159 kHz whether L and C are in series or parallel. In series, the resonant effect is to produce maximum current and maximum voltage across L or C at 159 kHz. The effect is desirable for the input circuit of an RF amplifier tuned to f_r because of the maximum signal. In parallel, the resonant effect at 159 kHz is minimum line current and maximum impedance across the generator. This effect is desirable for the output circuit of an RF amplifier, as the gain is maximum at f_r because of the high Z.

If we assume the 1-mH coil used for L has an internal resistance of 20 Ω, the Q of the coil is 1000 Ω/20 Ω, which equals 50. This value is also the Q of the series resonant circuit. If there is no shunt damping resistance across the parallel LC circuit, its Q is also 50. With a Q of 50 the bandwidth of the resonant circuit is 159 kHz/50, which equals 3.18 kHz for Δf.

Answers at end of chapter.
a. What is f_r for 1000 pF of C and 1 mH of L?
b. What is f_r for 250 pF of C and 1 mH of L?

Series and parallel resonance are compared in Table 26-5. The main difference is that series resonance produces maximum current and very low impedance at f_r, but with parallel resonance the line current is minimum to provide a very high impedance. Remember that these formulas for parallel resonance are very close approximations that can be used for circuits with a Q higher than 10. For series resonance, the formulas apply whether the Q is high or low.

TABLE 26-5 COMPARISON OF SERIES AND PARALLEL RESONANCE

SERIES RESONANCE	PARALLEL RESONANCE (HIGH Q)
$f_r = \dfrac{1}{2\pi\sqrt{LC}}$	$f_r = \dfrac{1}{2\pi\sqrt{LC}}$
I maximum at f_r with θ of $0°$	I_T minimum at f_r with θ of $0°$
Impedance Z minimum at f_r	Impedance Z maximum at f_r
$Q = X_L/r_S$, or	$Q = X_L/r_S$, or
$Q = V_{out}/V_{in}$	$Q = Z_{max}/X_L$
Q rise in voltage $= Q \times V_{gen}$	Q rise in impedance $= Q \times X_L$
Bandwidth $\Delta f = f_r/Q$	Bandwidth $\Delta f = f_r/Q$
Circuit capacitive below f_r, but inductive above f_r	Circuit inductive below f_r, but capacitive above f_r
Needs low-resistance source for low r_S, high Q, and sharp tuning	Needs high-resistance source for high R_P, high Q, and sharp tuning
Source is inside LC circuit	Source is outside LC circuit

SELF-TEST

ANSWERS AT BACK OF BOOK.

Choose (*a*), (*b*), (*c*), or (*d*).

1. For a series or parallel LC circuit, resonance occurs when (*a*) X_L is 10 times X_C or more; (*b*) X_C is 10 times X_L or more; (*c*) $X_L = X_C$; (*d*) the phase angle of the circuit is $90°$.
2. When either L or C is increased, the resonant frequency of the LC circuit (*a*) increases; (*b*) decreases; (*c*) remains the same; (*d*) is determined by the shunt resistance.

3. The resonant frequency of an *LC* circuit is 1000 kHz. If *L* is doubled but *C* is reduced to one-eighth of its original value, the resonant frequency then is (*a*) 250 kHz; (*b*) 500 kHz; (*c*) 1000 kHz; (*d*) 2000 kHz.

4. A coil has a 1000-Ω X_L and a 5-Ω internal resistance. Its *Q* equals (*a*) 0.005; (*b*) 5; (*c*) 200; (*d*) 1000.

5. In a parallel *LC* circuit, at the resonant frequency, the (*a*) line current is maximum; (*b*) inductive branch current is minimum; (*c*) total impedance is minimum; (*d*) total impedance is maximum.

6. At resonance, the phase angle equals (*a*) 0°; (*b*) 90°; (*c*) 180°; (*d*) 270°.

7. In a series *LC* circuit, at the resonant frequency, the (*a*) current is minimum; (*b*) voltage across *C* is minimum; (*c*) impedance is maximum; (*d*) current is maximum.

8. A series *LC* circuit has a *Q* of 100 at resonance. When 5 mV is applied at the resonant frequency, the voltage across *C* equals (*a*) 5 mV; (*b*) 20 mV; (*c*) 100 mV; (*d*) 500 mV.

9. An *LC* circuit resonant at 1000 kHz has a *Q* of 100. The bandwidth between half-power points equals (*a*) 10 kHz between 995 and 1005 kHz; (*b*) 10 kHz between 1000 and 1010 kHz; (*c*) 5 kHz between 995 and 1000 kHz; (*d*) 200 kHz between 900 and 1100 kHz.

10. In a low-*Q* parallel resonant circuit, when $X_L = X_C$, (*a*) I_L equals I_C; (*b*) I_L is less than I_C; (*c*) I_L is more than I_C; (*d*) the phase angle is 0°.

QUESTIONS

1. (**a**) State two characteristics of series resonance. (**b**) With a microammeter measuring current in the series *LC* circuit of Fig. 26-2, describe the meter readings for the different frequencies from 600 to 1400 kHz.

2. (**a**) State two characteristics of parallel resonance. (**b**) With a microammeter measuring current in the main line for the parallel *LC* circuit in Fig. 26-6*a*, describe the meter readings for frequencies from 600 to 1400 kHz.

3. State the *Q* formula for the following *LC* circuits: (**a**) series resonant; (**b**) parallel resonant, with series resistance r_S in the inductive branch; (**c**) parallel resonant, with zero series resistance but shunt R_P.

4. Explain briefly why a parallel *LC* circuit is inductive but a series *LC* circuit is capacitive below f_r.

5. What is the effect on *Q* and bandwidth of a parallel resonant circuit if its shunt damping resistance is decreased from 50,000 to 10,000 Ω?

6. Describe briefly how you would use an ac meter to measure the bandwidth of a series resonant circuit in order to calculate the circuits *Q*.

7. Why is a low-resistance generator good for a high *Q* in series resonance, while a high-resistance generator is needed for a high *Q* in parallel resonance?

8. Referring to Fig. 26-13, why is it that the middle frequency of 1080 kHz does not correspond to the middle capacitance value of 200 pF?

9. (**a**) Give three criteria for parallel resonance. (**b**) Why is the antiresonant frequency f_a different from f_r with a low-*Q* circuit? (**c**) Why are they the same for a high-*Q* circuit?

10. Show how Formula (26-10) reduces to R_P/X_L when r_S is zero.
11. (a) Specify the edge frequencies f_1 and f_2 for each of the three response curves in Fig. 26-11. (b) Why does lower Q allow more bandwidth?
12. (a) Why does maximum Z for a parallel resonant circuit correspond to minimum line current? (b) Why does zero phase angle for a resonant circuit correspond to unity power factor?
13. Explain how manual tuning of an LC circuit can be done with a capacitor or a coil.
14. What is meant by electronic tuning?
15. Suppose it is desired to tune an LC circuit from 540 to 1600 kHz by varying either L or C. Explain how the bandwidth Δf is affected by: (a) varying L to tune the LC circuit; (b) varying C to tune the LC circuit.

PROBLEMS

ANSWERS TO ODD-NUMBERED PROBLEMS AT BACK OF BOOK.

1. Calculate f_r for a series LC circuit with $L = 5$ μH and $C = 202.64$ pF.
2. Calculate f_r for a series LC circuit with $L = 33$ μH and $C = 7.67$ pF.
3. Calculate f_r for a parallel LC circuit with $L = 25.48$ μH and $C = 500$ pF.
4. Calculate f_r for a parallel LC circuit with $L = 2.2$ μH and $C = 58.74$ pF.
5. What value of inductance L must be connected in series with a 50-pF capacitance for an f_r of 3.8 MHz?
6. What value of capacitance C must be connected in parallel with a 100-μH inductance for an f_r of 1.9 MHz?
7. Calculate f_r in Fig. 26-16.

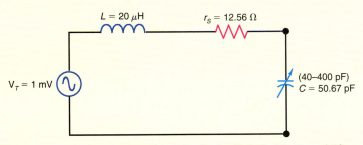

FIG. 26-16 Series resonant circuit for Probs. 7, 8, 9, 10, 11, and 19.

8. In Fig. 26-16, calculate the following values at f_r: X_L, X_C, Z_T, I, V_L, V_C, and θ_Z.
9. In Fig. 26-16, calculate Q, Δf, and the edge frequencies f_1 and f_2.
10. In Fig. 26-16, calculate Z_T, I, and θ_Z at: (a) f_1; (b) f_2.
11. Refer to Fig. 26-16. (a) To what value must the capacitance, C be adjusted to provide an f_r of 2.5 MHz? (b) What are the values for X_L, X_C, Q, and Δf for an f_r of 2.5 MHz?
12. Calculate f_r in Fig. 26-17.

13. In Fig. 26-17, calculate the following values at f_r: X_L, X_C, I_L, I_C, Q, Z_{EQ}, and I_T.

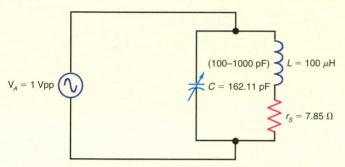

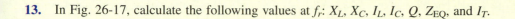

FIG. 26-17 Parallel resonant circuit for Probs. 12, 13, 14, 15, 16, and 17.

14. In Fig. 26-17, calculate Δf, f_1, and f_2.

15. In Fig. 26-17, determine Z_{EQ}, I_T, and θ_I at: (**a**) f_1; (**b**) f_2.

16. In Fig. 26-17, calculate Q and Δf if a 100-kΩ resistance R_P is placed across the tank circuit.

17. Refer to Fig. 26-17. (**a**) To what value must the capacitance C be adjusted to provide an f_r of 1 MHz? (**b**) What are the values for X_L, X_C, Q, Δf, Z_{EQ}, and I_T at 1 MHz?

18. Calculate the lowest and highest values of capacitance C needed to tune an LC circuit from 3.5 MHz to 4 MHz if a 25-μH inductor is used.

19. Using the original values in Fig. 26-16, calculate the power factor (PF), real power, and apparent power at: (**a**) f_r; (**b**) f_1; (**c**) f_2.

20. A series resonant circuit whose f_r is 10 MHz has a bandwidth Δf of 100 kHz. If the circuit's L/C ratio is doubled, calculate the edge frequencies f_1 and f_2.

21. Calculate f_r in Fig. 26-18.

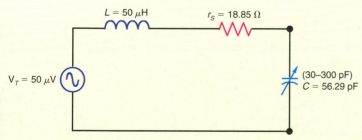

FIG. 26-18 Series resonant circuit for Probs. 21, 22, 23, 24, 25, and 26.

22. In Fig. 26-18, calculate the following values at f_r: X_L, X_C, Z_T, I, Q, V_L, V_C, and θ_Z.

23. In Fig. 26-18, calculate Δf, f_1, and f_2.

24. In Fig. 26-18, calculate Z_T, I, and θ_Z at f_1 and f_2.

25. In Fig. 26-18, what value of series resistance R_S must be added to double the bandwidth Δf when $C = 56.29$ pF?

26. Refer to Fig. 26-18. **(a)** To what value must the capacitance C be adjusted to provide an f_r of 1.5 MHz? **(b)** What are the values for X_L, X_C, Q, and Δf at 1.5 MHz?
27. Calculate f_r in Fig. 26-19.
28. In Fig. 26-19, calculate the following values at f_r: X_L, X_C, I_L, I_C, Q, Z_{EQ}, and θ_I.
29. In Fig. 26-19, calculate Δf, f_1, and f_2.
30. In Fig. 26-19, calculate Q and Δf if a 1-MΩ R_P is placed across the tank.

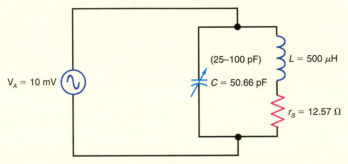

FIG. 26-19 Parallel resonant circuit for Probs. 27, 28, 29, and 30.

CRITICAL THINKING

1. Prove that

$$X_L = \sqrt{\frac{L}{C}}$$

for an LC circuit at f_r.

2. Suppose you are an engineer designing a coil to be used in a resonant LC circuit. Besides obtaining the required inductance L, your main concern is in reducing skin effect so as to obtain as high a Q as possible for the LC circuit. List three design techniques which would reduce or minimize skin effect in the coil windings.

26-1 **a.** 1000 kHz
     b. 1000 kHz

26-2 **a.** F
     b. T
     c. T

26-3 **a.** F
     b. T
     c. T

26-4 **a.** decreased
     b. 200 μH
     c. 39.48

26-5 **a.** $Q = 100$
     b. $Z_{EQ} = 25\ k\Omega$

26-6 **a.** $\Delta f = 0.25\ \text{MHz}$
     b. $Q = 50$

26-7 **a.** lowest
     b. 1:4

26-8 **a.** capacitive
     b. inductive

26-9 **a.** low
     b. less

26-10 **a.** $Q = 50$
     b. $Q = 50$
     c. $Z_{EQ} = 50\ k\Omega$

26-11 **a.** $f_r = 159\ \text{kHz}$
     b. $f_r = 318\ \text{kHz}$

CHAPTER 27

FILTERS

A filter separates different components that are mixed together. For instance, a mechanical filter can separate particles from liquid, or small particles from large particles. An electrical filter can separate different frequency components.

Generally, inductors and capacitors are used for filtering because of their opposite frequency characteristics. Inductive reactance X_L increases but capacitive reactance X_C decreases with higher frequencies. In addition, their filtering action depends on whether L and C are in series or parallel with the load.

The amount of attenuation offered by a filter is usually specified in decibels (dB). The decibel is a logarithmic expression which compares two power levels. The frequency response of a filter is usually drawn as a graph of frequency versus decibel attenuation.

The most common filtering applications are separating audio from radio frequencies, or vice versa, and separating ac variations from the average dc level. There are many of these applications in electronic circuits.

CHAPTER OBJECTIVES

Upon completion of this chapter, you should be able to:

- *State* the difference between a low-pass and a high-pass filter.
- *Explain* what is meant by *pulsating direct current*.
- *Understand* how a transformer acts as a high-pass filter.
- *Understand* how an *RC* coupling circuit couples ac but blocks dc.
- *Understand* the function of a bypass capacitor.
- *Calculate* the cutoff frequency, output voltage, and phase angle of basic *RL* and *RC* filters.
- *Explain* the operation of bandpass and bandstop filters.
- *Explain* why log-log graph paper or semilog graph paper is used to plot a frequency response.
- *Define* the term *decibel*.
- *Explain* how resonant circuits can be used as bandpass or bandstop filters.
- *Describe* the function of a power-line filter and a television antenna filter.

IMPORTANT TERMS IN THIS CHAPTER

ac component	cycle	octave
active filter	dc component	π-type filter
attenuation of filter	decade	piezoelectric effect
bandpass filter	decibels	pulsating dc values
bandstop filter	fluctuating dc values	ripple
bypass capacitor	graph cycles	semilog graph
capacitive coupling	high-pass filter	T-type filter
crystal filter	low-pass filter	transformer coupling
cutoff frequency	L-type filter	

TOPICS COVERED IN THIS CHAPTER

27-1 EXAMPLES OF FILTERING

Electronic circuits often have currents of different frequencies corresponding to voltages of different frequencies. The reason is that a source produces current with the same frequency as the applied voltage. As examples, the ac signal input to an audio circuit can have high and low audio frequencies; an RF circuit can have a wide range of radio frequencies in its input; the audio detector in a radio has both radio frequencies and audio frequencies in the output. Finally, the rectifier in a power supply produces dc output with an ac ripple superimposed on the average dc level.

In such applications where the current has different frequency components, it is usually necessary either to favor or to reject one frequency or a band of frequencies. Then an electrical filter is used to separate higher or lower frequencies.

The electrical filter can pass the higher-frequency component to the load resistance, which is the case of a high-pass filter, or a low-pass filter can be used to favor the lower frequencies. In Fig. 27-1a, the high-pass filter allows 10 kHz to produce output, while rejecting or attenuating the lower frequency of 100 Hz. In Fig. 27-1b, the filtering action is reversed to pass the lower frequency of 100 Hz, while attenuating 10 kHz. These examples are for high and low audio frequencies.

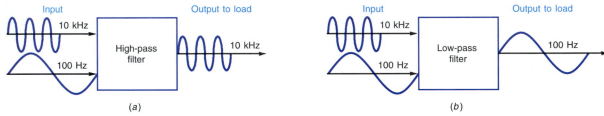

(a)   (b)

FIG. 27-1 Function of electrical filters. (a) High-pass filter couples higher frequencies out to the load. (b) Low-pass filter couples lower frequencies out to the load.

For the case of audio mixed with radio frequencies, a low-pass filter allows the audio frequencies in the output. Or a high-pass filter allows the radio frequencies to be passed to the load.

27-2 DIRECT CURRENT COMBINED WITH ALTERNATING CURRENT

Current that varies in amplitude but does not reverse in polarity is considered *pulsating* or *fluctuating* direct current. It is not a steady direct current because its value fluctuates. However, it is not alternating current because the polarity remains the same, either positive or negative. The same idea applies to voltages.

Figure 27-2 illustrates how a circuit can have pulsating direct current or voltage. Here, the steady dc voltage of the battery V_B is in series with the ac voltage V_A. Since the two series generators add, the voltage across R_L is the sum of the two applied voltages, as shown by the waveshape of v_R in Fig. 27-2b.

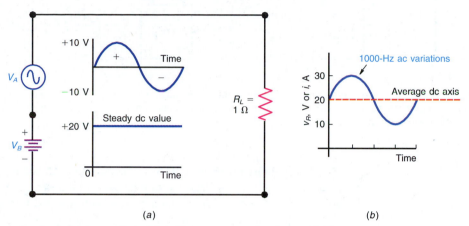

(a)   (b)

FIG. 27-2 An example of a pulsating or fluctuating direct current and voltage. (*a*) Circuit. (*b*) Graph of voltage across R_L. This V equals V_B of the battery plus V_A of the ac source, with frequency of 1000 Hz.

If values are taken at opposite peaks of the ac variation, when V_A is at $+10$ V, it adds to the $+20$ V of the battery to provide $+30$ V across R_L; when the ac voltage is -10 V, it bucks the battery voltage of $+20$ V to provide $+10$ V across R_L. When the ac voltage is at zero, the voltage across R_L equals the battery voltage of $+20$ V.

The combined voltage v_R then consists of the ac variations fluctuating above and below the battery voltage as the axis, instead of the zero axis for ac voltage. The result is a pulsating dc voltage, since it is fluctuating but always has positive polarity with respect to zero.

The pulsating direct current i through R_L has the same waveform, fluctuating above and below the steady dc level of 20 A. The i and v values are the same because R_L is 1 Ω.

Another example is shown in Fig. 27-3. If the 100-Ω R_L is connected across 120 V 60 Hz as in Fig. 27-3a, the current in R_L will be V/R_L. This is an ac sine wave, with an rms value of 120/100 or 1.2 A.

Also, if you connect the same R_L across the 200-Vdc source in Fig. 27-3b, instead of using the ac source, the steady direct current in R_L will be 200/100, or 2 A. The battery source voltage and its current are considered steady dc values because there are no variations.

However, suppose that the ac source V_A and dc source V_B are connected in series with R_L, as in Fig. 27-3c. What will happen to the current and voltage for R_L? Will V_A or V_B supply the current? The answer is that both sources will. Each voltage source produces current as though the other were not there, assuming the sources have negligibly small internal impedance. The result then is the fluctuating dc voltage or current shown, with the ac variations of V_A superimposed on the average dc level of V_B.

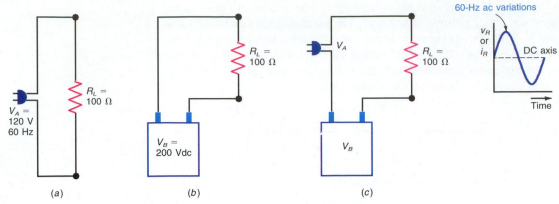

FIG. 27-3 A combination of ac and dc voltage to provide fluctuating dc voltage across R_L. (*a*) An ac source alone. (*b*) A dc source alone. (*c*) The ac source and dc source in series for the fluctuating voltage across R_L.

DC AND AC COMPONENTS The pulsating dc voltage v_R in Fig. 27-3*c* is just the original ac voltage V_A with its axis shifted to a dc level by the battery voltage V_B. In effect, a dc component has been inserted into the ac variations. This effect is called *dc insertion.*

Referring back to Fig. 27-2, if you measure across R_L with a dc voltmeter, it will read the dc level of 20 V. An ac-coupled oscilloscope* will show only the peak-to-peak variations of ± 10 V.

It is convenient, therefore, to consider the pulsating or fluctuating voltage and current in two parts. One is the steady dc component, which is the axis or average level of the variations; the other is the ac component, consisting of the variations above and below the dc axis. Here the dc level for V_T is $+20$ V, while the ac component equals 10 V peak or 7.07 V rms value. The ac component is also called ac *ripple.*

It should be noted that with respect to the dc level the fluctuations represent alternating voltage or current that actually reverses in polarity. For example, the change of v_R from $+20$ to $+10$ V is just a decrease in positive voltage compared with zero. However, compared with the dc level of $+20$ V, the value of $+10$ V is 10 V more negative than the axis.

TYPICAL EXAMPLES OF DC LEVEL WITH AC COMPONENT As a common application, transistors always have fluctuating dc voltage or current when used for amplifying an ac signal. The transistor amplifier needs steady dc voltages to operate. The signal input is an ac variation, usually with a dc axis to establish the desired operating level. The amplified output is also an ac variation superimposed on a dc supply voltage that supplies the required power output. Therefore, the input and output circuits have fluctuating dc voltage.

The examples in Fig. 27-4 illustrate two possibilities, in terms of polarities with respect to chassis ground. In Fig. 27-4*a*, the waveform is always positive, as in the previous examples. This example could apply to collector voltage on an NPN transistor amplifier. Note the specific values. The average dc axis is the steady dc level. The positive peak equals the dc level plus the peak ac value. The

---

*See App. D for an explanation of how to use the oscilloscope.

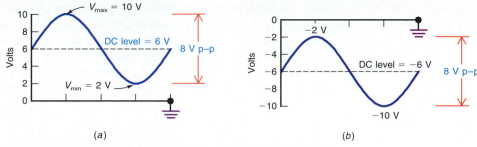

FIG. 27-4 Typical examples of a dc voltage access for an ac component. (*a*) Positive fluctuating dc values because of a large positive dc component. (*b*) Negative fluctuating dc values because of a large negative dc component.

minimum point equals the dc level minus the peak ac value. The peak-to-peak value of the ac component and its rms value are the same as for the ac signal alone. However, it is better to subtract the minimum from the maximum for the peak-to-peak value, in case the waveform is unsymmetrical.

In Fig. 27-4*b*, all the values are negative. Notice that here the positive peak of the ac component subtracts from the dc level because of the opposite polarities. Now the negative peak adds to the negative dc level to provide a maximum point of negative voltage.

SEPARATING THE AC COMPONENT   In many applications, the circuit has pulsating dc voltage, but only the ac component is desired. Then the ac component can be passed to the load, while the steady dc component is blocked, either with transformer coupling or with capacitive coupling. A transformer with a separate secondary winding isolates or blocks steady direct current in the primary. A capacitor isolates or blocks a steady dc voltage.

TEST-POINT QUESTION 27-2

Answers at end of chapter.

For the fluctuating dc waveform in Fig. 27-4*a*, specify the following voltages:

a. Average dc level.
b. Maximum and minimum values.
c. Peak-to-peak of ac component.
d. Peak and rms of ac component.

27-3 TRANSFORMER COUPLING

Remember that a transformer produces induced secondary voltage just for variations in primary current. With pulsating direct current in the primary, the secondary has output voltage only for the ac variations, therefore. The steady dc component in the primary has no effect in the secondary.

In Fig. 27-5, the pulsating dc voltage in the primary produces pulsating primary current. The dc axis corresponds to a steady value of primary current that has a constant magnetic field, but only when the field changes can secondary voltage be induced. Therefore, only the fluctuations in the primary can produce output in the secondary. Since there is no output for the steady primary current, this dc level corresponds to the zero level for the ac output in the secondary.

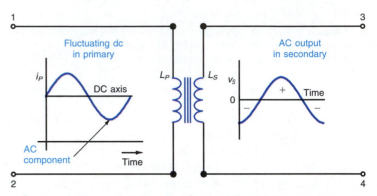

FIG. 27-5 Transformer coupling blocks the dc component. With fluctuating direct current in the primary L_P, only the ac component produces induced voltage in the secondary L_S.

When the primary current increases above the steady level, this increase produces one polarity for the secondary voltage as the field expands; when the primary current decreases below the steady level, the secondary voltage has reverse polarity as the field contracts. The result in the secondary is an ac variation having opposite polarities with respect to the zero level.

The phase of the ac secondary voltage may be as shown or 180° opposite, depending on the connections and direction of the windings. Also, the ac secondary output may be more or less than the ac component in the primary, depending on the turns ratio. This ability to isolate the steady dc component in the primary while providing ac output in the secondary applies to all transformers with a separate secondary winding, whether iron-core or air-core.

TEST-POINT QUESTION 27-3

Answers at end of chapter.
a. Is transformer coupling an example of a high-pass or low-pass filter?
b. In Fig. 27-5, what is the level of v_S for the average dc level of i_P?

27-4 CAPACITIVE COUPLING

Capacitive coupling is probably the most common type of coupling in amplifier circuits. The coupling means connecting the output of one circuit to the input of the next. The requirements are to include all frequencies in the desired signal, while rejecting undesired components. Usually, the dc component must be blocked from the input to ac amplifiers. The purpose is to maintain a specific dc level for the amplifier operation.

In Fig. 27-6, the pulsating dc voltage across input terminals 1 and 2 is applied to the RC coupling circuit. Capacitance C_C will charge to the steady dc level, which is the average charging voltage. The steady dc component is blocked, therefore, since it cannot produce voltage across R. However, the ac component is developed across R, between the output terminals 3 and 4. The reason is that the ac voltage allows C to produce charge and discharge current through R. Note that the zero axis of the ac voltage output corresponds to the average level of the pulsating dc voltage input.

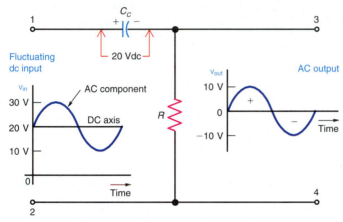

FIG. 27-6 The RC coupling blocks the dc component. With fluctuating dc voltage applied, only the ac component produces charge and discharge current for the output voltage across R.

THE DC COMPONENT ACROSS C The voltage across C_C is the steady dc component of the input voltage because the variations of the ac component are symmetrical above and below the average level. Furthermore, the series resistance is the same for charge and discharge. As a result, any increase in charging voltage above the average level is counteracted by an equal discharge below the average.

In Fig. 27-6, for example, when v_{in} increases from 20 to 30 V, this effect on charging C_C is nullified by the discharge when v_{in} decreases from 20 to 10 V. At all times, however, v_{in} has a positive value that charges C_C in the polarity shown.

The net result is that only the average level is effective in charging C_C, since the variations from the axis neutralize each other. After a period of time, depending on the RC time constant, C_C will charge to the average value of the pulsating dc voltage applied, which is 20 V here.

THE AC COMPONENT ACROSS R Although C_C is charged to the average dc level, when the pulsating input voltage varies above and below this level, the charge and discharge current produces IR voltage corresponding to the fluctuations of the input. When v_{in} increases above the average level, C_C takes on charge, producing charging current through R. Even though the charging current may be too small to affect the voltage across C_C appreciably, the IR drop across

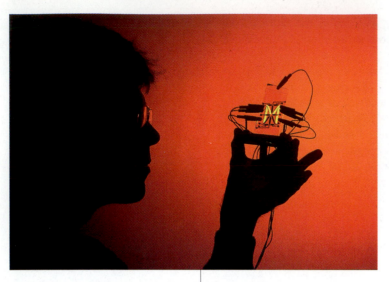

a large value of resistance can be practically equal to the ac component of the input voltage. In summary, a long RC time constant is needed for good coupling.

If the polarity is considered, in Fig. 27-6, the charging current produced for an increase of v_{in} produces electron flow from the low side of R to the top, adding electrons to the negative side of C_C. The voltage at the top of R is then positive with respect to the line below.

When v_{in} decreases below the average level, C loses charge. The discharge current then is in the opposite direction through R. The result is negative polarity for the ac voltage output across R.

When the input voltage is at its average level, there is no charge or discharge current, resulting in zero voltage across R. The zero level in the ac voltage across R corresponds to the average level of the pulsating dc voltage applied to the RC circuit.

The end result is that with positive pulsating dc voltage applied, the values above the average produce the positive half-cycle of the ac voltage across R; the values below the average produce the negative half-cycle. Only this ac voltage across R is coupled to the next circuit, as terminals 3 and 4 provide the output from the RC coupling circuit.

It is important to note that there is practically no phase shift. This rule applies to all RC coupling circuits, since R must be ten or more times X_C. Then the reactance is negligible compared with the series resistance, and the phase angle of less than 5.7° is practically zero.

VOLTAGES AROUND THE RC COUPLING CIRCUIT   If you measure the fluctuating dc voltage across the input terminals 1 and 2 in Fig. 27-6 with a dc voltmeter, it will read the average dc level of 20 V. Across the same two points, if you connect an ac-coupled oscilloscope, it will show only the fluctuating ac component. These voltage variations have a peak value of 10 V, or a peak-to-peak value of 20 V, or an rms value of $0.707 \times 10 = 7.07$ V.

Across points 1 and 3 for V_C in Fig. 27-6, a dc voltmeter reads the steady dc value of 20 V. An ac voltmeter across points 1 and 3 reads practically zero.

However, an ac voltmeter across the output R between points 3 and 4 will read the ac voltage of 7 V, approximately, for V_R. Furthermore, a dc voltmeter across R reads zero. The dc component of the input voltage is across C_C but is blocked from the output across R.

TYPICAL COUPLING CAPACITORS   Common values of RF and AF coupling capacitors for different sizes of series R are listed in Table 27-1. In all cases the coupling capacitor blocks the steady dc component of the input voltage, while the ac component is passed to the resistance.

The size of C_C required depends on the frequency of the ac component. At each frequency listed at the left in Table 27-1, the values of capacitance in the horizontal row have an X_C equal to one-tenth the resistance value for each column. The R increases from 1.6 to 16 to 160 kΩ for the three columns, allowing

| FREQUENCY | VALUES OF C_C | | | FREQUENCY BAND |
	$R = 1.6\ k\Omega$	$R = 16\ k\Omega$	$R = 160\ k\Omega$	
100 Hz	10 μF	1 μF	0.1 μF	Audio frequency
1000 Hz	1 μF	0.1 μf	0.01 μF	Audio frequency
10 kHz	0.1 μF	0.01 μF	0.001 μF	Audio frequency
100 kHz	0.01 μF	0.001 μF	100 pF	Radio frequency
1 MHz	0.001 μF	100 pF	10 pF	Radio frequency
10 MHz	100 pF	10 pF	1 pF	Radio frequency
100 MHz	10 pF	1 pF	0.1 pF	Very High frequency

*For coupling circuit in Fig. 27-6; $X_{C_C} = \frac{1}{10}\ R$.

smaller values of C_C. Typical audio coupling capacitors, then, are about 0.1 to 10 μF, depending on the lowest audio frequency to be coupled and the size of the series resistance. Typical RF coupling capacitors are about 1 to 100 pF.

Values of C_C more than about 1 μF are usually electrolytic capacitors, which must be connected in the correct polarity. These can be very small, many being ½ in. long, with a low voltage rating of 3 to 25 V for transistor circuits. Also, the small leakage current of electrolytic capacitors is not a serious problem in this application because of the low voltage and small series resistance for transistor coupling circuits.

TEST-POINT QUESTION 27-4

Answers at end of chapter.
a. In Fig. 27-6, what is the level of v_{out} across R corresponding to the average dc level of v_{in}?
b. Which of the following is a typical audio coupling capacitor with a 1-kΩ R: 1 pF; 0.001 μF; or 5 μF?

27-5 BYPASS CAPACITORS

A bypass is a path around a component. In circuits, the bypass is a parallel or shunt path. Capacitors are often used in parallel with resistance to bypass the ac component of a pulsating dc voltage. The result, then, is steady dc voltage across the RC parallel combination, if the bypass capacitance is large enough to have little reactance for the lowest frequency of the ac variations.

As illustrated in Fig. 27-7, the capacitance C_1 in parallel with R_1 is an ac bypass capacitor for R_1. For any frequency at which X_{C_1} is one-tenth of R_1, or less, the ac component is bypassed around R_1 through the low reactance in the shunt path. The result is practically zero ac voltage across the bypass capacitor because of its low reactance.

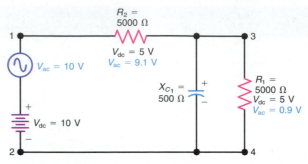

FIG. 27-7 Low reactance of bypass capacitor C_1 short-circuits R_1 for an ac component of fluctuating dc input voltage.

Since the voltage is the same across R_1 and C_1 because they are in parallel, there is also no ac voltage across R_1 for the frequency at which C_1 is a bypass capacitor. We can say that R is bypassed for the frequency at which X_C is one-tenth of R. The bypassing also applies to higher frequencies where X_C is less than one-tenth of R. Then the ac voltage across the bypass capacitor is even closer to zero because of its lower reactance.

BYPASSING THE AC COMPONENT OF A PULSATING DC VOLTAGE

The voltages in Fig. 27-7 are calculated by considering the effect of C_1 separately for V_{dc} and for V_{ac}. For direct current, C_1 is practically an open circuit. Then its reactance is so high compared with the 5000-Ω R_1 that X_{C_1} can be ignored as a parallel branch. Therefore, R_1 can be considered as a voltage divider in series with R_2. Since R_1 and R_2 are equal, each has 5 V, equal to one-half V_{dc}. Although this dc voltage division depends on R_1 and R_2, the dc voltage across C_1 is the same 5 V as across its parallel R_1.

For the ac component of the applied voltage, however, the bypass capacitor has very low reactance. In fact, X_{C_1} must be one-tenth of R_1, or less. Then the 5000-Ω R_1 is so high compared with the low value of X_{C_1} that R_1 can be ignored as a parallel branch. Therefore, the 500-Ω X_{C_1} can be considered as a voltage divider in series with R_2.

With an X_{C_1} of 500 Ω, this value in series with the 5000-Ω R_2 allows approximately one-eleventh of V_{ac} to be developed across C_1. This ac voltage, equal to 0.9 V here, is the same across R_1 and C_1 in parallel. The remainder of the ac applied voltage, equal to approximately 9.1 V, is across R_2. In summary, then, the bypass capacitor provides an ac short circuit across its shunt resistance, so that little or no ac voltage can be developed, without affecting the dc voltages.

Measuring voltages around the circuit in Fig. 27-7, a dc voltmeter reads 5 V across R_1 and 5 V across R_2. An ac voltmeter across R_2 reads 9.1 V, which is almost all the ac input voltage. Across the bypass capacitor C_1 the ac voltage is only 0.9 V.

In Table 27-2, typical sizes for RF and AF bypass capacitors are listed. The values of C have been calculated at different frequencies for an X_C one-tenth the shunt resistance given in each column. The R decreases for the three columns, from 16 to 1.6 kΩ and 160 Ω. Note that smaller values of R require larger values of C for bypassing. Also, when X_C equals one-tenth of R at one frequency, X_C will be even less for higher frequencies, improving the bypassing action.

DID YOU KNOW?

Crucial influences on the American electric power industry come from three sources: the need for reliable power to control sensitive electronics in business; regulations forcing transmission networks to be open to competitors; and the practice of bulk power sales.

| | VALUES OF C | | | |
FREQUENCY	$R = 16\text{ k}\Omega$	$R = 1.6\text{ k}\Omega$	$R = 160\ \Omega$	FREQUENCY BAND
100 Hz	$1\ \mu\text{F}$	$10\ \mu\text{F}$	$100\ \mu\text{F}$	Audio frequency
1000 Hz	$0.1\ \mu\text{F}$	$1\ \mu\text{F}$	$10\ \mu\text{F}$	Audio frequency
10 kHz	$0.01\ \mu\text{F}$	$0.1\ \mu\text{F}$	$1\ \mu\text{F}$	Audio frequency
100 kHz	$0.001\ \mu\text{F}$	$0.01\ \mu\text{F}$	$0.1\ \mu\text{F}$	Radio frequency
1 MHz	100 pF	$0.001\ \mu\text{F}$	$0.01\ \mu\text{F}$	Radio frequency
10 MHz	10 pF	100 pF	$0.001\ \mu\text{F}$	Radio frequency
100 MHz	1 pF	10 pF	100 pF	Very high frequency

*For RC bypass circuit in Fig. 27-7; $X_{C_1} = \frac{1}{10}\,R$.

Therefore, the size of bypass capacitors should be considered on the basis of the lowest frequency to be bypassed.

It should be noted that the applications of coupling and bypassing for C are really the same, except that C_C is in series with R and the bypass C is in parallel with R. In both cases X_C must be one-tenth or less of R. Then C_C couples the ac signal to R. Or the shunt bypass short-circuits R for the ac signal.

BYPASSING RADIO FREQUENCIES BUT NOT AUDIO FREQUENCIES

See Fig. 27-8. At the audio frequency of 1000 Hz, C_1 has a reactance of 1.6 MΩ. This reactance is so much higher than R_1 that the impedance of the parallel combination is essentially equal to the 16,000 Ω of R_1. Then R_1 and R_2 serve as a voltage divider for the applied AF voltage of 10 V. Each of the equal resistances has one-half the applied voltage, equal to the 5 V across R_2 and 5 V across R_1. This 5 V at 1000 Hz is also present across C_1, since it is in parallel with R_1.

For the RF voltage at 1 MHz, however, the reactance of the bypass capacitor is only 1600 Ω. This is one-tenth of R_1. Then X_{C_1} and R_1 in parallel have a combined impedance equal to approximately 1600 Ω.

Now, with a 1600-Ω impedance for the R_1C_1 bank in series with the 16,000 Ω of R_2, the voltage across R_1 and C_1 is one-eleventh the applied RF voltage. Then there is 0.9 V across the lower impedance of R_1 and C_1, with 9.1 V across the larger resistance of R_2. As a result, the RF component of the applied voltage can be considered bypassed. The capacitor C_1 is the RF bypass across R_1.

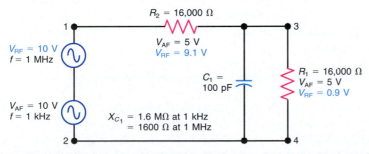

FIG. 27-8 Capacitor C_1 bypasses R_1 for the radio frequencies but not for the audio frequencies.

Answers at end of chapter.
a. In Fig. 27-8, is C_1 an AF or RF bypass?
b. Which of the following is a typical audio bypass capacitor across a 1-kΩ R: 1 pF; 0.001 μF; or 5 μF?

27-6 FILTER CIRCUITS

In terms of their function, filters can be classified as either low-pass or high-pass. A low-pass filter allows the lower-frequency components of the applied voltage to develop output voltage across the load resistance, while the higher-frequency components are attenuated, or reduced, in the output. A high-pass filter does the opposite, allowing the higher-frequency components of the applied voltage to develop voltage across the output load resistance.

The case of an RC coupling circuit is an example of a high-pass filter because the ac component of the input voltage is developed across R while the dc voltage is blocked by the series capacitor. Furthermore, with higher frequencies in the ac component, more ac voltage is coupled. For the opposite case, a bypass capacitor is an example of a low-pass filter. The higher frequencies are bypassed, but the lower the frequency, the less the bypassing action. Then lower frequencies can develop output voltage across the shunt bypass capacitor.

In order to make the filtering more selective in terms of which frequencies are passed to produce output voltage across the load, filter circuits generally combine inductance and capacitance. Since inductive reactance increases with higher frequencies, while capacitive reactance decreases, the two opposite effects improve the filtering action.

With combinations of L and C, filters are named to correspond to the circuit configuration. Most common types of filters are the L, T, and π. Any one of the three can function as either a low-pass filter or a high-pass filter.

For either low-pass or high-pass filters with L and C the reactance X_L increases with higher frequencies, while X_C decreases. The frequency characteristics of X_L and X_C cannot be changed. However, the circuit connections are opposite to reverse the filtering action.

In general, high-pass filters use:

1. Coupling capacitance C in series with the load. Then X_C can be low for high frequencies to be passed to R_L, while low frequencies are blocked.
2. Choke inductance L in parallel across R_L. Then the shunt X_L can be high for high frequencies to prevent a short circuit across R_L, while low frequencies are bypassed.

The opposite characteristics for low-pass filters are:

1. Inductance L in series with the load. The high X_L for high frequencies can serve as a choke, while low frequencies can be passed to R_L.
2. Bypass capacitance C in parallel across R_L. Then high frequencies are bypassed by a small X_C, while low frequencies are not affected by the shunt path.

For any filter, the ability to reduce the amplitude of undesired frequencies is called the *attenuation* of the filter. The frequency at which the attenuation reduces the output to 70.7 percent is the *cutoff frequency*, usually designated f_c.

TEST-POINT QUESTION 27-6

Answers at end of chapter.
a. Does high-pass or low-pass filtering require series C?
b. Which filtering requires parallel C?

27-7 LOW-PASS FILTERS

Figure 27-9 illustrates low-pass circuits from a single filter element with a shunt bypass capacitor in Fig. 27-9a or a series choke in b, to the more elaborate combinations of an L-type filter in c, a T type in d, and a π type in e and f. With an applied input voltage having different frequency components, the low-pass filter action results in maximum low-frequency voltage across R_L, while most of the high-frequency voltage is developed across the series choke or resistance.

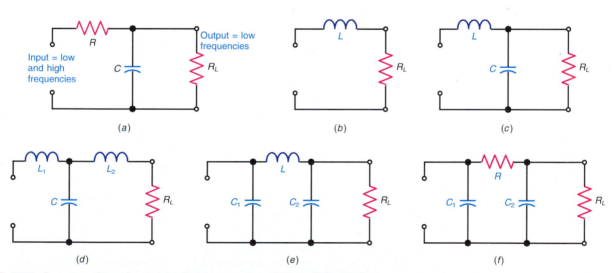

FIG. 27-9 Low-pass filter circuits. (*a*) Bypass capacitor C in parallel with R_L. (*b*) Choke L in series with R_L. (*c*) Inverted-L type with choke and bypass capacitor. (*d*) The T type with two chokes and one bypass capacitor. (*e*) The π type with one choke and bypass capacitors at both ends. (*f*) The π type with a series resistor instead of a choke.

In Fig. 27-9a, the shunt capacitor C bypasses R_L for high frequencies. In Fig. 27-9b, the choke L acts as a voltage divider in series with R_L. Since L has maximum reactance for the highest frequencies, this component of the input voltage is developed across L, with little across R_L. For lower frequencies, L has low reactance, and most of the input voltage can be developed across R_L.

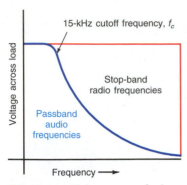

FIG. 27-10 The response of a low-pass filter with cutoff at 15 kHz. The filter passes the audio signal but attenuates radio frequencies.

In Fig. 27-9c, the use of both the series choke and bypass capacitor improves the filtering by providing sharper cutoff between the low frequencies that can develop voltage across R_L and the higher frequencies stopped from the load by producing maximum voltage across L. Similarly, the T-type circuit in Fig. 27-9d and the π-type circuits in e and f improve filtering.

Using the series resistance in Fig. 27-9f instead of a choke provides an economical π filter needing less space.

PASSBAND AND STOP BAND   As illustrated in Fig. 27-10, a low-pass filter attenuates frequencies above the cutoff frequency f_c of 15 kHz in this example. Any component of the input voltage having a frequency lower than 15 kHz can produce output voltage across the load. These frequencies are in the *passband*. Frequencies of 15 kHz or more are in the *stop band*. The sharpness of filtering between the passband and the stop band depends on the type of circuit. In general, the more L and C components, the sharper the response of the filter can be. Therefore, π and T types are better filters than the L type and the bypass or choke alone.

The response curve in Fig. 27-10 is illustrated for the application of a low-pass filter attenuating RF voltages while passing audio frequencies to the load. This is necessary where the input voltage has RF and AF components but only the audio voltage is desired for the AF circuits that follow the filter.

A good example is filtering the audio output of the detector circuit in a radio receiver, after the RF-modulated carrier signal has been rectified. Another common application of low-pass filtering is where the steady dc component of pulsating dc input must be separated from the higher frequency 60-Hz ac component, as in the pulsating dc output of the rectifier in a power supply.

CIRCUIT VARIATIONS   The choice between the T-type filter with a series input choke and the π type with a shunt input capacitor depends upon the internal resistance of the generator supplying input voltage to the filter. A low-resistance generator needs the T filter so that the choke can provide a high series impedance for the bypass capacitor. Otherwise, the bypass capacitor must have extremely large values to short-circuit the low-resistance generator for high frequencies.

The π filter is more suitable with a high-resistance generator where the input capacitor can be effective as a bypass. For the same reasons, the L filter can have the shunt bypass either in the input for a high-resistance generator or across the output for a low-resistance generator.

For all the filter circuits, the series choke can be connected either in the high side of the line, as in Fig. 27-9, or in series in the opposite side of the line, without any effect on the filtering action. Also, the series components can be connected in both sides of the line for a *balanced filter* circuit.

PASSIVE AND ACTIVE FILTERS   All the circuits here are passive filters, as they use only capacitors, inductors, and resistors, which are passive components. An active filter, however, uses the operational amplifier (op amp) on an IC chip, with R and C. The purpose is to eliminate the need for inductance L. This feature is important in filters for audio frequencies when large coils would be necessary. The operational amplifier is described in Chap. 32, "Integrated Circuits."

Answers at end of chapter.

a. Which diagrams in Fig. 27-9 show a π-type filter?

b. Does the response curve in Fig. 27-10 show low-pass or high-pass filtering?

27-8 HIGH-PASS FILTERS

As illustrated in Fig. 27-11, the high-pass filter passes to the load all frequencies higher than the cutoff frequency f_c, while lower frequencies cannot develop appreciable voltage across the load. The graph in Fig. 27-11a shows the response of a high-pass filter with a stopband of 0 to 50 Hz. Above the cutoff frequency of 50 Hz, the higher audio frequencies in the passband can produce AF voltage across the output load resistance.

The high-pass filtering action results from using C_C as a coupling capacitor in series with the load, as in Fig. 27-11b. The L, T, and π types use the inductance for a high-reactance choke across the line. In this way the higher-frequency components of the input voltage can develop very little voltage across the series

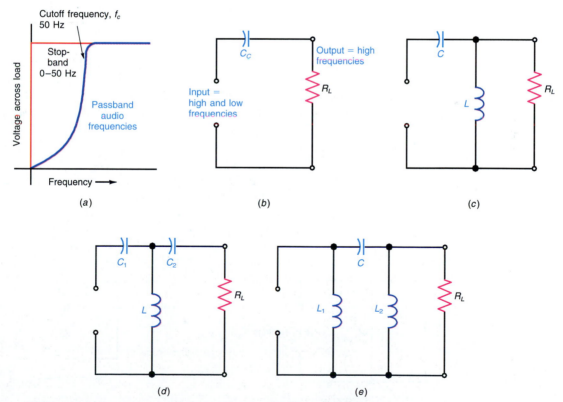

FIG. 27-11 High-pass filters. (a) The response curve for an audio-frequency filter cutting off at 50 Hz. (b) An *RC* coupling circuit. (c) Inverted-L type. (d) The T type. (e) The π type.

capacitance, allowing most of this voltage to be produced across R_L. The inductance across the line has higher reactance with increasing frequencies, allowing the shunt impedance to be no lower than the value of R_L.

For low frequencies, however, R_L is effectively short-circuited by the low inductive reactance across the line. Also, C_C has high reactance and develops most of the voltage at low frequencies, stopping these frequencies from developing voltage across the load.

TEST-POINT QUESTION 27-8

Answers at end of chapter.
a. Which diagram in Fig. 27-11 shows a T-type filter?
b. Does the response curve in Fig. 27-11*a* show high-pass or low-pass filtering?

27-9 ANALYZING FILTER CIRCUITS

Any low-pass or high-pass filter can be thought of as a frequency-dependent voltage divider, since the amount of output voltage is a function of frequency. Special formulas can be used to calculate the output voltage for any frequency of the applied voltage. What follows is a more mathematical approach in analyzing the operation of the most basic low-pass and high-pass filter circuits.

***RC* LOW-PASS FILTER**   Figure 27-12*a* shows a simple *RC* low-pass filter, while Fig. 27-12*b* shows how its output voltage, V_{out}, varies with frequency. Let's examine how the *RC* low-pass filter responds when $f = 0$ Hz (dc) and $f = \infty$ Hz. At $f = 0$ Hz, the capacitor C has infinite capacitive reactance X_C, calculated as:

$$X_C = \frac{1}{2\pi f C}$$

$$= \frac{1}{2 \times \pi \times 0 \text{ Hz} \times 0.01 \ \mu\text{F}}$$

$$= \infty \, \Omega$$

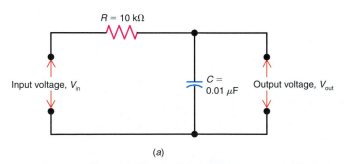

(a)   (b)

FIG. 27-12   *RC* low-pass filter. (*a*) Circuit. (*b*) Graph of V_{out} versus frequency.

Figure 27-13a shows the equivalent circuit for this condition. Notice that C appears as an open. Since all of the input voltage appears across the open in a series circuit, V_{out} must equal V_{in} when $f = 0$ Hz.

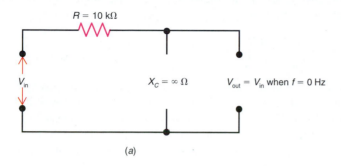

(a)

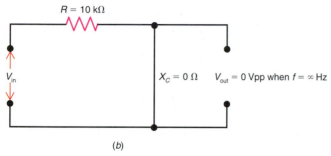

(b)

FIG. 27-13 *RC* low-pass equivalent circuits. (*a*) Equivalent circuit for $f = 0$ Hz. (*b*) Equivalent circuit for very high frequencies, or $f = \infty$ Hz.

At the other extreme, consider the circuit when the frequency f is very high or infinitely high. Then $X_C = 0\ \Omega$, calculated as:

$$X_C = \frac{1}{2\pi f C}$$

$$= \frac{1}{2 \times \pi \times \infty \text{ Hz} \times 0.01\ \mu\text{F}}$$

$$= 0\ \Omega$$

Figure 27-13b shows the equivalent circuit for this condition. Notice that C appears as a short. Since the voltage across a short is zero, the output voltage for very high frequencies must be zero.

When the frequency of the input voltage is somewhere between zero and infinity, the output voltage can be determined by using Formula (27-1):

▶ $$V_{out} = \frac{X_C}{Z_T} \times V_{in} \qquad\qquad (27\text{-}1)$$

where

$$Z_T = \sqrt{R^2 + X_C^2}$$

For very low frequencies, where X_C approaches infinity, V_{out} is approximately equal to V_{in}. This is true because the ratio X_C/Z_T approaches one as X_C and Z_T become approximately the same value. For very high frequencies, where X_C approaches zero, the ratio X_C/Z_T becomes very small, and V_{out} is approximately zero.

With respect to the input voltage V_{in}, the phase angle θ of the output voltage V_{out} can be calculated as:

$$\blacktriangleright \qquad \theta = \arctan\left(-\frac{R}{X_C}\right) \qquad\qquad \text{(27-2)}$$

At very low frequencies, X_C is very large and θ is approximately 0°. For very high frequencies, however, X_C is nearly zero and θ approaches $-90°$.

The frequency where $X_C = R$ is the *cutoff frequency*, designated f_c. At f_c the series current I is at 70.7 percent of its maximum value, because the total impedance Z_T is 1.41 times larger than the resistance of R. The formula for the cutoff frequency f_c of an RC low-pass filter is derived as follows. Because $X_C = R$ at f_c, we have

$$\frac{1}{2\pi f_c C} = R$$

Solving for f_c gives

$$\blacktriangleright \qquad f_c = \frac{1}{2\pi RC} \qquad\qquad \text{(27-3)}$$

The response curve in Fig. 27-12b shows that $V_{out} = 0.707V_{in}$ at the cutoff frequency f_c.

Example

EXAMPLE 1 In Fig. 27-12a, calculate: **(a)** the cutoff frequency f_c; **(b)** V_{out} at f_c; **(c)** θ at f_c. (Assume $V_{in} = 10$ Vpp for all frequencies.)

ANSWER

a. To calculate f_c, use Formula (27-3).

$$f_c = \frac{1}{2\pi RC}$$

$$= \frac{1}{2 \times \pi \times 10 \text{ k}\Omega \times 0.01 \text{ } \mu F}$$

$$= 1.592 \text{ kHz}$$

b. To calculate V_{out} at f_c, use Formula (27-1). First, however, calculate Z_T at f_c.

$$X_C = \frac{1}{2\pi f_c C}$$

$$= \frac{1}{2 \times \pi \times 1.592 \text{ kHz} \times 0.01 \text{ } \mu F}$$

$$= 10 \text{ k}\Omega$$

$$Z_T = \sqrt{R^2 + X_C^2}$$
$$= \sqrt{10^2 \text{ k}\Omega + 10^2 \text{ k}\Omega}$$
$$= 14.14 \text{ k}\Omega$$

Next,

$$V_{out} = \frac{X_C}{Z_T} \times V_{in}$$
$$= \frac{10 \text{ k}\Omega}{14.14 \text{ k}\Omega} \times 10 \text{ Vpp}$$
$$= 7.07 \text{ Vpp}$$

c. To calculate θ, use Formula (27-2).

$$\theta = \arctan\left(-\frac{R}{X_C}\right)$$
$$= \arctan -\frac{10 \text{ k}\Omega}{10 \text{ k}\Omega}$$
$$= \arctan(-1)$$
$$= -45°$$

The phase angle of $-45°$ tells us that V_{out} lags V_{in} by 45° at the cutoff frequency f_c.

RL LOW-PASS FILTER   Figure 27-14a shows a simple RL low-pass filter, while Fig. 27-14b shows how its output voltage V_{out} varies with frequency. For the analysis that follows, it is assumed that the coil's dc resistance r_s is negligible in comparison with the series resistance R.

Figure 27-15a shows the equivalent circuit when $f = 0$ Hz (dc). Notice that the inductor L acts as a short, since X_L must equal 0 Ω when $f = 0$ Hz. As a result, $V_{out} = V_{in}$ for very low frequencies and dc (0 Hz). For very high frequencies, X_L approaches infinity and the equivalent circuit appears as in Fig. 27-15b. Since L is basically equivalent to an open for very high frequencies, all of the

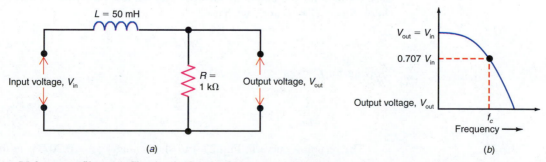

(a)   (b)

FIG. 27-14   *RL* low-pass filter. (*a*) Circuit. (*b*) Graph of V_{out} versus frequency.

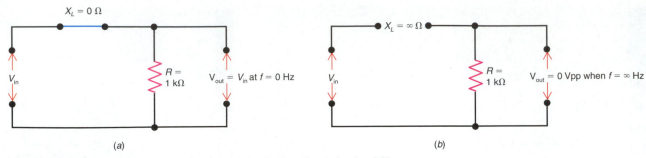

FIG. 27-15 *RL* low-pass equivalent circuits. (*a*) Equivalent circuit for *f* = 0 Hz. (*b*) Equivalent circuit for very high frequencies, or *f* = ∞ Hz.

input voltage will be dropped across *L* rather than *R*. Therefore, $V_{\text{out}} = 0$ Vpp for very high frequencies.

To calculate the output voltage at any frequency in Fig. 27-14*a*, use Formula (27-4).

$$\blacktriangleright \qquad V_{\text{out}} = \frac{R}{Z_T} \times V_{\text{in}} \qquad\qquad \textbf{(27-4)}$$

where

$$Z_T = \sqrt{R^2 + X_L^2}$$

For very low frequencies, where X_L is very small, V_{out} is approximately equal to V_{in}. This is true because the ratio R/Z_T approaches one as Z_T and R become approximately the same value. For very high frequencies the output voltage is approximately zero, because the ratio R/Z_T becomes very small as X_L and thus Z_T approach infinity.

The phase angle θ between V_{in} and V_{out} can be determined using Formula (27-5).

$$\blacktriangleright \qquad \theta = \arctan\left(-\frac{X_L}{R}\right) \qquad\qquad \textbf{(27-5)}$$

At very low frequencies, X_L approaches zero and θ is approximately 0°. For very high frequencies, X_L approaches infinity and θ is approximately −90°.

The frequency where $X_L = R$ is the cutoff frequency f_c. At f_c the series current I is at 70.7 percent of its maximum value, since $Z_T = 1.41R$ when $X_L = R$. The formula for the cutoff frequency of an *RL* low-pass filter is derived as follows. Since $X_L = R$ at f_c, we have

$$2\pi f_c L = R$$

Solving for f_c gives

$$\blacktriangleright \qquad f_c = \frac{R}{2\pi L} \qquad\qquad \textbf{(27-6)}$$

The response curve in Fig. 27-14*b* shows that $V_{\text{out}} = 0.707 V_{\text{in}}$ at the cutoff frequency f_c.

Example

EXAMPLE 2   In Fig. 27-14a, calculate: **(a)** the cutoff frequency f_c; **(b)** V_{out} at 1 kHz; **(c)** θ at 1 kHz. (Assume $V_{in} = 10$ Vpp for all frequencies.)

ANSWER

a.   To calculate f_c, use Formula (27-6).

$$f_c = \frac{R}{2\pi L}$$

$$= \frac{1\ \text{k}\Omega}{2 \times \pi \times 50\ \text{mH}}$$

$$= 3.183\ \text{kHz}$$

b.   To calculate V_{out} at 1 kHz, use Formula (27-4). First, however, calculate X_L and Z_T at 1 kHz.

$$X_L = 2\pi f L$$

$$= 2 \times \pi \times 1\ \text{kHz} \times 50\ \text{mH}$$

$$= 314\ \Omega$$

$$Z_T = \sqrt{R^2 + X_L^2}$$

$$= \sqrt{1^2\ \text{k}\Omega + 314^2\ \Omega}$$

$$= 1.05\ \text{k}\Omega$$

Next,

$$V_{out} = \frac{R}{Z_T} \times V_{in}$$

$$= \frac{1\ \text{k}\Omega}{1.05\ \text{k}\Omega} \times 10\ \text{Vpp}$$

$$= 9.52\ \text{Vpp}$$

Notice that $V_{out} \cong V_{in}$, since 1 kHz is in the passband of the low-pass filter.

c.   To calculate θ at 1 kHz, use Formula (27-5). Recall that $X_L = 314\ \Omega$ at 1 kHz.

$$\theta = \arctan\left(-\frac{X_L}{R}\right)$$

$$= \arctan\left(-\frac{314\ \Omega}{1\ \text{k}\Omega}\right)$$

$$= \arctan(-0.314)$$

$$= -17.4°$$

The phase angle of $-17.4°$ tells us that V_{out} lags V_{in} by $17.4°$ at a frequency of 1 kHz.

RC HIGH-PASS FILTER Figure 27-16a shows an RC high-pass filter. Notice that the output is taken across the resistor R rather than across the capacitor C. Figure 27-16b shows how the output voltage varies with frequency. To calculate the output voltage V_{out} at any frequency, use Formula (27-7).

$$\blacktriangleright \qquad V_{out} = \frac{R}{Z_T} \times V_{in} \qquad \textbf{(27-7)}$$

where

$$Z_T = \sqrt{R^2 + X_C^2}$$

For very low frequencies the output voltage approaches zero, because the ratio R/Z_T becomes very small as X_C and thus Z_T approach infinity. For very high frequencies V_{out} is approximately equal to V_{in}, because the ratio R/Z_T approaches one as Z_T and R become approximately the same value.

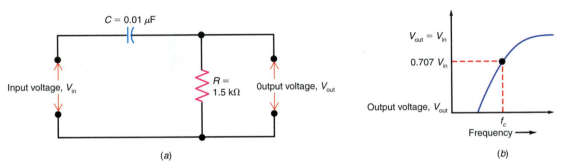

FIG. 27-16 RC high-pass filter. (a) Circuit. (b) Graph of V_{out} versus frequency.

The phase angle of V_{out} with respect to V_{in} for an RC high-pass filter can be calculated using Formula (27-8).

$$\blacktriangleright \qquad \theta = \arctan\left(\frac{X_C}{R}\right) \qquad \textbf{(27-8)}$$

For very low frequencies where X_C is very large, θ is approximately 90°. For very high frequencies where X_C approaches zero, θ is approximately 0°.

To calculate the cutoff frequency f_c for an RC high-pass filter, use Formula (27-3). Although this formula is used to calculate f_c for an RC low-pass filter, it can also be used to calculate f_c for an RC high-pass filter. The reason is that, for both circuits, $X_C = R$ at the cutoff frequency. In Fig. 27-16b, notice that $V_{out} = 0.707V_{in}$ at f_c.

RL HIGH-PASS FILTER An RL high-pass filter is shown in Fig. 27-17a, while its response curve is shown in Fig. 27-17b. In Fig. 27-17a, notice that the output is taken across the inductor L rather than across the resistance R.

To calculate the output voltage V_{out} at any frequency, use Formula (27-9).

$$\blacktriangleright \qquad V_{out} = \frac{X_L}{Z_T} \times V_{in} \qquad \textbf{(27-9)}$$

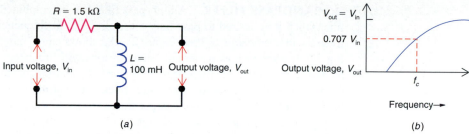

(a)

(b)

FIG. 27-17 *RL* high-pass filter. (*a*) Circuit. (*b*) Graph of V_{out} versus frequency.

where

$$Z_T = \sqrt{R^2 + X_L{}^2}$$

For very low frequencies, where X_L is very small, V_{out} is approximately zero. For very high frequencies, $V_{out} = V_{in}$ because the ratio X_L/Z_T is approximately one.

The phase angle θ of the output voltage V_{out} with respect to the input voltage V_{in} is

▶ $$\theta = \arctan\left(\frac{R}{X_L}\right) \tag{27-10}$$

For very low frequencies θ approaches 90°, because the ratio R/X_L becomes very large when X_L approaches zero. For very high frequencies θ approaches 0°, because the ratio R/X_L becomes approximately zero as X_L approaches infinity. To calculate the cutoff frequency of an *RL* high-pass filter, use Formula (27-6).

Example

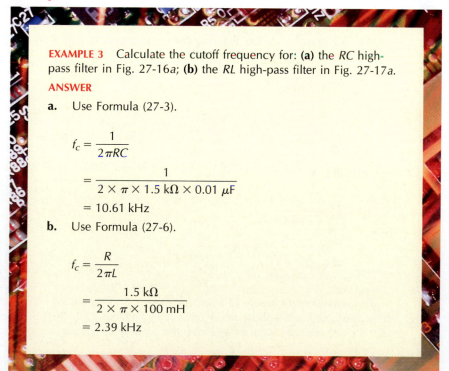

EXAMPLE 3   Calculate the cutoff frequency for: **(a)** the *RC* high-pass filter in Fig. 27-16*a*; **(b)** the *RL* high-pass filter in Fig. 27-17*a*.

ANSWER

a.   Use Formula (27-3).

$$f_c = \frac{1}{2\pi RC}$$

$$= \frac{1}{2 \times \pi \times 1.5 \text{ k}\Omega \times 0.01 \text{ }\mu F}$$

$$= 10.61 \text{ kHz}$$

b.   Use Formula (27-6).

$$f_c = \frac{R}{2\pi L}$$

$$= \frac{1.5 \text{ k}\Omega}{2 \times \pi \times 100 \text{ mH}}$$

$$= 2.39 \text{ kHz}$$

RC BANDPASS FILTER A high-pass filter can be combined with a low-pass filter when it is desired to pass only a certain band of frequencies. This type of filter is called a *bandpass filter*. Figure 27-18a shows an *RC* bandpass filter, while Fig. 27-18b shows how its output voltage varies with frequency. In Fig. 27-18a, R_1 and C_1 constitute the high-pass filter, while R_2 and C_2 constitute the low-pass filter. To ensure that the low-pass filter does not load the high-pass filter, R_2 is usually made 10 or more times larger than the resistance of R_1. The cutoff frequency of the high-pass filter is designated f_{c_1}, while the cutoff frequency of the low-pass filter is designed f_{c_2}. These two frequencies can be found on the response curve in Fig. 27-18b. To calculate the values for f_{c_1} and f_{c_2}, use the formulas given earlier for individual *RC* low-pass and *RC* high-pass filter circuits.

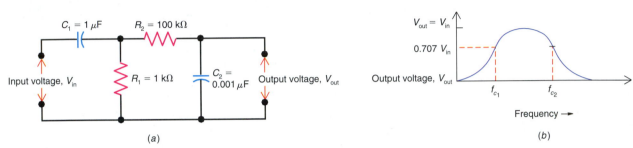

FIG. 27-18 *RC* bandpass filter. (a) Circuit. (b) Graph of V_{out} versus frequency.

Example

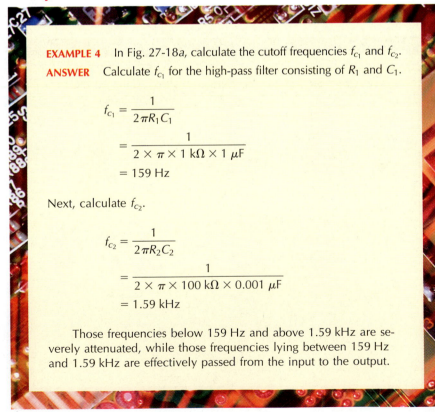

EXAMPLE 4 In Fig. 27-18a, calculate the cutoff frequencies f_{c_1} and f_{c_2}.

ANSWER Calculate f_{c_1} for the high-pass filter consisting of R_1 and C_1.

$$f_{c_1} = \frac{1}{2\pi R_1 C_1}$$

$$= \frac{1}{2 \times \pi \times 1\ \text{k}\Omega \times 1\ \mu\text{F}}$$

$$= 159\ \text{Hz}$$

Next, calculate f_{c_2}.

$$f_{c_2} = \frac{1}{2\pi R_2 C_2}$$

$$= \frac{1}{2 \times \pi \times 100\ \text{k}\Omega \times 0.001\ \mu\text{F}}$$

$$= 1.59\ \text{kHz}$$

Those frequencies below 159 Hz and above 1.59 kHz are severely attenuated, while those frequencies lying between 159 Hz and 1.59 kHz are effectively passed from the input to the output.

RC BANDSTOP FILTER A high-pass filter can also be combined with a low-pass filter when it is desired to block or severely attenuate a certain band of frequencies. Such a filter is called a *bandstop* or *notch filter*. Figure 27-19a shows an *RC* bandstop filter, while Fig. 27-19b shows how its output voltage varies with frequency. In Fig. 27-19a, the components identified as $2R_1$ and $2C_1$ constitute the low-pass filter section, while the components identified as R_1 and C_1 constitute the high-pass filter section. Notice that the individual filters are actually in parallel with each other. The frequency of maximum attenuation is called the *notch frequency* and is identified as f_N in Fig. 27-19b. Notice that the maximum value of V_{out} below f_N is less than the maximum value of V_{out} above f_N. The reason for this is that the series resistances ($2R_1$) in the low-pass filter provide greater circuit losses than do the series capacitors (C_1) in the high-pass filter.

To calculate the notch frequency f_N in Fig. 27-19a, use Formula (27-11).

▶ $$f_N = \frac{1}{4\pi R_1 C_1} \qquad\qquad\qquad \textbf{(27-11)}$$

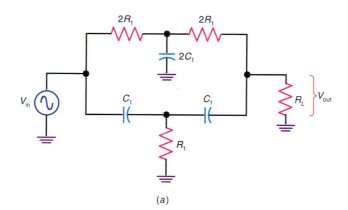

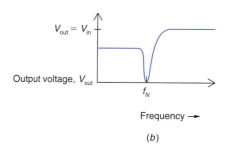

FIG. 27-19 Notch filter. (*a*) Circuit. (*b*) Graph of V_{out} versus frequency.

Example

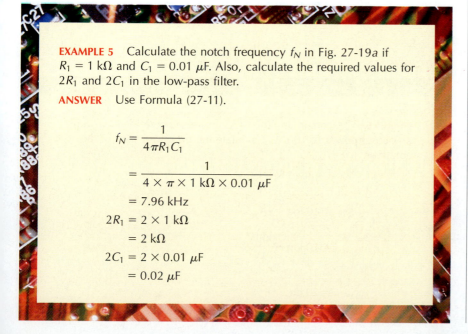

EXAMPLE 5 Calculate the notch frequency f_N in Fig. 27-19a if $R_1 = 1\ k\Omega$ and $C_1 = 0.01\ \mu F$. Also, calculate the required values for $2R_1$ and $2C_1$ in the low-pass filter.

ANSWER Use Formula (27-11).

$$f_N = \frac{1}{4\pi R_1 C_1}$$

$$= \frac{1}{4 \times \pi \times 1\ k\Omega \times 0.01\ \mu F}$$

$$= 7.96\ kHz$$

$$2R_1 = 2 \times 1\ k\Omega$$
$$= 2\ k\Omega$$
$$2C_1 = 2 \times 0.01\ \mu F$$
$$= 0.02\ \mu F$$

Answers at end of chapter.

Answer True or False.
a. Increasing the capacitance C in Fig. 27-12a raises the cutoff frequency f_c.
b. Decreasing the inductance L in Fig. 27-14a raises the cutoff frequency f_c.
c. Increasing the value of C_2 in Fig. 27-18a reduces the passband.
d. In Fig. 27-17a, V_{out} is approximately zero for very low frequencies.

27-10 DECIBELS AND FREQUENCY RESPONSE CURVES

In analyzing filters, the decibel (db) unit is often used to describe the amount of attenuation offered by the filter. In basic terms, the *decibel* is a logarithmic expression that compares two power levels. Expressed mathematically,

$$N_{db} = 10 \log \frac{P_{out}}{P_{in}} \tag{27-12}$$

where

$$N_{db} = \text{gain or loss in decibels}$$
$$P_{in} = \text{input power}$$
$$P_{out} = \text{output power}$$

If the ratio P_{out}/P_{in} is greater than one, the N_{db} value is positive, indicating an increase in power from input to output. If the ratio P_{out}/P_{in} is less than one, the N_{db} value is negative, indicating a loss or reduction in power from input to output. A reduction in power, corresponding to a negative N_{db} value, is referred to as *attenuation*.

Example

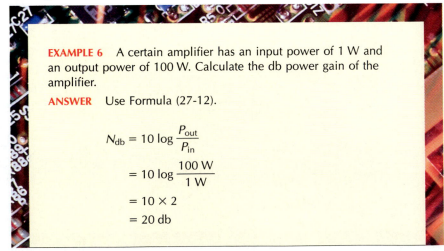

EXAMPLE 6   A certain amplifier has an input power of 1 W and an output power of 100 W. Calculate the db power gain of the amplifier.

ANSWER   Use Formula (27-12).

$$N_{db} = 10 \log \frac{P_{out}}{P_{in}}$$
$$= 10 \log \frac{100 \text{ W}}{1 \text{ W}}$$
$$= 10 \times 2$$
$$= 20 \text{ db}$$

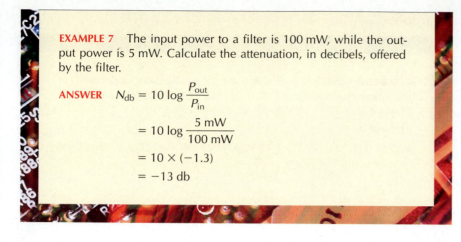

EXAMPLE 7   The input power to a filter is 100 mW, while the output power is 5 mW. Calculate the attenuation, in decibels, offered by the filter.

ANSWER $N_{db} = 10 \log \dfrac{P_{out}}{P_{in}}$

$$= 10 \log \dfrac{5 \text{ mW}}{100 \text{ mW}}$$

$$= 10 \times (-1.3)$$

$$= -13 \text{ db}$$

The power gain or loss in decibels can also be computed from a voltage ratio if the measurements are made across equal resistances.

▶ $$N_{db} = 20 \log \dfrac{V_{out}}{V_{in}}$$ (27-13)

where

$$N_{db} = \text{gain or loss in decibels}$$
$$V_{in} = \text{input voltage}$$
$$V_{out} = \text{output voltage}$$

For the passive filters discussed in this chapter, the N_{db} value can never be positive, because V_{out} can never be greater than V_{in}.

Consider the RC low-pass filter in Fig. 27-20. The cutoff frequency f_c for this circuit is 1.592 kHz, as determined by Formula (27-1). Recall that the formula for V_{out} at any frequency is

$$V_{out} = \dfrac{X_C}{Z_T} \times V_{in}$$

Dividing both sides of the equation by V_{in} gives

$$\dfrac{V_{out}}{V_{in}} = \dfrac{X_C}{Z_T}$$

Substituting X_C/Z_T for V_{out}/V_{in} in Formula (27-13) gives

$$N_{db} = 20 \log \dfrac{X_C}{Z_T}$$

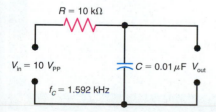

FIG. 27-20 RC low-pass filter.

Example

EXAMPLE 8   In Fig. 27-20, calculate the attenuation, in decibels, at the following frequencies: **(a)** 0 Hz; **(b)** 1.592 kHz; **(c)** 15.92 kHz. (Assume that $V_{in} = 10$ Vpp for all frequencies.)

ANSWER

a. At 0 Hz, $V_{out} = V_{in} = 10$ Vpp, since the capacitor C appears as an open. Therefore,

$$N_{db} = 20 \log \frac{V_{out}}{V_{in}}$$

$$= 20 \log \frac{10 \text{ Vpp}}{10 \text{ Vpp}}$$

$$= 20 \log 1$$

$$= 20 \times 0$$

$$= 0 \text{ db}$$

b. Since 1.592 kHz is the cutoff frequency f_c, V_{out} will be $0.707 \times V_{in}$ or 7.07 Vpp. Therefore,

$$N_{db} = 20 \log \frac{V_{out}}{V_{in}}$$

$$= 20 \log \frac{7.07 \text{ Vpp}}{10 \text{ Vpp}}$$

$$= 20 \log 0.707$$

$$= 20 \times (-0.15)$$

$$= -3 \text{ db}$$

c. To calculate N_{db} at 15.92 kHz, X_C and Z_T must first be determined.

$$X_C = \frac{1}{2\pi f C}$$

$$= \frac{1}{2 \times \pi \times 15.92 \text{ kHz} \times 0.01 \ \mu\text{F}}$$

$$= 1 \text{ k}\Omega$$

$$Z_T = \sqrt{R^2 + X_C^2}$$

$$= \sqrt{10^2 \text{ k}\Omega + 1^2 \text{ k}\Omega}$$

$$= 10.05 \text{ k}\Omega$$

Next,

$$N_{db} = 20 \log \frac{X_C}{Z_T}$$

$$= 10 \log \frac{1 \text{ k}\Omega}{10.05 \text{ k}\Omega}$$

$$= 20 \log 0.0995$$

$$= 20(-1)$$

$$= -20 \text{ db}$$

In Example 8, notice that N_{db} is 0 db at a frequency of 0 Hz, which is in the filter's passband. This may seem unusual, but the 0-db value simply indicates that there is no attenuation at this frequency. For an ideal passive filter, $N_{db} = 0$ db in the passband. As another point of interest from Example 8, N_{db} was determined to be -3 db at the cutoff frequency of 1.592 kHz. Since for any passive filter $V_{out} = 0.707V_{in}$ at f_c, N_{db} is always -3 db at the cutoff frequency for a passive filter.

The N_{db} value of loss can be determined for any filter if the values of V_{in} and V_{out} are known. Figure 27-21 shows the basic RC and RL low-pass and high-pass filters. For each filter the formula for calculating the N_{db} loss is provided.

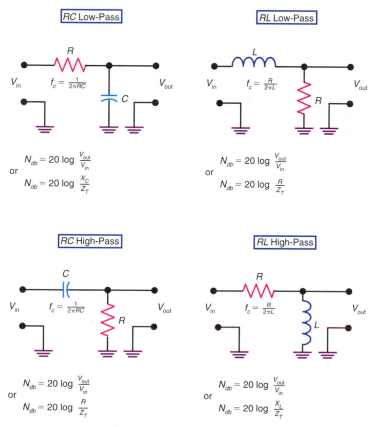

FIG. 27-21   *RC* and *RL* filter circuits, showing formulas for calculating decibel attenuation.

FREQUENCY RESPONSE CURVES

The frequency response of a filter is typically shown by plotting its gain (or loss) versus frequency on logarithmic graph paper. The two types of logarithmic graph paper are log-log and semilog. On *semilog graph paper,* the divisions along one axis are spaced logarithmically, while the other axis has conventional linear spacing between divisions. On *log-log graph paper,* both axes have logarithmic spacing between divisions. Logarithmic spacing results in a scale that expands the display of smaller values and compresses the display of larger values. On logarithmic graph paper, a 2-to-1 range of frequencies is called an *octave,* and a 10-to-1 range of values is called a *decade.*

DID YOU KNOW?

Yttrium is a strange material that can change from being a mirror to a window and back again. When it is a mirror, its electrons absorb light. When yttrium is exposed to hydrogen, the electrons stop absorbing or radiating light, and it becomes a window.

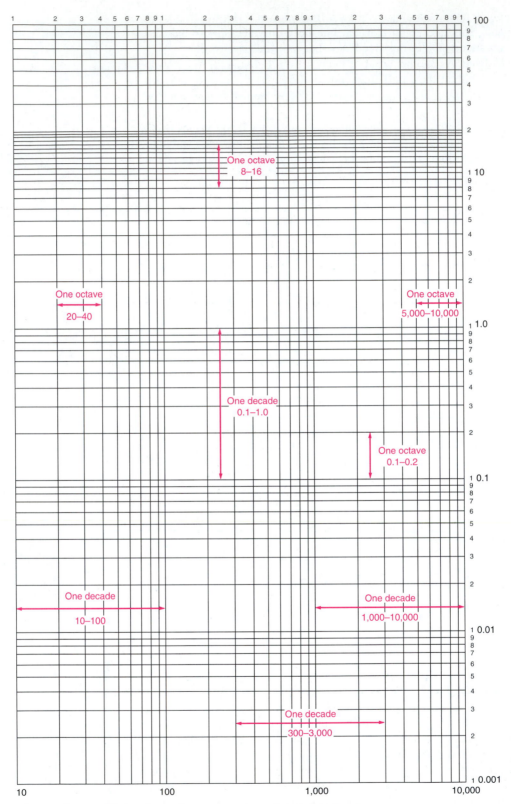

FIG. 27-22 Log-log graph paper. Notice that each octave corresponds to a 2-to-1 range of values and each decade corresponds to a 10-to-1 range of values.

One advantage of logarithmic spacing is that a larger range of values can be shown in one plot without losing resolution in the smaller values. For example, if frequency values between 10 Hz and 100 kHz were plotted on 100 divisions of linear graph paper, each division would represent approximately 1000 Hz and it would be impossible to plot values in the decade between 10 Hz and 100 Hz. On the other hand, by using logarithmic graph paper, the decade between 10 Hz and 100 Hz would occupy the same space on the graph as the decade between 10 kHz and 100 kHz.

Log-log or semilog graph paper is specified by the number of decades it contains. Each decade is a *graph cycle*. For example, 2-cycle by 4-cycle log-log paper has two decades on one axis and four on the other. The number of cycles must be adequate for the range of data being plotted. For example, if the frequency response extends from 25 Hz to 40 kHz, 4 cycles are necessary to plot the frequency values corresponding to the decades 10 Hz to 100 Hz, 100 Hz to 1 kHz, 1 kHz to 10 kHz, and 10 kHz to 100 kHz. A typical sheet of log-log graph paper is shown in Fig. 27-22. Because there are three decades on the horizontal axis and five decades on the vertical axis, this graph paper is called 3-cycle by 5-cycle log-log paper. Notice that each octave corresponds to a 2-to-1 range in values and each decade corresponds to a 10-to-1 range in values. For clarity, several octaves and decades are shown in Fig. 27-22.

When semilog graph paper is used to plot a frequency response, the observed or calculated values of gain (or loss) must first be converted to decibels before plotting. On the other hand, since decibel voltage gain is a logarithmic function, the gain or loss values can be plotted on log-log paper without first converting to decibels.

RC LOW-PASS FREQUENCY RESPONSE CURVE

Figure 27-23*a* shows an *RC* low-pass filter whose cutoff frequency f_c is 1.592 kHz as determined by Formula (27-1). Figure 27-23*b* shows its frequency response curve plotted on semilog graph paper. Notice there are 6 cycles on the horizontal axis, which spans a frequency range extending from 1 Hz to 1 MHz. Notice the vertical axis specifies the N_{db} loss, which is the amount of attenuation offered by the filter in decibels. Notice that $N_{db} = -3$ db at the cutoff frequency of 1.592 kHz. Above f_c, N_{db} decreases at the rate of approximately 6 db/octave, which is equivalent to 20 db/decade.

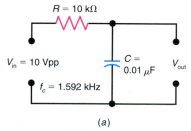

(a)

FIG. 27-23 *RC* low-pass filter frequency response curve. (*a*) Circuit. Continue on page 790 for part (*b*).

Example

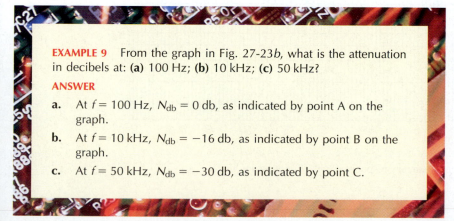

EXAMPLE 9 From the graph in Fig. 27-23*b*, what is the attenuation in decibels at: **(a)** 100 Hz; **(b)** 10 kHz; **(c)** 50 kHz?

ANSWER

a. At $f = 100$ Hz, $N_{db} = 0$ db, as indicated by point A on the graph.

b. At $f = 10$ kHz, $N_{db} = -16$ db, as indicated by point B on the graph.

c. At $f = 50$ kHz, $N_{db} = -30$ db, as indicated by point C.

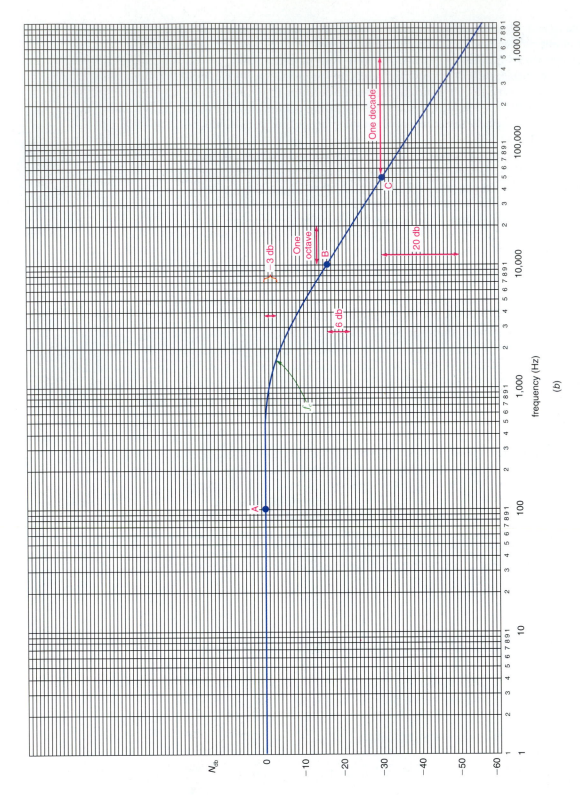

FIG. 27-23 (*continued*) *RC* low-pass filter frequency response curve. (*b*) Frequency response curve.

For filters such as the inverted-L, T, or π type, the response curve rolloff is much steeper beyond the cutoff frequency f_c. For example, a low-pass filter with a series inductor and a shunt capacitor has a rate of rolloff of 12 db/octave or 40 db/decade above the cutoff frequency f_c. To increase the rate of rolloff, more inductors and capacitors must be used in the filter design. Filters are available whose rolloff rates exceed 36 db/octave.

TEST-POINT QUESTIONS 27-10

Answers at end of chapter.

Answer True or False
a. For very low frequencies, a low-pass filter provides an attenuation of 0 db.
b. At the cutoff frequency, a low-pass filter has an N_{db} loss of -3 db.
c. On logarithmic graph paper, one cycle is the same as one octave.
d. The advantage of semilog and log-log graph paper is that a larger range of values can be shown in one plot without losing resolution in the smaller values.

27-11 RESONANT FILTERS

Tuned circuits provide a convenient method of filtering a band of radio frequencies because relatively small values of L and C are necessary for resonance. A tuned circuit provides filtering action by means of its maximum response at the resonant frequency.

The width of the band of frequencies affected by resonance depends on the Q of the tuned circuit, a higher Q providing narrower bandwidth. Because resonance is effective for a band of frequencies below and above f_r, resonant filters are called *bandstop* or *bandpass* filters. Series or parallel LC circuits can be used for either function, depending on the connections with respect to R_L. In the application of a bandstop filter to suppress certain frequencies, the LC circuit is often called a *wavetrap*.

SERIES RESONANCE FILTERS   A series resonant circuit has maximum current and minimum impedance at the resonant frequency. Connected in series with R_L, as in Fig. 27-24a, the series-tuned LC circuit allows frequencies at and near resonance to produce maximum output across R_L. Therefore, this is a case of bandpass filtering.

When the series LC circuit is connected across R_L as in Fig. 27-24b, however, the resonant circuit provides a low-impedance shunt path that short-circuits R_L. Then there is minimum output. This action corresponds to a shunt bypass capacitor, but the resonant circuit is more selective, short-circuiting R_L just for frequencies at and near resonance. For the bandwidth of the tuned circuit, therefore, the series resonant circuit in shunt with R_L provides band-stop filtering.

The series resistor R_S in Fig. 27-24b is used to isolate the low resistance of the LC filter from the input source. At the resonant frequency, practically all of the input voltage is across R_S, with little across R_L, because the LC tuned circuit then has very low resistance due to series resonance.

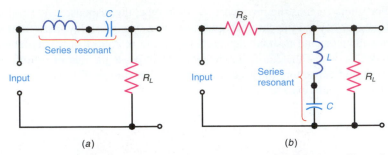

FIG. 27-24 The filtering action of a series resonant circuit. (*a*) Bandpass filter when L and C are in series with R_L. (*b*) Bandstop filter when LC circuit is in shunt with R_L.

PARALLEL RESONANCE FILTERS A parallel resonant circuit has maximum impedance at the resonant frequency. Connected in series with R_L, as in Fig. 27-25*a*, the parallel-tuned LC circuit provides maximum impedance in series with R_L, at and near the resonant frequency. Then these frequencies produce maximum voltage across the LC circuit but minimum output voltage across R_L. This is a bandstop filter, therefore, for the bandwidth of the tuned circuit.

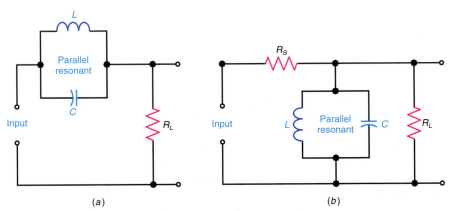

FIG. 27-25 The filtering action of a parallel resonant circuit. (*a*) Bandstop filter when LC bank is in series with R_L. (*b*) Bandpass filter when LC bank is in shunt with R_L.

The parallel LC circuit connected across R_L, however, as in Fig. 27-25*b*, provides a bandpass filter. At resonance, the high impedance of the parallel LC circuit allows R_L to develop its output voltage. Below resonance, R_L is short-circuited by the low reactance of L; above resonance, R_L is short-circuited by the low reactance of C. For frequencies at or near resonance, though, R_L is shunted by a high impedance, resulting in maximum output voltage.

The series resistor R_S in Fig. 27-25*b* is used to improve the filtering effect. Note that the parallel LC combination and R_S divide the input voltage. At the resonant frequency, though, the LC circuit has very high resistance for parallel resonance. Then most of the input voltage is across the LC circuit and R_L, with little across R_S.

L-TYPE RESONANT FILTER Series and parallel resonant circuits can be combined in L, T, or π sections for sharper discrimination in the frequencies to be filtered. Examples for an L-type filter are shown in Fig. 27-26.

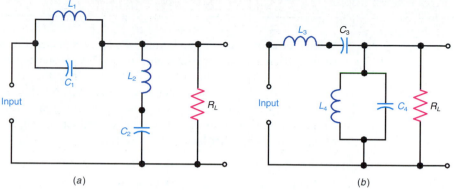

FIG. 27-26 Inverted-L filter with resonant circuits. (*a*) Bandstop filtering action. (*b*) Bandpass filtering action.

The circuit in Fig. 27-26*a* is a bandstop filter. The reason is that the parallel resonant L_1C_1 circuit is in series with the load, while the series-resonant L_2C_2 circuit is in shunt with R_L. There is a dual effect as a voltage divider across the input source voltage. The high resistance of L_1C_1 reduces voltage output to the load. Also, the low resistance of L_2C_2 reduces the output voltage.

For the opposite effect, the circuit in Fig. 27-26*b* is a bandpass filter. Now the series-resonant L_3C_3 circuit is in series with the load. Here the low resistance of L_3C_3 allows more output for R_L at resonance. Also, the high resistance of L_4C_4 allows maximum output voltage.

CRYSTAL FILTERS   A thin slice of quartz provides a resonance effect by mechanical vibrations at a particular frequency, like an *LC* circuit. The quartz crystal can be made to vibrate by a voltage input or produce voltage output when it is compressed, expanded, or twisted. This characteristic of some crystals is known as the *piezoelectric effect*. As a result, crystals are often used in place of resonant circuits. In fact, the *Q* of a resonant crystal is much higher than that of *LC* circuits. However, the crystal has a specific frequency that cannot be varied because of its stability. Crystals are used for radio frequencies, in the range of about 0.5 to 30 MHz. Figure 27-27 shows a crystal in its housing for the frequency of 3.579545 MHz, for use in the color oscillator circuit of a television receiver. Note the exact frequency. More details of the piezoelectric effect and crystal resonators are explained in Chap. 29.

FIG. 27-27   Quartz crystal in holder. Size is ½ in wide.

Special ceramic materials, such as leads titanate, can also be used for crystal filters. They have the piezoelectric effect like quartz crystals. Ceramic crystals are smaller in size and cost less but they have lower *Q* than quartz crystals.

TEST-POINT QUESTION 27-11

Answers at end of chapter.

Answer True or False.
a. A parallel-resonant *LC* circuit in series with the load is a bandstop filter.
b. A series resonant *LC* circuit in series with the load is a bandpass filter.
c. Quartz crystals can be used as resonant filters.

27-12 INTERFERENCE FILTERS

Voltage or current not at the desired frequency represents interference. Usually, such interference can be eliminated by a filter. Some typical applications are (1) low-pass filter to eliminate RF interference from the 60-Hz power-line input to a receiver, (2) high-pass filter to eliminate RF interference from the signal picked up by a television receiving antenna, and (3) resonant filter to eliminate an interfering radio frequency from the desired RF signal. As noted earlier, the resonant bandstop filter is called a wavetrap.

POWER-LINE FILTER   Although the power line is a source of 60-Hz voltage, it is also a conductor for interfering RF currents produced by motors, fluorescent lighting circuits, and RF equipment. When a receiver is connected to the power line, the RF interference can produce noise and whistles in the receiver output. To minimize this interference, the filter shown in Fig. 27-28 can be used. The filter is plugged into the wall outlet for 60-Hz power, while the receiver is plugged into the filter. An RF bypass capacitor across the line with two series RF chokes forms a low-pass balanced L-type filter. Using a choke in each side of the line makes the circuit balanced to ground.

The chokes provide high impedance for interfering RF current but not for 60 Hz, isolating the receiver input connections from RF interference in the power line. Also, the bypass capacitor short-circuits the receiver input for radio frequencies but not for 60 Hz. The unit then is a low-pass filter for 60-Hz power applied to the receiver while rejecting higher frequencies. The current rating means the filter can be used for equipment that draws 3 A or less from the power line without excessive heat in the chokes.

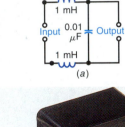

FIG. 27-28   Power-line filter unit. (*a*) Circuit of balanced L-type low-pass filter. (*b*) Filter unit.

TELEVISION ANTENNA FILTER   When a television receiver has interference in the picture resulting from radio frequencies below the television broadcast band, picked up by the receiving antenna, this RF interference can be reduced by the high-pass filter shown in Fig. 27-29. The filter attenuates frequencies below 54 MHz, which is the lowest frequency for channel 2.

At frequencies lower than 54 MHz the series capacitances provide increasing reactance with a larger voltage drop, while the shunt inductances have less reactance and short-circuit the load. Higher frequencies are passed to the load as the series capacitive reactance decreases and the shunt inductive reactance increases.

Connections to the filter unit are made at the receiver end of the line from the antenna. Either end of the filter is connected to the antenna terminals on the receiver, with the opposite end connected to the antenna line.

FIG. 27-29   A television antenna filter to pass TV channel frequencies above 54 MHz but attenuate lower frequencies that can cause interference.

- A filter separates high and low frequencies. With input of different frequencies, the high-pass filter allows the higher frequencies to produce output voltage across the load; a low-pass filter provides output voltage for the lower frequencies.
- Pulsating or fluctuating direct current varies in amplitude but does not reverse its direction. Similarly, a pulsating or fluctuating dc voltage varies in amplitude but maintains one polarity, either positive or negative.
- The pulsating direct current or voltage consists of a steady dc level, equal to the average value, and an ac component that reverses in polarity with respect to the average level. The dc and ac can be separated by filters.
- An RC coupling circuit is effectively a high-pass filter for pulsating direct current. Capacitance C_C blocks the steady dc voltage but passes the ac component.
- A transformer with an isolated secondary winding also is effectively a high-pass filter. With pulsating direct current in the primary, only the ac component produces output voltage in the secondary.
- A bypass capacitor in parallel with R provides a low-pass filter.
- Combinations of L, C, and R can be arranged as L, T, or π filters for more selective filtering. All three arrangements can be used for either low-pass or high-pass action. See Figs. 27-9 and 27-11.
- In high-pass filters, the capacitance must be in series with the load as a coupling capacitor, with shunt R or L across the line.
- For low-pass filters, the capacitance is across the line as a bypass capacitor, while R or L then must be in series with the load.
- The cutoff frequency f_c of a filter is the frequency at which the output voltage is reduced to 70.7 percent of its maximum value.
- For an RC low-pass or high-pass filter, $X_C = R$ at the cutoff frequency. Similarly, for an RL low-pass or high-pass filter, $X_L = R$ at the cutoff frequency. To calculate f_c for an RC low-pass or high-pass filter, use the formula $f_c = 1/2\pi RC$. To calculate f_c for an RL low-pass or high-pass filter, use the formula $f_c = R/2\pi L$.
- For an RC or RL filter, either low-pass or high-pass, the phase angle θ between V_{in} and V_{out} is approximately $0°$ in the passband. In the stop band, $\theta = \pm 90°$. The sign of θ depends on the type of filter.
- RC low-pass filters can be combined with RC high-pass filters when it is desired to either pass or block only a certain band of frequencies. These types of filters are called bandpass and bandstop filters, respectively.
- The decibel (dB) unit of measurement is used to compare two power levels. A passive filter has an attenuation of -3 db at the cutoff frequency.
- Semilog and log-log graph paper are typically used to show the frequency response of a filter. On semilog graph paper the vertical axis uses conventional linear spacing, while the horizontal axis uses logarithmically spaced divisions.

- The advantage of using semilog or log-log graph paper is that a larger range of values can be shown in one plot without losing resolution in the smaller values.
- A bandpass or bandstop filter has in effect two cutoff frequencies. The bandpass filter passes to the load those frequencies in the band between the cutoff frequencies, while attenuating all other frequencies higher and lower than the passband. A bandstop filter does the opposite, attenuating the band between the cutoff frequencies, while passing to the load all other frequencies higher and lower than the stop band.
- Resonant circuits are generally used for bandpass or bandstop filtering with radio frequencies.
- For bandpass filtering, the series resonant LC circuit must be in series with the load, for minimum series opposition, while the high impedance of parallel resonance is across the load.
- For bandstop filtering, the circuit is reversed, with the parallel resonant LC circuit in series with the load, while the series resonant circuit is in shunt across the load.
- A wavetrap is an application of the resonant bandstop filter.

SELF-TEST

ANSWERS AT BACK OF BOOK.

Choose (a), (b), (c), or (d).

1. With input frequencies from direct current up to 15 kHz, a high-pass filter allows the most output voltage to be developed across the load resistance for which of the following frequencies? (a) Direct current; (b) 15 Hz; (c) 150 Hz; (d) 15,000 Hz.
2. With input frequencies from direct current up to 15 kHz a low-pass filter allows the most output voltage to be developed across the load resistance for which of the following frequencies? (a) Direct current; (b) 15 Hz; (c) 150 Hz; (d) 15,000 Hz.
3. An $R_C C_C$ coupling circuit is a high-pass filter for pulsating dc voltage because: (a) C_C has high reactance for high frequencies; (b) C_C blocks dc voltage; (c) C_C has low reactance for low frequencies; (d) R_C has minimum opposition for low frequencies.
4. A transformer with an isolated secondary winding is a high-pass filter for pulsating direct primary current because: (a) the steady primary current has no magnetic field; (b) the ac component of the primary current has the strongest field; (c) only variations in primary current can induce secondary voltage; (d) the secondary voltage is maximum for steady direct current in the primary.
5. Which of the following is a low-pass filter? (a) L type with series C and shunt L; (b) π type with series C and shunt L; (c) T type with series C and shunt L; (d) L type with series L and shunt C.
6. A bypass capacitor C_b across R_b provides low-pass filtering because: (a) current in the C_b branch is maximum for low frequencies; (b) voltage across C_b is minimum for high frequencies; (c) voltage across C_b is minimum for low frequencies; (d) voltage across R_b is minimum for low frequencies.
7. An ac voltmeter across R in Fig. 27-6 reads (a) practically zero; (b) 7.07 V; (c) 10 V; (d) 20 V.
8. Which of the following L-type filters is the best bandstop filter? (a) Series resonant LC circuit in series with the load and parallel resonant LC circuit in shunt; (b) parallel resonant

LC circuit in series with the load and series resonant *LC* circuit in shunt; (*c*) series resonant *LC* circuits in series and in parallel with the load; (*d*) parallel resonant *LC* circuits in series and in parallel with the load.

9. A 455-kHz wavetrap is a resonant *LC* circuit tuned to 455 kHz and connected as a (*a*) bandstop filter for frequencies at and near 455 kHz; (*b*) bandpass filter for frequencies at and near 455 kHz; (*c*) bandstop filter for frequencies from direct current up to 455 kHz; (*d*) bandpass filter for frequencies from 455 kHz up to 300 MHz.

10. A power-line filter for rejecting RF interference has (*a*) RF coupling capacitors in series with the power line; (*b*) RF chokes in shunt across the power line; (*c*) 60-Hz chokes in series with the power line; (*d*) RF bypass capacitors in shunt across the power line.

11. Which of the following will increase the cutoff frequency of an *RC* high-pass filter: (*a*) increasing *R*; (*b*) decreasing *C*; (*c*) increasing *C*; (*d*) both (*a*) and (*c*).

12. At the cutoff frequency of an *RL* low-pass filter: (*a*) V_{out} is reduced to 70.7 percent of its maximum value; (*b*) $N_{db} = -10$ db; (*c*) $V_{out} = V_{in}$; (*d*) $V_{out} = 0$ Vpp.

13. On logarithmic graph paper, a 10-to-1 range of values is called a(n): (*a*) octave; (*b*) cycle; (*c*) decade; (*d*) both (*b*) and (*c*).

14. In the passband, an *RC* low-pass filter has a phase angle θ of approximately: (*a*) 0°; (*b*) −45°; (*c*) +90°; (*d*) −90°.

15. The decibel attenuation at the cutoff frequency of an *RC* low-pass filter equals: (*a*) 0 db; (*b*) −20 db; (*c*) −3 db; (*d*) −6 db.

QUESTIONS

1. What is the function of an electrical filter?
2. Give two examples where the voltage has different frequency components.
3. (**a**) What is meant by pulsating direct current or voltage? (**b**) What are the two components of a pulsating dc voltage? (**c**) How can you measure the value of each of the two components?
4. Define the function of the following filters in terms of output voltage across the load resistance: (**a**) High-pass filter. Why is an $R_C C_C$ coupling circuit an example? (**b**) Low-pass filter. Why is an $R_b C_b$ bypass circuit an example? (**c**) Bandpass filter. How does it differ from a coupling circuit? (**d**) Bandstop filter. How does it differ from a bandpass filter?
5. Draw circuit diagrams for the following filter types. No values are necessary. (**a**) T-type high-pass and T-type low-pass; (**b**) π-type low-pass, balanced with a filter reactance in both sides of the line.
6. Draw the circuit diagrams for L-type bandpass and L-type bandstop filters. How do these two circuits differ from each other?
7. Draw the response curve for each of the following filters: (**a**) low-pass cutting off at 20,000 Hz; (**b**) high-pass cutting off at 20 Hz; (**c**) bandpass for 20 to 20,000 Hz; (**d**) bandpass for 450 to 460 kHz.
8. Give one similarity and one difference in comparing a coupling capacitor and a bypass capacitor.
9. Give two differences between a low-pass filter and a high-pass filter.

10. Explain briefly why the power-line filter in Fig. 27-28 passes 60-Hz alternating current but not 1-MHz RF current.

11. Explain the advantage of using semilog and log-log graph paper for plotting a frequency response curve.

12. Explain why an RC bandstop filter cannot be designed by interchanging the low-pass and high-pass filters in Fig. 27-18a.

PROBLEMS

ANSWERS TO ODD-NUMBERED PROBLEMS AT BACK OF BOOK.

1. Refer to the RC coupling circuit in Fig. 27-6, with R equal to 16,000 Ω. **(a)** Calculate the required value for C_C at 1000 Hz. **(b)** How much is the average dc voltage across C_C and across R? **(c)** How much is the ac voltage across C_C and across R?

2. Refer to the R_1C_1 bypass circuit in Fig. 27-8. **(a)** Why is 1 MHz bypassed but not 1 kHz? **(b)** If C_1 were doubled in capacitance, what is the lowest frequency that could be bypassed, maintaining a 10:1 ratio of R to X_C?

3. Calculate the C_C needed to couple frequencies of 50 to 15,000 Hz with a 50-kΩ R.

4. Show the fluctuating collector current i_c of a transistor that has an average dc axis of 24 mA and a square-wave ac component with a 10-mA peak value. Label the dc axis, maximum and minimum positive values, and the peak-to-peak alternating current.

5. Calculate the cutoff frequency f_c for the filter in: **(a)** Fig. 27-30a; **(b)** Fig. 27-30b; **(c)** Fig. 27-30c; **(d)** Fig. 27-30d.

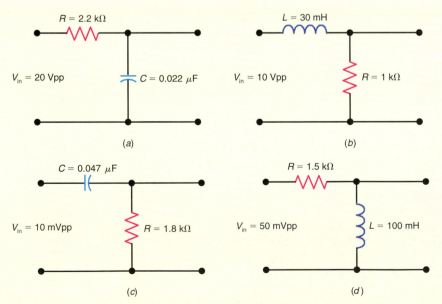

FIG. 27-30 Circuits for Probs. 5, 6, 7, 8, 9, 10, 11, and 15.

6. In Fig. 27-30a, calculate V_{out} and the phase angle θ between V_{in} and V_{out} at the following frequencies: **(a)** 0 Hz; **(b)** 100 Hz; **(c)** f_c; **(d)** 5 kHz; **(e)** 15 kHz; **(f)** 100 kHz.

7. In Fig. 27-30b, calculate V_{out} and the phase angle θ between V_{in} and V_{out} at the following frequencies: **(a)** 0 Hz; **(b)** 200 Hz; **(c)** f_c; **(d)** 10 kHz; **(e)** 53 kHz; **(f)** 1 MHz.

8. In Fig. 27-30c, calculate V_{out} and the phase angle θ between V_{in} and V_{out} at the following frequencies: **(a)** 10 Hz; **(b)** 500 Hz; **(c)** f_c; **(d)** 5 kHz; **(e)** 20 kHz; **(f)** 500 kHz.

9. In Fig. 27-30d, calculate V_{out} and the phase angle θ between V_{in} and V_{out} at the following frequencies: **(a)** 50 Hz; **(b)** 1.5 kHz; **(c)** f_c; **(d)** 6 kHz; **(e)** 50 kHz; **(f)** 1.5 MHz.

10. In Fig. 27-30a, calculate the attenuation in decibels (N_{db}) at the following frequencies: **(a)** 0 Hz; **(b)** f_c; **(c)** $10f_c$; **(d)** $20f_c$.

11. In Fig. 27-30d, calculate the attenuation in decibels (N_{db}) at the following frequencies: **(a)** 100 Hz; **(b)** f_c; **(c)** $5f_c$; **(d)** $100f_c$.

12. In Fig. 27-18a, calculate f_{c_1} and f_{c_2} for the following circuit values: $R_1 = 2.2$ kΩ, $C_1 = 0.068$ μF, $R_2 = 47$ kΩ, and $C_2 = 330$ pF.

13. In Fig. 27-19a, calculate the notch frequency f_N if $R_1 = 18$ kΩ and $C_1 = 0.001$ μF.

14. In Fig. 27-31, calculate: **(a)** the maximum output voltage at low frequencies; **(b)** the dc voltage across R_2; **(c)** the value of C_1 required to effectively bypass R_2 if the lowest frequency of the applied voltage is 15 kHz; **(d)** the value of C_1 required to effectively bypass R_2 if the lowest frequency of the applied voltage is 100 kHz.

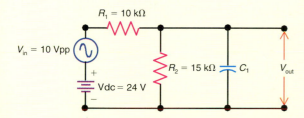

FIG. 27-31   Bypass circuit for Prob. 14.

15. The input power to an amplifier equals 1 W. Calculate the gain in decibels if the output power equals: **(a)** 2 W; **(b)** 10 W; **(c)** 20 W; **(d)** 100 W; **(e)** 1 kW; **(f)** 2 kW.

16. Using semilog graph paper, plot the frequency response curve for the filter in Fig. 27-30c. Mark the vertical axis in decibels. The frequency range on the horizontal axis should span the frequencies extending from 1 Hz to 100 kHz.

CRITICAL THINKING

1. In Fig. 27-32 on page 800, calculate: **(a)** the cutoff frequency f_c; **(b)** the output voltage at the cutoff frequency f_c; **(c)** the output voltage at 50 kHz.

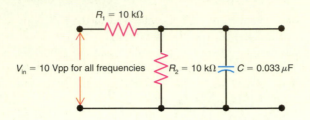

FIG. 27-32 Circuit for Critical Thinking Prob. 1.

2. In Fig. 27-33, calculate the values of L and C required to provide an f_r of 1 Mhz and a bandwidth, Δf, of 40 kHz.

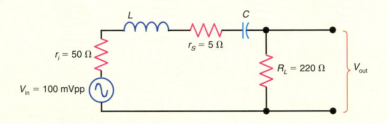

FIG. 27-33 Circuit for Critical Thinking Prob. 2.

3. In Fig. 27-34, calculate the values of L and C required to provide an f_r of 1 MHz and a bandwidth, Δf, of 20 kHz.

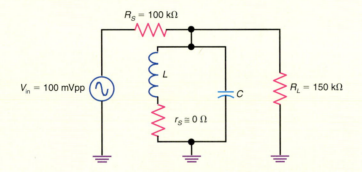

FIG. 27-34 Circuit for Critical Thinking Prob. 3.

ANSWERS TO TEST-POINT QUESTIONS

27-1	a. 500 kHz	27-4	a. 0 V	27-8	a. d	27-10	a. T
	b. 60 Hz		b. 5 μF		b. high-pass		b. T
27-2	a. 6 V	27-5	a. RF	27-9	a. F		c. F
	b. 10 and 2 V		b. 5 μF		b. T		d. T
	c. 8 V	27-6	a. high-pass		c. T	27-11	a. T
	d. 4 and 2.8 V		b. low-pass		d. T		b. T
27-3	a. high-pass	27-7	a. e and f				c. T
	b. 0 V		b. low-pass			27-12	a. T
							b. T

REVIEW: CHAPTERS 26 AND 27

SUMMARY

- Resonance results when the reactances X_L and X_C are equal. In series, the net reactance is zero. In parallel, the net reactive branch current is zero. The specific frequency that makes $X_L = X_C$ is the resonant frequency $f_r = 1/(2\pi\sqrt{LC})$.
- Larger values of L and C mean lower resonant frequencies, as f_r is inversely proportional to the square root of L and C. If the value of L or C is quadrupled, for instance, f_r will decrease by one-half.
- For a series resonant LC circuit, the current is maximum. The voltage drop across the reactances is equal and opposite; the phase angle is zero. The reactive voltage at resonance is Q times greater than the applied voltage.
- For a parallel resonant LC circuit, the impedance is maximum with minimum line current, since the reactive branch currents cancel. The impedance at resonance is Q times the X_L value, but it is resistive with a phase angle of zero.
- The Q of the resonant circuit equals X_L/r_S for resistance in series with X_L, or R_P/X_L for resistance in parallel with X_L.
- The bandwidth between half-power points if f_r/Q.
- A filter uses inductance and capacitance to separate high or low frequencies. A low-pass filter allows low frequencies to develop output voltage across the load; a high-pass filter does the same for high frequencies. Series inductance or shunt capacitance provides low-pass filtering; series capacitance or shunt inductance provides high-pass filtering.
- A fluctuating or pulsating dc is equivalent to an ac component varying in opposite directions around the average-value axis.
- An RC coupling circuit is effectively a high-pass filter for pulsating dc voltage, passing the ac component but blocking the dc component.
- A transformer with an isolated secondary is a high-pass filter for pulsating dc, allowing ac in the secondary but no dc output level.
- A bypass capacitor in parallel with R is effectively a low-pass filter, since its low reactance reduces the voltage across R for high frequencies.
- The main types of filter circuits are π type, L type, and T type. These can be high-pass or low-pass, depending on how L and C are connected.
- Resonant circuits can be used as bandpass or bandstop filters. For bandpass filtering, series resonant circuits are in series with the load or parallel resonant circuits are across the load. For bandstop filtering, parallel resonant circuits are in series with the load or series resonant circuits are across the load.
- A wavetrap is an application of a resonant bandstop filter.

- The cutoff frequency of a filter is the frequency at which the output voltage is reduced to 70.7 percent of its maximum value.
- The cutoff frequency of an RC low-pass or high-pass filter can be calculated as $f_c = 1/2\pi RC$. Similarly, the cutoff frequency of an RL low-pass or high-pass filter can be calculated as $f_c = R/2\pi L$.
- The decibel (db) is a logarithmic expression that compares two power levels. In the passband a passive filter provides 0 db of attenuation. At the cutoff frequency a passive filter provides an attenuation of -3 db.
- Semilog and log-log graph paper are typically used to show the frequency response of a filter. The advantage of using logarithmic graph paper is that a wide range of frequencies can be shown in one plot without losing resolution in the smaller values.

REVIEW SELF-TEST

ANSWERS AT BACK OF BOOK.

Fill in the numerical answer.

1. An L of 10 H and C of 40 μF has f_r of _____ Hz.
2. An L of 100 μH and C of 400 pF has f_r of _____ MHz.
3. In question 2, if $C = 400$ pF and L is increased to 400 μH, the f_r decreases to _____ MHz.
4. In a series resonant circuit with 10 mV applied across a 1-Ω R, a 1000-Ω X_L, and a 1000-Ω X_C, at resonance the current is _____ mA.
5. In a parallel resonant circuit with a 1-Ω r_S in series with a 1000-Ω X_L in one branch and a 1000-Ω X_C in the other branch, with 10 mV applied, the voltage across X_C equals _____ mV.
6. In question 5, the Z of the parallel resonant circuit equals _____ MΩ.
7. An LC circuit resonant at 500 kHz has a Q of 100. Its total bandwidth between half-power points equals _____ kHz.
8. A coupling capacitor for 40 to 15,000 Hz in series with a 0.5-MΩ resistor has the capacitance of _____ μF.
9. A bypass capacitor for 40 to 15,000 Hz in shunt with a 1000-Ω R has the capacitance of _____ μF.
10. A pulsating dc voltage varying in a symmetrical sine wave between 100 and 200 V has the average value of _____ V.
11. An RC low-pass filter has the following values: $R = 1$ kΩ, $C = 0.005$ μF. The cutoff frequency f_c is _____.
12. The input voltage to a filter is 10 Vpp and the output voltage is 100 μVpp. The amount of attenuation is _____ db.
13. On logarithmic graph paper, a 2-to-1 range of values is called a(n) _____, and a 10-to-1 range of values is called a(n) _____.
14. At the cutoff frequency, the output voltage is reduced to _____ percent of its maximum value.

Answer True or False.

15. A series resonant circuit has low I and high Z.
16. A steady direct current in the primary of a transformer cannot produce any ac output voltage in the secondary.
17. A π-type filter with shunt capacitances is a low-pass filter.
18. An L-type filter with a parallel resonant LC circuit in series with the load is a bandstop filter.
19. A resonant circuit can be used for a bandstop filter.
20. In the passband, an RC low-pass filter provides approximately 0 db of attenuation.
21. The frequency response of a filter is never shown on logarithmic graph paper.

REFERENCES

Bogart, T.: *Electric Circuits,* Glencoe/McGraw-Hill, Columbus, Ohio.

Grob, B.: *Electronic Circuits and Applications,* Glencoe/McGraw-Hill, Columbus, Ohio.

Kaufman, M., and A. H. Seidman: *Electronics Sourcebook: For Technicians and Engineers,* McGraw-Hill, New York.

Schuler, C., and R. Fowler: *Electric Circuit Analysis,* Glencoe/McGraw-Hill, Columbus, Ohio.

CHAPTER 28

ELECTRONIC DEVICES

For the most part, the subject of electronic devices means semiconductor components such as diodes, transistors, and integrated circuits. They are used for amplifiers, oscillators, rectifiers, and digital circuits, which include just about everything in electronics.

The semiconductors are a group of chemical elements with special electrical characteristics. Most common are silicon (Si) and germanium (Ge), with Si used for almost all semiconductor components. The semiconductors have a unique atomic structure that allows the addition of specific impurity elements to produce very useful features that can be applied in electronic circuits.

CHAPTER OBJECTIVES

Upon completion of this chapter, you should be able to:

- *Explain* the difference between an intrinsic and an extrinsic semiconductor.
- *Explain* what a hole charge is and describe the concept of hole current flow.
- *Describe* the physical construction of a diode.
- *List* the approximate values of forward voltage V_F for a silicon and germanium PN junction.
- *Name* the three terminals of a bipolar transistor.
- *State* the relationship among the three transistor currents.
- *Test* diodes and transistors using an ohmmeter.

IMPORTANT TERMS IN THIS CHAPTER

anode and cathode	field-effect transistor (FET)	metal-oxide-semiconductor FET (MOSFET)
bias, dc	forward voltage and current	NPN and PNP transistors
bipolar transistor	hole charge	peak inverse voltage
bridge rectifier	insulated-gate FET (IGFET)	rectifier
covalent bond	junction FET (JFET)	reverse voltage
diode	junction transistor	volt-ampere characteristic
doping	leakage current	

TOPICS COVERED IN THIS CHAPTER

28-1 SEMICONDUCTORS

The name *semiconductor* for materials such as silicon and germanium means that they are not as good as the metals as electrical conductors but they are not insulators. The reason is atomic structure.* All the semiconductor elements have atoms with an electron valence of ± 4. With this valence, the atom has four electrons in the outermost shell, which can have eight for a stable ring.

As an example, silicon has the atomic number 14. This atom has 14 positive protons in the nucleus, balanced electrically by 14 negative electrons in three outer shells. The electrons are distributed in shells or rings of 2, 8, and 4. For germanium with atomic number 32, the electrons are in shells of 2, 8, 18, and 4.

With four electrons in the outermost shell, halfway to the goal of eight for stability, the atom does not easily gain or lose individual electrons. Instead, the semiconductor atoms share the four valence electrons with other groups of four. Such a combination of atoms sharing groups of valence electrons is called a covalent bond. As a result, pure semiconductors have the following resistance characteristics: (1) medium R and (2) negative temperature coefficient α. The negative α means that R decreases with higher temperature. These characteristics apply just to the pure semiconductor, called *intrinsic,* which means without any added impurities.

Silicon itself is an element in most common rocks. Sand is silicon dioxide (SiO_2). The element Si was discovered in 1823 and Ge in 1886. Germanium is recovered from the ash of certain coals. The oxides of both Ge and Si are reduced chemically to produce the elements with almost perfect purity. Figure 28-1 shows a solid bar of silicon and the thin slices or disks that are used for semiconductor devices. The semiconductor must be very pure so that only the desired impurity elements are added. It can be noted that intrinsic Ge has only about $\frac{1}{1000}$ the resistance of Si, but silicon is generally used for semiconductor devices. It is the way that the semiconductors are altered electrically by added impurity elements that really makes it possible to have useful semiconductor devices.

COVALENT BONDS AND CRYSTAL STRUCTURE   Because of the covalent bonds between atoms, either in Si or Ge, the atoms form a network or lattice structure in a regular pattern that is characteristic of a crystal solid. This pattern is illustrated in Fig. 28-2 for Si atoms. A crystal has a definite geometrical form for the internal atoms. A diamond is an example of the crystalline structure for pure carbon which, incidentally, is also a semiconductor. When a crystal is broken into smaller segments, each has the same structure as the original crystal. With covalent bonds for the semiconductor elements Si and Ge and the crystal structure, it becomes possible to add impurity elements that result in the desired electrical characteristics. This process is called *doping* the pure semiconductor.

N-TYPE AND P-TYPE SEMICONDUCTORS   The production process involves vaporizing certain impurity elements under high heat in an oven with the semiconductor disk, such as the silicon disks shown in Fig. 28-1. For N-type material, the doping element is chosen to provide free electron charges. The P type

*More details of atomic structure for the chemical elements are described in Chap. 1.

FIG. 28-1   Pure silicon in the form of crystals, a solid bar, and wafer disks.

has free positive charges. The doped form is called an *extrinsic* semiconductor, as the opposite of the intrinsic or pure semiconductor.

For N-type, the doping element can be arsenic, antimony, or phosphorus. Each has an electron valence of 5. This value means one extra electron for each group of four in the outermost shell of the atom. As a result, each impurity atom provides an extra electron in the covalent bonds. Figure 28-3 illustrates silicon with atomic number 14 doped with phosphorus (P), which has atomic number 15. Note that P is the symbol for phosphorus and does not indicate positive polarity. The P atoms have five valence electrons. Four of these become part of co-valent bond structure. The extra electron, however, can be considered as a free negative charge, since it is not needed for a covalent bond. The result is N-type doped silicon. Since many phosphorus atoms are added, the doping provides many free electrons. Remember that free charges can move with relative ease, when voltage is applied, to produce electric current.

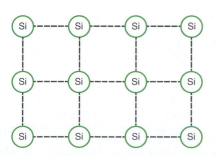

FIG. 28-2   Crystal lattice structure with covalent bonds between silicon (Si) atoms, for a pure semiconductor without doping.

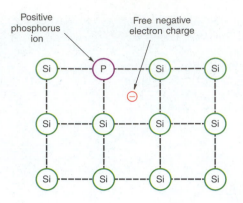

FIG. 28-3 Crystal lattice structure of silicon (Si) atoms doped with phosphorus (P). The covalent bonds have one free electron for each phosphorus atom.

For P type, the doping element can be aluminum, boron, gallium, or indium, with a valence of 3. Each atom has three electrons in the outermost ring. Figure 28-4 illustrates silicon (Si) doped with aluminum (Al). The element aluminum has atomic number 13, which means three outer electrons. In the covalent bonds of Al with Si atoms, there are seven electrons instead of eight. The one missing electron in such a covalent bond can be considered as a free positive charge, called a *hole*. Remember that taking away a negative electron is equivalent to adding the same amount of positive charge. With many aluminum atoms added, the doping provides many hole charges. The holes are free charges that can move with relative ease to produce electric current. However, the free hole charges have slightly less mobility than electrons.

It should be noted that the doping really does not add or subtract charges. The semiconductor is still neutral with equal positive and negative charges. However, the doping redistributes the valence electrons so that more free charges are available.

THE HOLE CHARGE   The hole is a new type of free charge that is present only in P-type semiconductors. The charge is the same amount as a proton, equal to that of an electron but with opposite polarity. A hole charge is not a proton, however. The proton is a stable, immobile charge in the nucleus where the proton is not free to move. The hole is a positive charge outside the nucleus, where it can be made to move by an applied voltage to produce electric current.

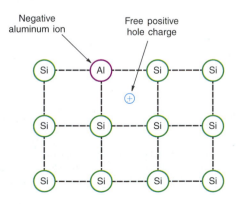

FIG. 28-4   Crystal lattice structure of silicon (Si) atoms doped with aluminum (Al). The covalent bonds have one free positive hole charge for each aluminum atom.

HOLE CURRENT   The idea of hole charges moving to produce current is illustrated in Fig. 28-5. Along the top row, a hole charge is shown at point 1, with some filled covalent bonds. Suppose that a valence electron from the filled bond at point 2 moves to point 1. Along the middle row, then, the bond at point 1 becomes filled and the hole charge moves to point 2. Similarly, along the bottom row, an electron can move from point 3 to point 2 to fill this bond and the hole charge is at point 3. With this sequence, the hole charge is moving from point 1 to point 6, from left to right in Fig. 28-5. To produce this current, a voltage could be applied across the semiconductor, with the positive terminal at point 1.

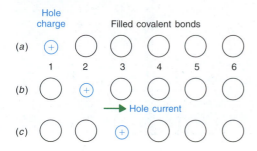

FIG. 28-5   An example of a hole charge moving in a P-type semiconductor to provide a hole current. Motion of the positive charge from location 1 to 2 to 3 is illustrated in (a), (b), and (c).

The direction of hole current is the same as conventional current, opposite from electron flow, because holes are positive charges. *All arrow symbols for the direction of current in semiconductor devices are shown for the direction of hole current.*

It should be noted, though, that hole current flows only in P-type semiconductors. The actual current is electron flow in N-type semiconductors and all wire conductors.

MAJORITY AND MINORITY CHARGES IN SEMICONDUCTORS

With doping, an N-type semiconductor has a large supply of free electrons, a result of the added impurity atoms. Then the electrons become the dominant or *majority charges*. Still, there are *minority* charges of holes. The reason is that thermal energy, even at room temperatures, always gives a random motion to free charges.

Similarly, a P-type semiconductor has majority hole charges. The minority charges here are electrons. In any one type of semiconductor, the majority and minority charge carriers always have opposite polarities.

When the majority charges are made to move in a semiconductor by an applied voltage, the result is a relatively large amount of *forward* current or *easy current.* The forward current is motion of electrons in N-type semiconductors or hole charges in P-type semiconductors. The amount of I is generally in the range of milliamperes to amperes.

When minority charges move, this current is in the reverse direction compared with the forward current of the majority charges. The reason is simply that the polarity of minority charges is the opposite of majority charges in the same conductor.

This very small current of minority charges is called *reverse current* or *leakage current.* The I is in the order of microamperes.

Furthermore, the reverse current increases with higher temperatures. More minority charges are produced by an increase of thermal energy. This increase in the reverse current of minority charges is the reason why temperature is very important in the operation of NPN and PNP transistors.

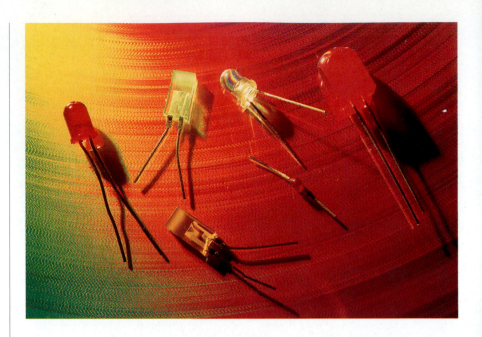

FIXED ION CHARGES IN THE DOPED SEMICONDUCTOR   The free charges in a doped semiconductor are balanced by ions of the impurity element. Two important features of ions in general are:

1. An ion is an atom, with its nucleus, where the atom has a net charge, either positive or negative.
2. Since an ion has the nucleus of the atom, the ions in a solid material are relatively immobile, meaning they cannot be moved as free charges.

As an example of ions resulting from the doping, refer to Fig. 28-3. Here the phosphorus ion is positive because it lacks one of its five valence electrons. As a result, the nucleus provides the charge of one proton in the phosphorus ion. For the example in Fig. 28-4, the aluminum ion is negative because effectively an extra electron has been added as part of its covalent bond structure.

The ions are fixed charges that are not easily moved in the solid crystal. Since the ions are impurity atoms, they are present only in the doped semiconductor. The fixed charges of the ions are important because they provide an internal contact potential that allows a junction to be formed where P and N materials meet. The PN or NP junction is the basis for the operation of all semiconductor devices.

TEST-POINT QUESTION 28-1

Answers at end of chapter.
a. What is the electron valence for Si and Ge?
b. Is the Si doped with phosphorus in Fig. 28-3 N type or P type?
c. True or False? Hole current is in the same direction as electron flow.
d. Is forward current a flow of majority or minority charge carriers?

28-2 THE PN JUNCTION

It is interesting that doping increases the free charges in a semiconductor material but the real usefulness comes from having a junction between successive layers of P and N types. The reason is that the junction provides an internal contact potential, which is 0.7 V for Si and 0.3 V for Ge. Controlling this potential across a PN junction makes it possible to control the current in the semiconductor. The P-type and N-type materials then serve just as electrodes for applying an external voltage to the junction. Although labeled here as a PN junction, it could just as well be NP. The opposite polarities provide the internal junction potential.

In Fig. 28-6, the semiconductor materials are shown with a thin junction between successive layers of P-type and N-type formed in a single crystal. The junction itself has a width of only 10^{-4} cm. In the magnified view of the junction, the impurity ions are shown at opposite edges. The ions provide the contact potential across the two sides of a junction.

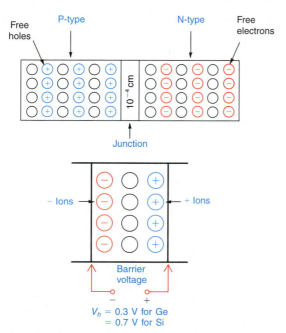

FIG. 28-6   A PN junction with magnified view below of depletion zone to show how ion charges produce internal barrier voltage V_b.

Some of the free electrons in the N material at the junction are attracted toward the P material. At the same time, some of the free hole charges at the P side move the other way to the junction. These opposite charges combine to produce *electron-hole pairs* that are neutral. However, this effect at the junction uncovers the charged ions of the impurity atoms in the junction. The ions do not move. As a result, the ions provide an internal contact potential labeled V_b. This potential is considered a barrier voltage because it prevents any more free electrons or hole charges from crossing the junction. In effect, V_b maintains the free electrons in the N-type semiconductor and the hole charges in the P-type to prevent opposite sides from neutralizing each other.

It should be noted that the junction is only an electrical boundary between the P and N semiconductors produced by alternate types of doping. There is actually no physical separation.

THE INTERNAL BARRIER POTENTIAL V_b Note the polarity of V_b at the junction in Fig. 28-6. The negative side of V_b is at the edge with the P-type semiconductor. Here there are negative ions as a result of the P-type doping. This side of V_b with negative polarity repels electrons from the N-type semiconductor, preventing them from crossing through the junction. Also, the positive side of V_b is at the edge with the N-type material. The positive ions are produced by the N-type doping. This side of V_b repels hole charges from the P-type semiconductor to prevent them from crossing through the junction.

Although V_b is an internal contact potential that cannot be measured directly, its effect can be overcome by applying an external voltage of 0.3 V for Ge or 0.7 V for Si, in the correct polarity. The V_b is higher for a silicon junction because its lower atomic number allows more stability in the covalent bonds.

The V_b of the junction is a characteristic of the element. Therefore, the values given for Ge and Si apply to all PN junctions for semiconductors, diodes, and transistors of any size or power rating. Remember the values of 0.7 V for a silicon junction and 0.3 V for a germanium junction.

The barrier voltage is what makes the junction useful because the effect of V_b can be controlled by applying an external voltage. With forward voltage applied in the polarity that neutralizes the effect of V_b, forward current can flow through the junction. For the opposite case, reverse voltage is applied in the polarity that does not cancel V_b.

DEPLETION ZONE Because of its neutral electron-hole pairs, the junction area is considered as a depletion zone. It has no free charge carriers that can be moved. However, the junction still has the ion charges anchored in position to produce V_b.

EFFECT OF TEMPERATURE The values of 0.3 V for Ge and 0.7 V for Si are at normal room temperature of 25°C. However, V_b decreases at higher temperatures. The reason is that more minority charge carriers are produced by increased thermal energy. The decrease in V_b is the reason why avoiding high temperature is an important precaution in the operation of circuits with NPN and PNP junction transistors.

POLARITY OF FORWARD VOLTAGE AND REVERSE VOLTAGE Refer to Fig. 28-7a. The forward voltage V_F is applied by wire conductors to the P and N electrodes. Such a connection without any barrier potential is called an *ohmic contact*. Then the external V_F is applied through the bulk materials to the PN junction. Forward current I_F flows as the forward voltage neutralizes the effect of the barrier voltage. The required polarity is

$$\left.\begin{array}{l} +V_F \text{ to the P electrode} \\ -V_F \text{ to the N electrode} \end{array}\right\} \text{ forward voltage}$$

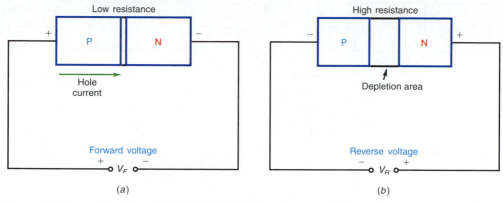

FIG. 28-7 (a) Forward voltage with $+V_F$ to the P side of the PN junction to produce forward current. (b) Reverse voltage with $+V_R$ to the N side of the junction. There is no forward current.

In other words, the polarity of V_F corresponds to the semiconductor types.

The $+V_F$ at the P electrode repels hole charges to the side of the junction that has the negative ions of V_b. Then the effect of the negative ions is canceled. Also, the $-V_F$ at the N electrode repels electrons to the side of the junction that has the positive ions of V_b. Then the effect of the positive ions is canceled. The overall effect then is that V_F cancels V_b.

The results of forward voltage are: (1) forward or easy current flows and (2) the junction has low resistance. All or part of V_b can be neutralized, depending on the amount of V_F.

For reverse voltage V_R, see Fig. 28-7b. The polarity of V_R is the opposite of V_F. Specifically,

$$\left.\begin{array}{l} -V_R \text{ to the P electrode} \\ +V_R \text{ to the N electrode} \end{array}\right\} \quad \text{reverse voltage}$$

In other words, the polarity of V_R is the reverse of the semiconductor types.

The $-V_R$ at the P electrode attracts hole charges away from the junction. Also, the $+V_R$ at the N electrode attracts electrons away from the junction. As a result, the ion charges at the junction are not neutralized but remain intact to maintain V_b.

Because of the reverse voltage, no forward current can flow. The junction R is practically infinite. A small reverse leakage current flows but this value is very low, in the order of 10^{-6} A for Ge and 10^{-9} A for Si.

Values of forward voltage are low, in millivolts or tenths of a volt. Too much V_F causes excessive I_F, producing heat which can destroy the electrical characteristics of the junction. Values of reverse voltage can be much higher, in the order of 100 V to 10 kV or higher, since there is no forward current.

FORWARD CURRENT I_F   An important feature of I_F is that it depends on the amount of V_F. Typical values of V_F across the junction itself are 0.5 to 0.7 V for silicon. In Fig. 28-8a, the V_F of 0.5 V is shown producing I_F of 20 mA, just as examples. In Fig. 28-8b, the higher V_F of 0.6 V increases I_F to 30 mA. This effect of varying V_F to control the amount of I_F is the basis of transistor amplifiers.

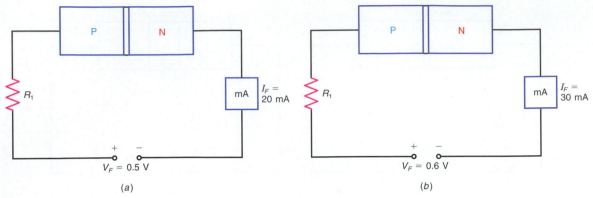

FIG. 28-8   Examples to show that more V_F increases I_F. In (a), 0.5 V produces 20 mA; in (b), 0.6 V produces 30 mA.

More details of the forward current are illustrated in Fig. 28-9 so that the direction can be considered. The PN junction has the symbol shown with an arrow at the P side into the bar at the N side. Actually, any PN junction is a semiconductor diode. For all semiconductor symbols, the arrow shows the direction of hole current. In the P electrode, then, hole charges are moving into the junction. However, in the N electrode the equivalent effect is electrons flowing toward the junction. In the wire conductors and R_1, the current is indicated by dashed arrows for electron flow. Resistor R_1 is inserted to limit the amount of current. Note that the hole current and electron flow are in opposite directions. However, there is only one current in the circuit. The net effect is the same as conventional I around the complete circuit in the direction shown. With the arrow symbol for hole current with semiconductors, it is usually simpler just to follow the current around the circuit in the direction of the arrow.

VOLT-AMPERE CHARACTERISTICS   A graphical plot of I and V for a PN junction is its volt-ampere characteristics. As shown in Fig. 28-10, the

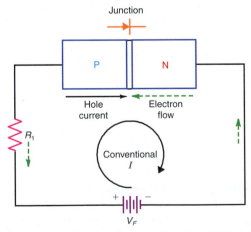

FIG. 28-9   Details of the forward current in a circuit with a PN junction. Note the symbol with an arrow in the direction of the hole current. The dashed arrow indicates the electron flow.

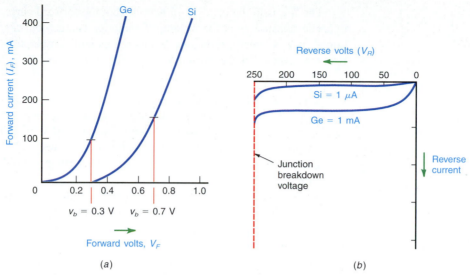

FIG. 28-10 Volt-ampere graphs of a PN junction for Ge and Si. (*a*) Forward characteristics in tenths of a volt and relatively high current. (*b*) Reverse voltage up to 250 V with very small reverse leakage current.

characteristic curves summarize the operation. Separate graphs are used here to illustrate forward voltage in Fig. 28-10*a* and reverse voltage in Fig. 28-10*b*.

In Fig. 28-10*a*, the graph plots low values of V_F, in tenths of 1 V, with relatively high values of I_F in units of mA, which could also be in amperes. The reverse characteristics in Fig. 28-10*b* need a separate graph because relatively high values of V_R up to 250 V are plotted. Also, the values of reverse leakage current are very small. Note that the polarities of V_F and V_R are opposite. Also, the forward current and reverse leakage current are in opposite directions.

In both Fig. 28-10*a* and *b*, separate curves are shown for germanium and silicon because they have different values for the internal barrier voltage V_b. Also, the reverse leakage current for Si is only 1 μA while Ge has 1 mA.

Consider the forward characteristic of Si in Fig. 28-10*a*, as it is used for most junction transistors. Forward current flows when V_F approaches 0.5 V. More V_F increases the I_F. At 0.7 V or more, the forward current increases sharply to its maximum value. A middle value is 0.6 V, which is a typical forward bias for Si junction transistors. Similarly, for a Ge junction, the range of V_F for forward current is 0.1 to 0.3 V. The values of forward current can be in the range of 1 mA to as high as 20 A.

REVERSE CURRENT With reverse voltage, only a small reverse current of minority charges can flow, as shown in Fig. 28-10*b*. The separate curves indicate typical leakage current of 1 mA for Ge and 1 μA for Si. Note the advantage of Si, with practically zero reverse current.

The reverse leakage current has the symbol I_{CO} to indicate a small cutoff current. With reverse voltage, the junction is practically an open circuit, which cuts off the current, except for the small leakage current. As a comparison of forward and reverse characteristics, the V_F makes the junction operate like a short circuit with high I and low V, but for V_R the junction is like an open circuit with just the small I_{CO} and higher values of voltage.

The I_{CO} is in the opposite direction from the forward current. They must be opposite because I_F is the motion of majority charges but I_{CO} consists of minority charges. Furthermore, the I_{CO} increases with temperature. For every 10°C rise, the reverse leakage current doubles for either Ge or Si.

JUNCTION BREAKDOWN VOLTAGE Refer to the reverse characteristics in Fig. 28-10b. The reverse current is constant at a very small value until V_R reaches the junction breakdown voltage shown as 250 V here. Then a relatively large reverse current can flow. Most important, the reverse voltage across the junction is constant at the breakdown value. This effect is used for voltage regulator diodes.

What happens with the breakdown voltage is that the electrical characteristics are changed by the strong electric field of the reverse voltage. The value of 250 V may not seem so high but the junction is very thin. For a width of 1×10^{-4} cm, the value of 25 V across the junction corresponds to 25×10^4 V/cm.

28-3 SEMICONDUCTOR DIODES

A diode is essentially a PN junction. The standard symbol is an arrow to indicate the direction of hole current and a bar, as shown in Fig. 28-11. The arrow is at the anode, which must be positive for current flow, while the bar is the cathode. Conventional current is in the direction of the arrow for the flow of hole charges. Electron flow is the opposite way, against the arrow. The practical use of diodes is to serve as a one-way valve. Current can flow only when positive voltage at the anode with respect to cathode, provides forward voltage. With the reverse polarity, no forward current can flow. This feature is the basis for the general use of the diode as a rectifier to change ac input to dc output.

Two small semiconductor diodes are shown in Fig. 28-11. In Fig. 28-11a, the symbol is on the diode to indicate anode and cathode. For the diode in Fig. 28-11b, the colored band at one end indicates the cathode side. Some diodes may have a + sign at the cathode end to show this is where positive dc output can be obtained in a rectifier circuit. Any mark at one end indicates the cathode side. The cathode has positive dc output when the ac input is applied to the anode.

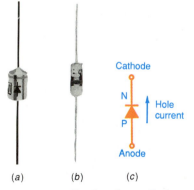

(a)   (b)   (c)

FIG. 28-11 Semiconductor diode rectifiers. Length is about ½ in. without leads. (a) Silicon power diode with current rating of 1 A. The arrow shows the direction of the hole current. (b) Germanium detector diode. The cathode end has a dark band. (c) Schematic symbol for diode.

TYPE NUMBERS   The numbering system for diodes uses the letter N for semiconductors, the prefix 1 before the N and numbers after the N for individual types. The 1 means one junction. As an example, the 1N3196 is a popular silicon diode. The 1N indicates a semiconductor diode, while the 3196 specifies the individual characteristics, which are listed in semiconductor handbooks, specification sheets, and application notes. There is no special indication for Si or Ge, but practically all rectifier diodes are made of silicon. In schematic diagrams, diodes are usually labeled D, CR, or X and Y. The CR stands for crystal rectifier.

DIODE APPLICATIONS   The following list summarizes the main uses for diodes:

1. *Power-supply rectifier.* This function is converting ac input from the 60-Hz power line to dc output. One diode is used for a half-wave rectifier. The full-wave rectifier needs two diodes. More details are described in Sec. 29-7.
2. *Signal detector.* The detector circuit uses a diode to rectify a modulated signal in order to recover the modulating signal. The circuit for an AM detector is explained in Chap. 29.
3. *Digital logic gates.* In these circuits, the diode functions as a switch. It is on when the diode conducts and off without conduction. More details of digital electronics are described in Chap. 31.

In addition, there are many different types of diodes for special applications. These include the capacitive diode for electronic tuning and the light-emitting diode used as a visual display. These diodes and other types are described in Sec. 28-8.

RECTIFIER PACKAGING   The plastic package shown in Fig. 28-11 is very common with rectifiers for currents of about 1 A. Even smaller diodes can be used for less current. Two other types of rectifiers are shown in Fig. 28-12. The metal can in Fig. 28-12*a* is called "top-hat" style. The type in Fig. 28-12*b* uses a stud mount that screws directly into a metal mounting for the cathode connection. Note the diode symbols printed directly on the unit to indicate the anode and cathode terminals. The stud mount types generally have high current ratings.

BRIDGE RECTIFIERS   This type of power supply is used very often, even though it requires four diodes. Connections for the bridge are shown in Fig. 28-13*a*, while typical packages are in Fig. 28-13*b* and *c*. Note that the bridge has four terminals. Two are connections for the ac input and two for the dc output.

RECTIFIER RATINGS   The two most important ratings are for maximum forward current I_F and maximum peak inverse voltage (PIV). Ratings for maximum I_F can be a fraction of one ampere up to 25 A or more. The PIV rating for popular diodes is typically about 1000 V. The peak inverse voltage is the value that can be used across the diode in reverse polarity, negative at the anode, without disrupting the electrical characteristics of the junction. The PIV rating is usually chosen to be at least double the value of the dc voltage output.

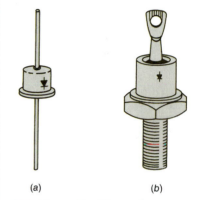

(a)   (b)

FIG. 28-12   Rectifier packages. (*a*) "Top-hat" style. Height is ⅜ in. without leads. (*b*) Heavy-duty rectifier for stud mounting. Height without stud is ½ in.

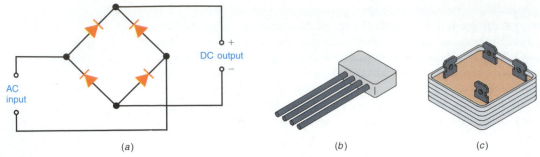

FIG. 28-13 The bridge rectifier. (*a*) Schematic with four diodes as a full-wave bridge. (*b*) Typical package. Length is ¾ in. (*c*) Heavy-duty package for mounting on heat sink. Size is 1 in. square.

28-4 PNP AND NPN TRANSISTORS

PNP and NPN transistors consist of a P or N semiconductor between opposite types, as shown in Fig. 28-14. The purpose is to have three electrodes with two junctions. In operation, the first section at one end supplies free charges, either holes or electrons, to be collected by the third section at the opposite end, through the middle section. The middle electrode controls the current. The names and functions of the electrodes are

Emitter—supplies free charges
Base—controls the flow of charges
Collector—collects the charges from the emitter

Note that the base electrode in the middle has a junction with the emitter and another junction with the collector.

With an N-type base, the transistor is PNP. A P-type base is used for an NPN transistor. The letters correspond to the polarities for emitter, base and collector. Both PNP and NPN transistors operate the same way but they take opposite polarities of dc supply voltage. Most small transistors are the NPN type and made of silicon.

In the schematic symbols, the emitter has an arrow at the junction with the base. The arrow indicates this electrode is the emitter. The third electrode is the

DID YOU KNOW?

A new microprocessor can sense a tornado's sound and alert a household 30 to 90 seconds before the twister hits.

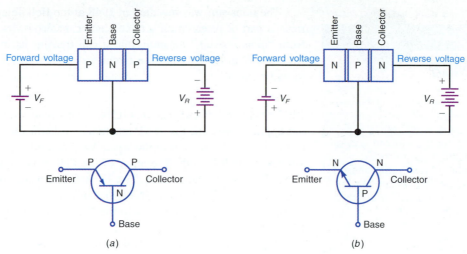

FIG. 28-14   Schematic symbols and dc bias voltages for junction transistors. (*a*) PNP. (*b*) NPN.

collector. As in all semiconductor symbols, the arrow is in the direction of hole current. In Fig. 28-14*a*, the emitter arrow shows that hole charges can move from the P emitter into the N base. Therefore, this transistor is PNP. In Fig. 28-14*b*, the arrow is in the opposite direction to show that hole charges can move from the P base to the N emitter. This transistor is NPN, therefore. Electron flow can be considered to be in the direction from the N emitter to the P base. In short, an arrow into the base means PNP; an arrow out from the base indicates an NPN transistor.

The NPN and PNP types are called *junction transistors.* Also, they can be considered *bipolar transistors* because they have the two opposite polarities of doped semiconductor.

With three electrodes, the transistor is a *triode.* Compared with the diode as a two-terminal device, a triode has one more electrode that can control the current. In a transistor, the base electrode controls the current from emitter to collector, which makes it possible to have amplification. Practically all amplifiers in electronic circuits use transistors. Figure 28-15 shows a simple transistor amplifier circuit as it would appear on a lab prototype board and on a surface-mount circuit board. Either discrete transistors can be used, or the transistor is part of an integrated circuit (IC). Transistors are the main components in IC chips.

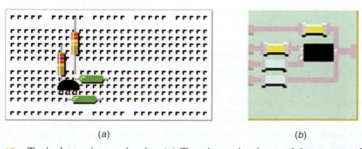

FIG. 28-15   Typical transistors circuits. (*a*) Transistor circuit on a lab prototype board. (*b*) Transistor circuit on a surface-mount circuit board.

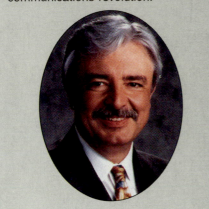
The transistor was invented in 1948 at the Bell Telephone Laboratories, as part of research on semiconductor materials. Its name is derived from "trans resistor," meaning that it can transfer its internal resistance from low R in the emitter-base circuit to a much higher R in the collector-base circuit.

THE EMITTER-BASE JUNCTION   This junction has forward voltage applied across the PN or NP junction in order to allow the free charges of the emitter to move into the base. As shown in Fig. 28-14a, positive voltage is applied to the P emitter of the PNP transistor, with the negative side of V_F to the N base. The V_F is forward bias for the transistor. These polarities are for the PNP type. In Fig. 28-14b, the NPN transistor has forward bias, with $-V_F$ to the N emitter, just the opposite of the PNP type.

Typical values of forward bias are 0.2 V for Ge or 0.6 V for Si in transistor amplifiers. The required bias voltage is the same for any size transistor.

It is important to realize that the forward bias is necessary for the transistor to operate. No current can come out of the collector unless V_F allows the emitter to inject free charges into the base. In short, the junction transistor is a normally off device. It needs forward voltage applied to start conducting.

THE COLLECTOR-BASE JUNCTION   The function of the collector-base junction is to remove charges from the base. Then current can flow in the collector return circuit to the emitter. However, to accomplish this result, the collector must have reverse voltage with respect to the base. As shown in Fig. 28-14a, the P collector of the PNP transistor is connected to the negative side of V_R, with $+V_R$ to the N base. These polarities for reverse voltage are opposite in Fig. 28-14b with the NPN transistor. In both Fig. 28-14a and b, though, the collector has reverse voltage.

In short, NPN transistors take positive dc supply voltage at the collector for reverse voltage; PNP transistors require negative dc collector voltage. Typical values are 9 to 100 V, depending on the power rating of the transistor.

The reverse voltage across the collector-base junction means that no majority charges can flow from collector to base. However, in the opposite direction from base to collector, the collector voltage attracts the charges in the base supplied by the emitter.

TRANSISTOR ACTION   The requirement is to have the collector current controlled by the emitter-base circuit. There are three factors:

1.   The emitter has heavy doping to supply free charges.
2.   The base has only light doping and is very thin.
3.   The collector voltage is relatively high.

As a result, practically all the charges supplied by the emitter to the base are made to flow in the collector circuit. Typically, 98 to 99 percent or more of the emitter charges provide collector current I_C. The remaining charges become the small base current I_B.

Consider the currents for an NPN transistor as illustrated in Fig. 28-16. The N emitter supplies electrons to the P base with forward voltage across the junction. In the P base, the electrons are minority charges. Because of the light doping in the base, though, very few of the electrons can recombine with hole charges.

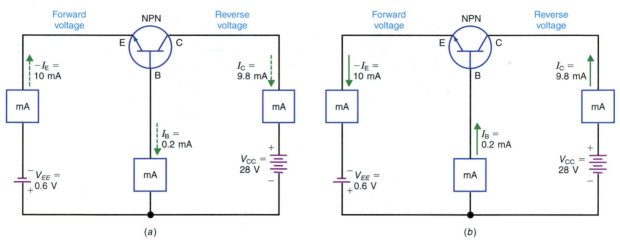

FIG. 28-16   Electrode currents I_E, I_B, and I_C for an NPN transistor. (a) Dashed arrows for I show the direction of the electron flow. (b) Solid arrows for I show the direction of the conventional current for the same NPN transistor as in (a).

Some electrons in the base return through the external base circuit to the emitter to provide the very small I_B. However, almost all the electrons move through the thin base to the collector junction. Here the N collector has reverse voltage of positive polarity. For electrons moving from the base, though, the positive collector voltage attracts these free electron charges. Therefore, the electrons supplied from the emitter side of the base move to the collector side of the base and are attracted into the collector electrode. The result is collector current in the external circuit, where I_C returns to the emitter.

For a PNP transistor, all voltage polarities are the opposite of Fig. 28-16, for the reverse and forward voltages. Also, the I_C at the collector electrode consists of hole charges.

ELECTRODE CURRENTS   For the example in Fig. 28-16, the emitter supplies 10 mA of forward current. This is I_E. Consider that 98 percent or 9.8 mA is injected into the collect circuit. This current is I_C. Only 2 percent, equal to 0.2 mA or 200 μA of I_B, flows through the base terminal to return to the emitter. As a formula,

$$I_E = I_C + I_B \qquad\qquad (28\text{-}1)$$

In Fig. 28-16a, the direction of the currents is shown with dashed arrows for electron flow. The opposite direction for conventional current is indicated by the solid arrows in Fig. 28-16b. For either direction of current, the values are still the same, with 10 mA of I_E supplying 9.8 mA for I_C and 0.2 mA for I_B.

Incidentally, note that the supply voltage for each electrode is labeled with double subscripts, as in V_{EE} and V_{CC}. This notation is standard practice. It is useful for typical amplifiers, where the actual electrode voltage may have a lower value than the supply voltage because of voltage drops in the circuit.

Note that I_E in Fig. 28-16 is marked negative, only to indicate that its direction is opposite from that of I_C and I_B. The I_E is into its electrode, while the I_C and I_B are out from their respective electrodes. *It is standard practice to consider hole current into a semiconductor as the positive direction of I.* This rule is only an arbitrary definition. As an application in Fig. 28-16a, the I_E is electron flow into the transistor. Electron flow moving in corresponds to hole charges moving out, which is the negative direction for I. Also I_C and I_B as electron flow are out from the transistor, which is the positive direction corresponding to hole charges moving in. The same signs for I apply in Fig. 28-16*b* where the currents are shown for the direction of hole charges or conventional current. Algebraically, the values of current in Fig. 28-16 can be stated as

$$-I_E + I_C + I_B = 0$$
$$-10 \text{ mA} + 9.8 \text{ mA} + 0.2 \text{ mA} = 0$$

In practical terms, the formula states that the collector and base currents must add to equal the emitter current, which is the source.

Example

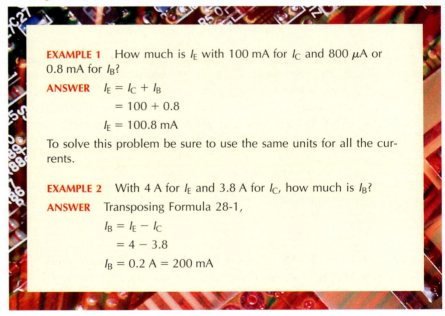

EXAMPLE 1 How much is I_E with 100 mA for I_C and 800 μA or 0.8 mA for I_B?

ANSWER $I_E = I_C + I_B$
$\qquad\qquad = 100 + 0.8$
$\qquad I_E = 100.8 \text{ mA}$

To solve this problem be sure to use the same units for all the currents.

EXAMPLE 2 With 4 A for I_E and 3.8 A for I_C, how much is I_B?

ANSWER Transposing Formula 28-1,
$\qquad I_B = I_E - I_C$
$\qquad\quad = 4 - 3.8$
$\qquad I_B = 0.2 \text{ A} = 200 \text{ mA}$

For most transistors, the I_B is in microamperes or milliamperes. The I_C and I_E are in milliamperes, usually, or in amperes for power transistors. In all cases, I_E and I_B return through the external circuit to the emitter, as the source of the free charges for all the currents.

THE BASE CURRENT CONTROLS THE COLLECTOR CURRENT
This factor is the reason why a transistor can amplify the signal input. When I_B is increased by more forward voltage, this effect means more majority charges are in the base to be injected into the collector. Therefore, increasing I_B means more I_C. For the opposite variation with less forward voltage and less I_B, the col-

lector current is reduced. As a result, the signal variations of amplitude in the base circuit produce equivalent variations in the collector circuit.

It is important to remember that there must be at least enough forward voltage across the emitter-base junction to produce current. For silicon, typical values for V_{BE} are 0.5 to 0.7 V. For lower voltages down to zero, the transistor is off, without any conduction. With enough V_{BE} to provide emitter current, then the transistor can produce amplification.

All these features apply to both PNP and NPN transistors. See Table 28-1 for a summary of the two types of junction transistors.

TABLE 28-1 JUNCTION TRANSISTORS

TYPE	SYMBOL	ELECTRODES	NOTES
NPN		C = collector B = base E = emitter	Needs $+V_C$ reverse voltage for collector; forward bias of 0.6 V for Si or 0.2 V for Ge typical values; Si more common for NPN Hole current out from base into emitter
PNP		C = collector B = base E = emitter	Needs $-V_C$ reverse voltage for collector; forward bias V_{BE} of 0.2 V for Ge or 0.6 V for Si typical values; Ge more common for PNP Hole current into base from emitter

TEST-POINT QUESTION 28-4

Answers at end of chapter.
a. Does the base-emitter junction have forward or reverse bias?
b. Does the collector-base junction have forward or reverse voltage?
c. If I_C is 1 mA and I_B is 50 μA, how much is I_E?
d. A Si transistor has 0.1-V forward bias. Is it conducting or cut off?
e. A transistor symbol shows the emitter arrow out from base. Is the transistor PNP or NPN?

28-5 FIELD-EFFECT TRANSISTOR (FET)

The field-effect transistor (FET) is an amplifier with the same function as a junction transistor, but FET construction is different. As a result, the FET has the following features:

1. The input resistance is very high. A typical value is 15 MΩ.
2. The input circuit can take several volts for the input signal, compared with tenths of a volt for junction transistors.

As illustrated in Fig. 28-17a, the FET operation depends on controlling current through a semiconductor channel of N or P polarity. One side of the channel is the *source electrode* and the other end is the *drain electrode.* An N channel is shown here, but a P channel can be used instead. For either type, current flows from the source to the drain. The voltage applied to the gate electrode controls the current through the channel. The bulk or substate material is neutral or lightly doped silicon. It only serves as a platform on which the other electrodes are diffused.

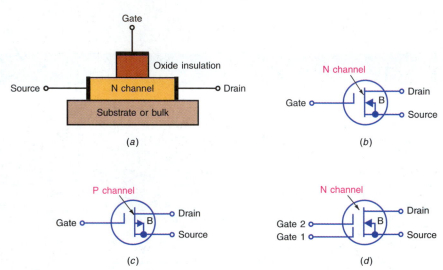

FIG. 28-17   Insulated-gate field-effect transistor, or IGFET. (*a*) Construction. (*b*) Schematic symbol for N-channel with arrow in. B is bulk or substrate connected internally to source. (*c*) Symbol for P-channel with arrow out. (*d*) N-channel IGFET with two-gate inputs.

The terms anode and cathode are not used for the FET because the channel can be either N or P. An FET is a unipolar device, as the charge carriers in the channel have only one polarity.

Compared with junction transistors, the FET has much higher input resistance, from gate to source, and can take more input signal voltage. Also, the FET is less sensitive to the effects of temperature, x-rays, and cosmic radiation. These energy sources can produce minority carriers in junction transistors. In addition, the FET has less internal noise as an amplifier. The disadvantages are less gain for a given bandwidth and smaller power ratings than the NPN or PNP type. Finally, the switching speed of the FET is slower.

In summary, the electrodes of an FET correspond to the emitter, base, and collector in a junction transistor as follows:

1. *Source.* This is the terminal where the charge carriers enter the channel bar to provide current through the channel. Source current is I_S. The source corresponds to the emitter.
2. *Drain.* This is the terminal where current leaves the channel. Drain current is I_D. The drain corresponds to the collector.
3. *Gate.* This electrode controls the conductance of the channel between the source and the drain. Input signal voltage is generally applied to the gate. The gate voltage is V_G. The gate corresponds to the base, but the gate voltage controls

the electric field in the channel, while the base current controls the collector current in a junction transistor.

In the schematic symbol, an arrow for hole charges into the channel indicates an N channel, as in Fig. 28-17b. For a P channel, the arrow direction is out, as in Fig. 28-17c. The source and drain have no polarity, since they are just ohmic contacts. An N channel takes positive drain voltage to provide drain current.

FUNCTION OF THE GATE In the input circuit, the gate and channel act like two plates of a capacitor. A charge of one polarity on the gate induces an equal and opposite charge in the channel. As a result, the conductivity of the channel can be increased or decreased by the gate voltage. With an N channel, positive voltage at the gate induces negative charges in the channel to allow more electron flow from source to drain.

IGFET This abbreviation is for the insulated-gate FET, with the construction illustrated in Fig. 28-17a. The IGFET consists of a metal electrode for the gate separated from the channel by a thin layer of silicon dioxide. This material is an insulator, like glass. However, by electrostatic induction a voltage applied to the gate can induce charges in the channel to control the current from source to drain. There is no PN junction. The IGFET construction is also used for transistors in integrated circuits. This type is a metal-oxide-semiconductor field-effect transistor (MOSFET). Either name, IGFET or MOSFET, is used for these transistors.

In the FET symbols in Fig. 28-17, an N channel is shown in b, and a P channel in c. For the FET in Fig. 28-17d, the symbol shows an N channel with dual gate electrodes for two input signals to control the drain current.

Because of its very high input resistance, an IGFET not connected in a circuit may require a shorting ring on the leads to protect against a buildup of static charge. Also, a grounded iron is used for soldering the leads. Excess charge can produce enough voltage to puncture the thin glass insulation of the gate electrode. However, many FETs have internal protective diodes for the gate.

DEPLETION OR ENHANCEMENT MODE IGFETs come as either depletion or enhancement types, depending on the amount of doping used for the channel construction. For the depletion type:

1. The channel has free charge carriers.
2. Current can be produced in the channel with voltage applied between drain and source, but no gate voltage.
3. The gate voltage can deplete the charge carriers in the channel to a greater or lesser extent to control the drain current.

As a result of these characteristics, the depletion type of IGFET is a normally on device. Drain current flows with zero gate voltage but usually a reverse voltage of about 1 V is used as dc bias for the desired operating characteristics as an amplifier.

For the enhancement type of IGFET, the channel has very little doping. Then gate voltage must be applied to enhance the amount of charge carriers in the channel to produce drain current. Therefore, the enhancement type is a normally off device. A forward voltage of about 5 to 7 V is applied to the gate as dc bias for the amplifier. With an N channel, forward voltage at the gate is positive and reverse voltage is negative.

A comparison between the on and off conditions is illustrated in Fig. 28-18 for these two types. The gate electrode is open for the condition of zero gate voltage. In Fig. 28-18a, the meter shows drain current produced by the voltage between drain and source for the depletion type. However, the enhancement type in Fig. 28-18b has no drain current.

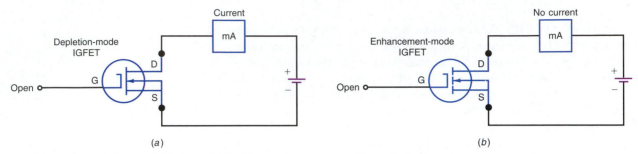

(a)   (b)

FIG. 28-18 (a) Depletion-mode N-channel IGFET operating as a normally on device. (b) Enhancement-mode N-channel IGFET operating as a normally off device.

Note the IGFET schematic symbols in Fig. 28-18a and b, which show an N channel with the arrow for hole current directed into the channel. In Fig. 28-18b, however, the channel is shown as a broken line in order to indicate that this is the enhancement type.

Another mode for the IGFET is the depletion-enhancement type, as a compromise between both characteristics. In this type, drain current flows with zero gate voltage, but less than the amount for a depletion type. As a result, the depletion-enhancement IGFET can be used in an amplifier circuit without any dc bias for the gate.

EIA TYPES The Electronic Industries Association (EIA) classifies the types of IGFET in three groups labeled A, B, and C. For an N channel, the depletion type A takes negative gate bias for a middle value of drain current I_D. With the same I_D, depletion-enhancement type B can operate with zero gate bias. Finally, enhancement type C requires positive gate bias.

JFET This abbreviation is for junction field-effect transistor. Instead of an insulated gate in an IGFET, the JFET uses a PN junction between the gate and channel. This construction is illustrated in Fig. 28-19. However, reverse bias is

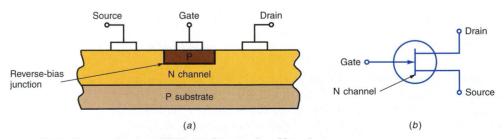

(a)   (b)

FIG. 28-19 Junction field-effect transistor, or JFET. (a) Construction. Note the reverse bias for junction of gate and source. (b) Schematic symbol for N-channel with arrow in.

used at the gate so that the input resistance is very high as in the IGFET. In the schematic symbol, the arrow is on the gate electrode. The arrow directed inward shows an N channel. All the different types of FETs, including the JFET, are summarized in Table 28-2, with symbols, electrodes, and characteristics.

TABLE 28-2	TYPES OF FIELD-EFFECT TRANSISTOR		
TYPE	**SYMBOL**	**ELECTRODES**	**NOTES**
JFET		D = drain G = gate S = source	Junction-type FET; arrow pointing to channel indicates P gate and N channel; reverse bias at junction
IGFET or MOSFET N channel		D = drain G = gate S = source	Insulated gate; depletion or depletion-enhancement type; B is bulk or substrate connected internally to source
IGFET or MOSFET P channel		D = drain G = gate S = source	Arrow pointing away from channel indicates P channel
IGFET or MOSFET, N channel, enhancement		D = drain G = gate S = source	Broken lines for channel show enhancement type
Dual-gate IGFET or MOSFET, N-channel		D = drain G_2 = gate 2 G_1 = gate 1 S = source	Either or both gates control amount of drain current I_D

TEST-POINT QUESTION 28-5

Answers at end of chapter.

Answer True or False.
a. The IGFET has high input resistance.
b. The gate electrode on an FET compares with the emitter electrode on a junction transistor.
c. The schematic symbol in Fig. 28-17b is for an N-channel depletion-mode IGFET.
d. The IGFET is unipolar while junction transistors are bipolar.

28-6 TESTING DIODES AND TRANSISTORS

An ohmmeter can be used to check a PN junction either for an open circuit or a short circuit. The short is indicated by R of practically zero ohms. A very high R of many megohms, in the direction of infinite ohms, means an open circuit. Power is off in the circuit for ohmmeter readings. Preferably, the device is out of the circuit to eliminate any parallel paths that can affect the resistance readings.

Figure 28-20 illustrates the idea of testing a diode with an ohmmeter. In Fig. 28-20a, the forward resistance should be low, perhaps 100 to 1000 Ω, with the positive battery lead of the ohmmeter at the anode side of the diode. Next, the ohmmeter leads are reversed to provide reverse voltage across the diode. Then the reverse resistance should be very high, near infinity for silicon diodes. These measurements also show which end of the diode is the anode. There are the following possibilities in the measurements:

1. When the ratio of reverse to forward R is very high, the diode is probably good.
2. When both the forward and reverse R are very low, close to zero, the diode junction is short-circuited.
3. When both the forward and reverse R are very high, close to infinity, the diode probably has an open at the terminal.

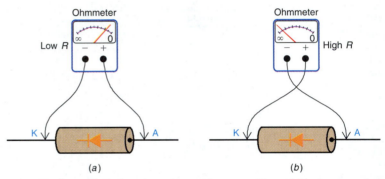

FIG. 28-20   Testing a diode junction with an ohmmeter. Normal R values are shown here. (a) Forward bias results in very low resistance. (b) Reverse leads for reverse bias and very high resistance.

These tests should not be made on the $R \times 1$ position of the ohmmeter because it may supply too much current. Use the $R \times 10$ or $R \times 100$ position.

When the ohmmeter has a low-ohms position, only about 0.3 V is used for the ohmmeter leads. This voltage is not high enough to turn on a silicon junction for testing. However, this type of meter usually has a *diode test* position. The meter then indicates the normal junction voltage of approximately 0.7 V for Si or 0.3 V for Ge. Current for testing the junction is provided by the meter.

For testing PNP and NPN transistors, the same diode test can be used for each junction. Check the emitter-base junction as a diode, in one polarity and then the opposite way. Similarly, the collector-base junction can be checked as a diode. Remember that an NPN transistor takes positive voltage at the base for forward bias. More detailed tests on transistors, including their electrical characteristics, can be checked with a transistor tester.

With junction transistors that are not marked, some practical points can help identify the type. In a circuit where the dc supply voltage is positive for the collector or negative for the emitter, the transistor is NPN. This polarity is needed as reverse voltage for V_{CE}. When the forward voltage for V_{BE} is 0.6 to 0.7 V, the transistor is silicon. Germanium transistors take 0.2 to 0.3 V.

Consider a transistor by itself, out of the circuit. Some ohmmeter tests can determine which leads are for the emitter, base, and collector. On the $R \times 100$ or $R \times 10,000$ scale for small transistors, measure the resistance for the three combinations between two pins. Then reverse the ohmmeter leads and repeat the measurements. The lowest reading is the resistance between base and emitter in the forward direction. With an NPN type, the base has the positive side of V_{BE}. The next-to-lowest reading is between collector and base, also in the forward direction. Incidentally, the reverse readings for silicon are practically infinite ohms.

To check a field-effect transistor, the resistance of the channel between source and drain should be about $10,000 \ \Omega$. The R between gate and channel should be infinitely high for an MOSFET, in either polarity. With a JFET check the gate-to-channel junction as a diode.

It should be noted that the red lead of the ohmmeter is usually the positive side of the internal battery. However, check the polarity by using another voltmeter to measure the voltage across the ohmmeter leads.

TEST-POINT QUESTION 28-6

Answers at end of chapter.

Answer True or False.
a. An ohmmeter test across a diode should show low R in one polarity and very high R in the opposite polarity.
b. The same test as in **a** across the base-collector junction of a transistor should show the same results.
c. The R of the channel between source and drain of an FET should measure infinitely high.

28   SUMMARY AND REVIEW

- The semiconductor elements used for electronic devices are silicon (Si) and germanium (Ge). Doped semiconductors are P type or N type. All semiconductor devices are solid-state components, generally using Si.
- A PN junction allows forward current with forward voltage V_F applied. The polarity of V_F is the same as the P and N electrodes. Reverse voltage of the opposite polarity prevents forward current.
- A semiconductor diode has a PN junction. The P side is the anode and the N side is the cathode. Positive voltage at the anode allows diode current. The diode is used as a one-way conductor for a rectifier to convert ac input to dc output.
- Both the NPN and the PNP transistors have two junctions for the three electrodes: emitter, base, and collector. The emitter supplies free charges through the base to be received at the collector. The collector requires reverse voltage, while the base-emitter junction needs forward bias. Typical forward bias is 0.6 V for Si or 0.2 V for Ge. The features of NPN and PNP transistors are summarized in Table 28-1.
- The field-effect transistor (FET) is an amplifier like junction transistors, but the FET has very high input resistance. The FET electrodes are a source corresponding to an emitter, a gate like the base electrode, and a drain corresponding to a collector. Features of the FET are summarized in Table 28-2.
- A semiconductor diode can be tested with an ohmmeter for low R in the forward direction and high R in the reverse direction. The same tests apply to the junctions in PNP and NPN transistors.

SELF-TEST

ANSWERS AT BACK OF BOOK.

1. Give the electron valence for the semiconductor elements.
2. How much is the internal barrier voltage for a silicon PN junction?
3. Is hole current in the same or opposite direction as electron flow?
4. Does an arrow on the emitter out from the base show a PNP or NPN transistor?
5. Which electrode controls the output current for junction transistors?
6. Is V_{BE} forward or reverse voltage?
7. Is V_{CB} forward or reverse voltage?
8. Do NPN transistors take positive or negative collector voltage?
9. In a rectifier circuit, is positive dc output taken from the diode anode or cathode?

10. Does the arrow in symbols for semiconductor devices show the direction of hole current or electron flow?
11. When I_C in a junction transistor is 99.9 mA and I_B is 0.1 mA, how much is I_E?
12. Which electrode in an FET corresponds to the emitter in junction transistors?
13. In schematic symbols for the FET, does an arrow directed into the channel show an N channel or P channel?
14. Give the abbreviations for three types of FET.
15. Does an FET have higher or lower input resistance than a junction transistor?
16. An ohmmeter test on a PN junction shows close to zero ohms with forward and reverse voltage. Is the junction shorted or open?
17. True or False? The normal resistance of the channel in an FET is infinitely high.

QUESTIONS

1. Define the following: (a) doping; (b) N-type silicon; (c) P-type silicon; (d) PN junction; (e) internal barrier potential; (f) depletion zone.
2. For a PN junction, show a battery applying: (a) forward voltage; (b) reverse voltage.
3. Give two comparisons between electron flow and hole current.
4. Compare hole charges to ion charges.
5. Draw the schematic symbols for NPN and PNP transistors and label the electrodes.
6. Give the functions for the three electrodes in a junction transistor.
7. Explain the biasing on the emitter-base junction and base-collector junctions of NPN and PNP transistors.
8. Draw the schematic symbols for an N-channel MOSFET, depletion type and enhancement type.
9. Compare the functions of the three electrodes in an FET with a junction transistor.
10. Give two comparisons between junction transistors and the FET.
11. Explain briefly how to test a PN junction with an ohmmeter for a short or open.
12. Identify the majority and minority charges in a(n): (a) N-type semiconductor; (b) P-type semiconductor.
13. Draw a schematic for a bridge rectifier which will produce a negative output voltage.
14. Explain what is meant when a depletion-type IGFET is said to be normally on.

PROBLEMS

ANSWERS TO ODD-NUMBERED PROBLEMS AT BACK OF BOOK.

1. Calculate the resistance of a silicon diode that has 0.5 A of current through it with a 0.8-V drop across the two terminals.

2. In a junction transistor, I_C is 1 mA and I_B is 20 μA. Calculate I_E.
3. In a junction transistor, I_E is 5.82 mA and I_B is 120 μA. Calculate I_C.
4. Refer to the graph in Fig. 28-10. How much is I_F with (a) V_F of 0.7 V for Si; (b) V_F of 0.3 V for Ge?
5. Refer to the electrode currents in Fig. 28-16. If I_B increases to 0.3 mA and I_C increases to 50 mA, how much is the higher value of I_E?
6. For each of the circuits in Fig. 28-21, indicate whether the diodes and transistors are cut off or conducting. Note the difference of potential for the voltage in each case.

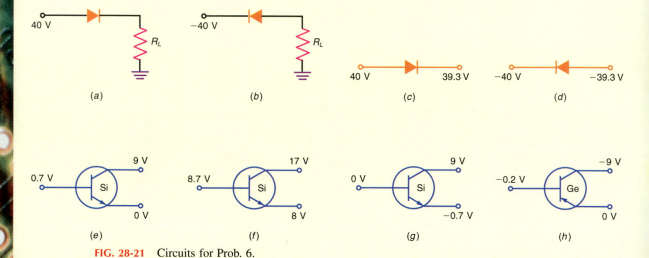

FIG. 28-21 Circuits for Prob. 6.

7. In Fig. 28-22, calculate the current I in the circuit. (Assume that $V_F = 0.6$ V.)
8. In Fig. 28-22, how much is the current I if the diode is reversed?

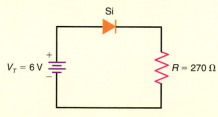

FIG. 28-22 Circuit for Probs. 7 and 8.

CRITICAL THINKING

1. In Fig. 28-23, calculate the forward current I_F in diode D_1. (Assume that $V_F = 0.6$ V.) Hint: Apply Thevenin's theorem.

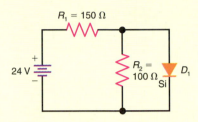

FIG. 28-23 Circuit for Critical Thinking Prob. 1.

ANSWERS TO TEST-POINT QUESTIONS

28-1 a. ±4
b. N-type
c. F
d. majority charges

28-2 a. 0.7 V
b. P
c. N
d. 1 μA
e. 150 mA

28-3 a. positive
b. hole current
c. cathode
d. one
e. cathode

28-4 a. forward
b. reverse
c. 1.05 mA
d. cut off
e. NPN

28-5 a. T
b. F
c. T
d. T

28-6 a. T
b. T
c. F

CHAPTER 29

ELECTRONIC CIRCUITS

Transistors and diodes are two common types of semiconductor devices that make it possible to have so many applications of electronics. One of the main applications of transistors is in amplifier circuits. An amplifier increases the magnitude or amplitude of signal variations to make the desired signal stronger.

Without electronic amplification, not much could be done with audio, radio, video, or control systems. An amplified audio signal makes the reproduced sound loud enough. In television, the amplified video signal makes the reproduced picture have enough contrast and brightness.

Another important function of electronic circuits is switching, with either diodes or transistors. When the device is conducting current, the internal resistance is low. The R can be low enough, less than $1\ \Omega$, to be considered a short circuit. This low resistance corresponds to a closed switch. For the opposite case, a diode or transistor can be cut off, meaning no current in the device. For this condition, the R is practically infinity. The high resistance corresponds to an open switch. The electronic switching, therefore, means making the diode or transistor change between conduction and cutoff.

The integrated circuit (IC) chip combines transistors and diodes in one unit, often with resistance and capacitance also. Inductors are not included, since they take up too much space. The semiconductor element silicon (Si) is used for most transistors, diodes, and IC chips. A transistor or diode not in an IC chip is called a *discrete component* because the part is complete in itself. Discrete transistors and diodes generally have higher power ratings than IC chips.

CHAPTER OBJECTIVES

Upon completion of this chapter, you should be able to:

- *Explain* the difference between an analog and a digital signal.
- *Give* examples of analog and digital signals.
- *Calculate* the voltage gain, current gain, and power gain of an amplifier or a chain of amplifiers.
- *Explain* the roles of resistors, capacitors, and inductors in an amplifier circuit.
- *Explain* what is meant by *positive feedback*.
- *Describe* the operation of an RF feedback oscillator.
- *List* three different types of multivibrators.
- *Define amplitude modulation* and *frequency modulation* and list the main characteristics of each.
- *Explain* the operation of a half-wave power supply.
- *Troubleshoot* a defective half-wave power supply.

IMPORTANT TERMS IN THIS CHAPTER

active device	detector circuit	integrated circuit
amplifier	dc supply	isolation transformer
amplitude modulation (AM)	digital signal	modulation
analog signal	discrete component	multivibrator
anode	feedback	oscillator circuit
baseband signal	flip-flop circuit	passive device
binary numbers	frequency modulation (FM)	power supply
bit	full-wave rectifier	rectifier
byte	gain of amplifier	semiconductor device
carrier wave	half-wave rectifier	sidebands
cascaded amplifiers	Hartley oscillator	signal variations
Colpitts oscillator	hum	transistor
crystal oscillator	IC chip	

TOPICS COVERED IN THIS CHAPTER

29-1 ANALOG AND DIGITAL SIGNALS

The world of electronics is divided into two broad areas—analog and digital—because they are so different. Analog circuits consist mainly of amplifiers for voltage or current variations that are smooth and continuous. Transistors and IC chips are generally used for the amplification. Digital circuits provide electronic switching of voltage pulses. A pulse has abrupt changes between two distinctly different amplitude levels. Diodes and transistors in digital IC chips are generally used. Just to indicate the division between the digital and analog areas, the manufacturers of IC chips have separate catalogs or handbooks for their digital and analog products. The analog form is generally called a *linear* type of IC unit because analog information deals with proportional values.

What do we mean by the signal? In general, the signal variations for electronic circuits are changes in voltage and current that correspond to the desired information. In an analog signal, the electrical variations have a direct relation to the changes that represent the information. Analog examples include audio signal for sound and video signal for the picture in television. As another practical example, a radio or television broadcast station transmits radio-frequency variations in an electromagnetic field that forms an RF signal. All analog signals have continuous variations, with smooth changes between many different values. For the opposite case, the pulses in a digital signal have abrupt changes between two levels, representing only two values. It is the combinations of pulses that provide the desired information.

As a specific example of analog signals, Fig. 29-1 shows sine-wave variations

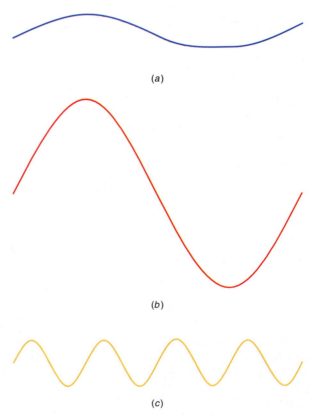

(a)

(b)

(c)

FIG. 29-1 Examples of analog signals. (*a*) Sine wave. (*b*) More amplitude. (*c*) Higher frequency.

in voltage or current. In Fig. 29-1a, note that the amplitude variations are smooth and continuous, without any abrupt transitions in amplitude. The sine wave is shown with higher amplitude in Fig. 29-1b, representing a stronger value of the corresponding voltage and current. In Fig. 29-1c, more cycles for the sine waves show a higher frequency. In all cases, these waveforms are analog signals, as the variations in the electrical values are similar to the changes in the desired information.

A digital signal consists of a train of pulses for the voltage or current, as illustrated in Fig. 29-2. The pulses all have the same amplitude with abrupt changes between the maximum level at 5 V here and the minimum level at 0 V. The voltages between the two extremes have no meaning in terms of information for the signal. The reason is that the pulses operate with a switching circuit, which is turned either on or off.

Since the digital signal in Fig. 29-2 has only two significant levels, either HIGH or LOW, it is useful to represent the pulses in a binary number system with the digits 1 and 0. The high level is generally indicated as binary 1 and the low level with 0. Then the binary numbers correspond to the digital pulse signal. For instance, the binary values in Fig. 29-2 represent the four pulses in each signal as follows:

$$1010 = \text{HIGH, LOW, HIGH, LOW (Fig. 29-2a)}$$
$$0101 = \text{LOW, HIGH, LOW, HIGH (Fig. 29-2b)}$$
$$1100 = \text{HIGH, HIGH, LOW, LOW (Fig. 29-2c)}$$
$$0011 = \text{LOW, LOW, HIGH, HIGH (Fig. 29-2d)}$$

The binary values are usually in groups of eight. Each pulse is a *bit* of information. The group of bits is a *word*. The words can have 4, 8, 16, or 32 bits. An 8-bit word is called a *byte*.

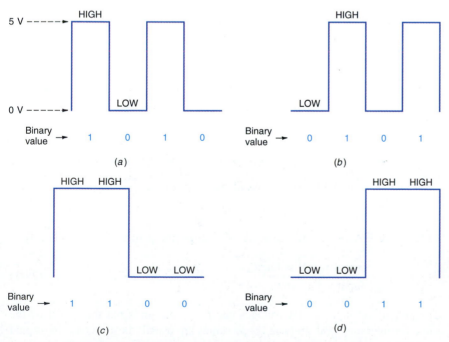

FIG. 29-2 Pulses for digital signals. HIGH voltage level considered as binary 1 with zero level as 0. Information is in the combinations of 1 and 0.

Answers at end of chapter.

Answer True or False.
a. The sine wave is an example of an analog signal.
b. The digits used in a binary number system are 1 and 2.
c. The pulses for the binary word 0111 are *not* shown in Fig. 29-2.

29-2 AMPLIFIER GAIN

The ability of an amplifier circuit to increase the amount of signal is measured by the gain, defined as the ratio of output signal to input signal. For example, when the output is 50 times more than the input, the gain of the amplifier equals 50.

General symbols for an amplifier circuit are shown in Fig. 29-3. The triangle is a general form that just shows terminals 1 and 2 for input and output signals. The square block indicates that two *pairs* of connections are needed. Note that one input terminal and one output terminal are tied together for a return path to the common terminal 3. The ground connection here does not necessarily mean earth ground but is just a common return. The block diagram is a simplified method of indicating an amplifier circuit.

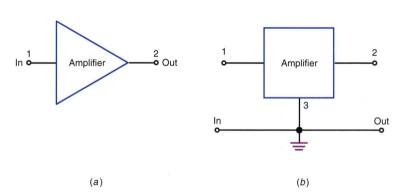

(a) *(b)*

FIG. 29-3 Block symbols for amplifier circuit. (*a*) Triangle as general symbol with input and output terminals. (*b*) Square block with details for two pairs of connections. Note common return for low side of input and output circuits.

VOLTAGE GAIN In terms of signal voltage, the gain with the symbol A_V is

$$A_V = \frac{\text{output signal voltage}}{\text{input signal voltage}} \tag{29-1}$$

The output and input signal voltages can be in any units but they must be the same for the ratio. Peak-to-peak (p–p) values are usually best, in case the signal waveform is not symmetrical. Typical values of voltage gain for transistor amplifier circuits are about 10 to 2000.

An example for a voltage gain of 40 is illustrated in Fig. 29-4. Here, the input signal equals 0.1 V, p–p. This signal is amplified to the value of 4 V p–p in the output. To calculate the voltage gain

$$A_V = \frac{4 \text{ V}}{0.1 \text{ V}} = 40$$

There are no units for A_V because it is a ratio of the same units, which cancel.

To do this problem on a calculator, just punch in 4 on the numerical keyband, press the division key ⊝ at the side, enter the .1, and press the ⊜ key for the answer of 40.

This example is for an audio circuit amplifying a signal with the audio frequency of 400 Hz, as noted in the diagram in Fig. 29-4. Note that the frequency of the output signal voltage is the same as the input signal. The 400 Hz is only one example. Almost any frequency can be amplified.

For the output signal voltage in Fig. 29-4, note that the phase is shown opposite from the input signal, meaning a phase inversion of 180°. The amplifier can then be considered as an *inverter* circuit also, because of the reversed polarity. The phase inversion does not affect the gain, however. Many amplifiers can invert the polarity of the signal, depending on the type of output circuit.

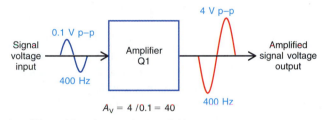

4 V p–p

0.1 V p–p

Signal
voltage
input

Amplifier
Q1

Amplified
signal voltage
output

400 Hz

400 Hz

$A_V = 4\,/0.1 = 40$

FIG. 29-4 Amplifier with voltage gain A_V of 40.

CURRENT GAIN With transistor amplifiers, the gain in current for the output signal compared with the input signal is probably more important than the voltage gain. The reason is that the amount of ac signal voltage for the input circuit is limited to ±0.1 V, approximately, but the input current can be in microamperes, milliamperes, or even as high as an ampere. This limitation applies to junction transistors such as the PNP and NPN types.

The amount of current gain, with the symbol A_I, is

▶ $$A_I = \frac{\text{output signal current}}{\text{input signal current}} \qquad\qquad \textbf{(29-2)}$$

For the example in Fig. 29-5 the input signal current of 200 μA is equal to 0.2 mA. The output is 6 mA. Then

$$A_I = \frac{\text{output } I}{\text{input } I}$$

$$A_I = \frac{6 \text{ mA}}{0.2 \text{ mA}} = 30$$

There are no units for A_I because it is a ratio of the same two units of current.

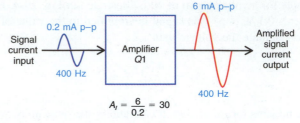

$$A_I = \frac{6}{0.2} = 30$$

FIG. 29-5 Amplifier with current gain A_I of 30.

Note that no phase inversion occurs with the signal current. Only the amplified signal voltage can be inverted in the output circuit. Typical values of current gain with transistors are about 1 to 500.

POWER GAIN   This quantity, with the symbol A_P, is the product of the voltage gain times the current gain, or

$$A_P = A_V \times A_I \qquad\qquad (29\text{-}3)$$

For the examples of A_V equal to 40 and A_I of 30, the power gain is

$$A_P = 40 \times 30 = 1200$$

There are still no units for A_P because it is the product of two gain values without any units. A high value of power gain for an amplifier means it can drive a load that requires appreciable current and voltage. It may be noted that discrete transistors generally have higher power ratings than IC chips for applications that require appreciable power gain.

OVERALL GAIN FOR AMPLIFIERS IN CASCADE   Most applications require more than one amplifier stage in order to provide enough gain. The reason is that the original signal to be amplified usually has low amplitude and a strong signal is needed for the desired output. As an example for audio signals, the output from a record player or magnetic tape may be just a few millivolts. However, a loudspeaker requires much more signal. Sufficient amplification is needed in order to hear the sound.

Actually, it is practical to have almost any amount of gain with transistors and IC chips because these semiconductor devices are so small. One IC chip can easily have two, three, or more transistor amplifiers.

A specific example is shown in Fig. 29-6 for the two amplifier stages $Q1$ and $Q2$ connected in cascade. Each amplifier circuit with one transistor is called

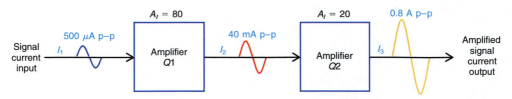

Cascaded $A_I = 80 \times 20 = 1600$

FIG. 29-6 Amplifiers in cascade to multiply the gain. Here, $A_I = 80 \times 20 = 1600$.

a *stage*. When the output terminal of one stage drives the input terminal of the next stage, the two stages are connected in *cascade*. Just to indicate how the circuit might be different, it should be noted that transistors can be connected in series or in parallel. When the amplifier stages are in cascade, the total gain equals the product of the individual gains for each stage.

Figure 29-6 shows cascaded values of gain to illustrate a practical example with transistors such as the PNP or NPN types. In the second amplifier $Q2$, this transistor needs signal input variations with an amplitude of 40 mA p–p in order to provide output of 0.8 A p–p. This current gain of $Q2$ is 0.8 A or 800 mA divided by 40 mA, for A_I of 20. However, the variation of current in the original signal has a magnitude of only 500 μA or 0.0005 A p–p. Therefore, this signal is fed into amplifier $Q1$, with a current gain of 80, to provide enough signal current for the input to $Q2$. The calculations for the current amplitudes in Fig. 29-6 are as follows:

$$I_2 = 80 \times I_1$$
$$= 80 \times 500 \ \mu\text{A}$$
$$I_2 = 40{,}000 \ \mu\text{A} = 40 \ \text{mA}$$
$$I_3 = 20 \times I_2$$
$$= 20 \times 40 \ \text{mA}$$
$$I_3 = 800 \ \text{mA} = 0.8 \ \text{A}$$

The overall gain for the two stages in cascade in Fig. 29-6 is $80 \times 20 = 1600$ for the total A_I. This value can also be calculated as the ratio of the 0.8 A output to the 500 μA input. Then converting 0.8 A to 800,000 μA gives

$$A_I = \frac{800{,}000 \ \mu\text{A}}{500 \ \mu\text{A}} = 1600$$

The multiplication of gain values for stages in cascade also applies to A_V and A_P.

TEST-POINT QUESTION 29-2

Answers at end of chapter.
a. Calculate the gain A_V for output of 10 V with input of 50 mV.
b. Calculate the gain A_I for output of 1 mA with input of 20 μA.
c. For a transistor amplifier stage, A_V equals 200 and A_I is 50. Calculate its power gain.
d. Three stages in cascade each have A_I of 30. Calculate the total current gain.

29-3 CHARACTERISTICS OF AMPLIFIER CIRCUITS

There must be hundreds of different kinds of amplifiers but a few basic features can reduce the problem of analysis. Referring to the amplifier circuit in Fig. 29-7, note the following:

1. The transistor $Q1$ is the amplifying device.

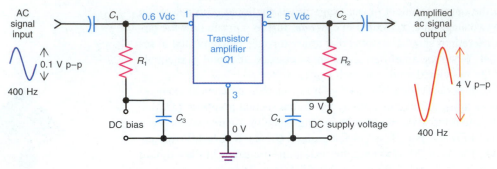

FIG. 29-7 Schematic diagram of an amplifier with a transistor as active device.

2. Resistors and capacitors are used. Inductors can also be included, although they are not shown here.

3. The frequency of 400 Hz for the ac signal indicates this circuit is for audio amplification.

4. Both dc and ac voltages are shown. The dc voltages are needed for electrode voltages on $Q1$, to make it conduct current. The ac voltages apply to the desired signal input and output, generally in peak-to-peak values.

These characteristics can be applied to all types of amplifier circuits.

The transistor is called an *active device* because it can amplify the signal. Diodes also are active devices since conduction depends on the polarity of the input voltage. The R, L, and C components are considered *passive devices*.

R, C, AND L COMPONENTS IN AN AMPLIFIER   Each of these passive components has its own characteristics, whether in an electric circuit or in an amplifier circuit. However, the special effects for dc and ac voltages become more important because an amplifier circuit has both.

Resistance reduces the amount of current. The R provides a voltage drop equal to IR. In terms of dc voltage, the voltage drop reduces the amount of voltage for $Q1$. Referring to Fig. 29-7, note that the dc voltage at terminal 2 of $Q1$ is 5 V, from the 9-Vdc supply, because of an IR drop of 4 V across R_2. The voltage drop across R_2 is even more important for ac signal voltage. Actually the ac voltage across R_2 in the amplifier output circuit provides the desired signal output. In Fig. 29-7, R_2 is the output load resistor for signal voltage from terminal 2 of $Q1$ to the next stage. Also, for the input circuit, R_1 is the load resistor that develops ac signal input to terminal 1 of $Q1$. It should be noted that a resistor has the same R for a steady current or a varying current. Also, R is constant for all frequencies in an ac signal.

Because a capacitor can charge and discharge with changes in voltage, it has capacitive reactance X_C. For steady dc voltage, however, a capacitor is practically an open circuit. Furthermore, X_C decreases for higher frequencies. Also, X_C is less with a higher value of C. The amount of X_C can easily be made less than 10 Ω, for a circuit path with very little opposition to the current. As a result, capacitors are often used in electronic circuits for the following important functions:

1. They block dc voltage but pass the ac signal. In this application, C is a coupling capacitor. In Fig. 29-7, C_1 and C_2 are coupling capacitors. C_2 passes

the ac signal output from terminal 2 of $Q1$ to the next stage, but blocks the dc voltage in the output circuit. For the input circuit, C_1 couples the ac signal input to terminal 1 of $Q1$ but blocks any dc voltage from the previous stage.

2. The C provides lower X_C for high frequencies compared with lower frequencies. This factor means more ac signal voltage can be passed for the higher frequencies. As an example, a division between radio frequencies for high values and audio frequencies for lower values is provided.

3. The same frequency characteristics of X_C can be applied to bypass capacitors as well as coupling capacitors. The only difference is that a coupling capacitor is a series component in the high side of the signal circuit to pass along the desired information, while a bypass capacitor is a parallel component to shunt the ac signal away from a part of the circuit where the signal is not desired. Usually, the shunt path of a bypass capacitor connects the ac signal to the common terminal of the return path for the low side of the circuit. In Fig. 29-7, C_4 at the low side of R_2 is a bypass capacitor to keep the ac signal from producing current in the source of the 9-Vdc supply voltage. This dc voltage is needed to produce current in the transistor. Similarly, C_3 at the low side of R_1 for the input current is a bypass capacitor for the 0.6-Vdc bias. The dc bias for the input circuit is needed to provide the desired operating characteristics for the transistor amplifier $Q1$.

In summary, the three main functions of capacitors in amplifier circuits are to block the dc supply voltage, couple the ac signal to the points in the circuit where the signal is needed, and bypass the ac signal around the components where the signal should not be. As typical values, 5 μF is commonly used for a coupling or bypass capacitor in audio amplifiers, while 100 pF is a typical size for RF circuits with frequencies of 0.5 to 10 MHz, approximately. Figure 29-8 shows how the electronic components look on a printed-circuit board.

Since an inductor is just a coil of wire, it allows direct current with dc voltage applied. The dc value depends on the resistance of the wire in the coil. However, because of the magnetic field around the coil with current, the inductor also has inductive reactance X_L for alternating current. The amount of X_L increases with higher frequencies and more L. These reactance effects for X_L are

FIG. 29-8 Components for electronic circuits mounted on a printed-circuit board.

the opposite of X_C. Inductors are used where it is desired to have high impedance for alternating current, with less opposition for lower frequencies down to 0 Hz for direct current. A common application is an RF or AF choke to isolate an ac component from the signal variations or to provide an ac load impedance that still allows direct current to flow. It should be noted that no L is shown in Fig. 29-7 because this circuit does not use any chokes.

Keep in mind that both L and C can be used for resonant circuits, particularly for RF signals, as in a tuned RF amplifier. The higher the resonant frequency the smaller the values of L and C required.

DC AND AC VOLTAGES IN AMPLIFIER CIRCUITS An important practical feature of amplifiers is that they generally use ac and dc voltages in the same circuit. This characteristic makes the circuits interesting but the combined effect may not be so easy to visualize. The reason for having the ac–dc combination is very specific. An active device such as a transistor needs operating voltages applied to its electrodes in order to operate at all. The operation means it must be conducting current. Furthermore, the dc voltage must have a specific polarity. For instance, with an NPN transistor the dc supply voltage must be positive for the output electrode. A PNP transistor takes negative dc supply voltage. In addition, as far as amplification is concerned, the desired variations are usually in the form of an ac signal.

The way that an amplifier circuit diagram looks in terms of dc electrode voltages is illustrated in Fig. 29-9a. The dc supply voltage is 9 V, and terminal 2 has 5 V because of the 4-V drop across R_2. The input terminal has 0.6 V for dc bias to provide the desired operating characteristic for the amplifier $Q1$. Without any ac signal input at all, the amplifier $Q1$ is conducting direct current. In other words, the amplifier $Q1$ is ready for amplification.

It should be noted that any path for dc operating voltage cannot have any series capacitor, since capacitors block the dc voltage. However, series chokes can be used because they pass direct current. A practical example of the need for dc voltage is the fact that in a small portable transistor radio, the 9-V battery is the dc supply for all the circuits.

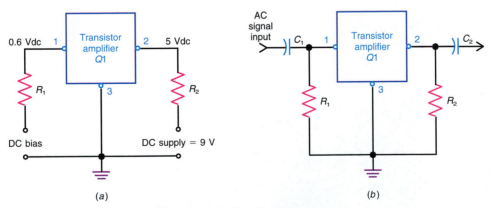

(a)   (b)

FIG. 29-9 The dc and ac voltages for the amplifier circuit in Fig. 29-7, shown separately in two diagrams. (a) The dc supply voltages without any ac signal. (b) Equivalent circuit for the ac signal without the dc supply voltages. The bypass C_3 and C_4 in Fig. 29-7 provide ac ground returns (not shown in diagram).

The circuit in Fig. 29-9*b* shows how the circuit looks for ac signal only, without any dc voltages indicated. The common ground return at terminal 3 of Q1 is only an equivalent ground for the ac signal. The dc voltages are really necessary but the low side of the actual circuit has bypass capacitors to ground. The bypass capacitors are chosen with enough *C* to have such low reactance at the operating frequency that they can be considered as a short circuit for the ac signal. In summary, the bypassing provides an effective ac ground for the signal without affecting the dc voltage. If ac signal were applied to the circuit without its dc operating voltages, the signal could not be amplified.

The complete amplifier circuit in Fig. 29-7 has ac signal input and the required dc operating voltages. The 0.1 V signal applied to the input is amplified up to 4 V for the output signal. However, the amplified output at terminal 2 is actually a fluctuating dc voltage, as shown in Fig. 29-10. The waveform has an ac component of 4 V p–p riding on the average dc level of 5 V. For instantaneous values, the output voltage is varying from 5 V down to 3 V and then from 5 V up to 7 V. These signal variations of ±2 V, or 4 V p–p, are much larger than the input signal of 0.1 V p–p. This amplification by transistor Q1 is accomplished by allowing the small signal at the input to control the output circuit with its dc supply, where current and voltage variations can be produced with much greater amplitude.

Consider the fluctuating dc voltage in Fig. 29-10 as the output at terminal 2 of the amplifier in Fig. 29-7. Then the dc and ac components are separated as follows:

1. The coupling capacitor C_2 blocks the 5-Vdc level by developing this voltage across the two terminals of the capacitor.

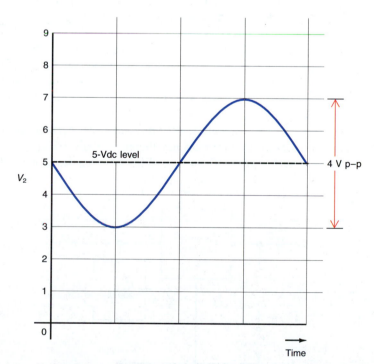

FIG. 29-10 Graph of fluctuating dc voltage with ac component at output terminal 2 of amplifier in Fig. 29-7.

2. The 4-Vac p–p signal is coupled to the next amplifier as the ac voltage across its input circuit.

SIGNAL FREQUENCIES IN AMPLIFIERS

An amplifier can be used for audio frequencies or radio frequencies and sometimes for both in wideband amplifiers. The range of audio frequencies is 20 to 20,000 Hz but 50 to 15,000 Hz is generally the frequency range for high-fidelity audio equipment. A narrower AF range can be used and still be suitable. In telephone applications, the range of audio frequencies is usually 100 to 3000 Hz.

Radio frequencies extend from 30 kHz up to the megahertz (10^6 Hz) and gigahertz (10^9 Hz) ranges. The values between 20 kHz at the top of the AF range and 30 kHz for RF can be considered ultrasonic audio frequencies.

The RF range includes the following four bands that are used for many RF applications, such as radio and television:

Medium frequencies (MF), 0.3 to 3 MHz
High frequencies (HF), 3 to 30 MHz
Very high frequencies (VHF), 30 to 300 MHz
Ultra-high frequencies (UHF), 300 to 3000 MHz

As examples, the AM radio broadcast service is 535 to 1605 kHz in the HF band; the FM commercial radio service is 88 to 108 MHz in the VHF band; television broadcast stations use 6 MHz channels in the VHF and UHF bands.

Amplifiers for AF signals are untuned, using a resistance load that is constant for all audio frequencies. The circuit shown before in Fig. 29-7 is for an audio amplifier. RF amplifiers are usually tuned with *LC* circuits resonant at the desired frequency. A wideband amplifier is a special case for amplifying audio and radio frequencies. An example is the video amplifier used in television for the broad range of video frequencies from 30 Hz up to 4 MHz.

An example of a tuned RF amplifier is shown in Fig. 29-11. This circuit is similar to the basic amplifier in Fig. 29-7, except that the parallel resonant circuit with L_1 and C_1 is used for the output load instead of a resistor. No bypass capacitors for the dc voltages are shown in Fig. 29-11 because this bypassing is not always necessary.

The values for L_1 and C_1 are chosen for resonance at 1 MHz in this example. The amplifier is tuned to this frequency therefore. Only 1 MHz and a narrow band of frequencies around it can be amplified. Other frequencies do not

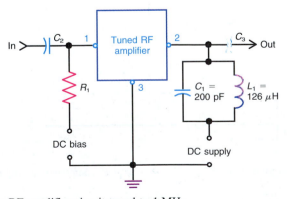

FIG. 29-11   An RF amplifier circuit tuned to 1 MHz.

have any gain. The reason for gain at 1 MHz is that the tuned circuit has high impedance for parallel resonance. A high impedance as the output load is necessary for the amplifier to provide high voltage gain. Frequencies off resonance have low impedance and little or no gain. The tuned RF amplifier then is able to select the frequency to be amplified. The amount of gain in a tuned RF amplifier is its *sensitivity*. How narrow the response is in terms of the band of frequencies that are amplified is the *selectivity*.

Answer True or False.
a. Coupling and bypass capacitors have low reactance at the frequency of the ac signal.
b. A resistor has the same R for dc or ac voltages.
c. The dc voltage in Fig. 29-10 has an ac component of 4 V p–p.
d. Audio amplifiers usually have resonant LC circuits for the load impedance.

29-4 OSCILLATORS

The process of oscillation means that variations in amplitude are repeated continuously at a specific frequency. A mechanical example is a swinging pendulum. In electronic circuits, an oscillator generates ac signal output without any ac signal input from an external source. Essentially, the oscillator is an ac generator for audio or radio frequencies, with many useful applications. This important function is accomplished by using an amplifier in a circuit where part of the output is fed back to the input. The feedback then corresponds to input signal. An example is shown in Fig. 29-12.

The oscillator output cannot be generated without using energy. This energy comes from the dc voltage for the circuit. In effect, the oscillator circuit converts the energy of the dc power supply into ac signal variations.

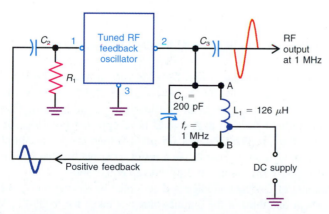

FIG. 29-12   An RF feedback oscillator circuit generating output at 1 MHz.

POSITIVE FEEDBACK   The positive polarity means that the ac signal for oscillator feedback must be in the same phase that an ac input signal would have for amplification. Usually, the positive feedback results from two phase reversals of 180° each. The amplifier itself has one phase inversion of 180°. Then another 180° phase reversal is provided by the circuit that feeds signal from the output back to the input.

TUNED RF FEEDBACK OSCILLATORS   This name is given to the type of circuit in Fig. 29-12. It is tuned to the radio frequency of 1 MHz in this example by the resonant circuit with C_1 and L_1. Note that C_1 is shown variable to serve as a tuning capacitor that sets the oscillator frequency. Actually, the oscillating part of an oscillator circuit is the L_1C_1 circuit. Any tuned circuit can produce sine-wave oscillations of current and voltage at its natural resonant frequency f_r. The oscillator ac output signal is at the f_r of the tuned circuit. The only function of the amplifier in an oscillator circuit is to provide feedback that supplies energy to prevent the oscillations from decaying to zero. The tuned circuit is often called a *tank circuit* because it stores energy.

The functions of the components in the oscillator circuit of Fig. 29-12 can be summarized as follows:

C_2 is a coupling capacitor for the feedback signal into terminal 1.
R_1 provides the ac feedback signal coupled by C_2.
C_3 is a coupling capacitor for the output signal to the next stage.
C_1 is the tuning capacitor to set the frequency of the oscillator output.
L_1 is the inductance for the tuned circuit.

The resonant frequency of the tuned circuit can be calculated as

$$f_r = \frac{1}{2\pi\sqrt{LC}}$$

For the values of 126 μH and 200 pF in Fig. 29-12,

$$f_r = \frac{1}{2\pi\sqrt{126 \times 10^{-6} \times 200 \times 10^{-12}}} = \frac{0.159}{\sqrt{126 \times 200}} \times 10^9$$

$$= \frac{0.159}{1590} \times 10^9$$

$$f_r = 0.001 \times 10^9 = 1 \times 10^6 = 1 \text{ MHz}$$

How to do a problem like this with a calculator is explained in Chap. 26 on resonant circuits.

The RF oscillator circuit in Fig. 29-12 is very similar to the RF amplifier in Fig. 29-11, except for the feedback. Note that the coil L_1 is tapped for the connection to the dc supply. This point is effectively an ac ground. Therefore, the two opposite ends of the coil at A and B have opposite polarities of ac signal, with respect to the tap. The feedback for point B then is opposite in phase to the amplified output signal for point A. As a result, positive feedback is provided because the two phase inversions of 180° add to equal 360°.

The amount of feedback voltage is determined by the tap on L_1 in Fig. 29-12. Generally, about one-third of the output voltage is taken for feedback. Most of the oscillator voltage is used for the ac output coupled by C_3 to the next stage.

Note that no dc bias source is shown for the input circuit at terminal 1 where the feedback signal is applied. The reason is that an oscillator circuit usually makes its own bias by rectifying the feedback signal at the input. In this way, the bias regulates itself since the amount of bias depends on the amount of feedback. In fact, a practical method of testing whether an oscillator circuit is oscillating is to measure its dc bias. No dc bias means no ac output from the oscillator.

HARTLEY AND COLPITTS OSCILLATORS   These are named for the inventors of the two main types of circuits for an RF feedback oscillator. In the Hartley circuit, the feedback is provided by a tapped coil as in Fig. 29-12. Then the coil serves as an ac voltage divider for the output voltage and feedback signal. In a Colpitts circuit, similar results are obtained with a capacitive voltage divider, as shown in Fig. 29-13. Here the voltage across C_2 is the feedback. The coil L_T is made variable for tuning the oscillator. There are many modifications in the circuits, but all RF feedback oscillators can fit into these two classifications.

CRYSTAL OSCILLATORS   In this type, a piezoelectric crystal is used as a resonant circuit, replacing an LC circuit. The crystal is a thin slice of natural quartz or a synthetic material. An example is shown in Fig. 29-14. The piezoelectric effect means the crystal can vibrate mechanically when excited electrically and produce ac voltage output. The resonant frequency is fixed by the size of the crystal. Typical values of f_r are 0.5 to 30 MHz. The advantages of the crystal is its very high Q as a resonant circuit, which results in good frequency stability for the oscillator.

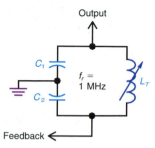

FIG. 29-13   Tuned circuit with capacitive voltage divider to provide feedback for Colpitts oscillator circuit.

FIG. 29-14   A 3.58-MHz crystal in its housing, commonly used for the color oscillator in television receivers.

RC **FEEDBACK OSCILLATORS**   This type of circuit, shown in Fig. 29-15, is used for audio oscillators. At audio frequencies, the LC values for a tuned circuit would be too large. In Fig. 29-15 on the next page, three RC networks are used to provide feedback. Each can provide a phase shift of 60°. Then $3 \times 60° = 180°$, which when added to the 180° for phase inversion in $Q1$ is able to provide positive feedback. Resistor R_3 is made variable to tune the oscillator for the required frequency. The circuit oscillates at the frequency that provides the required phase shift for positive feedback. Typical frequencies for the oscillator output are 20 Hz to 200 kHz.

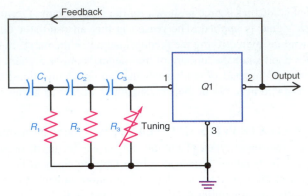

FIG. 29-15   The ac equivalent circuit of an *RC* feedback oscillator for AF, with three *RC* networks in feedback circuit. The dc supply voltage is not shown.

TEST-POINT QUESTION 29-4

Answers at end of chapter.
a. Is a tapped coil for feedback used in the Hartley or the Colpitts oscillator?
b. Does an oscillator circuit use positive or negative feedback?
c. For the circuit in Fig. 29-12, if C_1 is reduced to 50 pF, what will the oscillator frequency be?

29-5 MULTIVIBRATORS

The multivibrator (MV) circuit is in a class by itself as an oscillator because it is so important as a pulse generator in digital electronics. In this application, the MV serves as a *reference clock* to synchronize the timing in a digital system for the switching of pulses. Instead of generating sine waves, the MV produces oscillations between the HIGH and LOW voltage levels at the output electrode. One cycle includes the time for a HIGH and a LOW. When each level takes the same time, the output is a symmetrical square wave. With unequal times, the circuit produces unsymmetrical pulses. The voltage levels oscillate between the HIGH and LOW levels because of the changes between conduction and cutoff in the MV circuit. This type is sometimes called a *relaxation oscillator* because of the periods of cutoff.

Figure 29-16 shows the block diagram of an ac equivalent circuit, without dc supply voltages, for a free-running MV circuit. This MV operates as an oscillator without the need for any input signal. The same circuit with more details is shown in Fig. 29-17. Referring to the block diagram in Fig. 29-16, note the following:

1. This MV circuit consists of two cross-coupled amplifier stages $Q1$ and $Q2$.
2. The MV oscillator is a pulse generator.
3. Operation can be over a wide range of audio or radio frequencies. The output is shown here at 1 MHz only as an example. The frequency is determined by the time constants in the *RC* coupling circuits.

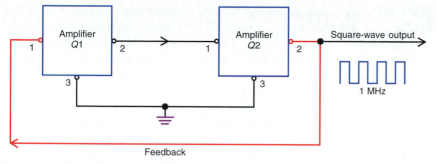

FIG. 29-16 Block diagram of MV circuit with two cross-coupled amplifier stages. The dc supply voltage is not shown.

The multivibrator is a very compact circuit, using transistors, diodes, resistors, and capacitors but without the need for any coils or transformers for feedback.

The cross-coupling means that the output of each stage drives the input of the other. Since each stage produces a phase inversion of 180°, the two inversions result in the positive feedback needed for oscillations.

The oscillations are in the conducting and cutoff conditions for Q1 and Q2 in Fig. 29-16. Conduction in a stage means it is turned ON with driving voltage at the input; the OFF means that the stage is not conducting because of cutoff voltage at the input. When one stage is ON, the other is OFF. In fact, it is the conduction in one stage that makes the other stage cut off because of the feedback. The ON–OFF times depend on the stage that is cut off. How long it is OFF is controlled by the *RC* time constant in the input circuit. When it starts to conduct, the drop in its output voltage cuts off the other stage. The action is almost instantaneous, resulting in sharp changes in voltage for the output pulses.

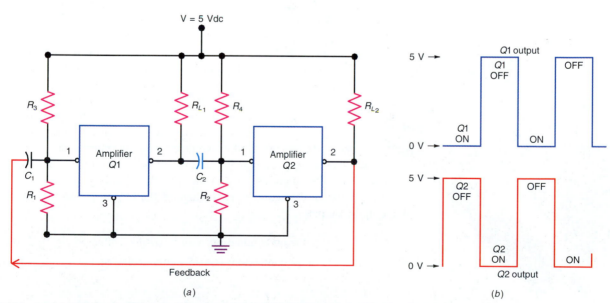

FIG. 29-17 (*a*) Details of a free-running MV circuit with dc operating voltage. (*b*) Waveform at output terminal 2 for *Q*1 above and *Q*2 below.

MV CIRCUIT More details can be seen from the schematic diagram in Fig. 29-17a. Note that the output electrode 2 of $Q1$ drives the input terminal of $Q2$ through the R_2C_2 coupling circuit. Also, in the feedback path shown at the bottom of the diagram, output terminal 2 of $Q2$ drives input terminal 1 of $Q1$ through the R_1C_1 coupling circuit. For the other components in the circuit, R_{L_1} is the output load resistor for $Q1$ while R_{L_2} has the same function for $Q2$. The output pulses are obtained from terminal 2 of either $Q1$ or $Q2$ but with opposite polarities, as shown in Fig. 29-17b. Resistors R_3 and R_4 provide dc bias for the input circuit of $Q1$ and $Q2$.

The drop in voltage at terminal 2 for the output from either $Q1$ or $Q2$ depends on conduction of current. When either stage is cut off, the output voltage at terminal 2 equals the dc supply of 5 V. The reason is that without any current, no IR drop is produced across the output load resistor. When the stage conducts, the result is an IR drop of 5 V across the load resistor, resulting in zero volts at terminal 2. Then the output drops from 5 V to 0 V.

For either $Q1$ or $Q2$, the input voltage cuts off the stage because of the drop in voltage at the output of the previous stage. How long the cutoff stage remains nonconducting depends on the RC time constant of the input coupling circuit. When the cutoff stage starts conducting, it cuts off the other stage.

When the two stages have the same values of components, they are cut off for the same amount of time. The output voltage then is a symmetrical square wave, as shown in Fig. 29-17b. One cycle includes the period of cutoff and conduction for either stage. Output can be obtained from terminal 2 of either $Q1$ or $Q2$, but with opposite polarities. The reason is that one stage has HIGH output voltage when the other is LOW. When the two stages are not symmetrical, one is HIGH for a longer time than the other. Then the output consists of unsymmetrical pulses.

To summarize the action of each stage of the MV in producing pulses of voltage in the output:

OFF = no conduction = HIGH output voltage

ON = conduction = LOW output voltage

TYPES OF MULTIVIBRATORS The circuits in Figs. 29-16 and 29-17 show the type called an *astable MV*, meaning it is not stable in terms of the ON and

OFF states for either stage. This circuit is a free-running oscillator. It does not need any input signal.

Another type is the *bistable MV*. By the choice of component values, the MV circuit can be made to remain stable with either stage OFF and the other ON. It has two stable states. For instance $Q1$ can remain OFF with $Q2$ conducting, or for the opposite condition, $Q2$ can be OFF with $Q1$ conducting. The circuit stays in one of these states until an input pulse is applied to the OFF stage to make it conduct. This function of forcing the stage into conduction is called *triggering*. Then the ON stage drives the other stage into the OFF condition. The circuit then stays in this condition until another input pulse reverses the states. The name *flip-flop* is used for the bistable MV circuit to describe this idea of switching the ON–OFF states one way and then the opposite way by means of input trigger pulses.

A third type is the *monostable* or *one-shot MV*. This circuit has only one stable state. An input pulse is needed to trigger the OFF stage into conduction. Then the MV goes through one cycle of changes and back to its original condition. The original OFF stage is again in the nonconducting condition, ready for another trigger pulse.

TEST-POINT QUESTION 29-5

Answers at end of chapter.

Answer True or False.
a. An astable MV circuit is a free-running oscillator.
b. In Figs. 29-16 and 29-17, the amplifiers $Q1$ and $Q2$ must be conducting at the same time.
c. The flip-flop circuit is a bistable MV.
d. The output electrode has high output voltage from the dc supply when the stage is not conducting.

29-6 MODULATION

The process of modulation in electronic circuits can be defined as modifying the characteristics of one waveform with the variations in another signal. The purpose is to transmit the information in a desired signal as the variations of another waveform that is better for transmission. Common examples are amplitude modulation and frequency modulation used in radio and television broadcasting. In these applications, a basic RF waveform serves as the *carrier wave* for the modulating information. The frequency of the carrier wave must be much higher than the modulating frequencies for minimum distortion in the modulating process. In AM radio broadcasting, AF signals modulate an RF carrier wave.

The lower-frequency signal for the modulation is called the *baseband signal*. In AM and FM radio broadcasting, the baseband modulation is an audio signal. For television, a video signal is used as the baseband modulation.

The baseband modulation can vary three characteristics of the carrier wave. These are peak-to-peak amplitude, instantaneous frequency, and phase angle.

AMPLITUDE MODULATION (AM)   Figure 29-18 illustrates this process. The carrier input to the modulator comes from an RF oscillator. For modulation, the baseband signal is from an audio amplifier. The modulator stage can be a diode or a nonlinear amplifier. Nonlinear amplification means that the output amplitudes are not exactly proportional to the input signal. Otherwise the two input signals would just be mixed in the output. In Fig. 29-18, the modulator stage amplifies the 1-MHz RF input, but the amount of gain is controlled by the audio modulating signal. Then the amplitudes of the RF output cycles vary in step with the variations in the audio modulating signal. The result is the amplitude-modulated signal shown for the output.

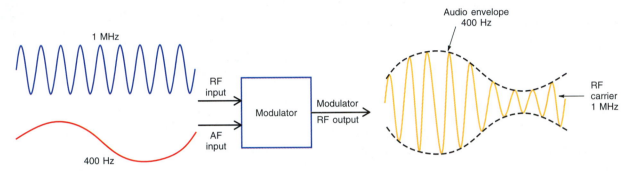

FIG. 29-18   Illustrating the process of amplitude modulation (AM). The 1-MHz RF carrier wave is modulated by the 400-Hz audio signal.

In the AM output signal, the RF peak-to-peak amplitudes have variations that correspond to the audio modulation. As shown in Fig. 29-18, the outline of varying amplitudes is the *modulation envelope.* The envelope is the same top and bottom or positive and negative, since the RF amplitude variations are symmetrical around the zero axis. Note that the envelope is shown as a dotted waveform because it is not actually a separate signal in itself. The reason is that the AM carrier signal has equal audio variations in opposite directions. The modulated signal must be detected by rectification to eliminate the symmetry and filtered in order to recover the audio modulation.

Important applications of amplitude modulation are the standard AM radio broadcast service and the picture signal for television. The AM radio service uses RF carrier frequencies spaced every 10 kHz in the band of 535 to 1605 kHz. Examples of the carrier frequencies for AM radio broadcast stations are 540, 880, 1010, and 1560 kHz. The 10-kHz spacing is needed for the bandwidth of ±5 kHz with an AF baseband signal of 50 to 5000 Hz.

In the television service, 6-MHz channels are used for broadcasting, starting with 54 to 60 MHz for channel 2. The 6-MHz bandwidth is needed for a video baseband signal of 0 to 4 MHz. The sound associated with the picture is broadcast as a separate FM signal.

SIDEBANDS   A modulated signal needs more bandwidth than the carrier wave itself, in order to allow for the variations produced by the modulation. The necessary bandwidth is at least equal to the frequencies in the baseband signal, in a band just above and below the carrier frequency. As an example, for a 1000-kHz carrier modulated by 5 kHz, the extra bandwidth is ±5 kHz for a total of 10 kHz.

Then the modulated signal includes frequencies from 995 to 1005 kHz. The 10-kHz range of frequencies around 1000 kHz then become the sidebands of the 1000-kHz carrier wave. This example is illustrated in Fig. 29-19.

PERCENT MODULATION This figure measures how much the carrier is changed by the baseband signal. In Fig. 29-18, the amplitudes are shown with approximately 50 percent modulation. With 100 percent modulation, the RF amplitudes go up to double the unmodulated level and down to zero. More than 100 percent modulation cannot be used in an AM signal because then part of the baseband signal would be missing while the carrier amplitude is zero.

FREQUENCY MODULATION (FM) In this method, the instantaneous *frequency* of the carrier wave is made to vary in step with the variation of *voltage* in the baseband signal. There are many types of FM modulator circuits for this process. An example of an FM signal is shown in Fig. 29-20. Here the RF carrier frequency is 1 MHz or 1000 kHz. This value is the *center frequency*. Because of the modulation, though, the instantaneous frequency is made to deviate 75 kHz by the audio baseband signal. This change from center is the *frequency deviation*. In this example, the deviation is 75 kHz and the total swing is ±75 kHz or 150 kHz.

The peak values of the audio modulation produce the maximum frequency deviation of ±75 kHz. Smaller AF values produce less frequency deviation in the RF carrier wave. When the audio modulating voltage is at its zero value, the FM signal is at its center frequency. In summary, the amplitude values in the audio baseband signal are indicated by the instantaneous frequency values of the FM carrier signal. The information about the frequencies in the baseband signal is indicated by the rate at which the frequency swings are produced in the FM signal.

The frequency deviation of 75 kHz is chosen in Fig. 29-20 to illustrate the value that is the maximum allowed for 100 percent modulation in the commercial FM radio broadcast band. In an FM signal, the percentage of modulation is

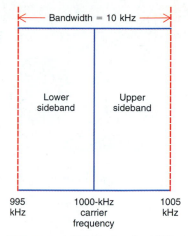

FIG. 29-19 Sidebands of a 1000-kHz carrier wave.

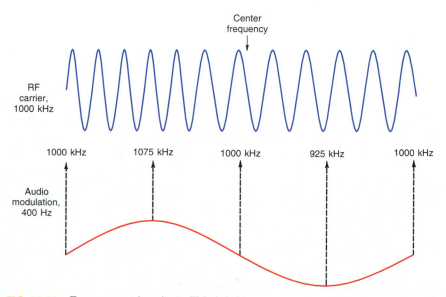

FIG. 29-20 Frequency swings in an FM signal.

in the amount of frequency change for the carrier wave. The 75-kHz deviation is for the loudest audio signal with the greatest amplitude in the baseband modulating signal. Lower values of audio amplitude result in less frequency deviation.

The value of 75 kHz for 100 percent modulation applies only to the FM radio broadcast band. In the FM sound signal in television broadcasting, the value is 25-kHz maximum deviation for 100 percent modulation. The FM range for the baseband signal is 50 to 15,000 Hz for FM radio and TV sound. It should be noted that in narrowband FM systems for communications radio, 5 kHz can be used for the maximum deviation, with a baseband audio signal of 300 to 3000 Hz.

The FM radio broadcast band is 88 to 108 MHz, with stations spaced every 200 kHz or 0.2 MHz for the center frequency of the transmitted carrier. Examples of carrier frequencies for FM broadcast stations are 92.1, 96.3, and 104.5 MHz. The 200-kHz spacing between carrier frequencies is needed to allow for a total frequency swing of 150 kHz, with a guard band of 25 kHz on each side to prevent interference between adjacent stations.

PHASE MODULATION In this method, the instantaneous phase angle of the RF carrier wave is made to vary in step with the modulating voltage. Actually, the PM is similar to FM, since any change in frequency affects the phase. Therefore, phase modulation produces *equivalent FM* or *indirect FM*. However, one important factor is that a change in phase angle produces a larger change in the RF carrier frequency for higher audio modulating frequencies. This relation can be corrected, though, by a predistortion network for the audio modulation. Many FM transmitters use a phase-modulator circuit with a crystal-controlled oscillator for good frequency stability of the center frequency for the carrier wave.

PULSE MODULATION This method is necessary with the pulses representing digital information. Typical systems of pulse modulation are:

PAM or pulse-amplitude modulation
PFM or pulse-frequency modulation
PWM or pulse-width modulation
PCM or pulse-code modulation

Pulse modulation is efficient because the carrier power is on for only the time of the pulses. However, greater bandwidth may be needed for the harmonic frequency components of sharp pulses.

TEST-POINT QUESTION 29-6

Answers at end of chapter.
a. In Fig. 29-20, is 400 Hz the baseband signal or the carrier frequency?
b. Is the carrier frequency of 880 kHz in the AM or FM radio band?
c. In an FM signal, the carrier frequency deviates from a center frequency of 96,000 kHz to 96,050 kHz. Give the frequency deviation.
d. Do the sidebands of a modulated signal increase or decrease the necessary bandwidth?
e. Is phase modulation similar to AM or FM?

29-7 DIODE RECTIFIERS

In Fig. 29-21, the ac input to the anode (terminal 1) of a diode has positive and negative half-cycles. However, current can flow in a diode only during the time when the anode is positive. Therefore, only the positive half-cycles of the input at terminal 1 can produce output at terminal 2. The negative half-cycles of the input are just not used. The nonconducting diode then is practically an open circuit, without any conduction because the anode is negative.

Although the output at terminal 2 in Fig. 29-21 has fluctuations in the positive half-cycles, it is still a dc waveform. In terms of voltage, it has only one polarity. For the current, there is only one direction, whether we consider conventional current or electron flow. Actually, the variations in the fluctuating dc voltage can be smoothed out by filter capacitors and chokes.

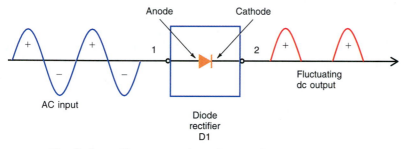

FIG. 29-21 The diode rectifier converts its ac input to fluctuating dc output.

THE DIODE AS A SWITCH   A diode can operate in such a way that it really serves as an electronic switch. It can be ON with very low resistance because of conduction. Then the diode switch is closed. The resistance may be less than 1 Ω. Or the diode can be OFF because no current flows. Then the diode switch is open with infinitely high resistance. The ON–OFF conditions can be summarized as follows:

1.   Anode positive. Current flows. Diode is ON as a closed switch with very low R.
2.   Anode negative. No current. Diode is OFF as an open switch with very high R.

Even as a rectifier, the diode can be considered as a switch. The diode connects the output to the input only for the polarity that makes the anode positive. As a result, the output has only one polarity. This action is called *commutating,* meaning a process of switching in accordance with a specific polarity.

HALF-WAVE POWER SUPPLY   One of the most common applications of the diode rectifier is to provide dc supply voltage from the 60-Hz ac power line. The circuit is a dc power supply. Its function can very well be to supply the dc operating voltages needed for the amplifiers and oscillators on an electronic circuit board. In portable electronic equipment, the dc supply voltage is provided by batteries. For electronic equipment that operates from the ac power line, though, a dc power supply is necessary.

In Fig. 29-22, D1 is the diode rectifier. One diode is always a half-wave rectifier because only one-half of the ac input cycles are used for dc output. These waveforms were shown before in Fig. 29-21. However, the fluctuations in dc output are not shown in Fig. 29-22 because the ac ripple is almost completely eliminated by the filter capacitors C_1 and C_2 with the smoothing choke L_1. The frequency of the ac ripple component in the dc output is 60 Hz, with ac input at this frequency.

Operation of the power supply depends on the diode. When the anode is positive, current flows and the ac input voltage charges C_1 and C_2 for dc output voltage. Capacitor C_1 is at the input side of the filter. It can charge through D1 but without L_1. Capacitor C_2 is the output filter for R_L. When the anode of the diode is negative, no current flows.

In the diode input circuit, R_1 is a surge-limiting resistor. It prevents excessive current through the diode when the input filter capacitor C_1 is charging. The output resistor R_L represents the combined resistance of all the load currents connected to the output of the dc power supply.

The dc output voltage may be higher than the rms value of the ac input voltage because the input filter capacitor C_1 can charge to the peak value. In Fig. 29-22, the dc output of 140 V is between 120 V as the rms value and 170 V as the peak value of the ac input. Remember that the peak of a sine wave equals 1.414 times the rms value. Then 120 V × 1.414 = 169.68 or approximately 170 V.

The value of dc output voltage, compared to the ac input voltage, depends on the amount of dc load current, represented by the 500-Ω value of R_L in Fig. 29-22. Since the load current represents discharge of C_1 and C_2, less current means the capacitors can charge to a higher voltage. The value of 140 Vdc with R_L of 500 Ω is typical for a load current I_L of 280 mA. This I_L can be calculated as

$$I_L = \frac{V_{dc}}{R_L} = \frac{140 \text{ V}}{500 \text{ } \Omega}$$

$$I_L = 0.28 \text{ A} = 280 \text{ mA}$$

In Fig. 29-22, the dc output voltage has positive polarity because it is taken from the diode cathode. In general, ac input at the anode produces positive dc

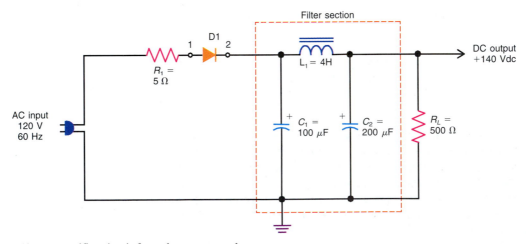

FIG. 29-22 Half-wave rectifier circuit for a dc power supply.

output at the cathode. Even though the dc voltage output makes the cathode positive, the ac input voltage makes the anode more positive to allow conduction. Basically, the cathode terminal in Fig. 29-22 must provide positive output because the output at terminal 2 is connected to the input at terminal 1 of the diode only when the source voltage at the input is positive.

For the opposite polarity of dc output, the ac input can be connected to the cathode for negative dc output at the anode. This circuit is sometimes called an *inverted power supply*. Both positive and negative polarities are commonly used as dc supply voltage for transistor circuits.

FILTER SECTION   Note that C_1 and C_2 in Fig. 29-22 are marked with a polarity sign, indicating they are electrolytic capacitors. The filter capacitors in a 60-Hz power supply are always electrolytics because of the high capacitance required. Typical values are 80 to 1000 μF. The filters are essentially shunt bypass capacitors that must have very low reactance at the frequency of the ac ripple, usually at 60 Hz or 120 Hz. Larger filter capacitors are needed with larger values of load current, corresponding to a smaller value of R_L. The filter L_1 is an iron-core choke used as a series component. It must have high reactance at the frequency of the ac ripple. In many cases, a series filter resistor of 100 to 1000 Ω may be used instead of L_1, for economy and to save space.

TYPES OF POWER SUPPLIES   The half-wave rectifier with one diode is the basic power supply but combinations are used with more diodes. Two diodes can be arranged in a *full-wave rectifier* circuit. It provides dc output for both cycles of the ac input. The advantage is that more load current can be supplied. Also, the frequency of the ac ripple on the output becomes 120 Hz instead of 60 Hz. The higher ripple frequency is easier to filter, allowing smaller values of C.

Other combinations include the voltage doubler with two diodes and the voltage tripler with three diodes. These circuits are used for higher dc output voltage with small values of load current.

DIODE DETECTOR CIRCUIT   Another important function of the diode rectifier is detecting a modulated signal to recover the modulation. Half-wave rectification is generally used. A detector is just a rectifier circuit for small values of signal voltage, usually about 1 to 10 V. Diodes for detectors are made of silicon (Si) or germanium (Ge). The Ge diodes have less resistance.

In Fig. 29-23, an AM signal is shown for input to the diode detector D1. The signal must be rectified for detection, in order to eliminate the symmetry in the modulation with positive and negative envelopes that are essentially the same. The load resistor R_L has the rectified output voltage, which consists of half-cycles of the rectified carrier signal. The polarity of the rectified output does not matter in this application for an audio signal. In the rectified output across R_L, the amplitudes vary in accordance with the changes in the envelope of the AM signal. The individual cycles of rectified carrier signal are not shown, however, because they are eliminated by the bypass capacitor C_1. It has very low reactance for the RF carrier frequency but not for the lower frequencies in the audio modulation. As a result, the output of the AM detector circuit is the desired audio signal, coupled by C_2 to the next stage.

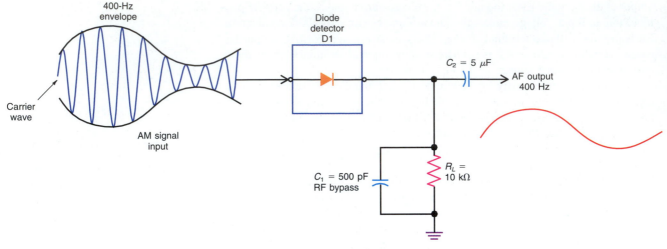

FIG. 29-23 Diode detector circuit to recover audio modulation in an AM signal.

TEST-POINT QUESTION 29-7

Answers at end of chapter.

Answer True or False.
a. The anode must be positive for current in a diode.
b. The diode symbol shows the direction of electron flow.
c. One diode can be used as a half-wave rectifier.
d. An ac input at the anode of a diode results in positive dc output voltage.
e. Filter capacitors in a power supply are generally electrolytic capacitors.
f. A diode can be used for detection of an AM signal.

29-8 TROUBLESHOOTING THE DC SUPPLY VOLTAGE

Although electronic circuits are used mainly for ac signals, most troubles in operation result from incorrect dc supply voltages. Remember that transistors and IC chips need dc electrode voltages so that they can conduct current and function. No dc supply voltage means no ac output, even with ac signal input. Low values of dc voltage also can cause troubles in the operation of the circuit.

Very common is the problem of no dc supply voltage. If a supply voltage problem appears in battery-powered portable equipment, first try a new battery. An old battery may appear to have its normal voltage when measured with a dc voltmeter, but the output drops drastically with load current.

When a power supply is used, as illustrated in Fig. 29-24, the first step in troubleshooting is to check the dc output voltage to see if it is normal. The measurement can be made between point C in the diagram to chassis ground. A dc voltmeter should read 170 V at C and 140 V at D. Normal values are generally within ±10 percent. If the dc output voltage is not correct, then additional voltage and resistance tests can be made to isolate the trouble. The procedure may

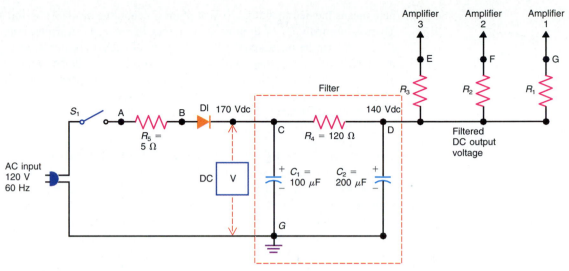

FIG. 29-24 Half-wave rectifier power supply illustrated for troubleshooting procedures explained in the text.

seem complicated, but usually the purpose is simply to find a component that is either open or shorted.

Use a multimeter, either the digital or analog type, in its three functions:

1. *AC voltmeter.* To check the 60-Hz ac voltage input.
2. *Ohmmeter.* To check for an open circuit, short circuit, or incorrect R value. A short circuit reads 0 Ω. With an open circuit, the ohms reading is infinitely high.
3. *DC voltmeter.* Measure dc voltages in the output of the rectifier, from different points in the circuit to chassis ground. The negative lead of the voltmeter stays connected to ground and the positive lead is moved to the check points, assuming positive dc supply voltage.

CHECKING THE AC VOLTAGE INPUT   The meter function is set for ac voltages, on a range for 120 V. With power turned on by the switch S_1, measure from point A in Fig. 29-24 to G, across the ac power plug. Zero voltage here means no input power, probably because of a blown fuse or open circuit breaker in the power line. The switch S_1 could be defective, but this is not a common trouble. Also measure the ac input voltage at point B to G. If there is ac voltage at A but not at B, then R_5 must be open.

When the diode anode at point B has ac input voltage but there is no dc output voltage at point C at the diode cathode, either the diode is defective or the output circuit has a trouble.

OHMMETER MEASUREMENTS   Power is off, preferably with the plug disconnected. First measure the resistance of the dc output circuit, from point C to G, to see if it is shorted. A power supply with a short circuit in the dc output will produce excessive current in the ac input, if the power is turned on.

The ohmmeter is really the meter that finds the defective component that needs replacement. Measure across R_5 to see if it is 5 Ω. An open here means no ac input voltage for the diode rectifier. Measure across R_4 for 120 Ω. An open here means the dc output of 170 V at point C cannot be available as 140 V at D. There is normally an IR drop of 40 V across R_4. Also the resistance across R_1, R_2, and R_3 can be measured for the dc supply of each amplifier. If one of these resistors is open, this stage will not have dc operating voltage while the other stages do. All the amplifiers are in parallel for dc voltage from the power supply. Finally, and maybe most important, the ohmmeter can be used to check the diode rectifier D1 and the filter capacitors C_1 and C_2.

CHECKING FILTER CAPACITORS WITH AN OHMMETER   These capacitors are the electrolytic type with large values of C. First, discharge the capacitor with a jumper lead across its two terminals. Then check the resistance with the ohmmeter leads. The capacitor charging action from the battery in the ohmmeter should be very definite, starting with a very low R and backing off to a high value for its normal R. Typical values are 100 kΩ to 500 kΩ, which is much lower than for other capacitor types. Check an electrolytic capacitor with the ohmmeter leads in one polarity and then reverse the leads. Use the higher readings. The following capacitor troubles can be indicated:

1.   No charging action with zero ohms of R means the capacitor is shorted.
2.   No charging action with a very high R means the capacitor is open.
3.   Very little charging action may mean lower C than normal.
4.   Too small a reading for the highest R may indicate the capacitor has leakage across its terminals.

For all these cases, the capacitor is defective and should be replaced. Install the new electrolytic in the same polarity as the old one. Also, with electrolytic capacitors, the replacement should have approximately the same voltage rating.

TESTING DIODES WITH AN OHMMETER   When the internal battery of the ohmmeter makes the diode anode positive, this polarity is called forward voltage because current can flow easily. The result is a low value of forward R. Typical values may be 10 to 1000 Ω. For the opposite polarity, when the ohmmeter leads are reversed, the resistance of a silicon diode is almost infinitely high, practically an open circuit. Therefore, a good diode has very much more R with one polarity of the ohmmeter leads, compared with the opposite polarity. The ratio of reverse to forward R should be at least 100:1 and usually 1000:1. A defective diode is indicated by the following:

1.   The diode resistance is low in both directions. This diode has an internal short circuit.
2.   The diode resistance is very high in both directions. This diode has an open circuit, probably at the electrode leads.

In either case, the diode is defective and must be replaced. Install the new diode in the same polarity as the old one.

Several points about ohmmeters should be kept in mind. Many ohmmeters have a *low-ohms* position with very low battery voltage. Do *not* use this position for testing diodes because there is not enough voltage to produce forward current. Also, many multimeters have a *diode test* position. This test is not for R but provides normal forward voltage across the diode terminals, typically 0.6 to 0.7 V for silicon diodes. A higher or lower value indicates a defective diode.

HUM When the filters in a power supply do not have enough capacitance, the result is too much of the ac ripple component in the dc output voltage. The effect is excessive *hum* in the sound output from an audio system. With a half-wave power supply, the hum is at 60 Hz. In a full-wave supply, the sound of the hum at 120 Hz has a little higher pitch. The problem with hum voltage is that it becomes part of the ac signal in the amplifier circuits.

With a hum component in the video signal for a television picture, the effect is horizontal dark and light bars across the screen. They usually appear to roll slowly upward. With 60 Hz, the screen has one pair of hum bars; for 120 Hz two pairs of bars are produced.

ISOLATION TRANSFORMER A power-supply circuit like the one in Fig. 29-24 is *line-connected* or *nonisolated.* This type has one side of the dc output connected directly to the 60-Hz power line. The result can be dangerous; it can present the hazard of a severe electrical shock. For safety in testing such a supply, an isolation transformer should be used (Fig. 29-25). It has separate primary and secondary windings to isolate the ac input and dc output circuits. The turns ratio may be 1:1, or the transformer can have adjustable values of ac input for testing with different values of input voltage.

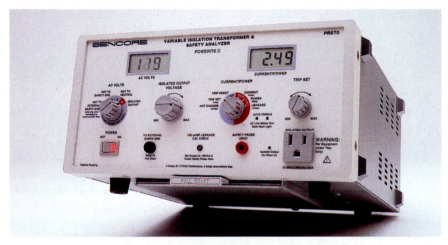

FIG. 29-25 Isolation transformer for safe bench work.

TEST-POINT QUESTION 29-8

Answers at end of chapter.

Refer to the diagram in Fig. 29-24.
a. Which open resistor can cause no ac input voltage?
b. Which open resistor can cause no filtered dc output voltage?
c. Which components can cause excessive hum?
d. Which component converts the ac input to dc output?
e. Which open resistor can cause no dc supply voltage for amplifier 3?

29   SUMMARY AND REVIEW

- A sine wave of *V* or *I* is an analog signal with continuous variations that are proportional to changes in the information of the signal.
- Amplification means increasing the magnitude or amplitude. The voltage gain A_V of an amplifier is given in Formula 29-1; current gain A_I is defined by Formula 29-2, and power gain A_P by Formula 29-3.
- Amplifiers in cascade have signal from the output of one stage driving the input of the next stage. The overall gain of the amplifiers equals the product of the individual gain values.
- Amplifiers use transistors and integrated circuit chips. These active electronic devices need dc supply voltage to provide current so that the ac signal can be amplified.
- Passive components in electronic circuits are resistors, capacitors, and inductors. Resistance is used to provide an *IR* voltage drop. Capacitance is used for coupling and bypass capacitors. Inductance is used for chokes and transformers. Inductance and capacitance are combined for resonant circuits at radio frequencies.
- An electronic oscillator circuit is an amplifier with positive feedback to sustain oscillations. No input signal is necessary because the feedback from the output circuit provides the input. The oscillator generates its own ac output, converted from the power of the dc voltage supply. For special applications, AF oscillators and RF oscillators are used.
- RF feedback oscillators use tuned circuits to determine the oscillator frequency. Two main types are the Hartley and the Colpitts. A Hartley oscillator uses a tapped coil as an ac voltage divider for feedback. In the Colpitts circuit, a capacitive voltage divider is used.
- The multivibrator is a circuit to produce voltage output that oscillates between HIGH and LOW states. The astable MV is a free-running oscillator that serves as a pulse generator. The bistable MV is stable in either the HIGH or LOW state but can be triggered into the reverse state.
- Modulation means modifying the characteristics of one waveform in accordance with the variations in another signal that has a lower frequency. Common applications are in radio and television broadcasting where an RF carrier wave is modulated by audio or video signal. For amplitude modulation, the amplitude of the RF carrier varies with the modulation. In frequency modulation, the frequency of the RF carrier varies with the modulation.
- In a power supply, the 60-Hz ac input from the power line is converted to dc output, generally for dc supply voltage to amplifier circuits. Electrolytic filter capacitors are used to eliminate the ac ripple in the dc fluctuations of the rectified output. The basic half-wave rectifier circuit uses one diode. A full-wave rectifier uses two diodes.
- An ac ripple in the fluctuating dc output of a power supply is called hum, from the way it sounds in an audio system. In a television picture, the ripple produces horizontal hum bars across the screen. The hum frequency is 60 Hz for a half-wave rectifier or 120 Hz in a full-wave rectifier. Hum is generally caused by defective filter capacitors in the power supply.
- Troubles in a power supply are generally no dc output, insufficient dc output, and hum.

1. Is the transistor an active or inactive device?
2. Does a coupling capacitor have high or low reactance at the frequency of the signal?
3. Is the sine wave a digital or analog waveform?
4. With 0.05 V input and 5 V output, calculate the voltage gain.
5. For 600-μA input and 30-mA output, how much is the current gain?
6. Which component blocks dc voltage, R, L, or C?
7. Give the radio frequencies in the very high frequency (VHF) band.
8. Is a resistance load in the output circuit more likely to be used in an AF or RF amplifier?
9. Is a tapped coil for feedback used in the Hartley or Colpitts oscillator?
10. Which circuit can be used as a pulse generator, the multivibrator or the Hartley oscillator?
11. Which circuit is a free-running oscillator, the astable MV or the flip-flop?
12. True or False? The sidebands in a modulated signal increase the bandwidth?
13. Does the standard radio band of 535 to 1605 kHz use AM or FM?
14. Is the baseband signal for the picture in television a video or audio signal?
15. Is phase modulation similar to AM or FM?
16. In the symbol for a diode, does the arrow point to the anode or cathode?
17. In order for a diode to conduct, which of its electrodes must be more positive?
18. In a half-wave rectifier circuit, is the hum frequency 60 or 120 Hz?
19. How many diodes would be necessary for a full-wave rectifier circuit?
20. True or False? In Fig. 29-24, if the diode is open, the trouble will be no dc output.
21. In Fig. 29-24, if filter capacitor C_2 is open, will the trouble be no dc output or excessive hum?
22. True or False? In testing a diode with an ohmmeter, the reverse R should be much higher than the forward R.
23. True or False? A filter capacitor that shows no charging action when tested with an ohmmeter should be replaced.

QUESTIONS

1. Define active device and passive device for electronic circuits. Give two examples of each.
2. Give two differences between analog and digital signals. Would you consider the audio signal that drives a loudspeaker an analog or digital signal?
3. Draw the waveform of a pulse signal representing the binary values 1010.
4. Define the terms bit, byte, and word for digital signals.
5. What is meant by amplification of an ac signal?
6. Why is dc supply voltage needed in a transistor amplifier circuit for ac signals?
7. Compare the terms voltage gain and current gain.
8. Give the definitions for the gain values A_V, A_I, and A_P.
9. What does the term *in cascade* mean for amplifiers?
10. Give two methods of calculating the gain for amplifiers in cascade.
11. Give one difference and one similarity in a comparison of coupling and bypass capacitors.

12. State the frequencies in the VHF and UHF bands.
13. Show a sine wave for the signal waveform as a fluctuating dc voltage. Label the average dc axis and the peak-to-peak value of the ac component.
14. Which of the following components can block dc voltage: RF choke; AF coupling capacitor; RF bypass capacitor; load resistor?
15. Compare the frequencies in AF and RF signals. Why are LC tuned circuits generally used for RF signals rather than AF signals?
16. Compare a tuned RF amplifier stage with an audio amplifier.
17. Give a definition of an oscillator circuit.
18. What is meant by feedback? How would positive feedback be different from negative feedback?
19. What is the main difference between the Hartley and Colpitts oscillator circuits?
20. What is meant by an RC feedback oscillator?
21. Give a definition for the multivibrator (MV) oscillator.
22. What is the difference between the astable and bistable MV?
23. What is meant by modulation of an RF carrier wave?
24. Define AM, FM, and PM. Give two commercial applications.
25. What is meant by the baseband signal in the modulation process? Give two examples of baseband signals.
26. Define center frequency and the frequency deviation for an FM signal.
27. Draw the schematic symbol for a silicon diode and label the anode and cathode. Indicate the direction of hole current and electron flow.
28. What is meant by a rectifier? Why is the diode a good example? Give two applications of diode rectifiers.
29. Describe briefly how you can test a diode with an ohmmeter.
30. What is the purpose of a filter capacitor in a power supply?
31. What is the function of a dc power supply in electronic circuits?
32. What is a flip-flop circuit?
33. Describe briefly how you would test an electrolytic filter capacitor with an ohmmeter.
34. Draw the circuit of a power supply using a half-wave rectifier with filter. Give the function of each component.
35. For the circuit in Question 34, list three possible troubles caused by defective components.

PROBLEMS

ANSWERS TO ODD-NUMBERED PROBLEMS AT BACK OF BOOK.

1. A transistor amplifier has input signal of 0.02 V and 400 μA p–p. The amplified output signal is 6 V and 36 mA, also p–p. Calculate the voltage gain A_V, current gain A_I, and power gain A_p.
2. With input signal of 600 μA p–p and a current gain of 80, calculate the peak-to-peak output current.
3. Three amplifier stages in cascade have current gain values of 50, 20, and 10 respectively. Calculate the total current gain.
4. In Prob. 3, when the input signal to the first stage is 200 μA p–p, how much is the output from the last stage?

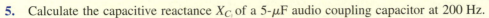

5. Calculate the capacitive reactance X_C of a 5-μF audio coupling capacitor at 200 Hz.
6. What size C is needed for an RF bypass capacitor at 88 MHz?
7. For a filter capacitor in a power supply calculate C needed for 8 Ω of X_C at: (a) 60 Hz; (b) 120 Hz.
8. What L is needed for an RF choke with 10 kΩ of X_L at 1.6 MHz?
9. An RF feedback oscillator has an LC circuit with 30 pF for C and 320 μH for L. Calculate the oscillator frequency.
10. In Prob. 9, what C is needed for an oscillator frequency of 995 kHz?
11. Refer to Fig. 29-26. What value of R_L is needed for the 7 V at terminal 2 of $Q1$?
12. What value of R_L would be needed in Fig. 29-26 if I were 4 mA?

CRITICAL THINKING

1. In Fig. 29-27, how much dc voltage exists at point A with respect to ground?
2. In Fig. 29-27, how much dc voltage exists at point A if the ground connection is removed from the center tap (C.T.) connection on the transformer secondary? Why?

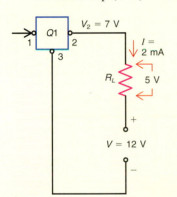

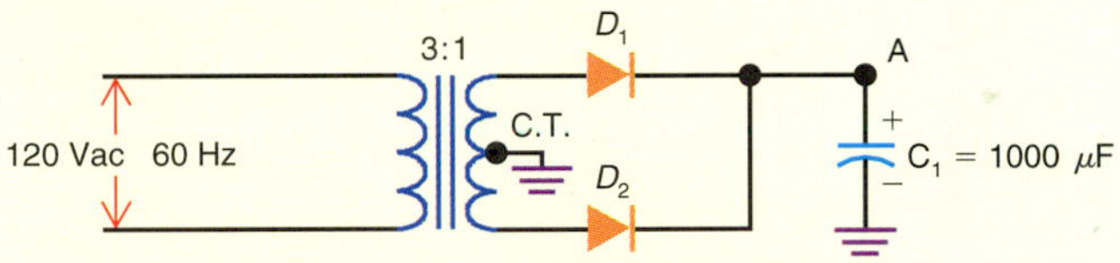

FIG. 29-27 Circuit for Critical Thinking Probs. 1 and 2.

FIG. 29-26 Circuit for Probs. 11 and 12.

ANSWERS TO TEST-POINT QUESTIONS

29-1 **a.** T
 b. F
 c. T

29-2 **a.** 200
 b. 50
 c. 10,000
 d. 27,000

29-3 **a.** T
 b. T
 c. T
 d. F

29-4 **a.** Hartley
 b. positive
 c. 2 MHz

29-5 **a.** T
 b. F
 c. T
 d. T

29-6 **a.** baseband
 b. AM
 c. 50 kHz
 d. increase
 e. FM

29-7 **a.** T
 b. F
 c. T
 d. T
 e. T
 f. T

29-8 **a.** R_5
 b. R_4
 c. C_1 and C_2
 d. D1
 e. R_3

CHAPTER 30

TRANSISTOR AMPLIFIERS

More details of exactly how a transistor amplifies its input signal are analyzed in this chapter. In general, any amplifier operates by having a small input able to control more power in the output circuit. The dc supply for electrode voltages provides the power. As a result, amplifier circuits for ac signals operate with a combination of ac and dc values.

Another question in amplifiers is which electrodes are used for input and output signals. These connections determine the circuit configuration for the amplifier. The features of amplifiers are described here for NPN and PNP transistors and the field-effect transistor (FET). The amplifiers can be discrete units or part of an integrated circuit (IC) chip.

CHAPTER OBJECTIVES

Upon completion of this chapter, you should be able to:

- *List* the three different amplifier configurations for bipolar transistors.
- *Define* class A, class B, and class C operation of a transistor amplifier and list the characteristics of each.
- *Analyze* a common-emitter amplifier in terms of biasing voltages and signal variations.
- *Define* the terms *alpha* (α) and *beta* (β).
- *Explain* the term *transconductance* (g_m) as it relates to an FET amplifier.
- *Troubleshoot* a transistor amplifier circuit.

IMPORTANT TERMS IN THIS CHAPTER

alpha characteristic (α)
beta characteristic (β)
bias current
bias stabilization
common-base (CB) circuit
common-collector (CC) circuit

common-drain (CD) circuit
common-emitter (CE) circuit
common-gate (CG) circuit
common-source (CS) circuit
Darlington circuit

emitter bias
emitter follower
self-bias
source-follower
transconductance (g_m)

TOPICS COVERED IN THIS CHAPTER

30-1 CIRCUIT CONFIGURATIONS

The circuit configuration specifies which electrodes in the amplifier are used for input and output signals. Actually, though, the configuration is named for the electrode that is the common return connection. The common electrode usually is the one that does not have any signal.

Since a PNP or NPN transistor is a triode with only three electrodes, there are only three possible configurations: the base common; the emitter common; or a common collector. Figure 30-1 shows these circuits. The general example in Fig. 30-1a would apply to any amplifier device with three terminals. For transistors specifically, the common-base circuit is in Fig. 30-1b, the common-emitter circuit is in Fig. 30-1c, and the common-collector circuit is in Fig. 30-1d. Although the common electrode is shown grounded here, it need not be connected to chassis ground. The main characteristics of these circuits are compared in Table 30-1. All the circuits have reverse voltage for the collector and forward bias for the emitter-base junction.

The common-emitter (CE) circuit is the one generally used for amplifiers because it has the best combination of current gain and voltage gain. However, each type of circuit has special features.

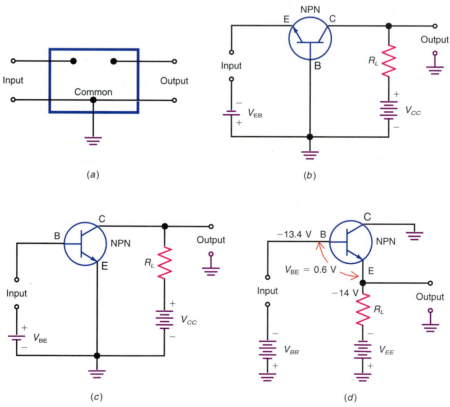

FIG. 30-1 Circuit configurations for amplifiers shown with NPN transistors. All polarities are reversed for PNP. (*a*) General case of a common terminal for two pairs of connections. (*b*) Common-base circuit. (*c*) Common-emitter circuit. (*d*) Common-collector circuit or emitter-follower.

TABLE 30-1 COMPARISON OF CIRCUITS FOR JUNCTION TRANSISTORS

CHARACTERISTIC	COMMON BASE (CB)	COMMON EMITTER (CE)	COMMON COLLECTOR (CC)
Signal into	Emitter	Base	Base
Signal out of	Collector	Collector	Emitter
Advantage	Stability	High gain	High r_i
Phase inversion	No	Yes	No
Input resistance* r_i	20 Ω	1000 Ω	150 kΩ
Output resistance* r_o	1 MΩ	50 kΩ	80 Ω

*Typical values of r for small-signal transistor with I_E of 1.5 mA.

COMMON-BASE (CB) CIRCUIT

In Fig. 30-1b, the input voltage is applied to the emitter, with respect to the grounded base. The amplified output is taken from the collector. Resistance R_L is in series with the collector supply V_{CC}.

In the CB circuit, the emitter input has low resistance r_i because I_E is high. The output resistance r_o for the collector is high because of the reverse voltage. Typical values for a small-signal transistor with I_E of 1.5 mA are 20 Ω for r_i and 1 MΩ for r_o. These values are for the internal resistance of each electrode to the common base.

The CB circuit is seldom used. It has no current gain from input to output because I_C must be less than I_E. The voltage gain can be high, but the output is shunted by the low input resistance of the next stage. The only advantage of the CB circuit is that it has the best stability with an increase in temperature. The reason is that reverse leakage current from collector to base is not amplified in the CB circuit.

COMMON-EMITTER (CE) CIRCUIT

See Fig. 30-1c. Input voltage is applied to the base instead of to the emitter, which is now the grounded electrode. Note that the emitter is shown at the bottom of the schematic symbol. The input circuit here involves the small I_B instead of I_E. As a result, the r_i for the CE circuit is much higher than for the CB circuit. A typical value is 1000 Ω for r_i.

The output voltage is still taken from the collector with its R_L. A typical value for r_o in the collector output circuit is 50 kΩ.

In the input circuit, the forward bias V_{BE} is applied to the base instead of the emitter. Note the polarity. Positive V_{BE} to the P base corresponds to negative V_{EB} at the negative emitter. Both are forward-bias voltages, with the polarity the same as the N or P electrode.

Furthermore, the positive V_{BE} at the base for forward bias uses the same voltage polarity as positive V_{CC} for reverse voltage at the N collector. This feature allows the practical convenience of using one voltage supply for both forward bias in the input and reverse bias in the output.

The CE circuit has current gain because I_C is much larger than I_B. The voltage gain is the same as for the CB circuit. With a higher r_i, however, the CE circuit can be used in cascaded amplifiers where the collector output of one stage drives the base input of the next. The CE circuit is the amplifier generally used

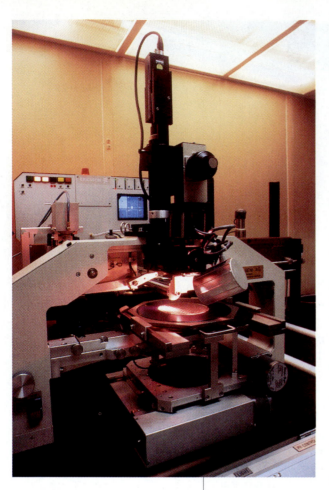

for transistors, therefore, because it has the best combination of voltage gain and current gain. The disadvantage is that reverse leakage current is amplified in the CE circuit, but bias stabilization methods can be used.

Only the CE amplifier inverts the polarity of signal voltage. This phase inversion of 180° is neither good nor bad, but just a result of the circuit connections. When the base input signal increases the forward voltage, the collector voltage of the same polarity decreases because of the voltage drop across R_L.

COMMON-COLLECTOR (CC) CIRCUIT   See Fig. 30-1d. Signal input is applied to the base, as in the CE circuit. However, the collector is grounded, instead of the emitter. Therefore, the emitter has the R_L for the output signal.

Now there are two questions for the electrode voltages: how to apply reverse voltage for the grounded collector and forward bias for the base. Note that the emitter is at -14 V with respect to chassis ground. The collector is at chassis ground. In effect, the collector is connected to the positive side of the emitter supply voltage. This makes the N collector positive with respect to the emitter, as reverse voltage for V_{CE}.

For forward voltage, the base must be more positive than the emitter. In this example V_E is -14 V. Therefore V_B is made -13.4 V. The actual bias V_{BE} then is $+0.6$ V.

In the CC circuit, the input circuit has high r_i. A typical value is 150 kΩ. The output in the emitter circuit has low r_o of about 80 Ω. Note that for the CC circuit, the input resistance is high and output resistance is low, compared with low r_i and high r_o for the other circuits.

There is no voltage gain in the CC circuit because the output signal across R_L in the emitter circuit provides negative feedback to the base input circuit. However, there is appreciable current gain.

The name *emitter-follower* is generally used for the CC circuit. The output signal at the emitter follows the polarity of the input signal at the base. The emitter-follower circuit is often used for impedance matching, from a high-impedance source to a low-impedance load. The circuit also provides isolation between a load in the emitter circuit and a source in the base circuit.

DARLINGTON PAIR   This circuit consists of two emitter followers connected in cascade. The two stages are usually in one package, with dc coupling internally and just three external leads. The package provides higher input resistance and more current gain than just one stage.

TEST-POINT QUESTION 30-1

Answers at end of chapter.
a. Which circuit has the most gain?
b. Which circuit has input signal to the base and output from the collector?
c. Give another name for the CC circuit.

30-2 CLASS A, B, OR C OPERATION

The amplifier class of operation is defined by the percentage of the input signal that is able to produce output current. In other words, is any part of the input cycle cut off in the output? The class of operation depends on two amplitudes: (1) the dc bias compared with the cutoff value and (2) the peak ac signal compared with the dc bias.

The input and output waveforms for class A, B, and C amplifiers are illustrated in Fig. 30-2. The output current is labeled I_O. This is collector current for junction transistors, or drain current for the FET. The class of operation determines the power efficiency and how much distortion of the signal may be produced by the amplifier.

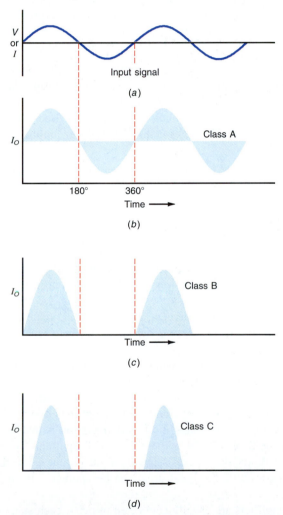

FIG. 30-2 Class of operation for amplifiers in terms of output current I_O. (a) Cycles of sine-wave ac input signal. (b) Class A with I_O for full cycle of 360°. (c) Class B with I_O for half-cycle of 180°. (d) Class C with I_O for less than 180°, typically 120°.

CLASS A   The output current I_O flows for the full cycle of 360° of input signal. This operation is shown by the waveform in Fig. 30-2b. An audio amplifier stage operates this way to follow the signal variations without too much distortion.

For class A operation, the dc bias allows an average I_O of about one-half the maximum value. Then the ac signal swings I_O around that middle value. The output current I_O can vary up to its maximum value, and close to zero, but is never cut off.

CLASS B   The I_O flows for 180°, or approximately one-half of the input cycle, as shown in Fig. 30-2c. The dc bias is at or near the cutoff value. I_O is at or close to zero, then, without any signal input. However, the positive half-cycle of signal can swing I_O up to its maximum value. The negative half-cycles of input signal are cut off in the output, because I_O then is zero. Class B operation requires more dc bias and more ac signal drive than class A.

Class B operation with a single stage corresponds to half-wave rectification of the ac signal input. However, two stages can be used to provide opposite half-cycles of the signal in the output.

CLASS C   In Class C operation the output current I_O flows for less than one-half the input cycle. Typical operation is 120° of I_O during the positive half-cycle of input, as shown in Fig. 30-2d. This result is produced by doubling the class B bias and using twice as much ac signal drive. Because of its high efficiency, class C operation is used for tuned RF power amplifiers.

CHARACTERISTICS OF EACH CLASS   Which class of operation is used for an amplifier depends on the requirements for minimum distortion, maximum ac power output, and efficiency. The degree by which the output signal waveshape differs from the input signal waveshape is known as distortion. The ac power output is signal output. Efficiency is the ratio of ac power output to the dc power dissipated at the output electrode of the amplifier.

In class A operation, distortion is lowest, but so also are ac power output and efficiency. Typical values are less than 1 to 5 percent distortion and an efficiency of 5 to 25 percent. At the opposite extreme, class C operation offers the highest efficiency, of about 90 percent, and allows the greatest ac power output but with the most distortion. Class B operation lies between A and C in distortion, power, and efficiency.

With audio amplifiers, class A must be used in a single stage for minimum distortion. Otherwise, the sound would be garbled. Also, an RF stage amplifying an amplitude-modulated signal must operate class A for minimum distortion of the modulation. In general, most small-signal amplifiers operate class A.

The reason for low efficiency in class A operation is that the middle value of I_C flows all the time, with or without ac signal input and for weak or strong signals. As a result, the dc power dissipation at the output electrode is high. Furthermore, relatively little ac signal drive can be applied without exceeding the cutoff voltage.

Class B amplifiers are usually connected in pairs, each stage of which supplies opposite half-cycles of the signal input. Such a circuit is called a *push-pull amplifier*. The results approximate the low distortion of class A, but more drive can be used for more ac power output with higher efficiency. The push-pull circuit arrangement is often used for audio power output to a loudspeaker.

Class C operation is generally used for RF amplifiers with a tuned circuit in the output. Then the *LC* circuit can provide a full sine-wave cycle of output for each pulse of I_O. Class C amplifiers have high efficiency because the average I_O is very low compared with the peak signal amplitude. The result is relatively low dc power dissipated at the output electrode compared with the amount of ac power output. In addition, a pulse clipper circuit operates as a class C amplifier.

30-3 ANALYSIS OF COMMON-EMITTER (CE) AMPLIFIER

Typical values of V_{BE} are in tenths of a volt for a junction transistor. The required bias at the base for a class A amplifier is 0.6 to 0.7 V for Si or 0.2 to 0.3 V for Ge. Class A operation means that the amplifier conducts current for 360° of the signal cycle, for minimum distortion. Furthermore, the maximum ac input signal without overload distortion is ±0.1 V. These values are summarized in Table 30-2. Note that 0.1 V is 100 mV.

Without any forward bias a junction transistor is cut off by its own internal barrier potential. The cut-in voltage in the first column of Table 30-2 is the lowest V_{BE} that allows appreciable I_C.

The saturation voltage in the second column of Table 30-2 is the highest V_{BE} that allows it to produce proportional changes in I_C. At saturation, the maximum I_C does not increase with an increase of forward voltage.

The input voltages listed in Table 30-2 apply to all junction transistors, regardless of size and power rating. However, the difference is in the amount of base current I_B and collector current I_C. As an example, a small-signal transistor may have I_B of 60 μA with V_{BE} of 0.6 V at the input, for I_C output of 3 mA. In a medium-power transistor I_B could be 20 mA for the same V_{BE} of 0.6 V, to produce I_C of 1 A.

TABLE 30-2	INPUT VOLTAGES V_{BE} AT 25°C FOR JUNCTION TRANSISTORS			
	CUT-IN VOLTAGE	SATURATION VOLTAGE	ACTIVE REGION	AVERAGE BIAS VOLTAGE
Ge	0.1	0.4	0.1–0.4	0.2–0.3
Si	0.5	0.8	0.5–0.8	0.6–0.7

CIRCUIT COMPONENTS The transistor amplifier itself is usually labeled Q, as for $Q1$ in Fig. 30-3. The 2.2-kΩ R_L is the collector load. It is in series with the positive V_{CC} of 12 V for reverse collector voltage on the NPN transistor.

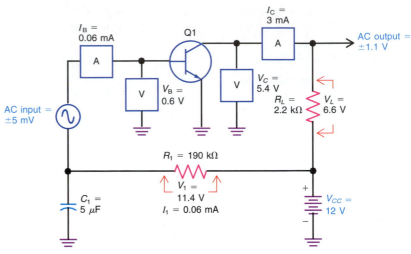

FIG. 30-3 Example of a circuit for the CE amplifier. The meters shown for V and I indicate average dc values. This circuit is analyzed in the text.

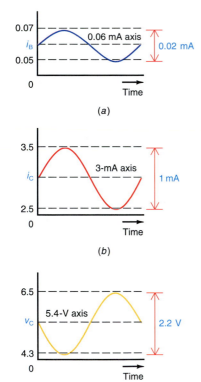

FIG. 30-4 Ladder diagram of signal waveforms for the circuit in Fig. 30-3. (a) Base current i_B. (b) Collector current i_C. (c) Collector output voltage v_C.

One supply voltage is used for both collector and base in the CE circuit, as positive base bias and positive collector voltage are needed. However, the collector voltage is too high for base bias. Therefore, the 190-kΩ R_1 is connected in series as a voltage-dropping resistor for the base.

The required forward-bias voltage for Si is 0.6 V. The corresponding bias current for this transistor is taken as 60 μA or 0.06 mA, as an example. The voltage drop across R_1 is 0.06 mA $\times$ 190 kΩ = 11.4 V. Then $12 - 11.4 = 0.6$ V remains for V_{BE} as forward voltage for the base.

The bypass capacitor C_1 allows the ac input signal voltage to vary the base current without the series resistance of R_1. Then very small changes of input voltage can produce appreciable changes in base current. We are assuming an ac input of ±5 mV for the base input signal, or 10 mV p–p.

SIGNAL VARIATIONS The input signal of base current i_B is shown in Fig. 30-4a. This waveform shows i_B varies by 10 μA above the 60-μA bias axis, up to the peak of 70 μA. On the down side, i_B decreases by 10 μA, from 60 to 50 μA. Positive signal voltage in the forward direction increases i_B, while negative signal voltage reduces i_B. The peak-to-peak signal in i_B then is $70 - 50 = 20$ μA, or 0.02 mA.

The variations in i_B cause corresponding variations in i_C, as shown in the waveform in Fig. 30-4b. Let the current transfer ratio be 50, meaning that this is the ratio of collector output current to base input current. Then the i_B variations of ±10 μA swing i_C by $50 \times 10 = 500$ μA or 0.5 mA.

The average dc level for I_C is taken as 3 mA for a small-signal low-power transistor. Then the ac signal swing of i_C is ±0.5 mA above and below the axis of 3 mA. In i_C, the p–p signal for the output current is $3.5 - 2.5 = 1$ mA.

The signal changes in i_C produce variations in the voltage drop $i_C R_L$ across the collector load for the output circuit. As a result, V_C varies because it is the difference between V_{CC} of the supply and the voltage drop across R_L. As a formula,

▶ $$V_C = V_{CC} - i_C R_L \qquad \text{(30-1)}$$

For example, with an average level of 3-mA I_C through the 2.2-kΩ R_L, this voltage drop is $0.003 \times 2200 = 6.6$ V. Subtracting 6.6 V from 12 V, the difference is 5.4 V for the average V_C.

The variations of V_C are in the waveform in Fig. 30-4c. This shows the amplified signal output voltage. The average dc level or axis is 5.4 V. When i_B increases, the v_C decreases to 4.3 V because of a larger voltage drop across R_L. On the next half-cycle i_C decreases. Then less voltage across R_L allows v_C to rise to 6.5 V. Then peak-to-peak signal voltage is $6.5 - 4.3 = 2.2$ V for v_C at the collector. This amplified output voltage is 180° out of phase with the signal input voltage at the base.

The basis for Formula (30-1) is just the fact that R_L and the collector-emitter circuit of the transistor are in series with each other as a voltage divider from the high side of V_{CC} to chassis ground. The equivalent collector circuit is shown in Fig. 30-5 with R_L as the external load resistor. The R_Q is the internal resistance of the transistor conducting current from emitter to collector. In this example, Q_1 is conducting 3 mA, resulting in 5.4 V for V_{CE} in the divider with V_L. Note that V_Q is the same as V_{CE}. These values are $12 - 6.6 = 5.4$ V for V_Q. Voltages V_C and V_{CE} are the same here because the emitter is grounded.

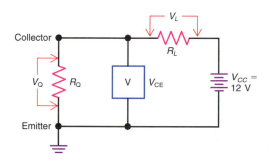

FIG. 30-5 Equivalent voltage divider circuit for R_Q in series with R_L across the supply voltage V_{CC}.

TYPICAL BASE BIAS VOLTAGE In a class A amplifier with collector output current for the full signal cycle, the general requirements are to have the value of I_C that allows the voltage drop $I_C R_L$ to equal about one-half the dc supply voltage. Then the average dc collector voltage V_C is also approximately $V_{CC}/2$. *The base bias V_{BE} is made the value, around 0.6 to 0.7 V, that produces the required I_C with the specified R_L.*

GAIN To calculate the voltage amplification, or gain, with peak-to-peak values for Fig. 30-3,

▶ $$A_V = \frac{V_{\text{out}}}{V_{\text{in}}} \qquad \text{(30-2)}$$

$$A_V = \frac{2.2\ \text{V}}{0.01\ \text{V}} = 220$$

The current amplification is

$$A_I = \frac{i_{\text{out}}}{i_{\text{in}}} \qquad \text{(30-3)}$$

$$A_I = \frac{i_C}{i_B} = \frac{1 \text{ mA p–p}}{0.02 \text{ mA p–p}} = 50$$

The power amplification is

$$A_P = A_V \times A_I \qquad \text{(30-4)}$$

$$A_P = 220 \times 50 = 11,000$$

Since the CE circuit has both voltage and current amplification, the power gain is high. Power gain is desirable because it means the voltage output can drive a low-impedance circuit without too much loss of voltage amplification.

CASCADED STAGES   As shown in Fig. 30-6, the amplifiers $Q1$, $Q2$, $Q3$, and $Q4$ are in cascade. With CE amplifiers, the collector output of one stage drives the base input of the next. It is important to realize that the voltage amplification need not build up the signal to a level greater than ± 0.1 V. This is the maximum signal swing without distortion for junction transistors. However, the cascaded stages provide enough signal current to drive the base of the output stage $Q4$.

FIG. 30-6   Amplifier stages in cascade. Drive increases for each successive stage.

As an example, suppose that $Q4$ is a power output stage to drive a loudspeaker that needs 5 A as the load. With an average level of 5 A for I_C in $Q4$, its I_B would be of the order of 250 mA or 5 A/20, with a current transfer ratio of 20. This I_B of 250 mA can be supplied by the collector output of the driver stage $Q3$. Similarly, $Q3$ with an I_C of 250 mA would have I_B of 12.5 mA with a current ratio of 20. Also, with a current ratio of 50 for $Q2$, its I_B would be 12.5 mA/50 = 0.25 mA, or 250 μA. This drive for base current in $Q3$ can be provided by the collector current of the input stage $Q1$. In summary, each of the cascaded amplifiers increases the signal current enough to drive the next stage.

TEST-POINT QUESTION 30-3

Answers at end of chapter.
a. In Fig. 30-3, how much is the base-bias V_{BE}?
b. In Fig. 30-5, how much is V_Q with V_L of 5.5 V?
c. In Fig. 30-4, how much is the peak-to-peak signal output in i_C?
d. In Fig. 30-4, how much is the peak-to-peak signal input for i_B?

30-4 COLLECTOR CHARACTERISTIC CURVES

The curves in Fig. 30-7 show the volt-ampere characteristics for the collector. I_C on the vertical axis is plotted against V_{CE} on the horizontal axis. Each curve is for a specific I_B. The different curves specify how I_C increases with increases in I_B.

The characteristic curves are provided by the manufacturer in a transistor manual or application notes. For the CE circuit, the collector curves are for different values of I_B; for a CB circuit, they would be for different values of I_E.

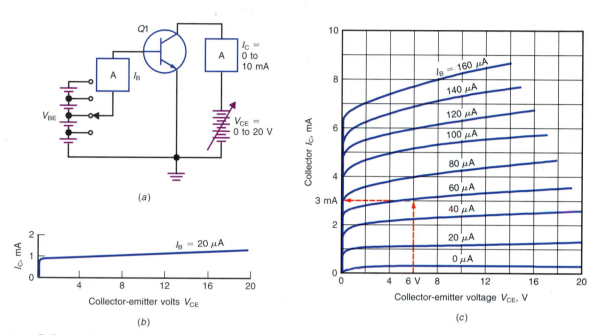

FIG. 30-7   Collector characteristic curves showing how I_C varies with V_{BE}. (a) Experimental CE circuit to determine values. The circuit does not have any load. (b) A single collector curve for a single value of 20 μA for I_B. (c) Family of collector curves for different values of I_B. The dashed arrows indicate the example in the text.

AN EXPERIMENTAL CIRCUIT FOR I_C   Figure 30-7a is a circuit in which the transistor voltages can be varied experimentally in order to determine the effect on I_C. One value of base voltage is used for a specific I_B while V_{CE} is varied to see how much I_C changes. Then the base voltage is set for another value of I_B and V_{CE} is varied again while I_C is measured. The V_{CE} and I_C values for each I_B setting are used for one curve.

The collector and base voltages are labeled V_{CE} and V_{BE} for the general case of a potential difference with respect to emitter. However, they are the same as V_C and V_B with the emitter grounded.

Note that no load resistance is used for the experimental circuit of Fig. 30-7a. Also, there is no signal input or output. The circuit is not an amplifier for signal voltage; it is only an experimental arrangement for measuring the volt-ampere characteristics of the transistor itself without any load.

ONE TYPICAL CURVE   The results for one value of I_B at 20 μA are shown by the curve in Fig. 30-7b. I_C rises from 0 to 1 mA when V_C is increased from 0 to 1 V, approximately, but there is only a slight rise in I_C to about 1.3 mA when V_C is further increased to 20 V. The reason for the small second increase in I_C is that the increase is limited by the value of I_B permitted by the forward voltage at the base junction. For more I_C, the transistor needs more base current. As an example of reading values from the curve, I_C is exactly 1 mA when V_{CE} is 4 V.

THE FAMILY OF CURVES   The results of different values of base current are shown by the family of collector characteristics in Fig. 30-7c. Each curve represents the relationship of I_C and V_C for a specific I_B. Note that the curve for 20 μA of I_B is the same as Fig. 30-7b.

There are several curves, but only one is read at a time. As an example, arrows are shown for the fourth curve up from the bottom, for I_B of 60 μA. The values there are 3 mA of collector current with 6 V at the collector.

The family of curves really shows how the values of I_C and I_B are related for the same collector voltage. If the vertical arrow in Fig. 30-7c is extended up to the next curve, for 80 μA of I_B, the I_C reading will increase to slightly less than 4 mA with the same 6 V at the collector. For the opposite change, when I_B decreases to 40 μA, the I_C is reduced to 2.2 mA, all with 6 V for V_{CE}.

BETA (β) CHARACTERISTIC   This specification for a junction transistor indicates the amount of current gain in the common-emitter circuit, since β compares the collector current I_C to the base current I_B. Specifically,

$$\beta = \frac{I_C}{I_B} \qquad\qquad (30\text{-}5)$$

As an example, when 60 μA of I_B produces 3 mA of I_C,

$$\beta = \frac{I_C}{I_B} = \frac{3 \text{ mA}}{60 \text{ } \mu\text{A}} = \frac{3000 \text{ } \mu\text{A}}{60 \text{ } \mu\text{A}}$$

$$\beta = 50$$

There are no units for β because it is a ratio of two currents. Note that 3 mA here is converted to 3000 μA. This example gives the static or dc value of β. The dynamic or ac value is calculated for changes in the current values. Then the ac value of beta gives the current gain in the CE circuit.

The β value can be used two ways to relate I_C and I_B:

$$I_C = \beta \times I_B$$

or

$$I_B = \frac{I_C}{\beta}$$

For instance, when I_B is 2 mA and β equals 90, then the I_C is 2 × 90 = 180 mA. Typical values of β are 40 to 300 for small-signal transistors and 10 to 30 for power transistors.

DID YOU KNOW?

Power blackouts are caused by failed technology and geomagnetic storms. A task force from commerce, research, and the military is devising a method to provide warnings of predictable space disturbances.

ALPHA (α) CHARACTERISTIC This ratio compares collector current to emitter current:

$$\blacktriangleright \quad \alpha = \frac{I_C}{I_E} \qquad \qquad \textbf{(30-6)}$$

As an example, for a transistor with 3 mA of I_C and 3.06 mA of I_E, the calculations are

$$\alpha = \frac{I_C}{I_E} = \frac{3 \text{ mA}}{3.06 \text{ mA}}$$

$$\alpha = 0.98 \qquad \text{approximately}$$

There are no units for α because it is a ratio of two currents. The α must always be less than 1 because I_C cannot be larger than I_E. The value of 0.98 is typical for α.

TEST-POINT QUESTION 30-4

Answers at end of chapter.
a. What is the I_B for the curve in Fig. 30-7b?
b. From the family of curves in Fig. 30-7c, what is I_C for 8 V of V_{CE} and 140 mA of I_B?
c. Calculate β for 80 μA of I_B and 4 mA of I_C.
d. The I_B is 80 μA and β is 100. Calculate I_C.

30-5 LETTER SYMBOLS FOR TRANSISTORS

Because of the combination of an ac component on a dc axis, it is important to distinguish between the different voltages and currents in a transistor amplifier. In general, there are letter symbols for three kinds of values:

1. Average dc values
2. Instantaneous values of the fluctuating dc waveform
3. Values for the ac signal variations alone

All these are summarized in Table 30-3, which shows how the various letters are used to indicate the different voltages or currents.

The capital letters V and I and their subscripts are used for average dc values. The subscript also is a capital letter. An example is V_C for average dc collector voltage.

Double subscripts that are repeated, as in V_{CC}, indicate the supply voltage that does not change. Also, V_{EE} is used to denote the dc supply voltage for the emitter.

The small letters v and i are used for instantaneous values that vary with the fluctuating dc waveform. As an example, v_C is an instantaneous value of the varying dc collector voltage.

A small letter in the subscript indicates the ac component.

TABLE 30-3 LETTER SYMBOLS FOR TRANSISTORS

SYMBOL	DEFINITION	NOTES
V_{CC}	Collector supply voltage	Same system for collector currents; also for base or emitter voltages and currents. Also applies to drain, gate, and source of field-effect transistors
V_C	Average dc voltage	
v_c	AC component	
v_C	Instantaneous value	
V_c	RMS value of ac component	
I_{CBO}	Collector cutoff current, emitter open	Reverse leakage current
BV_{CBO}	Breakdown voltage, collector to base, emitter open	Ambient temperature T_A is 25°C
h_{fe}	Small-signal forward-current transfer ratio in CE circuit	Same as ac β for CE circuit

The rms value, or effective value, of the ac component is a capital letter. However, its subscript is a small letter. As an example, V_c is the rms value of the ac component of collector voltage.

At the bottom of Table 30-3, additional letter symbols for transistors are listed. For example, I_{CBO} denotes reverse leakage current. The letter O shows which electrode is open when leakage current between the other two electrodes is measured. Therefore, I_{CBO} is leakage current between collector and base with the emitter open.

In the symbol h_{fe}, the h stands for *hybrid parameters*, which are combinations of voltage and current ratios in the forward and reverse directions. In the subscripts, f indicates a forward characteristic from the base input to collector output. The e indicates the common-emitter circuit. The symbol h_{fe} is used often, therefore, because its forward current-transfer ratio is the same as the small-signal or ac β of the transistor in the CE circuit.

TEST-POINT QUESTION 30-5

Answers at end of chapter.

Give the letter symbol for the following:
a. Collector dc supply voltage.
b. Instantaneous value of collector voltage.
c. The ac component of base current.

30-6 FET AMPLIFIERS

Just like the configurations with junction transistors, there are three types of circuit connections for the FET. As shown in Fig. 30-8, these are:

1. *Common-source (CS) circuit.* This circuit corresponds to the common-emitter with junction transistors. With an FET, input signal is applied to

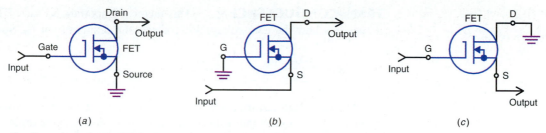

FIG. 30-8 Configurations for the FET. Shown for N-channel depletion type. (*a*) Common-source circuit. (*b*) Common-gate circuit. (*c*) Common-drain circuit.

the gate, which is the control electrode (Fig. 30-8*a*). Amplified output signal is taken from the drain. The source is the common electrode.

2. *Common-gate (CG) circuit.* As shown in Fig. 30-8*b*, the input signal is applied to the source, with output from the drain. The gate is the common electrode.

3. *Common-drain (CD) circuit.* As shown in Fig. 30-8*c*, input signal is applied to the gate, with output from the source. The drain is the common electrode. With the output load impedance in the source circuit, this configuration is a *source follower,* corresponding to the emitter follower with junction transistors.

Of the three types of circuits, the common-source configuration is probably used most often for FET amplifiers, just as the common-emitter circuit is the main type for junction transistors.

FET AMPLIFIER Refer to Fig. 30-9 for a common-source circuit using an N-channel depletion-type FET. The input signal is applied to the gate by the R_1C_1 coupling circuit. In the output circuit, R_L is the load resistor for amplified output signal. The R_L is in series with V_{DD} of 20 V for the drain supply voltage.

The source is the common electrode, even though it is not directly grounded. Actually, the bypass capacitor C_2 provides an effective ac ground for the source electrode.

The combination of R_2 with C_2 provides source bias for the amplifier. Drain current I_D returning to the source through R_2 produces a dc voltage drop that biases the source with respect to the gate. The bypass C_2 provides a steady value for the dc bias. This method is called *self-bias* because it is produced by the I_D of the transistor itself.

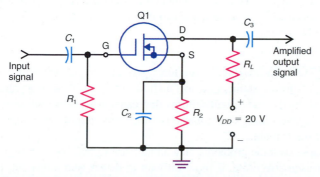

FIG. 30-9 Common-source amplifier circuit with N-channel depletion type of FET.

TRANSCONDUCTANCE g_m   This factor is important for the FET because it specifies how the gate voltage V_G controls the drain current I_D. Specifically,

$$g_m = \frac{\Delta I_D}{\Delta V_G} \tag{30-7}$$

The delta sign (Δ) means a change in value. As an example, suppose that a change of 0.5 V in the gate voltage can increase or decrease the drain current by 4 mA. Then

$$g_m = \frac{\Delta I_D}{\Delta V_G} = \frac{4 \times 10^{-3}\ A}{0.5\ V}$$
$$= 8 \times 10^{-3} = 8000 \times 10^{-6}\ S$$
$$g_m = 8000\ \mu S$$

Note that the unit for g_m is the seimens for conductance because it is a ratio of I/V. Since the conductance is less than 1 S, the g_m values are usually given in microsiemens. The g_m symbol indicates a mutual conductance relation of how the effect of the input voltage at the gate is transferred to the output current in the drain circuit.

TEST-POINT QUESTION 30-6

Answers at end of chapter.
a. Which FET circuit corresponds to the CE amplifier?
b. Which FET circuit is a source follower?
c. Calculate g_m in microsiemens when V_G of 1 V changes I_D by 10 mA.

30-7 TROUBLESHOOTING AMPLIFIER CIRCUITS

With all the advances in modern electronic equipment, troubles are still caused most of the time by a short circuit, open circuit, or low dc supply voltage. The open circuit can be in the transistor itself, resistors, capacitors, coils, or the wiring, especially at plugs, sockets, and connectors. A short circuit can occur in transistors and capacitors. It is a good idea to check the dc supply voltage first, especially with batteries.

TRANSISTOR TROUBLES   Failures generally result from an open weld at the wire leads to the semiconductor, a short circuit at a junction caused by momentary overloads, and circuit failures that cause transistor overheating. In most cases, a defective transistor is internally short-circuited or open, and simple tests will reveal the trouble.

The cause of some problems, like an increase in leakage, a drop in breakdown voltage, or excessive noise, is more difficult to detect, and direct substitution may be the easiest way to localize the fault. Transistor testers are available to check the transistor in or out of the circuit for an open circuit, short circuit, leakage, and β. However, open and short circuits in a transistor can also be checked by using a multimeter to test each junction as a diode. If a DMM has a diode test position, it can be used for a dynamic test of the junction.

IN-CIRCUIT TESTS These tests are helpful because transistors are usually soldered into the circuit board. Measurements for dc voltage can determine if the junctions are intact and if the transistor is conducting normal current.

Refer to Fig. 30-10 for examples in a CE amplifier. This circuit is typical because of the method of obtaining base-emitter bias V_{BE}. In the emitter circuit, self-bias of 2 V is obtained as the voltage drop of I_E through the 470-Ω R_E. This voltage is *self-bias* because V_E depends on the transistor's own emitter current. However, V_E biases $Q1$ in the reverse direction, as this voltage is positive at the N emitter. The purpose of R_E is to provide *bias stabilization,* by preventing I_C from increasing with more leakage current.

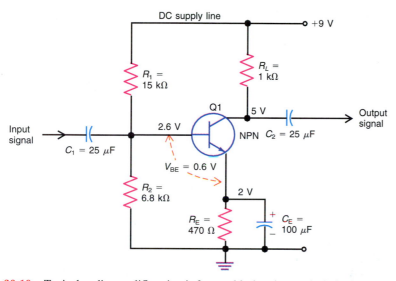

FIG. 30-10 Typical audio amplifier circuit for troubleshooting analysis in the text.

In order to provide forward bias on $Q1$, the voltage divider with R_1 and R_2 is used. The divider provides 2.6 V for V_B. It should be noted that I_B returns to the emitter through R_1. Both V_B and V_E are dc bias voltages, positive with respect to chassis ground. The net voltage for V_{BE}, then, is $2.6 - 2.0 = 0.6$ V, which provides forward bias for the transistor amplifier.

In the collector circuit of $Q1$, the 1-kΩ R_L is the collector load resistor connected to the V^+ supply of 9 V. R_L has a dc voltage drop of 4 V, which allows 5 V for the collector voltage to chassis ground. The actual V_{CE}, though, is $5 - 2 = 3$ V because the emitter has 2 V.

The collector has a positive dc supply for reverse voltage on the NPN transistor. As a CE amplifier, $Q1$ has input signal from the coupling capacitor C_1, with amplified output signal coupled by C_2.

DID YOU KNOW?

Information storage is getting smaller and faster. Today, researchers are going beyond the limits of silicon to find ways to transmit, store, and retrieve data with molecules. The first step in meshing molecular systems with electronics is to turn molecules into magnets and back again.

CHECKING FORWARD BIAS Measure the base-to-emitter voltage by putting the dc voltmeter leads directly across these terminals. In Fig. 30-10, as an example, V_{BE} should read 0.6 V.

If the reading for V_{BE} is zero, the base-emitter junction is short-circuited. For the opposite case, if V_{BE} is 0.8 V or higher, the junction is probably open. One word of caution, though. These voltage readings apply to class A amplifiers, which require a middle value of forward bias. However, in pulse circuits it may be normal to have reverse bias of several volts for V_{BE} in order to cut off I_C until the input pulse drives the transistor into conduction.

CHECKING $I_C R_L$ VOLTAGE To check for correct collector current, put the voltmeter between the collector and the supply voltage to read the voltage drop across the load. In Fig. 30-10, a VOM across the 1-kΩ R_L should read 4 V. Then divide this reading by the dc resistance in the collector circuit to calculate the current I_C. For this example, the normal I_C value is 4 V/1 kΩ = 4 mA.

If there is no voltage drop across R_L, then I_C must be zero. Then V_C at the collector will have the same 9-V value as the supply voltage V_{CC}. For the opposite trouble, excessive I_C can cause excessive voltage across R_L, resulting in zero or very low V_C.

If I_C is zero or very low, the transistor may be open. However, check for opens in the emitter circuit before making a replacement.

If excessive I_C is flowing, short-circuit the base-to-emitter voltage and repeat the test. Under these conditions only the small leakage current should flow. Remember that a junction transistor is cut off without any forward bias. If I_C is still high, the transistor is probably short-circuited.

MEASURING I_C Calculation of collector current may be difficult if the collector circuit contains little resistance or an unknown value, such as a transformer primary. In this case, current measurements may be better. You can open the collector circuit by cutting the foil of the printed-circuit board with a razor blade. Then put the leads of your VOM, set to read milliamperes or amperes, across the cut. You can bridge the cut with solder when the test is finished.

CHECKING THE EMITTER CIRCUIT You can check current by measuring the voltage drop across the emitter resistance and dividing by the value of the resistor. Be careful in your analysis. This voltage will read almost normal even if the emitter resistor is open because the voltmeter resistance then completes the emitter circuit. So check the value of the emitter resistor first.

TEST-POINT QUESTION 30-7

Answers at end of chapter.

Refer to the CE amplifier in Fig. 30-10.
a. How much is the normal dc voltage across R_E to ground?
b. How much is the normal forward bias V_{BE}?
c. How much is V_C when R_L is open?

30   SUMMARY AND REVIEW

- The three types of circuits for junction transistors are the common-emitter, common-collector or emitter follower, and common-base configurations. See Table 30-1.
- In class A operation, output current flows for 360° of the cycle, class B for 180°, and class C for less than 180°. Class A operation has the least distortion, while class C has the best efficiency.
- Typical forward bias for V_{BE} with NPN Si transistors as a class A amplifier is 0.6 V, with higher reverse voltage at the collector. The transistor is cut off without the forward bias. See Table 30-2.
- For junction transistors, β is the ratio of I_C to I_B. The β determines the current gain in a CE circuit. Typical values of β are 10 to 300. The α is the ratio of I_C to I_E. Typical values are 0.98 or 0.99; it must be less than 1.
- In letter symbols for transistors, double subscripts are used for dc supply voltage, as in V_{CC} or V_{DD}. Small letters are used for instantaneous values. See Table 30-3.
- In FET amplifiers, the common-source circuit corresponds to the common-emitter circuit for junction transistors; the common drain circuit corresponds to the common-collector circuit; the source follower corresponds to the emitter follower; the common-gate circuit corresponds to the CB circuit. See Fig. 30-8 for this comparison. The circuit of a common-source FET amplifier circuit is shown in Fig. 30-9.

SELF-TEST

ANSWERS AT BACK OF BOOK.

1. Which is the most common type of amplifier circuit, CB, CE, or CC?
2. Does the CB, CE, or CC circuit have the most gain?
3. Is the current gain for cascaded stages multiplied or added?
4. Which is an emitter follower, the CB, CE, or CC circuit?
5. Which FET circuit corresponds to the CE circuit?
6. Is dc bias of 0.6 V typical for an Si or Ge transistor?
7. Which type of amplifier operation has the least distortion, class A, B, or C?
8. In a class A amplifier, is V_{BE} forward or reverse voltage?
9. Is forward or reverse voltage needed at the collector?
10. Does the N-channel FET take positive or negative drain supply voltage?
11. Is positive collector supply voltage used for a PNP or NPN transistor?
12. Is the letter symbol for collector supply voltage V_{CC} or V_C?
13. With an FET, is the CG, CS, or CD circuit a source follower?
14. Is the current gain in the CE circuit indicated by the α or β of the transistor?

15. Is the value of α or β always less than 1?
16. With I_B of 50 μA and β of 50, how much is I_C?
17. Would class A or C operation be used for an audio amplifier?
18. Without forward bias, is a junction transistor off or on?
19. Does zero ohms across the base-emitter junction indicate a short circuit or an open circuit?
20. In an amplifier with R_L in the collector circuit, should V_C be more or less than V_{CC}?
21. Would a stage for impedence matching use the CB, CC, or CE circuit?
22. Is the forward bias for a CE amplifier a dc or ac voltage?
23. In Fig. 30-10, is self-bias provided by R_E or R_L?

QUESTIONS

1. Make three drawings to illustrate the CB, CE, and CC circuits.
2. Give an application for the CE and CC circuits.
3. Compare the emitter-follower and source-follower circuits.
4. Which FET circuits correspond to the CB, CE, and CC circuits?
5. Show a circuit diagram for an FET amplifier.
6. Specify what the following letter symbols indicate: V_{CC}, V_{DD}, V_{EE}, V_C, h_{fe}, and I_{CBO}.
7. Give two conditions that would cause a junction transistor to be cut off without any current.
8. Define the following classes of operation for amplifiers: A, B, and C.
9. Why would you say that the characteristic curves in Fig. 30-7 apply to the common-emitter circuit?
10. Why are the waveshapes in Fig. 30-4 called a *ladder diagram.*
11. How could a transistor amplifier circuit be used as a switch to operate in the on-off conditions?
12. What is meant by the cutoff condition of a transistor?
13. Refer to the waveform of i_c in Fig. 30-4b. How much current would a dc milliammeter read in the collector circuit?
14. Refer to the collector curves in Fig. 30-7c. Approximately how much bias current I_B would be used for a class A amplifier?
15. For the CE class A amplifier in Fig. 30-10 give the following values: (a) V_E; (b) V_B; (c) V_{BE}; (d) V_{CC}; (e) V_C; (f) V_{CE}.

PROBLEMS

ANSWERS TO ODD-NUMBERED PROBLEMS AT BACK OF BOOK.

1. A CE amplifier has the following values for ac signal: $v_b = \pm 60$ mV and $i_b = \pm 40$ μA; $v_c = \pm 3$ V and $i_c = \pm 3.2$ mA. Calculate (a) voltage gain A_V. (b) Current gain A_I. (c) Power gain A_P.
2. Three cascaded stages have A_I values of 70, 50, and 15. Calculate the total current gain.

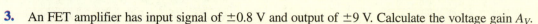

3. An FET amplifier has input signal of ± 0.8 V and output of ± 9 V. Calculate the voltage gain A_V.
4. Base current is 220 mA. Collector current is 3.4 A. Calculate β.
5. A CE amplifier with an R_L of 2 kΩ has the following dc voltages: $V_{CC} = 9$ V, $V_C = 5$ V. Calculate **(a)** Voltage across R_L. **(b)** Value of I_C.
6. $I_E = 10.2$ mA, $I_C = 10.098$ mA, and $I_B = 102$ μA. Calculate α and β.
7. With an FET in the CS circuit, V_{DD} is 20 V, V_D is 12 V, and R_L is 4 kΩ. **(a)** How much is V_{RL}? **(b)** Calculate I_D. **(c)** What value of R_S would be needed for a source bias voltage of 2 V?
8. In an emitter-bias circuit, V_E is 0.8 V and I_E is 0.6 A. Calculate the required value of R_E.
9. Refer to the collector characteristic curves in Fig. 30-7c. For V_{CE} of 8 V and I_B variations of ± 40 μA around a bias current of 80 μA: **(a)** How much is the variation in i_c? **(b)** Calculate β.

CRITICAL THINKING

1. Derive the formula

$$\beta = \frac{\alpha}{1 - \alpha}$$

2. Derive the formula

$$\alpha = \frac{\beta}{\beta + 1}$$

ANSWERS TO TEST-POINT QUESTIONS

30-1 a. CE
 b. CE
 c. emitter follower

30-2 a. class A
 b. class B
 c. class C
 d. class A

30-3 a. 0.6 V
 b. 6.5 V
 c. 1 mA
 d. 0.02 mA

30-4 a. 20 μA
 b. 7 mA
 c. 50
 d. 8 mA

30-5 a. V_{CC}
 b. v_C
 c. i_b

30-6 a. CS
 b. CD
 c. 10,000

30-7 a. 2V
 b. 0.6 V
 c. zero

REVIEW: CHAPTERS 28 TO 30

SUMMARY

- Analog signals have continuous variations, as in a sine wave of V or I. Digital signals consist of combinations of pulses.
- An amplifier circuit increases the amplitude of the signal. The ratio of output signal to input signal is the gain of the amplifier.
- An oscillator is an amplifier circuit with positive feedback to generate ac output from the dc power supply. The oscillator can produce audio or radio frequencies.
- In tuned RF feedback oscillators, the two main types are the Hartley and Colpitts circuits. The former has a tapped coil for feedback; the latter uses a capacitive voltage divider.
- The multivibrator (MV) circuit is used as a pulse generator. The astable MV is a free-running oscillator often used to produce timing pulses for digital circuits.
- Modulation means varying the characteristics of a carrier wave. The modulation is the base-band signal. The main types are amplitude, frequency, and phase modulation (PM).
- A semiconductor diode is a PN junction. The P terminal is the anode, the N side is the cathode. Electron flow is in the direction from cathode to anode, when the anode is positive. Therefore, the diode is used as a rectifier to convert ac input to dc output.
- One diode can serve as a half-wave rectifier. Two diodes are needed for the full-wave rectifier circuit. A bridge rectifier uses four diodes.
- An ac ripple in the fluctuating dc output of a rectifier causes hum interference. With 60-Hz ac input, the hum frequency is 60 Hz for a half-wave rectifier or 120 Hz for a full-wave circuit.
- Digital logic gates are used to control the flow of binary pulses in digital circuits.
- Junction transistors are the NPN and PNP types. There are three electrodes: emitter, base, and collector. The emitter supplies charges to the base to be received by the collector. Forward bias is necessary at the emitter-base junction. The collector-base junction needs reverse voltage. Positive collector voltage is used for NPN transistors.
- The field-effect transistor (FET) has three electrodes: source, gate, and drain, corresponding to emitter, base, and collector in junction transistors. Both types are commonly used for transistor amplifiers. See Table 28-2 for a summary of FET types.
- A semiconductor diode can be tested with an ohmmeter. The reverse R should be much higher than the forward R. This test also applies to the junctions for NPN and PNP transistors.
- The three circuits for junction transistors are common base, common emitter, and common collector. The CE circuit is used most often because it has the most gain. See Table 30-1.
- The corresponding circuits for the FET are common gate, common source, and common drain, also called a source follower.
- In class A operation of amplifiers, the output current flows for the complete input cycle of 360°; class B is 180°; class C is 120°.

- The β characteristic for junction transistors compares collector current to base current. Typical values of β are 10 to 300.
- The α characteristic for junction transistors compares collector current to emitter current. Typical values are 0.98 to 0.99. It must be less than 1.
- The transconductance g_m for the FET compares drain current to gate voltage. Typical values of g_m are about 8000 μmhos.

REVIEW SELF-TEST

ANSWERS AT BACK OF BOOK.

Answer True or False.

1. Transistors are active devices because they can amplify signals.
2. A coupling capacitor has low reactance but a bypass capacitor needs high reactance.
3. With 200-μA input signal and 8-mA output signal, the current gain is 200.
4. An RF choke also blocks direct current.
5. The VHF band of radio frequencies is 30 to 300 MHz.
6. The astable multivibrator circuit is a pulse generator.
7. The reverse resistance of a silicon diode is very high.
8. In a class A amplifier the output current flows for the full 360° of the input cycle.
9. In a class B amplifier the output current flows for 180° of the input cycle.
10. An electrolytic filter capacitor that shows no charging action with an ohmmeter is open.
11. Hole current is a motion of positive charges.
12. An N-channel FET takes positive drain supply voltage.
13. Hole current and electron flow are in opposite directions.
14. Normal voltage across a conducting silicon diode is about 0.7 V.
15. An NPN transistor takes a negative collector voltage.
16. The emitter supplies charges to the base electrode.
17. The emitter-base junction must have forward voltage to have the transistor conduct current.
18. The gate electrode in the FET corresponds to the emitter electrode in junction transistors.
19. A PNP transistor symbol has the emitter arrow into the base.
20. The emitter-follower circuit has high input resistance and low output resistance for impedance matching.
21. The common-source circuit with an FET corresponds to an emitter-follower.
22. With β of 60 and I_B of 400 μA, the I_C equals 2.4 mA.
23. Class A operation in an amplifier provides maximum efficiency and distortion.
24. The unit for transconductance g_m is the ohm.
25. The α characteristic of a transistor must be less than 1.

REFERENCES

Malvino, A.: *Electronic Principles,* Glencoe/McGraw-Hill, Columbus, Ohio.

Schuler, C.: *Electronics: Principles and Applications,* Glencoe/McGraw-Hill, Columbus, Ohio.

CHAPTER 31

DIGITAL ELECTRONICS

Digital electronic circuits operate using only two voltage levels for all of their input and output signals. The two voltage levels most commonly used are 0 V and +5 V. Regardless of the voltage levels used, all input and output signals encountered in digital circuits will be at one of two distinctly different voltage levels. This two-state design allows us to use the binary number system when working with digital circuits. The binary number system uses only two digits, which are 0 and 1. In most digital circuits, binary 0 is used to represent 0 V, and binary 1 is used to represent +5 V.

Modern calculators and computers that process binary numbers use decision-making elements called *logic gates.* There are several different types, such as the AND gate and the OR gate. Logic gates can have many input signals, but they have only one output signal.

Digital logic circuits are often classified into two very broad categories: combinational logic circuits and sequential logic circuits. Generally, a circuit is considered a combinational logic circuit if its output goes either LOW or HIGH with a specified combination of input signals. The order or sequence in which the inputs are applied is not important. What is important, though, is that the correct combination of inputs exists for the desired output. Sequential logic circuits, on the other hand, must have a definite order or sequence for their inputs before the desired output is obtained. The basic building block in combinational logic circuits is the logic gate; the basic building block of the sequential logic circuit is the flip-flop.

CHAPTER OBJECTIVES

Upon completion of this chapter, you should be able to:

- *Count* using the binary and hexadecimal number systems.
- *Convert* from the binary and hexadecimal number systems to the decimal number system and vice versa.
- *Understand* the BCD system and ASCII code.
- *Describe* the operation of and construct truth tables for the inverter, AND, OR, NAND, NOR, XOR, and XNOR logic gates.
- *Understand* boolean algebra and DeMorgan's theorem.
- *Define* what is meant by the terms *active high* and *active low*.
- *Explain* how to handle unused inputs on logic gates.
- *Derive* a boolean expression from a truth table.
- *Simplify* boolean expressions.
- *Describe* the operation of *RS* flip-flops, *D*-type flip-flops, and *JK* flip-flops.
- *Understand* binary counters.
- *Identify* the new rectangular logic symbols.

IMPORTANT TERMS IN THIS CHAPTER

active HIGH	counter	nibble
active LOW	DeMorgan's theorem	NOR gate
AND gate	flip-flop	OR gate
base	hexadecimal	radix
binary	inverter	reset
bit	least significant bit (LSB)	set
boolean algebra	minterm	truth table
byte	most significant bit (MSB)	XNOR gate
chunking	NAND gate	XOR gate

TOPICS COVERED IN THIS CHAPTER

31-1 COMPARING BINARY AND DECIMAL NUMBERS

In the binary number system there are only two digits, 0 and 1. Binary numbers, then, are just strings of 0s and 1s. In a moment we will see how to determine the decimal equivalent of a binary number, such as 1001. How to count with binary numbers will be explained later in this section.

All number systems have a *base* or *radix,* which specifies how many digits can be used in each place count. For binary numbers, the base is 2, with 0 and 1 as the only two digits. In the decimal system, the base is 10, so there are ten digits that can be used for each place count. The digits are 0, 1, 2, 3, 4, 5, 6, 7, 8, and 9. The decimal number system is familiar because it is used by all of us in our everyday world.

Each digit position in both the binary and decimal number systems has a specified weight in the value of the number. For binary numbers, the position represents a power of 2, such as 2, 4, 8, and 16. For decimal numbers, each digit position represents a power of 10, such as 100 and 1000.

The weight distribution for digit positions in the binary number system is illustrated in Fig. 31-1. Notice that the value, or weight, in each position doubles as we move left, because the base is 2. We know that $2^0 = 1$ is true, because any number raised to the zero power* equals 1. For the digit positions in the binary number 1001, the decimal value is $8 + 0 + 0 + 1 = 9$. Notice that the procedure is to add only those bit positions that contain a 1 in the original binary number.

FIG. 31-1   Weight distribution for binary numbers.

Typical binary numbers are often written in groups of four or eight digits. Examples are 1001 and 10010110. Each digit, either 0 or 1, is referred to as a *bit.* A string of four bits is called a *nibble,* and eight bits make a *byte.* Thus, 1001 is a nibble and 10010110 is a byte.

The weight distribution for digit places in the decimal system is shown in Fig. 31-2. Each positional value, or weight, increases by a factor of 10. Notice that as we move to the left, the place values are 10, 100, 1000, 10,000, and 100,000. Consider 2367 as a typical decimal number. This value is determined as

$$(2 \times 1000) + (3 \times 100) + (6 \times 10) + (7 \times 1) = 2367$$

etc. ←	10^5	10^4	10^3	10^2	10^1	10^0
	100,000	10,000	1000	100	10	1

FIG. 31-2   Weight distribution for decimal numbers.

For either binary or decimal numbers, the rightmost digit is referred to as the least significant digit, or LSD, because its positional value, or weight, is the

---

*See Grob, *Mathematics for Basic Electronics* for an explanation of exponents.

lowest. For the decimal number 2367, the 7 is the LSD. In the binary number 1001, the 1 at the right is the LSD.

The left-most digit is the most significant digit, or MSD, because its positional value, or weight, is the highest. For the decimal number 2367 the 2 is the MSD with a value of 2000. In the binary number 1001, the 1 at the left is the MSD with the value of 8 in decimal terms.

The method for counting with binary numbers is illustrated in Fig. 31-3. Only four places are shown here, but the same idea applies to more places, with the positional values shown in Fig. 31-1. For the numbers in the bottom row of Fig. 31-3 the values are

$$(1111)_2 = 8 + 4 + 2 + 1 = (15)_{10}$$

Note that the subscripts indicate that 1111 is to base 2 (and therefore a binary number), and 15 is to base 10 (so it's a decimal number).

Positional values

2^3	2^2	2^1	2^0	
8	4	2	1	**Decimal count**
0	0	0	0	$= 0 + 0 + 0 + 0 = 0$
0	0	0	1	$= 0 + 0 + 0 + 1 = 1$
0	0	1	0	$= 0 + 0 + 2 + 0 = 2$
0	0	1	1	$= 0 + 0 + 2 + 1 = 3$
0	1	0	0	$= 0 + 4 + 0 + 0 = 4$
0	1	0	1	$= 0 + 4 + 0 + 1 = 5$
0	1	1	0	$= 0 + 4 + 2 + 0 = 6$
0	1	1	1	$= 0 + 4 + 2 + 1 = 7$
1	0	0	0	$= 8 + 0 + 0 + 0 = 8$
1	0	0	1	$= 8 + 0 + 0 + 1 = 9$
1	0	1	0	$= 8 + 0 + 2 + 0 = 10$
1	0	1	1	$= 8 + 0 + 2 + 1 = 11$
1	1	0	0	$= 8 + 4 + 0 + 0 = 12$
1	1	0	1	$= 8 + 4 + 0 + 1 = 13$
1	1	1	0	$= 8 + 4 + 2 + 0 = 14$
1	1	1	1	$= 8 + 4 + 2 + 1 = 15$

FIG. 31-3 Counting in the binary system.

TEST-POINT QUESTION 31-1

Answers at end of chapter.
a. What digits are used in the binary number system?
b. What digits are used in the decimal number system?
c. How many bits are in a byte?
d. What binary number follows 0111?

31-2 DECIMAL TO BINARY CONVERSION

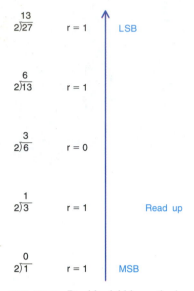

FIG. 31-4 Double-dabble method of converting a decimal number to its binary equivalent, using the remainders (r) 0 or 1.

It may be desirable or necessary to convert a decimal number to its binary equivalent. The method described here is called *double-dabble,* because it requires successive divisions by 2. When converting a decimal number to its binary equivalent, be sure to write down each quotient and its remainder, as shown for the example (decimal 27) in Fig. 31-4. First, divide 27 by 2, which is 13 with a remainder of 1. This bit of 1 is the least significant bit (LSB) in the resultant binary number. Next, divide the quotient of 13 by 2 to obtain a new quotient of 6 with a remainder of 1 again. This 1 is the next bit in the resultant binary number. Notice that you read the bits upward in Fig. 31-4.

Continue the process of dividing the new quotient by 2 and noting whether the remainder is 0 or 1 until you have performed as many divisions as possible (i.e., until you have obtained a quotient of 0 with a remainder of 1). The final remainder of 1 is the most significant bit (MSB) in the binary equivalent number. Reading the remainders in Fig. 31-4 from bottom to top we have 11011 as the binary equivalent of decimal 27. The values are $16 + 8 + 0 + 2 + 1$. Remember that when looking at the divisions and their remainders, the LSB appears at the top and the MSB is at the bottom.

31-3 HEXADECIMAL NUMBERS

Hexadecimal numbers are used extensively in the microcomputer field. As we will see, binary numbers start to get very lengthy and therefore become quite cumbersome to work with. However, hexadecimal numbers, with a base of 16, are much shorter and, therefore, much easier to work with. The first ten digits in the hexidecimal system are represented by the numbers 0 through 9, and the letters A through F are used to represent the numbers 10, 11, 12, 13, 14, and 15 respectively.

As is true for binary and decimal numbers, each digit in the hexidecimal system has a positional value or weight. For the right-most digit, the positional value, or weight, corresponds to 16^0 or 1, the next digit to the left corresponds to 16^1 or 16, and so on. Each digit to the left has a positional value, or weight, that increases in ascending powers of 16. The weight distribution for the hexadecimal number system is shown in Fig. 31-5.

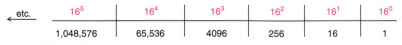

← etc.	16^5	16^4	16^3	16^2	16^1	16^0
	1,048,576	65,536	4096	256	16	1

FIG. 31-5 Weight distribution for a hexadecimal number system.

HEXADECIMAL COUNTING   Figure 31-6 shows the counting sequence. Start with 0 in the 16^0, or 1s, column and proceed down until the digit "F" appears. Then, for the next count, the digit in the 1s column returns to 0, and the digit in the 16^1, or 16s, column advances by 1. This process continues until the digits "FF" are reached. For the next count we start over with 0s in the 16^1 and 16^0 columns and add 1 to the 16^2 column.

Positional values				Decimal count
16^3	16^2	16^1	16^0	
4096	256	16	1	
			0	= 0
			1	= 1
			2	= 2
			3	= 3
			4	= 4
			5	= 5
			6	= 6
			7	= 7
			8	= 8
			9	= 9
			A	= 10
			B	= 11
			C	= 12
			D	= 13
			E	= 14
			F	= 15
		1	0	16 + 0 = 16
		F	F	240 + 15 = 255
	1	0	0	256 + 0 + 0 = 256
	F	F	F	3840 + 240 + 15 = 4095
1	0	0	0	4096 + 0 + 0 + 0 = 4096

FIG. 31-6   Counting in the hexadecimal system.

HEXADECIMAL TO DECIMAL CONVERSIONS   It is sometimes necessary or desirable to convert a hexadecimal number to its decimal equivalent. To convert hexadecimal number $B49F_{16}$, for example, start by writing the positional values, or weights, above each digit in the hexadecimal number:

$$16^3 \qquad 16^2 \qquad 16^1 \qquad 16^0$$
$$B \qquad\quad 4 \qquad\quad 9 \qquad\quad F$$

Next, multiply each digit by its positional value and add them all together. The resultant number is the decimal equivalent of the hexadecimal number. This can be shown as

$$(11 \times 4096) + (4 \times 256) + (9 \times 16) + (15 \times 1)$$
$$45{,}046 \quad + \quad 1024 \quad + \quad 144 \quad + \quad 15 \quad = 46{,}239$$

Thus, $B49F_{16} = 46{,}239_{10}$.

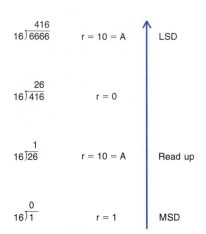

$$\begin{array}{ll} \dfrac{416}{16\overline{)6666}} & r = 10 = A \end{array}$$ LSD

$$\begin{array}{ll} \dfrac{26}{16\overline{)416}} & r = 0 \end{array}$$

$$\begin{array}{ll} \dfrac{1}{16\overline{)26}} & r = 10 = A \end{array}$$ Read up

$$\begin{array}{ll} \dfrac{0}{16\overline{)1}} & r = 1 \end{array}$$ MSD

FIG. 31-7 Hex-dabble method of converting a decimal number to its hexadecimal equivalent.

DECIMAL TO HEXADECIMAL CONVERSIONS　To convert from a decimal number to a hexadecimal number, we can use hex-dabble. The idea is the same as double-dabble, but with hex-dabble we have successive divisions by 16 rather than 2. You know the process is complete when you obtain a quotient of 0 and a remainder ranging anywhere from 1 to F. The example in Fig. 31-7 illustrates how the hex-dabble technique is used to obtain the hexadecimal equivalent of decimal number 6666_{10}.

　The first step is to divide 6666 by 16 to obtain a quotient of 416 with a remainder of 10. Next, we divide the quotient of 416 by 16 to obtain a quotient of 26 with a remainder of 0. Next 26 is divided by 16 to obtain a quotient of 1 with a remainder of 10. Finally 1 is divided by 16 to obtain a quotient of 0 with a remainder of 1. Therefore $6666_{10} = 1A0A_{16}$. Notice that the last indicated remainder is the MSD in the final answer, while the first remainder obtained is the LSD. Notice also that when remainders of 10, 11, 12, 13, 14, and 15 are obtained, we substitute the equivalent hexadecimal digit (A, B, C, D, E, and F, respectively) as the remainder.

HEXADECIMAL TO BINARY CONVERSIONS　To convert a hexadecimal number to a binary number, change each hexadecimal digit to its 4-bit binary equivalent. As an example, let's convert FAF_{16} to its binary equivalent.

$$F \qquad\qquad A \qquad\qquad F$$
$$1111 \qquad 1010 \qquad 1111$$

Therefore, we can say that $FAF_{16} = 111110101111_2$. However for easier reading, it is appropriate to show the binary number with spaces between each 4-bit nibble, such as:

$$FAF_{16} = 1111\ 1010\ 1111_2$$

BINARY TO HEXADECIMAL CONVERSIONS　To go from a binary to a hexadecimal number, simply reverse the process. For example, to convert binary 10100001001_2 to its hexadecimal equivalent, break the binary number into 4-bit groups. You must begin grouping from the far *right*. Then convert each 4-bit group to its hexidecimal equivalent. That is:

$$0101 \qquad\quad 0000 \qquad\quad 1001$$
$$5 \qquad\qquad\quad 0 \qquad\qquad\quad 9$$

(Notice that the 4-bit group on the left begins with a zero. When the MSB of the binary equivalent is 0, it can be dropped without affecting the value of the binary number.) Thus, $10100001001_2 = 509_{16}$.

You can see how much shorter it is to say, think, and write 509_{16} than one, zero, one, zero, zero, zero, zero, one, zero, zero, one. Hexadecimal numbers are simply much easier for technicians and engineers to deal with. The process of replacing long strings of data such as 10100001001 with a much shorter string, such as 509_{16}, is known as *chunking*. See Table 31-1 for a comparison of binary, decimal, and hexadecimal numbers.

TEST-POINT QUESTION 31-3

Answers at end of chapter.
a. List the digits used in the hexadecimal number system.
b. Convert hexadecimal number D104 to its decimal equivalent.
c. Find the binary equivalent for hexadecimal 2C14.
d. Find the hexadecimal equivalent of binary 101110011.

TABLE 31-1   BINARY, DECIMAL, AND HEXADECIMAL NUMBERS COMPARED

BINARY	DECIMAL	HEXADECIMAL
0000	0	0
0001	1	1
0010	2	2
0011	3	3
0100	4	4
0101	5	5
0110	6	6
0111	7	7
1000	8	8
1001	9	9
1010	10	A
1011	11	B
1100	12	C
1101	13	D
1110	14	E
1111	15	F

31-4 BINARY CODED DECIMAL SYSTEM

Another very commonly used number system in the field of digital electronics is the binary coded decimal (BCD) system. This system is different from the ordinary number system in that it expresses each decimal digit as a 4-bit nibble. For example, it may be desirable or necessary to convert 489_{10} to a binary coded decimal. It is done as follows:

4	8	9
0100	1000	1001

Note that the highest BCD value that a 4-bit nibble could represent is 9, which would be 1001_2 in binary. See Table 31-2 for decimal numbers and their equivalent BCD values. When using the BCD number system, remember that all zeros must be retained, unlike a binary number where leading zeros can be dropped.

The BCD number system is used when it is necessary to transfer decimal information into or out of a digital machine. Examples of digital machines include digital clocks, calculators, digital voltmeters, and frequency counters.

TABLE 31-2 BINARY CODED DECIMAL (BCD) VALUES

DECIMAL	BINARY CODED DECIMAL
0	0000
1	0001
2	0010
3	0011
4	0100
5	0101
6	0110
7	0111
8	1000
9	1001
10	0001 0000
11	0001 0001
12	0001 0010
↓	↓
128	0001 0010 1000
129	0001 0010 1001

Answers at end of chapter.
a. Convert the decimal number 245 to a binary coded decimal (BCD).
b. Write the BCD equivalent of decimal number 1056.
c. What is the decimal value for the BCD number 0101 1000 0111 0100?

31-5 THE ASCII CODE

For information to be transferred into or out of a computer, numbers, letters, and several other symbols must be translated into binary code. The system used is the American Standard Code for Information Interchange, or ASCII (pronounced "ask-key"). The ASCII code is an alphanumeric code; it has binary values for each letter, number, and symbol. The ASCII code has been used to standardize codes for numbers, letters, and symbols in equipment such as printers, keyboards, and computer displays. Each keystroke on an ASCII keyboard produces a corresponding binary code for the designated character.

The breakdown of the ASCII code is shown in Table 31-3. Each number, letter, and symbol is represented by a 7-bit binary word in the form of X_6, X_5, X_4, X_3, X_2, X_1, X_0, where X_6 is the first bit. As an example, the ASCII code for the

TABLE 31-3   BINARY VALUES IN THE ASCII CODE

$X_3X_2X_1X_0$	010	011	100	101	110	111
			$X_6X_5X_4$			
0000	SP	0	@	P		p
0001	!	1	A	Q	a	q
0010	"	2	B	R	b	r
0011	#	3	C	S	c	s
0100	$	4	D	T	d	t
0101	%	5	E	U	e	u
0110	&	6	F	V	f	v
0111	'	7	G	W	g	w
1000	(	8	H	X	h	x
1001	)	9	I	Y	i	y
1010	*	:	J	Z	j	z
1011	+	;	K		k	
1100	,	<	L		l	
1101	—	=	M		m	
1110	.	>	N		n	
1111	/	?	O		o	

capital letter "W" is 1010111. Table 31-3 shows that the X_6, X_5, X_4 bits are 101 and the X_3, X_2, X_1, X_0 bits are 0111.

TEST-POINT QUESTION 31-5

Answers at end of chapter.
   a. What capital letter corresponds to 1000101 in the ASCII code?
   b. What is the binary ASCII code for a question mark?

31-6 LOGIC GATES, SYMBOLS, AND TRUTH TABLES

A logic gate is a circuit that has one or more input signals but only one output signal. All logic gates can be analyzed by constructing a truth table. Truth tables list all input possibilities and the corresponding output for each input.

INVERTERS   The inverter is the simplest of all logic gates. It has only one input and one output, where the output is the opposite of the input. The schematic symbol for a logic inverter is shown in Fig. 31-8a and b. (The small bubble on the inverter diagram represents inversion. Notice that the bubble can be shown on either the input or output side without affecting the way the inverter operates. The reason why the bubble is shown on one or the other side for logic diagrams will be discussed in Sec. 31-9.)

A binary 0 represents 0 V, and a binary 1 represents +5 V. It is common to refer to a binary 0 as a LOW input or output, and a binary 1 as a HIGH input or output. The logic inverter works like this: When the input A is LOW, or a 0 V, the output X is HIGH, or at +5 V. Also, when the input A is HIGH, or at +5 V, the output X is LOW, or at 0 V.

The number of input possibilities for the truth table in Fig. 31-8c is 2^1 or 2, because the logic gates in Fig. 31-8a and b have only one input. We know the input can be either 0 or 1. In general, the number of possibilities listed in the truth table is 2^N, where N is the number of inputs to the logic gate.

OR GATES   Another commonly used logic gate is the OR gate. An OR gate has two or more inputs but only one output. The logic symbol for a 2-input OR gate and its truth table are shown in Fig. 31-9.

For any OR gate, the output X is LOW when *all* inputs are LOW. However, the output X of any OR gate is HIGH if any or all inputs are HIGH. The OR gate in Fig. 31-9 will have a HIGH X output if either or both inputs A and B are HIGH. The output X will be LOW when both inputs A and B are LOW. Notice that for a 2-input truth table there are 2^2 (or 4) input combinations of 0s and 1s.

Now look at Fig. 31-10, which shows a 3-input OR gate and its corresponding truth table. Notice in the truth table that the number of different input combinations equals 2^3, or 8. The output X in the truth table of Fig. 31-10 is LOW only when all inputs A, B, and C are LOW. When any or all inputs A, B, and C are HIGH, the output X is HIGH. The logic symbol for a multiple-input OR gate is basically the OR gate symbol drawn with the required number of inputs.

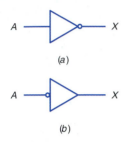

A	X
0	1
1	0

(c)

FIG. 31-8   Inverter logic gates. (a) Logic symbol with bubble at output. (b) Bubble at input. (c) Truth table.

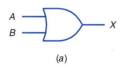

(a)

A	B	X
0	0	0
0	1	1
1	0	1
1	1	1

(b)

FIG. 31-9   Two-input OR gate. (a) Logic symbol. (b) Truth table.

AND GATES   The AND gate is another logic gate. An AND gate has two or more inputs but only one output. The logic symbol for a 2-input AND gate and its truth table are shown in Fig. 31-11. For any AND gate, the output X is HIGH when *all* inputs are HIGH. However, the output X of any AND gate is LOW if any or all inputs are LOW. For the 2-input AND gate in Fig. 31-11, the output X is HIGH only when both inputs A and B are HIGH. The output X is LOW if either or both inputs A and B are LOW.

Figure 31-12 shows the logic symbol and truth table for a 3-input AND gate. Notice from the truth table that the output is HIGH only when all inputs A, B, and C are HIGH. Also notice that the output is LOW for all other input combinations because at least one of the inputs A, B, or C is LOW.

The logic symbol for a multiple-input AND gate uses a basic AND gate symbol drawn with the required number of inputs.

NOR GATES   The NOR gate has two or more inputs but only one output. The logic symbol for a 2-input NOR gate and its truth table are shown in Fig. 31-13a on page 904. The output X is LOW if either or both inputs A and B are HIGH. The output X is HIGH only when both inputs A and B are LOW.

The NOR gate actually performs a logic function identical to that of an OR gate followed by an inverter. This is shown in Fig. 31-13b. Notice that the output of the OR gate is connected to the input of the inverter. If the output of the OR gate is HIGH, then the inverter output X is LOW. If the output of the OR gate is LOW, then the output X is HIGH. Therefore, the truth table for the logic circuit in Fig. 31-13b is the same as for the NOR gate in Fig. 31-13a.

A NOR gate with three or more inputs reacts the same way as a 2-input NOR gate. That is, the output X is LOW if any or all inputs are HIGH, and the output X is HIGH only when *all* inputs are LOW.

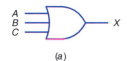

(a)

A	B	C	X
0	0	0	0
0	0	1	1
0	1	0	1
0	1	1	1
1	0	0	1
1	0	1	1
1	1	0	1
1	1	1	1

(b)

FIG. 31-10   Three-input OR gate. (a) Logic symbol. (b) Truth table.

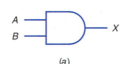

(a)

A	B	X
0	0	0
0	1	0
1	0	0
1	1	1

(b)

FIG. 31-11   Two-input AND gate. (a) Logic symbol. (b) Truth table.

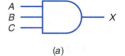

(a)

A	B	C	X
0	0	0	0
0	0	1	0
0	1	0	0
0	1	1	0
1	0	0	0
1	0	1	0
1	1	0	0
1	1	1	1

(b)

FIG. 31-12   Three-input AND gate. (a) Logic symbol. (b) Truth table.

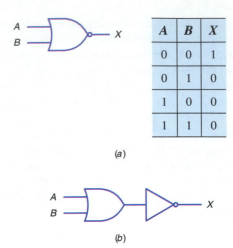

A	B	X
0	0	1
0	1	0
1	0	0
1	1	0

(a)

(b)

FIG. 31-13 The NOR gate. (*a*) Logic symbol with truth table. (*b*) Equivalent logic circuit for a NOR gate.

NAND **GATES**   The NAND gate also has two or more inputs but only one output. The logic symbol for a 2-input NAND gate and its truth table are shown in Fig. 31-14*a*. Notice that the output *X* is HIGH if either or both inputs *A* and *B* are LOW. The output *X* is LOW only when both inputs *A* and *B* are HIGH. The NAND gate actually performs a logic function identical to that of an AND gate followed by an inverter. This is shown in Fig. 31-14*b*.

Figure 31-14*b* shows that the output *X* of the AND gate is connected to the input of the inverter. If the output of the AND gate is HIGH, then the output *X* is LOW. If the output of the AND gate is LOW, then the output *X* is HIGH. Therefore, the truth table for the logic circuit in Fig. 31-14*b* is the same as for the NAND gate in Fig. 31-14*a*.

A NAND gate with three or more inputs reacts the same way as a 2-input NAND gate. That is, the output *X* is HIGH when any or all inputs are LOW, and the output *X* is LOW only when *all* inputs are HIGH.

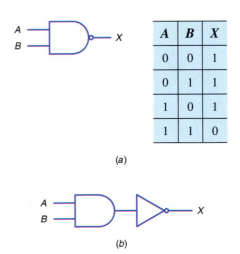

A	B	X
0	0	1
0	1	1
1	0	1
1	1	0

(a)

(b)

FIG. 31-14 The NAND gate. (*a*) Logic symbol with truth table. (*b*) Equivalent logic circuit for a NAND gate.

EXCLUSIVE OR (XOR) AND EXCLUSIVE NOR (XNOR) GATES Two other logic gates are exclusive OR and exclusive NOR gates. The logic symbols and truth tables for each are shown in Fig. 31-15. Notice that for the XOR gate in Fig. 31-15a the output X is HIGH only when the inputs A and B are different. For a multiple-input XOR gate with three or more inputs, the output X will be HIGH only if an odd number of 1s is applied to the inputs. If an even number of 1s is applied to the inputs of an XOR gate, then the output X is LOW.

For the XNOR gate in Fig. 31-15b the output X is HIGH only when both inputs are the same. The XNOR gate is equivalent to an XOR gate followed by an inverter. For a multiple-input XNOR gate with three or more inputs, the output x will be HIGH only when an even number of 1s is applied to the inputs. If an odd number of 1s is applied to the inputs of an XNOR gate, the output x will be LOW. It is important to note that zero 1s is an even number when looking at the truth tables.

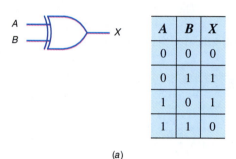

A	B	X
0	0	0
0	1	1
1	0	1
1	1	0

(a)

A	B	X
0	0	1
0	1	0
1	0	0
1	1	1

(b)

FIG. 31-15 The XOR and NXOR gates. (a) Logic symbol and truth table for a 2-input XOR gate. Note the extra curved line at the front of the symbol. (b) Logic symbol and truth table for a 2-input XNOR gate.

TEST-POINT QUESTION 31-6

Answers at end of chapter.

Answer True or False.
a. A HIGH input applied to an inverter produces a LOW output.
b. A 4-input AND gate will have a HIGH output only when all inputs are HIGH.
c. The output of an OR gate will be LOW if any or all inputs are HIGH.
d. A 2-input NAND gate has a LOW output when both inputs are HIGH.
e. A 3-input NOR gate has a HIGH output when all inputs are LOW.
f. A XNOR gate with two inputs has a HIGH output when both inputs are the same.
g. A XOR gate with two inputs has a LOW output when both inputs are the same.

31-7 BOOLEAN ALGEBRA

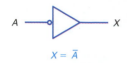

FIG. 31-16 Inverter logic gate. Boolean expression is $X = \bar{A}$.

It is very common to express the operation of a logic gate, or combination of logic gates, using boolean algebra. For the basic logic inverter in Fig. 31-16, the boolean expression would be $X = \bar{A}$. The overbar or "not sign" above the input variable A represents inversion, or complementing. To *invert,* or *complement,* a binary number means to change it to the opposite state, such as changing a 0 to a 1, or a 1 to a 0. For the inverter in Fig. 31-16, X = not A, or $X = \bar{A}$. Then,

$$\text{If } A = 0, \text{ then } X = \bar{0}, \text{ or } X = 1 \quad \text{and}$$
$$\text{If } A = 1, \text{ then } X = \bar{1}, \text{ or } X = 0$$

BOOLEAN ALGEBRA FOR OR GATES   The boolean expression for the OR gate in Fig. 31-17a is $A + B = X$. The "+" sign stands for OR addition. The following truth table shows all of the possibilities for the inputs.

$A + B = X$
$0 + 0 = 0$
$0 + 1 = 1$
$1 + 0 = 1$
$1 + 1 = 1$

According to the truth table, when 0 is ORed with 0, the result equals 0. Also, any variable ORed with 1 equals 1.

In Fig. 31-17b we see that when A is ORed with 0, the output will be whatever A is. If $A = 0$, then $0 + 0 = 0$. If $A = 1$, then $1 + 0 = 1$.

We see in Fig. 31-17c that when A is ORed with 1, the output will be 1. If $A = 0$, then $0 + 1 = 1$. If $A = 1$, then $1 + 1 = 1$.

In Fig. 31-17d we see that when A is ORed with A, then the output will be whatever A is. If $A = 0$, then $0 + 0 = 0$. If $A = 1$, then $1 + 1 = 1$.

In Fig. 31-17e we see that if A is ORed with its complement, then the output is 1. If $A = 0$, then $\bar{A} = 1$. If $A = 1$, then $\bar{A} = 0$. The output must be 1 because any variable ORed with 1 equals 1.

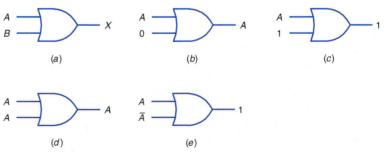

FIG. 31-17   Boolean algebra for OR gates. (*a*) $A + B = X$. (*b*) $A + 0 = A$. (*c*) $A + 1 = 1$. (*d*) $A + A = A$. (*e*) $A + \bar{A} = 1$.

To summarize Fig. 31-17:

0 ored with 0 equals 0

0 ored with 1 equals 1

1 ored with 1 equals 1

$$A + 0 = A$$

$$A + 1 = 1$$

$$A + A = A$$

$$A + \overline{A} = 1$$

BOOLEAN ALGEBRA FOR AND GATES

The boolean expression for the AND gate in Fig. 31-18a is $A \cdot B = X$, where the multiplication dot stands for the AND operation. (Note that the multiplication dot is often omitted, so the expression may appear as $AB = X$.) For Fig. 31-18a we have the following input possibilities:

$A \cdot B$	X
$0 \cdot 0$	$= 0$
$0 \cdot 1$	$= 0$
$1 \cdot 0$	$= 0$
$1 \cdot 1$	$= 1$

The truth table for an AND gate tells us that 0 ANDed with any variable equals 0. Also, 1 ANDed with 1 equals 1.

In Fig. 31-18b we see that when A is ANDed with 0, the output will be 0. If $A = 0$, then $0 \cdot 0 = 0$. If $A = 1$, then $1 \cdot 0 = 0$.

We see in Fig. 31-18c that when A is ANDed with 1, the output will be whatever A is. If $A = 0$, then $0 \cdot 1 = 0$. If $A = 1$, then $1 \cdot 1 = 1$.

In Fig. 31-18d we see that when A is ANDed with A, then the output will be whatever A is. If $A = 0$, then $0 \cdot 0 = 0$. If $A = 1$, then $1 \cdot 1 = 1$.

In Fig. 31-18e we see that if A is ANDed with its complement, then the output is 0. If $A = 0$, then $\overline{A} = 1$. If $A = 1$, then $\overline{A} = 0$. The output must be 0 because any variable ANDed with 0 equals 0.

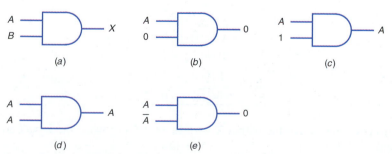

FIG. 31-18 Boolean algebra for AND gates. (a) $AB = X$. (b) $A \cdot 0 = 0$. (c) $A \cdot 1 = A$. (d) $A \cdot A = A$. (e) $A \cdot \overline{A} = 0$.

To summarize Fig. 31-18:

$$0 \text{ ANDed with } 0 \text{ equals } 0$$
$$0 \text{ ANDed with } 1 \text{ equals } 0$$
$$1 \text{ ANDed with } 1 \text{ equals } 1$$
$$A \cdot 0 = 0$$
$$A \cdot 1 = A$$
$$A \cdot A = A$$
$$A \cdot \overline{A} = 0$$

BOOLEAN ALGEBRA FOR OTHER LOGIC GATES

The boolean expression for the NOR gate in Fig. 31-19 is $\overline{A + B} = X$, which can be read as $X =$ not A or B. To perform the boolean algebra operation, it is important to perform the OR portion first, and then invert the OR sum. For Fig. 31-19 we have:

$$\overline{A + B} = X$$

FIG. 31-19   The NOR gate. $\overline{A + B} = X$.

$\overline{A + B}$	=	X
$\overline{0 + 0} = \overline{0}$	=	1
$\overline{0 + 1} = \overline{1}$	=	0
$\overline{1 + 0} = \overline{1}$	=	0
$\overline{1 + 1} = \overline{1}$	=	0

The boolean expression for the NAND gate in Fig. 31-20 is $\overline{A \cdot B} = X$, which can be read as $X =$ not $A \cdot B$. To perform the boolean algebra operation, first the inputs must be ANDed, and then the inversion is performed. For Fig. 31-20 we have:

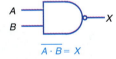

$$\overline{A \cdot B} = X$$

FIG. 31-20   The NAND gate. $\overline{A \cdot B} = X$.

$\overline{A \cdot B}$	=	X
$\overline{0 \cdot 0} = \overline{0}$	=	1
$\overline{0 \cdot 1} = \overline{0}$	=	1
$\overline{1 \cdot 0} = \overline{0}$	=	1
$\overline{1 \cdot 1} = \overline{1}$	=	0

The boolean expression for the exclusive OR gate in Fig. 31-21 is $A \oplus B = X$, which can be read as A XOR $B = X$. For Fig. 31-21 we have:

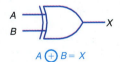

$$A \oplus B = X$$

FIG. 31-21   The XOR gate. $A \oplus B = X$. Note the $\oplus$ symbol used for the XOR and NXOR gates.

$A \oplus B = X$
$0 \oplus 0 = 0$
$0 \oplus 1 = 1$
$1 \oplus 0 = 1$
$1 \oplus 1 = 0$

This table tells us that the output X is HIGH, or 1, only when the inputs are different.

For XOR addition with more than 2-input variables, simply obtain the XOR sum of two variables and then XOR this sum with the other input variable. As an example: $1 \oplus 0 \oplus 1 = 0$. First find the XOR sum of $0 \oplus 1$, which is shown as $0 \oplus 1 = 1$. Then take this XOR sum of 1 and XOR it with 1 to obtain the following: $1 \oplus 1 = 0$.

The boolean expression for the exclusive NOR gate shown in Fig. 31-22 is $\overline{A \oplus B} = X$, which can be read as $X = $ not A XOR B. For Fig. 31-22 we have:

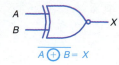

FIG. 31-22   The NXOR gate.
$\overline{A \oplus B} = X$.

$\overline{A \oplus B}$	X
$\overline{0 \oplus 0} = \overline{0} = 1$	
$\overline{0 \oplus 1} = \overline{1} = 0$	
$\overline{1 \oplus 0} = \overline{1} = 0$	
$\overline{1 \oplus 1} = \overline{0} = 1$	

Looking at this table we can see that the output X is HIGH only when the inputs are the same. Regardless of the number of input variables, first the XOR operation must be performed, followed by the inversion.

TEST-POINT QUESTION 31-7

Answers at end of chapter.
a. Write the boolean expression for a 3-input OR gate.
b. Write the boolean expression for a 3-input AND gate.
c. Write the boolean expression for a 2-input NOR gate.
d. Write the boolean expression for a 2-input NAND gate.
e. Write the output for each of the following:
$A + 0$
$A \cdot 0$
$A + A$
$A \cdot 1$
$A + \overline{A}$
$A \cdot \overline{A}$
$1 \oplus 1 \oplus 1$
$0 \oplus 1 \oplus 1$

31-8 DEMORGAN'S THEOREM

Two very important principles of boolean algebra, known as DeMorgan's theorem, can help to greatly simplify expressions in which a product or sum is inverted. The first theorem states that the complement of a sum equals the product of the complements. Expressed as an equation, we have:

$$\overline{A + B} = \overline{A} \cdot \overline{B}$$

In terms of the algebra itself, it is important to realize that if the overbar is broken directly above the OR operation sign for $\overline{A + B}$, then each variable remains complemented, but the operation sign changes from OR ($+$) to AND ($\cdot$). This basically means that the output of a NOR gate and the output of an AND gate with inverted inputs will be the same, provided both gates have the same inputs.

The logic symbols and truth tables for the NOR and the AND gate with inverted inputs are shown in Fig. 31-23. The truth tables prove the equivalency for the statement $\overline{A + B} = \overline{A} \cdot \overline{B}$.

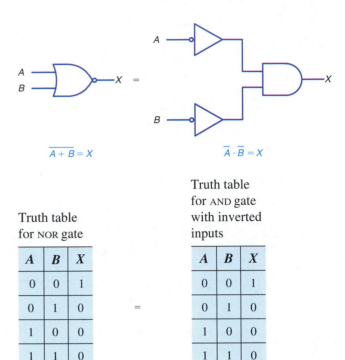

$\overline{A + B} = X$ $\overline{A} \cdot \overline{B} = X$

Truth table
for NOR gate

Truth table
for AND gate
with inverted
inputs

A	B	X		A	B	X
0	0	1		0	0	1
0	1	0	=	0	1	0
1	0	0		1	0	0
1	1	0		1	1	0

FIG. 31-23 Logic symbols and truth tables for a NOR gate, and for an AND gate with inverted inputs.

$\overline{A} \cdot \overline{B} = X$

FIG. 31-24 Bubbled AND gate.

The AND gate with inverted inputs is usually shown with just the bubbles on the input leads. Figure 31-24 shows the schematic symbol of a bubbled AND gate. This symbol is actually an alternate way of showing a NOR gate.

DeMorgan's second theorem states that the complement of a product equals the sum of the complements. Expressed as an equation, we have:

$$\overline{A \cdot B} = \overline{A} + \overline{B}$$

As far as the algebra is concerned, when the overbar is broken directly above the AND operation sign for $\overline{A \cdot B}$ the individual variables remain complemented, but the operation sign changes from AND ($\cdot$) to OR ($+$). This basically means that the output of a NAND gate is the same as for an OR gate with inverted inputs, provided both gates have the same inputs. The logic symbols and truth tables for the NAND gate and the OR gate with inverted inputs are shown in Fig. 31-25. The truth tables prove the equivalency for the statement $\overline{A \cdot B} = \overline{A} + \overline{B}$.

The OR gate with inverted inputs is usually shown with just the bubbles on the input leads, as in Fig. 31-26. This symbol is actually an alternate symbol for a NAND gate. (Note that there are many times when the alternate symbols for both NAND gates and NOR gates will be used to help simplify interpretation of a given logic circuit.)

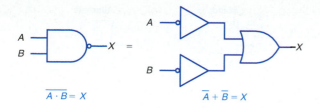

$$\overline{A \cdot B} = X \qquad \overline{A} + \overline{B} = X$$

Truth table for NAND gate

A	B	X
0	0	1
0	1	1
1	0	1
1	1	0

Truth table for OR gate with inverted inputs

A	B	X
0	0	1
0	1	1
1	0	1
1	1	0

FIG. 31-25 Logic inputs and truth tables for a NAND gate, and for an OR gate with inverted inputs.

Two other important concepts in boolean algebra should be understood. First, if a variable is double complemented, the variable appears uncomplemented at the output. As examples, $\overline{\overline{A}} = A$, $\overline{\overline{AB}} = AB$, $\overline{\overline{A + B}} = A + B$, and so on. Second, complementing both sides of an equation does not destroy its equality. As an example, if we complement $A + B = X$, we have $\overline{A + B} = \overline{X}$, and this simplifies to $A + B = \overline{\overline{X}}$. As another example, we can complement $AB = X$ to get $\overline{AB} = \overline{X}$, which simplifies to $AB = \overline{\overline{X}}$.

Figure 31-27 shows the standard and alternate logic symbols used for the various logic gates. The way a logic gate is drawn can help simplify interpretation of a given logic circuit when troubleshooting. For Fig. 31-27, look at the boolean expression for each logic gate in the left column and confirm that the logic gate and boolean expression in the right column is correct. Remember that when the overbar is broken directly above the operation sign, the individual variables remain complemented but the operation sign changes.

$$\overline{A} + \overline{B} = X$$

FIG. 31-26 Bubbled OR gate.

Answers at end of chapter.

Answer True or False
a. $\overline{A + B} = \overline{A} \cdot \overline{B}$
b. $\overline{A \cdot B} = \overline{A} + \overline{B}$
c. $\overline{\overline{A}} = A$
d. $\overline{\overline{A + B}} = A + B$
e. $A \cdot B = \overline{\overline{A \cdot B}}$

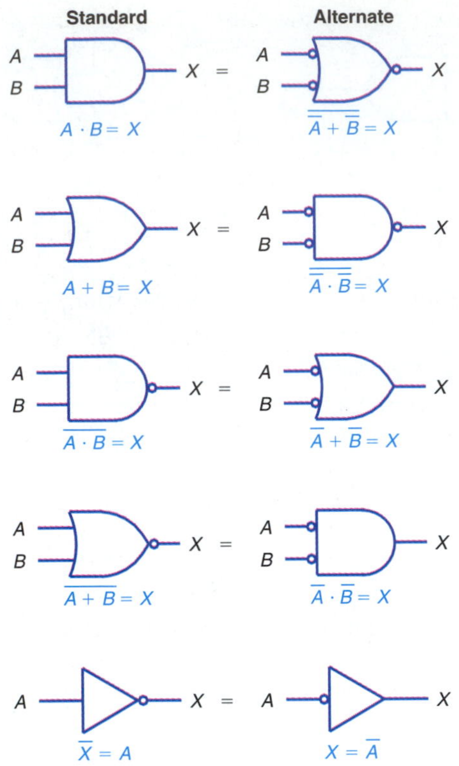

Standard   **Alternate**

$$A \cdot B = X$$ $$\overline{\overline{A} + \overline{B}} = X$$

$$A + B = X$$ $$\overline{\overline{A} \cdot \overline{B}} = X$$

$$\overline{A \cdot B} = X$$ $$\overline{A} + \overline{B} = X$$

$$\overline{A + B} = X$$ $$\overline{A} \cdot \overline{B} = X$$

$$\overline{X} = A$$ $$X = \overline{A}$$

FIG. 31-27   Standard and alternate logic symbols used for the common logic gates.

31-9 ACTIVE HIGH/ACTIVE LOW TERMINOLOGY

When an input or output line on a logic gate symbol does not show a bubble, it indicates that these lines are active HIGH. When an input or output line does show a bubble, these lines are said to be active LOW. Therefore, the presence or absence of a bubble on the inputs and outputs of logic gates indicates whether a line is considered to be active HIGH or active LOW. When an input variable or an output in a boolean expression has no overbar (or "not sign"), it means that the input variables or outputs are active HIGH. However, if an input or an output in a boolean expression does have an overbar, it means that the input variable or the output variable is active LOW.

Now that we can identify active HIGH and active LOW inputs and outputs on logic gates and in boolean expressions, let's examine what is really meant by the terminology "active HIGH" and "active LOW." An example (see Fig. 31-28) will best illustrate the idea. In this circuit, the LED turns on when the output X goes HIGH. The reason is that the X output is connected to the LED anode, which needs positive voltage to turn on. Also, the output X will only go HIGH if the input A is LOW. Because the bubble appears on the input line of the inverter, the input A is said to be active LOW. Likewise, because there is no bubble on the output line, it is said to be active HIGH. For Fig. 31-28, we can say that the output X is active HIGH when the input A is active LOW. We call the output X active HIGH because it is this HIGH signal that causes action in the circuitry being driven by the output X. Likewise we call the input A active LOW because when this input signal is LOW it causes action in the circuitry driven by the inverter output. For Fig. 31-28 the

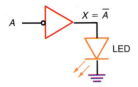

$$X = \overline{A}$$

LED

FIG. 31-28   Inverter logic gate with output X connected to the anode of the LED. The input A must be active LOW to produce an active HIGH output. The circuit is active when the LED comes on.

action is the LED being turned on. The logic gate symbol describes how the circuit functions.

Notice that the boolean expression $X = \overline{A}$ describes how the inverter functions. Because X has no overbar it is said to be active HIGH. The input variable A is complemented and is said to be active LOW. Technicians and engineers looking at the logic circuit in Fig. 31-28 and its corresponding boolean expression would realize that the output X is active HIGH only when the input A is active LOW. They would also know that the output X is inactive LOW when the input A is inactive HIGH. This condition is called inactive, because the LED does not light. The logic symbol for an inverter with the bubble on the output would not be appropriate, as it does not represent how the circuit operates.

For another example, see Fig. 31-29. Here we see there is no bubble on the input line, which means the input A is active HIGH. The bubble on the output indicates that X is active LOW. We can say that the output X is active LOW when the input A is active HIGH because now the X output is connected to the LED cathode.

It is common to see the boolean expression for this circuit as $A = \overline{X}$, rather than $X = \overline{A}$. (Here we have simply complemented both sides of the equation $X = \overline{A}$. Remember, complementing does not destroy the equality of the equation.) Notice how the boolean expression matches the logic symbol. For Fig. 31-29 the LOW output X causes the LED to turn on; this is the action. Likewise, when the input A is HIGH, the output X is LOW, which causes action in the circuit by lighting the LED. Thus, it is common practice to write the boolean expression for the circuit in Fig. 31-29 as $A = \overline{X}$, because the expression matches the inverter logic symbol.

Another example is shown in Fig. 31-30. Here we see both inputs are active HIGH and the output X is active LOW. Either, or both inputs must be active HIGH in order for the output X to be active LOW. The most appropriate boolean expression is $A + B = \overline{X}$. It is obtained by complementing both sides of the equation $\overline{A + B} = X$. Notice that the boolean expression and the logic diagram clearly describe how the logic circuit is functioning. Remember, that the turned on LED is the action.

Suppose we have a circuit like the one shown in Fig. 31-31. For this logic circuit we can say that both inputs A and B must be active LOW to produce an active HIGH output. It is important to realize that the logic gate in Fig. 31-31 is a NOR gate. The alternate logic symbol, rather than the standard logic symbol, is used to clarify circuit operation. The boolean expression that is most appropriate is $\overline{A} \cdot \overline{B} = X$, not $\overline{A + B} = X$. Notice how the boolean expression matches the logic diagram. The standard logic symbol for a NOR gate is not used here because it does not clearly represent how the circuit operates.

Any logic gate can be shown or represented by either of two symbols: standard or alternate. The logic symbol and boolean expression used should provide the clearest description of how the circuit operates.

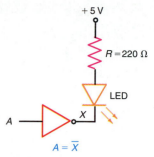

FIG. 31-29 Inverter logic gate with output X connected to the cathode of the LED. The input A must be active HIGH to produce an active LOW output.

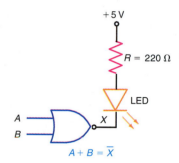

FIG. 31-30 A NOR logic gate with output X connected to the cathode of the LED. Either or both inputs A and B must be active HIGH to produce an active LOW output.

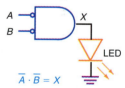

FIG. 31-31 A NOR logic gate drawn using the alternate symbol. Output X is connected to the anode of the LED. Inputs A and B must be active LOW to produce an active HIGH output.

Answers at end of chapter.
a. True or False? A bubble on the output of a logic gate indicates that the output is active HIGH.
b. True or False? Bubbles on the input lines of logic gates indicate that these inputs are active LOW.

c. True or False? The boolean expression for a logic circuit is $\overline{A} + \overline{B} = X$. This tells us both inputs are active LOW and the output is active HIGH.

d. Draw the logic gate symbol that would match the following boolean expression: $\overline{A} + \overline{B} = X$.

e. Draw the logic gate symbol that would match the following boolean expression: $AB = \overline{X}$.

31-10 TREATING UNUSED INPUTS ON LOGIC GATES

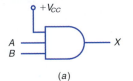

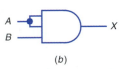

FIG. 31-32 Treating unused inputs on AND gates. (*a*) Tie unused input to V_{CC}. (*b*) Connect unused input to another used input.

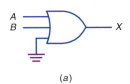

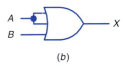

FIG. 31-33 Treating unused inputs on OR gates. (*a*) Tie unused input to ground. (*b*) Connect unused input to another used input.

Digital circuits often include inputs that are not needed in the final wiring. Unused inputs on logic gates should never be left disconnected or floating, because they can act as small antennas and pick up noise, causing erratic or unpredictable operation of logic circuits.

Unused inputs on AND gates should be connected directly to V_{CC} (+5 V), as in Fig. 31-32*a*, or tied to another input that is being used. Note that since the unused input is always HIGH, inputs *A* and *B* must then be made HIGH to obtain a HIGH output at *X*. Furthermore, the unused input should never be tied to ground for an AND gate because then the output *X* would always be LOW even if all other inputs are HIGH. Figure 31-32*b* shows the unused input connected to another input, which is also acceptable.

Unused inputs on OR gates should be tied to ground, as in Fig. 31-33*a*, or tied to another input that is being used. In Fig. 31-33*a*, the output *X* will be HIGH if either or both inputs *A* and *B* are HIGH. Figure 31-33*b* shows the unused input connected to another input, which is also acceptable. The unused input should never be tied to V_{CC} for an OR gate because this would cause the output *X* to remain HIGH regardless of the conditions of the other inputs.

Figure 31-34 shows how to handle unused inputs on NAND gates. Notice that this is the same method used with AND gates. The output *X* will be LOW only when inputs *A* and *B* are HIGH. The output *X* will be HIGH if either or both inputs *A* and *B* are LOW.

Figure 31-35 shows how to handle unused inputs on NOR gates, which is the same method used with OR gates. The output *X* will be HIGH only when inputs *A* and *B* are LOW. The output *X* will be LOW if either or both inputs *A* and *B* are HIGH.

It is also possible to use NAND and NOR gates as inverters, in which case, all inputs are tied together and used as one input. To summarize:

1. All unused inputs of AND gates and NAND gates should be tied to V_{CC} or to other used inputs.
2. All unused inputs of OR gates and NOR gates should be tied to ground or to other used inputs.
3. NAND and NOR gates can be used as inverters if all inputs are tied together.

TEST-POINT QUESTION 31-10

Answers at end of chapter.

a. A 3-input AND gate is to be used as a 2-input AND gate. Show how to connect the unused input.

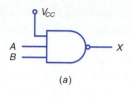

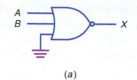

(a)

(a)

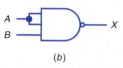

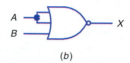

(b)

(b)

FIG. 31-34 Treating unused inputs on NAND gates. (*a*) Tie unused input to V_{CC}. (*b*) Connect unused input to another used input.

FIG. 31-35 Treating unused inputs on NOR gates. (*a*) Tie unused input to ground. (*b*) Connect unused input to another used input.

b. A 4-input NOR gate is to be used as a 2-input NOR gate. Show how to connect the unused inputs.

c. A 4-input NAND gate is to be used as an inverter. Show how to wire the inputs.

31-11 COMBINATIONAL LOGIC CIRCUITS

Digital circuits often consist of several different logic gates, interconnected in such a way as to perform a specific logic function. Figure 31-36*a* shows a logic circuit that uses one OR gate and one AND gate. The boolean expression for this logic circuit is $A \cdot (B + C) = X$, or just $A(B + C) = X$. The parentheses around the $B + C$ sum indicates that the input variable A is being ANDed with the $B + C$ sum. The truth table in Fig. 31-36*b* lists all input possibilities and the corresponding output for each input condition. According to the truth table, the output X is LOW when A is LOW. This must be true because any variable ANDed with 0 equals 0. The output X is HIGH only when A is HIGH *and* either or both inputs B and C are HIGH.

As another example, see Fig. 31-37*a*. This logic circuit also uses one OR gate and one AND gate. The boolean expression for this logic circuit is $A + BC = X$. The truth table in Fig. 31-37*b* lists all input possibilities and the corresponding output for each input condition. From the truth table we see that the output X is HIGH if A is HIGH *or* if B and C are both HIGH. This must be true because any variable ORed with 1 equals 1.

DERIVING A BOOLEAN EXPRESSION FROM A TRUTH TABLE

When designing digital circuits, we begin by constructing a truth table. For example, suppose we have to construct a logic circuit that will produce an output in accordance with the truth table in Fig. 31-38. The boolean expression for this truth table would be

$$\overline{A}\,\overline{B}C + A\overline{B}C + AB\overline{C} + ABC = X$$

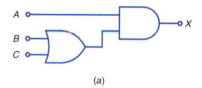

(a)

A	B	C	X
0	0	0	0
0	0	1	0
0	1	0	0
0	1	1	0
1	0	0	0
1	0	1	1
1	1	0	1
1	1	1	1

(b)

FIG. 31-36 Combinational logic circuit. (*a*) Logic diagram with OR gate feeding AND gate. (*b*) Truth table.

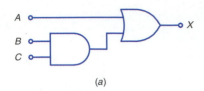

(a)

A	B	C	X
0	0	0	0
0	0	1	0
0	1	0	0
0	1	1	1
1	0	0	1
1	0	1	1
1	1	0	1
1	1	1	1

(b)

FIG. 31-37 Combinational logic circuit. (*a*) Logic diagram with AND gate feeding OR gate. (*b*) Truth table.

A	B	C	X
0	0	0	0
0	0	1	1
0	1	0	0
0	1	1	0
1	0	0	0
1	0	1	1
1	1	0	1
1	1	1	1

FIG. 31-38 Truth table used for obtaining boolean expression.

This is called a sum of products, or *minterm* boolean expression. Each AND product is ORed together. Let's examine each term in the minterm boolean expression and see how it is obtained from the truth table.

From the second row in the truth table we see that when $A = 0$, $B = 0$, and $C = 1$, the output X is HIGH. The boolean expression for this condition is $\overline{A}\,\overline{B}C$, which is the first term shown in the minterm boolean expression. Using the rules of boolean algebra we have $\overline{0} \cdot \overline{0} \cdot 1 = 1$, or $1 \cdot 1 \cdot 1 = 1$. Then from the sixth row in the truth table we see that X is HIGH when $A = 1$, $B = 0$, and $C = 1$. The boolean expression for this term is $A\overline{B}C = X$, the second term in the minterm boolean expression. Using the rules of boolean algebra we have $1 \cdot \overline{0} \cdot 1 = 1$, or $1 \cdot 1 \cdot 1 = 1$. The other two terms in the minterm boolean expression are obtained using the same method.

As a general rule, when designing a digital circuit from a truth table, write each of the AND products that will produce a HIGH output, and OR these AND products together. When a variable is 0 for an AND product that produces a HIGH output, indicate the variable as complemented. When a variable is 1 for an AND product that produces a HIGH output, the variable should appear uncomplemented.

SIMPLIFYING BOOLEAN EXPRESSIONS   Before building the logic circuit that would implement the boolean expression $\overline{A}\,\overline{B}C + A\overline{B}C + AB\overline{C} + ABC = X$, it should be determined whether or not the boolean expression can be simplified. Here we can apply the rules of boolean algebra. Remember that a simpler boolean expression means a simpler overall logic circuit that requires fewer ICs.

One way to simplify the boolean expression $\overline{A}\,\overline{B}C + A\overline{B}C + AB\overline{C} + ABC = X$ is to factor the boolean expression if possible. Notice that the first two terms have $\overline{B}C$ as common factors. The factoring can be shown as:

$$\overline{A}\,\overline{B}C + A\overline{B}C$$
$$\overline{B}C(\overline{A} + A)$$

$A + \overline{A} = 1$, and $\overline{B}C \cdot 1$ simplifies to $\overline{B}C$. Therefore, the first two terms $\overline{A}\,\overline{B}C + A\overline{B}C$ simplify to $\overline{B}C$. The second two terms $AB\overline{C} + ABC$ have AB as common factors. The factoring can be shown as:

$$AB\overline{C} + ABC$$
$$AB(\overline{C} + C)$$

$C + \overline{C} = 1$, and $AB \cdot 1$ simplifies to AB. Therefore, the second two terms can be simplified to AB. Thus,

$$\overline{A}\,\overline{B}C + A\overline{B}C + AB\overline{C} + ABC = AB + \overline{B}C$$

Again, not only is the boolean expression simplified, but so is the resultant logic circuit.

To build the logic circuit for $AB + \overline{B}C = X$, we use the configuration shown in Fig. 31-39*a*. Notice how the boolean expression and logic circuit match. Inputs A and B are active HIGH inputs for the upper AND gate, while the lower AND gate uses an active LOW input B and an active HIGH input C. The output X is also active HIGH.

REDUCING THE NUMBER OF ICs REQUIRED TO IMPLEMENT A BOOLEAN EXPRESSION

The compactness of an electronic device is extremely important in today's world. In digital electronic circuitry it is desirable to perform a specific logic function using as few ICs as possible. In digital electronics, ICs known as *TTL* (transistor-transistor logic) *circuits* are often used. For a complete listing of all TTL devices, refer to the TTL data book published by Texas Instruments. The logic circuit in Fig. 31-39a will require three different ICs, using one 7404 hex inverter (IC1), one 7408 quad 2-input AND gate (IC2), and one 7432 quad 2-input OR gate (IC3). IC pin numbers are shown for each logic gate. Although not shown in Fig. 31-39a, each IC is connected to V_{CC} and ground through the appropriate pins.

The same logic function that would implement the boolean expression $AB + \overline{B}C = X$ could be built using only one 7400 quad 2-input NAND gate. This is shown in Fig. 31-39b. IC pin numbers are shown for each gate inside the 7400 device. The inverter symbol in Fig. 31-39b is a NAND gate wired to work as an inverter. The boolean expression for this circuit would be $\overline{\overline{AB}} + \overline{\overline{\overline{B}C}} = X$, which simplifies to $AB + \overline{B}C = X$. The idea shown in Fig. 31-39b is that the logic circuit uses only one IC instead of three to perform the exact same logic function. This, of course, means less cost, less space, and less weight for the electronic device.

In Fig. 31-39b the circuit uses all NAND gates. NAND gates are often referred to as the universal logic gate because they can be used to implement any boolean expression. The right most NAND gate (numbered 4 in the diagram) is performing the OR function and is, therefore, drawn using the alternate symbol. The middle NAND gates (numbered 2 and 3) perform the AND function and are, therefore, drawn using the standard symbol. Notice the double inversion from the outputs of NAND gates 2 and 3 to the inputs of NAND gate 4. Whenever possible, connect bubbled outputs to bubbled inputs and nonbubbled outputs to nonbubbled inputs. This makes it much easier for technicians and engineers to follow the signal flow through the circuit and to determine the input conditions that are required to produce an active output, as explained in Sec. 31-9. Double inversion is the same as no inversion at all!

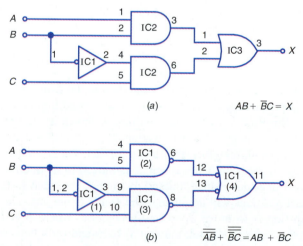

(a)

$AB + \overline{B}C = X$

(b)

$\overline{\overline{AB}} + \overline{\overline{\overline{B}C}} = AB + \overline{B}C$

FIG. 31-39 Constructing a logic diagram from the simplified boolean expression. (*a*) Circuit built using three ICs. (*b*) Circuit built using one 7400 IC. Both figures *a* and *b* implement the boolean expression $AB + \overline{B}C$.

Any logic circuit of AND gates and OR gates can be replaced with equivalent NAND gates. This can be accomplished by adding circles at the outputs of each AND gate and circles at the inputs of each OR gate. This converts all AND gates and OR gates to NAND gates. If two bubbles appear in series, they have no effect on the overall operation of the circuit. If converting AND gates and OR gates to NAND gates causes a single bubble to appear at an input or output, then an inverter must be added to the input or output to produce an equivalent boolean expression like that of the original AND/OR logic circuit. Using these rules, the boolean expression of the resultant logic circuit, consisting of NAND gates, will always simplify to the same equation shown for the original logic circuit consisting of AND's and OR's.

<hr>

TEST-POINT QUESTION 31-11

Answers at end of chapter.

Answer True or False.
a. The NAND gate is considered a universal logic gate.
b. Boolean expressions can be simplified using boolean algebra.
c. Two bubbles in series provide inversion.

31-12 FLIP-FLOPS

A *flip-flop* is a digital circuit that has two stable states. A flip-flop will stay in one of its two stable states until an input pulse forces the flip-flop to switch to its other stable state. The flip-flop can remain in either stable state indefinitely. Flip-flops are used to store binary information. Digital memory circuits that can store bits of data are an essential part of any computer system.

RS **FLIP-FLOPS**   The most basic type of flip-flop is the reset/set type, hence the name *RS* flip-flop. The basic *RS* flip-flop can be built using either two NOR gates, as shown in Fig. 31-40a, or two NAND gates, as shown in Fig. 31-40b. Notice that for either type, the inputs are labeled R for reset, and S for set. Also each flip-flop has two outputs labeled Q and $\overline{Q}$. The flip-flop is said to be set when $Q = 1$ and $\overline{Q} = 0$. When $Q = 0$ and $\overline{Q} = 1$, the flip-flop is said to be reset.

Let's analyze the NOR gate flip-flop and its truth table in Fig. 31-40a. When S_1 is in the up position, $R = 1$ and $S = 0$. For this condition $Q = 0$ and $\overline{Q} = 1$; therefore, the flip-flop is reset. This condition is shown in the third row of the truth table. When S_1 is in the down position, $R = 0$ and $S = 1$. For this condition $Q = 1$ and $\overline{Q} = 0$; therefore, the flip-flop is set. This condition is shown in the second row of the truth table. When S_1 is moved to its middle position, the flip-flop output does not change. In fact, moving S_1 back and forth between its middle and set positions will not change the Q and $\overline{Q}$ outputs. The only way to reset the flip-flop is to move S_1 to its up position, where $R = 1$ and $S = 0$. Again, once the flip-flop is reset, moving S_1 back and forth between its middle and reset positions does not change the condition of the Q and $\overline{Q}$ outputs. When S_1 is in its middle position, $R = 0$ and $S = 0$. This condition is shown in the first row of the truth table. Notice that for this condition, no change (NC) occurs in the flip-flop outputs. Resistors R_1 and R_2 are called *pull-down resistors*.

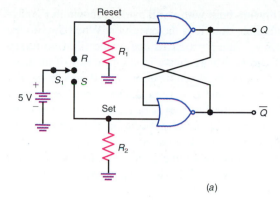

R	S	Q	Comment
0	0	NC	No change
0	1	1	Set
1	0	0	Reset
1	1	*	Illegal

(a)

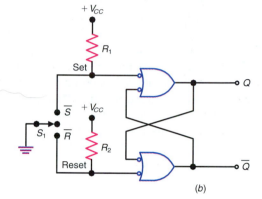

R	S	Q	Comment
0	0	*	Illegal
0	1	0	Reset
1	0	1	Set
1	1	NC	No change

(b)

FIG. 31-40 (a) Set/reset flip-flop with active HIGH inputs. (b) Set/reset flip-flop with active LOW inputs.

These resistors pull the TTL inputs to ground for binary 0. The maximum resistance value for either R_1 or R_2 should not exceed 500 Ω. This is determined by the following relationship:

$$R_{\text{pull-down (max)}} = V_{IL}/I_{IL}$$
$$= 0.8 \text{ V}/1.6 \text{ mA}$$
$$R_{\text{pull-down (max)}} = 500 \text{ Ω}$$

This basically says that R must not be so large that its voltage drop exceeds 0.8 V. Remember, logic gates will recognize an input as LOW if the voltage is 0.8 V or less. Also remember that a low TTL input could draw as much as 1.6 mA of current.

One more thing. The inputs without bubbles in Fig. 31-40a are active HIGH. This flip-flop will only respond to HIGH inputs. For example, in Fig. 31-40a assume that the flip-flop is reset, where $Q = 0$ and $\overline{Q} = 1$. If we move S_1 to its set position, the flip-flop outputs change to $Q = 1$ and $\overline{Q} = 0$. It is the HIGH S input, not the LOW R input, that causes the flip-flop outputs to change, hence the term active HIGH. To reset the flip-flop, we move S_1 to its reset position to obtain $Q = 0$ and $\overline{Q} = 1$. It is the HIGH R input, not the LOW S input, that causes the flip-flop outputs to change.

The bottom row of the truth table in Fig. 31-40a shows that $R = 1$ and $S = 1$. If this were possible (it is not with the circuit in Fig. 31-40a), both Q and $\overline{Q}$ would equal 0. This is commonly referred to as an illegal or prohibited state for the outputs.

An *RS* flip-flop built with NAND gates is shown in Fig. 31-40b. The truth table describes the operation. When S_1 is in the up position, $\overline{R} = 1$ and $\overline{S} = 0$. This causes the flip-flop to set, where $Q = 1$ and $\overline{Q} = 0$, as shown in the third row of the truth table. When S_1 is in the down position, $\overline{R} = 0$ and $\overline{S} = 1$. For this condition $Q = 0$ and $\overline{Q} = 1$, as shown in the second row of the truth table. If the flip-flop is reset, moving S_1 back and forth between the middle and reset positions will not affect the condition of the outputs. Likewise, if the flip-flop is set, moving S_1 back and forth between the set and middle positions will not change the condition of the flip-flop outputs. When S_1 is in the middle position, $\overline{R} = 1$ and $\overline{S} = 1$. This condition is shown in the last row of the truth table. Notice for this condition that no change occurs in the flip-flop outputs. Resistors R_1 and R_2 are called *pull-up resistors*. When S_1 is in the middle position, R_1 and R_2 pull the TTL inputs up to $+V_{CC}$ for a binary 1.

One more thing. The alternate logic symbol is used for the NAND gates because the $\overline{R}$ and $\overline{S}$ inputs are active LOW. For example, in Fig. 31-40b assume that the flip-flop is reset where $Q = 0$ and $\overline{Q} = 1$. If we move S_1 to the set position where $\overline{R} = 1$ and $\overline{S} = 0$, the flip-flop outputs change to $Q = 1$ and $\overline{Q} = 0$. It is the LOW $\overline{S}$ input, not the HIGH $\overline{R}$ input, that causes the flip-flop to set, hence the name active LOW. To reset the flip-flop, we move S_1 to the reset position where $Q = 0$ and $\overline{Q} = 1$. It is the LOW $\overline{R}$ input, not the HIGH $\overline{S}$ input, that resets the flip-flop.

RS flip-flops are often shown using the schematic symbols in Fig. 31-41a and b. If the inputs are active HIGH, then Fig. 31-41a is used, where no bubbles appear on the input lines. If the inputs are active LOW, then the symbol shown with bubbles in Fig. 31-41b is used to indicate the active LOW characteristic. From now on *RS* flip-flops will be shown using these symbols.

CLOCKED *RS* FLIP-FLOPS Figure 31-42a shows an *RS* flip-flop that has a clock (CLK) input. The clock voltage is a square wave that has a maximum value of +5 V and a minimum value of 0 V. The truth table in Fig. 31-42b explains the operation. When the clock is LOW, the outputs will not change regardless of the conditions of the R and S inputs. When the clock input is HIGH, the flip-flop will set if $R = 0$ and $S = 1$. When the clock input is HIGH and $R = 1$ and $S = 0$, the flip-flop will reset. Once the flip-flop is set, or perhaps reset, the inputs R and S can be pulled LOW and the flip-flop will remain in its last stable state even though the clock is HIGH. The last row in the truth table shows CLK $= 1$, $R = 1$, and $S = 1$. This is an illegal, or prohibited, state because both Q and $\overline{Q}$ would equal 1.

Figure 31-43 shows a clocked *RS* flip-flop with a timing diagram depicting how the Q output responds to inputs applied to the R, S, and clock inputs. When analyzing the Q output in the timing diagram, refer back to the truth table in Fig. 31-42b.

It should be pointed out that computers use literally thousands of flip-flops. To coordinate the overall circuit action inside of a computer, a clock signal is sent to each and every flip-flop. This clock input signal prevents the flip-flop outputs from changing until exactly the right time.

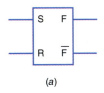

(a)

(b)

FIG. 31-41 Logic symbols for set/reset flip-flops. (a) *RS* flip-flop with active HIGH inputs. (b) *RS* flip-flop with active LOW inputs.

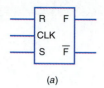

(a)

CLK	R	S	Q
0	0	0	NC
0	0	1	NC
0	1	0	NC
0	1	1	NC
1	0	0	NC
1	0	1	1
1	1	0	0
1	1	1	Illegal*

(b)

FIG. 31-42 Clocked *RS* flip-flop. (a) Logic diagram. (b) Truth table.

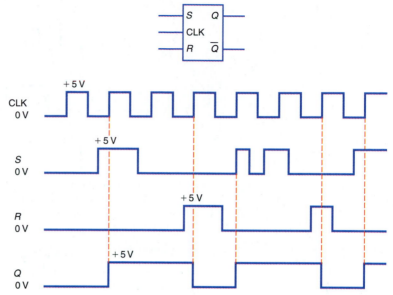

FIG. 31-43 Clocked *RS* flip-flop with timing diagram to show how the *Q* output responds to *R*, *S*, and CLK inputs.

D-TYPE FLIP-FLOPS Another commonly used flip-flop is the edge-triggered *D*-type flip-flop. The logic symbol and truth table for the *D*-type flip-flop are shown in Fig. 31-44a and b. The presence of a small triangle on the clock input indicates that the flip-flop is edge-triggered. Also the *X*s in the truth table of Fig. 31-44b are called "don't care's" because if the clock is LOW, HIGH, or on it's negative edge, the flip-flop is inactive. Therefore, we "don't care" what the data input value is for these conditions of the clock. The flip-flop is edge-triggered because the flip-flop only responds when the clock is changing states.

For the *D*-type flip-flop in Fig. 31-44a, the *Q* and $\overline{Q}$ outputs change only on the positive-going edge of the incoming clock pulse. In the truth table, arrows are used to indicate whether we have a positive- or negative-going clock edge. The up arrow (↑) indicates a positive-going clock edge, and the down arrow (↓) indicates a negative-going clock edge. A positive edge-triggered flip-flop is inactive if the clock is LOW, HIGH, or on its negative-going edge. This is shown in the first three rows of the truth table. The last two rows indicate an output change on the positive-going edge of the incoming clock pulse. If *D* = 0 when the positive-going clock edge appears, then *Q* = 0 and $\overline{Q}$ = 1. If *D* = 1 when the positive-going clock edge appears, then *Q* = 1 and $\overline{Q}$ = 0. The data input and output are the same after a positive-going pulse. In other words, the input data *D* is stored only on the positive-going edge of the incoming clock pulse.

Figure 31-45 shows a clocked *D*-type flip-flop with a timing diagram depicting how the *Q* output responds to the *D* and clock inputs. When analyzing the *Q* output in the timing diagram, refer to the truth table in Fig. 31-44b.

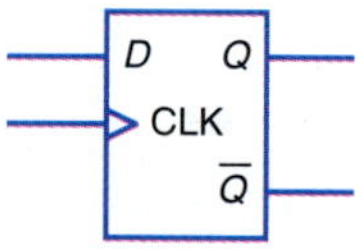

(a)

CLK	D	Q
0	X	NC
1	X	NC
↓	X	NC
↑	0	0
↑	1	1

(b)

FIG. 31-44 Positive edge-triggered *D*-type flip-flop. (a) Logic diagram. (b) Truth table.

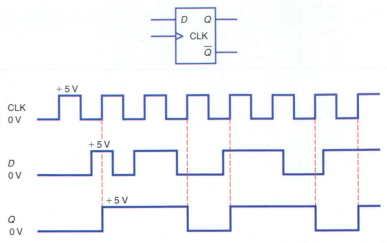

FIG. 31-45 Positive edge-triggered D-type flip-flop with timing diagram to show how the Q output responds to R, S, and CLK inputs.

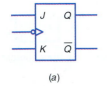

(a)

CLK	J	K	Q
0	X	X	NC
1	X	X	NC
↑	X	X	NC
X	0	0	NC
↓	0	1	0
↓	1	0	1
↓	1	1	Toggle

(b)

FIG. 31-46 Negative edge-triggered JK flip-flop. (a) Logic diagram. (b) Truth table.

JK **FLIP-FLOPS**   Another edge-triggered flip-flop is the negative edge-triggered flip-flop shown in Fig. 31-46a. This flip-flop will respond only to a negative-going clock pulse. The J and K inputs are control inputs; that is, they determine what the flip-flop will do when it receives a negative-going clock edge. The truth table in Fig. 31-46b summarizes the action. The flip-flop is inactive when the clock is LOW, HIGH, or on its positive-going edge. Likewise, the circuit is inactive when both the J and K inputs are LOW.

When $J = 0$ and $K = 1$, the negative-going edge of the clock pulse puts the outputs at $Q = 0$ and $\overline{Q} = 1$. When $J = 1$ and $K = 0$, the negative-going edge of the clock pulse puts the Q outputs at $Q = 1$ and $\overline{Q} = 0$. When $J = 1$ and $K = 1$, the Q and $\overline{Q}$ outputs toggle, or alternate, with each negative-going clock edge. Figure 31-47 shows a clocked JK flip-flop with the diagram depicting how the Q outputs respond to the J, K, and clock inputs.

TEST-POINT QUESTION 31-12

Answers at end of chapter.

Answer True or False.
a. A flip-flop is set when $Q = 0$ and $\overline{Q} = 1$.
b. For a positive edge-triggered D-type flip-flop, the Q output will equal the D input after a positive-going clock edge is applied.
c. A JK flip-flop will operate in the toggle mode when $J = 1$ and $K = 1$.
d. Pull-down resistors for standard TTL inputs should not exceed 500 Ω.

31-13 BINARY COUNTERS

JK flip-flops can also be connected together to form a binary counter. Binary counters are used when it is necessary to count the number of clock pulses that

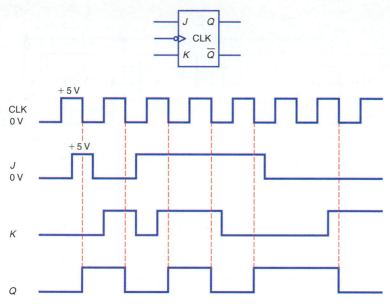

FIG. 31-47 Negative edge-triggered *JK* flip-flop with timing diagram to show how the Q output responds to R, S, and CLK inputs.

arrive at a clock input. Figure 31-48*a* shows a binary counter built using four *JK* flip-flops. Notice that for each flip-flop, the J and K inputs are tied HIGH. This means that each flip-flop will toggle when its clock input receives a negative-going clock pulse. The MSB of the counter is Q_3, and the LSB is Q_0.

Although not mentioned earlier, all D and JK flip-flops have clear and preset inputs. These inputs override all other inputs on the flip-flop. The flip-flops in Fig. 31-48 have an active LOW "clear input." When $\overline{CLR}$ is pulled LOW, all flip-flop Q outputs return to 0. Notice the overbar above $\overline{CLR}$. This is to indicate the active LOW condition. When the flip-flops are counting, the CLR input is pulled HIGH to its inactive state.

Figure 31-48*b* shows how the Q outputs of each flip-flop respond to each negative-going clock edge. Notice that with each negative-going clock edge, the count increases by 1. Initially the count is $Q_3Q_2Q_1Q_0 = 0000$. As shown in the diagram, the LSB flip-flop sets on the first negative-going edge of the clock input, increasing the count to 0001. The HIGH-going Q_0 output is fed to the clock input of the next most significant flip-flop. This HIGH-going clock pulse does not cause the Q_1 output to change. However, the second negative-going clock edge applied to the LSB flip-flop causes the Q_0 output to toggle from 1 to 0. This negative-going clock edge causes the Q_1 output to go from 0 to 1, changing the count to 0010. The count continues until 1111 is reached. Then on the next negative-going clock edge, all flip-flop outputs toggle back to 0 for a count of 0000.

Notice that each flip-flop divides the incoming clock pulse frequency by a factor of 2. The frequency of the Q_0 output is one-half that of the clock input. Likewise the Q_1 output is one-fourth the frequency of the clock input. The Q_2 output is one-eighth the frequency of the clock input, and the Q_3 output is one-sixteenth the frequency of the clock input. The counter in Fig. 31-48 is called a *ripple counter* because the output of one flip-flop is fed to the clock input of another. When the clock input to the counter receives a negative-going clock pulse, the count changes. In effect, pulses are fed down the line to the left like a ripple in water.

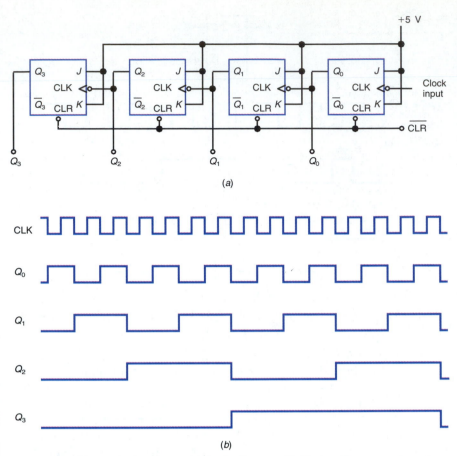

FIG. 31-48 Binary ripple counter. (a) Logic diagram. (b) Timing diagram to show the count for each negative-going clock pulse.

The modulus of a counter is the number of output states it has. A 4-bit ripple counter has a modulus of 16 because there are 16 different states ranging from 0000 to 1111. The modulus of a counter can be altered by connecting additional logic gates to the appropriate Q and $\overline{Q}$ outputs, and to the preset and clear inputs of the flip-flops.

Several different counter chips are available in the 7400 series. The 7490 decade counter is a popular one. As the name implies (decade means 10), it divides the clock input by 10. Another popular counter is the 74193 presettable up/down counter. As its name implies, the 74193 can count up or down. It is called presettable because there are provisions for starting the count from some number other than 0000.

Answers at end of chapter.

Answer True or False.
a. *JK* flip-flops could be used individually to divide the clock input frequency by a factor of 2.
b. A 3-bit ripple counter has eight different output states.

31-14 NEW LOGIC SYMBOLS

The logic symbols used earlier in this chapter are the traditional logic symbols used by industry and educators for many years. In 1984 a new set of standard symbols was introduced by the Institute of Electrical and Electronics Engineers and the American National Standards Institute, abbreviated IEEE/ANSI. The new logic symbols are becoming accepted by the electronic industries and have been appearing in the literature published by the manufacturers of new electronic equipment. U.S. military contracts for the manufacture of electronic equipment require the use of the new symbols. Figure 31-49 shows both the traditional and new rectangular logic gate symbols. The rectangular symbols use a small right triangle to indicate inversion instead of the small bubble used on the traditional symbols. The presence or absence of a triangle on the new logic symbols indicates whether an input or output line is active HIGH or active LOW.

A special symbol inside the rectangle of each logic gate describes how the gate functions. The "1" inside the inverter gate rectangle denotes a gate with only one input. For the logic inverter in Fig. 31-49, the output X will be active LOW when the input A is active HIGH. The "&" symbol inside the AND gate rectangle means the output X will go active HIGH only when all inputs are active HIGH. The "$\geq$" symbol inside the OR gate rectangle means that the output X will go active HIGH when one or more inputs are active HIGH.

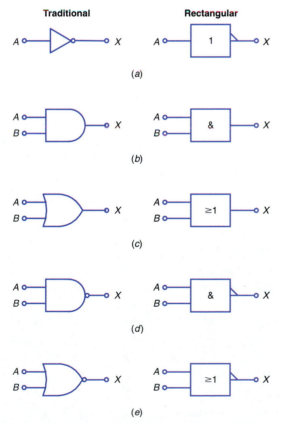

FIG. 31-49 Traditional and new rectangular logic symbols. (*a*) Inverter. (*b*) AND gate. (*c*) OR gate. (*d*) NAND gate. (*e*) NOR gate.

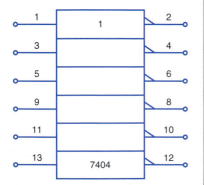

The rectangular symbols for the NAND and NOR gates are the same as for the AND and OR gates with the addition of the small triangle for inversion. The rectangular symbol for an XOR gate uses the "=1" symbol. This is shown in Fig. 31-50a. This indicates that the output will be active HIGH when only one input is HIGH. Notice that the rectangular symbol for the XNOR gate in Fig. 31-50b is the same as the XOR gate with a small triangle on the output.

Throughout this textbook and the rest of this chapter the traditional logic symbols, rather than the new rectangular logic symbols, are used. Most of the electronics industry is still using the traditional logic symbols. Also, most of the digital equipment already in the field still uses the standard traditional symbols. It is going to take several years before all electronics engineers and technicians, as well as manufacturers, convert to the new rectangular symbol. You should, however, be aware that the new symbols will become more and more commonplace in logic diagrams.

Complete ICs can be represented using the new rectangular symbol. Figure 31-51 shows the 7404 hex inverter IC. Notice that the notation "1" appears only in the top rectangle, but applies to all the inverters in the blocks below.

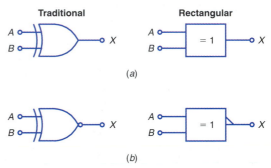

FIG. 31-50 (a) Traditional and new rectangular logic symbols for XOR gate. (b) Traditional and new rectangular logic symbols for XNOR gate.

TEST-POINT QUESTION 31-14

Answers at end of chapter.

Answer True or False.
a. The new rectangular symbols use a small triangle instead of a bubble to indicate inversion.
b. Complete ICs can be represented using the new rectangular symbols.

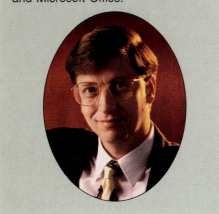

FIG. 31-51 New rectangular symbol for 7404 hex inverter.

31   SUMMARY AND REVIEW

Table 31-4 summarizes the basic logic gates used in digital circuits. Both the traditional and new rectangular symbols are shown, along with the boolean expression for each logic gate. Table 31-5 summarizes some of the different types of flip-flops found in digital machines.

The terminology used in the field of digital electronics is summarized in the following alphabetical list:

AND gate A logic circuit with two or more inputs. The output is LOW if any or all inputs are LOW. The output is HIGH only when all inputs are HIGH.

BCD Binary coded decimal. Each decimal digit is represented by a 4-bit nibble.

binary A number system that is used in digital electronics. The digits used are 0 and 1. Binary 0 is used to represent a LOW voltage, and binary 1 is used to represent a HIGH voltage.

bit A binary digit.

byte A string of 8 bits.

complement To change a binary digit to its opposite level, such as 0 to 1 or 1 to 0.

decimal The most commonly used number system. The digits used are 0, 1, 2, 3, 4, 5, 6, 7, 8, 9.

TABLE 31-4   SUMMARY OF LOGIC GATES

TRADITIONAL	RECTANGULAR	BOOLEAN EXPRESSION
A — ▷ — X	A — [1] — X	$X = \overline{A}$
A, B — D — X	A, B — [&] — X	$AB = X$
A, B — D — X	A, B — [≥1] — X	$A + B = X$
A, B — D○ — X	A, B — [&] — X	$\overline{AB} = X$
A, B — D○ — X	A, B — [≥] — X	$\overline{A + B} = X$

TABLE 31-5   SUMMARY OF FLIP-FLOPS

LOGIC DIAGRAM	DESCRIPTION
S Q / R $\overline{Q}$	Set/reset flip-flop with active HIGH set, and reset inputs.
$\overline{S}$ Q / $\overline{R}$ $\overline{Q}$	Set/reset flip-flop with active LOW set and reset inputs.
S Q / CLK / R $\overline{Q}$	Clocked set/reset flip-flop. Flip-flop is active only when clock input is HIGH.
D Q / CLK / $\overline{Q}$	Edge-triggered D-type flip-flop. The data input D is stored on the positive-going clock edge.
J Q / CLK / K $\overline{Q}$	Edge-triggered JK flip-flop. The J and K inputs determine what the flip-flop does when it receives a negative-going clock edge.

flip-flop A digital circuit that can hold or store digital data.

hexadecimal A number system used in the microcomputer field. The digits used are 0, 1, 2, 3, 4, 5, 6, 7, 8, 9, A, B, C, D, E, and F.

inverter A logic gate with one input and one output. For a LOW input, the output will be HIGH, and for a HIGH input, the output will be LOW.

NAND gate A logic circuit with two or more inputs but only one output. The output is HIGH if any or all inputs are LOW. The output is LOW only when all inputs are HIGH.

NOR gate A logic circuit with two or more inputs but only one output. The output is LOW if any or all inputs are HIGH. The output is HIGH only when all inputs are LOW.

OR gate A logic circuit with two or more inputs but only one output. The output is HIGH if any or all inputs are HIGH. The output is LOW only when all inputs are LOW.

radix The number of digits used by a number system.

truth table A listing of input possibilities for a logic gate and the corresponding output for each input condition.

XNOR gate A logic circuit with two or more inputs but only one output. The output is HIGH when an even number of 1s is applied to its inputs. The output if LOW if an odd number of 1s is applied to its inputs.

XOR gate A logic circuit with two or more inputs but only one output. The output is HIGH when an odd number of 1s is applied to its inputs. The output is LOW when an even number of 1s is applied to its inputs.

SELF-TEST

ANSWERS AT BACK OF BOOK.

1. How many digits are used in the binary number system?
2. What number system has a base, or radix, of 16?
3. What is the decimal equivalent of binary 1011?
4. Which logic gate has a LOW output only when all inputs are HIGH?
5. Which logic gate has a HIGH output only when all inputs are LOW?
6. Which logic gate has a LOW output only when all inputs are LOW?
7. Which logic gate has a HIGH output only when an even number of 1s is applied to its inputs?
8. Which logic gate has a HIGH output when any or all inputs are LOW?
9. Which logic gate has a LOW output when any or all inputs are HIGH?
10. A logic gate has four inputs. When constructing the truth table, how many input combinations of 0s and 1s will there be?
11. Is a flip-flop set or reset when $Q = 0$ and $\overline{Q} = 1$?
12. Is a flip-flop set or reset when $Q = 1$ and $\overline{Q} = 0$?
13. A negative edge-triggered flip-flop has the following inputs: $J = 0$ and $K = 1$. What are the Q and $\overline{Q}$ outputs after the clock input receives a negative-going clock pulse?

QUESTIONS

1. List the next 10 counts after the hexadecimal number OAFC.
2. Define active LOW and active HIGH terminology for logic diagrams and boolean expressions.
3. With an ASCII keyboard, each keystroke produces the ASCII equivalent of the designated character. What is the output from the ASCII keyboard if you type "TWINS"?
4. Using boolean algebra express DeMorgan's theorems.
5. Explain the purpose of having standard and alternate logic symbols.
6. Explain what is meant by the toggle mode of a JK flip-flop.

PROBLEMS

ANSWERS TO ODD-NUMBERED PROBLEMS AT BACK OF BOOK.

1. Find the decimal equivalent for the following binary numbers: (**a**) 1011; (**b**) 10001; (**c**) 10000; (**d**) 10110101.
2. Find the binary equivalent for the following decimal numbers: (**a**) 15; (**b**) 20; (**c**) 54; (**d**) 63.
3. Convert the following hexadecimal numbers to their binary equivalents: (**a**) 100; (**b**) F06; (**c**) 75; (**d**) C33A.
4. Find the hexadecimal equivalent for the decimal number 894.
5. Convert the following binary numbers to their hexadecimal equivalents: (**a**) 1011111; (**b**) 01011111; (**c**) 110101; (**d**) 101100101001.

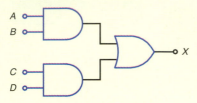

FIG. 31-52 Logic circuit for Probs. 10, 11, and 12.

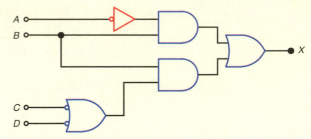

FIG. 31-53 Logic circuit for Prob. 17.

6. Find the decimal equivalent for the following BCD numbers: (**a**) 0010 0101; (**b**) 1001 1000 0111; (**c**) 0001 1001 1000 0010.

7. Draw the logic symbol and construct a truth table for a 2-input AND gate.

8. Draw the logic symbol and construct a truth table for a 3-input NAND gate.

9. Draw the logic symbol and construct a truth table for a 3-input NOR gate.

10. Write the boolean expression for the logic circuit shown in Fig. 31-52.

11. Construct a truth table for the logic circuit in Fig. 31-52.

12. Convert the logic circuit in Fig. 31-52 to a circuit that uses only NAND gates.

13. Draw the logic diagram that most clearly represents the boolean expression $A + B + C = \overline{X}$.

14. Simplify the boolean expression $\overline{A}\overline{B}\overline{C} + \overline{A}BC + A\overline{B}C + ABC = X$. Draw the logic diagram (using only NAND gates) that will implement the simplified boolean expression.

15. Draw the new rectangular logic gate symbols for the following: (**a**) inverter; (**b**) AND; (**c**) OR; (**d**) NAND; (**e**) NOR; (**f**) XOR; and (**g**) XNOR gates.

16. Draw the logic diagram for the expression $AB + BCD$. Use the new rectangular logic symbols.

17. Write the boolean expression and construct a truth table for the logic circuit in Fig. 31-53.

18. Construct a truth table and logic diagram for the boolean expression $(A + \overline{B})(C + \overline{D}) = X$.

19. Using JK flip-flops, draw a binary ripple counter whose natural modulus is 32. Be sure to identify each Q output and where the clock input is applied.

20. In Fig. 31-54, complete the Q output column in the truth table using 0 and 1. Notice in the first row that $Q = 0$ when the clock input is receiving a positive clock edge and $D = 0$.

CLK	D	Q
↑	0	0
1	1	
↓	1	
0	0	
↑	1	
1	0	
↓	0	
0	0	

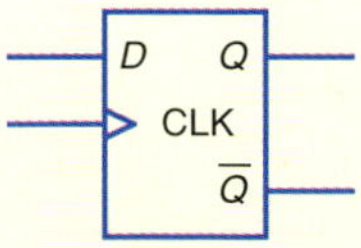

FIG. 31-54 Logic diagram and truth table for Prob. 20.

CRITICAL THINKING

1. Using only NOR gates, draw the logic diagram for the boolean expression $(A + B)(\overline{C} + D) = X$.
2. Referring to Fig. 31-48, show how a 2-input NAND gate can be connected to alter the natural count sequence so that the modulus of the counter becomes 10.
3. Simplify the boolean expression $(\overline{A} + B)(A + B + D)\overline{D}$.

ANSWERS TO TEST-POINT QUESTIONS

31-1 **a.** 0 and 1
      b. 0, 1, 2, 3, 4,
         5, 6, 7, 8, 9
      c. eight
      d. $(1000)_2$

31-2 **a.** $(11110)_2$
      b. $(30)_{10}$
      c. $(111\ 1111)_2$

31-3 **a.** 0, 1, 2, 3, 4, 5,
         6, 7, 8, 9,
         A, B, C, D, E, F
      b. 53,508
      c. 10110000010100
      d. 173

31-4 **a.** 0010 0100 0101
      b. 0001 0000 0101 0110
      c. 5874

31-5 **a.** E
      b. 0111111

31-6 **a.** T
      b. T
      c. F
      d. T
      e. T
      f. T
      g. T

31-7 **a.** $A + B + C = X$
      b. $\overline{ABC} = X$
      c. $\overline{A + B} = X$
      d. $\overline{AB} = X$
      e. $A + 0 = A$
         $A \cdot 0 = 0$
         $A + A = A$
         $A \cdot 1 = A$
         $A + \overline{A} = 1$
         $A \cdot \overline{A} = 0$
         $1 \oplus 1 \oplus 1 = 1$
         $0 \oplus 1 \oplus 1 = 1$

31-8 **a.** T
      b. T
      c. T
      d. T
      e. T

31-9 **a.** F
      b. T
      c. T
      d.
      e.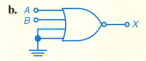

31-10 **a.**
      b.
      c.

31-11 **a.** T
      b. T
      c. F

31-12 **a.** F
      b. T
      c. T
      d. T

31-13 **a.** T
      b. T

31-14 **a.** T
      b. T

CHAPTER 32

INTEGRATED CIRCUITS

Integrated circuits (ICs) have reduced the size, weight, and power requirements of today's electronic equipment. They are replacing transistors in electronic circuits just as transistors once replaced vacuum tubes. ICs are actually
microelectronic circuits. Contained within the IC itself are microscopically
small electronic components such as diodes, transistors, resistors, and capacitors. The actual IC is formed on a single
piece of silicon about the size of a pin head.

ICs are classified as either digital or linear. Digital ICs
are used in computers, calculators, and digital clocks as well
as many other digital devices. Linear ICs are used in analog-
type circuits such as audio amplifiers, voltage regulators, operational amplifiers, and radio frequency circuits to name just
a few. Most linear ICs are low-power devices with power dissipation ratings less than 1 W. There are, however, some chips
available that are capable of handling larger amounts of power
such as 5 W or more. Most linear ICs manufactured today are
designed so that they can be used in a wide variety of applications.

CHAPTER OBJECTIVES

Upon completion of this chapter, you should be able to:

- *Understand* what is meant by the open-loop voltage gain of an op amp.
- *Define* the terms *slew rate* and *power bandwidth*.
- *Calculate* the voltage gain of an inverting amplifier.
- *Calculate* the voltage gain of a noninverting amplifier.
- *Explain* the usefulness of an op amp voltage-follower circuit.
- *Understand* the operation of an op amp summing amplifier.
- *Understand* the operation of an op amp differential amplifier.
- *Understand* the operation of a unity-gain active filter.

IMPORTANT TERMS IN THIS CHAPTER

closed-loop voltage gain	input bias current	slew rate
common mode input	input offset current	summing amplifier
common mode rejection ratio (CMRR)	inverting amplifier	unity-gain active filter
comparator	noninverting amplifier	virtual ground
differential amplifiers	open-loop voltage gain	zero-crossing detector
differential input voltage	power bandwidth	
	saturation	

TOPICS COVERED IN THIS CHAPTER

32-1 Operational Amplifiers and Their Characteristics

32-2 Op Amp Circuits

32-1 OPERATIONAL AMPLIFIERS AND THEIR CHARACTERISTICS

(a)

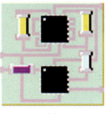

(b)

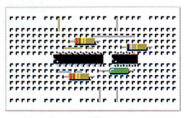

(c)

FIG. 32-1 Typical integrated circuit (IC). (*a*) Sample of 14-pin dual in-line package IC. (*b*) Two IC packages used in a circuit that is mounted on a lab prototype board. (*c*) Two surface-mount IC packages used in an SMT circuit.

Operational amplifiers (op amps) are the most commonly used type of linear integrated circuit (IC). (A typical IC package is shown in Fig. 32-1*a*.) By definition, an op amp is a high-gain, direct-coupled, differential amplifier. An op amp referred to as the 741 has become an industry standard. This op amp, which is contained in an 8-pin IC, is made by several different manufacturers. They are, however, all equivalent because the specifications are nearly identical from one manufacturer to another. Figure 32-1*b* shows two IC packages mounted on a lab prototype board. Figure 32-1*c* shows two surface-mounted IC packages on a printed circuit board.

Figure 32-2*a* shows the internal diodes, transistors, resistors, and capacitors for a 741 op amp. The base leads of *Q*1 and *Q*2 connect to pins on the IC unit and serve as the two inputs for the op amp. *Q*1 and *Q*2 form a differential amplifier circuit. This circuit is used because it can amplify the difference in voltage between the two input signals.

The output of the op amp is taken at the emitters of transistors *Q*8 and *Q*9. These transistors are connected in a push-pull configuration. As far as the output waveform is concerned, *Q*8 conducts during the positive half cycle, and *Q*9 conducts during the negative half cycle. This push-pull configuration allows the op amp to have a very low output impedance, which is analogous to a voltage source having a very low internal resistance.

When viewing the circuit in Fig. 32-2*a*, it is important to note that direct coupling is used between all stages. For this reason, the op amp can amplify signals all the way down to dc. Capacitor C_C affects the operation of the op amp at higher frequencies. This capacitor is called a *compensating capacitor,* with a value of about 30 pF. C_C is used to prevent undesirable oscillations from occurring within the op amp.

The schematic symbol commonly used for op amps is shown in Fig. 32-2*b*. Notice that the triangular schematic symbol shows only the pin connections to different points inside the op amp. Pin 7 connects to $+V_{CC}$, and pin 4 connects to $-V_{CC}$. Also, pins 2 and 3 connect to the op amp inputs, while pin 6 connects to the op amp output.

OPEN-LOOP VOLTAGE GAIN A_{VOL} The open-loop voltage gain A_{VOL} of an op amp is its voltage gain when there is no negative feedback. The open-loop voltage gain of an op amp is the ratio of its output voltage V_{out} to its differential input voltage V_{id}. The open-loop voltage gain A_{VOL} of an op amp is expressed as:

$$A_{VOL} = \frac{V_{out}}{V_{id}}$$

(32-1)

where

$$A_{VOL} = \text{open-loop voltage gain of op amp}$$
$$V_{out} = \text{output voltage}$$
$$V_{id} = \text{differential input voltage}$$

The typical value of A_{VOL} for a 741 op amp is 200,000.

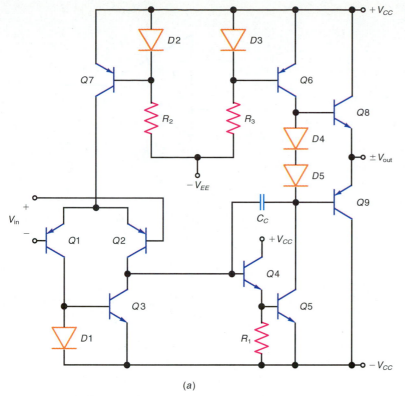

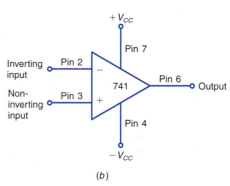

(a)   (b)

FIG. 32-2 (a) Simplified schematic of a 741 op amp. (b) Schematic symbol for op amp.

Figure 32-3a shows the idea. Notice that V_{id} is equal to $V_1 - V_2$, and that V_{out} equals $A_{VOL} \times V_{id}$. Only the differential voltage $V_1 - V_2$ is amplified, not the individual values of V_1 and V_2.

As a numerical example, assume in Fig. 32-3b that the differential input $V_{id} = \pm 50 \ \mu V$ and that $A_{VOL} = 200{,}000$. Then

$$V_{out} = A_{VOL} \times V_{id}$$
$$= 200{,}000 \times (\pm 50 \ \mu V)$$
$$V_{out} = \pm 10 \ V$$

The answer is shown as ± 10 V in Fig. 32-3b because the polarity of V_{id} has not been specified. The polarity of output voltage V_{out} for an op amp is determined using the following two rules.

1. When the voltage at the noninverting ($+$) input is made positive with respect to its inverting ($-$) input, the output is positive.
2. When the voltage at the noninverting ($+$) input is made negative with respect to its inverting ($-$) input, the output is negative.

Assume in Fig. 32-3c that $V_1 = 1$ V and $V_2 = 999.95$ mV. What will the output be? Simply multiply V_{id} by A_{VOL}. But first find V_{id}.

$$V_{id} = V_1 - V_2$$
$$= 1 \ V - 999.95 \ mV$$
$$V_{id} = 50 \ \mu V$$

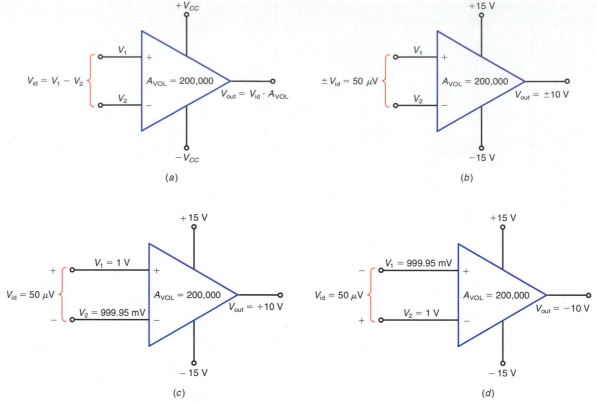

FIG. 32-3 Op amp circuits used to amplify the small value of V_{id} by the HIGH value of A_{VOL}. See text for analysis.

To calculate V_{out}, proceed as follows:

$$V_{out} = A_{VOL} \times V_{id}$$
$$= 200{,}000 \times 50 \ \mu V$$
$$V_{out} = +10 \ V$$

Notice again that the actual values of V_1 and V_2 are not amplified, only the difference between them is. Notice also in Fig. 32-3c that V_1 is connected to the noninverting (+) input terminal and V_2 is connected to the inverting input terminal. Because V_1 is more positive than V_2, the output is positive.

In Fig. 32-3d, V_1 and V_2 are reversed. Here V_1 is negative with respect to V_2. V_{id} is calculated as:

$$V_{id} = V_1 - V_2$$
$$= 999.95 \ mV - 1 \ V$$
$$V_{id} = -50 \ \mu V$$

To find V_{out}, multiply V_{id} by A_{VOL}:

$$V_{out} = A_{VOL} \times V_{id}$$
$$= 200{,}000 \times -50 \ \mu V$$
$$V_{out} = -10 \ V$$

The key point demonstrated in all the circuits in Fig. 32-3 is that only the differential input voltage V_{id} is amplified by the op amps HIGH value of open-loop voltage gain.

It should be pointed out that there are upper and lower limits for the output voltage V_{out}. The upper limit of V_{out} is called the *positive saturation voltage* and is designated $+V_{sat}$. The lower limit of V_{out} is called the *negative saturation voltage* and is designated $-V_{sat}$. For the 741, $\pm V_{sat}$ is usually within a couple volts of $\pm V_{CC}$. For example, if $\pm V_{CC} = \pm15$ V, then $\pm V_{sat} = \pm13$ V. Incidentally, the amount of differential input voltage V_{id} required to produce positive or negative saturation in Fig. 32-3 is found as follows:

$$\pm V_{id} = \frac{\pm V_{sat}}{A_{VOL}} = \frac{\pm 13 \text{ V}}{200,000}$$
$$\pm V_{id} = \pm 65 \ \mu\text{V}$$

Remember that V_{out} will be positive if the noninverting input $(+)$ is made positive with respect to the inverting $(-)$ input. Likewise, V_{out} will be negative if the noninverting input $(+)$ is made negative with respect to the inverting $(-)$ input.

One more point. If the output voltage of any op amp lies between $\pm V_{sat}$, then V_{id} will be so small that it can be considered zero. Realistically it is very difficult to measure a V_{id} of 65 μV in the laboratory because of the induced noise voltages present. Therefore, V_{id} can be considered zero, or $V_{id} = 0$ V, in most cases.

INPUT BIAS CURRENTS

In Fig. 32-2a, the base leads of $Q1$ and $Q2$ serve as the inputs to the op amp. These transistors must be biased correctly before any signal voltage can be amplified. In other words $Q1$ and $Q2$ must have external dc return paths to the power supply ground. Figure 32-4 shows current flowing from the noninverting and inverting input terminals when they're grounded. For a 741, these currents are very, very small, usually 80 nA (80×10^{-9} A) or less. In Fig. 32-4 the I_{B+} designates current flowing from the noninverting input terminal, and I_{B-} designates the current flowing from the inverting input terminal. Manufacturers specify I_B as the average of the two currents I_{B+} and I_{B-}. This can be shown as:

▶ $$I_B = \frac{|I_{B+}| + |I_{B-}|}{2} \tag{32-2}$$

where $|\ |$ means magnitude without regard to polarity. I_{B+} and I_{B-} may be different because it is difficult to match $Q1$ and $Q2$ exactly. The difference between these two currents is designated as I_{OS} for input offset current. I_{OS} can be expressed as:

▶ $$I_{OS} = |I_{B+}| - |I_{B-}| \tag{32-3}$$

For a 741, I_{OS} is typically 20 nA.

For most of our analysis, assume that the value of I_{B+} and I_{B-} are zero because they are so small. However, in some cases of high precision, the effects of I_{B+} and I_{B-} on circuit operation must be taken into account.

FREQUENCY RESPONSE

Figure 32-5 shows the frequency response curve for a typical 741 op amp. Notice that for frequencies below 10 Hz, $A_{VOL} =$

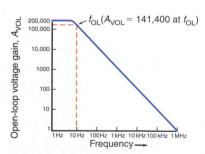

FIG. 32-4 Input bias currents flowing from op amp input terminals.

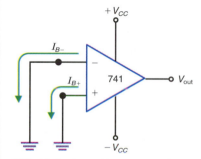

FIG. 32-5 Typical frequency response curve for a 741 op amp without negative feedback.

200,000, and at 10 Hz, A_{VOL} is down to 70.7 percent of its maximum value. That is, at 10 Hz, $A_{VOL} = 141,400$. This frequency is designated f_{OL}, for open-loop cut-off frequency.

Beyond f_{OL}, the voltage gain decreases by a factor of 10 for each decade (factor of 10) increase in frequency. This drop in A_{VOL} at higher frequencies is caused by capacitor C_C inside the op amp. The frequency where A_{VOL} equals 1 is designated f_{unity}. For a 741, f_{unity} is approximately 1 MHz.

SLEW RATE Another very important op amp specification is its slew rate, usually designated S_R. The slew-rate specification of an op amp tells how fast the output voltage can change. The slew rate of an op amp is specified in volts per microsecond, or V/µs. For a 741, the S_R is 0.5 V/µs. This means that no matter how fast the input voltage to a 741 op amp changes, the output voltage can only change as fast as its slew rate allows—0.5 V/µs. Figure 32-6 illustrates the idea. Here the op amp's output waveform should be an amplified version of the sinusoidal input V_{id}. Waveform A would be the expected output. However, if the slope of the output sine wave exceeds the S_R rating of the op amp, the waveform would appear triangular; that is, as waveform B. Therefore, slew-rate distortion of a sine wave produces a triangular wave.

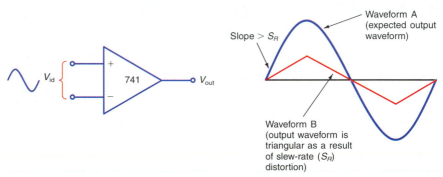

FIG. 32-6 Slew-rate distortion occurs when the initial slope of the output waveform exceeds the S_R rating of the op amp.

POWER BANDWIDTH There are two ways to avoid slew-rate distortion of a sine wave: either use an op amp with a faster slew rate or accept an output waveform with a lower peak voltage. Using an op amp with a faster slew rate seems like a logical solution because then the output waveform will be able to follow the sinusoidal input voltage V_{id}. But why would less peak voltage for the output waveform solve the problem? The answer is best illustrated in Fig. 32-7. Here, both waveforms A and B have exactly the same frequency but different peak values. Waveform A has a peak value of 1 V, while waveform B has a peak value of 10 V. If we compare the voltage change during the first 30° of each waveform, we see that the change in voltage ΔV for waveform A is 0.5 V during the first 30°, while ΔV for waveform B is 5 V during the same interval. Notice that the rate of voltage change for waveform B is 10 times that of waveform A during the same time interval even though both waveforms have exactly the same frequency! Therefore, it is true to say that two waveforms having identical frequencies, but different peak values have significantly different slopes during their positive and negative alternations.

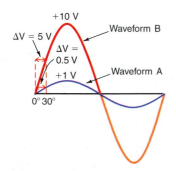

FIG. 32-7 Waveforms A and B have identical frequencies but different slopes during their positive and negative alternations.

The higher the peak voltage of a sine wave for a given frequency, the greater its initial slope. If the initial slope of the output waveform exceeds the S_R rating of the op amp, slew-rate distortion will occur. The following formula shows the highest undistorted frequency out of an op amp for a given S_R and peak voltage:

$$\blacktriangleright \quad f_{\text{max}} = \frac{S_R}{2\pi V_{\text{pk}}} \qquad\qquad (32\text{-}4)$$

where

$$f_{\text{max}} = \text{Highest undistorted frequency}$$
$$S_R = \text{Slew rate}$$
$$V_{\text{pk}} = \text{Peak value of output sine wave}$$

Notice that f_{max} can be increased by using an op amp with a higher slew rate or by accepting an output wave form with less peak voltage.

Example

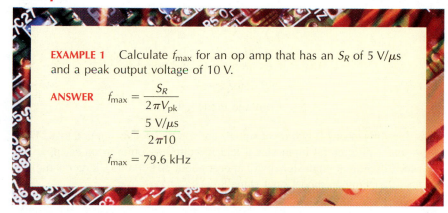

EXAMPLE 1   Calculate f_{max} for an op amp that has an S_R of 5 V/μs and a peak output voltage of 10 V.

ANSWER $f_{\text{max}} = \dfrac{S_R}{2\pi V_{\text{pk}}}$

$$= \frac{5 \text{ V/}\mu\text{s}}{2\pi 10}$$

$$f_{\text{max}} = 79.6 \text{ kHz}$$

The frequency f_{max} of 79.6 kHz is commonly called the 10-V power bandwidth. This means that slew-rate distortion for a 10-V peak sine wave will not occur for frequencies at or below 79.6 kHz.

If we try to amplify higher frequencies with the same peak value of 10 V, we get slew-rate distortion, meaning the output waveform gets triangular as described earlier.

OUTPUT SHORT-CIRCUIT CURRENT   An op amp, such as the 741, has short-circuit output protection. For a 741, the output short-circuit current is approximately 25 mA. This means that if the op amp output (pin 6) is tied directly to ground, the output current will not exceed 25 mA. It should be noted that small load resistances connected to the op amp output will usually have lower amplitudes of output voltage because the output voltage cannot exceed 25 mA $\times R_L$.

COMMON MODE REJECTION RATIO   As mentioned earlier, an op amp amplifies only the difference in voltage between its two inputs. This means that

if two identical signals are applied to the inputs of an op amp, each with exactly the same phasoral relationship and voltage values, the output will be zero. Such a signal is called a common mode signal because identical waveforms are applied to both inputs. In other words, the signal is common to both inputs. Unfortunately, even with a perfect common mode–type signal, the output from the op amp will not be zero. This is because op amps are not ideal. The rejection of the common mode signal, however, is very high. For a typical 741, the common mode rejection ratio (CMRR) is 90 db, which corresponds to a ratio of about 30,000. The CMRR of an op amp is defined as its ability to amplify differential input signals while attenuating or rejecting common mode signals. Expressed as an equation:

$$\text{CMRR} = \frac{A_d}{A_{cm}} \qquad\qquad (32\text{-}5)$$

where

$$\text{CMRR} = \text{Common mode rejection ratio}$$
$$A_d = \text{Differential gain}$$
$$A_{cm} = \text{Common mode gain}$$

What does this mean? If two input signals are simultaneously applied to a 741 op amp—one a differential input signal and the other a common mode input signal—the differential input signal will appear about 30,000 times larger at the output than the common mode input signal.

TEST-POINT QUESTION 32-1

Answers at end of chapter.

Answer True or False.

a. The open-loop voltage gain A_{VOL} of an op amp is very large for low frequencies.

b. For an op amp, $\pm V_{sat}$ is usually within a couple volts of $\pm V_{CC}$.

c. For a 741, the open-loop cut-off frequency f_{OL} is 10 Hz.

d. The slew rate S_R of an op amp tells how well the op amp can reject common mode signals.

e. In order to increase f_{max} (the highest undistorted frequency out of an op amp), either accept less peak voltage for the output waveform or use an op amp with a higher S_R.

f. The output of an op amp will be positive if its noninverting (+) input is positive with respect to its inverting (−) input.

g. The output of an op amp will be negative if its inverting (−) input is positive with respect to its noninverting (+) input.

h. When the output voltage of an op amp is less than $\pm V_{sat}$, then V_{id} is so small it can be considered zero.

32-2 OP AMP CIRCUITS

Most op amp circuits use negative feedback. As a general rule, op amp circuits without negative feedback are too unstable to be useful. Negative feedback reduces the overall voltage gain of the op amp circuit. In return for this loss of gain, however, we have a tremendously stable voltage gain over a very wide range of frequencies.

INVERTING AMPLIFIER   Figure 32-8 shows an op amp connected to work as an inverting amplifier. It is called an inverting amplifier because the input and output signals are 180° out of phase when V_{in} is applied to the inverting (−) input terminal, as shown.

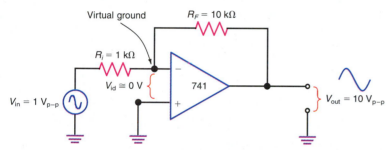

FIG. 32-8   Op amp connected to work as an inverting amplifier.

Resistors R_F and R_i provide the negative feedback, which in turn controls the circuit's overall voltage gain. The output signal V_{out} is fed back to the inverting input of the op amp through resistors R_F and R_i. The voltage between the inverting input and ground is V_{id}. The exact value of V_{id} is determined by the values A_{VOL} and V_{out}. For all practical purposes, V_{id} is so small that it can be considered zero. This introduces little or no error in our circuit analysis. Because V_{id} is so small (practically zero), the inverting input terminal of the op amp in Fig. 32-8 is said to be at virtual ground, because the voltage at the op amps inverting input is nearly the same as at ground, yet it can sink no current.

Since the inverting input is at virtual ground, V_{in} is dropped across R_i. Therefore,

$$I = \frac{V_{in}}{R_i} \qquad\qquad (32\text{-}6)$$

This says that all input current produced by V_{in} flows through resistor R_i. Rearranging this equation, we have:

$$V_{in} = I \times R_i \qquad\qquad (32\text{-}7)$$

Practically no current flows from the inverting input terminal, so all the current produced by V_{in} must flow through R_F to the op amp output terminal. Therefore,

▶ $V_{\text{out}} = I \times R_F$ **(32-8)**

Since $V_{\text{in}} = I \times R_i$ and $V_{\text{out}} = I \times R_F$, we can calculate the voltage gain of the circuit as follows:

▶ $$A_{\text{CL}} = \frac{V_{\text{out}}}{V_{\text{in}}}$$ **(32-9)**

$$= \frac{I \times R_F}{I \times R_i}$$

$$A_{\text{CL}} = \frac{R_F}{R_i}$$

where

$$A_{\text{CL}} = \text{Closed-loop voltage gain}$$

The closed-loop voltage gain of the circuit is the voltage gain with negative feedback. In Fig. 32-8, A_{CL} is found as:

$$A_{\text{CL}} = \frac{R_F}{R_i}$$

$$= \frac{10 \text{ k}\Omega}{1 \text{ k}\Omega}$$

$$A_{\text{CL}} = 10$$

In Fig. 32-8, V_{out} can be found by multiplying A_{CL} by V_{in}:

$$V_{\text{out}} = A_{\text{CL}} \times V_{\text{in}} = 10 \times 1 \text{ V}_{\text{p–p}}$$

$$V_{\text{out}} = 10 \text{ V}_{\text{p–p}}$$

Remember that for an inverting amplifier, V_{out} is 180° out of phase with V_{in}.

NONINVERTING AMPLIFIER   Figure 32-9 shows an op amp circuit connected as a noninverting amplifier. It is a noninverting amplifier because the input signal V_{in} drives the noninverting input terminal of the op amp. For this circuit, the input and output signals are always in phase. In Fig. 32-9, V_{id} is again considered to be approximately 0 V. This means that the voltage across resistor R_i must equal V_{in}. The current through R_i is found as follows:

▶ $I = \dfrac{V_{\text{in}}}{R_i}$ or   **(32-10)**

$V_{\text{in}} = I \times R_i$

All this current must flow through R_F, since virtually no current flows from the inverting input terminal. Therefore, the voltage across R_F is:

▶ $V_{R_F} = I \times R_F$ **(32-11)**

DID YOU KNOW?

Current researchers are going beyond the limits of silicon to find ways to transmit, store, and retrieve data with molecules (as it is done biologically). The first step is to turn molecules into magnets, and vice versa, at low temperatures and under colored light.

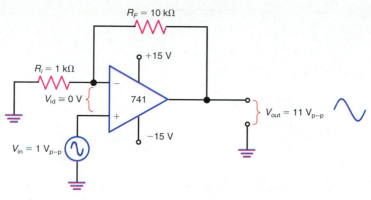

FIG. 32-9 Op amp connected to work as a noninverting amplifier.

Since the output voltage V_{out} is taken with respect to ground, V_{out} must be the sum of V_{R_i} and V_{R_F}:

▶ $$V_{\text{out}} = IR_F + IR_i \quad \text{or} \qquad\qquad\qquad \text{(32-12)}$$
$$V_{\text{out}} = I(R_F + R_i)$$

Therefore, the closed-loop voltage gain A_{CL} can be expressed as:

▶ $$A_{\text{CL}} = \frac{V_{\text{out}}}{V_{\text{in}}} \qquad\qquad\qquad \text{(32-13)}$$

$$= \frac{I(R_F + R_i)}{IR_i}$$

$$A_{\text{CL}} = \frac{R_F + R_i}{R_i} \quad \text{or}$$

$$A_{\text{CL}} = \frac{R_F}{R_i} + 1$$

For Fig. 32-9, A_{CL} is:

$$A_{\text{CL}} = \frac{10 \text{ k}\Omega}{1 \text{ k}\Omega} + 1$$

$$A_{\text{CL}} = 11$$

And V_{out} can be found by multiplying A_{CL} by V_{in}:

$$V_{\text{out}} = A_{\text{CL}} \times V_{\text{in}}$$
$$= 11 \times 1 \text{ V}_{\text{p-p}}$$
$$V_{\text{out}} = 11 \text{ V}_{\text{p-p}}$$

Here is an important point. Since the voltage source has to supply virtually no current to the op amp's noninverting input, the voltage source V_{in} sees a very high value of input impedance. For this reason there is virtually no loading of the voltage source V_{in}.

VOLTAGE FOLLOWER   Figure 32-10 shows the op amp connected to provide a voltage gain of one, or unity. Notice that the op amp output is connected

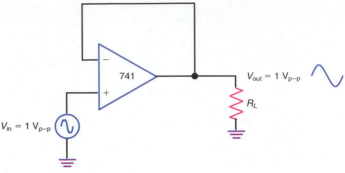

FIG. 32-10 Op amp connected to work as a voltage follower. $A_{CL} = 1$; $V_{out} = V_{in} = 1$ Vpp.

directly to the inverting input terminal to obtain the maximum amount of negative feedback possible. For this circuit, $A_{CL} = 1$. Because $V_{id} = 0$ V, V_{out} must equal V_{in}. But why use such a circuit if it provides no voltage gain? It is used because the op amp circuit will buffer, or isolate, the voltage source V_{in} from the load R_L. This means that rather than connect a relatively low value of load resistance across the terminals of V_{in}, the op amp can be used to eliminate any loading that might occur. Since the voltage source V_{in} is connected to the noninverting (+) input terminal of the op amp, it has to supply virtually no current to the circuit. Thus, the voltage source won't be loaded down. Also, because of negative feedback, the op amp circuit will have an output impedance that is close to zero. In effect, then, the load believes it is being driven by an ideal voltage source with zero internal impedance.

SUMMING AMPLIFIER The circuit shown in Fig. 32-11 is called a summing amplifier or summer. When $R_1 = R_2 = R_3 = R_F$, the output voltage V_{out} equals the negative sum of the input voltages. Because the right end of resistors R_1, R_2, and R_3 are at virtual ground, the input currents are calculated as

$$I_1 = \frac{V_1}{R_1}$$

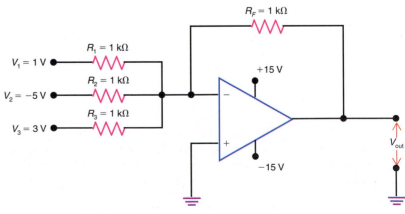

FIG. 32-11 Summing amplifier.

$$I_2 = \frac{V_2}{R_2}$$

$$I_3 = \frac{V_3}{R_3}$$

Because the inverting input has practically zero current, each of the input currents flow through the feedback resistor R_F. Because $R_1 = R_2 = R_3 = R_F$, the voltage gain of the circuit is 1 and the output voltage is calculated as

▶ $V_{\text{out}} = -(V_1 + V_2 + V_3)$ **(32-14)**

If more inputs are necessary, additional resistors can be added at the input. This is possible because the inverting input is at virtual ground, thus effectively isolating each input from the other. Because of the virtual ground, each input sees its own input resistance and nothing else.

If each input is amplified by a different amount, the output voltage will equal the negative of the amplified sum of the inputs. When the voltage gain is different for each input, the formula for the output voltage becomes

▶ $V_{\text{out}} = -\left(\dfrac{R_F}{R_1}V_1 + \dfrac{R_F}{R_2}V_2 + \dfrac{R_F}{R_3}V_3 \right)$ **(32-15)**

Example

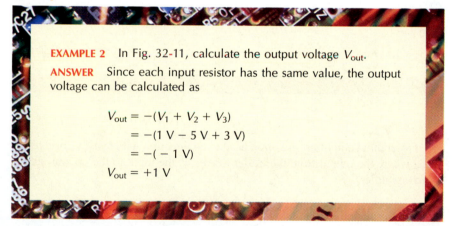

EXAMPLE 2   In Fig. 32-11, calculate the output voltage V_{out}.

ANSWER   Since each input resistor has the same value, the output voltage can be calculated as

$$V_{\text{out}} = -(V_1 + V_2 + V_3)$$
$$= -(1\ V - 5\ V + 3\ V)$$
$$= -(-1\ V)$$
$$V_{\text{out}} = +1\ V$$

DIFFERENTIAL AMPLIFIERS   Differential amplifiers are circuits which have the ability to amplify differential input signals and severely attenuate common-mode signals. A typical op amp differential amplifier is shown in Fig. 32-12. The formula for calculating the output voltage is

▶ $V_{\text{out}} = -\dfrac{R_F}{R_1}(V_X - V_Y)$ **(32-16)**

where V_X and V_Y represent the individual inputs being applied to the circuit. Note that if $V_X = V_Y$, the output voltage will be zero. Therefore, this circuit will amplify only the difference in voltage that exists between the inputs V_X and V_Y.

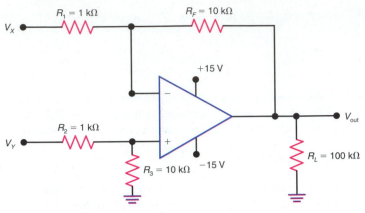

FIG. 32-12 Op amp differential amplifier.

Example

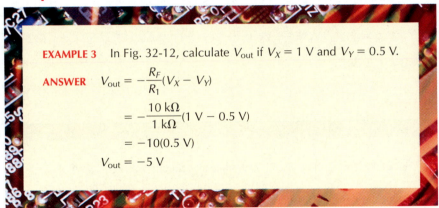

EXAMPLE 3 In Fig. 32-12, calculate V_{out} if $V_X = 1$ V and $V_Y = 0.5$ V.

ANSWER

$$V_{out} = -\frac{R_F}{R_1}(V_X - V_Y)$$

$$= -\frac{10 \text{ k}\Omega}{1 \text{ k}\Omega}(1 \text{ V} - 0.5 \text{ V})$$

$$= -10(0.5 \text{ V})$$

$$V_{out} = -5 \text{ V}$$

Differential amplifiers are often used in conjunction with resistive bridge circuits, where the output from the bridge serves as the input to the op amp differential amplifier.

UNITY-GAIN ACTIVE FILTERS   An active filter is a type of filter that uses components or devices that have the ability to amplify, such as transistors or op amps. A passive filter is one that uses only passive components such as resistors, capacitors, and inductors.

Figure 32-13 shows a type of active filter that has unity gain in the passband. Figure 32-13*a* shows an active low-pass filter, whereas Fig. 32-12*b* shows an active high-pass filter. It is important to notice that, in each case, the op amp is wired to work as a voltage follower. The main reason for using the op amp is that the RC filter can be isolated from the load R_L, which may have a very low impedance. More specifically, because Z_{in} is so high looking into the op amp and because Z_{out} is approximately zero, the RC filter is effectively isolated from the load R_L.

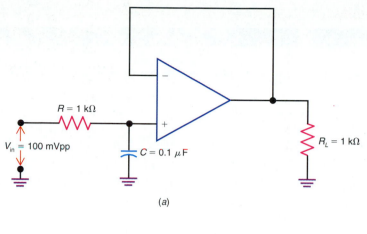

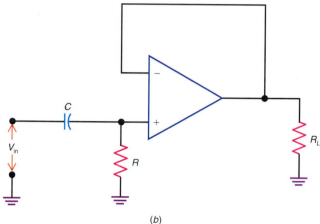

FIG. 32-13 Unity-gain active filters. (*a*) Active low-pass filter with unity gain. (*b*) Active high-pass filter with unity gain.

For both circuits in Fig. 32-13, the cutoff frequency f_c is calculated as

$$f_c = \frac{1}{2\pi RC} \qquad\qquad (32\text{-}17)$$

For both filters, the voltage gain is 1 in the passband, which corresponds to a voltage gain of 0 db.

In Fig. 32-13*a*, the output voltage at any frequency can be calculated as

$$V_{\text{out}} = \frac{X_C}{Z_T} \times V_{\text{in}} \qquad\qquad (32\text{-}18)$$

In Fig. 32-13*b*, V_{out} can be calculated as

$$V_{\text{out}} = \frac{R}{Z_T} \times V_{\text{in}} \qquad\qquad (32\text{-}19)$$

Example

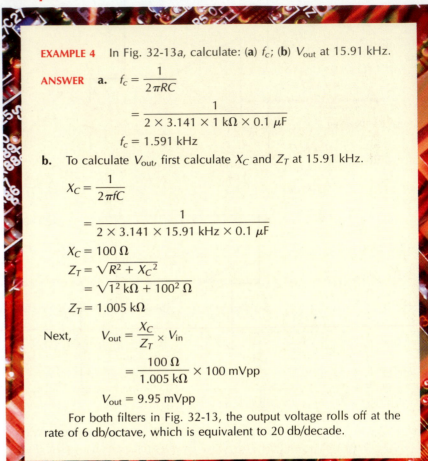

EXAMPLE 4 In Fig. 32-13a, calculate: (a) f_c; (b) V_{out} at 15.91 kHz.

ANSWER **a.** $f_c = \dfrac{1}{2\pi RC}$

$$= \dfrac{1}{2 \times 3.141 \times 1 \text{ k}\Omega \times 0.1 \text{ } \mu\text{F}}$$

$$f_c = 1.591 \text{ kHz}$$

b. To calculate V_{out}, first calculate X_C and Z_T at 15.91 kHz.

$$X_C = \dfrac{1}{2\pi fC}$$

$$= \dfrac{1}{2 \times 3.141 \times 15.91 \text{ kHz} \times 0.1 \text{ } \mu\text{F}}$$

$$X_C = 100 \text{ } \Omega$$

$$Z_T = \sqrt{R^2 + X_C^2}$$

$$= \sqrt{1^2 \text{ k}\Omega + 100^2} \text{ } \Omega$$

$$Z_T = 1.005 \text{ k}\Omega$$

Next, $V_{out} = \dfrac{X_C}{Z_T} \times V_{in}$

$$= \dfrac{100 \text{ } \Omega}{1.005 \text{ k}\Omega} \times 100 \text{ mVpp}$$

$$V_{out} = 9.95 \text{ mVpp}$$

For both filters in Fig. 32-13, the output voltage rolls off at the rate of 6 db/octave, which is equivalent to 20 db/decade.

COMPARATORS   A comparator is a circuit that compares the signal voltage on one input with a reference voltage on the other. An op amp comparator is shown in Fig. 32-14a. Notice that there is no feedback resistor to provide negative feedback. As a result, the op amp is running in the open-loop mode with a voltage gain equal to A_{VOL}.

In Fig. 32-14a, the inverting input of the op amp is grounded and the input signal is applied to the noninverting input. The comparator compares V_{in} to the 0-V reference on the inverting input. When V_{in} goes positive, the output is driven to $+V_{sat}$. When V_{in} goes negative, the output is driven to $-V_{sat}$. Usually, $\pm V_{sat}$ is within a couple of volts of $\pm V_{CC}$.

Because of the op amp's extremely high open-loop voltage gain, even the slightest input voltage forces the output voltage to $\pm V_{sat}$. Figure 32-14b shows the transfer characteristic for the op amp comparator. Notice that V_{out} switches to $+V_{sat}$ when V_{in} is positive, and to $-V_{sat}$ when V_{in} is negative. Because V_{out} switches when V_{in} crosses zero, the comparator in Fig. 32-14a is sometimes called a *zero crossing detector*.

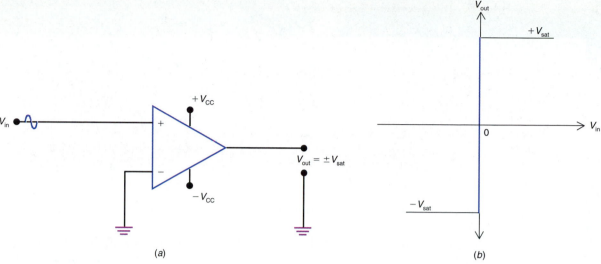

FIG. 32-14 Op amp comparator. (a) Circuit. (b) Transfer characteristic.

It should be noted that a comparator can use a reference voltage other than zero. For example, if the inverting input was a reference voltage of $+1$ V, then V_{out} would switch between $\pm V_{sat}$ when V_{in} crossed the $+1$-V reference value.

One final point. In some cases the noninverting input may be grounded or connected to a reference voltage rather than the inverting input. In this case, the output voltage switches to $+V_{sat}$ when V_{in} is less than the reference voltage, and to $-V_{sat}$ when V_{in} is greater than the reference voltage.

TEST-POINT QUESTION 32-2

Answers at end of chapter.
  a. In Fig. 32-8, calculate A_{CL} if $R_F = 15$ kΩ and $R_i = 1.2$ kΩ.
  b. In Fig. 32-8, calculate V_{out} if $R_F = 15$ kΩ, $R_i = 1$ kΩ, and $V_{in} = 0.5$ V$_{p-p}$.
  c. In Fig. 32-9, calculate A_{CL} if $R_F = 18$ kΩ and $R_i = 1.2$ kΩ.
  d. In Fig. 32-9, calculate V_{out} if $R_F = 24$ kΩ, $R_i = 1$ kΩ, and $V_{in} = 250$ mV$_{p-p}$.
  e. In Fig. 32-10, $V_{in} = 5$ V$_{p-p}$. What is V_{out}?
  f. What is the approximate value of V_{id} in Figs. 32-8, 32-9, and 32-10?
  g. In Fig. 32-11, what is V_{out} if $V_1 = -2$ V, $V_2 = +2$ V, and $V_3 = -4$ V?
  h. In Fig. 32-12, what is V_{out} if $V_X = +0.1$ V and $V_Y = +0.1$ V?

32 SUMMARY AND REVIEW

- Integrated circuits (ICs) are classified into two groups: digital or linear. Digital ICs process digital signals, and linear ICs process analog signals.
- Operational amplifiers (op amps) are the most commonly used type of linear IC. An op amp is a high-gain, direct-coupled, differential amplifier. The 741 op amp has become an industry standard.
- The open-loop voltage gain A_{VOL} of an op amp is its voltage gain without negative feedback. For a 741, the typical value for A_{VOL} is 200,000.
- For all practical purposes, the differential input voltage V_{id} applied to an op amp is considered zero when V_{out} is less than $\pm V_{sat}$.
- The input bias currents of a 741 are typically 80 nA. The input bias currents for the noninverting and inverting inputs are designated I_{B+} and I_{B-} respectively.
- The open-loop cut-off frequency of an op amp is designated f_{OL}. At this frequency, the op amps open-loop voltage gain A_{VOL} is down to 70.7 percent of its maximum value. For a 741, $f_{OL} = $ 10 Hz.
- The slew rate of an op amp is designated S_R. The S_R specification of an op amp tells how fast the output voltage can change. The S_R of a 741 op amp is 0.5 V/µs.
- The power bandwidth (designated f_{max}) of an op amp circuit is the highest undistorted frequency out of an op amp for a given S_R and peak voltage.
- The 741 has an output short-circuit current of approximately 25 mA.
- Most op amp circuits use negative feedback. When the op amp is connected to work as an inverting amplifier, the closed-loop voltage gain $A_{CL} = R_F/R_i$. For an inverting amplifier, the input and output signals are 180° out of phase.
- The op amp can also be connected to work as a noninverting amplifier. The closed-loop voltage gain $A_{CL} = R_F/R_i + 1$. For this circuit the input and output signals will be in phase with each other.
- An op amp circuit that has its output tied directly to the inverting input terminal is called a *voltage follower*. This circuit has a voltage gain of one, or unity.
- An op amp summer is a circuit whose output voltage is the negative sum of the input voltages.
- An op amp differential amplifier is a circuit which amplifies differential input signals and rejects common-mode signals.
- A unity-gain active filter has the advantage of isolating the RC filter from a low-impedance load, R_L.
- An op amp comparator is a circuit that compares the signal voltage on one of its inputs with a reference voltage on the other.

SELF-TEST

ANSWERS AT BACK OF BOOK.

Answer True or False.

1. Op amps can amplify both dc and ac signals.
2. An op amp comparator uses no negative feedback.
3. The open-loop voltage gain of an op amp is very high.
4. The $\pm V_{sat}$ for an op amp is usually 2 to 3 V higher than $\pm V_{CC}$.
5. To increase the power bandwidth of an op amp circuit, either accept less peak voltage V_{pk} or use an op amp with a higher slew rate.
6. For the op amp circuit in Fig. 32-8, the voltage gain A_{CL} could be increased by reducing the value of R_i.
7. For the op amp circuit in Fig. 32-8, $V_{id} = 1$ V$_{p-p}$.
8. Noninverting amplifier circuits have a very high input impedance.
9. A voltage follower has an input of 0.5 V$_{p-p}$. The output is also 0.5 V$_{p-p}$.
10. The slew rate of an op amp is of no concern when amplifying dc input signals.
11. Op amps will attenuate common mode input signals.

QUESTIONS

1. Define what is meant by a common mode input signal.
2. What type of circuit is used for the input stage of an op amp?
3. What is meant by f_{unity}?
4. Define the term "input offset current" (I_{OS}).
5. Describe how slew-rate distortion affects the operation of an op amp circuit used to amplify sinusoidal signals.
6. What is meant by virtual ground?
7. Why are op amp circuits without negative feedback so unstable?

PROBLEMS

ANSWERS TO ODD NUMBERED PROBLEMS AT BACK OF BOOK.

1. Refer to Fig. 32-8. If $R_F = 10$ kΩ, $R_i = 1.0$ kΩ, and $V_{in} = -0.5$ Vdc, then calculate (**a**) A_{CL}; (**b**) V_{out}.
2. Refer to Fig. 32-9. If $R_F = 24$ kΩ, $R_i = 1$ kΩ, and $V_{in} = 0.25$ V$_{pp}$, then calculate (**a**) A_{CL}; (**b**) V_{out}.
3. An op amp has a slew rate of 0.5 V/μs. Calculate the 5-V power bandwidth.

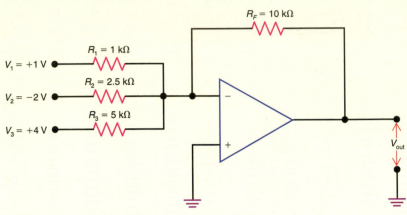

FIG. 32-15 Circuit for Prob. 9.

4. Refer to Fig. 32-11. If $V_1 = 2$ V, $V_2 = -3$ V, and $V_3 = -2$ V, calculate V_{out}.
5. In Fig. 32-12, calculate V_{out} for the following values of V_X and V_Y: (**a**) $V_X = +1$ V and $V_Y = -0.25$ V; (**b**) $V_X = -0.5$ V and $V_Y = +0.5$ V; (**c**) $V_X = +0.6$ V and $V_Y = +0.6$ V.
6. In Fig. 32-13a, calculate the cutoff frequency f_c for $R = 1.5$ kΩ and $C = 0.047$ μF. Also, calculate V_{out} at f_c and $10f_c$.
7. In Fig. 32-13b, calculate the cutoff frequency f_c for $R = 10$ kΩ and $C = 0.015$ μF. Also, calculate V_{out} at $0.25f_c$. ($V_{in} = 100$ mVpp.)
8. In Fig. 32-14a, assume $\pm V_{sat} = \pm13$ V and $A_{VOL} = 200,000$. Calculate the minimum value for V_{id} which will drive the output to $\pm V_{sat}$.
9. In Fig. 32-15, calculate V_{out}.

CRITICAL THINKING

1. In Fig. 32-12, assume $V_X = 100$ mVpp and $V_Y = 100$ mVpp. If the two waveforms are 180° out of phase, what is the value of V_{out}, and is V_{out} in phase with V_X or V_Y?

ANSWERS TO TEST-POINT QUESTIONS

32-1 **a.** T
 b. T
 c. T
 d. F
 e. T
 f. T
 g. T
 h. T

32-2 **a.** 12.5
 b. 7.5 V_{p-p}
 c. 16
 d. 6.25 V_{p-p}
 e. 5 V_{p-p}
 f. 0 V
 g. +4 V
 h. 0 V

REVIEW: CHAPTERS 31 AND 32

SUMMARY

- The basic logic gates used in digital electronic circuits are summarized in Table 31-4. Both the traditional and new rectangular symbols are shown, along with the boolean expression for each logic gate.
- The basic types of flip-flops used in digital electronic circuits are shown in Table 31-5.
- Digital circuits use a two-state design. For this reason, the binary number system is used when working with digital circuits. The binary digit 0 is used to represent a low voltage such as 0 V, whereas the binary digit 1 is used to represent a high voltage such as +5 V or +12 V.
- Truth tables are used to analyze the operation of a logic gate. The truth table lists all input possibilities and the corresponding output for each input.
- Table 31-1 compares the binary, decimal, and hexadecimal number systems. Memorize the binary and hexadecimal numbers that correspond to the decimal digits 0 through 15.
- DeMorgan's first theorem states that the complement of a sum equals the product of the complements. DeMorgan's second theorem states that the complement of a product equals the sum of the complements. The first and second theorems can be expressed as shown:

First Theorem: $\overline{A + B} = \overline{A} \cdot \overline{B}$

Second Theorem: $\overline{AB} = \overline{A} + \overline{B}$

- Operational amplifiers (op amps) are the most widely used type of linear IC. Op amps are high-gain, direct-coupled, differential amplifiers. The 741 is a very popular op amp and has become an industry standard.
- The open-loop voltage gain of an op amp is its voltage gain with no negative feedback. It is designated as A_{VOL}. If the output voltage of any op amp lies between $\pm V_{sat}$, then the differential input voltage V_{id} will be so small that it can be approximated as zero. Typical values of A_{VOL} are 200,000 or more at low frequencies. It should be realized, then, that V_{id} will be very small for a given output voltage owing to the op amp's high value of A_{VOL}.
- Most op amp circuits use negative feedback. This is because op amp circuits without negative feedback are usually too unstable. When an op amp is connected to work as an inverting amplifier, $A_{CL} = R_F/R_i$. When the op amp is connected to work as a noninverting amplifier, $A_{CL} = R_F/R_i + 1$.
- The slew rate (S_R) specification of an op amp tells how fast the op amp's output voltage can change. For a 741, $S_R = 0.5$ V/μs. Slew-rate distortion of a sine wave produces a triangular wave.

REVIEW SELF-TEST

ANSWER TRUE OR FALSE.

1. The binary number system uses the digits 0 and 1.
2. The binary equivalent of decimal 21 is 10101.
3. The hexadecimal number F0 has a binary equivalent of 11110000.
4. An AND gate will have a high output only when all of its inputs are HIGH.
5. The output of an OR gate will be low if any or all inputs are LOW.
6. The output of an XOR gate is high only when an even number of input 1s are applied.
7. A NOR gate will have a high output only when all inputs are LOW.
8. A NAND gate will have a low output only when all inputs are HIGH.
9. The boolean expression for a 4-input XNOR gate would be $\overline{A \oplus B \oplus C \oplus D} = X$.
10. An open, or disconnected, input on a logic gate acts like a HIGH input.
11. The absence of a bubble on the output line of a logic gate indicates that it is an active HIGH output.
12. A flip-flop is set when $Q = 0$ and $\overline{Q} = 1$.
13. The differential input voltage V_{id} for an op amp is usually 2 to 3 V in most cases.
14. Slew-rate distortion of a sine wave can be eliminated if we use an op amp with a higher S_R rating or accept less peak voltage for the output waveform.
15. An op amp's output will be negative if the inverting ($-$) input terminal is positive with respect to the noninverting input terminal.
16. A voltage follower is often used to amplify common mode signals.

REFERENCES

Malvino, A. P.: *Digital Computer Electronics,* Glencoe/McGraw-Hill, Columbus, Ohio.

Tokheim, R. L.: *Digital Principles,* Glencoe/McGraw-Hill, Columbus, Ohio.

Couglin, R. F., and F. F. Driscoll: *Operational Amplifiers and Linear Integrated Circuits,* Prentice-Hall, Englewood Cliffs, New Jersey.

APPENDIX A

ELECTRICAL SYMBOLS AND ABBREVIATIONS

Table A-1 summarizes the letter symbols used as abbreviations for electrical characteristics and their basic units. All the metric prefixes for multiple and fractional values are listed in Table A-2. In addition, Table A-3 shows electronic symbols from the Greek alphabet. Table A-4 shows the preferred values for resistors having tolerances of ±20 percent, ±10 percent, and ±5 percent.

TABLE A-1   ELECTRICAL CHARACTERISTICS

QUANTITY	SYMBOL*	BASIC UNIT
Current	I or i	ampere (A)
Charge	Q or q	coulomb (C)
Power	P	watt (W)
Voltage	V or v	volt (V)
Resistance	R	ohm (Ω)
Reactance	X	ohm (Ω)
Impedance	Z	ohm (Ω)
Conductance	G	siemens (S)
Admittance	Y	siemens (S)
Susceptance	B	siemens (S)
Capacitance	C	farad (F)
Inductance	L	henry (H)
Frequency	f	hertz (Hz)
Period	T	second (s)

*Capital letters for I, Q, and V are generally used for peak, rms, or dc value; small letters are used for instantaneous values. Small r and g are used for internal values, such as r_p and g_m of a tube.

TABLE A-2   MULTIPLES AND SUBMULTIPLES OF UNITS*

VALUE	PREFIX	SYMBOL	EXAMPLE
$1\ 000\ 000\ 000\ 000 = 10^{12}$	tera	T	$THz = 10^{12}\ Hz$
$1\ 000\ 000\ 000 = 10^{9}$	giga	G	$GHz = 10^{9}\ Hz$
$1\ 000\ 000 = 10^{6}$	mega	M	$MHz = 10^{6}\ Hz$
$1\ 000 = 10^{3}$	kilo	k	$kV = 10^{3}\ V$
$100 = 10^{2}$	hecto	h	$hm = 10^{2}\ m$
$10 = 10$	deka	da	$dam = 10\ m$
$0.1 = 10^{-1}$	deci	d	$dm = 10^{-1}\ m$
$0.01 = 10^{-2}$	centi	c	$cm = 10^{-2}\ m$
$0.001 = 10^{-3}$	milli	m	$mA = 10^{-3}\ A$
$0.000\ 001 = 10^{-6}$	micro	μ	$\mu V = 10^{-6}\ V$
$0.000\ 000\ 001 = 10^{-9}$	nano	n	$ns = 10^{-9}\ s$
$0.000\ 000\ 000\ 001 = 10^{-12}$	pico	p	$pF = 10^{-12}\ F$

*Additional prefixes are exa = 10^{18}, peta = 10^{15}, femto = 10^{-15}, and atto = 10^{-18}.

TABLE A-3 GREEK LETTER SYMBOLS*

NAME	LETTER CAPITAL	SMALL	USES
Alpha	A	α	α for angles, transistor characteristic
Beta	B	β	β for angles, transistor characteristic
Gamma	Γ	γ	
Delta	Δ	δ	Small change in value
Epsilon	E	ϵ	ϵ for permittivity; also base of natural logarithms
Zeta	Z	ζ	
Eta	H	η	
Theta	Θ	θ	Phase angle
Iota	I	ι	
Kappa	K	κ	
Lambda	Λ	λ	λ for wavelength
Mu	M	μ	μ for prefix micro, permeability, amplification factor
Nu	N	ν	
Xi	Ξ	ξ	
Omicron	O	o	
Pi	Π	π	π is 3.1416 for ratio of circumference to diameter of a circle
Rho	P	ρ	ρ for resistivity
Sigma	Σ	σ	Summation
Tau	T	τ	Time constant
Upsilon	Υ	υ	
Phi	Φ	ϕ	Magnetic flux, angles
Chi	X	χ	
Psi	Ψ	ψ	Electric flux
Omega	Ω	ω	Ω for ohms; ω for angular velocity

*This table includes the complete Greek alphabet, although some letters are not used for electronic symbols.

TABLE A-4 PREFERRED VALUES* FOR RESISTORS

TOLERANCE			TOLERANCE		
20%	10%	5%	20%	10%	5%
10	10	10			36
		11		39	39
	12	12			43
		13	47	47	47
15	15	15			51
		16		56	56
	18	18			62
		20	68	68	68
22	22	22			75
		24		82	82
	27	27			91
		30	100	100	100
33	33	33			

*Numbers and decimal multiples for ohms.

FROM SIMPLE TASK TO FINE ART

Soldering is the process of joining two metals together by the use of a low-temperature melting alloy. Soldering is one of the oldest known joining techniques, first developed by the Egyptians in making weapons such as spears and swords. Since then, it has evolved into what is now used in the manufacturing of electronic assemblies. Soldering is far from the simple task it once was; it is now a fine art, one that requires care, experience, and a thorough knowledge of the fundamentals.

The importance of having high standards of workmanship cannot be overemphasized. Faulty solder joints remain a cause of equipment failure, and because of that, soldering has become a *critical skill*.

The material contained in this appendix is designed to provide the student with both the fundamental knowledge and the practical skills needed to perform many of the high-reliability soldering operations encountered in today's electronics.

Covered here the fundamentals of the soldering process, the proper selection, and the use of the soldering station.

The key concept in this appendix is *high-reliability soldering*. Much of our present technology is vitally dependent on the reliability of countless, individual soldered connections. High-reliability soldering was developed in response to early failures with space equipment. Since then the concept and practice have spread into military and medical equipment. We have now come to expect it in everyday electronics as well.

THE ADVANTAGE OF SOLDERING

Soldering is the process of connecting two pieces of metal together to form a reliable electrical path. Why solder them in the first place? The two pieces of metal could be put together with nuts and bolts, or some other kind of mechanical fastening. The disadvantages of these methods are twofold. First, the reliability of the connection cannot be assured because of vibration and shock. Second, because oxidation and corrosion are continually occurring on the metal surfaces, electrical conductivity between the two surfaces would progressively decrease.

A soldered connection does away with both of these problems. There is no movement in the joint and no interfacing surfaces to oxidize. A continuous conductive path is formed, made possible by the characteristics of the solder itself.

---

*This material is provided courtesy of PACE, Inc., Laurel, Maryland.

THE NATURE OF SOLDER

Solder used in electronics is a low-temperature melting alloy made by combining various metals in different proportions. The most common types of solder are made from tin and lead. When the proportions are equal, it is known as 50/50 solder—50 percent tin and 50 percent lead. Similarly, 60/40 solder consists of 60 percent tin and 40 percent lead. The percentages are usually marked on the various types of solder available; sometimes only the tin percentage is shown. The chemical symbol for tin is Sn; thus Sn 63 indicates a solder which contains 63 percent tin.

Pure lead (Pb) has a melting point of 327°C (621°F); pure tin, a melting point of 232°C (450°F). But when they are combined into a 60/40 solder, the melting point drops to 190°C (374°F)—lower than either of the two metals alone.

Melting generally does not take place all at once. As illustrated in Fig. B-1, 60/40 solder begins to melt at 183°C (361°F), but it has not fully melted until the temperature reaches 190°C (374°F). Between these two temperatures, the solder exists in a plastic (semiliquid) state—some, but not all, of the solder has melted.

The plastic range of solder will vary, depending on the ratio of tin to lead, as shown in Fig. B-2. Various ratios of tin to lead are shown across the top of this figure. With most ratios, melting begins at 183°C (361°F), but the full melting temperatures vary dramatically. There is one ratio of tin to lead that has no plastic state and is known as *eutectic solder*. This ratio is 63/37 (Sn 63) and it fully melts and solidifies at 183°C (361°F).

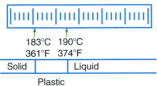

FIG. B-1 Plastic range of 60/40 solder. Melt begins at 183°C (361°F) and is complete at 190°C (374°F).

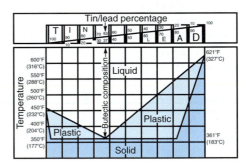

FIG. B-2 Fusion characteristics of tin/lead solders.

The solder most commonly used for hand soldering in electronics is the 60/40 type, but because of its plastic range, care must be taken not to move any elements of the joint during the cool-down period. Movement may cause a disturbed joint. Characteristically, this type of joint has a rough, irregular appearance and looks dull instead of bright and shiny. It is unreliable and therefore one of the types of joints that is unacceptable in high-reliability soldering.

In some situations, it is difficult to maintain a stable joint during cooling, for example, when wave soldering is used with a moving conveyor line of circuit boards during the manufacturing process. In other cases it may be necessary to use minimal heat to avoid damage to heat-sensitive components. In both of these situations, eutectic solder is the preferred choice, since it changes from a liquid to a solid during cooling with no plastic range.

THE WETTING ACTION

To someone watching the soldering process for the first time, it looks as though the solder simply sticks the metals together like a hot-melt glue, but what actually happens is far different.

A chemical reaction takes place when the hot solder comes into contact with the copper surface. The solder dissolves and penetrates the surface. The molecules of solder and copper blend together to form a new metal alloy, one that is part copper and part solder and that has characteristics all its own. This reaction is called *wetting* and forms the intermetallic bond between the solder and copper (Fig. B-3).

Proper wetting can occur only if the surface of the copper is free of contamination and from oxide films that form when the metal is exposed to air. Also, the solder and copper surfaces need to have reached the proper temperature.

Even though the surface may look clean before soldering, there may still be a thin film of oxide covering it. When solder is applied, it acts like a drop of water on an oily surface because the oxide coating prevents the solder from coming into contact with the copper. No reaction takes place, and the solder can be easily scraped off. For a good solder bond, surface oxides must be removed during the soldering process.

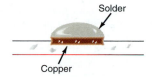

FIG. B-3 The wetting action. Molten solder dissolves and penetrates a clean copper surface, forming an intermetallic bond.

THE ROLE OF FLUX

Reliable solder connections can be accomplished only on clean surfaces. Some sort of cleaning process is essential in achieving successful soldered connections, but in most cases it is insufficient. This is due to the extremely rapid rate at which oxides form on the surfaces of heated metals, thus creating oxide films which prevent proper soldering. To overcome these oxide films, it is necessary to utilize materials, called *fluxes,* which consist of natural or synthetic rosins and sometimes additives called activators.

It is the function of flux to remove surface oxides and keep them removed during the soldering operation. This is accomplished because the flux action is very corrosive at or near solder melt temperatures and accounts for the flux's ability to rapidly remove metal oxides. It is the fluxing action of removing oxides and carrying them away, as well as preventing the formation of new oxides, that allows the solder to form the desired intermetallic bond.

Flux must activate at a temperature lower than solder so that it can do its job prior to the solder flowing. It volatilizes very rapidly; thus it is mandatory that the flux be activated to flow onto the work surface and not simply be volatilized by the hot iron tip if it is to provide the full benefit of the fluxing action.

There are varieties of fluxes available for many applications. For example, in soldering sheet metal, acid fluxes are used; silver brazing (which requires a much higher temperature for melting than that required by tin/lead alloys) uses a borax paste. Each of these fluxes removes oxides and, in many cases, serves additional purposes. The fluxes used in electronic hand soldering are the pure rosins, rosins combined with mild activators to accelerate the rosin's fluxing

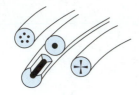

FIG. B-4 Types of cored solder, with varying solder-flux percentages.

capability, low-residue/no-clean fluxes, or water-soluble fluxes. Acid fluxes or highly activated fluxes should never be used in electronic work. Various types of flux-cored solder are now in common use. They provide a convenient way to apply and control the amount of flux used at the joint (Fig. B-4).

SOLDERING IRONS

In any kind of soldering, the primary requirement, beyond the solder itself, is heat. Heat can be applied in a number of ways—conductive (e.g., soldering iron, wave, vapor phase), convective (hot air), or radiant (IR). We are mainly concerned with the conductive method, which uses a soldering iron.

Soldering stations come in a variety of sizes and shapes, but consist basically of three main elements: a resistance heating unit; a heater block, which acts as a heat reservoir; and the tip, or bit, for transferring heat to the work. The standard production station is a variable-temperature, closed-loop system with interchangeable tips and is made with ESD-safe plastics.

CONTROLLING HEAT AT THE JOINT

Controlling tip temperature is not the real challenge in soldering; the real challenge is to control the *heat cycle* of the work—how fast the work gets hot, how hot it gets, and how long it stays that way. This is affected by so many factors that, in reality, tip temperature is not that critical.

The first factor that needs to be considered is the *relative thermal mass* of the area to be soldered. This mass may vary over a wide range.

Consider a single land on a single-sided circuit board. There is relatively little mass, so the land heats up quickly. But on a double-sided board with plated-through holes, the mass is more than doubled. Multilayered boards may have an even greater mass, and that's before the mass of the component lead is taken into consideration. Lead mass may vary greatly, since some leads are much larger than others.

Further, there may be terminals (e.g., turret or bifurcated) mounted on the board. Again, the thermal mass is increased, and will further increase as connecting wires are added.

Each connection, then, has its particular thermal mass. How this combined mass compares with the mass of the iron tip, the "relative" thermal mass, determines the time and temperature rise of the work.

With a large work mass and a small iron tip, the temperature rise will be slow. With the situation reversed, using a large iron tip on a small work mass, the temperature rise of the work will be much more rapid—even though the *temperature of the tip is the same.*

Now consider the capacity of the iron itself and its ability to sustain a given flow of heat. Essentially, irons are instruments for generating and storing heat, and the reservoir is made up of both the heater block and the tip. The tip comes in various sizes and shapes; it's the *pipeline* for heat flowing into the work. For small work, a conical (pointed) tip is used, so that only a small flow of heat occurs. For large work, a large chisel tip is used, providing greater flow.

The reservoir is replenished by the heating element, but when an iron with a large tip is used to heat massive work, the reservoir may lose heat faster than it can be replenished. Thus the *size* of the reservoir becomes important: a large heating block can sustain a larger outflow longer than a small one.

An iron's capacity can be increased by using a larger heating element, thereby increasing the wattage of the iron. These two factors, block size and wattage, are what determine the iron's recovery rate.

If a great deal of heat is needed at a particular connection, the correct temperature with the right size tip is required, as is an iron with a large enough capacity and an ability to recover fast enough. *Relative thermal mass,* then, is a major consideration for controlling the heat cycle of the work.

A second factor of importance is the *surface condition* of the area to be soldered. If there are any oxides or other contaminants covering the lands or leads, there will be a barrier to the flow of heat. Then, even though the iron tip is the right size and has the correct temperature, it may not supply enough heat to the connection to melt the solder. In soldering, a cardinal rule is that a good solder connection cannot be created on a dirty surface. Before attempting to solder, the work should always be cleaned with an approved solvent to remove any grease or oil film from the surface. In some cases pretinning may be required to enhance solderability and remove heavy oxidation of the surfaces prior to soldering.

A third factor to consider is *thermal linkage*—the area of contact between the iron tip and the work.

Figure B-5 shows a cross-sectional view of an iron tip touching a round lead. The contact occurs only at the point indicated by the "X," so the linkage area is very small, not much more than a straight line along the lead.

The contact area can be greatly increased by applying a small amount of solder to the point of contact between the tip and workpiece. This solder heat bridge provides the thermal linkage and assures rapid heat transfer into the work.

From the aforementioned, it should now be apparent that there are many more factors than just the temperature of the iron tip that affect how quickly any particular connection is going to heat up. In reality, soldering is a very complex control problem, with a number of variables to it, each influencing the other. And what makes it so critical is *time.* The general rule for high-reliability soldering on printed circuit boards is to apply heat for no more than 2 s from the time solder starts to melt (wetting). Applying heat for longer than 2 s after wetting may cause damage to the component or board.

With all these factors to consider, the soldering process would appear to be too complex to accurately control in so short a time, but there is a simple solution—the *workpiece indicator* (WPI). This is defined as the reaction of the workpiece to the work being performed on it—a reaction that is discernible to the human senses of sight, touch, smell, sound, and taste.

Put simply, workpiece indicators are the way the work talks back to you—the way it tells you what effect you are having and how to control it so that you accomplish what you want.

In any kind of work, you become part of a closed-loop system. It begins when you take some action on the workpiece; then the workpiece reacts to what you did; you sense the change, and then modify your action to accomplish the result. It is in the sensing of the change, by sight, sound, smell, taste, or touch, that the workpiece indicators come in (Fig. B-6).

For soldering and desoldering, a primary workpiece indicator is *heat rate recognition*—observing how fast heat flows into the connection. In practice, this means observing the rate at which the solder melts, which should be within 1 to 2 s.

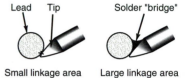

Lead   Tip   Solder "bridge"

Small linkage area   Large linkage area

FIG. B-5 Cross-sectional view (left) of iron tip on a round lead. The "X" shows point of contact. Use of a solder bridge (right) increases the linkage area and speeds the transfer of heat.

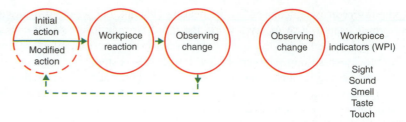

FIG. B-6 Work can be viewed as a closed-loop system (left). Feedback comes from the reaction of the workpiece and is used to modify the action. Workpiece indicators (right)—changes discernible to the human senses—are the way the "work talks back to you."

This indicator encompasses all the variables involved in making a satisfactory solder connection with minimum heating effects, including the capacity of the iron and its tip temperature, the surface conditions, the thermal linkage between tip and workpiece, and the relative thermal masses involved.

If the iron tip is too large for the work, the heating rate may be too fast to be controlled. If the tip is too small, it may produce a "mush" kind of melt; the heating rate will be too slow, even though the temperature at the tip is the same.

A general rule for preventing overheating is "Get in and get out as fast as you can." That means using a heated iron you can react to—one giving a 1- to 2-s dwell time on the particular connection being soldered.

SELECTING THE SOLDERING IRON AND TIP

A good all-around soldering station for electronic soldering is a variable-temperature, ESD-safe station with a pencil-type iron and tips that are easily interchangeable, even when hot (Fig. B-7).

The soldering iron tip should always be fully inserted into the heating element and tightened. This will allow for maximum heat transfer from the heater to the tip.

The tip should be removed daily to prevent an oxidation scale from accumulating between the heating element and the tip. A bright, thin tinned surface must be maintained on the tip's working surface to ensure proper heat transfer and to avoid contaminating the solder connection.

The plated tip is initially prepared by holding a piece of flux-cored solder to the face so that it will tin the surface when it reaches the lowest temperature at which solder will melt. Once the tip is up to operating temperature, it will usually be too hot for good tinning, because of the rapidity of oxidation at elevated temperatures. The hot tinned tip is maintained by wiping it lightly on a damp sponge to shock off the oxides. When the iron is not being used, the tip should be coated with a layer of solder.

MAKING THE SOLDER CONNECTION

The soldering iron tip should be applied to the area of maximum thermal mass of the connection being made. This will permit the rapid thermal elevation of the parts being soldered. Molten solder always flows toward the heat of a properly prepared connection.

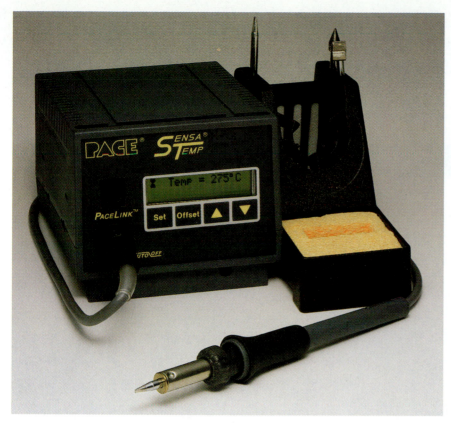

FIG. B-7   Pencil-type iron with changeable tips.

When the solder connection is heated, a small amount of solder is applied to the tip to increase the thermal linkage to the area being heated. The solder is then applied to the opposite side of the connection so that the work surfaces, not the iron, melt the solder. Never melt the solder against the iron tip and allow it to flow onto a surface cooler than the solder melting temperature.

Solder, with flux, applied to a cleaned and properly heated surface will melt and flow without direct contact with the heat source and provide a smooth, even surface, feathering out to a thin edge (Fig. B-8). Improper soldering will exhibit a built-up, irregular appearance and poor filleting. The parts being soldered must be held rigidly in place until the temperature decreases to solidify the solder. This will prevent a disturbed or fractured solder joint.

Selecting cored solder of the proper diameter will aid in controlling the amount of solder being applied to the connection (e.g., a small-gauge solder for a small connection; a large-gauge solder for a large connection).

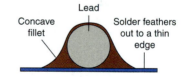

FIG. B-8   Cross-sectional view of a round lead on a flat surface.

REMOVAL OF FLUX

Cleaning may be required to remove certain types of fluxes after soldering. If cleaning is required, the flux residue should be removed as soon as possible, preferably within 1 hour after soldering.

APPENDIX C

SCHEMATIC SYMBOLS

TABLE C-1   GENERAL SYMBOLS

DEVICE	SYMBOL	DEVICE	SYMBOL
AC voltage source		Coil or inductance Air-core	
Amplifier		Iron-core	
Antenna General		Variable	
Dipole		Powdered iron or ferrite slug	
Loop		Conductor General	
		Connection	
		No connection	
Battery, cell, or dc voltage source		Current source	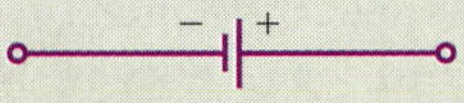
Longer line positive			
Capacitor General, fixed Curved electrode is outside foil, negative, or low-potential side		Crystal, piezoelectric	
		Fuse	
Electrolytic	or	Ground, earth or metal frame Chassis or common return connected to one side of voltage source	
Variable	or	Chassis or common return not connected to voltage source	
		Common return	
Ganged			

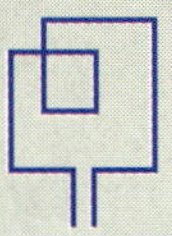

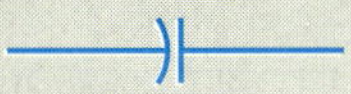

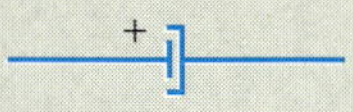

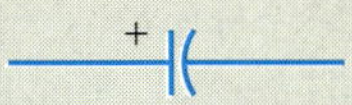

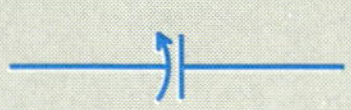

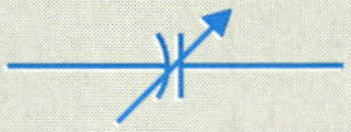

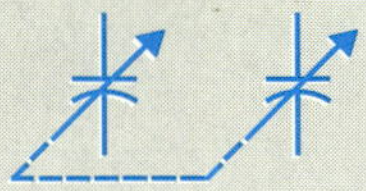

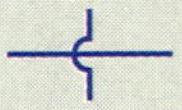

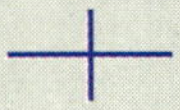

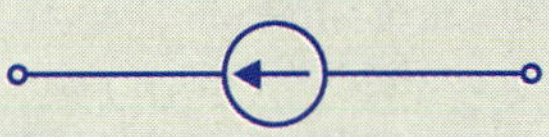

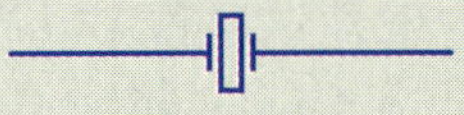

TABLE C-1 GENERAL SYMBOLS (CONTINUED)

DEVICE	SYMBOL	DEVICE	SYMBOL
Jack		Resistor, fixed	
Plug for jack	Tip Sleeve	Tapped	
		Variable	
Key telegraph		Switch, SPST SPDT	
Loudspeaker, general		2-pole (DPDT)	
Phones or headset		Shielding	
Magnet Permanent	PM	Shielded conductor	
Electromagnet		Transformer Air-core	
Microphone		Iron-core	
Meters; letter or symbol to indicate range or function	A mA V	Autotransformer	
Motor	M		
Neon bulb		Link coupling	
Relay, coil			
Contacts			

APPENDIX D

USING THE OSCILLOSCOPE

MAIN SECTIONS OF AN OSCILLOSCOPE

An oscilloscope is one of the most important types of test equipment for checking electronic circuits because the "scope" shows the waveform of an applied voltage (Fig. D-1). The fluorescent screen, usually green, of the cathode-ray tube (CRT) displays a graph of voltage amplitudes with respect to time. Examples of the oscilloscope display are shown in Fig. D-2 with sine waves in Fig. D-2*a* and square waves in Fig. D-2*b*. Actually, the pattern on the screen is produced by a small spot of light that is deflected vertically and horizontally to trace out the waveform.

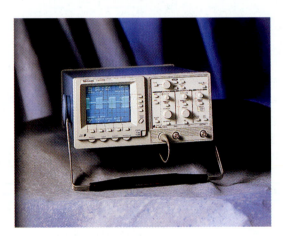

FIG. D-1   Oscilloscope.

The vertical axis represents voltage amplitudes. More voltage applied to the vertical input terminals of the oscilloscope results in a bigger trace pattern. For horizontal deflection, the oscilloscope has its own internal sawtooth voltage generator. No external connections are used for this function. A sawtooth waveform is needed for the internal horizontal sweep because of its linear increases in amplitude. Then the horizontal axis of the display is proportional to time. The screen pattern, therefore, displays a graph of the variations of the signal input voltage, with time on the horizontal axis and voltage on the vertical axis. It should be noted that for most measurements, only the vertical input terminal is used for connections to the circuit being tested.

The number of cycles you see depends on the frequency of the input signal and the timing of the internal sweep for horizontal deflection. Usually, the horizontal sweep frequency is set for a display of two or three complete cycles of the waveform, just for convenience.

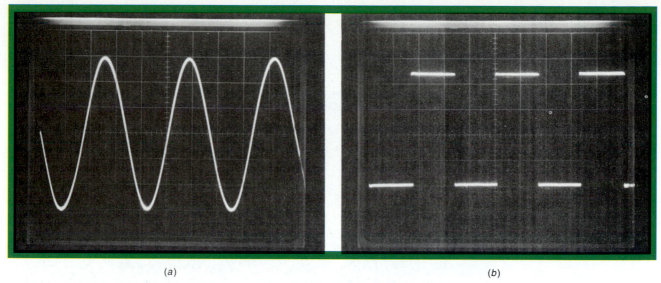

(a)   (b)

FIG. D-2   Examples of oscilloscope screen patterns. (*a*) Sine wave. (*b*) Square wave.
Vertical lines are not visible because the spot moves too fast.

Not only does the oscilloscope show the amount of voltage, as the height of
the trace pattern, but also the frequency can be checked in terms of the horizontal
time base. Furthermore, any distortion of the signal is shown in the trace pattern.

The oscilloscope is a voltmeter. You connect the cable lead from the vertical input terminal *across* a component to check the voltage. Although the oscilloscope is generally used for ac voltages, it can also indicate dc values in terms
of a steady displacement of the trace from the center position.

Figure D-3 shows the front panel of a typical dual-trace oscilloscope. The
key functional controls are indicated with an emphasis on three basic functions:

1.   The cathode-ray tube (or CRT) at the left, as the visual display, shows the
     signal waveform. The CRT provides a spot of light that can be deflected up
     or down and left or right.
2.   The vertical (V) deflection sections (one for each of the two channels) am-
     plify the signal applied to the V input terminals to provide enough signal
     voltage to the vertical deflection plates in the CRT.
3.   The horizontal (H) deflection section has the internal sawtooth voltage gen-
     erator that is used as a time base for horizontal deflection. The H deflection
     voltage is applied to the horizontal deflection plates in the CRT.

The combined functions of these three sections result in a trace pattern on the
oscilloscope screen. From the waveform, you can measure voltage amplitude of
the V input signal, determine the period or frequency, and observe any distortion.

CRT SECTION   In Fig. D-3, note the power on-off switch for the oscilloscope. The intensity control varies the brightness of the trace. Adjust the focus
control for a sharp pattern at the desired brightness. Do not keep the brightness
high with a stationary pattern for too long, as the screen phosphor may be
"burned," which produces a permanent brown discoloration of the screen. In some
cases, the power on-off switch may be part of the brightness control.

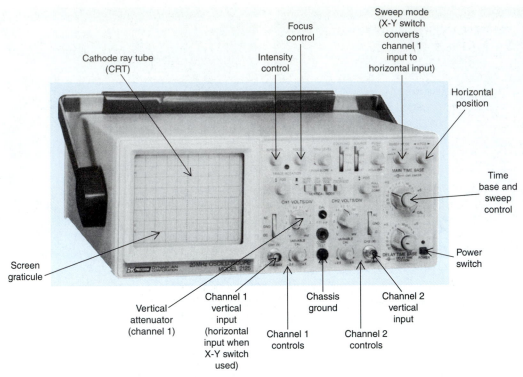

Cathode ray tube (CRT)

Intensity control

Focus control

Sweep mode (X-Y switch converts channel 1 input to horizontal input)

Horizontal position

Time base and sweep control

Power switch

Channel 2 vertical input

Channel 2 controls

Chassis ground

Channel 1 controls

Channel 1 vertical input (horizontal input when X-Y switch used)

Vertical attenuator (channel 1)

Screen graticule

FIG. D-3 The main functional controls of a typical dual-trace oscilloscope.

The screen *graticule* is a plastic sheet over the face of the CRT to make measurements more convenient. A typical size for the graticule is 10 cm × 8 cm, with each box 1 cm high and wide. Calibrated values for the oscilloscope are often specified per centimeter.

Although not labeled in Fig. D-3, the oscilloscope also has provisions for centering the electron beam in the CRT, which positions the trace vertically and horizontally. The centering controls adjust dc voltages to leave the trace in a reference position for ac deflection by the signal voltage. If the beam should be off the screen, there will be no trace visible, but it can be returned to the screen with the V and H, or X and Y, positioning controls.

V DEFLECTION Figure D-3 shows a step attenuator and a variable gain control to adjust the height of the trace pattern. The trace can be adjusted to fit most of the screen height in order to see the waveform better. The step attenuator cuts down the amount of signal applied to the V input terminals, often in multiples of ½, ¹⁄₂.₅, or ¹⁄₁₀. The gain control is a variable adjustment for in-between values of vertical deflection. In many oscilloscopes, the step attenuator is calibrated in units of V/cm of vertical deflection at the maximum setting of the variable gain control.

H DEFLECTION The H deflection section is more complicated than the V section, because horizontal deflection involves timing of the internal H sweep, synchronizing the H timing to make the pattern remain still and the H amplification for the desired width of the trace pattern. For oscilloscopes with a calibrated time base, the step control is marked in μs/cm, ms/cm, or s/cm for the maximum setting of the variable timing control. Without the calibration, these controls just indicate approximate frequency for the internal time base.

The timing of the internal time base for horizontal deflection must be synchronized with the vertical input signal; otherwise, the pattern on the screen drifts to the left or right across the screen. The general procedure is to adjust the timing for two or three cycles, with as little drift as possible. Then adjust the synchronization (sync) control to make the pattern remain stationary. Always use as little sync as possible to avoid distortion of the pattern. Many oscilloscopes have an automatic sync or *trigger* level.

The pattern on the screen represents a spot of light that is moving so fast that you see the complete trace. Actually, the trace patterns are repeated one over the other, but it looks like a steady picture with the correct synchronization. Below about 30 Hz for the H sweep frequency, the brightness of the trace may flicker. At lower sweep frequencies, it is possible to see the spot moving.

In summary, then, for the general operation of using the oscilloscope to observe a signal waveform, connect the signal to the V input terminals and adjust the horizontal timing of the internal sweep generator to obtain the desired trace pattern. The H input terminals for an external input signal are only used for some special functions, such as the Lissajous patterns described at the end of this appendix. In some oscilloscopes, one of the two V inputs can be made an H input through the operation of an X-Y switch.

Y AXIS FOR VOLTAGE AND *X* AXIS FOR TIME

Refer to Fig. D-4, which illustrates how oscilloscope waveforms are developed. In Fig. D-4*a*, the spot of light at the center of the screen is produced by the CRT. It has the required electrode voltages, including high voltage for the anode. The spot is focused for the smallest size and the intensity is set for the desired brightness. Also, the spot can be centered with the positioning controls. There is no vertical or horizontal deflection as yet.

In Fig. D-4*b*, horizontal deflection voltage is applied to the CRT to move the spot back and forth across the screen along the *x* axis. There is no vertical deflection. A horizontal line is displayed because the spot repeats its motion over the same area too fast for you to see movement. Horizontal deflection is provided by the internal sweep voltage, for most applications of the oscilloscope.

With vertical deflection, but without horizontal deflection, the spot moves up and down along the *y* axis. This action is shown in Fig. D-4*c*, without any horizontal deflection. You can obtain this display by turning off the internal horizontal sweep. The vertical deflection is provided by signal voltage at the vertical input terminal of the oscilloscope.

Both vertical and horizontal deflection are used for the trace pattern in Fig. D-4*d*. Two cycles of sine waves show the waveform of the vertical input signal. More cycles or less cycles can be displayed according to the frequency setting of the internal horizontal sweep voltage. The complete trace pattern results from the spot of light being deflected vertically by the input signal while the spot is deflected across the screen at a uniform rate of speed.

An example of oscilloscope waveforms for the video signal in a television receiver is shown in Fig. D-5. The oscilloscope is an important instrument for adjusting and troubleshooting television receivers.

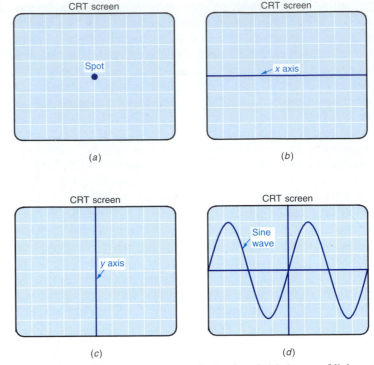

CRT screen

Spot

(a)

CRT screen

x axis

(b)

CRT screen

y axis

(c)

CRT screen

Sine wave

(d)

FIG. D-4 How an oscilloscope trace pattern is developed. (*a*) A spot of light at the center of the screen without V and H deflection. (*b*) A horizontal line produced by H deflection but no V deflection. (*c*) A vertical line produced by V deflection but no H deflection. (*d*) A sine-wave vertical input signal with V and H deflection.

CALIBRATED VOLTAGES ON THE VERTICAL *Y* AXIS In many oscilloscopes, the step attenuator that adjusts the amount of vertical input signal is calibrated so that you can read voltage amplitudes directly from the height of the trace on the oscilloscope screen. The variable gain control must stay set at its

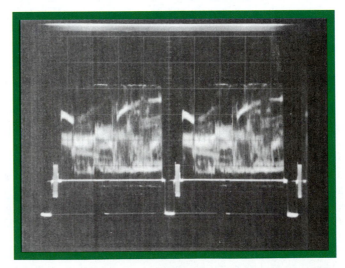

FIG. D-5 Oscilloscope pattern for a video signal in a television receiver. Two cycles shown.

maximum value. An example is illustrated in Fig. D-6 for a pattern with a height of six boxes equal to 6 cm. In Fig. D-6a, the horizontal deflection is turned off to show the vertical amplitude alone, whereas Fig. D-6b shows the complete trace pattern. Either way, the height of 6 cm represents the peak-to-peak voltage of the waveform. The measurement is peak-to-peak because it is between the two opposite peaks.

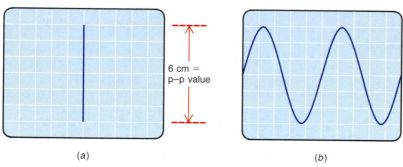

(a)   (b)

FIG. D-6 Voltage values on the vertical axis. (a) H deflection removed to show just the peak-to-peak amplitude on the y axis. (b) Complete waveform with V and H deflection.

Assume that the vertical attenuator is at 1 V/cm. This setting means that each 1-cm-high box represents 1 V of vertical input signal. For the six boxes equal to 6 cm, the amplitude is

$$6 \text{ cm} \times \frac{1 \text{ V}}{\text{cm}} = 6 \text{ V}$$

For this example, the amplitude of the vertical input signal is 6 V p-p.

Suppose that the height of the trace is 4 cm, instead of 6 cm. For the same 1-V/cm setting, the trace amplitude is

$$4 \times 1 = 4 \text{ V p-p}$$

As another example, let the V/cm setting be changed to 0.005 V/cm. For 6 cm of deflection, the V input signal then is

$$6 \text{ cm} \times \frac{0.005 \text{ V}}{\text{cm}} = 0.030 \text{ V or } 30 \text{ mV}$$

With 4 cm of vertical deflection, the amplitude is 20 mV. The 30 and 20 mV amplitudes are peak-to-peak values.

CALIBRATED TIME VALUES ON THE HORIZONTAL X AXIS

Refer to Fig. D-7 with the x axis shown 10 cm wide. Also, the calibrated H time setting is taken here as 1 ms/cm. Each box horizontally represents 1 ms of time, as shown in Fig. D-7a. The two sine-wave cycles in Fig. D-7b, therefore, take 10 ms of time. For one cycle that takes 5 boxes or 5 cm, the period T for one cycle is 5 ms.

Remember that the frequency f is equal to 1/T. Therefore, the frequency for the sine waves in Fig. D-7b is as follows.

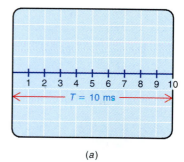

(a)

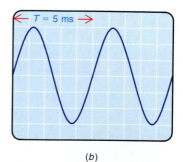

(b)

FIG. D-7 Time and frequency values on the x axis. (a) The period T of the horizontal sweep time is 10 ms. (b) One cycle of the vertical input signal has a period T of 5 ms, or f = 0.2 kHz = 200 Hz.

$$\blacktriangleright \quad f = \frac{1}{T} = \frac{1}{5 \text{ ms}} = 0.02 \text{ kHz or } 200 \text{ Hz}$$

This value is the frequency of the vertical input signal.

Suppose that five cycles with the same time base of 10 ms were shown in Fig. D-7. One cycle would take $10/5 = 2$ ms. With T of 2 ms, the f is equal to 0.5 kHz or 500 Hz.

As another example, let the H time setting be 20 μs/cm. Then the width of 10 cm in Fig. D-7 represents $20 \times 10 = 200$ μs. For two cycles across the entire width of the screen, one cycle takes 100 μs for T of the signal input. The frequency then is

$$f = \frac{1}{100 \text{ } \mu\text{s}} = 0.01 \text{ MHz or } 10 \text{ kHz}$$

These calculations give the frequency of the signal applied to the vertical input terminals of the oscilloscope.

For a quick, simple check of the calibrated time values, you can apply a sample of the 60-Hz voltage from the ac power line, which is an accurate reference for frequency. With the horizontal T/cm control set on 5 ms/cm there should be exactly three cycles on the screen.

COMPARISON OF H SYNC AND H SELECTOR SWITCHES

One of the most important problems in learning how to use the oscilloscope is to appreciate the difference between setting the type of deflection with the horizontal selector switch and choosing the type of sync with the sync selector switch to be used with internal sweep. The two switches have the same names for their three positions, but their functions are entirely different. In order to emphasize the comparison, Fig. D-8 shows the two switches with the name and function for each position.

HORIZONTAL SELECTOR SWITCH
In Fig. D-8*a* the horizontal selector switch turns on the internal sweep at the top position. This method supplies dc electrode voltages for the sawtooth generator used for H deflection. Then the sawtooth voltage output is fed to the H deflection amplifier for the horizontal deflection plates. This setting for internal sweep is the type of operation generally used for the oscilloscope to display the waveform of the vertical input signal.

On the 60-Hz position of the H selector switch, a sample of the 60-Hz ac line voltage is taken internally to be used for H deflection. The 60-Hz voltage is a sine wave.

On the "external" position of the H selector switch, the internal sweep is disconnected from the H amplifier. Without any signal connected to the H input terminals on the front panel of the oscilloscope, there is no horizontal deflection. However, any type of signal input can be connected to the external H terminals to provide signal for the H deflection amplifier.

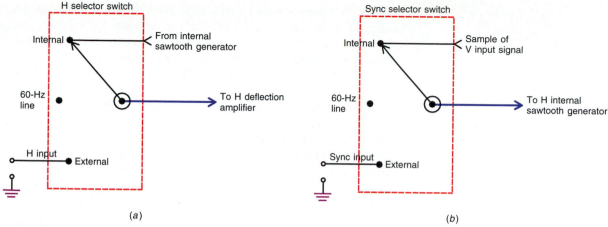

FIG. D-8 Comparison of a horizontal selector switch, shown in (a), with a sync selector switch, shown in (b).

HOW A SAWTOOTH GENERATOR IS SYNCHRONIZED Before we consider the different types of sync that can be chosen with the sync selector switch, an illustration of synchronization with the required sync pulses is shown in Fig. D-9. Keep in mind that the sync function only applies when the internal sawtooth generator is being used to provide the internal sweep. In general, any sawtooth generator is a pulse oscillator, like the bistable multivibrator (MV). An asymmetrical type of MV circuit is used with a waveshaping capacitor to provide sawtooth voltage in the output. This type of circuit is easily synchronized by injecting pulses at the input electrode to force the oscillator to run at the frequency of the synchronizing pulses. This value is the *forced frequency,* which is locked in by the sync pulses, compared with the *free-running frequency* without sync.

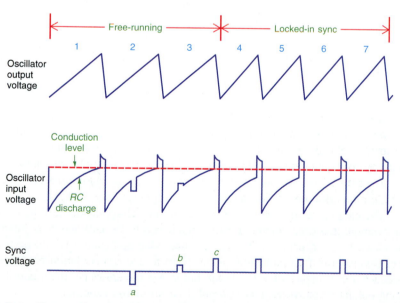

FIG. D-9 How synchronizing (sync) voltage is used to trigger a free-running pulse oscillator to lock it in at sync frequency.

The free frequency must be set a little lower than the sync frequency, thus allowing the input voltage of the oscillator to be ready for the sync voltage to produce the lock-in effect. Incidentally, synchronization of the deflection oscillator is also used in television receivers to hold the picture steady.

Refer to the waveforms in Fig. D-9. At the top is the sawtooth voltage output of the H deflection oscillator used for internal sweep. Just under the sawtooth waves are the waveforms of input voltage from the MV type of pulse oscillator. At the bottom of the figure, the pulses toward the end of the pulse train are able to lock in the oscillator at the sync frequency.

The trigger effect for controlling the oscillator occurs with the input voltage shown below the sawtooth waves in Fig. D-9. In the RC discharge curve, this voltage is cutting off one stage of the MV, until the input voltage approaches the conduction level. Then when one stage conducts, it cuts off the other stage.

The sync voltage is able to control when the input voltage can reach the conduction level. Consider the sync pulses at the bottom of Fig. D-9, in sequence, from left to right. First, there is no sync. Now the oscillator is free-running. Next, is the sync pulse marked a, but it has the wrong polarity. It cannot force the input voltage to the conduction level, and the oscillator cannot be affected. Then comes sync pulse b of the correct polarity but not enough amplitude. The oscillator is still free-running. During this time, the oscillator is free-running for the cycles marked 1, 2, and 3 at the top of the sawtooth waveform.

Finally, sync pulse c in Fig. D-9 has the correct polarity and enough amplitude to force the oscillator input voltage to the conduction level. The start of conduction corresponds to the start of retrace on the sawtooth wave.

All the sync pulses that follow c in Fig. D-9 also trigger the oscillator at the sync frequency, as shown for the sawtooth waves labeled 4, 5, 6, and 7 at the top of the figure. As long as the sync voltage is applied, it is able to hold the oscillator at the sync frequency.

SYNC SELECTOR SWITCH   We can now consider the different types of sync for the internal sawtooth generator. The sync section of the oscilloscope has waveshaping circuits to provide sharp pulses for exact synchronization. In many oscilloscopes, the polarity and slope of the sync pulses can be varied.

Most important is internal sync. It is used practically always with the internal sweep for horizontal deflection. The reason is that internal sync is automatically at the correct frequency for synchronizing the vertical input signal because this sync is a sample of the signal from the vertical deflection amplifier. In order to see complete cycles in the trace pattern, the sweep frequency must be at an exact submultiple of the signal frequency. As examples, for two cycles, the V input signal is double the H scanning frequency, and for three cycles the ratio is $3:1$.

The 60-Hz position on the sync selector switch is seldom needed. One application is in the use of the oscilloscope in aligning tuned circuits.

The external sync also is used only for special applications. When necessary, though, the sync voltage must be connected to the external sync input terminal on the front panel of the oscilloscope. One application would be to display the pattern of an AM radio signal with an audio modulation envelope. In this case, the V input is an RF signal, but to see the envelope, the audio modulating voltage must be used as external sync for the oscilloscope.

To summarize the functions for the selector switches in Fig. D-9, remember that the sync voltage does not produce deflection. Therefore, internal sync does not mean internal sawtooth scanning. The H selector switch determines when the internal sawtooth generator is used. Furthermore, selecting the type of sync has no meaning unless the internal sweep is on. The sync has its use only for the internal sawtooth generator. In other words, the H selector switch must be set for internal sweep in order to use any type of sync.

OSCILLOSCOPE PROBES

Oscilloscope probes are the test leads used for connecting the vertical input signal to the oscilloscope. There are three types: a direct lead that is just a shielded cable, the low-capacitance probe (LCP) with a series-isolating resistor, and a demodulator probe. Figure D-10 shows a circuit for an LCP for an oscilloscope. The LCP usually has a switch to short out the isolating resistor so that the same probe can be used either as a direct lead or with low-capacitance.

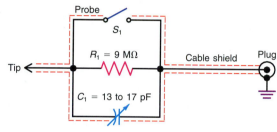

FIG. D-10   Circuit for low-capacitance probe (LCP) for an oscilloscope.

DIRECT PROBE   The direct probe is just a shielded wire without any isolating resistor. A shielded cable is necessary to prevent any pickup of interfering signals, especially with the high resistance at the vertical input terminals of the oscilloscope. The higher the resistance, the more voltage that can be developed by induction. Any interfering signals in the test lead produce distortion of the trace pattern. The main sources of interference are 60-Hz magnetic fields from the power line and stray RF signals.

The direct probe as a shielded lead has relatively high capacitance. A typical value is 90 pF for 3 ft (0.9 m) of 50-Ω coaxial cable. Also, the vertical input terminals of the oscilloscope have a shunt capacitance of about 40 pF. The total C then is $90 + 40 = 130$ pF. This much capacitance can have a big effect on the circuit being tested. For example, it could detune a resonant circuit. Also, nonsinusoidal waveshapes are distorted. Therefore, the direct probe can be used only when the added C has little or no effect. These applications include voltages for the 60-Hz power line or sine-wave audio signals in a circuit with a relatively low resistance of several kilohms or less. The advantage of the direct probe is that it does not divide down the amount of input signal, since there is no series isolating resistance.

LOW-CAPACITANCE PROBE (LCP)   Refer to the diagram in Fig. D-10. The 9-MΩ resistor in the probe isolates the capacitance of the cable and the oscilloscope from the circuit connected to the probe tip. With an LCP, the input capacitance of the probe is only about 10 pF. The LCP must be used for oscilloscope measurements when

1. The signal frequency is above audio frequencies.
2. The circuit being tested has R higher than about 50 kΩ.
3. The waveshape is nonsinusoidal, especially with square waves and sharp pulses.

Without the LCP, the observed waveform can be distorted. The reason is that too much capacitance changes the circuit while it is being tested.

THE 1:10 VOLTAGE DIVISION OF THE LCP

Refer to the voltage divider circuit in Fig. D-11. The 9-MΩ of R_P is a series resistor in the probe. Also, R_S of 1 MΩ is a typical value for the shunt resistance at the vertical terminals of the oscilloscope. Then $R_T = 9 + 1 = 10$ MΩ. The voltage across R_S for the scope equals R_S/R_T or 1/10 of the input voltage. For the example in Fig. D-11 with 10 V at the tip of the LCP, 1 V is applied to the oscilloscope.

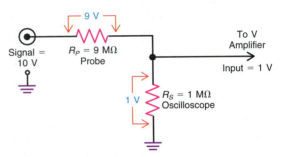

FIG. D-11   Voltage division of 1:10 with a low-capacitance probe.

Remember, when using the LCP, multiply by 10 for the actual signal amplitude. As an example, for a trace pattern on the screen that measures 2.4 V, the actual signal input at the probe is 24 V. For this reason, the LCP is generally called the "10 times" probe. Check to see whether or not the switch on the probe is on the direct or LCP position. Even though the scope trace is reduced by the factor of 1/10, it is preferable to use the LCP for almost all oscilloscope measurements to minimize distortion of the waveshapes.

TRIMMER CAPACITOR OF THE LCP

Referring back to Fig. D-10, note that the LCP has an internal variable capacitor C_1 across the isolating resistor R_1. The purpose of C_1 is to compensate the LCP for high frequencies. Its time constant with R_1 should equal the RC time constant of the circuit at the vertical input terminals of the oscilloscope. When necessary, C_1 is adjusted for minimum tilt on a square-wave signal.

DEMODULATOR PROBE

The demodulator probe has an internal diode to detect an amplitude-modulated RF signal. The output of the probe is the envelope or modulation. Polarity of the rectified dc output is usually negative. The demodulator probe can be used for signal tracing in the RF circuits of an AM receiver, where the signal has the modulation envelope. There is usually a problem, however, in very low signal amplitudes.

CURRENT MEASUREMENTS WITH OSCILLOSCOPE Although it serves as an ac voltmeter, the oscilloscope can also be used for measuring current values indirectly. The technique is to insert a low R in series where the current is to be checked. Use the oscilloscope to measure the voltage across R. Then the current is $I = V/R$. Keep the value of the inserted R much lower than the resistance of the circuit being tested to prevent any appreciable change in the actual I. Besides measuring the current this way, the waveform of V on the screen is the same as I because R does not affect the waveshape.

SPECIAL OSCILLOSCOPE FEATURES

Many oscilloscopes have special features that make operation more convenient. Actually, most oscilloscopes are the dual-trace type and have triggered sweep with a horizontal time base calibrated in ms and μs. Some oscilloscopes even have three or four vertical input channels.

DUAL-TRACE OSCILLOSCOPE The dual-trace oscilloscope can show two traces at the same time, one above the other, for two vertical input signals. An example of the two traces is shown in Fig. D-1.

For dual-trace operation, the oscilloscope has two vertical amplifier channels, labeled either A and B or 1 and 2. An internal electronic switch changes the signals alternately from each vertical amplifier to the deflection plates. The switching is accomplished with a square-wave generator. The switching rate is fast enough to make the changes invisible.

The advantages of a dual trace is that it permits observation of two signals at the same time. They both have the same linear time base. As a result, time and amplitude comparisons can be viewed directly.

Also, many dual-trace oscilloscopes have provision for either adding or subtracting the two vertical input signals with one resultant trace pattern. This operation is labeled A + B or A − B. The subtraction is accomplished by inverting the trace for B and adding to A.

Furthermore, one channel can also be used as a horizontal deflection amplifier when the internal horizontal sweep is turned off. This is usually the B channel. Then it becomes an amplifier for the x axis, whereas the A channel is for the y axis. The x-y operation is used for an external horizontal input signal, without the internal horizontal sweep.

DUAL-BEAM OSCILLOSCOPE The dual-beam oscilloscope can show two trace patterns also, but a special CRT is used that has two separate electron beams. One application of the dual-beam oscilloscope is in medical electronic equipment.

TV POSITIONS FOR INTERNAL SWEEP On the time or frequency switch for the internal horizontal sweep, many oscilloscopes have two positions marked V and H for television. At the H position, the internal sweep is set for two cycles of the horizontal scanning voltage in television receivers. This H frequency is exactly 5,734.26 Hz or the nominal value of 15,750 Hz. At the V position on the selector switch, the oscilloscope internal sweep is set for two cycles of 60-Hz signal at the vertical input terminal. This frequency is for the vertical scanning voltage in television receivers, which is exactly 59.94, but nominally 60 Hz. An example is the pattern in Fig. D-5, which shows the video signal for two horizontal scanning lines in a television receiver.

Do not confuse the abbreviation V for vertical deflection in television with V for the vertical input signal to an oscilloscope. The television V is 60 Hz for vertical deflection in the picture tube. The oscilloscope vertical input signal can have almost any frequency.

TRIGGERED SWEEP   The comparison here is with recurrent sweep, which uses a free-running sawtooth oscillator for internal sweep in the oscilloscope. In this method, horizontal deflection is produced by the internal sweep with or without the injection of pulses. With triggered sweep, though, the internal sawtooth generator produces one cycle of output only for each sync pulse as a trigger voltage at the input. As a result, the triggered sweep for internal horizontal deflection is produced only when the sawtooth generator has sync. The method of triggered sweep uses a monostable or one-shot multivibrator circuit to produce the sawtooth output voltage.

The advantages of triggered sweep are better synchronization and more exact control of the horizontal sweep time. Oscilloscopes with triggered sweep usually have the H sweep time calibrated. The time, rather than frequency, is calibrated because it applies for any number of cycles of the trace pattern on the screen.

Z AXIS FOR INTENSITY MODULATION   The y and x axes are for vertical and horizontal deflection in a CRT. In addition, though, the intensity of the electron beam can be varied by different values of the control grid voltage. The result is *intensity modulation* of the electron beam, which varies the light output from the screen. Such control is considered as z-axis modulation, because the effect is not vertical or horizontal.

In oscilloscopes, a separate z-axis external terminal may be provided for a connection to the control grid of the CRT. However, it is not used for the normal display of the vertical input signal.

When z-axis modulation is used in oscilloscopes, an amplitude of about 15 V p–p can vary the beam intensity between maximum light and zero. No light output is considered black, compared with bright illumination. The black level for control-grid voltage that cuts off the beam current can also be considered a blanking level. Any time the control-grid voltage is at the black level for zero beam current, the trace on the screen is blanked out, meaning it is not visible.

In television receivers, the picture is reproduced by z-axis modulation. The control-grid voltage of the picture tube is varied by the video signal, which corresponds to the picture information. Typical video signal amplitude is about 100 V p–p. More control-grid voltage is needed for TV picture tubes than for oscilloscopes because the anode voltage of 15 to 30 kV is much higher. It should be noted that for a television picture, horizontal and vertical deflection is used to fill the screen with scanning lines, whereas the video signal provides the visual information by varying the intensity of the electron beam.

DIRECT CONNECTIONS TO DEFLECTION PLATES   Direct connections to the deflection plates may be provided at the back of the oscilloscope, in order to bypass frequency limitations of the vertical and horizontal deflection amplifiers. However, appreciable deflection voltage is needed. Typically 30 V of potential difference between a pair of plates can deflect the beam 1 in.

H TRACE MAGNIFIER   The H trace magnifier expands the horizontal deflection to make the trace wider than the screen size. Then more details can be seen of the trace that is on the screen. The magnification is usually five times larger.

LISSAJOUS PATTERNS FOR PHASE AND FREQUENCY COMPARISONS

Examples of Lissajous patterns are shown in Fig. D-12 for phase angles and in Fig. D-13 for frequency comparisons. In Fig. D-12, the phase comparisons are for two sine waves that have the same frequency but different phase angles. The patterns in Fig. D-13 can be used to check a sine wave of unknown frequency with another sine wave that has a frequency known to be accurate, like the 60-Hz power line.

The Lissajous patterns are only for sine-wave signals. One is applied to the oscilloscope vertical input and the other to the horizontal input. Turn off the oscilloscope internal sweep, as it is not used for this application. Lissajous patterns are named after the man who first used them, Jules A. Lissajous.

Both the vertical and horizontal deflection signals should have equal amplitude. This requirement can be checked by adjusting the gain until the same height and width are obtained for each signal alone without the other.

PHASE-ANGLE COMPARISONS   Assume two sine waves have the same frequency. The combined trace looks like one of the patterns in Fig. D-12. Consider the diagonal line for 0°. The two sine waves are in phase. At the start, the spot is at the center without any deflection. When the V signal increases in a positive direction to deflect the spot upward, the H signal also is positive and moves the spot the same amount to the right. Halfway to the peak voltage, the spot is halfway to the top, as shown by a dot in the figure. At the peak value for both the

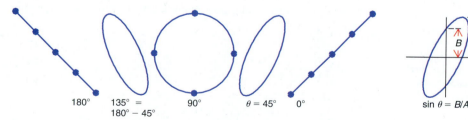

FIG. D-12 Lissajous patterns for phase angles on an oscilloscope screen. These patterns compare with the phases of two sine waves at the same frequency.

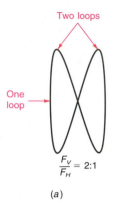

$$\frac{F_V}{F_H} = 2:1$$

(a)

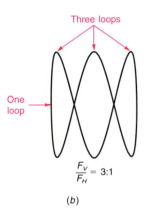

$$\frac{F_V}{F_H} = 3:1$$

(b)

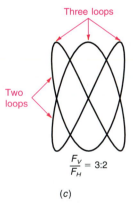

$$\frac{F_V}{F_H} = 3:2$$

(c)

FIG. D-13 Lissajous patterns for frequency comparisons on an oscilloscope screen. These patterns compare two sine waves of different frequencies. The frequency of the vertical signal is F_V. The frequency of the horizontal signal is F_H. The closed loops are counted to determine the frequency ratio of F_V/F_H.

V and H signals, the spot is at its extreme top right position shown at the end of the diagonal line. That action occurs during the first quarter-cycle for both signals.

On the next quarter-cycle the spot repeats the same positions on the way back to the center. In the same way, during the negative half-cycle the spot moves diagonally down to the bottom left. The spot repeating over this path produces a diagonal line. When the two waves are 180° out of phase, the line slopes in the opposite direction.

Consider the circle pattern produced by two waves 90° out of phase. One signal is at maximum when the other is at zero. When the vertical signal forces the spot to the extreme top or bottom position, the spot is in the center horizontally. Also, when the horizontal signal is maximum for the extreme left and right positions, the spot is in the center vertically. Then the spot traces a circle pattern for all the V and H values 90° out of phase.

For the pattern of an ellipse, the phase angle θ can be determined by calculating the ratio of the two lengths B and A shown at the right in Fig. D-12. As an example, when the B intercept is seven-tenths of A, the ratio is 0.7. Since sine θ equals 0.707 for 45°, the phase angle is 45°.

FREQUENCY COMPARISONS   In practical terms, the patterns in Fig. D-12 show that the frequency is the same for the vertical and horizontal input signals. The pattern may drift between a diagonal line and the circle as the phase changes slowly. Even so, the pattern indicates a 1:1 frequency ratio.

When the vertical input signal has a higher frequency than the horizontal input signal, the patterns in Fig. D-13 are produced. To determine the frequency ratio, count the loops across either the top or bottom of the trace for F_V. Count only the closed loops; an open loop, such as a half-loop, is not counted at all. Similarly, count the closed loops at either side for F_H. The frequency ratio is then equal to F_V/F_H.

As an example, let the horizontal input be a 60-Hz ac voltage from the power line as a reference frequency. The vertical-input signal is from an audio signal generator. Its frequency calibration can be checked at the dial setting of 60 Hz. With a circle or line pattern, the frequency is exactly 60 Hz for the signal generator.

Next change the frequency dial to 120 Hz. Where the generator produces the pattern in Fig. D-13a, the frequency is exactly $2 \times 60 = 120$ Hz. The 3:1 pattern in Fig. D-13b shows the generator frequency is 180 Hz. Frequencies that are not exact multiples can also be compared, as in Fig. D-13c. The 3:2 ratio shows the generator frequency is $3/2 \times 60 = 90$ Hz. In that way the generator frequency can be checked for patterns up to about 10 loops which would represent 600 Hz.

After the generator is calibrated, it can be used as the reference for checking an unknown frequency. Use the generator for the horizontal-input signal and connect the other to the oscilloscope vertical input. The unknown frequency can be determined by seeing how many loops it produces compared with the generator frequency.

ac Abbreviation for alternating current.

active device One that can control voltage or current. Examples are transistor amplifier and diode rectifier.

acute angle Less than 90°.

A/D converter A device that converts analog input signals to digital output.

admittance (Y) Reciprocal of impedance Z in ac circuits. $Y = 1/Z$.

air gap Air space between poles of a magnet.

alkaline cell or battery One that uses alkaline electrolyte.

alpha (α) Characteristic of junction transistors. Ratio of collector current to emitter current. Value is 0.98 to 0.99.

alternating current (ac) Current that reverses direction at a regular rate. Alternating voltage reverses in polarity. The rate of reversals is the frequency.

alternator AC generator.

ampere (A) Basic unit of electric current. Value of one ampere flows when one volt of potential difference is applied across one ohm of resistance.

ampere-turn Unit of magnetizing force equal to 1 A × 1 turn.

amplifier A device that increases the amplitude of a signal.

amplitude modulation (AM) Changing the amplitude of an RF carrier wave in step with a lower-frequency signal that has the desired information.

analog circuits Circuits that use continuous variations in voltage or current, compared with digital pulse circuits.

AND gate Digital logic circuit. Produces HIGH output of 1 only when all inputs are at 1.

antiresonance Term sometimes used for parallel resonance.

apparent power The product of voltage and current VA when V and I are out of phase.

armature The part of a generator in which the voltage is produced. In a motor it is commonly the rotating member. Also, the movable part of a relay.

astable MV Multivibrator that has no stable state. Used as an oscillator to generate clock timing pulses.

audio frequency (AF) Within the range of hearing, approximately 16 to 16,000 Hz.

autotransformer A single, tapped, winding used to step up or step down voltage.

average value In sine-wave ac voltage or current, is 0.637 of peak value.

back-off scale Ohmmeter readings from right to left.

bandpass Filter that allows a band of frequencies to be coupled to the load.

bandstop Filter that prevents a band of frequencies from being coupled to the load.

bandwidth A range of frequencies that have a resonant effect in LC circuits.

bank Components connected in parallel.

battery Group of cells connected in series or parallel.

BCD Binary coded decimal. Converts a decimal number to a binary equivalent.

beta (β) Current-gain characteristic of junction transistors. Ratio of collector current to base current.

bias Average dc level of amplifier voltage or current to set operating characteristic.

binary number system Uses only two digits, 0 and 1.

bipolar transistor NPN or PNP type.

bistable MV Multivibrator that has two stable states. Used as flip-flop circuit.

bit One unit of information, either 0 or 1, in the binary number system.

bleeder current Steady current from source, used to stabilize output voltage with changes in load current.

boolean algebra Logical system of using binary information in digital circuits.

branch Part of a parallel circuit.

bridge Circuit in which voltages or currents can be balanced for a net effect of zero.

brushes In a motor or generator, devices that provide stationary connections to the rotor.

bypass capacitor One that has very low reactance in a parallel path.

byte Digital word with a string of eight bits of 0 and 1.

C Symbol for capacitance.

C Abbreviation for coulomb, the unit of electric charge.

calorie Amount of heat energy needed to raise the temperature of one gram of water by 1°C.

capacitor Device used to store electric charge.

capacitance The ability to store electric charge.

carbon composition resistors Resistors made of finely divided carbon or graphite mixed with a powdered insulating material.

carbon film resistors Resistors made by depositing a thin layer of carbon on an insulated substrate. The carbon film is cut in the form of a spiral.

cascaded amplifiers Output terminal of one stage drives input terminal of next stage.

CB circuit Common-base amplifier for junction transistors. Signal into emitter and output from collector.

CB radio Citizen's band radio, 26.965 to 27.405 MHz.

CC circuit Common-collector amplifier for junction transistors. Signal into base and output from emitter. Is emitter-follower stage.

CD circuit Common-drain circuit for field-effect transistors. Is source-follower stage.

CE circuit Common-emitter amplifier for junction transistors. Signal into base and output from collector.

Celsius scale (°C) Temperature scale that uses 0° for the freezing point of water and 100° for the boiling point. Formerly called centigrade.

ceramic Insulator with a high dielectric constant.

chassis ground Common return for all electronic circuits mounted on one metal chassis or PC board. Usually connects to one side of dc supply voltage.

CG circuit Common-gate amplifier for FETs.

cgs Centimeter-gram-second system of units.

chip Miniature semiconductor for integrated circuit.

chip capacitor A surface-mounted capacitor.

choke Inductance with high X_L compared with the R of the circuit.

circuit breaker A protective device that opens when excessive current flows in circuit. Can be reset.

circular mil Cross-sectional area of round wire with diameter of 1 mil or 0.001 in.

clamp probe Measures current without opening the circuit.

clear Same as reset on a flip-flop. Puts Q output at logic 0.

clock A device that provides timing pulses for digital circuits; it is usually a multivibrator (MV) oscillator.

closed circuit A continuous path for current.

coaxial cable An inner conductor surrounded by an outer conductor that serves as a shield.

coding of capacitors The methods used to indicate the value of a capacitor.

coil Turns of wire conductor to concentrate the magnetic field.

color code System in which colors are used to indicate values in resistors and capacitors.

commutator Converts reversing polarities to one polarity.

comparator An op amp circuit that compares the signal voltage on one input with a reference voltage on the other.

complex number Has real and j terms; uses form $A + jB$.

conductance (G) Ability to conduct current. It is the reciprocal of resistance, $G = 1/R$. The unit is the siemens (S).

constant-current source One that has high r_i to supply constant I with variations in R_L.

constant-voltage source One that has low r_i to supply constant V with variations in R_L.

continuity Continuous path for current. Reading of zero ohms with an ohmmeter.

conventional current Direction of flow of positive charges, opposite from electron flow.

corona Effect of ionization of air around a point at high potential.

cosine A trigonometric function of an angle, equal to the ratio of the adjacent side to the hypotenuse in a right triangle.

cosine wave One whose amplitudes vary as the cosine function of an angle. It is 90° out of phase with the sine wave.

coulomb (C) Unit of electric charge. $1\ C = 6.25 \times 10^{18}$ electrons.

counter Digital circuit using a flip-flop to accumulate the count of pulses.

coupling capacitor Has very low X_C in series path.

covalent bond Pairing of atoms with electrical valence of ± 4.

cps Cycles per second. Formerly used as unit of frequency. Replaced by hertz (Hz) unit where $1\ Hz = 1\ cps$.

CRT Cathode-ray tube. A device that converts electric signals to a visual display on a fluorescent screen.

CS circuit Common-source amplifier for field-effect transistors.

current divider A parallel circuit to provide branch I less than the main-line current.

current source Supplies $I = V/r_i$ to load, with r_i in parallel.

cutoff No current in an active device such as a transistor.

cycle One complete set of values for a repetitive waveform.

D/A converter Converts digital input to analog output.

damping Reducing the Q of a resonant circuit to increase the bandwidth.

Darlington pair Combination of two transistor stages in cascade.

D'Arsonval meter A dc analog meter movement commonly used in ammeters and voltmeters.

dB Abbreviation for decibel. Equals 10 times the logarithm of the ratio of two power levels.

dc Abbreviation for direct current.

decade A 10:1 range of values.

decade resistance box A unit for providing any resistance within a wide range of values.

decibels A logarithmic expression that compares two power levels.

degaussing Demagnetizing by applying an ac field and gradually reducing it to zero.

delta (Δ) network Three components connected in series in a closed loop. Same as pi (π) network.

detector diode A device that rectifies a modulated signal to recover information in the modulation.

diamagnetic Material that can be weakly magnetized in the opposite direction from the magnetizing field.

dicing Cutting a slice of semiconductor material into tiny chips.

dielectric Insulating material. It cannot conduct current but does store charge.

dielectric constant (k) Ability to concentrate the electric field in a dielectric.

differential amplifiers An op amp circuit that amplifies differential signals but attenuates common-mode signals.

differentiating circuit An RC circuit with a short time constant for pulses across R.

digital circuit One that uses only two amplitudes for a pulse of voltage or current, either HIGH at 1 or LOW at 0.

digital IC Abbreviation for digital integrated circuit.

diode Electronic device with two electrodes. Allows current flow in only one direction.

DIP Dual inline package for the pins of IC chip.

direct current (dc) Current that flows in only one direction. Dc voltage has a steady polarity that does not reverse.

discrete component A single individually packaged component usually with two or three leads.

distributor Digital circuit to convert serial data to parallel data. Also called demultiplexer.

DMM Digital multimeter. A piece of test equipment used to measure voltage, current, and resistance in an electronic circuit.

doping   Adding impurities to pure semiconductor material to provide free positive and negative charges.

double subscripts   An example is V_{BA} to indicate voltage at point B with respect to point A.

DPDT   Double-pole double-throw switch or relay contacts.

DPST   Double-pole single-throw switch or relay contacts.

dynamometer   Type of ac meter, generally for 60 Hz.

eddy current   Circulating current induced in the iron core of an inductor by ac variations of magnetic flux.

effective value   For sine-wave ac waveform, 0.707 of peak value. Corresponds to heating effect of same dc value. Also called rms value.

efficiency   Ratio of power output to power input $\times$ 100%.

EIA   Electronic Industries Association.

electricity   Dynamic electricity is the effect of voltage in producing current in conductors. Static electricity is accumulation of charge.

electrolyte   Solution that forms ion charges.

electrolytic capacitor   Type with very high C because electrolyte is used to form very thin dielectric. Must be connected with correct polarity in a circuit.

electromagnet   Magnet whose magnetic field is associated with electric current in a coil.

electron   Basic particle of negative charge, in orbital rings around the nucleus in an atom.

electron flow   Current of negative charges in motion. Direction is from the negative terminal of the voltage source, through the external circuit, and returning to the positive side of the source. Opposite to the direction of conventional current.

electron volt   Unit of energy equal to the work done in moving a charge of 1 electron through a potential difference of 1 V.

electronics   Based on electrical effects of the electron. Includes applications for amplifiers, oscillators, rectifiers, control circuits, and digital pulse circuits.

emf   Electromotive force, voltage to produce current in a circuit.

emitter follower   Circuit in which signal input is to base and output is from emitter. Same as common-collector circuit.

F connector   Solderless plug for coaxial cable.

Fahrenheit scale (°F)   Temperature scale that uses 32° for the freezing point of water and 212° for the boiling point.

farad (F)   Unit of capacitance. Value of one Farad stores one coulomb of charge with one volt applied.

Faraday's law   For magnetic induction, the generated voltage is proportional to the flux and its rate of change.

FCC   Federal Communications Commission.

ferrite   Magnetic material that is not a metal conductor.

ferromagnetic   Magnetic properties of iron and other metals that can be strongly magnetized in the same direction as the magnetizing field.

FET   Field-effect transistor.

field   Group of lines of force; magnetic or an electric field.

field-effect transistor (FET)   A device that depends on an electric field to control the current in a silicon channel.

field winding   The part of a motor or generator that supplies the magnetic field cut by the armature.

film capacitor   A capacitor which uses a plastic film for its dielectric.

filter   Circuit to separate different frequencies.

fluctuating dc   Varying voltage and current but no change in polarity.

flux (ϕ)   Magnetic lines of force.

flux density   Amount of flux per unit area.

flywheel effect   Ability of an LC circuit to continue oscillating after the energy source has been removed.

forward voltage   Polarity that allows current of majority carriers through a semiconductor junction.

frequency (f)   Number of cycles per second for a waveform with periodic variations. The unit is hertz (Hz).

frequency modulation (FM)   Changing the frequency of an RF carrier wave in step with a lower-frequency signal that has the desired information.

function generator   A piece of test equipment which produces sine, square, and triangular waveforms. It is used when designing or testing electronic circuitry.

fuse   Metal link that melts from excessive current and opens circuit.

gain (A)   Also amplification. Ratio of amplified output to input.

galvanic cell   Electrochemical type of voltage source.

galvanometer   Measures electric charge or current.

gate   Logic circuit with two or more inputs but one HIGH or LOW output for specific combinations of input pulses.

gauss (G)   Unit of flux density in cgs system, equal to one magnetic line of force per square centimeter.

generator   A device that produces voltage output. Is a source for either dc or ac V and I.

germanium (Ge)   Semiconductor element used for transistors and diodes.

giga (G)   Metric prefix for 10^9.

gilbert (Gb)   Unit of magnetomotive force in cgs system. One gilbert equals 0.794 ampere-turn.

graph cycle   A 10:1 range of values on logarithmic graph paper.

ground   Common return to earth for ac power lines. Chassis ground in electronic equipment is the common return to one side of the internal power supply.

half-power frequencies   Define bandwidth with 70.7 percent response for resonant LC circuit.

Hall effect   Small voltage generated by a conductor with current in an external magnetic field.

harmonic   Exact multiple of fundamental frequency.

henry (H)   Unit of inductance. Current change of one ampere per second induces one volt across an inductance of one henry.

hertz (Hz)   Unit of frequency. One hertz equals one cycle per second.

hexadecimal   Number system with radix of 16.

h_{FE}   Hybrid parameter for junction transistors that specifies current gain for common-emitter circuit.

holding current   The minimum amount of current required to keep a relay energized.

hole   Positive charge that exists only in doped semiconductors because of covalent bonds between atoms. Amount of hole charge is the same as a proton and an electron.

hole current   Motion of hole charges. Direction is the same as that of conventional current, opposite from electron flow.

hot resistance   The R of a component with its normal load current. Determined by V/I.

hot-wire meter   Type of ac meter.

hybrid IC   A device that has discrete components with an integrated circuit.

hypotenuse   Side of a right triangle opposite the 90° angle.

hysteresis   In electromagnets, the effect of magnetic induction lagging in time behind the applied magnetizing force.

Hz   Hertz unit of frequency, equal to one cycle per second.

IC   Abbreviation for integrated circuit.

I_{CB_O} Leakage current from collector to base with emitter open.

IGFET   Insulated-gate field-effect transistor.

imaginary number   Value at 90°, indicated by j operator, as in the form jA.

impedance matching   Occurs when a transformer is utilized for its impedance transformation properties. With impedance matching, maximum power is delivered to the load, R_L.

inductance (L)   Ability to produce induced voltage when cut by magnetic flux. Unit of inductance is the henry (H).

induction   Ability to generate V or I without physical contact. Electromagnetic induction by magnetic field; electrostatic induction by electric field.

inductor   Coil of wire with inductance.

insulator   A material that does not allow current to flow when voltage is applied, because of its high resistance.

integrated circuit   Contains transistors, diodes, resistors, and capacitors in one miniaturized package. Can use bipolar transistor or FET technology.

integration circuit   An RC circuit with a long time constant. Voltage output across C.

internal resistance r_i   Limits the current supplied by the voltage source to $I = V/r_i$.

inverse relation   Same as reciprocal function. As one variable increases, the other decreases.

ion   Atom or group of atoms with net charge. Can be produced in liquids, gases, and doped semiconductors.

IR drop   Voltage across a resistor.

iron-vane meter   Type of ac meter, generally for 60 Hz.

j operator   Indicates 90° phase angle, as in $j8\ \Omega$ for X_L. Also, $-j8\ \Omega$ is at $-90°$ for X_C.

JFET   Junction field-effect transistor.

JK flip-flop   Type that has a clock input to toggle outputs between logic 1 and 0 when the J and K terminals are held HIGH.

joule (J)   Practical unit of work or energy. One joule equals one watt-second of work.

k   Coefficient of coupling between coils.

keeper   Magnetic material placed across the poles of a magnet to form a complete magnetic circuit. Used to maintain strength of magnetic field.

Kelvin (K) scale   Absolute temperature scale, 273° below values on Celsius scale.

kilo (k)   Metric prefix for 10^3.

Kirchhoff's current law (KCL)   The phasor sum of all currents into and out of any branch point in a circuit must equal zero.

Kirchhoff's voltage law (KVL)   The phasor sum of all voltages around any closed path must equal zero.

laminations   Thin sheets of steel insulated from one another to reduce eddy-current losses in inductors, motors, etc.

latch   A device that remains in one stable state until activated to the opposite state. Can store binary information as logic 1 or 0. Is flip-flop in digital circuits.

leakage current   Small reverse current of minority carriers across a PN junction.

Leclanché cell   Carbon-zinc primary cell.

LED   Light-emitting diode.

Lenz's law   Induced current has magnetic field that opposes the change causing the induction.

linear IC   A device that contains analog circuits, such as amplifiers, oscillators, rectifiers, and control circuits, rather than digital circuits.

linear relation   Straight-line graph between two variables. As one increases, the other increases in direct proportion.

load   Takes current from the voltage source, resulting in load current.

loading effect   Source voltage is decreased as amount of load current increases.

loop   In a circuit, any closed path.

LSI   Large-scale integration for IC chips.

magnetic pole   Concentrated point of magnetic flux.

magnetism   Effects of attraction and repulsion by iron and similar materials without the need for an external force. Electromagnetism includes the effects of a magnetic field associated with an electric current.

magnetomotive force (MMF)   Ability to produce magnetic lines of force. Measured in units of ampere-turns.

magnitude   Value of a quantity regardless of phase angle.

make and break   Occurs when contacts close and open.

maxwell (Mx)   Unit of magnetic flux, equal to one line of force in the magnetic field.

mega (M)   Metric prefix for 10^6.

memory device   Digital circuit that can store information as bits of logic 1 or 0. Often is flip-flop circuit.

mesh current   Assumed current in a closed path, without any current division, for application of Kirchhoff's current law.

metal film resistors   Resistors made by spraying a thin film of metal onto a ceramic substrate. The metal film is cut in the form of a spiral.

micro (μ)   Metric prefix for 10^{-6}.

microelectronics   Microscopic components used for IC chips to miniaturize size of equipment.

milli (m)   Metric prefix for 10^{-3}.

miniDIP   Miniature IC package with eight pins in dual inline form.

mks   Meter-kilogram-second system of units.

monostable   Having one stable state, as in one-shot multivibrator.

MOSFET   Metal-oxide semiconductor FET.

motor   A device that produces mechanical motion from electric energy.

multiplier   Resistor in series with a meter movement for voltage ranges.

multivibrator (MV)   Astable type of oscillator circuit to produce pulses as a clock generator for timing in digital circuits.

mutual induction (L_M)   Ability of one coil to induce voltage in another coil.

NAND gate   Logic circuit that produces a LOW output of 0 only when all inputs are HIGH at 1.

nano (n)   Metric prefix for 10^{-9}.

NC   Normally closed for relay contacts, or no connection for pinout diagrams.

neutron   Particle without electric charge in the nucleus of an atom.

nibble   Binary word with four bits, equal to one-half byte.

node   A common connection for two or more branch currents.

NOR gate   Logic circuit that produces a LOW output of 0 when any of the inputs is HIGH at 1.

Norton's theorem   Method of reducing a complicated network to one current source with shunt resistance.

NOT gate   Circuit to change binary 1 to 0 or 0 to 1. Same as inverter.

obtuse angle   More than 90°.

octal base   Eight pins for vacuum tubes. Eight digits for octal number system.

octave   A 2:1 range of values.

oersted (Oe)   Unit of magnetic field intensity; 1 Oe = 1 Gb/cm.

ohm (Ω)   Unit of resistance. Value of one ohm allows current of one ampere with potential difference of one volt.

Ohm's law   In electric circuits, $I = V/R$.

ohms per volt   Sensitivity rating for a voltmeter. High rating means less meter loading.

open circuit   One that has infinitely high resistance, resulting in zero current.

operational amplifier (op amp)   High-gain amplifier commonly used in linear IC chips for analog circuits.

OR gate   Digital logic circuit that produces a HIGH output of 1 when any of the inputs is HIGH at 1.

oscillator   Circuit that generates ac output from dc power input, without any ac signal input.

oscilloscope   A piece of test equipment used to view and measure a variety of different ac waveforms.

parallel circuit   One that has two or more branches for separate currents from one voltage source.

paramagnetic   Material that can be weakly magnetized in the same direction from the magnetizing force.

passive device   Components such as resistors, capacitors, and inductors. They do not generate voltage or control current.

PC board   A device that has printed circuits.

peak-to-peak value (p-p)   Amplitude between opposite peaks.

peak value   Maximum amplitude, in either polarity; 1.414 times rms value for sine-wave V or I.

permanent magnet (PM)   It has magnetic poles produced by internal atomic structure. No external current needed.

permeability   Ability to concentrate magnetic lines of force.

permeance   Reciprocal of magnetic reluctance.

phase angle   Angle between two phasors; denotes time shift.

phasing dots   Used on transformer windings to identify those leads having the same instantaneous polarity.

phasor   A line representing magnitude and direction of a quantity, such as voltage or current, with respect to time.

pickup current   The minimum amount of current required to energize a relay.

pico (p)   Metric prefix for 10^{-12}.

piezoelectric effect   Vibrations produced by some crystals compressed, expanded, or twisted, or when voltage is applied.

pinout   Pin numbers for IC package.

polar form   Form of complex numbers that gives magnitude and phase angle in the form $A \angle \theta°$.

polarity   Property of electric charge and voltage. Negative polarity is excess of electrons. Positive polarity means deficiency of electrons.

potential   Ability of electric charge to do work in moving another charge. Measured in volt units.

potentiometer   Variable resistor with three terminals connected as a voltage divider.

power (P)   Rate of doing work. The unit of electric power is the watt.

power factor   Cosine of the phase angle for a sine-wave ac circuit. Value is between 1 and 0.

power supply   A piece of test equipment used to supply dc voltage and current to electronic circuits under test.

preferred values   Common values of resistors and capacitors generally available.

primary cell or battery   Type that cannot be recharged.

primary winding   Transformer coil connected to the source voltage.

printed wiring   Conducting paths printed on plastic board.

proton   Particle with positive charge in the nucleus of an atom.

pulsating dc value   Includes ac component on average dc axis.

pulse   A sharp rise and decay of voltage or current of a specific peak value for a brief period of time.

Q   Figure of quality or merit, in terms of reactance compared with resistance. The Q of a coil is X_L/r_i. For an LC circuit, Q indicates sharpness of resonance. Also used as the symbol for charge: $Q = CV$.

quadrature   A 90° phase angle.

R   Symbol for resistance.

radian (rad)   Angle of 57.3°. Complete circle includes 2π rad.

radio   Wireless communication by electromagnetic waves.

radio frequencies (RF)   Those high enough to be radiated efficiently as electromagnetic waves, generally above 30 kHz. Usually much higher.

radix   Base for a number system; 10 for decimal numbers and 2 for binary numbers.

ramp   Sawtooth waveform with linear change in V or I.

reactance   Property of L and C to oppose flow of I that is varying. Symbol is X_C or X_L. Unit is the ohm.

read   Take out digital information from a memory device.

real number   Any positive or negative number not containing j. $(A + jB)$ is a complex number but A and B by themselves are real numbers.

real power   The net power consumed by resistance. Measured in watts.

reciprocal relation   As one variable increases, the other decreases.

rectangular form   Representation of a complex number in the form $A + jB$.

rectifier   A device that allows current in only one direction.

reflected impedance   The value of impedance reflected back into the primary from the secondary.

relay   Automatic switch operated by current in a coil.

relay chatter   Describes the vibrating of relay contacts.

reluctance (R)   Opposition to magnetic flux. Corresponds to resistance for current.

reset   Put Q output of flip-flop to logic 0.

resistance (R)   Opposition to current. Unit is the ohm (Ω).

resistance wire   A conductor having a high resistance value.

resonance   Condition of $X_L = X_C$ in an LC circuit to favor the resonant frequency for a maximum in V, I, or Z.

reverse voltage   Polarity that prevents forward current through a PN junction.

rheostat   Variable resistor with two terminals to vary I.

ringing   Ability of an LC circuit to oscillate after a sharp change in V or I.

rms value   For sine-wave ac waveform, 0.707 of peak value. Also called effective value.

rotor   Rotating part of generator or motor.

saturation   Maximum limit at which changes of input have no control in changing the output.

sawtooth wave   One in which amplitude values have a slow linear rise or fall and a sharp change back to the starting value. Same as a linear ramp.

secondary winding   Transformer coil connected to the load.

secondary cell or battery   Type that can be recharged.

self-inductance (L)   Inductance produced in a coil by current in the coil itself.

series circuit   One that has only one path for current.

set   Put Q output of flip-flop at logic 1.

shield   Metal enclosure to prevent interference of radio waves.

short-circuit   Has zero resistance, resulting in excessive current.

shunt   A parallel connection. Also a device used to increase the range of an ammeter.

SI   Abbreviation for *Système International,* a system of practical units based on the meter, kilogram, second, ampere, kelvin, mol, and candela.

siemens (S)   Unit of conductance. Reciprocal of ohms unit.

silicon (Si)   Semiconductor chemical element used for transistors, diodes, and integrated circuits.

sine   Trigonometric function of an angle, equal to the ratio of the opposite side to the hypotenuse in a right triangle.

sine wave   One in which amplitudes vary in proportion to the sine function of an angle.

slip rings   In an ac generator, devices that provide connections to the rotor.

solder   Alloy of tin and lead used for fusing wire connections.

solenoid   Coil used for electromagnetic devices.

source follower   FET amplifier circuit in which input is to the gate and output from the source electrode. Same as common-drain circuit. Corresponds to emitter follower.

spade lug   A type of wire connector.

SPDT   Single-pole double-throw switch or relay contacts.

specific gravity   Ratio of weight of a substance with that of an equal volume of water.

specific resistance   The R for a unit length, area, or volume.

SPST   Single-pole single-throw switch or relay contacts.

square wave   An almost instantaneous rise and decay of voltage or current in a periodic pattern with time and with a constant peak value. The V or I is on and off for equal times and at constant values.

static electricity   Electric charges not in motion.

stator   Stationary part of a generator or motor.

steady-state value   The V or I produced by a source without any sudden changes. Can be dc or ac value. Final value of V or I after transient.

storage cell or battery   Type that can be recharged.

string   Components connected in series.

summing amplifier   An op amp circuit whose output equals the negative sum of the inputs.

superconductivity   Very low R at extremely low temperatures.

superposition theorem   Method of analyzing a network with multiple sources by using one at a time and combining their effects.

supersonic   Frequency above the range of hearing, generally above 16,000 Hz.

surface-mount resistors   Resistors made by depositing a thick carbon film on a ceramic base. Electrical connection to the resistive element is made by means of two leadless solder end electrodes which are C-shaped.

surface-mount technology   Components soldered directly to the copper traces of a printed circuit board. No holes need to be drilled with surface-mounted components.

susceptance (B)   Reciprocal of reactance in sine-wave ac circuits; $B = 1/X$.

switch   Device used to open or close connections of a voltage source to a load circuit.

switching contacts   The contacts which open and close when a relay is energized.

tangent   Trigonometric function of an angle, equal to the ratio of the opposite side to the adjacent side in a right triangle.

tank circuit   An LC tuned circuit. Store energy in L and C.

tantalum   Chemical element used for electrolytic capacitors.

taper   How R of a variable resistor changes with the angle of shaft rotation.

tapered control   The manner in which the resistance of a potentiometer varies with shaft rotation. For a linear taper, one-half shaft rotation corresponds to a resistance change of one-half its maximum value. For a nonlinear taper, the resistance change is more gradual at one end, with larger changes at the other end.

taut-band meter   Type of construction for meter movement often used in VOM.

temperature coefficient   For resistance, how R varies with a change in temperature.

tesla (T)   Unit of flux density, equal to 10^8 lines of force per square meter.

Thevenin's theorem   Method of reducing a complicated network to one voltage source with series resistance.

three-phase power   AC voltage generated with three components differing in phase by 120°.

time constant   Time required to change by 63 percent after a sudden rise or fall in V and I. Results from the ability of L and C to store energy. Equals RC or L/R.

toggle   For digital circuits, changing between HIGH at logic 1 and LOW at logic 0.

toroid   Electromagnet with its core in the form of a closed magnetic ring.

transconductance   Ratio of current output to voltage input.

transformer   A device that has two or more coil windings used to step up or step down ac voltage.

transient   Temporary value of V or I in capacitive or inductive circuits caused by abrupt change.

transistor   Semiconductor device used for amplifiers. Includes NPN and PNP junction types and FETs.

trigonometry   Analysis of angles and triangles.

truth table   Listing of all possible combinations of inputs and outputs for a digital logic circuit.

tuning   Varying the resonant frequency of an LC circuit.

turns ratio   Comparison of turns in primary and secondary for a transformer.

twin lead   Transmission line with two conductors in plastic insulator.

UHF   Ultra high frequencies in band of 30 to 300 MHz.

unity-gain amplifier   An op amp circuit whose voltage gain is 1 or unity. It is used for buffering or isolating a low-impedance load from a high-impedance source.

VAR   Unit for voltamperes of reactive power, 90° out of phase with real power.

Variac   Transformer with variable turns ratio to provide different amounts of secondary voltage.

vector   A line representing magnitude and direction in space.

VHF   Very high frequencies, in band of 30 to 300 MHz.

volatile memory   Memory that loses its stored information when the power is turned off.

volt (V)   Practical unit of potential difference. One volt produces one ampere of current in a resistance of one ohm.

voltage divider   A series circuit to provide V less than the source voltage.

voltage drop   Voltage across each component in a series circuit. The proportional part of total applied V.

voltage regulator   A device that maintains a constant output voltage with changes of input voltage or output load current.

voltage source   Supplies potential difference across two terminals. Has internal series r_i.

voltampere (VA)   Unit of apparent power, equal to $V \times I$.

volt-ampere characteristic   Graph to show how I varies with V.

voltmeter loading   The amount of current taken by the voltmeter acting as a load. As a result the measured voltage is less than the actual value.

VOM   Volt-ohm-milliammeter.

watt (W)   Unit of real power. Equal to I^2R or $VI \cos \theta$.

watt hour   Unit of electric energy, as power $\times$ time.

wattmeter   Measures real power as instantaneous value of $V \times I$.

wavelength (λ)   Distance in space between two points with the same magnitude and direction in a propagated wave.

wavetrap   An LC circuit tuned to reject the resonant frequency.

weber (Wb)   Unit of magnetic flux, equal to 10^8 lines of force.

Wheatstone bridge   Balanced circuit used for precise measurements of resistance.

wire gage   A system of wire sizes based on the diameter of the wire. Also, the tool used to measure wire size.

wirewound resistors   Resistors made with wire known as *resistance wire* which is wrapped around an insulating core.

word   In digital circuits, a group of bits of 0 and 1. Usually written in groups of four, eight, or sixteen bits.

work   Corresponds to energy. Equal to power $\times$ time, as in kilowatthour unit. Basic unit is one joule, equal to one volt-coulomb, or one watt-second.

wye network   Three components connected with one end in a common connection and the other ends to three lines. Same as T network.

X_C Capacitive reactance, equal to $1/(2\pi f C)$.

X_L Inductive reactance, equal to $2\pi f L$.

XNOR gate   Digital logic circuit for exclusive NOR gate.

XOR gate   Digital logic circuit for exclusive OR gate.

Y Symbol for admittance in an ac circuit. Reciprocal of impedance Z; the $Y = 1/Z$.

Y network   Another way of denoting a wye network.

Z Symbol for ac impedance. Includes resistance with capacitive and inductive reactance.

zero-crossing detector   An op amp comparator whose output switches between $\pm V_{\text{sat}}$ when V_{in} crosses zero.

zero-ohm resistors   A resistor whose value is practically 0 Ω. The 0-Ω value is denoted by a single black band around the center of the resistor body.

zero-ohms adjustment   Used with ohmmeter of VOM to set the correct reading at zero ohms.

ANSWERS TO SELF-TESTS

CHAPTER 1	1. T	5. T	9. T	13. T	17. T
	2. T	6. T	10. T	14. T	18. T
	3. T	7. T	11. T	15. T	19. F
	4. T	8. T	12. T	16. T	20. F

CHAPTER 2	1. *b*	3. *a*	5. *c*	7. *a*	9. *c*
	2. *d*	4. *c*	6. *a*	8. *d*	10. *c*

CHAPTER 3	1. 2	6. 25	10. 72	14. 0.83	18. 3
	2. 4	7. 25	11. 8	15. 144	19. 0.2
	3. 16	8. 10	12. 2	16. 2	20. 0.12
	4. 0.5	9. 0.4	13. 2	17. 1.2	21. *d*
	5. 2				

CHAPTER 4	1. *d*	3. *d*	5. *c*	7. *c*	9. *b*
	2. *c*	4. *b*	6. *d*	8. *b*	10. *d*

CHAPTER 5	1. *b*	3. *a*	5. *a*	7. *c*	9. *c*
	2. *a*	4. *d*	6. *c*	8. *b*	10. *b*

CHAPTER 6	1. *c*	3. *c*	5. *d*	7. *d*	9. *d*
	2. *c*	4. *c*	6. *b*	8. *a*	10. *d*

REVIEW: CHAPTERS 1 TO 6	1. *a*	6. *c*	10. *d*	14. *b*	18. *a*
	2. *c*	7. *b*	11. *b*	15. *a*	19. *b*
	3. *b*	8. *c*	12. *a*	16. *a*	20. *a*
	4. *c*	9. *b*	13. *c*	17. *a*	21. *b*
	5. *c*				

CHAPTER 7	1. T	3. T	5. T	7. T	9. F
	2. T	4. F	6. T	8. T	10. T

CHAPTER 8	1. *a*	3. *a*	5. *c*	7. *a*	9. *d*
	2. *c*	4. *a*	6. *c*	8. *c*	10. *c*

REVIEW: CHAPTERS 7 AND 8	1. T	4. T	7. T	9. T	11. F
	2. T	5. F	8. T	10. F	12. T
	3. T	6. F			

CHAPTER 9	1. T	3. T	5. T	7. F	9. T
	2. F	4. T	6. T	8. T	10. T

CHAPTER 10	1. T	3. T	5. T	7. T	
	2. T	4. T	6. T	8. F	

REVIEW: CHAPTERS 9 AND 10	1. T	4. T	7. F	10. T	13. T
	2. T	5. T	8. T	11. T	14. T
	3. T	6. F	9. T	12. T	15. T

CHAPTER 11	1. *a*	3. *d*	5. *b*	7. *b*	9. *c*
	2. *d*	4. *b*	6. *a*	8. *c*	10. *c*

CHAPTER 12	1. *d*	3. *b*	5. *d*	7. *a*	9. *a*
	2. *c*	4. *a*	6. *d*	8. *c*	10. *d*

REVIEW: CHAPTERS 11 AND 12	1. *d*	3. *a*	5. *b*	7. *b*	9. *a*
	2. *c*	4. *c*	6. *d*	8. *b*	10. *d*

CHAPTER 13	1. T	4. F	7. T	10. T	13. T
	2. T	5. T	8. T	11. F	14. T
	3. T	6. T	9. T	12. T	15. F

CHAPTER 14	1. F	4. F	7. T	9. T	11. T
	2. T	5. T	8. T	10. T	12. T
	3. T	6. T			

CHAPTER 15	1. T	5. T	9. T	13. T	17. F
	2. T	6. T	10. T	14. T	18. T
	3. T	7. T	11. T	15. T	19. F
	4. T	8. T	12. T	16. F	

CHAPTER 16	1. T		11. 28.28 V		21. 1000 Hz
	2. T		12. 1.2 A		22. 180 Hz
	3. T		13. 70.7 V		23. 11.1 Hz
	4. T		14. 3×10^4 cm		24. 120 V
	5. T		15. 0.001 ms		25. 240 Hz
	6. T		16. 60 Hz		26. 240 V
	7. F		17. 0.01 μs		27. 120°
	8. F		18. 0.25 MHz		28. 208 V
	9. T		19. 7.07 V		
	10. T		20. 40 V		

REVIEW: CHAPTERS 13 TO 16	1. *b*	3. *c*	5. *b*	7. *a*	9. *c*
	2. *a*	4. *d*	6. *d*	8. *d*	10. *a*

CHAPTER 17	1. *a*	3. *b*	5. *c*	7. *c*	9. *c*
	2. *b*	4. *c*	6. *c*	8. *b*	10. *b*

CHAPTER 18	1. *b*	3. *c*	5. *d*	7. *a*	9. *b*
	2. *c*	4. *b*	6. *a*	8. *d*	10. *a*

CHAPTER 19	1. *d*	3. *a*	5. *b*	7. *c*	9. *b*
	2. *b*	4. *b*	6. *b*	8. *c*	10. *b*

REVIEW: CHAPTERS 17 TO 19	1. T	7. T	13. F	19. T	25. T
	2. T	8. T	14. T	20. T	26. T
	3. T	9. F	15. T	21. T	27. F
	4. T	10. T	16. T	22. T	28. F
	5. T	11. T	17. F	23. T	29. T
	6. T	12. F	18. T	24. T	30. T

CHAPTER 20	1. *b*	3. *c*	5. *c*	7. *b*	9. *b*
	2. *c*	4. *d*	6. *d*	8. *d*	10. *a*

CHAPTER 21	1. *a*	3. *c*	5. *a*	7. *c*	9. *d*
	2. *c*	4. *d*	6. *c*	8. *b*	10. *c*

CHAPTER 22	1. *c*	3. *c*	5. *b*	7. *c*	9. *d*
	2. *c*	4. *c*	6. *b*	8. *c*	

CHAPTER 23	1. *a*	3. *d*	5. *d*	7. *d*	9. *d*
	2. *c*	4. *b*	6. *c*	8. *c*	10. *c*

REVIEW: CHAPTERS 20 TO 23	1. *c*	5. *d*	8. *a*	11. *c*	14. *d*
	2. *b*	6. *d*	9. *b*	12. *a*	15. *b*
	3. *d*	7. *c*	10. *c*	13. *c*	16. *a*
	4. *d*				

CHAPTER 24	1. *b*	3. *a*	5. *c*	7. *b*	9. *c*
	2. *c*	4. *c*	6. *a*	8. *c*	10. *a*

CHAPTER 25	1. d	4. j	7. c	10. a	12. f
	2. m	5. h	8. k	11. b	13. g
	3. i	6. l	9. e		

REVIEW: CHAPTERS 24 AND 25	1. 300	9. 14.1	17. 5.66 $\angle 45°$
	2. 300	10. 1	18. 4 $\angle 10°$
	3. 300	11. 45°	19. T
	4. 250	12. −45°	20. T
	5. 250	13. 1	21. T
	6. 200	14. 1.41	22. F
	7. 200	15. 7.07	
	8. 14.1	16. 600	

CHAPTER 26	1. *c*	3. *d*	5. *d*	7. *d*	9. *a*
	2. *b*	4. *c*	6. *a*	8. *d*	10. *b*

CHAPTER 27

1. *d*	**4.** *c*	**7.** *b*	**10.** *d*	**13.** *d*
2. *a*	**5.** *d*	**8.** *b*	**11.** *b*	**14.** *a*
3. *b*	**6.** *b*	**9.** *a*	**12.** *a*	**15.** *c*

REVIEW: CHAPTERS 26 AND 27

1. 8	**7.** 5	**12.** -100 db	**17.** T
2. 0.8	**8.** 0.08	**13.** octave, decade	**18.** T
3. 0.4	**9.** 40	**14.** 70.7	**19.** T
4. 10	**10.** 150	**15.** F	**20.** T
5. 10	**11.** $f_c = 31.83$ kHz	**16.** T	**21.** F
6. 1			

CHAPTER 28

1. ± 4	**6.** forward	**11.** 100 mA	**15.** higher
2. 0.7 V	**7.** reverse	**12.** source	**16.** shorted
3. opposite	**8.** positive	**13.** N channel	**17.** F
4. NPN	**9.** cathode	**14.** JFET, IGFET,	
5. base	**10.** hole current	and MOSFET	

CHAPTER 29

1. active	**7.** 30 to 300 MHz	**13.** AM	**19.** two
2. low	**8.** AF	**14.** video	**20.** T
3. analog	**9.** Hartley	**15.** FM	**21.** hum
4. $A_V = 100$	**10.** Multivibrator	**16.** cathode	**22.** T
5. $A_I = 50$	**11.** astable MV	**17.** anode	**23.** T
6. C	**12.** T	**18.** 60 Hz	

CHAPTER 30

1. CE	**7.** A	**13.** CD	**19.** short-circuit
2. CE	**8.** forward	**14.** β	**20.** less
3. multiplied	**9.** reverse	**15.** α	**21.** CC
4. CC	**10.** positive	**16.** 2.5 mA	**22.** dc
5. CS	**11.** NPN	**17.** class A	**23.** R_E
6. Si	**12.** V_{CC}	**18.** off	

REVIEW: CHAPTERS 28 TO 30

1. T	**6.** T	**11.** T	**16.** T	**21.** F
2. F	**7.** T	**12.** T	**17.** T	**22.** F
3. F	**8.** T	**13.** T	**18.** F	**23.** F
4. F	**9.** T	**14.** T	**19.** T	**24.** F
5. T	**10.** T	**15.** F	**20.** T	**25.** T

CHAPTER 31

1. 2	**5.** NOR	**8.** NAND	**11.** reset
2. hexadecimal	**6.** OR	**9.** NOR	**12.** set
3. 11	**7.** XNOR	**10.** 2^4 or 16	**13.** $Q = 0, \overline{Q} = 1$
4. NAND			

CHAPTER 32

1. T	**4.** F	**7.** F	**10.** T
2. T	**5.** T	**8.** T	**11.** T
3. T	**6.** T	**9.** T	

REVIEW: CHAPTERS 31 AND 32

1. T	**5.** F	**9.** T	**13.** F
2. T	**6.** F	**10.** T	**14.** T
3. T	**7.** T	**11.** T	**15.** T
4. T	**8.** T	**12.** F	**16.** F

ANSWERS TO ODD-NUMBERED PROBLEMS AND CRITICAL THINKING PROBLEMS

ANSWERS TO ODD-NUMBERED CHAPTER PROBLEMS

CHAPTER 1

1. $I = 5$ A
3. $Q = 9$ C
5. $Q = -3$ C
7. Since there are four electrons in the valence shell, the valence is ± 4.
9. $T = 8$ s
11. (a) $R = 500\ \Omega$
(b) $R = 250\ \Omega$
(c) $R = 120\ \Omega$
(d) $R = 4\ \Omega$
13. (a) $G = 1$ S
(b) $G = 0.0001$ S
(c) $G = 0.025$ S
(d) $G = 2$ S
15. $V = 12$ V

CHAPTER 2

1. (a) $1.5\ \text{k}\Omega \pm 10\%$
(b) $27\ \Omega \pm 5\%$
(c) $470\ \text{k}\Omega \pm 5\%$
(d) $6.2\ \Omega \pm 5\%$
(e) $91\ \text{k}\Omega \pm 10\%$
(f) $10\ \Omega \pm 5\%$
(g) $1.8\ \text{M}\Omega \pm 10\%$
(h) $1.5\ \text{k}\Omega \pm 20\%$
3. (a) $470\ \text{k}\Omega$
(b) $1.2\ \text{k}\Omega$
(c) $330\ \Omega$
(d) $10\ \text{k}\Omega$
5. Reading from left to right, the colors are:
(a) brown, black, orange, and gold
(b) red, violet, gold, and gold
(c) green, blue, red, and silver
(d) brown, green, green, and gold
(e) red, red, silver, and gold
7. (a) $680,225\ \Omega$
(b) $8250\ \Omega$
(c) $18,503\ \Omega$
(d) $275,060\ \Omega$
(e) $62,984\ \Omega$
9. See Fig. 2-19*b*

CHAPTER 3

1. (b) $I = 15$ mA
(c) the same, 15 mA
(d) $I = 30$ mA
3. (a) $V = 36$ V
(b) $I = 150$ mA
5. (a) $V = 24$ V
(b) $P = 48$ W
(c) the same, 48 W
7. (a) $I = 300 \times 10^{-6}$ A $= 300\ \mu$A
(b) $I = 5 \times 10^{-6}$ A $= 5\ \mu$A
(c) $I = 200 \times 10^{-3}$ A $= 200$ mA
(d) $I = 300 \times 10^{-6}$ A $= 300\ \mu$A
(e) $I = 1 \times 10^{-3}$ A $= 1$ mA
(f) $I = 15 \times 10^{-3}$ A $= 15$ mA
9. (a) $R = 12 \times 10^3\ \Omega = 12\ \text{k}\Omega$
(b) $R = 3.75 \times 10^3\ \Omega = 3.75\ \text{k}\Omega$
(c) $R = 8.64 \times 10^3\ \Omega = 8.64\ \text{k}\Omega$
(d) $R = 18 \times 10^3\ \Omega = 18\ \text{k}\Omega$
11. (a) $R = 2.5\ \Omega$
(b) $V = 25$ V
(c) $P = 250$ W
15. $R = 240\ \Omega$
17. $I_{max} = 19.17$ mA
19. $I = 16.67$ mA
21. $P = 1.125$ W
23. Choose a resistor whose R and P values are $200\ \Omega$ and ¼ W respectively.
25. $R = 100\ \Omega$

CHAPTER 4

1. $I = 2$ A, $R_2 = 10\ \Omega$
3. $V_2 = 1.2$ V
5. $I = 4$ mA
$V_1 = 8$ V
$V_2 = 32$ V
$P_1 = 32$ mW
$P_2 = 128$ mW
$P_T = 160$ mW $= 0.16$ W
7. $R_T = 2,573,470\ \Omega$
9. Each $R = 6\ \text{k}\Omega$
11. 20.7 V
13. $R_1 = 300\ \Omega$
$P_1 = 27$ W
$P_2 = 9$ W
15. $R_2 = 25\ \Omega$
17. $V_2 = 13$ V
19. (a) $V = 18$ V
(b) $V = 0$ V
21. $R = 60\ \Omega$
23. $V_1 = 0$ V
$V_2 = 36$ V
$V_3 = 0$ V
$V_4 = 0$ V
25. $R_T = 4.5\ \text{k}\Omega$
$I = 12$ mA
$V_1 = 1.2$ V
$V_2 = 2.64$ V
$V_3 = 8.16$ V
$V_4 = 14.4$ V
$V_5 = 21.6$ V
$V_6 = 1.32$ V
$V_7 = 4.68$ V
$P_T = 648$ mW
$P_1 = 14.4$ mW
$P_2 = 31.68$ mW
$P_3 = 97.92$ mW
$P_4 = 172.8$ mW
$P_5 = 259.2$ mW
$P_6 = 15.84$ mW
$P_7 = 56.16$ mW
27. $I = 20$ mA
$V_1 = 2.4$ V
$V_2 = 2$ V
$V_3 = 13.6$ V
$V_T = 18$ V
$R_3 = 680\ \Omega$
$P_T = 360$ mW
$P_2 = 40$ mW
$P_3 = 272$ mW

CHAPTER 5

1. (b) 12 V
   (c) $I_1 = 2$ A, $I_2 = 1$ A
   (d) $I_T = 3$ A
   (e) $R_{EQ} = 4\ \Omega$
3. (b) 20 V
   (c) $I_2 = 2$ A
      $I_3 = 4$ A
5. (a) $I_2 = 0$ A
   (b) $I_1 = 1$ A
   (c) $I_T = 1$ A
   (d) $R_{EQ} = 10\ \Omega$
   (e) $P_T = 10$ W
7. (a) $7.14\ \Omega$
   (b) $2\ k\Omega$
   (c) $250\ \Omega$
   (d) $54.6\ \Omega$
   (e) $714\ \Omega$
   (f) $5\ k\Omega$
9. $G_T = 0.039$ S

11. (a) $R = 100\ k\Omega$
    (b) $R = 33.3\ k\Omega$
    (c) $R = 11.1\ k\Omega$
13. (a) 8.8 V
    (b) 8.8 V
    (c) $I_2 = 2.26$ mA
15. $G_T = 0.5$ S
    $R_{EQ} = 2\ \Omega$
17. $I_1 = 200$ mA
    $I_2 = 400$ mA
    $I_3 = 600$ mA
    $I_T = 1.2$ A
    $R_{EQ} = 10\ \Omega$
    $P_1 = 2.4$ W
    $P_2 = 4.8$ W
    $P_3 = 7.2$ W
    $P_T = 14.4$ W
19. $I_1 = 50$ mA
    $I_2 = 25$ mA

$I_3 = 75$ mA
$I_T = 150$ mA
$R_{EQ} = 60\ \Omega$
$P_1 = 450$ mW
$P_2 = 225$ mW
$P_3 = 675$ mW
$P_T = 1.35$ W
21. I_1 remains the same
    (50 mA)
23. $R_{EQ} = 25\ \Omega$
25. (a) 6 mA
    (b) 30 mA
27. (a) 42.5 mA
    (b) 157.5 mA
29. $I_T = 1.2$ A
    $I_1 = 100$ mA
    $I_2 = 900$ mA
    $R_1 = 1.08\ k\Omega$
    $R_2 = 120\ \Omega$

$R_3 = 540\ \Omega$
$P_T = 129.6$ W
$P_2 = 97.2$ W
$P_3 = 21.6$ W
31. $R_{EQ} = 1.105\ k\Omega$
    $I_T = 190$ mA
    $I_2 = 30$ mA
    $I_3 = 50$ mA
    $I_4 = 100$ mA
    $R_1 = 21\ k\Omega$
    $R_3 = 4.2\ k\Omega$
    $P_T = 39.9$ W
    $P_1 = 2.1$ W
    $P_2 = 6.3$ W
    $P_3 = 10.5$ W
    $P_4 = 21$ W

CHAPTER 6

1. (a) $R_T = 25\ \Omega$
   (b) $I_T = 4$ A
3. (b) $R_T = 15\ \Omega$
5. (a) $R = 6\ \Omega$
   (b) $R = 24\ \Omega$
7. (a) $V_1 = 2.23$ V
      $V_2 = 0.74$ V
      $V_3 = 6.7$ V
      $V_4 = 22.3$ V
   (b) $P_1 = 204$ mW
      $P_2 = 69$ mW
      $P_3 = 620$ mW
      $P_4 = 2.08$ mW

9. $V_1 = V_X = 1$ V
   $V_2 = V_S = 10$ V
   $R_X = 4.2\ \Omega$
11. $R_T = 10.45\ \Omega$
13. (a) $V_2 = 20$ V
    (b) $V_1 = V_2 = 22.5$ V
15. $R_T = 1\ k\Omega$
    $I_1 = 60$ mA
    $I_2 = 44$ mA
    $I_3 = 16$ mA
    $V_1 = 7.2$ V
    $V_2 = 52.8$ V
    $V_3 = 52.8$ V

$V_{AB} = -52.8$ V
17. V_{AB} increases. I_T and I_2
    both decrease, and I_3 is
    zero.
19. V_{AB} decreases to zero.
    I_T and I_3 both increase,
    and I_2 is zero.
21. $R_T = 300\ \Omega$
    $I_1 = 120$ mA
    $I_2 = 60$ mA
    $I_3 = 60$ mA
    $I_4 = 30$ mA
    $I_5 = 30$ mA

$I_6 = 15$ mA
$I_7 = 15$ mA
$I_8 = 7.5$ mA
$I_9 = 7.5$ mA
$V_1 = 18$ V
$V_2 = 18$ V
$V_3 = 9$ V
$V_4 = 9$ V
$V_5 = 4.5$ V
$V_6 = 4.5$ V
$V_7 = 2.25$ V
$V_8 = 2.25$ V
$V_9 = 2.25$ V

CHAPTER 7

1. $V_1 = 3$ V
   $V_2 = 6$ V
   $V_3 = 9$ V
3. $I_1 = 1$ A
   $I_2 = 2$ A
5. $V_1 = 12$ V
   $V_2 = 6$ V
   $V_3 = 10.8$ V

$V_4 = 7.2$ V
$V_{CG} = 7.2$ V
$V_{BG} = 18$ V
$V_{AG} = 24$ V
7. $I_1 = 16$ mA
   $I_2 = 8$ mA
9. $I_1 = 25\ \mu$A
   $I_2 = 37.5\ \mu$A

$I_3 = 12.5\ \mu$A
$I_4 = 75\ \mu$A
11. $R_1 = 236.8\ \Omega$
    $R_2 = 333\ \Omega$
    $R_3 = 1.5\ k\Omega$
    $P_1 = 342$ mW
    $P_2 = 108$ mW
    $P_3 = 54$ mW

13. $R_1 = 500\ \Omega$
    $R_2 = 800\ \Omega$
    $R_3 = 1.6\ k\Omega$
    $P_1 = 1.8$ W
    $P_2 = 1.62$ W
    $P_3 = 360$ mW
15. $V_{\text{load } C} = 11.21$ V

CHAPTER 8

1. (a) $R_S = 50\ \Omega$
   (b) $R_S = 5.55\ \Omega$
   (c) $R_S = 2.083\ \Omega$
   (d) $R_S = 0.505\ \Omega$
      (approximately)
   (e) Half-scale current is
      1 mA in (a), 5 mA
      in (b), 12.5 mA in

      (c), and 50 mA in
      (d).
3. (a) $R_{\text{mult}} = 2.95\ k\Omega$
   (b) $R_{\text{mult}} = 9.95\ k\Omega$
   (c) $R_{\text{mult}} = 29.95\ k\Omega$
   (d) $R_{\text{mult}} = 99.95\ k\Omega$
   (e) $R_{\text{mult}} = 299.95\ k\Omega$
5. (a) $V = 4.8$ V

(b) $V = 3$ V
(c) $V = 4.77$ V
7. $\dfrac{10\ k\Omega}{V}$
9. (a) $R_X = 0\ \Omega$
   (b) $R_X = 4.5\ k\Omega$
   (c) $R_X = 3\ k\Omega$

(d) $R_X = 1.5\ k\Omega$
(e) $R_X = 750\ \Omega$
(f) $R_X = 500\ \Omega$
11. $R_1 = 145\ \Omega$
13. $10\ \Omega$
15. 1.1 V

CHAPTER 9

1. (a) $I_3 = 8$ A
   (b) $I_5 = 16$ A
   (c) $I_6 = 11$ A
3. (a) $V_T = 36$ V
   $R_T = 4.5$ kΩ
   $I = 8$ mA
   $V_{R_1} = 8$ V
   $V_{R_2} = 12$ V
   $V_{R_3} = 16$ V
   (b) $V_{AG} = +12$ V
   $V_{BG} = +4$ V
   $V_{CG} = -8$ V

$V_{DG} = -24$ V
5. (a) $V_T = 30$ V
   $R_T = 25$ kΩ
   $I = 1.2$ mA
   $V_{R_1} = 12$ V
   $V_{R_2} = 6$ V
   $V_{R_3} = 12$ V
   (b) $V = +3$ V
   (c) $V = -3$ V
   (d) $V = 0$ V
7. $I_1 = 160$ mA
   $I_2 = 440$ mA

$I_3 = 600$ mA
   $V_{R_1} = 19.2$ V
   $V_{R_2} = 79.2$ V
   $V_{R_3} = 10.8$ V
9. Answers are the same as for Prob. 7.
11. Left loop: CCW from negative terminal of V_1: -10 V $+ 16.875$ V $- 15$ V $+ 8.125$ V $= 0$. Right loop: CCW from positive terminal of V_2:

15 V $- 16.875$ V $+ 1.875$ V $= 0$. Outside loop: CCW from negative terminal of V_1: -10 V $+ 1.875$ V $+ 8.125$ V $= 0$.
13. $I_1 = 2$ A
    $I_2 = 3$ A
    $I_3 = 1$ A
    $V_{R_1} = 30$ V
    $V_{R_2} = 30$ V
    $V_{R_3} = 10$ V

CHAPTER 10

1. $V_{TH} = 15$ V
   $R_{TH} = 3$ Ω
   $V_L = 6$ V
3. $I_S = 5$ A
   $R_S = 4$ Ω
   $I_L = 3$ A
5. R_L not open
7. $V_P = 4.2$ V
9. $V_{R_2} = 16.8$ V

11. $V_{R_2} = 16.8$ V
13. $V_{R_3} = 10.6$ V
15. See Fig. 10-31
17. (a) $V_1 = 54$ V,
    $R_1 = 6$ Ω,
    $V_2 = 72$ V,
    $R_2 = 12$ Ω
    (b) $V_{AB} = V_{TH} = -18$ V,
    $R_{TH} = 18$ Ω

(c) $I_L = 500$ mA
    (d) $I_N = 1$ A
    $R_N = 18$ Ω
19. $V_{AB} = V_{TH} = 36$ V
    $R_{TH} = 600$ Ω
    $V_L = 24$ V
21. $V_{TH} = 36$ V
    $R_{TH} = 600$ Ω
    $I_L = 20$ mA

$V_L = 24$ V
23. $I_N = 33.3$ mA
    $R_N = 300$ Ω
25. $I_1 = 890$ mA
    $I_2 = 522.5$ mA
    $I_3 = 1.41$ A
    $V_{R_1} = 8.9$ V
    $V_{R_2} = 20.9$ V
    $V_{R_3} = 21.1$ V

CHAPTER 11

1. (a) 1024 cmil
   (b) gage no. 20
   (c) $R = 1.035$ Ω
3. (a) 1-A fuse
   (b) 0 V

(c) 120 V
5. $R = 96$ Ω
7. 10,000 ft
9. (a) 4.8 Ω
   (b) 4000 ft

11. 3 V
13. $I = 30$ A
15. (a) No. 14 gage
    (b) 0.25 Ω
    (c) 3.25 Ω

17. (a) 200 ft
    (b) 0.3185 Ω
    (c) 115 V approx.
    (d) 71.7 W approx.

CHAPTER 12

1. 1.5 mA
3. 600 A
5. (a) 2.88×10^5 C
   (b) 40 h
7. 20 kΩ
9. 6 Ω
11. $R_L = 1$ Ω:
    $I = 2.5$ A
    $V_L = 2.5$ V
    $P_L = 6.25$ W
    $P_T = 37.5$ W
    % efficiency = 16.67
    $R_L = 3$ Ω:
    $I = 1.875$ A

$V_L = 5.625$ V
    $P_L = 10.547$ W
    $P_T = 28.125$ W
    % efficiency = 37.5
    $R_L = 5$ Ω:
    $I = 1.5$ A
    $V_L = 7.5$ V
    $P_L = 11.25$ W
    $P_T = 22.5$ W
    % efficiency = 50
    $R_L = 7$ Ω:
    $I = 1.25$ A
    $V_L = 8.75$ V
    $P_L = 10.938$ W

$P_T = 18.75$ W
    % efficiency = 58.3
    $R_L = 10$ Ω:
    $I = 1$ A
    $V_L = 10$ V
    $P_L = 10$ W
    $P_T = 15$ W
    % efficiency = 66.7
    $R_L = 15$ Ω:
    $I = 750$ mA
    $V_L = 11.25$ V
    $P_L = 8.438$ W
    $P_T = 11.25$ W
    % efficiency = 75

$R_L = 45$ Ω:
    $I = 300$ mA
    $V_L = 13.5$ V
    $P_L = 4.05$ W
    $P_T = 4.5$ W
    % efficiency = 90
    $R_L = 100$ Ω:
    $I = 142.9$ mA
    $V_L = 14.29$ V
    $P_L = 2.04$ W
    $P_T = 2.14$ W
    % efficiency = 95.3
13. $r_i = 25$ Ω
15. $R_L = 150$ Ω

CHAPTER 13

1. 5×10^3 Mx
   5×10^{-5} Wb
3. 0.4 T
5. 24×10^3 Mx

7. 300
9. 1μWb $= 10^{-6} \times 10^8$ Mx
11. (a) 500 kG
    (b) 600 Mx

(c) 0.25 T
    (d) 150 μWb
    (e) 40 G

13. $B = 500$ G
15. $B = 80 \times 10^{-3}$ T or 80 mT

CHAPTER 14

1. (a) 200
  (b) 500
3. (a) 300 G/Oe
  (b) 378×10^{-6} T/(A · t/m)
  (c) 300
5. (a) 126×10^{-6}

  (b) 88.2×10^{-6}
7. (b) 40 V
  (c) 1000 A · t/m
  (d) 0.378 T
  (e) 3.02×10^{-4} Wb
  (f) 66×10^{4} A · t/Wb

9. 14.4
11. $\mu_r = 133.3$
13. $B = 0.00189$ T or 1.89 mT
15. $\phi = 0.01$ Wb or 10 mWb

CHAPTER 15

1. 8 kV
3. (a) 2 Wb/s
  (b) -2 Wb/s
5. (a) 0.2 A

  (b) 80 ampere-turns
  (c) 400 ampere-turns/m
  (d) 0.252 T
  (e) 1.512×10^{-4} Wb

  (f) See Fig. 15-7 for an example.
7. $v_{ind} = 1$ V
9. $N = 1600$ turns

CHAPTER 16

1. (a) $I = 6$ A
  (b) $f = 60$ Hz
  (c) $0°$
  (d) 120 V
3. (a) $t = 0.25$ ms
  (b) $t = 0.0625$ μs
5. (a) $f = 20$ Hz
  (b) $f = 200$ Hz
  (c) $f = 0.2$ Hz
  (b) $f = 0.2$ Hz
7. (a) $+10$ and -10 V
  (b) $+10$ and -10 V
  (c) $+10$ and -10 V

  (d) $+15$ and -5 V
9. $I_1 = 40$ μA
  $I_2 = 20$ μA
  $V_1 = V_2 = 200$ V
  $P_1 = 8$ mW
  $P_2 = 4$ mW
11. $I = 2.5$ A
13. (a) 27.15 V
  (b) 20.8 V
15. $I_1 = 2.553$ A
  $I_2 = 1.765$ A
  $I_3 = 5.455$ A
  $I_T = 9.773$ A

17. (a) 462.5 μV
  (b) 9.84 mV
  (c) 35.19 mV
19. (a) 7.93 Ω
  (b) 1.55 W
21. (a) 42.42 V_{rms}
  (b) 60 V
  (c) 120 V
23. 7.5 MHz, first harmonic, first odd; 15 MHz, second harmonic, first even; 22.5 MHz, third harmonic, second odd;

30 MHz, fourth harmonic, second even.
25. (a) $\lambda = 113$ ft
  (b) $\lambda = 22.6$ ft
  (c) $\lambda = 4.52$ ft
  (d) $\lambda = 1.13$ ft
  (e) $\lambda = 0.0753$ ft
  (f) $\lambda = 0.0565$ ft
27. 1.6 kHz
29. $22.5°$
31. $T = 6.67$ nS

CHAPTER 17

1. $Q = 400$ μC
3. (a) $Q = 18$ μC
  (b) 9 V
5. $C = 1062$ pF
7. (a) 200 V
  (b) $Q = 200$ μC
  (c) $C = 1$ μF
9. (a) 2.5×10^{-2} J
  (b) 12.5 J
  (c) 3.2 J

11. (a) 0.06 μF
  (b) 74.2 pF
13. (a) $C_T = 0.01334$ μF
  (b) $C_T = 0.047$ μF
15. (a) 4700 pF $\pm 20\%$
  (b) 10,000 pF $\pm 5\%$
  (c) 220,000 pF $\pm 10\%$
  (d) 820 pF $\pm 20\%$
  (e) 0.0033 μF $+ 80\%$, -20%
  (f) 0.022 μF $+ 100\%$, -0%

  (g) 1800 pF $\pm 10\%$
  (h) 0.0027 μF $+ 80\%$, -20%
17. (a) 56 pF
  (b) 12,000 pF
  (c) 560,000 pF
  (d) 22 pF
19. (a) 470,000 pF or 0.47 μF, $\pm 10\%$
  (b) 6,200,000 pF or 6.2 μF, $\pm 5\%$
  (c) 15,000,000 pF or 15 μF, $\pm 10\%$
  (d) 820,000,000 pF or 820 μF, $\pm 5\%$

CHAPTER 18

1. 80 pF at 1 MHz
3. (b) $I = 4.5$ mA
  (c) $f = 1$ kHz
5. (b) $I = 2$ mA
  (c) $V_{C_1} = 2$ V
  (d) $V_{C_2} = 8$ V

7. $f = 3183$ Hz
9. (a) $X_{C_T} = 200$ Ω
  (b) $C = 333.3$ pF
  $C_T = 1000$ pF
11. (a) $X_C = 300$ Ω
  $C = 8.85$ μF

  (b) $C = 17.7$ μF
13. $X_C = 206.5$ Ω
15. $I = 0.78$ mA
17. $C = 422$ pF
19. See values in Prob. 21
21. See values in Prob. 19

23. (a) X_{C_T} is halved
  (b) I doubles
25. (a) I_T is halved
  (b) $X_{C_{EQ}}$ doubles
27. $C = 100$ pF

CHAPTER 19

1. (b) $Z = 50$ Ω
  (c) $I = 2$ A
  (d) $V_R = 80$ V
  $V_C = 60$ V

  (e) $\theta_Z = -37°$
3. $C = 0.08$ μF at 100 Hz
  $C = 80$ pF at 100 kHz
5. At 60 Hz, $C = 26.59$ μF

At 1 kHz, $C = 1.59$ μF
At 1 MHz, $C = 1590$ pF
7. $I_C = 15$ mA
  $I_R = 20$ mA

$I_T = 25$ mA
$Z_T = 1.2$ kΩ
$\theta_I = 37°$
$V_R = V_C = 30$ V

CHAPTER 19 (*cont.*)

9. For dc or ac,
   $V_1 = 400$ V
   $V_2 = 200$ V
   $V_3 = 100$ V
11. $Z_T = 583$ Ω
    $I = 0.2$ A
    $\theta_Z = -31°$
13. $I_T = 0.466$ A

$Z_{EQ} = 258$ Ω
$\theta_Z = 59°$
15. $C_1 = 66$ μF approx.
    $C_2 = 33$ μF approx.
17. $\theta_Z = -5.7°$
19. $C = 765$ pF
21. $X_C = 31.83$ kΩ
    $Z_T = 33.37$ kΩ

$I = 3.6$ mA
$V_C = 114.6$ V
$V_R = 35.6$ V
$\theta_Z = -72.6°$
23. $X_C = 965$ Ω
    $I_C = 10.36$ mA
    $I_R = 10$ mA
    $I_T = 14.4$ mA

$Z_{EQ} = 694$ Ω
$\theta_I = 55.2°$
25. $V_{AG} = 60$ V
    $V_{BG} = 12$ V
    $V_{CG} = 1.09$ V
27. (a) $C = 1.59$ μF
    (b) 0.0106 μF or
        10.61 nF

CHAPTER 20

1. (a) 1.5 A/s
   (b) 10,000 A/s
   (c) 10,000 A/s
   (d) $-10,000$ A/s
3. (a) 7.5 mV
   (b) 50 V
   (c) 50 V
   (d) -50 V
5. (a) 60 Hz
   (b) 960 V
   (c) 96 mA

(d) 0.768 A
7. (a) 300 μH
   (b) 66.7 μH
   (c) 320 and 280 μH
   (d) 0.0707
9. $R = 10.52$ Ω
11. 0.243×10^{-3} J
13. (a) 80 percent
    (b) 500 W
15. 1.26 mH
17. (a) 12 V

(b) 2.4 mA
(c) 28.8 mW
(d) 28.8 mW
19. $V_{S_1} = 120$ V
    $I_{S_1} = 50$ mA
    $P_{S_1} = 6$ W
    $V_{S_2} = 24$ V
    $I_{S_2} = 1$ A
    $P_{S_2} = 24$ W
    $P_P = 30$ W
    $I_P = 250$ mA

21. (a) $R_L = 3$ Ω
    (b) $I_P = 2.5$ A
23. (a) $Z_P = 200$ Ω
    (b) $Z_P = 12.5$ Ω
    (c) $Z_P = 6.25$ kΩ
    (d) $Z_P = 5$ kΩ
    (e) $Z_P = 5$ Ω
25. 11.18:1

CHAPTER 21

1. At 100 Hz, $X_L = 314$ Ω
   At 200 Hz, $X_L = 628$ Ω
   At 1000 Hz, $X_L = 3140$ Ω
3. (b) $I = 21.4$ mA
   (c) $V_L = 16$ V
5. $X_L = 1.2$ kΩ
   $L = 3.18$ H
7. $L = 0.159$ H
   $X_L = 10$ kΩ
9. (a) $X_{L_T} = 5$ kΩ
   (b) $I = 2$ mA
   (c) $V_{L_1} = 2$ V

$V_{L_2} = 8$ V
(d) $L_1 = 2.65$ H
    $L_2 = 10.6$ H
11. (a) $f = 0.16$ kHz
    (b) $f = 1.27$ kHz
    (c) $f = 0.4$ MHz
    (d) $f = 1.6$ MHz
    (e) $f = 16$ MHz
13. $X_L = 1628.6$ Ω
15. $X_L = 754$ Ω
17. (d) At 500 Hz,
        $X_L = 785$ Ω

$I = 12.7$ mA
19. (a) 250 μH
    (b) 125 μH
    (c) 500 μH
    (d) 25 μH
    (e) 2.5 mH
21. $X_{L_1} = 5$ kΩ
    $X_{L_2} = 10$ kΩ
    $X_{L_3} = 15$ kΩ
    $X_{L_T} = 30$ kΩ
    $I = 4$ mA
    $V_{L_1} = 20$ V

$V_{L_2} = 40$ V
$V_{L_3} = 60$ V
23. $X_{L_1} = 1.6$ kΩ
    $X_{L_2} = 6.4$ kΩ
    $X_{L_3} = 1.28$ kΩ
    $I_1 = 20$ mA
    $I_2 = 5$ mA
    $I_3 = 25$ mA
    $I_T = 50$ mA
    $X_{L_{EQ}} = 640$ Ω
25. (a) $X_{LT} = 5.34$ kΩ
    (b) $X_{LT} = 4.08$ kΩ

CHAPTER 22

1. (a) $Z = 100$ Ω
       $I = 1$ A
       $\theta = 0°$
   (b) $Z = 100$ Ω
       $I = 1$ A
       $\theta = 90°$
   (c) $Z = 70.7$ Ω
       $I = 1.41$ A
       $\theta = 45°$
3. (b) $X_L = 377$ Ω
   (c) $Z = 390$ Ω
   (d) $I = 0.3$ A
   (e) $I = 47.8$ mA
5. $Z = 400$ Ω
   $X_L = 400$ Ω

7. $R_e = 94$ Ω
9. $Z = 566$ Ω
   $I = 0.177$ A
   $V_L = 70.7$ V
   $V_R = 70.7$ V
   $\theta_Z = 45°$
11. At 800 Hz,
    $I_R = 0.25$ A
    $I_L = 0.125$ A
    $\theta_I = -26.6°$
13. $X_L = 500$ Ω
    $L = 15.9$ mH
15. $Z_T = 583$ Ω
    $I = 0.2$ A
    $\theta_Z = 59°$

17. ν_L is a square wave,
    ±160 V p-p
19. (a) 45°
    (b) 63.4°
    (c) 84.3°
    (b) 63.4°
21. (a) $-45°$
    (b) $-26.6°$
    (c) $-5.7°$
    (b) $-26.6°$
23. $X_L = 6$ kΩ
    $Z_T = 7.62$ kΩ
    $I = 3.15$ mA
    $V_L = 18.9$ V
    $V_R = 14.8$ V

$\theta_Z = 51.93°$
25. $X_L = 2$ kΩ
    $I_L = 12$ mA
    $I_R = 16$ mA
    $I_T = 20$ mA
    $Z_{EQ} = 1.2$ kΩ
    $\theta_I = -36.87°$
27. $L = 26.53$ mH
    $Z_T = 3.33$ kΩ
    $I = 4.51$ mA
    $V_L = 11.26$ V
    $V_R = 9.92$ V
    $\theta_Z = 48.6°$
29. (a) $L = 7.96$ mH
    (b) $L = 159.1$ μH

CHAPTER 23

1. (a) 0.05 s
   (b) 0.05 μs
   (c) 1 ms
   (b) 20 μs
3. (a) 4 s
   (b) 100 V

5. $v_C = 86$ V
7. $v_C = 15$ V
9. 1.4 ms
11. (a) short
    (b) short
    (c) long

13. $C = 10$ μF
15. 1.96 V
17. 0.05×10^6 V/s
19. Answer in instructor's manual.
21. (a) $V_C = 5$ V

(b) $V_C = 10$ V
(c) $V_C = 11.32$ V
(d) $V_C = 13.65$ V
(e) $V_C = 14.7$ V
(f) $V_C = 14.93$ V

CHAPTER 24

1. (a) 100 W
   (b) no reactance
   (c) 1
3. (b) $I = 10$ A, approx.
   (c) $Z = 10$ Ω
   (d) $\theta = 0°$
5. (c) $Z_T = 500$ Ω
   $I = 0.8$ A
   $\theta_Z = 53°$

7. (a) $X_L = 0$, approx.
   $X_C = 665$ Ω
   (b) $Z_T = 890$ Ω
   $I = 135$ mA
   $\theta_Z = -47.9°$
9. (a) $180°$
11. $R = 320$ Ω
13. $C = 300$ pF
15. $R = 9704$ Ω

17. $Z_T = 143$ Ω
    $I = 0.7$ A
    $\theta_Z = -36.5°$
    $\theta = 36.5°$
19. (a) $L_1 = 3$ mH
    $C_1 = 0.025$ μF
    (b) same reactances
21. $I_L = 60$ mA
    $I_C = 120$ mA

$I_R = 80$ mA
$I_T = 100$ mA
$Z_{EQ} = 120$ Ω
$\theta_I = 36.87°$
Real power = 960 mW;
apparent power =
1.2 W; power factor =
0.8.

CHAPTER 25

1. (a) $4 - j3$
   (b) $4 + j3$
   (c) $3 + j6$
   (d) $3 - j3$
3. (a) $5 \angle{-37°}$
   (b) $5 \angle{37°}$
   (c) $3.18 \angle{18.5°}$
   (d) $4.24 \angle{-45°}$
5. $Z_T = 65.36 + j23.48$
7. (a) $4.5 \angle{14°}$
   (b) $4.5 \angle{34°}$

(c) $100 \angle{-84°}$
(d) $100 \angle{-60°}$
9. $Z_T = 12.65 \angle{18.5°}$
11. $Z_T = 5.25 \angle{-13.7°}$
13. $R = 5.08$ Ω
    $X_C = 1.27$ Ω
15. $R = 21.4$ Ω
    $X_L = 10.2$ Ω
17. $Z_T = 50 \angle{-37°} = 40 - j30$ Ω
    $I = 2 \angle{37°} = 1.6 + j1.2$ A
    $V_R = 80 \angle{37°} = 64 + j48$ V

$V_L = 120 \angle{127°} = -72 + j96$ V
$V_C = 180 \angle{-53°} = 108 - j144$ V
19. $Z_T = 2.07$ kΩ $\angle{14.6°}$ kΩ
    $I = 3.88$ mA $\angle{-14.6°}$ mA
21. $Z_T = 13.4 \angle{46.5°}$
23. $Z_T = 1.28 \angle{-11.3°}$ kΩ
    $I_T = 18.75 \angle{11.3°}$ mA
    $V_{R_1} = 13.31 \angle{56.3°}$ V
    $V_{C_1} = 19.97 \angle{-33.7°}$ V
    $V_{L_1} = 13.31 \angle{56.3°}$ V
    $V_{R_2} = 19.97 \angle{33.7°}$ V

CHAPTER 26

1. $f_r = 5$ MHz
3. $f_r = 1.41$ MHz
5. $L = 35.08$ μH
7. $f_r = 5$ MHz
9. $Q = 50$
   $\Delta f = 100$ kHz
   $f_1 = 4.95$ MHz
   $f_2 = 5.05$ MHz
11. (a) $C = 202.68$ pF
    (b) $X_L = 314.2$ Ω
    $X_C = 314.2$ Ω
    $Q = 25$
    $\Delta f = 100$ kHz
13. $X_L = 785.4$ Ω

$X_C = 785.4$ Ω
$I_L = 1.273$ mA
$I_C = 1.273$ mA
$Q = 100$
$Z_{EQ} = 78.58$ kΩ
$I_T = 12.73$ μA
15. (a) $Z_{EQ} = 55.55$ kΩ
    $I_T = 18$ μA
    $\theta_I = -45°$
    (b) $Z_{EQ} = 55.55$ kΩ
    $I_T = 18$ μA
    $\theta_I = 45°$
17. (a) $C = 253.5$ pF
    (b) $X_L = 628.3$ Ω

$X_C = 628.3$ Ω
$Q = 80$
$\Delta f = 12.5$ kHz
$Z_{EQ} = 50.26$ kΩ
$I_T = 19.9$ μA
19. (a) Real power =
    79.62 nW; apparent
    power = 79.62 nW;
    power factor = 1.
    (b) Real power =
    39.8 nW; apparent
    power = 56.3 nW;
    power factor =
    0.707.

(c) Real power =
39.8 nW; apparent
power = 56.3 nW;
power factor =
0.707.
21. $f_r = 3$ MHz
23. $\Delta f = 60$ kHz
    $f_1 = 2.97$ MHz
    $f_2 = 3.03$ MHz
25. $R_S = 18.85$ Ω
27. $f_r = 1$ MHz
29. $\Delta f = 4$ kHz
    $f_1 = 998$ kHz
    $f_2 = 1.002$ MHz

CHAPTER 27

1. (a) $C = 0.1$ μF
   (b) $V_R \cong 0$ V
   $V_C = 20$ V
   (c) $V_R = 7.07$ V rms value

$V_C \cong 0$ V
3. $C = 0.64$ μF
5. (a) $f_c = 3.29$ kHz
   (b) $f_c = 5.305$ kHz

(c) $f_c = 1.88$ kHz
(d) $f_c = 2.39$ kHz
7. (a) $V_{out} = 10 \angle{0°}$ V
   (b) $V_{out} = 9.99 \angle{-2.16°}$ V

CHAPTER 27 (*cont.*)

(c) $V_{out} = 7.07 \; \underline{/-45°}$ V
(d) $V_{out} = 2.2 \; \underline{/-62°}$ V
(e) $V_{out} = 1 \; \underline{/-84°}$ V
(f) $V_{out} = 53 \; \underline{/-89.7°}$ mV
9. (a) $V_{out} = 1.05 \; \underline{/88.8°}$ mV
(b) $V_{out} = 26.58 \; \underline{/57.9°}$ mV
(c) $V_{out} = 35.35 \; \underline{/45°}$ mV

(d) $V_{out} = 46.45 \; \underline{/21.7°}$ mV
(e) $V_{out} = 49.94 \; \underline{/2.74°}$ mV
(f) $V_{out} = 50 \; \underline{/0°}$ mV
11. (a) $N_{db} = -27.6$ db
(b) $N_{db} = -3$ db
(c) $N_{db} = -0.17$ db
(d) $N_{db} = 0$ db (approximately)

13. $f_N = 4.42$ kHz
15. (a) $N_{db} = 3$ db
(b) $N_{db} = 10$ db
(c) $N_{db} = 13$ db
(d) $N_{db} = 20$ db
(e) $N_{db} = 30$ db
(f) $N_{db} = 33$ db

CHAPTER 28

1. $R = 1.6 \; \Omega$
3. $I_C = 5.7$ mA
5. $I_E = 50.3$ mA
7. $I = 20$ mA

CHAPTER 29

1. $A_V = 300$
$A_I = 90$
$A_P = 27{,}000$
3. 10,000
5. $159 \; \Omega$
7. (a) $C = 332 \; \mu F$
(b) $C = 166 \; \mu F$
9. $f_r = 1.625$ MHz
11. $R_L = 2.5$ kΩ

CHAPTER 30

1. (a) 50
(b) 80
(c) 4000
3. 11.25
5. (a) 4 V
(b) 2 mA
7. (a) 8 V
(b) 2 mA
(c) 1 kΩ
9. (a) ±2 mA
(b) 50

CHAPTER 31

1. (a) 11
(b) 17
(c) 16
(d) 181
3. (a) 100000000
(b) 111100000110
(c) 1110101
(d) 1100001100111010
5. (a) 5F
(b) 5F
(c) 35
(d) B29

7.

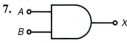

**Logic symbol for
2-input AND gate**

A	B	X
0	0	0
0	1	0
1	0	0
1	1	1

**Truth table for
2-input AND gate**

9.

**Logic symbol for
3-input NOR gate**

A	B	C	X
0	0	0	1
0	0	1	0
0	1	0	0
0	1	1	0
1	0	0	0
1	0	1	0
1	1	0	0
1	1	1	0

**Truth table for
3-input NOR gate**

11.

A	B	C	D	X
0	0	0	0	0
0	0	0	1	0
0	0	1	0	0
0	0	1	1	1
0	1	0	0	0
0	1	0	1	0
0	1	1	0	0
0	1	1	1	1
1	0	0	0	0
1	0	0	1	0
1	0	1	0	0
1	0	1	1	1
1	1	0	0	1
1	1	0	1	1
1	1	1	0	1
1	1	1	1	1

**Truth table for the boolean
expression: $AB + CD = X$**

13.

$A + B + C = \overline{X}$

15. See Fig. 31-49.

17.

A	B	C	D	X
0	0	0	0	0
0	0	0	1	0
0	0	1	0	0
0	0	1	1	0
0	1	0	0	1
0	1	0	1	1
0	1	1	0	1
0	1	1	1	1
1	0	0	0	0
1	0	0	1	0
1	0	1	0	0
1	0	1	1	0
1	1	0	0	1
1	1	0	1	1
1	1	1	0	1
1	1	1	1	0

**Truth table for the boolean
expression:** $B(\overline{A} + \overline{C} + \overline{D}) = X$

19.

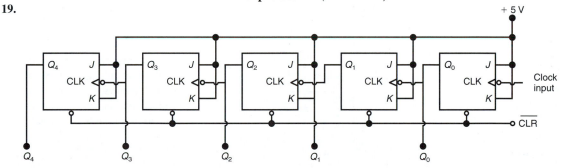

CHAPTER 32

1. (a) $A_{CL} = 10$
   (b) $V_{out} = 5$ V
3. $f_{max} = 15.92$ kHz

5. (a) $V_{out} = -12.5$ V
   (b) $V_{out} = +10$ V
   (c) $V_{out} = 0$ V

7. $f_c = 1.06$ kHz
   V_{out} at $0.25f_c = 24.25$ mVpp
9. $V_{out} = -10$ V

ANSWERS TO ODD-NUMBERED CRITICAL THINKING PROBLEMS

CHAPTER 1
1. $Q = 1.6 \times 10^{-16}$ C
3. $I = 0.0001$ A or 1×10^{-4} A

CHAPTER 2
1. $R = 250$ kΩ

CHAPTER 3
1. $I = 21.59$ A
3. Cost = \$7.52
5. I_{max} (120°C) = 13.69 mA

CHAPTER 4
1. $R_1 = 300$ Ω
  $R_2 = 600$ Ω
  $R_3 = 1.8$ kΩ
3. $I = 35.36$ mA
5. $R_1 = 250$ Ω
  $V_T = 1.25$ V

CHAPTER 5
1. $I_T = 44.53$ mA
3. $R_1 = 2$ kΩ
  $R_2 = 6$ kΩ
  $R_3 = 3$ kΩ
5. $R_1 = 15$ kΩ
  $R_2 = 7.5$ kΩ
  $R_3 = 3.75$ kΩ
  $R_4 = 1.875$ kΩ

CHAPTER 6
1. **(a)** $R_1 = 657$ Ω
  (b) 2×12.9 W = 25.8 W, approx.,
    for a safety factor of 2.
  (c) $R_T = 857$ Ω
3. **(b)** $A(B + C) = X$

CHAPTER 7
1. $R_1 = 1$ kΩ
  $R_3 = 667$ Ω

CHAPTER 8
1. $R_1 = 40$ Ω
  $R_2 = 9$ Ω
  $R_3 = 1$ Ω
3. 10 kΩ/V

CHAPTER 9
1. $R_1 = 1.5$ kΩ
  $R_3 = 1$ kΩ

CHAPTER 10
1. $V_{TH} = 20$ V
  $R_{TH} = 12$ Ω
  $I_L = 250$ mA
  $V_L = 17$ V
3. $V_{TH} = 0$ V
  $R_{TH} = 25$ Ω

CHAPTER 11
1. Drawing provided in instructor's manual.

CHAPTER 12
1. **(a)** $V_L = 11.45$ V
  (b) $I_L = 1.91$ A approx.
  (c) V_1 supplies 550 mA approx.
    V_2 supplies 275 mA approx.
    V_3 supplies 1.1 A approx.

CHAPTER 13
1. **(a)** $B = 1550$ G or 1.55 kG
  (b) $B = 0.155$ T or 155 mT

CHAPTER 14
1. $\mu_0 = B/H$:1 = 1 G/10e

Since 1 G = $\dfrac{1 \times 10^{-8} \text{ Wb}}{1 \times 10^{-4} \, m^2}$

$= 1 \times 10^{-4}$ T

and 10e = $79.36 \dfrac{A \cdot t}{m^2}$ then

$\mu_0 = \dfrac{1 \times 10^{-4} \text{ T}}{79.36 \dfrac{A \cdot t}{m^2}}$

$= 1.26 \times 10^{-6} \dfrac{T}{A \cdot t/m^2}$

CHAPTER 15
1. **(a)** $R_W = 1.593$ Ω
  (b) $R_T = 17.593$ Ω
  (c) $V_L = 218.3$ V
  (d) I^2R power loss = 296.5 W
  (e) $P_L = 2.98$ kW
  (f) $P_T = 3.27$ kW
  (g) % efficiency = 91.1
3. With a relay, the 1000-ft length of wire does not carry the load current I_L and thus the circuit losses are reduced significantly.

CHAPTER 16
1. **(a)** 65 ft
  (b) 1981 cm
3. $f = 3.9$ MHz

CHAPTER 17
1. $C_1 = 8.08$ nF
  $C_2 = 2.02$ nF
  $C_3 = 161.6$ nF
3. **(a)** $\mathcal{E} = 500$ mJ
  (b) $\mathcal{E} = 250$ mJ
  (c) Yes. 250 mJ of energy was lost as heat energy (I^2R) in the wire conductors when the second 100 μF was connected in part **(b)**.

CHAPTER 18
1. Connect the unmarked capacitor in series with an ac voltage source whose frequency and output voltage are known. Measure the ac current through the capacitor. Calculate X_C as V_C/I. Next, solve for C as $C = 1/2\pi f X_C$.

CHAPTER 19

1. $X_C = 300\ \Omega$
   $Z_T = 671\ \Omega$
   $I = 26.83$ mA
   $f = 19.649$ kHz
   $V_T = 18$ V
   $V_R = 16.1$ V
3. $I_C = 12$ mA
   $I_R = 20.79$ mA
   $I_T = 24$ mA
   $X_C = 2\ k\Omega$
   $R = 1.154\ k\Omega$
   $C = 7960$ pF approx.

CHAPTER 20

1. Since $V_P = V_S \times N_S/N_P$ and $I_P = I_S \times N_S/N_P$ then $Z_P = V_P/I_P =$
   $$\frac{V_S \times N_S/N_P}{I_S \times N_S/N_P}$$
   since $Z_S = V_S/I_S$ then $Z_P = (N_P/N_S)^2 \times Z_S$.
3. (a) $Z_P = 800\ \Omega$
   (b) $Z_P = 200\ \Omega$

CHAPTER 21

1. $L_1 = 60$ mH
   $L_2 = 40$ mH
   $L_3 = 120$ mH
   $L_T = 90$ mH
   $X_{L_1} = 1.2\ k\Omega$
   $X_{L_2} = 800\ \Omega$
   $X_{L_T} = 1.8\ k\Omega$
   $V_{L_1} = 24$ V
   $V_{L_3} = 12$ V
   $I_{L_2} = 15$ mA
   $I_{L_3} = 5$ mA
3. $L_1 = 10$ mH
   $L_2 = 120$ mH
   $L_3 = 40$ mH

CHAPTER 22

1. $X_L = 1.2\ k\Omega$
   $R = 2.08\ k\Omega$
   $L = 120$ mH
   $I = 15$ mA
   $V_L = 18$ V
   $V_R = 31.2$ V
3. $V_R = 6$ Vpp
   $V_{R_1} = 5$ Vpp
   $X_{L_1} = 833\ \Omega$
   $X_{L_2} = 500\ \Omega$
   $I = 6$ mApp
   $Z_T = 1.67\ k\Omega$
   $L_1 = 41.67$ mH
   $L_2 = 25$ mH
   $\theta_Z = 53.13°$

CHAPTER 23

1. (a) $V_C = 30$ V
   (b) 7.5 ms
3. (a) 100 V
   (b) 0 V
   (c) 26.4 V
   (d) 55.37 V
   (e) 72.93 V

CHAPTER 24

1. $L = 2.98$ mH

CHAPTER 25

1. $V_{in} = 24\ \angle 0°$ V

CHAPTER 26

1. $Q = 2\pi f_r L/r_S$
   $Q r_S = 2\pi L \times 1/2\pi\sqrt{LC}$
   $Q r_S = L/\sqrt{LC}$
   $Q^2 r_S^2 = L^2/LC$
   $Q^2 r_S^2 = L/C$
   $$\frac{X_L^2}{r_s^2} \times r_s^2 = L/C$$
   $X_L^2 = L/C$
   $X_L = \sqrt{L/C}$

CHAPTER 27

1. (a) $f_c = 965$ Hz
   (b) $V_{out} = 3.535$ Vpp
   (c) $V_{out} = 68.2$ mVpp
3. $L = 191\ \mu H$
   $C = 132.63$ pF

CHAPTER 28

1. $I_F = 150$ mA

CHAPTER 29

1. $Vdc = 27.68$ V

CHAPTER 30

1. $B = I_C/I_B$
   $= \alpha I_E/I_E - \alpha I_E$
   $= \alpha I_E/I_E\ (1 - \alpha)$
   $B = \alpha/1 - \alpha$

CHAPTER 31

1.

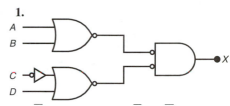

3. $(\overline{A} + B)(A + B + D)\overline{D} = B\overline{D}$

CHAPTER 32

1. $V_{out} = 2$ Vpp
   V_{out} is in phase with V_Y.

INDEX

Photo Credits

Cover and title page photos: Foreground: Larry Kennan Associates/The Image Bank; background: Simon Fraser/Welwyn Electronics/Science Photo Library/Photo Researchers

Pages iii, xiv: Stephen Simpson/FGP International; **Page 2:** Stephen Simpson/FPG International; **Page 3:** Bettmann Archive (Heinrich Hertz), Mark Steinmetz (Fig. S-1); **Page 5:** Science Photo Library/Photo Researchers; **Page 6:** Don Mason/The Stock Market (top), Bettmann Archive (Faraday), Courtesy of the Royal Institution (Faraday in lab); **Page 7:** Mark Steinmetz; **Page 9:** Mark Steinmetz; **Page 10:** Mark Steinmetz (Figs. S-5, S-6, and S-7); **Page 11:** Mark Steinmetz; **Page 12:** Mark Steinmetz; **Page 13:** Courtesy Fluke Corporation (Fig. S-10a), Courtesy of MCM Electronics (Fig. S-10b), Reproduced by permission of Tektronix, Inc. (Fig. S-11); **Page 14:** Phillip Hayson/Photo Researchers; **Page 18:** Bettmann Archive (Volta portrait), J-L Charmet/Science Photo Library/Photo Researchers (Volta with battery); **Page 21:** Sheila Terry/Science Photo Library/ Photo Researchers; **Page 26:** File photo; **Page 28:** Bettmann Archive (Millikan portrait and Millikan ray machine); **Page 33:** Mark Steinmetz; **Pages iv, 35:** Phillip Hayson/Photo Researchers; **Page 40:** Peter Aprahamian/Science Photo Library/Photo Researchers; **Page 45:** Mark Segal/Tony Stone Images; **Page 46:** Courtesy Harris Corporation; **Page 48:** Charles Krebs/Tony Stone Images (Fig. 2-1b), Mark Steinmetz (Fig. 2-2), P. R. Mallory (Fig. 2-3); **Page 49:** Bettmann Archive (Watt), Stackpole Corporation (Figs. 2-4 and 2-5); **Page 50:** Mark Steinmetz; **Page 55:** Mark Steinmetz; **Page 57:** Mark Steinmetz; **Page 60:** © Telegraph Colour Library/FPG International; **Page 66:** Courtesy Harris Corporation; **Page 67:** Charles Gupton/Stock Boston; **Pages v, 68:** Tom Tracy/The Stock Market; **Page 70:** Telegraph Colour Library/FPG International; **Page 77:** Bettmann Archive; **Page 79:** Bettmann Archive; **Page 80:** Bettmann Archive; **Page 82:** Telegraph Colour Library/FPG International; **Page 94:** Gabriel M. Covian/The Image Bank; **Page 95:** Matthew Borkoski/Stock Boston; **Page 96:** John Madere/The Stock Market; **Page 98:** Ben Swedowsky/The Image Bank; **Page 101:** File photo; **Page 103:** Metropolitan Museum of Art, Michael Friedsam Collection, 1931; **Page 111:** John Madere/The Stock Market; **Pages vi, 120:** Art Montes de Oca/FPG International; **Page 125:** Art Montes de Oca/FPG International; **Page 130:** Richard Pasley/Stock Boston; **Pages vii, 146:** Richard Nowitz/The Stock Market; **Page 151:** John Madere/The Stock Market; **Page 152:** Telegraph Colour Library/FPG International; **Page 157:** Richard Nowitz/The Stock Market; **Pages viii, 176:** Blair Seitz/Photo Researchers; **Page 188:** Geoff Tompkinson/Science Photo Library/Photo Researchers; **Pages xi, 194:** Joe Bator/The Stock Market; **Page 196:** Courtesy of MCM Electronics (Fig. 8-1a), Courtesy Fluke Corporation (Fig. 8-1a); **Page 197:** Weston Instrument Corp.; **Page 206:** Simpson Electric Company; **Page 215:** Courtesy of Simpson Electric Company; **Page 216:** Courtesy of Tektronix, Inc. (Fig. 8-20), Courtesy Fluke Corporation (Fig. 8-21); **Page 217:** Courtesy Fluke Corporation; **Page 223:** Alvis Upitis/The Image Bank; **Page 232:** Gary Gladstone/The Image Bank; **Page 236:** Bettmann Archive; **Page 239:** Rick Altman/Nawrock Stock Photo; **Page 240:** Charles Gupton/Stock Boston; **Page 243:** Gary Gladstone/The Image Bank; **Page 247:** Crown Studio; **Page 252:** Dick Luria/FPG International; **Page 261:** Dick Luria/FPG International; **Page 264:** SuperStock; **Pages xii, 288:** Gabe Palmer/The Stock Market; **Page 292:** L. S. Starrett; **Page 293:** Mark Steinmetz; **Page 294:** Hank Morgan/VHSID Lab/ECE Dept U of MA/Science Source/Photo Researchers; **Page 296:** Mark Steinmetz (Fig. 11-7a and b); **Page 299:** Mark Steinmetz; **Page 300:** Mark Steinmetz (Figs. 11-13, 11-14, and 11-15); **Page 305:** Bettmann Archive; **Page 306:** Gabe Palmer/The Stock Market; **Page 312:** Reproduced by permission of Tektronix, Inc.; **Page 317:** © Ken Biggs/Photo Researchers; **Page 318:** Mark Steinmetz; **Page 320:** Mark Steinmetz; **Page 321:** Mark Steinmetz; **Page 324:** Eveready Union Carbide Corporation; **Page 326:** Eveready Union Carbide Corporation; **Page 327:**

Eveready Union Carbide Corporation; **Page 329:** Mark Steinmetz; **Page 330:** Exide Corporation; **Page 332:** © Ken Biggs/Photo Researchers; **Page 333:** Exide Corporation; **Page 334:** Doug Martin; **Page 335:** Mark Steinmetz; **Page 345:** Bob Daemmrick/Stock Boston; **Page 352:** Toshiba America Consumer Products, Inc.; **Page 363:** Mark Steinmetz; **Page 364:** © 1990 David A. Wagner/The Stock Market; **Page 365:** Toshiba America Consumer Products, Inc.; **Page 366:** Courtesy F. W. Bell; **Page 370:** Charles Thatcher/Tony Stone Images; **Page 375:** KS Studios; **Page 378:** Michael Gilbert/Science Photo Library/Photo Researchers; **Page 385:** Charles Thatcher/Tony Stone Images; **Page 389:** Gregory MacNicol/Photo Researchers; **Page 390:** Ralph Mercer/Tony Stone Images; **Page 402:** © Alvis Upitis/The Image Bank; **Page 405:** Mark Steinmetz; **Page 413:** Michael Rosenfeld/Tony Stone Images; **Page 414:** Tony Craddock/SPL/Photo Researchers; **Page 419:** Kim Steele/The Image Bank; **Page 445:** Mark Steinmetz (Fig. 16-22a and b; Fig. 16-23); **Page 449:** Courtesy Meteor Communications Corp., Kent, Washington; **Page 458:** Greg Pease/Tony Stone Images; **Page 469:** Mark Steinmetz (Figs. 17-4b and 17-5b); **Page 470:** Mark Steinmetz (Figs. 17-6 and 17-7); **Page 471:** Mark Steinmetz; **Page 472:** Mark Steinmetz; **Page 474:** Mark Steinmetz; **Page 483:** Bob Johnston/Texas State Technical College; **Page 489:** Bob Daemmrich/Uniphoto; **Pages x, 496:** Ken Cooper/The Image Bank; **Page 500:** Peter Scholey/Nawrocki Stock Photo; **Page 507:** Ken Cooper/The Image Bank; **Pages ix, 516:** Larry Keenan Associates/The Image Bank; **Page 523:** Larry Keenan Associates/The Image Bank; **Page 526:** © 1990 Joe Robbins/FGP International; **Page 540:** © 1990 Stephen Hunt/The Image Bank; **Page 543:** Bettmann Archive; **Page 545:** Mark Steinmetz; **Page 545:** Mark Steinmetz; **Page 553:** Mark Steinmetz; **Page 559:** Mark Steinmetz; **Page 570:** Sencore, Inc.; **Page 575:** Gabe Palmer/The Stock Market; **Page 576:** Hewlett-Packard/Peter Arnold; **Page 583:** Scott Eklund/Gamma Liaison; **Page 584:** American Honda Motor Company; **Page 597:** American Honda Motor Company; **Page 597:** American Honda Motor Company; **Page 604:** Stephen Ferry/Gamma Liaison; **Page 618:** Stephen Ferry/Gamma Liaison; **Page 619:** Mark Steinmetz; **Page 630:** Eric Sander/Gamma Liaison; **Page 637:** Eric Sander/Gamma Liaison; **Page 660:** Alfred Pasieka/Science Photo Library/Photo Researchers; **Page 664:** Yvonne Hemsey/Gamma Liaison; **Page 679:** Spencer Grant/Gamma Liaison; **Page 679:** Spencer Grant/Gamma Liaison; **Page 690:** Uniphoto; **Page 705:** Courtesy of Metatec; **Page 713:** Uniphoto; **Page 719:** File photo (top left), Fred Wilson/FPG International (top right), Courtesy Apple Computers (bottom left), Davies & Starr/Gamma Liaison (bottom right); **Page 722:** Remi Benali/Gamma Liaison; **Page 730:** Remi Benali/Gamma Liaison; **Page 757:** Robert Severi/Gamma Liaison; **Page 758:** Volker Steger/Peter Arnold; **Page 766:** Volker Steger/Peter Arnold; **Page 793:** Mark Steinmetz; **Page 794:** Mark Steinmetz; **Page 794:** Mark Steinmetz; **Page 804:** Eduardo Garcia/FPG International; **Page 807:** Texas Instruments, Inc.; **Page 810:** Eduardo Garcia/FPG International; **Page 812:** Leonard Lessin/Peter Arnold; **Page 820:** Hayes Microcomputer Products; **Page 829:** DISH Network; **Page 834:** Studiohio; **Page 843:** Mark Steinmetz; **Page 849:** Mark Steinmetz; **Page 852:** Studiohio; **Page 863:** Sencore, Inc.; **Page 868:** Robert Nickelsberg/Gamma Liaison; **Page 872:** Uniphoto; **Page 892:** C. Falco/Photo Researchers; **Page 920:** C. Falco/Photo Researchers; **Page 926:** Microsoft Corporation; **Page 932:** Michael Fairchild/Peter Arnold; **Page 934:** Mark Steinmetz; **Page 940:** Michael Fairchild/Peter Arnold; **Page 967:** Reproduced by permission of Tektronix, Inc.; **Page 969:** BK Precision

Photo for Did You Know? feature: Pete Saloutos/The Stock Market; photo for About Electronics feature: Dale O'Dell/The Stock Market

Circuit board frame: Steve Allen/The Image Bank

Appendix B: Courtesy of PACE, Inc., Laurel, Maryland